Lecture Notes in Computer Science 8156

Commenced Publication in 1973
Founding and Former Series Editors:
Gerhard Goos, Juris Hartmanis, and Jan van Leeuwen

Alfredo Petrosino (Ed.)

Image Analysis and Processing – ICIAP 2013

17th International Conference
Naples, Italy, September 9-13, 2013
Proceedings, Part I

 Springer

Volume Editor

Alfredo Petrosino
University of Naples Parthenope
Department of Science and Technology
Naples, Italy
E-mail: alfredo.petrosino@uniparthenope.it

Cover Illustration: "ICIAP 2013" by Laura Zoé (2013)

ISSN 0302-9743   e-ISSN 1611-3349
ISBN 978-3-642-41180-9   ISBN 978-3-642-41181-6 (eBook)
DOI 10.1007/978-3-642-41181-6
Springer Heidelberg New York Dordrecht London

Library of Congress Control Number: 2013948577

CR Subject Classification (1998): I.4, I.5, I.2.10, H.3, F.2, I.3

LNCS Sublibrary: SL 6 – Image Processing, Computer Vision, Pattern Recognition, and Graphics

Typesetting: Camera-ready by author, data conversion by Scientific Publishing Services, Chennai, India

Printed on acid-free paper

Springer is part of Springer Science+Business Media (www.springer.com)

Preface

The International Conference on Image Analysis and Processing (ICIAP) is an established biennial scientific meeting promoted by the Italian Group of Researchers in Pattern Recognition (GIRPR), which is the Italian IAPR Member Society, and covers topics related to theoretical and experimental areas of image analysis and pattern recognition with emphasis on different applications.

The 17th International Conference on Image Analysis and Processing (ICIAP 2013), held in Naples, Italy, September 9–13, 2013, in the magnificent Castel dell'Ovo, (www.iciap2013-naples.org), was organized by the CVPR Lab of the University of Naples Parthenope (cvprlab.uniparthenope.it).

ICIAP 2013 was endorsed by the International Association for Pattern Recognition (IAPR), the IEEE Computer Society's Technical Committee on Pattern Analysis and Machine Intelligence (TCPAMI), and the IEEE Computational Intelligence Society (CIS).

The central aim of ICIAP 2013 was to highlight connections and synergies of image processing and analysis with pattern recognition and machine learning, human computer systems, biomedical imaging and applications, multimedia interaction and processing, 3D computer vision, and understanding objects and scene, providing to researchers, as well as people from industry, students, and interested newcomers, a forum for discussing current developments and applications. To this aim, Mei Han from Google Inc. USA was invited to give an industrial talk and participate in a panel discussion together with several high-tech companies.

ICIAP 2013 received 354 paper submissions from all over the world, a substantial increase on previous years, including Algeria, Argentina, Austria, Belgium, Brazil, Bulgaria, Canada, China, Colombia, Czech Republic, Denmark, Egypt, Finland, France, Germany, Greece, India, Israel, Italy, Japan, Korea, Marocco, Mexico, Pakistan, Poland, Romania, Russia, Saudi Arabia, Slovak Republic, Spain, Sweden, Switzerland, Tunisia, Turkey, United Kingdom, USA and, Vietnam. To select papers from these submissions, 10 expert researchers were invited to act as areas chairs, together with the International Program Committee and an expert team of reviewers. The rigorous peer-review selection process, carried out by three distinct reviewers and including a rebuttal phase, ultimately led to the selection 162 high-quality manuscripts, with an overall acceptance rate of 45.76%, the lowest in the ICIAP series.

The program included five invited talks by experts in computer vision and pattern recognition, Fei-Fei Li, Stanford University (USA), Jiri Matas, Czech Technical University (Czech Republic), Sankar K. Pal, Indian Statistical Institute (India), Ching Y. Suen, Concordia University (Canada), and Antonio Torralba, Massachusetts Institute of Technology (USA), who covered established approaches, recent results, and directions of future works of different topics like handwriting signature recognition, soft computing in computer vision, scene understanding, and tracking.

ICIAP 2013 also included several tutorials on topics of great relevance with respect to the state of the art: "Bio-Inspired Attention Methods in Computer Vision: Theory, Models and Biological Realities" (John Tsotsos, Canada); "Discrete Optimization in Computer Vision" (Nikos Komodakis and Pawan Kumar, France); "Digital Camera Images: Captured Scene Information vs. Engineered Errors" (Alessandro Rizzi, Italy and John McCann, USA); "Hands on Advanced Bag-of-Words Models for Visual Recognition" (Lamberto Ballan and Lorenzo Seidenari, Italy); "Non-rigid 3D Reconstruction from Images" (Alessio Del Bue, Italy); and "Artificial Consciousness: Theoretical and Empirical Issues" (Riccardo Manzotti, Italy).

ICIAP 2013 also hosted five satellite workshops: "First International Workshop on Assistive Computer Vision and Robotics" (*ACVR*), organized by Marco Leo and Danilo P. Mandic; "Emerging Aspects on Handwritten Signature Processing" (*EAHSP*), organized by Michael Fairhurst, Donato Impedovo, and Giuseppe Pirlo; "Multimedia for Cultural Heritage (*MM4CH*), organized by Costantino Grana, Johan Oomen, and Giuseppe Serra; "Pattern Recognition in Proteomics, Structural Biology, and Bioinformatics (*PR PS BB*), organized by Virginio Cantoni, Michele Ceccarelli, and Robert Murphy; "Social Behaviour Analysis" (*SBA*), organized by Alberto del Bimbo, Pietro Pala, and Maja Pantic. The workshop papers were all collected in a separate volume of the LNCS series by Springer.

Several awards were attributed. Five student support awards were provided by Google USA to cover conference/travel expenses. From a scientific standpoint, the International Association for Pattern Recognition (IAPR) sponsored two IAPR Best Paper awards, while the Caianiello Best Student Award, promoted by GIRPR, was attributed by the CVPR Lab of the University of Naples Parthenope to the best paper presented by a young researcher. All the authors of the awarded papers were invited to submit extended versions of their own papers to a special issue of the Pattern Recognition Letters journal.

The success of ICIAP 2013 is to be credited to the contribution of many people. Special thanks should be given to the area chairs, who did a truly good job. We wish to thank the International Program Committee and the additional Reviewers for the immense amount of hard work and professionalism that has gone into making ICIAP 2013. Our thanks also go to the Organizing Committee for their unstinting advice and support.

I would like to dedicate this book and the entire event to the memory of my great Teacher, Eduardo R. Caianiello. He was one of the most prominent scientists in the field of neural networks, other than quantum field theory, and primarily an inspiring teacher for most researchers in pattern recognition. I always feel that remembrance is the best way to meet a person. I am proud to have met him once again.

Hoping that ICIAP 2013 will serve as an inspiration as we venture towards other frontiers.

September 2013   Alfredo Petrosino

The most typical and distinctive characteristic of the human mind is, in our opinion, its ability to abstract what is 'common' to two, or more, situations or patterns, and to retain of this operation as a new pattern, which is entrusted to the memory as if learnt from the outside.

E.R. Caianiello

(From "Outline of a theory of thought-processes and thinking machines", Journal of Theoretical Biology, 1, 204–235, 1961.)

Organization

Organizing Institution

CVPR Lab of the University of Naples Parthenope, Italy
http://cvprlab.uniparthenope.it

General Chair

Alfredo Petrosino   University of Naples Parthenope, Italy

Area Chairs

Pattern Recognition and Machine Learning:

Marco Gori	University of Siena, Italy
Kai Yu	Baidu Inc., Germany

Human Recognition Systems:

Paola Campadelli	University of Milan, Italy
Caroline Pantofaru	Google, USA

BioMedical Imaging Applications:

Joan Martì	Universitat de Girona, Spain
Francesco Tortorella	University of Cassino, Italy

Multimedia Interaction and Processing:

Rita Cucchiara	University of Modena and Reggio Emilia, Italy
Fatih Porikli	MERL, USA

3D Computer Vision:

Shaogang Gong	Queen Mary University of London, UK
Vittorio Murino	University of Verona and IIT, Italy

Understanding Objects and Scene:

Silvio Savarese	University of Michigan, USA
Jiambo Shi	University of Pennsylvania, USA

Steering Committee

Virginio Cantoni	University of Pavia, Italy
Luigi Cordella	University of Naples Federico II, Italy
Alberto Del Bimbo	University of Florence, Italy
Marco Ferretti	University of Pavia, Italy

Fabio Roli   University of Cagliari, Italy
Gabriella Sanniti di Baja   ICIB-CNR, Italy

Local Committee Chairs

Alessio Ferone   University of Naples Parthenope, Italy
Maria Frucci   ICIB-CNR, Italy

Workshop Chairs

Lucia Maddalena   ICAR-CNR, Italy
Pietro Pala   University of Florence, Italy

Tutorial Chairs

Francesco Isgrò   University of Naples Federico II, Italy
Giosuè Lo Bosco   University of Palermo, Italy

Industrial Liason Chairs

Michele Nappi   University of Salerno, Italy
Francesco Camastra   University of Naples Parthenope, Italy

International Program Committee

Jake Aggarwal, USA   Kalman Palagyi, Hungary
Marco Andreetto, USA   Witold Pedrycz, Canada
Edoardo Ardizzone, Italy   Marcello Pelillo, Italy
Isabelle Bloch, France   Fatih Porikli, USA
Gunilla Borgefors, Sweden   Carlo Sansone, Italy
Alfred Bruckstein, Israel   Raimondo Schettini, Italy
Rama Chellappa, USA   Mubarak Shah, USA
Leila De Floriani, Italy   Josè Ruiz Shulcloper, Cuba
Aytul Ercil, Turkey   Stefano Soatto, USA
Gianluca Foresti, Italy   Arnold Smeulders, The Netherlands
Ashish Ghosh, India   Steven Tanimoto, USA
Edwin Hancock, UK   Massimo Tistarelli, Italy
Xiaoyi Jiang, Germany   John Tsotsos, Canada
Etienne Kerre, Belgium   Shimon Ullman, Israel
Walter Kropatsch, Austria   Mario Vento, Italy
Yanxi Liu, USA   Alessandro Verri, Italy
Gerard Medioni, USA   Hezy Yeshurun, Israel
Alain Merigot, France   Ramin Zabih, USA
Ram Nevatia, USA   Bertrand Zavidovique, France
Sankar Kumar Pal, India   Jacek Zurada, USA

Additional Reviewers

Alessia Albanese
Maria Grazia Albanesi
Antonis Argyros
Ira Assent
Lamberto Ballan
Yingze Bao
Sebastiano Battiato
Massimo Bertozzi
Giuseppe Boccignone
Nunzio Alberto Borghese
Nikolaos Bourbakis
Thierry Bouwmans
Koen Buys
Simone Calderara
Francesco Camastra
Paola Campadelli
Virginio Cantoni
Elena Casiraghi
Giovanna Castellano
Michele Ceccarelli
Yu-Wei Chao
Guillaume Chiron
Wongun Choi
Angelo Ciaramella
Sonya Coleman
Carlo Colombo
Luigi Pietro Cordella
Marco Cristani
Rita Cucchiara
Maria De Marsico
Claudio De Stefano
Alberto Del Bimbo
Matteo Dellepiane
Cecilia Di Ruberto
Giovanni Maria Farinella
Alessio Ferone
Marco Ferretti

Pasquale Foggia
Maria Frucci
Andrea Fusiello
Giorgio Giacinto
Costantino Grana
Edwin R. Hancock
Giulio Iannello
Francesco Isgro
Anne Jorsdstad
Frank Klawonn
Shripad Kondra
Malay K. Kundu
Takio Kurita
Marco La Cascia
Oswald Lanz
Marco Leo
Giosué Lo Bosco
Luca Lombardi
Lucia Maddalena
Marco Maggini
Pradipta Maji
Davide Maltoni
Antonio Maratea
Angelo Marcelli
Gian Luca Marcialis
Francesco Masulli
Alain Merigot
Christian Micheloni
Krystian Mikolajczyk
Mario Molinara
Vittorio Murino
Michele Nappi
Sankar Pal
Pietro Pala
Marco Paladini
Francesco A.N. Palmieri
Marco Pedersoli

Claudio Piciarelli
Giuseppe Pirlo
Roberto Pirrone
Clara Pizzuti
Giovanni Poggi
Moshe Porat
Andrea Prati
Enrico Puppo
Giuliana Ramella
Elisa Ricci
Daniele Riccio
Daniel Riccio
Alessandro Rizzi
Vito Roberto
Fabio Roli
Albert Rothenstein
Stefano Rovetta
Alessandro Rozza
Gabriella Sanniti di Baja
Nicu Sebe
Lorenzo Seidenari
Giuseppe Serra
Antonino Staiano
Katsumi Tadamura
Domenico Tegolo
Ryota Tomioka
Andrea Torsello
Athanasios Tsitsoulis
Cesare Valenti
Giorgio Valentini
Domenico Vitulano
Yu Xiang
Kai Yu
Junsong Yuan
Primo Zingaretti

Additional Sub-reviewers

Andy Bagdanov
Stefano Berretti

Marco Bertini
Battista Biggio

Alessandro Bria
Arcangelo Bruna

Dalia Coppi
Antonio Della Cioppa
Dario Di Fina
Michele Fornaciari
Giorgio Gemignani
Luca Ghiani
Marco Lippi
Carmen Alina Lupascu
Paola Magillo

Marco Manfredi
Mario Manzo
Claudio Marrocco
Iacopo Masi
Mario Molinara
Nicoletta Noceti
Antonio Parziale
Ignazio Pillai
Paolo Piro

Giovanni Puglisi
Paolo Santinelli
Adolfo Santoro
Patricio Simari
Daniele Ravì
Tiberio Uricchio
Roberto Vezzani

Endorsing Institutions

International Association for Pattern Recognition (IAPR)
IEEE Computer Society's Technical Committee on Pattern Analysis and
Machine Intelligence (IEEE-TCPAMI)
IEEE Computational Intelligence Society (IEEE-CIS)
Italian Group of Researchers in Pattern Recognition (GIRPR)
National Group for Scientific Computing (GNCS)

Institutional Patronage

Universitá di Napoli Parthenope, Italy
Campania Regional Board, Italy
National Research Council of Italy (CNR), Italy

Sponsoring Institutions

Italian Ministry of Education, University and Research (MIUR)
Italian Ministry of Economic Development (MiSE)
Comune di Napoli
Google Inc., USA
Ansaldo STS
Italian Aerospace Research Center (CIRA)
Selex ES
ST-Microelectronics
Unlimited Software srl

Acknowledgments

We acknowledge the support of the Project PT2LOG, National Operational
Programme for "Research and Competitiveness" 2007-2013, made available by
the Italian Ministry of Education, University and Research (MIUR) and the
Ministry of Economic Development (MiSE).

Table of Contents – Part I

Speeding Up Local Patch Dissimilarity............................. 1
   Radu Tudor Ionescu and Marius Popescu

A Graph-Based Hierarchical Image Segmentation Method Based on a
Statistical Merging Predicate 11
   Silvio Jamil F. Guimarães and Zenilton K.G. Patrocínio Jr.

Application of Local Binary Pattern to Windowed Nonlocal Means
Image Denoising 21
   Fakhry Khellah

Integrating Color Sampling into Depth Based Bilayer Segmentation 31
   Lorenzo Sorgi and Markus Schlosser

Local Intrinsic Dimensionality Based Features for Clustering 41
   Paola Campadelli, Elena Casiraghi, Claudio Ceruti,
   Gabriele Lombardi, and Alessandro Rozza

Deeply Optimized Hough Transform: Application to Action
Segmentation.. 51
   Adrien Chan-Hon-Tong, Catherine Achard, and Laurent Lucat

Layout-Based Document-Retrieval System by Radon Transform Using
Dynamic Time Warping 61
   Giuseppe Pirlo, Michela Chimienti, Michele Dassisti,
   Donato Impedovo, and Angelo Galiano

Evaluation of Low-Level Image Representations for Illumination-
Insensitive Recognition of Textureless Objects 71
   Sebastian Zambanini and Martin Kampel

Kernels for Visual Words Histograms 81
   Radu Tudor Ionescu and Marius Popescu

A New Adaptive Zoning Technique for Handwritten Digit
Recognition 91
   Sebastiano Impedovo, Francesco Maurizio Mangini, and
   Giuseppe Pirlo

Image Annotation by Learning Label-Specific Distance Metrics 101
   Xing Xu, Atsushi Shimada, and Rin-ichiro Taniguchi

Approximating the Skeleton for Fine-to-Coarse Shape Representation... 111
   Luca Serino, Carlo Arcelli, and Gabriella Sanniti di Baja

Learning Iterative Strategies in Multi-Expert Systems Using SVMs for
Digit Recognition 121
    Donato Barbuzzi, Donato Impedovo,
    Francesco Maurizio Mangini, and Giuseppe Pirlo

Learning Precise Local Boundaries in Images from Human Tracings 131
    Martin Horn and Michael R. Berthold

Age Estimation Using Local Binary Pattern Kernel Density Estimate ... 141
    Juha Ylioinas, Abdenour Hadid, Xiaopeng Hong, and
    Matti Pietikäinen

Improving the Quality of Color Image Segmentation Using Genetic
Algorithm.. 151
    Aniceto C. Andrade Jr., Zenilton K.G. Patrocínio Jr., and
    Silvio Jamil F. Guimarães

Detection of the Vanishing Line of the Ocean Surface from Pairs of
Scale-Invariant Keypoints 161
    Sergiy Fefilatyev, Matthew Shreve, and Dmitry Goldgof

Average Common Submatrix: A New Image Distance Measure 170
    Alessia Amelio and Clara Pizzuti

A Fast Jensen-Shannon Subgraph Kernel............................ 181
    Lu Bai and Edwin R. Hancock

Evaluation of Interactive Segmentation Algorithms Using Densely
Sampled Correct Interactions 191
    S.M. Rafizul Haque, Mark G. Eramian, and Kevin A. Schneider

Estimating Complex Refractive Index Using Ellipsometry 201
    Gul e Saman and Edwin R. Hancock

Lazy Nonlinear Diffusion Parameter Estimation..................... 211
    Daniel Thuerck and Arjan Kuijper

Analysis of WD Face Dictionary for Sparse Coding Based Face
Recognition 221
    Shejin Thavalengal and Anil Kumar Sao

Fast and Robust Edge-Guided Exemplar-Based Image Inpainting 231
    Yun Wu and Chun Yuan

A Watershed-Based Segmentation Technique for Multiresolution
Data 241
    Giuseppe Masi, Giuseppe Scarpa, Raffaele Gaetano, and
    Giovanni Poggi

Database for Arabic Printed Text Recognition Research.............. 251
Faten Kallel Jaiem, Slim Kanoun, Maher Khemakhem,
Haikal El Abed, and Jihain Kardoun

On the Stability of Ranks to Low Image Quality in Biometric
Identification Systems 260
Emanuela Marasco and Ayman Abaza

Approximated Overlap Error for the Evaluation of Feature Descriptors
on 3D Scenes.. 270
Fabio Bellavia, Cesare Valenti, Carmen Alina Lupascu, and
Domenico Tegolo

Exploiting the Golden Ratio on Human Faces for Head-Pose
Estimation 280
Gianluca Fadda, Gian Luca Marcialis, Fabio Roli, and Luca Ghiani

An Interactive Video Retrieval Approach Based on Latent Topics 290
Rubén Fernández-Beltran and Filiberto Pla

Performance Study of a Regularization-Based Deformable Handwritten
Recognition Approach 300
Yoshiki Mizukami, Shinya Nakanishi, and Katsumi Tadamura

Layered Self-Organizing Map for Image Classification in Unrestricted
Domains.. 310
Christian O'Connell, Andrea Kutics, and Akihiko Nakagawa

Wide Area Camera Localization 320
Valeria Garro, Maurizio Galassi, and Andrea Fusiello

Arabic Printed Word Recognition Using Windowed Bernoulli HMMs ... 330
Ihab Khoury, Adrià Giménez, Alfons Juan, and Jesús Andrés-Ferrer

Head Direction Estimation from Silhouette 340
Amina Bensebaa, Slimane Larabi, and Neil M. Robertson

Combined Supervised / Unsupervised Algorithm for Skin Detection:
A Preliminary Phase for Face Detection........................... 351
Eyal Braunstain and Isak Gath

Conic Based Camera Re-calibration after Zooming 361
Iuri Frosio, Cristina Turrini, and Alberto Alzati

Dynamic Hierarchical Segmentation of Remote Sensing Images......... 371
Giuseppe Scarpa, Giuseppe Masi, Raffaele Gaetano,
Luisa Verdoliva, and Giovanni Poggi

Road Traffic Conflict Analysis from Geo-referenced Stereo Sequences ... 381
Sebastiano Battiato, Stefano Cafiso, Alessandro Di Graziano,
Giovanni M. Farinella, and Oliver Giudice

Adaptive Compression of Stereoscopic Images 391
Alessandro Ortis, Francesco Rundo, Giuseppe Di Giore, and
Sebastiano Battiato

Trajectory Similarity Measures Using Minimal Paths 400
Brais Cancela, Marcos Ortega, Alba Fernández, and
Manuel G. Penedo

Structured Multi-class Feature Selection for Effective Face
Recognition 410
Giovanni Fusco, Luca Zini, Nicoletta Noceti, and Francesca Odone

Measuring Sandy Bottom Dynamics by Exploiting Depth from Stereo
Video Sequences 420
Rosaria E. Musumeci, Giovanni M. Farinella, Enrico Foti,
Sebastiano Battiato, Thor U. Petersen, and B. Mutlu Sumer

Daily Living Activities Recognition via Efficient High and Low Level
Cues Combination and Fisher Kernel Representation 431
Negar Rostamzadeh, Gloria Zen, Ionuţ Mironică,
Jasper Uijlings, and Nicu Sebe

What Epipolar Geometry Can Do for Video-Surveillance 442
Nicoletta Noceti, Luigi Balduzzi, and Francesca Odone

Class Representative Computation Using Graph Embedding 452
Fahri Aydos, Ahmet Soran, and M. Fatih Demirci

Robust Selective Stereo SLAM without Loop Closure and Bundle
Adjustment 462
Fabio Bellavia, Marco Fanfani, Fabio Pazzaglia, and Carlo Colombo

Demographics versus Biometric Automatic Interoperability 472
Maria De Marsico, Michele Nappi, Daniel Riccio, and
Harry Wechsler

Edge Detection on Polynomial Texture Maps 482
Cristian Brognara, Massimiliano Corsini, Matteo Dellepiane, and
Andrea Giachetti

A Ripplet Transform Based Statistical Framework for Natural Color
Image Retrieval 492
Manish Chowdhury, Sudeb Das, and Malay K. Kundu

A Fully Automatic Approach for the Accurate Localization of the
Pupils 503
Marco Leo, Dario Cazzato, Tommaso De Marco, and
Cosimo Distante

Problems in Distortion Corrected Texture Classification and the Impact
of Scale and Interpolation   513
  *Michael Gadermayr, Michael Liedlgruber, Andreas Uhl, and
  Andreas Vécsei*

On Optimized Color Image Coding Using Correlation of Primary
Colors   523
  Eyal Braunstain and Moshe Porat

Towards Learning Hierarchical Compositional Models in the Presence
of Clutter   532
  Jan Mačák and Ondřej Drbohlav

Social Groups Detection in Crowd through Shape-Augmented
Structured Learning   542
  Francesco Solera and Simone Calderara

Evaluation of Statistical Features for Medical Image Retrieval   552
  Cecilia Di Ruberto and Giuseppe Fodde

Hierarchical Image Representation Simplification Driven by Region
Complexity   562
  Petra Bosilj, Sébastien Lefèvre, and Ewa Kijak

Anisotropic Diffusion and Curve Evolution for Segmentation of Color
Images in Cultural Heritage   572
  Luigi Cinque and Rossella Cossu

Comparison of Three Approaches for Scenario Classification for the
Automotive Field   582
  *Nicola Bernini, Massimo Bertozzi, Luca Devincenzi, and
  Luca Mazzei*

Adverse Driving Conditions Alert: Investigations on the SWIR
Bandwidth for Road Status Monitoring   592
  Massimo Bertozzi, Rean Isabella Fedriga, and Carlo D'Ambrosio

Simple and Robust Facial Portraits Recognition under Variable Lighting
Conditions Based on Two-Dimensional Orthogonal Transformations   602
  Paweł Forczmański, Georgy Kukharev, and Nadezdha Shchegoleva

Investigation of Different Classification Models to Determine the
Presence of Leukemia in Peripheral Blood Image   612
  Lorenzo Putzu and Cecilia Di Ruberto

Estimating the Serial Combination's Performance from That of
Individual Base Classifiers   622
  Gian Luca Marcialis, Luca Didaci, and Fabio Roli

3D Interest Points Detection Using Symmetric Surround-Based Surface
Saliency 632
    Yitian Zhao and Yonghuai Liu

Robust Silhouette Extraction from Kinect Data 642
    *Michele Pirovano, Carl Yuheng Ren, Iuri Frosio, Pier Luca Lanzi,
    Victor Prisacariu, David W. Murray, and N. Alberto Borghese*

A Modified SIFT Descriptor for Image Matching under Spectral Variations 652
    Sajid Saleem and Robert Sablatnig

A New Fuzzy Skeletonization Algorithm and Its Applications to
Medical Imaging 662
    Dakai Jin and Punam K. Saha

A Subunit-Based Dynamic Time Warping Approach for Hand
Movement Recognition 672
    *Yanrung Wang, Atsushi Shimada, Takayoshi Yamashita, and
    Rin-ichiro Taniguchi*

Softmax Regression for ECOC Reconstruction 682
    Roberto D'Ambrosio, Giulio Iannello, and Paolo Soda

Multisubjects Tracking by Time-of-Flight Camera 692
    Piercarlo Dondi, Luca Lombardi, and Luigi Cinque

3D Tracking of Honeybees Enhanced by Environmental Context 702
    *Guillaume Chiron, Petra Gomez-Krämer, Michel Ménard, and
    Fabrice Requier*

Classification of Pollen Apertures Using Bag of Words 712
    *Gildardo Lozano-Vega, Yannick Benezeth, Franck Marzani, and
    Frank Boochs*

Evaluating Temporal Information for Social Image Annotation and
Retrieval 722
    *Tiberio Uricchio, Lamberto Ballan, Marco Bertini, and
    Alberto Del Bimbo*

VSCAN: An Enhanced Video Summarization Using Density-Based
Spatial Clustering 733
    Karim M. Mahmoud, Mohamed A. Ismail, and Nagia M. Ghanem

On the Impact of Alterations on Face Photo Recognition Accuracy 743
    Matteo Ferrara, Annalisa Franco, Davide Maltoni, and Yunlian Sun

An Efficient Indexing Scheme Based on Linked-Node m-Ary Tree
Structure 752
    *The-Anh Pham, Sabine Barrat, Mathieu Delalandre, and
    Jean-Yves Ramel*

A Natural Interface for the Training of Medical Personnel in an
Immersive and Virtual Reality System   763
   *Alberto Del Bimbo, Andrea Ferracani, Daniele Pezzatini, and
   Lorenzo Seidenari*

Saliency Based Image Cropping   773
   Edoardo Ardizzone, Alessandro Bruno, and Giuseppe Mazzola

First Quantization Coefficient Extraction from Double Compressed
JPEG Images   783
   *Fausto Galvan, Giovanni Puglisi, Arcangelo R. Bruna, and
   Sebastiano Battiato*

Soccer Ball Detection with Isophotes Curvature Analysis   793
   Tommaso De Marco, Marco Leo, and Cosimo Distante

A Bayesian Approach to Tracking Learning Detection   803
   *Giorgio Gemignani, Wongun Choi, Alessio Ferone,
   Alfredo Petrosino, and Silvio Savarese*

Blind Invisible Watermarking Technique in DT-CWT Domain Using
Visual Cryptography   813
   *Meryem Benyoussef, Samira Mabtoul, Mohamed El Marraki, and
   Driss Aboutajdine*

An Ensemble Algorithm Framework for Automated Stereology of
Cervical Cancer   823
   *Baishali Chaudhury, Hady Ahmady Phoulady, Dmitry Goldgof,
   Lawrence O. Hall, Peter R. Mouton, Ardeshir Hakam, and
   Erin M. Siegel*

Attributed Relational SIFT-Based Regions Graph for Art Painting
Retrieval   833
   Mario Manzo and Alfredo Petrosino

Video Segmentation Framework by Dynamic Background Modelling   843
   *Santiago Molina-Giraldo, Andres M. Álvarez-Meza,
   Julio C. García-Álvarez, and Cesar G. Castellanos-Domínguez*

Author Index   853

Table of Contents – Part II

Improving Gait Biometrics under Spoofing Attacks   1
   Abdenour Hadid, Mohammad Ghahramani, John Bustard, and
   Mark Nixon

Extracting Compact Information from Image Benchmarking Tools:
The SAR Despeckling Case   11
   Gerardo Di Martino, Giovanni Pecoraro, Giovanni Poggi,
   Daniele Riccio, and Luisa Verdoliva

Automatic Aesthetic Photo Composition   21
   Roberto Gallea, Edoardo Ardizzone, and Roberto Pirrone

Face Recognition in Uncontrolled Conditions Using Sparse
Representation and Local Features   31
   Alessandro Adamo, Giuliano Grossi, and Raffaella Lanzarotti

Eigenvector Sign Correction for Spectral Correspondence Matching   41
   Muhammad Haseeb and Edwin R. Hancock

An Interactive Image Rectification Method Using Quadrangle
Hypothesis   51
   Satoshi Yonemoto

MRF Based Image Segmentation Augmented with Domain Specific
Information   61
   Özge Öztimur Karadağ and Fatoş T. Yarman Vural

Segmentation of Time-Lapse Images with Focus on Microscopic Images
of Cells   71
   Jindřich Soukup, Petr Císař, and Filip Šroubek

Segmentation with Incremental Classifiers   81
   Guillaume Bernard, Michel Verleysen, and John A. Lee

3D Femur Reconstruction Using X-Ray Stereo Pairs   91
   Sonia Akkoul, Adel Hafiane, Eric Lespessailles, and Rachid Jennane

Information-Based Learning of Deep Architectures for Feature
Extraction   101
   Stefano Melacci, Marco Lippi, Marco Gori, and Marco Maggini

Image Classification with Multivariate Gaussian Descriptors   111
   Costantino Grana, Giuseppe Serra, Marco Manfredi, and
   Rita Cucchiara

Non–referenced Quality Assessment of Image Processing Methods in
Infrared Non-destructive Testing   121
   Tomas J. Ramírez-Rozo, Hern D. Benítez-Restrepo,
   Julio C. García-Álvarez, and Cesar G. Castellanos-Domínguez

Using Dominant Sets for k-NN Prototype Selection   131
   Sebastiano Vascon, Marco Cristani, Marcello Pelillo, and
   Vittorio Murino

Feature Extraction for Iris Recognition Based on Optimized
Convolution Kernels   141
   Lubos Omelina, Bart Jansen, Milos Oravec, and Jan Cornelis

Saliency Based Aesthetic Cut of Digital Images   151
   Luca Greco and Marco La Cascia

A Plant Recognition Approach Using Shape and Color Features in Leaf
Images   161
   Ali Caglayan, Oguzhan Guclu, and Ahmet Burak Can

Robust Coarse-to-Fine Sparse Representation for Face Recognition   171
   Yunlian Sun and Massimo Tistarelli

SketchSPORE: A Sketch Based Domain Separation and Recognition
System for Interactive Interfaces   181
   Danilo Avola, Luigi Cinque, and Giuseppe Placidi

Ontology-Assisted Object Detection: Towards the Automatic Learning
with Internet   191
   Francesco Setti, Dong-Seon Cheng, Sami Abduljalil Abdulhak,
   Roberta Ferrario, and Marco Cristani

Epithelial Cell Segmentation in Histological Images of Testicular Tissue
Using Graph-Cut   201
   Azadeh Fakhrzadeh, Ellinor Spörndly-Nees, Lena Holm, and
   Cris L. Luengo Hendriks

Urban Road Network Extraction Based on Fuzzy ART, Radon
Transform and Morphological Operations on DSM Data   209
   Darlis Herumurti, Keiichi Uchimura, Gou Koutaki, and
   Takumi Uemura

A Weighted Majority Vote Strategy Using Bayesian Networks   219
   Luigi P. Cordella, Claudio De Stefano, Francesco Fontanella, and
   Alessandra Scotto di Freca

Pedestrian Detection in Poor Visibility Conditions: Would SWIR
Help?   229
   Massimo Bertozzi, Rean Isabella Fedriga, Alina Miron, and
   Jean-Luc Reverchon

Multi-target Data Association Using Sparse Reconstruction 239
Andrew D. Bagdanov, Alberto Del Bimbo, Dario Di Fina,
Svebor Karaman, Giuseppe Lisanti, and Iacopo Masi

The Recognition of Polynomial Position and Orientation through the
Finite Polynomial Discrete Radon Transform 249
Ines Elouedi, Régis Fournier, Amine Naït-Ali, and Atef Hamouda

Multiple Classifier Systems for Image Forgery Detection 259
Davide Cozzolino, Francesco Gargiulo, Carlo Sansone, and
Luisa Verdoliva

Using the Watershed Transform for Iris Detection 269
Maria Frucci, Michele Nappi, Daniel Riccio, and
Gabriella Sanniti di Baja

Outdoor Environment Monitoring with Unmanned Aerial Vehicles 279
Claudio Piciarelli, Christian Micheloni, Niki Martinel,
Marco Vernier, and Gian Luca Foresti

Training Binary Descriptors for Improved Robustness and Efficiency in
Real-Time Matching 288
Sharat Saurabh Akhoury and Robert Laganière

Towards Semantic KinectFusion 299
Nicola Fioraio, Gregorio Cerri, and Luigi Di Stefano

Face Recognition under Ageing Effect: A Comparative Analysis 309
Zahid Akhtar, Ajita Rattani, Abdenour Hadid, and
Massimo Tistarelli

A Slightly Supervised Approach for Positive/Negative Classification of
Fluorescence Intensity in HEp-2 Images 319
Giulio Iannello, Leonardo Onofri, and Paolo Soda

Landmarks-SIFT Face Representation for Gender Classification 329
Yomna Safaa El-Din, Mohamed N. Moustafa, and Hani Mahdi

Discrete Morse versus Watershed Decompositions of Tessellated
Manifolds 339
Leila De Floriani, Federico Iuricich, Paola Magillo, and
Patricio Simari

A New Algorithm for Cortical Bone Segmentation with Its Validation
and Applications to In Vivo Imaging 349
Cheng Li, Dakai Jin, Trudy L. Burns, James C. Torner,
Steven M. Levy, and Punam K. Saha

Automatic Lesion Detection in Breast DCE-MRI   359
  Stefano Marrone, Gabriele Piantadosi, Roberta Fusco,
  Antonella Petrillo, Mario Sansone, and Carlo Sansone

Invariants to Symmetrical Convolution with Application to Dihedral
Kernel Symmetry   369
  Jiří Boldyš and Jan Flusser

Observing Dynamic Urban Environment through Stereo-Vision Based
Dynamic Occupancy Grid Mapping   379
  You Li and Yassine Ruichek

A Multiple Classifier Approach for Detecting Naked Human Bodies in
Images   389
  Luca Giangiuseppe Esposito and Carlo Sansone

Diversity in Ensembles of Codebooks for Visual Concept Detection   399
  Luca Piras, Roberto Tronci, and Giorgio Giacinto

A Novel Method for Fast Processing of Large Remote Sensed Image   409
  Adriano Mancini, Anna Nora Tassetti, Alessandro Cinnirella,
  Emanuele Frontoni, and Primo Zingaretti

MATRIOSKA: A Multi-level Approach to Fast Tracking by Learning . . .   419
  Mario Edoardo Maresca and Alfredo Petrosino

Towards a Realistic Distribution of Cells in Synthetically Generated 3D
Cell Populations   429
  David Svoboda and Vladimír Ulman

Single Textual Image Super-Resolution Using Multiple Learned
Dictionaries Based Sparse Coding   439
  Rim Walha, Fadoua Drira, Franck Lebourgeois,
  Christophe Garcia, and Adel M. Alimi

Mixed Kernel Function SVM for Pulmonary Nodule Recognition   449
  Yang Li, Dunwei Wen, Ke Wang, and A'lin Hou

Fast and Accurate Tree-Based Clustering for Japanese/Chinese
Character Recognition   459
  Yuichi Abe, Takahiro Sasaki, and Hideaki Goto

Towards Automatic Hands Detection in Single Images   469
  Athanasios Tsitsoulis and Nikolaos Bourbakis

Precise 3D Angle Measurements in CT Wrist Images   479
  Johan Nysjö, Albert Christersson, Ida-Maria Sintorn,
  Ingela Nyström, Sune Larsson, and Filip Malmberg

Layout Estimation of Highly Cluttered Indoor Scenes Using Geometric
and Semantic Cues   489
     *Yu-Wei Chao, Wongun Choi, Caroline Pantofaru, and
     Silvio Savarese*

Dissimilarity Measures for the Identification of Earthquake Focal
Mechanisms   500
     Francesco Benvegna, Giosué Lo Bosco, and Domenico Tegolo

Texture Classification Based on Co-occurrence Matrix and
Neuro-Morphological Approach   510
     Mohammed Talibi Alaoui and Abderrahmane Sbihi

A Virtually Continuous Representation of the Deep Structure of
Scale-Space   522
     Luigi Rocca and Enrico Puppo

Integral Spiral Image for Fast Hexagonal Image Processing   532
     Sonya Coleman, Bryan Scotney, and Bryan Gardiner

Rough Set Based Homogeneous Unsharp Masking for Bias Field
Correction in MRI   542
     Abhirup Banerjee and Pradipta Maji

Real-Time Estimation of Planar Surfaces in Arbitrary Environments
Using Microsoft Kinect Sensor   552
     *Francesco Castaldo, Vincenzo Lippiello,
     Francesco A.N. Palmieri, and Bruno Siciliano*

Data Ranking and Clustering via Normalized Graph Cut Based on
Asymmetric Affinity   562
     *Olexiy Kyrgyzov, Isabelle Bloch, Yuan Yang, Joe Wiart, and Antoine
     Souloumiac*

A Boosting-Based Approach to Refine the Segmentation of Masses in
Mammography   572
     Mario Molinara, Claudio Marrocco, and Francesco Tortorella

Visual Concept Detection and Annotation via Multiple Kernel Learning
of Multiple Models   581
     Yu Zhang, Stephane Bres, and Liming Chen

Facial Expression Recognition Based on Perceived Facial Images and
Local Feature Matching   591
     *Hayet Boughrara, Liming Chen, Chokri Ben Amar, and
     Mohamed Chtourou*

Real-Time 2DHoG-2DPCA Algorithm for Hand Gesture Recognition . . .   601
     Omnia S. El Saadany and Moataz M. Abdelwahab

Shearlet Network-based Sparse Coding Augmented by Facial Texture
Features for Face Recognition 611
    Mohamed Anouar Borgi, Demetrio Labate, Maher El'Arbi, and
    Chokri Ben Amar

Fuzzy Analysis of Classifier Handshapes from 3D Sign Language
Data 621
    Kabil Jaballah and Mohamed Jemni

Cooking Action Recognition with *i*VAT: An *Interactive* Video
Annotation Tool 631
    Simone Bianco, Gianluigi Ciocca, Paolo Napoletano,
    Raimondo Schettini, Roberto Margherita,
    Gianluca Marini, and Giuseppe Pantaleo

Spatial Resolution and Distance Information for Color Quantization 642
    Giuliana Ramella and Gabriella Sanniti di Baja

On the Robustness of Color Texture Descriptors across Illuminants 652
    Simone Bianco, Claudio Cusano, Paolo Napoletano, and
    Raimondo Schettini

Semiotic-based Conceptual Modelling of Hypermedia 663
    Elio Toppano and Vito Roberto

Modelling Visual Appearance of Handwriting 673
    Angelo Marcelli, Antonio Parziale, and Adolfo Santoro

Learning the Scene Illumination for Color-Based People Tracking in
Dynamic Environment 683
    Sinan Mutlu, Tao Hu, and Oswald Lanz

Multicamera People Tracking Using a Locus-Based Probabilistic
Occupancy Map 693
    Tao Hu, Sinan Mutlu, and Oswald Lanz

Construction and Application of Marine Oil Spill Gravity Vector
Differences Detection Model 703
    Weiguang Su, Bo Ping, and Fenzhen Su

A Graph-Based Method for PET Image Segmentation in Radiotherapy
Planning: A Pilot Study 711
    Alessandro Stefano, Salvatore Vitabile, Giorgio Russo,
    Massimo Ippolito, Daniele Sardina, Maria G. Sabini,
    Francesca Gallivanone, Isabella Castiglioni, and Maria C. Gilardi

White Paper on Industrial Applications of Computer Vision and
Pattern Recognition   721
  Giovanni Garibotto, Pierpaolo Murrieri, Alessandro Capra,
  Stefano De Muro, Ugo Petillo, Francesco Flammini,
  Mariana Esposito, Concetta Pragliola, Giuseppe Di Leo, Roald Lengu,
  Nadia Mazzino, Alfredo Paolillo, Michele D'Urso,
  Raffaele Vertucci, Fabio Narducci, Stefano Ricciardi,
  Andrea Casanova, Gianni Fenu, Marco De Mizio, Mario Savastano,
  Michele Di Capua, and Alessio Ferone

Empty Vehicle Detection with Video Analytics   731
  Francesco Buemi, Mariana Esposito, Francesco Flammini,
  Nicola Mazzocca, Concetta Pragliola, and Marcella Spirito

Stock Control through Video Surveillance in Logistics   740
  Mariarosaria Carullo, Gianluca Cavaliere, Aniello De Prisco,
  Michele Di Capua, Alfredo Petrosino, Donatella Padovano,
  Gennaro Nave, and Daniele Ruggeri

H.264 Sensor Aided Video Encoder for UAV BLOS Missions   749
  Cesario Vincenzo Angelino, Luca Cicala, Marco De Mizio,
  Paolo Leoncini, E. Baccaglini, M. Gavelli, N. Raimondo, and
  R. Scopigno

Pattern Recognition for Defect Detection in Uncontrolled Environment
Railway Applications   753
  Giuseppe Di Leo, Roald Lengu, Nadia Mazzino, and Alfredo Paolillo

Author Index   759

Speeding Up Local Patch Dissimilarity

Radu Tudor Ionescu and Marius Popescu

Faculty of Mathematics and Computer Science
University of Bucharest, No. 14 Academiei Street, Bucharest, Romania
{raducu.ionescu,popescunmarius}@gmail.com

Abstract. There are many patch-based techniques used in image processing, but most of them are heavy to compute with current machines. A dissimilarity measure for images based on patches, inspired from rank distance, called Local Patch Dissimilarity (LPD), was recently introduced. It has very promising results in optical character recognition, but, as other patch-based methods, it is computationally heavy.

This work aims at showing that LPD can be improved in terms of efficiency. Several ways of optimizing the LPD algorithm are presented, such as using a hash table to store precomputed patch distances or skipping the comparison of overlapping patches. Another way to avoid the problem of the higher computational time on large sets of images is to turn to local learning methods.

Several experiments are conducted on two datasets using both standard machine learning methods and local learning methods. All methods are based on LPD. The obtained results come to support the fact that LPD is a very good dissimilarity measure for images. In this paper, LPD is also used with success for classifying images other than handwritten digits.

Keywords: image dissimilarity, image classification, handwritten digit recognition, patches, patch-based technique.

1 Introduction

Researchers have developed sophisticated methods for analyzing and editing digital images and video. Indeed, many of the most powerful of these methods are patch-based [2] and they divide the image into many small patches and then manipulate or analyze the image based on its patches. Patch-based sampling methods have become a popular tool for image and video synthesis and analysis. Applications include texture synthesis, image and video completion, summarization and retargeting, image recomposition and editing, object detection, noise removal, super-resolution and more.

Local Patch Dissimilarity (LPD) is a dissimilarity measure for images based on patches that was recently introduced in [7]. Patch-based algorithms are known to be computationally heavy because they must search and manipulate millions of patches. Local Patch Dissimilarity is no exception, but it gives a very high accuracy in optical character recognition. This work investigates methods of speeding up the LPD algorithm in order to make it more efficient. Using a hash

A. Petrosino (Ed.): ICIAP 2013, Part I, LNCS 8156, pp. 1–10, 2013.
© Springer-Verlag Berlin Heidelberg 2013

table to store precomputed patch distances or skipping the comparison step of some patches are the proposed methods for time optimization.

Finding a way of using patch-based algorithms on large datasets or for real-time applications is not always possible. But in some particular cases such as object recognition, image categorization, and image retrieval, there are possible solutions to overcome the time barrier. The aim of this paper is to show that LPD can also be used on large datasets of images. Local learning methods are preferred instead of standard machine learning algorithms to achieve this goal.

The paper is organized as follows. Related work about local learning methods and patch-based techniques is discussed in section 2. LPD is shortly described in section 3. Section 4 presents some optimizations of the LPD algorithm. Experiments with both standard and local learning methods based on LPD are presented in section 5. Finally, the conclusions are drawn in section 6.

2   Related Work

Patches belong to the category of local features, which means that they describe properties of a certain region of an image. In contrast to that, global features provide information about an image as a whole. Beside patches, another popular class of local image features are image descriptors, such as SIFT [12].

For numerous computer vision applications, the image can be analyzed at the patch level rather than at the individual pixel level or global level. Patches contain contextual information and have advantages in terms of computation and generalization. For example, patch-based methods produce better results and are much faster than pixel-based methods for texture synthesis [8]. However, patch-based techniques are still heavy to compute with current machines [2].

A paper that describes a patch-based approach for rapid image correlation or template matching is [9]. An approach to object recognition was proposed by [6], where image patches are clustered using the EM algorithm for Gaussian mixture densities and images are represented as histograms of the patches over the (discrete) membership to the clusters. The authors of [1] propose a method where images are represented by binary feature vectors that encode which patches from a codebook appear in the images and which spatial relationship they have. In [17] the authors propose a novel texture classification method via patch-based sparse texton learning. The patch transform, proposed in [4], represents an image as bag of overlapping patches sampled on a regular grid. In [2] the authors present a new randomized algorithm for quickly finding approximate nearest neighbor matches between image patches.

The development of unconventional learning formulations and non-inductive types of inference was studied in [15]. The author argues in favor of introducing and developing unconventional learning methods, as opposed to algorithmic improvements of existing learning methods. This view is consistent with the main principle of VC theory, suggesting that one should always use direct learning formulations for finite sample estimation problems, rather than more general settings (such as density estimation). In [3] the idea of local algorithms for pattern recognition was used, where local linear rules (instead of local constant rules) and VC bounds (instead of the distance to the k-th nearest neighbor) were used. The local linear rules demonstrated an improvement in accuracy on the popular MNIST dataset (from $4, 1\%$ to $3, 2\%$).

Local learning methods attempt to locally adjust the performance of the training system to the properties of the training set in each area of the input space. A simple local learning algorithm would work like this: for each testing example, select a few training examples located in the vicinity of the testing example, train a classifier with only these few examples and apply the resulting classifier to the testing example. The k-Nearest Neighbor algorithm (k-NN) is a method for classifying objects based on the closest training examples in the feature space. It is part of the family of local learning algorithms, but it has the advantage that no training time is required. In this paper, a k-NN with filtering approach is used. For the k-NN with filtering approach, the idea is to filter (or select) the nearest K neighbors (where K is larger than k) using a distance measure that is much faster and easy to compute. Instead of training a classifier, the next step is to select the nearest k neighbors from those filtered K examples using a distance measure that is able to capture much finer differences. This latter distance measure is allowed to consume more time in order to determine a better similarity (or dissimilarity) between training examples. This approach is appropriate when it is unreasonable, from the perspective of time, to compute the latter distance for all training examples. This two-step selection (or filtering) process is much faster to compute than a standard k-NN.

3   Local Patch Dissimilarity

This section gives an overview of LPD. To compute LPD between two gray-scale images, the idea is to sum up all the offsets of similar patches between the two images. The LPD algorithm is briefly described next. For every patch in one image the idea is to search for a similar patch in the other image. First, look for similar patches in the same position in both images. If those patches are similar, sum up 0 since there is no offset between patches. If the patches are not similar, start looking around the initial patch position in the second image to find a patch similar to the one in the first image. If a similar patch is found during this process, sum up the offset between the two patches. The search goes on until a similar patch is found or until a maximum offset is reached. Note that the maximum patch offset must be set a priori. The time complexity of LPD is $O(h^2 \times w^2)$, where h and w represent minimum height and width of the two compared images, respectively. For a more detailed description of the LPD algorithm and a better understanding of section 4, please refer to [7].

LPD is based on another distance between patches. Any image distance can be used to compute the similarity between patches. Note that the LPD algorithm [7] determines patch similarity using the mean squared euclidean distance that corresponds to the L_2-norm. Another version of the LPD algorithm [7] is also used in the experiments, that determines patch similarity using the mean euclidean distance that corresponds to the L_1-norm. Both algorithms show good results.

4   LPD Algorithm Optimization

As stated in [2], patch-based algorithms are heavy to compute with current computers because these algorithms must deal with millions of patches. Some optimizations that improve the LPD algorithm in terms of speed are discussed here. It may not occur at first look, but LPD needs to compute the similarity between many pairs of patches and for some of them, even several times. Recomputation of patch similarities can be avoided by storing the precomputed values in a hash table. The tests presented in section 5 are performed using the hash table optimization which brings an $8 - 10\%$ speed improvement.

Another optimization is to stop the search for similar patches earlier. Recall that the search goes on until a similar patch is found or until a maximum offset is reached. It is very unlikely to find similar patches at high offset from each other. Even if two similar patches are found at a great distance from each other, it does not mean the images are similar. In fact, this phenomenon may bring noise into the computation of LPD. To avoid this extensive search that can potentially harm the dissimilarity measure, setting a maximum offset radius much lower than the image diagonal size is a good choice. This search limitation was included in all the experiments, which resulted in a great improvement in terms of speed and accuracy. However, one must be careful not to reduce the maximum offset by too much, which can badly alter the performance and strength of the dissimilarity measure. To stop the search too early would mean to disregard some similar patches that bring important information in the dissimilarity computation. In the experiments, an offset radius of $25 - 50\%$ of the image diagonal size works very well. This also brings a speed improvement of $25 - 30\%$.

The last proposed algorithm optimization comes from the fact that LPD computes the similarity between many overlapping patches. A fast version of the LPD algorithm [7] is to skip the comparison of overlapping patches. This can be achieved by increasing the offset by more than one unit, each time a similar patch is not found at the current offset. The results presented in section 5.4 are obtained by increasing the offset with two units at every step, thus, skipping half of the comparisons between patches and speeding the LPD algorithm by a factor of two. The proposed speed improvements do not affect the time complexity.

5   Experiments

5.1   Datasets

The first dataset used for testing the dissimilarity presented in this paper is the MNIST set, which is described in detail in [11]. Comparative experiments are also reported in [11]. The MNIST database, containing 60.000 training examples and 10.000 test examples, is available at **http://yann.lecun.com/exdb/mnist/**. The second dataset was collected from the Web by the authors of [10] and consists of 100 images each of six different classes of birds: egrets, mandarin ducks, snowy owls, puffins, toucans, and wood ducks. The Birds dataset is used in the last experiment presented in this paper.

Table 1. Error and time of 3-NN classifier based on LPD with filtering

K	Error	Avg. time per image	Overall time
3	2, 73%	2, 2 seconds	6 hours
10	1, 78%	2, 6 seconds	7 hours
30	1, 45%	3, 7 seconds	10 hours
50	1, 38%	4, 8 seconds	13 hours
100	1, 26%	7, 6 seconds	21 hours
200	1, 15%	13, 2 seconds	36 hours
500	1, 09%	30, 5 seconds	84 hours
1000	1, 05%	58, 8 seconds	162 hours

5.2 k-NN with Filtering

In [7] LPD was tested on small subsets of the MNIST dataset using a k-NN based on LPD model. The obtained error rates are 7, 89% on 300 images and 4, 59% on 1000 images, improving the baseline k-NN error rates by 8, 92% and 8, 41%, respectively. These results look promising, but LPD should be tested on the entire MINST dataset for a strong conclusion. The problem of the k-NN classifier based on LPD is that it is not feasible, from a time perspective, for very large datasets. To avoid this problem, local learning methods that speed-up the learning algorithm are used. Therefore, LPD is plugged into different local learning methods and used for image classification.

The first local learning classifier tested here is the k-NN with filtering model. This classifier was chosen because it reflects the characteristics of the LPD measure. For the k-NN with filtering approach, the idea is to filter the nearest K neighbors using the standard euclidean distance measure (that is much faster to compute). Next, select the nearest k neighbors from those K images using LPD. The two-step selection process is much faster to compute on a large dataset (such as the MNIST dataset) than a standard k-NN based entirely on LPD. Results show that this method can improve accuracy and is among the top 4 k-NN models that reported results on the MNIST dataset.

The k-NN based on LPD with filtering is compared with two other k-NN classifiers. One is the k-NN based on the euclidean distance measure (L^2-norm) between input images. This is the baseline classifier. In [11], an error rate of 5, 00% was reported on the regular test set with $k = 3$ for this classifier. Other studies [16] report an error rate of 3, 09% on the same experiment. Results obtained in this work also show an error rate of 3, 09% on this baseline experiment. The second classifier is the k-NN based on Tangent Distance [14]. Tangent distance (TD) is insensitive to small distortions and translations of the input image. The error rate reported for this classifier in [11] is 1, 1%, but it was necessary to preprocess the images by subsampling to 16×16 pixels to obtain this error rate.

The accuracy of the k-NN based on LPD depends very much on the filtering. Take into account that the nearest K images are selected using the euclidean distance in the filtering phase. If K is close to k, a very fast classifier is obtained, but its accuracy will be near the baseline k-NN. As K increases the accuracy improves, but the method also becomes slower (since it has to compute LPD between more images than it was before). If K is equal to the number of training examples, there is no filtering at all and the highest accuracy is obtained. However, the time to compute k-NN based on LPD with no filtering is too high to be considered. The idea is to choose an optimal K in order to obtain a trade-off between accuracy and time. Table 1 shows the error rate and the execution time of the 3-NN classifier based on LPD with filtering for several K values. The time was measured on a computer with Intel Core Duo 2.26 GHz processor and 4 GB of RAM memory using a single Core. Empirical results show that an optimal K would be somewhere between 100 and 500. When K is 100, approximately 8 seconds are needed to assign a label to a test image. This is reasonable for a real-time application, since adding parallel processing into assigning a label can improve the time even further.

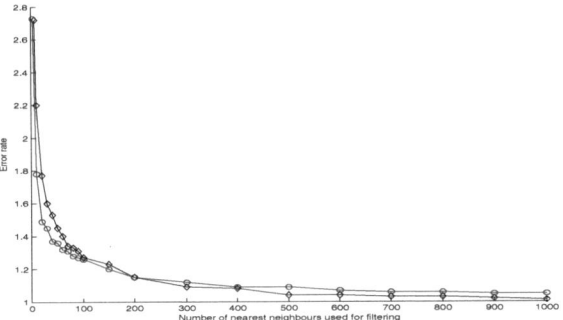

Fig. 1. Error rate drops as K increases for 3-NN (∘) and 6-NN (⋄) classifiers based on LPD with filtering

Looking at how the error gets lower as K increases, one can observe that the error tends to stabilize at some point. In other words, the error rate will not drop anymore after a certain K value. That error rate represents the actual error rate of a 3-NN classifier based on LPD (without filtering). The limitation of LPD induces this error, but what exactly is this error rate? Figure 1 gives an overview of this phenomenon and a hint about the point where the error stabilizes for both 3-NN and 6-NN classifiers. In order to make a prediction about the stabilization point of the error, the stability of the k-NN with filtering needs to be studied as K varies. It is clear from Figure 1 that the error drops with no variation when K increases. But increasing K may induce a misclassification on some test images (even if, overall, more images are classified correctly). For $K = 100, 200, ..., 1000$, the set of misclassified images for a certain value of K always includes the set of misclassified images for a greater value of K. Thus, the method has very stable behavior as K varies. In these circumstances, the error rate of the 3-NN and 6-NN classifiers based on LPD (without filtering) can be obtained by testing it only on the previously misclassified images. Doing so, an error rate of $1,03\%$ is obtained for the 3-NN classifier and an error rate of $0,98\%$ is obtained for the 6-NN classifier. These error rates are based only on a statistical proof. But as stated before, the real proof (that of testing the 3-NN and 6-NN classifiers with no filtering on all test images) is not practical from the time perspective.

Table 2. Error on MNIST dataset for baseline 3-NN, k-NN based on tangent distance (TD) and k-NN based on LPD with filtering

Method	baseline 3-NN	k-NN + TD	3-NN + LPD + Filter	6-NN + LPD + Filter
Error	$3,09\%$	$1,1\%$	$1,05\%$	$\mathbf{1,01\%}$

The obtained error rates only give a good indication of the actual ones, and they are not compared with the error rates of the other k-NN classifiers based on euclidean and tangent distance, respectively.

Table 2 compares error rates of the three k-NN classification methods (distinct only by the metric used) using the MNIST test set of 10.000 examples. For the k-NN classifier based on LPD with filtering, $k = 3$ and $k = 6$ are used. The reported results use the nearest $K = 1000$ images filtered with the euclidean distance. Note that the error rates of the k-NN models based on LPD are obtained with patches of 4×4 pixels and a similarity threshold of $0, 125$. These parameters are obtain through 10-fold cross validation on 1000 images selected from the MNIST dataset. The k-NN based on LPD with filtering model has a better accuracy than the other k-NN models. In fact, it is among the top 4 k-NN models that reported results on the MNIST dataset. The 6-NN classifier based on LPD with filtering has an error rate of only $1,01\%$. Note that unlike the k-NN with filtering based on LPD, the other three methods from top 4 need additional preprocessing steps.

5.3   Local Learning

The idea of combining LPD with kernel methods (such as SVM or kernel Ridge Regression) was presented in [7]. The reported error rates for the SVM based on LPD are $5, 67\%$ on 300 images and $3, 10\%$ on 1000 images, improving the standard SVM error rates by $8, 00\%$ and $5, 60\%$, respectively. The kernel Ridge Regression (kRR) based on LPD has an error rate of $6, 96\%$ on 300 images and an error rate of $3, 84\%$ on 1000 images, improving again the standard kRR error rates by $6, 85\%$ and $4, 58\%$, respectively. The accuracy improves when there are more training samples.

Kernel methods are based on similarity. LPD can be transformed into a similarity measure using the Gaussian-like kernel [13]:

$$K_{LPD}(img^1, img^2) = e^{-\frac{LPD(img^1, img^2)}{2\sigma^2}},$$

where img^1 and img^2 are two gray-scale images. The parameter σ is usually chosen to match the number of features (pixels) so that values of K_{LPD} are well scaled. Because LPD is computationally heavy, it is not feasible to compute a kernel matrix with high dimensions. Therefore, classifiers such as SVM and kRR based on LPD cannot be used directly on the entire MINST dataset, for example. Instead, kernel methods can be integrated into a local learning algorithm. The proposed approach is very similar to the k-NN with filtering model. The first step is to filter the nearest K neighbors using the standard euclidean distance. The second step is to train a kernel classifier using only the filtered K neighbors.

A new classifier will be trained for each test image. Each classifier will predict only the label of the test image that was build for.

Before training the classifier, it is necessary to build a kernel matrix using LPD. It is not feasible for a real-time application to build a $K \times K$ kernel matrix for each test image, when K is larger than 100. But, a large number of K neighbors is needed in order to improve the accuracy of the kernel method. A feasible solution is to train a kernel classifier only when there is not a majority of images with the same label greater than 60% among the filtered K neighbors, and use a k-NN model otherwise. There are two reasons for this approach. First, if there is such a majority, the kernel method will be biased towards choosing the majority class. Second, it is likely that a simple k-NN would also choose this majority class that is probably the right one.

This local learning algorithm was tested on the MNIST dataset. Using $K = 200$, it gives an error rate of $1,07\%$. From the 10.000 test cases, 983 of them had a majority class of less than 60% of the total 200 neighbors. For each of these 983 cases, a kRR classifier based on LPD was trained. The other test labels were predicted using the 3-NN based on LPD. Note that the 3-NN with filtering approach has an error rate of $1,15\%$ for $K = 200$ and the local learning algorithm is able to improve it (to $1,07\%$). Another local learning algorithm, that uses the SVM classifier instead of kRR classifier, was tested without being able to improve the accuracy.

In conclusion, using kernel methods in a local learning context doesn't bring a significant improvement in accuracy to the k-NN with filtering approach. However, the local learning algorithm proposed here can be successfully used for handwritten digit recognition. It also benefits from a faster computational time compared to standard kernel methods based on LPD.

5.4   Skip Overlapping Patches

In this experiment, LPD is used to classify a more general type of images available in the Birds dataset. Several k-NN models based on different distance measures are compared. The first k-NN model uses the Bhattacharyya coefficient to compare spatiograms of HSV values extracted from images. This improved measure of comparing spatiograms is proposed in [5]. The second k-NN model is based on the mean euclidean distance measure (L_1-norm) between input images. Another two k-NN classifiers based on LPD are tested. One of them uses a slightly modified version of the LPD algorithm [7]. It determines similar patches using the mean euclidean distance corresponding to the L_1-norm. The other one uses a fast version of the LPD algorithm that skips half of the comparisons between patches.

For both the euclidean distance measure and the LPD measure, images from the Birds dataset need to have the same size in order to be compared. Thus, images are resampled to 100×100 pixels. To observe the difference between original and resampled images, the k-NN model based on the Bhattacharyya coefficient is tested on both types of images. The other k-NN models are tested only on the resampled images, that are also converted to gray-scale. Table 3 compares error rates of all k-NN models.

Table 3. Error on Birds dataset for different k-NN models

Method	Preprocessing	Error
5-NN + Bhattacharyya	none	$57, 33\%$
5-NN + Bhattacharyya	resize	$54, 67\%$
3-NN + euclidean	resize, gray-scale	$51, 00\%$
3-NN + LPD	resize, gray-scale	$\mathbf{30, 33\%}$
3-NN + LPD + skip	resize, gray-scale	$\mathbf{30, 33\%}$

Table 4. Error on Birds dataset for texton learning methods and kernel methods based on LPD

Method	Naive Bayes	Exp+parts	Exp+rel	Exp+parts+rel	SVM+LPD+skip	kRR+LPD+skip
Error	$21, 33\%$	$\mathbf{7, 67\%}$	$24, 67\%$	$8, 33\%$	$27, 33\%$	$26, 00\%$

Next, two kernel methods based on the fast version of LPD (that skips half of the comparisons between patches) are compared with the state of the art texton learning methods from [10]. The proposed kernel methods based on LPD are the SVM and the kernel Ridge Regression. Table 4 compares error rates of kernel methods based on LPD and texton learning methods. The kernel methods based on LPD have an accuracy similar to some of the state of the art methods. However, the proposed kernel methods are more time efficient. The authors of [10] report a time of about 7 days for a single experiment, while the kernel methods based on LPD need about 4 days for the same experiment on an Intel Core Duo 2.26 GHz processor and 4 GB of RAM memory. Note that error rates of all models based on LPD, used in this experiment, are obtained with patches of 4×4 pixels and a similarity threshold of $0, 15$. In conclusion, the fast version of LPD can successfully be used as a kernel for image classification.

6   Conclusion and Further Work

This work presented several methods of speeding up the LPD algorithm, showing that LPD can be used for real-time image classification, especially when local learning methods are preferred instead of standard machine learning algorithms. Local learning methods based on LPD were proposed and tested on the popular MNIST dataset. The experiments show that local learning algorithms perform very well in terms of accuracy and time. The error rate achieved by the k-NN based on LPD with filtering approach is only $1, 01\%$ on the MNIST dataset.

Another important achievement of this work is that LPD has successfully been used for classifying images other than handwritten digits. The kRR based on LPD obtained an error rate of $26, 00\%$ on the Birds dataset, which is comparable to the results obtained in [10]. In conclusion, LPD is a powerful dissimilarity measure that can be used with good results in important problems in image processing, such as object classification, object recognition or similar tasks.

In future work, a very fast version of LPD based on image descriptors can be proposed. Instead of comparing the similarity between many patches, LPD can compare the similarity of a few SIFT descriptors, for example. This approach would bring a major improvement in terms of speed. Until further investigation, it remains uncertain if such an approach can have similar or more accurate results than the current formulation of LPD. A hint could be that LPD somehow measures the difference of spatial information between images. Can this difference still be measured using fewer image descriptors?

References

1. Agarwal, S., Roth, D.: Learning a sparse representation for object detection. In: Heyden, A., Sparr, G., Nielsen, M., Johansen, P. (eds.) ECCV 2002, Part IV. LNCS, vol. 2353, pp. 113–127. Springer, Heidelberg (2002)
2. Barnes, C., Goldman, D.B., Shechtman, E., Finkelstein, A.: The PatchMatch Randomized Matching Algorithm for Image Manipulation. Communications of the ACM 54(11), 103–110 (2011)
3. Bottou, L., Vapnik, V.: Local Learning Algorithms. Neural Computation 4, 888–900 (1992)
4. Cho, T.S., Avidan, S., Freeman, W.T.: The patch transform. PAMI 32(8), 1489–1501 (2010)
5. Conaire, C.O., O'Connor, N.E., Smeaton, A.F.: An Improved Spatiogram Similarity Measure for Robust Object Localisation. In: ICASSP, pp. 1069–1072 (2007)
6. Deselaers, T., Keyser, D., Ney, H.: Discriminative Training for Object Recognition using Image Patches. In: CVPR, pp. 157–162 (2005)
7. Dinu, L.P., Ionescu, R.-T., Popescu, M.: Local patch dissimilarity for images. In: Huang, T., Zeng, Z., Li, C., Leung, C.S. (eds.) ICONIP 2012, Part I. LNCS, vol. 7663, pp. 117–126. Springer, Heidelberg (2012)
8. Efros, A.A., Freeman, W.T.: Image quilting for texture synthesis and transfer. In: SIGGRAPH 2001, pp. 341–346. ACM (2001)
9. Guo, G., Dyer, C.R.: Patch-based Image Correlation with Rapid Filtering. In: CVPR (2007)
10. Lazebnik, S., Schmid, C., Ponce, J.: A Maximum Entropy Framework for Part-Based Texture and Object Recognition. In: ICCV, pp. 832–838 (2005)
11. LeCun, Y., Bottou, L., Bengio, Y., Haffner, P.: Gradient-based learning applied to document recognition. Proceedings of the IEEE 86(11), 2278–2324 (1998)
12. Lowe, D.G.: Object Recognition from Local Scale-Invariant Features. In: ICCV, vol. 2, pp. 1150–1157. IEEE Computer Society, Washington, DC (1999)
13. Shawe-Taylor, J., Cristianini, N.: Kernel Methods for Pattern Analysis. Cambridge University Press (2004)
14. Simard, P., LeCun, Y., Denker, J.S., Victorri, B.: Transformation Invariance in Pattern Recognition, Tangent Distance and Tangent Propagation. Neural Networks: Tricks of the Trade (1996)
15. Vapnik, V.: Estimation of dependencies based on empirical data (Information Science and Statistics), 2nd edn. Springer (2006)
16. Wilder, K.J.: Decision tree algorithms for handwritten digit recognition. Electronic Doctoral Dissertations for UMass Amherst (January 1998), http://scholarworks.umass.edu/dissertations/AAI9823791
17. Xie, J., Zhang, L., You, J., Zhang, D.: Texture classification via patch-based sparse texton learning. In: ICIP, Hong Kong, China, pp. 2737–2740 (2010)

A Graph-Based Hierarchical Image Segmentation Method Based on a Statistical Merging Predicate⋆

Silvio Jamil F. Guimarães and Zenilton K.G. Patrocínio Jr.

Audio-Visual Information Proc. Lab. (VIPLAB)
Computer Science Department – ICEI – PUC Minas
{sjamil,zenilton}@pucminas.br

Abstract. Hierarchical image segmentation provides a set of image segmentations at different detail levels in which coaser details levels can be produced by simple merges of regions from segmentations at finer detail levels. Most image segmentation algorithms, such as region merging algorithms, rely on a criterion for merging that does not lead to a hierarchy. In addition, for image segmentation, the tuning of the parameters can be difficult. In this work, we propose a hierarchical graph-based image segmentation relying on a statistical region merging. Furthermore, we study how the inclusion of hierarchical property have influenced the computation of quality measures in the original method. Quantitative and qualitative assessments of the method on two image databases show efficiency and ease of use of our method.

Keywords: hierarchical segmentation, vertex-edge-weighted graph, statistical region merging predicate.

1 Introduction

Image segmentation is the process of grouping perceptually similar pixels into regions. A hierarchical image segmentation is a set of image segmentations at different detail levels in which the segmentations at coarser detail levels can be produced from simple merges of regions from segmentations at finer detail levels. Therefore, the segmentations at finer levels are nested with respect to those at coarser levels. Hierarchical methods have the interesting property of preserving spatial and neighboring information among segmented regions. Here, we propose a hierarchical image segmentation in the framework of vertex-edge-weighted graphs, where the image is equipped with an adjacency graph, the cost of an edge is given by a dissimilarity between two points of the image and the cost of a vertex is the color information of the associated point. Therefore, the adjacency graph is represented by data structures in order to efficiently compute this hierarchy.

⋆ The authors are grateful to PUC Minas – Pontifícia Universidade Católica de Minas Gerais, CNPq, CAPES and FAPEMIG for the financial support of this work.

A. Petrosino (Ed.): ICIAP 2013, Part I, LNCS 8156, pp. 11–20, 2013.

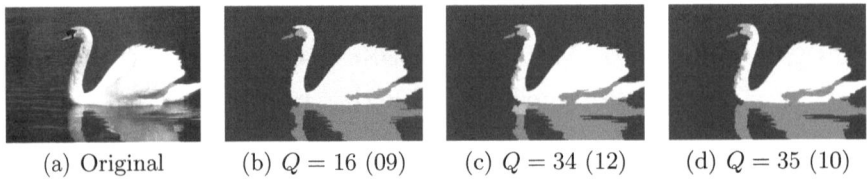

| (a) Original | (b) $Q = 16$ (09) | (c) $Q = 34$ (12) | (d) $Q = 35$ (10) |

Fig. 1. A real example illustrating the violation of the causality principle by [12]: the number of regions (in parentheses) is not monotonic, when the so-called "segmentation scale" increases

Any hierarchy can be represented with a minimum spanning tree. The first appearance of this tree in pattern recognition dates back to the seminal work of Zahn [13]. Lately, its use for image segmentation was introduced by Morris *et al.* [10] in 1986 and popularized in 2004 by Felzenszwalb and Huttenlocher [5], Noch and Nielsen [12] proposed a statistical method in which the merging order is similar to the creation of a MST. However the region-merging method [5, 12] does not provide a hierarchy. In [4, 11], it was studied some optimality properties of hierarchical segmentations. Considering that, for a given image, one can tune the parameters of the well-known method [5] for obtaining a reasonable segmentation of this image. In [8], we presented a framework to transform the non-hierarchical method proposed by [5] into its hierarchical version, the scale is computed by a maximum between two independent scales. Unfortunately, this methodology can not be directly applied to other methods. For example, the scales computed by the method proposed in [12], here-after called SRG, are dependent on a relation between two regions. Here, we provide a hierarchical version of this method that removes the need for parameter tuning.

Even if the image segmentation results obtained by the method proposed in [12] are interesting, the user faces two major issues:

- first, the number of regions is not monotonic when the parameter k increases. This should not be possible if k was a true scale: indeed, it violates the *causality principle* of multi-scale analysis, that states in our case [6] that a contour present at a scale k_1 should be present at any scale $k_2 < k_1$. Such unexpected behavior of missing causality principle is demonstrated on Fig. 1.
- Second, even when the number of regions decreases, contours are not stable: they can move when the parameter k varies, violating a *location principle*. Such a situation is illustrated on Fig. 2.

Given these two issues, the tuning of the parameters of [5, 12] is a difficult task.

Following [6], we believe that, in order for k to be a true scale-parameter, we have to satisfy both the causality principle and the location principle, which leads to work with a hierarchy of segmentations. Reference [9] is the first to propose an algorithm producing a hierarchy of segmentations based on [5]. However, this method is an iterative version of [5] that uses a threshold function, and requires a tuning of the threshold parameter, therefore this methodology can easily be extended to other methods. Reference [8] is the first to propose a method which

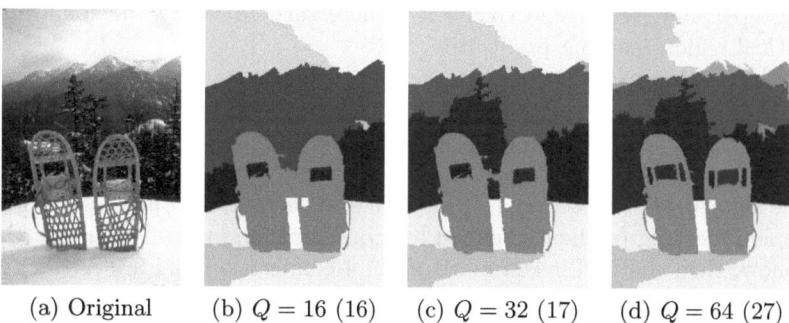

(a) Original   (b) $Q = 16$ (16)   (c) $Q = 32$ (17)   (d) $Q = 64$ (27)

Fig. 2. A real example illustrating the violation of the location principle by [12]: the contour location is not stable for two different "segmentation scales"

computing all hierarchy using the same dissimilarity measure of [5]. However, the extension of this methodology for other methods is not a trivial task.

The main result of this paper is an efficient hierarchical image segmentation algorithm based on the dissimilarity measure of [12]. Our algorithm has a computational cost similar to [12], but provides all statistical scales instead of only one segmentation level. As it is a hierarchy, the result of our algorithm satisfies both the locality principle and the causality principle. Namely, and in contrast with [12], the number of regions is decreasing when the scale parameter increases, and the contours do not move from one scale to another. Therefore, [12] produces under-segmented regions and our approach produces over-segmented regions since small regions may have high hierarchical values.

This work is organized as follows. In Section 2, we present the statistical predicate for merging two regions. In Section 3, we present our hierarchical method for color image segmentation. Some experimental results performed on a image database are given in Section 4. Finally, in Section 5, some conclusions are drawn and further works are discussed.

2   A Statistical Region Merging Predicate

Noch and Nielsen [12] proposed an efficient method for image segmentation based on a statistical predicate. Even if the statistical theory is not easy, the method is simple coping with noise corruption and handle occlusion. To follow, we describe this method.

Let I be an image, $|I|$ be its number of pixels, Q and g be the new and old number of expected different color values. Let $\mathcal{R}_l$ be the set of regions with l pixels and $b(R) = g\sqrt{(1/(Q|R|))ln(|\mathcal{R}_{|\mathcal{R}|}|/\delta)}$ in which $\delta = \frac{1}{(6|I|^2)}$. Thus,

$$b(R, R') \leq \sqrt{b^2(R) + b^2(R')} < b(R) + b(R') \tag{1}$$

According to [12], a region is an unordered bag of pixels, thus we may fix $|\mathcal{R}_l| \leq (l+1)^{\min\{l,g\}}$. For merging two adjacent regions R and R' with color average $|R|$ and $|R'|$, respectively, it is necessary to verify the following the statistical region merging predicate

$$\mathcal{P}(R, R') = \begin{cases} \text{true} & \text{if } |\overline{R} - \overline{R'}| \leq \sqrt{b^2(R) + b^2(R')} \\ \text{false} & \text{otherwise} \end{cases} \tag{2}$$

The algorithm proposed by [12] for image segmentation considers the statistical region merging predicate defined by Eq. 2 for merging two adjacent regions. To follow, their method is summarized. In 4-connexity, there are $N < 2|I|$ couples of adjacent pixels. Let S_f be the set of these couples. Let $f(p, p')$ be a real-valued function, with p and p' pixels of f. First, the couples of S_f are sorted in increasing order of $f(.,.)$, and then traverse this order only once. For any current couple of pixels p and p' that does not in the same region, the test $\mathcal{P}(R, R')$, and merge R and R' iff it returns true.

Considering that the parameter Q controls the coarseness (or "scale") of a segmentation, it is important to understand the influence of this parameter in the process. Thus, we re-write the Eq. 2 in order to explicit the parameter Q:

$$(\overline{R} - \overline{R'})^2 \leq b^2(R) + b^2(R') \tag{3}$$

$$(\overline{R} - \overline{R'})^2 \leq g^2 \frac{1}{(Q|R|)} ln \frac{|\mathcal{R}_{|R|}|}{\delta} + g^2 \frac{1}{(Q|R'|)} ln \frac{|\mathcal{R}_{|R'|}|}{\delta} \tag{4}$$

$$(\overline{R} - \overline{R'})^2 \leq g^2 \frac{|R'|}{(Q|R||R'|)} ln \frac{|\mathcal{R}_{|R|}|}{\delta} + g^2 \frac{|R|}{(Q|R'||R|)} ln \frac{|\mathcal{R}_{|R'|}|}{\delta} \tag{5}$$

$$Q \leq \frac{g^2}{(\overline{R} - \overline{R'})^2} \left(\frac{|R'|}{(|R||R'|)} ln \frac{|\mathcal{R}_{|R|}|}{\delta} + \frac{|R|}{(|R'||R|)} ln \frac{|\mathcal{R}_{|R'|}|}{\delta} \right) \tag{6}$$

Thanks to upperbound of $|\mathcal{R}_l|$, the Eq. 6 can be re-written by

$$Q \leq \frac{g^2}{(\overline{R} - \overline{R'})^2 |R||R'|} (|R'|\ m\ (l + d) + |R|\ m'\ (l' + d)) \tag{7}$$

in which $m = \min\{g, |R|\}$, $m' = \min\{g, |R'|\}$, $l = ln(1 + |R|)$, $l' = ln(1 + |R'|)$ and $d = 2 \times ln(6|I|)$.

We can see in Fig. 2 and Fig. 1 the segmentation results for fixed values of parameter Q (which controls the so-called *segmentation scale*), however the location and causality principle are missing, and this shows the absence of hierarchical properties of this method. In fact, we will look for to adapt the values of Q according to the regions analyzed for guaranteeing that two regions are correctly merged. First, since Q could easily be limited by the number of pixels in the image, i.e., $Q \leq |I|$, we consider in our analysis $Q' = |I| - Q$ because minimizing Q' will correspond to maximizing Q (we search for the smallest scale value for which two regions can be merged).

Then, the *scale* $Q(R, R')$ is defined as:

$$Q(R, R') = |I| - \frac{g^2}{(\overline{R} - \overline{R'})^2 |R||R'|} (|R'|\ m\ (l + d) + |R|\ m'\ (l' + d)). \tag{8}$$

Thanks to this notion of a scale, Eq. (7) can be written as:

$$Q' \geq Q(R, R'). \tag{9}$$

3   A Hierarchical Graph Based Image Segmentation

In this section, we describe our method to compute a hierarchy of partitions based on scales, so-called here *statistical scale*, as defined by Eq. 7. This method is based on our previous work [7, 8], however here it is necessary to compute all candidate values to represent a scale once, differently of [8], the merging depends on both, the information of the regions and the distance between these regions. Let us first recall some important notions for handling hierarchies [4, 10, 11].

To every tree T spanning the set V of the image pixels, to every map $w : E \rightarrow \mathbb{N}$ that weights the edges of T and to every threshold $\lambda \in \mathbb{N}$, one may associate the partition $\mathcal{P}_\lambda^w$ of V induced by the connected components of the graph made from V and the edges of weight below λ. It is well known [4, 10] that for any two values λ_1 and λ_2 such that $\lambda_1 \geq \lambda_2$, the partitions $\mathcal{P}_{\lambda_1}^w$ and $\mathcal{P}_{\lambda_2}^w$ are *nested* and $\mathcal{P}_{\lambda_1}^w$ is *coarser* than $\mathcal{P}_{\lambda_2}^w$. Hence, the set $\mathcal{H}^w = \{\mathcal{P}_\lambda^w \mid \lambda \in \mathbb{N}\}$ is a *hierarchy of partitions induced by the weight map w*.

Our algorithm does not explicitly produce a hierarchy of partitions, but instead produces a weight map L (scales of statistical values) from which the desired hierarchy $\mathcal{H}^L$ can be inferred on a given T. It starts from a minimum spanning tree T of the vertex-edge-weighted graph built from the image. In order to compute the scale $L(e)$ associated with each edge of T, our method iteratively considers the edges of T in a non-decreasing order of their original weights w. For every edge e, the new weight map $L(e)$ is initialized to $|I|$; then for each edge e linking two vertices p and p' the following steps are performed:

(i) Find the region R of $\mathcal{P}_{w(e)}^w$ that contains p.
(ii) Find the region R' of $\mathcal{P}_{w(e)}^w$ that contains p'.
(iii) Compute the hierarchical statistical scale $L(e) = Q(R, R')$.

At step (iii), the *hierarchical statistical scale* $L(e)$ is computed based on a minimization of the function which relates some sub-region of R to some sub-region of R'. More precisely, using an internal parameter v, this scale is computed as follows:

(1) Initialize the value of v to $|I|$.
(2) Decrement the value of v by 1.
(3) Find the region R_* of $\mathcal{P}_v^L$ that contains p.
(4) Find the region R'_* of $\mathcal{P}_v^L$ that contains p'.
(5) Repeat steps 2, 3 and 4 while $Q(R_*, R'_*) < v$
(6) Set $Q(R, R') = v$.

As the image is represented by a vertex-edge-weighted graph and the dissimilarity measures is based on color information of the regions, the computation of color average for each region can be done by computation of average cost of the region vertices.

To efficiently implement our method, we use some data structures similar to the ones proposed in [4, 8]; in particular, the management of the collection of partitions is due to Tarjan's union find and Fredman and Tarjan's Fibonnacci heaps. Furthermore, we made some algorithmic optimizations to speed up the computations of the statistical scales.

4   Experiments

To provide a basis of comparison for our hierarchical method with respect to its non-hierarchical version, we consider the dataset used in [1, 2] which contains clearly one or two objects. This dataset is divided into two groups containing 100 images each one: single and two objects, this dataset is so-called **Object data set**. According to [1, 2] the database was designed to contain a variety of images with objects that differ from their surroundings by either intensity, texture, or other low level cues. To avoid potential ambiguities it was selected images that clearly depict one object or two objects in the foreground. Some experiments are also given using Berkeley Segmentation DataSet [3], so-called **BSDS500**. In this database, a semantic segmentation is done, and sometimes, some images are under-segmented. With respect to the quality measures, we assess our method using measures presented in [3] (Ground-Truth Segmentation, Probabilistic Rand Index and Variation Information) and measures used in [1, 2] (F1-measure). The aim here is to show how the inclusion of hierarchical property will influence on the results obtained by the method without this property.

In Table 1 and 2, we present some results, represented by F-measures, obtained when we apply the method to Object Data set. As we can observe, the F-measure calculated for the results obtained by hSRG are much better than the

Table 1. Performances of our methods, called hSRG and hFH, when compared to their non-hierarchical method, SRG and FH, respectively. The performances showed in [2] are also presented. The performances are measures by F-measure for the Single-Object Data Set: (i) single segment coverage; and (ii) fragmented Coverage. See [2] for more details on the evaluation method.

Algorithm	Single	Multi	
	F-measure	F-measure	Fragmentation
Proba	0.86	0.87	2.66
SWA V1	0.83	0.89	3.92
SWA V2	0.76	0.86	3.71
Gpb	0.54	0.88	8.20
Mean Shift	0.57	0.88	12.08
NCut	0.72	0.84	3.12
SRG	0.58	0.56	8.53
hSRG	0.64	0.88	7.83
FH	0.73	0.85	5.96
hFH	0.63	0.85	5.12

one obtained by SRG. For hFH and FH, the measures are quite similar. Thus, these results illustrate that the inclusion of hierarchical property has no prejudice with respect to the computed measures.

Table 2. Performances of our methods, called hSRG and hFH, when compared to their non-hierarchical method, SRG and FH, respectively. The performances showed in [2] are also presented. The performances are measures by F-measure for fragmented coverage for the Two-Object Data Set: (i) fragmented coverage for smaller object; and (ii) fragmented coverage for larger object. See [2] for more details on the evaluation method.

(a) Fragmented coverage

Algorithm	2 Objects		Larger object		Smaller object	
	Aver. F-measure	Aver. Frag.	Aver. F-measure	Aver. Frag.	Aver. F-measure	Aver. Frag.
Proba	0.85	1.67	0.87	2.00	0.84	1.33
SWA V1	0.88	3.13	0.91	3.88	0.84	2.37
SWA V2	0.85	2.27	0.88	2.76	0.82	1.77
Gpb	0.84	2.95	0.87	3.60	0.81	2.30
Mean Shift	0.78	3.65	0.85	4.49	0.71	2.81
NCut	0.84	2.64	0.88	3.34	0.80	1.93
SRG	0.60	1.90	0.60	1.68	0.60	2.12
hSRG	0.87	4.96	0.88	4.77	0.87	5.15
FH	0.88	5.86	0.89	6.91	0.87	3.81
hFH	0.86	4.20	0.89	5.60	0.84	2.81

(b) Single coverage

Algorithm	2 Objects	Larger object	Smaller object
Proba	0.68	0.70	0.65
SWA V1	0.66	0.74	0.57
SWA V2	0.61	0.71	0.50
Gpb	0.72	0.70	0.75
Mean Shift	0.61	0.65	0.58
NCut	0.58	0.66	0.49
SRG	0.58	0.62	0.55
hSRG	0.75	0.74	0.76
FH	0.79	0.78	0.80
hFH	0.75	0.76	0.75

In Table 3, we present the quality measures when we apply four segmentation methods to the object dataset: two non-hierarchical methods (SRG and FH) and their hierarchical versions, called hSRG and hFH). In this experiment, we compute all measures for the datasets containing one object and two objects, we also compute the quality measures considering that we search only the smaller or larger objects into the two-object dataset. As we can see, these results are quite similar.

Table 3. Performances of hierarchical methods, called hSRG and hFH [8], and the non-hierarchical methods SRG [2] and FH [5] using three different measures applied to **Object Data Set**: (a) Ground-truth Covering (GT Covering), (b) Probabilistic Rand Index and (c) Variation Information. The presented scores are optimal considering a constant scale parameter for the whole dataset (ODS) and a scale parameter varying for each image (OIS), and a scale parameter varying for each region of the ground-truth (Best). See [3] for more details on the evaluation method.

(a) Ground-truth Covering (GT Covering)

	Smaller			Large			2 obj			1 obj		
	ODS	OIS	Best	ODS	OIS	Best	ODS	OIS	Best	ODS	OIS	Best
SRG	0.82	0.83	0.85	0.80	0.83	0.85	0.78	0.85	0.87	0.67	0.73	0.75
hSRG	0.81	0.84	0.88	0.80	0.84	0.88	0.73	0.80	0.84	0.65	0.73	0.76
FH	0.81	0.84	0.88	0.80	0.83	0.87	0.73	0.83	0.87	0.66	0.73	0.77
hFH	0.82	0.84	0.89	0.82	0.85	0.88	0.75	0.83	0.86	0.63	0.67	0.70

(b) Probabilistic Rand Index

	Smaller		Large		2 obj		1 obj	
	ODS	OIS	ODS	OIS	ODS	OIS	ODS	OIS
SRG	0.80	0.81	0.79	0.83	0.80	0.86	0.70	0.77
hSRG	0.80	0.83	0.80	0.83	0.74	0.83	0.68	0.78
FH	0.80	0.83	0.78	0.82	0.73	0.84	0.68	0.77
hFH	0.81	0.84	0.81	0.84	0.76	0.86	0.65	0.75

(c) Variation Information

	Smaller		Large		2 obj		1 obj	
	ODS	OIS	ODS	OIS	ODS	OIS	ODS	OIS
SRG	0.60	0.57	0.63	0.60	0.69	0.57	0.98	0.91
hSRG	0.57	0.54	0.58	0.55	0.84	0.72	0.85	0.80
FH	0.55	0.52	0.59	0.55	0.77	0.60	0.96	0.87
hFH	0.56	0.56	0.54	0.52	0.77	0.64	0.92	0.89

(a)   (b)

Fig. 3. Examples of application of SRG and hSRG methods to Object Dataset containing one object and two objects. In left column of (a) and (b), we present the original images, in middle column, we illustrate the results obtained from SRG method. In right column, we illustrate the results obtained from hSRG.

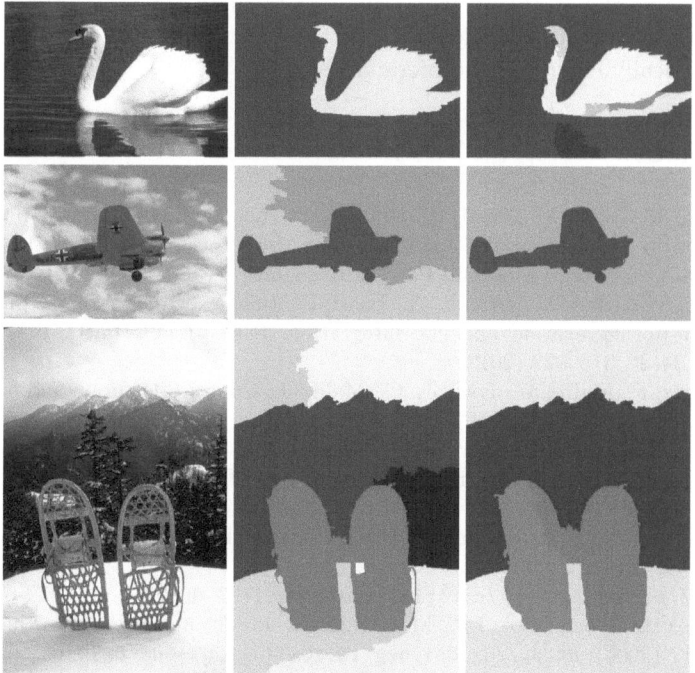

Fig. 4. Examples of application of SRG and hSRG methods to BSDS. In left column, we present the original images, in middle column, we illustrate the results obtained from SRG method. In right column, we illustrate the results obtained from hSRG.

In Fig. 3 and 4 we present some results of applications of SRG method and our hierarchical version. The two formers were extracted from Object Data Set, and the last one, is obtained from BSDS500. As we can see, the results are quite similar, however there are some differences, mainly, in second and third rows of Fig. 4, in second and third rows in Fig 3(b) and in first row of Fig. 3(a).

5   Conclusion and Further Works

In this work, we propose a method for transforming a non-hierarchical method into a hierarchical method preserving the merging criterium, *i.e.*, all regions are merged according to the same statistical criterium. Differently of the method that iteratively computes the hierarchies, our method produces a weight map L (scales of statistical values) from which the desired hierarchy can be easily inferred. According to our results, the inclusion of the hierarchical property on this region merging approach solves the causality and location problems which are missing in SRG method, and does not prejudice the quality of results, in fact, our method present higher f-measure values than SRG method. However, due to inclusion of hierarchical property, some small regions are present in high scale. To solve this problem, it is necessary to filter out this regions using an area opening. For future works, we will study the robustness to noise and how to automatically choice a good hierarchical scale.

References

1. Alpert, S., Galun, M., Basri, R., Brandt, A.: Image segmentation by probabilistic bottom-up aggregation and cue integration. In: Proceedings of the IEEE Conference on Computer Vision and Pattern Recognition (June 2007)
2. Alpert, S., Galun, M., Brandt, A., Basri, R.: Image segmentation by probabilistic bottom-up aggregation and cue integration. IEEE Trans. Pattern Anal. Mach. Intell. 34(2), 315–327 (2012)
3. Arbelaez, P., Maire, M., Fowlkes, C., Malik, J.: Contour detection and hierarchical image segmentation. PAMI 33, 898–916 (2011)
4. Cousty, J., Najman, L.: Incremental algorithm for hierarchical minimum spanning forests and saliency of watershed cuts. In: Soille, P., Pesaresi, M., Ouzounis, G.K. (eds.) ISMM 2011. LNCS, vol. 6671, pp. 272–283. Springer, Heidelberg (2011)
5. Felzenszwalb, P.F., Huttenlocher, D.P.: Efficient graph-based image segmentation. IJCV 59, 167–181 (2004),
   http://portal.acm.org/citation.cfm?id=981793.981796
6. Guigues, L., Cocquerez, J.P., Men, H.L.: Scale-sets image analysis. IJCV 68(3), 289–317 (2006), http://dx.doi.org/10.1007/s11263-005-6299-0
7. Guimarães, S.J.F., Cousty, J., Kenmochi, Y., Najman, L.: An efficient hierarchical graph based image segmentation. CoRR abs/1206.2807 (2012)
8. Guimarães, S.J.F., Cousty, J., Kenmochi, Y., Najman, L.: A hierarchical image segmentation algorithm based on an observation scale. In: Gimel'farb, G., Hancock, E., Imiya, A., Kuijper, A., Kudo, M., Omachi, S., Windeatt, T., Yamada, K. (eds.) SSPR & SPR 2012. LNCS, vol. 7626, pp. 116–125. Springer, Heidelberg (2012)
9. Haxhimusa, Y., Kropatsch, W.: Segmentation graph hierarchies. In: Fred, A., Caelli, T.M., Duin, R.P.W., Campilho, A.C., de Ridder, D. (eds.) SSPR&SPR 2004. LNCS, vol. 3138, pp. 343–351. Springer, Heidelberg (2004)
10. Morris, O., Lee, M.J., Constantinides, A.: Graph theory for image analysis: an approach based on the shortest spanning tree. Communications, Radar and Signal Processing IEE Proceedings F 133(2), 146–152 (1986)
11. Najman, L.: On the equivalence between hierarchical segmentations and ultrametric watersheds. JMIV 40, 231–247 (2011)
12. Nock, R., Nielsen, F.: Statistical region merging. IEEE Transactions on Pattern Analysis and Machine Intelligence 26(11), 1452–1458 (2004)
13. Zahn, C.T.: Graph-theoretical methods for detecting and describing gestalt clusters. IEEE Trans. Comput. 20, 68–86 (1971),
    http://dx.doi.org/10.1109/T-C.1971.223083

Application of Local Binary Pattern to Windowed Nonlocal Means Image Denoising

Fakhry Khellah

Prince Sultan University, Riyadh - 11586, Saudi Arabia
`fkhellah@cis.psu.edu.sa`

Abstract. This paper presents a new technique to image denoising that mainly addresses the incurred high blurring when the windowed nonlocal means is applied to images corrupted by high noise levels. The proposed method is based on an enhanced weighting function that computes patches similarity based on both their intensities and structural features. The structural features are encoded using Local Binary Pattern (LBP) a well known texture descriptors. A new LBP based weighting function is proposed that has properties complementing the intensity based weighting function. The LBP based weighting function is used to modulate the intensity based weighting function. The modulated weights are noise independent and reflect the actual patch similarity. The method is found to be quantitatively and qualitatively effective in denoising images when corrupted by high noise levels. It suppresses image noise while preserving significant image characteristics.

Keywords: Local Binary Patterns, Nonlocal means, Image Denoising.

1 Introduction

Image denoising is still one of the most fundamental areas of research in the field of image processing. Some of the spatial domain methods restore the intensity value of each image pixel by averaging in some way the intensities of its nearby neighbors. The major drawback of such techniques is that they blur the small scale structures in the image such as edges or textures. One of the introduced spatial domain methods that has been shown to produce high quality results is the nonlocal means filtering method proposed by Buades [2]. The nonlocal means algorithm is based on the idea that natural and textured images have redundancy and any image pixel has similar pixels that are not necessarily located in a nearby spatial neighborhood. In the classical definition of the nonlocal means filter, for each pixel being processed in the image, the whole image is searched and differences between corresponding neighborhoods (patches) are computed. In [2], the weights computation was limited to a local subdomain called search window centered at the pixel being processed. The method is called windowed nonlocal means filter. One main drawback of the nonlocal means filter is the obvious blurring of edges and fine texture details incurred to denoised images

A. Petrosino (Ed.): ICIAP 2013, Part I, LNCS 8156, pp. 21–30, 2013.

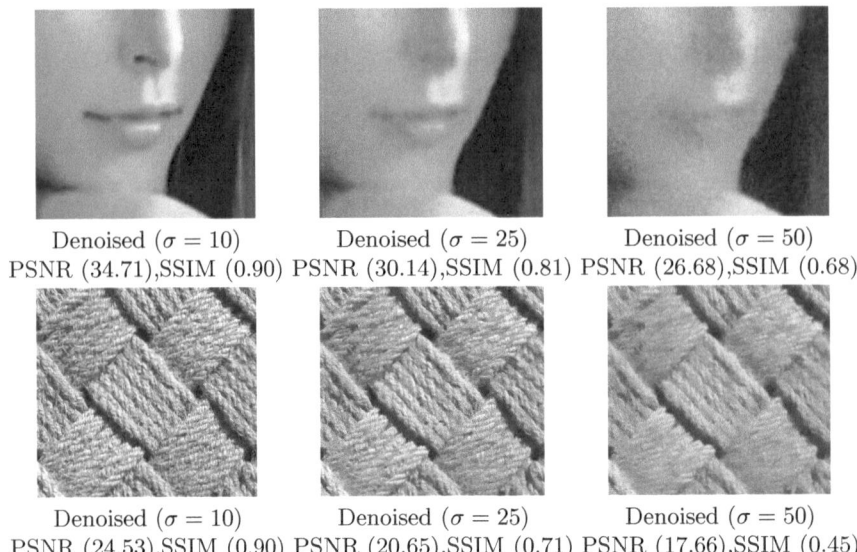

Denoised ($\sigma = 10$)   Denoised ($\sigma = 25$)   Denoised ($\sigma = 50$)

PSNR (34.71),SSIM (0.90) PSNR (30.14),SSIM (0.81) PSNR (26.68),SSIM (0.68)

Denoised ($\sigma = 10$)   Denoised ($\sigma = 25$)   Denoised ($\sigma = 50$)

PSNR (24.53),SSIM (0.90) PSNR (20.65),SSIM (0.71) PSNR (17.66),SSIM (0.45)

Fig. 1. Illustration of effect of the noise level on the windowed nonlocal means filter denoising quality. The smoothness increases as the noise level increases. The top row is zoomed Lena image, and the bottom row is zoomed D1 image.

especially at high noise levels. This is due to incorporating exaggerated weights that do not reliably reflect the actual patch similarity in the averaging of a given pixel. As a result, at high noise levels, the restored pixel will be approximately the average of the intensity values of almost all the pixels in the search window leading to strong smoothing of the image, and therefore, blurring of small scale structures of the image. Particularly, in textured regions of an image, small scale structures are not necessarily equivalent to noise. This is shown in Fig. 1 where the windowed nonlocal means filter is used to denoise two different types of images: a natural image; the standard Lena image, depicted in the top row of Fig. 1 and a highly textured image, (D1) depicted in the bottom row of the figure. The images were corrupted by three noise levels: ($\sigma = 10, 25, 50$). Observe the high blurring effect as the noise level increases.

In this work, we address the above drawback of the windowed nonlocal means filter by incorporating the structural features of the region where the restored pixel is located in the computation of its averaging weights.

The rest of the paper is organized as follows. In Section 2 an overview of the windowed nonlocal means image denoising method is presented. The proposed filtering method using is described in Section 3. Experimental results are presented and discussed in Section 4. Finally, conclusions are given in Section 5.

2 Overview of the Windowed Nonlocal Means Filter

In this section, we present a brief overview of the nonlocal means method introduced in [2]. Let the observed noisy intensity of pixel i be defined as: $I_n(i) = I_o(i) + n(i)$, where $I_o(i)$ is the original intensity of the noise-free pixel i and $n \sim \mathcal{N}(0, \sigma^2)$ is the observation noise. In the original formulation of the nonlocal means filter, the restored intensity $\hat{I}(i)$, is the weighted average of all pixel intensities in the noisy image $I_n(i)$ defined as:

$$\hat{I}(i) = \sum_{j \in I_n} w(i,j) I_n(j) \tag{1}$$

The weight $w(i,j)$ quantifies the similarity between two square local neighborhoods (patches) N_i and N_j centered on pixels i and j of predefined size $r \times r$ with typical sizes of 7×7 or 9×9. In order to reduce the computational complexity, the sum in (1) is restricted to pixels within a local subdomain around pixel i called search window S_i of size $M \times M$ with a typical size of 21×21 [1],[2],[3],[7]. This version of the nonlocal means filter is usually called semilocal or windowed nonlocal means. For each pixel i the weight in (1) is computed by:

$$w(i,j) = \frac{1}{Z(i)} e^{-\frac{\|v(N_i) - v(N_j)\|^2_{2,a}}{h^2}} \tag{2}$$

Where Z_i is a normalizing term ensuring that all computed weights will add to one, i.e., $\Sigma_j w(i,j) = 1$. The nonlocal means filter computes the intensity similarity between any two image pixels i and j based on their local neighborhoods N_i and N_j rather than their individual values. In the above equation, $v(N_i)$ is the vector of neighborhood pixel values around pixel i. The vector norm in (2) computes Euclidean distance weighted by a Gaussian kernel of standard deviation a between neighborhood intensities N_i and N_j.

The exponential weighting function in (1) is controlled mainly by two parameters: the intensity based similarity criterion and the smoothing kernel width parameter h; that controls the extent of averaging. Both parameters contributes directly to the filter performance degradation as the noise level increases.

The similarity criterion measures the similarity between two patches based on their intensities regardless of their underlying textured or structural characteristics such as curves, edges, corners, and spots. As a fact, two neighboring patches that are similar with respect to their measured intensities according to (1) are not necessarily similar with respect to their primitive textured structures. This is evident in texture images and in textured regions of natural images. Consequently, image details and fine texture can be oversmoothed due to incorporating dissimilar patches in the denoising of a given pixel.

The smoothing parameter h of the weighting function controls the extent of averaging and is usually set to the noise standard deviation $h = \sigma$ [2] regardless of the smoothness degree of the region where the search window S_i is located. Consequently, when the noise standard deviation is high, the smoothing parameter h will be also high and therefore, the exponential weighting function (2) is

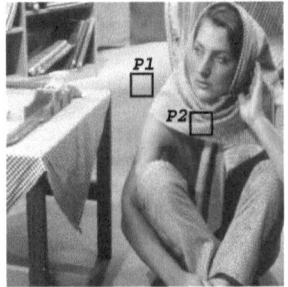

(a) Barbara image showing the location of two pixels $P1$ and $P2$ located in flat and textured regions, respectively.

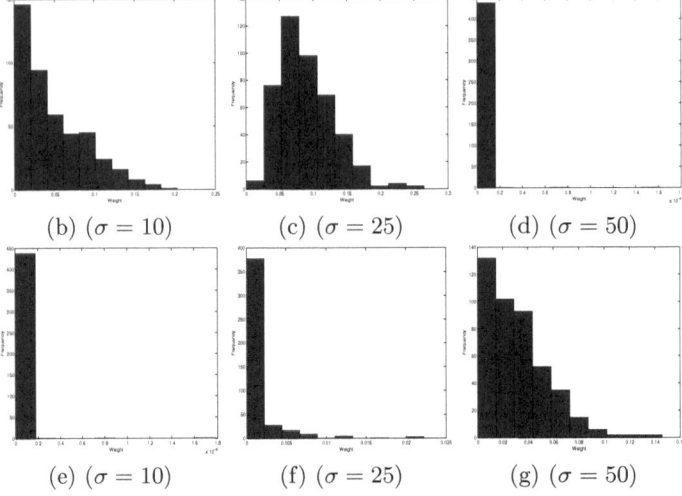

(b) $(\sigma = 10)$ (c) $(\sigma = 25)$ (d) $(\sigma = 50)$

(e) $(\sigma = 10)$ (f) $(\sigma = 25)$ (g) $(\sigma = 50)$

Fig. 2. Histograms of the intensity based weights for two Barbara image pixels $P1$ (top row), and $P2$ (bottom row) located in smooth and textured regions, respectively

mainly dominated by h. Therefore, the computed weights will not exactly reflect the similarity of the involved patches. Consequently, exaggerated weights will be assigned to almost all patches within the search window S_i. As a result, the restored value $\hat{I}(i)$ will be approximately the average of the intensity values of all the pixels in the search window S_i leading to strong smoothing of the image, and therefore, blurring of small scale structures of the image.

The above issues related to the degradation in the filter performance are illustrated in Fig. 2 which depicts the histograms (distributions) of the computed intensity based similarity weights for all patches within a 21×21 search window centered at two pixels of the standard Barbara image Fig. 2(a): $P1$ located in a flat region (the top row of Fig. 2) and $P2$ located at textured region (the bottom of Fig. 2). The weights are computed for three different noise standard deviations ($\sigma = 10, 25, 50$). First, observe that for the two regions, as the noise level increases, more dissimilar patches are incorporated and considered to be significant. This is clearly reflected in the computed histograms as they become flatter. For example, compared to the histograms in Figs. 2(b,e) when ($\sigma = 10$) for both pixels, the histograms in Figs. 2(d,g) ($\sigma = 50$) are much wider and there is a clear reduction in number of eliminated patches (the one assigned approximately zero weights) and increase in the number of patches that are considered significant (assigned high weights). More dissimilar patches to the reference patch are incorrectly incorporated while there has been no change in the structure of the underlying patches to be considered similar to the central patch.

In order to address the aforementioned issues that cause high blurring in the denoised image and becomes highly apparent as the noise level increase, we propose refining the filter intensity based weighting function using a new weighting function that is independent of the noise and utilizes a similarity selection criterion based on the patches structural characteristics. This is detailed in the following section.

3   LBP Based Nonlocal Means Filter

The proposed approach for enhancing the nonlocal mean filter is based on modulating the intensity based weights computed in (2) using another weighting function that computes the similarity between any two patches using their textured and structural features obtained by the LBP framework. The modulated weights $w_m(i,j)$ are obtained as a *product* of both the intensity based weights and the proposed LBP based weights:

$$w_m(i,j) = w(i,j)_{\text{intensity}} * w(i,j)_{\text{LBP}} \qquad (3)$$

A brief overview of the LBP feature extraction method is presented in the following section as detailed in [5], [6],.

3.1   LBP Overview

LBP is a gray-scale texture operator which characterizes the spatial structure of the local image texture. At a center pixel, each neighboring pixel is assigned a binary label, which can be either 0 or 1, depending on whether the center pixel has higher intensity value than the neighboring pixel. The neighboring pixels are the angularly evenly distributed sample points over a circle with radius R centered at the center pixel. The LBP label for that center pixel is given by [5]:

$$\text{LBP}_{P,R} = \sum_{p=0}^{P-1} s(g_p - g_c)2^p \qquad (4)$$

where g_c is the gray value of the central pixel, g_p is the value of its pth neighboring pixel, $p = 0, \cdots, P-1$, P is the total number of neighbors, and R is the radius of the neighborhood which determines how far the neighboring pixels are located away from the center pixel. $s(x)$ is a step function given by:

$$s(x) = \begin{cases} 1, x \geq 0 \\ 0, x < 0 \end{cases} \tag{5}$$

The value of P is assigned according to the value of R as suggested in [6]. For example, $P = 8$ when $R = 1$, $P = 16$ when $R = 2$, and $P = 24$ when $R = 3$. The total number of possible LBP features depends on P and is equal to 2^P. For the purpose of image denoising, we adopt certain class of local binary texture patterns termed "uniform" which have limited number of transitions or discontinuities in the circular presentation of the pattern. The "uniform" binary patterns correspond to primitive microfeatures, such as edges, corners, and spots [6]. The total number of possible "uniform" LBP features is $P * (P - 1) + 2$ [6]. In texture classification framework, given an image I of size $N \times M$, the LBP pattern is identified for each pixel i. Then, the whole image is represented by building a histogram H of the generated LBP features:

$$H_k(I) = \sum_{i=1}^{N} \sum_{j=1}^{M} f(\text{LBP}(i), k), k \in [0, K] \tag{6}$$

$$f(x, y) = \begin{cases} 1, x = y \\ 0, \text{otherwise} \end{cases} \tag{7}$$

Where K is the maximal LBP pattern value.

In this work, the LBP histogram is computed for a square patch L of size $l \times l$ surrounding each individual image pixels.

Therefore, the proposed LBP based weighting function is given by:

$$w(i, j)_{\text{LBP}} = \frac{1}{Z(i)} e^{-\frac{D(H(L_i) - H(L_j))}{h(S_i)}} \tag{8}$$

Where Z_i is a normalizing term, and $H(L_i)$ and $H(L_j)$ are the histograms of the extracted LBP features of the two patches L_i and L_j. D refers to the Chi-square dissimilarity metric used to compare histograms and given by [6]:

$$D(H(L_i), H(L_j)) - \sum_{n=1}^{B} \frac{(H(L_i)_n - H(L_j)_n)^2}{H(L_i)_n + H(L_j)_n} \tag{9}$$

Where B is the number of bins and $H(L_i)_n$ and $H(L_j)_n$ are, respectively, the values at the nth bin. Unlike the smoothing parameter h of the intensity based weighting function which is set globally (i.e., for the whole image), the decay parameter $h(S_i)$ of the LBP based exponential weighting function in (8) varies according to the smoothness of the underlying region where the search window S_i is located and centered by the reference pixel i. The region smoothness is estimated from the sampled standard deviation of all the LBP similarity distances obtained according to (9) between the reference patch centered by the noisy pixel i and all other patches within the search window S_i.

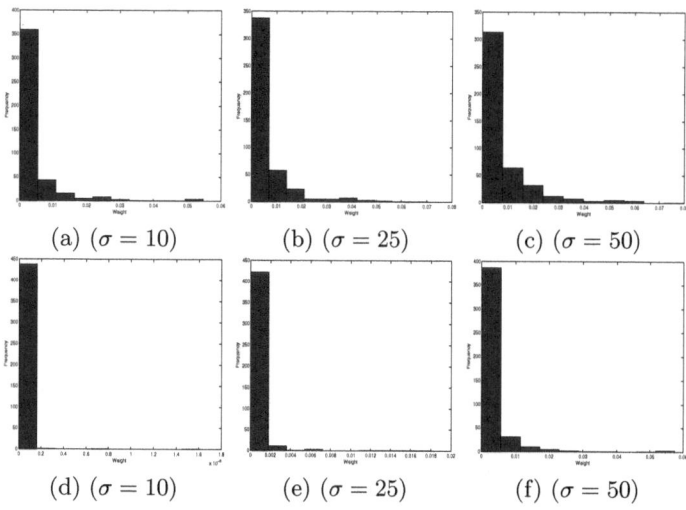

(a) $(\sigma = 10)$ (b) $(\sigma = 25)$ (c) $(\sigma = 50)$

(d) $(\sigma = 10)$ (e) $(\sigma = 25)$ (f) $(\sigma = 50)$

Fig. 3. Histograms of the modulated weights for the two Barbara image pixels $P1$ (top row), and $P2$ (bottom row) located in smooth and detailed regions, respectively

   The proposed LBP based weighting function in (8) poses two important properties: being noise independent and producing weights that vary according to structural characteristics of the underlying region. The two properties of the LBP based weighting function complement the intensity based weighting function once combined together as in (3). This is illustrated in Fig. 3 which shows the distribution of the modulated weights $w_m(i,j)$ for the same two pixels from the Barbara image Fig. 2(a): $P1$ and $P2$. Observe that the distribution of the modulated weights does not vary as the noise level increases. In other words, the number of patches that are considered significant and assigned high modulated weights stay the same regardless of the noise level. When comparing the modulated weight histograms Fig. 3 for both pixels, we found that they are almost identical to the intensity based histograms generated when the noise level is low ($\sigma = 10$) Fig. 2(b,e). That is, at all noise levels, the modulated weights will always be equivalent to the most reliable intensity based weights generated when the noise level is low. This shows that the generated weights are noise independent and, hence, are more reliable in reflecting the actual patch similarity regardless of the noise level. The effect of having region based smoothing parameter $h(S_i)$, can be observed when comparing the intensity based weights histogram of pixel $P1$ Fig. 2(b)($\sigma = 10$) and the modulated weights histograms, for all noise levels depicted at the top row of Fig. 3. In case of pixels located in flat regions, $h(S_i)$ will be low. Therefore, the LBP based weighting function will be more selective and will have the effect of scaling down the intensity based weights. Consequently, only few patches in flat regions will be considered significant and assigned high modulated weights. This has the effect of reducing the resulting blurring especially in flat regions.

3.2   Implementation Details

The proposed LBP based nonlocal means filter is performed in two main stages. In the first stage, histograms for "uniform" LBP features using $R = 2$ and $P = 16$ are computed for all image pixels and stored in a lookup table. In this work, the sufficient size of the LBP patch L is determined experimentally and found to be 13×13. During the filtering stage of a given image pixel i, both intensity based weights and LBP based weights are computed according to (2),(8), respectively. Then, modulated weights are computed according to (3).

4   Experimental Results

A qualitative comparison using the peak signal-to-noise ratio (PSNR) between the proposed method and the intensity based windowed nonlocal means filter is presented in Table 1. The experiments are conducted on five standard test images using three noise levels ($\sigma = 10, 25, 50$). The results show that the proposed method produces higher PSNR than the intensity based windowed nonlocal means for all test cases. Visual comparisons between the intensity based windowed nonlocal means filter and the proposed method are shown in Fig. 4 and Fig. 5. Comparing the proposed method results depicted in Fig. 4 to the denoised images by the intensity based windowed nonlocal means filter shown in Fig. 1, it is obvious that the proposed method produces images with noticeably better visual quality. This is supported by the higher structural similarity index (SSIM) values obtained by the proposed method. The blurring effect of the filter is highly reduced especially at high noise levels (i.e., $\geq= 25$). In case of

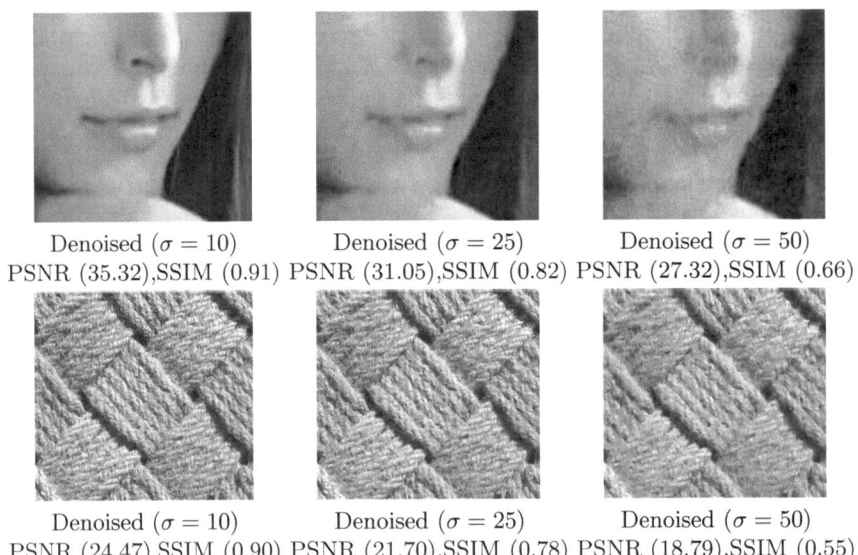

Denoised ($\sigma = 10$)   Denoised ($\sigma = 25$)   Denoised ($\sigma = 50$)
PSNR (35.32),SSIM (0.91) PSNR (31.05),SSIM (0.82) PSNR (27.32),SSIM (0.66)

Denoised ($\sigma = 10$)   Denoised ($\sigma = 25$)   Denoised ($\sigma = 50$)
PSNR (24.47),SSIM (0.90) PSNR (21.70),SSIM (0.78) PSNR (18.79),SSIM (0.55)

Fig. 4. PSNR and SSIM for Denoised images using the proposed method

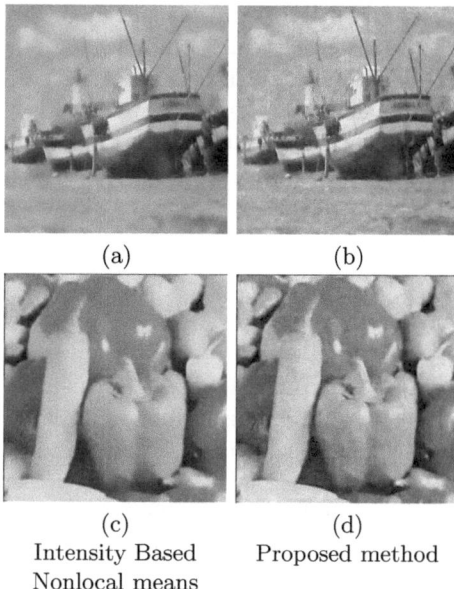

(a)	(b)
(c)	(d)
Intensity Based	Proposed method
Nonlocal means	

Fig. 5. Comparison of the denoised images using the proposed method and the intensity based Windowed nonlocal means filter. The noise standard deviation is 50.

Table 1. Comparison of the denoising quality in terms of PSNR for images denoised with the proposed method and the intensity based windowed nonlocal means for five standard test images. For each image, the three rows correspond to noise levels ($\sigma = 10, 25, 50$).

	Proposed Method PSNR (dB)	Windowed Nonlocal Means PSNR (dB)	PSNR Gain
Barbara	34.00	33.69	+0.31
	28.93	28.34	+0.59
	24.75	24.24	+0.51
Boat	33.13	32.38	+0.75
	29.02	27.74	+1.28
	25.46	24.45	+1.01
House	35.66	35.29	+0.37
	31.46	30.85	+0.61
	27.12	26.22	+0.90
Lena	35.32	34.71	+0.61
	31.05	30.13	+0.92
	27.32	26.68	+0.64
Peppers	33.55	33.05	+0.50
	29.51	28.51	+1.00
	25.60	24.43	+1.20

texture image (D1) depicted in the second row of Fig. 4, the proposed method is able to preserve fine texture details while removing undesirable noisy artifacts. The effect of having region dependent filter smoothing parameter on the visual quality of denoised flat image regions is illustrated in Fig. 5. The denoised flat regions using the proposed method (such as the sea and the pepper surfaces) Fig. 5(b,d) are not oversmoothed and have even clearer fine details than the denoising results of the intensity based windowed nonlocal means filter (Fig. 5(a,c)).

5   Conclusions

This paper introduced a novel approach to nonlocal means image denoising that computes the filter averaging weights using not only patch intensities but also their structural features extracted by the LBP method. Experimental results show that the proposed method has the advantage of preserving high frequency components and suppressing undesirable artifacts such as noise even when the noise level is high. The proposed LBP based windowed nonlocal means outperforms the intensity based windowed nonlocal means algorithm both quantitatively and qualitatively.

References

1. Brox, T., Kleinschmidt, O., Cremers, D.: Efficient nonlocal means for denoising of textural patterns. IEEE Trans. Image Process. 17(7), 1083–1092 (2008)
2. Buades, A., Coll, B., Morel, J.M.: A non-local algorithm for image denoising. In: International Conference on Computer Vision and Pattern Recognition, pp. 60–65 (2005)
3. Kervrann, C., Boulanger, J.: Optimal spatial adaptation for patch-based image denoising. IEEE Trans. Image Process. 15, 2866–2878 (2006)
4. Liao, S., Law, M.W.K., Chung, A.C.S.: Dominant local binary patterns for texture classification. IEEE Trans. Image Process. 18, 1107–1118 (2009)
5. Liao, S., Law, M.W.K., Chung, A.C.S.: Dominant local binary patterns for texture classification. IEEE Trans. Image Process. 18(5), 1107–1118 (2009)
6. Ojala, T., Pietikainen, M., Maenpaa, T.: Multiresolution gray-scale and rotation invariant texture classification with local binary patterns. IEEE Trans. Pattern Anal. Mach. Intell. 24(7), 971–987 (2002)
7. Tasdizen, T.: Principal neighborhood dictionaries for nonlocal image denoising. IEEE Trans. Image Process. 18(12), 2649–2660 (2009)

Integrating Color Sampling into Depth Based Bilayer Segmentation

Lorenzo Sorgi and Markus Schlosser

Technicolor R&I, 30625 Hannover, Germany
{lorenzo.sorgi,markus.schlosser}@technicolor.com
http://www.technicolor.com

Abstract. A trend in the computer vision community is observed towards the simultaneous exploitation of color images and depth maps. In this context we propose a novel approach for bi-layer segmentation, whose main contribution is given by the integration of a color classifier based on color sampling, within a depth-based segmentation framework. We have run tests on datasets available online and the outcoming results pointed out the effectiveness of this approach and its suitability for integration in automatic segmentation systems.

1 Introduction

Image segmentation is a fundamental brick of many video editing applications and despite the large volume of literature, it can be still considered an open research area. In the simplest scenario a single foreground subject needs to be extracted from the background scene. Techniques exploiting the assumption of a moving foreground and a static background are usually classified as Background Subtraction [3]. Recently active depth sensors have become available as off-the-shelf components and depth maps have been successfully integrated into the segmentation pipeline, taking advantage of their invariance to lighting conditions. However due to their reduced resolution, they are mostly used for the robust automatic initialization of the foreground mask [10]. Furthermore the misalignment with color images due to differences in viewpoints and spatial resolutions, implies an additional non trivial calibration process. Depth maps extracted from stereo or structure from motion do not suffer from these issues but provide a much lower reliability. A sort of ideal setup composed by a time-of-flight camera and a stereo camera, which combines robustness and resolution from both has been proposed in [16]. In some case depth maps have been used to prepare a trimap for a following alpha matting stage, and some matting techniques also integrate the depth information into their core matte estimation [5]. However, in our opinion their results still suffer from the broad initial trimap [11].

Binary segmentation systems using color and depth may be grouped in feature-level fusion and decision-level fusion. Approaches in the first group typically use a k-means clustering performed on feature vectors consisting of the color components and the spatial position [4,6]. The decision-level fusion systems instead, employ a graph-cuts framework, where the depth is integrated into the data term

A. Petrosino (Ed.): ICIAP 2013, Part I, LNCS 8156, pp. 31–40, 2013.

as a statistically independent variable [9,13,8,1]. Only in [15] the authors chose a voting scheme to combine the output of three classifiers based on background subtraction, color statistics and depth/motion consistency, and this approach is the closest to our work, even though we integrate different classifiers. In this paper we propose a statistical framework for video bilayer segmentation, which uses two independent classifiers based depth and color data. An objective function is defined as a weighted sum of their scores, and space-time regularization terms. All these terms are embedded in a segmentation graph as normalized probability measures. We believe that the numerical homogeneity of the different costs guaranteed by the probabilistic framework, allows for a more accurate treatment of the critical case where the pixel data cannot be explained by any of the distribution models and no useful information is provided by the classifiers. A novelty aspect is provided by the exploitation of the color data only in terms of color distance, whose statistical distribution model can be safely assumed content independent and learnt offline, providing a significant relief to the overall system. Furthermore differently from other techniques we include the color clue only in the second stage of a multiresolution framework, using color sampling. The latter has been recently proposed for alpha matting [12,7], but we are not aware of any attempt to exploitation for binary segmentation. The algorithm also is intrinsically suitable for video segmentation as the smoothness terms are homogeneously formulated both in space and time.

2   Graph Based Segmentation

Let us denote with $\mathbb{L} = \{F, B\}$ the space of binary segmentation labels and with $\alpha_i \in \mathbb{L}$ the i-th pixel label. Using this notation the segmentation of a videosequence is represented by the vector $\alpha = [\alpha_0, \ldots, \alpha_{N-1}]^T$ collecting the labels of the N unclassified pixels. The corresponding estimation problem is formulated as the minimization of an objective function comprised of three weighted terms:

$$\alpha = \min_{\alpha \in \mathbb{L}^N} \left\{ \sum_{i=0}^{N-1} e_{d,i}\left(\alpha_i\right) + w_s \sum_{(i,j) \in \Phi_s} e_{s,i,j}\left(\alpha_i, \alpha_j\right) + w_t \sum_{(i,j) \in \Phi_t} e_{t,i,j}\left(\alpha_i, \alpha_j\right) \right\}.$$

$$(1)$$

The first contribution in (1), denoted as data term, takes into account the coherence between the segmentation labels and the pixel data. As each pixel carries a 3D color vector and a depth measurement, the data term can be further expanded as a sum of two independent contributions, identified by the indexes c and z:

$$e_{d,i}\left(\alpha_i\right) = w_c e_{c,i}\left(\alpha_i\right) + w_z e_{z,i}\left(\alpha_i\right) \qquad . \qquad (2)$$

The other terms in (1), denoted as smoothness terms, provide the penalties for each pair of neighbor pixels differently labeled, within the sets Φ_s and Φ_t that collect the spatial and temporal cliques of the video sequence. The first is formed using the 4-connected neighborhood, while the temporal pairs are established using the background registration homographies. (Sec. 2.3).

The minimization of the objective function (1) can be modelled using the max-flow/min-cut optimization and solved via graph cut[18]. Video sequences normally lead to a massive graph representation and the problem must be tacked using technique ad hoc designed for this purpose [17]. We propose instead a more straightforward multiresolution framework. The sequence is initially processed at low resolution without the contribution of the color data terms. A trimap is then expanded from the contour of initialized segmentation mask and a second processing stage is performed at native resolution on the unknown area of the trimap. Furthermore this approach allows for the exploitation of color sampling in the computation color data term (Sec. 2.1).

2.1   Color Data Term

The data terms $e_{c,i}$ in (2) provide a measure of congruence between colors and segmentation labels. The color likelihood is widely used for this porpose, however we claim that this may not be the ideal way to tackle the task, as the color distribution within a video is strongly content dependent, and therefore an a-priori specified model could be inappropriate for a specific sequence. Futhermore the cumulative nature of global statistical models implies the loss of spatial accuracy, whereas it is not unrealistic that the color has an similar global distribution both in the background and foreground areas, making such models necessarily lose any classification power. On the other hand local models may suffer numerical instability which lead to a poor data representation.

For these reasons we opted for a different approach, inspired by the Alpha Matting technique proposed in [12], but implemented with a more aggressive sampling model (Fig. 2). In the second segmentation stage the unlabeled pixels are contained within the trimap stripe and for each of these two sample sets, denoted as $\mathbb{S}_{\{F,B\}}$, are randomly drawn from near foreground/background areas. We opted for a low set cardinality (3 samples). The two best samples are then selected and the color data terms are computed as a probability measure using the cumulative distribution of their distances in color space to the reference pixel:

$$e_{c,i}\left(\alpha_i\right) = P_c\left(d \leq \min_{k \in \mathbb{S}_{(F,B)}} \{d_{i,k}\} \,|\alpha_i = \alpha_k\right) \qquad , \qquad (3)$$

The remarkable aspect of the cost formulation (3) is that it does not need the inference of any specific color statistics for the video sequence.

Indeed, the color feature used in our segmentation framework is given by the pixel distance in Lab color space under the condition of homogeneous segmentation labeling. It is has empirically prooved that the corresponding probability density function, $f_c\left(d_{i,j}|\alpha_i = \alpha_j\right)$, which is also used as kernel for the integration of data terms, is roughly content independent and therefore the corresponding model can be selected and characterized offline. Eventually we claim that using only the color distance as classification feature we can successfully process any video sequence using a general statistical model. This approach at the same time overcomes the typical drawbacks of the color likelihood inference and is a remarkable advantage both in terms of performance and computational cost.

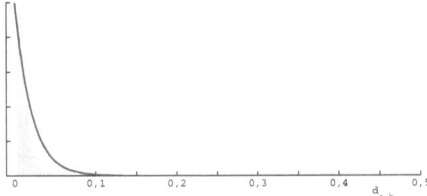

Fig. 1. Statistical distribution of the color distance in smooth areas. In green is represented the normalized histogram computed from the sample set and in red the estimated exponential model ($\lambda = 48.64$).

In our system the color distance likelihood have been estimated from 1e6 pixel pairs sampled from 50 web images with different content. The constraint on homogeneous segmentation has been approximated by two simple conditions, namely a spatial distance lower than 10 pixel and no edge across the connecting segment. These conditions do not guarantee the homogeneous labeling of the selected pixel pairs, however, we observed that the overall quality of the sample set is sufficiently high for our purpose. By visual inspection of the histogram of the color distance data obtained from the sample set we opted for an Exponential model, $f(d_{a,b}|\alpha_a = \alpha_b) = \lambda e^{-\lambda d_{a,b}}$, (Fig. 1).

The estimated model is then used to compute the data term (3) and the selected samples provide the function argument. As color sampling needs the availability of the sampling areas, in our multiresolution framework we can exploit the color data term only in the second segmentation round, when the trimap expanded from the initialized segmentation mask provides them. A second issue is related to the optimal setting for the trimap width. A narrow stripe boosts the sampling accuracy as the spatial nearness increases the color correlation between reference pixels and drawn samples. On the other hand a broad stripe allows the system to recover in case of poor initialization. Eventually it is extremely difficult to find a satisfying tradeoff. To overcome this problem, we propose an aggressive sampling which relaxes the role of the trimap, namely not only its foreground/background areas are sampled, rather the regions identified by the initial segmentation instead are considered eligible as sampling domains. In other words the samples can be drawn also inside the unlabeled stripe of the trimap (Fig. 2.a,b). The sampling model is shown in more details in Fig. 2.c. We denote with ω the half-width of the trimap and with δ the shortest distance between an unclassified pixel x and the initialized segmentation contour. The two sample set are drawn at $s_i = x + (\delta \pm dx_i) \cdot \eta_i$, where η is a random direction aiming at the contour, dx is a random variable uniformly distributed in the interval $[0, 2\omega]$ and the sign is negative if the initialized label of x and the sample label are equal, positive otherwise.

The proposed sampling strategy allows our system to cope in a natural way with a typical problem of trimap-based processing. Large portions of the foreground or background areas disappear when their linear extent is smaller than

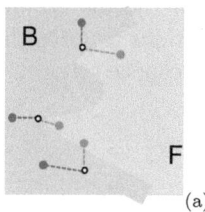 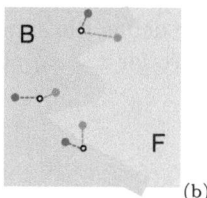 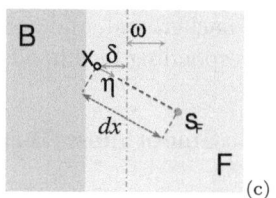

(a)   (b)   (c)

Fig. 2. *Sampling model.* The colors green, orange and grey identify the foreground, background and the unlabeled area of the trimap, and in yellow, blue and red are represented three reference pixels and the background and foreground samples. The conservative sampling draws the samples outside the grey stripe (a), in our approach instead, the initialized mask defines the sampling regions and the samples may belong to the unlabeled stripe (b). (c) Shows the sampling model.

size of the morphology operator used to expand the trimap. This happens for example when a segment has small internal holes, thin in- or out-ward lobes, or it is too close to the image border. In all these circumstances the sampling scheme shown in Fig.2.a is unfeasible as the nearby sampling areas have been absorbed into the trimap. Our approach instead, naturally tackles this problem as the trimap identifies only the unclassified pixels whereas the sampling domains are provided by the initial segmentation.

2.2   Depth Data Term

Similarly to (3) the depth data terms are computed from the depth distribution:

$$e_{z,i}(\alpha) = P_z \left(z \underset{\alpha=B}{\overset{\alpha=F}{\lessgtr}} z_i | \alpha \right) \quad . \tag{4}$$

We observed a general trend to use a Gaussian Model (GM) to capture depth likelihoods. According to our experience this representation is error prone in those areas where the segmentation inference should be easier. When the background contains multiple depth layers a GM polarizes towards one of them and the likelihood drops very low for high depth values. It is evident however, that pixels with very high depth measurements most likely belong to the background. Furthermore it is non-sporadic the case that the probability densities cross in the far depth zone making the depth classifier unreliable. Therefore we suggest a Single Side GM which better describe the natural distribution of depth data:

$$\begin{cases} f(z|\alpha = F) = C_F \cdot e^{-\frac{1}{2}\left(\frac{3z}{z_B}\right)^2} \\ f(z|\alpha = B) = C_B \cdot e^{-\frac{1}{2}\left(\frac{3(z-1)}{1-z_F}\right)^2} \end{cases} \quad , \quad 0 \le z \le 1 \tag{5}$$

The parameters C_F and C_B are estimated by constraining the probability densities intergation, and the values z_F and z_B are given by the first two peaks detected in the depth histogram. The underlying assumption is that the first two peaks in the depth histograms provide a rough localization of the foreground and background, and accordingly we define the standard deviations to force the corresponding likelihoods near to zero for $z \geq z_B$ and $z \leq z_F$ respectively.

2.3   Smoothness Terms

The space smoothness terms in eq.(1) are related to the image contrast, in order to penalize a labeling swap inside untextured areas. Almost each graph-based segmentation technique uses the color distance between neighbor pixels as a simple approximation of the local contrast. We propose to increase the dimension of the measurement domain up to four pixels, in order to make the measure more robust against image blur. If we denote with (l, i, j, k) a quadruplet of consecutive pixels, then the (i, j) contrast is defined as $\gamma_{i,j} = \max\{d_{l,j}, d_{i,k}\}$. This is still a distance in color space, therefore we suggest to compute the space smoothness term again using the precomputed color distance likelihood:

$$e_{s,i,j}\left(\alpha_i, \alpha_j\right) = \delta_{\alpha_i, \alpha_j} \cdot P_c\left(d \geq \gamma_{i,j} | \alpha_i = \alpha_j\right) \qquad, \tag{6}$$

where $\delta_{i,j}$ denotes the delta Kronecker function.

The underlying idea of the smoothness penalties is similar in the temporal or spatial domain, therefore we opted for reusing the cost expression (6) on the temporal cliques as well. The temporal smoothness terms are then computed as

$$e_{t,i,j}\left(\alpha_i, \alpha_j\right) = \delta_{\alpha_i, \alpha_j} \cdot P_c\left(d \geq \tau_{i,j} | \alpha_i = \alpha_j\right) \qquad. \tag{7}$$

where $\tau_{i,j}$ is the color distance of temporal clique (i, j). These are created by means of the background registration homographies between consecutive frames. At each frame t the registration homography is estimated using the Inverse Compositional Alignment [2] to minimize the photometric registration error:

$$H_t = \min_{H \in \mathcal{H}} \left\{ \sum_{i \in \Omega} f\left(z_i | \alpha = B\right) \left\| I_{t,i} - \tilde{I}_{t+1,i} \right\|^2 \right\} \tag{8}$$

where $\mathcal{H}$ is the space of 2D projective transformations, Ω is the registration support, $I_{t,i}$ is the color of i-th pixel at time t and the notation $\tilde{I}$ means the homography warping. The depth likelihood (5) is used as weighting mask in order to polarize the homography estimation towards the background alignment, and the registration support Ω is defined over highly textured areas.

3   Results

In our test we used the weight set $\{w_c = 0.6, w_z = 0.4, w_s = 1.0, w_t = 1.0\}$ to build the objective (1). Different configurations have been also tested and the results do not change significantly. Furthermore we point out that the assignment of meaningful values to the cost weights is an easy task, as each cost term ranges in the interval $[0, 1]$. The image resolution for the initialization round has been set to 160x120 pixels.

We performed a preliminary evaluation on a data set freely available online with the corresponding ground truth [1]. In Table 1 we present our results compared with those obtained by TofCut [14], the technique proposed by the dataset provider. In the same paper the performance of other three algorithms challenged with the same dataset have been measured. For simplicity we chose not to report these additional results also here, as TofCut is anyway able to outperform all of them. In Fig. 3 some sample frames from the dataset are shown with the corresponding segmentation mask. These results show that the proposed method can outperform TofCut.

Table 1. Comparison between TofCut and the proposed Color Sampling Segmentation with and without temporal smoothness (CSS, CSS_t), on four test video sequences. As performance indicator the average percentage of misclassified pixels over the whole sequence is computed.

Seq.ID	# Frames	% Err		
		TofCut	CSS	CSS_t
WL	200	1.35	0.84	0.59
MS	400	0.51	0.27	0.21
MC	300	0.15	0.14	0.12
CW	300	0.38	0.25	0.22

Although the proposed technique turned out to be very accurate, however it is worth further discussing its weak aspects, which will be object of the future refinement work. In this system we did not introduce any adaptive weighting to control the contribution of depth and color clues across the time according to their reliability. Unlike TofCut we opted for a set of constant weights as our main focus was the evaluation of the performance level when color sampling is integrated into a binary segmentation framework. Nevertheless, we consider the adaptive weighting an important feature which will be restored in the future development. The effect of this choice affects only the sequence CW, as a second moving subject is included by our algorithm in the foreground layer whereas in the ground truth it is labeled as background. The error rate provided in Table 1 for this sequence, is computed without considering the frames where this subject is present, otherwise by including the all sequence the error rate raise over %2. However we still consider this result relevant. Indeed, the moving subject is perceived by the depth sensor at the same distance as the static foreground, therefore the misclassification is mostly due to the limitations of the sensor rather than to the segmentation algorithm itself. We do not believe that it is relevant for the algorithm design to accurately model the sensor limitation, many sensors are indeed available on the market which may not suffer of this

---

[1] http://vis.uky.edu/~gravity/Research/ToFMatting/ToFMatting.htm

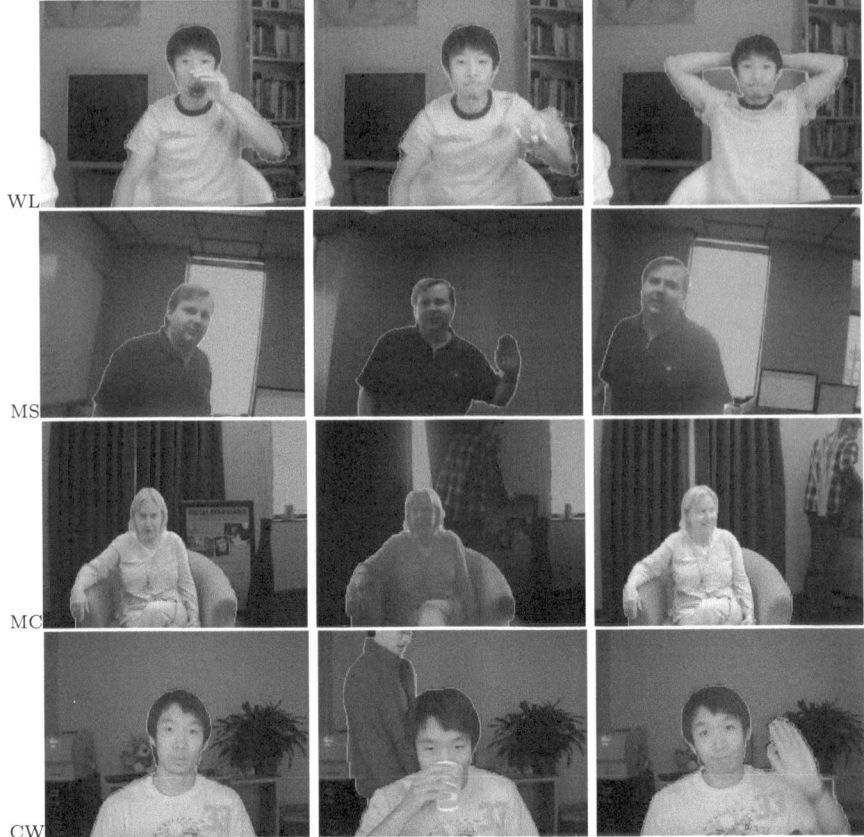

Fig. 3. Segmentation samples from each of the four dataset

problems. Furthermore for some applications, like video segmentation for 2D/3D conversion, distinct objects with so similar depth can be considered as part of the same foreground layer even if spatially distant. It is also interesting to notice that after the moving subject disappeared, the algorithm does not suffer from any error drift and it is able to recover correctly the segmentation mask corresponding to the main foreground subject.

Further tests have been run on two stereo-sequences at 1920x1080 HD resolution, with a moving foreground and a static but highly textured background. The sequences have been pre-processed using the graph-based matching software available online[2], to extract the depth maps, and the occluded pixels produced by the disparity estimator have been roughly filled using a median filter. In Fig. 4 a sample frame from each of the sequence is shown.

---

[2] http://pub.ist.ac.at/~vnk/software.html

Fig. 4. Segmentation samples from HD stereosequences. The color frames and the disparity maps are shown on the left and the segmented foreground on the right.

4 Conclusion

We presented a novel approach to bi-layer segmentation which integrates color sampling within a depth based segmentation framework. The results obtained from real video sequences, in different formats and with challenging content have confirmed the effectiveness of the approach. Our segmentation system does not require at runtime the estimation of any statistical model for the color data, this is a remarkable advantage as one of the weakest part of almost every Bayesian segmentation framework is therefore completely dropped. Furthermore the proposed objective function encapsulate depth, color and smoothness clues only in terms of probability measures, leading to a consistent normalized range for the edge capacities. This is also a simple but remarkable aspect since it greatly simplifies the choice of the weighting factors for non-expert users. We believe that this work has produced interesting results and it is a promising starting point for the design of a really operative video editing tool.

References

1. Arif, O., et al.: Visual tracking and segmentation using Time-of-Flight Sensor. In: IEEE Int. Conf. on Image Proc., pp. 2241–2244 (2010)
2. Baker, S., Matthews, I.: Lucas-Kanade 20 Years On: A Unifying Framework. Int. J. Comput. Vision 56(3), 221–255 (2004)
3. Benezeth, Y., Jodoin, P.M., Emile, B., Laurent, H., Rosenberger, C.: Review and evaluation of commonly-implemented background subtraction algorithms. In: 19th International Conference on Pattern Recognition, pp. 1–4 (2008)
4. Bleiweiss, A., Werman, M.: Fusing Time-of-Flight Depth and Color for Real-Time Segmentation and Tracking. In: Kolb, A., Koch, R. (eds.) Dyn3D 2009. LNCS, vol. 5742, pp. 58–69. Springer, Heidelberg (2009)
5. Cho, J., Ziegler, R., Gross, M., Lee, K.: Improving alpha matte with depth information. IEICE Electronics Express 6(22), 1602–1607 (2009)
6. Dal Mutto, C., Zanuttigh, P., Cortelazzo, G.M.: Scene Segmentation by Color and Depth Information and its Application. In: STreaming Day (2010)
7. Gastal, E., Oliveira, M.: Shared Sampling for Real-Time Alpha Matting. Computer Graphics Forum 2(29), 575–584 (2010)
8. He, H., McKinnon, et al.: Graphcut-based interactive segmentation using colour and depth cues. In: Australasian Conf. on Robotics and Automation ACRA (2010)
9. Kolmogorov, V., Criminisi, A., Blake, A., Cross, G., Rother, C.: Bi-Layer Segmentation of Binocular Stereo Video. In: Int. Conf. on Computer Vision and Pattern Recognition, vol. 2, pp. 407–414 (2005)
10. Leens, J., Piérard, S., Barnich, O., Van Droogenbroeck, M., Wagner, J.-M.: Combining Color, Depth, and Motion for Video Segmentation. In: Fritz, M., Schiele, B., Piater, J.H. (eds.) ICVS 2009. LNCS, vol. 5815, pp. 104–113. Springer, Heidelberg (2009)
11. Wang, J., Cohen, M.F.: Image and video matting: a survey. Found. Trends. Comput. Graph. Vis. 3(2), 97–175 (2007)
12. Wang, J., Cohen, M.: Optimized Color Sampling for Robust Matting. In: Int. Conf. on Computer Vision and Pattern Recognition (2007)
13. Wang, L., Gong, M., Zhang, C., Yang, R., Zhang, C., Yang, Y.: Automatic Real-Time Video Matting Using Time-of-Flight Camera and Multichannel Poisson Equations. Int. Journal of Computer Vision 97(1), 104–121 (2012)
14. Wang, L., Zhang, C., Yang, R., Zhang, C.: TofCut: Towards Robust Real-time Foreground Extraction Using a Time-of-Flight Camera. In: 3DPVT Conf. (2010)
15. Zhang, G., Jia, J., Hua, W., Bao, H.: Robust Bilayer Segmentation and Motion/Depth Estimation with a Handheld Camera. IEEE Trans. on Pattern Anal. Mach. Intell. 33(3) (2011)
16. Zhu, J., Liao, M., Yang, R., Pan, Z.: Joint depth and alpha matte optimization via fusion of stereo and time-of-flight sensor. In: Int. Conf. on Computer Vision and Pattern Recognition, pp. 453–460 (2009)
17. Lin, F., Cohen, W.: Power Iteration Clustering. In: 27th Int. Conf. on Machine Learning (2010)
18. Kolmogorov, V., Zabin, R.: What energy functions can be minimized via graph cuts? IEEE Trans. on Pattern Anal. Mach. Intell. 26(2), 147–159 (2004)

Local Intrinsic Dimensionality Based Features for Clustering

Paola Campadelli[1], Elena Casiraghi[1],
Claudio Ceruti[2], Gabriele Lombardi[1], and Alessandro Rozza[3]

[1] Dipartimento di Informatica, Università degli Studi di Milano,
Via Comelico 39-41, Milano, Italy
{campadelli,casiraghi,lombardi}@di.unimi.it
[2] Dipartimento di Matematica, Università degli Studi di Milano,
Via Saldini 50, Milano, Italy
claudio.ceruti@unimi.it
[3] Dipartimento di Scienze Applicate, Università degli Studi di Napoli Parthenope
Centro Direzionale di Napoli - Isola C4, Napoli, Italy
alessandro.rozza@uniparthenope.it

Abstract. One of the fundamental tasks of unsupervised learning is dataset clustering, to partition the input dataset into clusters composed by somehow "similar" objects that "differ" from the objects belonging to other classes. To this end, in this paper we assume that the different clusters are drawn from different, possibly intersecting, geometrical structures represented by manifolds embedded into a possibly higher dimensional space. Under these assumptions, and considering that each manifold is typified by a geometrical structure characterized by its intrinsic dimensionality, which (possibly) differs from the intrinsic dimensionalities of other manifolds, we code the input data by means of local intrinsic dimensionality estimates and features related to them, and we subsequently apply simple and basic clustering algorithms, since our interest is specifically aimed at assessing the discriminative power of the proposed features. Indeed, their encouraging discriminative quality is shown by a feature relevance test, by the clustering results achieved on both synthetic and real datasets, and by their comparison to those obtained by related and classical state-of-the-art clustering approaches.

Keywords: Local features, Intrinsic dimensionality, Dataset clustering, Multi-manifold structures.

1 Introduction

At the present, continuous technological advances allow to work on multiple sources to collect increasing amounts of informations, which are coded as vectors of numerical values usually called features. Therefore, a real dataset $\boldsymbol{X}_N$ usually comprises an high number N of D-dimensional feature vectors, that is $\boldsymbol{X}_N \equiv \{\boldsymbol{x}_i\}_{i=1}^{N} \equiv \{[x_1, \ldots, x_D]_i\}_{i=1}^{N} \subset \Re^D$. In the pattern recognition field

A. Petrosino (Ed.): ICIAP 2013, Part I, LNCS 8156, pp. 41–50, 2013.

it is often useful to partition $\boldsymbol{X}_N$ in C disjoint classes (generally called clusters) having somehow different peculiarities, thus obtaining: $\boldsymbol{X}_N \equiv \bigcup_{c=1}^{C} \boldsymbol{X}_c$ and $\forall i \neq j, i, j \in \{1, \cdots, C\}, \boldsymbol{X}_i \cap \boldsymbol{X}_j = \varnothing$.

For this reason, unsupervised clustering techniques usually aim to create compact neighborhoods (clusters), by considering the clusters' internal homogeneity (similarity) and their external separation (dissimilarity). However, choosing the proper function to measure the similarity and the dissimilarity is often difficult. For this reason, and since clustering approaches are employed in a wide variety of fields, many techniques have been presented that differ for the employed similarity criteria and for the automatic algorithm used to identify the best partition. At the state-of-the-art, among the several recent surveys and comparative researches describing unsupervised biclustering or clustering methods, those reported in [18,7,25] are notable since they deeply consider the problem of clustering high dimensional data belonging to sets eventually having high cardinalities.

To cope with this kind of data, relevant literature works generally compute their lower dimensional projections through classical techniques, e.g. Principal Component Analysis (PCA) and its variants [2], and then apply either classical clustering techniques, such as K-means and its (fuzzy) variants [21], the Expectation Maximization (EM, [8]) algorithm, or algorithms specifically designed for the clustering problem to be handled. Furthermore, the similarity between points is often computed by employing the common Euclidean distance; yet, as explained in [1], this is a quite limited methodology that might not properly capture and express the typifying geometrical structure underlying each cluster, specially in case of high dimensional data. Consequently, though promising results have been obtained, the problem of high dimensional dataset clustering is still open.

For this reason, and based on the aforementioned considerations, recent clustering approaches change their perspective and view the cth cluster as a set of points $\boldsymbol{X}_c = \{\phi_c(\boldsymbol{z}_{i,c})\}_{i=1}^{n_c} \subset \Re^D$ drawn from a low d_c-dimensional space (manifold) $\mathcal{M}_c \subseteq \Re^{d_c}$ and embedded into an higher $D-$dimensional space $\Re^D$ $(d_c \leq D)$ through a map $\phi_c(\cdot)$. Under this framework the dimensionality d_c of $\mathcal{M}_c$, generally called intrinsic dimensionality (id), becomes a distinctive feature.

This conceptual framework guarantees that each cluster is strongly typified by the geometrical structure characterizing the cluster as a whole. This intrinsic structure is the one inherited by the feature space $\mathcal{M}_c$ in $\Re^{d_c}$ from which the points of the cluster are supposed to be drawn. Therefore, the clustering of $\boldsymbol{X}_N$ can be achieved by multi-manifold clustering techniques, aimed at identifying the C intersecting manifolds underlying $\boldsymbol{X}_N$, and being (possibly) uniquely identified by their distinctive id (where $d_1 \neq \ldots \neq d_C$). To this aim, most of the few works proposed in literature [11,3,23,24,12] code each point as a vector of local id estimates[1], or local features related to them, with the aim of capturing the geometrical structure underlying the neighborhood of the coded point. Indeed, the discriminative power of these features allows to obtain promising results by employing classical clustering algorithms.

---

[1] The local id relative to one point is the id estimate computed on its neighborhood.

Considering the works in literature, one of the most related to ours is that described in [3], where the proposed clustering method (hereinafter referred to as NS) is based on local id estimates obtained by exploiting a modified version of the id estimator described in [6]. Precisely, in [6] the authors work on the k-nearest neighbors ($k << N$) of each point to estimate its local id, and then cluster the dataset by employing all the estimated local ids. Note that, as highlighted in Section 4 of [20], when dealing with high id datasets a strong underestimation problem affects most of the neighborhood based id estimators, causing unreliable id estimations and consequent inaccurate clusterings. To reduce this problem in [3] an effective "Neighborhood Smoothing" procedure is employed.

In this paper we describe some features that can be conceptually viewed as local id estimates and local characteristics of the underlying manifold portion (see Section 2); note that these features have been also exploited by effective global id estimators [17,5,4]. In Section 3 we show how they can be effectively exploited by classical clustering algorithms; indeed, a feature relevance test, promising clustering results on both synthetic and real datasets, and their comparison with those achieved by state-of-the-art clustering techniques (see Section 4), show the discriminative quality of the proposed features also when applied to high dimensional points characterized by both high and low ids.

2   Local id-Based Features

In this section we describe the three local features we exploited for dataset clustering; they are derived from two relevant global id estimators [17,5,4], that compute the global id estimate that characterizes the manifold from which the dataset X_N is assumed to be drawn.

The first local feature is successfully employed when estimating the (global) id by the well-known **Maximum Likelihood Estimator** for id (MLE [17]). The rationale of our choice is that this feature, referred as $\hat{d}(x_i, k)$ in the following, can be theoretically viewed as a local id estimate computed in the k-neighborhood of each point $x_i \in X_N$. More precisely, it is computed by treating the neighbors of each point $x_i \in X_N$ as events in a Poisson process and considering the Euclidean distance $r^{(j)}(p_i)$ between the query point x_i and its jth nearest neighbor as the event's arrival time. Since this process depends on the id that characterizes the underlying manifold's portion, MLE estimates it by maximizing the log-likelihood of the observed process. In practice $\hat{d}(x_i, k)$ is computed as:

$$\hat{d}(x_i, k) = \left(\frac{1}{k} \sum_{j=1}^{k} \log \frac{r^{(k+1)}(x_i)}{r^{(j)}(x_i)} \right)^{-1}$$

The other two local features used by our clustering approaches, referred as $\hat{\nu}(x_i, k)$ and $\hat{\tau}(x_i, k)$ in the following, collect informations related to the distribution of the pairwise angles in the k-neighborhood of each $x_i \in X_N$. They have been introduced and used by the global id estimators proposed in [5,4] since they express further, and different, informations about the local geometry of the unknown manifold's portions underlying each data neighborhood. Indeed, their exploitation has shown to improve the reliability of the computed id estimates since they allow to reduce the underestimation problem [20] that affects most of the id estimators when applied to high id data (for details see [5,4]).

The theoretical basis of these features is expressed by the following theorem proved in [5,4]:

Theorem 1. *Given two independent random unit vectors (x_1, x_2) in $\Re^d$, drawn from a uniform distribution on S^{d-1}, for increasing values of d the concentration parameter τ of the von Mises (VM) distribution describing the angle θ between x_1 and x_2 converges asymptotically to the dimensionality d.*

Taking into account this theorem, it is possible to consider local neighborhoods in X_N to capture the information provided by the concentration of pairwise angles. Practically, for each point $x_i \in X_N$ its k nearest neighbors $\bar{X}_k^i$ are identified, they are centered to obtain $\hat{X}_k^i = \{x_j - x_i : \forall\, x_j \in \bar{X}_k^i\}$, and then used to compute:

$$\theta(x_z, x_j) = \arccos \frac{x_z \cdot x_j}{\|x_z\|\|x_j\|}. \tag{1}$$

Employing Equation (1) the $\binom{k}{2}$ angles of all the possible pairs of vectors in $\hat{X}_k^i$ are computed to compose the vector $\hat{\theta}_i = \{\theta(x_z, x_j) : \forall\, x_z,\, x_j \in \hat{X}_k^i\}_{1 \le z < j \le k}$. Considering Theorem 1, each component of $\hat{\theta}_i = \left[\theta_1, \cdots, \theta_{\binom{k}{2}}\right]$ follows a VM pdf of parameters $\nu(x_i, k)$ and $\tau(x_i, k)$; therefore, the Maximum Likelihood (ML) of the population direction $\nu(x_i, k)$ equals the sample mean direction:

$$\hat{\nu}(x_i, k) = \text{atan}_2\left(\sum_{j=1}^{\binom{k}{2}} \sin\theta_j, \sum_{j=1}^{\binom{k}{2}} \cos\theta_j\right)$$

where $\text{atan}_2(x, y)$ is the arc tangent of y/x.

Likewise, the ML of the concentration parameter $\tau(x_i, k)$ equals the estimate $\hat{\tau}(x_i, k)$ calculated as a solution of $\eta(x_i, k) = \frac{I_1(\tau(x_i, k))}{I_0(\tau(x_i, k))} \equiv A(\tau(x_i, k))$, where I_v is the modified Bessel function of the first kind with order v, and $\eta(x_i, k)$ is the norm of the sample mean vector defined in [22] as:

$$\eta(x_i, k) = \sqrt{\left(\frac{1}{\binom{k}{2}}\sum_{j=1}^{\binom{k}{2}} \cos\theta_j\right)^2 + \left(\frac{1}{\binom{k}{2}}\sum_{j=1}^{\binom{k}{2}} \sin\theta_j\right)^2}$$

Being A a non invertible function, in [5,4] $A^{-1}(\eta(x_i, k))$ is approximated by:

$$\hat{\tau}(x_i, k) = \tilde{A}^{-1}(\eta(x_i, k)) = \begin{cases} 2\eta(x_i, k) + \eta(x_i, k)^3 + \frac{5\eta(x_i, k)^5}{6} & \eta(x_i) < 0.53 \\ -0.4 + 1.39\eta(x_i, k) + \frac{0.43}{1 - \eta(x_i, k)} & 0.53 \le \eta(x_i, k) < 0.85 \\ \frac{1}{\eta(x_i, k)^3 - 4\eta(x_i, k)^2 + 3\eta(x_i, k)} & \eta(x_i, k) \ge 0.85 \end{cases}$$

3   The Clustering Approaches

In this section we describe how we employ either the classical Expectation Maximization algorithm (EM, [8]) or a simple variant of the Label Propagation Algorithm (LPA, [19]) to cluster input datasets coded by the geometrical features described in Section 2; we underline that our choice of using classical and simple clustering techniques is motivated by the fact that our aim in this research work is to assess the discriminative ability of the features we are proposing.

Precisely, we consider input sets composed of C clusters, that is $X_N = \{x_i\}_{i=1}^{N} = \{\{\phi_c(z_{i,c})\}_{i=1}^{n_c}\}_{c=1}^{C} \subset \Re^D$ (n_c is the cardinality of the cth cluster, and $N = \sum_{c=1}^{C} n_c$), and we assume that the C clusters are composed of independent identically distributed points $z_{i,c}$ drawn from C different low-dimensional manifolds (possibly characterized by different ids) embedded in the higher dimensional space $\Re^D$ by (possibly different) proper maps ϕ_c ($\phi_1 \neq \phi_2 \neq \cdots \neq \phi_C$). Our aim is to exploit $\hat{d}(x_i, k)$, $\hat{\nu}(x_i, k)$, and $\hat{\tau}(x_i, k)$ to cluster X_N.

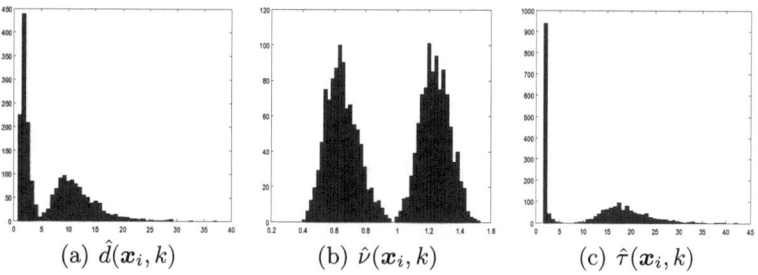

$$\text{(a) } \hat{d}(x_i, k) \qquad \text{(b) } \hat{\nu}(x_i, k) \qquad \text{(c) } \hat{\tau}(x_i, k)$$

Fig. 1. The three local parameters computed on points drawn by two manifolds having id $= 2$ and id $= 14$ embedded in $\Re^{30}$, and using $k = 30$

The first three clustering approaches we experimented employ only one of the aforementioned features. Precisely, the algorithm transforms each point $x_i \in X_N \subset \Re^D$ in a unique real value $y_i \in \Re$ by computing the local feature being used; this allows to obtain a 1-dimensional set $\{y_i\}_{i=1}^{N} = Y_N \subset \Re$. If we consider points belonging to different manifolds, each feature tends to be distributed as a mixture of gaussian distributions (see Figure 1). For this reason we chose to employ EM on Y_N. Precisely, being $h(s)$ the pdf of the event y_i and assuming that h is the sum of C normal distributions $h(s) = \sum_{c=1}^{C} w_c \mathcal{N}(y_i|\mu_c, \sigma_c)$, $w_c = n_c/N$, EM estimates from Y_N the means $\{\mu_1, ..., \mu_C\}$ and the standard deviations $\{\sigma_1, ..., \sigma_C\}$ of the C normal distributions. Employing the estimated values, each point x_i (coded as y_i) is associated to the cluster $c \in \{1, ..., C\}$ maximizing the probability $g_c(y_i, \mu_c, \sigma_c)$, i.e. the probability of a given y_i to belong to the cth cluster. $g_c(y_i, \mu_c, \sigma_c)$ is defined as follows:

$$g_c(y_i, \mu_c, \sigma_c) = \frac{1}{\sigma_c \sqrt{2\pi}} e^{-\frac{1}{2}\left(\frac{y_i - \mu_c}{\sigma_c}\right)^2}$$

The three algorithms obtained by employing $\hat{d}(\boldsymbol{x}_i, k)$, $\hat{\nu}(\boldsymbol{x}_i, k)$, and $\hat{\tau}(\boldsymbol{x}_i, k)$ are called $\text{EM}_{\hat{d}}$, $\text{EM}_{\hat{\nu}}$, and $\text{EM}_{\hat{\tau}}$, respectively. These techniques depend on the number k of the neighbors considered when computing the features, and on the parameter C of the EM algorithm (the number of clusters).

To improve the results obtained by the aforementioned clustering methods, we combine the three features $\hat{d}(\boldsymbol{x}_i, k)$, $\hat{\nu}(\boldsymbol{x}_i, k)$, and $\hat{\tau}(\boldsymbol{x}_i, k)$, to obtain, for each $\boldsymbol{x}_i \in \boldsymbol{X}_N \subset \Re^D$, a 3-dimensional vector $\boldsymbol{y}_i \in \Re^3$. We can then apply the same procedure described above to assign each point to the cluster that maximizes the probability $g_c(\boldsymbol{y}_i, \boldsymbol{\mu}_c, \boldsymbol{\Sigma}_c)$. This approach, called $\text{EM}_{\hat{d}, \hat{\nu}, \hat{\tau}}$, still depends on the parameters k and C.

Finally, to relax the dependence with respect to the parameter C we substitute EM with the LPA variant proposed in [19], obtaining $\text{LPA}_{\hat{d}}$, $\text{LPA}_{\hat{\nu}}$, $\text{LPA}_{\hat{\tau}}$, and $\text{LPA}_{\hat{d}, \hat{\nu}, \hat{\tau}}$. Briefly, this version of LPA automatically determines the number of clusters by an iterative process that assigns each $\boldsymbol{x}_i \in \boldsymbol{X}_N$ to the cluster to which the maximum number of its k nearest neighbors belong. To this aim, every point is initialized with a unique label and the labels are left to propagate through the network of points; as the labels propagate, densely connected clusters are formed, and they continue to expand until it is possible to do so.

4   Experimental Results

To obtain a preliminary assessment of the discriminative quality of the proposed features, we initially employed them to augment the dimensionality of points in the real datasets described below, and we apply a classical feature selection technique provided by WEKA [13].

The considered real datasets are: the Yeast dataset [15], which is composed of 1484 points in $\Re^8$ representing yeast proteins organized into 10 classes according to their positions in cells; the Segmentation dataset [10], which is composed of 2310 points in $\Re^{19}$ describing pixels randomly drawn from a dataset of 7 (classes) outdoor images; the MNIST test dataset [16] containing 10000 grey-level images of size 28×28 representing hand-written digits from 0 to 9 (10 classes). To use the MNIST dataset, we downsampled its images to the size 12×12, we vectorized them, and we appended the 576 gradient and curvature features described in [9], thus obtaining a dataset composed of 10000 points in $\Re^{720}$.

Note that, according to [14] and to the results we obtained by employing the id estimators described in [20], the MNIST clusters are characterized by id values higher than those of Segmentation and Yeast; indeed, we obtained id values between 2 and 4 for the Segmentation clusters, between 2 and 8 for the Yeast clusters, whilst the id values of the MNIST clusters are in the range $[7, 14]$.

As mentioned before, after augmenting the dimensionality of each vector in each dataset by appending the proposed local features, each dataset is processed by using the RankSearch feature selector (with GainRatio as a feature evaluator[2]) proposed in WEKA. We evaluated the relevance of each feature by separately applying RankSearch on 10 disjoint sets (folds) randomly generated from each dataset, and counting the number of times each feature is selected. According to the obtained results, $\hat{\nu}(\boldsymbol{x}_i, k)$ seems to be not relevant since it is never selected, while both $\hat{d}(\boldsymbol{x}_i, k)$ and $\hat{\tau}(\boldsymbol{x}_i, k)$ seem to be relevant (see Table 1); specifically, $\hat{\tau}(\boldsymbol{x}_i, k)$ seems important when considering high id datasets, whilst $\hat{d}(\boldsymbol{x}_i, k)$ is always selected for the low id ones. Nevertheless, as can be noticed in the following experiments, when coupled with $\hat{\tau}(\boldsymbol{x}_i, k)$ and $\hat{d}(\boldsymbol{x}_i, k)$, $\hat{\nu}(\boldsymbol{x}_i, k)$ allows to improve the clustering results.

Table 1. Percentage of times each feature has been selected as relevant in the 10 fold experiments

Dataset	id	$\hat{d}(\boldsymbol{x}_i, k)$	$\hat{\nu}(\boldsymbol{x}_i, k)$	$\hat{\tau}(\boldsymbol{x}_i, k)$
MNIST	$[7, 14]$	0%	0%	100%
Segmentation	$[2, 4]$	100%	0%	70%
Yeast	$[2, 8]$	100%	0%	100%

At this stage, we proceeded with tests on datasets generated by composing two or three clusters, which belong to either the MNIST test dataset, or to the synthetic datasets generated by the tool proposed in [14] (see Table 2). Precisely, to use the samples in the MNIST test dataset, we simply vectorized the 28×28 digit images obtaining samples in $\Re^{784}$. Though these points belong to 10 clusters (one cluster per digit) each being typified by a (probably) specific id, all the samples are already embedded in $\Re^{784}$; therefore, no embedding procedure is needed. To reduce the number of possible combinations, we run our clustering tests by randomly choosing one cluster (digit 1 images), and combining it with the other clusters to form all the possible cluster couples and triplets (obtaining 9 datasets of cluster couples and 36 of triplets).

Similarly, the synthetic datasets were created by merging two or three synthetic point sets (clusters), each comprising 1000 samples generated by the tool proposed in [14]. Note that each cluster is linearly embedded in a 40-dimensional space and the points belonging to the different clusters are concatenated, thus producing a point set containing either 2000 or 3000 points representing, respectively, 2 or 3 intersecting clusters. Furthermore, to reduce the possible combinations we randomly selected one of the synthetic sets (the $\mathcal{M}_{10}$ with $D = 15$), and we intersected it with 1 or 2 synthetic sets to obtain all the possible couples and triplets (12 datasets of cluster couples and 66 of triplets). Note that, when the employed generator requires to set a dimensionality D (see Table 2), this parameter was set to 10 for the cluster selected as the second, and 20 for the third one.

To objectively evaluate the effectiveness of employing the proposed features for clustering, we compared the results achieved on both synthetic and real datasets to those obtained on raw data by well-known clustering techniques

---

[2] GainRatio evaluates the worth of a single feature by measuring the gain ratio with respect to the class, where the gain ratio is defined as: $\frac{H(Class) - H(Class|Feature)}{H(Feature)}$, being H the relative entropy function.

Table 2. Brief description of the 13 synthetic datasets employed in our experiments, where d is the id and D is the embedding space dimension. Note that for some datasets the parameter D should be selected by the user.

Name	d	D	Description
$\mathcal{M}_1$	$D-1$	D	Uniformly sampled sphere linearly embedded.
$\mathcal{M}_2$	3	5	Affine space.
$\mathcal{M}_3$	4	6	Concentrated figure, confusable with a $3d$ one.
$\mathcal{M}_4$	4	8	Nonlinear manifold.
$\mathcal{M}_5$	2	3	2-d Helix
$\mathcal{M}_6$	6	36	Nonlinear manifold.
$\mathcal{M}_7$	2	3	Swiss-Roll.
$\mathcal{M}_8$	12	72	Nonlinear manifold.
$\mathcal{M}_9$	20	20	Affine space.
$\mathcal{M}_{10}$	$D-1$	D	Uniformly sampled hypercube.
$\mathcal{M}_{11}$	2	3	Möebius band 10-times twisted.
$\mathcal{M}_{12}$	D	D	Isotropic multivariate Gaussian.
$\mathcal{M}_{13}$	1	13	Curve.

(EM and LPA variant) and by the multi-manifold clustering approach (NS) presented in [3]. Note that, for each method, the parameter settings were chosen to obtain the best mean results (see Table 3). To assess the compared clustering techniques we employed the following measure: $accuracy = \frac{\sum_{i=1}^{N}(\chi(l_i=\hat{l}_i))}{N}$, where χ is the indicator function, $\hat{l}_i$ is the label associated to the sample point x_i by the employed clustering approach, and l_i is the correct label for that point.

Table 3. The methods used in our experiments and the chosen parameters. k is the number of neighbors for each point, k_{LPA} is the number of neighbors considered for Label Propagation, C is the number of clusters in the dataset, γ is the edge weighting factor, M is the number of Least Square (LS) runs, N is the number of re-sampling trials per LS iteration, Q is the number of different re-sampling values to be considered by NS.

Method	Parameters
Neighborhood Smoothing (NS)	$k = 20, \gamma = 1, M = 1, N = 10, Q = 10$
EM	$C = \{2,3\}$
$\text{EM}_{\hat{d}}$	$k = 30,\ C = \{2,3\}$
$\text{EM}_{\hat{\nu}}$	$k = 30,\ C = \{2,3\}$
$\text{EM}_{\hat{\tau}}$	$k = 30,\ C = \{2,3\}$
$\text{EM}_{\hat{d},\hat{\tau}}$	$k = 30,\ C = \{2,3\}$
$\text{EM}_{\hat{d},\hat{\nu},\hat{\tau}}$	$k = 30,\ C = \{2,3\}$
LPA	$k_{\text{LPA}} = 15$
$\text{LPA}_{\hat{d}}$	$k_{\text{LPA}} = 15$
$\text{LPA}_{\hat{\nu}}$	$k_{\text{LPA}} = 15$
$\text{LPA}_{\hat{\tau}}$	$k_{\text{LPA}} = 15$
$\text{LPA}_{\hat{d},\hat{\tau}}$	$k_{\text{LPA}} = 15$
$\text{LPA}_{\hat{d},\hat{\nu},\hat{\tau}}$	$k = 30,\ k_{\text{LPA}} = 15$

Table 4 shows the mean accuracies achieved on the synthetic and real datasets composed by two clusters, and those obtained on the datasets composed by three clusters[3]. It is possible to notice that $\text{EM}_{\hat{d},\hat{\nu},\hat{\tau}}$, which combines the information captured by the proposed local features, generally outperforms the other methods. Moreover, it is important to highlight that $\hat{\tau}(x_i, k)$ is a very discriminative

---

[3] Note that the dataset points are coded by employing only the proposed features.

information, especially when facing datasets with high id, but it needs to be combined with $\hat{d}(\boldsymbol{x}_i, k)$ and $\hat{\nu}(\boldsymbol{x}_i, k)$ to effectively cope with datasets composed by clusters characterized by both high and low ids and eventually embedded in high dimensional spaces. This consideration is further shown by the lower accuracies achieved by coding the points combining $\hat{\tau}(\boldsymbol{x}_i, k)$ and $\hat{d}(\boldsymbol{x}_i, k)$ to obtain LPA$_{\hat{d}, \hat{\tau}}$ and EM$_{\hat{d}, \hat{\tau}}$. Note that we run these further tests since the feature selection approach has highlighted that these features are the most discriminative ones.

Table 4. Mean accuracies computed on synthetic cluster couples ($\mathcal{M}_{10} + \mathcal{M}_*$), real cluster couples (MNIST$_1$ + MNIST$_*$), synthetic cluster triplets ($\mathcal{M}_{10} + 2\mathcal{M}_*$), and real cluster triplets (MNIST$_1$ + 2MNIST$_*$). In boldface the best results have been highlighted.

Dataset	Measure	NS	EM	EM$_{\hat{d}}$	EM$_{\hat{\nu}}$	EM$_{\hat{\tau}}$	EM$_{\hat{d},\hat{\tau}}$	EM$_{\hat{d},\hat{\nu},\hat{\tau}}$	LPA	LPA$_{\hat{d}}$	LPA$_{\hat{\nu}}$	LPA$_{\hat{\tau}}$	LPA$_{\hat{d},\hat{\tau}}$	LPA$_{\hat{d},\hat{\nu},\hat{\tau}}$
$\mathcal{M}_{10}+\mathcal{M}_*$	mean	0.74	0.64	0.78	0.64	0.87	0.86	**0.88**	0.75	0.18	0.16	0.20	0.78	0.87
	std	0.12	0.17	0.19	0.07	0.16	0.17	0.15	0.20	0.05	0.03	0.07	0.25	0.18
MNIST$_1$+MNIST$_*$	mean	0.77	0.53	0.68	0.51	0.93	0.88	**0.95**	0.61	0.21	0.14	0.18	0.83	0.89
	std	0.08	0.01	0.06	0.01	0.02	0.12	0.03	0.17	0.04	0.04	0.05	0.17	0.14
$\mathcal{M}_{10}+2\mathcal{M}_*$	mean	0.34	0.46	0.63	0.52	0.71	0.71	**0.72**	0.63	0.06	0.05	0.06	0.64	0.65
	std	0.22	0.16	0.17	0.10	0.17	0.19	0.17	0.16	0.02	0.02	0.02	0.22	0.22
MNIST$_1$+2MNIST$_*$	mean	0.36	0.37	0.52	0.37	0.72	0.70	**0.74**	0.39	0.08	0.04	0.07	0.57	0.60
	std	0.22	0.01	0.03	0.03	0.08	0.10	0.08	0.08	0.04	0.02	0.03	0.12	0.08

5   Conclusions

In this paper we show that effective clustering results can be obtained by viewing dataset clustering as a multi-manifold clustering problem, where the dataset to be clustered is assumed to be drawn from a geometrical structure composed of several, eventually intersecting, clusters drawn from manifolds embedded into a higher dimensional space, and being characterized by (possibly) different ids. Under this assumption, we achieve promising clustering results by coding the input data by means of local id estimates and features related to them. The promising discriminative quality of the proposed features is shown by a feature relevance test, by the clustering results achieved on both synthetic and real datasets, and by their comparison to those obtained by related and classical state-of-the-art clustering approaches. Note that the proposed features have shown their discriminative power also when applied to a difficult problem such as the clustering of high dimensional datasets characterized by high and low ids.

To further assess the quality of the proposed features, our future works will be focused at their usage to deal with supervised classifion problems.

References

1. Bennett, R.S.: The Intrinsic Dimensionality of Signal Collections. IEEE Trans. on Information Theory IT-15(5), 517–525 (1969)
2. Bishop, C.M.: Bayesian PCA. In: Proc. of NIPS 11, pp. 382–388 (1998)
3. Carter, K.M., Raich, R., Hero, A.O.: On local intrinsic dimension estimation and its applications. IEEE Trans. on Signal Processing 58(2), 650–663 (2010)

4. Ceruti, C., Bassis, S., Rozza, A., Lombardi, G., Casiraghi, E., Campadelli, P.: DANCo: Dimensionality from Angle and Norm Concentration. ArXiv e-prints (June 2012)
5. Ceruti, C., Rozza, A., Bassis, S., Lombardi, G., Casiraghi, E., Campadelli, P.: DANCo: an intrinsic Dimensionalty estimator exploiting Angle and Norm Concentration. Submitted to Pattern Recognition Letters (2013)
6. Costa, J.A., Hero, A.O.: Learning intrinsic dimension and entropy of high-dimensional shape spaces. In: Proc. of EUSIPCO, pp. 231–252 (2004)
7. Jiang, D., Tang, C., Zhang, A.: Cluster analysis for gene expression data: A survey. IEEE Trans. Knowl. Data Eng. 16(11), 1370–1386 (2004)
8. Dempster, A.P., Laird, N.M., Rubin, D.B.: Maximum likelihood for incomplete data via the EM algorithm. J. of the Royal Statistical Soc.: Series B 39(1), 1–38 (1977)
9. Dong, J., Krzyzak, A., Suen, C.: Fast svm training algorithm with decomposition on very large data sets. IEEE Trans. on PAMI 27(4), 603–618 (2005)
10. Frank, A., Asuncion, A.: UCI machine learning repository (2010), http://archive.ics.uci.edu/ml
11. Goldberg, A.B., Zhu, X., Singh, A., Xu, Z., Nowak, R.: Multi-manifold semi-supervised learning – learning when data lives on multiple, intersecting manifolds. In: Proc. of 12th International Conference on Artificial Intelligence and Statistics (2009)
12. Gong, D., Zhao, X., Medioni, G.: Robust multiple manifolds structure learning. In: Proceedings of the 29th International Conference on Machine Learning, Edinburgh, Scotland, UK (2012)
13. Hall, M., Eibe, F., Holmes, G., Pfahringer, B., Reutemann, P., Witten, I.: The weka data mining software: An update. SIGKDD Explorations 11 (2009)
14. Hein, M., Audibert, J.: Intrinsic dimensionality estimation of submanifolds in Rd. In: Proceedings of the ICML, pp. 289–296. ACM (2005)
15. Horton, P., Nakai, K.: A probablistic classification system for predicting the cellular localization sites of proteins. In: Intelligent Systems in Molecular Biology, pp. 109–115 (1996)
16. LeCun, Y., Bottou, L., Bengio, Y., Haffner, P.: Gradient-based learning applied to document recognition. Proc. of IEEE 86, 2278–2324 (1998)
17. Levina, E., Bickel, P.J.: Maximum likelihood estimation of intrinsic dimension. In: Proc. of NIPS, vol. 17(1), pp. 777–784 (2005)
18. Madeira, S., Oliveira, A.: Biclustering algorithms for biological data analysis: A survey. IEEE/ACM Trans. Comp. Biol. Bioinfor. 1(1), 24–45 (2004)
19. Raghavan, U.N., Albert, R., Kumara, S.: Near linear time algorithm to detect community structures in large-scale networks. Physical Review E 76 (2007)
20. Rozza, A., Lombardi, G., Ceruti, C., Casiraghi, E., Campadelli, P.: Novel high intrinsic dimensionality estimators. Machine Learning Journal (May 2012)
21. Steinbach, M., Karypis, G., Kumar, V.: A comparison of document clustering techniques. Tech. Rep. 00034 (2000)
22. Upton, G.J.G.: Approximate confidence intervals for the mean direction of a von Mises distribution. Biometrika 73(2), 525–527 (1986)
23. Wang, Y., Jiang, Y., Wu, Y., Zhou, Z.: Local and structural consistency for multi-manifold clustering. In: Proceedings of IJCAI 2011. AAAI Press (2011)
24. Xiao, Y., Yu, J., Gong, S.: Intrinsic dimension induced similarity measure for clustering. In: Tang, J., King, I., Chen, L., Wang, J. (eds.) ADMA 2011, Part II. LNCS, vol. 7121, pp. 110–123. Springer, Heidelberg (2011)
25. Xu, R., Wunsch, D.: Survey of Clustering Algorithms. IEEE Trans. on Neural Networks 16(3), 645–678 (2005)

Deeply Optimized Hough Transform: Application to Action Segmentation

Adrien Chan-Hon-Tong[1], Catherine Achard[2], and Laurent Lucat[1]

[1] CEA, LIST, DIASI, Laboratoire Vision et Ingénierie des Contenus, France
{adrien.chan-hon-tong,laurent.lucat}@cea.fr
[2] Institut des Systèmes Intelligents et Robotique, UPMC, France
catherine.achard@upmc.fr

Abstract. Hough-like methods like *Implicit Shape Model (ISM)* and Hough forest have been successfully applied in multiple computer vision fields like object detection, tracking, skeleton extraction or human action detection. However, these methods are known to generate false positives. To handle this issue, several works like *Max-Margin Hough Transform (MMHT)* or *Implicit Shape Kernel (ISK)* have reported significant performance improvements by adding discriminative parameters to the generative ones introduced by *ISM*. In this paper, we offer to use only discriminative parameters that are globally optimized according to all the variables of the Hough transform. To this end, we abstract the common vote process of all Hough methods into linear equations, leading to a training formulation that can be solved using linear programming solvers. Our new Hough Transform significantly outperforms the previous ones on *HoneyBee* and *TUM* datasets, two public databases of action and behaviour segmentation.

Keywords: Hough Transform, Learning, Action Segmentation.

1 Introduction

The Hough Transform [9] has first been introduced to detect lines in picture. The main idea of this method is to perform the detection not directly in the picture space but in the line parameter space (Hough space) where each line in the image is mapped into a single point. This method has subsequently been extended to parametric objects [1], and non-parametric objects [11] (e.g. car, pedestrian, sport activities, ...). For non parametric objects, the Hough Transform first learns a probabilistic-like parametrization of the objects on a training database, and, then performs the detections as a local problem in the corresponding Hough space.

Due to this property of local detection, Hough Transform is a very fast process both in theory (time complexity theory) and practice. For this reason, it has been applied in context of real-time systems like [7] for skeleton extraction and more generally in multiple computer vision fields like tracking [6], object detection [5], human action detection [19]. Actually, this method can be used to perform

A. Petrosino (Ed.): ICIAP 2013, Part I, LNCS 8156, pp. 51–60, 2013.

multi-class segmentation in multi-dimensional data e.g. pixel segmentation in image [12], temporal human action segmentation [20].

In this paper, we focus on temporal action segmentation in video: for each frame, the goal is to determine automatically which action is performed by an actor among a set of actions. The Hough Transform, used in this context, is composed of three main steps:

1: feature extraction and quantization to form codewords.
2: each of these codewords ω extracted at frame t votes for the action l at frame $t + \Delta_t$ with a specific weight $\theta(\omega, l, \Delta_t)$. During this step, the same codeword ω votes for all actions l and all relative time displacements Δ_t. The weights have been learned previously, during a learning step.
3: All votes, for all codewords and all frames, are agglomerated to build the Hough score which is the basis for segmentation decisions.

More formally, the Hough Transform (step 2-3) is based on the function $\theta()$ (traditionally positive) that links codewords, time displacements and actions to vote weights. Thus, a codeword w extracted at time t votes with a weight $\theta(w, l, \Delta_t)$ for the hypothesis that an action l is present at time $t + \Delta_t$ (this weight does not depend on the time t but only on l, Δ_t and w). Hence, given a set of localized codewords $W = \{w, t\}$ (w for codewords, t for the extraction time), the Hough score $\mathcal{H}$ for the action l at the time $\bar{t}$ is:

$$\mathcal{H}(\bar{t}, l) = \sum_{(w,t) \in W} \theta(w, l, \bar{t} - t) \tag{1}$$

and, the decision about the action at time $\bar{t}$ is given by:

$$\hat{l}(\bar{t}) = \arg\max_l \left(\mathcal{H}(\bar{t}, l) \right) \tag{2}$$

Thus an action is decided for each frame of the video.

Hence, all the purpose of the training is to select values for $\theta(w, l, \Delta_t)$ that will provide correct decisions with respect to the equations (1) (2) at testing time. Several works, recalled in section 2, offer to improve the generative votes introduced by the *ISM* method by introducing a partial discriminative optimization process during the vote estimation step. In section 3, we offer to extend these methods by optimizing globally all the votes in a discriminative way. With this new learning process, our Hough method significantly outperforms previous ones on two public datasets of action segmentation (the *Honeybee* dataset [14] and the *TUM* dataset [16]) as reported in section 4, before the conclusion in section 5.

2   State of the Art

In this section, we present the different published methods to select the vote weight during the training step of Hough Transform.

2.1   Implicit Shape Model

In the *ISM* [11], the Hough Transform (the values of θ ()) is based on generative weights. Let $\mathcal{P}(l, \Delta_t | w)$ be the probability that the action at time $t + \Delta_t$ is l, knowing that a codeword w has been extracted at time t. This probability is estimated with statistics on the training dataset and is supposed to be independent of t (it just depends on l, Δ_t and w). Then, the weights are given by:

$$\theta_{ISM}(w, l, \Delta_t) = \mathcal{P}(l, \Delta_t | w) \tag{3}$$

In practice, the probability $\mathcal{P}(l, \Delta_t | w)$ is estimated by:

$$\mathcal{P}(l, \Delta_t | w) \approx \frac{N(l, \Delta_t, w)}{N(w)} \tag{4}$$

where $N(l, \Delta_t, w)$ is the number of occurrences of an action l observed with a displacement Δ_t from a codeword w and $N(w)$ is the number of occurrences of the codeword w.

These *ISM*-based weights have several advantages (e.g. parameter-free training, robustness to over-training), but they suffer from several drawbacks. In particular, all codewords and training examples have the same importance and are considered independently from each other. Two methods, *MMHT* [13] and *ISK* [21] have been introduced to address these drawbacks.

2.2   Max-Margin Hough Transform

In *MMHT* [13], a coefficient is introduced for each codeword to ponderate the *ISM* values, resulting in:

$$\theta_{MMHT}(w, l, \Delta_t) = \lambda_w \times \theta_{ISM}(w, l, \Delta_t) = \lambda_w \times \mathcal{P}(l, \Delta_t | w) \tag{5}$$

The weights λ_w give more or less importance to the different codewords w according to their discriminative power. They are learnt simultaneously in a discriminative way through an optimisation process similar to *support vector machine* (*SVM*) training [4].

2.3   Implicit Shape Kernel

In *ISK* [21], the votes are also based on the *ISM* generative ones, but some coefficients are introduced to weight the different training examples. Hence, *ISK* training leads to:

$$\theta_{ISK}(w, l, \Delta_t) = \sum_i \lambda_i \times \mathcal{P}_i(l, \Delta_t | w) \tag{6}$$

where $\mathcal{P}_i(l, \Delta_t | w)$ is an estimation of the probability $\mathcal{P}(l, \Delta_t | w)$ based only on the training example i. The weights λ_i are learnt simultaneously in a discriminative way using a specific *kernel-SVM* training [21].

MMHT and *ISK* report experimental improvements over *ISM* by adding discriminative parameters. This trend is also supported by [18].

2.4 *ISM + SVM*

In [18], it is highlighted that learning directly the Hough map $\mathcal{H}\left(\bar{t}, l\right)$ with a *SVM* is equivalent to the *ISM* with the introduction of a weighting coefficient for each displacement. This results in:

$$\theta_{ISM+SVM}\left(w, l, \Delta_t\right) = \lambda_{\Delta_t} \times \mathcal{P}\left(l, \Delta_t | w\right) = \lambda_{\Delta_t} \times \theta_{ISM}\left(w, l, \Delta_t\right) \qquad (7)$$

2.5 Hough Forest

To our knowledge *ISM* [11] and the presented extensions [13,18,21] are the only published methods to estimate the weights of Hough Transform. More precisely, these methods define links between codewords and votes. There are, of course, various ways to select the features and the codewords, like, the Hough forest [5] methods which are major methods of the state of the art. Hough forests use *ISM* votes, but the mapping between features (usually data patches) and codewords (a leaf in a weak binary classifier tree) is constructed such that all training features associated with the same codeword are expected to come from training examples with a same label. Several works, like [5,20], report that this automatic feature mapping process associated with *ISM* votes leads to significant experimental improvements against codewords obtained without learning, by K-means algorithm for example.

However, in this paper, we focus on the optimisation of the weights used during the vote process and so to the link between codewords and votes which is generic whatever the features and codewords used. Thus, the offered method can be employed in the Hough forest context by substituting the weights estimated by *ISM* by the weights optimized by our offered method.

The common point between *MMHT*, *ISK* and *ISM+SVM* is that they add discriminative parameters to the generative ones introduced by the *ISM*. In this paper, we offer to use only discriminative votes strongly optimized. We call this method Deeply Optimized Hough Transform (*DOHT*).

3 Deeply Optimized Hough Transform

The goal of the training process is to establish a mapping between codewords and weights. While *ISM* method only uses generative weights, *MMHT*, *ISK* and *ISM+SVM* introduce discriminative parameters optimized according to codewords, training examples or displacements. We offer in this paper to optimize all these weights in a global way, according to all parameters of $\theta\left(w, l, \Delta_t\right)$ in multi-class context. Hence, our set of variables is indexed by codewords w (as in *HHMT*), displacements Δ_t (as in *ISM+SVM*) and also by actions l. These differences are summarized in table 1. In this way, we do not use *ISM* values and the method becomes deeply discriminative.

The goal is to define a function $\theta()$ such that for all training examples W and all times $\bar{t}$, the predicted action $\widehat{l}$ is the right one l^* (known on training data).

Considering the definition of the predicted action $\widehat{l}$ (eq. (2)), our problem formulation is equivalent to: $\forall W, \overline{t}, l \neq l^* \left(\overline{t} \right)$

$$\mathcal{H} \left(\overline{t}, l \right) < \mathcal{H} \left(\overline{t}, l^* \left(\overline{t} \right) \right) \tag{8}$$

By dividing $\theta()$ by the minimal gap in eq. (8), this is equivalent to

$$\mathcal{H} \left(\overline{t}, l \right) + 1 \leq \mathcal{H} \left(\overline{t}, l^* \left(\overline{t} \right) \right) \tag{9}$$

and, using equation (1), to $\forall W, \overline{t}, l \neq l^* \left(\overline{t} \right)$

$$\left(\sum_{(w,t) \in W} \theta \left(w, l, \overline{t} - t \right) \right) + 1 \leq \left(\sum_{(w,t) \in W} \theta \left(w, l^* \left(\overline{t} \right), \overline{t} - t \right) \right) \tag{10}$$

In addition, as a codeword extracted at time t should not provide more information about the time $t \pm 1$ than about the time t, we constraint θ to be decreasing with the absolute value of Δ_t.

However, it is not sure that a function can satisfy these constraints. Hence, we introduce a soft margin as in [4]: some variables ξ are introduced in eq. (10) leading to: $\forall \overline{t}, l \neq l^* \left(\overline{t} \right), W$

$$\sum_{(w,t) \in W} \theta \left(w, l, \overline{t} - t \right) + 1 - \xi \left(\overline{t} \right) \leq \sum_{(w,t) \in W} \theta \left(w, l^* \left(\overline{t} \right), \overline{t} - t \right) \tag{11}$$

and these variables are minimized to reduce the number of not-satisfied constraints, leading to the objective function: $\min\limits_{\theta \geq 0, \xi \geq 0} \left(\sum_{\overline{t}} \xi \left(\overline{t} \right) \right)$.

To prevent over-fitting, a regularity term in added to the objective function as in [2]. A coefficient Υ regulates the trade-off between the attachment to data and the regularity as in [4,2].

Finally, the training problem is formulated as:

$$\min_{\theta \geq 0, \xi \geq 0} \left(\sum_{(w, l, \Delta_t)} \theta(w, l, \Delta_t) + \Upsilon \sum_{\overline{t}} \xi \left(\overline{t} \right) \right)$$
$$\text{under constraints: } \forall W, \overline{t}, l \neq l^* \left(\overline{t} \right), \tag{12}$$
$$\sum_{(w,t) \in W} \left(\theta \left(w, l^* \left(\overline{t} \right), \overline{t} - t \right) - \theta \left(w, l, \overline{t} - t \right) \right) + \xi \left(\overline{t} \right) \geq 1$$
$$\text{and: } \forall w, l, \Delta_t, \theta \left(w, a, \Delta_t + \text{sign} \left(\Delta_t \right) \right) \leq \theta \left(w, a, \Delta_t \right)$$

where sign is the sign function.

These formulation is a linear program which which is a well studied problem in literature (e.g. [10]), and, which can be solved efficiency (for example using the solver CPLEX[1], freely available for academic purposes).

In the next section, we evaluate *ISM, HHMT, ISM+SVM* and *DOHT* in action segmentation or behavior segmentation contexts. As *ISK* is only intended

---

[1] www-01.ibm.com/software/websphere/products/
optimization/academic-initiative/

Table 1. The different learning methods of the Hough Transform

methods	θ	variables
ISM [11]	$\theta_{ISM}(w, l, \Delta_t) = \mathcal{P}(l, \Delta_t|w)$	-
MMHT [13]	$\theta_{MMHT}(w, l, \Delta_t) = \lambda_w \times \mathcal{P}(l, \Delta_t|w)$	λ_w
ISK [21]	$\theta_{ISK}(w, l, \Delta_t) = \sum_i (\lambda_i \times \mathcal{P}_i(l, \Delta_t|w))$	λ_i
ISM+SVM [18]	$\theta_{ISM+SVM}(w, l, \Delta_t) = \lambda_{\Delta_t} \times \mathcal{P}(l, \Delta_t|w)$	λ_{Δ_t}
DOHT (our)	$\theta_{DOHT}(w, l, \Delta_t) = \lambda_{w,l,\Delta_t}$	λ_{w,l,Δ_t}

$P(l, \Delta_t|w)$ is the probability that the action at time $t + \Delta_t$ is l knowing that a codeword w has been extracted at time t. $P_i(l, \Delta_t|w)$ is the same probability estimated using only the training example i.

for detection and can not be straightforwardly extended to segmentation, we can not compare it to the others methods.

4   Experimental Results

Experiments have been conducted on the *TUM* [16] and *Honeybee* [14] datasets. These datasets are well designed for segmentation as each frame is associated to an action.

4.1   Application to Human Action Segmentation

TUM is a multi-sensor dataset and in particular it contains skeleton streams (fig. 1). It is composed of 19 sequences about 2 minutes each, containing 9 kinds of actions like *Lowering an object*, *Opening a drawer*, performed by 5 subjects. To provide results comparable to [20], the dataset is separated in training and testing sequences, all algorithms decide an action for each frame of each testing sequence and the performance is measured by the ratio of correctly decided actions.

example of action: *Lowering an object*   Provided skeleton

Fig. 1. TUM dataset [16]

As [20] reports better performances using skeleton features (than visual or visual plus skeleton ones), we decide to consider only skeleton based features. Hence, the input signal of our algorithm is the 3D positions of each articulation at each time.

We use the same preprocessing (features and codewords) than the bag-of-gestures from [3] which achieves the best published performance on this dataset (with a manual segmentation). First, the positions are normalized (positions are expressed in a system of coordinates linked to the subject in order to be invariant to camera point of view, global body position, rotation and size). Then, we consider short temporal series of 3D positions of each articulation as features: let the vector $(p_1, ..., p_T)$ be the normalized trajectory of one articulation, then, we consider the vector $(p_{t-\tau}, ..., p_{t+\tau})$ as a feature extracted at time t. Similar features are also considered in [15,20,17] which emphasize the efficiency of interest points trajectories for human action recognition. Finally, all these features are clustered by K-means. The cluster centers define the codebook and features are mapped to their nearest codeword. We consider the 8 main articulations: feet, hands, knees, elbows. The quantization with K-means is performed independently for each articulation.

The parameters of this set of experiments are τ, K, Υ (for the learning process). On this dataset, the maximal performances achieved by *ISM, MMHT* and *ISM+SVM* when empirically varying the parameters values are less than the mean performances of *DOHT*. A typical run is obtained with $\tau = 6$, $K = 10$, and $C = 1$. Results of this experiment are presented in table 2.

In this set of experiments, *DOHT* significantly outperforms *ISM, MMHT* and *ISM+SVM* and achieves equivalent performance than a *SVM* based on the same features and codeword applied on the optimal segmentation (obtained from the ground truth) from [3]. Hence, for this dataset, we achieve equivalent performance than the best published (82.6% against 84.3%) without using any manual segmentation.

4.2 Application to Behaviour Segmentation

Experiments have also been conducted on the *Honeybee* dataset [14]. The Honeybee dataset provides tracking output of honey bees having 3 kinds of behaviour correlated with their trajectories (figure 2). It composed of 6 large sequences. To provide results comparable to [14], all algorithms decided an action for each frame in a leave-one-out cross validation setting and the performance is measured by the ratio of correctly decided actions.

The input signals in this dataset are the sequences of bee 2D positions and orientations (x_t, y_t, α_t). As in the previous experiment, normalized short temporal series of (2D here) positions are considered as features. Let us call $R(\beta)$ the matrix of the 2D rotation of angle β and $p(t) = (x_t, y_t)$, then we consider the vector $(R(-\alpha_t)(p_{t-\tau} - p_t), ..., R(-\alpha_t)(p_{t+\tau} - p_t))$ as the feature extracted at time t. All these features are clustered using K-means. The cluster centers define the codebook and features are mapped to their nearest codeword. K-means is performed independently for each τ.

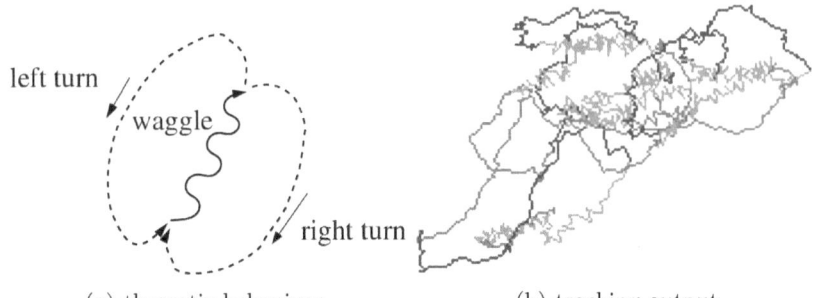

(a) theoretic behaviour   (b) tracking output
Green correspond to waggle, magenta to right turn and blue to left turn.

Fig. 2. Honeybee dataset [14]

The parameters of this set of experiments are τ, K, Υ. On this dataset, the maximal performances achieved by *ISM*, *MMHT* and *ISM+SVM* when empirically varying the parameters values are less than the mean performances of *DOHT*. A typical run is obtained with $\tau \in \{1, 3, 6\}$, $K = 10$, and $C = 1$. Results of this experiment are presented in table 2.

In this set of experiment, *DOHT* significantly outperforms *ISM*, *MMHT* and *ISM+SVM*. In addition, *DOHT* achieves equivalent performances than the latter best published results [8]. In [8], a multi-class *SVM* is applied on each temporal window (with similar kind of features and codewords). Then, segmentation is computed using dynamic programming. As scores are computed on each temporal window, this method is **quadratic** with respect to the maximal length of an activity while our is **linear**. This quadratic property is a common drawback caused by performing scoring as a global problem. Hence, for this dataset, we achieve equivalent performances (86.5% against 89.3%) than the best published results while being significantly faster.

Table 2. Global results on *TUM* [16] and *Honeybee* [14]

Method	Accuracy on *TUM*	Accuracy mean on *Honeybee*
ISM [11]	58.4	71.9
MMHT [13]	69.6	78.8
ISM+SVM [18]	68.5	77.5
DOHT (our)	**82.6**	**86.5**

5   Conclusion

In this paper, we offer to use Hough transform to segment temporal series. In a non parametric context, the training of Hough transform consists in properly selecting the weights used in the voting process. The simple way (Implicit Shape Model) consists in computing some probabilities on the training database, leading to a generative model. Some methods (Max-Margin Hough Transform, Implicit Shape Kernel) offer to add some parameters optimized on a training database in a discriminative way. In this article, we offer to skip the first step based on a generative model, and, to globally learn all the parameters of the Hough transform on the training database, resulting a deeply discriminative model. This required to reformulate the voting process to express it in a linear form in order to use linear programming solvers.

We performed several experiments on public datasets where the Hough transform trained with our method significantly outperforms other Hough transform methods and provides equivalent results than best published results for these datasets while avoiding some limitations of the corresponding algorithms (e.g. manual segmentation).

In future works, we will adapt and apply our method on other contexts e.g. object segmentation in image, video spatio-temporal segmentation, automatic speech segmentation, sign language segmentation.

References

1. Ballard, D.: Generalizing the Hough transform to detect arbitrary shapes. Pattern Recognition (1981)
2. Chakrabartty, S., Cauwenberghs, G.: Gini support vector machine: Quadratic entropy based robust multi-class probability regression. Journal of Machine Learning (2007)
3. Chan-Hon-Tong, A., Ballas, N., Achard, C., Delezoide, B., Lucat, L., Sayd, P., Coise Prêteux, F.: Skeleton point trajectories for human daily activity recognition. In: International Conference on Computer Vision Theory and Application (2013)
4. Cortes, C., Vapnik, V.: Support-vector networks. Journal of Machine Learning (1995)
5. Gall, J., Lempitsky, V.: Class-specific Hough forests for object detection. In: Internationnal Conference on Computer Vision and Pattern Recognition (2009)
6. Gall, J., Yao, A., Razavi, N., Van Gool, L., Lempitsky, V.: Hough forests for object detection, tracking, and action recognition. Pattern Analysis and Machine Intelligence (2011)
7. Girshick, R., Shotton, J., Kohli, P., Criminisi, A., Fitzgibbon, A.: Efficient regression of general-activity human poses from depth images. In: Internationnal Conference on Computer Vision and Pattern Recognition (2011)
8. Hoai, M., Lan, Z., De la Torre, F.: Joint segmentation and classification of human actions in video. In: Internationnal Conference on Computer Vision and Pattern Recognition (2011)
9. Hough, P.V.: Method and means for recognizing complex patterns. Tech. rep. (1962)
10. Karmarkar, N.: A new polynomial-time algorithm for linear programming. In: Theory of Computing (1984)
11. Leibe, B., Leonardis, A., Schiele, B.: Combined object categorization and segmentation with an implicit shape model. In: Workshop on Statistical Learning in Computer Vision (2004)
12. Sreedevi, M., Jeno Paul, P.: An efficient image segmentation using Hough transformation. Asian Journal of Information Technology (2011)

13. Maji, S., Malik, J.: Object detection using a max-margin Hough transform. In: Internationnal Conference on Computer Vision and Pattern Recognition (2009)

14. Oh, S., Rehg, J., Balch, T., Dellaert, F.: Learning and inferring motion patterns using parametric segmental switching linear dynamic systems. Journal of Computer Vision (2008)

15. Sun, J., Wu, X., Yan, S., Cheong, L., Chua, T., Li, J.: Hierarchical spatio-temporal context modeling for action recognition. In: Internationnal Conference on Computer Vision and Pattern Recognition (2009)

16. Tenorth, M., Bandouch, J., Beetz, M.: The tum kitchen data set of everyday manipulation activities for motion tracking and action recognition. In: International Conference on Computer Vision Workshops (2009)

17. Wang, H., Kläser, A., Schmid, C., Cheng-Lin, L.: Action Recognition by Dense Trajectories. In: Internationnal Conference on Computer Vision and Pattern Recognition (2011)

18. Wohlhart, P., Schulter, S., Kostinger, M., Roth, P., Bischof, H.: Discriminative Hough forests for object detection. In: British Machine Vision Conference (2012)

19. Yao, A., Gall, J., Van Gool, L.: A Hough transform-based voting framework for action recognition. In: Internationnal Conference on Computer Vision and Pattern Recognition (2010)

20. Yao, A., Gall, J., Fanelli, G., Van Gool, L.: Does human action recognition benefit from pose estimation? In: British Machine Vision Conference (2011)

21. Zhang, Y., Chen, T.: Implicit shape kernel for discriminative learning of the Hough transform detector. In: British Machine Vision Conference (2010)

Layout-Based Document-Retrieval System by Radon Transform Using Dynamic Time Warping

Giuseppe Pirlo[1,*], Michela Chimienti[2], Michele Dassisti[3],
Donato Impedovo[4], and Angelo Galiano[4]

[1] Dipartimento di Informatica, Università degli Studi di Bari "A. Moro",
via Orabona 4, 70125-Bari, Italy
[2] Laboratorio Kad3, C.da Baione, 70043 Monopoli (BA), Italy
[3] Dip. Meccanica, Management e Matematica, Politecnico di Bari,
viale Japigia 182, 70126 - Bari, Italy
[4] Dyrecta Lab, Via V. Simplicio 45, 70014 Conversano (BA), Italy
giuseppe.pirlo@uniba.it

Abstract. In the context of sustainability of document management technologies, this paper presents a new system for layout-based document retrieval specifically designed for commercial form retrieval. The system first uses a technique based on mathematical morphology to extract grid-based structural components from the document image. Successively, Radon Transform is used for document layout description. A document matching technique based on dynamic time warping is finally adopted. The experimental results carried out on real and simulated data set, demonstrate the effectiveness of the approach with respect to different classes of commercial forms.

Keywords: Document management, Document Image Retrieval, Sustainability, Mathematic Morphology, Radon Transform, Dynamic Time Warping.

1 Introduction

Information Retrieval (IR) is a critical task of document management systems as the number of documents available in databases and digital libraries exponentially grows. Quite often useless reprinting becomes a necessary activity in case of document loss or unavailability. This is also due to standard systems for document retrieval that use text data. They require a document to be present in text form and the querying method is based on a specific textual content in the document. Several advanced techniques have been proposed, based on set-theoretic, algebraic and probabilistic models [1, 2, 3]. Whatever the model used, one of the main drawback of text-based document retrieval systems is that they require a document in text form, since the search for similar documents is based on comparing the textual contents. As a consequence, a preliminary stage of image to text conversion by an Optical Character Recognizer (OCR) is required when a document is in image form. OCR is a time-consuming

---

* Corresponding author.

A. Petrosino (Ed.): ICIAP 2013, Part I, LNCS 8156, pp. 61–70, 2013.

error-prone process, specifically in the case of multi-lingual/multi-font documents and poor-quality document images [4, 5, 6]. The interested reader can refer also to two comprehensive surveys on this topic [7, 8].

In many cases, document search does not depend on the textual content while it is useful to search a document on the basis of its structure. In such cases, methods adopted for document retrieval is a feature vector, in which each feature is extracted from a specific region of the document image. For instance, Tzacheva et al. [9] used a static zoning strategy for document image decomposition to extact a fixed-size feature vector from the document image. In this approach, a regular grid is superimposed to the document image in order to extract regional characteristics. Duygulu et al. [10] proposed a hierarchical zoning strategy to overcome the problem of optimal grid selection, in order to face with the treatment of set of documents of different characteristics. Huang et al. [11] presented a system that extracts text lines and describes the layout by means of relationships between pairs of these lines. Erol et al. [12] used Brick Wall Coding Features (BWC) features, that are local features which represents bounding boxes of the words. Although the features are scale invariant and robust to slight perspective distortion, the accuracy of their system is very low. In addition the method does not work correctly when documents are written in languages such as Japanese and Chinese, in which words are not separated. The system of Liu and Liao [13] combines several approaches to identify a document, as for instance barcode, micro optical patterns, encoding hidden information, paper fingerprint, character recognition, local features etc. . Unfortunately, the retrieval process is time consuming and requires special equipment.

In this paper we deal with commercial forms, such as invoices, waybills, receipts, etc., where layout is strongly characterized by a grid-structure. In this particular cases, traditional document-image approaches are not effective since they are not able to describe documents on the basis of the grid-based structure. A layout-based Document Retrieval (LDR) system is proposed in this paper to handle automatically the documents. In a first step it uses a technique based on mathematical morphology for extracting the grid-based structure in the document layout, removing textual components. Subsequently Radon Transform is used to obtain the feature vector characterizing the specific grid structure of the document. A Dynamic Time Warping (DTW) - based technique finally performs document matching.

The paper is organized as follow. Section 2 presents overall structure of the system. In Section 3 the preprocessing phase is described, based on mathematical morphology operators. Feature extraction is presented in Section 4. Section 5 discusses the matching process based on DTW. The decision combination process is reported in Section 6. Section 7 presents the experimental results while Section 8 presents the conclusion of this work and highlights some directions for further research.

2   System Architecture

Figure 1 shows the structure of the LDR system presented in this paper. After document image acquisition, the document is preprocessed and transformed by Radon

Transform. The features extracted are then stored in the reference database in the enrollment stage. In the running stage, an unknown document is first scanned and preprocessed, successively the features are extracted compared to the those stored into the database. The matching module performs matching by Dynamic Time Warping (DTW) and outputs the ranked list of similar documents.

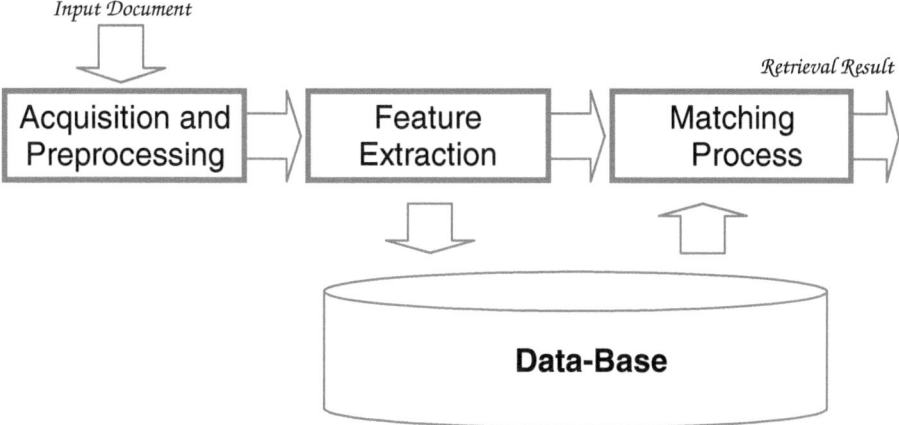

Fig. 1. The LDR system

Each processing step is performed by a well-defined software module. The following Sections will describe each module in detail.

3   Data Acquisition and Preprocessing

The data acquisition and pre-processing module is a key part of the system. It controls the acquisition of the input document as a standard 256 gray-level – 100dpi PDF file. Figures 2 shows an input document concerning a real invoice.

Fig. 2. Input document image I=I(x,y)

Successively, after noise removal, document is resampled to 100 dpi and grid-based structure is extracted by mathematical morphology [14, 15]. More precisely, let I=I(x,y) be the document image ($1 \leq x \leq X$, $1 \leq y \leq Y$) and let be

- B_{hor} the horizontal structure element defined as (see Figure 3a):
$$B=\{(-s,0), \dots, (-1,0), (0,0), (1,0), \dots , (s,0)\};$$

- B_{ver} the horizontal structure element defined as (see Figure 3b):
$$B=\{(0,-s), \dots, (0,-1), (0,0), (0,1), \dots , (0,s)\} ;$$

being s a small positive integer which determine the size of the structure element.

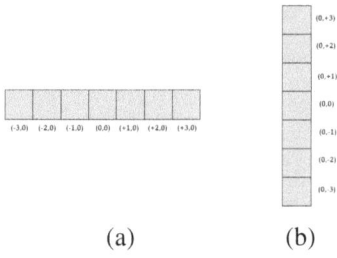

(a)   (b)

Fig. 3. Structure elements (s=3): (a) B_{hor} , (b) B_{ver}

In the preprocessing phase from the image I(x,y) two filtered images $I_{hor}=I_{hor}(x,y)$ and $I_{ver}=I_{ver}(x,y)$, which contains respectively horizontal and vertical segments, are obtained by a closure operator as follows:

$$I_{hor}= I \bullet B_{hor} = (I \oplus B_{hor}) \ominus B_{hor} \tag{1a}$$

$$I_{ver}= I \bullet B_{ver} = (I \oplus B_{ver}) \ominus B_{ver} \tag{1b}$$

being "•" the closure operator, while "⊕" and "⊖" indicate respectively Minkowski sum and difference.

(a)   (b)

Fig. 4. Example of filtered images: (a) I_{hor} , (b) I_{ver}

Finally, $I_{hor}(x,y)$ and $I_{ver}(x,y)$ are combined to reconstruct the preprocessed image I^* according to XOR operator:

$$I^* = I_{hor} \text{ XOR } I_{ver} \qquad (2)$$

Figure 5 shows an example of document image after preprocessing.

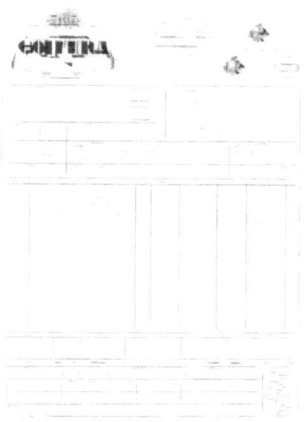

Fig. 5. The preprocessed image I^*

4   Feature Extraction

The feature extraction module, that has been specifically designed for grid-based layout document images, used the Radon Transform then has been extensively used in image analysis and has a number of important applications, like those related to MRI and computed tomography [16, 17]. The complete description of the Radon Transform is beyond the scope of this paper (see further details in [18, 19]). For the aim of this paper we only remind that the Radon Transform computes projection sum of the image intensity along a oriented at line $(\rho - x\cos \vartheta - y\sin \vartheta) = 0$, for each ϑ and ρ. More precisely the Radon Transform of a function $I^*(x,y)$ in an Euclidean space is defined by [20]:

$$S_{\vartheta,\rho} = \int_{-\infty}^{+\infty}\int_{-\infty}^{+\infty} I^*(x, y) \cdot \delta(\rho - x\cos \vartheta - y\sin \vartheta)dxdy \qquad (3)$$

where the d (r) is Dirac function, which is infinite for argument zero and zero for all other arguments (it integrates to one).

Therefore, reckoning of the Radon Transform of a two dimensional image intensity function $I^*(x,y)$ results in its projections across the image at arbitrary orientations ϑ and offsets ρ. Figure 6 presents the results of the Radon Transform applied to the preprocessed image I^* for the parameter values related to horizontal ($\vartheta=0$, $\rho=0$) and vertical ($\vartheta=\pi/2$, $\rho=0$) projections.

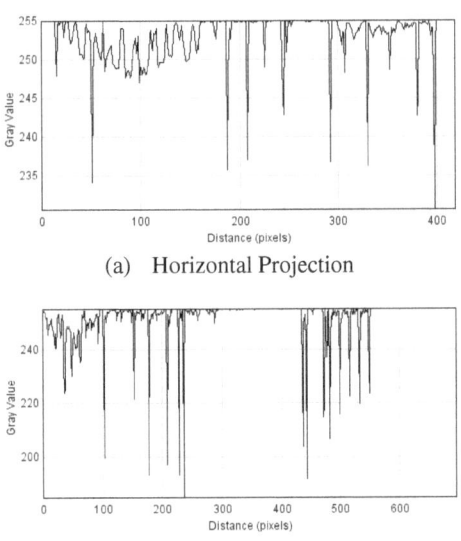

(a)   Horizontal Projection

(b) Vertical Projection

Fig. 6. Feature extraction by Radon Transform

5   Matching

The matching procedure is performed applying the Dinamic Time Warping (DTW) on the feature vectors extracted by the radon transform. In particular, let be F^r, S^t the feature vectors of M elements extracted from the document images I^{*r} and I^{*t}, a warping function between S^r and S^t is any sequence of couples of indexes identifying points of S^r and S^t to be joined [21, 22]:

$$W(S^r,S^t)=c_1,c_2,...,c_K, \qquad (4)$$

where $c_k=(i_k,j_k)$ (k,i_k,j_k integers, $1 \le k \le K$, $1 \le i_k \le M$, $1 \le j_k \le M$). Now, if we consider a distance measure $d(c_k)=d(z^r(i_k), z^t(i_k))$ between elements of S^r and S^t, we can associate to $W(S^r,S^t)$ the dissimilarity measure

$$D_{W(S^r,S^t)} = \sum_{k=1}^{K} d(c_k) . \qquad (5)$$

The DTW detects the warping function $W^*(S^r,S^t)= c^*_1,c^*_2,...,c^*_{K^*}$ which satisfies the condition of [21]:

- Monotonicity (i.e. $i_{k-1} \le i_k$, $j_{k-1} \le j_k$ for k=2,...K)   (6a)
- Continuity (i.e. $i_k - i_{k-1} \le 1$, $j_k - j_{k-1} \le 1$ for k=2,...K)   (6b)
- Boundary (i.e. $i_1 = 1$, $j_1 = 1$ and $i_K = M$, $j_K = M$)   (6c)

and which provides the distance value between S^r and S^t defined as [21, 22]:

$$D_{W^*(S^r,S^t)} = \min_{W(S^r,S^t)} D_{W(S^r,S^t)} . \qquad (7)$$

The value in eq. (7) represents the similarity between the document images I^{*r} and I^{*t}. Hence, according to this strategy, give a document image as input the matching module will outputs the ranked list of the k top similar document images retrieved from the database.

6   The Decision-Making Process

The matching procedure provides two distance-based ranked lists of documents, obtained respectively from horizontal and vertical projection. The decision making process combines the two ranked lists to let take the final decision. In this paper the Borda-count strategy was considered   [23, 24]. According to this strategy, let be $D=\{D_1, D_2,..., D_k,..., D_K\}$ the set of K documents enrolled into the system for reference and D" the unknown input document. Furthermore, let be:

- $L^h : < D^h_1, D^h_2, ..., D^h_k, ..., D^h_K >$ the ranked list of documents obtained from the match of the horizontal projection ($D^h_k \in D$, for k=1,2,,,,,K and $D^h_{k1} \neq D^h_{k2}$ for $k_1 \neq k_2$);
- $L^v : < D^v_1, D^v_2, ..., D^v_k, ..., D^v_K >$ the ranked list of documents obtained from the match of the vertical projection ($D^v_k \in D$, for k=1,2,,,,,K and $D^v_{k1} \neq D^v_{k2}$ for $k_1 \neq k_2$);

The Borda-count approach assigns to each reference document D_k a confidence score $S(D_k)$ defined as [25]:

$$S(D_k) = S^h(D_k) + S^v(D_k) \qquad (8)$$

being $S^h(D_k)=K-i$, if $D_k= D^h_i$; $S^v(D_k)=K-j$, if $D_k= D^v_j$.
Hence, the final list of ranked documents is

$$L^* : < D^*_1, D^*_2, ..., D^*_k, ..., D^*_K > \qquad (9)$$

so that D^*_{k1} precedes D^*_{k2} in L^* if and only is $S(D_{k1}) \geq S(D_{k2})$, and – of course - $_1$ is the top candidate document [25].

7   Experimental Results

For the experimental test, a set of 33 documents have been considered. They are authentic commercial forms belonging to 16 different categories, as Table 1 reports.

Table 1. Dataset

Category	1	2	3	4	5	6	7	8	9	10	11	12	13	14	15	16
Number of Documents	9	4	3	3	2	2	1	1	1	1	1	1	1	1	1	1

Documents were scanned (100dpi , 256 gray-level) and preprocessed. Finally they were stored into a database along with the values of the Radon Transform concerning the horizontal ($S_{0,0}$) and vertical ($S_{0,\pi/2}$) projection. In the testing phase each document has been considered for verifying the effectiveness of the system. In order to estimate the quality of the ranked list provided by the system for a given query, the Average Normalized Rank (ANR) was adopted, defined as in [26]:

$$ANR = \frac{1}{N \cdot N_w} \cdot \sum_{i=1}^{N_w} \left(R_i - \frac{N_w + 1}{2} \right) \tag{10}$$

being

- N the number of documents in the set,
- N_w the number of relevant documents (for the given query) in the set,
- R_i is the rank of each relevant document in the set.

It is worst noting that ANR ranges in [0,1]:

- ANR=0 means that relevant documents are at the top of the ranked list (right position);
- ANR=1 means that relevant documents are at the bottom the ranked list (wrong position).

Figure 7 shows the experimental results. They demonstrate that the proposed approach is very robust with respect to different categories of documents. On average the value of ANR is equal to 0.08. Furthermore, 26 cases out of 33 the ANR is less than 10%, whereas only in one case it is greater than 0.5. Of course, further experiments are in progress and will be shown at the conference.

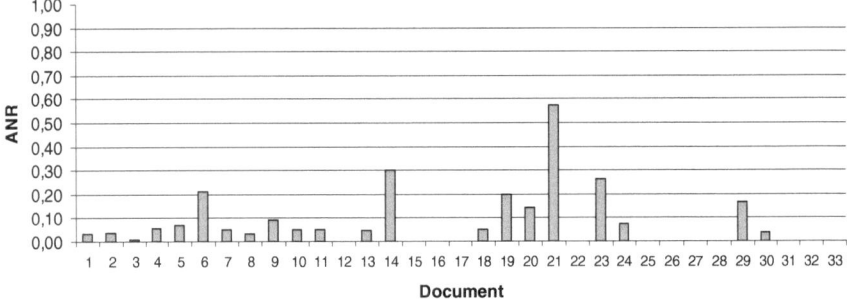

Fig. 7. Experimental Results: Average Normalized Rank (ANR) vs document type

8   Conclusion

A new system for layout-based document image retrieval was presented in this paper.

The system was specifically designed for retrieval of commercial forms as invoices, waybills and receipts, to optimize document management and sustainability.

It uses a morphologic filtering technique and the Radon Transform to obtain multiple document image descriptions. Document matching is then performed on each description by Dynamic Time Warping and a Borda-count decision combination strategy is finally used to combine multiple decisions.

The experimental results, carried out on a dataset of real commercial documents, demonstrate the effectiveness of the proposed solutions. Of course, further experiments are in progress to evaluate system robustness with respect to the size of the dataset as well as to document quality (i.e. faxed/photocopied documents) and image alterations in the acquisition process (i.e. document shift, rotation,   etc.). The results will be shown at the conference.

References

1. Manning, C.D., Raghavan, P., Schütze, H.: An Introduction to Information Retrieval. Cambridge Press (2009)
2. Doermann, D.: The Indexing and Retrieval of Document Images: A Survey. Computer Vision and Image Understanding 70(3), 287–298 (1998)
3. Ko, Y.: A study of term weighting schemes using class information for text classification. In: Proceedings of the 35th International ACM SIGIR Conference on Research and Development in Information Retrieval, pp. 1029–1030. ACM, NY (2012)
4. Marukawa, K., Hu, T., Fujisawa, H., Shima, Y.: Document retrieval tolerating character recognition errors - Evaluation and application. Pattern Recognition 30(8), 1361–1371 (1997)
5. Taghva, K., Borsack, J., Condit, A.: Evaluation of model-based retrieval effectiveness with OCR text. ACM TOIS 14(1), 64–93 (1996)
6. Lopresti, D.: Robust Retrieval of noisy text. In: Proceedings of the Third Forum on Research and Technology Advances in, pp.76–85 (1996)
7. Doermann, D.: The Indexing and Retrieval of Document Images: A Survey. Computer Vision and Image Understanding 70(3), 287–298 (1998)
8. Mitra, M., Chaudhuri, B.: Information retrieval from documents: A Survey. Information Retrieval 2(2/3), 141–163 (2000)
9. Tzacheva, A., El-Sonbaty, Y., El-Kwae, A.: Document Image Matching Using a Maximal Grid Approach. In: Proc. SPIE Document Recognition and Retrieval IX, pp. 121–128 (2002)
10. Duygulu, P., Atalay, V.: A Hierarchical Representation of Form Documents for Identification and Retrieval. International Journal on Document Analysis and Recognition 5(1), 17–27 (2002)
11. Huang, M., Dementhon, D., Doermann, D., Golebiowski, L.: Document ranking by layout relevance. In: Proc. Eighth International Conference on, vol. 1, pp. 362–366 (2005)
12. Erol, B., Antúnez, E., Hull, J.J.: Hotpaper: multimedia interaction with paper using mobile phones. In: Proceeding of the 16th ACM International Conference on Multimedia, pp. 399–408 (2008)
13. Liu, Q., Liao, C.: PaperUI. In: Iwamura, M., Shafait, F. (eds.) CBDAR 2011. LNCS, vol. 7139, pp. 83–100. Springer, Heidelberg (2012)
14. Serra, J.: Image Analysis and Mathematical Morphology. Academic Press (1982)

15. Pirlo, G.: Removing Underlines from Handwritten Text: An experimental investigation. In: Downton, C., et al. (eds.) Handwriting Recognition, pp. 497–502. World Scientific Publishing Co. Pte. Ltd., Singapore (1997) (in Progress)
16. Cormack, A.M.: Computed tomography: Some history and recent developments. In: Proc. Symposia in Applied Mathematics, vol. 27, pp. 35–42 (1983)
17. Deans, S.R.: The Radon Transform and Some of Its Applications. Wiley, NY (1983)
18. Jafari-Khouzani, K., Soltanian-Zadeh, H.: Radon Transform orientation estimation for rotation invariant texture analysis. IEEE Trans. Pattern Anal. Mach. Intell. 27(6), 1004–1008 (2005)
19. Seo, S., et al.: A robust image fingerprinting system using the Radon transforms. Signal Process. Image Commun. 19(4), 325–339 (2004)
20. Hjouj, F., Kammler, D.W.: Identification of Reflected, Scaled, Translated, and Rotated Objects From Their Radon Projections. IEEE Trans. Image Processing 17(3), 301–310 (2008)
21. Salvador, S., Chan, P.: Fast DTW: Toward Accurate Dynamic Time Warping in Linear Time and Space. In: Proc. KDD Workshop on Mining Temporal and Sequential Data, pp. 70–80 (2004)
22. Lemire, D.: Faster Retrieval with a Two-Pass Dynamic-Time-Warping Lower Bound. Pattern Recognition 42(9), 2169–2180 (2009)
23. Kittler, J., Hatef, M., Duin, R.P.W., Matias, J.: On combining classifiers. IEEE Trans. on Pattern Analysis Machine Intelligence 20(3), 226–239 (1998)
24. Xu, L., Krzyzak, A., Suen, C.Y.: Methods of Combining Multiple Classifiers and Their Applications to Handwriting Recognition. IEEE Transaction on Systems, Man and Cybernetics 22(3), 418–435 (1992)
25. Ho, T.K., Hull, J.J., Srihari, S.N.: Decision combination in multiple classifier systems. IEEE Trans. Pattern Anal. Mach. Intell. 16(1), 66–75 (1994)
26. Huang, M., Dementhon, D., Doermann, D., Golebiowski, L.: Document ranking by layout relevance. In: Proc. 8th ICDAR, pp. 362–366 (2005)

Evaluation of Low-Level Image Representations for Illumination-Insensitive Recognition of Textureless Objects

Sebastian Zambanini and Martin Kampel*

Computer Vision Lab, Vienna University of Technology, Austria

Abstract. In this paper the problem of recognizing textureless objects in unconstrained illumination and material conditions is investigated. We evaluate the discriminative power of various low-level image features for a pixelwise representation of the underlying surface characteristics of the object. For this purpose, a new dataset with rendered images of 3D models is used which allows to directly compare the influences of texture and material properties in an object recognition scenario. The results are further validated on a dataset of real object images and finally reveal that jets of single- and multi-scale even Gabor filter responses outperform other proposed features in scenarios with textureless objects and strong variations of illumination.

1   Introduction

Achieving invariance to illumination conditions is a major problem of many computer vision tasks. It has been heavily researched in the past in research areas like face recognition [11] or object recognition [14]. However, usually methods presented in these areas have the goal to extract the albedo information (a.k.a. reflectance or intrinsic image) and reduce the illumination effects. In the standard model [2], the image $I(x, y)$ is considered to be the product of the reflectance $R(x, y)$ (i.e. the albedo or texture) and the illumination effects $L(x, y)$, $I(x, y) = R(x, y) \cdot L(x, y)$. For textured objects, this decomposition makes sense because the albedo image $R(x, y)$ provides a discriminative basis for object comparison. For instance, for face images $R(x, y)$ describes the position and shapes of the lips, eyes, eyebrows etc. However, there is also a wide range of objects in the world with constant albedo (i.e. textureless objects), like coins, statues or facades. As for these objects $R(x, y)$ is constant over the entire image, such a decomposition does not provide any new information helpful for object recognition. In this paper we especially focus on this kind of objects. We address the problem of determining if *two aligned images of textureless objects or object parts show the same 3D surface*. We thereby restrict our study to low-level features in scenarios with *arbitrary, unknown illumination conditions* and *arbitrary, unknown*

---

* This research was supported by the Austrian Science Fund (FWF) under the grant TRP140-N23-2010 (ILAC).

A. Petrosino (Ed.): ICIAP 2013, Part I, LNCS 8156, pp. 71–80, 2013.

bidirectional reflectance distribution function (BRDF) of the objects' material.
Furthermore, we do not consider methods which exploit any other additional information, like 3D object models [3] or learned object appearance from training images [9].

Related Work. In fact, there is only few research related to this problem. Local image features for multi-view object matching have been evaluated with regard to their performance under lighting variations [14,16], but these studies were more general and did not especially evaluate the pixelwise low-level representations on textureless objects. Illumination-invariance in general has also been investigated by Chen et al. [5] concluding that without a priori information for general object classes a true illumination invariant does not exist. Nevertheless, there are representations which are less sensitive to illumination conditions than the original images. Chen et al. proposed to use image gradient directions as they are invariant to linear brightness transformations. Thus, gradient directions are a good choice for flat, textured objects or objects with surfaces of anisotropic depth (i.e. surfaces where the depth changes rapidly in one direction and slowly in another). Another study focusing on low-level features was performed by Osadchy et al. [15]. They showed that the decorrelation induced by a whitening filter for isotropic surfaces increases the distinctiveness of object images. As an approximation for whitening, the Laplacian of Gaussian (LOG) filter was proposed. LOG could be effectively combined with the gradient orientation to a jet of oriented second derivative filters for a distinctive representation for both isotropic and anisotropic surfaces.

Contribution. The problem with these existing studies is that they do not explicitly separate the cases of textured and textureless objects and thus can not give a well-founded statement about the performance of the investigated representations on textureless objects. It is also unclear how the performances of low-level features are related to the material properties or the amount of illumination changes. To explore these questions, we evaluate several low-level representations proposed by earlier works with respect to their performance on textureless and textured objects. This is achieved by a comprehensive evaluation on a new database of synthetically generated images allowing to directly investigate the effects of different material BRDFs and the texturedness of objects. Furthermore, we validate our results on real images of textureless objects. Thus, this paper helps to assess the discriminability and illumination-insensitivity of low-level representations which is helpful for researchers developing superior features and methods for textureless objects. The investigated representations are the basis for more sophisticated local features, e.g. gradient direction [13], gradient orientation [6] or steerable filter responses [4]. Therefore, one can also derive from the presented evaluation how these features act in scenarios with textureless objects.

2   Low-Level Image Representations

We compare eight image representations that have been chosen from literature:

Gradient Direction (GD): Image gradient directions have been identified by Chen et al. [5] as an illumination-insensitive image feature. In the circular domain, the distance between two gradient directions $GD(p)$ and $GD'(p)$ at the image point $p = (x, y)$ is computed as the minimum between $(|GD(p) - GD'(p)|)$ and $(2\pi - |GD(p) - GD'(p)|)$. Using this distance metric for the individual pixels, we take the Sum of Squared Differences (SSD) to compare two images.

Gradient Orientation (GO): Instead of representing image gradients in a *signed* version (directions between 0-360 degree), we can also use an *unsigned* version (orientations between 0-180 degree) of gradients. Gradient orientations are in theory less sensitive to the lighting directions than gradient directions, as opposite lighting directions tend to produce opposite gradient directions at depth discontinuities on the surface [15]. From the gradient direction $GD(p)$, the gradient orientation $GO(p)$ can be simply computed as $GO(p) = \mod(GD(p), \pi)$. To compare two images, the SSD is used where the pixel difference is defined as the minimum between $(|GO(p) - GO'(p)|)$ and $(\pi - |GO(p) - GO'(p)|)$.

Laplacian of Gaussian (LOG): The Laplacian of Gaussian is an approximation of the whitening filter tending to decorrelate the images which makes the filter appropriate for isotropic surfaces [15]. We use the LOG filter by convolving the image and normalizing the absolute responses to unit length. The distance between two images is then again determined by the SSD.

Jets of Gabor Filter Responses (JG): Gabor filters refer to the work of Dennis Gabor [8] in which he proposed to represent a signal as a combination of elementary functions. Gabor filters are widely mentioned to be insensitive against illumination conditions [1,15,12] due to their invariance against additive and multiplicative intensity changes which makes them a popular low-level feature for applications like face recognition [1]. A Gabor filter G has complex coefficients and can thus be defined in terms of a real/even part G_e and an imaginary/odd part G_o,

$$G_e(x,y) = e^{-\frac{x'^2 + \gamma^2 y'^2}{2\sigma^2}} \cos\left(2\pi\frac{x'}{\omega}\right), G_o(x,y) = e^{-\frac{x'^2 + \gamma^2 y'^2}{2\sigma^2}} \sin\left(2\pi\frac{x'}{\omega}\right) \quad (1)$$

with $x' = x\cos\theta + y\sin\theta$ and $y' = -x\sin\theta + y\cos\theta$. The parameter σ defines the standard deviation of the Gaussian envelope whereas ω represents the wavelength of the sinusoidal plane wave. To construct Gabor filters of different sizes but equal shapes one can define σ as a linear function of ω, $\sigma = c \cdot \omega$. θ defines the orientation of the filter and γ is the spatial aspect ratio. To construct a Gabor filter bank we keep the parameters σ, c and γ fixed, use N equally spaced orientations $\theta_1 \ldots \theta_N$ and filter the image with the corresponding N Gabor filters $G_e^{\theta_i}$ and $G_o^{\theta_i}$. The jet $\widetilde{JG}(p)$ is a vector of the magnitude responses of the filtered images $I_e^{\theta_i} = I \star G_e^{\theta_i}$ and $I_o^{\theta_i} = I \star G_o^{\theta_i}$,

$$\widetilde{JG}(p) = [\sqrt{(I_e^{\theta_1}(p))^2 + (I_o^{\theta_1}(p))^2}, \ldots, \sqrt{(I_e^{\theta_N}(p))^2 + (I_o^{\theta_N}(p))^2}] \quad (2)$$

In addition to complex shading patterns, illumination variations can also induce simple multiplicative changes of image intensities which can be compensated by normalizing the jet to unit length [12,15]. The final feature is thus given by the normalized jet $JG(p)$ and the distance between two jets $JG(p)$ and $JG'(p)$ is computed as the L2-norm of their vector difference. Image distances are computed by taking the SSD of $JG(p)$ and $JG'(p)$ for all image points p.

Jets of Even Gabor Filter Responses (JEG): Besides Gabor jets, jets of oriented second derivatives of Gaussians [7] have also been proposed as an effective way of combining LOG and GO to produce a representation which is appropriate for both isotropic and anisotropic surfaces [15]. Even Gabor filters have a very similar shape to oriented second derivatives of Gaussians if the cosine bandwidth is chosen such that the Gaussian envelope roughly covers the cosine range of $[-1.5\pi, 1.5\pi]$ (i.e., $c \approx 0.4$) [12,15]. In this study we use even Gabor filters as they provide a higher flexibility in the definition of the filter shape, due to a more general set of parameters. However, it is clear from the high similarity of the filters that substantially the same performance can be achieved by the use of second derivatives of Gaussians. In contrast to JG, the jet $\widetilde{JEG}$ is formed only from the absolute values of $I_e^{\theta_i}$,

$$\widetilde{JEG}(p) = [|I_e^{\theta_1}(p)|, \dots, |I_e^{\theta_N}(p)|]   \tag{3}$$

The final feature is again given by the normalized jet $JEG(p)$.

Jets of Multi-Scale Even Gabor Filter Responses (JMSEG): The optimal size of the filters depends on the surface characteristics, as for smoother surface parts a wider filter is needed than for less smooth surface parts [15]. One can learn the optimal filter size for a given application domain by means of training data as done in [15], but nonetheless the variation of surface smoothness is disregarded if only one single filter size is used. As in a general scenario the surface characteristics are usually unknown and varying, it is beneficial to extend the single-scale jet JEG towards a multi-scale representation JMSEG. For this jet the single-scale jets JEG_{ω_i}, obtained by filtering with Gabor filters of scales $\omega_1 \dots \omega_M$, are simply concatenated,

$$JMSEG(p) = [JEG_{\omega_1}, \dots, JEG_{\omega_M}]   \tag{4}$$

Self-Quotient Image (SQI): SQI was introduced by Wang et al. [17] as a method to separate the albedo information $R(x, y)$ from images. Similar to other works in this area, the idea is - based on the Lambertian assumption - that the illumination effects mainly appear in the low-frequency components of the image and that they can therefore be eliminated by dividing the image by a smoothed version of it. The method is intentionally designed for textured objects, but showed superior performance in a study of illumination invariance for face recognition [11] on the nearly textureless face parts cheek, chin and nose. Another motivation for including SQI in our evaluation is to assess the performance of the vast amount of methods dedicated to textured objects by evaluating one representative method.

For our experiments, we use the implementation of SQI provided by the INFace[1] toolbox and take the SSD of the SQIs as distance measure.

Gray Value (GV): In order to have a baseline performance, we also report results for simple image differencing, i.e., measuring the SSD between the original gray values of the two images.

3   Experiments

Experiments are conducted on synthetic image datasets built from 3D historical coin models as well as on real datasets of textureless and textured objects. We use synthetic images because this way the parameters of image formation can be freely changed to produce images with different illumination conditions and material properties with or without texture. In this manner, we are able to directly compare the performance of the features under different conditions without introducing a bias due to different objects used between datasets. The real dataset is used to validate our results for various real-world material properties and illumination conditions.

Synthetic Datasets. The synthetic datasets consist of images of 14 coin models which were rendered using the open-source graphics software *Blender*[2]. For each model, twelve sets of 500×500 images with 65 illumination directions were rendered where each set represents one out of four material BRDFs and one out of three texture density levels. Material BRDFs are intended to represent different levels of specularity starting from a Lambertian material with zero specularity up to specular intensity values of 0.25, 0.50 and 1.00. Three texture density levels were chosen to show the correlation of the features' performances to the amount of texture on the objects. The first level shows no texture and thus represents the set of textureless objects. For the remaining two levels we used synthetically generated textures. For each model and dataset, 65 images with varying illumination directions were rendered. The camera image plane is placed parallel to the coin and light source positions are defined by their azimuth angle ϕ and elevation angle λ. We used eight levels of λ with eight levels of ϕ each to produce 64 images. The 65th image is rendered with the light placed at the camera position (i.e. $\lambda = 90°$). Fig. 1(a) shows images of one model rendered with the same illumination parameters for the twelve synthetic datasets. Please note that the coin models have, on a local level, smooth isotropic as well as non-isotropic surface parts and thus cover the wide range of surface characteristics desired for our purpose. The dataset is available for download[3].

Amsterdam Library of Object Images. The Amsterdam Library of Object Images (ALOI) [10] is an image database of 1000 objects that were photographed from three viewpoints and with eight illumination configurations each.

---

[1] http://luks.fe.uni-lj.si/sl/osebje/vitomir/face_tools/INFace/

[2] http://www.blender.org/

[3] http://www.caa.tuwien.ac.at/cvl/people/zamba/sidire/

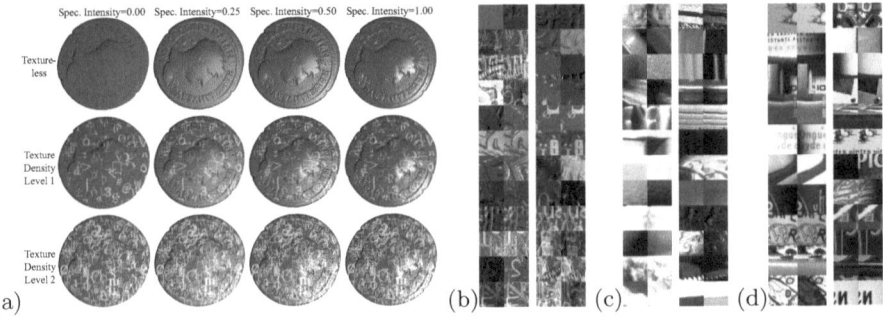

Fig. 1. (a) Coin model rendered with different material properties and texture densities; Patch pairs of size 64×64 from (b) the synthetic datasets, (c) ALOI textureless dataset and (d) ALOI textured dataset

The database contains a wide variety of textureless objects (e.g., a nut, a sponge, white cotton, a metal elephant, a plastic cup...) as well as textured objects (e.g., labeled boxes, an alarm clock, a calendar, a cream tube, a shoe ...). Therefore, the ALOI images provide a realistic and challenging database due to the high variation of material BRDF and surface smoothness among the objects.

3.1   Evaluation Scheme

The performance evaluation scheme is inspired by [4]: as a "good" feature will minimize the distance between image patches showing the same object part and maximize the distance between image patches showing different object parts, we measure these two groups of distances for a given feature and set of true and false image patch pairs. True patch pairs show the same object patch but with different illumination conditions, whereas false patch pairs show different object patches. False positive and true positive rates are sampled from these two groups of distances by means of a varying threshold and used to build a ROC curve of which the area under curve (AUC) is computed as performance measure. For patch pair generation we randomly extracted the same amount of true patch pairs and false patch pairs from the images of a dataset. We use a patch size of 16×16 pixels, but in general the patch size has no significant impact on the results, as has been observed in initial tests. Figure 1(b)-(d) show examples of true patch pairs from the synthetic datasets, the ALOI textureless dataset and the ALOI textured dataset, respectively (64×64 patches for better illustration). To generate patch pairs from the ALOI datasets, we manually identified 80 textureless objects and 80 textured objects in the dataset and randomly picked non-overlapping true and false patch pairs (12000 from the textureless objects and 18000 from the textured objects, as the textureless objects in the dataset are smaller on average).

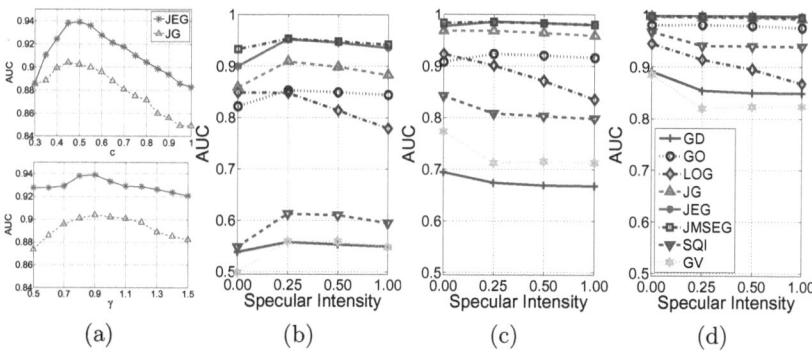

Fig. 2. (a) Recognition performance dependent on parameters c and γ; Recognition performance for different levels of material specularity on (b) textureless objects, (c) objects with texture density level 1 and (d) objects with texture density level 2

3.2   Results and Discussion

Parameter Selection. As our main purpose is the study of the features' behavior on textureless objects with varying material properties, tests for parameter selection were conducted on a mixed set consisting of 50000 true and false patch pairs extracted from the four synthetic datasets of textureless objects. Parameter selection was then achieved by an exhaustive search over the parameter space.

For GD and GO, we tested if a presmoothing of the patches or a bigger Sobel filter than the standard 3×3 one are beneficial in terms of recognition performance, but no improvement could be detected. For LOG we used an exhaustive search to find the optimal standard deviation of the Gaussian. For SQI, no exhaustive parameter selection was conducted as this method is intended for textured objects and initial tests with several parameter settings were not successful in substantially improving the generally bad performance of SQI. Therefore, we used the standard settings defined in the INFace Toolbox. For the features GJ and GEJ, the parameters defining the shape of the Gabor filters (c and γ) as well as the number N of orientations are of interest. Figure 2(a) shows the maximum AUC achieved over the parameter space for various fixed values of c and γ. We can derive from these results that the best performance is achieved when c is set in a range of $0.45 - 0.50$, i.e. the filters have a shape close to second derivatives of Gaussians. The optimal value for the aspect ratio of the filters defined by the parameter γ is around 0.9. Our experiments also revealed that the number of orientations does only have a minor influence on the overall performance for $N > 4$. Based on our results, for the further experiments we used parameter values of $\gamma = 0.9$ and $N = 6$ for JG, JEG and JMSEG, as well as $c = 0.50$ for JEG and JMSEG and $c = 0.45$ for JG. We also identified optimal filter sizes $\omega_1 \ldots \omega_M$ for JMSEG as $\omega_j = \omega_1 2^{(j-1)/2}$ with $\omega_1 = 1$ and $M = 8$.

Recognition Performance Depending on Object Specularity and Texturedness. To evaluate the recognition performance of the features for the twelve synthetic datasets, we randomly extracted 50000 true and false patch pairs for each dataset. Patch pairs contained in the mixed set for parameter selection are not contained in the four textureless datasets used for this evaluation. The results are plotted in Figure 2(b)-(d). It can it can be clearly seen that on textureless objects the representations based on even Gabor responses (JEG and JMSEG) perform best. The multi-scale representation of JMSEG is beneficial especially on Lambertian surfaces where it shows a significant improvement of recognition performance over JEG (AUC of 0.933 against 0.899). Complex Gabor filter responses (JG) are better than the other remaining features but it can be concluded from the worse performance compared to JEG that the phase invariance of the complex filter decreases its recognition power. For image gradients, there is a large discrepancy between the use of gradient orientations (GO) and gradient directions (GD). GD is much less stable than GO as it is highly vulnerable to edge polarity changes induced by opposite lighting directions between true patch pairs. SQI is only slightly better and performs substantially worse than the top-performing features, as this representation is designed for textured objects and is thus highly affected by changes of the shading patterns on textureless objects. Therefore, the method achieves its best results on Lambertian objects with a high texture density (Figure 2(d)). Another conclusion from the results on textureless objects is that more specularity of the objects' material increases the performance. Although a specular BRDF causes more appearance variations from light source variations than a Lambertian BRDF, surface characteristics are also more accentuated by a specular surface, which in turn supports its recognition. The only exception of this effect is LOG which has been especially proposed for smooth, Lambertian objects [15].

The results on textured objects shown in Figure 2(c)-(d) demonstrate that texture increases the recognition performance of all features and in general that their performance is correlated to the degree of texture variation, since albedo discontinuities are less affected by lighting variations than discontinuities of object depth. However, the representations based on Gabor filters are the best performing features for all material types, regardless of the texture density of the objects.

Influence of the Amount of Light Source Change. An interesting question in the context of our evaluation is how the amount of light source difference between the images to be compared has an influence on the discriminative power of the features. Therefore, we took this issue into account for the ROC curve generation by subselecting patch pairs from the textureless objects with a given difference of light source azimuth or elevation. Hence, only true patch pairs with a specified azimuth difference and no elevation difference, and vice versa, are considered. The results of these tests are shown in Figure 3(a)-(b). The plotted curves demonstrate that for smaller light source changes the performances of the features are close together whereas for stronger changes there is also a higher difference in performance. GD is a competitive feature for small light source changes of $45°$ azimuth and $10° - 20°$ elevation, but its performance decreases stronger than that of other features for larger light source changes. GD, SQI and

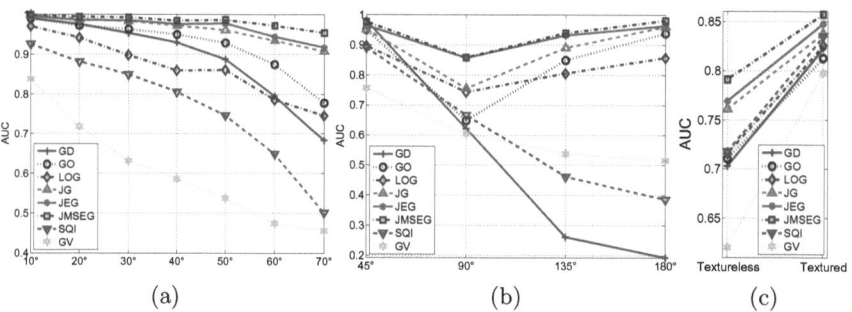

Fig. 3. Recognition performance in relation to (a) the difference of light elevation angle λ and (b) the difference of light azimuth angle φ; (c) Performance on real ALOI datasets

GV are especially vulnerable to changes of the light azimuth, as they are not invariant to edge polarities which tend to change on depth discontinuities for opposite lighting directions. An important aspect of these experiments is that the Gabor-based features show the top performance for all levels of light source changes, but their dominance is more pronounced for higher levels of change.

Recognition Performance on Real Datasets. As can be seen in Fig. 3(c), the results on the real datasets widely reflect the findings of the experiments on the synthetic datasets. JMSEG is again the best performing feature for textureless and textured objects, followed by JEG and JG. The generally lower performance on the real datasets is explained by image noise on the images as well as the acquisition setup used. There are more underexposed (i.e. completely black) and overexposed (i.e. completely white) objects parts which evidently hinders recognition. Nonetheless, the results show that the insights gained from our experiments on synthetic datasets can be transferred to the real world.

4   Conclusions

In this paper we addressed the problem of comparing images of textureless 3D objects in unconstrained conditions. Although in general invariance to illumination conditions is a central computer vision problem, we identified the subarea of textureless objects as an under-researched topic. Therefore, we evaluated several well-known pixelwise low-level features with respect to their recognition performance for textureless objects. Unlike previous studies [5,15], we separately evaluated textureless objects and demonstrated that features claimed to be insensitive to illumination conditions, like gradient direction or the self-quotient image, perform substantially worse on textureless surfaces than on textured surfaces. Our findings are based on a comprehensive evaluation on synthetic datasets with varying degrees of specularity and texturedness as well as real images of textureless and textured objects.

  Our experiments revealed that jets of even Gabor responses are the features of choice for capturing surface characteristics in an illumination-insensitive way, not only for textureless but also for textured objects under strong illumination changes. We also demonstrated that for improved performance one can extend the single-scale representation towards multiple scales by concatenating the single-scale jets. We think that this representation offers a powerful basis for more sophisticated methods that tackle computer vision problems involving textureless objects or heavily changing illumination conditions. For future work, we plan to intensively investigate their usage for higher-level features which will allow for the improved recognition, registration or reconstruction of objects in such scenarios.

References

1. Adini, Y., Moses, Y., Ullman, S.: Face recognition: The problem of compensating for changes in illumination direction. PAMI 19(7), 721–732 (1997)
2. Barrow, H., Tenenbaum, J.: Recovering intrinsic scene characteristics from images. Computer Vision Systems, 3–26 (1978)
3. Basri, R., Jacobs, D.: Lambertian reflectance and linear subspaces. PAMI 25(2), 218–233 (2003)
4. Brown, M., Hua, G., Winder, S.: Discriminative learning of local image descriptors. PAMI 33(1), 43–57 (2011)
5. Chen, H., Belhumeur, P., Jacobs, D.: In search of illumination invariants. In: CVPR, pp. 254–261 (2000)
6. Dalal, N., Triggs, B.: Histograms of oriented gradients for human detection. In: CVPR, pp. 886–893
7. Freeman, W., Adelson, E.: The design and use of steerable filters. PAMI 13(9), 891–906 (1991)
8. Gabor, D.: Theory of communication. Journal of the Institute of Electrical Engineers 93, 429–457 (1946)
9. Georghiades, A., Belhumeur, P., Kriegman, D.: From few to many: Illumination cone models for face recognition under variable lighting and pose. PAMI 23(6), 643–660 (2001)
10. Geusebroek, J., Burghouts, G., Smeulders, A.: The amsterdam library of object images. IJCV 61(1), 103–112 (2005)
11. Gopalan, R., Jacobs, D.: Comparing and combining lighting insensitive approaches for face recognition. CVIU 114(1), 135–145 (2010)
12. Kamarainen, J., Kyrki, V., Kalviainen, H.: Invariance properties of gabor filter-based features-overview and applications. TIP 15(5), 1088–1099 (2006)
13. Lowe, D.: Distinctive image features from scale-invariant keypoints. IJCV 60(2), 91–110 (2004)
14. Moreels, P., Perona, P.: Evaluation of features detectors and descriptors based on 3d objects. IJCV 73(3), 263–284 (2007)
15. Osadchy, M., Jacobs, D., Lindenbaum, M.: Surface dependent representations for illumination insensitive image comparison. PAMI 29(1), 98–111 (2007)
16. Van De Sande, K., Gevers, T., Snoek, C.: Evaluating color descriptors for object and scene recognition. PAMI 32(9), 1582–1596 (2010)
17. Wang, H., Li, S., Wang, Y.: Face recognition under varying lighting conditions using self quotient image. In: FG, pp. 819–824 (2004)

Kernels for Visual Words Histograms

Radu Tudor Ionescu and Marius Popescu

Faculty of Mathematics and Computer Science
University of Bucharest, No. 14 Academiei Street, Bucharest, Romania
{raducu.ionescu,popescunmarius}@gmail.com

Abstract. Computer vision researchers have developed several learning methods based on the bag-of-words model for image related tasks, such as image retrieval or image categorization. For such an approach, images are represented as histograms of visual words from a codebook that is usually obtained with a simple clustering method. Next, kernel methods are used to compare such histograms. Popular choices, besides the linear SVM, are the intersection, Hellinger's, χ^2 and Jensen-Shannon kernels.

This paper aims at introducing a kernel for histograms of visual words, namely the PQ kernel. This kernel is inspired from a class of similarity measures for ordinal variables, more precisely Goodman and Kruskals gamma and Kendalls tau. A proof that PQ is actually a kernel is also given in this work. The proof is based on building its feature map.

Object recognition experiments are conducted to compare the PQ kernel with other state of the art kernels on two benchmark datasets. The PQ kernel has the best mean average precision (AP) on both datasets. In one of the experiments, PQ and Jensen-Shannon kernels are combined to improve the mean AP score even further. In conclusion, the PQ kernel can be used with success, alone or in combination with other kernels, for image retrieval, image classification or other related tasks.

Keywords: kernel method, rank correlation measure, ordinal measure, ordinal data, visual words histograms, bag-of-words, BoW model.

1 Introduction

The classical problem in computer vision is that of determining whether or not the image data contains some specific object, feature, or activity. Particular formulations of this problem are image classification, object class recognition, object detection. Computer vision researchers have recently developed sophisticated methods for such image related tasks. Among the state of the art models are discriminative classifiers using bag-of-words (BoW) representation [13, 19] and spatial pyramid matching [8], generative models [5] or part-based models [7]. The BoW models, which represent an image as a histogram of local features, have demonstrated impressive levels of performance for image categorization [19], image retrieval [11], or related tasks.

This paper focuses on learning methods based on the BoW model. A vocabulary (or codebook) of visual words is obtained by clustering local image descriptors extracted from images. An image is then represented as a histogram of

A. Petrosino (Ed.): ICIAP 2013, Part I, LNCS 8156, pp. 81–90, 2013.

visual words (or bag-of-visual-words). Next, kernel methods are used to compare such histograms. Popular choices, besides the linear SVM, are the intersection, Hellinger's, χ^2 and Jensen-Shannon (JS) kernels. There is no reason to limit the choice of kernels to these options, when other kernels are available. The final goal, that is to improve the results for image related tasks, can be achieved by trying different kernels that could possibly work better.

In this work, a kernel for histograms of visual words, namely the PQ kernel, is introduced. The PQ kernel is inspired from a class of similarity measures for ordinal variables, more precisely Goodman and Kruskals gamma and Kendalls tau. The idea is to treat the visual words histograms as ordinal data, in which data is ordered but cannot be assumed to have equal distance between values. In this case, a histogram will be considered as a rank of visual words according to their frequencies in that histogram. Usage of the ranking of visual words instead of the actual values of the frequencies may seem as a loss of information, but the process of ranking can actually make PQ more robust, acting as a filter and eliminating the *noise* contained in the values of the frequencies. This work proves that PQ is a kernel and it also shows how to build its feature map.

Experiments are conducted in order to assess the performance of different kernels, including PQ, on two benchmark datasets of images. The idea behind the evaluation is to use the same framework and variate only the feature maps computed with different kernels. The experiments show that the PQ kernel has the best mean average precision on both datasets.

The paper is organized as follows. Section 2 presents the learning framework used for image retrieval, image categorization and related tasks. The PQ kernel for histograms of visual words is discussed in section 3. Experiments conducted on two benchmark datasets are presented in section 4. Finally, the conclusions are drawn in section 5.

2   Learning Model

In computer vision, the BoW model can be applied to image classification and related tasks, by treating image descriptors as words. A bag of visual words is a sparse vector of occurrence counts of a vocabulary of local image features. This representation can also be described as a histogram of visual words. The vocabulary is usually obtained by vector quantizing image features into visual words. The proposed learning model (framework) has two stages, one for training and one for testing. Each stage is divided in two important steps. The first step in both stages is for feature detection and representation. The second step is to train a kernel method (in the training stage) in order to predict the class label of new images (in the testing stage).

The feature detection and representation step in the training stage works as follows. Features are detected using a regular grid across the input image. At each interest point, a SIFT feature [10] is computed. This approach is known as dense SIFT [1,3]. Next, SIFT descriptors are vector quantized into visual words and a codebook of visual words is obtained. The vector quantization process is done by k-means clustering [9], and visual words are stored in a randomized forest of k-d trees [11] to reduce search cost. The frequency of each visual word is then recorded in a histogram which represents the final feature vector for the image. The histograms of visual words enter the training step. A kernel method is used for training. Several kernels can be used, such as the linear SVM, the intersection kernel, the Hellinger's kernel, the χ^2 kernel or the Jensen-Shannon kernel. In this paper, a novel approach is proposed, that of using the PQ kernel described in section 3.

Feature detection and representation is similar during the testing stage. The only difference is to use the same vocabulary that was already obtained in the training stage by vector quantization. The histogram of visual words that represents the test image is compared with the histograms learned in the training stage. The system can return either a label (or a score) for the test image or a ranked list of images similar to the test image, depending on the application. For image categorization a label (or a score) is enough, while for image retrieval a ranked list of images is more appropriate.

As expected for an image retrieval system, the training stage can be done offline. For this reason, the time that is necessary for vector quantization and learning is not of great importance. What matters most is to return the result for a new (test) image as quick as possible.

Performance level of the described model depends on the number of training images, but also on the number of visual words. The number of visual words must be set a priori. The accuracy gets better as the number of visual words is greater.

Note that the described model ignores spatial relationships between image features. Despite this fact, visual words showed a high discriminative power and have been used for region or image level classification [2, 6, 19]. Although most approaches are based on sparse descriptors, others have used dense descriptors [6, 17]. A good way to improve performance is to include spatial information [8]. This can be done by dividing the image into spatial bins. The frequency of each visual word is then recorded in a histogram for each bin. The final feature vector for the image is a concatenation of these histograms. The aim of this paper is to improve the performance of the learning model by trying different kernel methods. Therefore, other methods of improving the performance level are disregarded, since they are beyond the purpose of this work. However, one should be aware of all the possibilities of improving the described model for a real application, where the level of performance is of great importance.

3   PQ Kernel for Visual Words Histograms

All common kernels used in computer vision treat histograms of visual words either as finite probability distributions, for example, the Jensen-Shannon kernel, either as random variables whose values are the frequencies of different visual words in the respective images, for example, the Hellinger's kernel (Bhattacharyya's coefficient) and the χ^2 kernel. Even the linear kernel can be seen as the Pearson's correlation coefficient if the two histograms are standardized.

But the histograms of visual words can also be treated as ordinal data, in which data is ordered but cannot be assumed to have equal distance between values. In this case, the values of histograms will be the ranks of visual words according to their frequencies in image rather than of the actual values of these frequencies.

An entire set of correlation statistics for ordinal data are based on the number of concordant and discordant pairs among two variables. The number of concordant pairs among two variables (or histograms) $X, Y \in \mathbb{R}^n$ is:

$$P = |\{(i, j) : 1 \leq i < j \leq n, (x_i - x_j)(y_i - y_j) > 0\}|$$

In the same manner, the number of discordant pairs is:

$$Q = |\{(i, j) : 1 \leq i < j \leq n, (x_i - x_j)(y_i - y_j) < 0\}|$$

Goodman and Kruskal's gamma [14] is defined as:

$$\gamma = \frac{P - Q}{P + Q}$$

Kendall developed several slightly different types of ordinal correlation as alternatives to gamma. *Kendall's tau-a* [14] is based on the number of concordant versus discordant pairs, divided by a measure based on the total number of pairs (n is the sample size):

$$\tau_a = \frac{P - Q}{\frac{n(n-1)}{2}}$$

Kendall's tau-b [14] is a similar measure of association based on concordant and discordant pairs, adjusted for the number of ties in ranks. It is calculated as $(P - Q)$ divided by the geometric mean of the number of pairs not tied on X and the number of pairs not tied on Y, denoted by X_0 and Y_0, respectively:

$$\tau_b = \frac{P - Q}{\sqrt{(P + Q + X_0)(P + Q + Y_0)}}$$

All the above three correlation statistics are very related. If n is fixed and X and Y have no ties, then P, X_0 and Y_0 are completely determined by n and Q. Actually, all are based on the difference between P and Q, normalized differently.

The PQ kernel between two histograms X and Y is defined as:

$$k_{PQ}(X, Y) = 2(P - Q)$$

To prove that k_{PQ} is indeed a kernel, the explicit feature map induced by k_{PQ} is provided next.

Let $X, Y \in \mathbb{R}^n$ be two histograms of visual words. Let Ψ be defined as follows:

$$\Psi : \mathbb{R}^n \to \boldsymbol{M}_{n,n} \ \ \Psi(X) = (\Psi(X)_{i,j})_{1 \leq i \leq n, 1 \leq j \leq n}$$

with

$$\Psi(X) = \begin{cases} 1 & \text{if } x_i > x_j \\ -1 & \text{if } x_i < x_j \\ 0 & \text{if } x_i = x_j \end{cases}$$

Note that Ψ associates to each histogram a matrix that describes the order of its elements.

If matrices are treated as vectors, then the following equality is true:

$$k_{PQ}(X,Y) = 2(P - Q) = \langle \Psi(X), \Psi(Y) \rangle,$$

where $\langle \cdot, \cdot \rangle$ denotes the scalar product. This proves that k_{PQ} is a kernel and provides the explicit feature map induced by k_{PQ}.

Another approach inspired from rank correlation measures is the WTA hash proposed in [18]. For K=2, the WTA hash is closely related to the PQ kernel. However, there are two important differences. The first one is that WTA hash works with a random selection of pairs from the feature set. The second one is that, unlike PQ kernel, the WTA hash ignores equal pairs. In terms of feature representation, the PQ kernel represents a histogram with a feature vector containing $\{-1, 0, 1\}$ (0 for equal pairs), while the WTA hash with $K = 2$ uses only $\{1, 0\}$. In the experiments, one can observe that these differences have direct consequences to the performance level, creating an even greater gap between the two methods.

The authors of [16] state that histograms of γ-homogeneous kernels should be L_γ-normalized. Being linear in the feature space, PQ is a 2-homogeneous kernel and the histograms should be L_2-normalized. Therefore, in the experiments, the PQ kernel is based on the L_2-norm. An important advantage of PQ being linear is that it can be used with linear SVMs, such as the PEGASOS algorithm [12], that are much faster to learn and evaluate than the original non-linear SVMs.

Treating visual words frequencies as ordinal variables means that in the calculation of the distance (or similarity) measure, the ranks of visual words according to their frequencies in image will be used, rather than the actual values of these frequencies. Usage of the ranking of visual words in the calculation of the distance (or similarity) measure, instead of the actual values of the frequencies, may seem as a loss of information, but the process of ranking can actually make the measure more robust, acting as a filter and eliminating the *noise* contained in the values of the frequencies. For example, the fact that a specific visual word has the rank 2 (is the second most frequent feature) in one image, and the rank 4 (is the fourth most frequent feature) in another image can be more relevant than the fact that the respective feature appears 34 times in the first image, and only 29 times in the second. It is important to note that for big vocabularies (with more than 1.000 words), the kernel trick should be employed to obtain the kernel representation of PQ instead of computing its feature map, since there is a quadratic dependence between the feature map and the number of visual words.

4   Experiments

4.1   Datasets Description

The Pascal Visual Object Classes (VOC) challenge [4] is a benchmark in visual object category recognition and detection, providing the vision and machine learning communities with a standard dataset of images and annotation, and standard evaluation procedures. In the experiments of this work, the Pascal VOC 2007 dataset is used. The reason for this choice is that this is the latest dataset for which testing labels are available for download, and the experiments can be done offline.

The second dataset was collected from the Web by the authors of [7] and consists of 100 images each of 6 different classes of birds: egrets, mandarin ducks, snowy owls, puffins, toucans, and wood ducks. This dataset of 600 images is used in order to assess kernels behavior when less training data is available. The Birds dataset is available at http://www-cvr.ai.uiuc.edu/ponce_grp/data/.

4.2   Implementation and Evaluation Procedure

The framework described in section 2 is used for object class recognition. Details about the implementation of the model are given next. In the feature detection and representation step, a variant of dense SIFT descriptors extracted at multiple scales, called PHOW features [1], are used. The number of visual words used in the experiments is 500. For better accuracy, up to 10.000 visual words or more can be used.

Several state of the art kernel methods are compared with the PQ kernel in both experiments. The baseline method is the linear SVM, for which the histograms are L_2−normalized. One of the state of the art methods is based on the Hellinger's kernel. Two variants with different norms of this kernel are used. The first one is based on L_1−normalized feature vectors, and the second one is based on L_2−normalized feature vectors. Another state of the art kernel is Jensen-Shannon, which is L_1−normalized. Finally, these kernels are to be compared with the PQ kernel described in this paper. The PQ kernel is L_2−normalized. For all kernel methods, feature maps are computed from the visual words histograms. The training is always done using a linear SVM on the computed feature maps. The linear SVM is based on a implementation of the PEGASOS algorithm described in [12]. Note the feature map of the JS kernel cannot be computed directly. In order to use the same learning setting, its feature map has to be approximated using the method proposed in [16]. To approximate the JS kernel, 10.500 features are used. The idea behind the evaluation is to use the same framework and variate only the feature maps computed with different kernels, since the final goal of the experiments is to evaluate the difference between these kernels, in terms of performance. The implementation of both the feature detection and representation step, and the learning step, is mostly based on the VLFeat library [15].

The evaluation procedure for both experiments follows the Pascal VOC benchmark. The qualitative performance of the learning model is measured by using

Table 1. Mean AP on Pascal VOC 2007 dataset for machine learning methods based on visual words histograms with different kernels. The best AP on each class is highlighted with bold.

Class	Lin. L_2	Hel. L_1	Hel. L_2	WTA L_2	JS L_1	PQ L_2	JS+PQ
Aeroplane	0, 395%	0, 555%	0, 558%	0, 534%	0, 564%	0, 526%	**0, 574%**
Bicycle	0, 189%	0, 339%	0, 337%	0, 398%	0, 367%	**0, 409%**	0, 386%
Bird	0, 178%	0, 248%	0, 247%	0, 274%	0, 284%	0, 281%	**0, 305%**
Boat	0, 334%	0, 540%	0, 551%	0, 476%	0, 549%	0, 505%	**0, 553%**
Bottle	0, 122%	**0, 143%**	0, 139%	0, 139%	0, 127%	0, 140%	0, 129%
Bus	0, 239%	0, 334%	0, 336%	0, 404%	0, 379%	**0, 419%**	0, 406%
Car	0, 518%	0, 599%	0, 602%	0, 659%	0, 644%	**0, 670%**	0, 659%
Cat	0, 281%	0, 349%	0, 351%	0, 382%	0, 378%	**0, 402%**	0, 393%
Chair	0, 308%	0, 399%	0, 399%	0, 398%	**0, 414%**	0, 405%	**0, 414%**
Cow	0, 117%	0, 174%	0, 172%	**0, 209%**	0, 169%	**0, 209%**	0, 198%
Dining Table	0, 205%	0, 238%	0, 227%	0, 237%	0, 242%	0, 253%	**0, 255%**
Dog	0, 212%	0, 271%	0, 266%	0, 263%	0, 293%	0, 287%	**0, 299%**
Horse	0, 484%	0, 518%	0, 530%	0, 601%	0, 595%	0, 609%	**0, 614%**
Motorbike	0, 213%	0, 398%	0, 389%	0, 427%	0, 413%	**0, 451%**	0, 450%
Person	0, 639%	0, 715%	0, 717%	0, 756%	0, 759%	0, 773%	**0, 774%**
Potted Plant	0, 099%	**0, 125%**	0, 110%	0, 110%	0, 112%	0, 111%	0, 115%
Sheep	0, 220%	0, 217%	0, 237%	0, 219%	0, 222%	**0, 259%**	0, 243%
Sofa	0, 184%	0, 304%	0, 320%	0, 310%	0, 325%	0, 322%	**0, 333%**
Train	0, 363%	0, 534%	0, 528%	0, 547%	0, 554%	0, 570%	**0, 574%**
TV Monitor	0, 196%	0, 309%	0, 295%	0, 345%	0, 336%	**0, 351%**	0, 342%
Overall	0, 275%	0, 365%	0, 365%	0, 384%	0, 386%	0, 398%	**0, 401%**

the classifier score to rank all the test images. Next, the retrieval performance is measured by computing a precision–recall curve. In order to represent the retrieval performance by a single number (rather than a curve), the mean average precision (mAP) is often computed. The average precision as defined by TREC is used in the experiments. This is the average of the precision observed each time a new positive sample is recalled.

4.3   Pascal VOC Experiment

The first experiment is on the Pascal VOC 2007 dataset. There are 20 classes available in this dataset, and for each class the dataset provides a training set, a validation set and a test set. The training and validation sets have about 2.500 images each, while the test set has about 5.000 images. The validation set is used to validate the parameter C of the linear SVM algorithm. Table 1 presents the mean AP of the linear SVM, the Hellinger's kernel, the JS kernel, the WTA hash (with $K = 2$ and 10.000 random pairs) and the PQ kernel, on the Pascal VOC dataset. Looking at the results obtained by the JS kernel on one hand, and the PQ kernel on the other, one can observe that these methods are somehow complementary in terms of performance. This gives the idea of combing the two kernels to possibly obtain better results. Indeed, in this experiment another kernel based on the sum of JS and PQ kernels is presented. In order to obtain the feature map of this kernel combination, the feature maps of JS and PQ kernels are simply concatenated.

The accuracy of the state of the art kernels is well above the accuracy of the baseline linear SVM. In terms of AP, the state of the art kernels are about 10% better than the baseline method. The PQ kernel improves the accuracy of the

Table 2. The time for the second stage of the learning model and the number of features for each kernel. The time is measured in seconds.

	Lin. L_2	Hel. L_1	Hel. L_2	WTA L_2	JS L_1	PQ L_2	JS+PQ
Time	$1-2$	$2-3$	$2-3$	$15-16$	$15-16$	$830-860$	$850-880$
Features	500	500	500	10.000	10.500	250.000	260.500

learning model, when compared to the state of the art methods. The mAP of the PQ kernel is $3,3\%$ above the mAP of the Hellinger's kernels, $1,4\%$ above the mAP of the WTA hash, and $1,2\%$ above the mAP of the JS kernel. The combination of JS and PQ kernels improves the performance even further. The mAP of the JS+PQ kernel is $3,6\%$ above the mAP of the Hellinger's kernels, $1,7\%$ above the mAP of the WTA hash, and $1,5\%$ above the mean AP of the JS kernel. PQ kernel improves results over WTA hash by $1,4\%$, showing that the two methods are distinct. If the best AP per class is considered, the PQ kernel and the JS+PQ kernel win most of the classes (18 out of 20). The results presented in Table 1 come to support this statement. The L_1-normalized Hellinger's kernel seems to work best when classes are very difficult for all kernel methods.

The feature detection and representation stage, that builds a vocabulary of visual words and obtains histograms, takes a few hours on this dataset. The time for the second stage of the learning framework, that includes computing feature maps, training and testing, depends on the number of features in the feature space for each kernel. The time for the second stage and the number of features for each kernel is given in Table 2. The time was measured on a computer with Intel Core i7 2.3 GHz processor and 8 GB of RAM memory using a single Core. While the feature maps can be computed only once for the entire experiment along with the feature detection and representation stage, training and testing has to be repeated for each class. Despite the time for the PQ kernel ($14-15$ minutes) is higher than the time for other kernels ($2-15$ seconds), it doesn't add an overhead to the overall time of the learning framework, since the overall time is about $4-6$ hours. The PQ kernel and the JS+PQ kernel are constantly better than the other methods. In conclusion, the PQ kernel, used either alone or in combination with the JS kernel, has the best performance on this experiment.

4.4   Birds Experiment

The second experiment is on the Birds dataset. The training set consists of 300 images and the test set consists of another 300 images. There are 6 classes in this dataset. For each class, the dataset contains 50 positive train images and 50 positive test images. Since there is no validation set this time, the parameter C of the linear SVM algorithm is cross-validated on the training set.

Table 3 presents the mAP of the linear SVM, the Hellinger's kernel, the JS kernel, the WTA hash (with $K=2$ and 10.000 random pairs) and the PQ kernel, on the Birds dataset. A variant of the PQ kernel that ignores equal pairs (PQ ieq), which is more similar to the WTA hash, is also added to the experiment to emphase the difference between PQ kernel and WTA hash. The performance of the Hellinger's kernels is above the baseline linear SVM, as in the previous

Table 3. Mean AP on Birds dataset for machine learning methods based on visual words histograms with different mthods. The best AP on each class is highlighted with bold.

Class	Lin. L_2	Hel. L_1	Hel. L_2	JS L_1	WTA L_2	PQ ieq L_2	PQ L_2
Egret	$0,552\%$	$\mathbf{0,760\%}$	$0,747\%$	$0,416\%$	$0,735\%$	$0,738\%$	$0,753\%$
Mandarin Duck	$0,446\%$	$0,585\%$	$0,607\%$	$0,375\%$	$0,784\%$	$0,791\%$	$\mathbf{0,835\%}$
Owl	$0,815\%$	$0,895\%$	$0,887\%$	$0,490\%$	$0,879\%$	$0,889\%$	$\mathbf{0,915\%}$
Puffin	$0,427\%$	$0,696\%$	$0,730\%$	$0,369\%$	$0,708\%$	$0,703\%$	$\mathbf{0,764\%}$
Toucan	$0,572\%$	$0,715\%$	$0,747\%$	$0,558\%$	$0,776\%$	$0,787\%$	$\mathbf{0,845\%}$
Wood Duck	$0,608\%$	$0,795\%$	$0,816\%$	$0,361\%$	$0,767\%$	$0,769\%$	$\mathbf{0,849\%}$
Overall	$0,570\%$	$0,741\%$	$0,756\%$	$0,428\%$	$0,775\%$	$0,779\%$	$\mathbf{0,827\%}$

experiment. Both Hellinger's kernels are about 18% better than the baseline method. Unlike the previous experiment, the JS kernel has the worst accuracy on this dataset, when compared to the rest of the methods. The mAP of the JS kernel is $14,2\%$ below the baseline AP. The bad performance of the JS kernel on this dataset can be explained by the fact that it is based on an informational measure that uses an estimation of the distribution of the data. The number of training samples may not be enough for a good estimation.

The results of the PQ kernel on this experiment are consistent with the previous experiment. The PQ kernel improves the performance of the learning model, when compared to the state of the art kernels. The mean AP of the PQ kernel is $8,6\%$ above the mAP of the L_1-normalized Hellinger's kernel, $7,2\%$ above the mAP of the L_2-normalized Hellinger's kernel, and $5,2\%$ above the mAP of the WTA hash. Table 3 also shows that by ignoring equal pairs the mAP of the PQ kernel drops by $4,8\%$. By taking into account equal pairs and by considering the entire feature set, PQ has a significant improvement in terms of accuracy over WTA hash. There is no question that the two methods are distinct. If the best AP per class is considered, the PQ kernel wins most of the classes, again. The Hellinger's kernel based on the L_1-norm wins the *Egret* class. The PQ kernel wins the rest 5 classes. Note that the linear SVM, the L_2-normalized Hellinger's kernel and the JS kernel are not able to win any class. The PQ kernel is constantly better than the other methods. In conclusion, the PQ kernel has the best performance on the Birds dataset experiment.

5   Conclusion and Further Work

This paper discussed learning methods based on the BoW model. Usually, kernel methods, such as the linear SVM, the Hellinger's kernel or the JS kernel, are used to compare such histograms. This work showed that results for image classification, image retrieval or related tasks, can be improved by changing the kernel. Object recognition experiments compared the PQ kernel with other state of the art kernels on two benchmark datasets. The PQ kernel, used either alone or in combination with the JS kernel, has the best accuracy on the Pascal VOC 2007 experiment. The mAP of the JS+PQ kernel is at least $1,5\%$ above the best mAP of the state of the art kernels. On the Birds experiment, the PQ kernel improved the performance again. The mAP of the PQ kernel is at least $5,2\%$ above the best mAP of the state of the art kernels. The PQ kernel is constantly better than the other methods.

A possible way of improving the results for the PQ kernel may be that of using a TF-IDF measure for visual words as in [11]. Furthermore, eliminating visual words that have a low TF-IDF score can lead to an approximation of the PQ kernel that works faster and possibly better. In future work, other methods inspired from ordinal measures can be investigated.

References

1. Bosch, A., Zisserman, A., Munoz, X.: Image Classification using Random Forests and Ferns. In: ICCV, pp. 1–8. IEEE Computer Society Press (2007)
2. Csurka, G., Dance, C.R., Fan, L., Willamowski, J., Bray, C.: Visual categorization with bags of keypoints. In: Workshop on Statistical Learning in Computer Vision, ECCV, pp. 1–22 (2004)
3. Dalal, N., Triggs, B.: Histograms of Oriented Gradients for Human Detection. In: CVPR, vol. 1, pp. 886–893. IEEE Computer Society, Washington, DC (2005)
4. Everingham, M., van Gool, L., Williams, C.K., Winn, J., Zisserman, A.: The Pascal Visual Object Classes (VOC) Challenge. IJCV 88(2), 303–338 (2010)
5. Fei-Fei, L., Fergus, R., Perona, P.: Learning generative visual models from few training examples: An incremental Bayesian approach tested on 101 object categories. CVIU 106(1), 59–70 (2007)
6. Fei-Fei, L., Perona, P.: A Bayesian Hierarchical Model for Learning Natural Scene Categories. In: CVPR, vol. 2, pp. 524–531. IEEE Computer Society (2005)
7. Lazebnik, S., Schmid, C., Ponce, J.: A Maximum Entropy Framework for Part-Based Texture and Object Recognition. In: ICCV 2005, vol. 1, pp. 832–838. IEEE Computer Society, Washington, DC (2005)
8. Lazebnik, S., Schmid, C., Ponce, J.: Beyond Bags of Features: Spatial Pyramid Matching for Recognizing Natural Scene Categories. In: CVPR 2006, vol. 2, pp. 2169–2178. IEEE Computer Society, Washington, DC (2006)
9. Leung, T., Malik, J.: Representing and Recognizing the Visual Appearance of Materials using Three-dimensional Textons. IJCV 43(1), 29–44 (2001)
10. Lowe, D.G.: Object Recognition from Local Scale-Invariant Features. In: ICCV, vol. 2, pp. 1150–1157. IEEE Computer Society, Washington, DC (1999)
11. Philbin, J., Chum, O., Isard, M., Sivic, J., Zisserman, A.: Object retrieval with large vocabularies and fast spatial matching. In: CVPR 2007, pp. 1–8 (2007)
12. Shalev-Shwartz, S., Singer, Y., Srebro, N.: Pegasos: Primal Estimated sub-GrAdient SOlver for SVM. In: ICML, pp. 807–814. ACM (2007)
13. Sivic, J., Russell, B.C., Efros, A.A., Zisserman, A., Freeman, W.T.: Discovering Objects and their Localization in Images. In: Proceedings of ICCV, pp. 370–377 (2005)
14. Upton, G., Cook, I.: A Dictionary of Statistics. Oxford University Press (2004)
15. Vedaldi, A., Fulkerson, B.: VLFeat: An Open and Portable Library of Computer Vision Algorithms (2008), http://www.vlfeat.org/
16. Vedaldi, A., Zisserman, A.: Efficient additive kernels via explicit feature maps. In: CVPR, pp. 3539–3546. IEEE Computer Society, San Francisco (2010)
17. Winn, J., Criminisi, A., Minka, T.: Object Categorization by Learned Universal Visual Dictionary. In: ICCV, vol. 2, pp. 1800–1807. IEEE Computer Society (2005)
18. Yagnik, J., Strelow, D., Ross, D.A., Lin, R.S.: The power of comparative reasoning. In: ICCV, pp. 2431–2438. IEEE (2011)
19. Zhang, J., Marszalek, M., Lazebnik, S., Schmid, C.: Local Features and Kernels for Classification of Texture and Object Categories: A Comprehensive Study. IJCV 73(2), 213–238 (2007)

A New Adaptive Zoning Technique for Handwritten Digit Recognition

Sebastiano Impedovo, Francesco Maurizio Mangini, and Giuseppe Pirlo

Department of Computer Science, University of Bari "Aldo Moro"
Via Edoardo Orabona, 4 – 70125 Bari, Italy
francescomaurizio.mangini@uniba.it

Abstract. In this paper we present a new adaptive zoning technique based on Voronoi tessellation for the task of handwritten digit recognition. This technique extracts features according to an optimal zoning distribution, obtained by an evolutionary-strategy based search. Several experiments have been conducted on the MNIST and the USPS datasets to investigate the proposed approach. Comparisons with regular square zoning reveal that the presented zoning strategy achieves better results for any type of SVM classifier. Furthermore, the proposed zoning method shows that the combination of the adaptive zoning strategy with the Voronoi topology leads to find a distribution of zones able to improve accuracy significantly. As a matter of fact reached accuracies are close to the best algorithms.

Keywords: Zoning, Feature Extraction, Support Vector Machines.

1   Introduction

In handwritten character recognition, zoning is a well-known approach for feature extraction. It consists in dividing a pattern image into zones and collecting information from each zone. Whatever type of information is collected, the purpose of zoning is to allow to extract features from pattern images so that a given classifier can improve its accuracy. Many zoning methods have been proposed so far, based on different partitioning criteria [10].

Among the most recent approaches are those that attempt to adapt the distribution of zones depending on the patterns to be recognized. They have in common the idea to tailor zoning positions with the aim of meeting certain requirements. The most interesting works that fall into this category are briefly described below.

Radtke et al. [17] introduced an automatic approach to define zoning using Multi-Objective Evolutionary Algorithms (MOEAs). They provided a self adaptive methodology to find the best distribution of zones that obey to two optimality criteria: an error rate as low as possible and a minimal number of non-overlapping zones. They reported 95% accuracy with a nearest neighbor classifier on the NIST Special Database 19 (SD19).

Gatos et al. [8] adopted a zoning strategy that extracted features after adjusting the position of each zone according to local pattern information. For classification, they

A. Petrosino (Ed.): ICIAP 2013, Part I, LNCS 8156, pp. 91–100, 2013.

used the Euclidean distance between feature vectors combined with a minimum distance classifier. They achieved the best recognition accuracy of 88.35% on the CIL database of Greek handwritten characters, which consists of 28750 samples and has 46 classes.

Impedovo et al. [11] considered zoning design as a problem in which zone distribution had to be adapted in order to increase its discrimination capability. They reached an accuracy of 96% in handwritten digit recognition on the CEDAR dataset.

In this paper, inspired by the cited works, a new adaptive zoning technique based on Voronoi tessellation is presented. This approach tackles zoning design as an optimization problem and the optimal zoning distribution is obtained by an evolutionary strategy [1]. It differs from [11] in the problem definition, the optimization strategy, as well as the classifier.

Since SVM is a state-of-the-art classification method, accuracy performances of the proposed zoning approach have been experimentally investigated with several standard SVM classifiers. In addition, these performances have been compared with the ones obtained by regular square zoning.

Experimental tests have been conducted on the MNIST and the USPS databases of handwritten numerals. The results show that the proposed zoning method outperforms regular square zoning for any type of SVM classifier.

The remaining part of this paper is organized as follows. Section 2 formulates the zoning problem. Section 3 describes the proposed technique for zoning design. The experimental tests and the results are discussed in Section 4. The conclusions are drawn in Section 5.

2   Zoning Problem Definition

The introduced zoning strategy is based on Voronoi tessellation. It is a set of zones, called Voronoi polygons, built with respect to M generator points enclosed in the plane. Each generator p_i lies inside the boundaries of a Voronoi polygon Pol_i with the following property:

$$Pol_i = Pol(p_i) = \{p|d(p_i, p) \leq d(p_j, p), i{\neq}j\} \tag{1}$$

where $d(p_i, p_j)$ is the distance from point p_i to p_j.

Pol_i is the set of points closer to p_i than to any other p_j. The tessellation is $D_{Voronoi} = \{Pol_i\}_{i=1...M}$ (M being the number of zones).

For a given set of generator points, $D_{Voronoi}$ is unique and generates regions which are path connected. Hence, the Voronoi Tessellation can be a zoning method that partitions the image into a set of Voronoi polygons, given a set of distinct points into the image plane. In particular, if I is an image of WxH pixels, a Voronoi diagram refers to a tessellation of the domain $I = \{(x, y) | 0 \leq x \leq W, 0 \leq y \leq H\}$ by the Voronoi regions $\{Pol_i\}_{i=1...M}$ associated with a set of given generators $\{p_i\}_{i=1...M}$. Here, the Euclidean distance has been used in (1).

The aim of our approach is to find the image partition that improves the accuracy of a given classifier. This can be viewed as an optimization problem $\mathcal{P}(S, \mathcal{F})$, where S is the search space, whose elements are candidate solutions, and $\mathcal{F}: S \rightarrow R$ is the function to be minimized. In our case, S consists of all the possible M-permutations without repetition of all the n pixel points $p_i = (x_i, y_i)$ belonging to the image I, being M the number of zones and n the number of pixels in the image. According to our purpose, the objective function $\mathcal{F}$ is assumed equal to the misclassification error rate Err committed by a given classifier:

$$\mathcal{F}(S) = Err(S) \tag{2}$$

If S is the set of all the Voronoi generating points $\{p_i\}_{i=1...M}$, with $p_i \in I$, the minimization problem $\mathcal{P}(S, \mathcal{F})$ can be written as follows:
Find $s^* \in S, s^* = \{p^*_i\}_{i=1,...,M}$ so that:

$$\mathcal{F}(s^*) = \min{}_{s \in S}\, \mathcal{F}(s) \tag{3}$$

where p_i are the object variables of the optimization problem.

In this paper, for the optimization problem (3), an evolutionary approach has been used [6] because evolution strategies are generally known to be quite fast and efficient solvers.

3   The Evolutionary Strategy

Evolutionary strategies [1, 6] are iterative optimization techniques which try to discover an optimal solution through stochastic small changes of the object variables.

Figure 1 depicts the evolution cycle of the presented Evolutionary Strategy as a flowchart. This flowchart starts from an initial candidate solution s^0 that enters an evolution loop. The parent is then used for generating a set of λ offsprings by the mutation operator (4).

Thereafter, the population of the next generation is selected as the μ individuals with the best fitness, either from the offsprings and the parent (plus strategy $\mu+\lambda$). This loop is repeated until a termination criterion is met.

If p_i^g and p_i^{g+1} are respectively the point p_i at g-th and (g+1)-th generation, the mutation involves replacing each point $p_i^g = (x_i^g, y_i^g)$ with $p_i^{g+1} = (x_i^{g+1}, y_i^{g+1})$ as follows:

$$\begin{cases} x_i^{g+1} = x_i^g + \Delta_{x_i^g}(0, \sigma^g) \\ y_i^{g+1} = y_i^g + \Delta_{y_i^g}(0, \sigma^g) \end{cases} \quad \forall\, i = 1,..,M \tag{4}$$

where:

- $\Delta_{x_i^g}(0, \sigma^g)$ and $\Delta_{y_i^g}(0, \sigma^g)$ are random Gaussian numbers with a mean of zero and standard deviation of σ^g.
- σ^g is the standard deviation at g-th generation.

The standard deviation plays a key role as it allows to control the speed of convergence. It is often called step size or mutation strength and its adjustment is one of the most important factors in the algorithm. In this work, the control of standard deviation is performed by log-normal self-adaptation operator [1]:

$$\sigma^g = \sigma \, e^{\tau \, N(0, \, 1)} \tag{5}$$

Equation (5) contains a new fundamental parameter to be fixed: the learning rate τ. The general recommendation is to choose $\tau \propto 1/\sqrt{M}$, as it has been proven to be optimal with respect to the convergence speed in a theoretical case [6].

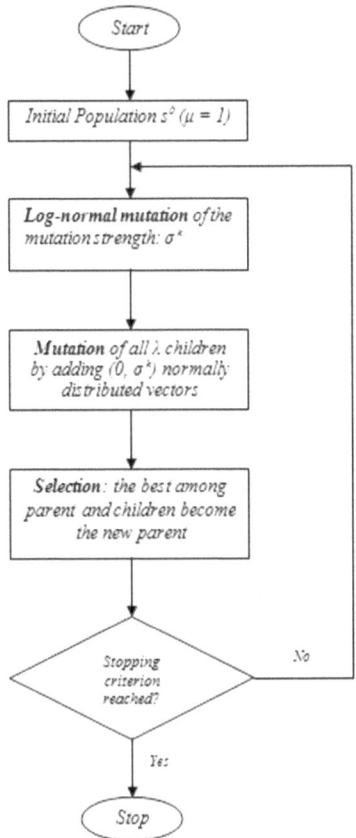

Fig. 1. Flow chart of the evolutionary approach

There is not an exact rule to decide the size of the parent population. Several theoretical studies [1] have been conducted and reported in the literature on the estimation of the optimal ratio λ/μ for $(1, \lambda)$ strategies. They showed that, although the optimal ratio depends on the objective function and its complexity, a ratio λ/μ equal to 5 can be a good starting point.

The convergence criterion is based on the evolution of the best value of the objective function along generations. The optimum is assumed to be reached as far as the best value has not significantly changed in the last generations. If F_{min}^{g} and F_{min}^{g+1} indicate the minimal values of the objective function inside the g-th and (g+1)-th generation, the iterative process will be stopped if:

$$| F_{min}^{g} - F_{min}^{g+1} | < \varepsilon \qquad (6)$$

Moreover, a maximum number of iterations N_{max} is defined, after which the iterative process is stopped even though the stopping rule (6) has not been satisfied.

4   Experimental Results

Experiments have been carried out using regular square zoning and the presented zoning technique (called "Voronoi-based zoning" in the following), combined with a feature set composed of gradient-based and pixel density features.

The MNIST and the USPS datasets have been used. The former contains 60000 images of digits 0 – 9 for training and 10000 images for testing. The latter contains 7291 images of digits 0 – 9 for training and 2007 images for testing. The MNIST images are 28 × 28 grayscale images, while the USPS images are 16 × 16 pixels sized grayscale images.

On a trial and error basis, the number of zones has been chosen equal to 25. A preprocessing phase has been performed to generate the input digit images. It has been consisted of binarization using ImageJ software [19], and size normalization into 30×30 images for the MNIST dataset and 20×20 images for the USPS dataset.

According to the mentioned schemes, gradient features complemented by information on pixel density have been extracted from each zone.

Gradient features consist of calculating gradient values and constructing local histograms along certain orientations. As each pixel of the image has an orientation and magnitude depending on the local gradient, histograms have been built by accumulating magnitude values over the same directions for all the pixels belonging to each zone of the image. The gradient computation has been performed by the Sobel operator. The input image has been convolved with Sobel masks to calculate the gradient components in the horizontal $g_x(p)$ and vertical direction $g_y(p)$ at each pixel p. The magnitude $G(p)$ and the angle $\Phi(p)$ of the pixel p have been given by :

$$G(p) = \sqrt{g_x^2(p) + g_y^2(p)} \qquad (7)$$

$$\Phi(p) = atan2\,(g_x(p), g_y(p)) \in [0, 2\pi[\qquad (8)$$

The gradient direction at each pixel has been decomposed into components according to the 8-directional chain code.

Pixel density has been computed as the number of black pixels that stand in each zone over the total number of pixels in the image. These complementary features have been added to the gradient ones, because they were able to further improve the accuracy. Therefore, we obtained a feature vector of 225 elements for each handwritten numeral.

All the experiments have been done in MATLAB version 7.8 (Release 2009a) 64 bits. It has been used an Intel Core i5 (64 bits), running on a Windows 7 Professional 64 bits Operating System, with 8 GB of RAM memory and 2 TB of disk space.

The first experiment has been dedicated to find the optimal zoning distribution for both the MNIST and the USPS datasets. To this purpose, the optimization problem (3) has been tackled adopting the *plus* strategy with $\mu = 1$. The initial parent population s^0 has been chosen as the set of Voronoi generating points $\{p^0_i\}_{i=1,...,M}$ to which corresponds the regular square zoning (see Figure 2).

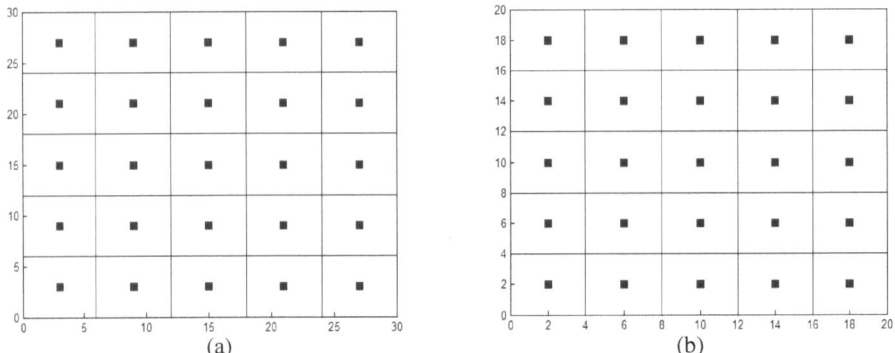

Fig. 2. Initial population s^0 for the MNIST (a) and the USPS (b) datasets

This choice has been taken because no assumptions can be done about the optimal distribution of zones. The objective function $F(s)$ (2) has been assumed as the k-fold cross validation error (k=5), computed by a linear SVM classifier. For this classifier, the value of the hyperparameter C has been set to 10.

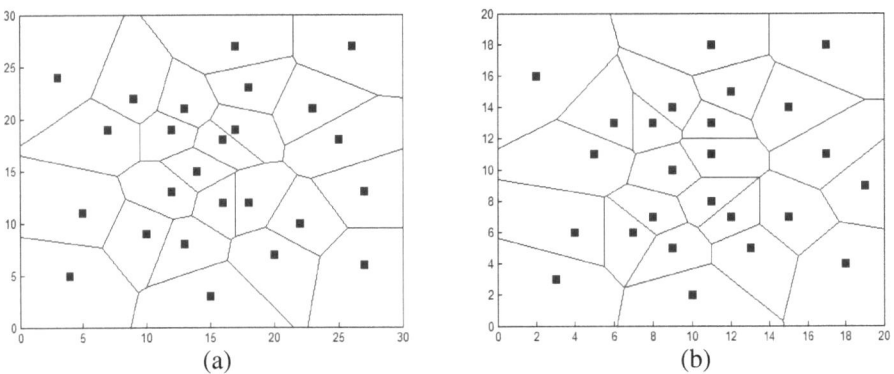

Fig. 3. Optimal zoning distribution for (a) the MNIST and (b) the USPS datasets

The following working parameters have been selected for the evolution strategy: $\lambda = 5$, $\sigma = 1$, $\tau = 0.8$, $N_{max} = 40$, $\varepsilon = 0.01$.

Figure 3 shows that the MNIST and the USPS optimal zoning configurations are quite similar. As a matter of fact, it can be observed that for both cases there is a higher concentration of Voronoi generator points in the center part of the image, while the edge points are less dense and have rather similar positions.

These zoning distributions are the ones which allow to separate in the most effective way patterns belonging to different classes, while bringing together those belonging to the same class.

From the previous figure, it comes out that there is a set of generating points able to discriminate effectively handwritten numeral shapes, according to the proposed zoning method and the selected features.

Intuitively, their fine positions depend on people's handwriting styles and most probably this is the reason why the two sets are not completely overlapped.

In order to evaluate the performances of the Voronoi-based zoning, several standard SVM classifiers were trained on the features extracted by these zoning distributions:

$$\Phi_{LIN}(\mathbf{x_i}, \mathbf{x_j}) = \mathbf{x_i} \cdot \mathbf{x_j} \tag{9}$$

$$\Phi_{POLY}(\mathbf{x_i}, \mathbf{x_j}) = (1 + \mathbf{x_i}^T \cdot \mathbf{x_j})^5 \tag{10}$$

$$\Phi_{RBF}(\mathbf{x_i}, \mathbf{x_j}) = \exp(-\gamma \| \mathbf{x_i} - \mathbf{x_j} \|^2) \tag{11}$$

They are, respectively, a linear, a degree 5 polynomial and a radial basis function kernel SVM. It has been chosen the polynomial degree $d = 5$, because it has been proven to have the best performance [14].

Using LIBSVM [3], one-against-one SVM classifiers have been trained for each possible pair of classes. Then, a test pattern has been assigned by evaluating all binary classifiers and performing voting among them [20].

The values of C and γ used to perform classification are summarized in the following table (Table 1).

Table 1. Parameters used to perform classification

	MNIST	USPS
C	10	10
γ	0.125	0.004

The combination of C and γ has been selected assuming $C = 10$ (default value) and calculating γ on a trial-and-error basis. So, it must be underlined that we did not perform any parameter optimization for these experiments.

Table 2 reveals that Voronoi-based zoning always outperforms regular square zoning for any type of SVM classifier.

In particular, even the performance of the linear kernel is remarkably improved for both datasets. It means that Voronoi-based zoning, combined with the feature we chose, is able to enhance the performance of SVM classifiers.

Besides, it can be observed that RBF and polynomial kernel SVM accuracy are competitive with state-of-the-art approaches.

Table 2. Accuracy on the test sets (%)

	MNIST		USPS	
	Regular Square Zoning	Voronoi - based Zoning	Regular Square Zoning	Voronoi - based Zoning
SVM$_{\text{LIN}}$	96.71	97.86	95.07	96.46
SVM$_{\text{POLY}}$	98.19	99.18	95.81	96.96
SVM$_{\text{RBF}}$	98.26	99.23	96.21	97.01

For the MNIST dataset, the best result available in the literature is 99.81% using a Convolutional Neural Network as a trainable feature extractor and a RBF kernel SVM as a recognizer [15]. Therefore, our results are not too far from [15] that achieves a better accuracy value, but uses a more sophisticated feature extraction algorithm.

For the USPS dataset, the human recognition rate estimated to be 97.5% [2] highlights that it is a quite hard database. In this case, some of the best results available in the literature are those obtained by Simard et al. [5, 22]. They reached 97.4% accuracy on the test set using a modified training set, that had been enhanced by adding machine-printed characters. Hence, once again, our results are noteworthy, since we employed the original training set.

Fig. 4. Examples of some MNIST misclassified digits

Fig. 5. Examples of some USPS misclassified digits

Anyway, our recognition results on the MNIST and the USPS test sets are a little bit lower than the best results reported in the literature. Notwithstanding this, our goal in this paper was not to reach the maximum recognition rate, but instead to find a zoning distribution capable of improving performance significantly. Thus, it can be concluded that the combination of the adaptive approach with the Voronoi topology leads to achieve excellent results. The adaptive approach can discover the distribution of zones that maximizes the interclass differences while minimizing the intraclass ones. Furthermore the Voronoi topology, allowing variations in the shape of the zones, introduces an additional degree of freedom useful to the same purpose.

The classification times of our method are shown in Table 3. They are measured in seconds.

Table 3. Classification times (s)

	MNIST	USPS
SVM_{LIN}	19	5
SVM_{POLY}	61	13
SVM_{RBF}	223	46

Hence, it can be observed that the polynomial SVM classifier with polynomial degree d = 5 represents a good trade-off in terms of both accuracy and classification speed.

5   Conclusions

In this paper, we presented a new zoning technique to address the handwritten digit recognition problem. Our technique considered zoning design as an optimization problem and used an evolutionary strategy to find the optimal zoning distribution. Our method is able to adaptively determine the optimal position of the Voronoi generating points.

The comparison of zoning method performance for handwritten numeral recognition is a difficult task since there are differences in experimental methodology and settings, as well as differences in the databases used. Here, in order to test the proposed approach, several standard SVM classifiers were trained on the features extracted by this zoning distribution. We chose SVM classifiers as they have been successfully used in character recognition.

The experiments show that this zoning method achieves a general improvement in performance, for any type of SVM classifier. Hence, this work demonstrates the utility of adaptive zoning and Voronoi topology for SVM classification of handwritten digits. In particular, our zoning strategy led to a recognition rate of 99.23% on the MNIST dataset and of 97.01% on the USPS dataset.

Future works include investigation on the effectiveness of our method by using larger features, in order to improve recognition rate without a significant increase of the classification time.

References

1. Beyer, H.-G.: The Theory of Evolution Strategy. Springer, New York (2001)
2. Bromley, J., Sackinger, E.: Neural-network and k-nearest-neighbor-classifiers. TechnicalReport 11359-910819-16TM, ATT (1991)
3. Chang, C.-C., Lin, C.-J.: LIBSVM: A library for support vector machines, http://www.csie.ntu.edu.tw/~cjlin/libsvm
4. de Berg, M., Cheong, O., van Kreveld, M., Overmars, M.: Computational Geometry: Algorithms and Applications. Springer, Berlin (2008)
5. Drucker, H., Schapire, R., Simard, P.: Boosting performance in neural networks. International Journalof Pattern Recognition and Artificial Intelligence 7, 705–719 (1993)

6. Fogel, D.B.: An introduction to simulated evolutionary optimization. IEEE Transactions on Neural Networks 5(1), 3–14 (1994)
7. Freitas, C.O.A., Oliveira, L.S., Bortolozzi, F.: Handwritten Character Recognition using nonsymmetrical perceptual zoning. IJPRAI 21(1), 1–21 (2007)
8. Gatos, B., Kesidis, A., Papandreou, A.: Adaptive Zoning Features for Character and Word Recognition. In: Proceedings of ICDAR 2011, pp. 1160–1164 (2011)
9. Hull, J.: A database for handwritten text recognition research. IEEE Transactions on Pattern Analysis and Machine Intelligence 16, 550–554 (1994)
10. Impedovo, D., Pirlo, G., Modugno, R.: New Advancements in Zoning-Based Recognition of Handwritten Characters. In: Proceedings of ICFHR 2012, pp. 661–665 (2012)
11. Impedovo, S., Modugno, R., Pirlo, G.: Analysis of membership Functions for Voronoi-based Classification. In: Proceedings of ICFHR 2010, pp. 220–225 (2010)
12. Impedovo, S., Modugno, R., Pirlo, G.: Membership Functions for Zoning-based Recognition of Handwritten Digits. In: Proc. International Conference on Pattern Recognition, Istanbul, Turkey, pp. 1879–1879 (2010)
13. LeCun, Y., et al.: Gradient-based learning applied to document recognition. In: Proceedings of the IEEE, vol. 86, pp. 2278–2324 (1998)
14. Liu, C.-L., Nakashima, K., Sako, H., Fujisawa, H.: Handwritten digit recognition: benchmarking of state-of-the-art techniques. Pattern Recognition 36, 2271–2285 (2003)
15. Niu, X.X., Suen, C.Y.: A novel hybrid CNN–SVM classifier for recognizing handwritten digits. Pattern Recognition 45, 1318–1325 (2012)
16. Oliveira, L.S., Sabourin, R., Bartolozzi, F., Suen, C.Y.: Automatic Recognition of Handwritten Numeral Strings: A Recognition and Verification Strategy. IEEE T-PAMI 24(11), 1438–1454 (2002)
17. Radtke, P.V.W., Wong, T., Sabourin, R.: A Multi-objective Memetic Algorithm for Intelligent Feature Extraction. In: Coello Coello, C.A., Hernández Aguirre, A., Zitzler, E. (eds.) EMO 2005. LNCS, vol. 3410, pp. 767–781. Springer, Heidelberg (2005)
18. Rajashekararadhya, S.V., Ranjan, P.V.: Support Vector Machine based Handwritten Numeral Recognition of Kannada Script. In: Proc. IEEE International Conference on Advance Computing 2009, pp. 381–386 (2009)
19. Rasband, W.S.: ImageJ, U. S. National Institutes of Health, Bethesda, Maryland, USA, http://imagej.nih.gov/ij/
20. Schölkopf, B., Smola, A.: Learning with Kernels: Support Vector Machines, Regularization, Optimization and Beyond. MIT Press, Cambridge (2002)
21. Sharma, D., Gupta, D.: Isolated Handwritten Digit Recognition using Adaptive Unsupervised Incremental Learning Technique. International Journal of Computer Applications 7(4), 27–33 (2010)
22. Simard, P., LeCun, Y., Denker, J.S.: Efficient pattern recognition using a new transformation distance. In: Advances in Neural Information Processing Systems 5, San Francisco, CA, USA, pp. 50–58 (1993)
23. Vamvakas, G., Gatos, B., Pratikakis, I., Stamatopoulos, N., Roniotis, A., Perantonis, S.J.: Hybrid off-line OCR for isolated handwritten Greek characters. In: Proc. 4th IASTED International Conference on Signal Processing, Pattern Recognition, and Applications (SPPRA 2007), Innsbruck, Austria, pp. 197–202 (2007)
24. Xiang, P., Xiuzi, Y., Sanyuan, Z.: A hybrid method for robust car plate character recognition. Engineering Applications of Artificial Intelligence 18(8), 963–972 (2005)

Image Annotation by Learning Label-Specific Distance Metrics*

Xing Xu, Atsushi Shimada, and Rin-ichiro Taniguchi

Department of Advanced Information Technology, Kyushu University, Japan
{xing,atsushi}@limu.ait.kyushu-u.ac.jp, ring@ait.kyushu-u.ac.jp

Abstract. Recently, weighted k nearest neighbor based label prediction
model combined with distance metric learning (KNN+ML) [10, 14, 17],
has become more attractive and showed exciting results on image annotation task. Usually, in KNN+ML framework, a uniform distance metric
is learned given a collection of similar/dissimilar image pairs from training data. Thus, for a couple of images, their distance is globally unique.
However, this might not be sufficient for label prediction on annotation
task because it is impossible to distinguish the multiple labels attached
to each image. In this paper, we are motivated to learn multiple label-
specific distance metrics, and measure the distance of an image pair under
different labels' distance metrics. We also propose a novel label specific
prediction model, in which the weight of each label is determined by its
specific distance value rather than previous global distance value. Compared with previous KNN+ML methods, our proposed method is able
to exactly discriminate each label in each neighbor, and efficiently reduce the prediction of *false positive* and *false negative* labels. Extensive
experimental results on three benchmark datasets demonstrate that proposed method achieves more accurate annotation results and competitive
overall performance.

1 Introduction

The task of image annotation is to automatically assign keywords to an image,
and it has become an active topic in computer vision and machine learning areas due to its potential useful applications, including image search and photo
management. Recently, k nearest neighbor (KNN) based methods has been successfully applied to image annotation problem, as this kind of local learning
technique has potentiality to capture the similarity graph of labeled and unlabeled images.

In order to extend KNN based methods to image annotation task, two primary
issues need to be considered. The first is how to select appropriate neighbors for
an unlabeled image. Metric learning (ML) methods [11,13,19] are often imported
to find optimal metric over feature space of provided pairs of labeled images,
and linear combination of base metrics for multiple high-dimensional features

---

* This work has been partly supported by Grant-in-Aid for Scientific Research (B),
Grant Number 24300074.

A. Petrosino (Ed.): ICIAP 2013, Part I, LNCS 8156, pp. 101–110, 2013.

is alternative to traditional low-dimensional Mahalanobis metric. The second is how to design efficient label transfer mechanism through learned distance metric. Some well known methodologies including greedy diffusion [14, 23], weighted nearest neighbor label prediction [10, 17, 21], are usually adopted.

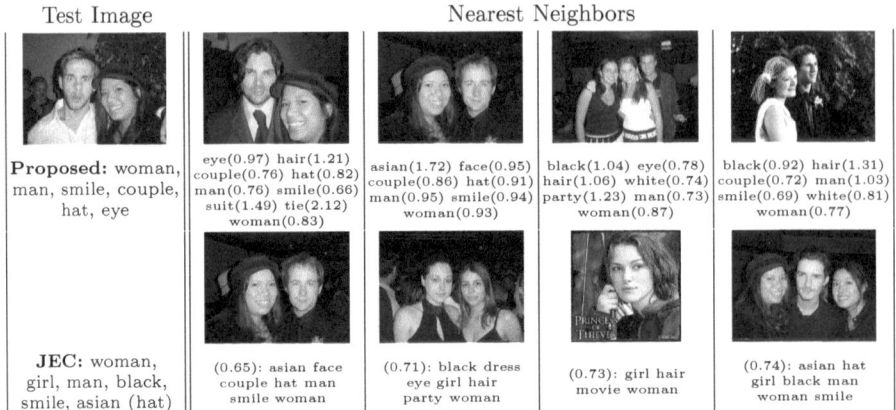

Fig. 1. For a test image from ESP Game dataset (first column), the first and second rows on the right section show its 4 nearest images and distance value of each label from proposed method (uses label-specific distance metrics) and JEC [17] (uses global distance metric). Predicted labels (with smallest distance values) in two methods can be compared with ground truth {couple, eye, hat, man, smile, woman}.

The main shortcomings of existing works based on KNN+ML are two folds. First, these works incline to use a single global distance metric to measure the similarity of an image pair, which is convincing to address traditional classification problem. However, in multi-label annotation condition, the degree of similarity ought to vary upon different label affiliated to the image pair. Second, in the celebrated weighted nearest neighbor label prediction model [10, 17], each label of one neighbor has identical weight $(\exp(-D(\cdot))$ in Equ. 1) since the distance is uniquely determined by global distance metric. Thus, during the final label prediction, some labels would get equal weight, such as {*asian, hat*} in JEC in Fig. 1, where we can only select these two labels {*asian, hat*} randomly and the accuracy of final annotation would be disturbed.

As the proverb goes, "there are a thousand Hamlets in a thousand people's eyes." Given an image pair, we are motivated to measure discriminative distance values ("*Hamlets*") by different label specific distance metrics ("*eyes*"). Then we can distinguish labels according to their specific distance values. In this paper, we propose a new weighted nearest neighbor type model that predicts each label's weight $(\exp(-D_{y_l}(\cdot))$ in Equ. 2) according to its specific distance value. As shown in Fig.1, unlike that different labels of one neighbor share same distance value in JEC, each label of one neighbor has its specific distance value in proposed method. This ensures proposed method to discriminate labels in one neighbor, select proper labels and avoid irrelevant labels simultaneously.

Our contributions are: 1) We propose a label-specific prediction model for annotation task, where the weight of each label in a neighbor is measured by its specific distance metric. 2) We extend [11] in a high dimensional multi-feature fusion setting to learn distance metric for each label. 3) We design a complete annotation framework of training and testing procedures, including learning label-specific distance metrics and predicting labels for new image.

In the next section, we review previous notable works in image annotation field. In Sect.3, we describe our label-specific prediction model, distance metric learning algorithm, and entire training and testing procedures. Experimental results compared with previous KNN+ML based methods are presented in Sect.4. Finally, we make conclusion in Sect.5.

2   Related Work

Large quantities of approaches have been proposed to address image annotation problem. One pipeline of research focuses on modeling medium sized image databases with fixed vocabularies. A common consensus has been reached from [3,8,10,14,17,23] that three main groups exist: 1) Generative models [2,5,22] aim to learn the joint probability of labels and image features, various relationships between semantic and visuality have been imagined, e.g. mixture of topics, and different hypotheses of probability distribution of labels and image features have been assumed, such as multinomials, separate Bernoullis, mixture of Gaussian. 2) Discriminative models [3, 12] treat each label as a semantic class of multi-class multi-label problem, and learn a separate classifier for each label, where balanced training data is required and correlation among the labels may be ignored. 3) Nearest neighbor based models [3,10,14,17,23] have become more attractive recently and shown state-of-the-art annotation performance. As visual close similar images possess certain semantic similarity, after selecting proper neighbors for unlabeled image, labels are then transferred from these neighbors.

On the other line of research, data-driven approaches [4,16,18] have demonstrated their capacity on large-scale web-based image databases with open vocabularies. These approaches usually firstly search a group of visually closely similar candidates for the query image, and then mine relevant tags from associated clues (such as image filename, URL and surrounding texts) available on the web. These approaches can be regarded as hybrid models which combine the generative/discriminative/nearest neighbor models in more practical circumstance.

3   Proposed Method

In this section, we first propose the label-specific prediction model, then describe the label-specific distance metric learning algorithm, finally depict our annotation framework of training and testing procedures.

3.1   Label-Specific Prediction Model

Consider a collection of labeled images $\mathcal{C} = \{\{I_1, Y_1\}, ..., \{I_D, Y_D\}\}$ with a fixed vocabulary of L labels $\mathcal{Y} = \{y_1, ...y_L\}$, where each $Y_i \subseteq \mathcal{Y}$ contains multiple labels. In weighted nearest neighbor based label prediction model [10, 17], each label's presence/absence of an unseen image J is a weighted sum over its K nearest neighbors $N_J = \{I_1, ...I_K\}$ in training set. We denote the weight of k-th neighbor I_k to image J as π_{J, I_k}, where $\pi_{J, I_k} \geq 0$ and $\sum_k \pi_{J, I_k} = 1$. In previous KNN+ML works, π_{J, I_k} is usually represented by a smooth exponential function over distance $D(J, I_k)$, as $\pi_{J, I_k} = \exp(-D(J, I_k))/\sum_{k'=1}^{K} \exp(-D(J, I_{k'}))$. Thus, the presence probability of l-th label y_l in J can be formulated as,

$$P(y_l = +1|J) = \sum_{(I_k, Y_k) \in N_J} \pi_{J, I_k} \cdot \delta(y_l \in Y_k|I_k)$$

$$= \sum_{(I_k, Y_k) \in N_J} \exp(-D(J, I_k)) \cdot \delta(y_l \in Y_k|I_k), \qquad (1)$$

where $\delta(y_l|I_k)$ is an indicator function that denotes the presence/absence of label y_l in I_k, with $\delta(\cdot) = 1$ when $y_l \in Y_k$ and $\delta(\cdot) = 0$ otherwise. Through ranking probabilities of all the labels $y_1 \rightarrow y_L$ based on Equ. 1, top-ranked labels can be assigned to the unseen image J.

Note that in Equ. 1, all labels in Y_k share the same weight value $\exp(-D(J, I_k))$ since the distance value $D(J, I_k)$ is unique given the learned global distance metric. This may lead to significant potential risk that, although we can correctly predict *true positive* labels of one neighbor for image J, it is still ambiguous to reject the *false positive* or *false negative* labels in that neighbor. Inspired by [7, 20] which learns local distance functions for every training image or image clusters in visual classification task, here we aims to learn local distance metric for each label. For one image pair, under distance metrics of different labels, the distances are various. Thus, based on these local distance metrics, our label-specific prediction model can be formulated as:

$$P(y_l = +1|J) = \sum_{(I_k, Y_k) \in N_J} \exp(-D_{y_l}(J, I_k)) \cdot \delta(y_l \in Y_k|I_k). \qquad (2)$$

Different from traditional prediction model in Equ. 1, here weight $\exp(-D_{y_l}(J, I_k))$ includes multiple values involved in different $y_k \in Y_k$. Intuitively, we can measure the distance between neighbor image I_k and J using the specific distance metrics of labels in I_k. This allows us to distinguish the importance of each label $y_k \in Y_k$ of I_k, reduce the prediction of irrelevant labels and preserve relevant labels from Y_k.

3.2   Learning Label-Specific Distance Metrics

Suppose we have generated a collection of similar and dissimilar image pairs for the l-th label, in the manner of learning distance metric [11, 13, 19, 20], our goal is to learn a distance metric for the l-th label to ensure distances of similar image pairs are smaller than dissimilar pairs. Following [11], we model the probability p_n that an pair $n = (A, B)$ is similar (positive), and the pair response r_n is 1, as:

$$p_n = p(r_n = 1 | A, B; D_{y_l}(A, B), b) = \sigma(b - D_{y_l}(A, B)), \qquad (3)$$

where $\sigma(z) = (1 + \exp(-z))^{-1}$ is the sigmoid function and b is a bias term. Notably, $D_{y_l}(A, B) = \sum_i u_l(i) \sum_j v_l(j) \cdot d^i_{A,B}(j)$ represents the distance of (A, B) on the view of l-th label y_l. Here in the multiple features fusion setting, $D_{y_l}(A, B)$ is a linear combination of base distances $d^i_{A,B}(\cdot)$ of multiple features [17]. Inter-feature weights $u_l(i)$ and intra-feature weights $v_l(j)$ are specific parameters we need to learn for the l-th label. We use maximum log-likelihood to optimize the parameters of the model. The log-likelihood $\mathcal{L}$ can be written as:

$$\mathcal{L} = \sum_n r_n \ln p_n + (1 - r_n) \ln(1 - p_n), \qquad (4)$$

and the gradient of $\mathcal{L}$ with respect to u_l and v_l equals

$$\frac{\partial \mathcal{L}}{\partial u_l} = \sum_n (r_n - p_n) v_l \cdot d_{A,B}, \qquad \frac{\partial \mathcal{L}}{\partial v_l} = \sum_n (r_n - p_n) u_l \cdot d_{A,B}, \qquad (5)$$

which are smooth and concave. Moreover, non-negative constraints are required to weights $\{u_l, v_l\}$ as the distance value should be non-negative. In practice, we use projected gradient ascend method to optimize $\{u_l, v_l\}$ in an alternating manner.

It is worth saying that our label-specific distance metric learning algorithm is quite different from "*word-specific logistic discriminate model (σML)*" proposed in [10]. Firstly, the target of our algorithm is to learn a metric to identify the similarity of a pair under a label, while σML aims to learn word-specific smooth factors for the weighted nearest neighbor prediction model. Secondly, our algorithm directly impacts the weight of each label (see Equ. 2, $\exp(-D_{y_l}(J, I_k))$), whereas σML does not effect weight factor, which implies similarity of a pair is still considered on a global viewpoint.

3.3   Training and Testing Procedures

To learn distance metric of each label, it's necessary to create a training set of similar/dissimilar image pairs for each label. Our scheme of obtaining label-specific image pairs is a modified version of [17], first we give some key definition:

Semantic cluster. For a label $y_l \in \mathcal{Y}$, its semantic cluster $S_{y_l} \subseteq \mathcal{C}$ contains all images annotated with label y_l in training set $\mathcal{C}$.

Semantic neighborhood. For an image T, its semantic neighborhood S_T has L subsets $\{S_{T,y_1} \bigcup ... \bigcup S_{T,y_L}\}$, where the l-th subset $S_{T,y_l} \subseteq S_{y_l}$ includes K_1 images that are most similar to T in semantic cluster S_{y_l}.

Similar/dissimilar pairs. For an labeled image (T, Y_T) with its semantic neighborhood S_T, its similar samples are images from S_{T,y_p}, $y_p \in Y_T$, the residual S_{T,y_q}, $y_q \in \mathcal{Y} \backslash Y_T$ are dissimilar samples.

Training

Input: A set of annotated training images $C = \{\{I_1, Y_1\}, ..., \{I_D, Y_D\}\}$, L semantic cluster $S_{Y_1}, ..., S_{y_L}$. For each label $l = \{1, ..., L\}$, do

1. Generate semantic neighborhood using base distance measure for each sample.
2. Generate similar/dissimilar pairs for each sample in S_{y_l}.
3. Learn parameters $\{b_l, u_l, v_l\}$ for l-th distance metric following Sect.3.2.

Output: L label-specific parameters $\{b_1, u_1, v_1\}, ..., \{b_L, u_L, v_L\}$.

Testing

Input: L label-specific parameters $\{b_1, u_1, v_1\}, ..., \{b_L, u_L, v_L\}$, L semantic clusters $S = \{S_{y_1}, ..., S_{Y_L}\}$ from training data, a test image J. For test image J, do

1. Generate its semantic neighborhood $S_J = \{S_{J,y_1} \bigcup ... \bigcup S_{J,y_L}\}$, where the l-th S_{J,y_1} contains K_1 training samples that are most similar to J measured by the l-th distance metric.
2. For each image $\forall \{I_t, Y_t\} \in S_J$, do
   • Calculate the presence probability (Equ. 2) of label $y_t \in Y_t$ by its label specific distance $D_{y_t}(J, I_t)$, using l-th distance metric.
   • Accumulate probability score of label y_t.
3. From probability scores of all labels in S_J, select top-5 labels with highest probability scores.

Output: predicted top-5 labels for test image J.

The training and testing procedures are incorporated into the complete annotation framework depicted above. Distance metrics are learned in training stage through the well organized semantic neighborhood of each training sample, which fully leverages image-image, image-label, label-label similarities. In testing stage, similarities between the test image and its neighbors are calculated according to different label's metric, finally labels that are most semantically similar to test image are dug out.

4   Experiments

In this section, we first present the experimental configuration: datasets, multiple features, evaluation measures and details, then we compare our proposed method with previous methods from different aspects.

4.1   Configuration

Datasets and Features. To keep coherence with previous works [10, 14, 17], we also consider three well-explored data sets: **Corel 5K** [6], **ESP Game** [1], **IAPR TC12** [9]. Table 1 summarizes the general statistics of the images and fused multiple features we use in our experiments.

Evaluation Measures. Following [10,17], we choose top-5 most relevant labels for each test image. Then we compute the mean precision **P**, mean recall **R** and

Table 1. General statistics for the three datasets and multiple features. In column 6 and 7, the items are in the format "mean, maximum." Total dimension of multiple features is 13,900.

Dataset	Num. of images	Num. of labels	Training images	Test images	Labels per image	Images per label
Corel 5K	5000	260	4500	500	3.4, 5	58.6, 1004
ESP Game	20,770	268	18,689	2,081	4.7, 15	326.7, 4553
IAPR TC12	19,627	291	17,665	1,962	5.7, 23	34.7, 4999
Feature	RGB	LAB	HSV	Gist	SIFT	hue
Dimension	4,096	4,096	4,096	512	1,000	100
Base metric	L_1	L_1	L_1	L_2	χ^2	χ^2

their trade-off **F1** score ($F1 = 2.P.R/(P+R)$). Moreover, number of words with non-zero recall **N+** is also taken into account.

Details. In Sect.3.3, to extract similar/dissimilar image pairs, we use entire training images to form semantic cluster of each label on entire Corel 5K dataset, subsets of 30% random training samples of both ESP Game and IAPR TC12 datasets. K_1 is set as 4 for Corel 5K, 3 for both ESP Game and IAPR TC12 datasets. Since for a labeled image, its dissimilar pairs is far more than similar pairs ($(L - K_1) \gg K_1$), following the advice in [23], we find that randomly selecting partial dissimilar pairs $4 \sim 10$ times larger than similar pairs (there are average $500 \sim 6000$ pairs (N) for each label in all three datasets) is sufficient to learn stable distance metrics.

The training and testing procedures are repeated three times on each dataset, and we choose models that perform best on testing sets for comparison. All experiments are executed using MATLAB 7.11 on a 3.4 GHz, 8GB RAM PC.

4.2   Comparison

Global *vs.* Label-Specific Distance Metric. First we compare our method with celebrated work of JEC [14] which uses global distance metric, and Fig. 2 (top row) shows performance in terms of **P**, **R** and **N+** with respect to changing neighborhood size. It is remarkable that our method makes significant improvement on these measures compared with global distance metric based method on all three datasets. In addition, unlike JEC which needs large numbers of neighbors (nearly 200) to improve performance, our proposed method achieves best performance using less neighbors (30, 100 and 80) on three datasets correspondingly. This is because in our testing procedure (in Sect.3.3), more semantically related neighbors are pulled nearer when generating semantic neighborhood.

Secondly, we follow σML in [10] and group labels according to their frequency in each dataset and explore which labels benefit most by using specific distance metrics. From Fig. 2 (bottom row), it is illustrated that our method could further care for rare labels and achieve significant improvements for these labels

Table 2. Comparison of annotation performance between proposed label-specific *vs.* previous global distance metric models

Method	Corel 5K				ESP Game				IAPR TC12			
	P	R	F1	N+	P	R	F1	N+	P	R	F1	N+
JEC [14]	27	32	29.3	139	22	25	23.4	224	28	29	28.5	250
GS [23]	30	33	31.4	146	-	-	-	-	32	29	30.4	252
TagProp (ML) [10]	31	37	33.7	146	49	20	28.4	213	48	25	32.9	227
TagProp (σML) [10]	33	42	37.0	160	39	27	31.9	239	46	35	39.8	266
2PKNN [17]	39	40	39.5	177	51	23	31.7	245	49	32	38.7	274
2PKNN+ML [17]	44	46	45.0	191	53	27	35.7	252	54	37	43.9	278
Proposed	40.5	44.7	42.5	185	44.1	26.2	32.9	247	47.6	36.1	41.1	264

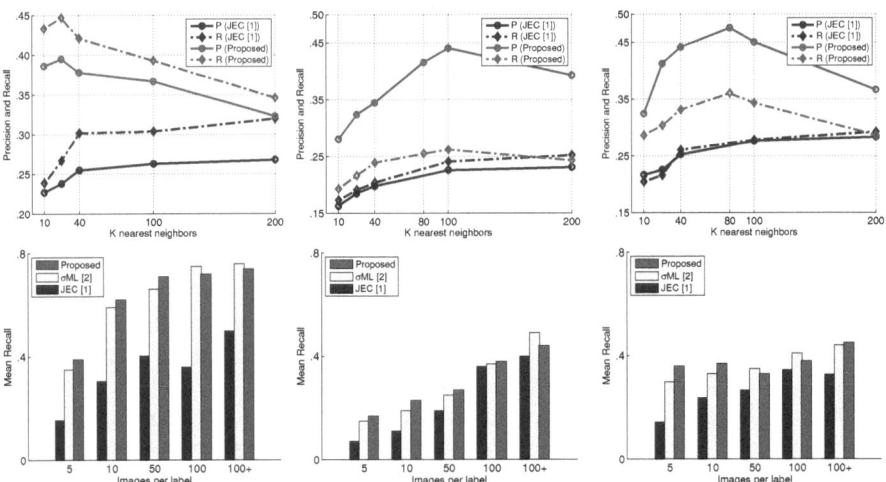

Fig. 2. Label-specific *vs.* global distance metric: performance in terms of **P**, **R** and **N+** with respect to neighborhood size changing (top row) and mean recall of labels (bottom row) on three datasets (from left to right: Corel 5K, ESP Game and IAPR TC12)

compared with σML. The reason is that smooth factor learned in σML could only change the weight for a rare label slightly, whereas the weight is directly decided and promoted by its specific distance metric in proposed method.

Comparison by Annotation Measures. Table 2 summarizes the overall evaluation from our results as well as those reported by previous KNN+ML methods on three datasets. It shows that our method outperforms previous global metric based methods, such as JEC, TagProp, and 2PKNN, but is worse than the prominent 2PKNN+ML. As for 2PKNN+ML, it utilizes sophisticated metric learning algorithm (LMNN [19]) to learn a global large marginalized distance metric, and it requires large quantities of seriously unbalanced similar/dissimilar pairs, e.g. for Corel 5K dataset, there are total 2 million training pairs, and the proportion

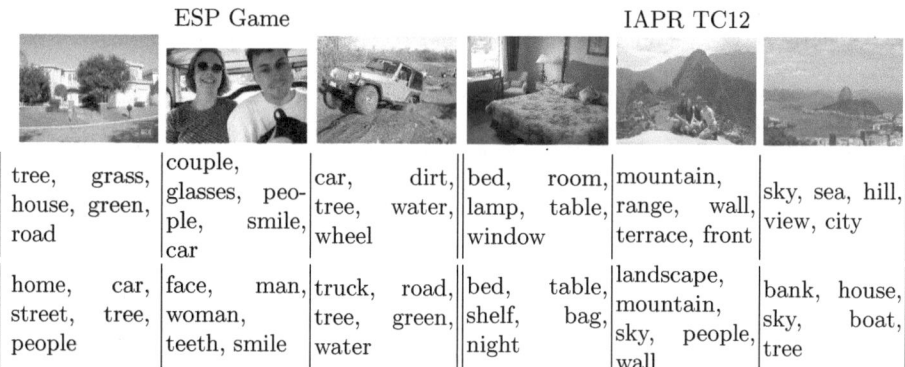

ESP Game			IAPR TC12		
tree, grass, house, green, road	couple, glasses, people, smile, car	car, dirt, tree, water, wheel	bed, room, lamp, table, window	mountain, range, wall, terrace, front	sky, sea, hill, view, city
home, car, street, tree, people	face, man, woman, teeth, smile	truck, road, tree, green, water	bed, table, shelf, bag, night	landscape, mountain, sky, people, wall	bank, house, sky, boat, tree

Fig. 3. Annotations of example images from ESP game (left three images) and IAPR TC12 (right three images). Since these exemplars have ground truth labels more than 5 words, we explicitly compare proposed method (second row) with JEC [14] (third row) on accuracy of predicted top-5 labels.

of similar/dissimilar pairs is about 1:50. The training procedure is complex and needs to be well designed for unbalanced setting. Our method requires much less (e.g. thousands pairs per label for Corel 5K) and fairly balanced pairs for each label and is scalable to larger vocabulary. On this point, we think our method is promising and competitive to the state-of-the-art method. Moreover, in Fig.3 we present some qualitative annotation results from our method compared with results from global metric based method JEC. It shows that proposed method is able to assign more accurate labels related to image content, whereas JEC might be ambiguous to distinguish the relevant/irrelevant labels, since some equal weighted labels are selected randomly.

5   Conclusion

In this paper, we have proposed a novel label prediction model for image annotation task. In proposed model, labels of one neighbor have different weights depending on their label-specific distance values. And we have extended [11] to high dimensional multiple-feature fusion setting, to learn the specific distance metric for each label. Moreover, we have also designed complete annotation framework of training and testing procedures. To further explore our annotation framework, it is feasible to import more sophisticated metric learning algorithms in high dimensional feature (distance) space, such as LMNN based methodology used in [17] [15]. This would be a primary issue to be tackled in our future work.

References

1. von Ahn, L., Dabbish, L.: Labeling images with a computer game. In: Proceedings of the SIGCHI Conference on Human Factors in Computing Systems (2004)
2. Blei, D.M., Jordan, M.I.: Modeling annotated data. In: ACM SIGIR 2003 (2003)

3. Carneiro, G., Chan, A., Moreno, P., Vasconcelos, N.: Supervised learning of semantic classes for image annotation and retrieval. IEEE Transactions on PAMI (2007)
4. Dai, L., Wang, X.J., Zhang, L., Yu, N.: Efficient tag mining via mixture modeling for real-time search-based image annotation. In: ICME (2012)
5. Putthividhya, D., Attias, H.T., Nagarajan, S.S.: Topic regression multi-modal latent dirichlet allocation for image annotation. In: CVPR (2010)
6. Duygulu, P., Barnard, K., de Freitas, J.F.G., Forsyth, D.: Object recognition as machine translation: Learning a lexicon for a fixed image vocabulary. In: Heyden, A., Sparr, G., Nielsen, M., Johansen, P. (eds.) ECCV 2002, Part IV. LNCS, vol. 2353, pp. 97–112. Springer, Heidelberg (2002)
7. Frome, A., Singer, Y., Sha, F., Malik, J.: Learning globally-consistent local distance functions for shape-based image retrieval and classification. In: ICCV (2007)
8. Fu, H., Zhang, Q., Qiu, G.: Random forest for image annotation. In: Fitzgibbon, A., Lazebnik, S., Perona, P., Sato, Y., Schmid, C. (eds.) ECCV 2012, Part VI. LNCS, vol. 7577, pp. 86–99. Springer, Heidelberg (2012)
9. Grubinger, M.: Analysis and Evaluation of Visual Information Systems Performance. Ph.D. thesis, Victoria University (2007)
10. Guillaumin, M., Mensink, T., Verbeek, J., Schmid, C.: Tagprop: Discriminative metric learning in nearest neighbor models for image auto-annotation. In: ICCV (2009)
11. Guillaumin, M., Verbeek, J., Schmid, C.: Is that you? metric learning approaches for face identification. In: ICCV (2009)
12. Huang, S.J., Zhou, Z.H.: Multi-label learning by exploiting label correlations locally. In: AAAI 2012 (2012)
13. Kostinger, M., Hirzer, M., Wohlhart, P., Roth, P., Bischof, H.: Large scale metric learning from equivalence constraints. In: CVPR (2012)
14. Makadia, A., Pavlovic, V., Kumar, S.: A new baseline for image annotation. In: Forsyth, D., Torr, P., Zisserman, A. (eds.) ECCV 2008, Part III. LNCS, vol. 5304, pp. 316–329. Springer, Heidelberg (2008)
15. Mensink, T., Verbeek, J., Perronnin, F., Csurka, G.: Metric learning for large scale image classification: Generalizing to new classes at near-zero cost. In: Fitzgibbon, A., Lazebnik, S., Perona, P., Sato, Y., Schmid, C. (eds.) ECCV 2012, Part II. LNCS, vol. 7573, pp. 488–501. Springer, Heidelberg (2012)
16. Torralba, A., Fergus, R., Freeman, W.: 80 million tiny images: A large data set for nonparametric object and scene recognition. IEEE Transactions on PAMI (2008)
17. Verma, Y., Jawahar, C.V.: Image annotation using metric learning in semantic neighbourhoods. In: Fitzgibbon, A., Lazebnik, S., Perona, P., Sato, Y., Schmid, C. (eds.) ECCV 2012, Part III. LNCS, vol. 7574, pp. 836–849. Springer, Heidelberg (2012)
18. Wang, X.J., Zhang, L., Liu, M., Li, Y., Ma, W.Y.: Arista - image search to annotation on billions of web photos. In: CVPR (2010)
19. Weinberger, K.Q., Blitzer, J., Saul, L.K.: Distance metric learning for large margin nearest neighbor classification. In: NIPS (2006)
20. Weinberger, K., Saul, L.: Fast solvers and efficient implementations for distance metric learning. In: ICML (2008)
21. Wu, P., Hoi, S.C.H., Zhao, P., He, Y.: Mining social images with distance metric learning for automated image tagging. In: WSDM (2011)
22. Xiang, Y., Zhou, X., Chua, T.S., Ngo, C.W.: A revisit of generative model for automatic image annotation using markov random fields. In: CVPR (2009)
23. Zhang, S., Huang, J., Huang, Y., Yu, Y., Li, H., Metaxas, D.: Automatic image annotation using group sparsity. In: CVPR (2010)

Approximating the Skeleton for Fine-to-Coarse Shape Representation

Luca Serino, Carlo Arcelli, and Gabriella Sanniti di Baja

Institute of Cybernetics "E. Caianiello", CNR, Naples, Italy
{l.serino,c.arcelli,g.sannitidibaja}@cib.na.cnr.it

Abstract. A method to generate a hierarchical skeleton structure is presented. The curve skeleton of a 3D object is used, where each voxel is labeled with the radius of the associated ball, i.e., with its distance from the complement of the object. Polygonal approximation is accomplished on all skeleton branches represented in a 4D space, where the coordinates are the (x,y,z) coordinates plus the radius r associated with each skeleton voxel. In this way, skeleton branches are divided into geometrically straight line segments, whose voxels are characterized by either constant or linearly increasing/decreasing radius. By increasing the threshold used for polygonal approximation the hierarchical skeleton structure is generated, which allows the user to get a fine-to-coarse shape representation.

Keywords: Skeleton, shape representation, polygonal approximation.

1 Introduction

The skeleton is a well known representation system and has been used in the framework of shape analysis for 2D and 3D objects. A wide literature is available as concerns different skeletonization methods, devised in the continuous and in the digital space, and the use of the skeleton in several application fields [1].

The skeleton is a homotopic subset of the object. It consists of the union of curves, termed skeleton branches, each of which can be seen as in correspondence with a part (main body or limb) of the object. Skeleton branches can be classified in two categories, respectively including peripheral and internal branches. The former branches are in correspondence with limbs of the object and are characterized by the presence of end points on their tips, i.e., points of the skeleton having only one neighbor in the skeleton. The latter branches are in correspondence with main bodies of the object from which the limbs protrude and are delimited by the so called branch points, i.e., points of the skeleton with more than two neighbors in the skeleton. Branch points are the skeleton points where different branches meet.

Skeleton points are symmetrically placed within the object. For 2D objects, skeleton points are all aligned along symmetry axes of the object. In turn, for 3D object this is the case only if the object consists exclusively of parts with tubular shape. In the general case, symmetry points of 3D objects are placed along planes and

A. Petrosino (Ed.): ICIAP 2013, Part I, LNCS 8156, pp. 111–120, 2013.

axes, so that only a linear subset of the set of the symmetry points can be included in the skeleton.

An important skeleton feature for shape analysis is related to the reconstruction ability of the skeleton. This means that the represented object can be fully reconstructed starting from its skeleton. To have such an important feature, each skeleton point should be associated with its distance from the complement of the object, i.e., with the radius of the ball that, centered on the skeleton point, touches the boundary of the object in at least two distinct parts. Moreover, the skeleton should include all the symmetry points of the object. Actually, the skeleton of a 3D object generally is not characterized by full reconstruction ability, since only a subset of the symmetry points of the object can be included in the skeleton. For example, refer to Fig. 1, where an input object, its skeleton and the object reconstructed from the skeleton are shown. Though full recovery is not possible, the reconstructed object is still a meaningful approximation of the input object. Thus, differences between input objects and reconstructed objects can be disregarded and in this paper we will consider the objects reconstructed by the skeletons as if they were our input objects.

Fig. 1. An input object, left, its skeleton, middle, and the reconstructed object, right

Another interesting skeleton feature, particularly useful in the framework of the structural approach to object recognition, is the possibility to represent shape in a hierarchical manner. In fact, different skeletons of the same object can be computed, each of which characterized by a different level of detail. Thus, the user can select the skeleton having enough detail to solve the current task.

Two main approaches exist to build a hierarchical skeleton structure. The most common approach is based on skeleton pruning. Given the initial skeleton, its branches are associated a parameter measuring the perceptual relevance of the corresponding object part. Then, peripheral branches can be pruned in increasing perceptual relevance order so generating a sequence of skeletons, each of which is a proper subset of the previous skeleton and whose structure becomes simpler and simpler. The second approach divides each branch of the skeleton into a number of parts by means of a polygonal approximation or by using an approximation with curves of higher order. By increasing the value of the threshold used during the approximation, rougher and rougher representations of the skeleton are obtained.

With the first approach (e.g., [2]), the objects recovered by the successively pruned skeletons differ from the input object for the absence of an increasing number of peripheral parts. Thus, the hierarchical skeleton structure represents an object characterized by simpler and simpler shape until a very rough shape is obtained, coinciding with what could be understood as the main body of the object. In turn, by following the second approach (e.g., [3]), no object's limbs are lost when approximating the skeleton with a higher value for the threshold, but the geometry of each object part is represented in a rougher and rougher manner.

The literature on multi-scale skeletonization driven by pruning includes contributions mainly for the 2D case. In the 3D case, pruning is mainly used as a necessary post-processing step to clean the skeleton. In fact, the skeleton is strongly conditioned by the presence of boundary noise, especially with continuous skeletonization methods such as those based on the use of the Voronoi Diagram. An extension of pruning to build a hierarchical skeleton structure also in the 3D space seems to be straightforward. In turn, as regards the second approach, to our knowledge only methods effective in the 2D space are available. In this paper we follow the second approach to provide a hierarchical skeleton structure for 3D objects.

2   Basic Notions

We use binary voxel images in cubic grids. The 3×3×3 neighborhood of a voxel p includes the six face- the twelve edge- and the eight vertex-neighbors of p.

The distance between two voxels p and q is given by the length of a minimal path from p to q. If the same weight is used for all unit moves, the D^{26} distance is obtained, which is the 3D version of the 2D chessboard distance. In turn, if the weights 3, 4 and 5 suggested in [4] are respectively used to measure the unit moves from p towards its face-, edge- and vertex-neighbors along the path, the <3,4,5>-distance is obtained. The <3,4,5>-distance provides a good approximation to the Euclidean distance.

The skeleton S of an object P is a homotopic subset of P, consisting of curves centered in P, whose voxels are labeled with their distance from the complement of P. We use the algorithm [5] to compute the <3,4,5>-distance labeled skeleton.

A voxel p of S is an end point if has only one neighbor in S, a normal point if has two neighbors in S, and a branch point if has more than two neighbors in S.

A skeleton branch is a connected subset of S entirely consisting of normal points, except for the two extremes of the branch that are end points or branch points. If both extremes are branch points, the skeleton branch is an internal branch. All other skeleton branches are peripheral branches.

The reverse distance transformation [6] is a process that associates to a set of voxels, labeled with integer distance values, the envelope of the balls centered on the voxels and having radii equal to the corresponding distance values. We use the <3,4,5>-distance when applying the reverse distance transformation to the skeleton.

3   Polygonal Approximation

Different approaches to compute the polygonal approximation of a digital curve are available in the literature. The split type approach [7] is particularly convenient when working with open curves, as it is the case for skeleton branches. In fact, in this case the two starting points from which the process recursively splits a curve are the extremes of the curve itself. This guarantees that the vertices identified along the curve are not influenced by the selection of the two starting points, as it would be the case for a closed curve.

Roughly speaking, a split type polygonal approximation algorithm can be described as follows. The two extremes of the input open curve are taken by all means as vertices of the polygonal approximation. The Euclidean distance of all points of the curve from the straight line joining the two vertices is computed. The point with the largest distance is taken as a new vertex, provided that such a distance overcomes an a priori fixed threshold θ (to be set depending on the desired approximation quality). Otherwise the process terminates. If a new vertex is detected, such a vertex divides the curve into two sub-curves, to each of which the above split type algorithm is applied. The splitting process is repeated as far as points are detected having distance larger than the threshold from the straight lines joining the extremes of the sub-curves to which the points belong. When recursion is completed, the curve results to be approximated by a number of segments that, in the limit of the adopted tolerance, are rectilinear. The curve is represented by the ordered sequence of all and only the points that have been detected as vertices.

For the sake of simplicity, let us consider the object shown in Fig. 2 left, whose skeleton consists of a unique branch. By comparing Fig. 2 left and Fig. 2 middle, we may observe that the geometry of the skeleton reflects the geometry of the object. In fact, curvature changes along the skeleton correspond to curvature changes along the boundary of the object. Thus, a polygonal approximation of the skeleton would divide it into straight line segments. Each segment would be in correspondence with a region of the object characterized by the absence of curvature changes along its boundary, and could be seen as the spine of that region.

Fig. 2. From left to right, an object, its skeleton (different colors denote different distance values), and the vertices (black) resulting in the skeleton after polygonal approximation in 4D

Still with reference to Fig. 2, we also note that the different distance values r_i of the voxels p_i successively met when tracing the skeleton from the first extreme to the second extreme (i.e., the different radii of the balls associated with the skeleton voxels) take into account the changes in width of the object. By plotting the ordered pairs (p_i, r_i) on a Cartesian plane, a 2D curve is obtained and a split type polygonal approximation can be performed to set vertices wherever the radii fail to be aligned in the limits of the adopted tolerance. The skeleton could then be divided into subsets such that the radii of the corresponding voxels are either constant, or change in a linear manner. Each subset could be interpreted as the spine of a region of the object that is characterized either by constant thickness or by thickness that linearly increases/decreases.

If changes in geometry and distance values along a skeleton branch are simultaneously considered, polygonal approximation will divide the skeleton branch into subsets characterized by linearity both in geometry and in the distribution of distance values. In this way, each segment of the skeleton branch will correspond to a simple region characterized by the following two properties: 1) absence of significant curvature changes along the boundary and 2) thickness that either is constant or linearly increases/decreases. To this purpose, we here perform polygonal approximation in a 4D space, where the four coordinates are the three Cartesian coordinates and the radii of the voxels of the 3D skeleton branches.

For the sake of completeness, we note that while a skeleton branch is obviously a connected set in the 3D space, this is no longer guaranteed when the skeleton branch is represented in the above 4D space. There, a skeleton branch may result in a sparse set of points. However, this does not influence the performance of polygonal approximation.

The computation of the Euclidean distance d of a point C from the straight line joining two points A and B in the 4D space is done by using the following expression:

$$d^2 = \|AC\|^2 - P_{ABC} * P_{ABC} / \|AB\|^2$$

where $\|AB\|$ is the norm of the vector AB, and P_{ABC} is the scalar product between vectors AB and AC. If the point C is the one at maximal distance from the straight line joining A and B, C is taken as a vertex of the polygonal approximation provided that its distance d satisfies the condition $d>\theta$.

Once all vertices have been detected in the 4D space, the skeleton voxels in the 3D space corresponding to them are marked as vertices. In the limits of the adopted tolerance, these vertices divide the skeleton branch into straight line segments, whose voxels are characterized by radii that either are constant or linearly increase/decrease. See Fig. 2 right, where the vertices are shown in black. Thus, a skeleton branch can be efficiently represented by orderly giving only the spatial coordinates and radii of the found vertices.

Each segment of a skeleton branch is interpreted as the spine of a simple region of the object part associated to the whole skeleton branch. A simple region is shaped as a cylinder (Fig. 3 middle left), when radii are all equal along the spine (Fig. 3 left, where all skeleton voxels, depicted with the same color, have the same radius), or as a truncated cone (Fig. 3 right), when radii linearly increase/decrease along the spine (Fig. 3 middle right, where different colors denote different radii). Of course, the simple region is delimited by two half balls centered on the two vertices of the spine.

Fig. 3. Straight line skeleton segments and their corresponding simple regions

4   The Hierarchical Skeleton Structure

The procedure described in the previous section is applied to all skeleton branches, so originating an approximated version of the entire skeleton. By using different values for the threshold θ, differently approximated skeletons are obtained. In particular, the more the threshold increases, the more the approximated skeleton actually represents a rough version of the object.

The smallest value that can be taken into account for the threshold is θ=1. In fact, due to the chord property [8], if the distance of each point of a digital curve from the straight line joining the extremes of the curve is at most equal to 1, then the digital curve is a digital straight line segment. To fix the maximum possible value for the threshold, we compute for each skeleton branch the maximal distance of its voxels from the straight line joining the extremes of the branch. Then, the maximum μ among all computed maximal distances is taken as the maximum value for the threshold θ. In fact, with this value for the threshold each skeleton branch is obviously approximated by the straight line segment joining the two extremes of the branch, since no skeleton voxel can be found characterized by distance larger than μ. It would be useless to consider a larger threshold.

Starting from the minimal value, the threshold is increased by a fixed incremental amount (set to 0.5 in this work), until μ is achieved. In principle, each obtained polygonal approximation of the skeleton should be recorded to build the hierarchical skeleton structure. However, the current polygonal approximation is skipped if it has the same number of vertices as the previously recorded polygonal approximation. In fact, it would be identical to the previous approximation.

Once all threshold values have been taken into account, the hierarchical skeleton structure is available. At a given level of the hierarchy, which is in correspondence with a given threshold value, each skeleton branch is represented by the sequence of vertices detected along that branch during the polygonal approximation by using that threshold.

To visualize the representations and to understand how the different polygonal approximations more or less faithfully represent the initial skeleton, we build the approximated skeletons starting from the available information, i.e., the ordered sequences of spatial coordinates and radii of the vertices.

Let (x_i,y_i,z_i) and $(x_{i+1},y_{i+1},z_{i+1})$ be the spatial coordinates of two successive vertices v_i and v_{i+1} and let r_i and r_{i+1} be the associated radii. Let Δx, Δy, Δz and Δr be the absolute values of the differences between homologous coordinates of the two vertices.

The number L of voxels in a minimal length path linking v_i and v_{i+1} is given by the maximum among Δx, Δy and Δz, i.e., by the D^{26} distance between v_i and v_{i+1}. We need to establish how many moves should be done via each of the three kinds of neighbors. In this respect, we note that, since a move via vertex-neighbors requires that the three spatial coordinates all change by 1, the total number of vertex-moves is conditioned by the minimum among Δx, Δy and Δz. For the sake of simplicity, let us suppose that the minimum is Δx. Then, Δx is the number of moves via vertex-neighbors. With similar arguments, the number of moves via edge-neighbors will be the minimum out of the remaining two differences of coordinates Δy and Δz, each of which diminished by Δx. Finally, the remaining moves, if any, are via face-neighbors. Thus, we can easily compute the number of moves necessary in total to link v_i and v_{i+1} and how many of these moves occur via each of the three types of neighbors.

However, several minimal length paths with the same number of moves for each kind of neighbors can be built. Thus, to build the minimal length path aligned along the digital straight line joining v_i and v_{i+1}, the different moves have to be suitably alternated along the path.

To this aim, let us denote by m_f, m_e and m_v the number of moves in face, edge and vertex direction respectively. Let us consider the minimum among m_f, m_e and m_v. For simplicity, we suppose that the minimum is m_v. Let us also consider the sum of m_f and m_e. The ratio $(m_f + m_e)/m_v$ counts the number of moves in face and edge directions, after which one move via a vertex-neighbor has to be done. To determine the way in which the moves via face- and edge-neighbors alternate, we compute the minimum between m_f and m_e. For simplicity, let us suppose that the minimum is m_e. Then the ratio m_f/m_e indicates after how many moves via face-neighbors a move via edge-neighbor should be done. Special cases where the ratios are not integer numbers or the denominator is zero are properly treated.

To select the first move to be accomplished starting from v_i, we compute the maximum among m_f, m_e and m_v. For simplicity, suppose that such a maximum is m_f. Then, the first move in the path is in face direction. Since six face-neighbors of v_i exist, the proper face-neighbor to be accepted in the path is the one at minimal distance from v_{i+1}. To select the remaining voxels constituting the path, it is enough to analyze only a subset of the neighborhood of the current voxel, which is determined based on the position of the previous voxel in the path. Among the neighbors of the current voxel, only those that can be reached with the proper move are considered and the one at minimal distance from the vertex v_{i+1} is selected as the next voxel in the path.

Fig. 4. Polygonal approximations of the skeleton with different threshold values

Since the hierarchical skeleton structure is recorded by the ordered sequences of spatial coordinates and radii of the vertices, the only available radii are those pertaining any two successive two vertices v_i and v_{i+1} of a given approximation. To distribute linearly distance values to all voxels in the path that we have just shown how to build, we make use of the number L of voxels in the path, i.e., the D^{26} distance between v_i and v_{i+1}, and of Δr, i.e., the absolute value of the difference between r_i and r_{i+1}. The ratio $\Delta r/L$ indicates the increment in radius that, starting from the smallest radius (say the radius r_i that pertains to vertex v_i), will allow us to ascribe the proper radius to each voxel in the segment until we reach the second extreme of the segment (v_{i+1} to which the radius r_{i+1} pertains). Then, the so computed radii are approximated by the closest integers, in order each representation in the hierarchical structure can be regarded as a digital skeleton.

As an example, see Fig. 4, where nine different polygonal approximations are shown. Colors are used to distinguish skeleton segments. Actually, 17 different approximations were available for the input skeleton, but for space restriction only a few approximations are shown, namely those corresponding to θ=2.5, θ=4.5, θ=6.5, θ=8.5, θ=10.5, θ=13.5, θ=24.5, θ=32.5 and θ=34.5. Since for θ=34.5 the nine branches of the skeletons are approximated by nine straight line segments, it would have no sense to consider larger values for the threshold.

To show how faithfully the object is represented when the skeleton is approximated by using an increasing threshold value, we apply the reverse distance transformation to the approximated skeletons. This is possible since we have assigned integer distance values to the voxels constituting the spines approximating the skeleton branches. In Fig. 5 the objects recovered starting from the skeletons of Fig. 4 are orderly shown.

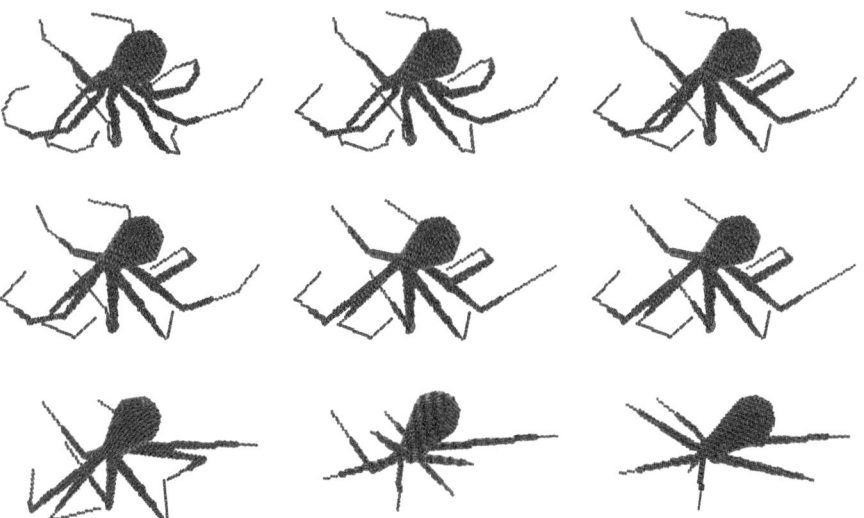

Fig. 5. Objects recovered by applying the reverse distance transformation to the input skeleton and its approximated versions

As expected, when the threshold increases the approximated skeleton becomes a more schematic representation of the input object. For example, since the roughest polygonal approximation of the skeleton (Fig. 4 bottom right) consists only of nine straight line segments, the corresponding recovered object (Fig. 5 bottom right) is the combination of nine simple parts (as many parts as many are the branches of the input skeleton) and information on the fact that, for example, the tentacles of the octopus are bent is no longer available. Actually, the straight line segments joining the extremes of the branches associated with the tentacles of the octopus are not at all centered within the actual tentacles of the input object.

To build a hierarchical skeleton structure able to represent the object's parts without excessive loss of shape information, one should use as maximum threshold a value definitely smaller than μ. Even better, one should fix a different maximum threshold value for each skeleton branch, which should obviously be smaller than the maximal distance of the skeleton voxels along that branch from the straight line joining the extremes of the branch. A possible solution is to use as maximum threshold value for each branch the largest radius associated with the skeleton voxels along the branch. In this way the risk of replacing the input skeleton, definitely centered within the object, by a skeleton whose voxels are likely to be placed outside the object is significantly reduced.

We work with a discrete skeleton whose voxels are labeled with the <3,4,5>-distance, while during polygonal approximation we detect vertices by computing the Euclidean distance. Thus, if we want to set the maximal threshold value in function of the maximal distance value of the skeleton voxels in a branch, we need to properly evaluate the Euclidean length of the radii. Roughly speaking, if we divide by 3 the distance value r_i of a skeleton voxel p_i, we obtain the number of moves in face direction to reach the complement of the object starting from p_i. Thus, a reasonable approximation of the Euclidean length of the radius of p_i is given by $(r_i/3)+0.5$, where the amount 0.5 is added to take into account that also the voxel p_i contributes to the length of the radius.

Fig. 6. Top, approximated versions of the skeleton, not all equally meaningful. Bottom, the corresponding reconstructed objects.

In Fig. 6, an example is given with reference to an object whose skeleton consists of a single branch, for which the largest distance of the skeleton voxels from the straight line joining the extremes of the branch μ equals 41.5. Eight different polygonal approximations are possible, respectively obtained for $\theta=1.5$, $\theta=2.5$, $\theta=3.5$

(not shown in Fig. 6 for space restrictions), θ=4.5, θ=5.5, θ=11.5, θ=12.5 (not shown in Fig. 6 for space restrictions) and θ=41.5. In Fig. 6 bottom, the corresponding reconstructed objects are shown. For this object, the largest radius along the branch equals 18, if measured with the <3,4,5>-distance. Its equivalent Euclidean length is 6.5. If such a value is taken as the maximal threshold value, the hierarchical skeleton structure would include only the first four approximated skeletons shown in Fig. 6 top (corresponding to the first four reconstructed objects in Fig. 6 bottom). In fact, θ=11.5 is the first threshold value for which a polygonal approximation is obtained with a smaller number of vertices with respect to the approximated version obtained for θ=5.5. Thus, θ=6.5 originates the same approximation as θ=5.5.

5   Conclusion

We have introduced a method to build a hierarchical skeleton structure, starting from the curve skeleton of a 3D object. Skeleton branches are represented in a 4D space, where the coordinates are the (x,y,z) coordinates plus the radius r associated with each skeleton voxel. Then, for each skeleton branch polygonal approximation is accomplished so as to divide skeleton branches into straight line segments, whose voxels have either constant or linearly increasing/decreasing radius. Different threshold values are used so as to obtain differently approximated versions of the skeleton. The hierarchical structure includes all the differently approximated skeletons, which allows the user to get a fine-to-coarse shape representation. The algorithm has been implemented on an Intel Core i7 (3.5 GHz, 8 GB RAM) personal computer and tested on images taken from publicly available shape repositories, e.g. [9].

References

1. Siddiqi, K., Pizer, S.M. (eds.): Medial Representations: Mathematics, Algorithms and Applications. Springer, Berlin (2008)
2. Shen, W., Bai, X., Hu, R., Wang, H., Latecki, L.J.: Skeleton growing and pruning with bending potential ratio. Pattern Recognition 44, 196–209 (2011)
3. Sanniti di Baja, G., Thiel, E.: A multiresolution shape description algorithm. In: Chetverikov, D., Kropatsch, W.G. (eds.) CAIP 1993. LNCS, vol. 719, pp. 208–215. Springer, Heidelberg (1993)
4. Borgefors, G.: On digital distance transform in three dimensions. CVIU 64, 368–376 (1996)
5. Arcelli, C., Sanniti di Baja, G., Serino, L.: Distance driven skeletonization in voxel images. IEEE Trans. PAMI 33, 709–720 (2011)
6. Nystrom, I., Borgefors, G.: Synthesising objects and scenes using the reverse distance transformation in 2D and 3D. In: Braccini, C., Vernazza, G., DeFloriani, L. (eds.) ICIAP 1995. LNCS, vol. 974, pp. 441–446. Springer, Heidelberg (1995)
7. Ramer, U.: An iterative procedure for the polygonal approximation of plane curves. CGIP 1, 244–256 (1972)
8. Rosenfeld, A.: Convex digital arcs. IEEE Trans. Computers C-23, 1264–1269 (1974)
9. Shilane, P., Min, P., Kazhdan, M., Funkhouser, T.: The Princeton Shape Benchmark. In: Shape Modeling International, Genova, Italy (2004)

Learning Iterative Strategies in Multi-Expert Systems Using SVMs for Digit Recognition

Donato Barbuzzi[1], Donato Impedovo[2],
Francesco Maurizio Mangini[1], and Giuseppe Pirlo[1]

[1] Department of Computer Science, University of Bari, Bari, Italy
{donato.barbuzzi,francescomaurizio.mangini,
giuseppe.pirlo}@uniba.it
[2] Department of Electrical and Electronic Engineering, Polytechnic of Bari, Bari, Italy
impedovo@deemail.poliba.it

Abstract. This paper presents three different learning iterative strategies, in a multi-expert system. In first strategy entire new dataset is used. In second strategy each single classifier selects new samples starting from those on which it performs a misclassification. Finally, the collective behavior of classifiers is studied to select the most profitable samples for knowledge base updating. The experimental results provide a comparison of three approaches under different operating conditions and feedback process. A classifier SVM and four different combination techniques were used by considering the CEDAR (handwritten digit) database. It is shown how results depend by the iterations on the feedback process, as well as by the specific combination decision schema and by data distribution.

Keywords: Feedback-based strategies, Instance Selection, Multi Expert Systems.

1   Introduction

While in the on-line handwriting recognition temporal and spatial information about each stroke is available, in off-line handwriting recognition only the image of the written character is used to perform the classification task. Indeed, off-line handwriting recognition is still considered a difficult problem that is only partially solved.

It has been observed, in the last few decades, that the accuracy of handwriting character recognition can be also improved by multiple expert fusion. The idea is not to rely on a single decision making scheme but to use several designs (experts) for decision making [1, 2, 3]. In fact, the collective behavior of a set of classifiers can convey more information that those of each classifier of the set, and this information can be exploited for classification aims [4, 5].

More specifically this paper proposes to select those samples, to be used for retraining specific experts of the set, misclassified by the multi-expert system [6, 7, 8, 9]. This approach is compared to situation in which the entire new dataset is used for

A. Petrosino (Ed.): ICIAP 2013, Part I, LNCS 8156, pp. 121–130, 2013.

learning as well as the case in which specific samples are selected by the individual classifier.

Furthermore at the three standard retraining rules is computed an iteration each on the feedback process. A significant decrease in the error rate is reported considering the approach feedback-based ME.

Tests have been performed on the task of handwritten digit recognition, on the CEDAR database, by considering different types of features and a state of the art classifier (Support Vector Machine). Four different combination techniques (Majority Vote, Weighted Majority vote, Sum Rule and Product Rule) have been used between abstract and measurement level. It is shown how results depend by the feedback process, as well as by the specific combination decision schema and by data distribution.

The paper is organized as follows: Section 2 presents an overview of retraining rules and the different strategies feedback-based. Experiments and results are in Section 3 and 4, respectively. Section 5 reports a discussion and the conclusion of the work.

2   Learning Strategies

2.1   Related Work

When new labeled data became available, the following fundamental question arises: "how to use new data?". The simplest thing is probably to use the entire new set to update the knowledge base of each expert in the system, already trained and in its working phase. On the other hand, many interesting algorithms can be adopted in order to select (or focus the attention on) specific samples.

In particular, the algorithm AdaBoost [10, 11] is able to improve performance of a classifier on a given data set by focusing the learner attention on difficult instances. Even if this approach is very powerful, it works well in the case of weak classifiers, moreover not all the learning algorithms accept weights for the incoming samples. Another interesting approach is the bagging one: a number of weak classifiers trained on different subset (random instance) of the entire dataset are combined by means of the simple majority voting [1, 2]. Bagging and AdaBoost algorithms are adapted when considering a single classifier but applied to a ME system, them performance are boosted [12].

From these observations, specific strategies are depicted in the next paragraph taking into account of a multi-expert system that works in supervised learning.

2.2   Selecting Instances

Let be:

- C_j, for j=1,2,...,M, the set of pattern classes;
- $P = \{x_k \mid k = 1,2,..., K\}$, a set of pattern to be feed to the Multi Expert (ME) system. P is considered to be partitioned into S subsets $P_1, P_2,...,P_s,...,P_S$, being

$P_s=\{x_k \in P \mid k \in [N_s \cdot (s-1)+1, N_s \cdot s]\}$ and $N_s=K/S$ (N_s integer), that are fed one after the other to the multi-expert system. In particular, P_1 is used for learning only, whereas $P_2, P_3,...,P_s,...,P_S$ are used both for classification and learning (when necessary);

- $y_s \in \Omega$, the label for the x_s pattern, $\Omega = \{C_1, C_2,...,C_M\}$;

- A_i the $i\text{-}th$ classifier for $i = 1,2,...,N$;

- $F_i(k) = (F_{i,1}(k), F_{i,2}(k), ..., F_{i,r}(k), ..., F_{i,R}(k))$ the feature vector used by A_i for representing the pattern $x_k \in P$ (for the sake of simplicity it is here assumed that each classifier uses R real values as features);

- $KB_i(k)$, the knowledge base of A_i after the processing of P_k. In particular $KB_i(k) = (KB_i^1(k), KB_i^2(k),..., KB_i^M(k))$;

- E the multi expert system which combines H_i hypothesis in order to obtain the final one.

In first stage (s=1), the classifier A_i is trained using the patterns $x_k \in P^*_i = P_1$. Therefore, the knowledge base $KB_i(s)$ of A_i is initially defined as:

$$KB_i(s)=(KB^1_i(s), KB^2_i(s),..., KB^j_i(s),..., KB^M_i(s)) \qquad (1a)$$

where, for j=1,2,...,M:

$$KB^j_i(s)=(F^j_{i,1}(s), F^j_{i,2}(s),..., F^j_{i,r}(s),..., F^j_{i,R}(s)) \qquad (1b)$$

being $F^j_{i,r}(s)$ the set of the $r\text{-}th$ feature of the $i\text{-}th$ classifier for the patterns of the class C_j that belongs to P^*_i.

Successively, the subsets $P_2, P_3,...,P_s,...,P_{S-1}$ are provided one after the other to the multi-classifier system both for classification and for learning. P_S is just considered to be the testing set in order to avoid biased or too optimistic results. When considering new labeled data (samples of $P_2, P_3,...,P_s,...,P_{S-1}$), a naïve and two not naïve strategies can be used.

The naïve strategy uses all the available new patterns to update the knowledge base of each individual classifier:

- $\forall x_t \in P_s : update\_KB_i$ $\qquad (2)$

Of course, in order to select patterns from Ps to train A_i, in the first strategy (not naïve) A_i is updated by considering all misclassified samples:

- $\forall x_t \in P_s \ni' A_i(x_t) \neq y_t : update\_KB_i$ $\qquad (3)$

The second approach (not naïve) is derived from AdaBoost and bagging. A_i is updated by considering all its misclassified samples if and only if these produce (or contribute to) a misclassification of the ME:

- $\forall x_t \in P_s \ \ni' (A_i(x_t) \neq y_t \wedge E(x_t) \neq y_t) : update\_KB_i$ (4)

In order to inspect and take advantage of the common behavior of the ensemble of classifiers, the following simple strategy is evaluated and compared to the previous two.

3   Experiments

3.1   CEDAR Database

In the experimental session, a multi-expert system for handwritten digit recognition has been considered [6, 8] and the CEDAR Database of handwritten digits has been used [13]. In this case P={x_k | k=1,2,…,20351} (classes from "0" to "9").

The DB has been initially partitioned into 6 subsets:

- P_1={$x_1,x_2,x_3,…, x_{12750}$},
- P_2={$x_{12751},…, x_{14119}$},
- P_3={$x_{14120},…, x_{15488}$},
- P_4={$x_{15489},…, x_{16857}$},
- P_5={$x_{16858},…, x_{18223}$},
- P_6={$x_{18224},…, x_{20351}$}.

In particular, $P_1 \cup P_2 \cup P_3 \cup P_4 \cup P_5$ represent the set usually adopted for training when considering the CEDAR DB. P_6 is the testing dataset. Each digit is zoned into 16 uniform (regular) regions [14, 15], successively, for each region, the following set of features have been considered [16, 17, 18, 19]:

F_1: features set 1 (geometric features): hole, up cavity, down cavity, left cavity, right cavity, up end point, down end point, left end point, right end point, crossing points, up extrema points, down extrema points, left extrema points, right extrema points;

F_2: features set 2 (contour profiles): max/min peaks, max/min profiles, max/min width, max/min height;

F_3: features set 3 (intersection with lines): 5 horizontal lines, 5 vertical lines, 5 slant -45° lines and 5 slant +45° lines.

3.2   Setup

The classifier used for the experimentation is a SVMs (Support Vector Machines). This is a classifier that separates a given set of binary labeled training data with a hyper-plane that is maximally distant from them. For no linear separation, SVM can work in combination with the technique of "kernels", that automatically realizes a non-linear mapping to a feature space. Here, for multi-class recognition, more binary SVM are performed and the kernel function adopted is the rbf gamma [3].

Also many approaches have been considered so far for classifiers combination. These approaches differ in terms of type of output they combine, system topology and degree of a-priori knowledge they use [1,2,3]. The combination technique plays a crucial role in the selection of new patterns to be feed to the classifier in the proposed approach. In this work the following decision combination techniques have been considered and compared: Majority Vote (MV), Weighted Majority Vote (WMV), Sum Rule (SR) and Product Rule (PR). MV just considers labels provided by the individual classifiers, it is generally adopted if no knowledge is available about performance of classifiers so that they are equal-considered. The second approach can be adopted by considering weights related to the performance of individual classifiers on a specific dataset. Given the case depicted in this work, it seems to be more realistic, in fact the behavior of classifiers can be evaluated, for instance, on the new available dataset. In particular, let ε_i be the error rate of the i-th classifier evaluated on the last available training set, the weight assigned to

$$\text{is, } w_i = \log(1/\beta_i) \text{ being } \beta_i = \varepsilon_i /(1 - \varepsilon_i) \tag{5}$$

Sum Rule (SR) and Product Rule (PR) take into account the confidence of each individual classifier given the input pattern and the different classes [1]. Before the combination, confidence values provided by different classifiers were normalized by means of Z-score [20, 21].

4   Experimental Results

This section presents the results in terms of both error rate percentage (ER) and number of selected samples (SS), for each different learning strategies, obtained using binary images on CEDAR database. We combined, adopting a multi-expert system, the three set of features (F_1, F_2 and F_3) and a classifier SVM. Values of the similarity index (SI) [22, 23] are reported in the last row of each table.

The label "X-feed" refers to the use of the X modality for the feedback training process: "All" is the feedback of the entire set, "C" is feedback at classifier level. "MV", "WMV", "SR", "PR" are feedback at ME level adopting, respectively, the majority vote, the weighted majority vote, the sum rule and the product rule schema.

Tables 1, 2 and 3 show results related to the use of SVM classifier. The three set of features F_1, F_2 and F_3 are represented here as: SVM_1, SVM_2 and SVM_3. P_1 is used for training and P_6 for testing. $P_2 \cup P_3 \cup P_4 \cup P_5$ is used for feedback learning. The first column (No-feed) reports results related to the use of P_1 for training and of P_6 for testing, without applying any feedback (0 Selected Samples), while the approach All-feed uses all samples belonging to the new set in order to update the knowledge base of each single classifier (All Selected Samples). Depending by the combination technique, a specific strategy can outperform the others. More specifically, applying two iterations in the feedback process, WMV-feed, SR-feed and PR-feed, respectively, an improvement of 0,05%, 0.09% and 0,05% compared to the use of the entire new data-set (All-feed), while for WMV-feed and SR-feed, respectively, an improvement of 0,10% and 0,09% compared to the use of the feedback at single expert level. While, iterating the feedback process in three steps, MV-feed, WMV-feed and SR-feed, respectively, improvement of 0.14%, 0.05% and 0.09% compared to All-feed and only SR-feed improvement of 0.04% compared to C-feed.

Finally using the feedback-based strategies at ME level, it is of interest the fact that a very restricted subset of samples are selected for retraining.

Table 1. SVM, Feedback - $P_2 \cup P_3 \cup P_4 \cup P_5$, One Iteration

	No-feed	C-feed		MV-feed		WMV-feed		SR-feed		PR-feed		All-feed	
	ER	ER	SS	ER	SS	ER	SS	ER	SS	ER	SS	ER	SS
SVM$_1$	2.94	2.82	335	3.01	143	3.01	93	2.96	76	2.96	64	2.92	5466
SVM$_2$	8.37	8.13	572	8.36	167	8.55	117	8.22	91	8.22	72	7.79	5466
SVM$_3$	4.09	4.35	225	4.46	124	4.46	124	4.18	69	4.18	52	4.23	5466
MV	2.54	2.49		**2.35**		X		X		X		2.58	
WMV	**1.69**	1.83		X		1.79		X		X		1.74	
SR	1.46	1.41		X		X		**1.36**		X		1.41	
PR	1.22	**1.17**		X		X		X		1.22		**1.17**	
SI	91.29	91.30		90.98		90.91		91.32		91.24		91.55	

Table 2. SVM, Feedback - $P_2 \cup P_3 \cup P_4 \cup P_5$, Two Iterations

	No-feed	C-feed		MV-feed		WMV-feed		SR-feed		PR-feed		All-feed	
	ER	ER	SS	ER	SS	ER	SS	ER	SS	ER	SS	ER	SS
SVM$_1$	2.94	2.82	642	3.05	327	2.87	327	3.01	197	3.01	159	2.68	10932
SVM$_2$	8.37	8.32	988	8.46	418	8.36	283	8.41	221	8.60	171	7.47	10932
SVM$_3$	4.09	4.23	1127	4.37	430	4.37	295	4.51	181	4.37	174	4.14	10932
MV	**2.54**	**2.54**		2.63		X		X		X		**2.54**	
WMV	1.69	1.74		X		**1.64**		X		X		1.69	
SR	1.46	1.50		X		X		**1.41**		X		1.50	
PR	**1.22**	**1.22**		X		X		X		**1.22**		1.27	
SI	91.29	91.29		91.02		91.12		90.99		90.85		92.04	

Table 3. SVM, Feedback - $P_2 \cup P_3 \cup P_4 \cup P_5$, Three Iterations

	No-feed	C-feed		MV-feed		WMV-feed		SR-feed		PR-feed		All-feed	
	ER	ER	SS	ER	SS	ER	SS	ER	SS	ER	SS	ER	SS
SVM$_1$	2.94	2.96	879	3.10	427	2.96	425	2.96	254	3.05	204	2.87	16398
SVM$_2$	8.37	8.04	1296	8.55	552	8.60	366	8.79	287	8.88	217	7.85	16398
SVM$_3$	4.09	4.23	1616	4.51	574	4.32	391	4.61	299	4.51	228	4.37	16398
MV	**2.54**	**2.54**		2.63		X		X		X		2.77	
WMV	**1.69**	1.74		X		1.74		X		X		1.79	
SR	**1.46**	1.50		X		X		**1.46**		X		1.55	
PR	1.22	1.22		X		X		X		1.27		**1.17**	
SI	91.29	91.35		90.85		91.02		90.68		90.66		91.57	

Figures 1, 2 and 3 represent the performances of the tables: Table 1, Table 2 and Table 3, respectively, show that in any case a combination technique exists that definitively outperforms the other two approaches, adopting an iteration on the feedback process.

In particular, the strategy WMV-feed, after two iterations on the feedback process, reduces the error rate both compared to one iteration and three iterations, respectively, of 0.15% and 0.10% and respect to other two feedback-based strategies C-feed and All-feed.

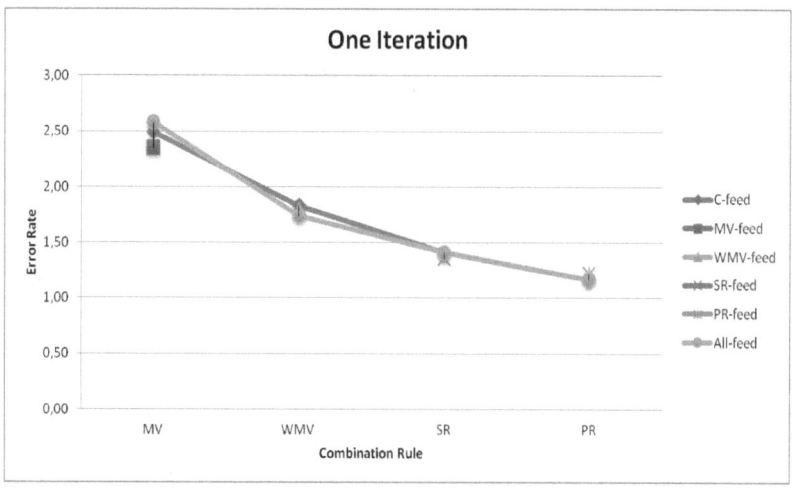

Fig. 1. SVM, Feedback - $P_2 \cup P_3 \cup P_4 \cup P_5$, One Iteration

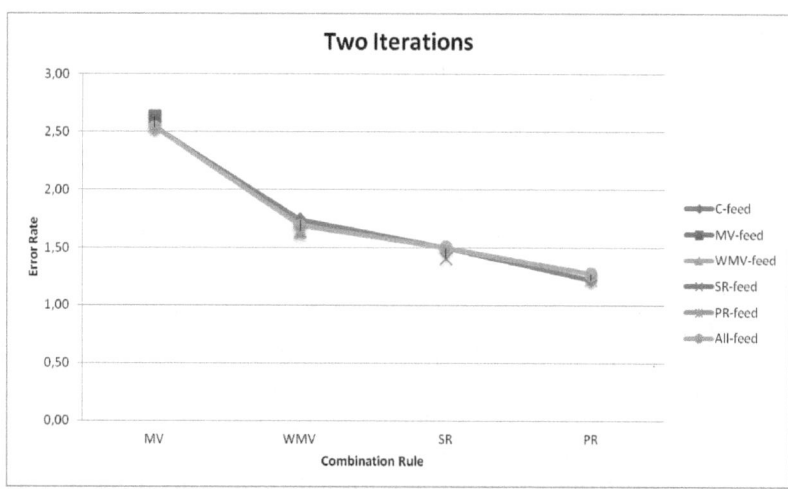

Fig. 2. SVM, Feedback - $P_2 \cup P_3 \cup P_4 \cup P_5$, Two Iterations

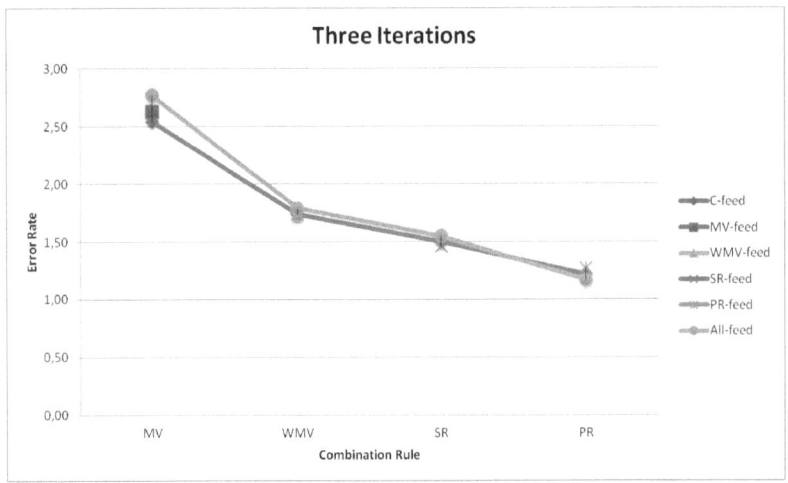

Fig. 3. SVM, Feedback - $P_2 \cup P_3 \cup P_4 \cup P_5$, Three Iterations

5   Discussion and Conclusion

This paper provides a comparison of three iterative learning strategies for multi-expert systems, when new labeled data become available. More precisely, the feedback-based strategies are used to the optimal way in which new selected samples must be used to update the knowledge base of the individual classifiers. For the purpose, four different combination techniques between abstract and measurement level strategies have been investigated under different operating conditions.

The experimental results have shown that performance of feedback-based training strictly depend by the iteration on the feedback process, by the combination strategy of the ME, but also by the data distribution and the similarity between samples in the feedback set and samples of the testing set. In particular, the not naïve strategy proposed in this paper (see eq. (4)) is able to select not only samples to be used for the updating process, but also the classifiers (see eq. (5)) to which those samples must be feed. Of course, considering initially trained classifiers, the multi-expert will return few instances for the retraining process if the classification performances on new data available are high. This can happen depending by performances of classifiers, by iterations on the feedback process as well as by the ratio new/old data. Especially, given a specific classifier, the difference between the confidence value in the case of misclassification and in the case of correct one could be imputed to the fact that the specific classifier (features, matching technique, etc.) is unable to represent it, and no improvements would be obtained by introducing the new sample in the knowledge base. This is particularly true under the assumption that strong (not weak) classifiers are used. Finally, the result shown that also when the cardinality of the new selected training set is negligible if compared to that of the initial training set, the feedback strategy is able to produce improvements.

Future work will inspect deeply the possibility of evaluate the approaches on the task of semi-supervised learning as well as in unsupervised learning [24, 25, 26].

References

1. Kittler, J., Hatef, M., Duin, R.P.W., Matias, J.: On combining classifiers. IEEE Trans. on PAMI 20(3), 226–239 (1998)
2. Suen, C.Y., Nadal, C., Legault, R., Mai, T.A., Lam, L.: Computer Recognition of unconstrained handwritten numerals. Proc. IEEE 80(7), 1162–1180 (1992)
3. Liu, C.L., Nakashima, K., Sako, H., Fujisawa, H.: Handwritten digit recognition: benchmarking of state-of-the-art techniques. Pattern Recognition 36(10), 2271–2285 (2003)
4. Pirlo, G., Impedovo, D.: Fuzzy-Zoning-Based Classification for Handwritten Characters. IEEE Trans. on Fuzzy Systems 19(4), 780–785 (2011)
5. Suen, C.Y., Tan, J.: Analysis of errors of handwritten digits made by a multitude of classifiers. Pattern Recognition Letters 26(3), 369–379 (2005)
6. Impedovo, D., Pirlo, G.: Updating Knowledge in Feedback-based Multi-Classifier Systems. In: Proc. of ICDAR, pp. 227–231 (2011)
7. Barbuzzi, D., Impedovo, D., Pirlo, G.: Feedback-Based Strategies In Multi-Expert Systems. In: Sesto Convegno del Gruppo Italiano Ricercatori in Pattern Recognition (2012)
8. Impedovo, D., Pirlo, G., Barbuzzi, D.: Supervised Learning Strategies in Multi-Classifier Systems. In: Proceedings of 11th International Conference on Information Science, Signal Processing and their Applications (ISSPA 2012), pp. 1215–1220 (2012)
9. Barbuzzi, D., Impedovo, D., Pirlo, G.: Benchmarking of Update Learning Strategies on Digit Classifier Systems. In: Proceedings of the 13th International Conference on Frontiers in Handwriting Recognition, pp. 35–40 (2012)
10. Freud, Y., Schapire, R.E.: Decision-theoretic generalization of on-line learning and an application to boosting. J. of Computer and System Sciences 55(1), 119–139 (1997)
11. Schapire, R.E.: The strength of weak learnability. Machine Learning 5(2), 197–227 (1990)
12. Polikar, R.: Bootstrap-Inspired Techniques in Computational Intelligence. IEEE Signal Processing Magazine 24(4), 59–72 (2007)
13. Hull, J.: A database for handwritten text recognition research. IEEE T-PAMI 16(5), 550–554 (1994)
14. Impedovo, S., Modugno, R., Ferrante, A., Pirlo, G.: Zoning Methods for Hand-written Character Recognition: An Overview. In: Proceedings of the 12th International Conference on Frontiers in Handwriting Recognition (ICFHR 2012), Kolkata, India, November 16-18, pp. 329–334 (2010)
15. Impedovo, D., Modugno, R., Pirlo, G.: New Advancements in Zoning-Based Recognition of Handwritten Characters. In: Proc. XIII International Conference on Frontiers in Handwriting Recognition (ICFHR 2012), Monopoli, Bari, Italy, September 18-20, pp. 661–665 (2012)
16. Impedovo, S., et al.: Feature Membership Functions in Voronoi-Based Zoning. In: Serra, R., Cucchiara, R. (eds.) AI*IA 2009. LNCS, vol. 5883, pp. 202–211. Springer, Heidelberg (2009)
17. Impedovo, S., Modugno, R., Pirlo, G.: Analysis of Membership Functions for Voronoi-based Classification. In: Proceedings of the 12th Interational Conference on Frontiers in Handwriting Recognition (ICFHR 2012), November 16-18, pp. 220–225. IEEE Computer Society Press, Kolkata (2010)

18. Pirlo, G., Impedovo, D.: Adaptive Membership Functions for Hand-Written Character Recognition by Voronoi-based Image Zoning. IEEE Transactions on Image Processing 21(9), 3827–3837 (2012)
19. Impedovo, S., Pirlo, G.: Tuning between Exponential Functions and Zones for Membership Functions Selection in Voronoi-based Zoning for Handwritten Character Recognition. In: Proc. of the 11th International Conference on Document Analysis and Recognition (ICDAR 2011), September 18-21, pp. 997–1001. IEEE Computer Society, Beijing (2011) ISBN: 978-0-7695-4520-2
20. Impedovo, D., Modugno, R., Pirlo, G.: Score Normalization by Dynamic Time Warping. In: Proceedings of the International Conference on Computational Intelligence for Measurement Systems and Applications (CIMSA), Taranto, Italy, September 6-8, pp. 82–85. IEEE Computer Society Press, Taranto (2010) ISBN: 978-1-4244-7229-1
21. Pirlo, G., Impedovo, D.: Adaptive Score Normalization for Multi-Classifier Systems. IEEE Signal Processing Letters 19(12), 837–840 (2012) ISSN: 1070-9908
22. Impedovo, D., Pirlo, G., Sarcinella, L., Stasolla, E.: Artificial Classifier Generation for Multi-Expert System Evaluation. In: Proceedings of the 12th International Conference on Frontiers in Handwriting Recognition (ICFHR 2012), November 16-18, pp. 42–426. IEEE Computer Society Press, Kolkata (2010) ISBN: 978-0-7695-4221-8
23. Bovino, L., Dimauro, G., Impedovo, S., Lucchese, M.G., Modugno, R., Pirlo, G., Salzo, A., Sarcinella, L.: On the Combination of Abstract-Level Classifiers. International Journal on Document Analysis and Recognition 6, 42–54 (2003) ISSN 1433-2833
24. Frinken, V., Bunke, H.: Evaluating Retraining Rules for Semi-Supervised Learning in Neural Network Based Cursive Word Recognition. In: Proc. of ICDAR, pp. 31–35 (2009)
25. Frinken, V., Fischer, A., Bunke, H., Fornes, A.: Co-Training for Handwritten Word Recognition. In: Proc. of ICDAR, pp. 314–318 (2011)
26. Blum, A., Mitchell, T.: Combining Labeled and Unlabeled Data with Co-Training. In: ACM Proc. of COLT, pp. 92–100 (1998)

Learning Precise Local Boundaries in Images from Human Tracings

Martin Horn and Michael R. Berthold

Nycomed Chair for Bioinformatics and Information Mining AND
Konstanz Research School Chemical Biology
University of Konstanz, Box 712, 78457 Konstanz, Germany
{martin.horn,michael.berthold}@uni-konstanz.de

Abstract. Boundaries are the key cue to differentiate objects from each other and the background. However whether boundaries can be regarded as such cannot be determined generally as this highly depends on specific questions that need to be answered. As humans are best able to answer these questions and provide the required knowledge, it is often necessary to learn task-specific boundary properties from user-provided examples. However, current approaches to learning boundaries from examples completely ignore the inherent inaccuracy of human boundary tracings and, hence, derive an imprecise boundary description. We therefore provide an alternative view on supervised boundary learning and propose an efficient and robust algorithm to derive a precise boundary model for boundary detection.

Keywords: boundary detection, supervised learning.

1 Introduction

A key cue in differentiating objects from each other and their background lies in the discovery of their boundaries. Hence, the detection of object boundaries is one of the most studied problems in computer vision and finds application in many different tasks like object detection/recognition, segmentation and tracking.

But characterizing boundaries can be quite a complex task and methods range from measuring abrupt changes in some low-level image features such as brightness or color (e.g. detected by looking for positions with high derivative) to considering texture gradients or other properties distinguishing the interior and exterior of a region (e.g. detected by comparing the distribution of some property of two halves of a disc [1]). Even more difficult, a high response of an unsupervised boundary detector alone does not tell us much in many real-world applications as high boundary strength and high boundary importance are considerably different concepts [2].

A popular way to deal with these difficulties is to combine different boundary cues by learning from user provided examples. To apply supervised learning to boundary detection, a set of features (boundary cues) is commonly calculated for an image patch of a certain size and orientation and an image patch classifier is

A. Petrosino (Ed.): ICIAP 2013, Part I, LNCS 8156, pp. 131–140, 2013.

trained by minimizing its pixel-level disagreement with the given human boundary tracing. Applied to new images it results in a boundary map estimating the probability of a pixel to be a boundary pixel.

2   Learning Boundaries - Previous Research

The methods for supervised boundary detection [1,3,4,5,6,7,8] mainly differ in the boundary features they calculate and classification models they use. Martin et al. [1] for instance proposed a small, carefully selected, hand-designed feature set and tried various classifiers. They conclude that the choice of the classifier is of minor importance. Dollar et al. [5] in turn learned boosted trees on a huge set of generic features. A recent and very successful approach is suggested by Ren and Bo [3] in which features are automatically learned from the data using sparse coding before being classified with a linear SVM. Most of the methods are ranked and compared by means of the popular Berkeley Segmentation Benchmark [9], a set of natural images with a human-marked groundtruth.

3   An Alternative View on Boundary Learning

Amongst others Martin et al. [1] diagnose that the boundary learning problem is characterized by a high degree of class overlap in any low-level feature if tackled as the straightforward minimization of the pixel-level disagreement with the human boundary tracings. This is mainly caused by the inherent and unavoidable inaccuracy of human tracings. Requiring the human to provide a perfect and consistent boundary tracing, even for a single region, is ambitious and not actually feasible. But due to the mere amount of available training samples given by the Berkeley Benchmark [9] the effect is negligible if evaluated on that basis, and has therefore not been considered problematic so far.

But if boundaries are learned from a sufficient *small* set of training samples, the problem gets severe and unacceptable for many tasks and has at least two serious implications:

First, the boundaries detected are become broader, less precise and locally ambiguous the more inaccuracy the user introduces. But respecting the fact that a boundary in the continuous space is actually infinitely thin, a detected boundary in digital images should preferably not exceed the width of one pixel. This problem is illustrated in Fig. 1.

The second implication, closely related to the first, is that the complexity of the model learned (complexity for instance in terms of the *minimum description length*) increases with the users inaccuracy. And it is common sense that less complex models or hypotheses should be preferred (*Occam's razor*).

Consequently, a human tracing cannot be considered to cover the true desired boundary and only roughly indicates that the true boundary is located somewhere near the boundary labeling. Instead of receiving a set of instances that are labeled either positive (boundary) or negative (non-boundary), the learner should receives a set of *bags* containing *potentially* positive instances. During the

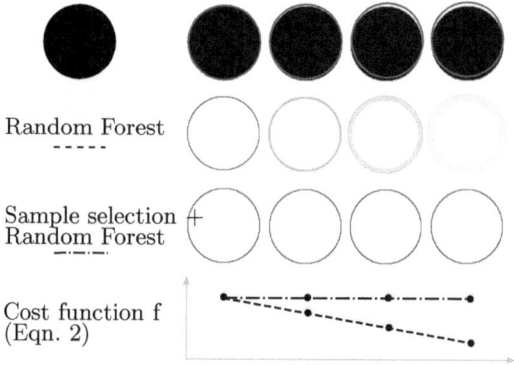

Fig. 1. Illustration of the problem when a boundary is learned by minimizing the pixel-level disagreement with the human boundary tracings. First row: An artificial image object and a labeling, which is increasingly disturbed (i.e. moved upwards). For the naive creation of the boundary model, the pixels (or rather the associated feature vector) underneath the red labels are the positive training samples, the negatives are randomly selected. Second row: If classifiers are naively trained on the training set, they will only perform well, if the label perfectly covers the right boundary positions. If not, the detected boundary will be broader and less precise. Third row: This is the boundary model that was learned by the alternative view on the supervised learning problem, by using the method proposed. Last row: The cost function indicating the homogeneity of the positive training samples used for classification.

inference of the model a maximum of one of these is selected. The non-boundary instances remain as single instances.

In the remainder of this section we will formally define the learning problem (Sect. 3.1) and present an approach to tackle it efficiently (Sect. 3.2).

3.1   Problem Definition

In order to learn boundaries from images a local boundary description (i.e. a feature vector for each pixel for a certain angle) needs to be derived and a set of closed contours (indicating where the boundaries to model are located, i.e. the human tracing) is required. Formally we define an oriented feature image and a closed contour as:

Definition 1 (Oriented Feature Image). *Let $\Omega \subset \mathbb{R}^2$ be the image domain (e.g. a pixel grid). Then the function I (oriented feature image) assigns to each pixel position (x, y) in the image and an angle α a feature vector (sample) $s \in \mathbb{R}^d$, i.e. $I : \Omega \times [0, 2\pi] \mapsto \mathbb{R}^d$.*

Definition 2 (Closed Contour). *A contour $\gamma : [0, 1] \mapsto \mathbb{R}^2$ is a differentiable function from the closed interval $[0, 1]$ to the plane, also known as Jordan Curves. A contour is closed, if $\gamma(0) = \gamma(1)$.*

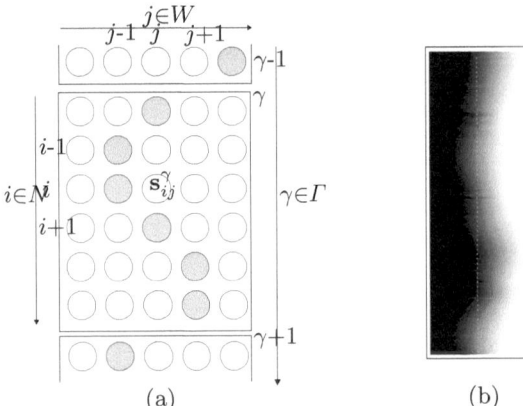

(a)   (b)

Fig. 2. (a) The grid, constructed curve-linear to the human-provided contours, with an exemplary boundary sample selection (shaded) which satisfies the constraints given in (3); (b) The grid curve-linear to the human trace (here as green dotted line) and a constraint-satisfying optimal solution (red dashed line).

Now let $\Omega \subset \mathbb{R}^2$ again be the image domain we are working on and Γ a set of closed contours (the given human tracings). Furthermore assume a feature image I according to Definition 1. Please note that for notational convenience the contour set as well as the feature image are continuous functions.

We now construct a finite set of feature vectors (samples) S by sampling along the contours in Γ as follows: Let $N = \{1, ..., n\}$ be the sampling points along a contour (n maybe chosen for each contour individually according to its length $l(\gamma) = \int_0^1 \|\gamma'(t)\| dt$, but w.l.o.g we choose n equal for each contour) and $W = \{0, ..., 2w\}$ the sampling points along the normal vectors of a contour with w being the number of points in one direction (width). Then we define a sampling function $s : N \times W \times \Gamma \mapsto \mathbb{R}^d$ for all contours

$$s(i, j, \gamma) = I(\gamma(i/n) + (j - w) * \gamma'(i/n), \angle\gamma'(i/n)) \tag{1}$$

where $\gamma'(i)$ denotes the normal vector of γ at position i and $\angle\gamma'(i)$ its angle. For notational convenience we will use s_{ij} instead of $s(i, j, \gamma)$ in the remainder and, therewith, ignore the presence of multiple contours. The extension of the process to multiple contours is straightforward.

Essentially we sampled within the feature image I along stripes, curve-linear to the given contour, and have constructed a grid with n rows and $(2w + 1)$ columns consisting of $n(2w + 1)$ feature vectors/samples. S is the union of all samples, i.e. $S = \{s_{ij} | \forall i \in N; \forall j \in W\}$. See Fig. 2 for illustration.

Now, we want to determine a subset $\hat{S} \subset S$ such that (i) in almost each row exactly one sample is selected (*thinness*); (ii) the selected samples in successive rows are close together with respect to their column coordinates, if they belong to the same closed contour (*continuity*); (iii) the selected samples are located around the middle column, where the true contour is expected to be, i.e. their column positions average to 0 (*centrality*); and (iv) the selected samples form dense clusters in the feature space $\mathbb{R}^d$ across all contours, i.e. maximizing $\sum_{s,s' \in S'} \exp\left(-\frac{\|s-s'\|^2}{2\theta^2}\right)$ (*homogeneity*), following the intuition that positions *on* the contour should look similar. A fifth condition, enforcing a large difference between samples of the *same* row (related to feature selection), could possibly be regarded as well, but is out of the scope of this paper and subject to future work.

More formally, it is a constraint discrete optimization problem (also known as a general nonlinear integer programming problem) which is generally defined as the maximization of a function $f(\mathbf{x})$ (soft constraint) subject to certain (hard) equality $h_i(\mathbf{x}) = 0$ and inequality constraints $g_i(\mathbf{x}) \leq 0$.

For the problem formulated above we want to maximize

$$f(\mathbf{x}) = \sum_{i \in N} \sum_{i' \in N} \exp\left(-\frac{\|s_{ix_i} - s_{i'x_{i'}}\|^2}{2\theta^2}\right) \tag{2}$$

subject to the constraints

$$h(\mathbf{x}) = \frac{1}{n} \sum_{i \in N} x_i = 0 \tag{3}$$

$$g(\mathbf{x}) = \max_{i \in N} |x_i - x_{i'}| \leq 1$$

$$\text{with } i' = (i+1) \mod n$$

whereas $\mathbf{x} = [x_1, ..., x_n] \in W^n$ represents a certain solution (samples selection) of the optimization problem such that each entry of the vector $\mathbf{x}$ is the column index in the grid for the respective row within a contour. Hence, the actual sample set to be determined is then $\hat{S}_\mathbf{x} = \{s_{ix_i} | \forall i \in N\}$. Please note that if multiple contours are considered, the cost function f has to be evaluated across all contours whereas the constraints (at least g) need to be satisfied within each contour individually.

The resulting sample subset $\hat{S}_\mathbf{x}$ serves as a positive training set and is used in conjunction with a negative training set (e.g. $S \backslash \hat{S}_\mathbf{x}$) to train an arbitrary classifier $c : \mathbb{R}^d \mapsto [0, 1]$ (e.g. random forest). The boundary detector, assigning each pixel a likelihood being a boundary pixel, is then defined as $P_b(x, y, \alpha) = c(I(x, y, \alpha))$.

As the number of feasible solutions is exponential $O(x^n)$ in the number of rows n, i.e. the total "length" of the given contour, it is almost impossible to enumerate all possible solutions to find the best one within a reasonable time (exhaustive search). Already the exact evaluation of the cost function (2) has a complexity of $O(n^2)$. Hence, approximations are inevitable. In the next section we will propose an algorithm which allows us to determine an approximate solution deterministically and very efficiently.

3.2   Boundary Sample Selection

Let Q_{ij} be the likelihood of the respective sample s_{ij} being selected as a boundary sample. Initially we consider all samples in the middle of the grid (the original human selection) as most likely to be selected, hence, all $Q_{iw} = 1$ and the rest is set to 0. Then we iteratively update each $\tilde{Q}_{ij}$ with

$$\tilde{Q}_{ij} = \sum_{i'j'} \exp\left(\frac{-\|s_{ij} - s_{i'j'}\|^2}{2\theta^2}\right) Q_{i'j'} \ . \tag{4}$$

That is, the samples with a likelihood of nonzero spread their values to nearby samples in the feature space and samples originally deemed as unlikely are increasingly likely to be selected if they had similar samples in their neighborhood, which, in turn, were defined as having high likelihoods in the previous iteration. After this for each row the sample with the maximum likelihood is determined and reset to 1, the rest of the same row again to 0. This process is repeated till convergence, i.e. the most likely sample for each row stops changing (see Algorithm 1).

Algorithm 1. *Basic* Sample Selection Algorithm

1: $Q_{ij} = \begin{cases} 1 \text{ if } j = w \\ 0 \text{ else} \end{cases}$

2: **while** not converged **do**

3: $\quad \tilde{Q}_{ij} \leftarrow \sum_{i'j'} \exp(\frac{-\|s_{ij}-s_{i'j'}\|^2}{2\theta^2})Q_{i'j'}$

4: $\quad \hat{Q}_{ij} \leftarrow \begin{cases} 1 \text{ if } j = \underset{j'}{\text{argmax}}\ \tilde{Q}_{ij'} \\ 0 \text{ else} \end{cases}$

5: **end while**

Algorithm 2. *Extended* Sample Selection Algorithm

1: $Q_{ij} = \begin{cases} 1 \text{ if } j = w \\ 0 \text{ else} \end{cases}$

2: $k(s) = \sum_{s' \in F} \exp(-\frac{\|s-s'\|^2}{2\theta^2})$

3: **while** not converged / not oscillating **do**

4: $\quad \tilde{Q}_{ij} \leftarrow$
$\quad (\sum_{i'j'} \exp(\frac{-\|s_{ij}-s_{i'j'}\|^2}{2\theta^2})Q_{i'j'})/k(s_{ij})$

5: $\quad \hat{Q}_{ij} \leftarrow \begin{cases} 1 \text{ if } j = \underset{j'}{\text{argmax}}\ \tilde{Q}_{ij'} \\ 0 \text{ else} \end{cases}$

6: $\quad$ Ensure $h(\mathbf{x}) = 0$ and $g(\mathbf{x}) \le 1$ (3) with $x_i = \text{argmax}_{j'}\ \tilde{Q}_{ij'}$

7: **end while**

If we consider

$$f = \sum_{ij} \sum_{i'j'} \exp\left(\frac{-\|s_{ij} - s_{i'j'}\|^2}{2\theta^2}\right) Q_{ij} Q_{i'j'} \tag{5}$$

then it is apparent that $f^1 \le f^2 \le \dots$ holds true after each iteration $1, 2, \dots$. Hence, it is converging towards a local maximum of the same function as given in (2).

In fact, the final solution we derive in this manner does not necessarily comply with the constraints (3) defining a feasible solution. Hence, after each iteration it might be necessary to alter the likelihoods of some dedicated samples such that the constraints are not violated anymore (e.g. by identifying the most probable line via dynamic programming). But changing some Q's might result in a temporary decrease of f (as the row-wise maxima are no longer considered for each row) and one has to be aware of that this can possibly lead to an oscillation

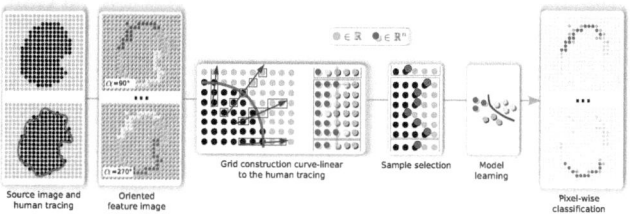

Fig. 3. The whole proposed process of boundary learning and detection: A feature image is derived from the source image by calculating boundary properties (e.g. gradients) for local image patches for different orientations. A grid, curve-linear to the (inconsistent) human tracing, is created (see Sect. 3.1 and Fig. 2) where each grid point is a feature vector. With the approach proposed in Sect. 3.2 the "true" boundary pixels can be identified efficiently. Taking these as positive training samples for a conventional classifier (e.g. random forest), a robust boundary model is derived and used to detect boundaries via pixel-wise classification.

rather than a convergence. Alternatively one can also disregard those samples that break ranks and consider only the constraint-satisfying samples in the next iteration or generally ignore the rows lacking a certain heterogeneity (which are likely to be non-boundary positions) completely.

Moreover, we observed that if true background samples accidentally have high likelihoods (e.g. due to the initial assignment) they negatively affect the convergence behavior and lead to undesired results (i.e. other background samples becoming successively most probable). That is due to the fact that background samples are usually more frequent and are therefore more likely to have more samples in the local neighborhood that contribute to the sum. But assuming that boundary samples are less frequent than background samples we simply convert the weighted sum into a weighted average, i.e. each sum is additionally normalized by the sample density. Very rare background samples are not selected either, as it is highly likely that they will not be similar to already-selected boundary samples. This little modification leads to much more robust results.

These insights are incorporated into the extended version of the proposed algorithm listed in 2.

3.3 Implementation

Both the maxima determination (line 5 in Algorithm 2) and assurance of the grid position constraints (line 6) run in linear time. The bottleneck is the update of the likelihoods (line 4). For each position in the grid the likelihoods have to be updated requiring the evaluation of the sum over all other variables and the estimation of the local density for normalization. A naive implementation would yield a quadratic complexity in the number of all samples, i.e. $O((2wn)^2)$, for each iteration. But looking more closely at the update equation in line 4 of the

Algorithm 2, one may recognize that this is actually a so-called Gauss transform, which is generally defined as

$$\hat{v}_i = \sum_j \exp\left(\frac{-|p_i - p_j|^2}{2}\right) v_j \ . \tag{6}$$

Various approaches have been proposed to overcome the computational complexity of Gauss transforms by approximate algorithms (e.g. [10]). They use the fact that the vector distances are faded out by a Gaussian kernel and only the distances of spatially close vectors have to be evaluated (those with big distances contribute very little to the overall sum and, hence, can be disregarded). This can be achieved by embedding the feature space into a special downsampled space (e.g. a Permutohedral lattice [10]) and reduces the complexity of the update step from quadratic to linear(!). Moreover, if we use a homogeneous vector representation the weighted sum in (6) can also perform a weighted average as required in Algorithm 2 (line 4). Then the input value is $[x, 1]$ and the output value of $[x, y]$ should be understood as $\left[\frac{x}{y}\right]$.

4   Results and Discussion

In our experiments we used two different feature sets to characterize boundary pixels. To describe simple edges (step or delta edges) we convolved the source images with various Gabor filters of different scales, each filter response representing an individual feature. To approximately capture the objects' interior and exterior we additionally regarded each pixel intensity along a line of certain length for different orientations as a feature (i.e. a line profile). These simplistic features have shown to be very powerful. None of the features we used measure more complex properties as those computational expensive texture features suggested by Martin et al. [1] or the learned feature set based on sparse codes introduced by Ren and Bo [3], because the aim is to understand and demonstrate the methodical need and success of the proposed boundary learning approach and not to drive an extensive evaluation of different hand designed feature sets, which, in fact, would have a strong impact on the results for a certain data set.

The examples in Fig. 4 compare the results obtained if the boundaries are learned naively or with the proposed boundary samples selection. The results clearly show that the alternative view on boundary learning significantly improves the outcome of the according boundary detector, i.e. the actual boundaries have almost the width of one pixel and they receive a significantly higher confidence than the background. A considerable increase of the cost function (2) and, hence, the altered curve-linear grid (Fig. 4 (d)) confirm this perception.

The generated angle-dependent probability map can subsequently be used to determine a segmentation by, for instance, performing an over-segmentation (e.g. with a watershed transform) followed by a region merging process, or, if the objects of interest are roundish (e.g. cells), by using dynamic programming applied on derived polar images at dedicated seeding points (see [11]).

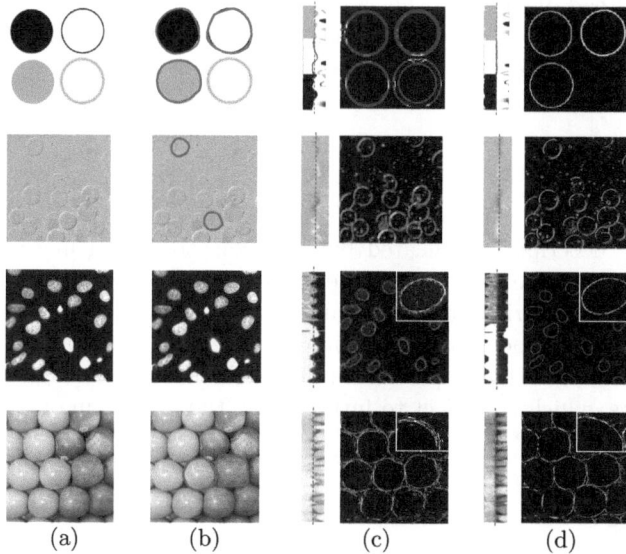

(a)   (b)   (c)   (d)

Fig. 4. Examples demonstrating the impact of the alternative boundary learning approach. (a) Source images: an artificial image, two cell images (taken from www.broadinstitute.org/bbbc) and a color image of apples; (b) Human selected examples; (c) The curve-linear grid without a correction via sample selection. The derived boundary detector is unsatisfactory; (d) The curve-linear grid corrected according to the selected instance identified with the proposed Algorithm 2. The boundary detection result is precise and as expected (the boundary orientation is known but not shown here).

These accurate segmentation results are, for instance, of major importance in an active segmentation framework where a segmentation is learned successively from as few examples as possible in a user-feedback-loop which is the ultimate goal of the work presented here.

5   Future Work

Essential for the suitability of the boundary model for a certain task is the choice of the right features calculated on the basis of the local pixel neighborhood. Although many combinations of available characteristics of a local image patch work quite well with our approach, it is still desirable to be able to determine the importance of individual features automatically for the supervised boundary learning problem at hand. It may, for instance, reduce the costs for the pixel-wise feature calculations and may lead to less complex boundary models.

Moreover, the choice of θ in (5) has a considerable impact on the quality of the results and an automatic determination would be desirable, possibly for subsection of the contours individually, which is essentially closely related to feature selection.

Finally, instead of selecting the samples after the boundary already has been traced by the human, a more consistent and homogeneous boundary might be suggested right during the actual manual delineation process.

6   Conclusions

We have proposed a simple and efficient algorithm to derive the "true" local boundary description indicated by an inconsistent and inaccurate human boundary tracing. The subsequently determined local boundaries via pixel-wise classification are invariant to disturbance and imprecision and examples have shown that the method significantly improves the accuracy and robustness of local boundary detection. Apart from the definition of the right local image patch features, which is subject to future work, the algorithm only requires one parameter θ (5).

References

1. Martin, D.R., Fowlkes, C.C., Malik, J.: Learning to detect natural image boundaries using local brightness, color, and texture cues. IEEE PAMI 26(5), 530–549 (2004)
2. Köthe, U.: Reliable Low-Level Image Analysis. Professorial dissertation, Department Informatik. University of Hamburg (2008)
3. Ren, X., Liefeng, B.: Discriminatively trained sparse code gradients for contour detection. In: Bartlett, P., Pereira, F., Burges, C., Bottou, L., Weinberger, K. (eds.) Advances in Neural Information Processing Systems 25, pp. 593–601 (2012)
4. Arbelaez, P., Maire, M., Fowlkes, C., Malik, J.: Contour detection and hierarchical image segmentation. Technical Report UCB/EECS-2010-17, EECS Department, University of California, Berkeley (February 2010)
5. Dollar, P., Tu, Z., Belongie, S.: Supervised learning of edges and object boundaries. In: CVPR, pp. 1964–1971 (2006)
6. Konishi, S., Yuille, A.L., Coughlan, J.M., Zhu, S.C.: Statistical edge detection: Learning and evaluating edge cues. IEEE PAMI 25, 57–74 (2003)
7. Meila, M., Shi, J.: Learning segmentation by random walks. In: Advances in Neural Information Processing Systems, pp. 873–879. MIT Press (2001)
8. Will, S., Hermes, L., Buhmann, J.M., Puzicha, J.: On learning texture edge detectors. In: Proc. Int. Conf. Image Processing, pp. 877–880 (2000)
9. Martin, D.R., Fowlkes, C., Tal, D., Malik, J.: A database of human segmented natural images and its application to evaluating segmentation algorithms and measuring ecological statistics. Technical Report UCB/CSD-01-1133, EECS Department, University of California, Berkeley (Jan 2001)
10. Adams, A., Baek, J., Davis, M.A.: Fast high-dimensional filtering using the permutohedral lattice. Computer Graphics Forum 29(2), 753–762 (2010)
11. Horn, M., Berthold, M.: Towards active segmentation of cell images. In: Biomedical Imaging: From Nano to Macro, March 30 -April 2, pp. 177–181 (2011)

Age Estimation Using Local Binary Pattern Kernel Density Estimate

Juha Ylioinas, Abdenour Hadid, Xiaopeng Hong, and Matti Pietikäinen

Center for Machine Vision Research, P.O. Box 4500,
FI-90014 University of Oulu, Finland

Abstract. We propose a novel kernel method for constructing local binary pattern statistics for facial representation in human age estimation. For age estimation, we make use of the *de facto* support vector regression technique. The main contributions of our work include (i) evaluation of a pose correction method based on simple image flipping and (ii) a comparison of two local binary pattern based facial representations, namely a spatially enhanced histogram and a novel kernel density estimate. Our single- and cross-database experiments indicate that the kernel density estimate based representation yields better estimation accuracy than the corresponding histogram one, which we regard as a very interesting finding. In overall, the constructed age estimation system provides comparable performance against the state-of-the-art methods. We are using a well-defined evaluation protocol allowing a fair comparison of our results.

1 Introduction

Determining the exact age of a person in a given image is a challenging task even for a human observer. For as long as possible, the assignment is done based on the overall hints available. Clearly, there are many factors affecting the final judgement including clothes, posture, and so on. What if the judgement must be based exclusively on target's face which is the case in the branch of face recognition? Naturally, the problem turns out to be even more troublesome.

Automatic estimation of human age based on facial images is an understudied problem. The lack of studies can be explained by many factors including the shortage of ground-truth age information in many existing face databases. As a result, investigation has mainly been focused on solving two to four class problems, where data has been roughly divided into groups containing child, young, adult and old human targets. Until recently, efforts have given birth to two well-known databases named as FG-NET [1] and Images of Groups [2]. The announcement of these two databases has challenged the research community to investigate the very complex recognition problem of human aging.

Automatic age classification or estimation aims to assign a label to a face regarding the exact age or the age group it belongs. The latter is the case of many early studies in age estimation, but the recent efforts in collecting new databases has seen the birth of interesting corpuses providing the labels of exact ages of the target's which has further driven the focus on more exact age estimation.

A. Petrosino (Ed.): ICIAP 2013, Part I, LNCS 8156, pp. 141–150, 2013.

As one of the recent dimensions of facial image analysis, automatic age estimation is useful in many applications such as more affective Human-Computer Interaction, video surveillance, forensics, audience measurement and reporting, and in age invariant face recognition [3].

Facial image based age estimation is challenging because of the appearance of a particular face varies due to changes in pose, expressions, illumination, and other factors such as make-up, occlusions (like eye glasses), image degradations caused by blur and noise, and so on. In addition to these there are variations that are due to, for example, living environment, lifestyle, and genes. Because of the diverse nature of facial aging process it is extremely difficult to find a model for this process.

Existing solutions for age estimation from facial images fundamentally differ in (i) the face representation and (ii) the classification scheme. Many face image representation methods have been studied such as anthropometric models, active appearance models (AAM), aging pattern subspace, and age manifold. An extensive review of age representation methods can be found in [3]. Regarding age classification schemes, the existing methods are based on either pure classification or regression analysis. Perhaps, among the pioneering studies on age classification are those proposed by Kwon and Vitoria Lobo [4], Lanitis et al. [5], and Guo et al. [6]. More recent methods include the ones proposed by Guo et al. [6] and Ruiz et al. [7] which treat age recognition as a regression problem.

The significance of face alignment is crucial in face recognition [8]. Recently, Vu and Caplier [9] noticed that by horizontally flipping the facial image the unpleasant effects due to pose variation can be mitigated in face matching. Considering the real-life nature simulated by the latest face databases and the inherent real-life-like set-up in face databases collected from the internet, the role of face alignment will remain important in face analysis.

In this paper, we propose a novel method for facial representation tackling the problem of human age estimation. Our proposal is an alternative to histograms and it is based on a kernel method that we use for constructing local binary pattern statistics for facial representation. To the best of our knowledge, we are the first ones to use the proposed kernel method for representing faces in age estimation. The outline of the paper is as follows. We first describe the modules of our facial age estimation pipeline. Then, in the experimental section we provide single- and cross-database evaluations proving the stated efficiency of the proposed kernel method against basic histograms. Finally, we provide some discussion about the advantages of our proposal and make the concluding remarks.

2   Our Age Estimation Pipeline

We built a system containing modules for face alignment and facial representation generated by statistics of local features. Our face alignment consists of facial shape and pose normalizations by a similarity transformation and a simple image flipping, respectively. For facial representation, we make use of an established spatially enhanced method based on local binary pattern (LBP) distributions. We compare two methods to estimate LBP distributions, namely the histogram method and kernel density estimation. Finally, we train a support vector regressor for age estimation.

We geometrically normalize faces based on both eyes and corners of the mouth by a similarity transformation. The motivation for using the similarity arises from the assumption that in the most usual cases the image to be aligned contains a subject directly facing the camera. Obviously, in that kind of case the input face may need some rotation, scaling, and perhaps, only a bit of translation. The face pose is subsequently corrected by flipping the image so that the facial normal is always directed to the left or right side relative to the view of a camera. From the age estimation point of view, the rigid similarity transformation and the image flipping operation are both pleasing as they retain the original shape and texture to large extent.

In our feature extraction and representation module we are using a local binary pattern variant called completed local binary pattern (CLBP) [10]. Compared to the conventional definition of LBP, the method provides two measures for local texture description, one for binary patterns and one for measuring the contrast of them. The CLBP method is explained in more detail in the next section. For the facial representation, we make use of a widely applied spatially enhanced method based on local feature distributions proposed by Ahonen et al. [12]. We compare two methods to estimate LBP distributions, namely the histogram method and the normal kernel method proposed by Aitchison and Aitken [13].

For estimating the age of a target person in a given image, we used the extracted facial representations as inputs to a support vector regressor (SVR) with a non-linear radial basis function kernel. The parameters of the SVR were determined using a grid search. We used the publicly available LIBSVM library [1].

3   Facial Representation by LBP Statistics

The local binary pattern (LBP) operator is a simple yet powerful gray-scale invariant texture primitive [14]. The original form of the operator works in a 3×3 neighborhood, using the center value as a threshold to label each pixel and considering the result as a binary number.

Later on, a more generic form of the operator was proposed using circular sampling and bilinear interpolation, which allowed to use any size of neighborhoods. Another extension addressed the importance of different binary patterns. This so called uniform patterns ($u2$) was inspired by the fact that some binary patterns occur more frequently than others. Uniform patterns are those that contain at most two bitwise transitions from 0 to 1 or vice versa. In general, the success of the method in image description can be seen in many variants and extensions proposed in the literature.

---

[1] http://www.csie.ntu.edu.tw/~cjlin/libsvm/

One of the most powerful extensions of LBP is so called completed modeling of local binary pattern (CLBP) [10]. In that local neighborhood is decomposed into two complementary components, the difference signs and the difference magnitudes. The sign component is coded using the conventional LBP operator defined as

$$
\text{CLBP\_S}_{P,R} = \sum_{p=0}^{P-1} t(g_p - g_c)2^p,
\tag{1}
$$

where g_c corresponds to the gray value of the center pixel (x_c, y_c), g_p refers to gray values of P equally spaced pixels on a circle of radius R, and t defines a thresholding function with $t(x) = 1$ if $x \geq 0$ and $t(x) = 0$ otherwise. The magitude component (CLBP_M$_{P,R}$) is coded replacing the threshold function in Eq.1 to $t(m_p, c)$, where m_p is is the magnitude of local pixel difference and c is a predetermined threshold value usually set as the mean value of local pixel differences in the whole image. As the magnitude operator encodes the difference in local pixel intensities, it gives a measure of contrast. The key idea of CLBP is to gain more comprehensive image representation by combining these two complementary descriptions.

After turning the input into two separate sign and magnitude labeled images, a histogram can be built by

$$
H(i) = \sum_{x,y} \delta_{l,i}, \quad i = 0 \ldots 2^P - 1,
\tag{2}
$$

where l is a labeled pixel at a position (x, y), 2^P is the number of different labels produced by the operator, and $\delta_{l,i}$ is the Kronecker delta function. Broadly speaking, a histogram is an estimate of the probability distribution. In the context of LBPs, it contains information about the distribution of the local micropatterns, such as edges, spots and flat areas, over the whole image [12].

In addition to a histogram, there is another widely used estimator for probability distributions, namely a kernel density estimator. In our case, however, common estimators in a continuous domain such as the Parzen-Rosenblatt window method is not applicable because LBPs essentially are variables in a multidimensional binary space. Fortunately, there is a kernel method, originally proposed by Aitchison and Aitken [13], that is suitable for estimating the probability distribution of LBP-like random variables. The kernel is given by

$$
K_h(l|l') = h^{P-d(l,l')}(1 - h)^{d(l,l')},
\tag{3}
$$

where l and l' are both P-dimensional binary variables, $d(l, l')$ is the Hamming distance between them, and $h \in [\frac{1}{2}, 1)$ is a bandwidth parameter. Finally, using the given kernel, instead of a histogram, one is then able to estimate the LBP probability distribution by

$$
\hat{f}_h(i) = \sum_{x,y} K_h(l|i), \quad i = 0 \ldots 2^P - 1.
\tag{4}
$$

Both the histogram H and the kernel density estimate $\hat{f}_h$ can be further normalized so that they sum up to one. Using the kernel function $K_h(l|l')$ one is able to distribute the same probability mass among several bins instead of putting a probability mass equal to one to a single bin, like in a histogram. The determining factor is the Hamming distance between the given label l and the possible entry i in the statistic, so that the smaller the Hamming distance the larger the probability mass given to the bin. Findings about the possible benefits using the kernel density estimate instead of a histogram are discussed in the experiments.

To retain spatial information, both the histogram and the kernel density estimate can be used to form a spatially enhanced statistic of the whole face as Ahonen et al. described in [12].

4   Experimental Results

To gain insight into the effectiveness of our age estimation system, we conducted experiments following the guidelines of the BeFIT (http://fipa.cs.kit.edu/befit/) standards for benchmarking age estimation methods. In addition, we performed a cross-database evaluation where two distinct face databases are used separately as a training and testing sets.

Setup. For the single-database benchmarks, we considered the FG-NET database which contains 1,002 uncontrolled images from 82 subjects. There are 6-18 samples per subject with ages from 0 to 69. The database also provides 68 landmarks on each face image. Fig. 1 shows some exemplars from the database, which further highlight the typical conditions of the images containing varying illumination, expression, and pose.

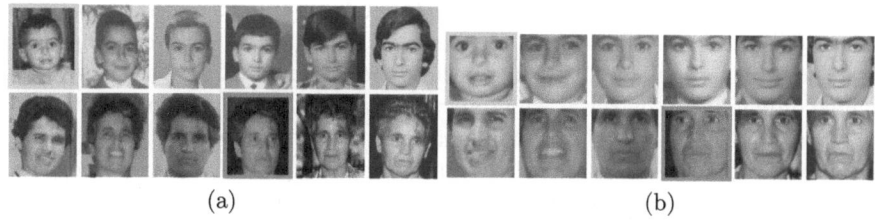

(a)   (b)

Fig. 1. (a) Original FG-NET images and (b) corresponding geometrically normalized and pose corrected faces samples. Framed samples illustrate pose correction by image flipping.

We use Leave-One-Person-Out (LOPO) for testing our proposals in the FG-NET database. This means that the images of one person are used as the test set and those of the others are used as the training set. After going through all 82 folds each subject has been used once as a the test set. The final result is then an average of the results of each fold.

As described earlier, the step before representing the faces is to first align them to reduce the effect of scale, rotation, translation variations. Hence, to get rid of the most of the variation, clearly visible in the images above, all face images are geometrically normalized by a similarity transformation with respect to both eyes and corners of the mouth using their coordinates provided by the database. We use a 76×76 pixels size of model to which all face images are fitted using the similarity calculated by point correspondences between the input and the model points. Before feature extraction, we further normalize all faces with respect to a facial normal by an image flipping operation. As visible in the faces above, they contain also severe pose variation which can be alleviated by forcing the pose to one of either right or left sides by flipping the image. The pose correction was performed by subjective conclusions of the target's facial normal based on, for example, if the right cheek is entirely visible (with respect to the camera), whereas the left shows less, it can be assumed that the facial normal is directed more to the left.

Once normalized, local features are extracted from uniformly distributed patches across the face. The face image is first divided into a set of L overlapping patches of a size 13×13 pixels, each patch overlapping its vertical and horizontal neighbors by 4 units. With a face image of size 76×76, this results in a total of 64 patches. The completed local binary pattern (CLBP) sign and magnitude operators are then used for feature extraction. CLBP has shown to be very powerful means for texture description, it has already been shown to suit well for facial representation in age group classification [11]. Our facial representation is based on local statistics of local features so we compared two sign and magnitude based statistics namely a histogram and a kernel density estimate using the normal kernel proposed in [13]. As the normal kernel requires to find a value for the smoothing bandwidth parameter h we were enforced to perform parameter tunings for both sign and magnitude component statistics.

After constructing the statistics for each patch we applied feature selection to reduce the patch-specific feature dimensions. Thus, for the sign component we used the $u2$-mapping and for the magnitude the ri-mapping. For the sign we considered only uniform patterns excluding the final 59th bin that is for non-uniform patterns. Further, as we believe that contrast is a rotation invariant property, it is well-founded to use rotation invariant magnitude component. For both measures, we operated on an $(8, 2)$-neighborhood. Using these settings the sign component yields a 58-dimensional, whereas the magnitude only a 36-dimensional feature vector. Finally, after concatenation of the sign and magnitude components we have 94-dimensional feature vectors for each of L patches. Thus, in our case, the final feature vector size is $L \times 94$, in the case of concatenated CLBP_S_M histogram $64 \times 94 = 6016$.

For estimating the age of a person in a given test image, we used the extracted facial representations as inputs to an SVR using a non-linear radial basis function kernel. The parameters of the SVR were determined using a grid search. According to the BeFIT standards, the performance of age estimation was then measured by the mean absolute error (MAE). The MAE is a mean of the absolute errors between the estimates and the true ages, $\text{MAE} = \sum_{k=1}^{N} |\hat{a}_k - a_k| / N$, where $\hat{a}_k$ and a_k are the estimate and the true age of the sample image k, and the N is the total number of samples.

Results. In our experiments we have two test variables namely facial representation and the face pose correction. The conditions of facial representation vary between six different methods, whereas for pose correction we compare the effect of using the original faces and manually pose corrected faces. If the pose was corrected then it was done both during training and testing. For each setting, we compare the following representations: spatially enhanced conventional LBP sign and magnitude histograms ($H + S$ and $H + M$), their kernel density versions ($K + S$ and $K + M$), combined spatially enhanced sign and magnitude histograms ($H + S\_M$), and its kernel density version ($K + S\_M$).

At first we went through finding an optimal value of the smoothing parameter h, for both sign and magnitude kernel density estimates separately, in both original and pose corrected settings. We plot the MAE measure against the h in Fig. 2. The h that gave the highest MAE for the representation in each setting was then selected for further evaluation.

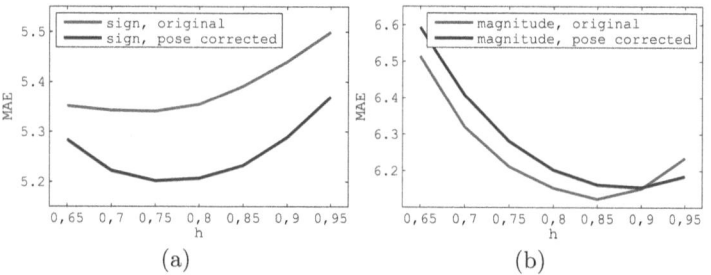

(a)   (b)

Fig. 2. The effect of h on MAE. (a) for sign and (b) for magnitude based representations in original and pose corrected settings.

After finding the optimal h, we continued by evaluating sign and magnitude, and their combination using face samples without pose correction. The results, shown in Table 1, indicate that the combined representation, i.e. using both sign and magnitude components, clearly outperforms the separated versions. Evidently, as hinted in [10], the sign component is more discriminative compared to the magnitude. The phenomenom is repeated using kernel estimator instead of a histogram, as the combination of sign and magnitude outperforms separated versions, and sign yields better result than magnitude. The most interesting is to highlight the effect of using the kernel density estimator as in that mode the sign component turns out to outperform the combination of sign and magnitude in the histogram mode.

To answer whether pose correction by image flipping would enhance the age estimation performance, we performed further evaluation. Based on the results, age estimation accuracy improves compared to the preceding setting. Evidently,

Table 1. MAE measures for different CLBP-based face representations in two different settings

representation	original faces	pose corrected
$H + S$	5.61	5.47
$H + M$	6.36	6.29
$H + S\_M$	5.39	5.23
$K + S$	5.34	5.20
$K + M$	6.12	6.16
$K + S\_M$	5.20	5.09

the used block-based facial representations are sensitive to the off-plane rotations of faces. Intuitively thinking the idea makes sense as the facial representation methods used here are based on local measures of texture regions. Therefore, in a such demanding task of human age estimation, it is important that faces used in training and testing are well aligned with respect to each other.

What makes the kernel method more powerful than the histogram, can be partly explained by the limited sample size scenario at hand. Given a patch size used here (13×13 pixels) and applying LBP and subsequent histogramming results in extremely sparse descriptors which may cause problems in the learning phase given also the limited number of training instances. By distributing the confidence, brought by one detected LBP label, among many bins instead of a single bin, one is able to produce more robust representation. Secondly, we believe that by distributing the confidence we are able to tackle the problems due to hard quantization, a component of the LBP label calculation, while there are degradations in the image such as noise.

Comparison to State-of-the-Art. We compared our methods against the most relevant works in the literature that report results using the same BeFIT benchmarking standard with FG-NET. The results, shown in Table 3, indicate fairly comparable performance.

Table 2. MAE measures for the most relevant works in the literature

method	QM [15]	MLP [15]	RUN [16]	BM [17]	LARR [18]	BGRM [7]	BIFs [6]	ours
MAE	6.55	6.98	5.78	5.33	5.07	4.96	4.77	**5.09**

The advantage of our method is computational lightness and its algorithmic simplicity comparing, for example, to the methods in [7] and in [6] where the facial representations are based on the outputs of Gaussian or Gabor-like filter bank responses. Besides, in [7] they also apply LBP operators to encode the resulting Gaussian receptive maps into histograms which hints that the method might gain even higher accuracy by using our proposed kernel density estimate instead of a histogram.

Cross-Database Evaluation. To investigate the generalization ability of the trained age regressors, we also performed a cross-database evaluation. For the testing benchmark, we ended up using the Images of Groups (IoG) database as it is the only available corpus containing face samples representative of the same age groups and ethnic origins present in FG-NET. However, as IoG only provides rough age group groundtruth, we explored solely how the age model trained using all FG-NET faces manages to categorize the given test set into age groups assigning the regressor output to the specified age range.

IoG consists of 28,231 facial images collected from Flickr images, taken in uncontrolled conditions. Each face is labeled with an age category defining seven age groups as follows: 0-2, 3-7, 8-12, 13-19, 20-36, 37-65, and 66+. The grouping roughly corresponds to different life stages [2]. In our evaluation, we considered only faces having interocular distance more than 40 pixels. By that way, we collected a subset of 1495 face images. We further relaxed the experiment by reorganizing the age labels into child, teen, and adult classes setting 0-12, 13-19, and 20+, respectively. The setting yielded the following amount of samples per each age group: 546, 250, and 699. Finally, we went through all of the IoG faces performing same normalizations than in the previous experiment. For facial representation, we only considered sign component statistics.

Based on the results, given in Table 3, the pose correction did not seem to provide any meaningful improvement, but comparing the facial representations we found a clear margin in performance between the histogram and the kernel density estimate.

Table 3. Age group classification performance on IoG using the sign component based histogram and kernel density estimate representations

representation	original faces (%)	pose corrected (%)
$H + S$	56.99	56.19
$K + S$	61.67	61.87

5   Conclusions

In this work we investigated human age estimation problem proposing a kernel method for constructing local binary pattern based statistical face representation. In our experiments, we compared our proposal to the widely used histogram based representation concluding that the kernel one yields much better accuracy. We validate our conclusion using single- and cross-database evaluations.

The motivation of our proposal arises from the limited-sample-size problem inherent to the spatially enhanced facial representation by histograms. While solving such a complex problem as human aging using local features, one is confined to very small image patches from which the aging trace might be possible to be captured. The problem with widely used local binary pattern histograms is then the resulting sparse nature of the representation. In our experiments we show that by using the proposed kernel estimator one is able to tackle this problem.

We also analysed the effect of performing face pose correction by image flipping. Based on the single-database experiments, pose correction provided significant performance improvement. However, in our cross-database experiment, pose correction did not provide any meaningful improvement which may be due to the simplified setting.

References

1. The FG-NET Aging Database, http://www.fgnet.rsunit.com/
2. Gallagher, A.C., Chen, T.: Understanding images of groups of people. In: CVPR 2009, pp. 256–263 (2009)
3. Fu, Y., Guo, G., Huang, T.S.: Age synthesis and estimation via faces: A survey. IEEE TPAMI 32(11), 1955–1976 (2010)
4. Kwon, H.Y., da Vitoria Lobo, N.: Age classification from facial images. In: CVPR 1994, pp. 762–767 (1994)
5. Lanitis, A., Taylor, C.J., Cootes, T.F.: Toward automatic simulation of aging effects on face images. IEEE TPAMI 24(4), 442–455 (2002)
6. Guo, G., Mu, G., Fu, Y., Huang, T.S.: Human age estimation using bio-inspired features. In: CVPR 2009, pp. 112–119 (2009)
7. Ruiz, J., Crowley, J., Lux, A.: "How old are you?": Age estimation with tensors of binary gaussian receptive maps. In: BMVC 2010, pp. 6.1–6.11 (2010)
8. Gross, R., Baker, S., Matthews, I., Kanade, T.: Face recognition across pose and illumination. In: Li, S.Z., Jain, A.K. (eds.) Handbook of face recognition, pp. 197–221. Springer, London (2011)
9. Vu, N.-S., Caplier, A.: Face recognition with patterns of oriented edge magnitudes. In: Daniilidis, K., Maragos, P., Paragios, N. (eds.) ECCV 2010, Part I. LNCS, vol. 6311, pp. 313–326. Springer, Heidelberg (2010)
10. Guo, Z., Zhang, L., Zhang, D.: A completed modeling of local binary pattern operator for texture classification. IEEE TIP 19(6), 1657–1663 (2010)
11. Ylioinas, J., Hadid, A., Pietikäinen, M.: Age classification in constrained conditions using LBP variants. In: ICPR 2012, pp. 1257–1260 (2012)
12. Ahonen, T., Hadid, A., Pietikäinen, M.: Face description with local binary patterns: Application to face recognition. IEEE TPAMI 28(12), 2037–2041 (2006)
13. Aitchison, J., Aitken, C.: Multivariate binary discrimination by the kernel method. Biometrika 63(3), 413–420 (1976)
14. Ojala, T., Pietikäinen, M., Mäenpää, T.: Multiresolution gray-scale and rotation invariant texture classification with local binary patterns. IEEE TPAMI 24(7), 971–987 (2002)
15. Lanitis, A., Draganova, C., Christodoulou, C.: Comparing different classifiers for automatic age estimation. IEEE TSMCB 34(1), 621–628 (2004)
16. Yan, S., Wang, H., Tang, X., Huang, T.S.: Learning auto-structured regressor from uncertain nonnegative labels. In: ICCV 2007, pp. 1–8 (2007)
17. Yan, S., Wang, H., Tang, X., Liu, J., Huang, T.S.: Regression from uncertain labels and its applications to soft biometrics. IEEE TIFS 3(4), 698–708 (2008)
18. Guo, G., Yun, F., Dyer, C.R., Huang, T.S.: Image-based human age estimation by manifold learning and locally adjusted robust regression. IEEE TIP 17(7), 1178–1188 (2008)

Improving the Quality of Color Image Segmentation Using Genetic Algorithm*

Aniceto C. Andrade Jr.,
Zenilton K.G. Patrocínio Jr., and Silvio Jamil F. Guimarães

Audio-Visual Information Proc. Lab. (VIPLAB)
Computer Science Department – ICEI – PUC Minas
anicetojunior@gmail.com, {zenilton,sjamil}@pucminas.br

Abstract. Color image segmentation is the process of grouping regions according to some criterium. In this work, we cope with this problem using a graph-based approach based on removal of minimum spanning tree edges, however the tuning of parameters is a difficult task. To better identify the set of parameters which optimizes the error producing good segmentations, we propose the use of genetic algorithm in order to establish the best set of parameters. According to test experiments, our proposed method presents better results when compared to other approaches from the literature.

Keywords: Color image segmentation, genetic algorithm.

1 Introduction

Color image segmentation aims to group image regions using some criterium, however the choice of a strategy for grouping is a difficult task, and it is dependent on the application domain. One approach to cope with this problem is to model an image as a grid graph whose vertices correspond to the pixels and the edges connect the nearest neighbor pixels and their labels represent a dissimilarity measure computed from the connected pixels [4, 13] followed by the computation of a minimum spanning tree (MST). The first appearance of this tree in pattern recognition dates back to the seminal work of Zahn [15]. Lately, its use for image segmentation was introduced by Morris *et al.* [11] in 1986 and it was popularized in 2004 by Felzenszwalb and Huttenlocher [4]. Considering that, for a given image, one can tune the parameters of the well-known method [4] for obtaining a reasonable segmentation of this image. In [8], it was presented a framework to transform the non-hierarchical method proposed by [4] into its hierarchical version. In [7] two methods were proposed for color image segmentation, these methods also are MST-based and are called *Hierarchical Euclidean Minimum Spanning Tree (HEMST)* and *Maximum Standard Deviation Reduction Clustering Algorithm (MSDR)*. In fact, image segmentation methods may

---

* The authors are grateful to PUC Minas – Pontifícia Universidade Católica de Minas Gerais, CNPq, CAPES and FAPEMIG for the financial support of this work.

A. Petrosino (Ed.): ICIAP 2013, Part I, LNCS 8156, pp. 151–160, 2013.

be considered as a clustering problem that are usually handled by MST and K-means approaches (see [3, 5, 10, 12, 13, 14] for more informations).

In computer vision and pattern recognition, Fuzzy C-means (FCM) has been extensively used to improve the compactness of segmented regions, but its implementations always suffer from initialization problems related to the specification of the cluster number and with the selection of initial cluster centroids [12]. In [14], a method based on ant colony optimization – named *Ant System Algorithm* (AS) – was proposed to overcome the sensitiveness of FCM to the initialization conditions. However, the solutions obtained by AS are not very compact in the feature space, so to improve that the *Ant Colony–Fuzzy C-means Hybrid Algorithm* (AFHA) was introduced by the same authors in [14], which incorporates the FCM algorithm to the AS in order to improve the compactness of the clustering results in the feature space. Moreover, the *Improved Ant Colony–Fuzzy C-means Hybrid Algorithm* (IAFHA) is also presented in [14] in order to reduce AFHZ computational complexity using a sub-sampling method. Although the efficiency of has been increased, it still suffers from high computational compexity. The *Histogram Thresholding–Fuzzy C-means Hybrid Algorithm* (HTFCM), presented in [12], uses a histogram thresholding method for setting the initial conditions for the FCM algorithm. This approach does not require high computational complexity when compared to the ant-based proposals.

In [13], a simple method was proposed for simplifying color image using a minimum spanning tree approach for improving fire pixel classification. Even if the results for this application were quite good, the tuning of parameters is a real challenge. To deal with that, we consider genetic algorithms to look for the best choice of parameters. Genetic algorithms (GA) are mathematical algorithms inspired by natural evolution and genetic recombination. This technique provides a search engine which is based on adaptive *Darwinian principle of natural selection* [6, 9]. According to [6], the basic principle of this theory is related to features transmission through generations of individual, and the individuals which are better adapted to the environment have the higher possibility to survive and reproduce, and consequently, perpetuating their features. In GA, a chromosome is a data structure that can represent a solution. Thus, chromosomes are subjects to an evolutionary process which involves evaluation, selection, sexual recombination (*crossover*) and mutation. A final solution is obtained after several evolution cycles of a population in which only the best individuals are preserved [6]. In order to identify the best individual, fitness function according to a specific metric is needed.

In [17], it was presented an analysis and comparison of some metrics using a methodology for evaluation of image segmentation methods. Figure 1 illustrates that methodology with its three main components: (i) analytical method; (ii) goodness method; and (iii) discrepancy method. Analytical method acts directly on the segmentation algorithms, and not over the segmented image, defining its principles, requirements, utility, complexity, and other features of the algorithm. A goodness method assesses the segmentation result without using any reference image, e.g., it could be made indirectly through goodness metrics according to

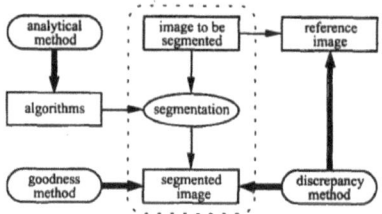

Fig. 1. The methodology for evaluation image segmentation methods proposed by [17]

the human intuition. It is also important to notice that a discrepancy method is used to quantify the difference between the segmented image and a reference one.

In this paper, we present a method for improving the quality of image segmentation using a genetic algorithm. Our proposal is based on the *Hierarchical Clustering Method* (HCM) presented in [13] and, therefore, we named it *Genetic Improvement of Hierarchical Clustering Method* (GHCM). The main contribution of this work is a graph-based approach for image segmentation, which uses a genetic algorithm to optimize the quality assessment according to some goodness method [17].

This paper is organized as follows. Section 2 presents our approach for image segmentation. Experimental results are presented in Section 3. And, finally in section 4, some considerations and future works are presented.

2   Proposed Approach

In this section, we present our proposal – GHCM – for improving the quality of image segmentation according to some metric using a genetic algorithm. Figure 2 shows the main steps of GHCM.

Our proposal uses the algorithm HCM presented in [13] (see Algorithm 1) for simplifying color images in order to decrease the number of colors. This method needs six parameters [13]: (i) original image; (ii) number of colors, which represents the number of representative colors that will be identified; (iii) outlier threshold, which represents the smaller permitted connected component size in terms of color frequency; (iv) a color space, which is the basis for graph creation values; (v) a distance measure, which is used to define the weight of graph edges; and (vi) number of nearest points considered during the graph creation. The computation time is directly related to the number of colors and, of course, to the adjacency relation of the graph, which will influence the graph size. In the adjacency relation, we will consider only the nK nearest points to each color c_1 in the color space. The strategy for the clustering process is presented in Algorithm 2 in which the best possible edge is identified to be eliminated and, consequently, its removal will divide a cluster into two others.

In order to apply a GA to a given problem, it is necessary to define the genotype required by the problem, i.e. the chromosome representation. In other

Algorithm 1. HCM – Hierarchical clustering method

input : An image – f; Number of colors – nC; Outlier threshold – T_n;
            Color space – cS; Distance measure – dM; Nearest points number – nK.
output: Color set of segmented image.
Color graph $G^\delta \leftarrow$ createGraph(f, cS, dM, nK);
$\text{MST}_{G^\delta} \leftarrow$ computeMinimumSpanningTree(G^δ);
for $i \leftarrow 1$ **to** nC **do**
  $e \leftarrow$ getBestViableEdge(MST_{G^δ}, T_n);
  if ($e = null$) **then** exit;
  $C_1 \leftarrow$ findConnectedComponentWithEdge(e);
  $(C_2, C_3) \leftarrow$ removeEdge(e);
  $\text{MST}_{G^\delta} \leftarrow (\text{MST}_{G^\delta} - C_1) \cup C_2 \cup C_3$;
end for
return MST_{G^δ};

words, a decision must be made on how the parameters of the problem will be mapped into a finite string of symbols (genes), encoding a possible solution in the problem space. In this work, a chromosome was coded by a decimal string which is divided in five parts that are responsible for coding the parameters of the algorithm presented in [13], except for the original image.

At initialization, our method generates an initial population – see *step 1* in Figure 2. Each one of those individuals represents a specific combination of parameters that could be used with Algorithm 1. The fitness function evaluation plays a very important role in guiding the GA to obtain the best solutions within a large search space. Good fitness functions help the GA to explore the search space more effectively and efficiently. On the other hand, inappropriate fitness functions can easily weaken the GA search ability and result in getting stuck into a local optimum solution. In *step 2*, the fitness of each individual is measured by applying the Algorithm 1 to an image (or a set of images) – *step*

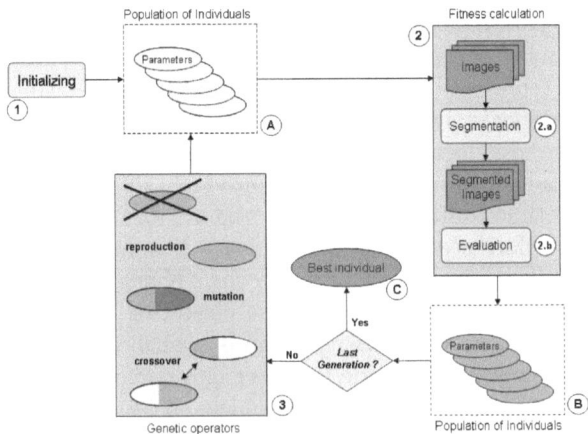

Fig. 2. Outline of the method of automatic segmentation

Algorithm 2. getBestViableEdge – Identifies the best edge to remove
  input : Minimum spanning tree – $\text{MST}_{G\delta}$; Outlier threshold – T_n.
  output: Best edge to remove – *best*.
  $best \leftarrow$ null;
  $freqTotal \leftarrow$ sumOfFrequencies($\text{MST}_{G\delta}$);
  while ($best = null$) **do**
  $\quad$ $w_e \leftarrow 0$;
  $\quad$ **foreach** *edge* $e \in \text{MST}_{G\delta}$ **do**
  $\quad\quad$ | **if** ($weight(e) \geq w_e$) **then** $w_e \leftarrow weight(e)$;
  $\quad$ **end foreach**
  $\quad$ **if** ($w_e = 0$) **then return** null;
  $\quad$ $(C_1, C_2) \leftarrow$ componentsConnectedByEdge(e);
  $\quad$ freqSum1 $\leftarrow$ sumOfFrequencies(C_1);
  $\quad$ freqSum2 $\leftarrow$ sumOfFrequencies(C_2);
  $\quad$ $freq \leftarrow$ min(freqSum1,freqSum2);
  $\quad$ **if** ($freq \geq (freqTotal \times T_n)$) **then** $best \leftarrow e$ **else** $weight(e) \leftarrow 0$;
  end while
  return $best$;

2a – and, then, the goodness of the segmented image(s) is(are) computed – *step 2b*, according to an unsupervised evaluation method (also known as empirical goodness methods) [16]. More details about specific evaluation method used in this work will be discussed later (in Section 3).

If the maximum number of generations is not reached, a set of genetic operators (selection, recombination, and mutation) is applied in order to produce a new generation of individuals. Then the process starts again. In Figure 2, the individuals of the population **A** have not been evaluated, while the fitness calculation has already be done for all individuals of the population **B**. Finally, when the number of generations reaches a limit, the best individual is found (see **C** in Figure 2). After that, it could be used to segment the image(s) in order to obtain an improved final result(s) according to a specified metric (the one used in fitness calculation).

3   Experimental Analysis

In this section, we discuss some implementation issues. And we also present quantitative and qualitative assessments to show the effectiveness of our method when compared to AS [14], AFHA [14], IAFHA [14], and HTFCM [12].

Our method was implemented in C♯ using a genetic algorithm library named GALib[1]. In order to compare our method to others, we used the same images considered in [12]: House (256 x 256), Football (256 x 256), Smarties (256 x 256), Capsicum (256 x 256), Gantry Crane (400 x 264), Beach (321 x 481) and Girl (321 x 481).

---

[1] `http://www.codeproject.com/Articles/54151/GALib`

Table 1. Genotype description: each gene represents a parameter for Algorithm 1

Gene Locus	Phenotypic feature (Coded parameter)	Alleles (Possible values for parameters)
1	Color space (cS)	RGB (1); YC_bC_r (2); L*u*v* (3); YIQ (4); L*a*b* (5)
2	Distance measure (dM)	Euclidean (1); Manhattan (2); Max (3)
3	Nearest points (nK)	Values in $[5, 10]$
4	Number of colors (nC)	Values in $[5, 15]$
5	Outlier threshold (T_n)	Values in $[1, 5]$

With respect to the our population, each individual is represented by a decimal string. Table 1 illustrates the genotype structure of each individual. Other parameters used in GA are: (i) population size: 30; (ii) the maximum number of generations: 10, 15, or 20; (iii) mutation probability: 0.10; (iv) crossover probability: 0.90; and (v) percentage of elitism: 6%. Those values were obtained after preliminary empirical tests. We have also adopt one-point crossover and fitness proportional selection.

Two different metrics were used as unsupervised evaluation to assess the goodness of segmented image(s) during fitness calculation: (i) V_{PC}; and (ii) $Q(I)$ – both were used in [12, 14]. These metrics allow us to assess the compactness and the homogeneity of pixel clusters based on a pixel representation. Let $\mathbf{x}_p$ be a vectorial representation of the pixel p in a color space, e.g., $\mathbf{x}_p = (r_p, g_p, b_p)$, where r_p, g_p, and b_p are the values of the red, green, and blue components of the pixel p in RGB color space. The first metric is Bezdek partition coefficient – V_{PC} [1] – and it measures the compactness of the generated clusters. A smaller V_{PC} value indicates a fuzzier result, thus the larger the V_{PC} value, the better the clustering result. This metric ranges from 0 (less compact) to 1 (more compact), and it is defined as follows:

$$V_{PC} = \frac{\sum_{i=1}^{N} \sum_{j=1}^{M} u_{ji}^2}{N}, \tag{1}$$

where N is the total pixel number in image, M is number of clusters, and u_{ji} is the membership degree of ith pixel to jth cluster whose value is calculated by:

$$u_{ji} = \frac{1}{\sum_{k=1}^{M} (d_{ji}/d_{ki})^{2/(m-1)}}, 1 \leq j \leq M, 1 \leq i \leq N \tag{2}$$

where d_{ji} is the (euclidean) distance between ith pixel and jth cluster centroid, and m is the exponential weight of membership degree (that was set to 2). Notice that if $d_{ji} = 0$, then $u_{ji} = 1$ and other membership degrees of this pixel are set to 0.

The second metric – $Q(I)$ [2] – is used to penalize a segmentation that forms too many regions and whose regions are non-homogeneous. This metric produces higher values for over-segmented images with non-homogeneous regions and smaller values otherwise, and is defined by:

$$Q(I) = \frac{1}{(1000 \times N)} \sqrt{R} \sum_{j=1}^{R} \left[\frac{e_j^2}{1 + \log N_j} + \left(\frac{S(N_j)}{N_j} \right)^2 \right], \qquad (3)$$

where I is the image and N is the total pixel number in I. The segmentation can be described as an assignment of pixels in image I into R regions. Let C_j denote the set of pixels in region j, $N_j = |C_j|$ denote the number of pixels in C_j. We also define $\mathbf{c}_j$ as the region centroid, $\mathbf{c}_j = (\sum_{p \in C_j} \mathbf{x}_p)/N_j$. The squared error of region j is defined as $e_j^2 = \sum_{p \in C_j} (\mathbf{x}_p - \mathbf{c}_j)^2$. Finally, $S(a)$ denotes the number of regions in image I that have an area of exactly a pixels [2].

3.1 Experimental Results Using V_{PC}

The number of color groups found by the five different methods is presented in Table 2(a) – for GHCM the best results found after 4 runs are shown. One can observe that GHCM produces the smaller number of groups using V_{PC} after 15 iterations, except for one image (these results correspond to the best individuals whose V_{PC} values are shown in Table 2(c)). Tables 2(b), 2(c) and 2(d) show the V_{PC} values obtained by GHCM after 10, 15, and 20 iterations, respectively, along with the values obtained by the other methods. One can easily see that GHCM presents a similar result when compared to HTFCM [12] after only 10 iterations, and, for 15 and 20 iterations, GHCM outperforms all tested methods. Therefore, all the remaining tests were made with a maximum number of iterations (generations) equals to 15.

3.2 Experimental Results Using $Q(I)$

In this experiment, the aim is to verify the behavior of GHCM in different runs using an empirical goodness method to estimate the segmentation quality with some human characterization about the properties of "ideal" segmentation without any prior knowledge of correct segmentation [16]. Therefore, metric $Q(I)$

Table 2. Results obtained using V_{PC} (Bold values represent the best results)

(a) Number of regions after 15 iterations

Image	AS	AFHA	IAFHA	HTFCM	GHCM
House	10	10	10	**07**	08
Football	10	10	10	07	**05**
Capsicum	15	15	17	08	**05**
Smarties	07	07	07	06	**05**
Beach	15	15	14	10	**06**
Gantry Crane	10	10	09	08	**05**
Girl	14	14	15	09	**06**

(b) Values of V_{PC} after 10 iterations

Image	AS	AFHA	IAFHA	HTFCM	GHCM
House	0.742	0.736	0.729	**0.804**	0.758
Football	0.628	0.633	0.634	**0.679**	0.673
Capsicum	0.474	0.464	0.449	0.593	**0.654**
Smarties	0.771	0.784	0.782	0.793	**0.827**
Beach	0.593	0.587	0.603	**0.689**	0.679
Gantry Crane	0.648	0.631	0.649	0.691	**0.772**
Girl	0.654	0.629	0.596	0.668	**0.706**

(c) Values of V_{PC} after 15 iterations

Image	AS	AFHA	IAFHA	HTFCM	GHCM
House	0.742	0.736	0.729	**0.804**	0.758
Football	0.628	0.633	0.634	0.679	**0.699**
Capsicum	0.474	0.464	0.449	0.593	**0.659**
Smarties	0.771	0.784	0.782	0.793	**0.828**
Beach	0.593	0.587	0.603	0.689	**0.696**
Gantry Crane	0.648	0.631	0.649	0.691	**0.772**
Girl	0.654	0.629	0.596	0.668	**0.733**

(d) Values of V_{PC} after 20 iterations

Image	AS	AFHA	IAFHA	HTFCM	GHCM
House	0.742	0.736	0.729	**0.804**	0.758
Football	0.628	0.633	0.634	0.679	**0.699**
Capsicum	0.474	0.464	0.449	0.593	**0.659**
Smarties	0.771	0.784	0.782	0.793	**0.828**
Beach	0.593	0.587	0.603	0.689	**0.696**
Gantry Crane	0.648	0.631	0.649	0.691	**0.800**
Girl	0.654	0.629	0.596	0.668	**0.767**

Table 3. Results obtained using $Q(I)$ after 15 iterations $(\times\ 10^3)$

Image	IAFHA	HTFCM	GHCM			
			Run 1	Run 2	Run 3	Run 4
House	0.4004	0.1237	0.0012	0.0017	0.0059	0.0022
Football	0.8403	0.1895	0.0039	0.0032	0.0029	0.0029
Capsicum	0.3806	0.3606	0.0065	0.3513	0.0193	0.0131
Smarties	0.1456	0.1442	0.0017	0.0022	0.0015	0.0017
Beach	0.1513	0.1495	0.0065	0.0074	0.0064	0.0042
Gantry Crane	0.3166	0.1717	0.0063	0.0021	0.0068	0.0089
Girl	0.2987	0.1864	0.0099	0.0095	0.0087	0.0090

was used for fitness calculation. Table 3 presents the results after 15 iterations. One can see that even for different runs, GHCM is always better than both – IAFHA [14] and HTFCM [12].

3.3 Qualitative Analysis

In our experiments, the color model L*u*v* was chosen in 39.28% for the best individual, followed by the model YIQ in 28.58% of best individuals, YC_bC_r in 25.00% and the last one is the RGB model in 7.14% of the cases. With respect to distance measure, Max was choosen in 42.86%, followed by Manhattan (32.14%) and Euclidean (25.00%). Table 4 shows some examples of the best individuals found after 15 iterations in 4 distinct runs using metric $Q(I)$ for fitness calculation.

Finally, Fig. 3 illustrates the segmented images obtained with the best individuals found by GHCM after 15 iterations using V_{pc} and $Q(I)$ for fitness calculation. It is important to notice that, even though these metrics are very

Table 4. Examples of best individuals found after 15 iteration in 4 runs using $Q(I)$

Image	Feature	Run 1	Run 2	Run 3	Run 4
	Color space	YIQ	YC_bC_r	L*u*v*	L*u*v*
	Distance measure	Euclidean	Manhattan	Euclidean	Manhattan
House	Nearest points	10	07	07	08
	Number of colors	05	07	08	07
	T_n	01	01	01	03
	Color space	YIQ	YIQ	YIQ	YIQ
	Distance measure	Manhattan	Euclidean	Manhattan	Manhattan
Football	Nearest points	09	05	06	06
	Number of colors	07	05	05	05
	T_n	02	01	01	02
	Color space	YC_bC_r	YIQ	YC_bC_r	YIQ
	Distance measure	Maxim	Maxim	Maxim	Maxim
Capsicum	Nearest points	09	05	06	06
	Number of colors	07	05	05	05
	T_n	01	01	01	01
	Color space	L*u*v*	L*u*v*	L*u*v*	YIQ
	Distance measure	Maxim	Euclidean	Maxim	Manhattan
Beach	Nearest points	09	06	07	05
	Number of colors	06	06	05	05
	T_n	05	02	05	05

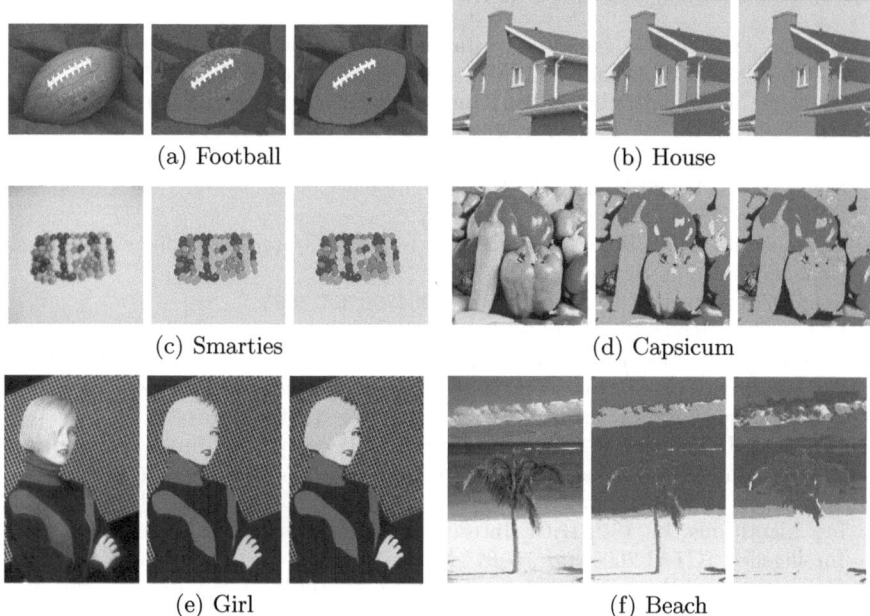

(a) Football

(b) House

(c) Smarties

(d) Capsicum

(e) Girl

(f) Beach

Fig. 3. Results for GHCM: original image (left), segmented image using V_{PC} (center); and segmented image using $Q(I)$ (right)

common in literature about unsupervised evaluation of image segmentation methods, they may produce some results (i.e., segmented images) which are not visually good/acceptable (specially for Fig. 3(d) and 3(f)). Despite of this, GHCM remains valid as a method adaptive selection of parameters since one can change the evaluation metric for a more suitable for a given task (or set of images).

4   Conclusion

In this paper, we present a method for improving the quality of segmentation results using a genetic algorithm to select the most suitable parameters in order to apply a graph-based approach for image segmentation. The main contribution of this work is a framework to optimize the quality assessment of the segmented image(s) according to some goodness method that could be selected according to a specific task (or image set).

Experimental results show that GHCM outperforms all other tested methods from the literature. But, in the future, more extensive experiments will be conducted to evaluate GHCM performance over different datasets. Another future line of work is to analyze the impact of the use of different metrics as empirical goodness methods.

References

1. Bezdek, J.C.: Cluster validity with fuzzy sets. Journal of Cybernetics 3(3), 58–73 (1973), http://www.tandfonline.com/doi/abs/10.1080/01969727308546047
2. Borsotti, M., Campadelli, P., Schettini, R.: Quantitative evaluation of color image segmentation results. Pattern Recogn. Lett. 19(8), 741–747 (1998)
3. Çiğla, C., Alatan, A.: Efficient graph-based image segmentation via speeded-up turbo pixels. In: 2010 17th IEEE International Conference on Image Processing (ICIP), pp. 3013–3016 (September 2010)
4. Felzenszwalb, P.F., Huttenlocher, D.P.: Efficient Graph-Based Image Segmentation. Int. J. Comput. Vision 59, 167–181 (2004), http://dl.acm.org/citation.cfm?id=981793.981796
5. Garcia, A., Vachier, C.: Simplification of color images using semi-flat morphological operators and statistical metrics. In: Proceedings of the 16th IEEE International Conference on Image Processing, ICIP 2009, pp. 469–472. IEEE Press, Piscataway (2009), http://dl.acm.org/citation.cfm?id=1818719.1818865
6. Goldberg, D.E.: Genetic Algorithms in Search, Optimization and Machine Learning, 1st edn. Addison-Wesley Longman Publishing Co., Inc., Boston (1989)
7. Grygorash, O., Zhou, Y., Jorgensen, Z.: Minimum Spanning Tree Based Clustering Algorithms. In: 18th IEEE International Conference on Tools with Artificial Intelligence, ICTAI 2006, pp. 73–81 (November 2006)
8. Guimarães, S.J.F., Cousty, J., Kenmochi, Y., Najman, L.: A hierarchical image segmentation algorithm based on an observation scale. In: Gimel'farb, G., Hancock, E., Imiya, A., Kuijper, A., Kudo, M., Omachi, S., Windeatt, T., Yamada, K. (eds.) SSPR&SPR 2012. LNCS, vol. 7626, pp. 116–125. Springer, Heidelberg (2012)
9. Holland, J.H.: Adaptation in natural and artificial systems. MIT Press, Cambridge (1992)
10. Mikolov, T.: Color Reduction Using K-Means Clustering. In: 11th Central European Seminar on Computer Graphics (April 2007)
11. Morris, O., Lee, M.J., Constantinides, A.: Graph theory for image analysis: an approach based on the shortest spanning tree. IEE Proceedings F Communications, Radar and Signal Processing 133(2), 146–152 (1986)
12. Siang Tan, K., Mat Isa, N.A.: Color image segmentation using histogram thresholding - Fuzzy C-means hybrid approach. Pattern Recogn. 44(1), 1–15 (2011), http://dx.doi.org/10.1016/j.patcog.2010.07.013
13. Souza, K.J.F., Guimarães, S.J.F., Patrocínio Jr., Z., Araújo, A.D.A., Cousty, J.: A Simple Hierarchical Clustering Method for Improving Flame Pixel Classification. In: 2011 23rd IEEE International Conference on Tools with Artificial Intelligence (ICTAI), pp. 110–117 (November 2011)
14. Yu, Z., Au, O.C., Zou, R., Yu, W., Tian, J.: An adaptive unsupervised approach toward pixel clustering and color image segmentation. Pattern Recogn. 43(5), 1889–1906 (2010), http://dx.doi.org/10.1016/j.patcog.2009.11.015
15. Zahn, C.T.: Graph-theoretical methods for detecting and describing gestalt clusters. IEEE Trans. Comput. 20, 68–86 (1971)
16. Zhang, H., Fritts, J.E., Goldman, S.A.: Image segmentation evaluation: A survey of unsupervised methods. Computer Vision and Image Understanding 110(2), 260–280 (2008), http://www.sciencedirect.com/science/article/pii/S1077314207001294
17. Zhang, Y.J.: A review of recent evaluation methods for image segmentation. In: Sixth International, Symposium on Signal Processing and its Applications, vol. 1, pp. 148–151 (2001)

Detection of the Vanishing Line of the Ocean Surface from Pairs of Scale-Invariant Keypoints

Sergiy Fefilatyev, Matthew Shreve, and Dmitry Goldgof

University of South Florida, 4202 E. Fowler Ave, Tampa, FL 33620
{sfefilatyev,mashreve,goldgof}@gmail.com

Abstract. In this paper, we propose an algorithm for estimating the vanishing line of a stochastically-textured plane in a single image taken by an uncalibrated perspective camera. As an example of such type of texture we take images of ocean surface for which existing methods of vanishing line detection from texture perform poorly. The proposed algorithm relies on finding pairs of similarly looking scale-invariant keypoints that are different in scale. The location of the vanishing line is estimated directly from those pairs of points by finding the vanishing line that represents the consensus of individually found vanishing points. We demonstrate the potential of the proposed method on a number of real images of ocean surface by estimating the horizon line using SIFT keypoints.

1 Introduction

We are investigating the method for estimating the affine geometry of a stochastically-textured plane from pairs of similarly looking scale-invariant keypoints. The motivation for this approach is the need for a method for inferring the attitude of a forward-looking camera attached to a highly-unstationary buoy platform surveying the ocean horizon [6]. It is not difficult to detect the horizon line in images when the horizon is indeed present in the field of view of the camera using traditional methods [4,12]. However, due to perspective effects on their appearance model, these methods fail to reliably identify the situation when camera points to the water regions only and no horizon in the image exists. By finding the orientation of the textured plane, and, thus, the camera's attitude, we are able resolve ambiguity of horizon line presence in maritime images.

Traditionally, methods for estimating the orientation of a plane are divided into two main categories: spectral [7,5] and structural [10,2] (although, the combinations of the two are also well represented [13,14]). The spectral category category of methods aims to estimate the orientation of a plane by finding the gradient of perspective distortion. Different spectral measurements are used to qualitatively estimate the gradient, making certain assumptions about the texture, such as homogeneity [3] or isotropy [15]. The methods from structural category rely on explicit identification of texture elements (textels) which then are used to find the geometric solution for the vanishing line.

In this paper, we chose the second path to find the orientation of a plane by identifying the vanishing line directly. However, in order to avoid a very fragile process of segmentation of textels, which are unique for every category of texture, we relied on pairs of similarly looking scale-invariant keypoints. Opposite to the traditional structural methods, the keypoints are only similarly looking within pairs, thus, each pair of

A. Petrosino (Ed.): ICIAP 2013, Part I, LNCS 8156, pp. 161–169, 2013.

keypoints may actually represent very different features of the texture. For example, in an image of the water surface one pair of such keypoints may come from tops of the ridges of sea waves, while another pair may come from hollow-looking depressions in the water. Thus, the homogeneity of appearance of structural elements is not required. Only the consensus of pairs, which are usually heterogeneous in appearance, is needed to find the orientation of the plane. In order to select the keypoints comprising a pair we chose Scale-Invariant Feature Transform (SIFT) [11].

The method works in several steps. First, the SIFT algorithm is used to detect candidate interest points within the image that are highly distinctive. Opposite to their original purpose of finding correspondences in *pairs* of images, these keypoints are, then, matched to other keypoints of different scales within the *same* image. Each potential pair consists of points that are similarly looking and are, necessarily, coming from different scales. Thus, we try to find similar features in the image that are located at different distances from the camera and, due to the foreshortening effect of perspective projection, are of different size in the image. We use the scales and positions of those keypoints in the pair to triangulate the position of an individual vanishing point in the image. Having identified a number of such vanishing points we estimate the position of the vanishing line (see Figure 1 (a)).

The appeal of such method is in its potential to estimate the plane's orientation in real images with very difficult stochastic textures. The method does not require prior knowledge of appearance of individual similar textels that comprise an image, avoiding the segmentation step of traditional structural methods.

2 Algorithm

The basic idea to identify the vanishing line of a plane under perspective projection comes from the properties of the intersection of two sets of parallel lines [9]. Each set of parallel lines intersects at the infinity point, which, under perspective projection, is a vanishing point, located on the vanishing line. In classical approaches of plane's orientation estimation, special structural elements (textels) are used with the assumption that the their sizes are the same and the lines drawn along their sides are parallel. We adopt a similar approach for vanishing point triangulation. However, instead of using pairs of textels we use pairs of scale-invariant keypoints. Such keypoints provide very intuitive locations, scale and orientations of keypoints which can be matched with high probabilities with other keypoints.

For geometrical computations, we make an assumption that each keypoint in the pair is of equal size under orthographic projection. Thus, under perspective projection, the position and sizes of two similar keypoints in the image can be used for computing a single vanishing point on a vanishing line. The size of matched keypoints is proportional to the scale where the keypoints are detected.

Two algorithms, SURF [1] and SIFT have been tested on a number of images of ocean surface. However, the results in this work are only reported for the SIFT algorithm.

The SIFT keypoints are only partially invariant to affine transformations [11] and the percentage of correctly matched points drops when significant affine transformation

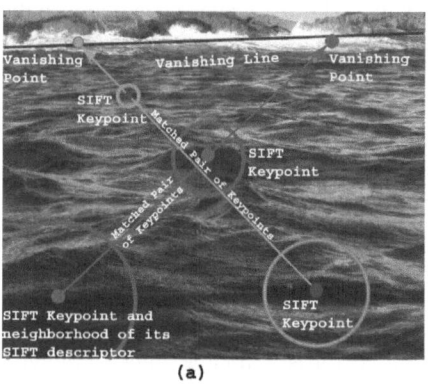
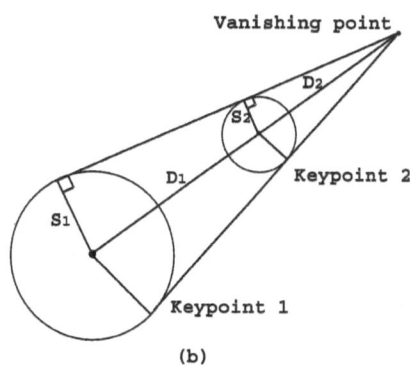

Fig. 1. Estimation of the vanishing line of the textured plane from pairs of matched SIFT keypoints. (a) Four SIFT keypoints with the regions (circles) under their descriptors are shown in the image. After matching by appearance of keypoints' descriptors two pairs are identified (each matching pair of keypoints is shown in its own color) and are used to estimate two vanishing points. The vanishing line is estimated through the locations of the vanishing points. (b) Geometry of vanishing point triangulation. Using the scales S_1 and S_2 on which the two matched keypoints were found and the distance D_1 between the keypoints it is possible to find the location of one vanishing point.

occurs. Thus, one may argue that it is impossible to compare keypoints of the same image features because they would be distorted by projective transformation. However, if keypoints in the matched pair are showing high similarity they are detected at locations in the image that are not distorted too much, otherwise the match would not have occurred.

Individual vanishing points computed from pairs of matched keypoints provide an imprecise location of true points on a line at infinity because of instability of the appearance of a stochastic texture. Another factor for imprecise localization of vanishing points comes from the fact that the scales on which the keypoints are found are selected from a limited number of scales in a scale pyramid (which depends on a number of layers and octaves), and, thus, do not represent a continuous range. However, a substantial number of such 'weak' vanishing points provides meaningful information about the vanishing line orientation. Some pairs of matched keypoints may not be pointing to the vanishing line at all, because the stochastic texture does provide the conditions for such pairs of keypoints to exist. However, these outliers can be removed from consideration using several geometrical constraints.

The following term is used to describe triangulation of a vanishing point. The vector that starts at the keypoints with bigger scale, goes through the keypoint with smaller scale, and ends at the vanishing point is called a *supporting vector* for the vanishing line (see Figure 1 (b)). The term reflects the idea of a feature in the image that supports the hypothesis of a specific vanishing line and is not related to *support vector* from the *Support Vector Machine* learning algorithm.

The constraints listed below are geometric rules used in order to estimate the locations of vanishing points and to filter out outliers:

1. A vanishing line of the plane can only be estimated if the matched keypoints in the image come from the texture of a surface plane. Thus, those keypoints that come from features of the image that do not belong to the textured plane should not be considered.
2. Keypoints in a matched pair should be of different scales. Difference in scale and the position of keypoints allows triangulation of a vanishing point.
3. Keypoints in a pair should be located far enough from each other that the regions under their descriptors do not intersect.
4. The vanishing point from a supporting vector with an angle to the gradient of perspective distortion (which is perpendicular to the vanishing line) above the maximal threshold, $A > T_A$, need to be disregarded from vanishing line computation. The reason behind it is that when the supporting vector is at a large angle to the gradient both keypoints from which the vanishing point is computed are located at approximately same distance from the camera. Thus, the scale change caused by perspective distortion is small and susceptible to error caused by the noise.
5. The supporting vector from a pair of matched keypoints should not point away from the estimated vanishing line. If, for a particular estimated vanishing line, the supporting vector points away, the vanishing point of such supporting vector should be considered an outlier.

The algorithm for estimating the vanishing line consists of two steps. First, the keypoints that are found in the image are matched to other keypoints in the image to generate vanishing points. The vanishing points are used to estimate the orientation of the vanishing line. Second, the estimated orientation of the vanishing line is used in a search for the position of the line that satisfies the geometrical constraints (4)-(5) and maximizes the optimization criterion.

2.1   Estimating Direction of Vanishing Line

The keypoints are selected by directly applying the SIFT keypoints detection algorithm on the raw image data. Once the keypoints are detected, their corresponding SIFT descriptors are computed and matched to the descriptors of other keypoints in that image. Since there are n^2 matches in the image with n keypoints, and most of them are of no interest, an effective constraint on the distance in the similarity space needs to be imposed to reduce the number of matches. In this work, we used a k-nearest neighbor in appearance space to select a 10 closest neighboring keypoints. The choice of selection of such a constraint as opposed to the radial distance in similarity space is based on the fact that some keypoints are very distinctive, and when unrestricted in number of neighbors, these points will dominate in the set of matches, skewing results for vanishing line.

In order to increase the accuracy of vanishing point localization, the number of scales in the scale pyramid is increased as compared to default values proposed in the description of SIFT algorithm [11] (see Table 1). The only indirect parameter during the SIFT keypoint localization used in our algorithm is the number of keypoints per image. This number drives the selection of other parameters for SIFT, such as the keypoint response threshold.

The matches selected from the previous step are checked against the constraints (1)-(3) listed above. The matches that are not consistent with the constraints are ruled out from further consideration. The remaining matches are sorted by the similarity measure and the top 10% of the matches are used to generate (triangulate) vanishing points.

Figure 1 (b) shows the geometrical interpretation of triangulation of a vanishing point from a pair of matched keypoints. Let the distance between the locations of matched keypoints in the image be D_1. The radiuses for regions under the SIFT descriptors are proportional to the scales where the matched keypoints were found. If S_1 is the scale for the first keypoint with the bigger radius for SIFT descriptor, and S_2 is the scale for the second keypoint with the smaller radius, the distance D_2 from the second keypoint to the vanishing point is found as the following:

$$D_2 = \frac{D_1 S_2}{S_1 - S_2} \tag{1}$$

Using (1) the positions of all vanishing points are computed (see Figure 2). Having a set of vanishing points it is possible to estimate the line that best models the set using the sum of least squares as a optimization function. The line is found using the linear square regression. However, since the position of a vanishing point is subject to error in both dimensions, the orthogonal (or total) [8] linear least square fit is more appropriate than the regular linear square regression. The implementation of the algorithm used singular value decomposition to find the total least square solution on which the orthogonal linear least square fit is based, as described in [8]. The fit of the vanishing line to the points is done twice. First, the fit is performed on all vanishing points found in the previous step. After such initial computation, 20% of the vanishing points that have the biggest distance from the computed line are removed as outliers and the line parameters are recomputed on the remaining points. The found line is represented in a slope-intercept form $y = mx + b$, where m is the slope and b is the intercept.

The computed line can be used as the final estimate of the vanishing line. However, if the orientation of such line may be close enough to the orientation of the real vanishing line in the image, the positional accuracy can be improved.

2.2   Estimating Position of Vanishing Line

The nature of stochastic texture allows for conditions where matched SIFT features in an image look very similar, however, do not correspond to the features of texture what would be of the same size when they are imaged orthographically. Although the number of such matches is small compared to the number of correct matches, it is still important to filter them out from consideration in order to improve accuracy of the vanishing line estimation. To remove such outliers the geometrical constraints (4)-(5) are applied as described below.

For each supporting vector the angle between between it and the vector of the gradient of perspective distortion is computed and compared with the allowed threshold. The vanishing points of those supporting vectors that make an angle of more than T_a degrees from the gradient are removed from consideration for vanishing line position determination. In order to obtain the vector of perspective gradient the following procedure is applied. First, since the vector of perspective gradient is perpendicular to the vanishing line (as shown in [3]) the slope of such perpendicular is found as $m_\perp = \frac{1}{m}$,

Fig. 2. Estimating direction of the vanishing line from a ensemble of vanishing points. The green line shows the ground truth. The red line shows the line estimated from the ensemble of vanishing points (red dots in the image) using total least squares algorithm. The vanishing points that are located far away from the estimated line are outliers.

where m is the slope of the vanishing line. Second, to convert the slope into a vector all the supporting vectors are projected onto the slope of the perpendicular and the projections are summed. The sign of such sum will indicate the direction along the slope as the direction of the gradient vector.

The orientation of the vanishing line found in the previous step is used in the search for the second parameter of the line - its position. The implementation of such search is done as following. The vanishing line in slope-intercept form is converted into normal representation of a line:

$$\rho = x \sin \Theta + y \cos \Theta \qquad (2)$$

where ρ is the distance from origin of coordinate system to the estimated line and Θ is the angle between the normal to the line and x-axis. During the search , the Θ parameter is fixed, and the value of ρ is varied.

A special criterion is used in order to find the optimal ρ. The criterion minimizes the sum of absolute distances from the estimated line to the set of vanishing points from supporting vectors that make an angle of less than T_a degrees:

$$\rho = \arg_\rho \min \sum_i D(L_{\Theta,\rho}, V_i) \qquad (3)$$

where $L_{\Theta,\rho}$ is the vanishing line in a normal form parameterized by Θ and ρ, V_i is the vanishing point i, supporting vector of which makes an angle with the perspective gradient less than T_a degrees.

Fig. 3. Examples of vanishing line detection on real images of water surface. The green line shows the groundtruth. The red line shows the output of the algorithm.

3 Experiments

A special dataset of 83 real images of ocean surface was collected using Google Images search engine to evaluate the performance of the algorithm. Each image in the set contained a scene of ocean surface with significant perspective effect present. The images in the dataset were assigned the groundtruth and preprocessed according to the following rules. First, each image forms approximately a square in its dimensions so there is no bias in orientation due to the aspect ratio of the image. Second, each image was mostly showing ocean surface to minimize the chance that the keypoints come from the features of the image not related to the texture. The dataset of original 83 images was split into the *development set* and the *test set*. The development set included 20 images and was used to tune the parameters of the algorithm. The test set included 53 images used to evaluate the performance.

The performance of the vanishing line estimation is evaluated in units directly related to the line parameters. The chosen metric is composed of average absolute error in the Θ angle and average absolute error in the normalized distance from the origin ρ_N. The normalized distance ρ_N is obtained for each image by dividing ρ of the evaluated line by the length of diagonal of the image. Such normalization allows comparison of errors in images of different sizes.

The Table 1 shows the values of parameters in the algorithm obtained experimentally in order to maximize performance on the development set. The performance of the algorithm with parameters from Table 1 was evaluated on the independent test set of 53 images. Table 2 shows the results of performance evaluation according to two setups. In the first setup, the vanishing line orientation and position are estimated in one step as described in Section 2.1 (thus, without correction of the position of the line). In the second setup, two step approach is applied: first, estimation of the orientation of the line, and then, correction of the position. Figure 3 shows some examples of images

Table 1. Values of parameters in the algorithm for vanishing line detection

Parameter name and description	Value
Number of layers in scale pyramid for SIFT	6
Number of octaves in each layer for SIFT	6
Number of SIFT keypoints	500
Length of SIFT descriptor vector	128
Number of nearest neighbors for each keypoints to consider when matching	10
Proportion of scales in a matched pair of keypoints to consider for supporting vector	1.5
Top percentage of matches by similarity distance to consider for vanishing line orientation	30%
Percentage of outliers in vanishing line orientation estimation	20%
Maximum angle deviation between the current gradient of perspective distortion and a valid supporting vector, T_a	45 degrees

Table 2. Performance of vanishing line estimation according to two setups

First Setup - Estimation of orientation and position of the vanishing line in one step

Average absolute error Θ, degrees	7.84
Standard deviation, error Θ, degrees	10.20
Average absolute error ρ_N	0.23
Standard deviation, error ρ_N	0.19

Second Setup - estimation of orientation and position in the first step, correction of position of the vanishing line in the second step

Error ρ_N	0.18
Standard deviation, Error ρ_N	0.15

with identified vanishing line. A good accuracy of vanishing line orientation estimation (average absolute error of 7.84 degrees) was achieved. The error in the line position estimation was 0.23 in the first setup, and 0.18 in the second. The images of the real stochastic texture are difficult to work with for all Shape From Texture methods, and, thus, the current performance is very promising. With structural methods, for example, it is impossible to define a textural element for such textures, which makes them inappropriate for the task. With spectral methods it is hard to define the scale which can be used to compare neighborhoods in the image to extract meaningful changes in the perspective gradient.

4   Conclusions

This paper describes a direct and simple method for estimating the vanishing line of a stochastically textured plane in real world images. As an example of such textured plane we used images of ocean surface. The described algorithm does not require pre-segmentation of textels, but uses the SIFT detector to find pairs of correspondent image features satisfying special properties, and, thus, can be applied directly to raw images. In our initial experiments on a number of real world images of the ocean surface, the proposed algorithm demonstrates promising performance and has a potential for further improvement. Images and groundtruth used in the experiments will be available on our web-site.

References

1. Bay, H., Tuytelaars, T., Van Gool, L.: SURF: Speeded up robust features. In: Leonardis, A., Bischof, H., Pinz, A. (eds.) ECCV 2006, Part I. LNCS, vol. 3951, pp. 404–417. Springer, Heidelberg (2006)
2. Blostein, D., Ahuja, N.: Shape from texture: Integrating texture-element extraction and surface estimation. IEEE Transactions on Pattern Analysis and Machine Intelligence 11(12), 1233–1251 (1989)
3. Criminisi, A., Zisserman, A.: Shape from texture: homogeneity revisited. In: Proceedings of British Machine Vision Conference, pp. 82–91 (2000)
4. Ettinger, S.M., Nechyba, M.C., Ifju, P.G., Waszak, M.: Vision-guided flight stability and control for micro air vehicles. In: International Conference on Intelligent Robots and Systems, vol. 3, pp. 2134–2140 (2002)
5. Farid, H., Kosecka, J.: Estimating planar surface orientation using bispectral analysis. IEEE Transactions on Image Processing 16(8), 2154–2160 (2007)
6. Fefilatyev, S., Goldgof, D., Shreve, M., Lembke, C.: Detection and tracking of ships in open sea with rapidly moving buoy-mounted camera system. Ocean Engineering 54, 1–12 (2012)
7. Galasso, F., Lasenby, J.: Shape from texture via fourier analysis. Advances in Visual Computing, 803–814 (2008)
8. Golub, G.H., Loan, C.F.V.: Matrix computations, vol. 3. Johns Hopkins Univ Press (1996)
9. Hartley, R., Zisserman, A.: Multiple view geometry in computer vision, vol. 2. Cambridge University Press (2000)
10. Kwon, J., Hong, H., Choi, J.: Obtaining a 3-d orientation of projective textures using a morphological method. Pattern Recognition 29(5), 725–732 (1996)
11. Lowe, D.: Distinctive image features from scale-invariant keypoints. International Journal of Computer Vision 60(2), 91–110 (2004)
12. McGee, T., Sengupta, R., Hedrick, K.: Obstacle detection for small autonomous aircraft using sky segmentation. In: Proceedings of the 2005 IEEE International Conference on Robotics and Automation, ICRA 2005, pp. 4679–4684 (2005)
13. Ribeiro, E., Hancock, E.R.: Estimating the 3d orientation of texture planes using local spectral analysis. Image and Vision Computing 18(8), 619–631 (2000)
14. Ribeiro, E., Hancock, E.: Estimating the perspective pose of texture planes using spectral analysis on the unit sphere. Pattern Recognition 35(10), 2141–2163 (2002)
15. Witkin, A.: Recovering surface shape and orientation from texture. Artificial Intelligence 17(1-3), 17–45 (1981)

Average Common Submatrix:
A New Image Distance Measure

Alessia Amelio and Clara Pizzuti

National Research Council of Italy (CNR)
Institute for High Performance Computing and Networking (ICAR)
{amelio,pizzuti}@icar.cnr.it

Abstract. A new information-theoretic distance measure for images is proposed. The measure is based on the concept of average common sub-matrix by considering the pixel matrices associated with the images. An algorithm to compute such a value is described, and its computational complexity analyzed. Experimental results show that the measure is able to discriminate images by correctly reflecting human perception. Furthermore, comparison with state-of-the-art information-theoretic measures, points out that the new measure outperforms these measures in terms of retrieval precision.

Keywords: image retrieval, similarity measure, pattern matching.

1 Introduction

Distance computation between images is an important and challenging problem in computer vision, image recognition, image registration, and, more in general, pattern recognition. Many different distance measures have been defined based on Euclidean distance among pixels [8], Hausdorff distance [4], cross correlation [5], and on the concept of entropy [1]. In particular, information-theoretic (dis)similarity measures rely on pixel intensity distributions and use the histograms of two images, i. e. the number of times each gray value occurs in an image, to determine the similarity between the images to be matched. Several information-theoretic measures have been defined and successfully applied in different contexts, such as medical imaging [6].

In this paper a new information-theoretic measure to compute the distance between two images I_A and I_B is proposed. The measure, named *Average Common Sub-Matrix* ($ACSM$), considers the pixel matrices A and B, defined on an alphabet Σ, associated with I_A and I_B respectively, and counts the number of square sub-matrices of matrix A that exactly occur in B, to quantify the distance between I_A and I_B. $ACSM$ is an extension in two dimensions of the average common substring (ACS) measure defined in [7] to measure the pairwise distances between sequences. Intuitively, if we have a matrix C on the same alphabet Σ, A is considered more similar to B than to C, if the average area of the sub-matrices of A that occur in B, is larger than the area of the sub-matrices of A occurring in C. In order to evaluate the $ACSM$ measure, two preliminary experimentations have been performed. The former computes distances among images containing similar objects, and shows that the $ACSM$ measure is able to reflect the concept of similar images as perceived by a human, i.e. it assigns smaller distance value to

A. Petrosino (Ed.): ICIAP 2013, Part I, LNCS 8156, pp. 170–180, 2013.

two images considered perceptually similar, and larger distance value to images deemed different. The second experimentation compares the $ACSM$ measure with other seven information-theoretic measures, and shows that $ACSM$ outperforms these measures in terms of retrieval precision.

The paper is organized as follows. In the next section the $ACSM$ measure is introduced, an algorithm to compute it described, and its complexity analyzed. In section 3 the experimental results are reported, showing that the new distance measure is very competitive with respect to the other information-theoretic measures. Section 4, finally, concludes the paper and gives some suggestion on future work.

2   Average Common Sub-matrix Measure

In this section we introduce a new (dis)similarity metric between matrices as extension in two dimensions of the average common substring (ACS) measure defined in [7], and we prove that it can be used to evaluate the distance among data in two dimensions, such as images. Intuitively, consider three matrices A, B and C defined on the same alphabet Σ. A can be considered more similar to B than to C if the average area of the sub-matrices in A that are also sub-matrices in B is larger than the same average area in C. This idea can be formalized as follows.

Let Σ be a finite alphabet, and A a square matrix over Σ of size $N \times N$.

Definition 1. *For any position (i, j) of A, let $A_{i,j}^n$ denote the set of all the square sub-matrices of size $n \times n$, for $1 \leq n \leq min\{i, j\}$, whose bottom right corner occurs at position (i, j). When $n = min\{i, j\}$, the sub-matrix P of A of size $n \times n$ is said maximal.*

Example 1. Figure 1 shows a 5×5 matrix A. Fixed position $(3, 4)$, being $min\{3, 4\} = 3$, the square sub-matrices starting at position $(3, 4)$ are $A_{3,4}^3$ of size 3×3, $A_{3,4}^2$ of size 2×2, and $A_{3,4}^1$ of size 1×1. Thus $A_{3,4}^n = \{A_{3,4}^3, A_{3,4}^2, A_{3,4}^1\}$, with $n = 1, 2, 3$. $A_{3,4}^3$ is the maximal sub-matrix.

Definition 2. *Given two matrices A and B over Σ, of size $N \times N$ and $M \times M$ respectively, for any position (i, j) of A, let $P \in A_{i,j}^n$ be the sub-matrix of A of greatest size $r \times r$, for $1 \leq r \leq min\{i, j\}$, that exactly matches a sub-matrix $Q \in B_{k,l}^m$ starting at some position (k, l) of B. The size $r \times r$ of such sub-matrix is called* area *of P and it is denoted as $W(i, j)$.*

Example 2. Figure 2 shows a 5×5 matrix A and a 4×4 matrix B. For position $(3, 5)$ in A, $P = A_{3,5}^2$ is the sub-matrix with the greatest n value that exactly matches a sub-matrix $Q = B_{3,3}^2$ starting at position $(3, 3)$ in B. Consequently, the greatest n value $r = 2$. In fact, $A_{3,5}^3$, that is greater than P and of maximal size ($n = 3$) for position $(3, 5)$, does not match with any sub-matrices in B. The area $W(3, 5)$ of P is thus equal to 2×2.

The average of all these areas of the sub-matrices of A that match sub-matrices of B can be used to define a similarity measure between A and B. Note that m is dependent from n. In fact, if an exact match of size $n \times n$, $n = min\{i, j\}$, does not exists, we consider $A_{i,j}^{n-1}$ and search for an exact match in B of size $m \times m$, where $m = n - 1$, and so on until $n = 1$.

$$A_{[5,5]} = \begin{vmatrix} 1 & 0 & 0 & 1 & 1 \\ 0 & 1 & 0 & 0 & 1 \\ 1 & 1 & \mathbf{1} & \mathbf{0} & \mathbf{0} \\ 0 & 0 & 1 & 1 & 1 \\ 0 & 1 & 0 & 1 & 0 \end{vmatrix} \quad A_{3,4}^3 = \begin{vmatrix} 0 & 0 & 1 \\ 1 & 0 & 0 \\ 1 & 1 & 0 \end{vmatrix} \quad A_{3,4}^2 = \begin{vmatrix} 0 & 0 \\ 1 & 0 \end{vmatrix} \quad A_{3,4}^1 = \begin{vmatrix} 0 \end{vmatrix}$$

Fig. 1. A matrix A of size 5×5, and the set of sub-matrices $A_{3,4}^n$, $n = 3, 2, 1$ whose bottom right corner occurs at $(3, 4)$

$$A_{[5,5]} = \begin{vmatrix} 1 & 0 & 0 & \mathbf{1} & \mathbf{1} \\ 0 & 1 & 0 & 0 & 1 \\ 1 & 1 & \mathbf{1} & \mathbf{0} & 0 \\ 0 & 0 & 1 & 1 & 1 \\ 0 & 1 & 0 & 1 & 0 \end{vmatrix} \quad B_{[4,4]} = \begin{vmatrix} 1 & 1 & 0 & 0 \\ 1 & 0 & 1 & 0 \\ 0 & 0 & 0 & 1 \\ 1 & 1 & 1 & 1 \end{vmatrix} \quad A_{3,5}^3 = \begin{vmatrix} 0 & 1 & 1 \\ 0 & 0 & 1 \\ 1 & 0 & 0 \end{vmatrix} \quad P = A_{3,5}^2 = \begin{vmatrix} 0 & 1 \\ 0 & 0 \end{vmatrix} \quad A_{3,5}^1 = \begin{vmatrix} 0 \end{vmatrix} \quad Q = B_{3,3}^2 = \begin{vmatrix} 0 & 1 \\ 0 & 0 \end{vmatrix}$$

Fig. 2. Two input matrices, A of size 5×5 and B of size 4×4. Fixed the position $(3, 5)$ in A, $P = A_{3,5}^2$ is the sub-matrix with the greatest n value that exactly matches a sub-matrix $Q = B_{3,3}^2$ starting at position $(3, 3)$ in B.

Definition 3. *Given two square matrices A and B, the* Average Common Sub-Matrix (ACSM) similarity *of A and B is defined as*

$$S(A, B) = \sum_{i=1}^{N} \sum_{j=1}^{N} W(i, j)/N^2 \tag{1}$$

Thus $S(A, B)$ computes the average of all the areas of the sub-matrices of A and B that match. Given another matrix C, if $S(A, B) > S(A, C)$, then A will be considered more similar to B than to C since the content of A is more embedded in B than in C.

In our basic measure, we need to identify, for any position (i, j) in A, the largest sub-matrix exactly matching some sub-matrix in B. Sometimes, this exact match is available only at a very small sub-matrix level. The problem is that, in some contexts, the similarity evaluation by using this thin granularity could be trivial due to the redundancy of very small patches in common between the two images. So, we introduce a parameterization in the size of the smallest sub-matrices considered in the similarity measure. A parameter α is introduced to fix a lower bound to the size of the sub-matrices.

More formally, the similarity measure is changed as,

$$S_\alpha(A, B) = \sum_{i=1}^{N} \sum_{j=1}^{N} W(i, j)/N^2 \ s.t. \ W(i, j) \geq \alpha \tag{2}$$

Now we derive a distance measure from this similarity measure. For a fixed α, since $S_\alpha(A, B)$ increases if B size increases, analogously to [7], we normalize $S_\alpha(A, B)$ with respect to the size of B by dividing it by $log(M^2)$. We take the inverse of the normalized similarity and then subtract a correction term in order to obtain zero if the two matrices are the same. The distance measure is then defined as,

$$d_\alpha(A, B) = \frac{log(M^2)}{S_\alpha(A, B)} - \frac{log(N^2)}{S_\alpha(A, A)} \tag{3}$$

Note that if $A = B$ the left part of the formula coincides with the right part, thus $d_\alpha(A, A) = 0$. This distance measure is not symmetric, thus we compute,

$$d_s(A, B) = d_s(B, A) = \frac{d_\alpha(A, B) + d_\alpha(B, A)}{2} \tag{4}$$

Input:
- two matrices A and B of size $N \times N$ and $M \times M$
- α

Output:
- the distance measure $d_s(A, B)$

begin
1. $d_\alpha(A, B) = computeACSM(A, B, \alpha)$
2. $d_\alpha(B, A) = computeACSM(B, A, \alpha)$
3. $d_s(A, B) = \frac{d_\alpha(A,B)+d_\alpha(B,A)}{2}$

end

computeACSM(A, B, α){
1. $W_\alpha(A, B) := 0, d := 0, W_\alpha(A, A) := 0, k := 0$
2. **for each** $i = 1 \ldots N$ in A
3. **for each** $j = 1 \ldots N$ in A
4. $d = min\{i, j\}, found = false, k = d$
5. **if** $d \geq \sqrt{\alpha}$
6. $W_\alpha(A, A) = W_\alpha(A, A) + d^2$
7. end
8. **while** ($k \geq \sqrt{\alpha}$ AND $\neg found$)
9. **if** $exactMatch(A^k_{i,j}, B)$
10. $W_\alpha(A, B) = W_\alpha(A, B) + W(i, j)$
11. $found = true$
12. end
13. $k = k - 1$
14. end
15. end
16. end
17. $S_\alpha(A, B) = W_\alpha(A, B)/N^2$
18. $S_\alpha(A, A) = W_\alpha(A, A)/N^2$
19. $d_\alpha(A, B) = \frac{log(M^2)}{S_\alpha(A,B)} - \frac{log(N^2)}{S_\alpha(A,A)}$
20. **return** $d_\alpha(A, B)$}

Fig. 3. The ACSM algorithm

that is the final distance measure in 2D. It is worth to note that the granularity is not lost in the definition of the symmetric version, as the same α is adopted for both $d_\alpha(A, B)$ and $d_\alpha(B, A)$.

2.1 The ACSM Algorithm

An algorithm to compute the distance between two matrices, by applying the concept of *Average Common Sub-Matrix*, introduced in the previous section, is shown in figure 3. The main procedure receives as input two matrices A and B, and the granularity parameter α. Then, it computes the $ACSM$ distance between A and B (step 1) and between B and A (step 2), given the α parameter. At the end of the procedure, the symmetric distance $d_s(A, B)$ is evaluated (step 3) by employing the computed $d_\alpha(A, B)$ and $d_\alpha(B, A)$, as in equation (4), and returned as output.

The $ACSM$ distance between two generic matrices A and B, given the α parameter, is calculated by the function *computeACSM(A,B,α)*. Step 1 of the function initializes some variables, including the cumulative area $W_\alpha(A, B)$ of the greatest common sub-matrices between A and B, and the cumulative area $W_\alpha(A, A)$ of the greatest common sub-matrices between A and itself. Observe that the similarity of the matrix A with itself $S_\alpha(A, A)$ can be computed simply by considering that, for each (i, j) in A, the greatest common sub-matrix matching inside A itself is exactly the sub-matrix of maximal extension at that position, whose size is $\geq \alpha$. Consequently, given a position (i, j) in A (steps 2-3), the function computes the area of the sub-matrix of maximal size $d \times d$, with $d = min\{i, j\}$ at that position and updates $W_\alpha(A, A)$ only if this area is larger than or equal to α, i.e. $d \geq \sqrt{\alpha}$ (steps 4-6). After that, the function finds the sub-matrix $A^k_{i,j}$ with the greatest size $k \times k$ that exactly matches a sub-matrix in B (steps 8-14). Searching starts from k equal to $d = min\{i, j\}$ and gradually decreases the k value (step 13). In any case, this value cannot be smaller than $\sqrt{\alpha}$, which is the lower bound for k (step 8). As soon as the greatest sub-matrix at position (i, j) in A matching inside B is detected, the cumulative area $W_\alpha(A, B)$ is augmented (step 10) with the area $W(i, j)$ of the found greatest common sub-matrix, and the next position in A is considered. Note that if no common sub-matrix is found at position (i, j) within the α bound, the current position of A does not contribute to the computation of the distance measure. Finally, equations (2) and (3) are evaluated (step 17 and step 19), by computing also the similarity of the A matrix with itself $S_\alpha(A, A)$ (step 18), and the value of $d_\alpha(A, B)$ is returned as output of the function.

Theorem 1. *The ACSM algorithm takes $O(M^2N^3)$ time. It can be reduced to $O(M^2N^2log(N))$ time by performing a binary search on d.*

Proof. The cost of the ACSM algorithm is mainly dependent on the $exactMatch$ procedure. Given the current pattern $A_{i,j}^k$ and the input matrix B, it searches $A_{i,j}^k$ in B, by verifying if the pattern exactly occurs into the input matrix. Consider the worst case where $\alpha = 1$ and the size $k \times k$ of the greatest common sub-matrix is equal to 1 for each position (i, j) in A. This means that, for each (i, j) in A, the algorithm will exactly match all the patterns with k varying from $d = min\{i, j\}$ to 1 with the input matrix B, and that the correspondence will be found between the pattern of size $k \times k = 1$ and B. The number of comparisons is:

$$nc = \sum_{i,j} \sum_{k=1}^{d} MC = \sum_{i,j} \sum_{k=1}^{d} M^2 = M^2 \sum_{i,j} \sum_{k=1}^{d} 1 \tag{5}$$

where MC is the execution time of $exactMatch$. A pattern matching procedure for searching a two dimensional pattern P inside the matrix B can take $O(M^2)$ time [2], independently from the size $k \times k$ of the pattern P, with k that varies from 1 to $d = min\{i, j\}$. Because the number of positions (i, j) is N^2 and d is at most equal to N, the overall cost is $O(M^2N^3)$.

However, for a given position (i, j) in A, the cost for searching the largest pattern P exactly matching a sub-matrix of B can be improved by employing a binary search strategy. In particular, starting from the pattern P of size $k \times k$ with $k = \frac{d}{2}$, the matching of P inside B is verified. If a correspondence is found, P occurs in B and eventually also a larger pattern containing P could be there. Consequently, the larger patterns of size $k \times k$, with $d \leq k \leq \frac{d}{2} + 1$ will be checked as possible expansion of P to match with B. Otherwise, looking at the patterns which are larger than P and that contain P is useless. In fact, if P doesn't match inside B, none of the larger patterns containing P can match inside B. Consequently, the smaller patterns of size $k \times k$, with $\frac{d}{2} - 1 \leq k \leq 1$ will be considered as possible reductions of P to match with B. In both cases, the process will start from the new middle points of the two intervals and it will continue, until the greatest common sub-matrix is found at that position. By performing this binary search along d, the number of patterns of size $k \times k$ to match with B for each position (i, j) in A, is reduced to $log(N)$. Consequently, nc is:

$$nc = \sum_{i,j} M^2 log(N) \tag{6}$$

Because the number of positions (i, j) is N^2 in A, the overall cost can be reduced to $O(M^2N^2log(N))$.

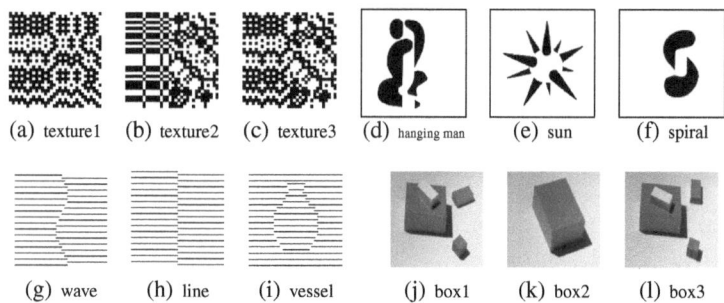

(a) texture1   (b) texture2   (c) texture3   (d) hanging man   (e) sun   (f) spiral

(g) wave   (h) line   (i) vessel   (j) box1   (k) box2   (l) box3

Fig. 4. The test images: the distance is computed for the triples (a-c), (d-f), (g-i) and (j-l)

2.2 Accelerating the ACSM Procedure

Recall that, for each position (i, j) in A, the ACSM algorithm finds the largest sub-matrix exactly matching some sub-matrix in B. In order to find, for each position (i, j) in A, the largest sub-matrix matching inside B, a generalized suffix tree in two dimensions can be constructed by employing the *Lsuffix tree* for a matrix [3], generalized for a set of matrices $\{A^1...A^s\}$, each of size $n_i \times n_i$, $1 \leq i \leq s$. In this case, the set of matrices is composed of A and B. The generalized Lsuffix tree is a compacted trie representing the set of all the square sub-matrices of both A and B matrices in a linearized form. Visiting properly this trie, the sub-matrices of maximal extension in A that are also sub-matrices in B, for each position (i, j) in A, can be discovered. This is mainly because a path from the root to a leaf node in the tree represents a sub-matrix starting at a given position inside A or B, and all the positions inside the two matrices are considered. This procedure would reduce the execution time from $O(M^2 N^2 log(N))$ to $O(N^2 + M^2)$, which is linear in the size of the input images, i.e. the area of the matrices.

3 Experimental Results

In this section preliminary tests to evaluate the proposed distance measure are presented. In particular, two kinds of experimentations have been performed. The former aims at assessing the correspondence between the distance values computed and the visual perception of a human. The latter quantitatively compares the capability of the $ACSM$ measure, with respect to state-of the art similarity measures, in finding images belonging to the same class of a given query image.

The test images used are extracted from the online database of the Computer Vision Group, University of Granada, freely available at http://decsai.ugr.es/cvg/dbimagenes/. This database contains gray level and color images of various size. For our experimentation, we used gray level illusory and color miscellaneous images of size 128×128. Without loss of generality, the size of the selected images is always the same. Gray level illusory images represent synthetic objects with some recurrent patterns inside, and relevant shapes useful for testing the effectiveness of the distance measure. Color miscellaneous images consist of real world objects under different poses, helpful for

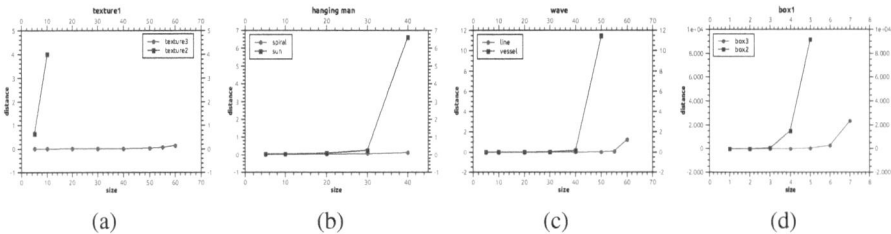

Fig. 5. $ACSM$ distance for different values of $\sqrt{\alpha}$ between (a) $texture1$ with $texture2$ and $texture3$, (b) $hanging$ man with sun and $spiral$, (c) $wave$ with $line$ and $vessel$, and (d) $box1$ with $box2$ and $box3$

evaluating the robustness of the distance measure. The used images have been manually grouped into five classes: *textures*, *symbols*, *lines*, *boxes* and *carafes*.

3.1 Human Perception Evaluation

An important characteristics of a distance measure between images is that it should reflect human perception, i.e. images deemed more similar than others by a human observer should have smaller distance among them. In order to evaluate the ability of the $ACSM$ measure to discriminate according to the visual similarity, we considered four out of the five different classes of objects, namely *textures*, *symbols*, *lines*, and *boxes* with three images each. Then we chose one target image out of the three images and computed the distance between the target image and the other two. Figure 4 shows the four image classes, the first of each is selected as the target image. A visual inspection of the figure clearly points out that *texture1* (Figure 4(a)) is more similar to *texture3* than to *texture2*, *hanging man* (Figure 4(d)) is more similar to *spiral* than to *sun*, *wave* (Figure 4(g)) is more similar to *line* than to *vessel*, and *box1* (Figure 4(j)) is more similar to *box3*, than to *box2*. This insight is confirmed by the computation of the $ACSM$ distance at different granularity values, as depicted in Figure 5. In all cases we can observe that the distance measure increases as the minimum size of the patches grows, until it goes to infinity. The motivation is that the distance between two images depends on the exact match of increasingly large patches. If such an exact match does not exists, then the value of formula (2) becomes zero, and consequently, the value of distance as computed in (3) is infinity. Figure 5(a) shows that the distance between $texture1$ and $texture3$, with α varying from 5×5 to 60×60, is lower than that between $texture1$ and $texture2$, and it slightly increases for larger values of α. This is due to the presence of recurrent regular and compact patterns shared by the first and the third image, even for sub-matrices of larger size. On the contrary, even if $texture1$ and $texture2$ have common small patches, if α is above 10×10, there is no overlap between the two objects, thus their similarity is zero, and, consequently, their distance grows to infinity. As regards the *hanging man* figure, the distance with *sun* and *spiral* becomes to be distinguished for sub-matrices of size larger than 20×20, and clearly returns a higher similarity between *hanging man* and *spiral* for $\alpha \geq 30 \times 30$. The necessity of higher values of α comes from the presence in both *sun* and *spiral* of many small pure black and white parts that overlap and that should not contribute to the similarity evaluation.

They are no more overlapping only if areas of larger size are considered. In such a case the value computed for the distance allows to correctly discriminate the shape of the inner objects and, consequently, the more similar object. An analogous behavior can be observed in Figure 5(c). In this case *wave*, *line*, and *vessel* are rather similar among them, thus a granularity of 40×40 is necessary in order to obtain a higher distance between *wave* and *vessel*. It's important to observe that, if small sub-matrices are also included, the difference in distance computed between *wave* and *vessel* and between *wave* and *line* is low. In fact, the background of all the three images is almost the same. However, visually *wave* and *line* contain a curved and straight line splitting the image area in two vertical parts, while in *vessel* two lines "draw" something similar to a vessel and split the image in two concentric regions. Finally, Figure 5(d) shows the distance values for *boxes*. These images are color images having an alphabet of very large size. The high variability in the pixel values drastically reduces the probability to have an exact match between two sub-matrices if their size is large. So, differently from the previous tests, we chose a range of small values of the α parameter for the computation of ACSM distance. In particular, because $\sqrt{\alpha}$ is fixed between 1 and 7, α is between 1 and 49. *box1* and *box3* images contain some details that are absent in *box2*, that consists of a single box. Furthermore, in *box1* and *box3* the same internal objects are placed in a different position, and some of them are rotated or located under a different perspective. Although the objects around the boxes are only small details, their presence influences the computation of the distance. In fact, *box1* appears as more distant from *box2* than from *box3*, for α values between 3×3 and 5×5.

3.2 Comparative Evaluation

In this section we compare the $ACSM$ measure with other seven standard similarity measures well known in Information Theory: Joint entropy, Conditional entropy, Mutual information, Normalized mutual information, Kullback-Leibler divergence, Arithmetic geometric mean divergence, Jensen divergence [6].

The performance index adopted to compare $ACSM$ and the above measures is the *retrieval precision* used in *content-based image retrieval*, and employed by Tourassi and Harrawood [6] in the medical context.

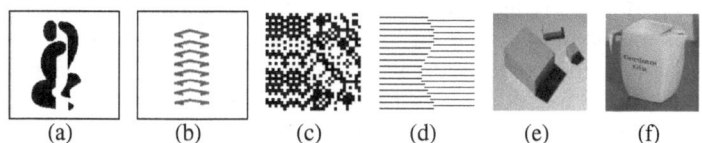

Fig. 6. Images representing the queries. Images (a) and (b) belong to *symbols* class, image (c) to *textures* class, image (d) to *lines* class, image (e) to *boxes* class and image (f) to *carafes* class.

As pointed out in [6], relevance in image retrieval can be of two types: $visual$ and $semantic$. A retrieved image I_R is considered relevant if it belongs to the same class of the query image I_Q. Since the precision depends on the query, precision results are averaged across multiple queries. In our case, the concept of relevant retrieved image that belongs to the correct class is interpreted as follows. We consider six query images belonging to the five different classes. The queries and the classes are reported in Figure

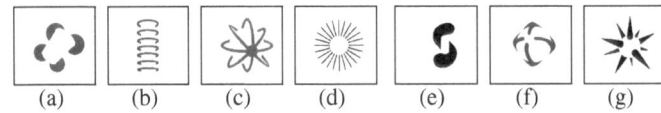

Fig. 7. Images belonging to the *symbols* class. They represent abstract objects, although some of them are similar to objects of real life, such as sun (d-g), spirals (e), columns (b).

Fig. 8. Images belonging to the *textures* class (a-e). They represent synthetic textures. Images belonging to the *boxes* class (f-j).

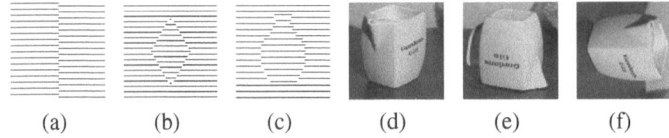

Fig. 9. Images belonging to the *lines* class (a-c). Two different figures in images (b) and (c) are depicted by the same lines background. Images belonging to the *carafes* class (d-f).

6. In particular, two query images (Figures 6(a-b)) belong to the class $symbols$, and there is one query image for each of the classes $textures$, $lines$, $boxes$, and $carafes$ (Figures 6(c-f)). For each class we examine a set of particularly significant images. For the class $symbols$ we have seven images (Figure 7), for $textures$ five images (Figure 8 (a-e)), for the classes $lines$ and $carafes$ three images each are considered (Figure 9), and for the class $boxes$ five images (Figure 8 (f-j)). As regards the total number of retrieved images, this number depends on the number of images contained in the query class. In order to detect the number of relevant images, for each query image I_Q, the similarity between I_Q and each image I of the five classes is computed. After that, the top K most similar images are selected. This procedure is performed for each similarity measure and all the similarity measures are evaluated by counting how many of the top K images belong to the query class.

Table 1 reports the average retrieval precision obtained by averaging on multiple query images for the top K retrieved images. Observe that the number of images in each class is different. Consequently, we computed the retrieval precision by averaging on all the queries, for $K = 1, 2, 3$, because some query classes don't contain more than 3 images. Then, for $K = 4, 5$, we considered queries (a), (b), (c) and (e) in Fig. 6, because their classes contain at least 5 images. Finally, for $K = 6, 7$, we averaged on queries (a) and (b) in Fig. 6, because they are the only queries whose number of images in the corresponding class is at least 7. The table points out that the $ACSM$ measure obtains the same precision of *Arithmetic-geometric mean divergence*, *Jensen divergence*, and *Kullback-Leibler divergence* for the first top 1 and 2 most similar images. In all the other cases $ACSM$ outperforms the other measures for increasing values of K. These results show that the new measure, based on the concept of average common sub-matrix is able to better discriminate (dis)similar images with respect to state-of-the-art measures.

Table 1. Average retrieval precision achieved by multiple similarity measures: Joint entropy (JE), Conditional entropy (CE), Mutual information (MI), Normalized mutual information (NMI), Arithmetic-geometric mean divergence (AGM), Jensen divergence (JD), KL divergence (KL), Average common submatrix distance (ACSM)

k	JE	CE	MI	NMI	AGM	JD	KL	ACSM
1	0.34	0.50	0.50	0.67	**1.00**	**1.00**	**1.00**	**1.00**
2	0.34	0.58	0.58	0.84	**1.00**	**1.00**	**1.00**	**1.00**
3	0.29	0.50	0.50	0.72	0.89	0.89	0.89	**0.94**
4	0.19	0.50	0.50	0.75	0.69	0.69	0.75	**0.94**
5	0.25	0.45	0.45	0.65	0.60	0.60	0.65	**0.80**
6	0.50	0.00	0.00	0.34	0.42	0.42	0.50	**0.67**
7	0.43	0.00	0.00	0.00	0.50	0.43	0.43	**0.58**

4   Conclusions and Future Work

A new information-theoretic distance measure for two dimensional matrices A and B of symbols has been proposed. The measure is based on the concept of average common sub-matrix, and considers the number of square sub-matrices of matrix A that exactly occur in B, to quantify the distance between the two matrices. The $ACSM$ measure has been applied to compute the similarity between images. Preliminary experimental results showed that $ACSM$ outperforms other information-theoretic similarity measures well known in the literature, by obtaining higher precision values in finding images belonging to the same class of query images. The $ACSM$ distance requires as input parameter the granularity level α. Experimentations have pointed out that, if images to compare are visually very different, small values of α are necessary to better discriminate similar images. Future work will extend the ACSM approach to rectangular matrices. Furthermore it will realize a more efficient implementation of the algorithm, by taking in account also an approximate matching between sub-matrices, and investigating more deeply the invariance of the distance measure to object rotation and scaling.

Acknowledgements. This work has been partially realized while the first author was visiting Georgia Institute of Technology, under the supervision of Prof. Alberto Apostolico. The work has been supported by the project *MERIT : MEdical Research in Italy*, funded by MIUR.

References

1. Cover, T.M., Thomas, J.A.: Elements of Information Theory. Wiley, New York (1991)
2. Crochemore, M., Gasieniec, L., Rytter, W., Plandowski, W.: Two-dimensional pattern matching in linear time and small space. In: Mayr, E.W., Puech, C. (eds.) STACS 1995. LNCS, vol. 900, pp. 181–192. Springer, Heidelberg (1995)
3. Giancarlo, R.: A generalization of the suffix tree to square matrices, with applications. SIAM Journal on Computing 24(3), 520–562 (1995)
4. Huttenlocher, D., Klanderman, G.A., Rucklidge, W.J.: Comparing images using the hausdorff distance. IEEE Trans. on Pattern Analysis and Machine Intelligence 15(9), 850–863 (1993)
5. Pratt, W.K.: Digital Image Processing. Wiley, New York (1991)

6. Tourassi, G.D., Harrawood, B.: Evaluation of information-theoretic similarity measures for content-based retrieval and detection of masses in mammograrams. Medical Physics 34(1), 140–150 (2007)
7. Ulitsky, I., Burstein, D., Tuller, T., Chor, B.: The average common substring approach to phylogenomic reconstruction. Journal of Computational Biology 13(2), 336–350 (2006)
8. Wang, L., Zhang, Y., Feng, J.: On the euclidean distance of images. IEEE Trans. on Pattern Analysis and Machine Intelligence 27(8), 1334–1339 (2005)

A Fast Jensen-Shannon Subgraph Kernel

Lu Bai and Edwin R. Hancock*

Department of Computer Science, University of York, York, YO10 5GH, UK

Abstract. In this paper we present a fast subgraph kernel based on Jensen-Shannon divergence and depth-based representations. For graphs with n vertices and m edges, the worst-case time complexity for our kernel is $O(n^3 + mn)$, in contrast to $O(n^6)$ for the classic graph kernel. Key to this efficiency is that we manage to compute the Jensen-Shannon divergence involved in our kernel with $O(n^2)$ operations. This computational strategy enables our subgraph kernel to easily scale up to graphs of reasonably large sizes and thus overcome the size limits arising in state of the art graph kernels. Experiments on standard bioinformatics graph datasets together with graph datasets extracted from images demonstrate the effectiveness and efficiency of our subgraph kernel.

1 Introduction

There has recently been an increased interest in learning graph structures using graph kernels[1,12]. In the research literature, most graph kernels are formulated as instances of the R-convolution kernel family defined by Haussler [8]. Here R-convolution is a generic way for defining graph kernels based on the similarities of decomposed subgraphs. In the light of this formulation, any new type of graph decompositions or new type of subgraph similarity measures could result in a new graph kernel. Accordingly, existing graph kernels can be generally categorized into three classes [15], i.e. graph kernels based on comparing all pairs of a) walks, b) paths and c) restricted subgraph and subtree structures. One major limitation of these existing graph kernels is that in practical computation they do not easily scale up to substructures of large sizes (e.g. (sub)graphs with hundreds or even thousands vertices). The resulting graph kernels are limited by the size of the substructures used, which only roughly capture the overall topological structure of a graph. Furthermore, even for relatively small subgraphs, most graph kernels require significant computational overheads. In this paper we propose a new graph kernel, which overcomes this problem and builds upon previous work by Bai and Hancock [2] to develop an information theoretic kernel using the Jensen-Shannon divergence and a novel depth-based graph representation of graph substructure. This results in a computationally efficient kernel.

The Jensen-Shannon divergence is a non-extensive mutual information theoretic measure based on non-extensive entropies. An extensive entropy is defined as the sum of the individual entropies of two probability distributions. The definition of non-extensive entropy generalizes the sum operation into composite actions. The classical Jensen-Shannon divergence is a function of probability distributions, and is related to the Shannon entropy [11]. The problem of establishing Jensen-Shannon divergence measures for

---

* Edwin R. Hancock is supported by a Royal Society Wolfson Research Merit Award.

A. Petrosino (Ed.): ICIAP 2013, Part I, LNCS 8156, pp. 181–190, 2013.

graphs can thus be posed as that of computing the required entropies for individual and composite graphs. Depth-based representations of undirected graphs have proved powerful for characterizing topological structures in terms of intrinsic complexities [4]. One approach for graph characterization is to gauge information content flow on K-layer subgraphs (e.g. subgraphs around a vertex having a maximum topological distance or minimal path length K) of increasing size and to use the flow as a structural signature. Unfortunately, existing methods for constructing such a depth-based representation of a graph always require a burdensome computational measure of the intrinsic structural complexity. Moreover, straightforward construction of the depth-based representation possessing global characteristics of a graph tends to be elusive, since it is difficult to determine a stable root vertex for graphs with potentially intricate vertex distributions.

The aim of this paper is to develop a novel subgraph kernel for efficient computation, even when a pair of fully sized subgraphs are compared. To this end, we investigate how to kernelize depth-based graph representations by measuring the information content for K-layer subgraphs using the Jensen-Shannon divergence. The contributions of this paper are threefold. First, we describe how to compute the Jensen-Shannon divergence for a pair of (sub)graphs based on the entropy difference between the original and composite (sub)graphs. The required entropies for the individual (sub)graphs are computed using the von Neumann and the Shannon entropy associated with the steady state random walk. Second, we develop a novel depth-based graph representation. We identify the centroid vertex of a graph by selecting the vertex with the minimum variance of shortest path lengths. From the centroid vertex, we derive a family of centroid expansion subgraphs with increasing layer size K. To avoid the burdensome subgraph enumeration of intrinsic complexity, we establish a depth-based representation for a graph by measuring the entropies of its centroid expansion subgraphs. Third, based on the first two contributions, we develop a fast Jensen-Shannon subgraph kernel. We empirically demonstrate the effectiveness and efficiency of our subgraph kernel on several challenging standard datasets furnished by bioinformatics and object recognition.

2   Jensen-Shannon Divergence on Graphs

In this section, we exploit the Jensen-Shannon divergence to develop a fast valid kernel measure for (sub)graphs. To commence, consider the graph $G(V, E)$ with vertex set V and edge set $E \subseteq V \times V$, the adjacency matrix A for $G(V, E)$ has elements

$$A(i, j) = \begin{cases} 1 \text{ if}(v_i, v_j) \in E; \\ 0 \text{ otherwise.} \end{cases} \tag{1}$$

The vertex degree matrix of $G(V, E)$ is a diagonal matrix D with diagonal elements given by $D(i, i) = d(i) = \sum_{j \in V} A(i, j)$. The Laplacian matrix L is defined as $L = D - A$. The normalized Laplacian matrix is given by $\hat{L} = D^{-1/2} L D^{-1/2}$. The spectral decomposition of the normalized Laplacian matrix is $\hat{L} = \hat{\Phi} \hat{\Lambda} \hat{\Phi}^T$ where $\hat{\Lambda} = diag(\hat{\lambda}_1, \hat{\lambda}_2, ..., \hat{\lambda}_{|V|})$ is a diagonal matrix with the ordered eigenvalues as elements $(0 = \hat{\lambda}_1 < \hat{\lambda}_2 < ... < \hat{\lambda}_{|V|})$ and $\hat{\Phi} = (\hat{\phi}_1|\hat{\phi}_2|...|\hat{\phi}_{|V|})$ is a matrix with the corresponding ordered orthonormal eigenvectors as columns. The normalized Laplacian matrix is positive semi-definite and so has all eigenvalues non-negative. The number of zero eigenvalues is the number of connected components in $G(V, E)$.

2.1   A Jensen-Shannon Divergence on Graphs

The classical Jensen-Shannon divergence is a non-extensive mutual information similarity measure defined on probability distributions. Assume $M_+^1(\chi)$ is a set of probability distributions where χ is a set provided with some $\sigma - algebra$ of measurable subsets, the Jensen-Shannon divergence $D_{JS} : M_+^1(\chi) \times M_+^1(\chi) \to R$ between the (discrete) probability distributions $P = (p_1, p_2, \ldots, p_M)$ and $Q = (q_1, q_2, \ldots, q_M)$, is negative definite (**nd**) with the following function

$$D_{JS}(P,Q) = H_S(\frac{P+Q}{2}) - \frac{H_S(P) + H_S(Q)}{2}. \tag{2}$$

where $H_S(P) = \sum_{m=1}^{M} p_m \log p_m$ is the entropy of the probability distribution P. In our work, we develop a Jensen-Shannon divergence for a pair of graphs. Given a pair of graphs $G_p(V_p, E_q)$ and $G_q(V_q, E_q)$, the Jensen-Shannon divergence is

$$D_{JS}(G_p, G_q) = H(G_p \oplus G_q) - \frac{H(G_p) + H(G_q)}{2}. \tag{3}$$

where $G_p \oplus G_q$ is a composite structure obtained from the two graphs $G_p(V_p, E_q)$ and $G_q(V_q, E_q)$. Here we use the disjoint union defined in Sec.2.3 as the composite structure, and explore the use of both the von Neumann entropy and Shannon entropy over the graphs. The Jensen-Shannon divergence D_{JS} for graphs defined in Eq.(3) is symmetric. With the Jensen-Shannon divergence for graphs defined in Eq.(2) to hand, we define a Jensen-Shannon diffusion graph kernel $k_{JS}: G_p \times G_q \to R$ with the kernel value

$$k_{JS}(G_p, G_q) = \exp(-\lambda D_{JS}(G_p, G_q)). \tag{4}$$

where λ is a decay factor and satisfies $0 < \lambda < 1$. The resulting diffusion kernel is positive definite.

2.2   Graph Entropies

To compute the Jensen-Shannon diffusion kernel, we require a means of evaluating the entropy of a graph. There are several alternative routes, but here we consider two computationally efficient strategies.

Von Neumann Entropy. The classical von Neumann entropy of $G(V, E)$ associated with the normalized Laplacian eigenspectrum is defined as $H_N = - \sum_{i=1}^{|V|} \frac{\hat{\lambda}_i}{|V|} \log \frac{\hat{\lambda}_i}{|V|}$. The computation of the von Neumann entropy requires cubic number of vertices operations. Han et al. [7] have shown how the computation can be rendered quadratic in the number of the vertices. By approximating the von Neumann entropy by its quadratic counterpart, the simplfied von Neumann entropy is

$$H_N(G) = 1 - \frac{1}{|V|} - \frac{1}{|V|^2} \sum_{(v_i, v_j) \in E} \frac{1}{d(i)d(j)}. \tag{5}$$

Shannon Entropy. An alternative entropy for $G(V, E)$ is to exploit steady state random walks on $G(V, E)$. The probability of the steady state random walks on $G(V, E)$ visiting vertex v_i is $P_G(i) = d(i)/\sum_{v_j \in V} d(j)$. And the Shannon entropy for $G(V, E)$ is

$$H_S(G) = -\sum_{i=1}^{|V|} P_G(i) \log P_G(i). \tag{6}$$

For the graph $G(V, E)$ with $n = |V|$ vertices, both the von Neumann entropy $H_N(G)$ and the Shannon entropy $H_S(G)$ require time complexity $O(n^2)$. This comes from the fact that the degree matrix D of $G(V, E)$ can be computed by visiting every entries $A(i, j)$ in the adjacency matrix A. Thus both entropies $H_N(G)$ and $H_S(G)$ can be directly computed by visiting all the n^2 pairs of vertices.   □

2.3   A Composite Entropy of a Pair of Graphs

We now turn our attention to computing the composite entropy $H(G_p \oplus G_q)$ for the pair of graphs $G_p(V_p, E_p)$ and $G_q(V_q, E_q)$. Since we aim to develop an efficient kernel, we seek a computationally inexpensive route and use the disjoint union $G_{DU} = G_p \cup G_q = \{V_p \cup V_q, E_p \cup E_q\}$ for constructing our composition of graphs $G_p(V_p, E_p)$ and $G_q(V_q, E_q)$. Let graphs G_p and G_q be the connected components of the disjoint union graph G_{DU}, then the composite entropy is $H(G_{DU}) = \rho_p H(G_p) + \rho_q H(G_q)$., where $\rho_p = |V(G_p)|/|V(G_{DU})|$ and $\rho_q = |V(G_q)|/|V(G_{DU})|$. Here the entropy function $H(\cdot)$ could be either the von Neumann entropy $H_N(\cdot)$ defined in Eq.(5) or the Shannon entropy $H_S(\cdot)$ defined in Eq.(6).

3   Fast Jensen-Shannon Subgraph Kernel

The idea of using Jensen-Shannon divergence to develop graph kernels has previously been explored by Lu and Hancock [2]. However, the kernel is slow since it requires a global computation of the divergence. In this section we overcome this problem and aim to develop a novel and fast subgraph kernel based on the Jensen-Shannon divergence (JSD). Our idea is to decompose a graph into substructures (i.e. subgraphs) spanned from a root vertex to the remaining vertices with respect to the minimal path length K, and measure the entropies on these subgraphs as a depth-based signature of the graph. To obtain a family of subgraphs capturing fine structures of a graph, we identify a centroid vertex as the root vertex. In the literature of graph theory, the centroid of a graph has been defined as a structure composed of vertices closest to all others [17]. Here we present a novel method to identify the centroid vertex of a graph by evaluating the shortest path length distribution around a vertex. We select the vertex with the minimum variance of shortest path lengths to the remaining vertices as the centroid vertex. The vertices surrounding the centroid vertex in a graph lie along the different shortest paths from the vertex, and the centroid vertex has a global view of the vertex path length distribution surrounding it. Finally, we establish an information theoretic subgraph kernel by applying the Jensen-Shannon diffusion kernel to subgraphs derived from the centroid vertices.

3.1 Subgraphs from the Centroid Vertex

We commence by identifying the centroid vertex of a graph. For an undirected graph $G(V, E)$, the shortest path $S_G(i, j)$ for a pair of vertices v_i and v_j can be computed by using Johnson's algorithm [9]. The matrix S_G whose element $S_G(i, j)$ represents the shortest path length between vertices v_i and v_j is referred to as the shortest path matrix for graph $G(V, E)$. The average-shortest-path vector S_V for $G(V, E)$ is a vector with the same vertex order as S_G and with element $S_V(i) = \sum_{j=1}^{|V|} S_G(i, j)/|V|$ representing the average shortest path length from vertex v_i to the remaining vertices. We then locate the centroid vertex $\hat{v}_i$ for $G(V, E)$ as follows

$$\hat{v}_i = \arg\min_i \sum_{j=1}^{|V|} [S_G(i, j) - S_V(i)]^2. \tag{7}$$

The centroid vertex $\hat{v}_i$ of $G(V, E)$ is identified by selecting the vertex with the minimum variance of shortest path lengths over all vertices in $G(V, E)$. Therefore, the shortest paths starting from the centroid vertex $\hat{v}_i$ form a *steady* path set that exhibits the least length variability compared with those path sets originating from the other vertices. Thus the centroid vertex has a global view of the path length distribution. Let $N_{\hat{v}_C}^K$ be a subset of V satisfying $N_{\hat{v}_C}^K = \{u \in V \mid S_G(\hat{v}_C, u) \leq K\}$. For $G(V, E)$ with the centroid vertex $\hat{v}_C$, the K-layer centroid expansion subgraph $\mathcal{G}_K(\mathcal{V}_K; \mathcal{E}_K)$ is

$$\begin{cases} \mathcal{V}_K = \{u \in N_{\hat{v}_C}^K\}; \\ \mathcal{E}_K = \{(u, v) \subset N_{\hat{v}_C}^K \mid (u, v) \in E\}. \end{cases} \tag{8}$$

The number of centroid expansion subgraphs is equal to the greatest length L of the shortest paths from the centroid vertex to the remaining vertices of the graph $G(V, E)$. The L-layer expansion subgraph is the graph $G(V, E)$ itself.

For graph $G(V, E)$ ($n = |V|$ and $m = |E|$), computing entropies for its centroid expansion subgraphs requires time complexity $O(n^2L + mn)$.

3.2 Depth-Based Representation

In this subsection, we develop a fast Jensen-Shannon subgraph kernel ($k_{\mathcal{JS}}$) as an information theoretic decomposition kernel. The proposed kernel $k_{\mathcal{JS}}$ is defined by kernelizing depth-based graph representations in terms of measuring information content similarities for K-layer subgraphs using the Jensen-Shannon divergence. For a sample graph $G(V, E)$, we commence by identifying the centroid vertex $\hat{v}_C$ using Eq.(7). Based on $\hat{v}_C$ we construct the K-layer centroid expansion subgraph $\mathcal{G}_K$ of $G(V, E)$ using Eq.(8). As we increase K from 1 to the greatest shortest path length L with respect to the centroid vertex $\hat{v}_C$, we obtain a family of centroid expansion subgraphs $\{\mathcal{G}_1, \cdots, \mathcal{G}_K, \cdots, \mathcal{G}_L\}$. We then measure the entropies of the subgraphs and establish the depth-based representation $D(G)$ of G(V,E) as $D(G) = \{H(\mathcal{G}_1), \cdots, H(\mathcal{G}_K), \cdots, \,$, where $H(\mathcal{G}_K)$ is the entropy for the K-layer subgraph $\mathcal{G}_K$ of $G(V, E)$. For a pair of graphs $G_p(V_p, E_p)$ and $G_q(V_q, E_q)$, we develop a similarity measure between their depth-based representations $D(G_p)$ and $D(G_q)$ as follows

$$s(D(G_p), D(G_q)) = \sum_{K=1}^{L} s_H(H(\mathcal{G}_{p;K}), H(\mathcal{G}_{q;K})). \tag{9}$$

where $s_H(H(\mathcal{G}_{p;K}), H(\mathcal{G}_{q;K}))$ is an entropy-based similarity measure for the K-layer subgraphs $\mathcal{G}_{p;K}$ and $\mathcal{G}_{q;K}$ of $G_p(V_p, E_p)$ and $G_q(V_q, E_q)$. By using the Jensen-Shannon diffusion kernel $k_{JS}(\cdot, \cdot)$ in Eq.(4) as the entropy-based similarity measure $s_H(\cdot, \cdot)$ in Eq.(9), the similarity between the depth-based representations $D(G_p)$ and $D(G_q)$ is formulated as the sum of the diffusion kernel measures for all the pairs of K-layer subgraphs of $G_p(V_p, E_p)$ and $G_q(V_q, E_q)$.

For the graphs $G_p(V_p, E_p)$ and $G_q(V_q, E_q)$ the Jensen-Shannon subgraph kernel $k_{JS}(G_p, G_q)$ is defined as

$$k_{JS}(G_p, G_q) = s(D(G_p), D(G_q)) = \sum_{K=1}^{L} k_{JS}(\mathcal{G}_{p;K}, \mathcal{G}_{q;K}). \tag{10}$$

where $\mathcal{G}_{p;K}(\mathcal{V}_{p;K}, \mathcal{E}_{p;K})$ and $\mathcal{G}_{q;K}(\mathcal{V}_{q;K}, \mathcal{E}_{q;K})$ are the K-layer centroid expansion subgraphs of $G_p(V_p, E_p)$ and $G_q(V_q, E_q)$ rooted from their centroid vertices $\hat{v}_{p;C}$ and $\hat{v}_{q;C}$, respectively, and $k_{JS}(\mathcal{G}_{p;K}, \mathcal{G}_{q;K})$ is the Jensen-Shannon diffusion kernel between $\mathcal{G}_{p;K}(\mathcal{V}_{p;K}, \mathcal{E}_{p;K})$ and $\mathcal{G}_{q;K}(\mathcal{V}_{q;K}, \mathcal{E}_{q;K})$. According to Eq.(2.3), Eq.(3) and Eq.(4), $k_{JS}(\mathcal{G}_{p;K}, \mathcal{G}_{q;K})$ is

$$k_{JS}(\mathcal{G}_{p;K}, \mathcal{G}_{q;K}) = \exp\left[-\lambda\left\{\frac{2|\mathcal{V}_{p;K}| - |\mathcal{V}_{q;K}|}{2|\mathcal{V}_{p;K}| + 2|\mathcal{V}_{q;K}|} H(\mathcal{G}_{p;K}) + \frac{2|\mathcal{V}_{q;K}| - |\mathcal{V}_{p;K}|}{2|\mathcal{V}_{p;K}| + 2|\mathcal{V}_{q;K}|} H(\mathcal{G}_{q;K})\right\}\right]. \tag{11}$$

The resulting Jensen-Shannon subgraph kernel k_{JS} is positive semidefinite. The computation of the proposed subgraph kernel between a pair of graphs, each of which has n vertices and m edges, requires time complexity $O(n^2 L + mn)$.

One extreme case for the time complexity $O(n^2 L + mn)$ is the chain structure, where the centroid vertex is the intermediate vertex along the chain and L is $\lfloor n/2 \rfloor$. Hence, the worst-case time complexity of our subgraph kernel is $O(n^3 + mn)$. But for most real-world graphs, the vertices surrounding the centroid vertex tend to be well distributed, and then our subgraph kernel usually only requires time complexity $O(n^2 + mn)$ for the practical kernel computation.

For a pair of graphs $G_p(V_p, E_p)$ and $G_q(V_q, E_q)$ with different sizes, the greatest sizes of the expansion subgraphs for the two graphs could be different. Suppose that $\hat{v}_{C;p}$ and $\hat{v}_{C;q}$ are the centroid vertices of $G_p(V_p, E_p)$ and $G_q(V_q, E_q)$, and the lengths of the greatest shortest paths from the centroid vertices $\hat{v}_{C;p}$ and $\hat{v}_{C;q}$ are L_p and L_q, respectively, where $L_p > L_q$. In practical computations, to balance the size difference between largest centroid expansion subgraphs of the two graphs, we use the graph $G_q(V_q, E_q)$ as the $(L_q + 1)$-layer to L_p-layer expansion subgraphs of $G_q(V_q, E_q)$.

Table 1. Information of the Graph-based Datasets

Datasets	MUTAG	D&D	ENZYMES	PPIs
Max # vertices	28	5748	126	232
Min # vertices	10	30	2	3
Mean # vertices	17.9	284.3	32.6	109.6
# graphs	188	1178	600	86
# disjoint graphs	0	21	31	0
# classes	2	2	6	2

4   Experimental Results

4.1   Standard Graph Datasets

We commence by demonstrating the performance of our Jensen-Shannon subgraph kernel and compare it to several state of the art graph kernels on four standard graph based datasets abstracted from bioinformatics databases. These datasets include: MUTAG, D&D, ENZYMES and PPIs. The MUTAG dataset contains graphs representing 188 chemical compounds to predict mutagenicity. The D&D dataset contains 1178 protein structures. Each protein is represented by a graph. The ENZYMES dataset contains graphs representing protein tertiary structures consisting of 600 enzymes from the BRENDA enzyme database. The PPIs dataset consists of protein-protein interaction networks (PPIs). Here we select Proteobacteria40 PPIs and Acidobacteria46 PPIs as the testing graphs. Details about the datasets are shown in Table.1.

We evaluate the performance of our proposed Jensen-Shannon subgraph kernel using the von Neumann entropy (JSSV) or Shannon entropy (JSSS) separately on the graph datasets abstracted from the bioinformatics databases, and then compare it with several alternative state of the art graph kernels. These graph kernels for comparison include 1) the Weisfeiler-Lehman subtree kernel (WL) [15], 2) the shortest path graph kernel (SPGK) [3], 3) the graphlet count graph kernel (GCGK) [16], 4) the random walk graph kernel (RWGK) [6], 5) the p-random walk graph kernel (PWGK) [10], and 6) the Ramon & Ganter graph kernel (RGGK) [5]. We compute the kernel matrix on each dataset by using our proposed method. We also compute the kernel matrices of the Weisfeiler-Lehman subtree kernel and shortest path graph kernel on the PPIs dataset. For each kernel matrix we perform Principle Component Analysis (PCA) to embed the graphs into feature space as vectors. We then perform 10-fold cross-validation using the Support Vector Machine (SVM) Classification associated with the Sequential Minimal Optimization (SMO) and the PUK kernel[13] to evaluate the performance of our method and those of the alternative methods. We use nine samples for training and one for testing. All the SMO-SVMs were performed along with their parameters optimized on each dataset. We report the average classification accuracies and runtime of establishing the kernel matrices of each method in Table.2, with the runtime measured under Matlab R2011a running on a 2.2GHz Intel 2-Core processor. Shervashidze et al. [15] have reported the accuracies of state of the alternative art graph kernels on each dataset excluding the PPIs dataset, by performing 10-fold cross-validation associated with the C-Support Vector Machine Classification (C-SVM). The runtime for establishing the kernel matrices using these methods was measured under Matlab R2008a running on

Table 2. Accuracy and CPU Runtime Comparisons on Graph Datasets

Datasets	MUTAG	D&D	ENZYMES	PPIs
JSSV	84.04	76.40	33.16	74.41
JSSS	85.10	78.50	38.00	80.01
WL	82.05	79.78	46.42	75.90
SPGK	87.28	78.45	41.68	70.93
GCGK	75.61	78.59	32.70	−
RWGK	80.72	71.70	21.68	−
PWGK	79.19	66.64	27.67	−
RGGK	85.72	57.27	13.35	−
Datasets	MUTAG	D&D	ENZYMES	PPIs
JSSV	1"	2'50"	4"	1"
JSSS	1"	2'49"	4"	1"
WL	1"	11'	20"	27"
SPGK	2"	23h17'2"	5"	47"
GCGK	3"	30'21"	25"	−
RWGK	12"	48days	12'19"	−
PWGK	4'42"	4days	10'	−
RGGK	40'60"	103days	38days	−

a 3.0GHz Intel 8-Core processor with 16GB RAM. We report these accuracies and runtime in Table.2.

There are a number of conclusions that can be drawn from Table 2. First, on the MUTAG, D&D, PPIs datasets, our subgraph kernel outperforms or is competitive with the alternatives. Secondly, on the ENZYMES dataset, the performance of our subgraph kernel is lower than that of the Weisfeiler-Lehman subtree kernel, is competitive with the shortest path graph kernel, and outperform the remaining methods. Thirdly, the runtime of our subgraph kernel is better than that of each of the alternatives on all datasets studied. This includes the challenging D&D dataset which contains graphs with thousands of vertices. Finally, the average greatest layer sizes of the subgraphs for the MUTAG, D&D, ENZYMES and PPIs datasets are 8.53, 24.14, 6.45 and 8.79, respectively.

The runtimes of our subgraph kernel using the von Neumann and Shannon entropies are similar. However the classification accuracies are quite dissimilar. The reason for this is that according to Eq.(5) the von Neumann entropy is more sensitive to variable graph size than the Shannon entropy. This implies that the Shannon entropy is better suited to graph datasets containing graphs of variable size. On the ENZYMES data-set, all methods perform relatively poorly. The reason for this is that there are six classes of graph in the ENZYMES dataset, and only two classes in the remaining datasets.

4.2   View-Based Object Recognition

In this subsection, we focus on graphs extracted from images and access the effectiveness of the proposed subgraph kernel in detecting object clusters. The first dataset (i.e. ALOI) consists of graphs extracted from images of three similar boxes in the ALOI database. The second dataset (i.e. MOVI-CMU) consists of graphs extracted from images of three similar toy houses in the MOVI and CMU databases. The third dataset (i.e. COIL) consists of graphs extracted from images of three similar cups in the COIL database. For each object there are 18 images captured from different viewpoints. The graphs are the Delaunay triangulations of feature points extracted from the different images. The minimum and maximum vertices of the three datasets are 295 (min) and 1288 (max) for ALOI, 249 (min) and 734 (max) for MOVI-CMU, and 13 (min) and 80 (max) for COIL.

For our subgraph kernel with the von Neumann entropy (JSSV) or the Shannon entropy (JSSS), we compute the kernel matrix on each dataset. We embed the testing graphs into feature space as vectors by performing the Principle Component Analysis (PCA) on each kernel matrix. Furthermore, we also compare our methods against a number of alternative graph characterisations, including a) the Ihara zeta function of graphs (CIZF) [14], and b) the pattern vectors from algebraic graph theory (PVAG) [18]. For these alternative methods we compute the feature vectors of graphs on each dataset. To evaluate the performance of all the methods, we perform the K-means clustering method to compute the classification accuracies for the three datasets. The experimental results are shown in Table 3

For the graphs in the COIL dataset of which the average number of vertices is less than 60, all the methods give good performance and can achieve 100% recognition accuracy. For the graphs in the ALOI and MOVI-CMU datasets of which the average number of vertices is between 200 to 700, the proposed methods outperform all the alternatives, only CIZF is competitive to our methods on the MOVI-CMU dataset. The CIZF and PVAG can not finish the computation on the ALOI dataset, since they generate infinite feature values for a graph of large size. The evaluation reveals that our subgraph kernel can easily scale to large graphs.

Table 3. Experimental Comparisons on Image Based Graph Dataset

Datasets	JSSV	JSSS	CIZF	PVAG
ALOI	87.03	90.73	–	–
MOVI-CMU	96.29	100	96.29	81.49
COIL	100	100	100	100

5 Conclusion

In this paper, we have shown how to construct a fast Jensen-Shannon subgraph kernel using depth-based representations of graphs. The proposed method is based on a fast Jensen-Shannon diffusion kernel measure defined in terms of the Jensen-Shannon divergence on (sub)graphs and an elegant graph decomposition through the depth-based representation. The proposed subgraph kernel overcomes the subgraph size restrictions arising in state of the art graph kernels, and also renders an efficient computation. The experimental results have demonstrated the effectiveness and efficiency of the proposed subgraph kernel.

Acknowledgments. We thank Prof. Francisco Escolano, Dr. Peng Ren and Dr. Boyan Bonev for the constructive discussion and suggestions.

References

1. Amizadeh, S., Wang, S., Hauskrecht, M.: An efficient framework for constructing generalized locally-induced text metrics. In: Proceedings of International Joint Conference on Artificial Intelligence, pp. 1159–1164 (2011)
2. Bai, L., Hancock, E.: Graph kernels from the Jensen-Shannon divergence. Journal of Mathematical Imaging and Vision (to appear 2013)
3. Borgwardt, K.M., Kriegel, H.-P.: Shortest-path kernels on graphs. In: Proceedings of the IEEE International Conference on Data Mining, pp. 74–81 (2005)
4. Crutchfield, J.P., Shalizi, C.R.: Thermodynamic depth of causal states: Objective complexity via minimal representations. Physical Review E 59, 275–283 (1999)
5. Gärtner, T., Driessens, K., Ramon, J.: Graph kernels and gaussian processes for relational reinforcement learning. In: Proceedings of International Conference on Inductive Logic Programming, pp. 146–163 (2003)
6. Gärtner, T., Flach, P.A., Wrobel, S.: On graph kernels: Hardness results and efficient alternatives. In: Schölkopf, B., Warmuth, M.K. (eds.) COLT/Kernel 2003. LNCS (LNAI), vol. 2777, pp. 129–143. Springer, Heidelberg (2003)
7. Han, L., Escolano, F., Hancock, E.R., Wilson, R.C.: Graph characterizations from von neumann entropy. Pattern Recognition Letters 33, 1958–1967 (2012)
8. Haussler, D.: Convolution kernels on discrete structures. In: Technical Report UCS-CRL-99-10 (1999)
9. Johnson, D.B.: Efficient algorithms for shortest paths in sparse networks. Journal of the ACM 24, 1–13 (1977)
10. Kashima, H., Tsuda, K., Inokuchi, A.: Marginalized kernels between labeled graphs. In: Proceedings of International Conference on Machine Learing, pp. 321–328 (2003)
11. Martins, A.F., Smith, N.A., Xing, E.P., Aguiar, P.M., Figueiredo, M.A.: Nonextensive information theoretic kernels on measures. Journal of Machine Learning Research 10, 935–975 (2009)
12. Nori, N., Bollegala, D., Ishizuka, M.: Interest prediction on multinomial, time-evolving social graph. In: Proceedings of International Joint Conference on Artificial Intelligence, pp. 2507–2512 (2011)
13. Platt, J.C.: Fast training of support vector machines ssing sequential minimal optimization. In: Schölkopf, B., Burges, C.J.C., Smola, A.J. (eds.) Advances in Kernel Methods, pp. 185–208 (1999)
14. Ren, P., Aleksic, T., Wilson, R.C., Hancock, E.R.: A polynomial characterization of hypergraphs using the ihara zeta function. Pattern Recognition 44, 1941–1957 (2011)
15. Shervashidze, N., Schweitzer, P., van Leeuwen, E.J., Mehlhorn, K., Borgwardt, K.M.: Weisfeiler-lehman graph kernels. Journal of Machine Learning Research 1, 1–48 (2010)
16. Shervashidze, N., Vishwanathan, S.V.N., Petri, T., Mehlhorn, K., Borgwardt, K.M.: Efficient graphlet kernels for large graph comparison. Journal of Machine Learning Research 5, 488–495 (2009)
17. Slater, P.J.: Centers to centroids in graphs. Journal of Graph Theory 2, 209–222 (1978)
18. Wilson, R.C., Hancock, E.R., Luo, B.: Pattern vectors from algebraic graph theory. IEEE Trans. Pattern Anal. Mach. Intell. 27, 1112–1124 (2005)

Evaluation of Interactive Segmentation Algorithms Using Densely Sampled Correct Interactions

S.M. Rafizul Haque, Mark G. Eramian, and Kevin A. Schneider

University of Saskatchewan, Saskatoon, SK, S7N 5C9, Canada
rafizul.cs@usask.ca,
{eramian,kas}@cs.usask.ca

Abstract. The accuracy and reproducibility of semiautomatic interactive segmentation algorithms are typically evaluated using only a small number of human observers which only considers a very small number of the possible *correct interactions* that an observer might provide. A *correct interaction* is one that provides contextual information that would be expected to result in a correct segmentation. In this paper, we demonstrate new evaluation methods for semiautomatic interactive segmentation algorithms that employ simulated observer models constructed from a large number of segmentations computed by uniformly sampling the entire set of possible correct interactions. The advantages of this method are that it is free of observer biases and the large number of segmentations produced for each object of interest to be segmented allow a range of statistical methods to be brought to bear on the analysis of segmentation algorithm performance. The methods are demonstrated using a semi-automated segmentation algorithm for ovarian follicles in ultrasonographic images.

Keywords: interactive segmentation, semiautomatic, correct interaction, reproducibility, performance, evaluation.

1   Introduction

Algorithms for segmentation of semantic objects from images are, ideally, desired to be fully automatic due to the tedious and time-consuming nature of manual segmentation. Some segmentation problems, however, are still very difficult to solve with fully automatic methods such as problems where an arbitrary number of objects of interest need to be both detected and segmented or where the imaging modality results in poorly resolved boundaries between neighbouring objects. This motivates the use of semiautomated segmentation algorithms in which a human operator provides high-level contextual information in an interactive phase, which is followed by an automatic phase where the segmentation is performed under the constraints of the operator-provided guidance.

Evaluation of accuracy and/or reproducibility of semiautomatic segmentation algorithms is typically performed by having a small number of experts in the

A. Petrosino (Ed.): ICIAP 2013, Part I, LNCS 8156, pp. 191–200, 2013.

problem domain who are well-trained in the use of the semiautomatic segmentation system segment a number of cases. Indeed, this has been the case with numerous recent studies that analyze intra- and/or inter-observer variability [5–8, 10, 13, 14]. Of these, only the studies of Stammberger et al. [14] and Claudia et al. [8] used more than 5 observers. Even in the simplest of situations, where the interaction is selecting a seed point somewhere within an object, it is not possible to robustly characterize the inherent variability in segmentation accuracy due to variations in seed point placement using only a small number of example interactions. Recently, some authors [11, 12] have turned to constructing simulated *observer models* to take into account more interactions per case, and to avoid observer bias but the number of seeds and brush strokes used were not sufficient to represent a very diverse set of examples of possible correct interactions.

The main deficiency, therefore, of existing evaluation methods is that an insufficiently diverse sampling of the set of correct interactions for each case are used to draw conclusions about overall segmentation accuracy and reproducibility. One must consider a diverse set of correct interactions in order to compare algorithms fairly and take into account the consequences of poor choices resulting from fatigue or lapses in judgement on the part of the operator. To this end we propose the use of observer models where the user interaction is generated programmatically in a similar way as in [11] and [12]. However, in contrast to these previous methods, our observer models are unbiased in the sense that we systematically and uniformly sample all possible correct interactions for each case to be segmented. Herein we consider a simple interaction mode where a user click is made to supply a seed point for each object to be segmented. For each object, all the segmentations generated from the uniformly sampled set of possible seed points are analyzed statistically to assess the impact of the seed point location on the quality of the resulting segmentation using four models. One of these four models is an overall unbiased one as it includes all the sampled seed points and the other three models are biased as each of these models includes only a specific subset of the sampled seed points based on the locations of these seed points. Using this synthetic interaction model, we have enough data to use statistical methods to test for significant differences in segmentation accuracy between observer models (subsets of correct interactions) and to quantify any differences found. As a vehicle for the demonstration of our methods, we consider an ovarian follicle segmentation algorithm based on binary graph cuts with a shape prior and evaluate the reproducibility of the results, and look for statistically significant differences between the aforementioned four observer models.

2   Methods

2.1   Interactions

We define a *correct interaction* to be the contextual information provided by the operator that would be expected to produce a correct segmentation, e.g. a seed point that is inside the object to be segmented, or a set of brush strokes that correctly indicate areas of foreground and background. Herein we demonstrate some new methods of analyzing the variability of segmentation accuracy over a wide range of uniformly sampled correct interactions using interactive graph cuts.

2.2   Binary Graph Cuts for Follicle Segmentation

Graph cut segmentation [3, 4] utilizes a max-flow/min-cut energy minimization algorithm which eventually generates the optimal segmentation with respect to the weights assigned on the edges of the graph. The minimum cut can be found using efficient solutions to the *maximum flow* problem since the set of saturated edges in a maximum flow solution coincide with the edges in the minimum cut. For details on binary graph cuts, the reader is referred to [4]. Instead of using general graph cut segmentation, a generic shape prior called "star shape" [15] was incorporated to enhance the segmentation accuracy. This shape prior is appropriate due to the roughly elliptical but low-eccentricity shape of ovarian follicles in normal ovaries. Within the Voronoi region of each seed point we added new edges to the graph radiating from each user supplied follicle seed point to encode the shape prior in the graph; see [15] for details.

For our follicle segmentation, asymmetric weights between non-terminal nodes were defined similarly to as in [4]. Terminal weights were derived from intensity distribution models (histograms) built from neighbourhoods defined by user-supplied clicks on example foreground and background regions.

Constraints were added based on the seed points given for each follicle. In our experimental methodology, seed points are selected from a set of correct interactions for each follicle (see Section 2.3).

2.3   Experimental Setup

A set of 32 ultrasound *in vivo* human ovarian images, obtained from a previous study [1], were used in this experiment. Size of each image is 640×480 pixels and the maximum number of follicles in an image was fourteen. The total number of follicles in all images was 132 among which 53 were very small, having a cross-sectional diameter in the image of less than 2.5mm; this is a significant size threshold because even human observers have difficulty correctly identifying follicles of this size or smaller. These follicles were not considered in our experiment, leaving 79 follicles of diameter greater than 2.5mm. Manually delineated ground truth segmentations of these follicles were provided by a single, highly experienced human operator. For each follicle to be analyzed we constructed a set of correct interactions (seed points). Seed points were sampled on a grid with a spacing of between 2 and 14 pixels depending on follicle size. Seed point locations were sampled more sparsely for larger follicles to maintain computational feasibility. Grid spacing was determined using the following procedure:

1. Determine number of seed points N to be used for the follicle as a function of its area using the piecewise cubic polynomial function in Figure 1(a). Interpolation points were selected empirically based on the data set.

2. Determine the grid spacing as $\lceil\sqrt{A/N}\rceil$ where A is the area (in pixels) of the follicle (determined from the ground truth).

This resulted in sets of correct interactions consisting of between 50 and 375 seed points for each follicle.

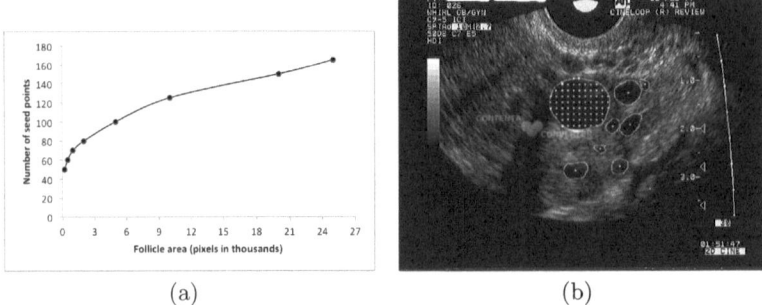

(a)   (b)

Fig. 1. (a) Piecewise cubic polynomial function used for determining the number of seed points for each follicle. (b)Segmenting a follicle with its set of correct interactions. The large follicle is segmented once with each of the seed points shown while the remaining follicles seed points (at the centroids of their regions) are held constant. This process is repeated for each other follicle in the image.

Each follicle was segmented using each seed point from its set of correct interactions exactly once while the seed points for any other follicles in the image were held constant (Figure 1(b)). These constant seed points were the centroid of the follicle region, determined from the ground truth. Thus each follicle was segmented with each of its correct interactions and the Dice coefficient, HD, and RMSD were determined for each interaction.

For each follicle, sampled seed points were categorized into three groups: central, intermediate and peripheral depending on their locations within the follicle relative to the follicle region's centroid. To determine the category of a seed point, a binary ground truth image was negated and then distance transform of that image was computed. From this transform, distance a of the seed point from the nearest boundary point was determined and distance from the seed point to the centroid b was calculated using the Euclidean distance metric. The seed point category $\mathbf{c}$ was then determined by a double threshold of the quantity $\frac{a}{a+b}$:

$$\mathbf{c} = \begin{cases} \text{peripheral}, & \frac{a}{a+b} \leq 1/3 \\ \text{intermediate}, & 1/3 < \frac{a}{a+b} \geq 2/3 \\ \text{central}, & 2/3 < \frac{a}{a+b} \end{cases} \tag{1}$$

Mean and standard deviation of Dice coefficient, RMSD and HD were computed over each follicle's set of correct interactions and over the central, intermediate and peripheral subsets of interactions for each follicle. Larger values of the Dice coefficient and lower values of RMSD and HD indicate a more accurate segmentation. Coefficients of variation of these segmentation accuracy measures were computed for each follicle and analyzed to evaluate segmentation reproducibility.

3   Analysis of Results

All of the 79 follicles with diameter > 2.5mm, were analyzed. Segmentation accuracy was measured in terms of Dice coefficient [9], root mean squared distance (RMSD) and Hausdorff Distance (HD) [2] of the segmented follicle contour from the manual contour. These measures were determined individually for each follicle.

3.1   Coefficient of Variation within Seed Point Categories

Coefficient of variation (CV) is the ratio of standard deviation of a sample to the mean of the sample and indicates the extent of variability in relation to mean of the population. CV of the Dice coefficient, RMSD, and HD were computed over all seed points of each follicle. Histograms of the resulting CV values for the 58 follicles are shown in Figure 2(a). The range of CV values have been divided into ten unequal intervals and have been represented along the horizontal axis; since most of the values are in the range of 0 to 0.2, this interval has been divided into 9 bins illustrate the distribution of CVs within this range. CV values greater than 0.2 have been included in a single bin. The vertical axis represents the number of follicles for which the CV fell into the specified interval. For Dice, RMSD and HD, 58% of the coefficients of variation are less than 5%. Figure 2(b) shows histograms of the Dice coefficient CV values from the 79 follicles calculated for the central, intermediate, and peripheral groups of seed points. The distribution of the these CVs follow the same general trends as the overall distributions; overall 86% of the Dice CV values over all three seed point groupings were less than 5%. The mean Dice CV for the central, intermediate, and peripheral seed point groups, computed over all follicles, were 2.11%, 3.04% and 6.90% respectively, while the overall mean Dice CV across all seed points and follicles was 5.48%. The central and intermediate seed point group's mean Dice CVs were found to be significantly different from the overall mean Dice CV (two-tailed Student's paired two-sample t-test, $p = .00024$ and $p = 0.005$, respectively).

The mean Dice CV for the peripheral region was not significantly different from the overall Mean Dice CV. The mean Dice CV for the central seed point group was significantly less than both the intermediate and the peripheral groups. The mean Dice CV for the intermediate seed point group was significantly less than the mean Dice CV for the peripheral seed point group. This is strong evidence that reproducibility is, on average, higher when the seed points are placed in either the central or intermediate regions of a follicle.

Figures 2(c) and 2(d) show the distributions of CV for the central, intermediate and peripheral seed point groups for RMSD and HD respectively, again showing similar trends. For RMSD, 68% follicles had a CV of less than 5% for all three seed point groups, while for HD, this number was 66%.

The overall mean RMSD CV was 13.5%, and the mean RMSD CV's for the central, intermediate, and peripheral seed point groups were 3.46%, 6.63% and 15.9%; all of these were significantly different from the overall RMSD CV. The mean RMSD CV for the central seed point group was significantly less than both the intermediate and the peripheral groups. The mean RMSD CV for the intermediate seed point group was significantly less than the mean Dice CV for the peripheral seed point group.

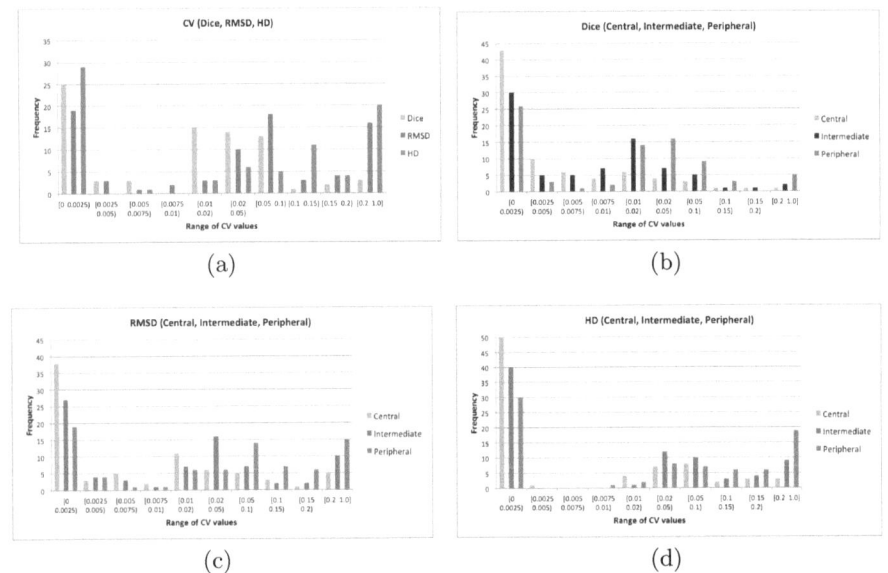

(a)   (b)

(c)   (d)

Fig. 2. (a)Coefficient of variation of overall Dice Coefficients, RMSD and HD. (b) Coefficient of variation of Dice Coefficients for three groups of seed points (c)Coefficient of variation of RMSD for three groups of seed points (d)Coefficient of variation of HD for three groups of seed points.

The overall mean HD CV was 14.3%, and the mean HD CV's for the central, intermediate, and peripheral seed point groups were 3.33%, 6.83% and 15.9%; the central and intermediate means were significantly different from the overall HD CV. The mean HD CV for the central seed point group was significantly less than both the intermediate and the peripheral groups. The mean HD CV for the intermediate seed point group was significantly less than the mean Dice CV for the peripheral seed point group.

Again we have very strong evidence that reproducibility is considerably greater, on average, when seed points are confined to the central and intermediate regions.

3.2   Mean Segmentation Accuracy within Seed Point Categories

Figure 3(top row) presents the mean and standard deviations (as error bars) of Dice, RMSD, and HD for each follicle. The follicles are positioned on the horizontal axis in decreasing order of their cross-sectional diameter. A linear regression line fit to this data shows that Dice coefficients generally decrease for smaller follicles; R^2 was significant ($p < 0.05$). There was no evidence of a linear trend for RMSD and HD with decreasing follicle size; the R^2 values for both of these regressions were not significant.

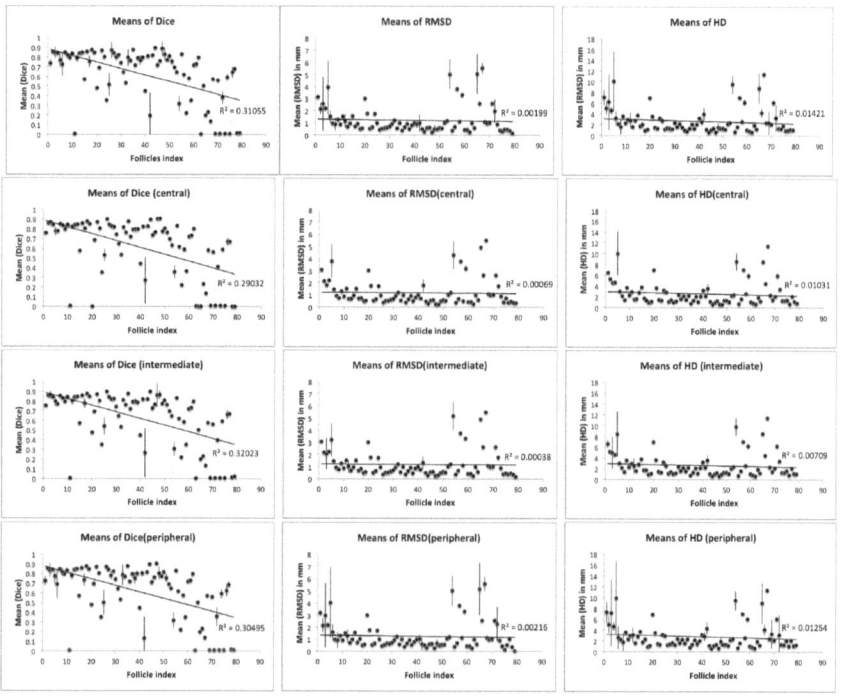

Fig. 3. Mean and Standard Deviation of Dice, RMSD and HD with error bar, 1st row: for overall, 2nd row: for central points, 3rd row: for intermediate points, 4th row: for peripheral points

On the whole, values of RMSD and HD generally remain stable as follicle size decreases, while Dice coefficient worsens. This can be explained by noting that Dice coefficient is a proportional error based on areas where as RMSD and HD are absolute errors based on distances. Consider two follicles region segmentation results with the same Dice coefficient, but where the follicles have vastly different area. Absolute deviations from the true boundary will be smaller for the small follicle than for the larger follicle simply because of the size difference. Thus, it is possible for RMSD and HD to remain stable as Dice coefficient and follicle size decrease. Figures 3 (2nd - 4th row) show the same information but subdivided into subgroups of central, intermediate, and peripheral seed points, respectively. It can be seen that the standard deviations (error bars) are generally smaller for the groups of central seed points and larger for the groups of peripheral seed points. Again, the linear regression line shows, in all cases, a trend of worsening Dice coefficient (R^2 significant, $p < 0.00001$), and no trend in RMSD and HD values (R^2 not significant) as follicle size decreases.

3.3 Results Summary

From our statistical study of groups of correct interactions for the follicle segmentation algorithm we have the following main results.

1. For many, but not all follicles, the overall CV of segmentation accuracy is low to moderate. Overall Dice CV was less than 5% for 76% of follicles; overall CV of RMSD and HD was less than 5% for 48% and 49% of follicles, respectively.
2. The mean Dice CV (respectively HD CV and RMSD CV) for the central seed point groups was significantly smaller than for the intermediate and peripheral groups. The mean Dice CV (respectively HD CV and RMSD CV) of the central and intermediate seed point groups were significantly smaller than the peripheral group, and the magnitude of the difference in means from the peripheral group were quite large.
3. There was a statistically significant trend of decreasing (worsening) mean Dice coefficient with decreasing follicle diameter. However, there was no evidence of a statistically significant trend in the values of mean HD and RMSD with respect to the diameter of the follicles.

3.4 Discussion

In this section we discuss the interpretation and implications of the results in the previous section.

Result #1 tells us, in the broadest of strokes, that the studied segmentation algorithm is not able to generate easily reproducible results on perhaps as many as half of all follicles. is a type of result that could have been obtained from the standard method of examining results from a small number of human observers. However, our methods result in many more samples from which to estimate the mean and variance of the segmentation accuracy measures, possibly resulting in better estimates of the true mean and variance over all correct interactions, and therefore more accurate estimates of coefficient of variation.

Result #2 is very strong evidence that reproducibility is, on average over all follicles, much higher for the central and intermediate seed point groupings than for the peripheral seed point group. This result could not have been obtained using standard evaluation with human observers. Even if there were controls to ensure an equal number of samples in each seed point group, with 10 or fewer observers there would be insufficiently many samples per grouping to obtain any reasonable level of statistical power for our methods. Traditional evaluation would only have provided an estimate of the overall reproducibility and would not have elucidated the magnitude of the degradation of reproducibility with increasing seed point distance from the centroid of the follicle region.

Result #3 indicates that, while the segmentation error with respect to the deviation of the segmented follicle boundary from the ground truth boundary was not related to follicle size, the mismatch between the segmented follicle region and the ground truth follicle region was larger for smaller follicles because said deviations from the true boundary result in a greater proportion of region mismatch for smaller follicles. This result might be obtained using standard evaluation with human observers if reproducibility of the algorithm is already very high, but otherwise, a larger number of samples are needed to get more accurate estimates of mean segmentation accuracy measures on a per-follicle basis.

By judiciously choosing subsets of correct interactions to analyze using our methods, one can potentially obtain evidence that can be used to make recommendations on how operators can best use the system to produce the most accurate and consistent results. In the case of our analysis of the follicle segmentation algorithm, one would likely recommend that observers avoid placing seed points within the follicle's periphery to reduce inter-observer variability.

As our methods sample the set of correct interactions uniformly, our methodology is free from observer bias, observer training effects, and other biases that might result from the instructions/protocols that human observers are instructed to use during data collection. Uniform sampling of correct interactions incorporates into the evaluation those correct interactions that might, under normal circumstances, never be used by human observers because they contradict established usage protocols, but which might nevertheless occur in cases of observer fatigue or a lapse in judgment, and allows us to study the effect and risk of such lapses by studying the appropriate subsets of correct interactions.

4   Conclusion

Instead of characterizing segmentation performance through the actions of a small number of observers, we constructed synthetic observers and characterized their behaviour using a much larger number of segmentations over the set of all correct interactions. Our methods allow for a much richer, statistically backed characterization of interactive segmentation algorithm performance, resulting in new kinds of information that elucidate the best practices for how an interactive algorithm should be used to avoid any inherent sources of error in the algorithm while better understanding those sources of error.

References

1. Baerwald, A.R., Adams, G.P., Pierson, R.A.: Characterization of ovarian follicular wave dynamics in women. Biology of Reproduction 69(3), 1023–1031 (2003)
2. Bowyer, K.W.: Validation of medical image analysis techniques. In: Sonka, M., Fitzpatrick, J.M. (eds.) Handbook of Medical Imaging: Medical Image Processing and Analysis, vol. 2, SPIE press (2000)

3. Boykov, Y., Jolly, M.P.: Interactive graph cuts for optimal boundary & region segmentation of objects in N-D images. In: Proceedings of the International Conference on Computer Vision (ICCV 2001), pp. 105–112 (2001)
4. Boykov, Y., Lea, G.F.: Graph cuts and efficient n-d image segmentation. International Journal of Computer Vision 70(2), 109–131 (2006)
5. Byrum, C.E., MacFall, J.R., Charles, H.C., Chitilla, V.R., Boyko, O.B., Upchurch, L., Smith, J.S., Rajagopalan, P., Passe, T., Kim, D., Xanthakos, S., Ranga, K., Krishnan, R.: Accuracy and reproducability of brain and tissue volumes using a magnetic resonance segmentation method. Psychiatry Research: Neuroimaging 67, 215–234 (1996)
6. Cates, J.E., Lefohn, A.E., Whitaker, R.T.: Gist: an interactive gpu-based level set segmentation tool for 3d medical images. Medical Image Analysis 8, 217–231 (2004)
7. Coehn, B.A., Barash, I., Kim, D.C., Sanger, M.D., Babb, J.S., Chandarana, H.: Intraobserve and interobserver variability in renal volume measurements in polycystic kidney disease using a semiautomated mr segmentation algorithm. American Journal of Roentgenology 199, 387–393 (2012)
8. Dach, C., Held, C., Wenzel, J., Gerlach, S., Lang, R., Palmisano, R., Wittenberg, T.: Evaluation of an interactive cell segmentation for flourescence microscopy based on the graph cut algorithm. In: Microscopic Image Analysis with Applications in Biology, Heidelberg, Germany (September 2, 2011)
9. Dice, L.R.: Measures of the amount of ecologic association between species. Ecology 26(3), 297–302 (1945)
10. McGuinness, K., O'Connor, N.E.: A comparative evaluation of interactive segmentation algorithms. Pattern Recognition 43(2), 434–444 (2010)
11. Moschidis, E., Graham, J.: A systematic performance evaluation of interactive image segmentation methods based on simulated user interaction. In: IEEE International Symposium on Biomedical Imaging: From Nano to Macro, pp. 928–931 (2010)
12. Nickisch, H., Rother, C., Kohli, P., Rhemann, C.: Learning an interactive segmentation system. In: Proceedings of the Seventh Indian Conference on Computer Vision, Graphics and Image Processing (ICVGIP 2010), pp. 274–281 (2010)
13. Schenk, A., Prause, G.P.M., Peitgen, H.-O.: Efficient semiautomatic segmentation of 3D objects in medical images. In: Delp, S.L., DiGoia, A.M., Jaramaz, B. (eds.) MICCAI 2000. LNCS, vol. 1935, pp. 186–195. Springer, Heidelberg (2000)
14. Stammberger, T., Eckstein, F., Michaelis, M., Englmeier, K.-H., Reiser, M.: Interobserver reproducibility of quantitative cartilage measurements: Comparison of b-spline snakes and manual segmentation. Magnetic Resonance Imaging 17(7), 1033–1042 (1999)
15. Veksler, O.: Star shape prior for graph-cut image segmentation. In: Forsyth, D., Torr, P., Zisserman, A. (eds.) ECCV 2008, Part III. LNCS, vol. 5304, pp. 454–467. Springer, Heidelberg (2008)

Estimating Complex Refractive Index
Using Ellipsometry

Gul e Saman and Edwin R. Hancock*

Department of Computer Science, University of York, YO10 5GH, UK

Abstract. The paper is about estimating complex refractive index images of surfaces. Complex refractive index allows both the transmission and absorption properties of a surface to be characterised. Although widely used in applied optics, there has been little work on its use in image processing and surface inspection. In this paper we describe how to compute the complex refractive index of a material using simple digital image measurements. Specifically, a novel technique for estimating complex refractive index using ellipsometry is reported. The method uses a simple experimental setup that relies on the use of raw intensity data. This represents an advantage over the methods used in applied optics. Rather than relying on the calibrated measurement of absorption, instead our method relies on measuring the relative change in reflection amplitude which is a dimensionless quantity.

Keywords: Complex Refractive Index, Absorption, Extinction Co-efficient, Naturally occurring surfaces, Manufactured surfaces, Photometric stereo, Fresnel Theory.

1 Introduction

When incident light is scattered from a surface, then there is always partial absorption. Classical refractive index does not fully capture the situation where scattering involves both absorption and transmission, since it measures only the ratio of the speed of light in material to that in a vacuum. Complex refractive index, on the other hand, can be used to describe both the transmission and absorption properties of naturally occurring surfaces. Such a characterisation is essential for diagnostic purposes in naturally occurring biological tissues.

Complex refractive index is a composite quantity comprised of a real part that gives the refractive index and the imaginary part which gives the extinction co-efficient for the material, and is given as $m = n + i\alpha$. where n is the refractive index of the scatterer and α is the extinction co-efficient which is used to determine absorption. The quantity can be directly measured using ellipsometry for planar samples. For naturally occurring materials, chemical suspensions must be prepared, and then ellipsometry applied.

The refractive index of a naturally occurring surface is a basic quantity that determines the interaction of light and its propagation in the medium. As most types of naturally occurring surface are inhomogeneous and optically turbid, the methods that

---

* Edwin Hancock is supported by the EU FET project SIMBAD and by a Royal Society Wolfson Research Merit Award.

A. Petrosino (Ed.): ICIAP 2013, Part I, LNCS 8156, pp. 201–210, 2013.

have been employed thus far to determine the refractive index of materials cannot be applied to such surfaces. Methods involving the use of refractometers for determining the refractive index require large optical contact areas between the sample and the instrument. The refractive index of many kinds of naturally occurring tissues are hence unmeasurable using refractometers. Refractive index is key to understanding how light propagates in a material and knowledge about the refractive index of a naturally occurring surface is important for optical diagnosis and treatment. The refractive index of a biological tissue can be measured by cladding an optic fiber with a tissue and measuring the cone angle of the transmitted light. In this way the refractive indices for different kinds of mammalian tissues have been measured at a wavelength of 632.8 nm [1]. Moreover, the refractive indices of thin samples of dehydrated human skin have been measured using a laser refractometers based on total internal reflection. To determine the refractive index of fresh hydrated tissues, the values obtained from the dehydrated tissues must be modified based on estimates of the water content [2].

The above methods can only determine the real part of the complex refractive index of a material and this does not incorporate the absorption of light by the material.

Confocal microscopy has also been used in order to measure the refractive index of tissues [3]. Using a layer of immersion fluid of exactly the same depth as the tissue, it takes into account the 3-D point spread function (PSF) of the microscope. Greenleaf et al. [4], have used computerised tomography to compute two images for the refractive index n and acoustic attenuation α representation. The values have been computed using profiles of propagation times and amplitudes of digitized acoustic pulses of tissues viewed at multiple angles from rectilinear transmission scans. It might be considered intrinsically more valuable to have quantitative data than qualitative data, as it gives more information about the variation in the absorption of the material and also in the refractive index.

There are many methods for computing the complex refractive index of a material such as ellipsometry and reflectance versus incidence angle methods [5]. Ding et al. [6], used the automated reflectometer system that was developed to determine the refractive index of a turbid sample by measuring the coherent reflectance (R) vs the angle of incidence (θ). The Fresnel equations have been used in conjunction with non-linear regression $R(\theta)$ to determine the complex refractive index of human skin. It has been determined for eight wavelengths of both the dermis and the epidermis tissues. The reflectance of p and s polarisation is measured at different angles of incidence or polarisation conditions. These are related to the variation of the light intensity. The results are affected by the scattering light, internal reflection and other factors.

The aim in this paper is to develop a more direct surface inspection method for determining complex refractive index using the type of equipment commonly found in a machine vision laboratory. Specifically our set up consists of an ellipsometry prism placed at the centre of an geodesic light dome. Polarised light sources placed at the nodes of the dome, illuminate a sample of material. Images collected using a digital camera and a polaroid analyser, can be used to make measurements of complex refractive index. Moreover, the method provides for the first time a means of producing a complex refractive index image of the surface, which can be used to segment surfaces into regions of different composition or to inspect for surface lesions or flaws.

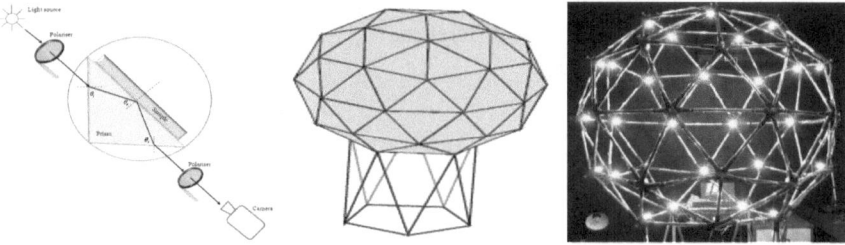

Fig. 1. The experimental setup for complex refractive index estimation using ellipsometry (left), light dome (center and right)

2   Polarised Light

Unpolarised light incident on a surface acquires partial polarisation due to the orientation of the dipoles in the scatterer. Polarisation information can be used for analysing small changes in a surface. Partially linearly polarised light has information that can be used for various purposes, for example leaves of different plants reflect light with a varying degree of polarisation [8]. The use of polarisation information by animals for navigation can be largely divided into two categories 1. macro-scale-orientation and 2. micro-scale-orientation. The macro-scale-orientation can be used to extract information about larger objects such as, trees and tree tops while, for the micro-scale-orientation information about smaller objects such as individual leaves [7].

3   Ellipsometry

Ellipsometry is the measurement of the effect of reflection on the state of polarisation of light. The method involves the measurement and interpretation of the change in the state of polarisation, when polarised light is reflected from a scattering surface. It is a method that is used in industry to measure the optical constants of materials and also for measuring the thickness and the optical constants for films placed on materials under study. It is important to characterise light waves incident on scattering surfaces. It is also of significance to understand how the light wave interacts with a reflecting surface and how it changes on contact with it and how can the change be related to the optical properties of a material.

A beam of incident light can be separated into electric field components that are perpendicular ($E_\perp$) and parallel ($E_\parallel$) to the plane of incidence. Similarly the perpendicular and parallel components of the reflected electric field are given by $R_\perp$ and $R_\parallel$, respectively.

The amplitude reflection co-efficient can be defined in terms of amplitude and phase change given as

$$r_\parallel = \frac{R_\parallel}{E_\parallel} \exp\left[i\left(\theta_{\parallel r} - \theta_{\parallel i}\right)\right] \tag{1}$$

where $\theta_{\|i}$ and $\theta_{\perp i}$ are the phase angles of the incident light, and $\theta_{\|r}$ and $\theta_{\perp r}$ the phase angles of reflected light. The phase differences $\theta_{\|r} - \theta_{\|i}$ and $\theta_{\perp r} - \theta_{\perp i}$ are not directly measurable. However, the ratio of

$$\frac{r_{\|}}{r_{\perp}} = \frac{\frac{R_{\perp}}{R_{\|}}}{\frac{E_{\|}}{E_{\perp}}} \exp\left(i\triangle\right) = \tan\phi \exp\left(i\triangle\right) \tag{2}$$

where
$$\phi = \arctan\left[\left(\frac{R_{\|}}{R_{\perp}}\right)\left(\frac{E_{\|}}{E_{\perp}}\right)\right] \tag{3}$$

Ellipsometry involves the measurement of $\tan\phi$ and $\triangle$ which are the relative change in amplitude and the phase difference on reflection. Here we deal with materials that absorb light, hence the use of the imaginary part of the refractive index.

3.1   Data Acquisition

We collect a succession of images of the different samples using a geodesic light dome constructed along the lines suggested in [9]. A schematic of the dome is shown in the centre of Figure (1) and a photograph of the completely lit dome is also shown in the figure. It consists of a geodesic grid of diameter 1.58 meter with 41 ultra bright white LEDs attached to the vertices of twice sub-divided icosahedron. A rotating polarising filter is placed in front of each LED. The geodesic dome has been used to accurately determine light source positions. The ellipsometry apparatus (essentially a prism) shown at the left of Figure (1) is placed at the centre of the dome. The prism is then imaged through a polaroid analyser using a digital camera, placed at a known viewing angle. Images are collected with the analyser at various angles. The LED polarisers are oriented and images collected as described below.

Various materials have been studied including chemical compound suspensions (industrially produced) and naturally occurring biological tissues. The chemical compounds that were used for creating suspensions for the experimentation are: Zinc Oxide, Graphite and Cobaltus Sulphate (coarse heterogeneous particle size and finely powdered to nearly homogeneous particle size). Painted samples using acrylic paints of different colours mixed with various materials forming new compounds. For the naturally occurring biological tissues, slices of potato, pepper, apple and orange (peel) have been used for estimating the complex refractive index. Slices have been taken due to the small surface area of the base of the prism (as the dimensions have to be comparable) and the requirement of the method for the use of planar samples, which is possible in the case of the chemical compounds, but a challenge in the case of the biological tissues. This is so because the sample has to be in complete contact with the surface of the prism.

3.2   Complex Refractive Index Using Ellipsometry

Here the complex refractive index is estimated by using a modification of the method proposed in [10]. This combines reflection ellipsometry with the principle of total reflection. Conventionally, a goniometric table has been used for the setup and recording of the observations with a He-Ne laser as a light source. However, in this paper, we have used the geodesic light dome where the LED acts as a light source and the ellipsometry setup (a prism) is placed at the approximate center of the dome. The orientation of the polariser placed in front of the incident light is set to $45°$, in order to equally divide the power into the perpendicular $\perp$ and parallel $\parallel$ polarisation components before it reaches the right angled glass prism (N-BK7) of refractive index 1.517. A camera (Nikon D200 digital SLR) has been used to measure the transmitted intensity I_t. A polaroid filter is placed in front of the camera, the analyser angle is rotated manually by increments of $10°$ or $5°$ (in some cases) to give 19 and 37 images per sample, respectively. The orientation of the second polariser is set such that the angle of separation between the two polarisers is $0°$. The degree of polarisation state can be computed by using Saman and Hancock's method of moments given in [11], while the angle of incidence is set to $34.12°$.

The composite quantity for the complex refractive index 'm' is defined as $m = n + i\alpha$ where α is obtained from the extinction co-efficient $k = \frac{4\pi\alpha}{\lambda}$. The complex refractive index can also be expressed as: $m = n - i\alpha$, and when $\alpha > 0$ light is absorbed by the scatterer while in case of $\alpha = 0$ light travels without any loss, for the case of $\alpha < 0$ the incident light is considered to be amplified. The convention followed for the expressing the complex refractive index during the course of this paper is: $m = n + i\alpha$.

Using the angles of each of the faces of the prism of AB, AC and BC with Snell's Law, the refractive index is given by

$$
n = \sqrt{\left(\left(\frac{q+1}{q-1}\right) n_0 \frac{\sin^2 \theta_r}{\cos \theta_r}\right)^2 + n_0 \sin^2 \theta_r}
\tag{4}
$$

where $q = -\frac{\alpha_e e^{-i\triangle}}{\cos^2\left(\theta_i - \theta_i'\right)}$, $\theta_i' = \sin^{-1}\left(\frac{\sin\theta_i}{n_0}\right)$, $\theta_r = \sin^{-1}\left(\frac{1}{n_0}\sin(\theta_i)\right) + 45°$ and n_0 is the refractive index of the prism while θ_i is the angle of incidence of the light source. The values for α_e and $\triangle$ once computed can be used to measure the complex refractive index of the sample. Ellipsometry is used to determine the values of α_e and $\triangle$. The transmitted light intensity I_t varies with the angle of the polariser placed in front of the camera, and follows the transmitted radiance sinusoid

$$
I_t = (I_0 - I_1)\left[T_0 + \sqrt{A_0^2 + B_0^2}\sin(2\phi + \beta)\right]
\tag{5}
$$

where $T_0 = \frac{|r_\parallel|^2 + |r_\perp|^2}{2}$, $A_0 = \frac{|r_\parallel|^2 - |r_\perp|^2}{2}$, $B_0 = |r_\parallel| \cdot |r_\perp|\cos\triangle$, and $\beta = \tan^{-1}\frac{B_0}{A_0} = \tan^{-1}\frac{B_1}{A_1}$. The extinction co-efficient α_e which is given by the dimensionless

$$
\alpha_e = \sqrt{\frac{1 - \frac{A_1}{T_1}}{1 + \frac{A_1}{T_1}}}
\tag{6}
$$

where $\triangle = \delta_\parallel - \delta_\perp = \cos^{-1}\left(\frac{1 - \alpha_e^2}{2\alpha_e}B_0 A_0\right) = \cos^{-1}\left(\frac{1 - \alpha_e^2}{2\alpha_e}B_1 A_1\right)$
$\tag{7}$

where $r_\perp$ and $r_\parallel$ are the amplitude transmission coefficients of the perpendicular $\perp$ and parallel $\parallel$ components of polarisation. θ_i and θ_i' are the angles of incidence on the AB face of the prism while the corresponding angles of refraction are, θ_r and θ_i', θ_s and θ_s' on the AC and BC faces of the prism.

The incident light intensity and loss of light due to scattering and absorption by the sample are given by I_0 and I_1. The polarisation angle, ϕ gives the difference between the transmission directions of the two polarisers.

3.3   Least Squares Estimation

The transmitted radiance sinusoid in Equation (5) has amplitude $\sqrt{A_1^2 + B_1^2} = \frac{I_{tmax} - I_{tmin}}{2}$, and offset (DC level) $T_1 = \frac{I_{tmax} + I_{tmin}}{2}$. The least squares fit for a sine function of known frequency [12] is used in order to compute T_1, $\left(A_1^2 + B_1^2\right)$ and β from the values of $I_t\left(\phi\right)$ acquired experimentally and α_e, $\triangle$. As has been mentioned the values for the transmitted light intensity are taken through the change in the polariser angle of the polaroid placed in front of the camera. It is evident from the above equations that the values of α_e and $\triangle$ are dependent on the ratios between A_1, T_1 and A_1, B_1, which show that the values for the complex refractive index are independent of the loss of light, due to the sample and the absolute intensity of the incident light.

4   Experiments

As mentioned earlier, the experiments were carried out in a dark room. The camera and the light sources are positioned at an angle to the object in the geodesic light dome using the experimental setup given in Figure (1). The samples are roughly divided into two categories 1. samples with unknown refractive indices comprising of naturally occurring surfaces and painted surfaces and 2. samples of known refractive indices comprising of chemical suspensions. For naturally occurring surfaces, the experiments were carried out on a wrinkled and fresh apple, orange, pepper and potato while, for industrially produced surfaces the experiments were carried out on painted pieces of paper. The same paper was painted with the same paints forming various compositions by mixing with other materials such as: 1. mineral blush, 2. talcum powder, 3. cumin powder and 4. powdered coal, the sample of plain paint was later coated with a layer of vaseline. The orange has been chosen in order to test the method for the effect of the presence of natural indentations in the surface. As shown in the Figure (1) the material whose refractive index is to be determined is placed/applied at the base of the prism. As for the chemical compounds, suspension were created by mixing 1g to 4g of the compound in vaseline and IPA (isopropanol) with increments of 1g. The chemical suspensions were applied directly to the base of the prism so that the estimates are only of the material under study while for the experiments in the category of unknown refractive index, the samples were mounted on a cardboard mount. The aim here is to determine the effectiveness of the method for material variation and also a sample was included to test its effectiveness for non-planar surface.

4.1   Results for Estimating the Complex Refractive Index Using Ellipsometry

The values for the real part of the complex refractive indices fall in the range $0 < n < 3$, where the non-physical outliers have been filtered. This corresponds to the parts of the surface in the case of rotting and wrinkled tissue that are not in contact with the surface of the prism. Errors arise due to the inter-reflection that occur because of the wrinkled surface resulting in unrealistic values and also to the non-physical values of the refractive index, which are less than unity. The potential sources of error are those that occur mainly due to the image acquisition process which result from camera jitter and human error in the positioning of the polariser filter, which also result in errors in the computation of the degree of polarisation. Additional sources of error in this experiment are due to the lack of contact between the material and the base of the prism also, in the case of the chemical suspensions created in IPA (which is volatile hence evaporates rapidly) when the IPA had significantly evaporated the coating started to crack and in case of the 4g a gap was created between the suspension and the prism surface leading to an error that is caused by non-contact.

The suspensions were used in order to verify how the method performs for materials of known refractive indices. Some experiment were also conducted in order to determine whether increasing concentrations of the compounds have an effect on the computations of the complex refractive index especially the extinction coefficient. The results for the estimations of the complex refractive index for chemical suspension are given in Table (2). The concentrations of the solutes were changed in order to study the variation of the complex refractive index especially the extinction co-efficient. It was found out by repeated experimentation that the change in the imaginary part of the composite quantity of the complex refractive index was not large hence the values were almost constant. Referring to results in existing literature such changes can be experienced when the sample has aged [13] while in this case the images were taken within the span of a few minutes or a day at the maximum. The variation in the values can be attributed to the noise in the image acquisition process.

The results for the estimations of the complex refractive index for mosaics are given in Table (1). The mosaics were created by combining two different materials as, 1. Zinc Oxide and Cobaltus Sulphate by using vaseline as a solvent, 2. Dormant yeast in cold water and activating the yeast by making a suspension with sugar and warm water, and 3. Cobaltus Sulphate by using IPA and vaseline. The mosaics were created and experimented on such that the method could be tested to verify its ability to compute the complex refractive index effectively for two different materials in the same image.

For the mosaics of 1. Zinc Oxide and Cobaltus Sulphate and 2. Cobaltus Sulphate, the histograms for the real and imaginary parts of the refractive index are give in Figure (2)and (3). It is evident from the bi-modal histogram for the imaginary values that it signifies the two materials in the mosaic. Also, for the same material, the histogram is bi-modal as the suspensions were created in a solvent whose characteristics are different from those of the chemical hence, the two peaks (as the absorption differs for the materials). The histograms for the real values of the same material have a single high peak which falls around the value of the refractive index of the material. The chemical suspensions that were created did not have uniform dimensions especially, near the edges of the prism where the material thinned and did not cover the base completely and

Table 1. The estimated Complex Refractive Indices for Mosaics using Ellipsometry

Sample Material	Estimated Complex Refractive index 'n'
Zinc Oxide and Cobaltus Sulphate	1.54-0.743i
Active and Dormant Yeast	1.90-1.481i
Cobaltus Sulphate	1.70-0.964i

Table 2. The estimated Complex Refractive Indices for Chemical Suspensions using Ellipsometry

Sample Material	Quantity (g)	Solvent	Estimated Complex Refractive index 'n'
Zinc Oxide	1	IPA	1.62-0.318i
Zinc Oxide	2	IPA	1.47-0.907i
Zinc Oxide	3	IPA	1.53-0.003i
Zinc Oxide	4	IPA	1.69-0.617i
Zinc Oxide	1	Vaseline	1.54-1.432i
Zinc Oxide	2	Vaseline	1.55-0.727i
Zinc Oxide	3	Vaseline	1.60-0.787i
Zinc Oxide	4	Vaseline	1.56-0.199i
Cobaltus Sulphate	1	IPA	1.58+0.858i
Cobaltus Sulphate	2	IPA	1.48-0.347i
Cobaltus Sulphate	3	IPA	1.48-0.372i
Cobaltus Sulphate	4	IPA	1.80-0.216i
Cobaltus Sulphate	1	Vaseline	2.01-0.572i
Cobaltus Sulphate	2	Vaseline	1.93-0.496i
Cobaltus Sulphate	3	Vaseline	1.95-0.359i
Cobaltus Sulphate	4	Vaseline	1.94-0.364i

Table 3. The estimated Complex Refractive indices using Ellipsometry

Sample Material	Estimated Complex Refractive index 'n'
Colour Patch	1.62 + 1.207i
Colour Patch Coated with Vaseline	1.42+1.308i
Cumin Powder	1.70+1.213i
Powdered Coal	1.79+1.176i
Talcum Powder	1.37 + 1.342i
Mineral Foundation	1.77 + 1.118i
Fresh Apple	1.66+1.162i
Wrinkled Apple	1.62 + 1.335i
Potato	1.40 + 1.952i
Orange	1.66 + 1.363i
Pepper	1.78 + 1.929i

with non-uniform thickness. For the histograms of real values of the refractive index, for the mosaic of Zinc Oxide and Cobaltus Sulphate two to three peaks can be observed. They can be attributed to the two materials comprising the mosaic, but it is not a very clear determinant of the refractive index. The average estimated values for the complex refractive index are in compliance with the existing published values of the materials.

4.2   Limitations

The limitations of the methods proposed thus far are: low accuracy due to the use of large amount of tissue and complicated setup or specific depth measurements of the tissue. In our experiments the method is simple and relatively easy to set up as the equipment being used is not specialised for the estimation of complex refractive index (the sample's images have been taken at intervals of $10°$ and in some cases at $5°$), neither is the tissue thickness consistent but, has varied with the different samples. The computed refractive indices of the experimental samples are given in the Table (3). It can be seen from the table that the computed value for the refractive index of vaseline coated paint, has an error of 4.15 %, while the values for the refractive index of apple are quite similar but, there is a difference in the computation of the extinction co-efficient hence more light is being absorbed by the wrinkled apple where the tissue water content is low as compared to the fresh apple.

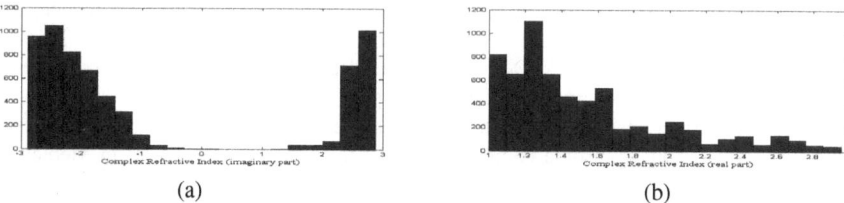

(a)   (b)

Fig. 2. The histogram of the estimation for the real and imaginary parts of the complex refractive index of the material mosaics using Ellipsometry of a. Zinc Oxide and Cobaltus Sulphate.

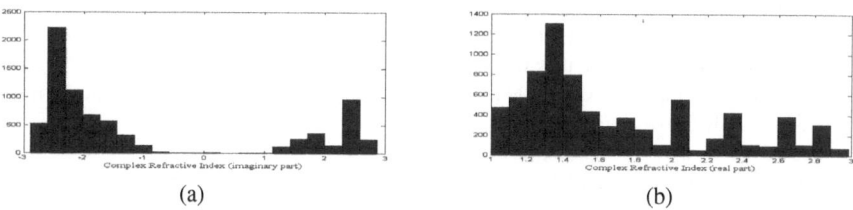

(a)   (b)

Fig. 3. The histogram of the estimation for the real and imaginary parts of the complex refractive index of the material mosaics using Ellipsometry of Cobaltus Sulphate in IPA and Vaseline

5   Conclusion

In this paper we show how complex refractive index can be computed by making use of ellipsometry. Rather than using a goniometer to perform the required calibrated photometric measurements, we use a geodesic light dome, using a digital camera together with polarisation measurements. Using this more simple experimental setup complex refractive index can be computed using dimensionless quantities that do not require photometric calibration. When used on mosaics of different materials, the method can be used to detect changes in material composition.

The method needs to be refined in various way. For instance in the data acquisition process, there is a problem associated with shadowing from the prism faces. The solution could be the use of a prism with a large base area so that the image can be captured without the shadowing effect. Another drawback of our method is that it requires a planar surface sample that has to be in complete contact with the base of the prism. The case of loss-of-contact leads to invalid complex refractive index values. However, the method that has been used in this paper does not require large tissue samples. The proposed method can be used to segment surfaces into regions of different composition or to inspect for surface lesions or flaws.

References

1. Bolin, F.P., Preuss, L.E., Taylor, R.C., Ference, R.J.: Refractive Index Estimation Using Polarisation and Photometric Stereo. Applied Optics, Optical Society of America (1989)
2. Tsenova, V., Stoykova, E.V.: Refractive index measurement in human tissue samples. In: Proceedings of SPIE (2003)
3. Dirckx, J.J.J., Kuypers, L.C., Decraemer, W.F.: Refractive index of tissue measured with confocal microscopy. Journal of Biomedical Optics (2005)
4. Greenleaf, J.F., Johnson, S.A., Lent, A.H.: Measurement of spatial distribution of refractive index in tissue by ultrasonic computer assisted tomography. Ultrasound in Medicine and Biology (1978)
5. Hunter, W.R.: Error in using the reflectance vs angle of incidence method for measuring optical constants. Journal of the Optical Society of America (1965)
6. Ding, H., Lu, J.Q., Wooden, W.A., Kragel, P.J., Hu, X.: Refractive indices of human skin tissues at eight wavelengths and estimated dispersion relations between 300 and 1600nm. Journal Physics in Medicine and Biology (2006)
7. Shashar, N., Cronin, T.W., Wolff, L.B., Condon, M.A.: The polarisation of light in a tropical rain forest. The association for Tropical Biology and Conservation (1998)
8. Land, M.F.: Old twist in a new tale. Nature (1993)
9. Cooper, P., Thomas, M.: Geodesic Light Dome. Department of Computer Science. University of York, UK (2010),
   http://www-users.cs.york.ac.uk/-pcc/Circuits/dome
10. Lai, J.C., Zhang, Y.Y., Li, Z.H., Jiang, H.J., He, A.Z.: Complex refractive index measurement of biological tissues by attenuated total reflection ellipsometry. Applied Optics (2010)
11. Saman, G., Hancock, E.R.: Robust Computation of the Polarisation Image. In: International Conference on Pattern Recognition (2010)
12. IEEE: IEEE Standard for Digitizing Waveform Recorders, IEEE Std 1057-1994 (1994)
13. Sato, K., Manabe, T., Polivka, J., Ihara, T., Kasashima, Y., Yamaki, K.: Measurement of the complex refractive index of concrete at 57.5 GHz. IEEE Transactions on Antennas and Propagation (1996)

Lazy Nonlinear Diffusion Parameter Estimation

Daniel Thuerck[1] and Arjan Kuijper[1,2]

[1] TU Darmstadt,
[2] Fraunhofer IGD, Darmstadt,
Germany

Abstract. Perona–Malik diffusion is a well-known type of nonlinear diffusion that can be used for image segmentation and denoising. The process itself needs an parameter k to decide which edges will be retained and which can be blurred and a stopping time t_S. Although there have been investigations on how to set these parameters, especially for regularized diffusion models, as well as different criteria for the optimal stopping time have been suggested, there is yet no quick and conclusive way to estimate both parameters – or to reduce the search space at least. In this paper, we show that Gaussian noise characteristics of an image and the diffusion parameters for an optimal optical result can be estimated based on the image histogram. We demonstrate the effectiveness of lazy learning in this area and develop a custom feature weighting algorithm.

1 Introduction

In 1990, Perona and Malik proposed nonlinear diffusion as a model mainly for image segmentation in a scale-space process [10]. It has been shown that this model can also be used for denoising. The model itself is expressed through a partial differential equation (PDE): $I_t = \text{div}(c(||\nabla I||)\nabla I)$ with $I(x,0) = I_0(x)$, where $\Omega \subset \mathbb{R}^2_+$ denotes the picture domain, $I : \Omega \rightarrow \mathbb{R}_+$ the intensity mapping for gray scale images, $x \in \Omega$, and appropriate boundary condition are taken. Furthermore, Perona and Malik introduced *conduction coefficient functions* expressed by c. The proposed explicit functions are

$$c_1(||\nabla I||) = \exp\left(-(||\nabla I||/k)^2\right) \text{ and } c_2(||\nabla I||) = \left(1 + (||\nabla I||/k)^2\right)^{-1}.$$

Of utmost importance is the parameter k. It can best be described as an *gradient threshold*. When we investigate the derivative of the *flux function*

$$\Phi(\nabla I) = \nabla I \cdot c(||\nabla I||), \tag{1}$$

we notice that with c_1 for edges with $||\nabla I|| > k$, we experience *backwards diffusion*, for $||\nabla I|| \leq k$ *forward diffusion*. On the one hand the backwards diffusion leads to edge enhancement [10,4,2,6,13], but also creates the problem of ill-posedness and image segmentation, which is not desirable for denoising.

2 Related Work

Different methods to estimate or determine k and t_S have been investigated. For k, Perona and Malik proposed in [10] a method: in each iteration, take the Quantile $p_{0.9}$ of the

A. Petrosino (Ed.): ICIAP 2013, Part I, LNCS 8156, pp. 211–220, 2013.

gradient histogram as k. Of course, this approach was meant for image segmentation, not denoising.

In the total variation image processing domain, the constrained optimization approach of Rudin et al. [12] avoids the problem of stopping time estimation – at the cost of a Lagrangian multiplier that has to be chosen. Convergence of the PDE then yields the desired minimal energy and optimized image [6,5].

Although the choice of k is such an important issue in nonlinear diffusion, not much research has been done in the area lately. Today, there are mostly simple histogram-based and morphological approaches. Histogram-based methods include the mentioned Canny algorithm which changes k in each iteration or a simple quadratic approximation of k by the image's average gradient as carried out in [14]. This is an interesting and yet simple approach; however, the formula is only based on two images. A quick experiment using our training set with ca. 130 images reveals that the method cannot be generalized.

In [16], Voci et al. proposed estimating the gradient threshold by the gradient histogram's p-Norm. For this, they introduced a new parameter σ for weighting the p-Norm. In effect, we get an histogram-based method again. Their morphological method, in contrast, compares the image opening and closing, thus estimating the average noise quantity and in the end, k.

An interesting approach, similar to the Canny method can be found in [7]. This method does not take the full histogram into account. Instead, it ranks blocks of a given window size w by a defined homogeneity and only regards the most inhomogeneous blocks for parameter estimation.

Weickert introduced his coherence enhancing diffusion and discussed the choice of the parameters - like a suitable stopping time t_S of the process [17]. Mrazek and Navara developed a time-selection strategy for iterative image restoration techniques: the stopping time is chosen such that the correlation of signal and noise in the filtered image is minimized [8]. Gilboa et al. used SNR Analysis [1] which was shown to be quite effective with respect to maximizing the SNR, but the SNR is per se not a good indicator for the subjective quality of an image as perceived by a human. This issue has been investigated by Wang et al. in [9] who developed the *structural similarity index* SSIM. The resulting index is a good approximation of the human visual system, as discussed by Ndajah et al. [9]. In this paper, we will concentrate on estimating parameters which will maximize the SSIM.

Different to the mentioned authors, we try to give *a priori* estimates for the parameters, not an *a posteriori* stopping criterion.

3   Parameter Estimation

In this section, we present for estimating the mentioned parameters. The first method is reducing the parameter space such that brute force parameter search becomes feasible. When an explicit model, on the other hand, is required, we show a simple machine learning method that is able to give reasonable estimates for a given input image.

3.1   Parameter Space Reduction

In [15], we present a new model for nonlinear diffusion that is constructed for efficient denoising and well-posedness. Although it requires again the parameter pair (k, t_S), we could show that the range of k values is significantly smaller than with the original Perona and Malik model. All images in our test set had the optimal parameter k in the range of $[6, 12]$ which results in acceptable denoising for a constant value $k = 9$.

3.2 Machine Learning Methods

Machine learning methods are usually categorized into *lazy learning* and *eager learning*. When applying lazy learning, an agent usually works with a training set of classified instances and compares them with a given, unclassified instance. The main methods in this field are nearest neighbour-based algorithms.

In eager learning, the learning algorithm usually gets the database as an input once and derives an algorithm for classifying from this data. Afterwards, a given example can be classified with respect to the learned model. While eager learning has the advantages of needing less memory and being able to reduce the influence of noise in the data, lazy learning techniques are especially useful when the data has no simple underlying model that can be learned by common machine learning algorithms.

In the remainder of this paper we show that for the estimation of k, no easy model can be derived. Hence we are restricted to lazy learning techniques. In contrast to that, when estimating the stopping time t_S methods from both classes work almost equally well.

3.3 Data Structure

For all machine learning experiments, we need to create input data from a set of images as training or test data. A prerequisite for this process is to define the feature set for the input training data set. In contrast to [17], we focus on a histogram approach: Each gray scale image is transformed in a *intensity histogram* and a *gradient magnitude histogram*. After that, we interpret both histograms as a statistical distribution and use these standard descriptors for characterizing the distribution:

- Quantiles $\left\{ p_{\frac{1}{8}}, p_{\frac{1}{4}}, p_{\frac{1}{2}}, p_{\frac{3}{4}} \right\}$
- Average value
- Median value
- Standard deviation

These result in a $(14 + 1)$-dimensional space. Hence, each dimension is called an *attribute* or *feature* of an *instance*, which is a n-tuple of values, one for each attribute and one for the parameter k or t_S. For later sections, we will denote the set of all attributes by $\mathcal{A}$. The 15th dimension is the dimension of the value that shall be estimated, e.g. k. Our used training set contains 128 images in different categories, mainly from the Berkeley Segmentation database. The test set contains 10 images (see Fig. 3) not in the training set.

Interestingly, in most publications, the authors conclude that structural information is required to estimate good parameter values. In contrast to that, our results show that the simple histogram approach does a reasonable job while being relatively simple.

3.4 Estimating Noise Variance

In this paper, we assume a Gaussian noise model with zero mean as underlying noise model for our images. For our numerical experiments, the noise variance will be discretized in a set of classes: $\{0.001, 0.005, 0.01, 0.05, 0.1\}$. To the human vision system, a higher noise variance is commonly perceived as *stronger noise*. For denoising purposes, this usually means that more iterations of the denoising process must be executed. Hence, the estimation of an image's noise variance is a prerequisite for estimating t_S.

For noise estimation, we used a simple regression tree algorithm with alternating growing and pruning, using randomly chosen $\frac{2}{3}$ of the data for growing in each iteration. The error measurement is carried out using the mean squared error metric. A leaf's value is then the average variance of all instances in the leaf.

When using the resulting model for classifying the test set presented in the next sections, we can estimate the correct value in 90% of all cases. Of course, this might be due to the discretized classes. This weakness can be improved by interpolating between classes or by using more complex regression tree algorithms like *M5P* [11] which create linear functions for each leaf. However, a 90% success rate is sufficient for our following tasks. The value can be used as a 16th dimension in the parameter estimation process or for inducing convergence in the denoising process (see [15]).

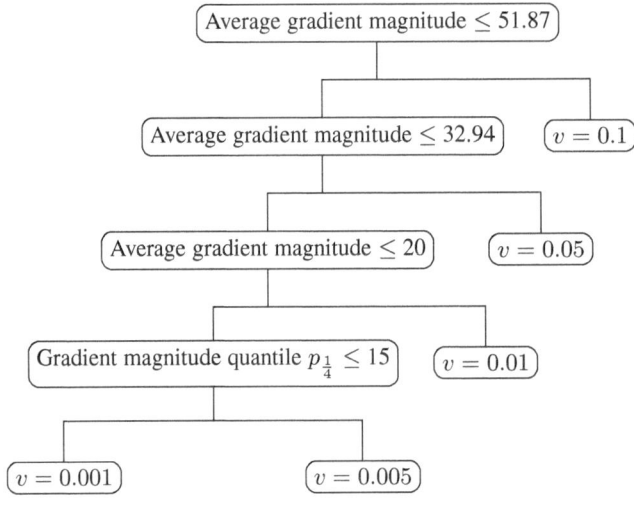

Fig. 1. Example regression tree for estimating noise variance v for a given image, trained on our training set

3.5 Estimating k

Unfortunately, the estimation of k is not quite as easy as the variance. If we plot k against each one of our features, most plots are likely to resemble normal distributions. Therefore, there is no usable information which could be used with regression algorithms. Experiments with our regression program verify that claim: The data is at most distributed in 2 leafs which would result in a binary choice for k. This is also the reason why the approach of Shao and Zou cannot work: There is no quadratic interpolation model for the correlation of average gradient, we can only recognize the different variance classes in a plot of k over the average gradient. This fact can be proven by using an regression algorithm based on a quadratic model. Additionally, additive regression or other ensemble techniques as boosting do not work. In fact, none of the classifiers provided by the WEKA [3] framework deliver a satisfying model.

This insight leads to the necessity of using a lazy learning technique, for example a simple nearest neighbor (NN) method. In this algorithm, all features or attributes $a \in \mathcal{A}$ for a given image are calculated. Afterwards, we compute the distance between two instances by an modified euclidean metric

$$d(x, y) = \sqrt{\sum_{a \in \mathcal{A}} (a_x - a_y)^2}. \tag{2}$$

as a combination of the distances of two instances' attributes. However, as most attributes usually have different orders of scale - e.g. the average gradient may be around 32, while the intensity Quantile $pi_{0.75}$ is 160 - thus leading to an unwanted implicit attribute weight in the naive approach. A first solution here is to normalize all attributes to $[0, 1]$. Normalized attributes will be denoted by $\hat{a}$, leading to the metric

$$d(x, y) = \sqrt{\sum_{a \in \mathcal{A}} (\hat{a}_x - \hat{a}_y)^2}. \tag{3}$$

where all attributes share the implicit weight 1. Equation (6) shows the metric that is used in simple nearest neighbor methods in our experiments („1NN" and „4NN" where x in xNN is the number of nearest neighbors whose values are included in calculating the estimation).

Up to here, our technique uses equal weights for all attributes of a histogram. As explained before, some attributes are normally distributed when plotting their relation to k. Hence, those attributes deliver less information in the sense of entropy than non-normally distributed attributes. This motivates our attribute weighting method: Attributes which contain more information about the distribution of k receive a higher weight than others. We will measure the amount of transferred information by the discrete entropy.

After Shannon, the entropy of a discrete random variable X over an alphabet $Z = \{z_1, z_2, ...\}$ is defined as $H(X) = -\sum_{z \in Z} P(X = z) \cdot \log_2 P(X = z)$. In the application, we discretize the values of each attribute in 256 values, resulting in the alphabet $\{0, 1, ..., 255\}$ and calculate the attribute's entropy. Afterwards, the entropy value is divided by the entropy of a normal distribution, which is in this case $\log_2 256 = 8$. The derived formula for the weights is now

$$w_a = 1 - \frac{H(a)}{8} \tag{4}$$

which results in our *entropic distance metric*

$$d(x, y) = \sqrt{\sum_{a \in \mathcal{A}} w_a (\hat{a}_x - \hat{a}_y)^2}. \tag{5}$$

The metric can be explained intuitively: An attribute that has a lower entropy value will usually reveal more regularities in contrast to the normal distribution. Hence, an attribute with lower entropy should be weighted stronger in the distance measure. The comparison with the normal distribution's entropy is only for normalizing purposes.

After calculating the distance to all images in the database, those images are ranked after their distance, low distances first. To estimate k, the values of the first n nearest neighbors is simply averaged. Here again, we can introduce weights in the average to penalize higher distances. In our experiments, a weight $\frac{1}{d^2}$ with d being the distance, delivered the best results.

This entropic measure will be called *nENN* with n as the number of instances that are involved in the averaging process.

3.6 Estimating t_S

Having k fixed, we usually get a development for the image quality as depicted in Fig. 2. The image quality (independent of used metric) has a single global maximum at a given time t_S which we call *optimal stopping time*. For all $t > t_S$, the image quality decreases; convergence is not guaranteed. Apart from estimating k, the next task is now estimating the ideal stopping time.

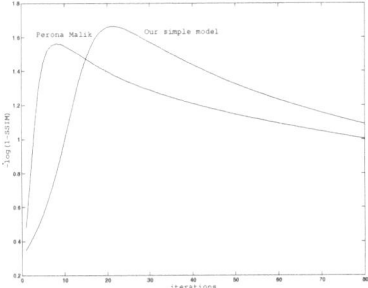

Fig. 2. Typical image quality development (SSIM [9]) in time when applying a diffusion model (here from [15])

The problem is very similar to the problem explained above: Scatter plots show that there also is no trivial model for t_S. Hence we need to use our learning techniques from above again; the technique can be applied without further changes. Interestingly, experiments show that the eager learner M5P delivers results of nearly the same quality for the test set as the entropic lazy learning techniques. For M5P, a linear model with an underlying regression tree of 2 classes was used. As a minimum occupancy per node, we chose $n = 100$; the model was created using the machine learning framework *WEKA* [3].

4   Numerical Experiments and Comparison

All learning algorithms mentioned above were tested on a test set of 10 images, none of those contained in the training set (see Fig. 3). We tried to concentrate on pictures which were not taken for testing purposes, as this shows the performance of our method in practical use. All images were scaled to 512 pixels on the larger side and saved as binary PGM.

The variance levels used for noise on the test set were $\{0.005, 0.05\}$. For each picture, we determined optimal parameters (k, t_S) for each image and variance levels using brute force to optimize the SSIM. The metric used for image quality measure is the SSIM [9]. The SSIM offers an human vision system-based quality metric that penalizes noise as well as blur [9]; just notice that the SSIM is normalized to $[0, 1]$ and equal images result in an SSIM of 1.

The Perona Malik filter itself is discretized as explained in [10] as a simple finite differences *lattice* model and implemented in pure C. The resulting programs are connected to MATLAB via the MEX interface. For testing purposes, we benchmarked the algorithms 1ENN, 4ENN, 1NN, 4NN, M5P and Shao [14], with abbreviations explained above.

Fig. 3. Images in test set in the order of their reference number

4.1   Estimation Performance for k and t_S

As the short statistical evaluation for k (Table 1) and t_S (Table 2) show, for the ks in an interval of [32,111], all methods deliver a quite good estimation. Although the average performance is comparable, the standard nearest neighbour methods 1NN and 4NN tend to have outliers, e.g. Fig. 5. The ENN methods, on the other hand, present a constant estimation performance, preventing extremely wrong results.

When comparing the number of data points used for estimation, the entropic method improves in performance with a higher n. Further experiments have shown that increasing n beyond 4 does not yield a significant improvement in the estimation performance. In the light of this result, we recommend two combinations for parameter setting:

1. 1ENN for k, 4ENN for t_S
2. 4ENN for k, 4ENN for t_S

as 1ENN and 4ENN are likewise suited for k estimation. Both methods and their SSIM results are compared in the next section.

4.2   Denoising with Estimated Parameters

After evaluating the estimated parameters in contrast to the optimised parameters, we shall now proceed to investigate the effects of estimation errors on the image denoising

Table 1. Statistical evaluation of k estimation methods

Quantity	1ENN	4ENN	1NN	4NN	M5P	Shao
average absolute error	11.15	11.19	17.55	21.6	51.45	35.57
average relative error	0.18	0.19	0.36	0.48	0.8	0.58
absolute error span	0-30	2-27	3-79	2-91	14-95	19-60
relative error span	0-0.3	0-0.39	0-1.6	0-2.9	0.65-1	0.53-0.65

Table 2. Statistical evaluation of t_S estimation methods

Quantity	1ENN	4ENN	1NN	4NN	M5P
average absolute error	5.3	4.5	7.7	5.79	5.67
average relative error	0.52	0.39	0.71	0.52	0.19
absolute error span	0-24	0-13	0-27	0-24	2-44
relative error span	0-3.5	0-1.43	0-3.5	0-1.4	0.34-0.56

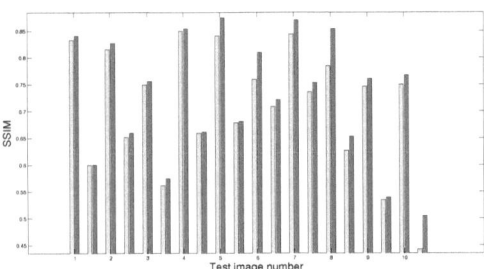

(a) 1ENN for k, 4ENN for t_S

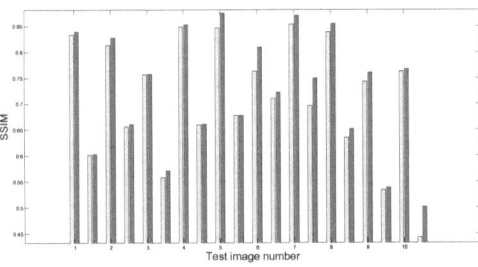

(b) 4ENN for both k and t_S

Fig. 4. SSIM results for denoising with 2 parameter estimation methods. For each test images, we present two bar groups: One for variance 0.005 and one for variance 0.05, where in turn each group consists two bars, denoting the result of the application of the estimated and optimal parameters.

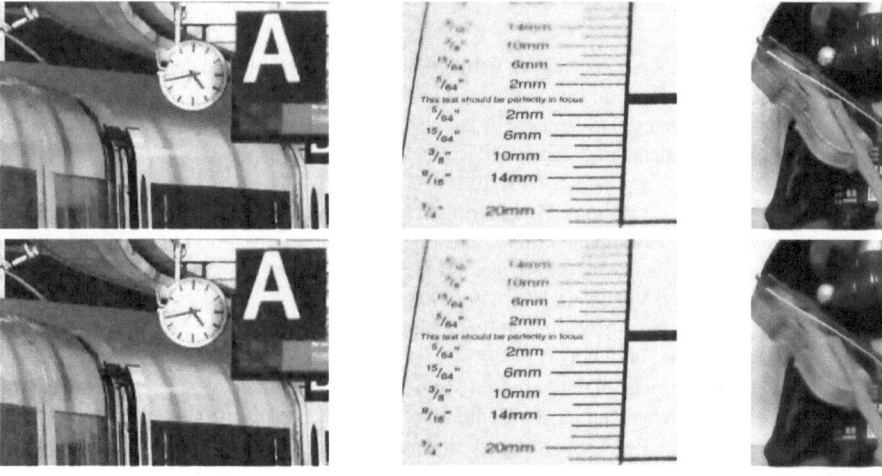

Fig. 5. Testing images #8,#7,#5: two typical cases and #5 as worst case, i.e. worst estimation. Top row: optimal parameters, bottom row: estimated parameters.

process. As mentioned above, the denoising success is measured by an increase of the denoised image's SSIM. When looking at bar charts Fig. 4 (a) for method 1 and (b) for method 2, we can see that the errors from estimation result in only small SSIM deficits. Altogether, the estimated values provided almost equal denoising as the optimised parameters. It is obvious that higher estimation errors lead to higher SSIM deficits, but for a quick estimation, our methods perform well. If one wants to optimise the SSIM, our estimates can be used as bounds for reducing the parameter search space by over 70%.

5 Conclusion

This paper has investigated the use of different machine learning techniques for Perona and Malik parameter estimation. An entropy-based distance measure was developed and used for estimation, which resulted in parameter combinations that led to denoised images of a quality comparable to the image denoised with optimal parameters. Different estimation methods were compared, with 4ENN winning the overall comparison. Its underlying database can be gradually improved by adding new images, making it possible to create even better approximations. The methods itself can also be applied to other denoising processes.

Acknowledgements. We thank Annemarie Sonnenberg for the clearance to use test pictures #1 and #5 as well as flickr users joachim_s_mueller, hotgear, mitch-in-wanderlust, sigsegv, 34053291@N05 and tuxdriver. We also thank Oren Halvani from CASED for his help in refining the manuscript.

References

1. Gilboa, G., Sochen, N.A., Zeevi, Y.Y.: Estimation of optimal pde-based denoising in the SNR sense. IEEE Transactions on Image Processing 15(8), 2269–2280 (2006)
2. Guo, Z., Sun, J., Zhang, D., Wu, B.: Adaptive Perona-Malik model based on the variable exponent for image denoising. IEEE TIP 21(3), 958–967 (2012), http://dx.doi.org/10.1109/TIP.2011.2169272
3. Halland, M., Frank, E., Holmes, G., Pfahringer, B., Reutemann, P., Witten, I.H.: The WEKA data mining software: An update. In: ACM SIGKDD Explorations, vol. 11, pp. 10–18 (2009)
4. Kichenassamy, S.: The Perona-Malik method as an edge pruning algorithm. Journal of Mathematical Imaging and Vision 30, 209–219 (2008)
5. Kuijper, A.: p-Laplacian driven image processing. In: 14th International Conference on Image Processing, ICIP 2007, vol. V, pp. 257–260 (2007)
6. Kuijper, A.: Geometrical PDEs based on second order derivatives of gauge coordinates in image processing. Image and Vision Computing 27(8), 1023–1034 (2009)
7. Monteil, J., Beghdadi, A.: A new interpretation and improvement of the nonlinear anisotropic diffusion for image enhancement. IEEE TPAMI 21(9), 940–946 (1999)
8. Mrázek, P., Navara, M.: Selection of optimal stopping time for nonlinear diffusion filtering. International Journal of Computer Vision 52(2-3), 189–203 (2003)
9. Ndajah, P., Kikuchi, H., Yukawa, M., Watanabe, H., Muramatsu, S.: An investigation on the quality of denoised images. Circuits, Systems and Signal Processing 5, 423–434 (2011)
10. Perona, P., Malik, J.: Scale-space and edge detection using anisotropic diffusion. IEEE Transactions on Pattern Analysis and Machine Intelligence 12, 629–639 (1990)
11. Quinlan, J.R.: Learning with continuous classes. In: Proceedings of the Australian Joint Conference on Artificial Intelligence, pp. 343–348 (1992)
12. Rudin, L.I., Osher, S., Fatemi, E.: Nonlinear total variation based noise removal algorithms. Physica D 60(259-268) (1992)
13. Schwarzkopf, A., Kalbe, T., Bajaj, C., Kuijper, A., Goesele, M.: Volumetric nonlinear anisotropic diffusion on gpus. In: Bruckstein, A.M., ter Haar Romeny, B.M., Bronstein, A.M., Bronstein, M.M. (eds.) SSVM 2011. LNCS, vol. 6667, pp. 62–73. Springer, Heidelberg (2012)
14. Shao, H., Zou, H.: Threshold estimation based on Perona-Malik model. In: Int. Conf. on Computational Intelligence and Software Engineering, pp. 1–4 (2009)
15. Thuerck, D., Kuijper, A.: Cosine-driven non-linear denoising. In: Kamel, M., Campilho, A. (eds.) ICIAR 2013. LNCS, vol. 7950, pp. 245–254. Springer, Heidelberg (2013)
16. Voci, F., Eiho, S., Sugimoto, N., Sekibuchi, H.: Estimating the gradient in the Perona-Malik equation. IEEE Signal Processing Magazine 21(3), 39–65 (2004), http://dx.doi.org/10.1109/MSP.2004.1296541
17. Weickert, J.: Coherence-enhancing diffusion of colour images. Image Vision Comput. 17(3-4), 201–212 (1999)

Analysis of WD Face Dictionary for Sparse Coding Based Face Recognition

Shejin Thavalengal and Anil Kumar Sao

School of Computing and Electrical Engineering
Indian Institute of Technology Mandi, India
s_t@students.iitmandi.ac.in, anil@iitmandi.ac.in

Abstract. This paper deals with the analysis of WD Face dictionary for sparse coding based face recognition. WD (weighted decomposition) Face dictionary emphasizes subject specific unique information of a person. This dictionary has an advantage to adapt to the nature of training images. In the resultant dictionary rows are uncorrelated, which is an essential criterion for dictionary to ensure sparse representation of coefficient vector. The range of sparsity determined by calculating the lower and upper bounds of sparse recovery of coefficient vector for WD Face dictionary exhibits its capability to sparsely represent a test image as a linear combination of training images, even when available training images are small in number. Experimental results solidify our proposal that sparse coding based face recognition with WD Face dictionary is preferable to the existing face recognition techniques.

Keywords: Sparse coding, dictionary, face recognition.

1   Introduction

Face recognition stands as one of the most interesting and promising area in computer vision and pattern recognition, inspired by the ability of human being to recognize and discriminate the faces of his fellow people. Automatic face recognition has been studied extensively for the last two decades owing to its significance in biometrics and other applications [20]. Recently sparse coding based approaches have been introduced and studied in face recognition [5, 18]. These methods allow to represent a test face image of a person ($\mathbf{y}_i$) as a linear combination of the training images of the same person, as

$$\mathbf{y}_i = \sum_{j=1}^{n_i} a_{i,j} \mathbf{v}_{i,j}, \tag{1}$$

where n_i is the number of training images available for the ith person, $\mathbf{v}_{i,j} \in \mathbb{R}^m$ is the jth training face image of the same person and corresponding coefficient is denoted by $a_{i,j}$. This equation can be rewritten as

$$\mathbf{y} = \mathbf{D}\mathbf{a}, \tag{2}$$

where $\mathbf{D} \in \mathbb{R}^{m \times n}$ denotes dictionary (sensing matrix), which contains all the available n training face images and $\mathbf{a} \in \mathbb{R}^n$ is denoted as coefficient vector, which contains zero

A. Petrosino (Ed.): ICIAP 2013, Part I, LNCS 8156, pp. 221–230, 2013.

corresponding to those face images in dictionary which do not belong to the same class of the test face image.

The choice of dictionary **D** plays a crucial role in any application in sparse coding framework. The available dictionary in literature can be classified in to two classes namely, (i) pre-constructed dictionaries, such as wavelets, random matrix and down sampling matrix and (ii) empirically-learned dictionary such as KSVD, MOD and WD Face dictionaries [7, 16]. The pre-constructed dictionaries are generic and do not exploit the nature of the signal, while empirically learned dictionaries can adapt to the family of signals.

Extensive details and information regarding sparse coding based face recognition and recovery of coefficient vector **a** can be found in [2, 5, 16, 18]. Reference [18] has shown that dictionary is not significant if very large number of training images are available. Assuming there are sufficient images available, this method uses severely downsampled face images (DS Face) or face images projected to a random matrix (Randomface) as dictionary. This work is extended to incorporate mis-alignment in reference [17]. Reference [5] attempted to overcome the issue of requirement of large number of training samples by generating an intra-class variant dictionary. This requires prior knowledge of all variations which may be present during face recognition and generation of all possible intra-class variant images. In reference [16] it is shown that when a very large number of training images is not available, dictionaries generated using DS Face and Randomface fail to deliver a good face recognition performance. WD Face dictionary was proposed in reference [16] to overcome this problem by emphasizing subject specific unique information of a person, which helps in discriminating face images of several people. The work presented in this paper focuses on demonstrating the characteristics of WD Face dictionary and how well this dictionary ensures recovery of sparse coefficient vector **a** via l_1 minimization.

The experiments carried out in this paper follow the algorithm described in [16]. Results are demonstrated using Extended Yale Face Database B [9], which contains 2432 images of 38 human subjects under 64 illumination conditions [13]. All images are resized to 100×100 prior to any operation. Experiments were conducted using four sets of training face images consisting of 1216 (32 face images per person), 570 (15 images per person), 380 (10 images per person), and 190 (5 images per person) face images. Half of the database is randomly chosen to generate first set of training images (1216 images, 32 images per person). The successive sets are randomly chosen subset of their predecessor. This approach is designed to remove the dependency of results on types of face images to maintain uniform nature in experiments.

Rest of the paper is organized as follows: Section 2 explains the necessary conditions to be satisfied by the dictionary (**D**) for good estimation of sparse coefficient vector (**a**). The deterministic construction of dictionary for face recognition which ensures sparse signal recovery is explained in Section 3. Evaluation of performance of WD Face dictionary is explained in Section 4, and Section 5 summarizes the paper.

2   Conditions for Sparse Signal Recovery

The fundamental concern in sparse coding based face recognition is how well the sensing matrix (dictionary) **D** is created. In reference [16] it was shown that it is necessary to create a dictionary which emphasizes the subject specific unique information of a person. Moreover, the dictionary $\mathbf{D}$ should ensure the reconstruction of sparse coefficient vector $\mathbf{a}$. Restricted Isometry Property (RIP) is a sufficient condition for unique and sparse reconstruction of $\mathbf{a}$ via l_1 minimization [3]. The dictionary $\mathbf{D} \in \mathbb{R}^{m \times n}$ follows RIP with parameters δ and $2s$, for all $\mathbf{a} \in \mathbb{R}^n$, if

$$\sqrt{1 - \delta}\|\mathbf{a}\|_2 \le \|\mathbf{Da}\|_2 \le \sqrt{1 + \delta}\|\mathbf{a}\|_2, \tag{3}$$

where $\delta \in (0, 1)$ and $s \in \mathbb{N}$. If $\delta < \sqrt{2} - 1$, then one can define upper bound on sparse recovery as explained in [3].

$$\|\bar{\mathbf{a}} - \mathbf{a}\|_1 \le 2(1 - \rho)^{-1}\left[\alpha\varepsilon\sqrt{s} + (1 + \rho)\|\mathbf{a} - \mathbf{a}^s\|_1\right], \tag{4}$$

where $\bar{\mathbf{a}} \in \mathbb{R}^n$ is the coefficient vector obtained via l_1 minimization, $\rho = \frac{\sqrt{2}\delta}{1-\delta}$, $\alpha = \frac{2\sqrt{1-\delta}}{1-\delta}$, ε is reconstruction error and $\mathbf{a}^s \in \mathbb{R}^n$ is a vector with largest s absolute values of $\mathbf{a}$ and remaining elements as zero. But it is an NP hard problem to check whether a given matrix satisfies RIP condition. In this context, reference [11] derived a verifiable condition for sparse recovery of a given sensing matrix. A parameter $\hat{\gamma}_s(\mathbf{D}, \beta)$ is defined as

$$\hat{\gamma}_s(\mathbf{D}, \beta) = \max_a\left[\|\mathbf{a}\|_{s,1} - \beta\|\mathbf{Da}\|_2 : \|\mathbf{a}\|_1 \le 1\right], \tag{5}$$

where $0 \le \beta \le \infty$ and $\|\mathbf{a}\|_{s,1}$ is the sum of largest s absolute values of $\mathbf{a}$. If $\hat{\gamma}_s(\mathbf{D}, \beta) \le \frac{1}{2}$ and $\beta \le \infty$, then the upper bound on sparse recovery is similar to the one obtained using equation (4), and written as in [11].

$$\|\bar{\mathbf{a}} - \mathbf{a}\|_1 \le (1 - 2\hat{\gamma}_s(\mathbf{D}, \beta))^{-1}[2\beta\varepsilon + 2\|\mathbf{a} - \mathbf{a}^s\|_1 + v], \tag{6}$$

where ε denotes reconstruction error and v is the inaccuracy in solving equation (6).

This method allows to obtain upper bound and lower bound of sparsity of the given dictionary. The lower bound is computed using linear programming of $(2n^2+n) \times (n(m+n+1))$ constrained matrix and upper bound is computed from lower bound by sequential convex approximation [11].

3   Analysis of WD Face Dictionary

The transformed face images such as Down-Sampling face (DS Face) where the face image is downsampled to 15×12 or Randomfaces where a face image is projected in to a random matrix of size $350 \times m$ whose values are drawn from independent and identically distributed Gaussian (where the image of size $\sqrt{m} \times \sqrt{m}$ is converted to a vector of $m \times 1$ before projection) can be used as dictionary for sparse coding based face recognition [18]. Even though the above mentioned random projection guarantees sparse recovery with high probability [4, 6], its performance in face recognition is not good since Randomfaces does not emphasize subject specific unique information of face image [16].

Dictionary derived using Weighted Decomposition face (WD Face) representation is given as

$$\hat{\mathbf{D}} = \mathbf{W}\mathbf{\Psi}^T\mathbf{D}, \tag{7}$$

where $\hat{\mathbf{D}}$ is the derived dictionary, $\mathbf{W}$ is a weight matrix and $\mathbf{\Psi}$ is the matrix consisting of eigenvectors obtained from the eigen analysis of the set of training images. WD Face is generated by assigning proper weights to those underlying components of a face image which emphasize subject specific unique information of the person. A physical inter-

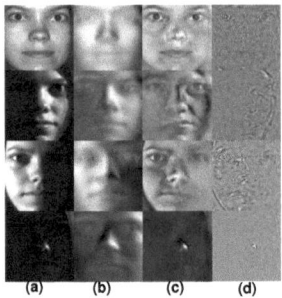

Fig. 1. Illustration of face decomposition in to different components. (a) Gray level image. The reconstructed face image using (b) first 10 eigenvectors, (c) 11-350 eigenvectors, (d) 351-remaining eigenvectors.

pretation of WD Face is given based on the assumption that a face image can be decomposed into three components [12] as shown in Fig. 1. The first component (Fig. 1(b)) corresponds to a component which is common to all the face images. The third component (Fig. 1(d)) can be considered as a noise component which contains negligible information about the identity of the person. The second component (Fig. 1(c)) carries all the person specific unique information which accounts for the discrimination of a person from the other. These components are derived from the eigenvectors of the covariance matrix of the dictionary $\mathbf{D}$. These eigenvectors are arranged according to the descending order of corresponding eigenvalues. Fig. 1(b) depicts the reconstruction of Fig. 1(a) using 10 eigenvectors which corresponds to the 10 largest eigenvalues. In a similar manner Fig. 1(c) and Fig. 1(d) are reconstructed using 11-350 eigenvectors and 351-10000 eigenvectors, respectively. WD Faces are generated by giving higher weightage to the component shown in Fig. 1(c). This can be done with the weight matrix $\mathbf{W} = \mathbf{\Lambda}^{-1/2}$ where $\mathbf{\Lambda}$ is a diagonal matrix containing the eigenvalues corresponding to the eigenvector matrix $\mathbf{\Psi}$. One has to take care of the situation where eigenvalues are very small (close to zero), which results in infinitely large weights for the less significant eigen components. This can be done by removing the least significant eigen components with the help of a thresholding operator. The eigen analysis incorporated in the computation of WD Face dicionary requires all the training images to be well-aligned [15].

Apart from just giving higher weights to more discriminative features of face image, this transformation projects the face images to an orthogonal subspace. This transformation take advantage of the nature of signal. Moreover, we can see that

$$E\{\hat{\mathbf{D}}\hat{\mathbf{D}}^T\} = \mathbf{W}\mathbf{\Psi}^T E\{\mathbf{D}\mathbf{D}^T\}(\mathbf{W}\mathbf{\Psi}^T)^T, \tag{8}$$

where $E\{.\}$ denotes the expectation operator. We can decompose $E\{\mathbf{D}\mathbf{D}^T\}$, which is the covariance matrix of all the training face images, in terms of eigenvectors and eigenvalues as $\mathbf{\Psi}\mathbf{\Lambda}\mathbf{\Psi}^T$. Then equation (8) can be rewritten as

$$E\{\hat{\mathbf{D}}\hat{\mathbf{D}}^T\} = \mathbf{W}\mathbf{\Psi}^T\mathbf{\Psi}\mathbf{\Lambda}\mathbf{\Psi}^T\mathbf{\Psi}\mathbf{W}^T$$
$$= \mathbf{\Lambda}^{-\frac{1}{2}T}\mathbf{\Psi}^T\mathbf{\Psi}\mathbf{\Lambda}\mathbf{\Psi}^T\mathbf{\Psi}\mathbf{\Lambda}^{-\frac{1}{2}T}$$
$$= \mathbf{I}, \tag{9}$$

where $\mathbf{I}$ is an identity matrix. That is, WD transform makes the rows of derived dictionary uncorrelated [10], which is one of the criteria for dictionary to ensure the sparse recovery of coefficient vector. One can observe the advantage of WD Face representation from Fig. 2.

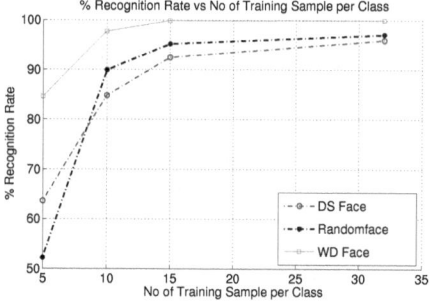

Fig. 2. Face recognition performance for different dictionaries on Extended Yale Face Database B as a function of number of training images per class

In addition, lower and upper bounds on sparse recovery (equations (5) and (6)) can be used to analyze the statistical RIP to check the efficiency of dictionary. The lower and upper bounds computed in this fashion are shown in Fig. 3. It has to be noted that the bounds are not calculated for the case where half of the data is used for training and remaining for testing (32 images per class, total 1216 Images for training), since it will result in solving linear programming of $(2n^2 + n) \times (n(m + n + 1)) = 5.6373 \times 10^{12}$ constrained matrix which is computationally intensive. Fig. 3(a) compares the lower bound for different dictionaries generated for sparse coding based face recognition. In the worst case scenario, where there are just five training images per class, the dictionaries generated using DS Face and Randomfaces return a lower bound of 10 and 17

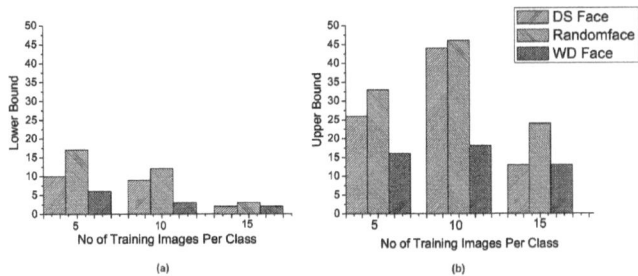

Fig. 3. Illustration of (a) lower bound and (b) upper bound on sparse recovery for different dictionaries as function of number of training images per class

respectively. It is higher than the number of actual available training images per class. Hence, these representations do not ensure a sparse recovery of the coefficient vector **a** with five coefficients, which explains their poor performance (Fig. 2). On the other hand, dictionary generated using WD Face provides a lower bound of 6 which is very close to the number of training images available for each class. Hence the probability to recover sparse coefficient vector **a** is higher as compared to vectors generated using DS Face and Randomface, which justifies the better performance of face recognition using WD Face based dictionaries. We can notice that as the number of training samples increases, the lower bound of sparsity is decreasing for all the dictionaries. This is because of the availability of new training faces which can represent the corresponding class efficiently. When number of training samples reaches 15 or more images per class, the lower bound of sparsity remains more or less the same for all the dictionaries, which explains the claim of reference [18] that when 'sufficient' training images are available, dictionary is no more significant.

Fig. 3(b) compares the upper bound of sparsity derived for the different dictionaries. The lower bound and upper bound together define the range of sparsity for the given dictionaries. It is proposed that the upper bound should not be very large as compared to the lower bound, and hence the range of sparsity. As the range of sparsity is small, it will help the recovery algorithm to ensure recovery of sparse coefficient vector **a**. Here, in the case of WD Face dictionary, we can note that the upper bound is 15 and lower bound is 6 for 5 training images per person. While, upper bound and lower bounds for DS Face dictionary are 26 and 10; and for Randomface dictionary are 32 and 16 respectively. Hence, the range of sparsity for WD Face dictionary is smaller as compared to the that for DS Face dictionary and Randomface dictionary. Also, as the number of training samples increases, the range of sparsity becomes narrower for all the three dictionaries because of the availability of new training faces which can represent the corresponding class efficiently. From Fig. 3, we can conclude that the dictionary generated using WD Faces is superior over its counterparts in sparse coding based face recognition, since it ensures the sparse recovery of the coefficient vector even when 'sufficient' training samples are not available and the difference between upper bound and lower bound is small as compared to its counterparts.

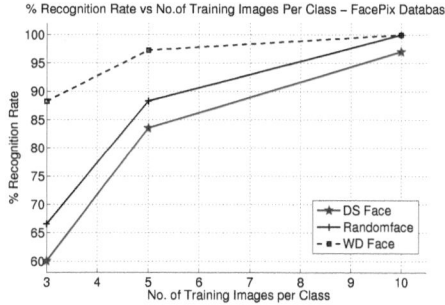

Fig. 4. Face recognition performance for different dictionaries on CUbiC FacePix Database as a function of number of training images per class

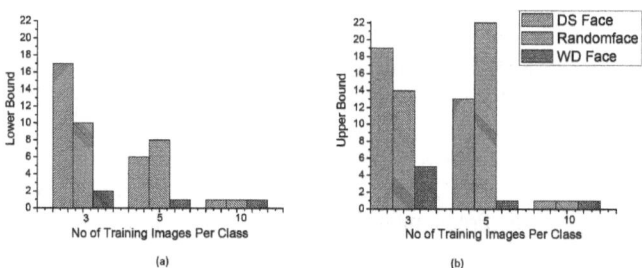

Fig. 5. Illustration of (a) lower bound and (b) upper bound on sparse recovery for different dictionaries as function of number of training images per class on CUbiC FacePix Database

4   Experiments and Results

In addition, analysis of WD Face dictionary is carried out on CUbiC FacePix database [1, 14]. A subset of CUbiC FacePix database corresponding to frontal face images captured with a point light source at various illumination angles without an ambient light source is used for our experiments. This subset consists of a total number of 5430 images of 30 human subjects. Three sets of training images are chosen in the same manner as explained in the previous section. Performance of various dictionaries is compared in Fig. 4. The lower bound and upper bound for these dictionaries are shown in Fig. 5. The results validate our claim that dictionary created using WD Face is superior to Randomfaces and DS Faces especially when the number of training images is very small.

In order to compare the results obtained by WD Face dictionary in sparse coding based face recognition with other studies on these databases, we extended our experiments as in references [8] and [1]. Table 1 compares the performance of WD Face dictionary with different classifiers as in [18] and [8]. These experiments were carried out in Extended Yale Face Database B, where half of the database was used for training and remaining half for testing. Face recognition performance of different methods with feature space dimensions 30, 56, 120, and 504, respectively is shown. We have to note that maximum number of valid Fisher faces is $k - 1$, where k refers total number of

Table 1. Face Recognition Results for Different Methods in Extended Yale Face Database B

Dimension	30	56	120	504
Eigen+NS	89.9%	91.1%	92.5%	93.2%
Eigen+SVM	70.6%	84.3%	93.1%	96.8%
Eigen+CRC [19]	67.97%	85.62%	94.41%	98.55%
Eigen+KSR-Polynomial [8]	91.74%	95.75%	97.42%	98.28%
Eigen+KSR-Gaussian [8]	89.01%	94.42%	97.49%	99.16%
Laplacian+NS	89.0%	90.4%	91.9%	93.4%
Laplacian+SVM	72.0%	85.0%	94.0%	97.7%
Laplacian+CRC [19]	72.25%	86.89%	94.54%	97.91%
Laplacian+KSR-Polynomial [8]	**92.25%**	95.14%	97.07%	99.11%
Laplacian+KSR-Gaussian [8]	88.86%	94.24%	97.11%	98.12%
DS Face+NS	80.8%	88.2%	91.1%	93.4%
DS Face+SVM	48.9%	69.5%	79.0%	91.6%
DS Face+CRC [19]	67.97%	82.84%	92.93%	97.23%
DS Face+KSR-Polynomial [8]	87.20%	93.09%	95.81%	97.47%
DS Face+KSR-Gaussian [8]	83.57%	91.65%	95.31%	97.80%
Fisher+NS	81.9%	NA	NA	NA
Fisher+SVM	86.7%	NA	NA	NA
Fisher+CRC [19]	68.10%	NA	NA	NA
Fisher+KSR-Polynomial [8]	91.74%	NA	NA	NA
Fisher+KSR-Gaussian [8]	88.93%	NA	NA	NA
WD Face+Sparse Coding	88.90%	**96.05%**	**98.44%**	**99.91%**

classes in the database, which is 38 in Extended Yale Face Database B. Hence recognition performance for the methods using Fisher face is limited to 30 dimensions. It can be noted that, sparse coding based face recognition, which incorporates the proposed WD Face dictionary in sparse coding based face recognition consistently outperforms all the other approaches (even the most recent kernel sparse representation method [8]) for all the dimensions except for the case of 30 dimensions. The drop-off in performance of WD Face dictionary may be because 30 dimensions won't be sufficient to represent person specific unique information.

Table 2. Face Recognition Results for Different Methods in FacePix Database

Number of Training Images	3	5
PCA	71.71%	90.33%
LDA	79.52%	94.92%
HMM	37.38%	59.37%
BIC	79.10%	93.54%
WD Face+Sparse Coding	**93.18%**	**98.24%**

Similarly, Table 2 shows the comparison of our method against existing approaches in CUbiC FacePix database as in [1]. Two different sizes of training image set are used; the first one contains 3 images per class for training (total 90 images) and the second set contains 5 images per class for training (total 150 images). We can observe from the table that, in both cases sparse coding based face recognition with WD Face dictionary significantly outperforms the other methods.

5   Summary

In this paper we analyzed the properties of WD Face dictionary for sparse coding based face recognition. WD Face dictionary, which is generated using WD face representation reflects person specific unique information, which is critical in face recognition. The WD Face transform makes the rows of the derived dictionary uncorrelated. Moreover, WD Face dictionary has the ability to adapt to the training face data. These properties of the dictionary make it a potential candidate for use in sparse coding based face recognition. Upper and lower bounds for sparsity computed experimentally serve to strengthen our proposal that a dictionary generated in this fashion ensures sparse representation of test image as a linear combination of training images even when training images are very small in number. Further experiments strengthen the proposal that sparse coding based face recognition using WD Face dictionary is favourable as compared to the existing approaches.

References

1. Black Jr., J.A., Gargesha, M., Kahol, K., Kuchi, P., Panchanathan, S.: A framework for performance evaluation of face recognition algorithms. ITCOM, Internet Multimedia Systems II, 163–174 (July 2002)
2. Candes, E., Romberg, J., Tao, T.: Robust uncertainty principles: exact signal reconstruction from highly incomplete frequency information. IEEE Trans. on Information Theory 52(2), 489–509 (2006)
3. Candes, E.J.: The restricted isometry property and its implications for compressed sensing. Comptes Rendus Mathematique 346(910), 589–592 (2008)
4. Candes, E.J., Romberg, J.K., Tao, T.: Stable signal recovery from incomplete and inaccurate measurements. Communications on Pure and Applied Mathematics 59(8), 1207–1223 (2006)
5. Deng, W., Hu, J., Guo, J.: Extended SRC: Undersampled face recognition via intraclass variant dictionary. IEEE Trans. on Pattern Analysis and Machine Intelligence 34(9), 1864–1870 (2012)
6. Donoho, D.L.: Compressed sensing. IEEE Trans. on Information Theory 52(4), 1289–1306 (2006)
7. Elad, M.: Sparse and Redundant Representations - From Theory to Applications in Signal and Image Processing. Springer, New York (2010)
8. Gao, S., Tsang, I.W.-H., Chia, L.-T.: Sparse representation with kernels. IEEE Trans. on Image Processing 22(2), 423–434 (2013)
9. Georghiades, A., Belhumeur, P., Kriegman, D.: From few to many: Illumination cone models for face recognition under variable lighting and pose. IEEE Trans. on Pattern Analysis and Machine Intelligence 23(6), 643–660 (2001)

10. Hyvärinen, A., Karhunen, J., Oja, E.: Independent Component Analysis. Wiley (2001)
11. Juditsky, A., Nemirovski, A.: On verifiable sufficient conditions for sparse signal recovery via l_1 minimization. Mathematical Programming 127, 57–88 (2011)
12. Sao, A.K., Yegnanarayana, B.: Analytic phase-based representation for face recognition. In: Seventh International Conference on Advances in Pattern Recognition, Kolkata, India, pp. 453–456 (February 2009)
13. Lee, K., Ho, J., Kriegman, D.: Acquiring linear subspaces for face recognition under variable lighting. IEEE Trans. on Pattern Analysis and Machine Intelligence 27(5), 684–698 (2005)
14. Little, D., Krishna, S., Black, J., Panchanathan, S.: A methodology for evaluating robustness of face recognition algorithms with respect to variations in pose angle and illumination angle. In: IEEE International Conference on Acoustics, Speech, and Signal Processing, Philadelphia, USA, vol. 2, pp. 89–92 (March 2005)
15. Shejin, T., Sao, A.K.: Dictionary for sparse coding based pose invariant face recognition. unpublished
16. Shejin, T., Sao, A.K.: Significance of dictionary for sparse coding based face recognition. In: 11th International Conference of the Biometrics Special Interest Group, pp. 1–6 (2012)
17. Wagner, A., Wright, J., Ganesh, A., Zhou, Z., Mobahi, H., Ma, Y.: Towards a practical face recognition system: Robust alignment and illumination by sparse representation. IEEE Trans. on Pattern Analysis and Machine Intelligence 34(2), 372–386 (2012)
18. Wright, J., Yang, A., Ganesh, A., Sastry, S., Ma, Y.: Robust face recognition via sparse representation. IEEE Trans. on Pattern Analysis and Machine Intelligence 31(2), 210–227 (2009)
19. Zhang, L., Yang, M., Feng, X.: Sparse representation or collaborative representation: Which helps face recognition? In: 2011 IEEE International Conference on Computer Vision, Barcelona, Spain, pp. 471–478 (November 2011)
20. Zhao, W., Chellappa, R., Phillips, P.J., Rosenfeld, A.: Face recognition: A literature survey. ACM Computer Survey 35(4), 399–458 (2003)

Fast and Robust Edge-Guided Exemplar-Based Image Inpainting

Yun Wu and Chun Yuan

Tsinghua-CUHK Joint Research Center for Media Sciences, Technologies
and Systems Graduate School at Shenzhen, Tsinghua University

Abstract. A fast and robust edge-guide exemplar-based method of image inpainting is proposed in this paper. Unlike traditional exemplar-based methods, we introduce an edge-reconstruction procedure before inpainting textures. The edge reconstruction procedure exploits different properties of edges, such as the curvature similarity, color similarity, and other estimate of how well two edges connect to each other. Guided by the reconstructed edge lines, the improved exemplar-based method is used to restore the textures and remaining structures. Moreover, we redesign the random search strategy to make it more suitable for our framework to solve the time-consuming problem caused by exhaustive search in most exemplar-based methods. After the match patch is chosen, color transfer is used to propagate the match patch information to further improve the visual quality and perceptual reasonability. Comprehensive experiments are performed to compare the proposed method with other well-known methods on synthetic images and natural images. The results show the proposed method can not only greatly reduce the computing time of exemplar-based methods, but also behave better on visual plausibility and continuity.

Keywords: image inpainting, edge, exemplar-based, PatchMatch, image completion.

1 Introduction

Image inpainting is this kind of technology whose goal is to remove some objects or restore the missing information from the source image in a way the observer can not notice the artificial trace.

In the literature, most inpainting methods can be categorized into two types, exem-plar-based and diffusion-based. The diffusion-based method is often called PDE (partial differential equation)-based method because the inpainting image is often modeled as a PDE. These methods source from the heat diffusion, it propagates the information of the known area into the missing area along some direction. Bertalmio et al. [1] took the isophote direction as the propagating direction, and iteratively propagated the Laplace smooth information into the missing region until the image didn't change. Chan et al. [2] developed the Total Variation (TV) model for the image inpainting. To solve the discontinuity

A. Petrosino (Ed.): ICIAP 2013, Part I, LNCS 8156, pp. 231–240, 2013.

problem in the TV model, curvature-driven diffusion (CDD) [3] model was introduced. These diffusion-based methods can deal the image of small damaged area without texture very well, but when it comes to the image of large damaged area or area with textures, they often result in some blurs.

On the contrary, the exemplar-based method can handle the image of large dam-aged area with texture well. The exemplar-based method copies the information from the known region by finding the most similar patch. However, directly copying and pasting the known information can't get a plausible result. Criminisi et al. [4] devel-oped the patch priority, which is determined by the flux of the isophote flowing into the inpainting contour. Using the patch priority to determine the filling order, the texture information and the linear structure information can be well reconstructed. Several variants were proposed after article [4]. Cheng [5] reformed the priority func-tion of article [4] to get more robust performance, Tae-o-sot et al. [6] introduced the patch shifting scheme. However, most of these exemplar-methods still result in global structure inconsistency and use exhaustive search to find the best match patch which is very time-consuming. These two points are the main points we focus on in this paper.

As to the time cost, Kwok et al. [7] used the DCT to decompose the patches into frequency coefficients to speed up the process of finding the best match patch. Barnes et al. [8] developed the PatchMatch method, which can effectively reduce the heavy time cost. Tae-o-sot et al. [9] redesigned the PatchMatch method to make it suit to their framework. But these acceleration algorithms don't consider about the inconsistency problem of the global structures. As we know, only article [10] did similar work as us, but they still use the method in article [4] without any change to fill the texture, that leads huge time cost and some visual discontinuity.

In this paper, we proposed a fast and robust edge-guided exemplar-based image inpainting method. To solve the global structure inconsistency problem, we introduce an edge-reconstruction procedure into the exemplar-based method. Exploiting differ-ent properties of edges, the fractured global edge lines are well connected. Guided by the reconstructed edge lines, the improved exemplar-based method is used to restore the textures and remaining structures. Moreover, we redesign the random search strategy to solve the time-consuming problems caused by exhaustive search. To fur-ther improve the visual quality and perceptual rea-sonability, we use color transfer to propagate the match patch information to the target patch. Comprehensive experi-ments are performed to compare the proposed method with other well-known meth-ods on synthetic images and natural images. The results show the proposed method can gain promising improvements both on visual plausibility and computation time.

The rest of the paper is organized as follows. In section 2, we introduce edge re-construction procedure in detail. The redesigned exemplar-based method is described in section 3. The comparison results over the typical inpainting method are presented in the section 4. We conclude this paper in section 5.

2   Edge Reconstruction Procedure

2.1   Overview of the Proposed Method

The proposed method mainly includes two important procedures, the global structure restoration procedure and the texture inpainting procedure. Given an image with a missing area, we first generate the edge line using the segmentation and edge detec-tion algorithm. As shown in Fig. 1(b), $L_1L_2...L_m$ are the generated interceptive edges. By analyzing the local and global features of these lines, the missing lines in the miss-ing region are reconstructed by our edge reconstruction method. Then by using the restored lines as guideline, the information near the restored lines is first inpainted by our improved exemplar-based method. Finally, the left region is inpainted by the exemplar-based method. Figure 1 shows the procedure demonstration of the proposed algorithm. The following will describe the first procedure-edge reconstruction in detail.

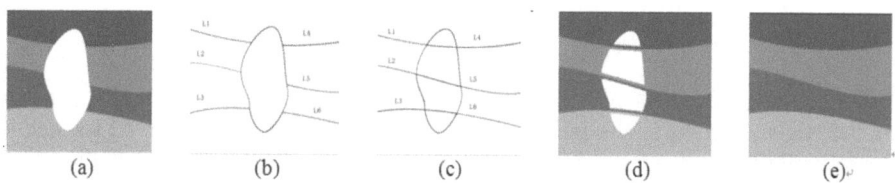

| | | | | |
| (a) | (b) | (c) | (d) | (e) |

Fig. 1. Overview of the proposed algorithm. (a) The source image with white mask. (b) Edge image after segmentation and edge detection. (c) Reconstructed edge-line image. (d) Intermediate result of inpainting guided by edge image. (e) Final result of the proposed method.

2.2   Edge Generation and Reconstruction

In Edge generationthe goal is to get the contour lines of the image, and item-ize the lines into each segmented part. In this paper, Mean Shift algorithm [11] is adopted to segment the source image because of its relatively fast speed and efficiency. As to the edge detection, we adopt the canny edge detection algorithm [12].

After getting the edge image, we extract all the edges hitting on the missing region contour. All these edges are defined as lines $L_1L_2...L_m$(e.g. Fig. 1(b)). Then we need to determine which two edge lines should connect to each other, which is the most im-portant part of edge reconstruction procedure. We define as the score to estimate how well two edge lines can connect to each other. The score is calculated accord-ing to the following equation:

$$C_{i,j} = CS_{i,j} + S_{i,j} + I_{i,j} + D_{i,j} \tag{1}$$

Where $CS_{i,j}$ represents the average curvature similarity of the pair of lines, $S_{i,j}$ repre-sents the direction consistency of the pair of lines, $I_{i,j}$ represents the color similarity surrounding the edge lines, $D_{i,j}$ represents the distance of the pair of lines. All the four parameters have values between 0 and 1.The better the two lines match, the closer the variants are to 1. The calculation details are described in the following.

1. The average curvature similarity $CS_{i,j}$:

$$CS_{i,j} = \left| \frac{\min(\text{average}(k_{i1}, k_{i2}...k_{in}), \text{average}(k_{j1}, k_{j2}...k_{jn}))}{\max(\text{average}(k_{i1}, k_{i2}...k_{in}), \text{average}(k_{j1}, k_{j2}...k_{jn}))} \right| \quad (2)$$

Where k_{it} represents the curvature t of point i on the edge line from the crossing point with the boundary to the other direction. n is the number of points to compute the average curvature on the edge line, we here set it as 5 to 10. When the average curva-tures of edge i and edge j are equivalent, the value is 1. The bigger difference the average curvatures have, the smaller the values is.

2. The color similarity surrounding the edge $I_{i,j}$:

$$I_{i,j} = \frac{\min(d_{1,2}, d_{1,3}...d_{2,3}...d_{m-1,m})}{d_{i,j}} \text{ with } d_{i,j} = \sum_{t=1}^{m}(patch_{it} - patch_{jt}) \quad (3)$$

Where the $d_{i,j}$ means the color difference of edge i and edge j, m is the number of the edges, $patch_{it}$ is the patch center on the point t on the edge i, its size is set to be 5 to 9. $patch_{it} - patch_{jt}$ is the sum of the corresponding color difference of the two patches. The smaller the difference is, the surrounding pixels of the edge are more alike, the bigger $I_{i,j}$ is.

3. The spatial distance of the two edges $D_{i,j}$

$$D_{i,j} = \frac{\min(dis_{1,2}, dis_{1,3}, ...dis_{2,3}, ...dis_{m-1,m})}{dis_{i,j}} \quad (4)$$

$$\text{with } dis_{i,j} = \sqrt{(p_{ix} - p_{jx})^2 + (p_{iy} - p_{jy})^2}$$

Where $dis_{i,j}$ is the coordinate distance of the cross point with the boundary of the edge i and edge j. The smaller the distance is, the bigger is.

As shown in Fig. 2(a), $L1L2L3L4$ have similar curvatures, and if textures surrounding these edges are also similar, then as the method in article [10] states, the spatial distance would decide which two edges should connect. Finally, the result of Fig. 2(b) is obtained. We can notice the unnaturalness very easily. So we introduce a new feature-direction consistency to solve the problem in Fig. 2(a). We first compute the distance of the centers of curvature circles. As shown in Fig. 2(c), the distance of $R1$ and $R4$ are much nearer, so we are more prone to connect $L1$ and $L4$, and so do with $L2$ and $L3$. The detail definitions are given in the following.

4. Direction Consistency $S_{i,j}$:

$$cdis_{i,j} = d(center_i, center_j), \ center_i = g(center_{i1}, center_{i2}...center_{in})$$

$$S_{i,j} = \frac{\min(cdis_{1,2}, cdis_{1,3}, ...cdis_{2,3}, ...cdis_{n-1,n})}{cdis_{i,j}} \quad (5)$$

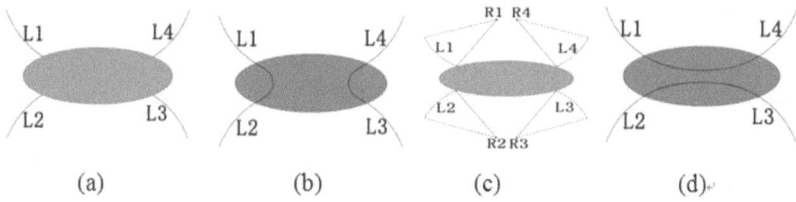

(a)	(b)	(c)	(d).

Fig. 2. (a) The situation method in article [10] can not deal with. (b) Unnatural result from method in article [10]. (c) Centers of curvature circles. (d) Natural result from the proposed method.

Where g returns the geometric center, $center_{ik}$ is the center of curvature circle of kth points on the edge i, n is the number of points, n is usually set to be 3 to 7, $cdis_{ij}$ is the coordinate distance of the two centers. The smaller the distance is, the bigger $S_{i,j}$ is.

After computing all the scores of the every pair of edge lines, we choose the pair with the highest score every time. Then we use $NURBS$ curve [13] to fit the two lines to connect the pair of edge lines. $NURBS$ curve is a powerful method to design curves. It is geometrically invariant and affine invariant. It is very convenient to joint two curves. Taking into consideration these advantages, we decide to use it to fit the broken edges.

Until all the lines are connected or all the pairs' scores are less than a threshold TH, we stop the connecting procedure. It would be appropriate to set the threshold TH as values between 0.3 and 1.5, we here set it as 0.7.

3   Improved Edge-Guided Exemplar-Based Inpainting Procedure

3.1   Redesigned Generalized Random Search Strategy

Here we first review the PatchMatch algorithm introduced by the Barnes [8], which is a very efficient method to solve problems like this: given an image I, for every square patch center on the pixel of I, find out the approximate best match patch in an image S, with SSD (sum-squared difference) between corresponding pixels as the measure.

The original PatchMatch algorithm includes three steps, initialization, propagation and random search. For patch T center on pixel (x, y), define $f : I \to R^2$ as the nearest neighbor function. Then $f(x, y)$ returns the corresponding patch offset of $T(x, y)$.

In the initialization step, all the patches are assigned to their corresponding patch randomly. Of course, you can adopt other strategies to initialize, such as basing on the prior knowledge or up-sampling the corresponding patch of the low resolution level.

The propagation procedure tries to improve the best match using the known best match above or the left. Let $f(x,y)$ denotes the patch difference between patch cen-ters on (x,y) and $(x,y)+v$. So the candidate best patch comes from $\min D(f(x,y)), D(f(x-1,y)+(1,0)), D(f(x,y-1)+(0,1))$,   Through   this propagation, if (x,y) has a good match, then all the patches below and right to (x,y) will have good matches. Besides, on the even iteration, scan the pixel on the reverse order, propagate the good match of down and right.

The propagation procedure converges very fast, but it may trap in the local mini-mal. So the random search strategy is used: candidate is sampled from an window randomly, and $f(x,y)$ is updated if any of the candidate has smaller distance. Let v_0 be the current best offset of $f(x,y)$. Then the candidate is randomly chosen by search around v_0 at an exponentially decrease radius: $u_i = v_0 + w\alpha^i R_i$, where w is the maximum search radius, α is the ratio decreasing the window size, R_i is a random in $[1,-1]\mathrm{x}[-1,1]$. The index i increases from 1 to n until the window size is less than 1.

Tae-o-sot et at. [9] changed the random search procedure. As the radius decreases, the search center no longer remains unchanged. Every time the radius decrease, it lets the previous best patch be the search center: $u_i = u_{i-1} + w\alpha^i R_i$. Obviously, this change makes it more accurately to find the better patch. But it brings another problem: if the final best patch is far from u_i, this change may end up with a local optimum, figure 3(a) demonstrates this situation. Suppose q_i is the match patch, and the smaller the index is, the better the patch matches. When $w=2$(the red box), if q_9 is not selected, for example, q_8 is selected, then the best match in the top right corner would be missed. But we here save k best match patches, then q_9 is very likely to be saved in the list, and the best match patch will be also more likely to be selected. The following is the details.

We save k best match patches $p_0, p_1, ...p_{k-1}$ in the search list. p_0 is initialized as v_0, then act as the best patch match, and p_{k-1} is the kth best match patch. Every time inserting one better patch into the list and delete the worst patch from the list, we changed the random search strategy as follows:

$$u_i = \min(p_0 + w\alpha^i R_i, p_1 + w\alpha^i R_i, ...p_{k-1} + w\alpha^i R_i) \qquad (6)$$

If u_i is better than p_k, we insert it into the list at appropriate position, and delete p_k from the list. That strategy makes the random search not be trapped into local optimum. And because the k is a constant and the most time-consuming procedure in the PatchMatch is not random search, so our redesigned scheme almost behaves as fast as the original scheme, and gets more consistent results than the original one. Figure 3(b) shows the running time of the original PatchMatch [8], kd-tree [14] and our scheme on different images with different number of pixels.

3.2   Color Transfer

In the typical exemplar-based methods, once the best match patch is selected, the pixels of the patch are copied to the target patch without any change.

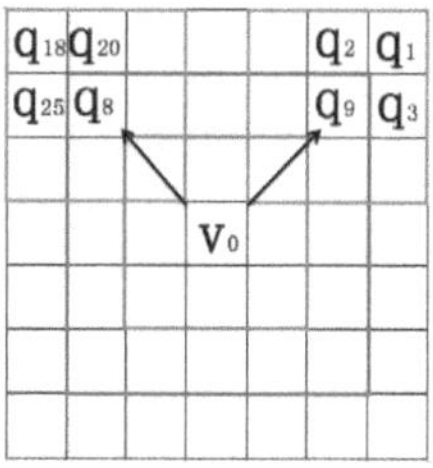

| | q_{18} | q_{20} | | | q_2 | q_1 |
| q_{25} | q_8 | | | q_9 | q_3 |

(a)

Megapixels	Running time(s)		
	Original PatchMatch	Kd-tree	Ours
0.1	0.656	16.237	0.675
0.2	1.578	38.675	1.623
0.4	2.874	89.984	2.896

(b)

Fig. 3. (a) The situation the method in [9] fails. (b) Running time contrast of different schemes.

Although the selected patch is the best match patch in the known region, it is not always in the same distribution with the target patch, which may lead some unnatural result in the end. Here we use some simple statistical feature to transfer the selected patch to fit the target patch.

As shown in articles [15] [16], image color characteristics affects human visual sen-sibility heavily. So we use the skewness and the kurtosis to transfer the selected patch to make it more similar with the target patch. Skewness is to measure uniformity of data distribution. Its value is M_3/σ^3, M_3 is the third moment of the distribution, σ is the standard deviation. We use the following equation [17] to transfer the skewness:

$$Y_i(\lambda) = \begin{cases} \frac{(x_i+t)^{\lambda}-1}{\lambda g_1^{\lambda}} & \lambda \neq 0 \\ g_1 \ln(x_i + t) & \lambda = 0 \end{cases} \tag{7}$$

g_1 is the geometric mean of source data, we should ensure $(x_i + t) > 0$, and adjust λ to change the skewness.

We use the following equation [17] to transfer the kurtosis, which equals to M_4/σ^4, M_4 is the fourth moment of the distribution:

$$Z_i(\lambda) = \begin{cases} \text{sign}(\frac{(|x_i-b|+1)^{\lambda}-1}{\lambda g_2^{\lambda-1}}) & \lambda \neq 0 \\ \text{sign}(g_2 \ln(|x_i - b| + 1)) & \lambda = 0 \end{cases}$$

$$g_2 = (\prod_1^n (|x_i - b| + 1))^{1/n}, \text{ sign=sgn}(x_i - b) \tag{8}$$

Where b is the geometric or arithmetic mean, we also transfer the target through ad-justing λ. When the difference of the skewness is bigger than that of kurtosis, we mainly transfer the skewness, else transfer the kurtosis. After the transfer, we copy the transferred data to the target patch. And that returns more robust results.

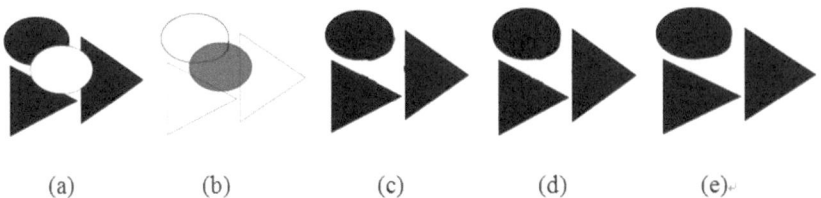

(a)   (b)   (c)   (d)   (e)

Fig. 4. Restore the synthesis image. (a) The source image with an ellipse missing region. (b) Reconstructed sketch line image. (c) The result of the method in article [4]. The method can only deal with linear structure, so the reconstructed curves are obviously inconsistent. (d) The result of method in article [7] (e) Final result of the proposed method. The curve of the ellipse and the truncated straight lines are well recon-structed.

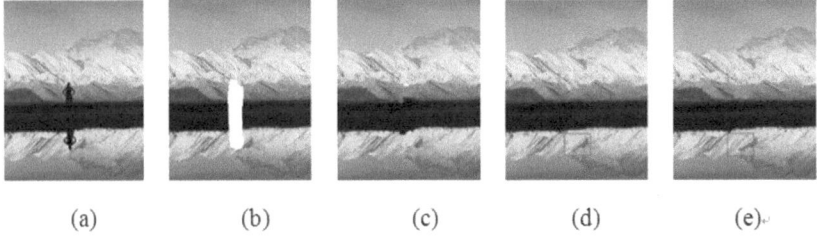

(a)   (b)   (c)   (d)   (e)

Fig. 5. Restore the natural scene image. (a) Source image with a man. (b) Image after removing the man and his reflection in the water. (c) The result from the method in article [4]. The grass lines have obvious artifact, the mountain lines are also not so well reconstructed. (d) The result of method in article [7] (e) The result of the proposed method. The structures and textures are well restored in a way human almost can't detect the artifact.

(a)   (b)   (c)   (d)   (e)   (f)

Fig. 6. Removing large object. (a) Source image. (b) Removing the man from source image. (c)Result of method in article [4]. (d)Result of method in article [7]. (e) Result of method in article [9]. (f) Result of the proposed method. Notice that the structures of the roof are well restored.

4 Experiment Results

The proposed method is experimented on many kinds of images, from the natural pictures to synthesis images. The final results show the robustness and the effective-ness of the proposed method.

We first test the proposed algorithm on a synthetic image with two triangles and one ellipse to show how the algorithm reconstructs the structure of the synthetic im-age. As shown in Fig.4, the curve can be well reconstructed; the texture also can be filled appropriately guided by the reconstructed edge lines. However, the method in article [4] can only restore some linear structure and may lead to some inconsistency. The method in article [7] also results in some artifact.

Figure 5 demonstrates the results to inpaint the natural scene pictures using our al-gorithm. There are a variety of textures and structures near the missing region. As shown in (e), both the structures and the textures are restored in a plausible way by our algorithm. By contrast, the method in article [4] behaves not so well in reconstruct-ing the boundary of the different textures and results in some artifacts. The proposed method also behaves better in structures restoration than the method in article [7] as shown in the red box.

The results of using different methods to remove large object are shown in Fig 6. Our method shows the superiority in restoring the structures and textures, especially in the roof (as marked in the red box), and returns comparative even better results over the typical and improved exemplar-based methods.

5 Conclusions

In this paper, we proposed a fast and robust edge-guided exemplar-based inpainting method basing on the basic truth that an image is a variety of textures organized by the global edge lines. Different from the method in article [4], we not only use the local structures to guide our texture inpainting, but also use the global structures restored by our algorithm, which makes both the structures and the textures be completed consistently. To solve the time-consuming problem of most exemplar-based meth-ods, we redesign the random search strategy to search the match patch. Besides, color transfer is used to further improve the visual consistency. In the future, we consider continuing to perfect our method of restoring edge lines, and test if we can apply the PDE method in our framework.

Acknowlegment. The work is supported by National Significant Science and Technology Projects of China under Grant No. 2013ZX01039001-002-003, National Natural Science Foundation of China Project under Grant No.61170253, Promotion Project of Shenzhen Key Laboratory on Information Science and Technology 2012.

References

1. Bertalmio, M., Saxpiro, G., Caselles, V., Ballester, C.: Image inpainting. In: ACM Comput. Graph (SIGGRAPH 2000), pp. 412–424 (2000)
2. Chan, T.F., Shen, J.: Mathematical models for local non-texture inpaintings. SIAM J. Appl. Math. 62, 1019–1043 (2001)
3. Chan, T.F., Shen, J.: Nontexture Inpainting by Curvature-Driven Diffusions. Journal of Visual Communication and Image Representaion 12, 436–449 (2001)
4. Criminisi, A., Perez, P., Toyama, K.: Region Filling and Object Removal by Exemplar-Based Image Inpainting. IEEE Transaction on Image Processing 13, 1200–1212 (2004)
5. Cheng, W.H., Hsieh, C.W., Lin, S.K., Wu, J.L.: Robust Algorithm for Exemplar-based Image Inpainting. In: Proc. Int. Conf. Comput. Graphics, Imaging Vis (CGIV), pp. 64–69 (2005)
6. Tae-o-sot, S., Nishhara, A.: Exemplar-based image inpainting with patch shifting scheme. In: 17th International Conference on Digital Signal Processing, pp. 1–5 (2011)
7. Kwok, T.H., Sheung, H., Wang, C.C.: Fast Query for Exemplar-Based Image Completion. IEEE Transaction on Image Processing 19(12), 3106–3115 (2010)
8. Barnes, C., Shechtman, E., Finkelstein, A., Goldman, D.B.: PatchMatch: A randomized correspondence algorithm for structural image editing. ACM Transaction on Graphics 28(24) (2009)
9. Tae-o-sot, S., Nishihara, A.: Iterative gradient-driven patch-based inpainting. In: Ho, Y.-S. (ed.) PSIVT 2011, Part II. LNCS, vol. 7088, pp. 71–81. Springer, Heidelberg (2011)
10. Chen, Y., Luan, Q., Li, H.Q., Au, O.: Sketch-Guided Texture-Based Image Inpainting. In: Proc. IEEE International Conference on Image Processing, pp. 1997–2000 (2006)
11. Cheng, Y.: Mean Shift, Mode Seeking, and Clustering. IEEE Transaction on Pattern Analysis and Machine Intelligence 17(8), 790–799 (1995)
12. Canny, J.: A Computational Approach to Edge Detection. IEEE Transaction on Pattern Analysis and Machine Intelligence 8(6), 679–698 (1986)
13. Piegl, L., Tiller, W.: The NURBS Book. Springer (1995-1997)
14. Wexler, Y., Shechtman, E., Irani, M.: Space-Time Completion of Video. IEEE Transaction on Pattern Analysis and Machine Intelligence 29(3), 463–476 (2007)
15. Reinhard, E., Adhikhmin, M., Gooch, B., Shirley, P.: Color transfer between images. IEEE Computer Graphics and Applications 21, 34–41 (2001)
16. Zhao, G.Y., Xiang, S.M., Li, H.: Application of Higher Moments in Color Transfer between Images. Journal of Computer-Aided Design & Computer Graphics 16(1) (2004)
17. Gao, H.X.: Statistic Computation [M]. Beijing Peking University Press (1995) (in Chinese)

A Watershed-Based Segmentation Technique for Multiresolution Data

Giuseppe Masi[1], Giuseppe Scarpa[1], Raffaele Gaetano[2], and Giovanni Poggi[1]

[1] DIETI, University Federico II of Naples
firstname.lastname@unina.it
[2] TSI Department,TELECOM-ParisTech
gaetano@telecom-paristech.fr

Abstract. A new watershed-based technique is proposed for the segmentation of multiresolution remote-sensing images. These images are composed by a high-resolution panchromatic band and a low-resolution multispectral set. To achieve a segmentation with the high resolution of the panchromatic image and the high accuracy granted by the spectral information, the two components are processed jointly, using both spectral and morphological properties. In addition, a fully automatic marker generation procedure is introduced to reduce the oversegmentation typical of watershed methods. Experiments on WorldView-2 multiresolution images demonstrate the potential of the technique.

1 Introduction

Image segmentation is a low-level processing task of critical importance in several applicative domains, like remote sensing [1–5] or medicine [6, 7], to mention just a few. In remote sensing, in particular, segmentation of multiresolution (MR) images is becoming a very relevant topic. In fact, because of technological limitations, satellite sensors cannot reach both high spatial and spectral (large number of bands) resolution. MR sensors, such as Ikonos, GeoEye, or World-View, overcome this problem by providing a single high-resolution (typically, below 50 cm) panchromatic (PAN) image complemented by a low-resolution multispectral (MS) image composed of 4-8 bands, relying then on signal processing, like pansharpening techniques [8], to recover a full multispectral data cube. Pansharpening, however, cannot really preserve all desired information, and high accuracy can be only guaranteed by direct processing of the original data. Here we propose a new segmentation technique specifically designed for multiresolution images and based on the watershed transform [9], a widespread technique particularly suited for the preservation of fine details.

To obtain a watershed segmentation, a topographic surface is first associated with the image. Then it is progressively filled with "water", and each time two basins meet a dam is built between them in order to avoid their merging. Once the surface is completely flooded the process stops and each water basin is regarded as an image segment. In the context of edge-based segmentation, one can apply watershed by using the distance transform to define the topographic

A. Petrosino (Ed.): ICIAP 2013, Part I, LNCS 8156, pp. 241–250, 2013.

surface. Watershed will therefore provide a full segmentation where all original edges are guaranteed to be part of the final region boundaries, thus completing the original edge map.

Unfortunately, watershed produces many more segments than desired, because of noise and intrinsic image textures, calling for the adoption of some suitable merging strategy to reduce them, a computation-intensive step, and a possible source of errors. Several approaches have been proposed in the literature to limit over-segmentation. One can identify shallow basins, likely due to noise, and merge them with some suitably chosen neighbor [5]. Inconsistent region separations can be singled out and removed by analyzing the image at a larger scale of interaction [10]. Better yet, one can intervene before running the watershed by introducing some "markers" in the topographic surface, which force separate basins to be treated as a single entity. Markers can be put manually from an operator, a tedious and low-precision task or, more interestingly, through a specific automatic procedure, e.g. [1, 11]. This latter approach was followed in [11], leading to the Edge, Mark and Fill (EMF) algorithm, suited for single-band images. In this work we generalize the EMF strategy to the case of color/multispectral, and in particular multiresolution images, providing new tools to further reduce oversegmentation.

After briefly recalling EMF in next Section, the proposed segmentation algorithm is described in Section 3, while Section 4 shows experimental results. Concluding remarks are given in Section 5.

2   Edge, Mark and Fill Segmentation

The Edge, Mark and Fill technique is a marker-controlled watershed applied to the distance transform, which embeds a suitable morphological process for a fully-automatic generation of markers.

To gain insight into the EMF algorithm consider the example of Fig.1. In (a) the image to be segmented is shown together with the contours, in red, produced by a suitable edge detector. The distance transform, D, is shown in (b): its opposite $-D$ is used as topographic surface (DEM) for the watershed. In order to produce effective markers, EMF computes the crest lines of D, shown in (c) together with the edge map, which represent a support for the skeletons of the closed regions singled out by the watershed. The local maxima of D on the crest lines are then extracted, the red points in (d), and used as marker seeds. These seeds, say s_k, are dilated with a circular morphological profile of radius $D(s_k) - \epsilon$, where the margin ϵ prevents crossing the edges. Overlapping dilated seeds are merged together as shown in (d). The segmentation provided by watershed with the DEM $-D$ and the above markers is shown in (e), while (f) shows the difference between the initial edge map (a) and the final (closed) contour map (e). Removed edges are represented in red, while additional edges needed to close open segments are shown in green. For further details about EMF the reader is referred to [11].

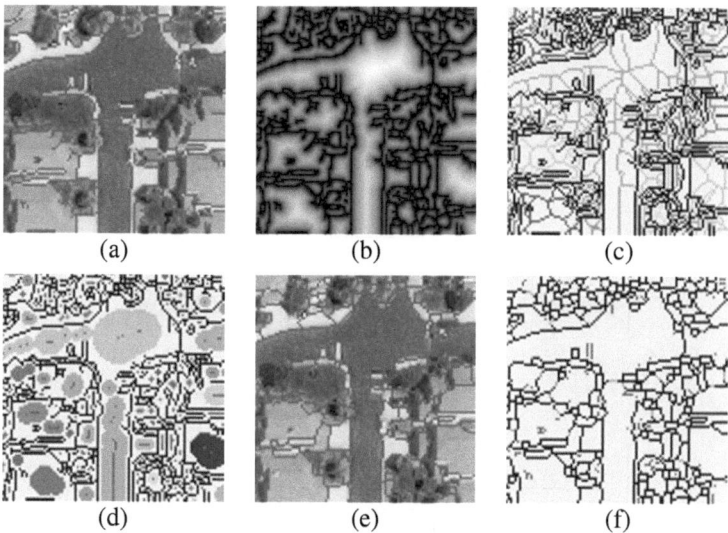

Fig. 1. Intermediate results of the EMF Algorithm

Experimental evidence shows that EMF reduces oversegmentation with no loss of valuable information, and preserving the high local accuracy of edge-based approaches. However, it keeps generating too many segments in textured and noisy areas, also because it works on a single-band image on the basis of purely morphological information. On one hand, this is a strength of EMF, which can deal with gray-level images or, more in general, with any given edge map, with no need for further information. On the other hand, when further information *is* available, as with MR images, it must be taken into account. Therefore, in next Section, we describe an enhanced version of EMF for MR images.

3   Multiresolution EMF

The oversegmentation typical of watershed is mostly due to edge map imperfections, with meaningless edges originated by minor texture variations, or edges which, because of noise, depart from their correct position. In both cases, the effect is a proliferation of small regions, with very similar spectral content, which should be merged together. Our primary goal, here, is to take into account the available spectral information to reduce oversegmentation from the beginning. In addition, with reference to MR images, we want to exploit effectively the different cues coming from the two image components, the single-band but high-resolution PAN, and the low-resolution spectrally rich MS.

In Fig.2 a high-level flowchart of the MR-EMF algorithm is shown. Unlike in basic EMF, the work-flow is split in two channels, in order to process the two image components. Edge detection is carried out in both channels, by means of

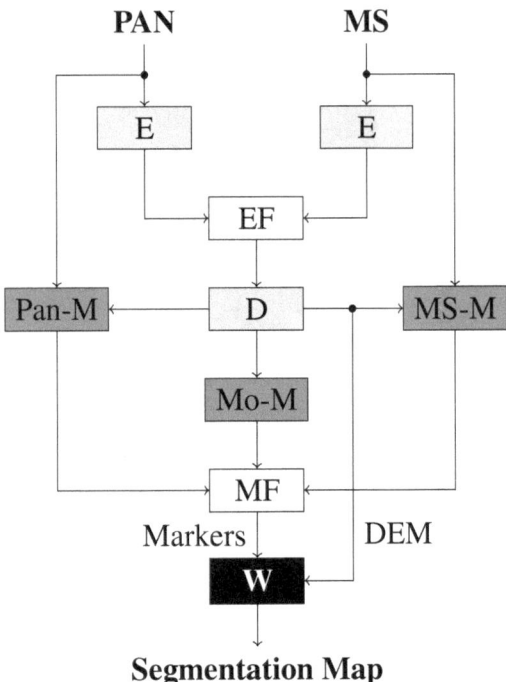

Segmentation Map

Fig. 2. Flowchart of MR-EMF. E: Edge Detection, EF: Edge Fusion, D: Distance transform, Pan-M: PAN Marker generation, MS-M: MS Marker generation, Mo-M: Morphological Marker generation, MF: Marker Fusion, W: Watershed.

a standard Canny detector [12], with the results combined to produce a single high-resolution edge map (here and in the following, for the sake of simplicity, we neglect all up/down-sampling and registration operations needed to combine data of the two channels). Two new sets of spectral markers are generated, one for each channel, besides the morphological markers used in EMF, and all markers are eventually combined at high resolution to guide the watershed transform.

Let us now analyze the algorithm structure in more detail. Intuitively one could choose to consider only the PAN data to generate geometrical markers, and only MS data for spectral markers, given the superior geometric resolution of the former and spectral resolution of the latter, thus renouncing to the full-symmetric approach depicted in Fig.2. However, there are good reasons to keep a balanced approach.

First of all, it can happen that edges visible in the MS component are not visible at all in the single-band PAN. Therefore, both edge maps must be taken into account. Of course, when contours are detected in both channels, preference must be given to the more accurate PAN data, neglecting the others. Experiments confirm that MS-edges help improving results. This is all the more true

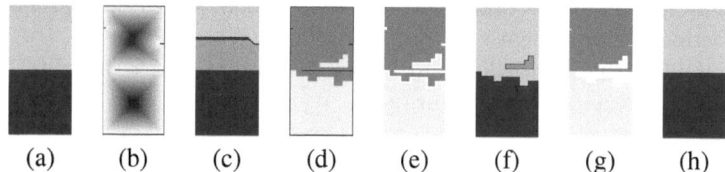

(a)   (b)   (c)   (d)   (e)   (f)   (g)   (h)

Fig. 3. Marker erosion and application: ground-truth (a); edge map (blue) with related distance transform (b); unmarked watershed (c); external spectral clustering (d) and its erosion, pre-markers, (e) by edge subtraction; (pre-)marker-controlled watershed (f); final markers (g) and related watershed (h)

for spectral marker generation, where both PAN and MS data are essential for a correct use of spectral information.

3.1 Generation of Spectral Markers

The generation of spectral markers is, in principle, very simple. Since our goal is to recognize and merge together homogeneous segments that share the same spectral properties, we only have to cluster data in the spectral domain and use the resulting connected components as markers. When a marker crosses the border between two segments, because pixels on both sides are spectrally homogeneous, the segments will be eventually merged. This vision, however, is overly simplistic as it relies on perfect segment homogeneity. In the actual cases, instead, within each segment, it can happen to find pixels assigned to different clusters, which risk to produce undesired results. This is basically due to the obvious inconsistency between the edge detection, which is a local process, and the clustering, which is a global one. Hence it is necessary to reshape the detected clusters before using them as markers.

Marker Erosion Process. In order to explain the proposed solution, we consider the toy example of Fig.3. Let (a) be the ideal segmentation (ground-truth), and (b) the detected edge map (blue) with corresponding distance function (gray scale). The given morphology gives rise to the wrong (over-) segmentation (c), if a simple unmarked watershed is applied. Now, assume to have a spectral-based clustering like that shown in (d), where three connected components can be distinguished. Edge map subtraction gives the markers in (e) which lead to an unsatisfactory marker-controlled segmentation (f) with both over- (additional region) and under-segmentation (lost edges). We therefore apply an additional erosion process to obtain more reliable markers. For each pixel p of a pre-marker M in (e) the attracting minimum in the DEM, say p_m, is localized. If $p_m \in M$ than p is kept, otherwise it is removed from M. As result, pre-markers which do not cover any minimum of the DEM will simply disappear because unreliable. Those who cover one or more minima will be constrained to the their attraction basins. The final markers obtained after the erosion, shown in (g), lead to the desired segmentation shown in (h).

In particular, the algorithm used in our method to single out the clustering (d) is a simple mean shift procedure [13].

PAN *vs* MS Image Domain Partition. To apply this tool to MR images, a preliminary key observation is necessary, that the reliability of the spectral information provided by the MS component depends on the size of the segments to characterize. In fact, since MS pixels are much larger than PAN pixels (typically 4×4 times) they can happen to average together high-resolution pixels with different spectral characteristics, typically near segment boundaries. As a consequence, "pure" MS pixels can be found only in the segment interior, provided this is large enough. Therefore, for small or elongated segments, the MS spectral characterization is highly unreliable and, below a given segment size, it is more appropriate to use the PAN data.

Based on this consideration, we perform a preliminary partition of the image in two regions where "thick" and "thin" segments are found, using MS-based spectral markers in the former region and PAN-based in the latter. To this end, a simple thresholding on the distance function may seem to work well. Given a suitable threshold d, objects less than d pixel wide (along the shortest dimension) are better characterized in the PAN domain, while the others are reliably featured by MS data. The region $\widetilde{M} = D < d$, which is nothing but a dilation of radius d of the edge map, will certainly cover entirely the thin objects. However, it will also overlap with the contours of thick elements, partially covering them. To avoid this inconvenient, we apply to $\widetilde{M}$ the same region erosion mechanism described above to reshape the spectral markers. In other words, elements of $\widetilde{M}$ whose attracting minimum does not belong to $\widetilde{M}$ are discarded. Once singled out the mask for thin object, the PAN domain, its complement is used as MS domain.

An example of this process for a real-world MR image is shown in Fig.4. In (a) is the PAN component of the image to segment, and in (b) the composite MS-PAN edge map. (c) shows the thin-thick mask obtained on the basis of the distance transform of (b). The final PAN and MS markers are shown in (d) in green and red, respectively.

4   Experimental Results

The segmentation and classification accuracy assessment in remote sensing is a challenging problem as testified by recently developed benchmarking systems, like [14], specifically thought for segmentation of remote sensing data. Unfortunately no one such solutions deals with multiresolution data.

The lack of multiresolution data with suitable ground-truth (not just the interior of large segments) prevents a reliable quantitative assessment of segmentation accuracy, as well as a meaningful comparison with competing algorithms, like the TFR [15, 16] or the technique proposed in [17]. Therefore, we confine

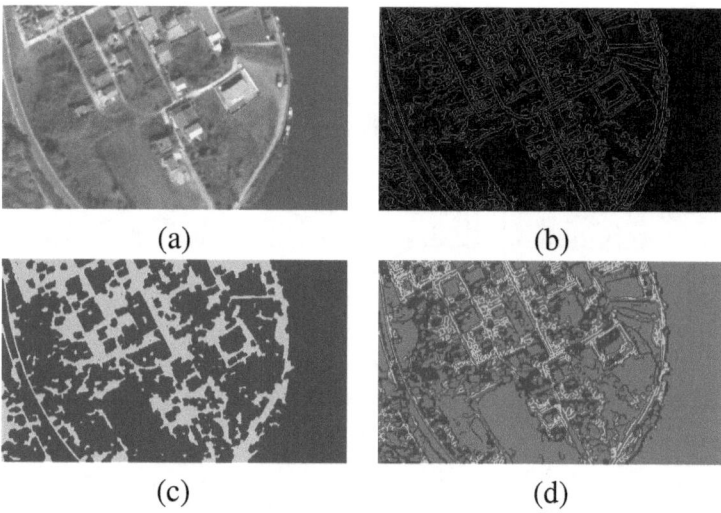

Fig. 4. Spectral markers generation: PAN component (a); edge map (b); PAN (light) and MS (dark) domains (c); PAN (green) and MS (red) markers (d)

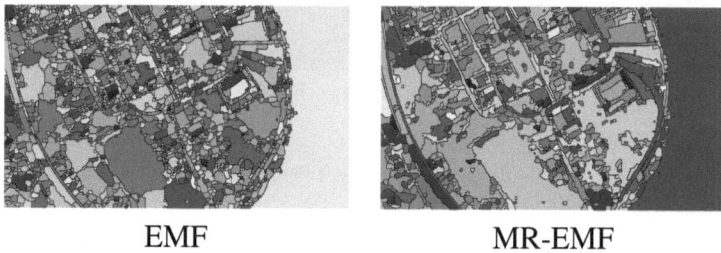

EMF   MR-EMF

Fig. 5. Segmentation maps obtained with EMF (left) and with MR-EMF (right)

our analysis, for the time being, to the visual inspection of results obtained on a large dataset of ©WorldView-2 images, comprising a 8-band MS component with 2-meter resolution and a PAN component with 50-cm resolution. We found experimentally that the best value for d, the threshold used to generate the PAN-MS domain partition, is in the range 6-8 (high-resolution) pixels, not far from the size of a MS pixel. In particular we selected $d = 6$ for the experiments.

The first experiment presented concerns the image used in the running example of Fig.4. In Fig.5 we show the marker-controlled watershed segmentation with morphological markers alone (EMF), and with superposition of spectral markers (proposed MR-EMF). Although a significant simplification of the map is evident, the achieved gain can be better appreciated by looking at the few enlarged details shown in Fig.6. Segmentation contours (in red) are superimposed to the RGB subset of the MS component. On the left column are the results of the unmarked

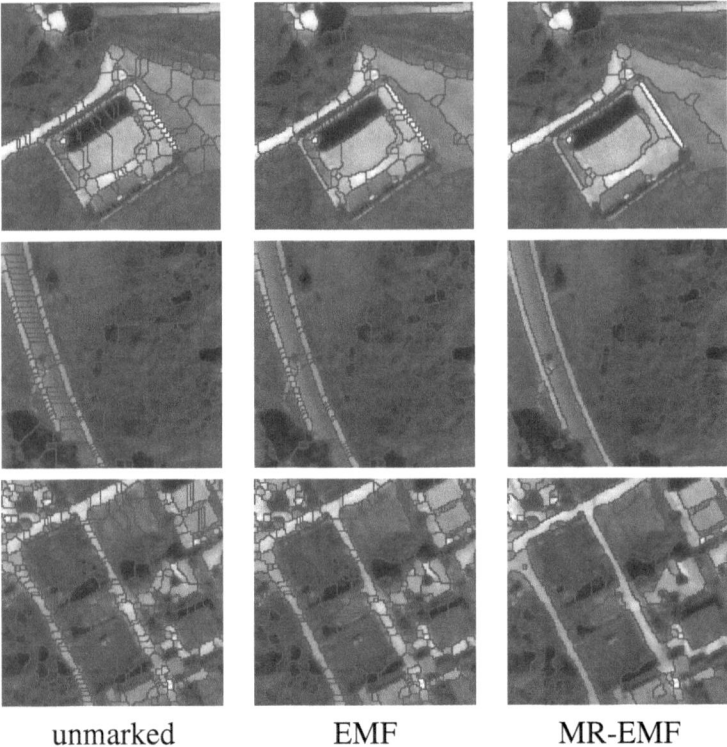

unmarked   EMF   MR-EMF

Fig. 6. Segmentation details for the running experiment of Fig.4: unmarked watershed (left), EMF (center), and MR-EMF (right)

EMF   MR-EMF

Fig. 7. Segmentation resultrs provided by EMF (left) and MR-EMF (right)

watershed. In the middle and right columns the results of EMF and MR-EMF, respectively. As can be easily appreciated by visual inspection the accuracy improves considerably moving toward the proposed solution. Over-segmentation is largely reduced, with no apparent loss in discriminating capabilities.

Similar considerations hold for the second experiments shown in Fig.7 which confirms a general improvement with respect to the basic EMF.

5 Conclusions

The proposed watershed-based segmentation algorithm for multiresolution images guarantees the high local accuracy of all edge-oriented techniques and a limited over-segmentation. A key step is the automatic generation of morphological and spectral markers, obtained by properly combining cues coming from both PAN and MS components and weighting their relative reliability.

Preliminary experiments have shown very promising results, with a large reduction of the over-segmentation and no apparent loss of details. In future work we will consider replacing the edge map fusion with a high resolution edge detector that works jointly on PAN and MS data, and designing an *ad hoc* clustering algorithm to improve upon the basic mean-shift.

References

1. Xiao, P., Feng, X., Zhao, S., She, J.: Multispectral ikonos image segmentation based on texture marker-controlled watershed algorithm. In: SPIE 6790, MIPPR (2007)
2. Cagnazzo, M., Poggi, G., Verdoliva, L.: Region-based transform coding of multispectral images. IEEE Transactions on Image Processing 16, 2916–2926 (2007)
3. Parrilli, S., Poderico, M., Angelino, C.V., Scarpa, G., Verdoliva, L.: A nonlocal approach for SAR image denoising. In: IEEE International Geoscience and Remote Sensing Symposium, IGARSS 2010, pp. 726–729 (2010)
4. Cagnazzo, M., Parrilli, S., Poggi, G., Verdoliva, L.: Improved Class-Based Coding of Multispectral Images With Shape-Adaptive Wavelet Transform. IEEE Geoscience and Remote Sensing Letters 4(4), 566–570 (2007)
5. Li, P., Guo, J., Song, B., Xiao, X.: A multilevel hierarchical image segmentation method for urban impervious surface mapping using very high resolution imagery. IEEE Journal of Selected Topics in Applied Earth Observations and Remote Sensing 4(1), 103–116 (2011)
6. Bova, N., Ibanez, O., Cordon, O.: Image Segmentation Using Extended Topological Active Nets Optimized by Scatter Search. IEEE Computational Intelligence Magazine 8(1), 16–32 (2013)
7. D'Elia, C., Marrocco, C., Molinara, M., Poggi, G., Scarpa, G., Tortorella, F.: Detection of microcalcifications clusters in mammograms through TS-MRF segmentation and SVM-based classification. In: 17th International Conference on Pattern Recognition, ICPR 2004, vol. 3, pp. 742–745 (2004)
8. Alparone, L., Wald, L., Chanussot, J., Thomas, C., Gamba, P., Bruce, L.M.: Comparison of Pansharpening Algorithms: Outcome of the 2006 GRS-S Data-Fusion Contest. IEEE Transactions on Geoscience and Remote Sensing 45(10), 3012–3021 (2007)

9. Beucher, S., Lantuejoul, C.: Use of Watersheds in Contour Detection. In: International Workshop on Image Processing: Real-time Edge and Motion Detection/Estimation, Rennes, France (September 1979)

10. Arbeláez, P., Maire, M., Fowlkes, C., Malik, J.: Contour detection and hierarchical image segmentation. IEEE Transactions on Pattern Analysis and Machine Intelligence 33(5), 898–916 (2011)

11. Gaetano, R., Masi, G., Scarpa, G., Poggi, G.: A marker-controlled watershed segmentation: Edge, mark and fill. In: IEEE International Geoscience and Remote Sensing Symposium, IGARSS 2012, pp. 4315–4318 (2012)

12. Canny, J.: A computational approach to edge detection. IEEE Transactions on Pattern Analysis and Machine Intelligence PAMI 8(6), 679–698 (1986)

13. Comaniciu, D., Meer, P.: Mean Shift: a robust approach toward feature space analysis. IEEE Transactions on Pattern Analysis and Machine Intelligence 24(5), 603–619 (2002)

14. Mikes, S., Haindl, M., Scarpa, G.: Remote sensing segmentation benchmark. In: 7th IAPR International Workshop on Pattern Recognition in Remote Sensing, PRRS 2012, Tsukuba Science City, Japan (November 2012)

15. Scarpa, G., Haindl, M.: Unsupervised texture segmentation by spectral-spatial-independent clustering. In: 18th International Conference on Pattern Recognition, ICPR 2006, vol. 2, pp. 151–154 (August 2006)

16. Gaetano, R., Scarpa, G., Poggi, G.: Hierarchical texture-based segmentation of multiresolution remote-sensing images. IEEE Transactions on Geoscience and Remote Sensing 47(7), 2129–2141 (2009)

17. Yuan, J., Wang, D.L., Li, R.: Remote Sensing Image Segmentation by Combining Spectral and Texture Features. IEEE Transactions on Geoscience and Remote Sensing (to appear)

Database for Arabic Printed Text Recognition Research

Faten Kallel Jaiem[1], Slim Kanoun[1], Maher Khemakhem[1],
Haikal El Abed[2], and Jihain Kardoun[3]

[1] MIRACL laboratory, ISIMS, University of Sfax, Tunisia
{kallelfaten,slim.kanoun}@gmail.com,
maher.khemakhem@fsegs.rnu.tn
[2] Institute for Communications Technology, Braunschweig University, Germany
elabed@tu-bs.de
[3] Department of Computer Engineering, ENIS, University of Sfax, Tunisia
jihen.kardoun@gmail.com

Abstract. This paper presents a real database for the Arabic printed text recognition, APTID / MF (Arabic Printed Text Image Database / Multi-Font).This database can be used to evaluate the system that recognizes Arabic printed texts with an open vocabulary. APTID / MF may be also used for research in word segmentation and font identification. APTID / MF is obtained from 387 pages of Arabic printed documents scanned with grayscale format and 300 dpi resolutions. From this documents, 1,845 text-blocks have been extracted. In addition ground truth file is provided for each texts-block. APTID / MF also includes an Arabic printed character image dataset made up of 27,402 samples. The database is freely available to interested researchers.

Keywords: Arabic printed text, APTID / MF database, Open vocabulary, Ground truth.

1 Introduction

A deep observation of the printed Arabic documents reveals that the number of font styles and sizes used is important. So far, there has been no standard corpus which includes multi-font, multi-style and multi-size printed Arabic writing that can be used to evaluate the score (recognition rate) of any given Arabic OCR system.

Indeed, intensive experiments achieved on some existing Arabic OCR systems reveal that their recognition rates is sometimes very sensitive to the variability of the font style and/or the font size.

Consequently, this paper presented a detailed description of Arabic multi-font, multi-style and multi-size printed text database. The first section presents the Arabic script complexity and the diversity of the fonts and styles of the Arabic printed script. The second section sheds some light on the existing databases for the printed Arabic script recognition. The third section gives a detailed description of the published databases. In the last section of this paper, we talk about the various prospects and future work for the database extension.

A. Petrosino (Ed.): ICIAP 2013, Part I, LNCS 8156, pp. 251–259, 2013.

2   Brief Description of Arabic Script Characteristics

The Arabic script has certain characteristics which cause the complexity of its recognition. These characteristics come from its cursive nature and the variation of its character forms according to their position in the word [1]. The Arabic characters change forms not only according to their position in the word but also according to the used calligraphic style. Calligraphy is very developed in the Arab-Muslim world. In fact, the Arabic writing can appear in different calligraphic styles such as Neskhi, Thoulthi, Diwani [2]. The analysis of the Arabic documents shows the existence of a diversity of writing fonts for the printed paper resulting from the Arabic calligraphic styles.

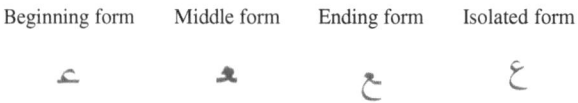

Fig. 1. An example of different shapes of an Arabic letter

3   Current Arabic Printed Database

In literature, the database for text recognition research is often organized in two classes. The first class represents the database of the printed documents; the second includes the handwritten documents [3] [4] [5]. This section presents an overview of the current Arabic Printed Database.

3.1   APTI

The APTI database [6] is made up of 45,313,600 Arabic Printed word images. These word images cover approximately 250,000,000 characters. These images result from 113,284 various word writings with 10 fonts, 10 sizes and 4 styles. The Xml files associated with each element of the database present the ground truth data.
The APTI database is a synthetic and it is designed for the evaluation of screen-based OCR systems . The images are generated at 72 dpi by an automatic program. This presents a disadvantage when evaluating the systems for recognizing scanned Arabic printed documents.

3.2   DARPA

The DARPA (Defense Advanced Research Projects Agency) Arabic corpus was created by Scientific Application International Company for the US Department of Defense [7]. DARPA Corpus data were collected from the books, the Magazines, the newspapers and the computer generated documents covering only 4 fonts. The DARPA corpus includes 345 Arabic printed pages with ground truth data. These pages were scanned with a 600 dpi resolution.

3.3   PATDB

The PATDB (Printed Arabic Text DataBase) was published by Al Hashim in 2010 [8]. This database was issued from the printed Arabic pages of texts, resulting from books, advertisements, magazines, newspapers and reports. These pages were scanned with 3 resolution levels 200, 300 and 600 dpi. The database is made up of 6,954 pages of Arabic printed texts. Ground truth data files were attributed to each page of the database.

4   Overview of the APTID / MF

In this section we present our database of Arabic printed text called APTID / MF (Arabic Printed Text Image Database / Multi-Font). In our work we initially aimed to create an Arabic multi-font and multi-size printed text database. The APTID / MF included an Arabic printed text image dataset, and an Arabic printed character image dataset.

4.1   Arabic Printed Text Image Dataset

The Arabic printed text image dataset was selected from the official site of the tunisien newspaper "El-chourouk". The set of document pages, included in the APTID / MF, is written in two writing styles (normal and bold).

The document pages are organized in 10 sets and we attributed a writing font for each set, The figure below present this fonts. The set of documents is written with 4 sizes (12, 14, 16, and 18 pts).

Andalus	المصارحة والمصالحة
Simplified Arabic	المصارحة والمصالحة
Tahoma	المصارحة والمصالحة
Traditional Arabic	المصارحة والمصالحة
Decotype Thuluth	المصارحة والمصالحة
Arabic transparent	المصارحة و المصالحة
Af-Diwani	المصارحة والمصالحة
Advertising Bold	المصارحة والمصالحة
Decotype Naskh	المصارحة والمصالحة
M-Unicode Sara	المصبارحة والمصبالحة

Fig. 2. An exemple of Arabic text written with the used fonts

Table 1. The distribution of document page

	Size 12	Size 14	Size 16	Size 18
The number of document pages	40	54	45	53

This set of document pages are printed with a laser printer and an inkjet printer. The set is stored in tow groups: the first includes the pages printed using a laser printer and the second includes those printed using an inkjet printer. Then, the pages are scanned. The first set of printed pages is scanned with an HP scanner while, the pages obtained by the second printer are scanned with an Epson one. All the page images are scanned with 300 dpi resolution in Grayscale format. The images are stored in "PNG" format. The APTID / MF contains 386 page images of Arabic printed texts. These page images are divided into text-blocks with a manual segmentation. This phase gave us 1,845 Arabic printed image text-blocks made up of 126,792 Arabic words, resulting from 6,989 distinct words.

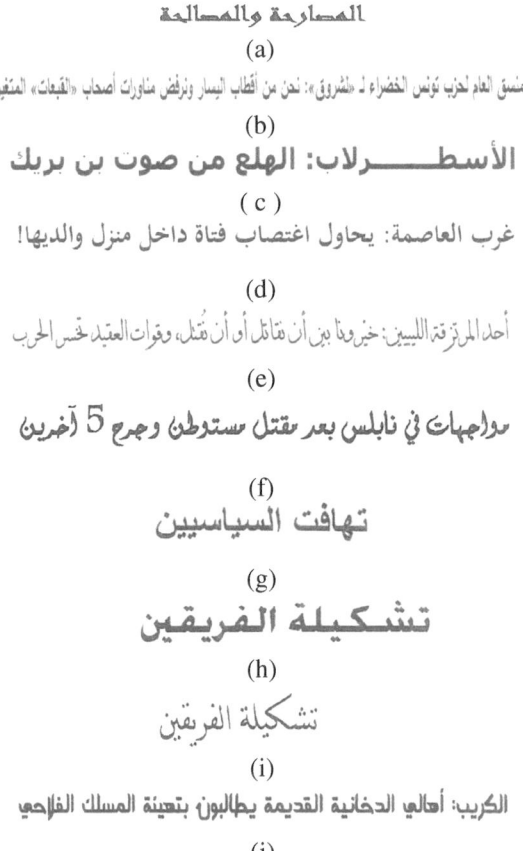

Fig. 3. Examples of image texts-blocks : (a)Andalus, (b) Simplified Arabic, (c) Tahoma, (d) Traditional Arabic, (e) Decotype Thuluth, (f) Af-Diwani, (g) Arabic transparent, (h)Advertising Bold, (i) Decotype Naskh and (j) M Unicode Sara

Statistics. The database APTID / MF included 1845 text-blocks images. The different text-block images in the APTID / MF are stored in different sets. TABLE 2 presents the distribution of APTID / MF text-blocks.

Table 2. The distribution of APTID / MF text-blocks

A laser printer and an HP scanner				
Font	size 12	size 14	size 16	size 18	Total
Andalus	26	26	24	21	97
ArabicTransparent	19	19	17	21	76
AdvertisingBold	24	24	23	26	97
Diwani Letter	21	21	20	21	83
DecoTypeThuluth	24	23	23	26	96
Simplified Arabic	20	20	20	20	80
Tahoma	19	19	19	21	78
Traditional Arabic	27	27	27	26	107
DecoType Naskh	20	20	21	17	78
M Unicode Sara	30	30	30	30	**120**
An inkjet printer and an Epson scanner					
Font	size 12	size 14	size 16	size 18	Total
Andalus	26	26	26	29	107
ArabicTransparent	19	19	17	25	80
AdvertisingBold	24	24	23	28	99
Diwani Letter	21	21	20	23	85
DecoTypeThuluth	24	23	23	28	98
Simplified Arabic	20	20	20	19	79
Tahoma	19	19	19	19	76
Traditional Arabic	27	27	27	26	101
DecoType Naskh	20	20	21	20	81
M Unicode Sara	30	30	30	31	121
TOTAL	**460**	**458**	**450**	**477**	**1845**

Ground Truth File Description. The database APTID / MF included 1,845 text-block images with their associated metadata files (XML file). These files present the ground-truth value of each sample of text image dataset, These files described at the text-block and line levels using XML file.

At the text-block level, these XML files include the following information: the text-block name (<TextImage Id = ...>) and the number of lines and words in text-block (<text nbligne=... nbword = ...>)

At the line level, these XML files include the following information: the row of line and the number of words on the line (<ligne Id= ... nbword=...> and the row and electronics of the word in the line (<word Id =... value=... />).

In addition these Xml files present the font (<Font name=... />), the style (<Style name=... />) and the size (<Size value=... />) of the text-block, the type of printer used(<Imprimant name=... />) and the name of the scanner used (<scanner name=.../>).

In the final structure of our text dataset, each folder that contains printed text-block samples is also provided with the ground truth data file for the sample. This ground truth is useful to evaluate the recognition results.

```xml
<?xml version="1.0" encoding="UTF-8" ?>
- <TextImage Id="text4_doc199">
  - <text nbligne="1" nbword="2">
    - <ligne Id="1" nbword="2">
        <word Id="1" value="تهافت" />
        <word Id="2" value="السياسيين" />
      </ligne>
    </text>
    <Font name="Arabic Transparent" />
    <Style name="Gras" />
    <Size value="16" />
    <Imprimant name="Laser" />
    <scanner name="Hp" />
  </TextImage>
```

Fig. 4. Example of XML files including ground truth

4.2   Arabic Printed Character Image dataset

The Arabic printed character image dataset was selected from various printing forms, magazines, Book chapters and newspapers... The construction process starts by scanning all the pages with a 300 dpi resolution in grayscale format scanner. After that, a binary copy of these pages is segmented in character images. The character dataset contains 27,402 binary images of characters. The different forms of the characters are used to store the character images in 32 classes.

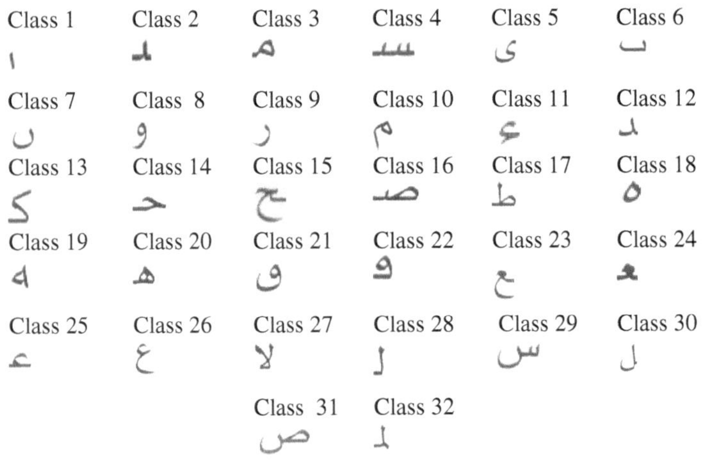

Class 1	Class 2	Class 3	Class 4	Class 5	Class 6
ا	ـلـ	ـه	ـلــ	ى	ـب
Class 7	**Class 8**	**Class 9**	**Class 10**	**Class 11**	**Class 12**
ن	و	ر	م	ـع	ـل
Class 13	**Class 14**	**Class 15**	**Class 16**	**Class 17**	**Class 18**
ک	ـحـ	ح	ـص	ط	ة
Class 19	**Class 20**	**Class 21**	**Class 22**	**Class 23**	**Class 24**
۹	۵	ف	۹	ح	٭
Class 25	**Class 26**	**Class 27**	**Class 28**	**Class 29**	**Class 30**
ـع	ع	لا	ل	س	ل
		Class 31	**Class 32**		
		ص	ٹ		

Fig. 5. Character class

The dataset of character images is divided into a training set that contains 18,404 samples and a test set which contains 8,998 samples.

Table 3. The distribution characters of training set

Class	Number image	Class	Number image
1	733	17	543
2	752	18	523
3	650	19	615
4	604	20	564
5	578	21	483
6	609	22	695
7	572	23	488
8	622	24	618
9	640	25	536
10	570	26	449
11	510	27	590
12	625	28	643
13	607	29	497
14	636	30	459
15	537	31	354
16	556	32	546

Table 4. The character distribution of the test set

Class	Number image	Class	Number image
1	361	17	267
2	371	18	258
3	316	19	301
4	296	20	277
5	282	21	238
6	295	22	339
7	279	23	242
8	301	24	304
9	311	25	263
10	278	26	218
11	248	27	285
12	306	28	313
13	297	29	245
14	310	30	223
15	263	31	171
16	273	32	267

Recognition Results. A character recognition system is developed . In our work we used the Hu's invariant moments [9], Affine invariant moments [10], Zernike's moments [11], Tsirikolias-Mertzios Moments [12], Fourier Mellin Transform [13], and Fourier Descriptor [14],which all represent the statistical features. In addition we referred to the Freemanchain codes [15] for the structural feature. Finally we chose the works of Heutte [16] for the topological feature to analyze the statistical and structural features existing in the literature. In this paper , ower aim is to present an Arabic characters dataset. So, as a first experiment of this dataset, we have chosen the K-nearest neighbor as classifier with K=1 and the training set and test set of the character image dataset. The table below presents the results.

Table 4. Recognitions Rate

Features	Recognitions Rate
Affine invariants Moment	65.40%
Hu's invariants Moments	84.07%
Zernike's Moments	76.77%
Tsirikolias–Mertzios Moments	78.04%
Fourier Mellin Transform	76.76%
Fourier Descriptors	67.70%
Topological features	96.81%
Freeman code chain	96.97%

5   Conclusions

In this paper, an Arabic printed text database is presented. This database may be used by the Arabic printed text recognition and font identification research community. The APTID / MF contains 1,845 image text-blocks that are scanned at 300 dpi resolution in grayscale format. The character dataset includes 27,402 image characters. For each piece of the text-block dataset, a corresponding ground truth file is available. APTID / MF was then extended to include a dataset of printed multi-font multi-style and multi-size Arabic image words with large vocabulary resulting from these text-block dataset. The APTID / MF is prepared to organize a competition for the large vocabulary Arabic printed text recognition.

References

1. Amara, N.B.: On the Problematic and Orientations in Recognition of the Arabic Writing. In: CIFED 2002, pp. 1–10 (2002)
2. Kanoun, S., Alimi, A.M., Lecourtier, Y.: Affixal Approach for Arabic Decom-posable Vocabulary Recognition: A Validation on Printed Word in Only One Font. In: ICDAR 2005, pp. 1025–1029 (2005)
3. Pechwitz, M., Maddouri, S., Margner, V., Ellouze, N., Amiri, H.: IFN/ENIT-Database of Handwritten Arabic Words. In: CIFED 2002, pp. 127–136 (2002)
4. Mozaffari, S., Faez, K., Faradji, F., Ziaratban, M., Golzan, M.: Isolated Far-si/Arabic character database for handwritten OCR research. In: International Work-shop on Frontiers of Handwriting Recognition, pp. 385–389 (2006)
5. Mozaffari, S., El Abed, H., Margner, V., Faez, K., Amirshahi, A.: IfN/Farsi-Database: A Database of Farsi Handwritten City Names. ICFHR (2008)
6. Slimane, F., Ingold, R., Kanoun, S., Alimi, A., Hennebert, J.: A New Arabic Printed Text Image Database and Evaluation Protocols. In: proc. of 10th IEEE International Conference on Document Analysis and Recognition, ICDAR 2009, pp. 946–950 (2009)
7. Davidson, R., Hopely, R.: Arabic and Persian OCR Training and Test Data Sets. In: Proceedings of Symposium. On Document Image Understanding Technology (1997)
8. AL-hashim, A.G., Mahmoud, S.A.: Benchmark Database and GUI Environment for Printed Arabic Text Recognition Research. Wseas Transactions Information Science and Applications 7(4), 10 (2010)

9. Hu, M.: Visual pattern recognition by moment invariants. IRE Trans. Information Theory, IT 8, 179–187 (1962)

10. Flusser, J., Suk, T.: Pattern recognition by affine moment invariants. Pattern Recognition 26(1), 167–174 (1993)

11. Zernike, F.: Diffraction theory of the cut procedure and its improved form, the phase contrast method. Physica 1, 689–704 (1934)

12. Tsirikolias, K., Mertzios, B.G.: Statistical pattern recognition using efficient two dimensional moments with applications to character recognition. Pattern Recognition 26, 877–882 (1993)

13. Derrode, S., Ghorbel, F.: Digital Fourier Mellin Transform- Reconstruction and es-timate of objects movement on levels of gray. In: Proc. of GRETSI conference, Grenoble, France, pp. 566–658 (1997)

14. Davis, C.B., Beecher, R., Beecher, M.: The statistical use of Fourier descriptors. Original Research Article Mathematical and Computer Modeling 11, 419–424 (1988)

15. Freeman, H.: On the encoding of arbitrary geometric configurations. IEEE Trans. Electronic Comp. EC-10, 260–268 (1968)

16. Heutte, L.: Reconnaissance de caractères manuscrits: Application a la lecture au-tomatique des chèques et des enveloppes postales. Doctorat Thesis, University of Rouen (1994)

On the Stability of Ranks to Low Image Quality in Biometric Identification Systems

Emanuela Marasco[1] and Ayman Abaza[2,3]

[1] Lane Department of Computer Science and Electrical Engineering,
West Virginia University
P.O. Box 6109 Morgantown, WV, USA
emanuela.marasco@mail.wvu.edu
[2] West Virginia High Technology Consortium Foundation
Fairmont, WV 26554, USA
[3] Biomedical Engineering and Systems,
Cairo University, Egypt

Abstract. The goal of a biometric identification system is to determine the identity of the input biometric probe. This is accomplished using a matcher which compares the input probe data against each labeled biometric data present in the gallery database. The output is a set of similarity scores that are ranked in decreasing order. The identity of the gallery entry corresponding to the highest similarity score (i.e., rank 1) is associated with that of the probe. In multibiometric systems, the outputs of multiple biometric matchers are combined. Such a combination, or fusion, can be accomplished at the score level or rank level (apart from other levels of fusion). In the literature, rank is believed to be a stable statistic. However, this belief has not been experimentally demonstrated. The contribution of this paper is to investigate the stability of ranks to the image quality degradation in both unimodal and multimodal scenarios. Experiments were carried out using two databases: 1) West Virginia University (WVU) dataset, composed of four fingerprints per subject for 240 subjects, 2) Face and Ocular Challenge Series (FOCS) collection, composed of three frontal faces per subject for 407 subjects. Experimental results show that, in a unimodal scenario when dealing with low quality data, ranks are more stable than scores. However, such a rank stability is not verified when fusing multiple matchers. Experiments demonstrate that, in the presence of low quality data, performance achieved by score-level fusion is better than that one achieved by rank-level fusion.

1 Introduction

In a generic biometric system, operating in identification mode, the input probe (e.g., a fingerprint image) is compared to the labeled biometric data in the gallery database (e.g., fingerprint database) and a set of similarity scores is generated. Scores are sorted in decreasing order and based on this ordering a set of integer values or *ranks* is assigned to these retrieved identities. The lowest rank indicates

A. Petrosino (Ed.): ICIAP 2013, Part I, LNCS 8156, pp. 260–269, 2013.

the best match; hence the corresponding identity is associated with that of the input probe. The identity of the gallery that corresponds to the true identity of the probe is known as the genuine identity; otherwise it is called impostor one.

The recognition accuracy of a biometric system generally decreases in the presence of low quality biometric data wherein the similarity between the probe and associated gallery image may be reduced [14] [6]. It has been observed that, in such a critical scenario, consolidating the evidence provided by multiple biometric sources can increase the recognition accuracy [8] [4]. Evidence can be integrated *before matching*, at sensor or feature level; or, *after matching* at decision, rank or score level [21] [7]. While the amount of information to integrate progressively decreases from the sensor level to the decision level, the degree of noise also decreases [20] [19]. Since match scores are easy to access and combine, score-level fusion has been widely used.

Recent research [13] [15] [22] has established the benefits of rank-level fusion in identification systems. Ranks only carry information about the relative ordering of the different identities in the gallery. However, there are cases where ranks are useful. First, when the output of commercial systems is only a list of candidate identities and no match scores are given [9]. Second, when conducting statistical parametric tests where distributions are assumed to be normal [5]. These tests may fail when considering match scores whose distributions are not normal. Using ranks instead of match scores, can lead to more robust results. Third, when applying monotonous transformation to match scores, the corresponding ranks are kept unchanged. Ranks do not change when the scale on which the corresponding numerical measurements changes [24]. Further, when combining multiple modalities, fusing ranks does not require a normalization phase as typically needed with heterogeneous match scores [16]. Ranks provided by multiple biometric matchers are consolidated and, for each identity in the gallery, a consensus rank is determined [18].

Monwar and Gavrilova presented a Markov chain approach for combining ranks from face, ear and iris [15]. Their experiments showed the superiority in accuracy and reliability over other biometric rank aggregation methods. They reported a rank-1 multimodal identification accuracy of 98.5% compared to the unimodal accuracies of 87%, 92% and 94% for ear, face and iris respectively. However, this improvement may be due to the presence of the iris modality. Abaza and Ross proposed a quality-based Borda Count scheme that is able to increase the robustness of the traditional Borda Count in the presence of low quality images without requiring a training phase [1]. Marasco *et al.* proposed a predictor-based approach to perform a reliable fusion at rank level. The predictor (classifier) was trained using both ranks and match scores and designed to operate before fusion [12]. Results demonstrated its effectiveness in detecting potential unimodal identification errors. An interesting analysis was conducted by Lee [10], who investigated the effect of using rank instead of similarity values when combining multiple evidence by Lee [10]. The study focuses on generating rank-similarity curves where the rank-similarity was computed by applying the function γ to the rank of the ith subject following:

$$\gamma_{(r_i)} = [1 - (r_i - 1)/N] \tag{1}$$

$i = 1 \ldots N$, where N indicates the number of enrolled subjects. The resulting value is used as the similarity value of the subject.

In the presence of low quality biometric data, the genuine match score is claimed to be low and it is expected to be an unreliable individual output, able to confuse a score level fusion algorithm and result in a potential identification error. Conversely, the rank assigned to that genuine identity is expected to remain stable even when using low quality biometric data [23]. However, the stability of ranks has been argued but not experimentally demonstrated. The contribution of this paper is to investigate the stability of ranks in the presence of low quality probes in both unimodal and multimodal scenarios. This paper is organized as follows: Section 2 defines the rank stability. Section 3 presents the approaches for fusion at rank level used to conduct this study. Section 4 describes the technique adopted to synthetically degrade the quality of the fingerprint images and the actual low quality face samples used in our experiments. Section 5 reports results and Section 6 summarizes the conclusions of this work.

2   Rank (and Score) Stability

The concept of stability is introduced here in order to have a method to evaluate the robustness of ranks and scores to low quality data. A biometric system is considered stable when small perturbations of its inputs do not alter its outputs [11]. The stability of ranks can be measured as a function of the difference between the rank assigned to the genuine identity using various low quality probes. A rank difference close to zero indicates that the system is rank stable. In this case the variation of the rank assigned to the genuine identity in the gallery when reducing the quality of the probe is limited. Similarly, the stability of match scores is based on the difference between the score assigned to the genuine identity using a high quality and a low quality probe, respectively. In this work, we consider sources of noisy input data that may arise during the image capture where the image quality can be impacted for example by an incorrect presentation of the biometric sample to the system.

Let $\mathbf{G} = [G_1, G_2, \ldots G_N]$ be the gallery set, composed by N biometric samples belonging to N different subjects. Given a single probe image, N comparisons of the probe against the gallery are performed and N similarity scores are generated. Let P denote a high quality probe image. Let P' denote the same probe image under degradation. Let s_i and s'_i indicate the score output by the matcher after comparing P and the ith gallery, and P' and the ith gallery, respectively. Let $r[s_i]$ and $r[s'_i]$ indicate the rank assigned to the scores s_i and s'_i, respectively. The score-stability statistic τ_S and rank-stability statistic τ_R can be measured as described in Eqn. (2) and Eqn. (3), respectively.

$$\tau_S = \mathbf{f}(s_i - s'_i) \tag{2}$$

$$\tau_R = \mathbf{f}(r[s_i] - r[s'_j]) \qquad (3)$$

Ranks (scores) are stable if the rank (score) assigned to the genuine identity would not change with respect to the probe quality. In other words, a small difference in ranks (scores) between using high and low quality probes indicates high stability, and viceversa. A biometric measure which measures the distance between two distributions is the relative entropy. The entropy measures the amount of information required to describe a random variable; however, as a functional of the distribution of the random variable, it does not depend on the actual values assumed by the random variable but only on the probabilities [2]. In order to measure the stability of ranks, it is important to keep the rank value. For the unimodal case, we develop statistical tests as non-parametric measure to estimate rank (and score) stability over low quality images. Tests based on an assumption of normality, like t-test, are not suitable to measure stability. We return this to the fact that these tests poorly approximates data under study. Also, the Wilcoxon and the Sign tests are not suitable to measure stability, since they assume that the distribution under the Null hypothesis is a standard normal. In this paper, we used the Kendall and the Spearman's rank correlation coefficient, whose inputs are two vectors composed by the ranks assigned to the genuine identity with high and low quality probes. For the later tests, higher correlation coefficient value indicates higher stability.

3 Experimental Results

This section discusses the used datasets and presents experiments to estimate the stability of ranks and scores for both unimodal and multimodal scenarios.

3.1 Datasets

The performance of the proposed strategy was evaluated using two databases. The first database was collected at West Virginia University (WVU). A subset of this database pertaining to the fingerprint (left thumb [FL1], right thumb [FR1], left index [FL2], right index [FR2]) was used [3]. Fingerprint images were collected using an optical sensor. The entire dataset was divided into five sets: one sample of each identity was used to compose the *gallery* and the remaining four samples of each identity were used as *probes* (P_1, P_2, P_3, P_4). VeriFinger[1] software was used for generating the fingerprint scores. Matching scenarios considered the gallery image of high quality and the probe image degraded to simulate low quality ones. The fingerprint image quality was quantified using the IQF software[17]. Degradation effects are simulated using a gray-scale saturation technique which converts fingerprint pixels corresponding to the ridges into background pixels [1]. The gray-scale saturation level (SL) indicates the gray level value above which pixels are saturated to white (255) (see Fig.1). Figure 2 illustrates the unimodal performance when using various levels of the image quality of probes.

---

[1] http://www.neurotechnology.com/verifinger.html

Fig. 1. Examples of low quality fingerprint images artificially degraded by using five different noise saturation levels ST = [128, 160, 192, 224, 240]

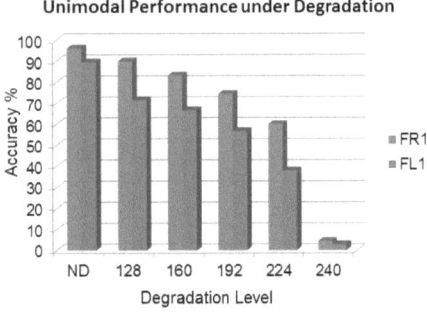

Fig. 2. The unimodal performance decreases when degrading the quality of the probe image at different levels. ND indicates the case with no degradation.

The second database is a subset of the Face and Ocular Challenge Series (FOCS) collection (the Good, Bad and Ugly database) composed by three frontal instances of faces, with two high quality images (from the Good dataset) and one actual low quality image (from the Ugly dataset in which images are taken under uncontrolled illumination, both indoors and outdoors). The partitions of interest are referred to as *Good* and *Ugly*, that have an average identification accuracy of 0.98 and 0.15 respectively [2]. PittPatt[3] software was used for generating the face match scores. Two different matching scenarios were considered: high quality gallery versus high quality probe, referred to as *Good-Good* and high quality gallery versus low quality probe, referred to as *Good-Ugly*. Table 1 provides the details of the database. Fig. 3 shows examples of actual low quality images.

3.2   Results and Discussion

We conducted experiments to estimate the stability of ranks (and match scores) in the presence of low quality input data for both unimodal and multimodal scenarios.

---

[2] http://www.nist.gov/itl/iad/ig/focs.cfm

[3] http://www.pittpatt.com/

Table 1. Details of the datasets used for the experiments

Database	Biometric	Subjects	Samples	Scores
WVU	Fingerprint (4 fingers)	240	5 per finger	Gen: $(1200 \times 4) \times 4$ Imp: $(240 \times 239 \times 25) \times 4$
FOCS	Face	407	3 per subject	Gen: 407×2 Imp: $407 \times 406 \times 2$

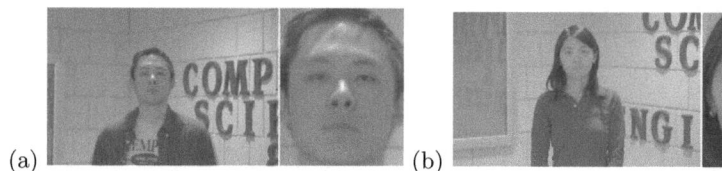

(a)   (b)

Fig. 3. Examples of face images taken from the Face and Ocular Challenge Series (FOCS) collection: (a) sample image from the Ugly partition; (b) sample image from the Good partition

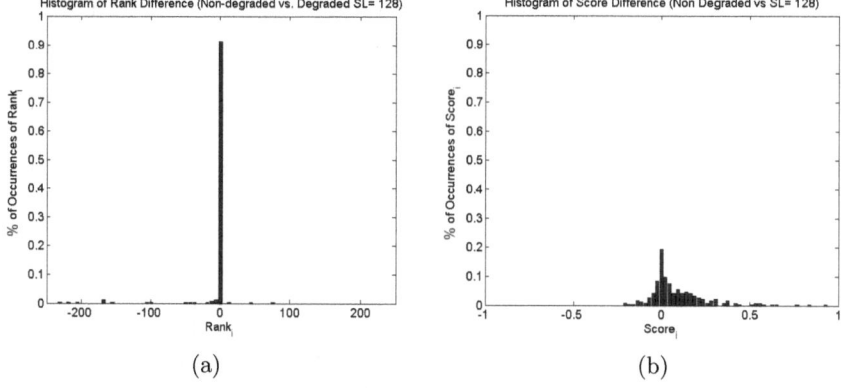

(a)   (b)

Fig. 4. Histograms of the difference between the rank (and the score) assigned to the genuine identity in the gallery before and after degradation of the probe image: (a) Rank difference: Non Degraded vs. degraded with SL= 128; (b) Score difference: Non Degraded vs. degraded with SL= 128.

Unimodal Rank/Score Stability. Ranks appears to be more stable than match scores, see Fig. 4 for fingerprints (a similar result is obtained for faces as well). Histograms of the difference between the rank (score) assigned to the genuine identity in the presence of a high and low quality probe image is shown in Fig. 4.

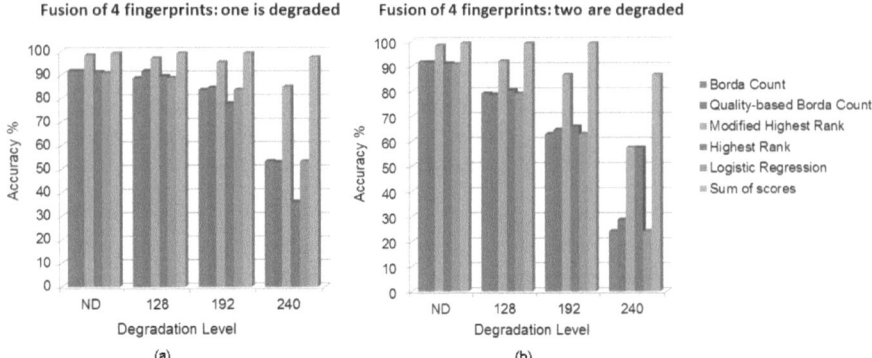

Fig. 5. (a) Fusion of four fingerprints when one of them is degraded: change in performance of different schemes at rank and score under different degradation levels. (b) Fusion of four fingerprints when two of them are degraded: change in performance of different schemes at rank and score under different levels of degradation of two fingerprint probe images.

Multimodal Rank/Score Stability. We integrated ranks in multimodal biometric systems, and compare them to the performance achieved using scores.

Fig. 5 (a) shows the accuracy achieved by rank- and score-level fusion schemes when combining four fingerprints where two are of low quality. In this scenario, the modified highest rank exhibits the best performance among the considered rank level fusion schemes; the achieved rank identification rate decreases from 92.08% to 57.5% when increasing the degradation factor. The score sum performance decreases only from 99.17% to 86.67%. Fig. 5 (b) shows the accuracy achieved when fusing four fingerprints where one is of low quality. The modified highest rank exhibits the best robustness to image quality degradation. It achieves a rank-1 identification rate of 97.08% when the noise saturation level applied to one fingerprint image in every pair is 128 and 85.00% when increasing the noise saturation level to 240. However, the performance of the score sum exceeds that obtained by rank level fusion by achieving a rank-1 identification rate of 99.17% in both non-degraded and degraded conditions. The experiments showed that rank is stable using low quality probes. The correlation coefficient value for ranks is higher than that for match scores, ranks are more stable than scores. When the level of quality degradation is significant, both ranks and scores are not stable (0.418 for ranks and 0.199 for scores).

Fig. 6 illustrates the performance when fusing one face and two fingerprints all of low quality. When combining ranks, the best accuracy is achieved by the Modified Highest Rank. For the traditional Borda Count scheme, the presence of only one incorrect identification where a high rank is assigned to the genuine identity to have a final error. The Highest Rank rule requires that only one of the combined biometric matchers assigns rank-1 to the genuine identity. Errors due to ties are solved with its modified version. When all the combined modalities

Fusion of face and two fingerprints

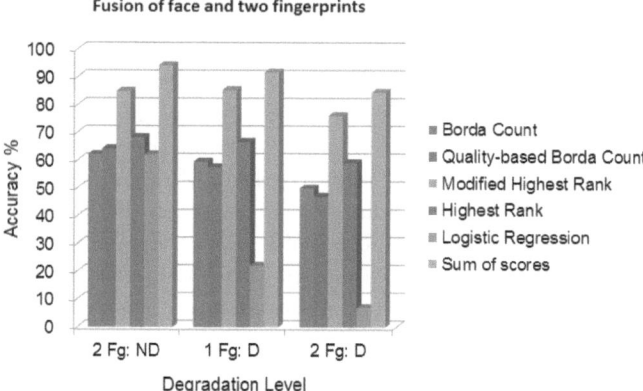

- Borda Count
- Quality-based Borda Count
- Modified Highest Rank
- Highest Rank
- Logistic Regression
- Sum of scores

Fig. 6. Fusion of one face and two fingerprints both are of low quality. The face modality is taken from the GBU data set of the FOCS database and both fingerprints from WVU database (FR1 and FL1 fingers).

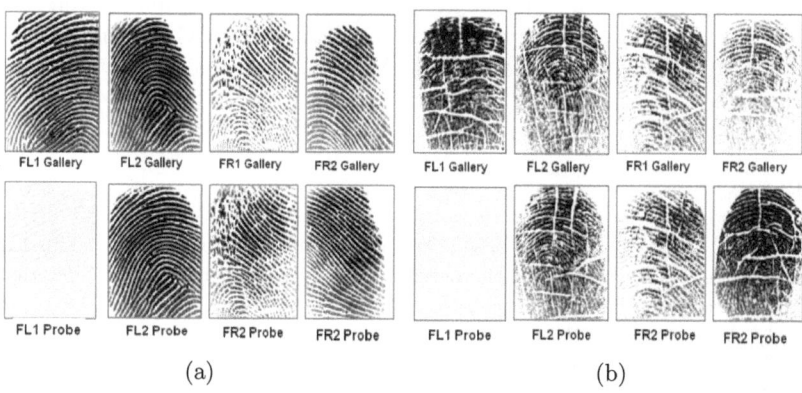

(a)   (b)

Fig. 7. (a)Fusion scenario where gallery is of high quality for all the matchers but one of the probe (FL1) is low quality (obtained with SL= 240). (b) Examples of a high quality (ND) and low quality fingerprint images where the low quality probe has been obtained with SL= 240. For this subject, when using the low quality probe, only the sum of scores is able to correctly classify that probe.

are all of low quality, the Quality-based Borda Count is the most effective fusion scheme. However, fusion at score level outperforms all the rank-level fusion schemes.

Fig. 7 (a) shows a fusion casework in which the sum of scores and the quality-based Borda Count assign a rank greater than 1, while all the other rank level fusion schemes performs well. In Fig. 7 (b) where only the sum of scores is able to output a correct decision.

4   Conclusion

This study carried out an investigation regarding the stability of the rank in the context of biometrics. Further, we analyzed different non learning-based rank level fusion schemes in the presence of both synthetically degraded fingerprint images and actual low quality face images. The experiments showed that rank is stable with low quality images, when the level of degradation is not significant; While both ranks and scores are not stable, when the level of degradation is significant. Further, ranks are more stable than scores since they present a higher rank correlation coefficient value. (However, the performed study may be dependent upon the matcher used). Conditions under which it is reasonable to use ranks can be expressed as follows:

- When match scores are not available, fusing ranks by applying the modified highest rank scheme leads to the best identification accuracy.
- When match scores are available, a better identification accuracy can be obtained by employing score level fusion.

Acknowledgments. The authors are grateful to West Virginia University and, in particular, Dr. Arun Ross. This material is supported by ONR under Contract No. N00014-09-C-0388 awarded to West Virginia High Technology Consortium Foundation.

References

1. Abaza, A., Ross, A.: Quality-based rank level fusion in biometrics. In: Third IEEE International Conference on Biometrics: Theory, Applications and Systems, pp. 1–6 (September)
2. Andy, A., Youmaran, R., Loyka, S.: Towards a measure of biometric information. In: Canadian Conference on Electrical and Computer Engineering, pp. 210–213 (2006)
3. Crihalmeanu, S., Ross, A., Schuckers, S., Hornak, L.: A protocol for multibiometric data acquisition, storage and dissemination. Technical Report, West Virginia University (2007)
4. Fiérrez-Aguilar, J., Chen, Y., Ortega-Garcia, J., K.Jain, A.: Incorporating image quality in multi-algorithm fingerprint verification. In: Zhang, D., Jain, A.K. (eds.) ICB 2005. LNCS, vol. 3832, pp. 213–220. Springer, Heidelberg (2005)
5. Friedman, M.: The use of ranks to avoid the assumption of normality implicit in the analysis of variance. Journal of the American Statistical Association 32(200), 675–701 (1937)
6. Grother, P., Tabassi, E.: Performance of biometric quality measures. IEEE Transaction On Pattern Analysis and Machine Intelligence 29(4), 531–543 (2007)
7. Kittler, J., Hatef, M., Duin, R.P., Matas, J.: On combining classifiers. IEEE Transaction on Pattern Analysis and Machine Intelligence 20(3), 226–239 (1998)
8. Kittler, J., Li, Y.P., Matas, J., Sanchez, M.U.R.: Combining evidence in multimodal personal identity recognition systems. In: Bigün, J., Borgefors, G., Chollet, G. (eds.) AVBPA 1997. LNCS, vol. 1206, pp. 327–334. Springer, Heidelberg (1997)

9. Labovitz, S.: The assignment of numbers to rank order categories. American Sociological Review 35(3), 515–524 (1970)
10. Lee, J.: Combining multiple evidence from different properties of weighting schemes. In: Proceedings of the 18th Annual International ACM SIGIR Conference on Research and Development in Information Retrieval, vol. 9, pp. 180–188. ACM, New York (1995)
11. Lempel, R., Moran, S.: Rank-stability and rank-similarity of link-based web ranking algorithms in authority-connected graphs. In: Third IEEE International Conference on Biometrics: Theory, Applications and Systems (September 2004)
12. Marasco, E., Ross, A., Sansone, C.: Predicting identification errors in a multibiometric system based on ranks and scores. In: Fourth IEEE International Conference on Biometrics: Theory, Applications and Systems, pp. 1–6 (September 2010)
13. Melnik, O., Vardi, Y., Zhang, C.: Mixed group ranks: Preference and confidence in classifier combination. IEEE Transaction on Pattern Analysis and Machine Intelligence 26(8), 973–981 (2004)
14. Monwar, M., Gavrilova, M.: Multimodal biometric system using rank-level fusion approach. IEEE Transactions on Systems, Man, and Cybernetics 39, 867–878 (2009)
15. Monwar, M., Gavrilova, M.: Markov chain model for multimodal biometric rank fusion. Signal, Image and Video Processing, 1863–1703 (2011)
16. Nandakumar, K., Jain, A., Ross, A.: Score normalization in multimodal biometric systems. Pattern Recognition 38(12), 2270–2285 (2005)
17. Nill, N.: Mitre technical report IQF (Image Quality of Fingerprint) software application (2007)
18. Ross, A., Jain, A.: Information fusion in biometrics. Pattern Recognition Letters 24, 2115–2125 (2003)
19. Ross, A., Jain, A.: Multimodal biometrics: an overview. In: 12th European Signal Processing Conference (EUSIPCO), pp. 1221–1224 (2004)
20. Ross, A., Jain, A., Nandakumar, K.: Introduction to Biometrics: A Textbook. Springer (2011)
21. Ross, A., Nandakumar, K., Jain, A.: Handbook in MultiBiometrics. Springer (2008)
22. Saranli, A., Demirekler, M.: A statistical unified framework for rank-based multiple classifier decision combination. Pattern Recognition 34, 865–884 (2001)
23. Tulyakov, S., Govindaraju, V.: Combining biometric scores in identification systems, 1–34 (2006)
24. Wilcoxon, F.: Individual comparisons by ranking methods. Biometrics Bulletin 1(6), 80–83 (1945)

Approximated Overlap Error for the Evaluation of Feature Descriptors on 3D Scenes

Fabio Bellavia[1], Cesare Valenti[2],
Carmen Alina Lupascu[2], and Domenico Tegolo[2]

[1] Università degli Studi di Firenze, Dipartimento di Sistemi Informatici,
Via di Santa Marta 3, 50139 FI, Italy
bellavia.fabio@gmail.com
[2] Università degli Studi di Palermo, Dipartimento di Matematica e Informatica,
Via Archirafi 34, 90123 PA, Italy
{cesare.valenti,carmen.lupascu,domenico.tegolo}@unipa.it

Abstract. This paper presents a new framework to evaluate feature descriptors on 3D datasets. The proposed method employs the approximated overlap error in order to conform with the reference planar evaluation case of the Oxford dataset based on the overlap error. The method takes into account not only the keypoint centre but also the feature shape and it does not require complex data setups, depth maps or an accurate camera calibration. Only a ground-truth fundamental matrix should be computed, so that the dataset can be freely extended by adding further images. The proposed approach is robust to false positives occurring in the evaluation process, which do not introduce any relevant changes in the results, so that the framework can be used unsupervised. Furthermore, the method has no loss in recall, which can be unsuitable for testing descriptors. The proposed evaluation compares on the SIFT and GLOH descriptors, used as references, and the recent state-of-the-art LIOP and MROGH descriptors, so that further insight on their behaviour in 3D scenes is provided as contribution too.

Keywords: keypoint descriptors, descriptor evaluation, epipolar geometry, SIFT, LIOP, MROGH.

1 Introduction

1.1 Related Works

Keypoint extracted from images have been adopted as primitive parts with good results in many computer vision tasks, such as recognition [10], tracking [9] and 3D reconstruction [16]. Detection and extraction of meaningful image regions, named keypoints or image features, is usually the first step of these methodologies. Numerical vectors that embody the image region properties are successively computed to compare the keypoints according to the particular computer vision task.

A. Petrosino (Ed.): ICIAP 2013, Part I, LNCS 8156, pp. 270–279, 2013.
© Springer-Verlag Berlin Heidelberg 2013

Different feature detectors have been proposed in the last decades including, but not limiting to, corners and blobs, invariant to affine transformations or scale and rotation only. Since this paper deals with feature descriptor, the reader may refers to [13] for a general overview.

After the keypoint is located, a meaningful descriptor vector to embody the characteristic properties of the keypoint support region is computed. Different descriptors have been developed, mainly divided in two categories: distribution-based descriptors and banks of filters [12]. In general, while the former kind of descriptors gives better results, the latter category provides more compact descriptors [12]. Banks of filters include complex filters, colour moments, the local jet of the keypoint and the differential operators, please refer to [12] for more details.

Distribution-based descriptors, also named histogram-based descriptors, divide the image patch into different areas and compute specific histograms for some image properties of each area. The final descriptor is given by the ordered concatenation of these histograms. The rank and the census transforms [20], which consider binary tests of the intensity of the central pixel against its neighbourhood, is the precursors of the histogram based descriptors. The recent BRIEF [4] descriptor can be considered as an extension of this kind of approach, obtained by the concatenation of binary tests on the intensity values between couples of pixels of the patch. To be mentioned are the spin image descriptor, the shape context and the geometric blur, please refer to [12].

One of the most popular histogram based descriptor is the SIFT (Scale Invariant Feature Transform) descriptor [11], given by the 3D histogram of gradient orientations on a Cartesian grid. SIFT has been extended in various ways since its first introduction, for instance GLOH (Gradient Local Orientation Histogram) [12] combines a log-polar grid with PCA (Principal Component Analysis). Overlapping regions using multiple support regions combined by intensity order pooling are used by MROGH (Multi Support Region Order Based Gradient Histogram) [5]. Recently, LIOP (Local Intensity Order Pattern) [19] uses the relative order of neighbour pixels to define the histogram.

Feature descriptors and detectors have been evaluated on different frameworks [3, 6, 7, 12–15, 17]. In the case of planar images, the repeatability index and the matching score [13] established de facto standards for evaluate feature detectors, while precision/recall curves are used in the case of descriptors [12]. The Oxford dataset [13] is considered the standard benchmark for evaluation on planar images [12–14], but an extension to the 3D space is not immediate [7] or is limited to planar object in the 3D environment [8]. Other evaluation methodologies use laser-scanner images [17] or structure from motion algorithms [3] to generate the ground-truth data, while epipolar reprojection on more than two images have been also applied to constrain the matches [15], as wells as indirect evaluations by object recognition tasks [21]. These methodologies may require complex and error prone setups [7] or would not apply to any 3D scenario [3, 8] or may not have a direct relation with the overlap error [6, 21].

1.2   Contributions

We propose a new framework based on the approximated overlap error described in [1] to evaluate feature descriptors on 3D datasets. The approximated overlap error is reported for clarity in Section 2. This measure represents an overlap between a planar approximation of the surfaces inside the feature patches, so that the evaluation in the 3D case can be made very similar to that obtained through the overlap error in the case of planar scenes [12]. The approximated overlap provides a meaningful measure which considers the feature shape and not only the feature centre as proposed in [15]. Only the fundamental matrix is required, so that no complex schemas [3,7,15,17], threshold setups [3,15], depth maps [3,7] or an accurate camera calibration [7,15] are needed. This allows to easily extend the dataset to further images.

The proposed framework is reported in Sect. 3. A minimal user interaction can be required to inspect some matches by hand, since false positive can occur as for other approaches based on the epipolar geometry [6,15], due to wrong matches lying on correct epipolar lines. An experimental analysis of the expected errors in the case of an unsupervised application of the method is reported in Sect. 3.2, which shows that results are equivalent in practice to the supervised case. Finally, an effective comparison of recent descriptors based on the proposed framework is reported in Sect. 3.3. We compared the SIFT [11] and GLOH descriptors [12], used as references, with the recent LIOP [19] and MROGH [5]. According to recent evaluations [5,14] these last descriptors outperform other novel descriptors. Conclusions and final discussions are reported in Sect. 4.

2   The Approximated Overlap Error

The working plan described by Mikolajczyk and Schmid [12] on the Oxford dataset can be considered a reference in the evaluation of descriptors in planar cases. For a stereo pair (I_1, I_2) in the dataset, precision/recall curves on correct matches are extracted through the overlap error ε

$$\varepsilon(R_w, R_z) = 1 - \frac{R_w \bigcap \mathcal{T}_{2\rightarrow1}(R_z)}{R_w \bigcup \mathcal{T}_{2\rightarrow1}(R_z)} \qquad (1)$$

where $R_w \in I_1$, $R_z \in I_2$ are the matched features and $\mathcal{T}_{2\rightarrow1}$ is the function that reprojects the feature R_z from I_2 to I_1.

Precision/recall curves for a fixed overlap error $\varepsilon = 50\%$ are extracted for increasing distance values by using a match criterion chosen between the NN (Nearest Neighbour) and the NNR (Nearest Neighbour Ratio) matching [11]. The precision is defined as the fraction of correct matches according to the overlap error with respect to all matches observed so far [12] and the recall is the fraction of correct matches detected with respect to all the possible correct matches [12]. A good descriptor requires that the precision/recall curve increase rapidly and attains high recall values when it become stable [12].

The approximated overlap error ε_q extends on planes the linear overlap error ε_l introduced by Forseén and Lowe [6], which is briefly described for the sake of

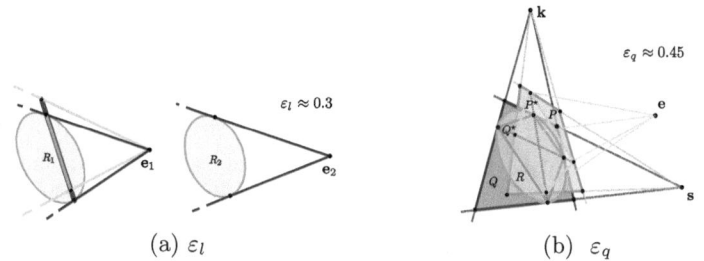

(a) ε_l (b) ε_q

Fig. 1. Overlap error approximations (best viewed in color)

clarity, see Fig. 1a. The tangency relation between epipoles and feature ellipses is preserved by perspective projection, since it is an incidence relation [18] (blue). By intersecting the line through the tangent points with the epipolar lines (yellow) of the corresponding tangent points on the respective ellipse in the other image, the linear overlap error ε_l between the small (azure) and the wider (red) segments is obtained.

In [1] an extension to the linear overlap error measure ε_l has been proposed, by observing that in order to compute the ground-truth fundamental matrix needed for ε_l, not only the correspondence between the epipoles is available, but also of fixed correspondences $(\mathbf{k}, \mathbf{s})$, $\mathbf{k}, \mathbf{s} \in C$, provided by additional hand taken points.

As shown in Fig. 1b, the tangent points of an ellipse patch R (orange) define an inscribed quadrilateral $Q^\star$ (azure), while the tangent lines provide a circumscribed quadrilateral Q (blue). The corresponding quadrilaterals $P^\star$ (light green) and P (dark green) are obtained by projecting from the other images through the fundamental matrix as done for ε_l. With an abuse of notation, the area of the ellipse R can be roughly approximated by the average area between Q and $Q^\star$

$$R \approx \frac{Q + Q^\star}{2} \tag{2}$$

The final approximate overlap error ε_q is defined as

$$\varepsilon_q = \frac{\varepsilon(Q, P) + \varepsilon(Q^\star, P^\star)}{2} \tag{3}$$

and it depends on the choice of the fixed correspondence pair $(\mathbf{k}, \mathbf{s})$. When epipolar and tangent lines are almost parallel the computation can suffer due to numerical instability. This issue can be resolved by introducing some heuristics [1].

The measure ε_q is not symmetric so the maximal value obtained for the ordered stereo pairs (I_1, I_2) and (I_2, I_1) is assigned to the match. It represents an overlap between a planar approximation of the surfaces inside the feature patches. As other approaches based on the stereo epipolar geometry, it fails for some configurations. In particular, all feature ellipses which share the same cone of epipolar lines with the correct matched feature are wrongly estimated as correct. This is however a stronger constraint with respect to consider only the epipolar distance from the keypoint centre. Furthermore, ε_q gives better results than ε_l when epipoles are inside the images [1] and can be directly related to the overlap error ε.

Since an exhaustive search should be done in order to find the best pair $(\mathbf{k}, \mathbf{s})$, the computational time $T(\varepsilon_q)$ depends quadratically on the cardinality $|C| = n$ of the fixed correspondence set C, i.e. $T(\varepsilon_q) = O(n^2)$. In order to limit n for practical purposes, we introduce a new a greedy sampling strategy on the redundant correspondences of C before the computation of ε_q described in [1]. The fixed correspondences $c \in C$ used to compute the fundamental matrix are sorted in increasing order according to their epipolar errors. Both images are divided into a $m \times m$ grid and the hand taken correspondences c are examined in turn. If any point of the correspondence c falls in a grid cell of any of the images not covered by a previous retained correspondence, the correspondence c is also retained. We experimentally verified that for a 24×24 grid size this strategy approximatively halves the number n of the hand taken fixed correspondences, thus reducing the time required to about 30% without variations in the value of ε_q. About 20 minutes on a standard PC are required to get ε_q on a stereo pair with roughly 800 matches and $n \approx 70$.

3   3D Evaluation

3.1   Setup

The evaluation on 3D images was done according to the freely available dataset[1] used in [1], see Fig. 3. The dataset consists of 10 different 3D scenes, with 3 images for each scenes so that a total of 30 stereo pairs are obtained. With respect to other evaluation methodologies, this comparison strategy does not require a complex setup [3,7,15,17], depth maps [3,7] or camera calibration [3,7,15]. Only the computation of a ground-truth fundamental matrix is required, so that the dataset can be easily extended by adding further images.

Fig. 3. Image dataset

Table 1. Evaluation strategy details

	SIFT	GLOH	LIOP	MROGH	Average	All
FP (%)	5.18	4.83	4.59	6.42	5.26	10.11
Error (%)	1.24	1.05	1.19	1.90	1.35	1.38
Positives (%)	24.02	21.67	26.02	29.52	25.30	13.66
Checked (%)	35.21	32.06	37.03	45.27	37.39	25.44
Positives	5712	5153	6188	7021	6018	10218
Checked	8375	7626	8807	10767	8893	19022
Total	23784	23784	23784	23784	23784	74784

The fundamental matrix is computed on more than 50 hand-taken correspondences for each stereo pair where the average epipolar error is of about 1 pixel.

---

[1] http://www.math.unipa.it/fbellavia/dl/3d_eval.tar.bz2

Note that the fundamental matrix computation does not require expert users, as small errors in the computation do not influence the shape area required by ε_q. This is due to the fact that not only the keypoint center is used, which would require an accurate estimation. We experienced that for higher epipolar reprojection errors in the test images, results on the approximated overlap error do not differ noticeably.

The HarrisZ detector [2], which selects robust and stable Harris corners in the affine scale-space, was used to extract keypoints. Previous evaluations [2] have shown that it is comparable with state-of-the-art detectors and provides better keypoints than Harris-affine. Furthermore, as noted in other work [5], although descriptors are influenced by detectors, the relative performances of the descriptors among different detectors are consistent [5, 7, 12, 15, 19]. The average number of keypoints extracted for the images of the dataset is about 800. For the evaluation of descriptors, SIFT and GLOH implementations by Mikolajczyk [12] are used, while the implementations available by the respective authors are used for LIOP [19] and MROGH [5].

The ε_q measure can extract correct matches with a recall close to 100% and a precision between 70-98% when considering an overlap error threshold $t_{\varepsilon_q} < 1$ [1], i.e. matches are considered correct if they share only a minimal overlap region. As the threshold t_{ε_q} decreases, the precision increases is similar way of the framework by Moreels and Perona [15].

The precision loss depends on the transformation applied to the scene as well as the uncertainly of the feature point on the epipolar line, as all the methodologies based on a ground-truth fundamental matrix. By using more than two images, as done in [15], this would be drastically reduced [1]. However, the recall of the correct matches is also reduced by about 30% [1,15] due to detector faults or occlusions, which can be unsuitable for testing descriptors. To deal with this issue, we compute ε_q on the candidate matches and a first selection is done automatically by thresholding with $t_{\varepsilon_q} < 1$. The reduced subset of matches is then inspected by hands, by simply discarding matches not sharing at least a common point. This step requires a few minutes for each stereo pair by using the tool freely provided[3]. Furthermore, time decreases as more descriptors are analysed, as shared matches between descriptors do not need to be checked again.

The remaining matches with an approximated overlap error less than a predefined threshold, set to 0.5 as for the planar test [12], are only retained to obtain the final set of correct matches used as ground-truth in the evaluation. As discussed in the next section, to threshold directly to 0.5 without user interaction gives an average expected error rate close to 5%, similar to that found in [15] and it does not change the overall results of the evaluation.

Note that the approximated overlap error threshold was fixed at 0.5 just to make the experiment consistent with those on the planar test [12], but can be further reduced to improve the point location accuracy as other approaches [3, 15]. As a positive side effect, the time required by the user to check matches will be reduced too, since the number of the estimated correct matches decreases accordingly.

3.2   Evaluation Strategy Accuracy

Table 1 shows the estimated accuracy of the unsupervised alternative, i.e. when matches are threshold by $t_{\varepsilon_q} < 0.5$ without user inspection, in the case of the proposed evaluation. Details for each stereo pair are reported in the additional material[2]. In particular, wrong matches are reported with respect to all positive matches retained for $t_{\varepsilon_q} < 0.5$ in terms of false positives (FP) and as error rate with respect to the total number of matches. These values are reported by considering all stereo pairs in the dataset for each descriptor separately, their average value and all unique matches gathered from all descriptors, since detectors share matches.

The false positives rate is about 5% on average for a single descriptor. Among the 30 stereo pairs in the dataset, the approximated overlap error gives less than 4% of false positives for 23 image pairs, i.e. for about 75% of the image pairs in the dataset, see the additional material. The same considerations hold for the error rate which is about 1.3%. The obtained false positive rate does not change the relative ranks between descriptors, so that results obtained by the unsupervised evaluation are still consistent with the ground-truth results, as reported in the additional material.

The false positive rate increases when considering all descriptors altogether. This accuracy decrease can be explained by considering that descriptors tend to share correct matches, which grow linear with the number of descriptors, while possible wrong matches increase quadratically. More wrong matches are present when the whole set of matches is considered, and the accuracy decreases accordingly.

The histograms of the false positives and of the error distribution reported in Fig. 4a-b show that the accuracy can considerably vary for different stereo pairs, in particular a considerable loss in the accuracy is present for the Kermit scene (rightmost image on the top row in Fig. 3), more details can be found in the additional material. This observation, should be taken into account in these kinds of evaluation.

Table 1 and Fig. 4c also report the number of matches that were checked for the supervised strategy in the proposed evaluation, i.e. all matches for which $t_{\varepsilon_q} < 1$. This value decreases from 37% for a single descriptor to 25% when we consider all descriptors since shared matches are taken into account. In the case of a direct threshold by $t_{\varepsilon_q} < 0.5$ to reduce the time spent, average values from 25% to 13% occur, corresponding to the positive matches in Table 1. Furthermore, matches to be checked are proportional to the correct matches, as the latter set is a subset of the former. For the same reasons, the matches to be checked proportionally decrease with the complexity of the stereo image pair.

Taking into account these observations, in order to add further new stereo pairs in the dataset, we suggest to check their accuracy for a reference descriptor and, if this value is reasonable, just follow the unsupervised evaluation for each descriptor to test.

---

[2] http://www.math.unipa.it/fbellavia/dl/3d_eval_additional_material.pdf

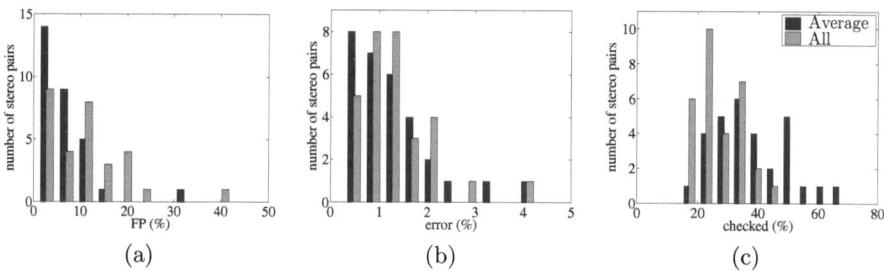

Fig. 4. Histograms of true positives (a), error rates (b) and checked matches (c) for the 30 stereo pairs in the dataset

3.3   Results

Fig. 5 shows the correct matches rate for each stereo pair and descriptor with respect to SIFT. We reported results for NNR matching with the L_1 distance, which is the best choice according to our experiments, see the additional material[4]. Histogram bars are clustered by scenes and then by stereo pairs. Inside a 3D scene, the stereo pair with the highest perspective distortion and occlusion achieves the lowest number of correct matches. The most challenging scene is represented by Kermit, where a high perspective distortion is present.

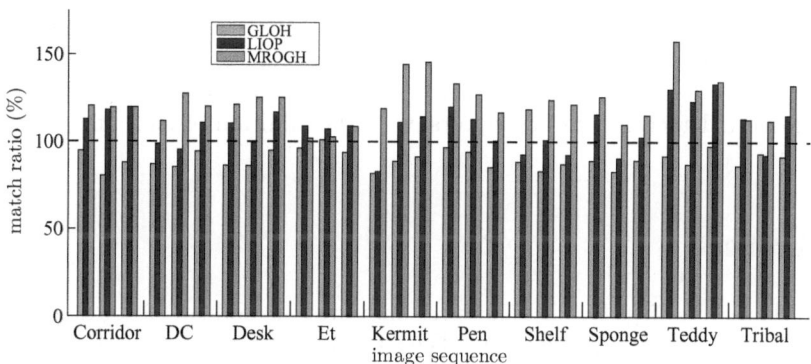

Fig. 5. Match ratio with respect to SIFT (best viewed in color)

GLOH performs slightly worse than SIFT with respect to other evaluations [12]. The difference in correct matches between SIFT, GLOH and rotation invariant descriptors is reduced from 50% in the planar case [5,19] to about 20% in 3D scenes. MROGH obtains in general the best results. Although LIOP does better than SIFT in most cases, it obtains a lower number of matches for some image pairs, because the local relative order of pixel intensities can suffer due to image discontinuities associated to perspective distortions in 3D image sequences.

Both the recently proposed rotational invariant LIOP and MROGH descriptors based on the intensity ordering pooling [5] are promising and can represent a new research direction toward the design of more efficient descriptors to avoid the bottleneck represented by the dominant orientation assignment [11] required by SIFT and GLOH.

4   Conclusions

This paper presents a new framework to evaluate feature descriptors on 3D datasets based on the recent approximated overlap error ε_q. This measure has a geometric mean which allows to uniform the evaluation on 3D scenes to the planar case.

The proposed framework, which does not require complex setups [3,6,7,12,13, 15,17], is freely available[1] and can be extended to include more images without relevant efforts. As other methods based on the epipolar stereo geometry [6,15], some false positive matches can occur, comparable to that obtained with other evaluation strategies, but no loss in recall is present since only stereo pairs are used, which makes this methodology appropriate to compare feature descriptors. The unsupervised use of the framework is still available, since the relative rank between the descriptors does not change.

By using the proposed evaluation strategy, a new comparison between the SIFT and GLOH descriptors with respect to novel rotational invariant approaches by intensity pooling, represented by LIOP and MROGH, is given. Descriptors based on the latter approach, especially MROGH, obtain better results, which show the validity on the rotational invariant method used.

Future works will include the extension of this evaluation scheme to further feature detectors and descriptors, with the addition of more images to the dataset.

References

1. Bellavia, F., Tegolo, D.: New error measures to evaluate features on three-dimensional scenes. In: Maino, G., Foresti, G.L. (eds.) ICIAP 2011, Part I. LNCS, vol. 6978, pp. 524–533. Springer, Heidelberg (2011)
2. Bellavia, F., Tegolo, D., Valenti, C.: Improving Harris corner selection strategy. IET Computer Vision 5(2) (2011)
3. Brown, M., Hua, G., Winder, S.: Discriminative learning of local image descriptors. IEEE Transactions on Pattern Analysis and Machine Intelligence 33(1), 43–57 (2011)
4. Calonder, M., Lepetit, V., Strecha, C., Fua, P.: BRIEF: Binary robust independent elementary features. In: Daniilidis, K., Maragos, P., Paragios, N. (eds.) ECCV 2010, Part IV. LNCS, vol. 6314, pp. 778–792. Springer, Heidelberg (2010)
5. Fan, B., Wu, F., Hu, Z.: Rotationally invariant descriptors using intensity order pooling. IEEE Transactions on Pattern Analysis and Machine Intelligence 34(10), 2031–2045 (2012)

6. Forssén, P., Lowe, D.G.: Shape descriptors for maximally stable extremal regions. In: International Conference on Computer Vision. IEEE Computer Society Press (2007)
7. Fraundorfer, F., Bischof, H.: A novel performance evaluation method of local detectors on non-planar scenes. In: IEEE Conference on Computer Vision and Pattern Recognition, p. 33. IEEE Computer Society Press (2005)
8. Gauglitz, S., Höllerer, T., Turk, M.: Evaluation of interest point detectors and feature descriptors for visual tracking. Int. J. Comput. Vision 94(3), 335–360 (2011)
9. Gil, A., Mozos, O.M., Ballesta, M., Reinoso, O.: A comparative evaluation of interest point detectors and local descriptors for visual SLAM. Machine Vision and Applications (MVA) 21(6), 905–920 (2010)
10. Lazebnik, S., Schmid, C., Ponce, J.: Beyond bags of features: spatial pyramid matching for recognizing natural scene categories. In: IEEE Conference on Computer Vision and Pattern Recognition, pp. 2169–2178. IEEE Computer Society Press (2006)
11. Lowe, D.G.: Distinctive image features from scale-invariant keypoints. International Journal of Computer Vision 60(2), 91–110 (2004)
12. Mikolajczyk, K., Schmid, C.: A performance evaluation of local descriptors. IEEE Transactions on Pattern Analysis and Machine Intelligence 27(10), 1615–1630 (2005)
13. Mikolajczyk, K., Tuytelaars, T., Schmid, C., Zisserman, A., Matas, J., Schaffalitzky, F., Kadir, T., Gool, L.V.: A comparison of affine region detectors. International Journal of Computer Vision 65(1-2), 43–72 (2005)
14. Miksik, O., Mikolajczyk, K.: Evaluation of local detectors and descriptors for fast feature matching. In: International Conference on Pattern Recognition (2012)
15. Moreels, P., Perona, P.: Evaluation of features detectors and descriptors based on 3d objects. International Journal of Computer Vision 73, 263–284 (2007)
16. Snavely, N., Seitz, S., Szeliski, R.: Modeling the world from internet photo collections. International Journal of Computer Vision 80(2), 189–210 (2008)
17. Strecha, C., von Hansen, W., Gool, L.V., Fua, P., Thoennessen, U.: On benchmarking camera calibration and multi-view stereo for high resolution imagery. In: Computer Vision and Pattern Recognition (2008)
18. Szeliski, R.: Computer Vision: Algorithms and Applications. Springer (2010)
19. Wang, Z., Fan, B., Wu, F.: Local intensity order pattern for feature description. In: IEEE International Conference on Computer Vision, pp. 603–610 (2011)
20. Zabih, R., Woodfill, J.: Non-parametric local transforms for computing visual correspondence. In: Eklundh, J.-O. (ed.) ECCV 1994. LNCS, vol. 801, Springer, Heidelberg (1994)
21. Zhang, J., Marszałek, M., Lazebnik, S., Schmid, C.: Local features and kernels for classification of texture and object categories: a comprehensive study. International Journal of Computer Vision 73(2), 213–238 (2007)

Exploiting the Golden Ratio on Human Faces for Head-Pose Estimation

Gianluca Fadda, Gian Luca Marcialis, Fabio Roli, and Luca Ghiani

Department of Electrical and Electronic Engineering,
University of Cagliari
Piazza d'Armi - 09123 Cagliari, Italy
{marcialis,roli,luca.ghiani}@diee.unica.it

Abstract. In this paper, a novel method for automatic head pose estimation is presented. This is based on a geometrical model of the head, in which basic features for estimating the pose consist in eyes and nose coordinates only. Worth noting, the majority of state-of-the-art approaches requires at least five features. The novelty of our work is the exploitation of the Vitruvian man's proportions and the related "Golden Ratio". The "Vitruvian man" is the well-known master-work by Leonardo Da Vinci, never used for automatic head pose estimation. Proposed method is compared by experiments with state-of-the-art ones, and shows a competitive performance although its simplicity and its low computational complexity.

1 Introduction

Head-pose estimation is based on the computation of three angles related to three reference axis, named yaw, pitch, and roll angles, which describe the rotation amount of the head along the three-dimensional space, with respect to an ideal frontal view.

In this case, brute-force approaches as machine learning-based ones, where a classifier is trained on several features from face images, is simply unsuitable. In fact, a lot of samples, covering different poses, are necessary. Moreover, head pose captured during system's operations can be significantly far from the ones used during training. This is due to the simple fact that, if no 3D information is available, estimating the direction according to 3D axes from 2D face images is a typical bad posed problem.

According to the taxonomy proposed in [1], head pose estimation approaches can be subdivided in: (1) appearance-template, where each face image is horizontally reverted and the symmetry degree with respect to the vertical axis along the xy reference plan must be assessed; (2) feature-tracking, where spatial features are extracted and tracked in a video-sequence. Experiments have shown that these methods are highly unreliable [2]; (3) moment-based, where a tessellation is projected on the face image. Among parts of the tessellation, three are selected which contain eyes and nose. From each part, a moment-based feature is extracted and used for detecting a discrete set of head poses [2]; (4)

A. Petrosino (Ed.): ICIAP 2013, Part I, LNCS 8156, pp. 280–289, 2013.

geometrical, which are based on the relative position of several features as eyes, nose and mouth.

The majority of head pose estimation methods relies on the computation of three angles named, respectively, Pitch, Yaw, Roll. Each one is related to the head rotation with respect of a reference axis in the 3D space. Thanks to these angles, it is possible to evaluate how much a certain estimated head pose is far from another one, by using appropriate distance functions [1].

Unfortunately, estimation of roll, yaw and pitch angles requires at least five features [1,3,6,7], for geometrical approaches, and this number strongly increases when considering other methods, as the ones based on neural networks or feature tracking [1]. To summarize current limits of state-of-the-art approaches [1]:

- Head pose is estimated by assuming an incremental variation of the facial position during video-sequence.
- Initial head pose in video-sequence is known.
- Scene calibration is required and several information must be extracted about the adopted camera, especially for the computation of the Pitch angle.
- All facial features are manually computed (e.g. eyes and nose position).
- The head is supposed to rotate around one axis at a time.

In this paper, we propose a novel method to automatic head pose estimation, where only three features, namely, eyes and nose locations are required. This is based on a novel model of the human head where geometrical relationships among above features are ruled by the concept of "Golden Ratio".

The "Golden Ratio" is the proportionality constant adopted by Leonardo Da Vinci in his master-work called "The Vitruvian Man". To the best of our knowledge, no work adopted the "Golden Ratio" for estimating the head pose. This method allows overcoming three limitations: (1) the use of a set of features smaller and easier to find, whose location cannot be necessarily exact; (2) no calibration of the scene is required, as well as camera parameters or characteristics; (3) it can be used for real-time applications, due to its low computational complexity.

Proposed method is tested on three benchmark data sets publicly available and adopted for comparing head pose estimation algorithms. All features are automatically computed, without human intervention. Reported results, in terms of well-known evaluation parameters, confirm that, although simpler, our approach is competitive with other state-of-the-art ones.

Paper is organized as follows. Section II describes the proposed model and algorithm. Section III reports some experimental results. Section IV concludes the paper.

2 The Proposed Model

The mathematical concept behind this work is the so-called "Golden Ratio". The "Golden Ratio" is the proportionality constant adopted by Leonardo Da Vinci in his master-work called "The Vitruvian Man" [4]. This is largely used in dentistry

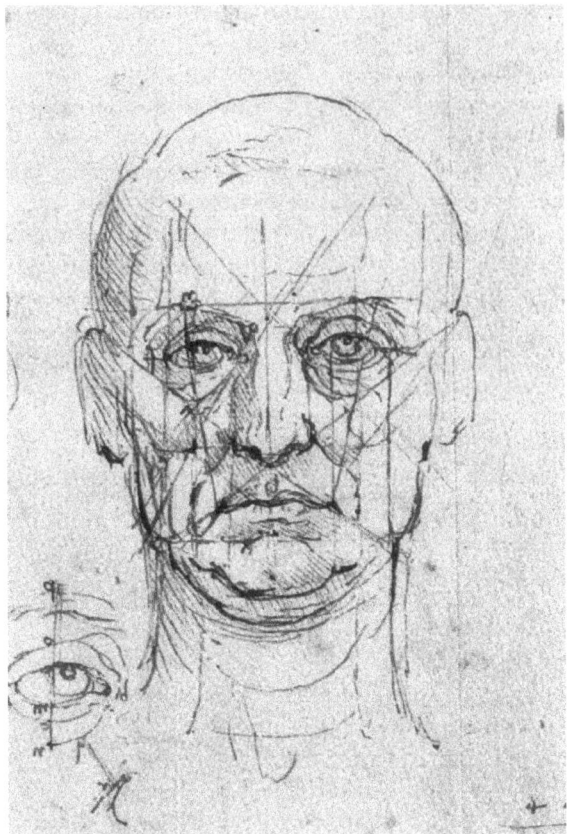

Fig. 1. Face and eyes proportions based on the "Golden Ratio", from a picture by
Leonardo da Vinci, 1488–9

and plastic surgery as it represents the ideal "perfection" and harmony of human
proportions [5]. It easy to find several approaches to assess the human face
beauty, based on the Vitruvian man's proportions. A detail on such proportions
is reported in Fig. 1.

In few words, the concept consists in recursively subdividing a certain line,
which may be represented by the duration/pitch of musical notes, or the size of
physical objects.

The "Golden ratio" (*phi*) is a value which represents the proportionality
constant of a line divided into two segments a and b, such that the whole
line is to the longer a segment as the a segment is to the shorter b segment:
$\phi = (a + b) : a = a : b$. By solving the related harmonic equation we obtain:

$$\phi = \frac{1 + \sqrt{5}}{2} = 1.61803399\ldots \tag{1}$$

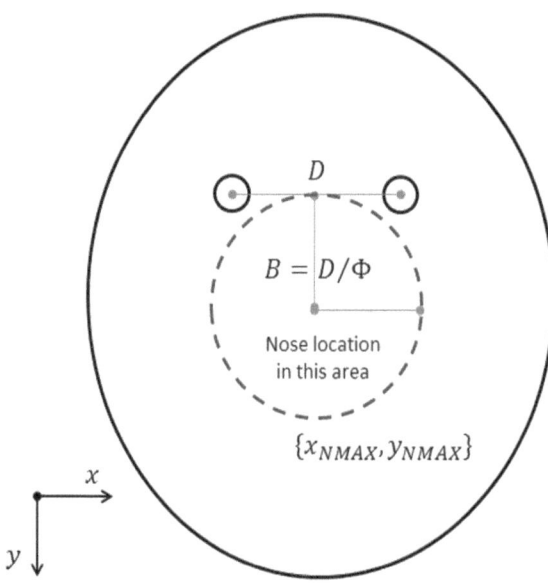

Fig. 2. Proposed model of a frontal head

This number, ϕ (*phi*), re-calling the initial letters of the sculptor Fidia, who used the "Golden Ratio" to create the Parthenon sculptures, is called "Golden Ratio".

Fig. 2 shows the proposed geometrical model of head. It is easy to see that this requires the eyes position (x-y coordinates). The "Golden Ratio" is exploited in locating the "ideal" nose location (x_{FN}, y_{FN}), according to the Vitruvian man's proportion. This point can be positioned along the line orthogonal with respect to the one jointing eyes, and passing on the middle point of that line (x', y'). The distance from this point is given by B, as shown in Fig. 2. Therefore, given the interocular distance D, we obtain $B = |y' - y_{FN}|$ that can be approximated by following Eq. (1):

$$\frac{D}{B} = \phi \Rightarrow B = \frac{D}{\phi} \approx 0.618D \tag{2}$$

Where: $\phi = 1.61803399\ldots$ is the "Golden Ratio".

By following the above assumption, three angles may be easily computed as follows.

Roll Angle Computation. According to Fig. 3, Roll angle is computed as:

$$Roll = arctg\left(\frac{dy}{dx}\right) \tag{3}$$

On the basis of dy sign, detectable roll angles are: $(-90° \div +90°)$.

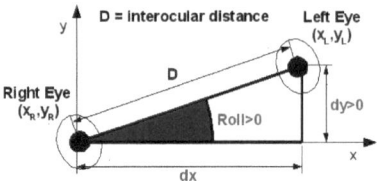

Fig. 3. Roll angle computation

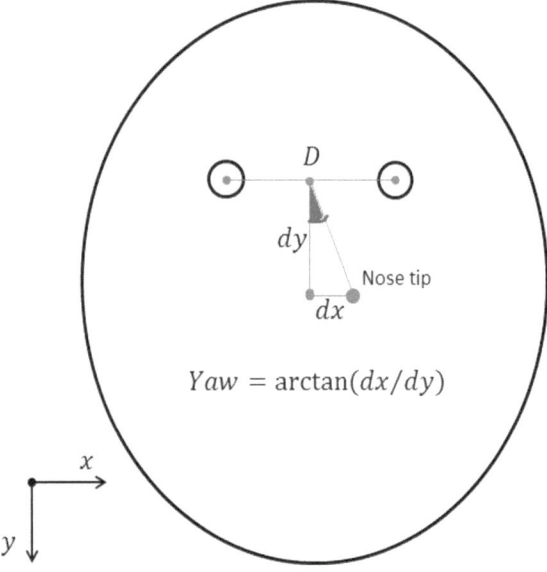

Fig. 4. Yaw angle computation

Yaw Angle Computation. Let (x_N, y_N) be the coordinates related to the nose location. According to Fig. 4, Yaw angle is computed as:

$$Yaw = arctg\left(\frac{dx}{dy}\right) \tag{4}$$

Pitch Angle Computation. Main innovation of this paper is the computation of Pitch angle, which is the most difficult to compute in the majority of approaches to head pose estimation, since it requires knowledge about the "depth" of the scene. In other words, 3D information, along the axis orthogonal to the image plan (xy).

State-of-the-art methods solve the problem of Pitch angle estimation by adding information about the used camera. Since it is necessary to know a reference distance between xy plan and another point along z axis, the usual solution is to consider the focal point of the camera. However, this approach makes the overall head pose estimation algorithm strongly dependent on the adopted hardware.

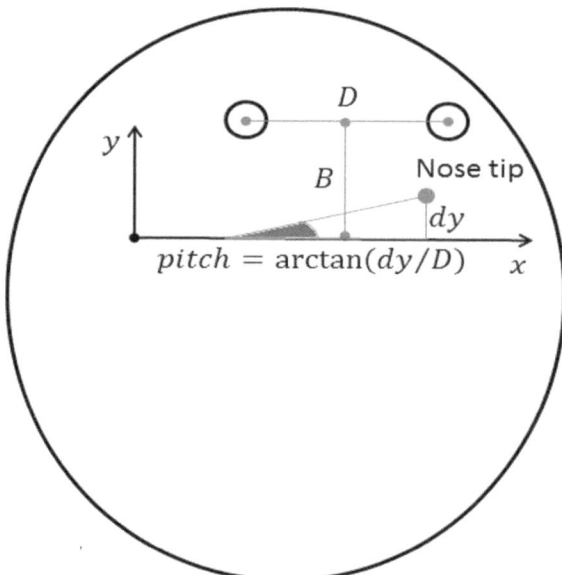

Fig. 5. Pitch angle computation

Solution proposed here overcome above limitations. Fig. 5 shows our face model, where the Pitch angle is given by:

$$Pitch = arctan\left(\frac{dy}{D}\right) \tag{5}$$

Maximum Angle Values. From equations above, it is possible to derive the maximum values of roll, pitch, and yaw angles which are tolerable by this approach.

- Roll angle. Depending on the sign of dy, detectable values are: $(-90° \div +90°)$.
- Yaw angle. Since the worst case corresponds to the nose location on the circumference of the circle depicted in Fig. 2, detectable angles are: $(-45° \div +45°)$.
- Pitch angle. As in the previous case, the maximum detectable values correspond to the nose location equal to (x_{NMAX}, y_{NMAX}). Accordingly, we obtain the interval: $(-31.716° \div +31.716°)$.

Since a non-frontal pose points out the loss of "harmony" with respect to the Golden ratio-based proportions, we expect that this method can be affected by an estimation error of the pose. In the next Section, we quantify this loss and compare it with some state-of-the-art algorithms.

3   Esperimental Results

Performance of the proposed system has been evaluated on three benchmark data sets:

- *Pointing '04 Head Pose Image Database* [10] is made up of 15 image galleries related to 15 different persons. Each gallery contains two sequences of 93 face images. Fifteen persons in the data set exhibit different characteristics in terms of age, eye glasses, skin colour. Since a quantitative performance of head pose estimation algorithms is necessary, a ground truth is given in terms of yaw and pitch angle, both discretized between $-90°$ and $+90°$ (thirteen values for yaw and nine values for pitch). The ground truth has been obtained by constraining captured subjects to look at several markers in the scene. Roll angle is not given.
- *Boston University (BU) Face tracking dataset* [3], is made up of 72 sequences taken from five subjects. First 45 sequences (9 per subject) have been captured under uniform lighting. Second 27 sequences (9 per a subset of 3 subjects) under different lighting variations. Each sequence is made up of 200 frames during which several free rotations and translations of the head have been taken. Ground truth is given in terms of Yaw, Roll and Pitch angles, evaluated through a magnetic sensor.
- *Head Pose and Eye Gaze (HPEG) Dataset* [8] is made up of 20 video-sequences subdivided in two different sessions. First one is studied for evaluate exactly the head pose; second one is aimed to the estimation of eye gaze. In each session, 10 subjects are captured. Different head rotations are allowed. Ground truth is given in terms of Yaw and Pitch angles, and computed thanks to a semi-automatic labelling process, based on three leds located around the subject's face.

All video-sequence frames are evaluated in order to assess the performance of the proposed method.

Systems we developed for head detection and facial feature localization is based on the Viola-Jones framework available on the OpenCV libraries [9]. Eyes and nose locations are evaluated using Viola-Jones classifiers esplicitly trained for the detection of these biometrics.

Starting from basic state-of-the-art algorithms, we assessed the estimation of the Yaw, Roll, and Pitch angles according to the formulas we showed in the previous Section.

Adopted evaluation parameters have been suggested in [1], and explained in the following.

Mean Absolute Error (MAE):

$$MAE = \frac{1}{N} \sum_{i=1}^{N} |a_i - GT_i] \tag{6}$$

Table 1. MAE of the proposed method on images of the Pointing'04 data set using manual and automatic localization of eyes and nose. Results refer to all 35 poses and 15 individual on the Pointing'04 data set.

	Mean Absolute Error	
	Yaw	Pitch
Automatic localization	9.6°	13.6°
Manual localization	12.6°	13.7°

And Root Mean Square Error (RMSE):

$$RMSE = \sqrt{\frac{1}{N}\sum_{i=1}^{N}(a_i - GT_i)^2} \tag{7}$$

Where a_i and GT_i are the considered angle (Yaw/Roll/Pitch) and the correspondent ground truth, and N is the overall number of frames.

Table 1 reports, first of all, an experiment aimed to see the difference of MAE parameter where nose and eyes are localized manually and automatically. Results are averaged on all images available in the Pointing'04 data set, that is, all 35 poses over 15 individuals are taken into account. It is possible to see that the Yaw angle is the most affected by a unaccurate estimation of eyes and nose, whilst no appreciable difference can be noticed on the Pitch angle. The Pointing'04 data set contains specific images for each range of possible poses, thus the assessment of results in simplified. Moreover, Pointing'04 data set is used as benchmark in the majority of papers on head pose estimation [1].

Tables 2-3 report the performance of our method and the one of other state-of-the-art approaches extracted from [1]. These evidences show that our simple method, based on a geometrical analysis of the face, and the Vitruvian man's proportions exploitation, is suitable for fine head pose estimation. Reported results clearly show that our method can be successfully used for this aim, and, in particular, allows an estimation degree more accurate than other state-of-the-art solutions [1]. It is worth to point out that state-of-the-art geometrical approaches perform much worse than ours, and this is the reason for which we did not report comparison with such approaches[1]. State-of-the-art methods reported here are based on much more complex approaches: in particular they exploit the video information by feature tracking, whilst out system estimates the head pose frame-by-frame. Therefore, whilst these methods can be effective only if a video-sequence is available, our method show superior or comparable performance by exploiting information extracted on still images. Since state-of-the-art algorithms are used to extract eyes positions, and that data sets used are made up of video-sequences, our method can be used in real time, as well as other reported ones.

Finally, it is worth to point out that adopted data sets exhibit a third-party ground truth. Thus, they allow the comparison and assessment of the methods

Table 2. Comparison of results on BU Face Tracking dataset

	Mean Absolute Error		
	Yaw	Pitch	Roll
Proposed Method	**5.5°**	**3.8°**	**3.4°**
La Cascia (Tracking) [3]	3.3°	6.1°	9.8°
Xiao (Tracking) [6]	3.8°	3.2°	1.4°

Table 3. Comparison of results on HPEG dataset

	Root Mean Square Error	
	Yaw	Pitch
Proposed Method	**7.45°**	**5.10°**
Asteriadis (Feat. Track. with Optical Flow) [7]	8.39°	5.51°
Asteriadis (Feat. Track. with Distance Vector Fields) [7]	6.65°	5.59°

performance. On the basis of reported results, estimated angles by our method are very near to this ground truth. With regard to the Yaw angle, error is less than 7 degrees, on average, and appears to be the most affected by a unaccurate estimation of eyes and nose. On the other hand, average error is less than for the Pitch angle, which is the most difficult to estimate and also the one requiring significant information especially for other state-of-the-art approaches. With regard to this point, it is worth remarking that our method does not require knowledge about the scene or the camera characteristics, thus reported results are noticeable.

4   Conclusions

In this paper, we presented a novel geometrical model for head pose estimation. Computation of Roll, Yaw and Pitch angles requires the location of eyes and nose, which is a number of feature much inferior than that required by other geometrical and non-geometrical methods. This low number of required features is due to the exploitation of the "Golden Ratio" among such features, which led, in particular, to a very effective estimation of the Pitch angle. In fact, estimating this angle requires 3D information which state-of-the-art approaches derive from the scene or the camera adopted. As a further advantage, no scene calibration is required.

Our method has been quantitatively evaluated on three benchmark data sets already used for fine head pose estimation. Method has been also compared with other non-geometrical approaches, resulting in a better performance on average, which is coupled with the low amount of computational complexity required.

Proposed algorithm suffers from the inexact computation of eyes and nose coordinates, as usual for all geometrical methods. However, on overall, it appears as quite stable, as shown here, where all feature have been automatically detected.

Finally, thanks to the our approach based on the "Golden Ratio", we have been able to derive the maximum estimation range allowable analytically, thus predicting the applications set for which our model can be suitable.

This preliminary set of experiments has shown that this geometrical approach is worth of further investigations. Future works will include extensive experiments on other data sets, and also possible countermeasures to correct estimation when nose and eyes can't be found reliably.

References

1. Murphy-Chutorian, E., Trivedi, M.M.: Head pose estimation in computer vision: A survey. IEEE Transactions on Pattern Analysis and Machine Intelligence 31(4), 607–626 (2009)
2. Moeslund, T.B., Mortensen, B.K., Hansen, D.M.: Detecting head orientation in low resolution surveillance video. Technical report, CVMT-06-02 ISSN 1601-3646 (2006)
3. Cascia, M.L., Sclaroff, S., Athitsos, V.: Fast, reliable head tracking under varying illumination: An approach based on registration of texture-mapped 3d models. IEEE Transactions on Pattern Analysis and Machine Intelligence 22, 322–336 (2000)
4. Leonardo da Vinci (1452–1519). Uomo Vitruviano, Gallerie dell'Accademia, Venice, ITALY (ca. 1490), http://brunelleschi.imss.fi.it/ stampa_leonardo/images/uomo_vitruviano_accademia_v.jpg
5. Baker, B.W., Woods, M.G.: The role of the divine proportion in the esthetic improvement of patients undergoing combined orthodontic/orthognathic surgical treatment. International Journal of Adult Orthodon Orthognath Surgery 16(2), 108–120 (2001)
6. Xiao, J., Moriyama, T., Kanade, T., Cohn, J.: Robust full-motion recovery of head by dynamic templates and re-registration techniques. International Journal on Imaging Systems and Technology 13(1), 85–94 (2003)
7. Asteriadis, S., Karpouzis, K., Kollias, S.: Head Pose Estimation with One Camera, in Uncalibrated Environments. In: International Workshop on Eye Gaze in Intelligent Human Machine Interaction (2010)
8. Asteriadis, S., Soufleros, D., Karpouzis, K., Kollias, S.: A natural head pose and eye gaze dataset. In: International Conference on Multimodal Interfaces (ICMI 2009), Boston, MA (November 2-6, 2009)
9. Viola, P., Jones, M.: Robust real-time object detection. International Journal of Computer Vision 57(2), 137–154 (2004)
10. Stiefelhagen, R.: Estimating Head Pose with Neural Networks - Results on the Pointing04. In: ICPR Workshop on Visual Observation of Deictic Gestures, Cambridge, UK (2004)

An Interactive Video Retrieval Approach Based on Latent Topics*

Rubén Fernández-Beltran and Filiberto Pla

Institute of New Imaging Technology, Jaume I University, Castellón, Spain
{rufernan,pla}@uji.es

Abstract. The huge collections of unconstrained videos have amplified the so-called semantic gap for content-based video retrieval. Therefore, new efficient approaches with higher generalisation power are needed. In this work, we present an interactive video retrieval approach based on latent topics to cope with the semantic gap in an efficient way. A supervised Symmetric extension of probabilistic Latent Semantic Analysis model is presented (sSpLSA). Then, this model is adapted to an on-line interactive information retrieval problem and it is applied to a video retrieval framework based on explicit short-term Relevance Feedback (RF) where queries are inside the database. Finally, several retrieval simulations using the Consumer Columbia Video (CCV) database are performed to compare the proposed approach with a distance-based RF baseline.

Keywords: Content-based video retrieval, relevance feedback, latent topics.

1 Introduction

In recent years the great expansion of video collections has boosted the video media as the biggest resource for information exchange on Internet. In this situation, one of the most important challenges is how to retrieve user relevant data from this huge amount of information in an effective way. Traditional search engines retrieved videos using only textual information. This limits the capabilities of the retrieval process due to the fact that they were devoid of the capacity to understand media contents. Over the last years, Content-Based Video Retrieval (CBVR) has become a very important area of research and several content-based video retrieval systems have been developed [1]. Recent works deal with video content at various levels: low-level descriptors and concepts detector [14]. However, recent results have shown that this approach does not scale adequately when the number of trained concepts increases [16]. Therefore, this information is not enough for discriminating across different multimedia content when unconstrained videos are considered [11].

---

* This work was partially supported by FPU-AP-2009-4435 from the Spanish Ministry of Education, PROMETEO/2010/028 project from Generalitat Valenciana and P1-1B2010-27 project from the Plan de Promoció de la Investigació UJI.

A. Petrosino (Ed.): ICIAP 2013, Part I, LNCS 8156, pp. 290–299, 2013.

The main problem in content-based video retrieval is the so-called semantic gap between computable low-level features and semantic concepts that users want to retrieve [15]. Most of current approaches use classifiers to attempt filling the semantic gap. Thus, they use features without semantic meaning to represent semantic concepts. For this reason, current approaches in video retrieval are only reliable under a specific domain (constrained videos) and they also need an extensive computation. One of the promising research directions that might improve video retrieval performance is based on using other kind of representation methods beyond conventional bag-of-words (BoW). In this field, topic models based techniques [3] can be used to characterize the samples in a higher level of semantic significance. Topic models have been used obtaining satisfactory results in many areas, such as text categorization [6], image recognition [13] or video classification [5]. Two of the most used algorithms are probabilistic Latent Semantic Analysis (pLSA) [9] and Latent Dirichlet Allocation (LDA) [4].

The presented work is focused on providing an interactive video retrieval approach based on latent topics, in order to cope with the semantic gap challenge using topic model representations. In this work, we propose a new point of view for the retrieval process. The problem becomes in a class discovery problem using a supervised latent topic model. The underlying goal is twofold, on the one hand, the use of latent topic representation is intended to express the data in a characterization space that is semantically nearer to the user's concepts. On the other hand, the use of the probabilistic ranking approach to be developed based on topic model representations is intended to be computationally effective for class discovering tasks in large scale information retrieval systems.

The rest of the paper is organized as follows. Section 2 discusses the background of the work. In Section 3, we present a supervised Symmetric extension of pLSA model (sSpLSA) in order to relate class labels and topic model characterisation. Subsequently, we adapt this supervised model to an on-line interactive information retrieval system where the problem becomes in a new class discovery problem by using the feedback provided by the user. Later, we apply this approach to a video retrieval framework based on explicit short-term Relevance Feedback (Section 4). In Section 5, we perform several experiments using one of the most challenging dataset for unconstrained video retrieval (Consumer Columbia Video database) and we compare proposed approach with respect a distance-based baseline. Finally, Section 6 draws the main conclusions arisen from this work and notes the future work.

2   Related Work: Content-Based Video Retrieval

Popular video retrieval approaches [1] have been focused on shots retrieval, that is, short pieces of video with homogeneous content. Each shot is summarized in a feature vector (descriptor) using combinations of low-level feature values or concepts. Then, these descriptors are used to rank the database and return the most relevant videos according to user's query. Therefore, two issues are essential for video retrieval performance: the descriptor and the ranking algorithm.

Regarding to the ranking algorithms, there are several methods to rank the database samples according to their relevance to the query. One of the most frequently used in multimedia retrieval has been distance-based ranking (e.g. [10]). However, the measures of distance or similarity are not able to work properly when the multimedia data are too complicated. Other ranking algorithms are based on inductive learning which typically uses a bank of classifiers to represent the set of possible events to test [14]. But, the performance of this approach depends on the training data and also the number of concepts to retrieve are very constrained for general applications. An alternative ranking methods are based on transductive ranking. That is, they use the data distribution of all the samples in order to improve the output ranking in the retrieval process. One of the most representative is Manifold Ranking (MR) [17] which rank the data with respect to the intrinsic data distribution.

Several of these approaches have shown to be successful in video retrieval process when they are used on edited videos with a specific number of concepts [2]. However, with the popularity of hand-held video recording devices, a huge amount of videos are captured by non-professional users under unconstrained conditions [11]. For unconstrained videos, the visual representation variability of a concept is so high that these approaches do not scale adequately when the number of trained concepts increases [16]. This amplifies the semantic gap challenge and demands for new capabilities in video retrieval with a higher generalization power. Moreover, video collections contain increasingly samples, therefore new efficient approaches are needed to deal with this amount of data in real life applications.

In order to bridge the semantic gap Relevance Feedback (RF) techniques can be used where the user collaborates with the retrieval system. The RF can be defined inside of the on-line learning paradigm [7]. The system learns from the user feedback to update its internal representation for the user preferences and the user interacts with the system until a prefixed number of relevant items are at the top of the system output. The feedback can be explicit where users are asked to assess the relevance of the videos, or implicit where the system is able to extract indicators of relevance according to the user's interaction (e.g. playing a video can be interpreted as implicit indicator of relevance because user is interested in corresponding sample). In addition, RF can be divided into two groups, short-term and long-term. Short-term only considers the information provided by the current user, and long-term also uses the feedback of previous users.

3   An Interactive Information Retrieval Approach Based on sSpLSA

Extracting hidden information in a data set to map it into a feature space that may fill the gap between sensory representation and higher level understanding is a key factor in many data analysis applications. In order to develop a probabilistic approach for video retrieval based on latent topic models, the model chosen to be explored is based on the unsupervised pLSA model [9], a relaxed version of LDA model [4], which will be extended to a supervised model by adding the observed random variable corresponding to class labels (y). In this case, the approach is directed to a similar scenario than the single author topic model, used by Fei-Fei and Perona [8] in the framework of a LDA-based model. The here proposed model is a supervised symmetric pLSA (sSpLSA) (figure 1). This model corresponds to an extension on the symmetric unsupervised pLSA parameterization introduced by Hoffman [9]. This supervised extension allows us to handle the query class, which is unknown, in order to develop a process to estimate the probability that samples belong to this unknown class.

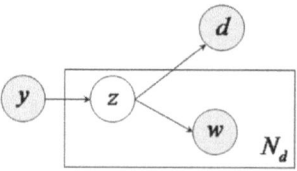

Fig. 1. Graphical representation of the sSpLSA model. y is the class, z the topic (hidden variable), w the word, d the document and N_d the number of words of d.

Based on the sSpLSA model, the next step is to extend it to an interactive on-line framework approach for information retrieval. An on-line user interactive problem can be formulated as follows: a user query at a given stage of the interactive process is represented by a query class y' of documents the user has in mind and he/she is looking for, and a set of query documents d' that represents instances of documents belonging to the query class. The query document set $d' = 1,...,D'$ is a dynamic set that changes during the interactive process, according to the positive documents the user provides as feedback from the iterations in the searching process.

Let us suppose that a learning process has been previously carried out with the documents d in the database, which will be expressed in a set of latent topics as $p(z|d)$. The latent topics may have been learned either in a supervised or unsupervised way. In the case of a supervised learning, this may have been done according to a pre-defined set of classes or concepts, different from the query class at each interactive user process, which is unknown.

The goal at each iteration in the interactive process is to provide a guess about the probability that a document of the database belongs to the query class, that is, we look for a estimation of the class conditional probability $p(y'|d)$. This probability will allow us to establish a retrieval ranking for the documents in the database. According to the sSpLSA model (figure 1), this probability could be estimated from the present user's query as follows. Let us express the conditional probability $p(y'|d)$ by marginalizing over topics, that is:

$$p(y'|d) = \frac{p(y',d)}{p(d)} = \frac{\sum_w \sum_z p(w,d,z,y')}{p(d)} = \frac{\sum_w \sum_z p(w|z)p(d|z)p(z|y')p(y')}{p(d)} \quad (1)$$

Where it has been assumed that the joint probability $p(w,d,z,y')$ is expressed according to the introduced sSpLSA model. Regarding the conditional topic probability of a given class $p(z|y')$, it can be estimated in function of the parameters of the model as follows by marginalizing over the query set $d' = 1,...,D'$:

$$p(z|y') = \sum_{d'} p(z,d'|y') = \sum_{d'} p(z|d',y')p(d'|y')$$
$$\approx \sum_{d'} p(z|d')p(d'|y') = \sum_{d'} \frac{p(z|d')p(y'|d')p(d')}{p(y')} \quad (2)$$

Topics have been calculated in a previous step using the documents of the database, and therefore we can consider that topic descriptions do not depend on the new class y' (query class) in order to estimate the conditional probability $p(z|d',y') = p(z|d')$. Inserting (2) in (1), applying Bayes rule and assuming the normalization constraint $\sum_w p(w|z) = 1$, the expression to estimate the class conditional probability $p(y'|d)$ can be expressed as follows:

$$p(y'|d) \approx \sum_z p(z|d) \left[\sum_{d'} p(d'|z) \right] \quad (3)$$

In this expression, $p(z|d)$ stands for the parameters of the database, and $p(d'|z)$ is the probability that a given topic z belongs to the query document d', which could be estimated from the same learning model used to estimate the database document description in topics, for instance from a maximum log-likelihood approach. Expression (3) would allow us to infer the class conditional probabilities $p(y'|d)$ in a simple and fast way from the document database described in topics. It has the advantage of being very efficient computationally and easy to update for subsequent iterations in the interactive process that dynamically changes the query documents set $d' = 1,...,D'$ that defines the user's query class y'. In addition, latent topics would allow a representation nearer to the user's semantic understanding.

The ranking process is made as follows. First of all, z topics are extracted from the database using some topic extraction method (like pLSA [10], LDA [11] or any other topic model algorithm). Then, each document d and also the query documents d' are represented according these topics. To represent the samples in topics given a set of topics, we have used a maximum log-likelihood approach as in [9]. Later, the database is sorted according the probability that a sample belongs to the query class using expression (3). This equation has two terms: the document d (which is expressed in z topics) and the query d' (sum of the probabilities of the query documents given the z topics, obtained by the maximum log-likelihood approach).

4   Application to Video Retrieval Framework

In the previous section, a model for interactive information retrieval has been introduced which can be applied to several information retrieval problems. The goal in this work is the application for unconstrained video retrieval. Specifically, the main objective is to adapt the general model to an interactive video retrieval framework based on explicit short-term relevance feedback. We are going to model a simplified user interaction scenario (simulation) to evaluate the effectiveness of proposed approach. The relevance feedback simulation can be divided into two parts: Firstly, an initial search query is triggered and a simulated user provides feedback on retrieved results. Secondly, the initial query is expanded with the items extracted from the feedback and a new query is triggered. The grade of feedback quality depends on the simulated user reliability. Therefore, the video retrieval framework based on sSpLSA has to include the following functional requirements:

- *Video representation*: The video collection has to be described in latent topics $p(z|d)$. First of all, the videos are represented in low level features which include static or spatio-temporal information. From this low level representation, K latent topics are extracted and each video sample is represented according these topics.
- *Query initialization*: The system has to be initialized with a query which contains Q video samples of the dataset. This set of videos defines the concept that user wants to retrieve (target) and it has to be expressed according the K topics extracted from the database.
- *Feedback information*: The video retrieval process is an iterative method of I iterations. In each one, the system shows a video ranking with S retrieved videos (scope). From these plausible videos, user selects the P positive samples according to query concept (target).
- *Propagate feedback*: The system uses the positive samples to enlarge the query $(Q + P)$. Therefore, the new query has more video samples and the system is able to refine the retrieved videos for the next iteration.

The considered target for the simulation has been each video class of the database (each one of the C classes), that is, the query is initialized with Q random videos of one class and the simulation has to be able to retrieve videos of this class. Queries can be initialized using samples inside the database or external samples which are not in the database. The only requirement is that they have to be expressed using the topics extracted from the database. In this simulation, queries are initialized with samples from the database. Moreover, this process is repeated R times per class to obtain relevant statistical results. The simulation video retrieval process is presented in Algorithm 1.

It should be noted that the S most probable videos would be inspected and checked by the user in a real video retrieval system. However, this process is automatically made in the proposed simulation which assumes user reliability of 100%.

Algorithm 1. Video Retrieval Simulation for $DATASET$

Require: C=classes, R=repetitions, Q=initSize, I=iterations, S=scope.

```
 1: for class c in C do
 2:    for repetition r in R do
 3:       Initialize QUERY with Q random samples of the class c
 4:       REST = DATASET − QUERY
 5:       for iteration i in I do
 6:          for video v in REST do
 7:             p(y′|v) = ∑z p(z|v)∑d′ p(d′|z)
 8:          end for
 9:          Rank REST in descending order of probability
10:          P = Videos which belong to class c from the S top ranking
11:          Enlarge QUERY adding P
12:          Update REST subtracting P
13:       end for
14:    end for
15: end for
```

5 Experiments

5.1 CCV DataSet

The video dataset selected for the experiments is Columbia Consumer Video Database (CCV) [12]. This recent video collection is a benchmark for consumer video analysis. It contains 9.317 YouTube videos over 20 semantic categories. The total number of video hours is 210 and the average length of the videos is 80 seconds. Also, different video descriptors are available to run experiments with CCV dataset (SIFT, STIP and MFCC). According to the classification accuracy for the CCV database [12], SIFT descriptor achieves better average precision than STIP and MFCC. Furthermore, the combination of them does not improve the accuracy in a significant way. For this reason, we have decided to use the SIFT descriptors provided by the authors of the dataset as a preliminary experiment to test the proposed approach. The vocabulary was defined as a Bag of Words (BoW) model from 500 clusters on SIFT descriptors over Hessian-Affine and DoG feature points extracted over the entire and 2x2 image blocks, which makes a total of 5000 words.

In this corpus, there are samples with null content that are been removed for the experiments. Furthermore, the samples with no annotation have been eliminated too. For the remaining samples, samples labelled with more than one category have been replicated one for each class. Therefore, we have considered a total of 7.846 video samples annotated in 20 classes.

5.2 Validation Set Up

As it has been mentioned in Section 4, the simulation of video retrieval framework has several functional requirements with parameters. These parameters have to be defined and established for the experiments. We have chosen the parameters of the simulation thinking about a real user, that is, the size of the initial query, the number of retrieval iterations and the number of selected videos for propagation (scope) have to be in an appropriate range for user comfort in a real video retrieval system. Therefore, the selected configuration for the simulation process according to the functional requirements are the following:

- *Video representation*: LDA [4] procedure has been applied to the SIFT descriptors provided by the CCV authors [12], in order to express the database in a set of latent topics as $P(Z|D)$. Several values of the number of topics have been considered for the experiments: $K = \{100,200,300,400,500\}$.
- *Query initialization*: For each simulation, an initial query has been initialized with $Q = \{1,2\}$ number of random video samples of the same class of the CCV database. These values have been selected due to the fact that users do not usually select more than one or two videos to initialize a search.
- *Feedback information*: A maximum of $I = \{5\}$ retrieving iterations has been executed using relevance feedback, which is acceptable to the user.
- *Propagate feedback*: In the query updating process, the number of the top selected items has been established to $S = \{20,40\}$. We have considered that user only examines one or two screens of videos to find items of the query class considering a screen size of 20 videos. Moreover, we have assumed that user marks all the positive videos of the query class in the inspected set.

The queries have been initialized with random samples of the same class and this process has been repeated $R = \{500\}$ times. To evaluate the retrieval results the precision over the five iterations have been calculated.

5.3 Relevance Feedback Based on Distance Ranking

In order to evaluate proposed approach we have implemented and executed a baseline method for simulation video retrieval. This baseline algorithm has a RF structure similar to the simulation showed in Section 4, but with two main differences. On the one hand, it uses the original BoW representation of the dataset (SIFT features). On the other hand, the database is sorted by the minimum Euclidean distance to the query (when the query has more than one sample, this distance is the average).

5.4 Experimental Results and Discussion

According to the validation setup (Section 5.2), four main configurations have been used to run the experiments. For these scenarios, several values of number of topics have been used. In order to evaluate the results, two metrics have been used: overall mean precision and execution time. Figure 2 shows the results for each scenario considering the baseline and the proposed approach.

Over the different scenarios, results show that the best retrieval performance has been obtained using 500 topics. It should be noted that the size of the

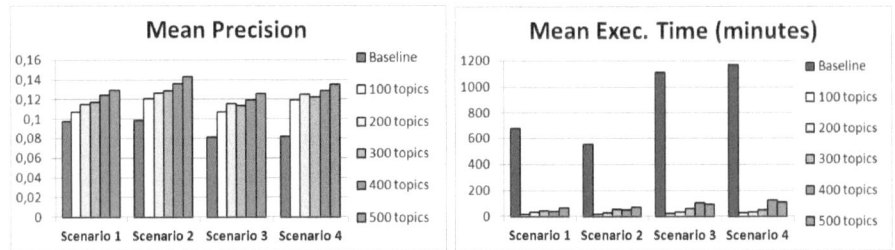

Fig. 2. Results for CCV database. Scenario 1 {Q=1, S=20}, Scenario 2 {Q=2, S=20}, Scenario 3 {Q=1, S=40} and Scenario 4 {Q=2, S=40}. Mean precision (right) and execution time in minutes (left).

initial query Q and the size of the scope S are important factors to the retrieval performance. With a bigger initial query the mean precision rises, but with a larger scope the precision drops. The best precision has been obtained in scenario 2 and the best mean of retrieved videos in scenario 4. Comparing the proposed approach with the baseline, the first one obtains a greater improvement with a bigger initial query and lesser loss with a larger scope.

Regarding to the computational time, results show a remarkable performance of the proposed topic-based approach. Note that the topic-model procedure is applied over the dataset as an off-line processing model, therefore it does not do any extra processing on queries. Furthermore, topic-model representation is able to significantly reduce the dimensionality of the samples. In other words, LDA model is estimated off-line once, and then the proposed approach can process queries much faster than baseline with a greater precision improvement.

6   Conclusions and Future Work

In this work, we have presented an alternative approach for interactive on-line video retrieval tasks based on latent topic models. The supervised latent topic model sSpLSA have been presented and adapted to a video retrieval framework based on relevance feedback. Several experiments have been done using a simulation-based methodology and a distance-based RF algorithm as baseline. The results provide evidences about the viability of proposed approach. The main conclusion that arises from the work is the importance of topic models to attempt filling the semantic gap for video retrieval. Topic models are able to extract hidden patterns of a data set and these patterns can be used to provide a higher level understanding. Besides, the proposed approach shows an important execution time reduction with respect the distance-based baseline. Although results are encouraging, much more experimental evidence is needed to really assess the properties and quality of the proposed approach. In particular, further work is directed to compare its performance with other recent information retrieval approaches and to implement a testing protocol in order to assess proposed approach using a user-based methodology.

References

1. Trecvid, http://trecvid.nist.gov/
2. The challenge problem for automated detection of 101 semantic concepts in multimedia (2006)
3. Blei, D.: Probabilistic topic models. Communications of the ACM 55(4), 77–84 (2012)
4. Blei, D., Ng, A., Jordan, M.: Latent dirichlet allocation. Journal of Machine Learning Research 3(4-5), 993–1022 (2003)
5. Bosch, A., Zisserman, A., Munoz, X.: Scene classification via 2009. In: European Conference on Computer Vision (2009)
6. Brants, T., Chen, F., Tsochantaridis, I.: Topic-based document segmentation with probabilistic latent semantic analysis. In: International Conference on Information and Knowledge Management (2002)
7. Chechik, G., Sharma, V., Shalit, U., Bengio, S.: Large scale online learning of image similarity through ranking. Journal of Machine Learning Research 11, 1109–1135 (2010)
8. Fei-Fei, L., Perona, P.: A bayesian hierarchical model for learning natural scene categories. In: IEEE Computer Society Conference on Computer Vision and Pattern Recognition, CVPR (2005)
9. Hofmann, T.: Unsupervised learning by probabilistic latent semantic analysis. Machine Learning 42(1-2), 177–196 (2001)
10. Huang, J., Kumar, S., Mitra, M., Zhu, W., Zabih, R.: Image indexing using color correlograms. In: IEEE Int. Conf. Computer Vision and Pattern Recognition (1997)
11. Jiang, Y., Bhattacharya, S., Chang, S., Shah, M.: High-level event recognition in unconstrained videos. International Journal of Multimedia Information Retrieval 2, 73–101 (2013)
12. Jiang, Y., Ye, G., Chang, S., Ellis, D., Loui, A.: Consumer video understanding: A benchmark database and an evaluation of human and machine performance. In: Proceedings ACM International Conference on Multimedia Retrieval, ICMR (2011)
13. Monay, F., Gatica-Perez, D.: On image auto-annotation with latent space models. In: ACM International Conference on Multimedia (2003)
14. Ren, W., Singh, S., Singh, M., Zhu, Y.: State-of-the-art on spatio-temporal information-based video retrieval. Pattern Recognition 42(2), 267–282 (2009)
15. Smeulders, A., Worring, M., Santini, S., Gupta, A., Jain, R.: Content-based image retrieval at the end of the early years. IEEE Transactions on Pattern Analysis and Machine Intelligence 22(12), 1349–1380 (2000)
16. Snoek, C., Worring, M.: Concept-based video retrieval. Foundations and Trends in Information Retrieval 4(2), 215–322 (2009)
17. Zhou, D., Weston, J., Gretton, A., Bousquet, O., Schölkopf, B.: Ranking on data manifolds. In: Advances in Neural Information Processing Systems. NIPS. MIT Press (2004)

Performance Study of a Regularization-Based Deformable Handwritten Recognition Approach

Yoshiki Mizukami, Shinya Nakanishi, and Katsumi Tadamura

Yamaguchi University, 2-16-1 Tokiwadai, Ube 755-8611, Japan
{mizu,s028vm,tadamura}@yamaguchi-u.ac.jp

Abstract. This study clarifies the accuracy performance of a deformable handwritten recognition approach (DHRA) for digit characters. The deformable approach consists of regularization-based displacement computation, coarse-to-fine strategy, distance measurement and k-nearest neighborhood method. We focus on several conditions for investigating the accuracy and the sensitivity, that is, the definition of averaging area in regularization process, regularization parameters and the number of k for k-nearest neighborhood method. According to the simulation results, it was shown that the proposed method has the error rate of 0.42% for MNIST handwritten digit database, resulting in the top-group of the performances reported until now.

Keywords: deformable handwritten recognition, MNIST digit database, regularization.

1 Introduction

Handwritten character recognition is one of the interesting topics in the field of computer vision and machine intelligence, since it does not only relate with the applications of optical character recognition but also the functionality of the human brain for reading many kinds of characters [1]. In order to compare the accuracy performances of different recognition approaches, several evaluation databases have been provided, and modified-NIST (MNIST) database is frequently employed [2].

Neural network approaches have been successfully applied to MNIST database. LeCun et al. proposed convolutional neural networks (CNN) [2–4], which are specifically designed to deal with the variability of 2D shapes. LeNet-5, a convolutional network, comprises 8 layers and has feature mapping planes where a set of the units shares identical weight values for an efficient learning. Under the hypothesis that more training data would improve the accuracy, they artificially generated more samples by randomly distorting the original training samples. For the distortion operation, planar affine transformation or elastic transformation was utilized. These ideas of deep convolutional neural network (DNN), earlier introduced by [5] and sophisticated by [2–4], were more developed by Ciresan et al. [6, 7]. They takes advantage of the recent parallel computation devices

A. Petrosino (Ed.): ICIAP 2013, Part I, LNCS 8156, pp. 300–310, 2013.

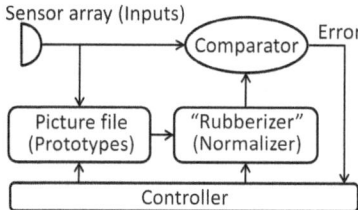

Fig. 1. Rubber mask technique for pattern recognition

(graphics processing units; GPUs) for their powerful learning process, and employed several effective strategies, that is, only winner-neurons are trained and several DNN columns were combined to form a multi-column DNN (MCDNN). The error rate of MCDNN was reported as 0.23%.

The support vector machine (SVM) is an extremely economical way of representing complex surfaces in high-dimensional spaces [8], which has been applied to various problems in pattern recognition. LeCun et al. investigated the application of SVM to MNIST [2], and then DeCoste and Scholkopf reported the error rate of their SVM as 0.56% [9]. Lin et al. reported their performance of 0.42% by using 8-direction gradient features and RBF kernel [10]. Recently, Niu and Suen presented a hybrid CNN-SVM classifier [11] and clarified the error rate of 0.19%, which is the top performance for MNIST database right now.

Deformable approach with k-nearest neighborhood method seems to be an alternative possibility. Compared with the above-mentioned approaches, deformable approach does not need the training process in advance, and the variability of 2D shapes is dealt with by deforming one of an input and prototype pattern into the other. The idea of the deformable approach can go back to Widrow's "rubber mask" technique [12], which was inspired by Gregory's insights [13],

> ··· perception is not determined simply by the stimulus patterns; rather it is a dynamic searching for the best interpretation of the available data ···

Figure 1 illustrates a recognition process of the rubber mask technique. An input is captured on the sensor array and the prototypes, which might be image features, are deformed in the "rubberizer" under the controller. Finally, both of them are compared in the comparator. The obtained error is referred by the controller to update the deformation. Many deformable approaches were proposed until now, and Belongie et al. applied their deformable approach to MNIST database earlier [14]. They defined shape context (image feature) at the contour pixels so as to represent the distribution of the surrounding contour pixels and utilized the shape context as image feature for elastically corresponding two character shapes with sub-pixel displacement. The error rate was reported as 0.63%. Keysers et al. proposed a novel approach [15], where the horizontal and vertical gradients with 3 x 3 pixel size are extracted as local context (image feature) and warping algorithm [16] is employed for the minimization in pixel-wise correspondence. They reported the error rate of 0.52%.

This study focuses on the accuracy performance of deformable approaches motivated by above-mentioned insights of Gregory [13] and Widrow [12]. A regularization-based deformable approach, which was originally proposed by Mizukami et al. [17–19], is employed for investigating the accuracy performance. Although they reported the error rate of 0.57% in 2010 [19], their study focused on reducing the computation time and did not succeed in investigating the fundamental accuracy performance. This study reviews their proposed deformable approach, conducts several simulations for clarifying the fundamental performance including the accuracy and the sensitivity to the parameters, and demonstrates how the approach deals with the variability of 2D shapes.

Section 2 describes the algorithm of the deformable recognition approach, Section 3 describes the simulation and Section 4 gives the conclusion.

2   Algorithm

The framework of regularization theory has been successfully applied to many early vision problems including optical flow, shape from shading and so on [20]. Inspired by a very simple but powerful regularization-based stereo correspondence method [21], Mizukami et al. proposed a deformable handwritten character recognition approach [17–19]. This section overviews the procedures in the proposed approach.

2.1   Regularization-Based Displacement Computation

The two-dimensional correspondence problem between pixels of an input pattern f and a prototype pattern g is ill-posed, and then it is necessary to introduce some adequate constraints to solve it. Figure 2 illustrates the input pattern f and the prototype g, where $u(x, y)$ and $v(x, y)$ indicates the horizontal and vertical displacements at the coordinate of (x, y) on the prototype g. In order to obtain the optimal displacement function for corresponding these two patterns, a cost function to be minimized is introduced,

$$E(u, v) = P(u, v) + \lambda S(u, v), \tag{1}$$

$$P(u, v) = \int\int (f(x + u, y + v) - g(x, y))^2 dxdy, \tag{2}$$

$$S(u, v) = \int\int (u_x^2 + u_y^2 + v_x^2 + v_y^2) dxdy, \tag{3}$$

where P is the Euclidean distance with considering the computed displacement, S is a stabilizing functional which imposes a smoothness constraint on (u, v), and λ is a so-called regularization parameter controlling the effect of S. The subscripts, x and y, are the derivative operation of horizontal and vertical directions, respectively.

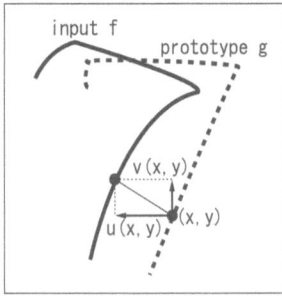

Fig. 2. Displacement function (u, v)

By applying calculus of variations to Eq. (1), the following iterative equations
are obtained,

$$u^{[t+1]}(x,y) = \bar{u}^{[t]} - \frac{1}{4\lambda}f_x(x+\bar{u}^{[t]}, y+\bar{v}^{[t]})(f(x+\bar{u}^{[t]}, y+\bar{v}^{[t]}) - g(x,y)), \quad (4)$$

$$v^{[t+1]}(x,y) = \bar{v}^{[t]} - \frac{1}{4\lambda}f_y(x+\bar{u}^{[t]}, y+\bar{v}^{[t]})(f(x+\bar{u}^{[t]}, y+\bar{v}^{[t]}) - g(x,y)), \quad (5)$$

where the superscript t is the number of iteration ($1 \leq t \leq T$), and $\bar{u}$ and $\bar{v}$ are
the average of u and v over the predefined neighborhoods of (x,y), respectively.
In order to make the computation more stable, $\bar{u}$ and $\bar{v}$ are employed not only
in the first term of the right hand but also in the second term instead of u and
v [17]. The first term smooths the displacement and the second term gives the
deforming force so as to overlap two images based on the derivatives of f and
the difference between the corresponding pixels on f and g.

2.2   Coarse-to-Fine Strategy with Distance Maps

For the fast convergence and preventing from being trapped in local minima, a
coarse-to-fine strategy with multi-resolution images is utilized. In this study, the
number of stage was set to 3. The original size of the pattern images is assumed
as 32×32 pixels and then the n-th stage deals with the size of $2^{n+2} \times 2^{n+2}$
($1 \leq n \leq 3$). The displacement function obtained at the n-th stage will be used
for preparing the initial displacement at the $n+1$-th stage.

Since the derivatives of the pattern images are zero in the background area,
most of the pixel area on the image will not have the deforming force, resulting in
an inefficient displacement computation. To overcome this problem, the binary
images are generated by a threshold processing and then they are translated
to the distance maps whose pixel value indicates the distance to the nearest
foreground pixel as shown in Fig. 3. Instead of the original pattern images, these
distance maps are used for computing the displacement [19].

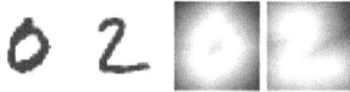

Fig. 3. Gray pattern images and distance maps

2.3   Prototype-Parallel Displacement Computation

Although the iterative equations seem to be efficiently implemented on GPUs due to its locally-parallel computation style, the sizes of the pattern images are too small to bring out the most latent strength of GPUs. To overcome this problem, prototype-parallel displacement computation (PPDC) is employed [18]. Figure 4 describes the diagram of PPDC. Three types of large plates, $(U_n^{[0]}, V_n^{[0]})$, F_n, and G_n, are generated by arranging displacement functions $(u_{n,l}^{[0]}, v_{n,l}^{[0]})$, input images f_n and prototypes $g_{n,l}$ at regular intervals respectively on the host memory, and then these plates are transferred collectively to the device memory, where l is the index of the prototype ($1 \leq l \leq L$). GPU computes the displacement for multiple pairs of input and prototype images, and updates the displacement function plate. Finally the plates of the computed displacement function $(U_n^{[T]}, V_n^{[T]})$ are transferred back from the device memory to the host memory.

2.4   Distance Measurement and Classification

Since the shape of a pattern will be deformed so as to be fitted to the other pattern shape by considering the computed displacement, there is a risk that the shapes of two patterns in different classes also become similar, resulting in very small distance. To avoid this problem, the local shape information of the original contours such as straight line or curvature is utilized in measuring the distance. Therefore, the distance between two pattern images is measured by applying the computed displacement function to eight-directional derivative images of two patterns, f^d and g^d ($1 \leq d \leq 8$; see Fig.5),

$$D(u, v) = \sum_{x,y} \sum_{d=1}^{8} (f^d(x + u, y + v) - g^d(x, y))^2. \tag{6}$$

After the distances of the input pattern to L prototypes $\{f_l^d\}$ are measured, the input pattern is classified to one of the classes by the terms of k-nearest neighborhood method. The previous studies [18, 19] employed gradual prototype elimination (GPE) for reducing the computation time, where the candidate prototypes are gradually eliminated through the coarse-to-fine strategy. On the other hands, this study did not employ GPE for the purpose of investigating the fundamental accuracy performance, since GPE might eliminate several prototypes which are helpful for classifying the given input pattern.

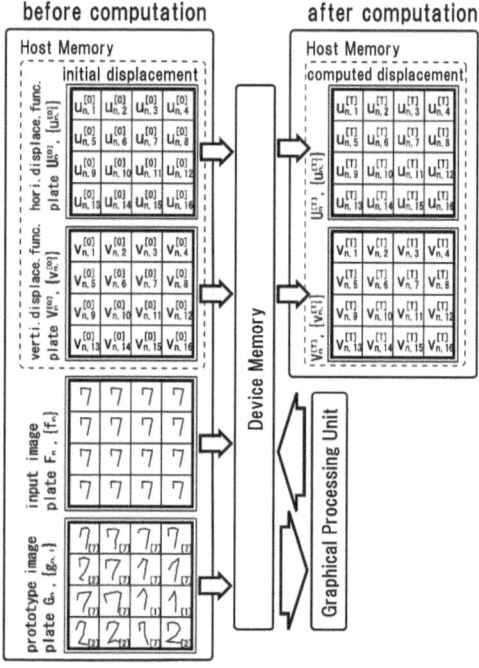

Fig. 4. Prototype-parallel displacement computation

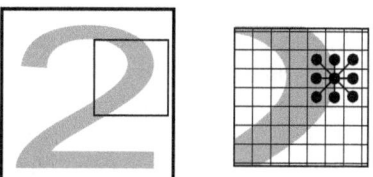

Fig. 5. Eight-directional derivatives

3   Simulation

This section investigates the effect of the regularization parameters, and the effect of k on the accuracy performance, and discusses how patterns are deformed. The deformable approach was implemented on a computer with Intel Core^{TM} i7 CPU-2600 (3.40GHz) and Nvidia Geforce GTX590. The programming environments were Microsoft Visual C++ 2010 and CUDA Toolkit 4.0 on Microsoft Windows 7. MNIST handwritten digit database consists of 10,000 input patterns and 60,000 prototypes. The regularization parameter λ_1 was set to 6.0, while λ_2 and λ_3 were set to [0.8:1.2]. The computation time for recognizing an input pattern was about 40 sec, where the displacement functions are computed for all of 60,000 prototypes.

Table 1. Error rates using 4-pixel average

λ_3	λ_2				
	0.8	0.9	1.0	1.1	1.2
0.8	<u>0.45</u>	<u>0.43</u>	<u>0.43</u>	<u>0.43</u>	<u>0.44</u>
0.9	<u>0.45</u>	<u>0.44</u>	<u>0.42</u>	<u>0.42</u>	<u>0.43</u>
1.0	<u>0.44</u>	<u>0.44</u>	<u>0.45</u>	<u>0.43</u>	<u>0.44</u>
1.1	<u>0.45</u>	0.47	0.47	<u>0.45</u>	<u>0.44</u>
1.2	0.48	0.50	0.49	0.46	0.46

Table 2. Error rates using 5-pixel average

λ_3	λ_2				
	0.8	0.9	1.0	1.1	1.2
0.8	0.49	0.49	0.49	0.48	0.48
0.9	0.46	<u>0.45</u>	<u>0.45</u>	<u>0.45</u>	<u>0.45</u>
1.0	<u>0.43</u>	<u>0.42</u>	<u>0.42</u>	<u>0.43</u>	<u>0.43</u>
1.1	0.46	0.46	<u>0.45</u>	<u>0.44</u>	<u>0.44</u>
1.2	0.46	<u>0.45</u>	<u>0.44</u>	<u>0.44</u>	<u>0.44</u>

Table 1 and 2 show the error rates of the deformable approach by using 4-pixel and 5-pixel averages, where 4-pixel average means that the four-pixel displacements surrounding the coordinate of (x,y) were used for obtaining $\bar{u}$ and $\bar{v}$, while 5-pixel average means that the surrounding four-pixel displacements and the displacement at (x, y) were used. The value of k was set to 3. In both case of 4 and 5-pixel averages, the minimum error rate was 0.42% as underlined by double line. The error rates, which are equal to 0.45% or less, are underlined by single line. It can be noticed that the accuracy is not so sensitive for averaging ways and the value of both λ_2 and λ_3.

Figure 6 illustrates the effect of k on the accuracy performance, where both λ_2 and λ_3 were set to 1.0 and 5-pixel average was used. The minimum error 0.42% was obtained by using $k = 3$. It was also confirmed that the use of $k = 3$ was adequate for other values of λ_2 and λ_3.

Figure 7 shows several input patterns and prototypes deformed to each input pattern. Fig.7(a) illustrates 10 prototypes, while Fig.7(b,c,d) show input patterns of '7', '2' and '1' and the 10 deformed prototypes. The numbers under the deformed prototypes mean the distance to the corresponding input pattern.

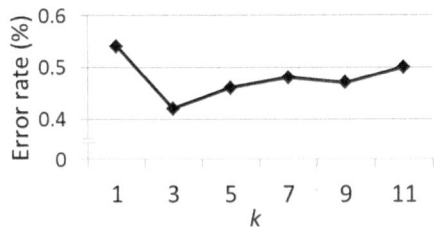

Fig. 6. Value of k versus accuracy

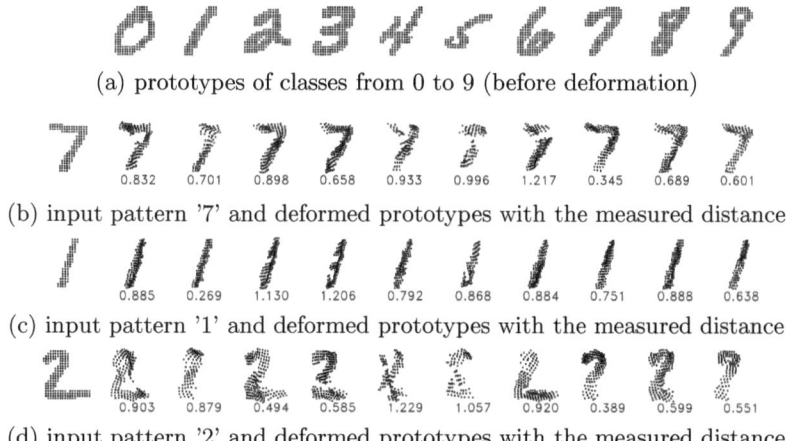

(a) prototypes of classes from 0 to 9 (before deformation)

0.832   0.701   0.898   0.658   0.933   0.996   1.217   0.345   0.689   0.601

(b) input pattern '7' and deformed prototypes with the measured distance

0.885   0.269   1.130   1.206   0.792   0.868   0.884   0.751   0.888   0.638

(c) input pattern '1' and deformed prototypes with the measured distance

0.903   0.879   0.494   0.585   1.229   1.057   0.920   0.389   0.599   0.551

(d) input pattern '2' and deformed prototypes with the measured distance

Fig. 7. Input patterns and deformed prototypes

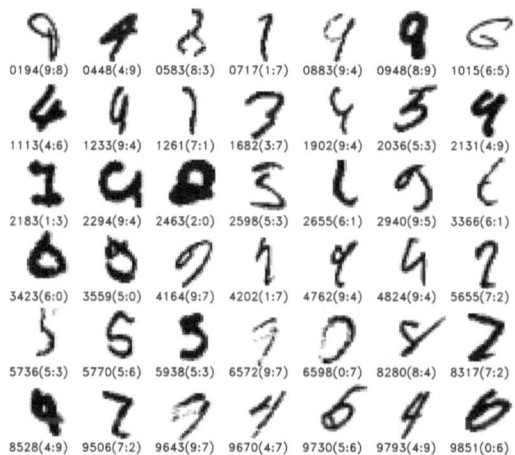

0194(9:8)   0448(4:9)   0583(8:3)   0717(1:7)   0883(9:4)   0948(8:9)   1015(6:5)

1113(4:6)   1233(9:4)   1261(7:1)   1682(3:7)   1902(9:4)   2036(5:3)   2131(4:9)

2183(1:3)   2294(9:4)   2463(2:0)   2598(5:3)   2655(6:1)   2940(9:5)   3366(6:1)

3423(6:0)   3559(5:0)   4164(9:7)   4202(1:7)   4762(9:4)   4824(9:4)   5655(7:2)

5736(5:3)   5770(5:6)   5938(5:3)   6572(9:7)   6598(0:7)   8280(8:4)   8317(7:2)

8528(4:9)   9506(7:2)   9643(9:7)   9670(4:7)   9730(5:6)   9793(4:9)   9851(0:6)

Fig. 8. Misclassified patterns with the index, the label and the misclassified label

It can be noticed that, in Fig.7(b), the deformation procedure also made the shapes of different-class prototypes of '1', '2', '3', '8' and '9' very similar to the shape of the input pattern '7', but the prototype '7' gave very smaller distance 0.345 than other prototypes. The same observation can be applied to the input pattern '1' in Fig.7(c). On the other hand, Fig.7(d) shows an interesting situation where the prototype '7' gave the smallest distance to the input pattern '2' due to the topological shape difference between the input pattern '2' and the prototype '2'. These results indicate that the quality or variation of the prepared prototypes should be important for the accurate recognition.

Figure 8 shows all the misclassified input patterns with their indices, labels and misclassified labels. These patterns seem to be very challenging since they were written in rough ways and some of them lack a part of the shape due to the pale pixel value.

4   Conclusion

This study clarified the accuracy performance of a deformable handwritten recognition approach for digit characters. The deformable approach consists of regularization-based displacement computation, coarse-to-fine strategy, distance measurement and k-nearest neighborhood method. We focused on several conditions for investigating the recognition accuracy and the sensitivity, that is, the definition of averaging area in regularization process, regularization parameters and the number of k for k-nearest neighborhood method. According to the simulation results, it was shown that the proposed method has the error rate of 0.42% for MNIST handwritten digit database, resulting in the top-group of the performances reported until now. It was also shown that the accuracy performance is not so sensitive to the condition including the parameter settings. The simulation results illustrated how the prototypes are deformed to the input pattern and explained that even though the prototype becomes deformed very similar to the input pattern, the proposed approach gives adequate distances, and that the topologically-different prototype will not help the accurate recognition.

Although the obtained error rate of 0.42% is still inferior to the previous record of 0.23% obtained by multi-column deep convolutional neural network (MCDNN) [6, 7] or the best record of 0.19% by a hybrid CNN-SVM classifier [11], it should be noticed that deformable approaches are totally different from both of them. They deal with the variability of 2D shapes by generating a lot of artificial training patterns and employing very sophisticated concepts of network learning or SVM, while the deformable approaches deal with the variability by flexibly fitting the prototypes in the memory to the input pattern. As pointed out by Gregory [13] and Widrow [12], deformable approaches have some aspects which resemble to the recognition model of human brain and then there seems to be still a possibility that the deformable approaches will give further performance in the future studies.

References

1. Plamondon, R., Srihari, S.N.: On-Line and off-Line handwriting recognition: a comprehensive survey. IEEE Trans. Pattern Anal. Mach. Intell. 22(1), 63–84 (2000)
2. LeCun, Y., Bottou, L., Bengio, Y., Haffner, P.: Gradient-based learning applied to document recognition. Proceedings of the IEEE 86(11), 2278–2324 (1998)
3. Simard, P.Y., Steinkraus, D., Platt, J.C.: Best practices for convolutional neural networks applied to visual document analysis. In: 7th International Conference on Document Analysis and Recognition, pp. 958–963 (2003)

4. Ranzato, M., Boureau, Y.-L., LeCun, Y.: Sparse feature learning for deep belief networks. In: Advances in Neural Information Processing Systems, vol. 20, pp. 1–8 (2007)
5. Fukushima, K.: Neocognitron: A self-organizing neural network model for a mechanism of pattern recognition unaffected by shift in position. Biological Cybernetics 36(4), 193–202 (1980)
6. Ciresan, D., Meier, U., Gambardella, L.M., Schmidhuber, J.: Convolutional neural network committees for handwritten character classification. In: 11th International Conference on Document Analysis and Recognition, pp. 1250–1254 (2011)
7. Ciresan, D., Meier, U., Schmidhuber, J.: Multi-column deep neural networks for image classification. In: International Conference on Computer Vision and Pattern Recognition, pp. 3642–3649 (2012)
8. Vapnik, V.N.: The Nature of Statistical Learning Theory. Springer (1995)
9. DeCoste, D., Scholkopf, B.: Training invariant support vector machines. Machine Learning 46, 161–190 (2002)
10. Liu, C., Nakashima, K., Sako, H., Fujisawa, H.: Handwritten digit recognition: benchmarking of state-of-the-art techniques. Pattern Recognition 36(10), 2271–2285 (2003)
11. Niu, X.X., Suen, Y.: A novel hybrid CNN-SVM classifier for recognizing handwritten digits. Pattern Recognition 45(4), 1318–1325 (2012)
12. Widrow, B.: The 'Rubber-Mask' Technique - I. Pattern Measurement and Analysis. Pattern Recognition 5(3), 175–198 (1973)
13. Gregory, R.L.: Eye and brain, the Psychology of Seeing. World University Library. McGraw-Hill, New York (1966)
14. Belongie, S., Malik, J., Puzicha, J.: Shape matching and object recognition using shape contexts. IEEE Trans. Pattern Anal. Mach. Intell. 24(4), 509–522 (2002)
15. Keysers, D., Deselaers, T., Gollan, C., Ney, H.: Deformation models for image recognition. IEEE Trans. Pattern Anal. Mach. Intell. 29(8), 1422–1435 (2007)
16. Uchida, S., Sakoe, H.: A monotonic and continuous two-dimensional warping based on dynamic programming. In: 14th International Conference on Pattern Recognition, vol. 1, pp. 521–524 (1998)
17. Mizukami, Y., Koga, K.: A handwritten Character recognition system using hierarchical displacement extraction algorithm. In: 13th International Conference on Pattern Recognition, vol. 3, pp. 160–164 (1996)
18. Mizukami, Y., Tadamura, K.: GPU implementation of deformable pattern recognition using prototype-parallel displacement computation. In: International Workshop on Image Registration in Deformable Environments - DEFORM 2006, vol. 1, pp. 71–80 (2006)
19. Mizukami, Y., Tadamura, K., Warrell, J., Li, P., Prince, S.: CUDA implementation of deformable pattern recognition and its application to MNIST handwritten digit database. In: 20th International Conference on Pattern Recognition, vol. 1, pp. 2001–2004 (2010)
20. Poggio, T., Torre, V., Koch, C.: Computational vision and regularization theory. Nature 317, 314–319 (1985)
21. March, R.: Computation of stereo disparity using regularization. Pattern Recognition Letters 8(3), 181–188 (1988)

Layered Self-Organizing Map for Image Classification in Unrestricted Domains

Christian O'Connell[1], Andrea Kutics[2], and Akihiko Nakagawa[2,3]

[1] University of Essex, Colchester, United Kingdom
cdocon@essex.ac.uk
[2] International Christian University (ICU), Tokyo, Japan
matz@icu.ac.jp
[3] University of Electro-Communications (UEC), Tokyo, Japan
ranaka@ppp.bekkoame.ne.jp

Abstract. The inherent difficultly in unrestricted image domain classification is due to the many different features exhibited by images. Efforts made toward classification of abstract features tend to focus on a single attribute. Without a method of unifying descriptors, it becomes very difficult to perform multi-feature analysis. Extending the concept of the Self-Organizing Feature Map to include multiple competitive layers, it has been possible to create a new type of Artificial Neural Network capable of analyzing image and signal datasets with multiple feature descriptors concurrently in a powerful yet computationally light manner. Compared to standard CBIR retrieval approach, a marked increase in the precision of clustering of 13 points has been achieved, along with a reduction in computation time.

Keywords: self-organizing map, image classification, features, image retrieval.

1 Introduction

The applications of image classification are vast, encompassing many fields from simple image searches to complex object detection. While great strides have been made to derive meaning from large datasets into a manner akin to a human observer, the complexity of the challenge means such a solution remains elusive. Fundamental differences between machine interpretation and the subjective nature of human perception – the so called "semantic gap" – have made it an extremely difficult task to which no clear solution for consistent meaningful results has arisen. Attempts to bridge this gap have led to numerous methods to derive meaning from image data being developed, utilizing most prominently the fields of artificial intelligence, data mining, cognitive psychology and learning from context. Progress toward achieving human-like reasoningwith image datasets are often inflexible, working only within predefined domains. Efforts such as the MPEG-7 multimedia content description standard [1], which provides a number of 'feature descriptors' to describe features such as color distribution and texture as a vector, have achieved a level of consistency

A. Petrosino (Ed.): ICIAP 2013, Part I, LNCS 8156, pp. 310–319, 2013.

for comparison, but little continuity between descriptors makes multi-descriptor analysis difficult. The non-linear statistical processing capabilities of Artificial Neural Networks (ANNs) have also generated interest in the field of image processing. By utilizing their architecture, it would be possible to unify their dynamicity with the accuracy offered by feature descriptors to improve image classification.This paper proposes a new architecture that is a derivation of the Self-Organizing Map (SOM) ANN [2] which employs an arbitrary number of competitive layers to perform unsupervised multi-descriptor analysis of datasets.

2 Related Work

Unsupervised clustering methods like SOM are frequently used for image classification, and the SOM has proved to be especially convenient for this task due to its 2D mapping capabilities making the resulting clusters easy to visualize. In SOMs, the neuron's weight often acts as a representation of the color domain such as RGB in vector format, and is altered during the training period to more closely match randomly selected pixels from the input image, "mapping" the SOM's topology to form a 2D representation of a higher dimensional vector space [3]. As in image classification the input data is represented as a composite high dimensional feature vector with variable component numbers in various individual features. Different types of SOM architectures are proposed, for example, satellite images can use different input layers according to the satellite specialization directive, and a GRID solution is proposed to be able to store satellite image codebook vectors and also different layers corresponding to soil and water in a weather forecast application [4]. The field of bioinformatics uses a Multi-SOM algorithm [5] for data mining large high-dimensional datasets with a number of small SOMs. Different data structures have been analyzed by SOMs using multiple kernels [6]. Another proposed text/image classification uses different font-types for grouping and presenting them to prototypes [7] and finally clustering them to three main clusters. SOMs can be considered as a solution to automatically estimate and linearly merge the weights of individual features with different distance measures for content-based image retrieval (CBIR) [8][9]. The problem can be traced back to [10] where the authors use a tree-structured type of SOM. Clustering is carried out on this basis and the contribution of each feature for the cluster structure is used for the selection of the proper weights. Although all of the above-mentioned contributions obtain relatively good results on restricted image domains, the problem of accurately classifying images of non-restricted domains and having specific, high dimensional and independent features, that cannot be combined linearly, remains unsolved.

3 Layered Self-Organizing Map

Analyzing more abstract image features such as an MPEG-7 feature can be performed effectively using an SOM to reduce the higher dimensional descriptors into a two dimensional network of relationships. Descriptor values are taken from an image

Algorithm 1. Layered SOM algorithm

$for\ t = 1\ to\ \lambda$
  $select\ X \in L_0$
  $N := \iota(X)$
  $while\ N \neq \{\emptyset\}$
    $\text{BMU} := N_i \mid \min_{\forall i \in N}(\|X_F - N_i\|)$
    $for\ each\ v \in \omega[BMU, \Omega(t)]$
      $v(t+1) = v(t) + \alpha(t)[X_F - v(t)]$
    end
    $N := \iota(BMU)$
  end
end

L_n – Layer n
$\iota(X)$ – Set of neurons inter-
      linked to X
X_F – Appropriate descriptor
      associated with X for
      distance calculation
$\omega(X, Y)$ – Set of neurons in
      the neighborhood of X
      with size Y
$\alpha(t)$ – Learning rate at t

dataset and fed into the SOM as input which maps itself to the feature space. Cluster-ing is discerned through analysis of the topographical and topological structure of the map. The primary limitation encountered is the maximum of one feature that can be analyzed at a time, significantly reducing the effectiveness of the map. A way of ex-tending the principle of SOMs to enable concurrent processing of multiple vector weights needed to be found, however descriptors can have different numerical repre-sentations of data components with varying probability distributions (number of di-mensions, range etc.), thus are unable to be compared directly. Initial attempts to adapt the original SOM algorithm to concurrently analyze multiple descriptors linear-ly proved unstable. Each neuron held all weights and used a calculated weighted measure to obtain the feature distance, however intrinsic differences between descrip-tor components based on their statistical distribution made it difficult to obtain fair comparison results, whilst quantization and scaling lead to an unacceptable loss of precision.This motivated the development the Layered SOM (LSOM), a new extended approach to the original SOM which delegates the task of analyzing each descriptor to individual competitive layers, interlinked sequentially to form a feed-forward network. This retained the simplicity and effectiveness of each neuron classifying a single weight, with each layer mapping a single descriptor, whilst main-taining a level of independence between descriptors so they need never be compared directly to one another; instead their topological position forming the basis for com-parison creating an implicit correlation between descriptors.

3.1   Feature Extraction

Before any classification can occur, the image features used in the LSOM need to be calculated via the appropriate methods for every image in the dataset. The features can be expressed in any format which allows comparison based on a scalar output i.e. distance; a vector or matrix representation being the most common. Feature extraction in this paper is performed mostly by using the methods defined by the MPEG-7 stan-dard. The Homogeneous Texture Descriptor (HTD) uses directional 2D Gabor filters to measure the statistical properties of the partitioned domain over a set of scales.

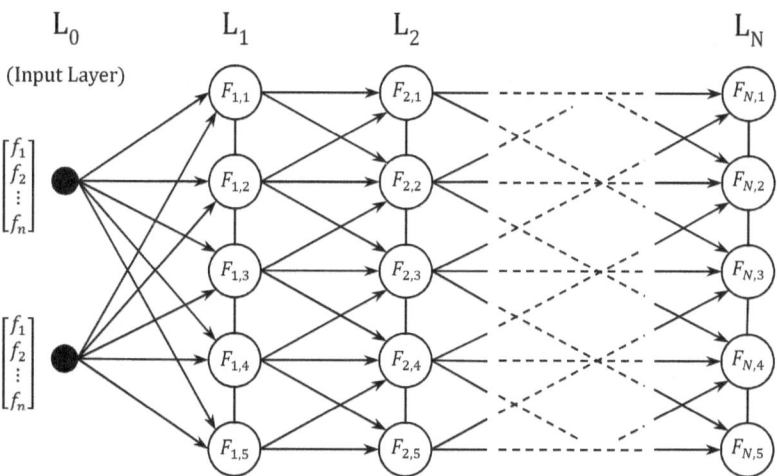

Fig. 1. LSOM Architecture

Its power comes from how the Gabor filter's operation closely models that of the human visual cortex, producing data in line with human understanding. The Edge Histogram Descriptor (EHD) is also used as a method to extract texture-like properties in a highly computationally efficient way by measuring direction of detected edges. Its functionality has been improved by incorporating Canny edge detection to create a new version more adept at texture detection through improved detection of relevant edges. Color analysis is handled by the Color Layout Descriptor (CLD), which segments an image into a spatially equal 8×8 grid, and the average of a cell's YCbCr color space is calculated. The final descriptor, although not a part of the MPEG-7 standard is the Color Correlogram [11], which documents the spatial relationship between pixels in an image. The Autocorrelogram is a variation which takes the main diagonal. Calculating the features for each image is the most computationally intensive part of the process; however is necessary only once as descriptors can be reused for repeated calculations. Each input to the LSOM is a collection of features to be analyzed by the LSOM. This collection of inputs is referred to as Layer 0 (L_0).

3.2   Network Initialization

- **Layer ordering.** Competitive layers in the LSOM are ordered in a feed-forward topology from L1 to LN, with neurons on each layer classifying a single descriptor. Descriptors positioned earlier in the map provide an approximate initialization of clusters for the next layer, thus descriptors which should have a higher influence on the results should be positioned later in the map.
- **Layer topology.** Each layer is identical in the number of neurons and topology, differing only in the type of descriptor held as the weight. More efficient topologies include hexagonal and square structures; the latter is used in this paper as it is effective for visualization, clearly showing the propagation of change evenly

throughout the network. Neurons are interlinked to neurons on adjacent layers using binary connections to create sets of neurons from which the training algorithm (Algorithm 1) selects the best matching units (BMUs). Inputs (Layer 0) are interlinked to every neuron on Layer 1 as with the regular SOM. For all other layers, neurons on Ln are interlinked to only a subset neurons on Ln+1. This is usually the neuron with the same coordinates and a 1-radius neighborhood (Figure 1). The effect is to apply an approximate regional sort based upon prior descriptors, which is then refined by subsequent layers.

- **Weights.** Neurons on L1 are the only ones which do not receive any form of real time initialization. At its simplest they can be initialized on a purely random basis. Better results can often be achieved by applying statistical analysis to the input domain (L0) to give an approximate mapping, which are then propagated throughout the layers during the training period. This paper calculates the initial weights using the properties obtained from the Gaussian type distribution of input features. Taking the central neuron of Layer 1 and assigning it the mean value of the appropriate input dataset descriptor, and using the standard deviation as the rate of change to set neurons in each neighborhood radius. Alternative methods include using Hidden Markov Models, although they are more computationally intensive.

3.3   Neighborhood Function

When a neuron is excited, its topological neighborhood, as defined by the neighborhood function, is also affected. Choice of implementation is critical to ensuring a correct propagation of inputs throughout the network, and best results were obtained using a retracting radius proportional to the learning restraint:

$$\Omega(t) = \max\left(1, \left[\max\left(w, h\right) \cdot \alpha(t)\right]\right) \tag{1}$$

Giving the number of concentric radii to include in the neighborhood, where w and h represent the width and height of the map respectively, α is the learning rate and t is the current epoch. The retracting radius ensures there is no variation with how the neighborhood is calculated between descriptors, increasing the systems modularity and adaptivity. This is as opposed to more feature dependent methods such as a Gaussian neighborhood, which requires a large sigma value to be effective as it has to describe the input dataset as a whole. The learning rate uses a logarithmic equation:

$$\alpha(t) = \left(0.9\left(\frac{\lambda}{100}\right)\right)\Big/\left(\left(\frac{\lambda}{100}\right) + t\right) \tag{2}$$

where t is the current epoch and λ is the maximum number of epochs the training period will run for. This starts high at 0.9 to quickly move neurons into position, but quickly dampens the map's ability to learn, entering the so called 'tuning phase'. If the map is to be updated with new inputs for an indeterminate period of time, as t tends toward infinity, the learning rate tends towards a minimum value above zero, effectively extending the tuning phase for all inputs past a certain epoch.

3.4   Algorithm

The LSOM algorithm (Algorithm 1) is a derived from the original SOM algorithm. The number of epochs varies according to the number of inputs and size of the map. However we found approximately 10,000 iterations gives good results for a map with 2,000 inputs and neurons on each layer. An input is taken at random from the input dataset and the BMU on L_1 by feature distance and its neighborhood are moved towards it using the learning restraint as a multiplier. The BMU for the next layer is selected from the neurons interlinked to the BMU on the previous layer.

4   Clustering and Visualization

4.1   Clustering

Clustering on the LSOM is performed by applying an adapted Watershed transformation [12] to the Unified Distance Matrix (U-Matrix) – a matrix of scalar values representing the average distance of each neuron from its immediate neighborhood. Clustering is performed on the N th layer of the LSOM as it represents the most ordered point, rendering explicit consideration of prior layers unnecessary. Clustering the trained map in a way that is semantically useful to humans is arguably the most difficult area to achieve a high level of accuracy since the algorithm will cluster based solely on differences between the descriptors leading to a high risk of under clustering. In an attempt to negate this, 'flooding' points, local U-Matrix minima that form the basis for each cluster are reduced prior to the application of the transformation by merging through differential smoothing of the U-Matrix. A reaction-diffusion model can be used to achieve nonlinear smoothing; this merges small clusters together whilst retaining the boundaries of larger ones. The transformation is applied to the unsmoothed matrix, with the minima equivalent to the remaining post-smoothed minima selected as the flooding points, ignoring all others. Neurons are incrementally added to neighboring clusters as if water were 'flooding' the structure from the selected minima. Where two clusters meet, the dividing neuron, known as a boundary neuron, is left unclustered until the algorithm has terminated where they are then sequentially assigned to the same cluster as their closest clustered topographical neighbor, starting with the smallest feature distance.

4.2   Relationship Graph

Discovered relationships by the LSOM can be used to create a graph detailing similar images for visualization and image retrieval in $O(1)$ time. The graph structure is identical to Layer N of the LSOM. Each node is assigned the image(s) closest to the equivalent neuron on the output layer by feature distance. It is often to be expected that a single image may be assigned to more than one node, however a large number of duplicates may be a sign of undertraining. A more useful relationship graph can be constructed by ranking images to each node according to their distance,

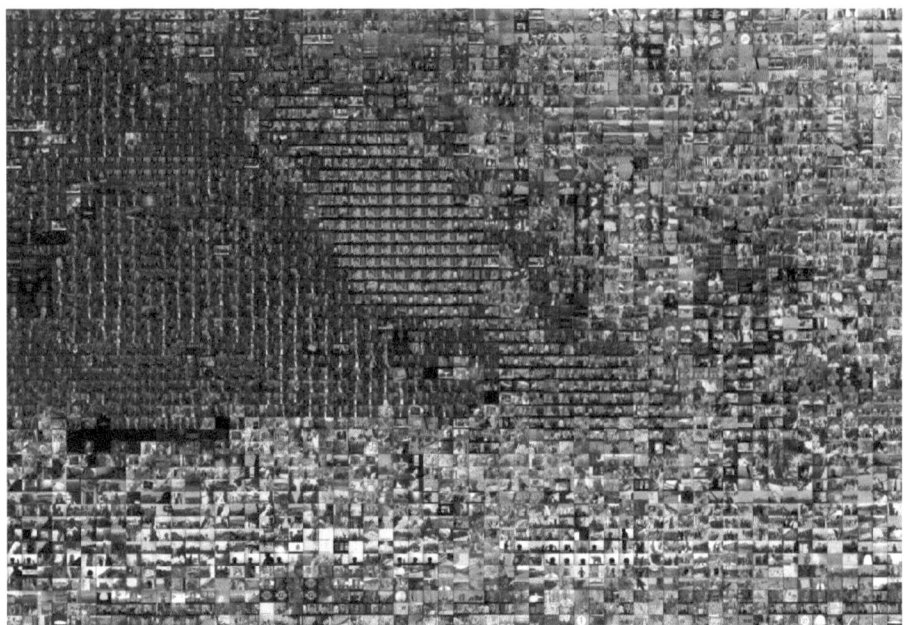

Fig. 1. Example visualization of relationship graph from LSOM with Homogenous Texture Descriptor (HTD) - Color Layout Descriptor (CLD) for 4,000 key frame images

then assigning the images according to their rank once only until all nodes have an associated image. An example visualization of 4,000 key frames is shown in Figure 2.

5   Experimental Results

Experimentation was performed using LSOMs with various layers and orderings, with each processing three image datasets (1,500 images from the Corel Gallery, 26,892 key frame images, and 2,000 color textures). The neurons in the resulting LSOMs were then clustered using the method described in section 4.1 and compared against the results from a CBIR search using the same descriptors combined using ranking. Comparison was made on the basis of how well a ground truth clustered using a rate of recall and precision metrics. The recall rate is given by the formula:

$$R_k = N/k \tag{3}$$

where N is the number of images in the ground truth in the main cluster(s), and k is the total number of images in the ground truth. The precision rate is given by:

$$P = (N + S)/L \tag{4}$$

where S is the number of similar images in the main cluster(s) of the ground-truth, but themselves are not part of the ground truth, and L is the number of images in the labeled area, including irrelevant images.

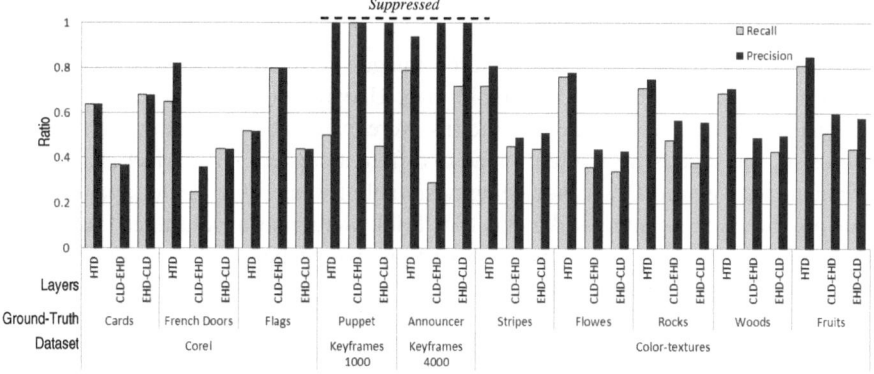

Fig. 2. Experimental results

Table 1. Comparison of Methods by Precision and Recall

Methods	LSOM	CBIR
Recall	56	55
Precision	67 (86*)	73

*Unsuppressed result   Ave. [%]

5.1 Corel Dataset

A set of 15 clusters were selected from the Corel Gallery dataset, with three selected as ground-truths to measure retrieval rate and precision: 'cards,' 'French doors' and 'flags.' The 1,500 images were sorted using a 50×30 LSOM. Expectedly, each ground-truth was categorized best by a different LSOM. Since the clusters were selected to represent a broad range of clusters with little overlap, it is unsurprising that the different features required different descriptor combinations to categorize their most prominent features. Further to this, the Corel Gallery also exhibited near identical precision and recall rates, as the clusters were too dissimilar to result in significant entropy.

5.2 Key Frames Dataset

The key frames dataset employed in [13] was formed from a group of television broadcast stills. As a test of the LSOM scalability, the results were run twice, once with a 1,000 images in a 40×25 LSOM, and again with a 4,000 image 80×50 map. While small clusters in the smaller map clustered very well, in one instance at 100% with the Color Layout-Canny EHD arrangement, they fared significantly less well in the larger LSOM. However this was compensated by retention of a high precision rate, for example for 'puppets', a 355% precision rate was obtained and suppressed to 100%, as seen in Figure 3, suggesting that the LSOM finds similarity amongst images on a macro level. Observation returned the notion that although many images were not connected in a way which satisfied the clustering condition, they were to be found in the same region.

5.3   Color Textures

Color textures consisted of 2,000 textured images selected from the Corel Gallery dataset and were classified with a 50x40 map. Unexpectedly, images containing approximately regular or directional patterns were well clustered. However, random textures such as clouds were often spread to different clusters. This effect is due to the Homogeneous Texture Descriptor being unsuitable for handling random patterns. On occasion, directional patterns like check or brick were also misclustered, suggesting the spatial frequency or directionality alterations would benefit from additional features such as steerable filters [14].

6   Conclusions

This paper proposes the Layered SOM (LSOM), a new form of ANN derived from the Self Organizing Feature Map for classifying images of arbitrary image domains. From the results of the conducted experiments on each domain containing thousands of images as well as from empirical observations, the findings can be summarized as follows:

On the basis of the achieved results we clearly attained a higher level of meaningful clustering and classification of the images. In many of the instances where falls in recall rates were observed, an increase in precision was also witnessed. Images in the ground truths may not have conformed to the clustering condition yet were still placed in the same region of the map, mixed with other images exhibiting similar properties, suggesting the LSOM forms clusters based on overriding themes rather than specifics. This is further supported during the scalability test where small ground truths clustered very well – in one case perfectly – for the smaller LSOM, yet represented a much smaller percentage of the larger map's area, and were scattered accordingly. Based on this the LSOM would lend itself well to data labeling. Compared to CBIR retrieval (Table 1), the LSOM exhibited only slightly better recall rates as the measurement was unable to account for when images gravitated towards similar clusters elsewhere in the map. While the same is true for similar images being clustered elsewhere, perhaps with greater accuracy, the LSOM precision still exhibits significantly better precision than CBIR supporting the notion that the LSOM clusters themes rather than specific features. The ordering of the layers and selection of descriptors is crucial to the success of the LSOM; and these choices depending very much on the type of data being analyzed. In several instances it was observed that while the measured ground truths did not cluster well, others clustered with a much greater degree of accuracy; this was particularly apparent with the Corel Gallery dataset, where a number of unrelated classes were categorizes together. The LSOM values in Table 1 have been impacted due to averaging including non-optimal layer arrangements. Automatic feature selection and layer ordering would make for interesting future work. The power of the LSOM lies in its flexibility to categorize large quantities of data implicitly in its topological structure, allowing new data to be incrementally added expending relatively few computational resources, as not all images need to be considered during the training period in order to be classified. In addition, the LSOM need only be trained once, yet can be queried multiple times, using it's topology to find similar data points, as opposed to distance retrieval where new distances need to be calculated or compared between every image in the dataset per query. As a framework, the LSOM approach can be applied to any general case where data features are compared using a distance function, with initial experimentation in the domain of one-dimensional signals exhibiting favorable results.

References

1. Majunath, B.S., Salembier, P.H., Sikora, T.: Introduction to MPEG-7. Wiley (2002)
2. Kohonen, T.: Self-organizing map, 3rd edn. Springer (2000)
3. Kirk, J.S., Chang, D.-J.: Zurada. J.M.: A self-organizing map with dynamic architecture for efficient color quantization. In: Proceedings of International Joint Conference on Neural Networks (IJCNN), vol. 3, pp. 2128–2132 (2001)
4. Arias, S.: Gomez. H.: Satellite Image Classification by Self-Organized Map on GRID Computing Infrastructures. In: Proceedings of the Second EELA-2 (2009)
5. Lu, S., Segall, R.S.: Multi-SOM: an Algorithm for High-Dimensional, Small Size Datasets. Journal of Systemics, Cybernetics and Informatics 11(2), 41–46 (2013)
6. Olteanu, M., Villa-Vialaneix, N., Cierco-Ayrolles, C.: Multiple Kernel Self-Organizing Maps. In: European Symposium on Artificial Neural Networks, Computational Intelligence and Machine Learning (2013)
7. Martín-Merino, M., Muñoz, A.: Extending the SOM Algorithm to Visualize Word Relationships. In: Famili, A.F., Kok, J.N., Peña, J.M., Siebes, A., Feelders, A. (eds.) IDA 2005. LNCS, vol. 3646, pp. 228–238. Springer, Heidelberg (2005)
8. Breiteneder, C., Merkl, D., Eidenberger, H.: Merging Image Features by Self-organizing Maps in Coats of Arms Retrieval. In: Proceedings of European Conference on Electronic Imaging and the Visual Arts, Berlin, Germany (1999)
9. Rahman, M.: Image Search in a Visual Concept Feature Space with SOM-Based Clustering and Modified Inverted Indexing. In: Application and Novel algorithm Design, pp. 173–188. Intech. (2011)
10. Oja, E., Laaksonen, J., Koskela, M., Brandt, S.: Self-Organizing Maps for Content-Based Image Database Retrieval. In: Kohonen Maps, pp. 349–362. Elsevier (1999)
11. Huang, J., Ravi Kumar, S., Mitra, M., Zhu, W.-J., Zabih, R.: Image Indexing Using Color Correlograms. In: Proceedings of IEEE Computer Vision and Pattern Recognition, pp. 762–768 (1997)
12. Vincent, L., Soille, P.: Watersheds in Digital Spaces: An Efficient Algorithm Based on Immersion Simulations. IEEE Transactions on Pattern Analysis and Machine Intelligence 3(6), 583–598 (1991)
13. Naito, M., Hoashi, K., Matsumoto, K., Shishibori, M., Kita, K., Kutics, A., Nakagawa, A., Sugaya, F., Nakajima, Y.: High-Level Feature Extraction Experiments for TRECVID 2007. In: TRECVID 2007 (2007)
14. Kondo, I., Kutics, A., Tanaka, H., Sakano, H.: A new texture descriptor using steerable filters, In: Technical Report of IEICE, Vol. 104, No. 573 (PRMU2004), pp. 13-18 (2004)

Wide Area Camera Localization

Valeria Garro[1], Maurizio Galassi[1], and Andrea Fusiello[2,*]

[1] Department of Computer Science,
University of Verona,
Strada Le Grazie 15, 37134 Verona, Italy
[2] Department of Electrical, Mechanical and Management Engineering,
University of Udine,
Via Delle Scienze 206, 33100 Udine, Italy

Abstract. In this paper we describe a mobile camera localization system that is able to accurately estimate the pose of an hand-held camera inside a known urban environment. The work leverages on a pre-computed 3D structure obtained by a hierarchical Structure from Motion pipeline to compute the 2D-3D correspondences needed to orient the camera. The hierarchical cluster structure, given by the SfM, guides the localization process providing accurate and reliable features matching. Experiments in outdoor challenging environments demonstrate the effectiveness of the method compared to a standard image retrieval approach.

Keywords: Localization, Camera pose, Structure from motion.

1 Introduction

The problem of providing a precise localization of a portable camera has been widely investigated in computer vision. A particular aspect of this issue is *image-based localization*, i.e., computing the camera pose estimation of the device using only the information given by the image or video itself. This topic is included in a wide range of applications such as video surveillance and robot localization, augmented reality application for eHeritage and gaming.

In these particular scenarios a very accurate level of localization is needed, hence positioning systems employing only GPS or Wi-Fi sensors are not sufficient. For example GPS signal is missing in indoor environment and even if available outside, its accuracy could be affected by atmospheric conditions and natural and artificial barriers, furthermore these type of sensors provide only the 3D position of the hand-held device but not the camera orientation. For these reasons further techniques based directly on image processing must be included in the system in order to provide a complete and precise camera pose estimation. We propose a complete image-based system that provides an accurate camera pose estimation of compact devices like smartphones, surveillance and consumer

---

* This work has been carried out while A.F. was with the University of Verona.

A. Petrosino (Ed.): ICIAP 2013, Part I, LNCS 8156, pp. 320–329, 2013.
© Springer-Verlag Berlin Heidelberg 2013

cameras in a urban environment. Image-based techniques usually request as input only a set of unordered images of the scene where one wants the positioning to take place. Thanks to the recent improvement in computer vision research on Structure from Motion (SfM) [1–5], in addition to this image archive we can also rely on the 3D reconstruction of the environment. Implementations of different SfM algorithms are available online[1], furthermore in the last years several new SfM techniques have been presented to improve scalability exploiting large scale photo collections [2, 3] and to augment efficiency and precision using hierarchical methods [4, 5].

A variety of approaches that exploit both 3D and 2D data for location recognition has been presented in the computer vision literature. In [6] the authors present a complete system integrating SfM and image-base technique for fast location recognition. They propose the creation of a set of synthetic views in addition to the initial dataset of images used for SfM reconstruction in order to cover as much as possible the corresponding area and be able to compute also the camera pose of query images taken far from the original dataset. In [7] a typical computer graphics approach for visibility estimation is applied in order to reduce the dataset of images to process during the retrieval step. The authors divide the 3D points cloud into view cells and pre-compute a cell-to-cell visibility data. These Potentially Visible Sets (PVS) determine the subset of 3D points and related descriptors that have to be considered according to the current cell.

Recent works focus on a direct 2D-to-3D registration that omits the conventional image retrieval step. In [8] a prioritized feature matching algorithm that matches a limited set of representative 3D scene features to features in the query image is proposed. In [9] the authors devised a direct matching procedure based on visual vocabulary quantization of the 3D features and a prioritized correspondence search. A further step has been introduced by [10] and [11], where a unified formulation of searching strategies has been explored that includes both 2D-to-3D and 3D-to-2D matching on large scale datasets.

In this paper we present a localization system leveraging on spatial 3D information, that combined with an efficient image retrieval technique, provides a fast and precise camera pose estimation of a single image or a video frame capture with an hand-held device. Our algorithm relies on a hierarchical SfM pipeline [4, 5] that besides the 3D points cloud creation provides a hierarchical cluster structure that guides the reconstruction process. It computes a sparse set of 3D points endowed with features descriptors (the "model") by processing a unordered set of images of the scene (the "images archive"). A set of 2D-3D point correspondences between the current frame and the model is needed in order to compute the camera pose estimation. Since typically the 2D points visible in one image are a small subset of the whole reconstruction, it is highly advisable to deploy pruning strategies to limit the set of 3D candidates. Our technique is based on retrieving a small set of the most similar images to the current frame from the archive and then limiting the candidates to those 3D points that are visible in the retrieved images. The retrieval procedure follows a Bag-of-Words

---

[1] http://homes.cs.washington.edu/~ccwu/vsfm/ or http://www.3dflow.net/

(BoW) approach with tf-idf weighting [12, 13] in order to give a compact representation of each image of the archive. Additionally this last step exploits also the hierarchical organization (called "dendrogram" or binary clustering tree) of the images archive produced by the clustering stage of the SfM algorithm.

More in details, the leaves of the binary cluster tree are associated with the single images of archive, the inner nodes represent a cluster of two or more images created during the reconstruction process of the SfM algorithm. The proposed approach leads to a fast and precise search algorithm. It is more efficient than the classic indirect image retrieval approach because the BoW vectors comparison is limited to a particular set of the inner nodes of the dendrogram avoiding a comparison procedure with the complete image archive. At the same time it guarantees more coherent retrieval results preventing the retrieval of single "outlier" image.

We test the performance of the algorithm in four different urban scenarios. In particular on the last more challenging dataset we have built also an handmade ground-truth in order to compute quantitative results on a video frame sequence.

2   System Overview

The system involves two main stages (see Fig.1):

- An "off-line" stage that runs the SfM pipeline and indexes each node of the dendrogram according to the Bag-of-words approach;
- An "on-line" stage during which the video stream captured from the mobile camera is transmitted over Wi-Fi connection to a server that processes each frame accordingly in order to orient the camera.

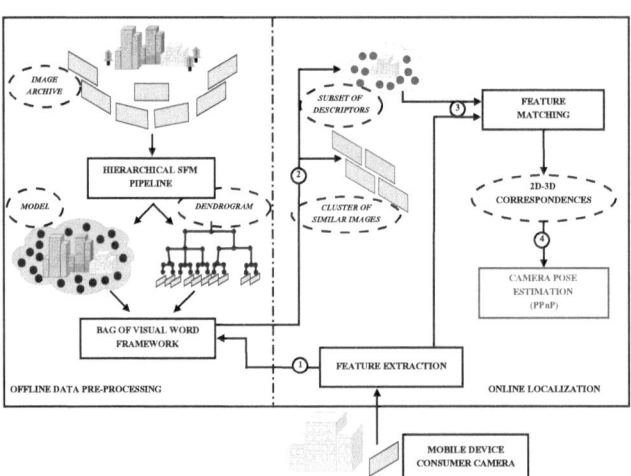

Fig. 1. System Overview. The "off-line" data pre-processing step are represented on the left of the image, the "on-line" stages are outlined on the right.

2.1 Offline Data Pre-processing

The off-line stage is devoted to compute a 3D reconstruction of the environment from the images archive. After running the SfM pipeline [4, 5] we obtain a 3D points cloud where each 3D point is endowed with a set of SURF [14] features descriptors, and a binary cluster tree (dendrogram) from which we can infer a hierarchical clustering of images.

Indexing and retrieval follows the well-known Bag-of-Words (BoW) framework. With the BoW approach images are represented by an histogram of occurrences of visual words from the codebook. These visual words are usually provided by quantization of the entire set of feature descriptors associated with the 3D points of the model. Additionally, we represent also each inner node of the dendrogram by a BoW histogram. Each inner node is a cluster of images and its BoW vector is computed using the feature descriptors related to the subset of 3D points visible from the images belonging to that cluster. The root of the binary cluster tree (level 0) identifies the whole 3D reconstruction. If the dendrogram is well balanced, its first levels are associated with big portions of the 3D points cloud, therefore their BoW histograms are not very discriminative and can be excluded from the retrieval step.

Different approaches can be employed for the descriptors quantization depending on the size of the dataset: for a relatively small dataset k-means clustering can be sufficient, however, in the pursue of scalability, more complex data structures like vocabulary tree [15] or random forest [16] must be used. We apply also the "term frequency - inverse document frequency" (tf-idf) weighting scheme: given a visual word t in an image d, its weight is given by: $\text{tf-idf}_{t,d} = \text{tf}_{t,d} \times \text{idf}_t$. The term frequency (tf) is simply the (normalized) occurrence count of a visual word in the image: $\text{tf}_{t,d} = \frac{n_{t,d}}{\sum_k n_{k,d}}$ where $n_{t,d}$ is the number of occurrences of the visual word t in the image d. The inverse document frequency (idf) evaluates the general importance (or rarity) of the visual term: $\text{idf}_t = \log \frac{|M|}{1+|\{i:n_{t,i}\neq 0\}|}$ where M is the set of all images and $\{i : n_{t,i} \neq 0\}$ is the set of images where the visual word t appears at least one time.

2.2 On-Line Processing

The on-line phase consists first in retrieving from the archive the most similar images to the current one, in order to limit the 3D matching candidates to those points that are visible in the retrieved images. Then the 2D-3D matches are used to orient the camera by solving an exterior orientation or Perspective-n-Point problem camera pose (PnP) problem. In particular, the on-line stage consists of the following steps, as illustrated in Fig. 1:

1. Fast-Hessian features detection and SURF descriptor extraction [14];
2. Retrieval of the most similar images exploiting the dendrogram and recover of SURF descriptors related to the 3D points viewed by the retrieved images;
3. Descriptors matching;
4. Camera orientation (or pose estimation) from 2D-3D correspondences.

First, keypoint features are detected and descriptors are extracted from the current frame (query image), in the specific case Fast-Hessian features and SURF descriptors [14] have been chosen, then each feature is assigned to a visual word of the codebook using a kd-tree structure, and the BoW histogram (H_q) of the query image is computed. Then, we run the retrieval step where the similarity between H_q and the BoW histograms related to the nodes of the dendrogram is computed by using the cosine similarity function: $\text{sim}(H_q, H_i) = \frac{H_q \cdot H_i}{\|H_q\|\|H_i\|}$ for each node i belonging to a specific level of the dendrogram D.

Suppose having a balanced dendrogram, the similarity check can be applied only to a particular subset of inner nodes D, reducing the number of comparisons. Nodes of the dendrogram with small depth (i.e. near the root of the tree) are associated to a large portion of the reconstruction and therefore their BoW histograms can be not so discriminative. For this reason we compare the BoW histogram of the query image only with the inner nodes whose subtrees contain a limited number of leaves (dataset images) (e.g. $6 - 10$). The most similar inner node $\tilde{D}$ is now determined, the leaves of the subtree with root $\tilde{D}$ are the subset of most similar images, defined $\tilde{M} \subset M$.

The second step consists in selecting the points of the 3D model visible from the cluster related to $\tilde{D}$ and $\tilde{M}$. Finally, a set of tentative correspondences between 2D query points and 3D model points is obtained with nearest-neighbor matching between the descriptors extracted from the query image and the descriptors of the 3D points just selected.

Given a number of 2D-3D point correspondences $\mathbf{m}_j \leftrightarrow \mathbf{M}_j$ and the intrinsic camera parameters K, the exterior image orientation problem requires to find a rotation matrix R and a translation vector $\mathbf{t}$ (which specify attitude and position of the camera) such that:

$$\zeta_j \tilde{\mathbf{m}}_j = K[R|\mathbf{t}]\tilde{\mathbf{M}}_j \quad \text{for all } j \tag{1}$$

where ζ_j denotes the depth of $\mathbf{M}_j$, and the $\tilde{\ }$ denotes homogeneous coordinates (with a trailing "1").

In literature there are many algorithms that solve this problem [17–19]; we adopted the PPnP approach [20], a simple and efficient solution that formulates it in terms of an instance of the anisotropic orthogonal Procrustes problem. In the remaining of this section we will briefly summarize this approach. After some rewriting, (1) becomes:

$$\underbrace{\begin{bmatrix} \mathbf{M}_1^T \\ \vdots \\ \mathbf{M}_n^T \end{bmatrix}}_{S} = \underbrace{\begin{bmatrix} \zeta_1 & 0 & \cdots & 0 \\ 0 & \zeta_2 & \cdots & 0 \\ \vdots & \vdots & \ddots & \vdots \\ 0 & 0 & \cdots & \zeta_n \end{bmatrix}}_{Z} \underbrace{\begin{bmatrix} \tilde{\mathbf{p}}_1^T \\ \vdots \\ \tilde{\mathbf{p}}_n^T \end{bmatrix}}_{P} R + \underbrace{\begin{bmatrix} \mathbf{c}^T \\ \vdots \\ \mathbf{c}^T \end{bmatrix}}_{\mathbf{1c}^T}. \tag{2}$$

where $\tilde{\mathbf{p}}_j = K^{-1}\tilde{\mathbf{m}}_j$, $\mathbf{c} = -R^T\mathbf{t}$, and $\mathbf{1}$ is the unit vector. Therefore, the previous equation can be written more compactly in matrix form:

$$S = ZPR + \mathbf{1c}^T \tag{3}$$

Fig. 2. Sample images of dataset *Piazza del Santo*(top) and *Piazza Brà* (bottom)

This is an instance of an *anisotropic* orthogonal Procrustes problem with *data* scaling [21]. The solution of this problem finds Z, R and $\mathbf{c}$ in such a way to minimize the sum of squares of the residual matrix $\Delta = S - ZPR - \mathbf{1c}^T$. This can be written as

$$\min\|\Delta\|_F^2 \text{ subject to } R^T R = I, \tag{4}$$

which can be solved with Lagrangian multipliers, yielding (the derivation of the formulae is reported in [20]):

$$R = U\text{diag}\left(1,1,\det(UV^T)\right)V^T \text{ with } UDV^T = P^T Z\left(I - \mathbf{1}\,\mathbf{1}^T/n\right) S \tag{5}$$

$$\mathbf{c} = (S - ZPR)^T \mathbf{1}/n \tag{6}$$

$$Z = \text{diag}\left(PR(S^T - \mathbf{c1}^T)\right)\text{diag}\left(PP^T\right)^{-1}. \tag{7}$$

It turns out that – as opposed to the isotropic case – here the unknowns are entangled in such a way that one must resort to a *block relaxation* scheme, where each variable is alternatively estimated while keeping the others fixed. Empirically, the procedure always converges to the correct solution starting from a random initialization. In order to cope with outliers we use PPnP as minimal solver ($n = 3$) and MSAC [22], as customary. A further processing of camera pose can be done applying a non-linear refinement on inliers that minimizes the reprojection error. This is however discretionary, subject to the time budget.

3   Experiments

In this section we describe the two different experiments performed to evaluate the proposed method. The first experiments has been run testing two outdoor scenarios, *Piazza del Santo* and *Piazza Brà*. The first archive is composed by 105 images (2592×1944) of a big city square outside an historical church, the second archive represents a bigger and more articulated square with much more repetitive structures, with 320 images (1504×1000).

Fig. 2 shows some image examples. A quantitative evaluation of camera pose estimation accuracy is based on leave-one-out tests.

Each camera provided by the SfM pipeline has been first removed from the image archive and consequently the related set of feature descriptors; then the proposed algorithm has been run on the updated archive. In this way we can consider the registered camera obtained with the SfM pipeline as our ground-truth data.

In order to test the performance of our retrieval method involving the dendrogram of the Structure from Motion reconstruction we compare it with the classic approach that measures the cosine distance between the BOW vector of the query image and the BOW vectors of each dataset image.

Table 1. Leave-one-out validation results on the two datasets. Values within the parenthesis indicate the errors after the non-linear refinement.

	# Images	# Features	# 3D points	Success Rate	Translation Error [m]	Rotation Error [deg]	Reprojection Error [px]
Piazza del Santo	105	45k	30k	95%	0.114	0.103	1.016
classic retrieval					(0.080)	(0.074)	(0.812)
Piazza del Santo	105	45k	30k	96%	0.050	0.091	0.951
dendrogram approach					(0.037)	(0.051)	(0.791)
Piazza Brà	320	233k	50k	92%	0.192	0.378	0.586
classic retrieval					(0.176)	(0.338)	(0.481)
Piazza Brà	320	233k	50k	92%	0.077	0.154	0.602
dendrogram approach					(0.058)	(0.114)	(0.486)

Table 1 shows the results for the leave-one-out tests, where the success rate is the percentage of images localized having a set of correspondences inliers larger than 20 after the camera pose estimation using MSAC. The accuracy of our algorithm is shown in terms of mean euclidean distance of the camera center with respect to the ground-truth, the mean reprojection error of the 3D points visible from the specific camera and the mean residual rotation angle given by the geodesic distance in $SO(3)$:

$$d_g(R_{gt}R_l) = \left\| \log R_{gt}^T R_l \right\|_F \qquad (8)$$

where R_{gt} is the rotation component of the camera matrix of the ground-truth data $P_{gt} = K_{gt}[R_{gt}|T_{gt}]$ and R_l is the rotation component of the camera matrix $P_l = K_l[R_l|T_l]$ computed by our algorithm. In both experiments our approach clearly outperforms the classic retrieval in terms of accuracy with comparable registration rate results.

Furthermore, for the *Piazza del Santo* dataset three different video sequences of a person walking on the area have been acquired with a simple consumer camera, each video sequence is formed by 900 frames out of which only 38 have not been successfully localized. A qualitative evaluation of the localization is reported in Fig 3.

The second test has been run on a challenging outdoor environment consisting of a parking space located in between several buildings with repetitive structures. The image archive is composed by 543 images (2048×1536), the 3D model of the

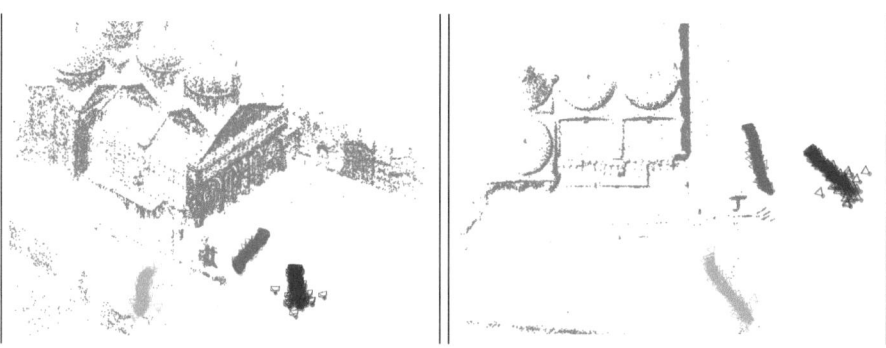

Fig. 3. The colored blobs represent the estimated camera pose for each frame of three different video sequences, perspective (left) and top (right) views

scene is described by a set of 32k 3D points and 203k SURF descriptors. Furthermore the experiment involved four static cameras slightly overlapped, installed on the parking area corners. These cameras were connected with a server that stored the 30 fps images from the cameras, synchronizing them and giving a common time stamp. The four cameras have been calibrated with respect to the reference system of the 3D model, achieving a coherent system. A sample of the four camera views can be seen in Fig. 4. The test consisted of an agent equipped with a proprietary device[2] fixed on the shoulder, walking in the area in a wide closed loop while recording the scene. Analyzing the video sequences recorded by the static cameras we computed the ground-truth for each frame, estimating the 3D position of the agent on the ground floor using a suitable homography transformation.

Fig. 4. Snapshots taken at the same time by the four static cameras

Due to the hardware setup the chain delays induced a variable frame rate transmission of the mobile device data, therefore it was not possible to couple frame by frame the agent and the static cameras views. In order to overcome this synchronization problem over the all frames we decided to evaluate the trajectory of the agent instead of comparing each single position. Each ground-truth path extracted by the four cameras has been approximated by fitting a polynomial curve to the data, generating a set of four segments representing

---

[2] This device have been specifically designed for mobile surveillance and provides high quality recording of time-stamped audio-video sequences. It has a ARM Cortex A8 core processor running at 720 MHz and an integrated Microsoft LifeCam Studio webcam with a resolution of 1280 × 720.

the path. The trajectory directions, the original ground-truth positions and the fitted segments are shown in Fig. 5. The average distance error is 2.82 m and the success rate is 45% We run our tests on a Intel QuadCore with 2.4Ghz, the C++ implementation of the algorithm takes less than 2 seconds. More in details, the feature and descriptor extraction requires 0.55 seconds, the retrieval 0.30 seconds, for the feature matching step the time is 0.65 seconds and finally the camera pose estimation with PPnP and MSAC takes 0.17 seconds.

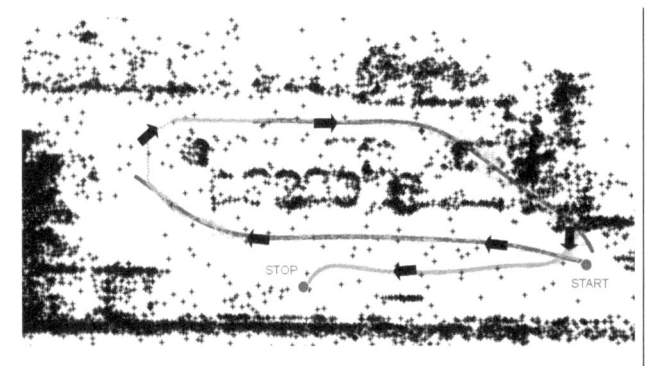

Fig. 5. Sequence trajectory. The ground-truth positions are marked in cyan, the four segments are indicated in alternated colors, green and red. Dashed lines indicate the absence of ground-truth data, as the path falls out of the field of view of the cameras.

4   Conclusions and Future Works

We described a mobile camera localization system in a known urban environment. Localization occurs via 2D keypoint matching against a 3D points cloud obtained by a hierarchical SfM pipeline, leveraging in the image retrieval step on the additional hierarchical structure given by the SfM. Future work will aim at achieving real-time processing of the video frames by GPU implementation and exploitation of motion constraints (presently, each frame is localized independently from the previous ones).

Acknowledgments. This work has been funded by the EU SAMURAI Project.

References

1. Snavely, N., Seitz, S.M., Szeliski, R.: Photo tourism: exploring photo collections in 3d. In: SIGGRAPH: International Conference on Computer Graphics and Interactive Techniques, pp. 835–846 (2006)
2. Frahm, J.-M., et al.: Building rome on a cloudless day. In: Daniilidis, K., Maragos, P., Paragios, N. (eds.) ECCV 2010, Part IV. LNCS, vol. 6314, pp. 368–381. Springer, Heidelberg (2010)
3. Snavely, N., Seitz, S., Szeliski, R.: Skeletal graphs for efficient structure from motion. In: IEEE Conference on Computer Vision and Pattern Recognition, pp. 1–8 (2008)

4. Farenzena, M., Fusiello, A., Gherardi, R.: Structure-and-motion pipeline on a hierarchical cluster tree. In: IEEE International Workshop on 3-D Digital Imaging and Modeling, pp. 1489–1496 (2009)
5. Gherardi, R., Farenzena, M., Fusiello, A.: Improving the efficiency of hierarchical structure-and-motion. In: IEEE Conference on Computer Vision and Pattern Recognition, pp. 1594–1600 (June 2010)
6. Irschara, A., Zach, C., Frahm, J., Bischof, H.: From structure-from-motion point clouds to fast location recognition. In: IEEE Conference on Computer Vision and Pattern Recognition, pp. 2599–2606 (2009)
7. Arth, C., Wagner, D., Klopschitz, M., Irschara, A., Schmalstieg, D.: Wide area localization on mobile phones. In: 8th IEEE International Symposium on Mixed and Augmented Reality, pp. 73–82 (2009)
8. Li, Y., Snavely, N., Huttenlocher, D.P.: Location recognition using prioritized feature matching. In: Daniilidis, K., Maragos, P., Paragios, N. (eds.) ECCV 2010, Part II. LNCS, vol. 6312, pp. 791–804. Springer, Heidelberg (2010)
9. Sattler, T., Leibe, B., Kobbelt, L.: Fast image-based localization using direct 2d-to-3d matching. In: International Conference on Computer Vision, pp. 667–674 (2011)
10. Li, Y., Snavely, N., Huttenlocher, D., Fua, P.: Worldwide pose estimation using 3D point clouds. In: Fitzgibbon, A., Lazebnik, S., Perona, P., Sato, Y., Schmid, C. (eds.) ECCV 2012, Part I. LNCS, vol. 7572, pp. 15–29. Springer, Heidelberg (2012)
11. Sattler, T., Leibe, B., Kobbelt, L.: Improving image-based localization by active correspondence search. In: Fitzgibbon, A., Lazebnik, S., Perona, P., Sato, Y., Schmid, C. (eds.) ECCV 2012, Part I. LNCS, vol. 7572, pp. 752–765. Springer, Heidelberg (2012)
12. Sivic, J., Zisserman, A.: Video Google: A text retrieval approach to object matching in videos. In: International Conference on Computer Vision, pp. 1470–1477 (2003)
13. Yang, J., Jiang, Y.G., Hauptmann, A.G., Ngo, C.W.: Evaluating bag-of-visual-words representations in scene classification. In: International Workshop on Multimedia Information Retrieval, pp. 197–206 (2007)
14. Bay, H., Ess, A., Tuytelaars, T., Van Gool, L.: Speeded-up robust features (surf). Computer Vision and Image Understanding 110(3), 346–359 (2008)
15. Nister, D., Stewenius, H.: Scalable recognition with a vocabulary tree. In: Conference on Computer Vision and Pattern Recognition, pp. 2161–2168 (2006)
16. Philbin, J., Chum, O., Isard, M., Sivic, J., Zisserman, A.: Object retrieval with large vocabularies and fast spatial matching. In: IEEE Conference on Computer Vision and Pattern Recognition (2007)
17. Moreno-Noguer, F., Lepetit, V., Fua, P.: Accurate non-iterative o(n) solution to the pnp problem. In: Proceeding of the IEEE International Conference on Computer Vision (October 2007)
18. Fiore, P.D.: Efficient linear solution of exterior orientation. IEEE Transactions on Pattern Analysis and Machine Intelligence 23(2), 140–148 (2001)
19. Schweighofer, G., Pinz, A.: Globally optimal o(n) solution to the pnp problem for general camera models. In: British Machine Vision Conference, pp. 55.1–55.10 (2008)
20. Garro, V., Crosilla, F., Fusiello, A.: Solving the pnp problem with anisotropic orthogonal procrustes analysis. In: Second Joint 3DIM/3DPVT Conference: 3D Imaging, Modeling, Processing, Visualization and Transmission (2012)
21. Bennani Dosse, M., Ten Berge, J.: Anisotropic orthogonal procrustes analysis. Journal of Classification 27(1), 111–128 (2010)
22. Torr, P.H.S., Zisserman, A.: MLESAC: A new robust estimator with application to estimating image geometry. In: Computer Vision and Image Understanding, vol. 78 (2000)

Arabic Printed Word Recognition Using Windowed Bernoulli HMMs

Ihab Khoury, Adrià Giménez, Alfons Juan, and Jesús Andrés-Ferrer

DSIC/ITI, Universitat Politècnica de València,
Camí de Vera s/n, 46022 València, Spain
{ialkhoury,agimenez,ajuan,jandres}@dsic.upv.es

Abstract. Hidden Markov Models (HMMs) are now widely used for off-line text recognition in many languages and, in particular, Arabic. In previous work, we proposed to directly use columns of raw, binary image pixels, which are directly fed into embedded Bernoulli (mixture) HMMs, that is, embedded HMMs in which the emission probabilities are modeled with Bernoulli mixtures. The idea was to by-pass feature extraction and to ensure that no discriminative information is filtered out during feature extraction, which in some sense is integrated into the recognition model. More recently, we extended the column bit vectors by means of a sliding window of adequate width to better capture image context at each horizontal position of the word image. However, these models might have limited capability to properly model vertical image distortions. In this paper, we have considered three methods of window repositioning after window extraction to overcome this limitation. Each sliding window is translated (repositioned) to align its center to the center of mass. Using this approach, state-of-art results are reported on the Arabic Printed Text Recognition (APTI) database.

Keywords: Bernoulli HMMs, APTI, Arabic Printed Recognition, Sliding Window, Repositioning.

1 Introduction

Hidden Markov Models (HMMs) are now widely used for off-line text recognition in many languages and, in particular, languages with Arabic scripts [1, 5–7, 4]. Following the conventional approach in speech recognition [9], HMMs at global (line or word) level are built from shared, *embedded* HMMs at character (sub-word) level, which are usually simple in terms of number of states and topology. In the common case of real-valued feature vectors, state-conditional probability (density) functions are modeled as Gaussian mixtures since, as with finite mixture models in general, their complexity can be easily adjusted to the available training data by simply varying the number of components.

After decades of research in speech recognition, the use of certain real-valued speech features and embedded Gaussian (mixture) HMMs is a de-facto standard [9]. However, in the case of text recognition, there is no such standard and,

A. Petrosino (Ed.): ICIAP 2013, Part I, LNCS 8156, pp. 330–339, 2013.

indeed, very different sets of features are in use today. In [2] we proposed to by-pass feature extraction and to directly feed columns of raw, binary pixels into *embedded Bernoulli (mixture) HMMs (BHMMs)*, that is, embedded HMMs in which the emission probabilities are modeled with Bernoulli mixtures. The basic idea was to ensure that no discriminative information is filtered out during feature extraction, which in some sense is integrated into the recognition model. In [3], we improved our basic approach by using a sliding window of adequate width to better capture image context at each horizontal position of the text image. This improvement, to which we refer as *windowed BHMMs*, achieved very competitive results on the well-known IfN/ENIT database of Arabic town names [8]. More recently, very good results on the Arabic Printed Text Image (APTI) database were achieved by using the same approach, which ranked first at the ICDAR 2011 - Arabic Recognition Competition for printed Arabic text [11].

Although windowed BHMMs achieved good results on IfN/ENIT and APTI, it was clear to us that text distortions are more difficult to model with wide windows than with narrow (e.g. one-column) windows. In order to circumvent this difficulty, we have considered new, adaptative window sampling techniques, as opposed to the conventional, direct strategy by which the sampling window center is applied at a constant height of the text image and moved horizontally one pixel at a time. More precisely, these adaptative techniques can be seen as an application of the direct strategy followed by a *repositioning* step by which the sampling window is repositioned to align its center to the center of gravity of the sampled image. This repositioning step can be done horizontally, vertically or in both directions. Although vertical repositioning was expected to have more influence on recognition results than horizontal repositioning, we decided to study both separately, and also in conjunction, so as to confirm this expectation.

In this paper, the repositioning techniques described above are introduced and extensively tested on an Arabic printed database. In particular, we provide new, state-of-art results on the Arabic Printed Text Image (APTI) database, which clearly outperform our previous results without repositioning [3, 11]. In what follows, we briefly review the Bernoulli mixtures HMMs, maximum likelihood parameter estimation and *windowed BHMMs* repositioning techniques. Then, empirical results are reported after a brief description of the APTI database.

2 Bernoulli Mixture HMMs

Let $O = (\mathbf{o}_1, \ldots, \mathbf{o}_T)$ be a sequence of feature vectors. An HMM is a probability (density) function of the form:

$$P(O \mid \Theta) = \sum_{q_1,\ldots,q_T} \prod_{t=0}^{T} a_{q_t q_{t+1}} \prod_{t=1}^{T} b_{q_t}(\mathbf{o}_t), \qquad (1)$$

where the sum is over all possible *paths* (state sequences) $q_0, \ldots, q_{T+1}$, such that $q_0 = I$ (special *initial* or *start* state), $q_{T+1} = F$ (special *final* or *stop* state), and

$q_1, \ldots, q_T \in \{1, \ldots, M\}$, being M the number of regular (non-special) states of the HMM. On the other hand, for any regular states i and j, a_{ij} denotes the *transition* probability from i to j, while b_j is the *observation* probability (density) function at j.

A Bernoulli (mixture) HMM (BHMM) is an HMM in which the probability of observing the binary feature vector $\mathbf{o}_t$, when $q_t = j$, follows a Bernoulli mixture distribution for the state j

$$b_j(\mathbf{o}_t) = \sum_{k=1}^{K} \pi_{jk} \prod_{d=1}^{D} p_{jkd}^{o_{td}} (1 - p_{jkd})^{1-o_{td}}, \tag{2}$$

where o_{td} is the d-th bit of $\mathbf{o}_t$, π_{jk} is the prior of the k-th mixture component in state j, and p_{jkd} is the probability that this component assigns to o_{td} to be 1.

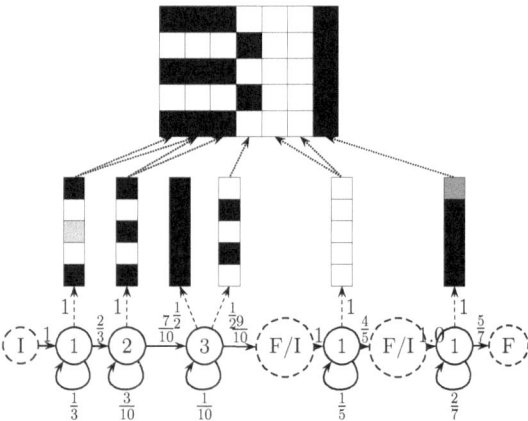

Fig. 1. BHMM examples for the numbers 31, together with binary images generated from it. Bernoulli prototype probabilities are represented using the following color scheme: black=1, white=0, gray=0.5 and light gray=0.1.

As discussed in the introduction, BHMMs at global (line or word) level are built from shared, embedded BHMMs at character level. More precisely, let C be the number of different characters (symbols) from which global BHMMs are built, and assume that each character c is modeled with a different BHMM of parameter vector $\mathbf{\Theta}_c$. Let $\mathbf{\Theta} = \{\mathbf{\Theta}_1, \ldots, \mathbf{\Theta}_C\}$, and let $O = (\mathbf{o}_1, \ldots, \mathbf{o}_T)$ be a sequence of feature vectors generated from a sequence of symbols $S = (s_1, \ldots, s_L)$, with $L \leq T$. The probability of O can be calculated, using embedded HMMs for its symbols, as:

$$P(O \mid S, \mathbf{\Theta}) = \sum_{i_1, \ldots, i_{L+1}} \prod_{l=1}^{L} P(\mathbf{o}_{i_l}, \ldots, \mathbf{o}_{i_{l+1}-1} \mid \mathbf{\Theta}_{s_l}), \tag{3}$$

where the sum is carried out over all possible segmentations of O into L segments, that is, all sequences of indices $i_1, \ldots, i_{L+1}$ such that

$$1 = i_1 < \cdots < i_L < i_{L+1} = T + 1;$$

and $P(\mathbf{o}_{i_l}, \ldots, \mathbf{o}_{i_{l+1}-1} \mid \boldsymbol{\Theta}_{s_l})$ refers to the probability (density) of the l-th segment, as given by (1) using the HMM associated with symbol s_l.

An embedded BHMM for the number 31 is shown in Fig. 1, which is the result of concatenating BHMMs for the digit 3, blank space and digit 1, in that order. Note that the BHMMs for blank space and digit 1 are simpler than that for digit 3. It is worth noting that prototypes do not account for the whole digits realizations, but only for single columns. This column-by-column emission of feature vectors attempts to better model horizontal distortions at character level and, indeed, it is the usual approach in both speech and handwriting recognition when continuous-density (Gaussian mixture) HMMs are used. The binary image of the number 31 shown above can only be generated from two paths, as indicated by the arrows connecting prototypes to image columns, which only differ in the state generating the second image column (either state 1 or 2 of the BHMM for the first symbol). It is straightforward to check that, according to (3), the probability of generating this image is 0.0004.

3   Maximum Likelihood Parameter Estimation

Maximum likelihood estimation (MLE) of the parameters governing an embedded BHMM does not differ significantly from the conventional Gaussian case, and it is also efficiently performed using the well-known EM (Baum-Welch) re-estimation formulae [9, 12]. Let $(O_1, S_1), \ldots, (O_N, S_N)$, be a collection of N training samples in which the n-th observation has length T_n, $O_n = (\mathbf{o}_{n1}, \ldots, \mathbf{o}_{nT_n})$, which corresponds to a sequence of L_n symbols $(L_n \leq T_n)$, $S_n = (s_{n1}, \ldots, s_{nL_n})$. At iteration r, the E step requires the computation, for each training sample n, of its corresponding forward (α) and backward (β) recurrences (see [9]), as well as

$$z_{nltk}^{(r)}(j) = \frac{\pi_{s_{nl}jk}^{(r)} \prod_{d=1}^{D} p_{s_{nl}jkd}^{(r)}{}^{o_{ntd}} \left(1 - p_{s_{nl}jkd}^{(r)}\right)^{1-o_{ntd}}}{b_{s_{nl}j}^{(r)}(\mathbf{o}_{nt})}, \tag{4}$$

for each t, k, j, l. In (4), $z_{nltk}^{(r)}(j)$ is the probability of $\mathbf{o}_{nt}$ to be generated in the k-th mixture component, given that $\mathbf{o}_{nt}$ has been generated in the j-th state of symbol s_l. The conditional probability function $b_{s_{nl}j}^{(r)}(\mathbf{o}_{nt})$ is analogous to that defined in (2).

In the M step, the Bernoulli prototype corresponding to the k-th component of the state j for character c has to be updated as

$$\mathbf{p}_{cjk}^{(r+1)} = \frac{1}{\gamma_{ck}(j)} \sum_n \frac{\sum_{l:s_{nl}=c} \sum_{t=1}^{T_n} \xi_{nltk}^{(r)}(j)\mathbf{o}_{nt}}{P(O_n \mid S_n, \boldsymbol{\Theta}^{(r)})}, \tag{5}$$

where $\gamma_{ck}(j)$ is a normalization factor

$$\gamma_{ck}(j) = \sum_n \frac{\sum_{l:s_{nl}=c} \sum_{t=1}^{T_n} \xi_{nltk}^{(r)}(j)}{P(O_n \mid S_n, \Theta^{(r)})}, \tag{6}$$

and $\xi_{nltk}^{(r)}(j)$ is the probability of O_n when the t-th feature vector of the n-th sample corresponds to symbol s_l and is generated by the k-th component of the state j,

$$\xi_{nltk}^{(r)}(j) = \alpha_{nlt}^{(r)}(j) z_{nltk}^{(r)}(j) \beta_{nlt}^{(r)}(j). \tag{7}$$

Similarly, the k-th component coefficient of the state j in the HMM for character c is updated by

$$\pi_{cjk}^{(r+1)} = \frac{1}{\gamma_c(j)} \sum_n \frac{\sum_{l:s_{nl}=c} \sum_{t=1}^{T_n} \xi_{nltk}^{(r)}(j)}{P(O_n \mid S_n, \Theta^{(r)})}, \tag{8}$$

where $\gamma_c(j)$ is a normalization factor

$$\gamma_c(j) = \sum_n \frac{\sum_{l:s_{nl}=c} \sum_{t=1}^{T_n} \alpha_{nlt}^{(r)}(j) \beta_{nlt}^{(r)}(j)}{P(O_n \mid S_n, \Theta^{(r)})}. \tag{9}$$

Finally, it is well-known that MLE tends to overtrain the models. In order to amend this problem Bernoulli prototypes are smoothed by linear interpolation with a flat (uniform) prototype, $\mathbf{0.5}$,

$$\tilde{\mathbf{p}} = (1 - \delta)\mathbf{p} + \delta\,\mathbf{0.5}, \tag{10}$$

where δ is usually optimized in a validation set or fixed to a sensible value such as $\delta = 10^{-6}$

4   Windowed BHMMs

Given a binary image normalized in height to H pixels, we may think of a feature vector $\mathbf{o}_t$ as its column at position t or, more generally, as a concatenation of columns in a window of W columns in width, centered at position t. This generalization has no effect neither on the definition of BHMM nor on its MLE, although it would be very helpful to better capture the image context at each horizontal position of the image. As an example, Fig. 2 shows a binary image of 4 columns and 5 rows, which is transformed into a sequence of four 15-dimensional feature vectors (first row) by application of a sliding window of width 3. For clarity, feature vectors are depicted as 3×5 subimages instead of 15-dimensional column vectors. Note that feature vectors at positions 2 and 4 would be indistinguishable if, as in our previous approach, they were extracted with no context ($W = 1$).

Although one-dimensional, "horizontal" HMMs for image modeling can properly capture non-linear horizontal image distortions, they are somewhat limited when dealing with vertical image distortions, and this limitation might be particularly strong in the case of feature vectors extracted with significant context. To overcome this limitation, we have considered three methods of window *repositioning* after window extraction: *vertical, horizontal,* and *both.* The basic idea is to first compute the center of mass of the extracted window, which is then repositioned (translated) to align its center to the center of mass. This is done in accordance with the chosen method, that is, horizontally, vertically, or in both directions. Obviously, the feature vector actually extracted is that obtained after repositioning. An example of feature extraction is shown in Fig. 2 in which the standard method (no repositioning) is compared with the three repositioning methods considered.

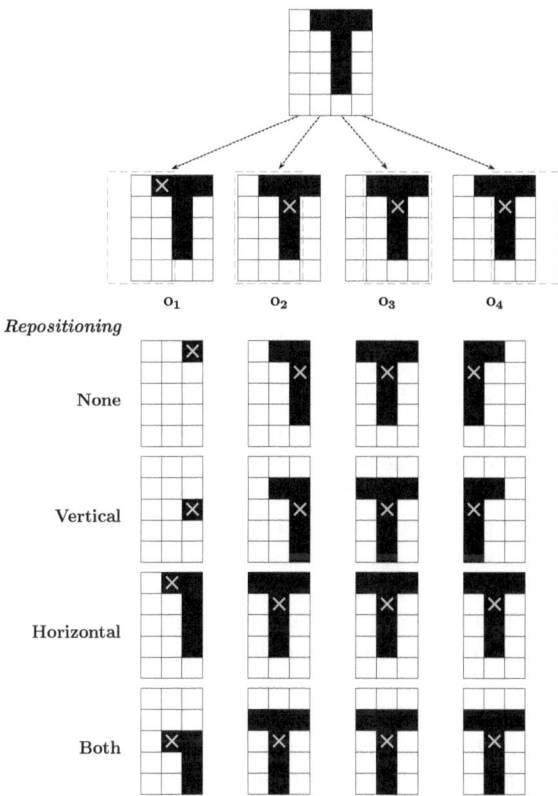

Fig. 2. Example of transformation of a 4×5 binary image (top) into a sequence of 4 15-dimensional binary feature vectors $O = (o_1, o_2, o_3, o_4)$ using a window of width 3. After window extraction (illustrated under the original image), the standard method (no repositioning) is compared with the three repositioning methods considered: vertical, horizontal, and both directions. Mass centers of extracted windows are also indicated.

5   APTI Database

The Arabic Printed Text Image (APTI) database is freely available for non-commercial research [10]. It is a multi-font, multi-size and multi-style database. It was used as a training data in the Arabic Recognition Competition held at ICDAR (Int. Conf. on Document Analysis and Recognition) in 2011 [11]. It comprises 113, 284 Arabic words generated in 10 different fonts, 10 different font sizes, and also 4 different styles. For the purpose of evaluating Arabic word recognition systems, APTI is divided into six equilibrated sets labeled as set_1, set_2, ..., set_6. The sixth set is unavailable for the public, and it was used as a testing data in the ICDAR 2011 competition. Each set has different words, but characters are distributed equally.

At the ICDAR 2011 competition, two protocols were defined which differ in the number of fonts used: APTIPC1 and APTIPC2. In APTIPC1, only the Arabic Transparent font was used. In APTIPC2, however, five different fonts were used: Arabic Transparent (Trans), Andalus (Anda), Diwani Letter (Diw), Simplified Arabic (Simp), and Traditional Arabic (Trad). In both protocols, only the *Plain* font style was used, with sizes of 6, 8, 10, 12, 18 and 24. Three systems participated: IPSAR, DIVA-REGIM and UPV-PRHLT (our system).

6   Experiments

As indicated above, experiments were carried out on the public part of the APTI database [10]. As the public part of APTI does not include set_6, we could not re-run exactly the experiments carried out at the ICDAR 2011 competition. Instead, we defined two new protocols: UPVPC1 and UPVPC2. UPVPC1 is similar to the APTIPC1 protocol described above though, as set_6 was not available, we randomly drew a number of images from all available sets for testing. More precisely, UPVPC1 uses 10000 images for training, 2000 for validation, and 3000 for testing. The UPVPC2 protocol was designed to approximate the ICDAR 2011 competition strategy reported in [11]. It uses sets set_1 to set_4 for training and set_5 for testing. It is worth noting that APTI was divided into six equilibrated sets to allow for flexibility in the composition of development and test sets [10]. Thus, we assume that the results on set_5 should not be very different from those on set_6.

For both protocols, UPVPC1 and UPVPC2, text images were scaled in height to a given dimension of D pixels (for 10 different values of D from 30 to 54) while keeping the aspect ratio. They were then binarized by means of the Otsu algorithm. For each tried value of D, we also tried different values of the sliding window width W (from 1 to 13).

The UPVPC1 protocol was first used to find out, for each font size, appropriate values for the dimension D, window width W, number of states per character Q, and number of mixture components per state K. To this end, a number of experiments were carried out using different values for these parameters. For all experiments, we used a 5-grams language model at character level instead of the conventional class priors, due to the huge amount of classes. Table 1 shows the best Character Error Rate (CER%) obtained for each size, together with the corresponding system values. From Table 1, it is clear that appropriate values for the parameters are $D \in \{40, 42\}$, $W = 9$, $Q = 7$ and $K = 32$. It is worth noting that the results with sliding windows ($W > 1$) are much better than those obtained with $W = 1$. For instance, for size 6, the best result we had with no window is a CER of 18.4% while, as can be seen in Table 1, it is reduced to a 3.5% with $W = 9$.

Table 1. Character Error Rate (CER%) for different font sizes, using the UPVPC1 protocol. For each CER% reported, the corresponding system values for D, W, Q and K are also provided.

Size	CER%	D	W	Q	K
6	3.5	38	9	7	32
8	2.4	42	9	7	32
10	2.6	40	9	6	32
12	2.3	40	9	6	32
18	1.9	40	11	6	32
24	1.8	42	11	7	32

In the experiments reported above, we did not try window repositioning after window extraction but, as previously mentioned, we think that many recognition errors of our BHMM-based recognizer might be due to its limited capability to properly model vertical image distortions. In order to study the effect of repositioning on the recognition accuracy, the standard method (no repositioning) was compared with the three repositioning methods considered in this work: vertical, horizontal, and both directions. This was done for font size 6, $D = 42$, $W = 9$ and $Q = 7$, with the data partition used in the previous experiments. The resulting CER% is shown in Fig 3, as a function of the grammar scale factor (a weight on the language model to adjust their importance with respect to word-conditional likelihoods). Clearly, vertical (or both) window repositioning outperforms the standard method or horizontal repositioning alone. It is worth noting that the CER% values obtained with no repositioning are approximately halved when using vertical repositioning. In particular, for $G = 70$, the standard method achieves a 3.8% of CER while vertical repositioning reduces this figure to 2.2%. These results are included in Table 2 together with the results obtained using font sizes other than 6. As can be seen in Table 2, the use of vertical repositioning leads to huge CER reductions in relative terms.

In order to compare our results with those reported in the ICDAR 2011 competition, a final experiment was carried out using the UPVPC2 protocol. In this experiment, we used a different recognizer for each font (font-dependent). Each test sample was recognized according to the font-dependent recognizer of higher maximum posterior probability. The results at character and word level

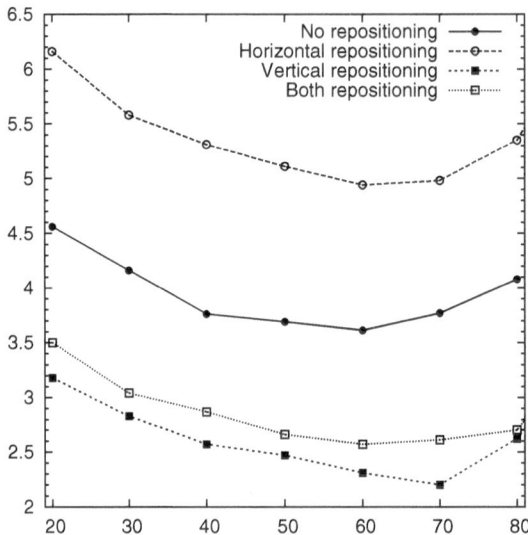

Fig. 3. CER(%) as a function of the grammar scale factor (G) for the different repositioning techniques

Table 2. Character Error Rate (CER%) using the standard method (no repositioning) and vertical repositioning, for different font sizes

Repositioning	6	8	10	12	18	24
No	3.8	2.4	2.6	2.3	1.9	1.8
Vertical	2.2	1.0	0.9	0.7	0.2	0.3
Relative reduction %	42	58	65	70	89	83

are shown in Table 3, together with the results obtained at the ICDAR 2011 competition. The system presented in this work would rank first.

Table 3. Character Error Rate (CER%) and Word Error Rate (WER%) for all participated systems in the ICDAR 2011 - Arabic Recognition Competition. Results are shown for different fonts.

System		Trans	Anda	Diw	Simp	Trad
IPSAR system	CER	9.8	13.0	22.5	10.2	25.1
	WER	41.2	46.2	63.0	41.0	64.5
UPV-PRHLT-REC	CER	3.6	3.3	17.2	2.8	15.4
	WER	16.6	16.2	56.1	13.1	54.6
DIVA-REGIM(APTIPC1)	CER	0.9	-	-	-	-
	WER	5.2	-	-	-	-
UPV (this work)	CER	**0.3**	**0.5**	**8.3**	**0.4**	**3.0**
	WER	**1.7**	**2.4**	**37.0**	**1.9**	**13.2**

7   Concluding Remarks

Windowed Bernoulli mixture HMMs (BHMMs) for Printed Arabic word recognition have been described and improved by the introduction of window *repositioning* techniques. In particular, we have considered three techniques of window *repositioning* after window extraction: *vertical, horizontal,* and *both.* The only differ in the way in which extracted windows are shifted to align mass and window centers (only in the vertical direction, horizontal direction or in both directions). These techniques have been carefully described and extensively tested on the APTI database for multi-font, multi-size and multi-style Arabic Printed text images. In all cases, state-of-art results were obtained with vertical repositioning.

Acknowledgment. The research leading to these results has received funding from the European Union Seventh Framework Programme (FP7/2007-2013) under grant agreement no 287755. Also supported by the Spanish Government (Plan E, iTrans2 TIN2009-14511 and AECID 2011/2012 grant).

References

1. Dehghan, M., et al.: Handwritten Farsi (Arabic) word recognition: a holistic approach using discrete HMM. Pattern Recognition 34(5), 1057–1065 (2001), http://www.sciencedirect.com/science/article/pii/S0031320300000510
2. Giménez, A., Juan, A.: Embedded Bernoulli Mixture HMMs for Handwritten Word Recognition. In: ICDAR 2009, Barcelona, Spain, pp. 896–900 (July 2009)
3. Giménez, A., Khoury, I., Juan, A.: Windowed Bernoulli Mixture HMMs for Arabic Handwritten Word Recognition. In: ICFHR 2010, Kolkata, India, pp. 533–538 (November 2010)
4. Grosicki, E., El Abed, H.: ICDAR 2009 Handwriting Recognition Competition. In: ICDAR 2009, Barcelona, Spain, pp. 1398–1402 (July 2009)
5. Günter, S., et al.: HMM-based handwritten word recognition: on the optimization of the number of states, training iterations and Gaussian components. Pattern Recognition 37, 2069–2079 (2004)
6. Märgner, V., El Abed, H.: ICDAR 2007 - Arabic Handwriting Recognition Competition. In: ICDAR 2007, Curitiba, Brazil, pp. 1274–1278 (September 2007)
7. Märgner, V., El Abed, H.: ICDAR 2009 Arabic Handwriting Recognition Competition. In: ICDAR 2009, Barcelona, Spain, pp. 1383–1387 (July 2009)
8. Pechwitz, M., et al.: IFN/ENIT - database of handwritten Arabic words. In: CIFED 2002, Hammamet, Tunis, pp. 21–23 (October 2002)
9. Rabiner, L., Juang, B.: Fundamentals of speech recognition. Prentice-Hall (1993)
10. Slimane, F., et al.: A new arabic printed text image database and evaluation protocols. In: ICDAR 2009, pp. 946–950 (July 2009)
11. Slimane, F., et al.: ICDAR 2011 - arabic recognition competition: Multi-font multi-size digitally represented text. In: ICDAR 2011 - Arabic Recognition Competition, pp. 1449–1453. IEEE (September 2011)
12. Young, S.: et al.: The HTK Book. Cambridge University Engineering Department (1995)

Head Direction Estimation from Silhouette

Amina Bensebaa[1], Slimane Larabi[1], and Neil M. Robertson[2]

[1] University of Sciences and Technology Houari Boumediene,
Computer Science Department, BP 32 El Alia, Algiers, Algeria
slarabi@usthb.dz

[2] Edinburgh Research Partnership in Engineering and Mathematics,
School of Engineering and Physical Sciences, Heriot-Watt University,
Edinburgh, EH14 4AS, UK
n.m.robertson@hw.ac.uk

Abstract. Due to the absence of features that may be extracted from face, heading direction estimation for low resolution images is a difficult task and requires the taking into account all information that may be inferred from human body in image, particularly its silhouette. We propose in this paper a set of geometric features extracted from shape head-shoulders, feet and knees shapes which jointly allow the estimation of body direction. Other features extracted from head-shoulders are proposed for heading direction estimation based on body direction. The constraint of camera position related to proposed features is discussed and results of conducted experiments are presented.

Keywords: Head direction, Body direction, Low resolution, Silhouette.

1 Introduction

Head direction estimation is one of challenging tasks for computer vision researchers especially in case of low resolution images. In case of high and medium resolution images, many approaches has been proposed to solve this problem. A survey may be found in [11]. All of these approaches try to find the most discriminate set of facial features which permit to estimate the pose. The objective to reach for any proposed technique is to verify a set of criteria such as: Accuracy, Monocular, Autonomous, Multi-person, Identity and Lighting invariant, Resolution independent, Full range of head motion and Real time [11].

The problem of estimating heading direction for low-resolution images without adding contextual information requires yet more contributions in order to deal with complex scenes where human are far from the camera. The performance of proposed methods are mainly limited because they are based on extracted features from the head which are very dependent on camera placement and the chosen texture and skin color models depend on the resolution of the head in the image and therefore doesn't work for lower resolution.

In this paper, we will investigate what it can be done from shoulders-head and legs shapes for heading direction estimation in case of low-resolution images. Firstly, a set of features are extracted from shoulders-head and legs shapes

A. Petrosino (Ed.): ICIAP 2013, Part I, LNCS 8156, pp. 340–350, 2013.

and used for inferring body direction. In the next, heading direction is esti-mated using body direction and features extracted from head-shoulders shape. Section 2 covers the theoretical aspects of body and heading direction estimation based on features extracted from shoulders-head and legs shapes. Experiments are conducted to validate our approach and obtained results are presented in section 3.

2   Prior Works

Face extraction in low-resolution images is an important task in the process of heading direction estimation. Few works have been devoted for this purpose and all present difficulties for detecting faces when the resolution of images decreases [18]. Labeled training examples of head images are used to train various types of classifiers such as support vector machines, neural networks, nearest neighbor and tree based classifiers [13], [3], [4]. The disadvantage of these methods is the requirement of all combinations of lighting conditions and skin/hair colour variations in order to estimate an accurate classification.

Contextual features have been used in addition to visual ones in order to improve the quality of heading direction estimation [9] [8] [1]. Using multiple views camera, Voit et al [17] estimate head pose for low resolution image by appearance-based method. The head size varies around 20×25 and the obtained results are satisfactory due to the use of multiple cameras. Additional contextual information: multiple calibrated cameras and a specific scene allows estimating of absolute coarse head pose for wide-angle overhead cameras by integrating 3D head position [16].

Head-shoulders shape has been studied and many methods have been pro-posed for the purpose of human detection in images using wavelet decomposition technique and support vector machine [14] or background subtraction algorithm [12]. In other side, head-shoulders shape has been used for human tracking and head pose estimation. In [12], the direction of head movements is detected and tracked throughout video frames. Templates are captured for a specific position of the camera (mounted sufficiently high above to provide a top-view of the scene) without using all head poses.

Legs shape may contribute for heading-direction estimation. Indeed, detectors on the lower parts of the body has been introduced in many works for human body pose calculation and human action recognition [15]. Legs shape has been also used for human segmentation where body parts particularly the legs are modeled in order to detect and segment human [10]. The proposed approach is based on the matching of part-template tree images hierarchically proposed and used initially in [6], [7].

Our main contributions in this work are:

- The processing of images of far-field video when there is a difficulty to locate some features used such as skin, face.

- A set of geometric features are proposed that allow the estimating of body's and head's directions,
- Such features may be extracted even for far-field video because they concern only the silhouette.
- Possibility to extract these features whatever the position of the camera except for the top view.

3   Basic Principle of the Method

Assuming that silhouettes of humans are extracted from images of low resolution, our aim is to estimate the directions of body and head. Geometric features are extracted from silhouette due to the absence of other features that may be extracted from head (face and hair) for such images. We will focus in this paper on shapes of head, shoulders, knees and feet which may be considered as a good features to achieve this task. Body direction is firstly estimated using features extracted from head-shoulders, knees and feet shapes. Secondly, knowing body direction, head direction is inferred from features of head-shoulders shape.

3.1   Geometric Features from Silhouette

A shape leg is a part of silhouette which plays a dominant role for body direction estimation from image. Indeed, our visual system is able to infer body direction seeing only the outline shape legs (and/or) head-shoulders shape (see figure 1). We propose three determinant cues of shapes legs and head-shoulders that allow inferring body direction when they are extracted from outline shape. These features can't be computed for a fixed top down camera giving blobs of persons where head-shoulders can't be extracted from the rest of the silhouette.

Fig. 1. Some shapes of legs for which it is easy to infer body direction

The **first one** is the bent knee. When a leg is well separated from the other and the knee is bent, a coarse body direction can be inferred without ambiguity. Figure 2.a illustrates an example of shape legs where feet are hidden. Our visual system can easily give an estimate of body direction because the feet have limited possibilities of poses. Figure 2.b illustrates the correct poses and the directions can be inferred using the feet shapes, however figure 2.c shows impossible situation. The directions of the lines joining inflexion points of the same leg are used to infer the body direction.

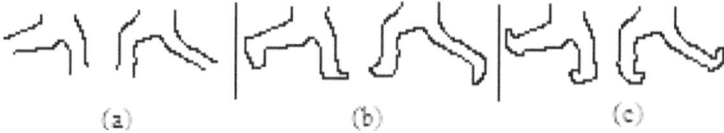

Fig. 2. Shapes of legs with inflected knee

The **second one** is the direction of shape foot. Indeed, our visual system encounters difficulties by looking at legs shapes without feet and can't estimate body direction for many configurations even if the body is moving and legs are well separated but without inflexion of knees. For example, seeing to the outlines of figure 3.a, without feet we can't recognize to what direction body is moving. This ambiguity is clear seeing at the original shapes (see figure 3.b) and at new shapes obtained drawing feet (see figure 3.c). The base lines of the feet are good features because they indicate the body direction. Their use is explained in the following subsection.

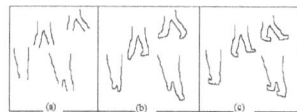

Fig. 3. Ambiguity in body direction estimation in case of missed shape feet

The **third feature** concerns the variation of silhouette's width along the shape head-shoulders. The ratio of the width of the upper part (head) and the lower part (shoulders) is related to the angle of rotation. We noticed that there's an opposite relationship between the ratio and the orientation angle.

3.2   Inferring Body Direction

Body Direction Estimation Using Feet's Features. This task consists to split the lower human shape into separated legs, separated lower legs or grouped legs (The two first cases include the case where the knee of one leg may be bent). This is done by the concavity points located on the outline in the lower part of the outline indicating the feet.

We associate to each foot a **base line** defined by two extremities of the foot located between the heel and the toes. The outline of lower part is processed in order to determine the baseline of the feet located between the heel and the toes. Firstly, high convexities points Cv_1 and Cv_2 characterizing the outline foot are located (see Figure 4(a)). Secondly, the last point of interest Cc representing a high concavity on this outline is located, such as the distances $CcCv_2$ is minimal. The convex point that represents toes, will be the closest point to the concave point of the feet outline, the other convex point will obviously correspond to the heel. Thus the base line joins the two convexities of the foot and the orientation of feet corresponds to the vector carried by the feet base line.

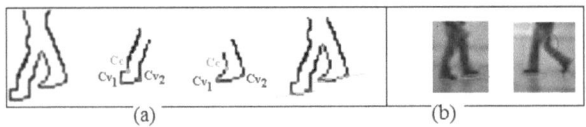

Fig. 4. (a) Steps of body direction estimation based on foot directions, (b) Body orientation from feet (In red color the feet orientations of foot and in blue the body orientation)

Applying the 2D quasi-invariant, the angle between the two vectors measured in 3D-space varies slowly in the image as viewpoint varies [2]. As in the scene the disposition of foot vectors is restricted by the human physic constraints, it will be the same case in image plane; the body direction is inferred as the average of foot directions. Once the base lines of feet are extracted, body orientation is computed as the resultant vector of the two orientations. When one foot isn't put on the ground, which correspond to a high curvature of the knee, the resultant vector will have the direction of the base line of the other foot (see figure 4(b)).

Body Direction Estimation Using Knee's Features. Extraction of curvature points consists to find the best concave or convex pixels of the lower part of the silhouette. Using the Chetverikov's algorithm [5], p is selected as curvature point if the angle defined by the three pixels: p^-, p, p^+ is lower that a given threshold α. Pixels p^-, p^+ are located pixels before and after the considered pixel p at d pixels (see figure 5). Default values of d, α are $7 pixels, 150$.

Fig. 5. Location of curvature points on outline legs

Many types of knees inflexion may be located (see figure 6) and body direction is considered as the direction defined by the concave point to the convex one of the bent knee. Only the direction left towards right and inversely will be considered.

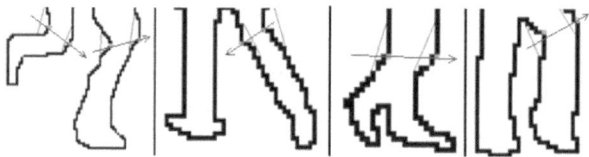

Fig. 6. Some cases of knee inflexion and the inferred direction of the knees

Body Direction Estimation Using Head-Shoulders Features. Body direction can inferred using the head-shoulders geometry. Indeed, when human body is rotating, the ratio R_w of the widths of head and shoulders are estimated as follow:

For the rotation angle in the intervals : $[0\,°, 15\,°]$, $[15\,°, 30\,°]$, $[30\,°, 45\,°]$, $[45\,°, 60\,°]$, $[60\,°, 75\,°]$, $[75\,°, 90\,°]$, the ratio R_w belongs respectively to the intervals: $[1.82, 4]$, $[1.70, 1.81]$, $[1.61, 1.69]$, $[1.51, 1.60]$, $[1.36, 1.5]$, $[1.4, 1.5]$.

Applying the algorithm of D. Chetverikov [5], the two concave points (left and right) delineating the head and the two convex points (left and right) extremities of shoulders are located. Head is separated by locating the pixels having the minimum angle among the selected candidate points. The second extremities of shoulders are located as convex pixels with high curvature along of the outline head-shoulders (see figure 7).

Fig. 7. The second extremities of shoulders

3.3   Inferring Head Direction from Head-Shoulders Shape

We assume that body direction is estimated based on the three features proposed above. In order to estimate head direction, two features are extracted from head-shoulders outline.

Features Extraction. The first feature concerns the lengths of shoulders S_L, S_R on shape head-shoulders. When the end of the neck isn't visible on one side due to shoulder's occlusion by head, the beginning of the shoulder is considered as the point of high curvature on head-shoulders outline.

The lengths of shoulders are important cues for both head and body directions estimation and the difference between lengths of S_L, S_R arises from one of the following configurations:

- Depending on the camera and body positions, the head can occludes a part of one shoulder and then decreases the shoulder length. For example, when the camera is on top at the right or at the left of the person (see figure 8).
- When human body is rotating, one of shoulders becomes less visible. This occurs for example when the camera is on top even if the person is in front of the camera. In this case, length of one shoulder decreases until that the two sides of the shape head-shoulders don't correspond to shoulders.

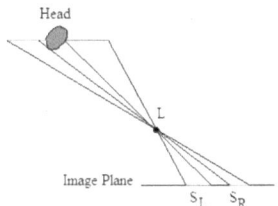

Fig. 8. Cases of occlusion of shoulder by head

Consequently, when body direction and head are in front to the camera, the lengths $L(S_L), L(S_R)$ of shoulders are identical. Otherwise, when the head is rotating or when body is at the lateral side of the camera, this equality isn't verified because in both cases the head occludes a part of one shoulder (see figure 8). We proved geometrically that without occlusion by head, the lengths of one shoulder decreases when body is rotating.

The second feature which completes the first one, concerns the occluded parts of shoulders that permit to estimate head rotation. Let I be the intersection point of the lines joining extremities of shoulders S_L and S_R (see figure 9). When body and head are in front to the camera, the distances d_L, d_R from I to shoulders are identical in the scene and in image. However, when head or body are rotating, these distances are different in image because a part of shoulder is occluded by head and thus in image the distance d_L or d_R includes the occluded segment of the shoulder and a part of the neck. The distances d_L, d_R will be used to infer the head direction.

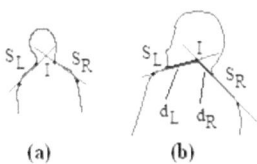

(a)   (b)

Fig. 9. Intersection of shoulders in case where (a) body and head are in front, (b) body and head rotating

Coarse Estimation of Head Direction. Head direction is estimated assuming body orientation, the difference ΔL between the images of (S_L), (S_R) and the difference Δd between the images of d_L, d_R are computed. When body is in the center of the field of view or at the left, table 1 gives the results obtained of head direction applying a geometric reasoning depending on the values of ΔL, δd and body direction. For example, figure 10 illustrates the variation of ΔL, δd in case where human in the center of the field of view of the camera. The case where body is at the right is symmetrical to the case where body is at the left.

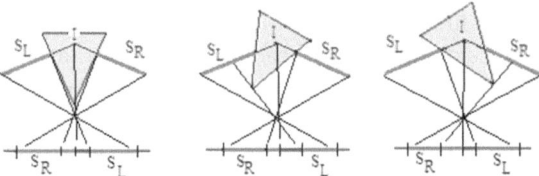

Fig. 10. Different poses of head where images of d_R, d_L are illustrated with blue and red color in case where human is in the center of the field of view

Table 1. Head direction inferred in cases where body is in front and at the left

Body in the center	$\delta d = 0$	$\delta d > 0$	$\delta d < 0$
$\Delta L = 0$	Head in Front	Not possible	Not possible
$\Delta L < 0$	Not possible	Rotation to left	Not possible
$\Delta L > 0$	Not possible	Not possible	rotation to right
Body at the left	$\delta d = 0$	$\delta d > 0$	$\delta d < 0$
$\Delta L = 0$	Low Rotation to right	Not possible	Not possible
$\Delta L < 0$	Not possible	Head in front or rotating to left	Not possible
$\Delta L > 0$	Not possible	Not possible	Hight Rotation to right

3.4   Study of the Camera Position Constraint

As we are interested in this work to images of low resolution which means a far field of view, the camera may be:
- Fixed at the top and far from the scene. In this case, none from the features: head, shoulders, legs and feet can't be located using the blob representing human.

- Fixed so as its optical axis is oblique or horizontal towards the scene. In this case, whatever the position of the camera relatively to human in the scene: in front or at the lateral position, its head-shoulders, legs and feet are viewed. Consequently, the availability of the proposed features depends only on the pose, which means that inflexion of knees or feet base lines may be missed, however the presence of the head-shoulders outline is required.

4   Results

We applied our method on some real images. Firstly silhouettes are extracted and body direction is firstly computed. In the next, head direction is estimated. We used all features extracted from head-shoulders, feet and knees outlines.

Figure 11 illustrates some poses, extracted silhouettes and computed body directions. Body direction is computed using the ratio R_w equal to $2.6, 2.89, 2.25,$ $1.33, 1.36, 2.27, 2.09$ giving the intervals of head direction equal to: $[0°, 15°], [0°, 15°], [15°, 30°], [75°, 90°], [75°, 90°], [15°, 30°], [0°, 15°]$. We can see that the computed body direction for the two last poses $(f), (g)$ are done using only the first feature which can't differentiate if the body in in front or of back with regard to the camera. The orientation of feet, when are located in the image, eliminate the ambiguity (in front or of back).

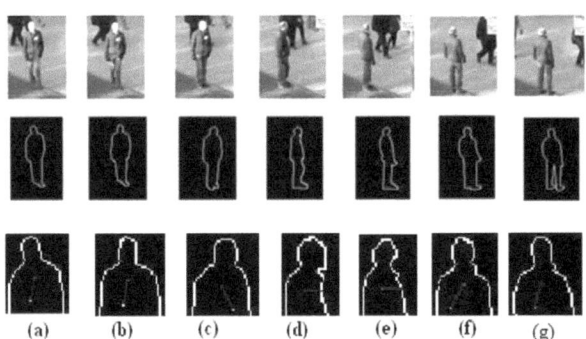

(a)   (b)   (c)   (d)   (e)   (f)   (g)

Fig. 11. Some poses and extracted silhouettes and the computed body directions based on R_w values

The combination of features used for body direction depends on what can be extracted. The features extracted from feet and knees are more strong than those extracted from head-shoulders. Figure 12 illustrates the results obtained when inflexion of knees are used in addition of the ratio R_w.

Head direction estimation is based on estimated body direction and the values of d_L, d_R computed using head-shoulders outline. Figure 13 summarizes this combination of features and shows that a good estimation is made even if the images are of low resolution.

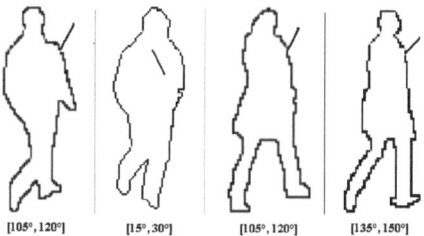

[105°,120°]   [15°,30°]   [105°,120°]   [135°,150°]

Fig. 12. Body Orientation using knee inflexion and R_w ratio

Fig. 13. Head and body directions from combined features

5   Conclusion

We proposed in this paper a method for head direction estimation based on geometric features extracted even if images are of low resolution. Body direction is inferred from features extracted from outlines of knees and feet and head-shoulders. This direction is used in addition to features extracted from outlines of head-shoulders for estimating head direction. The proposed method has been applied on real images and achieves good estimation of head direction. Also, the features extracted are independent from camera pose, except the top view where head-shoulders, knees and feet can't be located on human shape.

References

1. Ba, S.O., Odobez, J.M.: Multiperson Visual Focus of Attention from Head Pose and Meeting Contextual Cues. IEEE Transactions on Pattern Analysis and Machine Intelligence 33(1), 101–116 (2011)
2. Binford, T.O., Levitt, T.S.: Quasi-invariants: Theory and exploitation. In: Proceedings of DARPA Image Understanding Workshop, pp. 819–829 (1993)
3. Benfold, B., Reid, I.D.: Colour Invariant Head Pose Classification in Low Resolution Video. In: Proceedings of the 19th British Machine Vision Conference (2008)
4. Benfold, B., Reid, I.D.: Unsupervised Learning of a Scene-Specific Coarse Gaze Estimator. In: Proceedings of the International Conference on Computer Vision (ICCV), pp. 2344–2351 (2011)
5. Chetverikov, D.: A Simple and Efficient Algorithm for Detection of High Curvature Points in Planar Curves. In: Petkov, N., Westenberg, M.A. (eds.) CAIP 2003. LNCS, vol. 2756, pp. 746–753. Springer, Heidelberg (2003)

6. Gavrila, D.M.: The Visual Analysis of Human Movement: A Survey. Computer Vision and Image Understanding 73(1), 82–98 (1999)
7. Gavrila, D.M.: A Bayesian, Exemplar-Based Approach to Hierarchical Shape Matching. IEEE Transactions on Pattern Analysis and Machine Intelligence 29(8), 1408–1421 (2007)
8. Lanz, O., Brunelli, R.: Joint Bayesian Tracking of Head Location and Pose from Low-Resolution Video. In: Stiefelhagen, R., Bowers, R., Fiscus, J.G. (eds.) RT 2007 and CLEAR 2007. LNCS, vol. 4625, pp. 287–296. Springer, Heidelberg (2008)
9. Launila, A., Sullivan, J.: Contextual Features for Head Pose Estimation in Football Games. In: International Conference on Pattern Recognition (ICPR 2010), Turkey, pp. 340–343 (2010)
10. Lin, Z., Davis, L.S.: Shape-Based Human Detection and Segmentation via Hierarchical Part-Template Matching. IEEE Transactions on Pattern Analysis and Machine Intelligence 32(4), 604–618 (2010)
11. Murphy-Chutorian, E., Trivedi, M.M.: Head Pose Estimation in Computer Vision: A Survey. IEEE Transactions on Pattern Analysis and Machine Intelligence 31(4), 607–626 (2009)
12. Ozturk, O., Yamasaki, T., Aizawa, K.: Tracking of humans and estimation of body/head orientation from top-view single camera for visual focus of attention analysis. In: IEEE 12th International Conference on Computer Vision Workshops (ICCV Workshops), pp. 1020–1027 (2009)
13. Robertson, N., Reid, I.D.: Estimating Gaze Direction from Low-Resolution Faces in Video. In: Leonardis, A., Bischof, H., Pinz, A. (eds.) ECCV 2006. LNCS, vol. 3952, pp. 402–415. Springer, Heidelberg (2006)
14. Sun, Y., Wang, Y., He, Y., Hua, Y.: Head-and-shoulder detection in varying pose. In: Wang, L., Chen, K., S. Ong, Y. (eds.) ICNC 2005. LNCS, vol. 3611, pp. 12–20. Springer, Heidelberg (2005)
15. Singh, V.K., Nevatia, R., Huang, C.: Efficient Inference with Multiple Heterogeneous Part Detectors for Human Pose Estimation. In: Daniilidis, K., Maragos, P., Paragios, N. (eds.) ECCV 2010, Part III. LNCS, vol. 6313, pp. 314–327. Springer, Heidelberg (2010)
16. Tian, Y.L., Brown, L., Connell, C., Sharat, P., Arun, H., Senior, A., Bolle, R.: Absolute head pose estimation from overhead wide-angle cameras. In: IEEE International Workshop on Analysis and Modeling of Faces and Gestures, AMFG 2003, pp. 92–99 (2003)
17. Voit, M., Nickel, K., Stiefelhagen, R.: A Bayesian Approach for Multi-view Head Pose Estimation. In: IEEE International Conference on Multisensor Fusion and Integration for Intelligent Systems, pp. 31–34 (2006)
18. Zheng, J., Ramírez, G.A., Fuentes, O.: Face detection in low-resolution color images. In: Campilho, A., Kamel, M. (eds.) ICIAR 2010. LNCS, vol. 6111, pp. 454–463. Springer, Heidelberg (2010)

Combined Supervised / Unsupervised Algorithm for Skin Detection: A Preliminary Phase for Face Detection

Eyal Braunstain and Isak Gath

Department of Biomedical Engineering
Technion - Israel Institute of Technology, Haifa 32000, Israel
seyalbra@tx.technion.ac.il, isak@bm.technion.ac.il

Abstract. Skin detection in color images is of great importance for the computer vision research community. Many existing skin detection algorithms are characterized by high false detection rates. The proposed algorithm performs offline learning of skin color, but also of "false-skin colors", which may be misclassified as skin using regular histogram or Gaussian skin color models. This supervised learning of false-skin colors produces a significant reduction in false detection rates. Our aim is to extract skin blobs that are suspected to contain faces, which are usually ellipsoid-shaped.

Thus, to extract these blobs, an unsupervised optimal fuzzy clustering (UOFC) algorithm is applied in the spatial image space. Blobs segmented by the clustering procedure are then examined by specific features, e.g. geometrical, to classify them as face candidates.

Sample runs of the algorithm on a bank of images show high skin detection rates with reduced false detection rates.

Keywords: Skin Detection, Supervised Learning, Unsupervised Learning, Optimal Fuzzy Clustering, Likelihood Estimation.

1 Introduction

Skin color serves as a powerful cue for people detection in unconstrained environments, with applications ranging from detection and recognition of faces, of limbs gestures and movements for human-computer interaction systems, of naked people on the WWW etc. Skin color can be modeled in various color spaces (e.g. RGB, YCbCr, YES) and by various models, e.g. single Gaussian, mixture of Gaussians, or by histograms. The classification of a test-pixel can be made by various methods, e.g. by a probability threshold decision-making. Many skin detection techniques produce good detection rates, but may result in relatively high false detection rates [11].

The performance of histogram models was found to be superior to Gaussian mixture models, both in accuracy and in computational cost [3].

Much of previous research work on skin classification used the mixture of Gaussians model, comprised of a weighted sum of Gaussian kernels. The advantage of the model lies mainly in its ability to generalize well on a small training data set [3], [5].

A. Petrosino (Ed.): ICIAP 2013, Part I, LNCS 8156, pp. 351–360, 2013.

The authors of [6] suggest that normalization of RGB color values by the sum (R+G+B) yields a color space that is more efficient for skin color segmentation, since the sensitivity to variation of skin color is reduced. This is of high importance under variable illumination conditions. In [10], the skin color histograms of nine color spaces are examined, and it is stated that the distribution in r-g (Normalized R-G) space is compatible with a unimodal elliptical Gaussian model.

The authors of [1] suggest an image annotation scheme, based on supervised classification of colors, and learning classes' thresholds from ROC curves. In the present work, we suggest concomitant modeling of skin color and false-skin colors, i.e. colors that might be misclassified as skin by a skin color classifier, but do not actually represent skin.

The Unsupervised Optimal Fuzzy Clustering (UOFC) algorithm, developed by Gath and Geva [2], was tested for various applications, e.g. clustering of EEG signal recordings, in order to classify the signal into various sleep stages. In this paper, this algorithm was utilized to detect skin blobs that were suspected to contain faces.

The work is focused on white and asian skin-colored people. Dealing with more variable skin colors may require the construction of additional skin colors histograms for the different trained skin classes.

2   Skin Blobs Detection as Preliminary Phase to Face Detection

2.1   Training of the Skin Color and False-Skin Colors Classifiers

In the suggested algorithm, color is represented in CIE Lab space, which is perceptually more uniform than many other color spaces, e.g. RGB, HSV [7]. This enables the use of uniform color metric for color-related decision-making (e.g. clustering). For skin detection purposes, only the chroma components (a, b) are used, and luminance (L) is disposed. This is done since the intensity of the reflected light from skin of different people, or in variable lighting conditions, varies considerably, while the chroma remains relatively constant [7].

First, a single skin classifier was trained by manually labeling skin pixels and constructing a 2-D color histogram in CIE-Lab space, using a-b chroma values only.

We propose a supervised learning approach of skin color, as well as false-skin color, i.e. color values that allegedly may be classified as skin by a simple histogram classifier, but do not belong to skin. The suggested approach is aimed at lowering the false detection rate in skin classification. A skin histogram is constructed from a training set of 20 skin hand-labeled pictures. This histogram is used to classify skin on a validation set of 20 different images, in which it was observed that most false detection pixels belonged to 3 prime colors - brown, red and white. The training and validation sets are considered of appropriate sizes to avoid overfitting to the learning data.

The false detection pixels observed in the validation set were manually labeled to three new distinct training sets for brown, red and white, and three new histograms in CIE-Lab space (chroma components only, a-b) for these colors were constructed.

These histograms were smoothed by a Gaussian function; see fig. (1). It can be observed that there exists a significant overlap between different histograms, mainly between the skin to the brown and red histograms.

To train parameters for class W_i, let us assume that each color class can be represented by a single-Gaussian model.

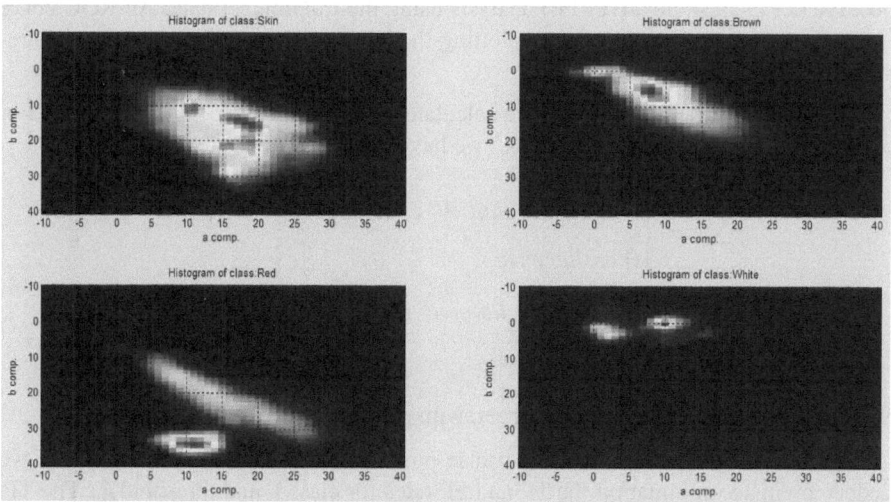

Fig. 1. Histograms of color classes in CIE-Lab space, by a-b chroma components. Upper left- skin color; upper right- brown; lower left- red; lower right- white.

This assumption is somewhat inaccurate, since certain color distributions are not unimodal; however, the discussed histograms display a distribution around a primary mode in a-b chroma space. Chroma values (a, b) are distributed in the range [-110, 110], and in fig. (1) are displayed in partial range [-10, 40].

Given the hand-labeled pixels of color class W_i, the mean chroma vector m_i and chroma covariance matrix K_i, for class W_i are:

$$m_i = [m_a, m_b]^T \qquad K_i = \begin{bmatrix} \sigma_a^2 & \sigma_{ab} \\ \sigma_{ab} & \sigma_b^2 \end{bmatrix} \tag{1}$$

Given a vector of chroma values for some pixel p, $x_p = [a_p, b_p]$, a unimodal Gaussian distribution is fitted to class W_i [1]:

$$p(x_p | W_i) = (2\pi)^{-1} |K_i|^{-\frac{1}{2}} \exp\{-\frac{1}{2}\lambda_p^i\} \tag{2}$$

Where λ_p^i is defined in Eq. (3). The contours of the Normal pdf of each class define an ellipse, centered at m_i, and whose principal axes are determined by K_i.

$$\lambda_p^i = [x_p - m_i]^T K_i^{-1} [x_p - m_i] \tag{3}$$

Given N color classes, N scalar values λ_p^i are to be calculated for pixel p, which are proportional to the probability that the pixel belongs to class W_i.

A specific threshold was learned for each class W_i, denoted t^i, by an optimum true positive (TP) - false positive (FP) analysis using the training set data. An ROC curve is generated for each color class i, plotting the TP vs. FP rates for a variety of threshold values λ_t [1].

The TP and FP rates for a class are calculated over the training set of that class. For pixel p, with some chosen threshold λ_t, we have the TP and FP rates for that pixel:

$$
\begin{aligned}
tp_p^{\lambda_t, I_j^i} &= \begin{cases} 1 & \text{if } p \text{ labeled } W_i \ \& \ \lambda_p^i \le \lambda_t \\ 0 & \text{otherwise} \end{cases} \\[2mm]
fp_p^{\lambda_t, I_j^i} &= \begin{cases} 1 & \text{if } p \text{ not labeled } W_i \ \& \ \lambda_p^i \le \lambda_t \\ 0 & \text{otherwise} \end{cases}
\end{aligned}
\tag{4}
$$

Where I_j^i denotes image j in training set, with pixels labeled as class i.

From the pixels' true positive and false positive indicators defined in Eq. (4), we deduce by image summation the TP and FP rates for class i, and threshold λ_t. The TP, FP rates are calculated for various threshold values, thus obtaining an ROC curve. The optimum threshold for class i, t^i, is computed as the intersection point of the ROC curve and the line $TP = 1 - FP$, [6]. This threshold simultaneously maximizes TP and minimizes FP, which serves the motivation of minimizing the false detection rate in skin classification.

2.2   Skin Color Classification

The classification of pixel p in a test image is performed by the application of N binary hypothesis tests based on Eq. (3), with thresholds t^i for each color class:

$$
c_p^i = \begin{cases} 1 & \lambda_p^i \le t^i \quad\quad \{p \in W_i\} \\ 0 & \lambda_p^i > t^i \quad\quad \{p \in \overline{W_i}\} \end{cases} \quad i = \{1,..,N\}
\tag{5}
$$

c_p^i is a binary classification of pixel p to class W_i, equal to 1 if p is in this color class.

This is followed by an MAP (Maximum a posteriori) decision rule, for pixels which are found to belong to more than one color class according to hypotheses.

Each pixel is classified to one of the 4 color classes, or to the "other" class ("rest of the world"), whose distribution is not modeled. The MAP absolute decision rule is defined by:

$$pixel \ \ p \in W_k \ \ iff \ \ \Pr(W_k \mid x_p) = \max\{\Pr(W_i \mid x_p)\}_{i=1..N} \tag{6}$$

Where, $\Pr(W_i \mid x_p)$ is attained by Bayes' theorem.

Finally, after the classification of the pixels to color classes for the test image, the pixels that were classified to the skin color class were extracted.

2.3   Extracting Skin Blobs Suspected as Faces by Optimal Fuzzy Clustering

The goal of the algorithm is to extract from a test image the skin blobs, which probably contain faces. Faces are often grossly described as elliptic-shaped, so we would like to extract blobs of ellipsoidal shape from the skin image produced so far. No prior knowledge of the number of faces in the image exists. A modified version of the Unsupervised Optimal Fuzzy Clustering algorithm was utilized. This algorithm, combining the fuzzy K-means and fuzzy maximum likelihood estimation (FMLE) has been shown to perform well in cases of clusters of variable hyper-ellipsoidal shapes. The data to be clustered is the spatial image points (x, y coordinates of the image). To reduce the computational load of the clustering algorithm, the skin points in the spatial image domain have been sub-sampled by a sampling grid. However, this required interpolation of the clustered data back to the original image grid. Let us denote by X the data matrix of skin image points (x, y), i.e. feature vectors, and assume grossly that the initial number of clusters is the number of adjoined components in the skin image. The number of ellipsoidal face aggregates does not necessarily correlate with the number of adjoined components, but considering the existence of noisy components in the image, it was found to usually lead to accurate clustering results. The UOFC algorithm has been applied to the features matrix, X. In its first stage, the fuzzy K-means algorithm was applied, minimizing the following objective function, with respect to U, the fuzzy membership matrix, and V, the prototypes - clusters centroids:

$$J_q(U,V) = \sum_{j=1}^{N}\sum_{i=1}^{K}(u_{ij})^q d^2(X_j,V_i) \quad K \le N \tag{7}$$

where N is the number of data points, K is the number of prototypes (assumed smaller than N), q, the weighting exponent for the fuzzy memberships, controls the "fuzziness" of the resulting clusters [8]. d is a defined metric (e.g. Euclidean).

Fuzzy K-means is run through an iterative optimization procedure. The end results of the fuzzy K-means clustering, V and U, serve as initial conditions to the FMLE algorithm, whose stages are given by:

1) Initial clusters centroids are given by the end results of fuzzy K-means stage.
2) To address the problem of clustering data that is not necessarily spherically distributed, a 2nd stage, based on maximum likelihood estimation (MLE) is carried out, with an "exponential" distance d_e [2]. It will be used to compute $h(i \mid X_j)$, the posterior probability of selecting cluster i, given feature vector X_j. $h(i \mid X_j)$ and the exponential distance d_e are given by:

$$h(i \mid X_j) = \frac{\dfrac{1}{d_e^2(X_j, V_i)}}{\sum_{k=1}^{K} \dfrac{1}{d_e^2(X_j, V_k)}} \qquad (8)$$

$$d_e^2(X_j, V_i) = \frac{[\det(F_i)]^{1/2}}{P_i} \exp\{(X_j - V_i)^T F_i^{-1}(X_j - V_i)/2\} \qquad (9)$$

At the first FMLE iteration, the posterior probabilities $h(i \mid X_j)$ can be inferred from the memberships U. To calculate d_e, F_i (the fuzzy covariance matrix of cluster i) and P_i (the a-priori probability) need to be computed:

$$F_i = \frac{\sum_{j=1}^{N} h(i \mid X_j)(X_j - V_i)(X_j - V_i)^T}{\sum_{j=1}^{N} h(i \mid X_j)}, \qquad P_i = \frac{1}{N}\sum_{j=1}^{N} h(i \mid X_j) \qquad (10)$$

3) Update clusters centroids by Eq. (11), and posterior probabilities by Eq. (8).

$$\hat{V}_i = \frac{\sum_{j=1}^{N} (h_{ij})^2 X_j}{\sum_{j=1}^{N} (h_{ij})^2} \qquad (11)$$

4) For a predefined threshold ε, if the following condition holds, than stop:

$\max_{ij}\{\mid h_{ij\_new} - h_{ij} \mid\} < \varepsilon.$ Otherwise, go to step 3, and iterate steps 3, 4 until convergence.

The end results of the UOFC algorithm are H, the posterior probabilities matrix (equivalent to the fuzzy membership matrix, U), and V, the prototype centroids. To determine, for each pixel, to which cluster it belongs, we utilize a hard-partition rule on the clusters' probabilities of H - each pixel will be attributed to the cluster to which its posterior probability is maximal. The end result is an absolute partition (segmentation) of the image. It is possible for pixels belonging to a specific cluster to be spatially disjoined, e.g. as two groups of pixels that are very close to each other in image space. However, disjoined pixels are to be regarded as separate clusters, so an analysis on the clusters map was performed, to separate disjoined components into different clusters. This might increase the number of clusters, but produces a more natural segmentation of the image. However, somewhat on the contrary, under certain geometrical conditions, neighboring clusters which are spatially adjoined in image space can be merged into one cluster.

2.4   Post-processing of Skin Blobs

Skin segments (clusters) have been extracted from the image. The post-processing of these skin blobs is comprised of morphological operations and geometrical tests. The validity of each cluster's fitting ellipse main axes ratio and cluster bounding box X-Y sizes ratio are checked to converge to pre-determined thresholds. Skin blobs can also be filtered by disposing of regions with high-amplitude variation in intensity [4]. In this work, a lower threshold is applied on the color variance of skin clusters, to filter skin segments, assuming that a face area is generally characterized by larger color variance than most other skin areas (e.g. hands, legs) due to the color variability of mouth, eyes and possibly facial hair, against skin color. This phase of post-processing enables further reduction of the number of detected skin blobs suspected as faces.

3   Skin Blobs Detection Experimental Results

The algorithm was tested on a bank of 50 self-picked images, mostly from the WWW, of white and Asian people, with variable background and lighting conditions.

A standard faces database was not used (e.g. Caltech faces database [12]), since the collected images were chosen to contain also skin parts other than faces, with indoor and outdoor variable backgrounds. Images were chosen to contain 2-6 individuals.

Sample results of the algorithm on an outdoor image of a group of 6 Asian people are presented (Fig. 2), with a distinct variability in visible skin color - from pale to brownish hue. The unified algorithm serves as more than a skin-pixels classifier, but is not a face detector. It would be wrong to define skin detection rates by simply counting skin pixels, since many skin patches (e.g. hands, legs) are deliberately filtered out. The percentage of undeleted faces on the test bank was ~97%. The false detection (FP) rate is intuitively related to pixels falsely detected as skin, and the average rate was 7%. The skin FN measure is high, since non-face skin patches were deleted. Thus, TP rate for skin may be low, but is not a valid performance measure. Table 1 provides a comparison to other skin detection techniques.

Fig. (2a) is the test image. Fig. (2b) is the skin classification image when training is performed on skin color class alone. Fig. (2c) is the skin classification image when the training is performed on skin class and false detection classes (false skin classes). It is apparent that the false detections existing in (2b) - some white background patches and brown hair of the figures in the image, are now diminished.

Fig. 2. Skin detection algorithm results. (a) Test image for skin blobs detection.
(b) Skin pixels classification; training of skin class only (apparent false detections).
(c) Skin pixels classification; skin class and false skin classes (red, brown and white).
(d) Running Optimal fuzzy clustering algorithm over image in (c), with post processing.
(e) Morphological filling holes in (d). The produced image can be a fed to a face detector.
(f) An RGB labeling map of the clusters computed by FMLE clustering.

Table 1. Comparison of TP and FP measures with other skin detection techniques

Method	Skin TP (%)	Skin FP (%)
Our algorithm	Not valid (See page 7) (Face detection rate - ~97)	7
Jones & Rehg [3]	93.9	8
Maximum Entropy Model [13]	80	8
Wu & Ai [14]	92.9	NA

The classification of (2c) caused the increase in miss-detection of some skin pixels, but the effect is somewhat negligible, and is mostly corrected by morphological operations performed at a later stage. Fig. (2d) presents the result of running Optimal Fuzzy Clustering algorithm on image in (2c), with post processing, thus erasing skin blobs that probably do not contain a face - hands and chest skin blobs in this case. This is done by some threshold criteria, e.g. x and y axes components ratio in bounding box, lower threshold on extent of cluster blob pixels to full cluster bounding box, lower threshold on standard deviation of cluster color values, etc.

Fig. (2e) presents the result of morphological filling of holes of the image in (2d), thus including eyes and mouth of the figures in the image. This image is comprised of skin blobs that are suspected to contain faces. It can be observed that besides some hands blobs, this mask image contains mostly "floating heads", which is very much desirable. Utilizing a face detection or recognition algorithm on this filtered skin image will require scanning of much less area than the full image. Fig. (2f) demonstrates the RGB labeling map of the clusters computed by Optimal FMLE clustering. Each color designates a different cluster in spatial image space. The advantage of FMLE clustering (segmentation) is apparent, especially in the middle of the image, in the area of the adult male and young child below him; we can clearly see in the clustering map that the UOFC algorithm succeeded in segmenting and distinguishing the two different heads of the adult and child, which are both ellipsoidal in shape and are adjoined components in the skin mask image. The algorithm was shown to be efficient on variable backgrounds and different shapes.

4   Summary and Conclusions

In the presented study, a new algorithm is described for the detection of skin blobs in color images, suspected to contain faces. Supervised learning of skin color histogram was performed, and was tested on a validation set. From the classification results on that set, three major false-skin color classes were recognized - red, brown and white, and thus three additional histograms were constructed, to model them. A Gaussian model was fitted to each class. Given a new test image, N hypothesis tests (for N classes) were implemented, followed by an MAP decision rule, to allocate a specific color class to each pixel. Pixels in the image that belonged to the skin class were then extracted. This skin mask image was segmented by a modified UOFC algorithm, followed by morphological and geometrical post-processing performed on the segmentation result, to yield the final result of filtered skin blobs image.

The algorithm was tested on an image bank containing 50 images, of variable background with different shapes, and variable lighting conditions. It produced very good detection performance, with reduced false detection rates.

The combination of supervised learning of skin color and false-skin colors models, in addition to the utilization of the UOFC algorithm to extract grossly ellipsoidal skin blobs that might contain faces, creates a skin mask that is drastically relieved of irrelevant segments, while preserving the blobs in the image that contain faces.

The suggested algorithm can serve as a preliminary phase to a face detection or recognition algorithm, by reducing significantly the image area that has to be scanned in the image to detect or recognize faces.

References

1. Saber, E., Murat Tekalp, A., Eschbach, R., Knox, K.: Automatic Image Annotation Using Adaptive Color Classification. Graphical Models and Image Processing 58(2), 115–126 (1996)
2. Gath, I., Geva, A.B.: Unsupervised Optimal Fuzzy Clustering. IEEE Transactions on Pattern Analysis and Machine Intelligence 11(7) (1989)
3. Jones, M.J., Rehg, J.M.: Statistical Color Models with Application to Skin Detection. International Journal of Computer Vision 46(1), 81–96 (2002)
4. Forsyth, D.A., Fleck, M., Bregler, C.: Finding naked people. In: Buxton, B.F., Cipolla, R. (eds.) ECCV 1996. LNCS, vol. 1065, Springer, Heidelberg (1996)
5. Schiele, B., Waibel, A.: Gaze tracking based on face-color. In: Proceedings of the International Workshop on Automatic Face- and Gesture-Recognition, Zurich, Switzerland, pp. 344–349 (1995)
6. Terrillon, J.-C., David, M., Akamatsu, S.: Automatic Detection of Human Faces in Natural Scene Images by Use of a Skin Color Model and of Invariant Moments. In: Proceedings of Third IEEE International Conference on Automatic Face and Gesture Recognition FG (1998)
7. Cai, J., Goshtasby, A.: Detecting human faces in color images. Image and Vision Computing 18, 63–75 (1999)
8. Bezdek, J.C.: Pattern Recognition with Fuzzy Objective Function Algorithms, New York, Plenum (1981)
9. Bezdek, J.C.: Fuzzy Mathematics in Pattern Classification. Ph.D. dissertation, Cornell Univ., NY (1973)
10. Terrillon, J.-C., Akamatsu, S.: Comparative Performance of Different Chrominance Spaces for Color Segmentation and Detection of Human Faces in Complex Scene Images. Vision Interface, Trois-Rivières, Canada (1999)
11. Vezhnevets, V., Sazonov, V., Andreeva, A.: A Survey on Pixel-Based Skin Color Detection Techniques. In: Proc. Graphicon (2003)
12. Caltech face database (1999), http://www.vision.caltech.edu/html-files/archive.html
13. Jedynak, B., Zheng, H., Daoudi, M., Barret, D.: Maximum entropy models for skin detection. Tech. report, Universite des Sciences et Technologies de Lille, France (2002)
14. Wu, Y.-W., Ai, X.-Y.: Face Detection in Color Images Using AdaBoost Algorithm Based on Skin Color Information. Knowledge Discovery and Data Mining (2008)

Conic Based Camera Re-calibration
after Zooming

Iuri Frosio[1], Cristina Turrini[2], and Alberto Alzati[2]

[1] Computer Science Dept., University of Milan, Italy
[2] Mathematics Dept., University of Milan, Italy
{iuri.frosio,cristina.turrini,alberto.alzati}@unimi.it

Abstract. We describe here a method to compute the internal parameters of a camera whose position and orientation are known. The method is based on the observation of at least three conics on a known plane; these can be easily extracted in a real scenario from a tiled floor or other regular structures. The method estimates the principal point and focal length using a unique image of the conics when these are observed by an additional calibrated camera. Differently from other methods, no assumption is made on the conics used for calibration. The experimental results demonstrate that the accuracy of the method is comparable to that of more traditional (and time consuming) approaches. It can find applications in systems of Pan-Zoom-Tilt (PZT) or traditional cameras, that are nowadays widely employed, for instance in the surveillance domain, and require frequent re-calibration.

Keywords: camera calibration, conics, computer vision, surveillance.

1 Introduction

Camera calibration is fundamental in computer vision to recover 3D information from the scene. Several approaches to calibration have been proposed, involving 3D objects of known shape, planar (2D) patterns and 1D objects [1]. Calibration generally involves the following steps: the calibration object is moved in the scene while the camera takes images; features are extracted from the images of the calibration object and the camera internal and external parameters are estimated from the positions of the features in the acquired images.

Lines and points have been widely employed to compose the calibration pattern because of the simplicity to identify them in the images [1, 2]. The Zhang's method [2, 3], one most diffused approach, uses a checkerboard pattern and it is based on the homographies relating the positions of the checkerboard corners in 3D space and in the acquired images. On the other hand, several authors noticed that conics convey more information and consequently produce a more accurate estimate of the camera parameters [4–8]. When conics are used for calibration, the common idea is that a planar conic in 3D space, described by a 3×3 matrix $\mathbf{C}$, is projected into a conic with matrix $\mathbf{H}^T \mathbf{C} \mathbf{H}$ onto the camera image plane, where $\mathbf{H}$ is the homography that relates the conic and the image

A. Petrosino (Ed.): ICIAP 2013, Part I, LNCS 8156, pp. 361–370, 2013.
© Springer-Verlag Berlin Heidelberg 2013

plane [4–6]. Putting in relation two ore more confocal or co-axial conics, or conics in known position, **H** is estimated and camera parameters are derived from its decomposition [1].

Because of the noise on the measured positions of the features, lots of images are necessary to get a reliable estimate of the camera parameters [1]. On the other hand, when systems composed by more than one camera are used (e.g. like in many surveillance systems, where frequent camera re-calibration is necessary) or only partial camera calibration is needed (e.g. after zooming), less parameters have to be estimated and/or information from other cameras can be profitably used to obtain a reliable calibration from few or even just one image[4, 9]. In this context, we describe a method to compute the internal parameters of a camera after zooming. The algorithm is based on the observation of at least three degenerate conics (i.e. pair of lines) from the camera itself and from a second, calibrated camera, but it can be used without any modification for any kind of conic. Differently from other approaches, our algorithm minimizes a cost function which involves directly the parameters of the observed conics. The experimental results demonstrate that only one image is sufficient to re-calibrate the camera with an accuracy similar to that of Zhang's method [2, 3], which requires a larger number of images and therefore more time to be performed.

2   Method

2.1   Preliminaries and Aim of the Method

The present method is aimed at estimating the internal parameters (principal point, focal length) of a camera after zooming, under the following hypothesis:

- two cameras observe at least three conics lying on a known plane (e.g. six lines extracted from a tiled floor can be used as a triplet of degenerate conics with no lines in common each other); the equation of the floor plane is supposed to be known, as this can be estimated from a set of features lying on the floor before camera zooming;
- the external parameters (camera orientation and position of the optical center) of the camera to be calibrated are known (e.g. from a previous calibration performed at a different zoom level, supposing that the camera position and orientation do not change after zooming);
- the internal (focal length, principal point) and external (position, orientation) parameters of the second camera are known.

No knowledge on the observed conics is necessary: they could be ellipses, hyperbolas or degenerate conics whose parameters are in any case unknown and unnecessary to the method described here.

2.2   Camera Model

For the i-th camera of the system, the projection of a 3D point $[X/U \; Y/U \; Z/U]^T$ onto the camera image plane in homogeneous coordinates is given by [10]:

$$[u \; v \; w]^T = \mathbf{K}_i \left[\mathbf{R}_i^T \; -\mathbf{R}_i^T \mathbf{t}_i \right] [X \; Y \; Z \; U]^T \tag{1}$$

$$\mathbf{K}_i = \begin{bmatrix} f_i & 0 & c_{x,i} \\ 0 & f_i & c_{y,i} \\ 0 & 0 & 1 \end{bmatrix} \tag{2}$$

where T is the transpose operator, $[u v w]^T$ are the homogeneous coordinates of the point in the image plane, f_i is the camera focal length, $[c_{x,i}\ c_{y,i}]^T$ is the camera principal point, $\mathbf{R}_i$ is a 3×3 rotation matrix describing the camera orientation and $\mathbf{t}_i$ is a 3×1 vector describing the position of the camera optical center in 3D space.

We will assume that no distortion is present on the images acquired by the cameras; this is justified with the fact that modern apparatuses generally included a lens model in their firmware: distortions are consequently automatically corrected [11, 12] and simplified distortion models can be adopted [13, 14].

2.3   Relation between Conics Observed by Two Cameras

Let us consider two cameras and a set of conics lying on the floor plane and observed by both cameras. The projection of a conic is again a conic; for the i-th camera, the j-th conic equation in homogeneous coordinates is given by:

$$\begin{bmatrix} u & v & w \end{bmatrix} \mathbf{C}_{i,j} \begin{bmatrix} u & v & w \end{bmatrix}^T = 0 \tag{3}$$

where $\mathbf{C}_{i,j}$ is a 3×3 symmetric matrix with the parameters of the conic. The cone associated to $\mathbf{C}_{i,j}$, whose vertex corresponds to the center of the i-th camera, is:

$$\begin{bmatrix} X & Y & Z & U \end{bmatrix} \mathbf{P}_i^T \mathbf{C}_{i,j} \mathbf{P}_i \begin{bmatrix} X & Y & Z & U \end{bmatrix}^T = 0 \tag{4}$$

where:

$$\mathbf{P}_i = \mathbf{K}_i \begin{bmatrix} \mathbf{R}_i^T & -\mathbf{R}_i^T \mathbf{t}_i \end{bmatrix} \tag{5}$$

is the i-th camera projection matrix. Let us suppose now that camera 2 has its optical center in $\mathbf{t}_2 = [0\ 0\ 0]^T$ and its orientation matrix $\mathbf{R}_2$ is the identity; this does not represent a limit, as it is always possible to change the reference frame such that these conditions are satisfied. Let us also suppose that the principal point of the camera 2 is in $[c_{x,2}\ c_{y,2}]^T = [0\ 0]^T$; this too can be achieved with a change of reference frame, without loosing generality. Under these hypotheses, the relation between the matrix of the conic observed by camera 1 and 2 is given by:

$$\mathbf{D}^T \mathbf{P}_1^T \mathbf{C}_{1,j} \mathbf{P}_1 \mathbf{D} = \lambda_j \mathbf{C}_{2,j} \tag{6}$$

where λ_j is an arbitrary scale factor, whereas:

$$\mathbf{D} = \begin{bmatrix} 1 & 0 & 0 \\ 0 & 1 & 0 \\ 0 & 0 & f_2 \\ \hline p_x & p_y & p_z f_2 \end{bmatrix} \tag{7}$$

and p_x, p_y and p_z are the parameters of the plane of the conic, given in the form $p_x Z + p_y Y + p_z Z = U$, in the reference of camera 2. In fact, if we eliminate

U from Eq. (4), by using the above equation of the plane, we get the equation of a cone, having vertex in $\mathbf{t}_2$, and projecting the observed conic from $\mathbf{t}_2$. By cutting such cone with the plane $Z = f_2 U$, we get the conic observed by camera 2. Hence, up to a scalar factor, the matrix of this conic is the first member of Eq. (6).

2.4   Estimation of Camera Internal Parameters

We will assume now that camera 1 has zoomed to a new focal length, thus changing both its focal length f_1 and its principal point $[c_{x,1}\ c_{y,1}]^T$ (notice that, because of the lens rotation during zooming, the position of the principal point may significantly change [13–15]). Eq. (6) poses a constraint on the parameters of the conics observed by camera 1 and 2 and it can be consequently used to estimate the internal parameters of camera 1. In fact, for the j-th observed conic, it can be rewritten as:

$$\mathbf{D}^T \left[\mathbf{R}_1^T - \mathbf{R}_1^T \mathbf{t}_1\right]^T \mathbf{K}_1^T \mathbf{C}_{1,j} \mathbf{K}_1 \left[\mathbf{R}_1^T - \mathbf{R}_1^T \mathbf{t}_1\right] \mathbf{D} = \lambda_j \mathbf{C}_{2,j} \tag{8}$$

where the unknowns f_1, $c_{x,1}$ and $c_{y,1}$ in $\mathbf{K}_1$, appear in the left side, whereas the unknown scale factor, λ_j, appears in the right side. The term $\mathbf{K}_1^T \mathbf{C}_{1,j} \mathbf{K}_1$ generates a quadratic polynomial including terms in f_1^2, $c_{x,1}^2$, $c_{y,1}^2$, $f_1 c_{x,1}$, $f_1 c_{y,1}$, $c_{x,1} c_{y,1}$, f_1, $c_{x,1}$, $c_{y,1}$ plus a scalar term.

Considering now a set of N conics, and properly rearranging the terms in Eq. (8) in a $6N \times (N + 9)$ matrix $\mathbf{A}$ and in a vector $\mathbf{b}$, the following non linear relation is derived:

$$\mathbf{A} \left[f_1^2\ c_{x,1}^2\ c_{y,1}^2\ f_1 c_{x,1}\ f_1 c_{y,1}\ c_{x,1} c_{y,1}\ f_1\ c_{x,1}\ c_{y,1}\ \lambda_1\ \lambda_2\ \dots\ \lambda_N\right]^T = \mathbf{b}; \tag{9}$$

which describes the equivalence (apart from the scale factors $\{\lambda_j\}_{j=1..N}$) of the parameters of the conics observed by camera 2 with those of the conics observed by camera 1, projected onto the conic plane (i.e. the floor) and then projected again onto the image plane of camera 2.

Eq. (9) can be solved in a least squares sense through an iterative optimization algorithm like the Levenberg-Marquardt method, implemented by *lsqnonlin* in Matlab. To this aim it is however necessary to provide an initial guess of the unknowns f_1, $c_{x,1}$, $c_{y,1}$, λ_1, $\dots$, λ_N. This is obtained solving the linear system $\mathbf{Ax} = \mathbf{b}$ in a least squares sense, without considering the non linear constraints (e.g., $x_1 = x_7^2$ should hold, where x_k is the k-th element of $\mathbf{x}$) that relate the elements of $\mathbf{x}$ [4]. In particular, $\mathbf{x}$ is obtained as:

$$\mathbf{x} = \left(\mathbf{A}^T \mathbf{A}\right)^\dagger \mathbf{A}^T \mathbf{b} \tag{10}$$

where † indicate the Moore-Penrose pseudo inverse, that has been adopted here since the rank of $\mathbf{A}$ can be not maximal when N is small [16]. We have verified experimentally that, at least for N small, the values in $\mathbf{x}$ provided by Eq. (10) are not reliable; on the other hand, a reliable estimate is achieved considering the following ratios between elements of $\mathbf{x}$:

$$\tilde{f}_1 = x_1/x_7; \quad \tilde{c}_{x,1} = x_4/x_7; \quad \tilde{c}_{y,1} = x_5/x_7. \tag{11}$$

Similarly, initial guesses for $\{\lambda_j\}_{j=1..N}$ are obtained. Eq. (9) is finally solved in a least squares sense, starting from this initial guess, through the *lsqnonlin* Matlab function.

3   Experimental Setup and Results

3.1   Experimental Setup and Line Fitting

The method has been tested using two reflex digital cameras, a Nikon D3100 equipped with a Nikkor 18-55mm zoom lens, acquiring images of 4608×3072 pixels, and a Pentax K-r equipped with a Pentax SMC 18-55mm zoom lens, acquiring images of 4288×2848 pixels. The cameras were positioned on two tripods, at a height of approximately $1.5m$ and at $3.5m$ of distance one from the other. The cameras were tilted about $20°$ low; their optical axes were convergent with an angle of approximately $70°$. Each camera looked at the floor, where squared tiles with a side of $0.6m$ were present; each camera could observe at the maximum focal length ($55mm$) an area of 3×3 tiles. The lines among the tiles were marked through a red pencil to highlight their visibility; moreover, six lines of white adhesive tape, approximately parallel to the red lines, were added to increase the number of available lines (see Fig. 1). Since each pair of lines represent a degenerate conic, a total of 7 degenerate conics with no lines in common were available to test the method (the tilted lines of white adhesive tape in Fig. 1 were not considered here).

We calibrated the cameras at different focal lengths using the Camera Calibration Toolbox for Matlab by J.-V. Bouguet [3], mainly inspired to the Zhang calibration procedure based on observation of a checkerboard planar pattern [2]. For each calibration, 30 images of the checkerboard were considered. The focal lengths considered were respectively {45mm, 38mm, 31mm, 24mm} and {25mm, 48mm} for cameras 1 and 2. Calibration parameters are reported in Table 1: notice that the principal point of camera 1 varies significantly (more than 100 pixels) for different focal lengths. Although the focal lengths changed, the position and orientation of both cameras remained unaltered between different experiments.

For each focal length and for each camera, we identified the red and white segments on the floor through the following procedure. For the red lines, we got an image $i_r(x, y)$ with enhanced red areas subtracting the green and blue channels from the red one and filtering then the image with a Gaussian filter of size 15×15 and $\sigma = 4$ pixels. For the white lines, we filtered in the same manner the sum of the three color channels to get $i_w(x, y)$. Then we clicked on the acquired images to roughly identify the area including the set of 4×4 red segments in Fig. 1; from this initial guess, the 8 red segments and the 6 white segments in the images were estimated by maximizing the average value of $i_r(x, y)$ and $i_w(x, y)$ along the segments. Typical results are shown in Fig. 1; the typical fitting errors observed on the available images were limited to 4-5 pixels in the worst cases (Fig. 1d).

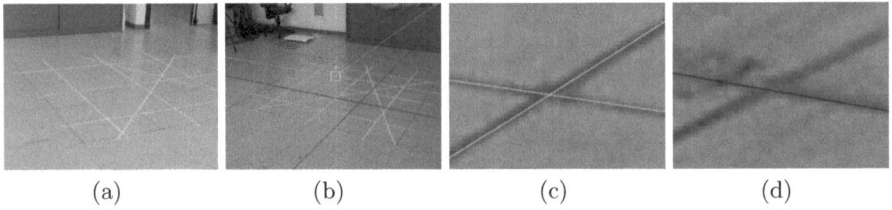

(a)   (b)   (c)   (d)

Fig. 1. (a) A typical image acquired by the Nikon camera at a focal length of 45mm, the longest one considered here. (b) Three degenerate conics (each with a different color) identified on the image of the Pentax camera. (c) Zoom of the area c in panel (b). (d) Zoom of the area d in panel (b): notice that in this case the red line is skewed by some pixels with respect to the red line on the floor.

3.2   Results

For each focal pair we built sets of 3 to 7 degenerate conics from the red and white lines identified in the images. Then we estimated the internal parameters of the Nikon and Pentax cameras for each possible pair of focal lengths. In a first experiment, we considered only degenerate conics constituted by two orthogonal lines, like those in Fig. 1; in a second experiment, we inserted two degenerate conics of parallel lines in the set of conics used for calibration. Tables 2 and 3 report the errors on the estimated focal lengths ($e_f = 100\frac{\tilde{f}_1 - f_1}{f_1}$, where $\tilde{f}_1$ is the estimated focal length and e_f defines the percent error) and principal point ($e_p = \| [(\tilde{c}_{x,1} - c_{x,1})\ (\tilde{c}_{y,1} - c_{y,1})]) \|$, where $[\tilde{c}_{x,1}\ \tilde{c}_{y,1}]$ is the estimated principal point and e_p is its distance from the ground true principal point in pixels) through the linear estimate provided by Eqs. (10) and (11) and after the non linear minimization of the quadratic cost function associated to Eq. (9). The same tables also report the rank of the matrix **A** in Eq. (8) with respect to its maximum admissible rank for the given number of conics.

When the conics are constituted of orthogonal lines and $N \leq 5$, the estimate of the camera parameters performed with the linear method (Eqs. (10) and (11)) is biased and poorly accurate, with an uncertainty always larger than 4.5% on f_1 (Table 2), which is significantly larger than the focal uncertainty obtained with the Zhang's method [2] (approximately equal to 1% in the worst case, see Table 1). Similar unsatisfying results are obtained for the estimate of the principal point, with typical uncertainties of hundreds of pixels. Notice that, for $N < 5$, the rank of **A** is not maximum and therefore the solution of the linear system in Eq. 10 is not uniquely defined.

On the other hand, when $N > 5$, even the simple linear procedure produces an estimate of the camera parameters that fit inside the confidence intervals provided by the Camera Calibration Toolbox for Matlab [3]. Finally, the least squares solution of Eq. (9) obtained with non linear optimization does produce a solution which is always in the confidence interval provided by the Camera Calibration Toolbox for Matlab [3], for any number of conics, with precision and accuracy that increase with N.

Table 1. Internal parameters (with uncertainties) of the cameras used during the experiments, computed through the Camera Calibration Toolbox for Matlab [3]

camera	focal	f_i [pixel]	$c_{x,i}$ [pixel]	$c_{y,i}$ [pixel]
1 (Nikon)	45mm	8711±65	2374±30	1624±38
1 (Nikon)	38mm	7743±51	2324±25	1505±26
1 (Nikon)	31mm	6089±28	2366±14	1605±18
1 (Nikon)	24mm	4723±19	2340±12	1533±15
2 (Pentax)	35mm	6314±64	2161±29	1481±33
2 (Pentax)	48mm	8219±68	2168±27	1485±25

Table 2. Errors (mean ± standard deviation) on the estimated focal and principal point for different set of N degenerate conics, each constituted by two orthogonal lines

	Linear estimate		Non linear estimate			
N	e_f [%]	e_p [pixels]	e_f [%]	e_p [pixels]	rank($\mathbf{A}$)	max rank
3	-5.66±16.11	336.20±722.22	0.16±0.69	4.98±2.58	10	12
4	-0.60±6.39	104.86±138.86	0.12±0.58	4.36±1.55	12	13
5	0.63±4.56	59.04±88.85	0.13±0.59	3.89±1.47	14	14
6	0.57±1.48	5.08±4.10	0.08±0.58	3.36±1.60	15	15
7	0.53±1.47	5.56±4.92	0.12±0.56	3.74±2.47	16	16

When two of the conics considered by the method are constituted of parallel lines (Table 3), the estimate of the camera parameters provided by the linear method becomes less stable, although it still represents a valid initial guess for the minimization of Eq. (9) through the Levemberg-Marquardt algorithm; an accurate result for f_1 and $[c_{x,1}\ c_{y,1}]$ is obtained in this case only for $N = 7$. After the non linear minimization, however, the estimate of the camera parameters have a precision and accuracy comparable to those reported in Table 2 and therefore to that typical of Zhang's method [2]. It is however to be noticed that, in this case, for $N = 3$ the present method fails to provide and accurate estimate of f_1 and $[c_{x,1}\ c_{y,1}]$ for all the cases considered (see Discussion).

Overall, the non linear method described in the previous section does provide an estimate of the camera internal parameters which generally fits into the confidence intervals provided by the Camera Calibration Toolbox for Matlab [3]. If the number of conics is sufficently high ($N > 6$), even the simple linear method associated to Eqs. (10) and (11) is capable to provide reliable results.

4   Discussion and Conclusion

The method described here permits to re-calibrate a camera after zooming, under the assumption that it does not change its position ($\mathbf{t}_1$) and orientation ($\mathbf{R}_1$) during zooming. This could be for instance the case of a surveillance camera installed with a fixed angle of view, or a reflex camera mounted on a tripod. Although assuming that the camera image plane does not rotate during zooming

Table 3. Errors (mean ± standard deviation) on the estimated focal and principal point for different set of N degenerate conics, each constituted by two orthogonal lines but two constituted by parallel lines

	Linear estimate		Non linear estimate			
N	e_f [%]	e_p [pixels]	e_f [%]	e_p [pixels]	rank($\mathbf{A}$)	max rank
3	-2.17±14.19	216.65±408.07	-9.62±31.19	123.09±372.50	10	12
4	-0.71±2.99	86.16±102.12	0.13±0.44	4.27±2.25	12	13
5	23.16±69.48	448.01±1158.01	0.12±0.48	3.98±2.09	14	14
6	-11.03±36.83	132.19±391.47	0.05±0.50	3.29±1.58	15	15
7	0.34±1.45	6.29±5.03	0.08±0.50	4.04±2.49	16	16

appears to be reasonable, assuming that $\mathbf{t}_1$ does not change could be more critical; in fact, in our camera model (Eq. (1)), $\mathbf{t}_1$ represents the center of the optical system of the camera [10], which actually goes forward and backward when the zoom lens elongates or shortens to modify its focal length. Despite this geometrical aspect was neglected in Eq. (9), the results reported in Tables 2 and 3 demonstrates that the accuracy achieved by the proposed method in estimating f_1, $c_{x,1}$ and $c_{y,1}$ is comparable to that of the Zhang's method [2, 3], that on the other hand does estimate also the external parameters ($\mathbf{R}_1$, $\mathbf{t}_1$) of the camera, as well as the distortion parameters of the lens.

The high accuracy achieved by the proposed method is even more noticeable considering that only one image (containing 3 ore more conics) is necessary for camera re-calibration. Such result is possible because the method does not utilize only data measured from a unique camera, like in the Zhang's method. Instead, it makes use of other information beyond $\mathbf{R}_1$ and $\mathbf{t}_1$, consisting in the parameters of the conic plane in the reference frame of camera 2 (p_x, p_y and p_z), that have to be estimated before the zooming of camera 1; and the matrices $\{\mathbf{C}_{2,j}\}_{j=1..N}$, measured by camera 2. The usage of such *a-priori* information makes the method also robust with respect to noise on the measured conics. Fig. 1 highlights in fact that the line fitting procedure adopted with our experimental setup is not characterized by a sub-pixel accuracy. Despite of this, the parameters estimated by the proposed method well fit in the interval of confidence provided by the Camera Calibration Toolbox [3] that represents our ground true.

A general approach to camera re-calibration in a real scenario would actually require to develop an algorithm to identify the conics in the images (for instance, the Hough transform could be used to identify the lines). Then, correspondence between the features observed by camera 1 and camera 2 should be solved. Since these problems have been widely studied and a wide literature exists on them [10], they have not been assessed here.

It is to be noticed that the present algorithm has been tested here only using pairs of degenerate conics, as these are naturally identifiable on any tiled floor; on the other hand, the approach can be applied to any kind of conic including circles, ellipses and hyperbolas without requiring any modification (furthermore, different kind of conics can actually appear at the same time in Eq. (9)). In particular, circles and ellipses could be interesting in practical applications, as this shapes occur quite frequently in real scenes [7].

Depending on the kind of conics appearing in Eq. (9), however, different kind of numerical issues could raise. For instance, for the case considered here of degenerate conics constituted of two lines with equations $xc\theta_1 + ys\theta_1 = \rho_1$ and $xc\theta_2 + ys\theta_2 = \rho_2$ (where $c\theta$ and $s\theta$ indicate the sin and cosine of θ), the corresponding conic matrix $\mathbf{C}$ is given by:

$$\mathbf{C} = \begin{bmatrix} c\theta_1 c\theta_2 & (c\theta_1 s\theta_2 + s\theta_1 c\theta_2)/2 & -(\rho_1 c\theta_2 + \rho_2 c\theta_1)/2 \\ (c\theta_1 s\theta_2 + s\theta_1 c\theta_2)/2 & s\theta_1 s\theta_2 & -(\rho_1 s\theta_2 + \rho_2 s\theta_1)/2 \\ -(\rho_1 c\theta_2 + \rho_2 c\theta_1)/2 & -(\rho_1 s\theta_2 + \rho_2 s\theta_1)/2 & \rho_1 \rho_2 \end{bmatrix}. \quad (12)$$

The elements of the 2×2 upper left block of $\mathbf{C}$ are always between -1 and 1; on the other hand, the third element of the diagonal of $\mathbf{C}$ is proportional to the product of the distances of the two lines from the point $[0\ 0]^T$ of the image and, given the size of the images considered here, it can easily exceed $1,000,000$, making the matrix $\mathbf{A}$ in Eq. (9) badly conditioned because of the presence of largely spread numbers. A different interpretation of badly conditioning in this situation is obtained considering that the least squares solution of Eq. (9) provides the estimate of f_1, $c_{x,1}$ and $c_{y,1}$ which minimizes the quadratic distances between the elements of any j-th conic matrix $\mathbf{D}^T \mathbf{P}_1^T \mathbf{C}_{1,j} \mathbf{P}_1 \mathbf{D}$ and $\lambda_j \mathbf{C}_{2,j}$ in Eq. (6). If one of the elements of the conic matrix is dramatically larger than the others, high importance is given in the quadratic cost function to this element whereas other elements appearing in the matrices of the conics are practically neglected. To avoid this, we have therefore normalized the pixel space dividing by $1,000$ the pixel coordinates (and therefore also the camera focal lengths appearing in $\mathbf{K}_1$ and $\mathbf{K}_2$) before applying the proposed method and finally multiplying by $1,000$ the estimate of $(f_1, c_{x,1}, c_{y,1})$ obtained with Eqs. (11) and (9). Some experimental results, not reported here, have demonstrated that such normalization procedure is actually fundamental to the success of the method: without it, the precision and accuracy decrease dramatically, making the obtained estimate of the internal parameters of camera 1 practically unusable.

An additional remark to the results presented in Tables 2 and 3 is to be done. We have in fact verified that large averages or standard deviations of the error are generally caused by only one very bad estimate of f_1, $c_{x,1}$ and $c_{y,1}$, among the ten experiments considered to compute each element of the tables. In practice, especially when N is low and the linear estimate is used, the method may fail to produce a reliable estimate of the camera parameters because of some particular geometrical configurations that induce numerical instability. We are going to investigate such conditions more in detail in the future.

Furthermore, we have obtained some analytic, preliminary results showing that Eq. (8) can be manipulated to obtain the estimate of f_1, $c_{x,1}$ and $c_{y,1}$ through the solution of a linear system. The accuracy and stability of such method still has to be evaluated and compared to the results obtained with the iterative solution of Eq. (9), but it promises to generate a more elegant framework where the linear estimate of the parameters can be analytically analyzed to identify geometrically critical situations; this would also permit to study noise propagation from the measured conics to the estimated parameters, for instance through covariance analysis.

Overall, the method described here permits to re-calibrate a camera after zooming using a unique image and features that are easily extracted from real world scenes, achieving an accuracy similar to that of more traditional and time consuming calibration procedures.

References

1. Zhang, Z.: Camera Calibration. In: Medioni, G., Kang, S.B. (eds.) Emerging Topics in Computer Vision, ch. 2, pp. 4–43. Prentice Hall Professional Technical Reference (2004)
2. Zhang, Z.: Flexible Camera Calibration by Viewing a Plane from Unknown Orientations. In: ICCV 1999 (1999)
3. Bouguet, J.-Y.: Camera Calibration Toolbox for Matlab,
   http://www.vision.caltech.edu/bouguetj/calib_doc/index.html/
4. Abado, F., Camahort, E., Vivó, R.: Camera Calibration Using Two Concentric Circles. In: International Conference on Image Analysis and Recognition, pp. 688–696 (2004)
5. Ying, X., Zha, H.: Camera calibration using principal-axes aligned conics. In: Yagi, Y., Kang, S.B., Kweon, I.S., Zha, H. (eds.) ACCV 2007, Part I. LNCS, vol. 4843, pp. 138–148. Springer, Heidelberg (2007)
6. Yang, C., Sun, F., Hu, Z.: Planar conic based camera calibration. In: The 15th International Conference on Pattern Recognition, vol. 1, pp. 555–558 (2000)
7. Frosio, I., Alzati, A., Bertolini, M., Turrini, C., Borghese, N.A.: Linear pose estimate from corresponding conics. Pattern Recognition 45, 4169–4181 (2012)
8. Kahl, F., Heyden, A.: Using Conic Correspondences in Two Images to Estimate the Epipolar Geometry. In: Int. Conf. on Computer Vision, pp. 761–766 (1998)
9. Borghese, N.A., Colombo, F.M., Alzati, A.: Computing Camera Focal Length by Zooming a Single Point. Pattern Recognition 39, 1522–1529 (2006)
10. Faugeras, O.: Three-dimensional computer vision: a geometric viewpoint. MIT Press (1993)
11. Nikon website, http://www.nikon.com/
12. Canon website, http://www.canon.com/
13. Fraser, C.S., Ajlouni, A.S.S.: Zoom-dependent camera calibration in digital close-range photogrammetry. Photogrammetric Engineering & Remote Sensing 72(9), 1017–1026 (2006)
14. Sun, X., Sun, J., Zhang, J., Li, M.: Simple zoom-lens digital camera calibration method based on exif. In: Three-Dimensional Image Capture and Applications VI, vol. 79 (2004)
15. Calore, E., Frosio, I.: Perspective correction in digital photography using on board accelerometer. Submitted to Image and Vision Computing (2013)
16. Adi, B.-I., Greville, T.N.E.: Generalized inverses. Springer, New York (2003)

Dynamic Hierarchical Segmentation
of Remote Sensing Images

Giuseppe Scarpa[1], Giuseppe Masi[1], Raffaele Gaetano[2],
Luisa Verdoliva[1], and Giovanni Poggi[1]

[1] DIETI, University Federico II of Naples
firstname.lastname@unina.it
[2] TELECOM-ParisTech
gaetano@telecom-paristech.fr

Abstract. Recursive tree-structured segmentation is a powerful tool to deal with
the non-stationary nature of images. By fitting model parameters to each re-
gion/class under analysis one can adapt the segmentation algorithm to the local
image statistics, thus improving accuracy. However, a single model/segmenter
cannot fit regions with wildly different nature, and one should be allowed to select in a suitable library the tool most suited to the local statistics. In this paper,
we implement this dynamic segmentation/classification paradigm, using two segmenters, based on spectral and textural properties, respectively, and defining suitable rules for switching model locally. Experiments on remote-sensing mosaics
show that the multiple-model dynamic algorithm largely outperforms comparable
single-model segmenters.

Keywords: Image segmentation, image model, hierarchical segmentation.

1 Introduction

Images are inherently non-stationary: most of the information they provide is related to
their non-stationary nature. Therefore, simple all-encompassing models cannot capture
the complexity of real-world images, and segmentation algorithms based on them are
bound to provide poor results. The recent literature testifies a considerable effort [1, 2]
towards the development of more complex models and segmentation algorithms for
images.

A natural approach is the use of tree-structured hierarchical models and segmentation algorithms, which adapt locally to each region and provide information at different
scales of observations of the image. Hierarchical segmentation can be obtained by recursive algorithms, where the image is segmented in a small number of regions, each
of which becomes the root of a new local segmentation process, going on until some
stopping criterion is met.

A tree-structured segmentation algorithm was proposed in [3], based on local Markov
random fields (MRF) models, and hence called TS-MRF. TS-MRF proved to deal quite
effectively with image non-stationarity [3, 4], thanks to the opportunity to adapt locally the parameters of the MRF model attached with each node/region. and was also
faster than algorithms based on "flat" models. Nonetheless, its flexibility was limited

A. Petrosino (Ed.): ICIAP 2013, Part I, LNCS 8156, pp. 371–380, 2013.
© Springer-Verlag Berlin Heidelberg 2013

to fitting locally the parameters of a given MRF model which could not deal equally well with homogeneous, textured, and strongly structured regions. Therefore, in [5] a dynamical segmentation paradigm with multiple models was proposed, but not implemented, providing just a proof of concept based on supervised experiments. In this work, following the paradigm of [5], we propose and implement a dynamical hierarchical segmentation/classification algorithm, DHC for short, based on two alternative models for homogeneous and textured regions. Here, the major challenge is not the segmentation itself, but the construction of the tree of regions that best describes the image, including thus the selection, at each node, of the model that best fits the region under analysis, and the definition of meaningful growth stopping criteria. To tackle these problems we define a suitable figure, attached with each node/region and each model, called split-gain, which allows us to decide which region to split first, with which model, and when to stop the algorithm.

2 Background

In this section we provide a short description of the basic segmentation algorithms, TS-MRF and TFR, necessary for understanding the dynamic algorithm proposed here. We assume the reader to have some familiarity with Bayesian segmentation and Markov random fields, referring to specific references [6] for a more thorough treatment.

Tree-Structured-MRF (TS-MRF) is a recursive segmentation algorithm. The whole image of interest, namely the set of sites S and the corresponding observables y is associated with the root of a tree. Its segmentation divides the set of sites in K disjoint subsets $\{S_1, \ldots, S_K\}$, each with its subset of observables, associated with the root children. Recursion on the newly generated nodes, together with some suitable stopping conditions, produces a tree of classes and the associated segmentation of the image. An arbitrary number of classes could be considered at each node, but we restrict attention to binary segmentation in order to avoid the challenging task of order selection. At each node, segmentation is carried out according to the MAP criterion

$$\widehat{x} = \arg\max_x p(x|y) = \arg\max_x p(y|x)p(x) \tag{1}$$

where $p(\cdot)$ indicates probability mass/density function (pmf or pdf) x is the label map, and $\widehat{x}$ is therefore the most probable map given the observables. The label map is modeled as an Ising MRF, where the labels are assumed to be equally likely, only cliques of two 8-connected sites are considered, $c = (i, j)$, and the potentials are defined as

$$V_c(x_i, x_j) = \begin{cases} \beta \text{ if } x_i \neq x_j \\ 0 \text{ otherwise.} \end{cases} \tag{2}$$

With this definition, we obtain the global pmf

$$p(x) = \frac{1}{Z} \exp[-\beta N_E(x)] \tag{3}$$

where $N_E(x)$ is the number of label transition or *edges* in the map or, under another point of view, the total length of region boundaries in the map. With this prior pmf, label

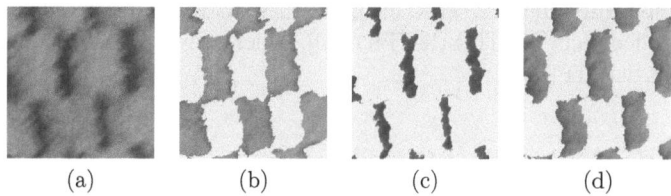

(a)   (b)   (c)   (d)

Fig. 1. An example texture with its three components (clusters of segments)

transitions are penalized, with a strength depending on the edge-penalty parameter β: the larger β the more regular the final segmentation map.

A key element of TS-MRF is the so called *split-gain* which, for each node, measures the convenience of actually proceeding with a further split of the associated class. For each leaf node we compute the associated split-gain, and proceed to split the node with the maximum gain, provided it is positive. Therefore, the segmentation tree is shaped by this measure, defined formally and analyzed in depth in the next Section, which also provides the necessary stopping condition.

The **Texture fragmentation and reconstruction (TFR)** algorithm [7–9] is oriented to the segmentation of textured images. Fig.1 shows a simple texture (a close view of a fabric) together with its decomposition in three basic components, each of which is, in turn, a collection of elementary segments with about the same color, size, and shape. This simple description fits very well a large variety of natural textures, and TFR translates it in three major processing steps

1. color-based segmentation;
2. segment clustering;
3. progressive cluster merging.

The first step detects all elementary connected regions with homogeneous color by means of any conventional region-based or edge-based segmenter. The second steps forms clusters of segments like those shown in Fig.1(b-d), which are the building blocks of textures. To this end, each segment is characterized by a vector of features. These include color, position, size, and shape, but the most discriminating features are related to spatial context, as natural for textures. For example, all black segments of Fig.1(a) have a red segment on their right, a distinctive contextual feature of this specific texture, not easily found in others. Once correctly featured the segment, it is relatively simple to retrieve meaningful clusters by means of standard data analysis techniques. The final step reconstructs the desired textures by progressive pair-wise merging of clusters. A suitable merging gain (called texture score) [9] is defined to decide which texture components should be merged at any step and when to stop the process.

Like TS-MRF, also TFR produces thus a hierarchy of nested segmentation maps. In the latter case, however, the process is a bottom-up merging, starting from the basic clusters of segments to obtain larger and larger textures. However, TFR can be trivially re-engineered to become a top-down process. By forcing the merging process to proceed until only two textures remain, we obtain a binary TFR (B-TFR) algorithm.

Recursive application of the B-TFR, then, gives rise to a tree-structured top-down segmentation, called recursive TFR (R-TFR) [10], which provides different (and typically better) results than TFR.

3   Dynamic Hierarchical Classification

DHC is a recursive algorithm, therefore it can be associated with a tree, where each node corresponds to a region (not necessarily connected) of the image. The root of the tree corresponds to the whole image while the leaves correspond to the elementary segments or classes produced by the segmentation process. We restrict attention to two-class segmentation and therefore deal with binary trees. A binary tree T is formally identified by a set of nodes $\{t \in T\}$ and their mutual relationships. Except for the root, each node t has one parent $u(t)$, and each internal node has two children $l(t)$ and $r(t)$, with $u[l(t)] = u[r(t)] = t$. In the following, to simplify notation, we will associate nodes with integers, with $t = 1$ corresponding to the root, $l(t) = 2t$, $r(t) = 2t + 1$ and $u(t) = \lfloor t/2 \rfloor$.

With each node t in T we associate

- a set of sites $\mathcal{S}^t \subseteq \mathcal{S}$, corresponding to a segment of the image;
- a set of observed data $y^t = \{y_i : i \in \mathcal{S}^t\}$, that is the restriction of y to $\mathcal{S}^t$.

A segmentation map x^t for node t is a field of labels defined on $\mathcal{S}^t$, with $x_i^t \in \{2t, 2t + 1\}$. When node t is split according to map x^t two new nodes are generated: node $2t$ with $\mathcal{S}^{2t} = \{i \in \mathcal{S}^t : x_i^t = 2t\}$, and $y^{2t} = \{y_i : i \in \mathcal{S}^{2t}\}$, and node $2t + 1$ with the associated items defined in a similar way.

Initially, the tree comprises only the root, $T = \{1\}$, which is also its unique leaf. We use the B-MRF segmentation algorithm to compute the segmentation map $x^{S,1}$, where the S superscript indicates spectral-based, with split gain $G^{S,1}$. Likewise, use the B-TFR segmentation algorithm to compute the segmentation map $x^{T,1}$, with split gain $G^{T,1}$, where T indicates texture-based. Then we set $G^1 = \max(G^{S,1}, G^{T,1})$. If $G^1 \leq 0$ the root is not split, because none of the two segmentation maps provides an improved representation of the data. Otherwise, it is split using the technique with the largest split gain, that is,

$$x^1 = \begin{cases} x^{S,1} \text{ if } G^{S,1} > G^{T,1} \\ x^{T,1} \text{ otherwise} \end{cases} \tag{4}$$

In the generic step of the algorithm, the leaf with the largest split gain is split using the same logic, provided the gain is positive, otherwise the algorithm stops.

Eventually, it all comes down to the definition of the spectral and textural split gains, which quantify the improvement in representation accuracy, if any, granted by segmentation. In [3] the split gain for TS-MRF is defined as the log likelihood ratio between the competing single-class (no split) and two-class (split) descriptions of the data. In the single-class description of node t, there is only one label available, and only one possible label field, with unit probability. The observables are assumed to be independent and identically distributed (i.i.d.) Gaussian with parameters estimated from the data. In the two-class description, instead, the label field is random, $x^t \in \{2t, 2t + 1\}^N$, and is described by the Ising MRF model of (3), while the observable are conditionally independent, given the label, following two Gaussian laws with different parameters.

Before the log, we therefore have

$$\exp(G^{S,t}) = \frac{p(x^t)p(y^t|x^t)}{p(y^t)} = p(x^t) \cdot \frac{g^{2t}(y^{2t})g^{2t+1}(y^{2t+1})}{g^t(y^t)} \tag{5}$$

where $g^t(\cdot)$ is the Gaussian distribution that best fits the observables y^t, with parameters estimated on the data, and $g^t(y^t) = \prod_{i \in S^t} g^t(y_i)$.

Equation (5) shows clearly the balance among two competing costs/gains. The last fraction makes clear that the two-class hypothesis can provide a better description of the observables as these can be suitably grouped and each group can be explained by a dedicated distribution, $g^{2t}(\cdot)$ or $g^{2t+1}(\cdot)$, while in the single-class case a unique distribution $g^t(\cdot)$ must explain all data. This obvious description gain is balanced by the cost of describing the grouping itself, represented by $p(x^t)$, which is absent in the single-class case. Therefore, the split gain balances the fitting gain, related to a better description of the data, with the segmentation cost, related to the number of edges $N_E(x^t)$ and hence to the map complexity.

We now want to define a split gain for B-TFR segmentation in order to compare the two solutions in a meaningful way. Therefore, we follow the same conceptual path, defining the split gain as a balance between the fitting gain of the data and the segmentation cost. In the case of textures, however, it makes no sense trying to describe data by means of a single Gaussian distribution. When we decide not to split a region, in the TFR framework, it is because we believe it is accurately represented by a single *texture*, not a single *homogeneous field*, therefore, its fitting must refer to some kind of mixture distribution. In other words, the region under analysis is considered under two complementary points of view: i) based on spatial properties, it is regarded as a single textured region, ii) based on color, it is divided again in two classes, each represented by its own distribution. Again, we choose the simplest possible mixture to describe data, that is, the mixture of two Gaussians. Therefore, even in the no-split hypothesis, the likelihood of the data is written as

$$p(y^t) = m(y^t) = \prod_{i \in \tilde{S}^{2t}} \tilde{g}^{2t}(y_i) \prod_{i \in \tilde{S}^{2t+1}} \tilde{g}^{2t+1}(y_i) \tag{6}$$

where $m(y^t)$ is the best mixture of Gaussians for the data of node t, $\tilde{S}^{2t}$ and $\tilde{S}^{2t+1}$ are the two segments in which S^t would be divided by the color-oriented clustering map c^t (not by the B-TFR map x^t), and $\tilde{g}^{2t}(\cdot), \tilde{g}^{2t+1}(\cdot)$ are the associated best fitting Gaussian distributions. Accordingly, in analogy with the former case, the splitting gain for B-TFR is written (again, before the log) as

$$\exp(G^{T,t}) = \frac{p(x^t)p(y^t|x^t)}{p(y^t)} = p(x^t) \cdot \frac{m^{2t}(y^{2t})m^{2t+1}(y^{2t+1})}{m^t(y^t)} \tag{7}$$

where the map description cost is computed with x^t regarded as an Ising MRF.

Given the estimate of mean vector μ and covariance matrix $\mathbf{C}$ for each set of observable data, we can write the Fitting cost as

$$F(y^t) = -\log[g^t(y^t)] = \alpha^t + \frac{N^t}{2}\log|\mathbf{C}^t| + \sum_{i \in S^t}\frac{1}{2}(y_i - \mu^t)^T[\mathbf{C}^t]^{-1}(y_i - \mu^t) \quad (8)$$

with $\alpha^t = N^t B \log(2\pi)/2$, by which, taking into account also (3), we can write (5) explicitly as

$$G^{S,t} = \left[F(y^t) - F(y^{2t}) - F(y^{2t+1})\right] - \left[\beta^t N_E(x^{S,t}) + \log Z^t\right] \quad (9)$$

where the first square brackets accounts for the fitting gain (cost reduction) granted by the split, and the second for the map description cost. Likewise, for a TFR split we can write

$$G^{T,t} = \left[\sum_{\tau=2t}^{2t+1} F(\tilde{y}^\tau) - \sum_{\tau=4t}^{4t+3} F(\tilde{y}^\tau)\right] - \left[\beta^t N_E(x^{T,t}) + \log Z^t\right] \quad (10)$$

Since all terms scale linearly with the segment dimension, N^t, we normalize the split gain to it. This has no influence on the comparison between MRF and TFR, but modifies the split order in favor of needy classes, rather than large ones.

A further suitable normalization concerns the data dimensionality. In fact, when $B \gg 1$ (think of hyperspectral images, with hundreds of bands), the regularization term related the map description cost becomes negligible w.r.t. the fitting gain, leading to a sure oversegmentation of the image. Our simple Gaussian model cannot automatically take into account this phenomenon. By normalizing the fitting gain term by the number of bands we increase the probability of stopping, splitting a node only when it is clearly necessary.

Finally, it is well known that the partition function Z^t cannot be computed exactly. Some good approximations have been proposed in the literature, especially for the Ising model, but only for the case of rectangular lattices, a constraint that we meet only for the segmentation of the root. Therefore we neglect it altogether, but compare the splitting gain with a small positive threshold, tuned by preliminary experiments, rather than zero.

Eventually, the split gains are defined as

$$G^{S,t} = \frac{1}{N^t}\left\{\frac{1}{B}\left[F(y^t) - F(y^{2t}) - F(y^{2t+1})\right] - \beta^t N_E(x^{T,t})\right\} \quad (11)$$

and

$$G^{T,t} = \frac{1}{N^t}\left\{\frac{1}{B}\left[\sum_{\tau=2t}^{2t+1} F(\tilde{y}^\tau) - \sum_{\tau=4t}^{4t+3} F(\tilde{y}^\tau)\right] - \beta^t N_E(x^{T,t})\right\} \quad (12)$$

4   Experiments and Discussion

To test the proposed algorithm, we carry out numerical experiments on multispectral remote-sensing images. Segmentation is especially relevant in this context not only as

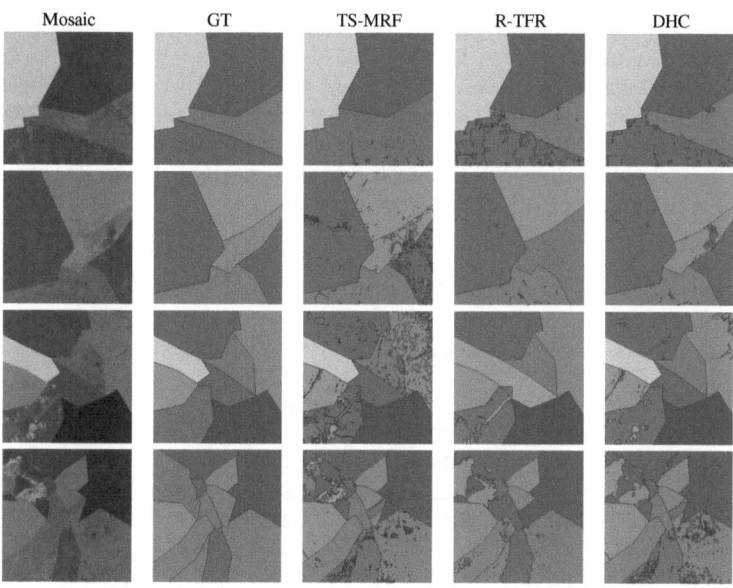

Fig. 2. Results on Prague benchmark. Left to right: mosaic, ground truth, TS-MRF, R-TFR, DHC.

an help for image analysis, but also in related applications, such as content-based image retrieval [11], or region-based image compression [12, 13] and denoising [14]. Objectivity and reproducibility of results are guaranteed by resorting to the publicly available Prague remote-sensing segmentation benchmark [15], which provides a number of multispectral mosaics with attached ground-truth, with many challenging cases. Results are compared under a large number of criteria with those of TS-MRF and R-TFR under two conditions *i)* fixed number of nodes *ii)* best number of nodes (with maximum correct segmentation). In any case, we keep the tree structure provided by the algorithm, with no user intervention.

Fig.2 shows some of the mosaics used in the experiments together with the associated ground truth and with the best segmentation provided by TS-MRF, R-TFR and DHC, respectively. Visual inspection testifies the superiority of the dynamic model approach: TS-MRF captures well the image structure, most of the times, but produces an obvious fragmentation of regions; on the other hand, TFR provides a more compact map, but happens to erroneously merge some segments. Most of these problems are overcome by DHC. Numerical results in Table 1, referring to a total of 20 mosaics, fully confirm this impression, with DHC almost always the best technique both for the fixed-number and best-number cases. In particular, DHC provides a 5 percent points improvement in overall Correct Segmentation (CS) w.r.t. the best fixed-model reference (R-TFR).

To gain insight into the reasons of such an improvement consider the segmentation tree of Fig.3, obtained by using exclusively the TS-MRF technique. In the box associated with each node we report the TS-MRF split gain (spectral gain, SG) and also, in smaller font, the TFR split gain (or texture gain, TG), which however is not taken into

Table 1. Numerical results on Prague benchmark for DHC, R-TFR and TS-MRF (with manual selection of the number of classes); (Benchmark criteria: CS = correct segmentation; OS = over-segmentation; US = under-segmentation; ME = missed error; NE = noise error; O = omission error; C = commission error; CA = class accuracy; CO = recall - correct assignment; CC = precision - object accuracy; I. = type I error; II. = type II error; EA = mean class accuracy estimate; MS = mapping score; RM = root mean square proportion estimation error; CI = comparison index; GCE = Global Consistency Error; LCE = Local Consistency Error; dD = Van Dongen metric; dM = Mirkin metric; dVI = variation of information).

	Benchmark – ALI					
	DHC Optimal # of classes	R-TFR Optimal # of classes	TS-MRF Optimal # of classes	DHC Given # of classes	R-TFR Given # of classes	TS-MRF Given # of classes
↑CS	**84.60**	78.45	66.45	75.05	70.33	55.49
↓OS	**6.78**	10.39	9.34	**7.56**	12.95	7.58
↓US	8.73	14.33	12.09	16.77	17.45	20.44
↓ME	4.70	**2.90**	14.26	4.07	8.28	15.33
↓NE	5.33	**1.28**	14.77	4.62	6.85	14.57
↓O	1.69	1.22	4.39	3.24	**1.07**	8.15
↓C	0.70	3.11	6.07	**0.52**	4.78	6.40
↑CA	**89.69**	84.74	80.82	83.36	80.52	73.11
↑CO	**92.80**	89.72	86.98	88.64	86.42	81.33
↑CC	**92.78**	88.98	87.88	87.19	88.30	81.47
↓I.	**7.20**	10.28	13.02	11.36	13.58	18.67
↓II.	**0.90**	1.25	2.64	2.25	1.98	4.23
↑EA	**92.29**	88.37	86.27	86.73	85.58	79.27
↑MS	**89.59**	85.45	81.60	82.95	79.63	72.00
↓RM	**1.94**	3.46	3.66	3.42	4.23	5.40
↑CI	**92.53**	88.84	86.83	87.24	86.44	80.27
↓GCE	4.38	**4.34**	10.26	5.74	8.45	12.31
↓LCE	2.67	**2.25**	6.72	2.80	2.84	6.52
↓dD	**4.61**	6.20	9.07	6.70	8.42	11.79
↓dM	**2.17**	3.06	5.66	4.33	4.60	8.38
↓dVI	14.51	14.43	14.77	**14.30**	14.57	14.44

account for the segmentation. The four-class mosaic associated with the root, node 1, is split along the red line with a large gain, SG=0.63. Both children nodes have a positive split gain, close to zero the left one, quite large the right one, which is therefore split first, generating nodes 6 and 7. Now look at node 6: although two different regions are clearly recognizable, the spectral gain is negative, because both regions are internally textured, and hence poorly represented by a single Gaussian fit. This node will no longer be split, with an unrecoverable undersegmentation. Going on we obtain only a wrong split of node 7 and possibly (depending on the decision threshold) of other nodes. In this case, instead, TFR would have split node 3 along the yellow line, correctly separating the two regions, with a relatively large texture gain TG=0.16.

Consider now the symmetric situation of Fig.4, where only the TFR segmenter is used. After splitting nodes 1, 3, and 2, in this order, along the red lines, the algorithm either stops, if the decision threshold exceeds 0.07, or splits unduly node 5 which corresponds to a homogeneous region. Node 6 is not split because the relatively large gain, TG=0.10, is obtained with a wrong split (not shown) which singles out an exceedingly

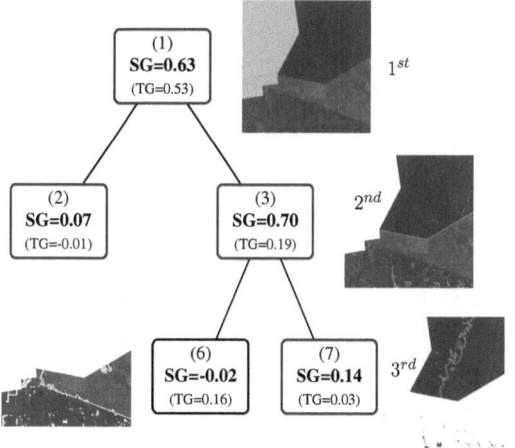

Fig. 3. Tree structured segmentation of mosaic # 2 by TS-MRF

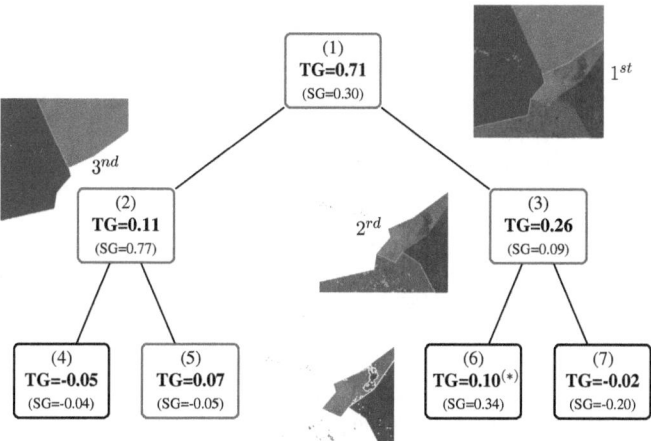

Fig. 4. Tree structured segmentation of mosaic # 2 by R-TFR

small region, and is hence rejected (marked with a *). The problem is that the two class region associated with node 6 is already well represented by a mixture of Gaussians, making a further textural split not convenient. For the same node, however, the spectral gain would be much larger, corresponding to the correct separations of the two large component regions along the yellow line.

In both cases, the use of the competing models would definitely improve the segmentation path and the final result. These results, however preliminary and limited, clearly show the potential of the dynamic hierarchical classification paradigm. Significant further improvements can be expected by including other image models and segmentation engines. The problem remains of how to define suitable metrics to switch between one model and the other.

References

1. Benboudjema, D., Pieczynski, W.: Unsupervised Statistical Segmentation of Nonstationary Images Using Triplet Markov Fields. IEEE Trans. on Pattern Analysis and Machine Intelligence 29, 1367–1378 (2007)
2. Liu, G., Qin, Q., Mei, T., Xie, W., Wang, L.: Supervised Image Segmentation Based on Tree-Structured MRF Model in Wavelet Domain. IEEE Geoscience and Remote Sensing Letters 6, 850–854 (2009)
3. D'Elia, C., Poggi, G., Scarpa, G.: A tree-structured Markov random field model for Bayesian image segmentation. IEEE Trans. on Image Processing 12, 1259–1273 (2003)
4. D'Elia, C., Marrocco, C., Molinara, M., Poggi, G., Scarpa, G., Tortorella, F.: Detection of microcalcifications clusters in mammograms through TS-MRF segmentation and SVM-based classification. In: The 17th International Conference on Pattern Recognition, ICPR 2004, vol. 3, pp. 742–745 (2004)
5. Masi, G., Gaetano, R., Scarpa, G., Poggi, G.: Dynamic segmentation for image information mining. In: IEEE International Geoscience and Remote Sensing Symposium, IGARSS 2010, pp. 1992–1995 (July 2010)
6. Li, S.Z.: Markov random field modeling in image analysis. Springer (2001)
7. Scarpa, G., Haindl, M.: Unsupervised texture segmentation by spectral-spatial-independent clustering. In: The 18th International Conference on Pattern Recognition, ICPR 2006, vol. 2, pp. 151–154 (August 2006)
8. Scarpa, G., Haindl, M., Zerubia, J.: A Hierarchical Finite-State Model for Texture Segmentation. In: IEEE International Conference on Acoustics, Speech and Signal Processing, ICASSP 2007, vol. 1, pp. I-1209–I-1212 (2007)
9. Gaetano, R., Scarpa, G., Poggi, G.: Hierarchical texture-based segmentation of multiresolution remote-sensing images. IEEE Trans. on Geoscience and Remote Sensing 47, 2129–2141 (2009)
10. Gaetano, R., Scarpa, G., Poggi, G.: Recursive Texture Fragmentation and Reconstruction segmentation algorithm applied to VHR images. In: IEEE International Geoscience and Remote Sensing Symposium, IGARSS 2009, pp. IV–101–104 (2009)
11. Li, Y., Bretschneider, T.R.: Semantic-Sensitive Satellite Image Retrieval. IEEE Trans. on Geoscience and Remote Sensing 45, 853–860 (2007)
12. Cagnazzo, M., Poggi, G., Verdoliva, L.: Region-based transform coding of multispectral images. IEEE Trans. on Image Processing 16, 2916–2926 (2007)
13. Cagnazzo, M., Parrilli, S., Poggi, G., Verdoliva, L.: Improved class-based coding of multispectral images with shape-adaptive wavelet transform. IEEE Geoscience and Remote Sensing Letters 4, 566–570 (2009)
14. Parrilli, S., Poderico, M., Angelino, C.V., Scarpa, G., Verdoliva, L.: A nonlocal approach for SAR image denoising. In: IEEE International Geoscience and Remote Sensing Symposium, IGARSS 2010, pp. 726–729 (2010)
15. Mikes, S., Haindl, M., Scarpa, G.: Remote sensing segmentation benchmark. In: The 7th IAPR International Workshop on Pattern Recognition in Remote Sensing, PRRS 2012, Tsukuba Science City, Japan (November 2012)

Road Traffic Conflict Analysis
from Geo-referenced Stereo Sequences

Sebastiano Battiato[1], Stefano Cafiso[2], Alessandro Di Graziano[2],
Giovanni M. Farinella[1], and Oliver Giudice[1]

[1] Image Processing Laboratory, Dipartimento di Matematica e Informatica
[2] Dipartimento di Ingegneria Civile e Ambientale
University of Catania, Italy

Abstract. In this paper an imaging system for road traffic conflict analysis is proposed. The system exploits geo-referenced stereo sequences and tracking procedure to compute traffic conflict measures which can be analysed by experts. Using the potentiality of the traffic conflict technique as a surrogate safety measure could constitute an effective tool in understanding how the driver interacts and adapts its behaviour with respect to the vehicle, the road characteristics, the traffic control devices and environment. Experiments performed on real data acquired in urban environment confirm the effectiveness of the system which makes simple and fast for the experts the understanding of the driver behaviour.

Keywords: Traffic conflict analysis, Stereo system, Tracking.

1 Introduction

With the increasing number of vehicles on the roads, accidents have become more and more frequent today. There are different causes of road accidents: distracted driving, speeding, unsafe lane changes, etc. Many people die as a result of road accidents, and many more have unwanted consequences, including serious injuries. The huge social, human and also economic costs induced by road accidents make the topic of road safety really important in nowadays society.

The objective of the traffic conflict techniques is the evaluation of traffic safety and the prediction of road accidents. By automatically understanding and predicting the risk of a road accident some actions can be taken in time by systems (e.g., Driver alerting) in order to avoid the conflict. Different studies have been done in the traffic safety context to build appropriate tools and measures for traffic conflict [1,2]. The major problem when using the Traffic Conflict Technique (TCT) is the complexity in data acquisition and processing: everything that happens in front of the vehicle needs to be taken into account in terms of time and space variables.

In this paper we propose a system which is useful for both, data acquisition and processing in the context of traffic conflict monitoring. The proposed system makes use of geo-referenced stereoscopic vision for the acquisition of data which are then analysed with a semi-automatic pipeline acting on the acquired stereoscopic video in order to obtain the traffic conflict risk measures.

A. Petrosino (Ed.): ICIAP 2013, Part I, LNCS 8156, pp. 381–390, 2013.
© Springer-Verlag Berlin Heidelberg 2013

The paper is organized as follows: we introduce the reader to the Traffic Conflict Technique fundamentals in Section 2. The hardware used in proposed system is summarized in Section 3, whereas the software components employed for data processing and traffic conflict analysis are described in Section 4. Experiments and results are reported in Section 5. Finally, conclusion and hints for future works are given in Section 6.

2   The Traffic Conflicts Technique

The Traffic Conflicts Techniques (TCT) are based on the "Heinrich Triangle" theory [3] founded on the relationship that "no-injury accidents" precede "minor injuries" (i.e., events closer to the base of the triangle precede events nearer the top). Application of this theory assumes that the appropriate traffic conflict factors can be defined as measures of near-crash events. A traffic conflict could be defined as "an observable situation in which two or more road users approach each other in space and time to such an extent that there is risk of collision if their movements remain unchanged" [4]. Traffic conflict measures, such as "time to collision" (TTC), address the first condition of surrogate measures, namely the common factors that are shared with safety highlighting advantages and limits of the TCT [5]. The TCT are based on the measurement of both spatial and temporal variables which describe the interactions between two road users involved in a critical event for safety. The traffic conflict measures considered in this paper are connected to the severity of the conflict that is a combination of accident probability and severity of injuries. The Time To Collision (TTC) is the indicator used by TCT and it is calculated for the vehicles (TTC_v) and an obstacles (TTC_o). The TTC is defined as follows:

$$TTC = \frac{D}{V} .$$
(1)

where D is the distance between the vehicle and the conflict point, and V is the velocity of a generic obstacle (e.g., another vehicle, a pedestrian, etc.).

In case the obstacle is a pedestrian crossing the street, the TTC measures at time i for the two actors could be defined as follows [6]:

$$TTC_v(i) = \frac{D_v(z_i)}{V_v(i)} ,$$
(2)

$$TTC_o(i) = \frac{D_v(x_i) - D_o(x_i)}{V_o(i)} .$$
(3)

where, by considering a 3D reference system centered on the vehicle, $D_v(z_i)$ represents the distance between the vehicle and the conflict point along the Z axis, and $D_v(x_i) - D_o(x_i)$ represents the distance between the vehicle and the obstacle along the X axis (see Fig. 1a).

Let be T_{f_i} the time required for the vehicle to stop at time i defined as:

$$T_f(i) = T_r + \frac{V_v(i)}{d} .$$
(4)

where T_r is the reaction time, $V_v(i)$ is the vehicle velocity at time i, and d is the deceleration during braking.

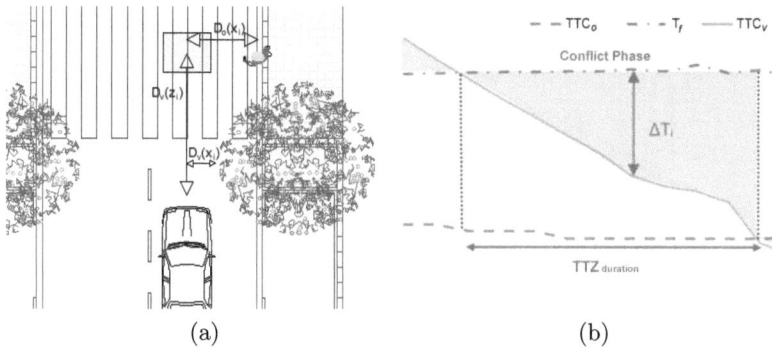

(a)   (b)

Fig. 1. (a) Distances involved in a TTC computation. (b)Temporal trend of the quantities involved in the TTC. The area where $TTC_o(i) < TTC_v(i) < T_f(i)$ is related to the conflict.

It is possible to consider the following cases, which determine the areas as in Fig. 1b:

i) $TTC_v(i) > T_f(i)$: vehicle may stop before conflict area,
ii) $TTC_o(i) > TTC_v(i)$: the vehicle passes the area of conflict before the pedestrian reaches it,
iii) $TTC_o(i) < TTC_v(i) < T_f(i)$: there is the conflict and it is possible to evaluate the $TTZ_{duration}$ that is the duration of the conflict.

Given the $TTZ_{duration}$ it is possible to compute the severity of the conflict as Risk Impact (RI) at time i according to the following equation:

$$RI(i) = \sum_{i \in TTZ_{duration}} V_v(i)^2 \times \Delta T(i) . \tag{5}$$

where $V_v(i)$ is the velocity of the vehicle at time i and $\Delta T(i) = T_f(i) - TTC_v(i)$. The RI for the entire conflict is given by the following formula:

$$RI = \frac{\frac{1}{N} \sum_i^N RI(i)}{TTZ_{duration}} . \tag{6}$$

In the case of an obstacle moving along the direction of the vehicle, as in a "car following" case, the TTC_o will be 0 and therefore the conflict will occur only if $TTC_v(i) < T_f(i)$.

3   Hardware Components

The hardware employed in the proposed system is composed by a number of devices connected together. Specifically, a **TYZX DeepSea G3 Embedded Vision System (EVS)** [7] is employed to acquire stereo images and to obtain a depth map of the scene in front of the vehicle. This piece of hardware is positioned in the front of the vehicle (i.e., a bus in our case) as depicted in Fig. 2.

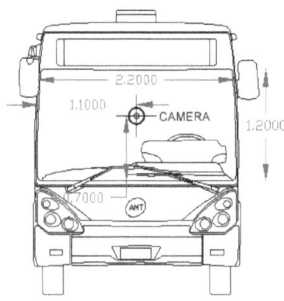
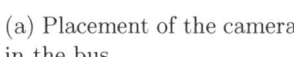

(a) Placement of the camera in the bus.

(b) Photo of the system installed on the BUS.

Fig. 2. The proposed system on a urban BUS during the experimental phase

The EVS implements a Census-Based stereo algorithm [8]. As the input pixels enter the EVS, the Census transform is computed at each pixel based on the local neighborhood, resulting in a stream of Census bit vectors. At every pixel a Hamming distance is used to compare the Census vectors around the pixel of one view (i.e., left image) to those at 52 locations in the other view (e.g., right image). These 52 comparisons are performed simultaneously making the stereoscopic system very fast at subpixels precision. The EVS processor converts the pixels disparity map to metric distance measurements using the stereo camera's calibration parameters and the depth units specified by the user [7]. In our setting a stereo cameras with a baseline of 33cm and an 83 degrees HFOV lens has been used. This configuration allows the overall system to work with distance in a range between 2.5m and 50m. All the specifications of the EVS used for the experiments are reported in Table 1.

Table 1. Datasheet of the EVS hardware used in our experiments

G3 Embedded Vision System Specifications	
Size	3.8cm x 18.7cm x 14.5cm
Power	11W typ.: 12vdc or PoE class III
Imagers	Aptina MT9V022 color, up to 60fps, 752x480 pixels
Lens	83 degrees horizontal FOV
Stereo Baseline	33 cm
CPU	Freescale PowerPC 8347 @400MHz
Memory	256 Mbytes
Operating System	Linux 2.6 Kernel

In the proposed system, the EVS is integrated with a series of components: a router, a GPS module and a notebook. The router solves the EVS data access problem. We used a gigabit Ethernet WLAN router to create a LAN in which the EVS can be accessed by a notebook. A client can access the EVS by connecting to a TCP service that sends each captured image and its computed depth map. For the application described in this paper, image and depth information must be geo-referenced. To achieve this requirement, we connected the EVS to a Bluetooth GPS receiver module. The choice of a wireless device was mandatory to solve the signal satellite problem: the specific shape of an urban BUS forced us to put the GPS module placed on the external roof of the bus to have the best possible number of satellites in sight. We connected the Bluetooth GPS module to the notebook communicating with the EVS through LAN network. Some ad-hoc routines have been developed to synchronize the GPS information with every single frame acquired by the EVS. Finally we have a notebook connected to the LAN running the software layer which coordinates the different hardware. The SW also performs the acquisition and analysis of the data. It gives commands to the EVS, receives data both from the EVS and the GPS module. It synchronizes every single EVS information making some computation on it and finally it gives a single output as archive files that can be opened and analyzed by a specific module, after the data acquisition phase.

After mounting the hardware on the vehicle (i.e., the bus), a dedicated calibration procedure must be done. This is because the measures of the indices of traffic conflicts depend on the reference system that must be integrated to the vehicle. The calibration procedure consists in placing outside and in front of the vehicle two targets orthogonal to the axis of the vehicle at 3 meters away from the camera. The targets are placed using an high precision laser rangefinder positioned on vehicles vertices and pointed to the targets. At this point, with the help of the data acquisition software in its initialization phase, the system evaluates the distances of the two targets and records the distance difference between the right target and the left one. This difference is due to the inclination of the camera related to the axis of the vehicle and the related values will be used for subsequent analysis as will be described in next section.

4   Software Components

As detailed in the following sub-sections, the software layer is composed by two main modules: one for data acquisition and the other for data analysis. The schema of the different components involved in the software layer are reported in Fig. 3.

4.1   Data Acquisition Module

This module takes in input the data coming from EVS and GPS as reported in Fig. 3. The EVS outputs are the color image sequences taken from the left camera, and the related depth-maps. The images are compressed according to the MPEG-4v3 codec and stored on a dedicated device. The depth-maps are represented with a 16 bits precision. From the depth-map stream a video file is obtained (the lossless compression codec FFV1 is used in this case). In this configuration, the software produces an amount of data of about 10GB every 2 hours of recording. Since the context of use of our system is related the urban buses traffic, we decided to stop the recording when the bus is not moving in order to reduce the amount of stored data of about 30%.

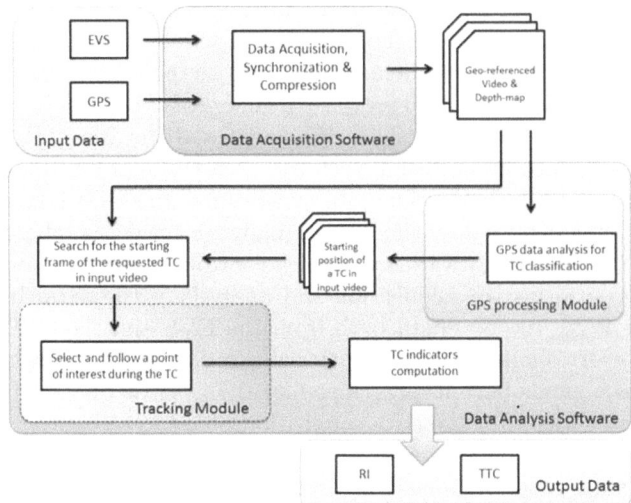

Fig. 3. The software layer of the proposed traffic conflict analysis system

As previously mentioned, the images (and the corresponding depth-maps from the EVS) need to be geo-referenced through an external GPS. The data acquisition module deals with this issue by solving the problem of the difference in the sampling frequency between the video information (20-30Hz) and the GPS (10Hz). For this purpose, the software operates as follows:

i) When a new video information arrives (i.e., a frame) the software operates a sample and hold mechanism on the GPS information by associating the last GPS information to each video frame. In this way there will be some frames corresponding to the same GPS information;

ii) When a new GPS information arrives, the acquisition module corrects all the previous informations with the same data interpolating the position and velocity values according to the data of the newly arrived GPS information.

Finally, the acquisition module produces a textual output that allows to bind each frame and depth-map to the corresponding GPS information. Since both the GPS and the EVS are connected to the computer via Bluetooth and LAN respectively, in the initialization phase, the module makes an evaluation of the latencies on a single packet transmission. This operation allows the acquisition module to know the arrival delay from the generation of each individual package to correct the output. In this way it's possible to obtain a more precise synchronization between each frame and the related GPS information.

4.2   Data Analysis Module

The data analysis module allows the expert to operate on the geo-referenced frames and depth-maps to obtain the traffic conflict indicators used for the analysis (see Section 2). The analysis starts from the first frame of a conflict which is selected

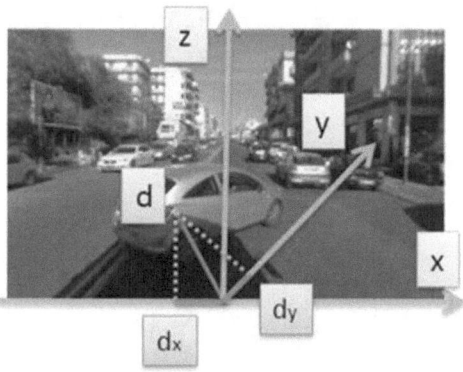

Fig. 4. Reference system used by the data analysis module

through a manual search in the video, or in an automatic way by processing the information provided by the GPS (i.e., vehicle speed, vehicle trajectory, acceleration, jerk and angular deviation rate) [11]. Once the initial frame of the traffic conflict is detected, the obstacles that may have generated the conflict is identified as the closest point to the vehicle. The obstacle is then tracked in the successive frames in order to obtain the traffic conflict measurements. For tracking the obstacle we have used a modified version of the well known mean-shift tracking algorithm [9]. The algorithm has been extended in order to take into account information provided by the depth-map and considering the speed information provided by GPS. Both depth and speed of the vehicle are useful to perform a fast binary segmentation and hence remove information which are not related to the obstacle, making the tracker more robust to clutter background (e.g., if the obstacle is a vehicle, the other vehicles or objects similar with respect to visual information are removed by using the distance given by the depth-map information).

Hence the closest point of the obstacle is tracked exploiting the data coming from the depth-map is possible to fully reconstruct the behavior of the vehicle and the target obstacle involved in the conflict in the space in front of the vehicle itself. The data analysis module hence computes speeds and distances of both vehicle and obstacle in X and Y axis components by considering the reference system synthetically sketched in Fig. 4. The quantities thus obtained are then used to compute the traffic conflict indicators as described in Section 2.

5 Experimental Phase

The described system has been used to analyze the behavior of bus drivers in urban context. The experiments have been performed on real traffic condition in the city of Catania, Italy. The system has been mounted on a urban bus and four different acquisition sessions of about 2 hours each have been carried. Using the acquired data, traffic conflict analysis was carried out. The results provided by the analysis are useful to understand the traffic behaviour and hence take

Fig. 5. Five frames and the related depth-maps of a pedestrian crossing the street hence generating the conflict

actions (i.e., installation of traffic lights) in order to prevent road accidents. In the following we present a case study, as well as the results obtained from the analysis performed with the system, related to a traffic conflict in which a pedestrian is involved. Among the other traffic conflicts (e.g., "car following") in the acquired data (about 8 hours of recording), the case here discussed is the most frequent and representative. Similar results have been found in the other traffic conflict cases. Due space limit, the other traffic conflicts examples (e.g., car following) will be presented and discussed in detail at conference time.

The conflict detailed in the following is generated from a pedestrian crossing the street. In this case the pedestrian decided to cross at the very last moment forcing the bus to make a stop and generating a conflict. Fig. 5 reports a sequence of frames and the related depth-maps which show the conflict tracked and analysed by the system. The data analysis module computes the behavior of the actors involved in the conflict, in terms of speed and distance on the two axes components, according to the reference system. The computed measures are reported in Fig. 6. The quantities extracted from the data analysis module are then further processed to product the measures of a TC. The obtained TTC for vehicle and pedestrian and the T_f are shown in Fig. 7. As shown in Fig. 7 the TC exists because there's an area in which $TTC_o < TTC_v < T_f$ (see Section 2). After obtaining traffic conflict indicators the data analysis module calculates the parameters indicating the severity of a conflict (RI) (see Equations (5) and (6)). The results of the RI computed in the case study presented here are reported in Table 2.

Given the variability and complexity of driver behaviors and performance, the random and rare nature of crashes and lack of adequate pre-crash data in the current crash record, it is becoming increasingly apparent that the necessary human factor data collection should be obtained from a "naturalistic" setting.

Table 2. Risk Impact results for the presented traffic conflict

	MAX RI	MEAN RI
Pedestrian	15.5613	9.2091

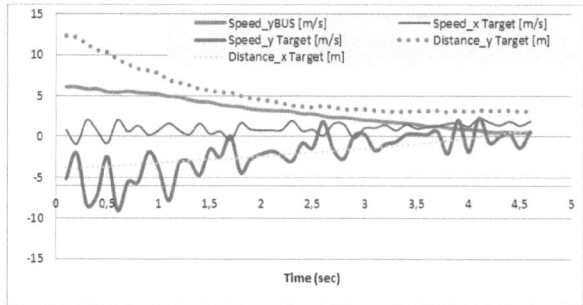

Fig. 6. Trend of speed and distance of vehicle and target in X and Y components

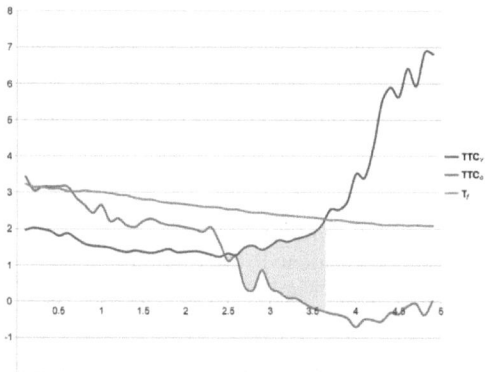

Fig. 7. Traffic conflict measures over time. The conflict phase is highlighted in yellow.

These approaches fill voids in the existing driving safety research methods. Specifically, the proposed method provides much greater information regarding pre-crash and crash events than what is currently available, even after a detailed crash investigation. Furthermore, the data provides much greater external validity relative to the larger context of driving than that of empirical methods, such as test tracks or simulators. For instance if the two types of conflicts are recurrent events in the same points a better regulation of the crosswalk/bus line could be analyzed.

6   Conclusions and Future Works

The main aim of traffic conflict analysis is the objective evaluation of traffic safety and the prediction of fatal crashes. In this paper we have presented a system able to make measurements of the traffic conflicts by exploiting geo-referenced stereo sequences and tracking procedure. The effectiveness of the system has been demonstrated on real urban environment. Future works will be devoted to exploit advanced tracking techniques (e.g., TLD [10]) combined with depth information to better deal with appearance changes (i.e., pose, occlusion, etc.) and to allow re-detection of obstacles which disappear in some frames of the conflict. Also, some prior information about the context could be used to refine the depth map [12]. Moreover, the ability to detect and discriminate some classes of obstacles (e.g., pedestrians, cars) will be included in order increase accuracy of the computation of the traffic conflict risk. Finally, we plan to build a labeled dataset involving different classes (e.g., crashes, near-crashes, other "accidents") and information coming from video, GPS, etc., as well as information about the driver behavior (e.g., head pose). The dataset will be used to understand the nature of the conflict events and to determine if the near-crashes could be quantitatively and automatically identified.

Acknowledgments. The authors wish to thank the Italian Ministry of Economic Development for the financial support of this research within the program "Industria 2015" and the Public Transport Company AMT Catania for the kind availability of the bus.

References

1. Chin, H.C., Quek, S.T.: Measurement of Traffic Conflicts. Safety Science 26(3), 169–187 (1997)
2. Migletz, D.J., Glauz, W.D., Bauer, K.M.: Relationships between Traffic Conflicts and Accidents. Report No: FHWA/RD-84/042. US Department of Transportation, Federal Highway Administration (1985)
3. Heinrich, H.W.: Industrial Accident Prevention. McGraw-Hill, New York (1932)
4. Hyden, C.: The Development of Method for Traffic Safety Evaluation: The Swedish Traffic Conflict Technique. Bulletin 70. Lund Ins. of Technology, Sweden (1987)
5. Songchitruksa, P., Tarko, A.P.: Extreme value theory approach to safety estimation. Accident Analysis & Prevention 38, 811–822 (2006)
6. Cafiso, S., Garcia, A.G., Cavarra, R., Romero Rojas, M.A.: Crosswalk safety evaluation using a pedestrian risk index as traffic conflict measure. In: The 3rd International Conference on Road safety and Simulation (2011)
7. Woodfill, J.I., Gordon, G., Buck, R.: Tyzx DeepSea High Speed Stereo Vision System. In: IEEE Workshop on Real Time 3-D Sensors and their Use, IEEE Conference on Computer Vision and Pattern Recognition (2004)
8. Zabih, R., Woodfill, R.: Non-parametric Local Transforms for Computing Visual Correspondence. In: Eklundh, J.-O. (ed.) ECCV 1994. LNCS, vol. 801, Springer, Heidelberg (1994)
9. Comaniciu, D., Ramesh, V., Meer, P.: Kernel-Based Object Tracking. IEEE Transactions on Pattern Analysis and Machine Intelligence 25(5) (2003)
10. Kalal, Z., Mikolajczyk, K., Matas, J.: Face-TLD: Tracking-Learning-Detection applied to faces. In: IEEE International Conference on Image Processing (2010)
11. Cafiso, S., Di Graziano, A.: Automated in-vehicle data collection and treatment for existing roadway alignment. In: Efficient Transportation and Pavement Systems: Characterization, Mechanisms, Simulation, and Modeling - International Gulf Conference on Roads, pp. 785–797 (2008)
12. Battiato, S., Curti, S., La Cascia, M., Scordato, E., Tortora, M.: Depth Map Generation by Image Classification. SPIE Electronic Imaging (2004)

Adaptive Compression of Stereoscopic Images

Alessandro Ortis[1], Francesco Rundo[2],
Giuseppe Di Giore[2], and Sebastiano Battiato[1]

[1] Università degli Studi di Catania - Dipartimento di Matematica ed Informatica
Viale A. Doria 6, 95125 – Catania, Italy
alessandro.ortis@gmail.com,
battiato@dmi.unict.it
[2] STMicroelectronics s.r.l. - Digital Convergence Group/CSP
Stradale Primosole 50, 95121 – Catania
{francesco.rundo,giuseppe.di-giore}@st.com

Abstract. Nowadays the growing availability of stereo cameras for common applications is becoming a commodity. This paper addresses the problem of stereoscopic images data compression proposing an innovative algorithm for compressing Multi Picture Object coded stereopairs. By means of self organizing reconstruction algorithm based on image redundancy we are able to reduce the size of the enclosed JPEG images. The overall perceived (and measured) quality is managed by considering that a stereoscopic image represents the same scene acquired from two different perspectives. In particular we achieve some compression gain just encoding the two images with different quality factors. The reported results and test benchmarks show the robustness and efficiency of the proposed algorithm.

1   Introduction

Stereoscopy (also called stereoscopics or 3D imaging) is a technique for representation and projection of images and videos for creating the illusion of depth in order to simulate the human binocular vision. The binocular vision is based on the principle that we can present two slightly different images separately to the left and the right eye whereas such images are then combined by the viewer's brain to give the perception of 3D vision. By comparing these two images the HVS (Human Vision System) is able to infer depth distances in the scene. Using this principle, the images are usually projected on the same screen with different polarization in order to present a different image to each eye. The viewer wears eyeglasses which contain a pair of opposite polarizing filters, each filter blocks the opposite polarized light and each eye sees only one image [1][2].

Multi Picture Object (MPO) is a file format specification to store multi-picture images [3]. The included object images are constituted by a chain of still JPEGs [4] merged together in a single file with specific tags allowing the images to be handled.

A. Petrosino (Ed.): ICIAP 2013, Part I, LNCS 8156, pp. 391–399, 2013.
© Springer-Verlag Berlin Heidelberg 2013

Fig. 1. Left and right shot taken as stored in an MPO file

One of the most used scenario is the so called "stereoscopic image representations". In that scenario, MPO tags contain both information about single images (independent JPEG images) and data about the global position of each image in the real environment. 3D viewers (e.g., 3D set top box, TV, etc.) make use of such peculiarities to properly render the stereoscopic image.

A stereoscopic image is composed by two images (left and right ones) of the same subject (called target object) captured from two different perspectives by using two cameras (or a 3D camera) where the reciprocal distance corresponds to the interocular distance (Figure1). The distance and the angle between these perspectives are called *baseline length* and *convergence angle* respectively and their measures are established in order to reproduce artificially the human binocular vision [3].

The needed storage for each MPO file is approximately the space needed to store the embedded JPEGs, so that, taking into account that the involved scene is the same for both images, we have a lot of common and redundant information.

In literature there are no other similar approaches in terms of adaptive coding denoted to attack such redundancy in MPO compliant way. Perkins's approach [5] requires a transmission format which could not replace established image formats. Unlike Perkins's approach, methods for achieve an improvement without departing from JPEG encoding are available [6][7]. In [8] a joint decoding scheme is proposed in order to enhance the quality of the reconstructed image pairs independently compressed with JPEG. But some regions of the processed image cannot possibly be reconstructed by this method and, with a middle quality of JPEG compression, there are PSNR [9] decrease and some ghosting artifacts appear seldom [8].

In the proposed method, during encoding, one of the two images is JPEG compressed with a low quality rate [4]. Then, in decoding process, we apply a block based reconstruction algorithm to enhance its quality by using the redundant information of the high level encoded image (where some block matching is exploited). If no reliable matching is achieved, i.e. the difference between the images is high, the algorithm

Fig. 2. An example of MPO file that contains two stereoscopic images. Differences between left and right images are located in the margin, in the left margin of the right image are shown some objects such a window with a tree in the background, a lamp and some details of a table which are not present in the left image.

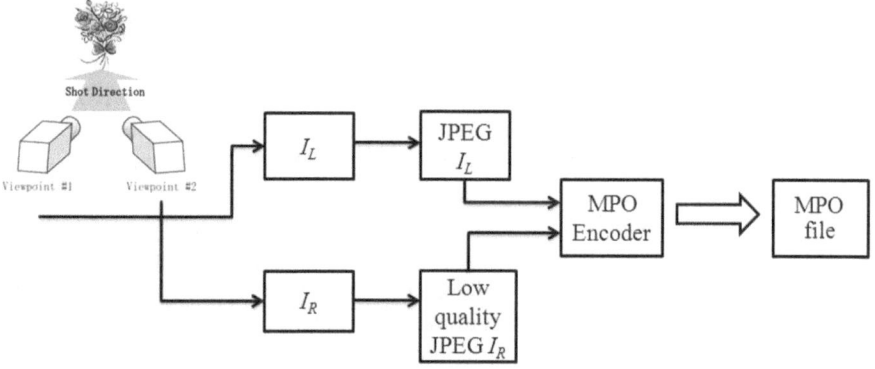

Fig. 3. Modified encoding pipeline

uses the low quality image information. This causes a losing rate which is not perceptible by the human vision system.

Experiments over a significant number of MPO images [10] confirm the effectiveness of the proposed approach. In addition, as further suggested in [11] our HVS can tolerate a slight amount of asymmetric image quality for stereo viewing (i.e. the perceived quality lies between that of the two views) thus the JPEG factor of the low quality image could be reduced even more than we've described allowing to obtain better results both in terms of bitrate saving and PSNR.

The paper is organized as follows. Section 2 describes the proposed algorithm, in Section 3 the results in terms of measured quality and bitrate saving are discussed, while a conclusion section gives direction for future works.

2   Proposed Method

In stereoscopic pair images the target object and all its main visual details appears in both cases, only a few differences are evident on the border (left and right), as shown in Figure 2. Figure 3 reports the encoding pipeline of the proposed approach where we split the processing according to two different encoding settings (Figure 3). As shown in Figure 3, we conventionally use the left image (denoted by I_L) as the high quality image and the right one (denoted by I_R) as the low quality image. Each block (of each channel) of the compressed image I_R is enhanced by using the image I_L as a reference; the same approach could be used inverting the role of the two involved images. This is a clear evidence of the robustness of the proposed method as it works without specific hypothesis related to the source images included in the original stereoscopic MPO. We limit the algorithm description to luminance plane only, but the same processing is applied also to the chromatic components.

During decoding, both images are subdivided in not overlapping blocks of equal dimensions (*NxMx3* with *N<W* and *M<H* where *W* and *H* are the dimensions of both images), the proposed scheme works block by block on each channel so it operates on blocks of *NxM* samples for each channel plane.

For the reconstruction of the generic i^{th} block of I_R, called b^R_i, the first step consists in finding the b^L_i block of I_L which best approximates the considered b^R_i block of I_R; to this purpose we consider two candidate blocks from the reference image: the first one is obtained by using a cross-correlation based template matching method and the other one is taken considering the b^L_i block which has the same position of b^R_i. This task is called *"Match similar"*. The first candidate block considered by *Match similar* is denoted by Y^{corr} and is obtained using a well known template matching method based on the computation of the normalized cross correlation (NCC) [12].

A common way to calculate the position of the pattern template t in the image f is to evaluate the NCC value $\gamma(u,v)$ at each point (u,v) for f and t, which has been shifted by u and v steps in the x and y direction respectively. The following equation gives a definition for the NCC coefficients:

$$\gamma(u,v) = \frac{\sum_{x,y}\left\{\left[f(x,y)-\bar{f}_{u,v}\right]\left[t(x-u,y-v)-\bar{t}\right]\right\}}{\left\{\sum_{x,y}\left[f(x,y)-\bar{f}_{u,v}\right]^2\sum_{x,y}\left[t(x-u,y-v)-\bar{t}\right]^2\right\}^{0.5}} \tag{1}$$

where $\bar{f}_{u,v}$ denotes the mean value of $f(x,y)$ within the area of template t shifted to (u,v) and $\bar{t}$ is the mean value of the template t. The normalization makes coefficients independent to changes in brightness or contrast of the image [12]. To reduce the computational cost of NCC we used an optimization approach which calculate the NCC using a subimage of the high quality image as search range instead of using all the image as reference image, this optimization permitted us to reduce the time required for reconstruction but it doesn't affected the results in terms of quality.

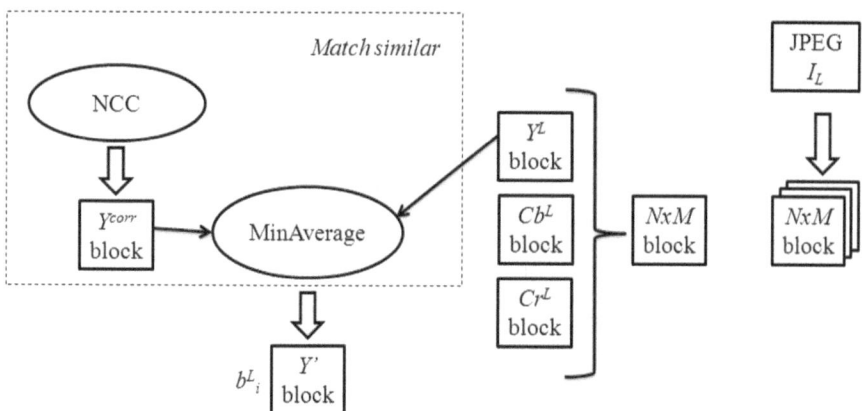

Fig. 4. Schematic view of the *Match similar* function where Y^{corr} and Y^L denotes the first and the second candidate block respectively, this function gets the candidate which minimize the average difference with respect to b^R_i.

The second candidate block, denoted by Y^L, is the *NxM* block in the luminance channel of I_L which is located in the same position of b^R_i. Figure 4 illustrates how *Match similar* works: for each candidate we calculate the average of the differences between candidate samples with respect to b^R_i samples (in the considered channel) with the following equations:

$$avg_{Y^{corr}} = \frac{1}{N \cdot M} \sum_{i=1}^{N} \sum_{j=1}^{M} \left| Y^{corr}_{i,j} - Y_{i,j} \right| \tag{2}$$

$$avg_{Y^L} = \frac{1}{N \cdot M} \sum_{i=1}^{N} \sum_{j=1}^{M} \left| Y^L_{i,j} - Y_{i,j} \right| \tag{3}$$

Match similar gets the candidate block which minimize this value (denoted by Y').

The second step of the reconstruction consists of enhancing the b^R_i block using the redundant information between the luminance block of b^R_i and the luminance block of b^L_i (i.e., the block Y' returned by *Match similar*).

This function is based on a simplified version of Kohonen reconstruction [13]. In fact, in this step we "pull" the value of some samples from b^R_i to b^L_i depending of the similarity between these two samples.

Our reconstruction uses the following equation:

$$\bar{b}^R_i(u,v) = \begin{cases} b^R_i(u,v) + \alpha \cdot \left(b^R_i(u,v) - b^L_i(u,v) \right) & if \left(b^R_i(u,v) - b^L_i(u,v) \right) < limit \\ b^R_i(u,v) & otherwise \end{cases} \tag{4}$$

where $\overline{b}_i^R(u,v)$ is the enhanced sample, $b_i^L(u,v)$ is the sample of Y' block chosen by *Match similar* and $b_i^R(u,v)$ is the sample which has to be enhanced (taken from Y block of $b^R{}_i$). The constants α and *limit* (in our cases $\alpha = 0,25$ and *limit* = 0,043) are two coefficients which determine the amount of "pulling" and the amount of "similarity" of the Kohonen reconstruction respectively [13]. This function works sample by sample and returns the enhanced block $\overline{Y}$ (Figure 5).

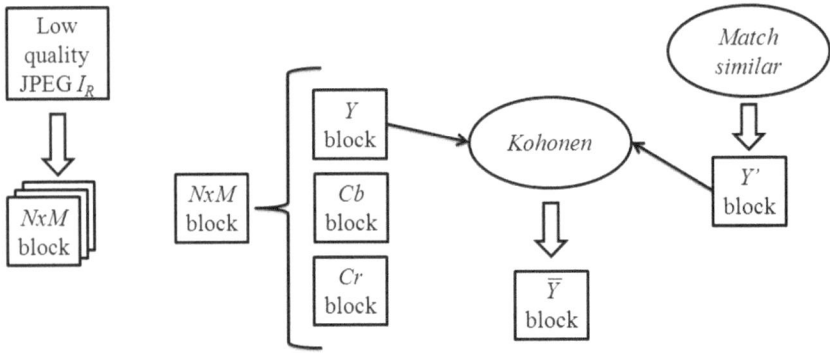

Fig. 5. $\overline{Y}$ block is the result of the reconstruction obtained enhancing Y through Kohonen reconstruction of Y using Y' as reference

3   Experimental Results

To evaluate the performances of the proposed MPO pipeline we have used for our experiments, 23 stereo MPO images conform with [3] specifications taken from [10] at various resolution sizes (1440x1080, 1444x1080, 1620x1080, 1920x1080 or 1924x1080). In our settings JPEG quality factor has been established to 85 for the high quality image and 65 (or 70) for the low quality image. Using a quality compression factor less than 65 causes an excessive degradation, viceversa using a quality compression factor greater than 70 involves a slight of bitrate saving. Reported results show compression gain in terms of total bitrate while the quality is measured considering the PSNR and the SSIM [14]. Table 1 list also the parameters used in our tests (e.g., the block size). For each MPO image and for each value of JPEG compression factor used to compress the low quality image (65 or 70) the following parameters are reported: file name, dimensions of the blocks used in decoding, the lossy (in terms of PSNR dB) measured after reconstruction and the rate of saving with respect to the same image codified using standard as high quality factor equal to 85.

Table 1. Comparison of the parameters and results of tests using 65 and 70 as JPEG quality factor for the low quality image

MPO image	NxM	Low quality 65		Low quality 70	
		Lossy (dB)	Bitrate saving	Lossy (dB)	Bitrate saving
Flowers1	360x360	2,17	40,7%	1,65	34,6%
Flowers2	360x481	2	40,6%	1,32	34,5%
Flowers3	360x481	2,76	48,7%	2,76	41,7%
Castle	360x481	2,62	38,3%	2,18	32,5%
Dorm	360x360	2,73	37,1%	2,64	31,1%
Pelion	360x481	2,34	37,6%	2,34	31,8%
Hallway	360x482	2,33	37,6%	2,33	32,1%
Statue	360x483	2,59	41,9%	2,60	35,7%
Library	360x270	1,98	38,7%	1,71	32,7%
Hall	360x360	1,66	41,1%	1,40	34,9%
Garden bridge	360x360	2,11	39,5%	1,82	33,5%
Autumn1	360x361	2,73	35,3%	2,53	29,8%
Autumn2	360x361	2,6	36,4%	2,40	30,6%
Autumn3	360x361	2,38	37,0%	2,15	31,2%
Autumn4	360x361	2,65	36,1%	2,44	30,4%
Animals1	360x240	2,16	38,8%	2,16	32,8%
Animals2	360x240	2,47	37,0%	2,18	31,3%
Cube	360x360	2,33	39,3%	2,00	33,3%
Covered	360x360	1,88	39,2%	1,73	33,4%
Garden	360x360	2,41	38,5%	2,15	32,5%
Snow	360x481	2,62	36,8%	2,45	31,2%
Tree	360x360	2,69	37,4%	2,52	31,4%
Zoo	360x240	2,67	36,9%	2,33	31,1%

Table 2. Comparison between overall results obtained using 65 or 70 as JPEG quality factor

Low quality	Average lossy (dB)	Average saving
65	2.39	38,7%
70	1.16	32,8%

Although the mean values reported in Table 2 are similar, the performances change for each image and the difference between using 65 than using 70 as JPEG quality factor appears more evident comparing this variability (see Table 1).

Compressing using 65 the bitrate saving is between 35% and 48,7% and the lossy is between 1,66 dB and 2,76 dB. Compressing using 70 the bitrate saving is between 30% and 42% and the lossy is between 1,32 dB and 2,76 dB. The visual degradation of the reconstructed images are so low that the SSIM values are all close to 1 (average value 0,93), therefore we considered the PSNR as the most significant quality metric.

4   Conclusion and Future Works

We have presented a reconstruction algorithm applied on a previous degraded image which use the redundant information between stereo pairs in order to reduce the bitrate of one view of a stereo image. The results show how this approach permits to reduce the size of MPOs with low visual degradation.

The proposed algorithm provides effective results both in terms of viewing and PSNR values, the only one disadvantage of this scheme is the computational cost caused by NCC. However other more powerful approaches [12] could be used instead of NCC in order to reduce the runtime of the reconstruction algorithm.

Furthermore, it would be interesting to evaluate the performance of the proposed algorithm by means of 3D quality metrics instead of PSNR or other 2D quality metrics, but very little research has been done to design a quality metric for 2-view case scenario. The HV3D metric [15] is a 3D video quality metric which takes into account the depth effect and the binocular properties of the HVS.

Future works will be focused on the improvement of the performances; this could be obtained using a different blocking approach or adding to the JPEG coding of the low quality image a previous compression scheme based on a different coding or a motion estimation compression based on the spatial parameters of the acquisition device or both. Moreover when the image has small fluctuation in depth it would be interesting to evaluate alternative linear template matching methods based on the asymmetry of the distortion. The encoding pipeline proposed by this scheme should be implemented directly in the digital cameras thus all alternative solutions must take into account that the latency of acquisition is a critical factor.

Acknowledgments. This work has been partially supported by STMicroelectronics - Digital Convergence Group/CSP - Catania. A related EU patent proposal has been currently submitted covering the main core of the involved processing pipeline.

References

1. Anderson, P.: Advanced Display Technologies JISC Technology and Standards Watch. JISC, Bristol (2005)
2. Wikipedia contributors, Polarized 3D system, Wikipedia The Free Encyclopedia (2013), http://en.wikipedia.org/wiki/Polarized_3D_system
3. Multi-Picture format, Camera & Imaging Products Association Standardization Committee, CIPA (2009)

4. Wallace, G.K.: The JPEG still picture compression standard. Communications of the ACM 34(4), 30–44 (1991)
5. Perkins, M.G.: Data compression of stereopairs. IEEE Transactions on Communications 40(4), 684–696 (1992)
6. Battiato, S., Mancuso, M., Bosco, A., Guarnera, M.: Psychovisual and Statistical Optimization of Quantization Tables for DCT Compression Engines. In: Proceedings of IEEE International Conference on Image Analysis and Processing, pp. 602–606 (2001)
7. Battiato, S., Bosco, C., Bruna, A., Di Blasi, G., Gallo, G.: Statistical Modeling of Huffman Tables Coding. In: Roli, F., Vitulano, S. (eds.) ICIAP 2005. LNCS, vol. 3617, pp. 711–718. Springer, Heidelberg (2005)
8. Schenkel, M.B., Luo, C., Frossard, P., Wu, F.: Joint decoding of stereo JPEG image Pairs. In: 2010 17th IEEE International Conference on Image Processing (ICIP), pp. 2633–2636 (2010)
9. Gonzalez, R.C., Woods, R.E.: Digital Image Processing, 3rd edn. Prentice Hall, Upper Saddle River (2008)
10. 3DMedia – 3D Technology and Software (2013),
    http://www.3dmedia.com/gallery
11. Pieter, S., Meesters, L., Ijsselsteijn, W.: Perceived quality of compressed stereoscopic images: Effects of symmetric and asymmetric JPEG coding and camera separation. ACM Transactions on Applied Perception (TAP) 3(2), 95–109 (2006)
12. Briechle, K., Hanebeck, U.D.: Template matching using fast normalized cross correlation. In: International Society for Optics and Photonics, Aerospace/Defense Sensing, Simulation, and Controls, pp. 95–102 (2001)
13. Kohonen, T.: The self-organizing map. Proceedings of the IEEE 78(9), 1464–1480 (1990)
14. Wang, Z., Bovik, A.C., Sheikh, H.R., Simoncelli, E.P.: Image quality assessment: From error visibility to structural similarity. IEEE Transactions on Image Processing 13(4), 600–612 (2004)
15. Banitalebi-Dehkordi, A., Pourazad, M.T., Nasiopoulos, P.: A human visual system based 3D video quality metric. In: The 2nd International Conference on 3D Imaging, IC3D, Liege, Belgium (2012)

Trajectory Similarity Measures Using Minimal Paths

Brais Cancela, Marcos Ortega, Alba Fernández, and Manuel G. Penedo

University of A Coruña,
Varpa Group, Department of Computer Science,
Campus de Elviña, s/n, Spain
{brais.cancela,mortega,alba.fernandez,mgpenedo}@udc.es
http://www.varpa.org

Abstract. Dealing with surveillance systems, large amount of distance measures are presented in order to classify both normal and abnormal behavior. Typically, techniques based in point-to-point distances are used. However, these techniques do not take into account information about the environment, like pits or restricted areas, for instance. Using a minimal path algorithm to model the usual paths, we develop new trajectory distance measures that are able to introduce information about the scene. The results obtained show promising results.

1 Introduction

Detecting human activities and behavior is a huge field of study in computer vision. One of the most active topics is related with the study of human behavior and their group relationships. This field has a special interest in surveillance systems. The idea of being able to detect abnormal behavior has being widely study. For instance, a strange movement could result in an abnormal behavior which has to be detected in order to throw an alarm.

The classical path classification methodologies are based in clustering techniques. Different configurations were used: direct [15], using techniques like k-means or fuzzy c means; agglomerative [5], where we merge clusters until we obtain the desired number; divisive [4], the top-down dual to agglomerative clustering; Hybrid [11], Graph-based [14] or Spectral [10]. All the techniques mentioned above are limited, since they require routes with the same number of samples to compute the clusters, and they are not easy to update along time. Suppose, for example, that an usual target is interrupted because of an object placed in the track. A new cluster is created with the new routes, but the previous cluster still remains in the system. A target which decides to jump that object, which clearly is an abnormal movement, will be declared as normal behavior because of the existing cluster. More recent techniques include the use of nonparametric Bayesian models [20], [19] or use models to predict the motion behavior [8].

To overcome the clustering issue, we developed a new system that can be easily updated [6]. We use a minimal path algorithm to model the abnormal behavior. In normal situations, a target tends to choose the route that costs the least time to reach its desired destination. Thus, trajectories that highly differ from this "ideal" route are marked as abnormal. A new metric were presented, based in a distance map algorithm, which requires a high computational time.

A. Petrosino (Ed.): ICIAP 2013, Part I, LNCS 8156, pp. 400–409, 2013.

In this work we present a new trajectory distance measure that can be used in surveillance systems to detect abnormal behavior. This technique uses the properties of the minimal path algorithm to obtain a metric without increasing the computational time, solving the distance map technique disadvantage. This paper is organized as follows: section 2 shows different trajectory measure algorithms; section 3 describes our method; section 4 shows some experimental results and section 5 offers conclusions and future work.

2   State of the Art

We define a trajectory as a collection of the positions a target reach along its way. So, we have a collection of N positions defining a target route. To compare it, there exist in the literature different approaches for trajectory distance metrics. Methods based in classical measure techniques were used, like euclidean distance [10] or principal component analysis (PCA) [3]. However, these methods obtain poor results, requiring trajectories with the same size to be compared. Other attempts, like the modified Hausdorff distance [2], does not take into account the order into the trajectory points.

Thus, distance measure techniques have to be able to compare unequal length trajectories, while taking into account the route orientation. In [12], Keogh et al. presented the Dynamic Time Warping (DTW) technique. Basically, this method tries to find a time warping that minimizes the distance between two different trajectories. It can be used with trajectories with different sizes. Buzan et al. [5] introduced a similar idea, the Longest Common Subsequence (LCSS). It can also be used with unequal length data, becoming more robust to noise. The reason is that not all the trajectory points need to be matched. Similar to these methods, Piciarelli and Foresti (PF) [17] uses a dynamic time warping window, which is increased along time, that is, the maximum error allowed is low at the starting trajectory point, becoming larger while we are reaching the end. The performance of these metrics were tested in [16].

Although these methods can deal with the problems mentioned before, they all present a major issue for real domains: they do not take into account information about the environment. For instance, in Fig. 1, we can see two examples of situations these techniques cannot correctly address. In the left, two parallel trajectories are defined. Using the similarity measure it is easy to conclude that both trajectories are similar. However, the red route is produced by a counterclockwise car, which clearly is an abnormal behavior. On the right image, both the red and the green route are similar to the blue one, but the red one crosses the central reservation.

To solve situations like the previously illustrated one, we develop a methodology that is able to introduce information about the environment [6], based in the geodesic active contours [7] and the level set theory [13]. A modification of the minimal path approach using geodesic active contours performed by Cohen and Kimmel [9] is provided. In this work, starting at any given point p_0, a minimal path map over an image is obtained, with a $\mathcal{O}(N \log N)$ complexity, being N the number of pixels in the image. To do that, a potential image P is created, which includes information about the environment. Later, the potential used in the algorithm is defined as

$$\tilde{P}(p) = \omega + P(p), \tag{1}$$

(a)   (b)

Fig. 1. These trajectories are defined as similar using classical distance measure techniques, while including the scene information the red routes are clearly abnormal

being ω the regularization term. This potential has to be defined so $\tilde{P} > 0$ and $P \geq 0$, meaning that $\omega > 0$. Values close to 0 implies pixels that are easy to reach, while higher values means the opposite.

Having this potential, the surface of minimal action U can be created. This surface assigns a value to each pixel in the image, which corresponds with the least cost that takes to reach that pixel, starting at the initial point p_0. To create that surface, the Fast Marching Method (FMM) proposed by Sethian is used [18]. An ordered upwind scheme is used, updating the cost of a given pixel $U_{i,j} = u$ following the equation

$$(\max\{u - U_{i-1,j}, u + U_{i+1,j}, 0\})^2 + (\max\{u - U_{i,j-1}, u + U_{i,j+1}, 0\})^2 = \tilde{P}_{i,j}^2. \quad (2)$$

Once the minimal path between the initial and the final point is obtained, we can evaluate the real path against this minimal cost approach. Note that the final point mentioned before is not necessarily the point where we stop to track a target. The minimal path can be computed each frame a target is detected, using the last position detected as final path. This is an advantage with respect to other methodologies, since the abnormal behavior can be detected as soon as it occurs.

So, in order to evaluate the real path against the minimal path approach, a distance map image is created. Having this minimal path algorithm in mind, it is easy to see that, if using as initial points all the points in a given trajectory $C(s)$ instead of the point p_0 in the original algorithm, the surface of minimal action U we obtain is a distance map, where the value $U_{i,j}$ of each pixel is the minimum cost that takes to reach that point, starting at any point that belongs to the trajectory $C(s)$. Note that, the distance between two consecutive pixels is defined by the potential $\tilde{P}$. In [6], a discussion about the potential image is provided.

The problem of this method is the computational cost needed to create the distance image. Since the algorithm is, in essence, equivalent to the minimal path approach, the cost to obtain the map is $\mathcal{O}(N \log N)$. Thus, our goal is to create a new trajectory distance measure that is able to include information about the environment, taking into account the properties of the minimal path approach, but avoiding unnecessary extra computations.

3   Minimal Path Metrics

Our goal is to use the properties of the Fast Marching Method used to create the minimal action surface U to detect if a trajectory is abnormal. That surface of minimal action has a convex like behavior, in the sense that, starting at any given point p, and following the gradient descent direction in U, we always converge to the initial point p_0. This means that the minimal action surface U has one local minimum, which is $U(p_0) = 0$. Furthermore, this geodesic active contour-based technique is consistent with the continuous problem, in the sense that the solution provided by the FMM becomes closer to the exact solution while reducing the grid. This property allows this algorithm to avoid the 'metrication error', which appears in the classical graph search algorithms, like A* or F*.

This property is crucial to present our methodology. If, for instance, we introduce the potential image $\tilde{P} = \tau > 0$, and we compute the minimal action surface U, starting at any given point within the grid p_0, we obtain a distance map, as the value at any point p, U_0, is the distance between this point with respect to the initial one p_0. And, contrary to the graph search algorithms, since the FMM is consistent with the continuous case, the solution we obtain is close enough to the euclidean distance. Note that this system is isotropic, having no information about directional forces. The cost to reach a point is always the same. no matter which direction the front-propagation come by.

However, using a potential like the mentioned before causes the system to lose the information about the environment, since this potential is constant all over the space. Fortunately, it is possible to compute the distance map regardless the type of potential used. We have to define another minimal action surface D, which is going to be updated using the equation

$$
D(p) = \begin{cases} \dfrac{D(p_a) + D(p_b) + \sqrt{2\tau^2 - (D(p_b) - D(p_a))^2}}{2} & \text{if } \tilde{P}(p) > (U(p_b) - U(p_a)) \\ D(p_a) + \tau & \text{otherwise} \end{cases} ,
$$

(3)

where p_a and p_b are the neighbors used in the Eq. 2 to update the surface of minimal action $U(p)$. satisfying that $U(p_a) \leq U(p_b)$, and τ is the distance between two neighbor pixels. Typically, $\tau = 1$.

Thus, while computing the minimal action surface U we can compute the distance map D without any substantial cost increment. In algorithm 1 a pseudocode of our FMM implementation is presented.

Once we have the distance map related to a trajectory, we are able to perform a similarity measure that can detect abnormal behavior. For notation, we have a trajectory

$$
Tr = \{p_0, p_1, \ldots, p_M\},
$$

(4)

where each point represents positions that are reached for the target. Note that this is an ordered sequence of events, where the position p_i is reached before the position p_{i+1}. These trajectories can be obtained using tracking techniques, being p_0 the first time a target is tracked. Since, as we demonstrate in [6], a target usually tends to follow the

Algorithm 1. Distance Surface Method

Definitions:

- *Alive* set: points of the grid for which U has been computed and it will not be modified.
- *Trial* set: next points in the grid to be examined (4-connectivity) for which an estimation of U is computed using the points in *alive* set.
- *Far* set: the remaining points of the grid for which there is not an estimate for U.

Initialization:

- For each point in the grid, let $U_{i,j} = \infty$ (large positive value). Put all points in the *far* set.
- Set the initial point p_0 to be zero: $U_{p_0} = 0$, $D_{p_0} = 0$, and put it in the *trial* set.

Marching loop:

- Select $p = (i_{min}, j_{min})$ from *trial* with the lowest value of U.
- If p is equal to p_1 being p_1 the final point then we finish.
- Else put p in *alive* and remove it from the *trial* set.
- If $\tilde{P}(i_{min}, j_{min}) < \tau$, for each of the 4 neighboring grid points (k, l) of (i_{min}, j_{min}):
  - If (k, l) belongs to *far* set, then put (k, l) in *trial* set.
  - If (k, l) is not in *alive* set, then set $U_{k,l}$ with Equation 2
  - and set $D_{k,l}$ with Equation 3.

path that cost less effort to reach the goal, we can conclude the relation between the minimal path distance with respect to a normal trajectory behavior is close to 1, that is,

$$\frac{\sum_{i=2}^{M} d(p_{i-1}, p_i)}{D(p_M)} \approx 1, \tag{5}$$

where $d(p_{i-1}, p_i)$ is the distance between two consecutive points in the trajectory. To compute this distance it is also possible to use the minimal path approach, starting in p_{i-1} instead of p_1. However, the points contained in the route are usually close together, meaning the euclidean distance often results in a good approach.

Having this in mind, different metrics are presented for detecting abnormal behavior. All these metrics are based in two different equations. The first one tries to obtain the relation between the target route and its associated minimal path. We called it Minpath Relation (MR), and is defined by

$$MR(p_N) = \left(\frac{\sum_{i=2}^{N} d(p_{i-1}, p_i)}{D(p_N)} - 1 \right)^2, \tag{6}$$

where $1 < N \leq M$. The second one tries to detect local variations in the MR metric. We called it Local Minpath Relation (LMR), and is defined by

$$LMR(p_N) = \left(\frac{d(p_{N-1}, p_N)}{D(p_N) - D(p_{N-1})} - 1 \right)^2. \tag{7}$$

In both metrics, values close to 0 mean the path is correct, while higher values could indicate an abnormal behavior.

4   Experimental Results

In our experiments we have used a dataset publicly available, BARD [1], in order to test the methodology. Several trajectories are developed over an intersection scene, resulting in more than 15000 trajectory points. In the experiments we decided to use a fixed potential image, that can be shown in Fig. 2-(a). We consider the grass areas as forbidden areas, using high values in the potential to model them.

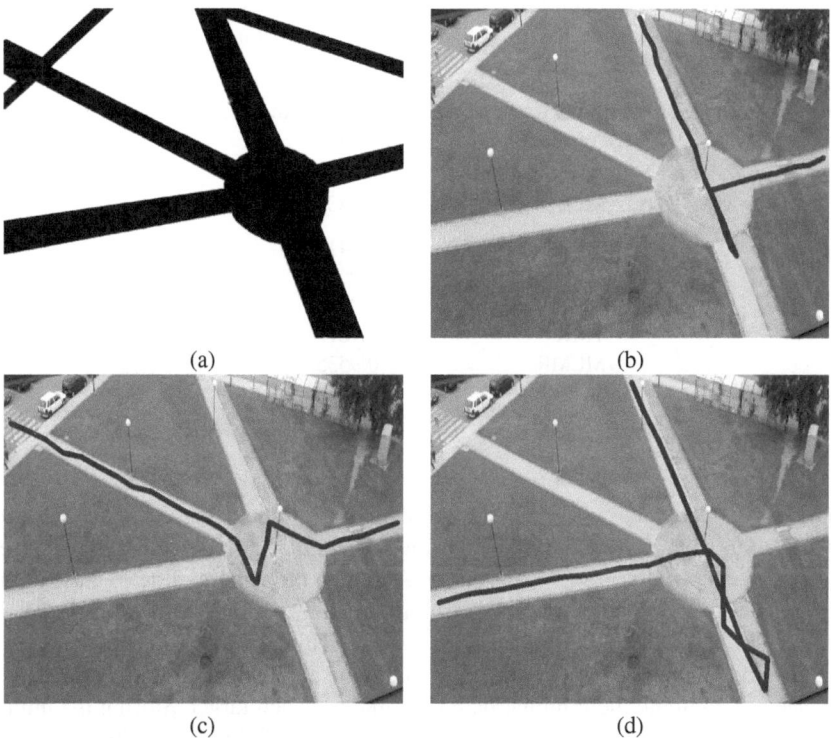

(a)   (b)

(c)   (d)

Fig. 2. Path trajectory examples. (a) Potential image used to computed the minimal path. (b, c, d) Abnormal behavior examples.

In first place, we decided to test different approaches of the metrics explained before. We decided to test, according to the MR equation, both MR, the Mean MR

$$MMR(p_N) = \frac{\sum_{i=2}^{N} MR(p_i)}{N}, \qquad (8)$$

its variance

$$VMR(p_N) = \frac{\sum_{i=2}^{N} MR(p_i)^2}{N} - MMR(p_N)^2 \tag{9}$$

and its top value

$$TMR(p_N) = \max_{i \in [2..N]} p_N. \tag{10}$$

When dealing with the LMR equation, the mean (MLMR), the variance (VLMR) and the maximum (TLMR) are used. As mentioned earlier, values close to 0 mean there is no abnormal behavior in the route. Thus, we can detect abnormal situations by simply introducing a threshold. In Table 1 we show the results obtained using these metric over the BARD dataset. Looking at the ROC Area Under the Curve values obtained, we can conclude the metrics exposed obtain good results, especially MMR and TMR measures. Most of the errors achieved by these metrics are related with the difficulty of annotate the correct moment where a normal route starts being erratic. As a result, it is possible that some errors may occur because of bad manually annotations.

Table 1. Minpath Metric Results. Using the ROC Area Under the Curve metric, we found that all the techniques achieve good results.

Metric	ROC AUC
MR	0.9417
MMR	0.9691
VMR	0.9578
TMR	0.9673
MLMR	0.9535
VLMR	0.9448
TLMR	0.9512

Having these results, we conclude the MMR technique achieve the better results. However, this comparison is made by using a potential image with the same size of the video frame, in this case, (576×720). Although the computation of these metrics is equivalent to the computation of the minimal path method, $\mathcal{O}(N \log N)$, the time needed is too high to be used in real-time systems. This is not a problem when dealing with a fixed potential, where the potential cannot be updated along frames, because we only need to compute the minimal path one time per each target. Storing the minimal action surface U, we can obtain the MMR metric with a $\mathcal{O}(1)$ complexity.

However, in more complex scenarios, we need to use a potential that is going to be modified along time. In this case, we need to compute the minimal action surface U every time we want to obtain the MMR metric. Thus, we need to reduce the potential image size in order to reduce the computational time. In Fig. 2 we can see the results obtained by reducing the image. As we can see, similar results are obtained if we reduce the potential image to (72×90), allowing our method to speed up the response without decreasing significantly the performance of the metric.

In order to compare the results of the metrics mentioned before against the baseline techniques, we use the ROC curve. In Fig. 3 we compare our metrics with the baseline

Table 2. Minpath Metric Results. Using the ROC Area Under the Curve metric, we found that all the techniques achieve good results.

Decreasing factor	ROC AUC						
	MR	**MMR**	**VMR**	**TMR**	**LMR**	**VLMR**	**TLMR**
f = 1 (576×720)	0.9417	**0.9691**	0.9578	0.9673	0.9535	0.9448	0.9512
f = 2 (288×360)	0.9488	**0.9646**	0.9519	0.9619	0.9493	0.9411	0.9471
f = 4 (144×180)	0.9574	**0.9581**	0.9424	0.9545	0.9453	0.9379	0.9438
f = 8 (72×90)	**0.9528**	0.9480	0.9224	0.9395	0.9308	0.9267	0.9326
f = 16 (36×45)	0.8965	**0.9124**	0.8678	0.8909	0.8646	0.8695	0.8712
f = 32 (18×22)	0.7912	**0.8598**	0.8179	0.8371	0.7214	0.7290	0.7327

methods. Since our method clearly outperforms methods that need samples with the same size to perform the computation, we decide to compare our method against more powerful techniques, like the previously mentioned PF, LCSS and DTW. Moreover, we introduce our previous distance map based techniques. We can conclude that the MMR metric clearly outperforms the baseline methods, except the Weighted Distance Map. However, the MMR can obtain similar results while avoiding the computation of the distance map image, which has a $\mathcal{O}(N \log N)$ complexity, meaning our new method is more suitable for being used in real-time applications.

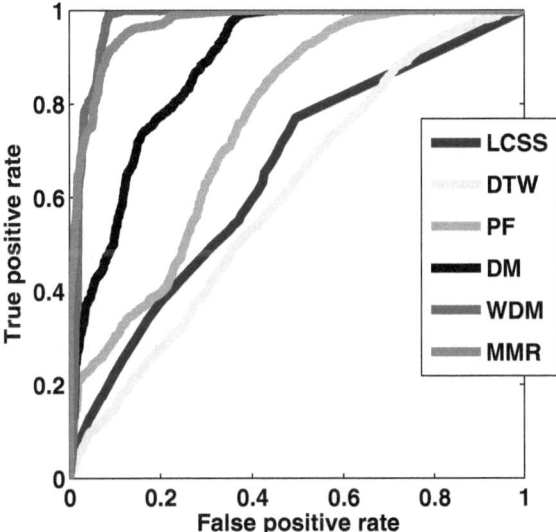

Fig. 3. ROC curve. Our new metric outperforms the baseline methods, with the exception of the Weighted Distance Map. However, the computational cost allows our new method to be more suitable to be used in real-time environments.

5   Conclusions

In this paper we present new metrics to detect abnormal behavior using target trajectories. Using minimal path techniques, which are proven to be useful detecting abnormal behavior, we develop new metrics that make use of the surface of minimal action properties, without increasing significantly the computation of such minimal paths. A comparison between different new metrics is performed, where the MMR obtains the better results. The results also show the potential image can be substantially reduced, having no significantly yield loss. Another comparison against baseline methods demonstrates our methods outperform classical abnormal behavior metrics, while obtains similar results to the recent Weighted Distance Map metric, which is proven to be computationally expensive.

In future works, we plan to introduce both direction and speed in our minimal path algorithm, allowing our system to have more information, in order to detect abnormal behavior that is not provided in our recent algorithm, like sudden speed changes or movements in the opposite direction of the usual routes.

Acknowledgments. This paper has been partly funded by the Consellería de Industria. Xunta de Galicia through grant contract 10TIC009CT, and by the Ministerio de Ciencia e Innovación through grant contract TIN2011-25476.

References

1. BARD, behavioral analysis and recognition dataset, http://www.varpa.org/bard/
2. Atev, S., Masoud, O., Papanikolopoulos, N.: Learning traffic patterns at intersections by spectral clustering of motion trajectories. In: 2006 IEEE/RSJ International Conference on Intelligent Robots and Systems, pp. 4851–4856. IEEE (2006)
3. Bashir, F., Khokhar, A., Schonfeld, D.: Object trajectory-based activity classification and recognition using hidden markov models. IEEE Transactions on Image Processing 16(7), 1912–1919 (2007)
4. Biliotti, D., Antonini, G., Thiran, J.P.: Multi-layer hierarchical clustering of pedestrian trajectories for automatic counting of people in video sequences. In: Proceedings of the IEEE Workshop on Motion and Video Computing (WACV/MOTION 2005), vol. 2, pp. 50–57. IEEE Computer Society, Washington, DC (2005)
5. Buzan, D., Sclaroff, S., Kollios, G.: Extraction and clustering of motion trajectories in video. In: Proceedings of the 17th International Conference on Pattern Recognition, ICPR 2004, vol. 2, pp. 521–524. IEEE (2004)
6. Cancela, B., Ortega, M., Penedo, M., Novo, J., Barreira, N.: On the use of a minimal path approach for target trajectory analysis. Pattern Recognition 46(7), 2015–2027 (2013)
7. Caselles, V., Kimmel, R., Sapiro, G.: Geodesic active contours. International Journal of Computer Vision 22(1), 61–79 (1997)
8. Chen, Z., Wang, L., Yung, N.H.: Adaptive human motion analysis and prediction. Pattern Recognition 44(12), 2902–2914 (2011)
9. Cohen, L., Kimmel, R.: Global minimum for active contour models: A minimal path approach. International Journal of Computer Vision 24(1), 57–78 (1997)
10. Hu, W., Xie, D., Fu, Z., Zeng, W., Maybank, S.: Semantic-based surveillance video retrieval. IEEE Transactions on Image Processing 16(4), 1168–1181 (2007)

11. Karypis, G., Han, E.H., Kumar, V.: Chameleon: Hierarchical clustering using dynamic modeling. Computer 32(8), 68–75 (1999)
12. Keogh, E., Pazzani, M.: Scaling up dynamic time warping for datamining applications. In: Proceedings of the Sixth ACM SIGKDD International Conference on Knowledge Discovery and Data Mining, pp. 285–289. ACM (2000)
13. Kimmel, R., Amir, A., Bruckstein, A.: Finding shortest paths on surfaces using level sets propagation. IEEE Transactions on Pattern Analysis and Machine Intelligence 17(6), 635–640 (1995)
14. Li, X., Hu, W., Hu, W.: A coarse-to-fine strategy for vehicle motion trajectory clustering. In: The 18th International Conference on Pattern Recognition, ICPR 2006, vol. 1, pp. 591–594 (2006)
15. Morris, B., Trivedi, M.M.: An adaptive scene description for activity analysis in surveillance video. In: ICPR, pp. 1–4. IEEE (2008)
16. Morris, B., Trivedi, M.M.: Learning trajectory patterns by clustering: Experimental studies and comparative evaluation. In: CVPR, pp. 312–319. IEEE (2009)
17. Piciarelli, C., Foresti, G.: On-line trajectory clustering for anomalous events detection. Pattern Recognition Letters 27(15), 1835–1842 (2006)
18. Sethian, J.: A fast marching level set method for monotonically advancing fronts. Proceedings of the National Academy of Sciences 93(4), 1591–1595 (1996)
19. Wang, X., Ma, K.T., Ng, G.W., Grimson, W.E.L.: Trajectory analysis and semantic region modeling using nonparametric hierarchical bayesian models. International Journal of Computer Vision 95(3), 287–312 (2011)
20. Wang, X., Ma, X., Grimson, W.E.L.: Unsupervised activity perception in crowded and complicated scenes using hierarchical bayesian models. IEEE Transactions on Pattern Analysis and Machine Intelligence 31(3), 539–555 (2009)

Structured Multi-class Feature Selection for Effective Face Recognition

Giovanni Fusco, Luca Zini, Nicoletta Noceti, and Francesca Odone

DIBRIS - Università di Genova
via Dodecaneso, 35
16146-IT, Italy
{Giovanni.Fusco,Luca.Zini,Nicoletta.Noceti,Francesca.Odone}@unige.it

Abstract. This paper addresses the problem of real time face recognition in unconstrained environments from the analysis of low quality video frames. It focuses in particular on finding an effective and fast to compute (that is, sparse) representation of faces, starting from classical Local Binary Patterns (LBPs). The two contributions of the paper are a new formulation of Group LASSO for structured feature selection (MCGroup LASSO) to cope directly with multi-class settings, and a face recognition pipeline based on a representation derived from MC-GrpLASSO. We present an extensive experimental analysis on two benchmark datasets, MOBO and Choke Point, and on a more complex dataset acquired in-house over a large temporal span. We compare our results with state-of-the-art approaches and show the superiority of our method in terms of both performances and sparseness of the obtained solution.

Keywords: Regularized feature selection, Group LASSO, multi-class feature selection, face recognition.

1 Introduction

With the diffusion of video-surveillance systems on one side and mobile devices equipped with video cameras on the other side, it is more and more common to address the problem of recognizing faces in video frames acquired in cluttered and unconstrained environments. These settings pose many challenges: low-quality images, relatively small faces, motion blur, variable lighting conditions, subjects in non standard poses. Moreover, these domains usually call for real-time performances, thus the computational efficiency of the methods plays a key role.

In this paper we present an efficient face recognition method, which learns a very sparse face representation from data by means of structured regularized feature selection. We are interested in unconstrained settings where people are free to move as they like and are captured by conventional video-surveillance cameras or by mobile devices. We start from an overcomplete description of face images based on Local Binary Patterns (LBP) [18,1], which have been shown effective for analyzing textures and patterns in low quality images. LBPs are

A. Petrosino (Ed.): ICIAP 2013, Part I, LNCS 8156, pp. 410–419, 2013.

computed at different scales and aspect ratios, then are subject to a feature selection step. In this way, not only we obtain an effective face representation, but we also improve the computational efficiency of face recognition at run time. With this aim, we propose a new formulation of the Group LASSO functional [17] – we call *MC-GrpLASSO* – to cope directly with a multi-class setting while capturing the block-structure of LBP. The new formulation allows us to select groups of features that simultaneously discriminate among all the identities. On top of this, we build a modular and efficient face recognition pipeline, analyzing a face image and associating with it one among N previously learnt identities.

Many works in the literature cope with the problem of selecting features in the face recognition domain. Often feature selection is based on Adaboost: in [18] the authors start off computing an overcomplete set of LBP histograms computed in different regions of face images. Then, they use Adaboost to select the most significant features. The work in [6], instead, introduces the Jensen-Shannon Boosting (JSBoost) algorithm, a modification of AdaBoost based on the Jensen-Shannon (JS) divergence, that is shown to provide a more appropriate measure of distance between examples belonging to two different classes. In [14], the authors consider Gabor features to extend Adaboost by incorporating mutual information. They achieve a lower training error rate with respect to the standard implementation. The Gabor features are also adopted in [11], where Real Adaboost is introduced to perform the selection wavelets. A genetic based approach is proposed in [13], where the selection of the LBP regions is based on Genetic & Evolutionary Computing (GEC). Many works look for identity-based descriptions [4,7] and are thus more specifically tailored for the face authentication problem. Here we focus on finding a universal description, common to all identities, with an advantage in terms of scalability. To this purpose our method is specifically designed for multi-class settings. Few works in the literature are, up to now, devoted to multi-class feature selection, in particular for face recognition. We mention [3] which extends Recursive Feature Elimination (RFE) to the multi-class case.

The remainder of the paper is organized as follows. Sec. 2 describes the multi-class feature selection method we propose, while Sec. 3 focuses on the entire face recognition pipeline. The experimental analysis on the different datasets is detailed on Sec. 4, while Sec. 5 is left to a final discussion.

2   Regularized Multi-class Feature Selection

In this section we describe the multi-class feature selection method we propose. Let us first formulate the *feature selection problem*: given a training set $(\mathbf{x}_i, y_i), i = 1, \ldots, n$, with $\mathbf{x}_i \in X \subseteq R^m$ $y_i \in R$, and a dictionary $\mathcal{D} = (\phi_j)_{j=1}^{p}$ which is a collection of atoms or features, we consider a generalized linear formulation of the input-output relationship

$$\sum_{j=1}^{p} \phi_j(\mathbf{x}_i)\beta_j = y_i \qquad (1)$$

or, in matrix form $\Phi\beta = \mathbf{y}$, where Φ is the feature matrix defined as $\Phi_{ij} = \phi_j(\mathbf{x}_i)$ of size $n \times p$, $\mathbf{y} \in \{-1, +1\}^{n \times 1}$ is the output vector, and $\beta \in \mathbb{R}^{p \times 1}$ is the vector of weights. The goal of feature selection is to find a *sparse* β which approximates well the input-output relationship.

Group LASSO. It is an extension of LASSO [15] to select groups of features, rather than considering each feature independently [17]. In several application domains, in fact, a prior on the structure of each input datum is available, and may be profitably inserted in the feature selection step. It can be formulated as a regularized minimization problem:

$$\beta^* = \arg\min_{\beta} \left(\|\mathbf{y} - \Phi\beta\|_2^2 + \tau \sum_{g=1}^{G} \|\beta_{I_g}\|_2 \right) \tag{2}$$

where $g = 1, \ldots, G$ is an index referring to a feature group, while $I_g \subseteq [1...p]$ is the set containing the positions in the input datum of the features of group g. The first term of the functional is a *data fidelity term* while the second one is a *penalty term* which enforces the selection of features in groups. The regularization parameter τ can be tuned to obtain solutions with a different level of sparsity.

Multi-class Group LASSO. We now re-write the functional (2) to deal with the simultaneous selection of groups of meaningful features for a multi-class problem. Let us consider N binary problems $\Phi\beta_c = \mathbf{y}_c$, $c = 1 \ldots N$. Similarly to above, Φ is the features matrix of size $n \times p$, $\mathbf{y}_c \in \{-1, +1\}^{n \times 1}$ is the output vector, containing 1 in correspondence of examples of class i and -1 elsewhere. Finally, $\beta_c \in \mathbb{R}^{p \times 1}$ is the weight vector for the class c. We formulate the Multi-Class Group LASSO (MC-GrpLASSO) as follows:

$$B^* = \arg\min_{B} \left(\|Y - \Phi B\|_2^2 + \tau \sum_{g=1}^{G} \|B_{I_g}\|_F \right). \tag{3}$$

Y is a $n \times N$ matrix composed as $[\mathbf{y}_1 \mathbf{y}_2 \ldots \mathbf{y}_N]$, and, similarly, B is a $p \times N$ such that $B = [\beta_1 \beta_2 \ldots \beta_N]$. F refers to the Frobenius norm. In this case sparsity enforces the selection of features in groups common to all the N classes. Roughly speaking, the method selects the features that best discriminate among all the N classes simultaneously.

3   The Proposed Pipeline

In this section we describe our approach to face recognition. We assume we have in input a training set $\{\mathbf{x}_i\}_{i=1}^n$ of n face images of N different identities. We first represent each image according to an appropriate dictionary of features based on Local Binary Patterns (LBP). Then, we apply MC-GrpLASSO to obtain a set of features groups meaningful for all the N classes. Finally, we train a multi-class classifier on the training set of images represented with the selected groups only.

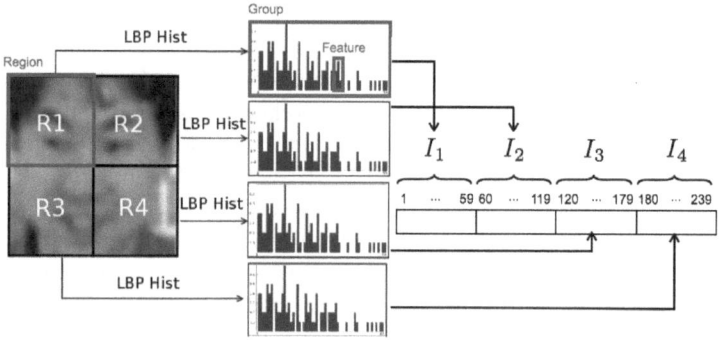

Fig. 1. A visualization of our face descriptors, a concatenation of LBPs

This allows us to obtain, at run time, a very fast representation and classification of each face image.

Construction of the Feature Matrix. In this work we adopt an overcomplete dictionary of Local Binary Pattern (LBP) descriptors, whose effectiveness in face recognition is well known. Similarly to [18], we consider a grid of overlapping regions having different scale and aspect ratio (faces images are all rescaled to the size 40×40 pixels). We then extract uniform 8-bits LBPs [10] and quantize their values with 59-bins histograms, accordingly to the original work. Figure 1 describes how to obtain the feature vector in a simplified situation, with no overlapping regions. The final overcomplete description consists of 841 LBPs, thus each training image $\mathbf{x}_i$ is described by $841 \times 59 = 49,619$ features (which is also the number of columns in the features matrix Φ). For numerical reasons matrix Φ needs to be normalized so that each column has mean 0 and variance 1.

Multi-class Feature Selection. We specialize the functional (3) to our case by grouping the features according to the image region they are associated with (see Figure 1). To find a solution to the optimization problem we adopt an iterative scheme as in [9] specialized for the multi-class case as follows:

$$\hat{B}^p_{\cdot,c} = B^{p-1}_{\cdot,c} + \frac{1}{\sigma} \Phi^T \left(Y_{\cdot,c} - \Phi B^{p-1}_{\cdot,c} \right) \quad \forall c \in [1, N] \tag{4}$$
$$B^p = S_{\frac{\tau}{\sigma}} \left(\hat{B}^p \right)$$

where σ is the iteration step size, which we set proportional to the highest eigenvalue of matrix $\Phi^\top \Phi$, to achieve an optimal convergence to the solution of (3) [9]. S is the soft-thresholding operator that acts on all features of a group $(S_{\frac{\tau}{\sigma}}(\hat{B}))_{i,c} = (\|\hat{B}_{I_g,\cdot}\|_F - \frac{\tau}{\sigma})_+ \hat{B}_{i,c}$. The iterative procedure can be initialized with a matrix B^0 of zeros. Once we have a solution B we select the regions associated with the rows without zero values: in this way the selected features are meaningful to *all* the N classes.

Recognition. Once we have found a sparse and meaningful representation for our multi-class problem, we train a multi-class classifier based on SVM and a *Winner-Takes-All* strategy [8]. Notice that it would have been possible to classify data directly via the solution of the MC-GrpLASSO, but the performance obtained with SVMs are usually higher because of the well known shrinkage effect of the weights of LASSO methods [2]. Thus, we train N One-vs-All binary SVM classifiers where each image $\mathbf{x}_i$ is represented by mapping it on the sub-set of selected regions, by a projector $\pi(\mathbf{x}_i) = \{\phi_j(\mathbf{x}_i)|B_{j,\cdot} \neq 0\}$. Given the binary classifier of class c, the negative examples are randomly sampled from all the classes $j, j = \ldots N, j \neq c$.

For a new datum $\mathbf{x}$, the resulting global classifier is then formulated as a combination of the N discriminant functions $h_1 \cdots, h_N$ of the binary classifiers as

$$h(\mathbf{x}) = \arg \max_{c=1\ldots N} \{h_c(\pi(\mathbf{x}))\}. \tag{5}$$

4 Experimental Analysis

In this section we present the experimental evaluations of the proposed pipeline, with particular emphasis on representation (feature selection) and recognition.

Datasets. The analysis is performed on three different datasets: two public benchmarks – *MOBO* [5] and *Choke Point* [16] – and a in-house dataset, *R309*. The *MOBO* dataset consists of low-resolution videos of 20 subjects, while they perform 4 different actions. It is characterized by small variations in the lighting condition and subject poses. The *Choke Point* dataset is characterized by a higher variability of image quality, appearance and lighting conditions. It includes videos of 29 subjects walking through two portals, acquired by an array of three cameras. The *R309* dataset (see Fig. 2) is a collection of indoor videos acquired in unconstrained conditions: the pose and the behavior of the volunteers are entirely unconstrained and very natural. The recordings have been made under various lighting conditions and in two very different environments. R309 includes 12 people with an average of 1200 frames for each identity acquired over a period of time of 7 months. It thus allows us to evaluate the robustness of the methods with respect to variations occurring over time.

Fig. 2. *R309* dataset: same subject under different poses and lighting conditions

Experimental Protocol. Each dataset has been divided in training, validation, and test, in the proportions specified later for each dataset. The validation
set is used for tuning the parameters of the classifiers (that is, the regularized
parameter of each SVM). The results reported in the tables have been obtained
on the test set.

Assessment of the Selected Features. The solution of the functional (3)
is a weight matrix B with a degree of sparsity that depends on τ. The choice
of the parameter is carried out on the validation set and guided by the aim of
obtaining a sparse solution producing a high recognition rate. The matrix has
a characteristic structure as the one shown in Fig. 3 (left): MC-GrpLASSO selects groups meaningful for all classes, producing a stripe structure (the selected
groups have non-zero weights along all the classes). It is interesting to compare
the structure of this matrix with the one derived by a 1vsALL architecture. In
this case we solve N binary structured feature selection problems as in Eq. (2)
obtaining N solutions β_i, $i = 1, \ldots, N$. Fig. 3 (right) shows a matrix obtained
by combining the N solutions (one per column). Notice how the β_i do not relate
to one another.

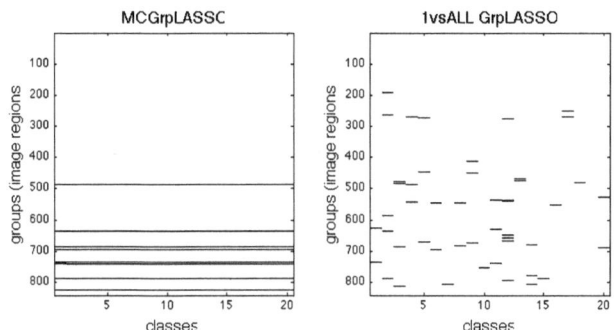

Fig. 3. (MOBO dataset) $G \times N$ binary maps: each entry is associated with an image
region (row) with respect to a given class (column). Black entries are the selected
groups.

With a 1vsALL approach we need to find a way to combine the N β_is; classical
choices are intersection (keep groups selected by all binary problems) and union
(keep groups selected by at least one problem). In this case parameter tuning
becomes more complex: we need to choose an appropriate τ for each binary
problem, while controlling the cardinality and the performances of the overall
problem. In particular, in the case of intersection, it is often the case that the
overall solution is empty; in this case a voting rule may be adopted. Fig. 4
highlights the features selected by MC-GrpLASSO (bottom) and 1vsALL with
union (top). The two representations achieve similar performances (see Table
1), but MC-GrpLASSO does it with less than 1/4 of the regions needed by the
1vsALL method.

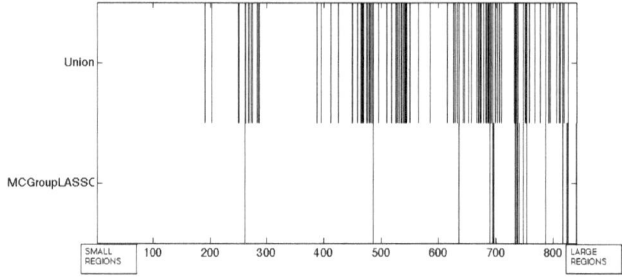

Fig. 4. (MOBO dataset) Binary maps of the obtained representations: in black are marked the indices of the groups selected by 1vsALL union (top) and MC-GrpLASSO (bottom). On the left smaller regions, on the right larger regions.

Finally, Fig. 5 gives a qualitative impression of the kind of features produced by our method, as it compares the 10 most meaningful features selected by different approaches to multi-class regularized feature selection: MCGroup LASSO, 1vsALL with union, and LASSO. The latter have been obtained by reformulating the N class problem with a simpler binary problem of unstructured features [7,4]: $\mathbf{x}_i$s are obtained by computing region-wise χ^2 similarities between LBP histograms of two different image pairs (positive examples are derived from image pairs of the same class). Then a region is associated with a single feature and unstructured feature selection (for instance with Adaboost or LASSO) may be applied. From the analysis of Fig. 5 it is apparent that structured feature selection methods better capture the complexity and the overall structure of the available data.

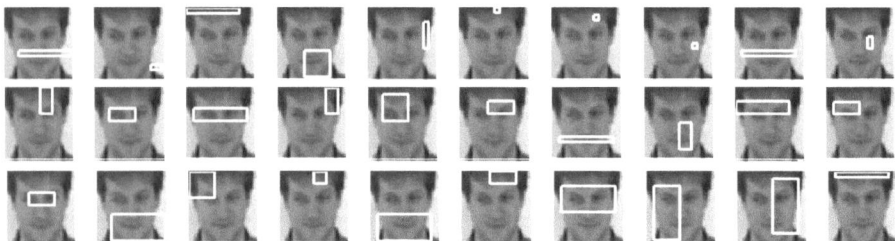

Fig. 5. MOBO dataset: the 10 most meaningful features (with respect to the corresponding value in β) selected by LASSO (I row), 1vsALL union (II row), MC-GrpLASSO (III row)

Face Recognition Assessment. On the **MOBO dataset** we consider a training set of 50 examples per identity, extracted from a pool of 5 videos (#training set = 1000). The validation set is collected following the same strategy on 5 other videos. The remaining videos are used to build the test set, made of 200 examples per subject (#test set = 4000). Table 1 reports the recognition

Table 1. Comparative analysis on the *MOBO* dataset

METHOD	# Features	TPR	FPR
fixed grid LBP [1]	49	95.6%	0.2%
LBP + LASSO	25	94.1%	0.3%
LBP + AdaBoost	25	90.1%	0,5%
LBP+ 1vsALL GrpLASSO (voting 50%.)	33	95.1 %	0.3 %
LBP+ 1vsALL GrpLASSO (union)	118	97.0%	0.1 %
LBP + MC-GrpLASSO	**26**	**96.9%**	**0.1%**
LPB + MC-GrpLASSO	13	96.1%	0.2%

Table 2. Comparative analysis on the *Choke Point* dataset (as in [16])

METHOD	REC. RATE
MRH + MSM on image subsets [16]	86.7%
LBP + MSM on image subsets [16]	75.8%
LBP + LASSO [4]	93.1%
LBP + MC-GrpLASSO	**96.9%**

Table 3. Comparative analysis on the R309 dataset

METHOD	# Features	TPR	FPR
fixed grid LBP	49	71,6%	3,5%
LBP+LASSO	19	71,4%	3,6%
LBP + GrpLASSO (union)	98	72.4%	0.3%
LBP + MC-GrpLASSO	**45**	**79,6%**	**2,6%**

performances we obtained starting from LBP features with different feature selection approaches. The goal is to achieve comparable or better performances with a sparser representation. MC-GrpLASSO achieves very good performances with a remarkably small representation.

On the **Choke Point dataset** we adopt the data groups and the protocol as proposed in [16]. In this paper the authors propose a video-based face recognition based on selecting the most meaningful faces in a video and then using those images to perform the recognition phase. Images are represented by Multi-Region Histograms (MRH) [12] and LBP [1]. Tab. 2 summarizes our results and also includes the best performances reported in [16], although the nature of the results is different. Tab. 3 shows how the overall performances on the **R309** dataset are significantly lower than for the other datasets, due to a consistent variability on the subjects appearances since acquisitions have been made on a temporal span of months. In any case our approach shows better performances than the others. Finally, Fig 6 reports the recognition results on the three dataset as a function of the selected regions. Notice how the results are quite stable with respect to this parameter.

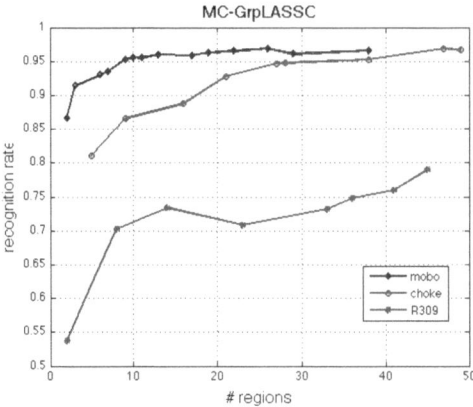

Fig. 6. Recognition rates function of the number of selected regions

5   Discussion

This paper presented an efficient face recognition method based on a compact and effective face representation derived by regularized structured feature selection. Starting from an overcomplete description of face images made of Local Binary Patterns (LBPs), we applied feature selection to select the discriminative information for the problem, and obtain a lower dimensional description. In particular, we proposed a new formulation of Group LASSO – the *MC-GrpLASSO* – to directly cope with a multi-class setting and exploiting the block-structure of features. Unlike previous approaches, the new formulation allows us to select groups that simultaneously discriminate among all the identities. We presented an extensive experimental analysis, both on benchmark datasets and on a complex set of data acquired in-house. The experimental analysis showed that our method outperforms other state-of-art approaches on all the three datasets. The future work will focus on two different aspects. On one hand, we will extend the pipeline from the standpoint of the description, by exploiting the temporal information we have available from the use of videos. On the other, the method we designed will be integrated on a prototype system for supporting blind users. It will be embedded on a device wore by the users, to help them in recognizing known people.

References

1. Ahonen, T., Hadid, A., Pietikäinen, M.: Face recognition with local binary patterns. In: Pajdla, T., Matas, J(G.) (eds.) ECCV 2004. LNCS, vol. 3021, pp. 469–481. Springer, Heidelberg (2004)
2. Candes, E., Tao, T.: The Dantzig selector: statistical estimation when p is much larger than n. The Annals of Statistics 35(6), 2313–2351 (2007)

3. Chapelle, O., Keerthi, S.: Multi-class feature selection with support vector machines. In: Proc. of the American Statistical Association (2008)
4. Destrero, A., Mol, C., Odone, F., Verri, A.: A regularized framework for feature selection in face detection and authentication. IJCV 83(2), 164–177 (2009)
5. Gross, R., Shi, J.: The cmu motion of body database. Tech. rep., Robotics Institute, Carnegie Mellon University (2001)
6. Li, H.X., Wang, S., Jensen-shannon, Y.: boosting learning for object recognition. CVPR 2, 144–149 (2005)
7. Hadid, A., Pietikäinen, M., Li, S.Z.: Learning personal specific facial dynamics for face recognition from videos. In: Zhou, S.K., Zhao, W., Tang, X., Gong, S. (eds.) AMFG 2007. LNCS, vol. 4778, pp. 1–15. Springer, Heidelberg (2007)
8. Jain, B.J., Wysotzki, F.: A competitive winner-takes-all architecture for classification and pattern recognition of structures. In: Hancock, E.R., Vento, M. (eds.) GbRPR 2003. LNCS, vol. 2726, pp. 259–270. Springer, Heidelberg (2003)
9. Mosci, S., Rosasco, L., Santoro, M., Verri, A., Villa, S.: Solving structured sparsity regularization with proximal methods. In: Balcázar, J.L., Bonchi, F., Gionis, A., Sebag, M. (eds.) ECML PKDD 2010, Part II. LNCS, vol. 6322, pp. 418–433. Springer, Heidelberg (2010)
10. Ojala, T., Pietikainen, M., Maenpaa, T.: Multiresolution gray-scale and rotation invariant texture classification with local binary patterns. PAMI (7) (2002)
11. Ruan, C., Ruan, Q., Li, X.: Real adaboost feature selection for face recognition. In: CSP, pp. 1402–1405 (2010)
12. Sanderson, C., Lovell, B.C.: Multi-region probabilistic histograms for robust and scalable identity inference. In: Tistarelli, M., Nixon, M.S. (eds.) ICB 2009. LNCS, vol. 5558, pp. 199–208. Springer, Heidelberg (2009)
13. Shelton, J., Dozier, G., Bryant, K., Adams, J., Popplewell, K., Abegaz, T., Purrington, K., Woodard, D., Ricanek, K.: Genetic based lbp feature extraction and selection for facial recognition. In: Southeast Regional Conf., pp. 197–200 (2011)
14. Shen, L., Bai, L., Bardsley, D., Wang, Y.: Gabor feature selection for face recognition using improved adaBoost learning. In: Li, S.Z., Sun, Z., Tan, T., Pankanti, S., Chollet, G., Zhang, D. (eds.) IWBRS 2005. LNCS, vol. 3781, pp. 39–49. Springer, Heidelberg (2005)
15. Tibshirani, R.: Regression shrinkage and selection via the lasso. Journal of the Royal Statistical Society. Series B (Methodological) 58(1) (1996)
16. Wong, Y., Chen, S., Mau, S., Sanderson, C., Lovell, B.C.: Patch-based probabilistic image quality assessment for face selection and improved video-based face recognition. In: CVPRW, pp. 74–81 (2011)
17. Yuan, M., Lin, Y.: Model selection and estimation in regression with grouped variables. J. R. Stat. Soc. Ser. B 68 (2006)
18. Zhang, G., Huang, X., Li, S.Z., Wang, Y., Wu, X.: Boosting local binary pattern (LBP)-based face recognition. In: Li, S.Z., Lai, J.-H., Tan, T., Feng, G.-C., Wang, Y. (eds.) Sinobiometrics 2004. LNCS, vol. 3338, pp. 179–186. Springer, Heidelberg (2004)

Measuring Sandy Bottom Dynamics by Exploiting Depth from Stereo Video Sequences

Rosaria E. Musumeci[1], Giovanni M. Farinella[2], Enrico Foti[1],
Sebastiano Battiato[2], Thor U. Petersen[3], and B. Mutlu Sumer[3]

[1] Dept. of Civil and Environmental Engineering, University of Catania, IT
[2] Dept. of Mathematics and Computer Science, University of Catania, IT
[3] DTU Mekanik, Section for Fluid Mechanics, Coastal and Maritime Engineering,
Technical University of Denmark, DK

Abstract. In this paper an imaging system for measuring sandy bottom dynamics is proposed. The system exploits stereo sequences and projected laser beams to build the 3D shape of the sandy bottom during time. The reconstruction is used by experts of the field to perform accurate measurements and analysis in the study of the final equilibrium conditions of sea bottoms in the presence of water flows. Results obtained by processing data acquired in hydraulic laboratory confirm the effectiveness of the system which makes simple and fast the understanding of the sandy bottom dynamics and the related equilibrium phenomena.

Keywords: Sandy bottom dynamics, Stereo system.

1 Introduction

The investigation of the three-dimensional morphological evolution of coastal sandy bottoms under the action of waves and currents is particularly relevant in the fields of physical oceanography and coastal engineering. For example, a detailed knowledge of scour processes at the foundation of coastal structures is fundamental to predict damages and failures potential of the structures themselves. Such a kind of studies are usually tackled by means of laboratory investigations, where all the complexity of the involved phenomena can be well controlled and reduced to simpler and reproducible conditions [1]. A typical experimental setup is shown in Figure 1.

In hydraulic wave flumes, mechanical instruments are often used to determine bottom evolution [2], however optical instruments are becoming more and more popular due to their noninvasiveness with respect to the experiments, i.e., the fact that it is possible to observe the dynamics of the bottom induced by the flow without generating spurious disturbance to the flow itself. This is one of the aims, for instance, of the approach proposed in [3] where the authors used a single mini waterproof camera coupled with a 45° mirror located within a transparent pile to monitor at a single point the maximum scour depth at the basis of a vertical cylinder.

A. Petrosino (Ed.): ICIAP 2013, Part I, LNCS 8156, pp. 420–430, 2013.

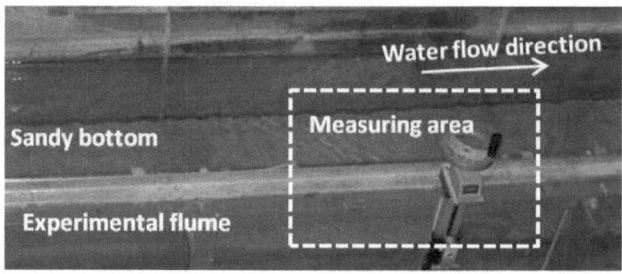

Fig. 1. View of the experimental setup

In order to gain knowledge of the three-dimensional characteristics of the coastal bottom morphology, several 3D based approaches have been recently proposed, ranging from commercial 3D laser scanners to more sophisticated stereo pairs techniques [4–7]. Unfortunately such optical methods are usually limited to measure the phenomena under consideration in dry condition, i.e., either the wave flume need to be stopped and drained either the measuring area is located in the dry part of the shore. Moreover the images of sandy bottom are structureless and it is difficult to find corresponding points in stereo pair images to be used to reconstruct the sandy bottom and to track interesting points and areas to understand their evolution in time. A recent study [4] have proposed a stereoscopic technique able to measure in the dry part of the swash zone (i.e., in the part of the beach which is alternately wet and dry) rapid variations of the sandy bed elevation based on the use of Particle Tracking Velocimetry (PTV) algorithms. A grid of light dots (each dot was 2 cm wide) projected on the dry part of the swash zone has been employed to add some recognizable structure to be tracked in the images.

Obtaining 3D measurements of the bed morphology undewater is much more difficult, due to problems both in image formation and image aquisition (see Figure 1). Indeed, due to the small size of the sand grains, $O(10^1 \mu m)$, and to the limited variability of colours, the images of sandy bottoms are "structure free", hence making difficult the matching of the points of the different point of view (i.e., correspondence problem in stereo systems). On the other hand it is preferable that the cameras are located below the water level and relatively far from the observing area, to avoid variable refraction and reflection disturbances from the moving water surfaces and undesired localized effects due to the presence of the optical equipment. Notice that classic computer vision techniques adopting interest feature points (like SIFT [8]) are not suitable to solve the matching problem in this context since the unstructured material observed into the scene (i.e., the sand) due to the high variability of the background induced by the water waves and the related reflection.

To deal with such problems, a stereoscopic technique with the cameras located in front of the glass wall of the flume below the still water level has been proposed in [5]. A grid of laser dots projected on the sandy bottom has been used to allow manual correspondences (made by the experts) between points observed by the stereo pairs and hence to perform 3D measurements. The main advantage of the technique is the possibility to perform 3D dynamic measurements of the bottom evolution (i.e., 3D depth is generated from stereo video) without disturbing the water flow. Such a technique was extended in [6] considering large scale applications in the swash zone region. In that case, not only a much greater number of dots was deemed necessary, but also longer time series were recorded up to equilibrium conditions. The technique has been also recently applied in [7] to investigate the backfilling process of a scour at the base of vertical cylinder. In this last work the possibility to measure the 3D dynamics of the bottom evolution has been fundamental to demonstrate that wave-induced scours and wave-induced backfillings of a current-scour lead to the same equilibrium conditions. However, although quite accurate from the measuring point of view, the procedure used in [5–7] to build the 3D maps is not easily applicable to the analysis of a large number of images (e.g. frames of long video pairs), required for an accurate analysis of the time evolution of the process. Indeed, both the automatic detection of hundreds of grid laser dots contained in each image and the matching of the stereo pairs are carried out manually by the experts. This process becomes impossible to do by considering hours of acquisition at 4 fps, as the experimental setting of this paper. Hence, in order to become an useful and versatile measurement instrument for sandy bottom evolution the aforementioned technique needs to be made automatic.

Building on previous works [5–7], in this paper an automatic system for measuring sandy bottom dynamics from stereo image sequences is proposed. After system calibration, the projected laser beams are detected on the stereo pair frames through an image processing pipeline. Exploiting the calibration parameters, the detected dots on the left image are matched with the ones on the right by projecting their coordinates and considering weak perspective assumption. Then the 3D maps are hence obtained by triangulation and the analysis is performed.

The paper is organized as follows: in Section 2 the proposed system is described by focusing on (i) the automatic detection of the laser dots onto the gathered images of the sandy bottom, and (ii) the automatic stereo matching of the dots on the stereo pairs. Section 3 discusses the experimental setting and summarize the results. Finally, conclusions and hints for future works are given in Section 4.

2   Proposed System

As it is standard in stereo vision approaches, the proposed technique can be schematized in the following three stages: (i) stereo image acquisition; (ii) matching of the corresponding points; (iii) 3D measurement and analysis.

The main objective of the instrumental setting used in this work is to obtain "suitable" images to make correspondences of points of the sandy bottom morphology throughout the experiment. Moreover, to make accurate measurements of the bottom dynamics the depth of specific points (e.g., evolution of points around a vertical pile) should be reconstructed in time. This is achieved by projecting a grid of laser points onto the sandy bottom. Figure 1 shows a typical

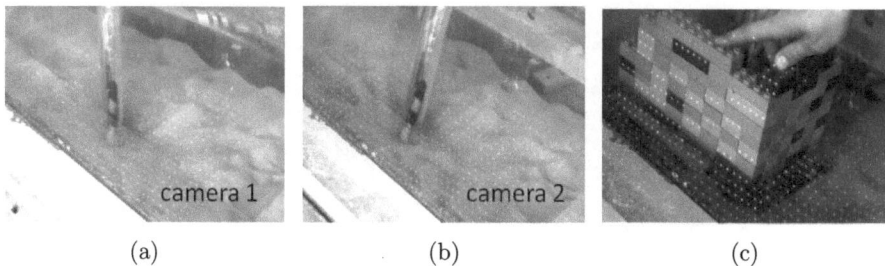

(a)   (b)   (c)

Fig. 2. (a) and (b): stereo pairs related to the experimental settings. The projected laser dots onto the sandy bottom and the the scour around a vertical pile are visible. (c) 3D rig used during calibration.

view of the experimental setup , while Figure 2(a) and b shows an example of a stereo images where the laser grid points are visible. The light dots become an optical well-defined point measuring stations on the sandy bottom, and also allows to add points to be matched between the stereo images.

In a previous work [5], Gaussian lenses have been used during acquisition. Here we follow the protocol of [6] where non-Gaussian lenses have been exploited to obtain an uniform density distribution of light with sharp ends. External lighting conditions of the hydraulic laboratory are also very important. In the ideal condition a dark room in which only the laser beams can be seen simplify the detection of the points. Unfortunately those conditions are difficult to be obtained in hydraulic laboratories, where for safety reason windows cannot usually be shaded to get complete darkness. Spurious reflections, due to the flume's glass walls or to the moving free surface of water, can also affect the quality of the images. It follows that the images must be carefully analyzed in order to reject points which do not belong to the laser grid projected onto the bed. As can be assessed by visual inspection of Figure 2(a) and (b), there are many source of variability in the images which make difficult both, dots detection (e.g., there are many false positive dots coming from reflection) and matching.

Being a stereoscopic approach, the image acquisition is obviously performed by using two cameras which simultaneously record videos of the sandy bottom evolution. In order to avoid any disturbance to the flow, the cameras are located outside of the flume, in front of glass walls in the correspondence of the measuring area. Due to the fact that in hydraulic laboratories several optical obstacles are usually present (e.g., steel frames, structural elements, other measuring instruments, etc.), the technique has been developed in such a way that there are no constraints on the relative position of the two cameras. The only requirement is that both cameras should have a "vision" as much complete as possible of the phenomenon under investigation.

About the camera calibration, while [5] and [6] used a direct linear transformation (DLT) [9, 10], in the present work a four-step camera calibration procedure is performed [11]. Traditionally, several views of a 2D planar rig are adopted during calibration [12]. To facilitate the operations within the flume filled with water, the

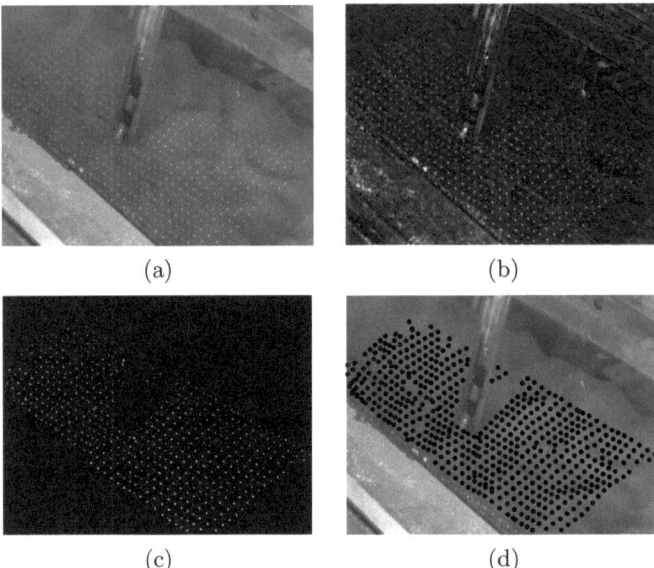

(a)

(b)

(c)

(d)

Fig. 3. Detection of laser dots. (a) Original image; (b) Filtered image; (c) Top-hat transform; d) Dots vs. other separation; e) Detected laser dots.

stereo images of a 3D calibration rig are used (see Figure 2 *(c)*). The rig has known dimensions, which can be modified according to the dimensions and shape of the investigated phenomenon to properly calibrate the stereo system in order to obtain the required accuracy in the measurements. Once, the calibration of the stereo system is done, the laser dots should be detected and matched in the image pairs to obtain the correspondences useful to build the 3D map of the sandy bottom at varying of time. The following subsections illustrate the techniques implemented for making automatic the laser dot detection and matching.

2.1   Detection of Laser Dots

The automatic detection of the light dots position on the images is obtained by means of an image processing pipeline involving a sequence of filtering, morphological operations and foreground (i.e., dots) versus background separation procedures. An example of the images obtained at the various steps of the pipeline is shown in Figure 3. Note that, due to the high variability and artifacts introduced by the reflection, noisy input, etc., this task is not so trivial as expected.

As first, each image is filtered by means of a Gaussian kernel to remove noise (see Figure 3 *(b)*). Then, in order to correct uneven illumination and to enhance the intensity of the laser dots, a morphological top-hat transform [13] is carried out by means of a disk structuring element (see Figure 3 *(c)*). The results of such an operation is hence binarized through a thresholding procedure to separate dots (i.e., foreground versus background). Therefore a flood-fill operation [14] of the obtained binary image is applied to fill holes, (i.e., to fill the background pixels that cannot be reached by filling in from the edge of the images). Finally, the

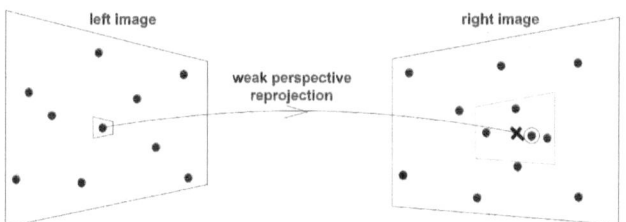

Fig. 4. Closest neighbour matching considering weak perspective

segmented spots (i.e., the connected elements on the image) are analyzed with respect to their shapes (through geometric constraints, such as area, etc.) in order to retain only the laser dots (see Figure 3 *(d)*). Then, the centroid coordinates of all the detected dots are determined and stored.

The parameters used in the aforementioned pipeline have been learned on the pair of images used for the stereo calibration procedure. Specifically, for those images a manual labeling has been performed and the parameters have been learned with a search procedure on a quantized version of the parameter space.

2.2 Stereo Matching

In a previous work [6], it was shown how standard methods used for stereo matching (e.g., correlation) fail to properly recover the sandy bottom geometry, due to the lack of structures useful to deal with the correspondence problem. As we have experimented in this context, the classic approach to match invariant feature points (e.g., SIFT [8]) in the pair images has problems due to the high variability and presence of artifacts generated by the reflection and to the periodic patterns and deformation present in the sandy texture and in the waves induced by the experiments.

At this stage we have the detected dots from the pipeline described in previous section. Also, since the cameras have been stereo calibrated, all the intrinsic and extrinsic parameters of the stereo system are known. Therefore, in the present work, the problem of stereo matching is solved by means of a Closest Neighbour Method (CNM), i.e., through a purely geometric approach, based on the distance of the reprojected points and making the assumption of weak-perspective model [15]. A sketch of the proposed procedure is shown in Figure 4. In our case, the approximation to a weak perspective model is possible and profitable since the dimension of the measuring area is relatively small compared to the distance of the cameras from it (i.e., the average variation of the depth of the measured area is very small with respect to the average distances of points). Hence, a constant size window is introduced to select a set of dots on the right image which are neighbours of the reprojection. Within such a set, the dot which is closest to the reprojection is chosen as the matching point. Subsequently, a series of checks are performed to guarantee a biunique correspondence of the matchings and hence remove false positive and wrong matchings.

2.3   3D Sandy Bottom Reconstruction and Analyses

Once the matching pairs are obtained the depth of the sandy bottom is reconstructed by using triangulation [15]. Moreover, in order to obtain a smooth 3D
depth maps over time, a linear interpolation of the 3D dots coordinates related
the sandy bed is carried out. The morphology of the sandy bottom surface is
enlightened by means of contour lines. Since the depth map is built for each
stereo pairs of the video streams, the 3D dynamics of the sandy bottom can be
analyzed following the dots of interest for the specific experiment (e.g, points
around the pile).

3   Experimental Setting and Results

In the present work, the proposed system has been applied to a laboratory
analysis of the evolution of the scour which occurs at the foundation of an
offshore structure due to marine currents and the refilling (or backfilling) of
such a scour hole generated by the subsequent action of wind waves. From the
engineering point of view, a throughout three-dimensional understanding of the
phenomenon is particularly important for protecting the foundations of off-shore
structures (e.g., wind turbines, oil platforms, etc.) from erosion.

The experiments have been carried out at the hydraulic laboratory of the
Section for Fluid Mechanics, Coastal and Maritime Engineering of the DTU
Mekanik of the Technical University of Denmark in a wave-current flume, which
is $28m$ long, $0.6m$ wide and $0.8m$ deep. Fine sand was used during the experiments, having a median diameter $d_{50} = 0.17mm$ and gradation equal to 1.3.
The water depth was $h = 0.4m$. The diameter of the plexiglass vertical circular
pile used to investigate scour and backfilling was $D = 40mm$. In Figure 3(a) is
possible to observe the aforementioned setting as seen from the right camera.

The evolution of the bottom morphology was recorded starting from flat bottom conditions up to current- or wave-induced equilibrium scour and backfilling
by means of two UI-2250RE-M monochrome cameras, located outside of the
wave flume (see Figure 2(a) and (b)). The grid of light points was obtained by
means of two 20 mW Lasiris SNF 660-nm laser generators (in order to cover the
measuring area) coupled with two GMN31614MNC-1 Goyo Optical Inc. diffrac-
tive lenses with 1.6 mm focal length. Each laser produced a square grid of 19×19
equally spaced laser beams, with a total of 722 dots projected onto the sandy
bottom. Some of the laser dots were not visible due to the pile shadow.

A total of 137 video pairs were recorded at 4 fps, using an image size of
1600×1200 pixels. The duration of each video was in the range of 2 to 15 minutes.
The longest video pairs were recorded at the beginning of the experiment to
monitor the initial stages of the evolution which are the fastest, then video pairs
were taken each 10 minutes up to an equilibrium condition was reached, after
$2 \div 4$ hours. One of the main motivation of the present work is that, the manual
treatment of the above data is extremely time consuming, since the analysis of
one image of a single stereo pair may take up to 2 hour. Moreover, due to the

small size of the light dots, the manual detection and matching are prone to material errors of the operator.

Here we report one case study related the investigation of the dynamics of sediment transport around a vertical cylinder [7] (see Figure 5). The considered images have been chosen in such a way to be representative of several stages of the bottom evolution.

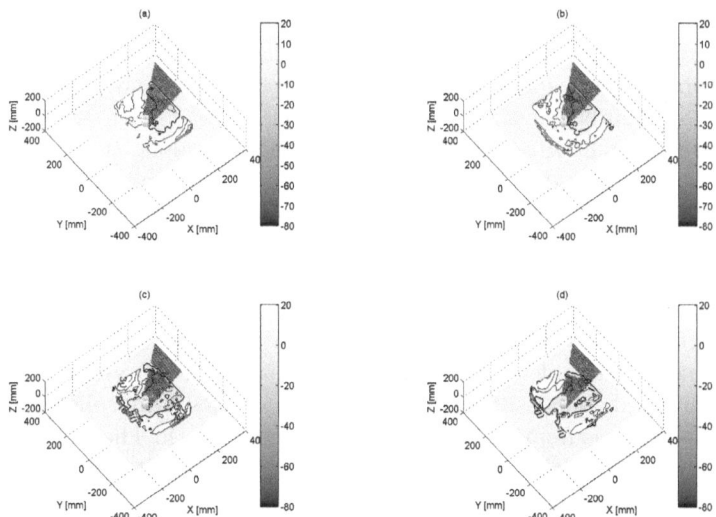

Fig. 5. 3D depth maps of the sandy bed: (a) flat bottom condition (Stereo Pair 1) as recovered manually [7]; (c) flat bottom condition (Stereo Pair 1) as recovered by the proposed automatic pipeline; (b) backfilling at equilibrium (Stereo Pair 5) as recovered manually [7]; (d) backfilling at equilibrium (Stereo Pair 5) as recovered by the proposed automatic pipeline

By defining T as the time scale of the backfilling process, the set of pairs related the case study refer to the following experimental conditions:
– Stereo Pair 1: initially flat bottom;
– Stereo Pair 2: current-induced scour at equilibrium;
– Stereo Pair 3: wave-induced backfilling after a time equal to $T/2$;
– Stereo Pair 4: wave-induced backfilling after a time equal to T;
– Stereo Pair 5: wave-induced backfilling at equilibrium;

To perform an objective evaluation of the proposed system, the dots in the images related the case study under evaluation were manually labeled and matched. The total number of manually labeled dots is 5610. The final number of manually labeled matchings is 2660. The overall detection rate obtained with the pipeline summarised in Section 2.1 was 0.83% with an average rate per image of 0.825%. The matching rate obtained considering the corrected detected points was 0.734% for the equilibrium condition state. The most important fact is that the interesting dots useful for the investigation of the dynamics of sediment transport around the pile were well detected and matched allowing the study of the flume at the equilibrium backfilling condition (Condition related Stereo Pair 5 in the list above). The results obtained can be considered satisfactory to perform the analysis, although further improvement is deemed necessary to reach better performances of the matching in the case of sandy bottom morphologies characterized by high steepness and high reflection noise and artifacts (e.g., condition related to the Stereo Pair 2 in the above list).

Once the matching is done, the 3D map of the sandy bottom is reconstructed. Examples of the 3D maps obtained from both manually labeled process [7] and from proposed method are reported in Figure 5 for Stereo Pair 1 and Stereo Pair 5, which corresponds respectively to the flat bed conditions and the wave-induced backfilling at equilibrium.

In Figure 6, the cross-sections of the sandy bottom along the central axis of the wave flume obtained by the manual reconstruction [7] and the one performed with the proposed automatic procedure are compared. The data are referred to the equilibrium condition of wave-induced backfilling. The independent measurement of the position of the sandy bottom (i.e., the ground-truth) as been recovered by a waterproof minicamera coupled with a 45° mirror [7] is also reported in the plots (i.e., the red dot in Figure 6). This result shows how the proposed method is able to accurately reconstruct the 3D sandy surface and hence can be used to measure the interesting quantities of morphological analyses, such as the measurement of the maximum scour depth. The results obtained with the proposed method outperforms the previous results obtained through manual approach [7].

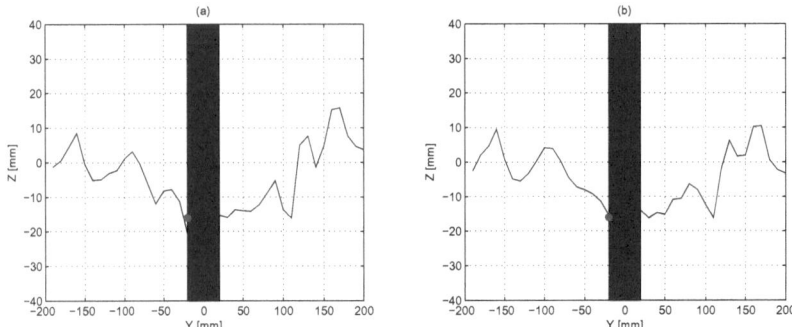

Fig. 6. Cross-section along the central axis of the flume at the equilibrium backfilling condition: (a) result recovered by using manual reconstruction [7]; (b) result obtained by the proposed automatic pipeline. The red dot indicates the ground-truth related the bed position[7]. The proposed method obtains the best result.

4   Conclusions and Future Works

In this paper a system for measuring quantities useful for sandy bottom morphological analyses over time has been presented. The system makes use of stereo vision and projected laser beams which are automatically detected and matched to find the correspondences useful for the 3D sandy surface reconstruction. The results obtained in measuring the final equilibrium conditions by processing data acquired in hydraulic laboratory confirm the effectiveness of the system which makes effortless for the expert the understanding of the sandy bottom dynamics phenomena. An improvement of the proposed method is under development to cope with the limits due to the weak perspective model assumption used here for reprojection. Moreover, future works will be devoted to test the performances of recent hardware (e.g., Kinect [16]) for the 3D sandy bottom reconstruction.

Acknowledgments. The research described in this manuscript was supported by the EC project HYDRALAB IV (contract no. 261520) and by the Italian Ministry for Education, University and Research funded PRIN 2010-2011 HYDROCAR (Prot. no. 20104J2Y8M_003).

References

1. Sumer, B.M., Fredsøe, J.: The mechanics of scour in the marine environement. World Scientific (2002)
2. Sánchez-Arcilla, A., Cáceres, I., van Rijn, L., Grüne, J.: Revisiting mobile bed tests for beach profile dynamics. Coastal Eng. 58(7), 583–593 (2011)
3. Sumer, B.M., Hatipoglu, F., Fredsöe, J.: Wave scour around a pile in sand, medium dense and dense silt. J. Waterway, Port, Coastal, Ocean Eng. 133(1), 14–27 (2007)
4. Astruc, D., Cazin, S., Cid, E., Eiff, O., Lacaze, L., Robin, P., Toublanc, F.: Cáceres: A stereoscopic method for rapid monitoring of the spatio-temporal evolution of the sand-bed elevation in the wash zone. Coastal Eng. 60(1), 11–20 (2012)
5. Baglio, S., Faraci, C., Foti, E., Musumeci, R.: Measurements of the 3-D scour process around a pile in an oscillating flow through a stereo vision approach. Measurement 30(2), 145–160 (2001)
6. Foti, E., Cáceres-Rabionet, I., Marini, A., Musumeci, R.E., Sánchez-Arcilla, A.: Experimental investigations of the bed evolution in wave flumes: Perrformances of 2D and 3D optical systems. Coastal Eng. 58(7), 606–622 (2011)
7. Sumer, B.M., Petersen, T.U., Locatelli, L., Fredsöe, J., Musumeci, R.E., Foti, E.: Backfilling of a scour hole around a pile in waves and current. J. Waterway, Port, Coastal, Ocean Eng. 139(1), 9–23 (2013)
8. Lowe, D.G.: Distinctive image features from scale-invariant key-points. International Journal of Computer Vision 60(2), 91–110 (2004)
9. Abdel-Aziz, Y.I., Karara, H.M.: Direct linear transformation into object space coordinates in close-range photogrammetry. In: Proc. Symposium in Close-Range Photogrammetry, Urbana, Illinois, pp. 1–18 (1971)
10. Faugeras, O.D., Toscani, G.: Camera calibration for 3D computer vision. In: Proc. International Workshop on Industrial Applications of Machine Vision and Machine Intelligence, Silken, Japan, pp. 240–247 (1987)
11. Heikkilä, J., Silvé, O.: A four-step camera calibration procedure with implicit image correction. In: IEEE Computer Society Conference on Computer Vision and Pattern Recognition, San Juan, Puerto Rico, pp. 1106–1112 (1997)
12. Tsai, R.Y.: A versatile camera calibration technique for hygh-accuracy 3D machine vision metrology using off-the shelf TV cameras and lenses. IEEE J. Robotics and Automation RA-3(4), 323–344 (1987)

13. Allan, G.H., Serra, J.: Morphological operators on the unit circle. IEEE Transactions on Image Processing 10(12), 1842–1850 (2001)
14. Soille, P.: Morphological image analysis: Principle and Applications. Springer (1999)
15. Hartley, R., Zisserman, R.: Multiple view geometry in computer vision. Cambridge University Press (2003)
16. Newcombe, R.A., Izadi, S., Hilliges, O., Molyneaux, D., Kim, D., Davison, A.J., Kohli, P., Shotton, J., Hodges, S., Fitzgibbon, A.: KinectFusion: Real-Time Dense Surface Mapping and Tracking. In: IEEE International Symposium on Mixed and Augmented Reality (2011)

Daily Living Activities Recognition via Efficient High and Low Level Cues Combination and Fisher Kernel Representation

Negar Rostamzadeh[1], Gloria Zen[1], Ionuţ Mironică[2], Jasper Uijlings[1], and Nicu Sebe[1]

[1] DISI, University of Trento, Trento, Italy
{rostamzadeh,zen,jrr,sebe}@disi.unitn.it
[2] LAPI, University Politehnica of Bucharest, Bucharest, Romania
imironica@imag.pub.ro

Abstract. In this work we propose an efficient method for activity recognition in a daily living scenario. At feature level, we propose a method to extract and combine low- and high-level information and we show that the performance of body pose estimation (and consequently of activity recognition) can be significantly improved. Particularly, we propose an approach extending the pictorial deformable models for the body pose estimation from the state-of-the-art. We show that including low level cues (*e.g.* optical flow and foreground) together with an *off-the-shelf* body part detector allows reaching better performance without the need to re-train the detectors. Finally, we apply the Fisher Kernel representation that takes the temporal variation into account and we show that we outperform state-of-the-art methods on a public dataset with daily living activities.

1 Introduction

Automatic video scene understanding and activity analysis are active research topics in computer vision. In this paper we focus on daily living activity scenarios. The interest in activity recognition in this scenario is motivated by the promise of important applications in areas such as patient monitoring and ambient assisted living.

Analyzing daily living scenarios is a challenging task. First of all, in such a scenario, different activities differ only slightly in motion and appearance. In some cases, the differences in appearance of the subjects performing the same task are more evident than the difference in activities. Moreover, one activity can be performed in many different ways, while two different activities may be performed in a very similar manner with respect to motion and appearance. For example, dialing and answering the phone are activities only slightly different in terms of hand movements. Particularly, if we consider the Activity of Daily Living (ADL) dataset[1], the difference between two activities in most cases is limited to taking *phone, banana* or *knife* from the *table, shelf*, or *refrigerator* and doing slightly different other activities (*e.g. eat snack* and *drink water*).

Recently, Bag-of-Words (BoW) models relying on local features have become popular in dynamic scene understanding due to their robustness to noise and occlusions.

---

[1] www.cs.rochester.edu/~rmessing/uradl/

A. Petrosino (Ed.): ICIAP 2013, Part I, LNCS 8156, pp. 431–441, 2013.

However, the traditional BoW representation that has typically been applied in activity recognition scenarios has some substantial restrictions [3], [13], [32]. First of all, a BoW representation based on low-level cues limits the access to the high-level information that may be more discriminative. Secondly, being a frequency histogram of quantized local appearances or motion, the relationships between temporal cues are totally ignored.

In this paper we address these two drawbacks in the BoW representation. First, we make an enriched descriptor by combining low- and high-level cues that are obtained from the local motion and a body pose detector. In this way, the source of motion (*e.g.* a hand) is taken into account. Second, we apply a Fisher Kernel representation of the combined low- and high-level features to model the temporal variation. Additionally, we provide an efficient method for body pose estimation which builds upon [33] and allows us to improve the detector performance on a new dataset by simply exploiting the information provided by easy-to-extract low level cues (thus saving the cost of creating the ground-truth and re-training the detector). Finally, we apply the popular non-linear SVM classification method and show that the obtained results outperform the state-of-the-art on the ADL dataset.

2   Related Work

Typical approaches for activity recognition rely on a two-steps paradigm. The first step concerns the generation of feature vectors: features are extracted and quantized according to a pre-defined codebook and are accumulated to form the so called bag-of-words. The second step takes these bags-of-words as input and learns how to classify the different actions. This phase is generally supervised and a training set is available for learning.

The first step is crucial for the good performance of the second one. In fact, the information discarded at this step can hardly be recovered afterwards. For example, if the codebook is defined based on local motion (*e.g.* tracklets or optical flow), all the information about the structure of the scene or about the entity involved in the motion is discarded. This causes a huge information loss, which heavily limits the capability of comprehending a scene in the learning step that follows. Over the past years, many works addressed this limitation and much effort has been devoted to enrich the descriptors with additional information beyond motion: (i) some works take into account the relationship between the spatio-temporal local features [8], [11], [16], [25]. Zhang *et al.* [35] enriched their descriptor not only with the relationship between neighboring local space-time features but also by considering the long-range relationship of local features. (ii) Malgireddy *et al.* [15] and Kovashka *et al.* [11] combined local features and made enriched descriptors. Others proposed taking the contextual features of interest points into account in a BoW representation [2], [30]. (iii) Lately, an increasing number of works exploited the information coming from detectors as a high level information about the observed scene [17], [22], [23], [34]. This is a step towards a higher level comprehension of the scene, w.r.t considering only low- or mid-level information represented by the local motion (*e.g.* optical flow, tracklets) or the local appearance (*e.g.* SIFT, HOG). In this way, the nature of the body parts involved in the observed motion is considered. In our daily living scenario, the person is monitored from a camera in a controlled environment and the body is clearly visible and mostly not occluded. We combine local motion with the high-level information coming from a body limbs detector. To do so, an efficient and accurate body pose estimator is required.

Over the past decade, many approaches have been proposed for capturing human body parts [6], [7], [10], [21], [29]. These works focused on generalizing and extending the pictorial model. Using a pictorial structure as a model to represent human body pose is a popular approach that tries to model an object by its parts arranged in a deformable configuration. The problems of the variety of body part appearances, different orientations, and different scales in which humans may appear were not well-investigated in the traditional pictorial structure. Felzenszwalb *et al.* [5] proposed an extension of the pictorial model to detect objects at different scales using a multi-scale HOG-pyramid. Yang and Ramanan [33] proposed a more general pictorial model covering a variety of body configurations. Their proposed approach is among the most efficient works that model the human body skeleton as a tree. They detect small bounding boxes around the body parts instead of complete body limbs. This makes their work more efficient because it prevents the problem of double counting. In their work, a local appearance template is obtained by a multi-scale HOG descriptor [4] that allows detection at different scales. Our human pose estimator is built upon their work [33].

Finally we investigate the use of the Fisher Kernel representation to model the temporal variation of videos. Involving the temporal variation is not very well-investigated yet [26]. Kuehne *et al.* [12] and Qi *et al.* [20] used Hidden Markov Models. Other works employed temporal rules with high-level concepts [14]. To the best of our knowledge the only work that used Fisher Kernel to model the temporal variation in videos is [18]. They employed a frame-based global feature descriptor for a movie-genre classification scenario. In our work, we use Fisher Kernel to model the temporal variation over local descriptors of individual body-parts that are detected in consequent frames of a video in an action recognition scenario.

3   Our Method

We propose a novel activity recognition method obtained by combining information taken from both the local motion and the body part detector. Combining low- and high-level cues exploits the advantages of both cues: on one side the robustness of low-level cues (*e.g.* optical flow) w.r.t occlusions, on the other side having the information about the body part involved in an activity increases the scene disambiguation.

In the case of body pose estimation, a significant drop in accuracy has been observed when a detector is trained on one dataset and it is evaluated on a different one [22]. The reason is that for some cases there are not enough samples in the training set. As the detector gives more priority to the positive samples of training set, the chance of detecting uncommon (w.r.t positive samples) body poses decreases. A possible solution to this is to obtain the body pose groundtruth for the new dataset and re-train the classifier. However, this procedure is very expensive and requires a considerable delay every time a new dataset has to be analyzed. Instead of training another classifier on the new dataset, we propose to use the already trained classifier, but we provide some additional information from the new dataset. Specifically, we used the classifier trained on the Buffy dataset [7], using the approach from [33]. Then we boost the classifier by exploiting the information of low-level cues from the ADL dataset. These low-level cues (*i.e.* optical flow and foreground pixels) can be easily extracted from a stationary webcam as in our case. To evaluate our contribution for pose estimation in a new dataset, we annotated the upper body poses for 371 frames obtained from different clips of the ADL dataset[2]. Then, we create our descriptors by combining our enhanced body-pose estimator with the local motion (*i.e.*, optical flow), that is already extracted for enhancing the pose estimator. Finally, we apply a Fisher Kernel representation to our descriptors to model the temporal variation in video, and apply a popular non-linear SVM classifier (SVM with RBF kernel) on our descriptors to classify the activities. The details of our approach are provided in the following sections.

3.1   Body Pose Estimation

Pictorial structures model the body as an ideal template represented as a graph, $G=(V,E)$, in which single body parts templates (V) are connected with springs (E) that represent the geometric constraints between them. The placement of these springs can change, while the structure of the model is preserved. These deformations present different possible configurations of body parts. Each possible body configuration is given a score that is based on the sum of local and pairwise scores [5,6]:

$$S(I,p,t) = \sum_{i \in V} w_i^{t_i} \phi(I,p_i) + \sum_{i,j \in E} w_{ij}^{t_i,t_j} \psi(p_i - p_j) + S(t) \qquad (1)$$

where $\phi(I,p_i)$ is a HoG descriptor extracted from the pixel location p_i in image I and $\psi(p_i - p_j)$ is the relative location of part i with respect to part j. The first term in Eq 1 represents the *local* score (also called *appearance model*) that indicates how likely is that a template $w_i^{t_i}$ for part $i \in \{1, ..., K\}$ of the body, tuned for type t_i, is located at position $p_i = (x,y)$ in the image I. The second term represents the *pairwise* score (also called *deformation model*) and controls the relative location of part i with respect to j. $S(t)$ is a *compatibility function* defined as,

$$S(t) = \sum_{i \in V} b_i^{t_i} + \sum_{i,j \in E} b_{ij}^{t_i,t_j} \qquad (2)$$

where $b_i^{t_i}$ represents the bias that favors particular type assignment for single part i and $b_{ij}^{t_i,t_j}$ represents the pairwise co-occurrence of parts i and j.

Our work builds upon [33] where the body relational graph is as a tree. The inference corresponds to maximizing the score function $S(I,p,t)$ over p and t and it can be efficiently solved with dynamic programming when the relational graph $G = (V,E)$ is modeled as a tree:

$$S_i(t_i,p_i) = b_i^{t_i} + w_{t_i}^i \phi(I,p_i) + \sum_{k \in kids(i)} m_k(t_i,p_i) \qquad (3)$$

---

[2] The groundtruth is available at:
https://sites.google.com/site/negarrostamzadeh/Ground-Truth.7z

where $m_k(t_i, p_i)$ collects the message from the children of part i (located at p_i for the type t_i). In Yang *et al* [33], the local score (the second term in Eq. 3) is based only on the appearance cues (*i.e.* HOG). Differently from them, in our work, we use a model that is trained on a dataset (Buffy dataset [7]) and we enrich the local score by including information provided by the local cues, such as *foreground* and *optical flow*, calculated for a new dataset (ADL dataset):

$$S(t_i, p_i) = b_i^{t_i} + w_{t_i}^i \phi(I, p_i) + \alpha\beta S_{FG}^i(p_i, \gamma) + (1-\alpha)\eta S_{OF}^i(p_i, \lambda) + \sum_{k \in kids(i)} m_k(t_i, p_i)$$

$$(4)$$

In Eq. 4, local *foreground* and *optical-flow* information are combined with the local appearance information at the testing level. S_{FG} and S_{OF} respectively present foreground and optical flow scores corresponding to the information that comes from these local cues. In our representation the impact of S_{FG} and S_{OF} is controlled respectively by parameters β and η. Moreover, the relative impact of the two added terms is controlled by the parameter α.

Computing the Foreground Score (S_{FG}). The foreground score S_{FG}^i is defined as the percentage of foreground pixels contained in the corresponding body part's bounding box, centered at location $p_i = (x, y)$. In order to extract foreground pixels, we applied the dynamic Gaussian Mixture background subtraction model [27]. For the foreground score, we consider a smaller bounding box w.r.t the one used for computing the HOG features, otherwise we would include some unnecessary portion of the background. In particular, we compute the number of foreground pixels $|pixels_{FG}^{\{p_i, \gamma\}}|$ in a bounding box of size $L_{FG} = \frac{1}{\gamma}L$, centered at p_i, where L is the size of the appearance bounding box. In the experimental section we report the effect of varying γ.

The foreground score S_{FG} is computed as follows:

$$S_{FG}^i(p_i, \gamma) = \frac{|pixels_{FG}^{\{p_i, \gamma\}}|}{|pixels^{\{p_i, \gamma\}}|}$$

$$(5)$$

where $|pixels_{FG}^{\{p_i, \gamma\}}|$ represents the number of foreground pixels that are present in a box centered at p_i with size L_{FG}, and $|pixels^{\{p_i, \gamma\}}|$ represents the total number of pixels in the foreground bounding box.

Computing the Optical Flow Score (S_{OF}). We use the Lucas-Kanade optical flow algorithm [28]. Similarly to the foreground score, we compute the number of optical flows $|pixels_{OF}^{\{p_i, \lambda\}}|$ in a bounding box of size $L_{OF} = \frac{1}{\lambda}L$, centered at p_i. The optical flow score is formulated as follows:

$$S_{OF}^i(p_i, \lambda) = \frac{|pixels_{OF}^{\{p_i, \lambda\}}|}{|pixels^{\{p_i, \lambda\}}|}$$

$$(6)$$

where $|pixels^{\{p_i, \lambda\}}|$ represents the number of pixels in the optical flow bounding box.

3.2 Activity Recognition

For the low-level cues, we quantize the motion vectors into 8 possible directions. For the high-level cues we apply our enhanced pose estimator and detect the placement of N_{bp} body-parts (in this experiment $N_{bp} = 18$). Then we make an 8 bin histogram for each body-part. Optical flows are assigned to the corresponding body part. Finally, we concatenate all of the 8 bin histograms and create an $8 \times N_{bp}$ bin histogram for each frame. Then we apply 2 representations and show how our approach outperforms the state-of-the-art. As the first representation, we simply accumulate all the histograms assigned to each clip in one histogram. For the second representation, we want to model the temporal variation within the video. We employ the Fisher Kernel to do so.

The Fisher Kernel representation was introduced recently to improve the BoW for representing sets of local appearance descriptors. The Fisher Kernel was designed to combine the benefits of both *generative* and *discriminative* approaches [9] and creates a fixed-length representation for a set of vectors. In this paper we use the Fisher Kernel to model the temporal variation in video. To do this, one can view a set of frame-based features (where we extract one feature from each frame) as a cloud of feature vectors. We can model this cloud with respect to a Gaussian Mixture Model (GMM) with diagonal covariance matrices. The resulting Fisher representation models the temporal variation in a generative way. Afterwards, we use the Fisher vector in a discriminative classifier (SVM).

The gradient vector is, by definition, the concatenation of the partial derivatives with respect to the model parameters. Let μ_i and σ_i be the mean and the standard deviation of i's Gaussian centroid, $\Gamma(i)$ be the soft assignment of descriptor x_t to Gaussian i, and let D denote the dimensionality of the descriptors x_t. $G^x_{\mu,i}$ is the D-dimensional gradient with respect to the mean μ_i and standard deviation σ_i of Gaussian i. Mathematical derivations lead to [19]:

$$G^x_{\mu,i} = \frac{1}{T\sqrt{\omega_i}} \sum_{t=1}^{T} \Gamma(i) \frac{x_t - \mu_i}{\sigma_i} \tag{7}$$

$$G^x_{\sigma,i} = \frac{1}{T\sqrt{2\omega_i}} \sum_{t=1}^{T} \Gamma(i) \left[\frac{(x_t - \mu_i)^2}{\sigma_i^2} - 1 \right] \tag{8}$$

where the division between vectors is a term-by-term operation. The final gradient vector G^x is the concatenation of the $G^x_{\mu,i}$ and $G^x_{\sigma,i}$ vectors, for $i = 1...K$. The final feature vector becomes a $2KD$ dimensional vector. At the end, we perform the normalization of the Fisher vectors since [19] has found this to significantly increase performance. The applied normalization is a combination of $L2$ and power normalization ($f(x) = sign(x)\sqrt{\alpha|x|}$) [19].

4 Results

4.1 Dataset

We present our pose estimation and activity recognition results on the ADL dataset. This dataset consists of 10 different activities: *answering a phone, dialing a phone,*

Fig. 1. A sample frame and its corresponding ground truth: (a) body pose tree showing the numbers in the correct positions (b) bounding boxes

looking up numbers in a phone book, writing on a white board, drinking water, eating a snack, peeling a banana, eating a banana, chopping a banana and *eating food with silverware.* Each of these activities is performed 3 times by 5 different people. These people have different genders, ethnicity, and appearance so sufficient appearance variation is available in the dataset. The original frame size is 1280×720. The frame-rate of the videos is 30 frames/s. Each clip is in the range of 3-50s and we extract features at a rate of one frame/s.

4.2 Groundtruth and Performance Evaluation

In this work, we provide qualitative analysis of our approach for body pose estimation and for activity recognition and we compare our results with related works. The ground truth for activity recognition comes with the dataset (*i.e.* each video clip contains a specific activity), but no groundtruth on the body pose is provided. Thus, we annotated 371 frames from different clips of ADL. For the annotation, we followed the procedure indicated in [33]. The example of an annotated frame is shown in Fig.1. Each of the 18 points in Fig.1(a) is the centroid of the bounding box of the corresponding body part of size L (as shown in Fig.1(b)).The accuracy of the body pose estimation is computed by comparing the positions of the groundtruth bounding box B_i^{GT} and of the estimated bounding box B_i^E, for each body part $i = 1, ..., 18$. If the overlap of B_i^E with B_i^{GT} is more than 80%, the body part is considered as being correctly detected. The accuracy of the body pose estimation is obtained by averaging over the accuracies of individual body parts.

4.3 Body Pose Estimation

In Eq. (4), α is a parameter controlling the relative importance of *foreground* and *optical flow* scores. To find the optimal values for different parameters, we tune parameters separately for S_{FG} and S_{OF}. To do so, we first set up $\alpha = 0$ and $\alpha = 1$ and find the optimum solution for (β, γ) and (η, λ), respectively. Then we tune α to get the best relative importance of S_{FG} and S_{OF}.

Varying Parameters of S_{FG} and S_{OF}. Fig. 2(a) and Fig. 2(b) show how varying the parameters γ, β and λ, η changes the detector's performance. We recall that increasing γ and λ respectively decreases the widths of the foreground window and the optical flow window. Choosing a too small value for the parameters γ and λ consequently increases the size of the foreground and optical flow windows which worsen the detection results by bringing background noise into account. Choosing too large values for γ and λ decreases the size of the windows and consequently some low-level information related to the foreground and optical flows is discarded and hence the performance will decrease. In our experiment we found $\gamma = \lambda = 5$ as the best values, and consequently the foreground and optical flow windows have the same size.

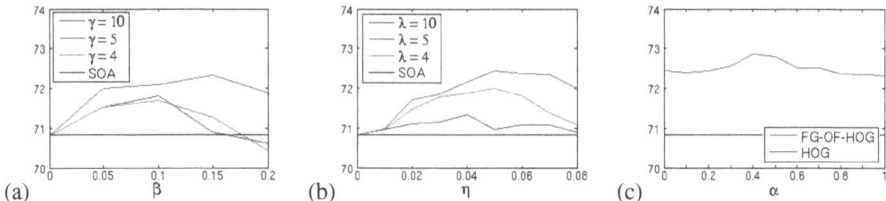

Fig. 2. Body parts detection accuracy at varying parameters (a) γ, β while $\alpha = 1$; (b) λ, η while $\alpha = 0$; (c) α. SOA refers to the results of the body-part detector in [33].

As we previously mentioned, β and η respectively represent the weights of the foreground and optical flow scores. Giving larger weights to the foreground or optical flow scores forces the detector to the ignore information that is obtained from HOG. In our experiments, we found that the best solution for these parameters is $\beta = 0.15$ and $\eta = 0.05$. In Fig. 2(c) we show how the relative pose estimation performance changes by giving different weights (α) to the foreground and optical flow scores. The highest performance is obtained by giving the weight 0.6 to the optical flow score and 0.4 to the foreground score (*i.e.* $\alpha = 0.4$).

Detection Performance on Different Body-Parts. Table 1 presents the best detection performance for different body parts using different local descriptors. Bolded numbers in Table 1 illustrate that applying the foreground descriptor improves the detection performance of the parts that are located in the subject's torso, while the optical flow score improves the detection performance mostly on the subject's hands as in the ADL dataset, usually the hands are moving more than the other parts.

Fig. 3(a) illustrates a sample in which using foreground information (Fig.3(c)) helps the detection of the right hand of the subject (Fig. 3(b)). Fig. 3(d) shows an example in which optical flow information (Fig. 3(f)) helps the body-pose estimator to detect the left hand of the subject in Fig. 3(e).

4.4   Activity Recognition

In Table 2(a) we present the performance of our activity recognition approach for the 2 different representations (we use leave-one-person-out cross-validation): (1) accumulate features descriptors over an entire video sequence and (2) use the Fisher-Kernel representation. The results show that even with the first representation that discards all the information about the temporal order and variation we obtain similar performance to some works in the literature that applied more expensive feature descriptors (see Table 2(b)). Additionally, by applying the Fisher Kernel representation we outperform all

Table 1. Accuracy of different parts of the body. For most of the cases, applying FG-OF-HOG local descriptor achieves a better detection accuracy. The last column represents the overall performance. Bold numbers show which single-descriptor works better on the correspondent part.

Body part	1	2	3	4	5	6	7	8	9	10	11	12	13	14	15	16	17	18	Average
HOG	83.3	89.0	92.2	84.9	67.7	60.9	46.1	84.4	60.9	56.3	89.2	84.4	65.8	63.1	51.2	80.6	60.7	54.5	70.8
FG-HOG	83.7	88.4	93.3	84.6	67.9	58.2	43.7	**87.9**	**64.4**	**62.5**	90.0	**87.6**	67.1	**68.5**	55.5	80.6	60.9	**57.4**	72.3
OF-HOG	**83.8**	**89.0**	93.3	**85.2**	**68.7**	60.7	**48.5**	86.5	63.9	59.8	89.7	86.5	67.4	66.9	55.8	**81.4**	60.9	56.3	72.5
FG-OF-HOG	83.8	89.0	93.3	85.4	70.1	62.0	49.1	86.5	63.9	60.7	89.2	85.4	67.4	68.7	55.5	83.0	61.5	57.1	**72.9**

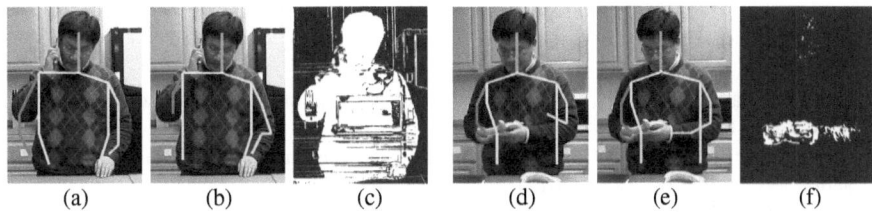

    (a)   (b)   (c)   (d)   (e)   (f)

Fig. 3. Body configuration obtained with (a) [33] and (b) our method, including the information of the foreground mask in the body pose estimation (c). (d) [33] and (e) our method, including the information of the optical flow in the body pose estimation (f).

Table 2. Activity recognition performance: (a) our approach: descriptor accumulation over a video sequence vs. Fisher Kernel representation for different body pose estimation methods (b) Performance comparison with the state-of-the-art on the ADL dataset.

Local descriptor in body part detector	Accumulation	Fisher-Kernel
HOG [33]	87.32	95.71
FG-HOG	88.93	98.57
OF-HOG	87.50	97.14
FG-OF-HOG	89.11	98.75

(a)

Method	Accuracy
Wang *et al.* [31]	96.0
Bilen *et al.* [1]	74.0
Matikainen *et al.* [16]	70.0
Satkin *et al.* [24]	80.0
Bilinski *et al.* [2]	93.33
Kuehne *et al.* [12]	82.0
Messing *et al.* [17]	89.0
Our approach	**98.75**

(b)

the state-of-the-art methods (Table 2(b)). The closest accuracy performance is reported by Wang *et al.* [31]. They applied Multi-Kernel-Learning, while our result is obtained using SVM with RBF kernel. The results with the Fisher Kernel representation are obtained with an optimized number of GMM centroids (the dictionary size), which in this case is equal to 20.

5 Conclusions and Future Work

In this paper we present an efficient method to recognize activities of daily living. We combine the cues obtained from a body pose detector and local motion. This step created a descriptor that uses the structure of located motion. In this way, we involve high-level information combined with the low-level cues. Moreover, we show that including low-level cues (*i.e.* optical flow and foreground) together with an *off-the-shelf* body part detector gives a better performance without the need to re-train the detectors. In fact, we generate optical flow information once, and then apply it for both *enriching the body-part detector* and *quantizing flows* for activity recognition task. We also model the temporal variation within the video using the Fisher Kernel representation. Finally, our novel descriptor with the Fisher Kernel representation achieved the best reporting results so far for the ADL dataset. In future work we plan to extend our approach for more challenging scenarios such as *fine-grained activities* [22].

References

1. Bilen, H., Namboodiri, V.P., Van Gool, L.: Action recognition: A region based approach. In: WACV (2011)
2. Bilinski, P., Bremond, F.: Contextual statistics of space-time ordered features for human action recognition. In: AVSS (2012)
3. Bilinski, P., Corvee, E., Bak, S., Bremond, F.: Relative dense tracklets for human action recognition. In: FG (2013)
4. Dalal, N., Triggs, B.: Histograms of oriented gradients for human detection. In: CVPR (2005)
5. Felzenszwalb, P., McAllester, D., Ramanan, D.: A discriminatively trained, multiscale, deformable part model. In: CVPR (2008)
6. Felzenszwalb, P., Huttenlocher, D.: Pictorial structures for object recognition. IJCV (2005)
7. Ferrari, V., Marin-Jimenez, M., Zisserman, A.: Progressive search space reduction for human pose estimation. In: CVPR (2008)
8. Gaur, U., Zhu, Y., Song, B., Roy-Chowdhury, A.: A string of feature graphs model for recognition of complex activities in natural videos. In: ICCV (2011)
9. Jaakkola, T.S., Haussler, D.: Exploiting generative models in discriminative classifiers. In: NIPS (1999)
10. Johnson, S., Everingham, M.: Clustered pose and nonlinear appearance models for human pose estimation. In: BMVC (2010)
11. Kovashka, A., Grauman, K.: Learning a hierarchy of discriminative space-time neighborhood features for human action recognition. In: CVPR (2010)
12. Kuehne, H., Gehrig, D., Schultz, T., Stiefelhagen, R.: On-line action recognition from sparce feature flow. In: VISAPP (2012)
13. Laptev, I., Marszalek, M., Schmid, C., Rozenfeld, B.: Learning realistic human actions from movies. In: CVPR (2008)
14. Liu, K.H., Weng, M.F., Tseng, C.Y., Chuang, Y.Y., Chen, M.S.: Association and temporal rule mining for post-filtering of semantic concept detection in video. IEEE Trans. MM (2008)
15. Malgireddy, M.R., Nwogu, I., Govindaraju, V.: A generative framework to investigate the underlying patterns in human activities. In: ICCV Workshops (2011)
16. Matikainen, P., Hebert, M., Sukthankar, R.: Representing pairwise spatial and temporal relations for action recognition. In: Daniilidis, K., Maragos, P., Paragios, N. (eds.) ECCV 2010, Part I. LNCS, vol. 6311, pp. 508–521. Springer, Heidelberg (2010)
17. Messing, R., Pal, C., Kautz, H.: Activity recognition using the velocity histories of tracked keypoints. In: CVPR (2009)
18. Mironica, I., Ionescu, B., Uijlings, J., Sebe, N.: Fisher kernel based relevance feedback for multimodal video retrieval. In: ICMR (2013)
19. Perronnin, F., Sánchez, J., Mensink, T.: Improving the fisher kernel for large-scale image classification. In: Daniilidis, K., Maragos, P., Paragios, N. (eds.) ECCV 2010, Part IV. LNCS, vol. 6314, pp. 143–156. Springer, Heidelberg (2010)

20. Qi, G.J., Hua, X.S., Rui, Y., Tang, J., Mei, T., Wang, M., Zhang, H.J.: Correlative multilabel video annotation with temporal kernels. ACM TOMCCAP (2008)
21. Ramanan, D., Sminchisescu, C.: Training deformable models for localization. In: CVPR (2006)
22. Rohrbach, M., Amin, S., Andriluka, M., Schiele, B.: A database for fine grained activity detection of cooking activities. In: CVPR (2012)
23. Sadanand, S., Corso, J.J.: Action bank: A high-level representation of activity in video. In: CVPR (2012)
24. Satkin, S., Hebert, M.: Modeling the temporal extent of actions. In: Daniilidis, K., Maragos, P., Paragios, N. (eds.) ECCV 2010, Part I. LNCS, vol. 6311, pp. 536–548. Springer, Heidelberg (2010)
25. Savarese, S., DelPozo, A., Niebles, J.C., Fei-Fei, L.: Spatial-temporal correlations for unsupervised action classification. In: IEEE Workshop on Motion and Video Computing (2008)
26. Snoek, C.G., Worring, M.: Concept-based video retrieval. FTIR 4(2), 215–322 (2009)
27. Stauffer, C., Grimson, W.E.L.: Adaptive background mixture models for real-time tracking. In: CVPR (1999)
28. Tomasi, C., Kanade, T.: Detection and tracking of point features. CMU-CS (1991)
29. Tran, D., Forsyth, D.: Improved human parsing with a full relational model. In: Daniilidis, K., Maragos, P., Paragios, N. (eds.) ECCV 2010, Part IV. LNCS, vol. 6314, pp. 227–240. Springer, Heidelberg (2010)
30. Wang, H., Ullah, M.M., Klaser, A., Laptev, I., Schmid, C.: Evaluation of local spatio-temporal features for action recognition. In: BMVC (2009)
31. Wang, J., Chen, Z., Wu, Y.: Action recognition with multiscale spatio-temporal contexts. In: CVPR (2011)
32. Willems, G., Tuytelaars, T., Van Gool, L.: An efficient dense and scale-invariant spatio-temporal interest point detector. In: Forsyth, D., Torr, P., Zisserman, A. (eds.) ECCV 2008, Part II. LNCS, vol. 5303, pp. 650–663. Springer, Heidelberg (2008)
33. Yang, Y., Ramanan, D.: Articulated human detection with flexible mixtures-of-parts. PAMI (2013)
34. Zen, G., Rostamzadeh, N., Staiano, J., Ricci, E., Sebe, N.: Enhanced semantic descriptors for functional scene categorization. In: ICPR (2012)
35. Zhang, Y., Liu, X., Chang, M.-C., Ge, W., Chen, T.: Spatio-temporal phrases for activity recognition. In: Fitzgibbon, A., Lazebnik, S., Perona, P., Sato, Y., Schmid, C. (eds.) ECCV 2012, Part III. LNCS, vol. 7574, pp. 707–721. Springer, Heidelberg (2012)

What Epipolar Geometry Can Do
for Video-Surveillance

Nicoletta Noceti, Luigi Balduzzi, and Francesca Odone

DIBRIS - Università degli Studi di Genova
via Dodecaneso, 35 - 16146-IT, Genova
{Nicoletta.Noceti,Luigi.Balduzzi,Francesca.Odone}@unige.it

Abstract. In this paper we deal with the problem of matching moving objects between multiple views using geometrical constraints. We consider systems of still, uncalibrated and partially overlapped cameras and design a method able to automatically learn the epipolar geometry of the scene. The matching step is based on a functional that computes the similarity between objects pairs jointly considering different contributions from the geometry. We obtain an efficient method for multi-view matching based on simple geometric tools, requiring a very limited human intervention, and characterized by a low computational load. We will discuss the potential of our approach for video-surveillance applications on real data, showing very good results. Also, we provide an example of application to the consistent labeling problem for multi-camera tracking, and report a comparative analysis with other methods from the state of the art on the PETS 2009 benchmark dataset.

Keywords: Epipolar geometry, multi-view object tracking, video-surveillance.

1 Introduction

State-of-the-art video-surveillance systems available on the market often adopt multiple cameras to be able to monitor large environments and tackle complex situations [1]. Quite surprisingly, the algorithms processing the acquired video streams rarely exploit prior information on the systems geometry. On one hand, in minimal configuration setups cameras have a small or null overlap, and thus system calibration becomes difficult and often not enough reliable. On the other hand, redundant setups are characterized by large field of views overlaps, making the calibration process more reliable but time-consuming. Also, all calibration procedures usually require a high degree of intervention of specialized users and may be not always accepted by surveillance systems installers.

In this work we consider systems of still, partially overlapped and uncalibrated cameras, observing generic environments with a moderate crowding level. Our goal is to build a model of the overall scene dynamics evolving over time. The method we propose is based on a coarse annotation of the scene, that identifies the main walkable components, as ground floor and stairs, that we approximate

A. Petrosino (Ed.): ICIAP 2013, Part I, LNCS 8156, pp. 442–451, 2013.

with planes. The annotation is the only part in the whole pipeline requiring human intervention. Given a pair of overlapped cameras, we relate the scenes at a global level – estimating the fundamental matrix – and at a more local one, building a homography relationship between each pair of homologous regions. Global and local geometrical constraints jointly contribute – possibly with different weights – to the evaluation of the similarity between objects observed in different views. Computing the pairwise similarity between all the objects at a given time t, we populate a matrix from which we deduce matching relationships and missing elements.

Over the last decades several methods adopting geometry within multi-camera systems have been proposed. Among the first attempts, the work in [12] addresses the problem of self-calibration of multiple cameras using feature correspondences to determine the camera geometry. It assumes planarity of the observed scene and sets the basis for working with an overhead view. In [14,17,19] calibration is used to model the 3D relationships between overlapped cameras. However, in most cases full calibration is not available, thus geometry is recovered estimating geometrical transformations between the views from image features. Multi-object matching has been addressed by imposing or learning geometrical constraints on the observed scene [13,2,6,4,16,3,9], often assuming planarity of the ground [16,3,9,13]. Some methods propose to precisely estimate the boundaries of the Field Of View (FOV) to disambiguate among the multiple possible objects associations [3,9], other methods tackle the same problem by combining geometry with appearance models [5,4].

The main contribution of this work is an analysis of what well-accepted geometrical tools can do to improve the reliability of real video-surveillance systems. The result is a method that, given a coarse annotation of the scene geometry, first provides a viable calibration procedure, and second builds a model of the scene dynamics which we use to match objects across the views. This model could be applied to deal with occlusions, tracking noise and consistent labeling. We do not add constraints on the environments and do not need to explicitly determine the common fields of view. Our method requires a very limited human intervention and is computationally very efficient, providing real-time performances.

We experimentally evaluate the multi-camera matching *per se* – addressing the so-called consistent labeling problem [9] – on both annotated data and observations obtained from a tracker to discuss the potential of our solution. Then, we also evaluate the accuracy of our method within multi-camera tracking on the benchmark data of *PETS 2009*, comparing them with the state of the art.

The rest of the paper is organized as follows. In Sec. 2 we discuss the estimation of the geometry, while in Sec. 3 we detail our approach. Sec. 4 and 5 are devoted to the experimental analysis, and, finally, Sec. 6 is left to discussions.

2   Estimation of the Geometry between Two Views

In this section we discuss a simple way to estimate the epipolar geometry between camera pairs, specifically designed to be practicable for video-surveillance systems installations.

The correspondences between the views can be easily established by considering videos with a single person walking, spanning all the walkable floor regions of the scene, similarly to [3]. We employ a motion segmentation algorithm to locate the moving objects and extract two points for each object: the head, or *upper point* (**up**), and the feet, or *lower point* (**lp**). We show in Fig. 1 an example of input for a given camera pair (upper points in green, lower points in blue). We thus finally collect the ordered sets $UP^c = \{\mathbf{up}_k^c\}_{k=1}^K$ and $LP^c = \{\mathbf{lp}_k^c\}_{k=1}^K$, where $c = \{1, 2\}$ from now on refers to the camera index and K is the number of corresponding points. We also call P^c the union of lower and upper points for each camera: $P^1 = UP^1 \cup LP^1$, $P^2 = UP^2 \cup LP^2$.

Fig. 1. An example of input corresponding points (upper points in green, lower points in blue). Two main regions are annotated, which correspond to ground floor and stairs.

To cope with possible non-planarities of the scene, each image plane is coarsely manually annotated to identify the main structural elements. We only consider walkable, not occluded regions (see Fig. 1) that may be approximated with a plane and characterized by a significant spatial extent.

Let us assume $\mathbf{R} = \{R_n\}_{n=1}^N$ is the set of regions globally annotated in the two cameras. We first consider the transformation between the image planes as a whole, that is the fundamental matrix F: $(\mathbf{p}^2)^T F \mathbf{p}^1 = 0$ with $\mathbf{p}^1 \in P^1$ and $\mathbf{p}^2 \in P^2$ corresponding points. Then we focus on the M *regions observed in both the views*, $M \leq N$. For each one of those, we estimate the homography H_m, such that $\mathbf{p}^2 = H_m \mathbf{p}^1$, where $p^1 \in LP^1$ lies on R_m^1 and $p^2 \in LP^2$ lies on R_m^2.

We solve all the obtained systems using the Direct Linear Transformation (DLT) algorithm with RANSAC [7], that allows us to cope with the presence of outliers (strongly affecting our input data) and avoid unstable solutions. At the end of the calibration procedure, we have obtained the matrices modeling the geometrical transformation from scene 1 to scene 2: F and H_m, $m = 1 \ldots M$.

3   Matching across Views

In this section we describe our approach to matching between multiple views. For the sake of clarity, in what follows we consider a single pair of overlapped cameras. In presence of more than two pairwise overlapped cameras, a graph modeling the transitions between cameras can be defined [16,3] to guide the

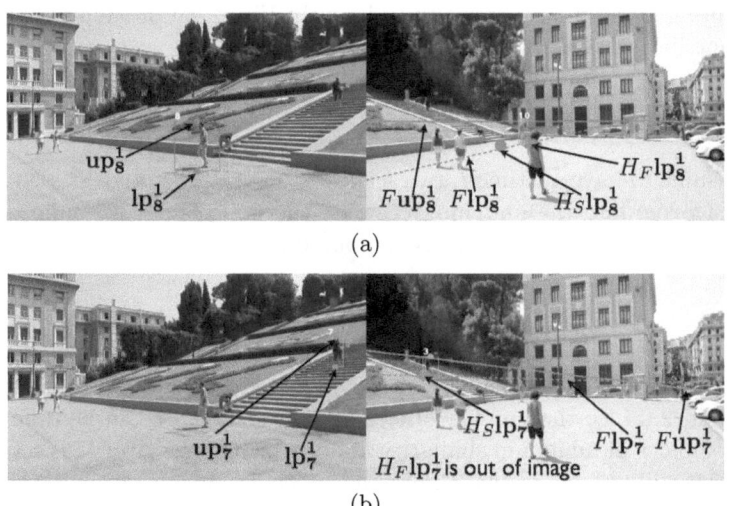

(a)

(b)

Fig. 2. The different contributions to the similarity measure of a person walking on the ground floor (Fig. 2(a)) and on the stairs (Fig. 2(b)). F is the fundamental matrix, while H_F and H_S refer to the homographies related to ground floor and stairs regions, respectively.

visit of the cameras network. The transitive property of subsequent assignments on different cameras guarantees the global consistency of the labeling.

3.1 Geometry-Based Objects Similarity

Let $O_t^1 = (\mathbf{up}_t^1, \mathbf{lp}_t^1)$ and $O_t^2 = (\mathbf{up}_t^2, \mathbf{lp}_t^2)$ be the descriptors of two objects observed in scene 1 and 2 respectively. We define $d_1 = d(F\mathbf{lp}_t^1, \mathbf{lp}_t^2)$ and $d_2 = d(F\mathbf{up}_t^1, \mathbf{up}_t^2)$, where d denotes the geometric distance between the epipolar lines and a point. Then we introduce the contributions of the regions R_m, $1 \le m \le M$ common to the views: $d_m = ||H_m\mathbf{lp}_t^1 - \mathbf{lp}_t^2||_2$.

The similarity between the objects is a combination of all the contributions:

$$S(O_t^1, O_t^2) = w_1 \exp\left(\frac{-d_1^2}{2\sigma^2}\right) + w_2 \exp\left(\frac{-d_2^2}{2\sigma^2}\right) + \sum_{m=1}^{M} w_{3+m-1} \exp\left(\frac{-d_m^2}{2\sigma^2}\right) \quad (1)$$

where σ controls the spatial region in which associations should be considered, while the ws weight the importance of each contribution to the final results. They might be chosen depending on prior information when available, or estimated from the data with an appropriate training procedure.

Notice that our method does not require a precise estimation of the common fields of view. Although we restrict the analysis to common regions only, the areas actually observed by each camera might only partially overlap. Thanks to the use of different geometrical contributions, we are able to automatically cope with points missing in one of the two views. In Fig. 2 we provide two examples of the contributions to the similarity measure.

3.2 Objects Matching

Let us assume to have, at time t, two scenes $S_t^1 = \{O_{i,t}^1\}_{i=1}^{N_1}$ and $S_t^2 = \{O_{j,t}^2\}_{j=1}^{N_2}$. In order to compute the matching between the two scenes, we build a matrix $M \in \mathbb{R}^{N_1 \times N_2}$ where each element is computed as

$$M(i,j) = \frac{S(O_{i,t}^1, O_{j,t}^2) + S(O_{j,t}^2, O_{i,t}^1)}{2}. \tag{2}$$

This matrix models the dynamics of the scene at time t as observed from the views 1 and 2. Since the number of elements in the scenes can be different, we fix a threshold τ of minimum similarity under which an entry of M is set to zero.

From M we may deduce which are the objects belonging to both the views (matches) and what objects are present in one view only (objects without a match). In the case of noiseless data, we could assume that when an object of scene 1 is viewed also in scene 2 then it will correspond to *exactly* one of its objects. In this case a match could be indicated by an entry in M maximum on its row and column (e.g. through the Hungarian algorithm [10]). Missing elements of scene 1 would be denoted by a null row, missing elements from scene 2 by a null column.

Fig. 3. An example of segmentation error: object 1 of scene 1 corresponds to objects 8 and 9 of scene 2

However in real world applications, the data intrinsically contain noise, due to error in the object segmentation and to objects partially overlapping. Thus, it is very likely to happen that an object of one view actually corresponds to more than one object in the other (see e.g. Fig. 3). We thus modify the matching rule to account for not univocal associations, by imposing that if $M(i,j) > 0$ then it is the only one on its row *or* on its column. If not so, we progressively simplify the matrix by rejecting the lowest values of a subregion of M until the condition holds. The subregion is identified by rows and columns of the elements in the i-row ($M(i,-)$) or in the j-column ($M(-,j)$) greater than zero.

4   Experiments on Multi-camera Matching

We validated our approach to multi-camera matching on a dataset acquired in-house (the dataset will be available for download at `http://slipguru.disi.unige.it`). It consists of 4 cameras with partially overlapped views (see Fig.4) that monitor a moderately crowded outdoor environment. All the cameras but *Cam* observe both planar and non-planar areas. A ground truth of the trajectories is available, with a common identifier between all the views for each person.

We annotated ground floor and stairs in each scene and estimated the projection matrices F, H_F and H_S. To evaluate the performance of our approach we interpret the data by considering matched objects as positives and objects without a match as negatives. We take into account both, estimating Positive Predictive Value, True Positive Rate, Negative Predictive Value, True Negative Rate, Accuracy and F-measure. In all the experiments, we fix $\tau = 0.5$ and $\sigma = 15$, while the weights ws are automatically selected on a training set using a grid-search approach (each w is sampled in the range $[0, 1]$ with sampling step 0.05 and such that the weights sum up to 1) and selecting the best performing combination in terms of matching correctness. We use the first minute of each video as training set, the remainder (about 2′) is adopted as a test set.

4.1   Assessment on Annotated Data

We first assess our approach on the annotated objects. In Tab. 1 are the weights learned automatically from the data, that reflect the peculiarities of the scenes pair. In the case of the pair *Ixus-Cam*, e.g., a great importance is done to H_F, since *Cam* only observes the floor. The homography is also rather significant for *Nikon-Cam* even if the distance between the common observed floor region and *Nikon* makes the contribution of H_F more noisy and thus the measures obtained with F gain a higher weight. When the cameras observe ground floor and stairs both the homographies are taken into account and enforced with at least one contribution from the epipolar geometry.

Tab. 2 reports the comparison of the methods with weights learning (last row) with a selection of other configurations. The case $H_F\mathbf{lp} + H_S\mathbf{lp}$ and $F\mathbf{lp} + F\mathbf{up}$ are considered as minimal configurations that explicitly account for scene with multiple planar regions. As shown, the weights learning allows us to reach the best results, showing the capability of adapting to different scene peculiarities.

| (a) *Ixus* | (b) *Nikon* | (c) *Cam* | (d) *Rx100* |

Fig. 4. Example frames of the dataset we adopted in the experimental analysis

Table 1. Weights selected via the training stage

Camera pair	W			
	Flp	Fup	$H_F$lp	$H_S$lp
Ixus-Nikon	0	0.6	0.2	0.2
Ixus-Cam	0	0.2	0.8	0
Ixus-Rx100	0	0.4	0.1	0.5
Nikon-Cam	0.2	0.3	0.5	0
Nikon-Rx100	0.3	0	0.4	0.3
Cam-Rx100	0	0.5	0.5	0

Table 2. Average performance of the matching procedure over all cameras pairs on annotated data

Configuration	PPV	TPR	NPV	TNR	ACC	F
$H_F$lp	**0.98**	0.74	0.67	0.99	0.80	0.79
$H_S$lp	0.94	0.56	0.40	**1.00**	0.67	0.65
Flp	0.76	0.72	0.68	0.93	0.81	0.78
Fup	0.62	0.57	0.54	0.86	0.69	0.64
$H_F$lp + $H_S$lp	0.96	0.74	0.54	0.99	0.80	0.79
Flp + Fup	0.79	0.73	0.68	0.92	0.82	0.79
Our approach	**0.98**	**0.96**	**0.91**	0.99	**0.97**	**0.97**

4.2   Evaluations on Measured Data

We now move to the analysis on real data. At each time instant we first apply a motion-based object segmentation and update a tracking [15] on each view independently, and then run the multi-view matching. We consider a match as correct if the two objects involved correspond to the same identity in the ground truth. The correspondence might be partial because of not univocal associations. Similarly, an object of one scene is considered correctly not matched if in the other view it is not observed or a detection is missing for it. The average performances (Tab. 3) show very accurate results although a decrease in the

Table 3. Average performance of the matching procedure over all cameras pairs on measured data

Configuration	PPV	TPR	NPV	TNR	ACC	F
$H_F$lp	0.61	0.58	0.76	0.79	0.71	0.71
$H_S$lp	0.36	0.21	0.67	**0.94**	**0.77**	0.73
Flp	0.45	0.80	0.88	0.54	0.60	0.64
Fup	0.36	0.77	0.86	0.40	0.49	0.53
$H_F$lp + $H_S$lp	**0.66**	0.57	0.78	0.82	0.75	0.75
Flp + Fup	0.42	**0.87**	**0.91**	0.43	0.54	0.58
Our approach	0.65	0.84	**0.91**	0.73	0.76	**0.79**

values due to the noise in the data. In this gap we can read an intrinsic limit of this approach: it is expected to increase as the crowd level in the scene grows, influencing the matching accuracy.

5   Experiments on Multi-camera Tracking

To show a possible application for video-surveillance, we apply our matching strategy to assign a common identifier to the same person observed from different views. This problem is commonly referred to as *consistent labeling*. At each time instant, we consider the tracking history from each camera and apply the multi-view matching. If for a given time interval a match has been continuously detected we assign a common identifier to the two objects and label the match as stable. Then, we exploit the association to recover the identities of objects whose trajectory has been cut in subparts, due to tracking failures.

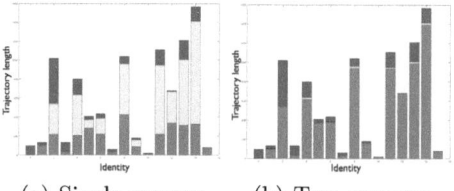

	Ixus	*Nikon*	*Cam*	*Rx100*
Ixus	0.61	0.98	0.97	0.97
Nikon	0.98	0.45	0.96	0.98
Cam	0.99	0.99	0.67	0.98
Rx100	0.95	0.95	0.94	0.48

(a) Single-camera   (b) Two-cameras

Fig. 5. Tracking results using 2 cameras observations to compute consistent labeling

We first consider our dataset and show in Fig. 5 the tracking performance for a cameras pair (*Ixus-Nikon*, the one in Fig. 1). Each bar corresponds to an identity in the ground truth, the height reflects the trajectory length. In red we denote the length of the *annotated* trajectory, while in yellow the length of the *measured* one. The latter might be lower due to tracking failures. The magenta bars give a visual impression of the spatio-temporal overlap between trajectories annotated and reconstructed with single-view (Fig. 5(a)) or multi-view (Fig. 5(b)) tracking. The latter clearly allows us to recover from tracking failures.

To show such capability in general, we evaluate the average percentage of spatio-temporal overlap between annotated and reconstructed trajectories. If more than one trajectories correspond to the same annotated object, we only consider the longest. We report the results in table of the right of Fig. 5 (1 means full overlap). The diagonal brings information on single view tracking, while the other values (i,j) tell us how much the i-th camera benefits from the mutual observations with the j-th camera. The table nicely show the gain of multi-view analysis.

We finally evaluate the performances of our approach on the benchmark dataset *PETS 2009*[1]. Due to the limited number of observations, we avoid the

---

[1] http://www.cvg.rdg.ac.uk/PETS2009/a.html

weights learning and instead force the geometrical constraints (F and a single H) to have the same importance. We compare our results on tracking accuracy and missing detections (evaluated following the *CLEAR* metric [8]) with the analysis reported in [11], where the authors propose a method to jointly track multiple objects in multiple views based on formulating the assignment problem as a min-cost problem. Tab. 4 reports the comparison: for camera pairs, our method performs comparably to [11], but differently from [11] we have a gain when increasing the number of cameras to three.

Table 4. Performance evaluation on *PETS 2009* benchmark data. In brackets, the number of cameras adopted for the evaluations.

Method	TA	Miss. Det.
Zhan et al.(1) [18]	**0.66**	0.28
Our approach [15] (1)	0.5	**0.26**
Proposed in [11] (2)	0.76	**0.17**
Our approach (2)	**0.79**	0.19
Proposed in [11] (3)	0.71	**0.13**
Our approach (3)	**0.8**	0.16
Berclaz et al. [1] (5)	0.75	–

6   Discussions

In this paper we showed how very simple geometrical tools can be profitably adopted within multi-camera surveillance setups. We considered systems of still, partially overlapped and uncalibrated cameras and proposed a multi-view matching strategy based on geometrical constraints. Our method estimates the epipolar geometry, and is based on a coarse annotation of the scene. We designed a similarity function making use of different geometry ingredients with variable importances, and showed in the experimental analysis – performed on real data – that learning the weights directly from the data allowed us to automatically adapt to general environment. We reported object matching performances on both annotated and measured data, validating our approach. We finally discussed the potential of our method to address the consistent labeling problem. We compared our method with other state-of-art approaches on the benchmark dataset *PETS 2009*, showing the benefit of increasing the number of cameras.

As a future development, we will integrate the matching module in a real surveillance setting. This will allow us to, on one hand, collect large amount of data, while, on the other, test the robustness of the method with respect to time.

References

1. Berclaz, J., Fleuret, F., Turetken, E., Fua, P.: Multiple object tracking using k-shortest paths optimization. PAMI 33(9), 1806–1819 (2011)
2. Black, J., Ellis, T.: Multi-camera image measurement and correspondence. Measurement 32(1), 61–71 (2002)

3. Calderara, S., Cucchiara, R., Prati, A.: Bayesian-competitive consistent labeling for people surveillance. PAMI 30(2), 354–360 (2008)
4. Chang, T., Gong, S., Ong, E.: Tracking multiple people under occlusion using multiple cameras. In: BMVC, pp. 566–576 (2000)
5. Chang, T.H., Gong, S.: Tracking multiple people with a multi-camera system. In: Work. on Multi-Object Tracking (2001)
6. Dockstader, S., Tekalp, A.: Multiple camera tracking of interacting and occluded human motion. Proc. of the IEEE 89(10), 1441–1455 (2001)
7. Fishler, M., Bolles, R.: Random sample consensus: a paradigm for model fitting with applications to image analtsis and automated cartography. In: Comm. ACM, vol. 24, pp. 381–395 (1981)
8. Kasturi, R., Goldgof, D., Soundararajan, P., Manohar, V., Garofolo, J., Bowers, R., Boonstra, M., Korzhova, V., Zhang, J.: Framework for performance evaluation of face, text, and vehicle detection and tracking in video: Data, metrics, and protocol. PAMI 31(2), 319–336 (2009)
9. Khan, S., Shah, M.: Consistent labeling of tracked objects in multiple cameras with overlapping fields of view. PAMI 25(10), 1355–1360 (2003)
10. Khun, H.: The hungarian method for the assignment problem. Naval Research Logistic Quarterly 2, 83–97 (1955)
11. Leal-Taixe, L., Pons-Moll, G., Rosenhahn, B.: Branch-and-price global optimization for multi-view multi-target tracking. In: CVPR, pp. 1987–1994 (2012)
12. Lee, L., Romano, R., Stein, G.: Monitoring activities from multiple video streams: Establishing a common coordinate frame. PAMI 22(8), 758–767 (2000)
13. Mittal, A., Davis, L.: Unified multi-camera detection and tracking using region-matching. In: Work. on Multi-Object Tracking, pp. 3–10 (2001)
14. Mittal, A., Davis, L.S.: M2Tracker: A multi-view approach to segmenting and tracking people in a cluttered scene using region-based stereo. In: Heyden, A., Sparr, G., Nielsen, M., Johansen, P. (eds.) ECCV 2002, Part I. LNCS, vol. 2350, pp. 18–33. Springer, Heidelberg (2002)
15. Noceti, N., Destrero, A., Lovato, A., Odone, F.: Combined motion and appearance models for robust object tracking in real-time. In: AVSS (2009)
16. Stauffer, C., Tieu, K.: Automated multi-camera planar tracking correspondence modeling. In: CVPR, vol. 1, pp. I–259 (2003)
17. Yue, Z., Zhou, S.K.: Robust two-camera tracking using homography. In: Int. Conf. on Acoustics, Speech, and Signal Processing, pp. 1–4 (2004)
18. Zhang, L., Li, Y., Nevatia, R.: Global data association for multi-object tracking using network flows. In: CVPR, pp. 1–8 (2008)
19. Zhou, Q., Aggarwal, J.K.: Object tracking in an outdoor environment using fusion of features and cameras. Image Vision Comput. 24(11), 1244–1255 (2006)

Class Representative Computation Using Graph Embedding

Fahri Aydos, Ahmet Soran, and M. Fatih Demirci

TOBB University of Economics and Technology,
Computer Engineering Department,
Sogutozu Cad. No:43, 06560 Ankara, Turkey
{aydos,asoran,mfdemirci}@etu.edu.tr

Abstract. Due to representative power of graphs, graph-based object recognition has received a great deal of research attention in literature. Given an object represented as a graph, performing graph matching with each member of the database in order to locate the graph which most resembles the query is inefficient especially when the size of the database is large. In this paper we propose an algorithm which represents the graphs belonging to a particular set as points through graph embedding and operates in the vector space to compute the representative of the set. We use the k-means clustering algorithm to learn centroids forming the representatives. Once the representative of each set is obtained, we embed the query into the vector space and compute the matching in this space. The query is classified into the most similar representative of a set. This way, we are able to overcome the complexity of graph matching and still perform the classification for the query effectively. Experimental evaluation of the proposed work demonstrates the efficiency, effectiveness, and stability of the overall approach.

Keywords: object recognition, graph embedding, clustering.

1   Introduction

Graph-based object recognition has received a great deal of research attention in literature. Once the objects are represented as graphs such that vertices correspond to features and edges encode some relations between the features, the problem of object recognition is reformulated as that of exact or inexact graph matching. Given a query represented as a graph, one can perform graph matching with each member of the database to locate the graph which most resembles the query. Considering the fact that graph matching is computationally expensive, this experimental setup is not practical especially when the size of the database is large.

To deal with this problem, some techniques concentrates on computing the representative graphs for a given set. Jiang et al. defines the median graph as the one whose sum distance to the other graphs of the set is minimized [11]. The median graph is drawn from the set, however, a more general concept of

A. Petrosino (Ed.): ICIAP 2013, Part I, LNCS 8156, pp. 452–461, 2013.

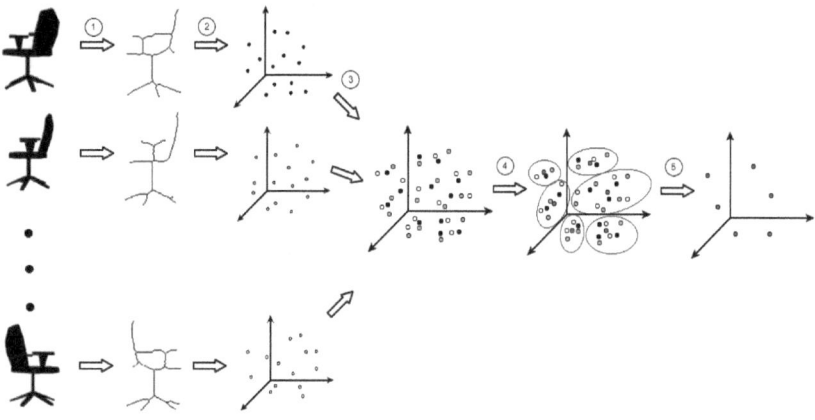

Fig. 1. Overview of the proposed algorithm. Each view of an object class is represented as a tree (transition 1). After embedding trees into the vector space (transition 2), we bring the embedded points to the same dimension (transition 3). The k-means clustering algorithm is then used to learn the centroids, which forms the class representatives (transitions 4 and 5).

the set median graph, which is not constrained to come from the set, is also introduced in this work. Both the median and set median graphs are used as the set representatives. This method has a number of applications where the graph representatives are needed. However, the cost of computing the such graphs are exponential in the number of database graphs [3].

The concept of graph embedding has been used by a number of approaches to map the graphs into the vector space, where the final matching is performed. The main advantage of moving from graph space to vector space is two-fold: using the representational power of graphs and approximating some computationally expensive algorithms in polynomial time. For instance, Demirci et al. [6,5] propose an algorithm using a graph embedding such that each vertex in the graph correspond to a point and the distance between a vertex pair is realized by its corresponding points. Once this transformation is computed, the matching and distance between two graphs can be easily obtained using their point sets. The reader is referred to [10,1] for an extensive review of graph embedding.

When graphs are embedded, the median and mean of these point sets can be easily computed in the vector space. Ferrer et al. [7] propose such an algorithm, which first embeds the graphs into the vector space and computes the median for the corresponding point sets using the Weiszfeld algorithm [17]. Obtaining the median in the vector domain requires simpler operations than those in the graph domain. Given a median in the vector space, the algorithm then uses a triangulation procedure that enables to go back to the graph domain in order to obtain an approximation of the median graph.

Lozano and Escolano [12] study the related problem of learning a probabilistic graph from a set of input graphs, in which node and edge probabilities show relative frequencies of node and edges in the input graphs. Torsello and Hancock [15,14] use a union tree model to capture the within-class variation. This model is learned from the class members by optimizing a minimum description length criterion. The authors' more recent approaches on this subject concentrates on tree edit-distance and tree clustering [16,18].

Although powerful, after computing the representative graph for a given set, one still faces the computational complexity of graph matching in order to obtain the most similar representative graph to the query. In this paper we propose an algorithm which represents the graphs belonging to a particular set (or, an object class) as points through graph embedding and operates in the vector space to compute the representative of the set. We use the k-means clustering algorithm to learn centroids forming the representatives. Once the representative of each set is obtained, we do not go back to the graph domain as done by the previous work. Instead, we embed the query into the vector space and compute the matching in the vector domain. The query is classified into the most similar representative of a set. This way, we are able to overcome the complexity of graph matching and still perform the classification for the query. Experimental evaluation of the proposed work shows that our approach is more efficient and effective than exhaustive search of the query in the original database without the representatives. Figure 1 shows an overview of the proposed approach.

The rest of the paper is organized as follows. In Section 2, we describe the graph embedding procedure, which takes a set of input graphs (in particular, trees) into the vector space and presents how the representative points are computed. Since the embedding procedure we used in this paper works with trees, we also introduce the process of representing input objects as trees in this section. In Section 3 we evaluate the proposed framework by conducting experiments in the domain of object recognition. Finally, we finish the paper with the conclusion and future work in Section 4.

2   Embedding into Vector Spaces and Computing Representative Points

Graph embedding represents an input graph as a set of points in a vector space such that the distance between a pair of vertices in the graph space realized by the distance between their corresponding points in the vector space. In this paper we follow the embedding procedure used in [5]. Since this embedding process is designed for trees, our algorithm starts by representing input images as trees through skeleton points. After defining this process in Section 2.1, we present the embedding procedure and for embedded points, we describe how we compute their representatives in Section 2.2 and Section 2.3, respectively.

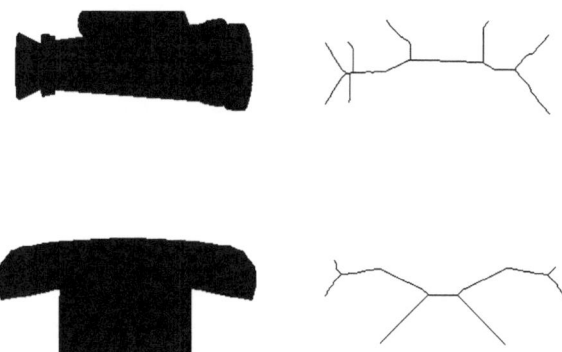

Fig. 2. Skeleton points of two silhouette images in left are shown in right

2.1   Tree Representations for Input Images

A skeleton point (or, shock point) is defined in [8] as the dynamic view of the medial axis where the propagation of waves from the shape boundary results in the formation of singularities. In [2] medial axis is described as the locus of centers of circles inside the region which are bi-tangent to the boundary in at least two places. Each skeleton point p is associated with a 3-dimensional vector $v(p) = (x, y, r)$, where (x, y) are the Euclidean coordinates of the point and r is the radius of the maximal bi-tangent circle centered at the point. Each shock point represented as a vertex in the skeleton graph, which takes over significant role especially on structural and statistical pattern recognition. Each pair of skeleton points in the graph is connected by an edge whose weight reflects the Euclidean distance between them. We convert the graph to a tree by computing its minimum spanning tree. As a result, nodes correspond to skeleton points, and edges connect nearby skeleton points. The root of the tree is the node that minimizes the sum of the shortest path distances to all other nodes in the tree. Figure 2 shows two silhouette images and their skeleton points.

2.2   Embedding through Caterpillar Decomposition

The embedding procedure used in this paper is based on a graph-theoretical concept, caterpillar decomposition, which can be defined as a way to represent a tree by its edge-disjoint root-leaf paths. This concept has been used before by Demirci [4] for efficient retrieval of database graphs, which are similar to a given query. The concept of the caterpillar decomposition is described in a sample tree shown in Figure 3. The two edge-disjoint root-leaf paths between a and h and a and i are called level 1 paths and represent first two paths in caterpillar decomposition. If we remove these two level 1 paths from the tree, we are left with the subtree rooted at c in which two edge-disjoint root-leaf paths exist.

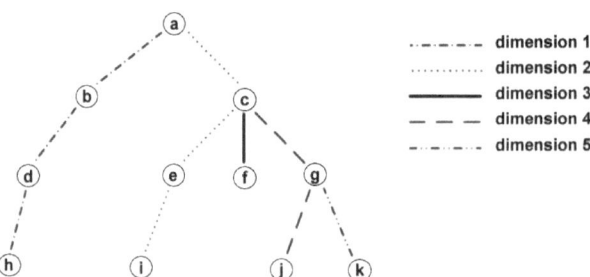

Fig. 3. The union of each edge-disjoint (sub)root-leaf paths forms the caterpillar decomposition of the tree. Each path in the caterpillar decomposition corresponds to a dimension in the vector space.Thus, the tree in this example is represented in a 5-dimensional vector space.

These are the paths between c and f and c and j, called level 2 paths, which represent the other two paths in caterpillar decomposition. Upon removal of these paths, only one path, between g and k, remains and this path is the level 3 path. If removing the level 3 path had left additional connected components, the process would be repeated until all the edges in the tree had been removed. The union of the paths is called the caterpillar decomposition of the tree. The total number of paths in the caterpillar decomposition specifies the dimensionality of the vector space into which the tree is embedded, i.e., each path corresponds to a different dimension. Thus, the tree shown in Figure 3 is embedded into 5 dimensional vector space. The reader is referred to [13] for a detailed description of the caterpillar decomposition and its properties.

To compute the coordinate of a vertex in the vector space, we follow the path from this vertex up to the root (the root is always embedded in the origin). The weight of the potion of the path in each level specifies the value of the coordinate for the corresponding dimension. As an example, the path between k and a consists of three edges; one between k and g corresponding to dimension 5, the other one between g and c corresponding to dimension 4, and the last one between c and a corresponding to dimension 2. Assuming that each edge has a weight of 1.0, the coordinate of vertex k is (0,1,0,1,1). It is easy to see that this embedding is distortion-free under the ℓ_1 norm. That is, the weight of the unique path between any pair of vertices in the tree equals the ℓ_1 distance of their corresponding points in the vector space.

One may notice that this embedding procedure is not unique and depends on the selection of the root and the root-leaf paths. To be consistent in our procedure, the vertex with total minimum distance to all other vertices is chosen as the root and the paths in the caterpillar decomposition is selected based on their weights. This way we ensure that the vertices of one tree is always represented by the same points in the vector space.

Note that since the dimension of the vector space into which an input tree is embedded is defined by its caterpillar decomposition, two trees with different structures are likely to be embedded into different dimensions. Thus before we proceed with computing the representative points of the set, we should perform a post-processing step whose objective is to bring the embedded points into the same dimension. There are two possible ways to do this. One is to bring the higher dimensional embeddings into the lower dimensions by some dimensionality reduction techniques, such as PCA. In general this is not possible without introducing some distortion. The other one is to bring the lower dimensional embeddings up to the higher dimensions by padding them with zeros. As used in [5], we employed the latter method since it does not result in distortion.

2.3   Computing Representative Points

Having embedded the input trees of one class into the same vector space as a set of points, we compute the representative points based on the k-means clustering algorithm. Given a set of n points in $d-$dimensional space $\Re^d$, the objective here is to find k representative points in $\Re^d$ such that the distance between each point to its closest representative point is minimized. More specifically, the k-means algorithm starts by initializing the cluster representatives. After assigning each point to its nearest cluster representative, the representatives are recomputed based on the current cluster memberships. This last step is repeated until no further change in the assignment of the data to new cluster representatives occurs. Although k-means is an effective clustering approach, it is well-known that it may converge to a local optimum. In order to avoid this, we start the algorithm from several initial cluster representatives, increasing the likelihood of obtaining a global optimum.

In the original formulation of k-means, each point has an equal importance in computing the representative of the cluster. However, in our case each point has a different importance reflected by its weight. Therefore, to adapt the original formulation of k-means to our application, we follow the following process. For each point $v(p) = (x, y, r)$ in the set, we insert as many points to location (x, y) as the value of r. This process moves cluster representatives to points with higher weights.

Given a set of images belonging to the same object class, we apply the above process to obtain their representative points. In order to classify the query, it is first embedded into the vector space and the distance with the representative points of each object class in the vector space is computed. As mentioned before, the proposed framework does not require an extra step to go back to the graph domain avoiding the computational complexity of graph matching.

3   Experiments

In this section we evaluate the proposed approach in the context of an object recognition experiment. We use a silhouette dataset, consisting of 9 different

Fig. 4. The top two row presents sample silhouettes from the dataset, while the bottom row shows some sample views for the same object class

objects with 180 views for each, i.e., objects are rotated 2 degrees in depth and for each rotation its 2D view is captured. The top two rows of Figure 4 presents sample silhouettes, while the bottom row shows some sample views for the same object. To test the proposed approach, we selected 19 equidistant views of each object. Each silhouette in the dataset is represented as a rooted tree using the method described in Section 2.1.

To evaluate the proposed approach on this database, we first compute the representative points for each class in the database using the proposed framework. Each view in the database is then used as a query. We say that the proposed approach is correct if the closest representative set to the query has been formed using the views of the query class. In our experiments we use the Hausdorff distance [9] to compute the dissimilarity between two point sets. Given two point sets, $P = \{p_1, \ldots, p_m\}$ and $Q = \{q_1, \ldots, q_n\}$, the Hausdorff distance is defined as follows:

$$d_H(P, Q) = \max\{\sup_{p \in P} \inf_{q \in Q} d(p, q), \sup_{q \in Q} \inf_{p \in P} d(p, q)\}, \tag{1}$$

where $d(p, q)$ is the distance between points p and q. Since our embedding is isometric under ℓ_1, we used the ℓ_1 norm (or, Manhattan distance) in the vector domain. It is well-known that the Hausdorff distance is sensitive to outliers. Thus, instead of using the original formulation, for each point in the first set, we compute its closest distance to the second set and took the average of all distances,i.e., the modified Hausdorff distance we used in this paper is computed as follows:

$$d'_H(P, Q) = \max\{\frac{1}{m} \sum_{p \in P} \inf_{q \in Q} d(p, q), \frac{1}{n} \sum_{q \in Q} \inf_{p \in P} d(p, q)\}, \tag{2}$$

where m is the size of set P, and n is the size of set Q.

The dataset used in the experiments has been employed before in [6], where the distance between each view and each of the remaining database entries is computed. This approach reports the average distance between a pair of objects.

Table 1. Recognition results for proposed, median, and exhaustive search algorithms for different sampling resolutions. The proposed algorithm outperforms the two algorithms in both experiments. The experiment in the bottom row is performed with higher sampling resolution than the top row.

Experiment	Proposed	Median	Exhaustive
RECOGNITION	97.4%	40.1%	67.3%
RECOGNITION	97.8%	42.0%	88.2%

That is given two objects A and B, for each 19 view of object A , its closest view of object B is found and the average over the resulting distances is computed. Since we aim at computing the class representatives, the result of this experiment is not appropriate for comparison purposes. However, the performance of our algorithm in the above experiment is still compared with two alternative approaches. In the first approach, we replace representative points of each class with the median view drawn from this class, i.e., from the set of all views used to construct the representative points, we choose the view whose sum distance to all other views is minimum. In the second approach, we eliminate the representative points and compare the query directly to each of the database views except itself.

The results are shown in the top row of Table 1 and reveal that the proposed framework outperforms both median, and exhaustive search algorithms. Specifically, the proposed framework, median, and exhaustive search algorithms achieve %97.4, %40.1, and %67.3 recognition rates, respectively. Since the proposed approach takes into consideration all views of a class for generating its representative points, the recognition results of our approach are better than the median algorithm. Upon taking a closer look at the results, the exhaustive search algorithm could have been expected to achieve a higher recognition rate. However, we should note that the skeletons are very sensitive to noise, thus even skeletons of neighboring views can be quite different in the database. When the sampling resolution increases, we expect the performance of exhaustive search to improve.

To measure how the sampling resolution of the database effects these results, we take all views in the database and repeat the above experiment. The retrieval performance of the proposed, median, and the exhaustive search algorithms are recorded as %97.8, %42.0, and %88.2, respectively (bottom row of Table 1). This demonstrates that the proposed algorithm is stable under the sampling resolution compared to the other algorithms. In fact, the results suggest that even only 19 equidistant views of each object are sufficient to generate a good representative for the class. The main advantage of our algorithm, on the other hand, lies in the number of comparisons it performs. For a single query, while the exhaustive search performs $n \times m$ comparisons, where n is the number of classes and m is the number of views per class, our approach needs only n comparisons.

To test the sensitivity of the proposed algorithm to perturbation, we perturbed the representative points of each class by first deleting a randomly selected 5%, 10%, and 20% of points for all its views in the vector space and applying the k-means technique to the resulting points. The recognition rates were recorded as 97.4%, 97.2%, and 96.9%, respectively for the database of 19 views per object. Note that a number of embedded points are placed nearby in the vector space. Thus, removing some of these points do not change the position of the cluster representatives much. Although not a true occlusion experiment, these results reflect the algorithm's stability to missing data, which is required for occlusion experiments.

4   Conclusions

In this paper we have proposed a new framework for graph-based object recognition based on graph embedding. Given a set of objects grouped in different classes, our algorithm first represents the objects as points through graph embedding and generates one representative for each class in the vector space. After embedding the query into the vector space, its classification is performed by comparing to each class representative only. This process precludes the need for comparing the query against each database view. Experimental evaluation of the framework, including a comparison with the two approaches demonstrates the efficacy of the overall algorithm. In addition, a set of perturbation experiments show the stability of the algorithm against missing data. We plan to extend the proposed approach in a number of ways. Performing a more comprehensive experimental test using a larger dataset with different image formats, measuring the recognition performance as a function of increasing database size using different alternative approaches, employing different clustering algorithms in the vector space, and applying our framework to different domains are our future plans.

References

1. Babilon, R., Matousek, J., Maxová, J., Valtr, P.: Low-distortion embeddings of trees. Journal of Graph Algorithms and Applications 7(4), 399–409 (2003)
2. Blum, H.: Biological shape and visual science (part i). Journal of Theoretical Biology 38(2), 205–287 (1973), http://www.sciencedirect.com/science/article/B6WMD-4F1Y9M7-D5/2/1b17959a78e759a89f524d9f3eae0938
3. Bunke, H., Münger, A., Jiang, X.: Combinatorial search versus genetic algorithms: a case study based on the generalized median graph problem. Pattern Recognition Letters 20(11-13), 1271–1277 (1999), http://dx.doi.org/10.1016/S0167-86559900094-X
4. Demirci, M.: Retrieving 2D shapes using caterpillar decomposition. Machine Vision and Applications 24(2), 435–445 (2013)
5. Demirci, M., Osmanlioglu, Y., Shokoufandeh, A., Dickinson, S.: Efficient many-to-many feature matching under the ℓ_1 norm. Computer Vision and Image Understanding 115(7), 976–983 (2011), http://dx.doi.org/10.1016/j.cviu.2010.12.012
6. Demirci, M., Shokoufandeh, A., Keselman, Y., Bretzner, L., Dickinson, S.: Object recognition as many-to-many feature matching. International Journal of Computer Vision 69(2), 203–222 (2006)

7. Ferrer, M., Valveny, E., Serratosa, F., Riesen, K., Bunke, H.: An approximate algorithm for median graph computation using graph embedding. In: 19th International Conference on Pattern Recognition, pp. 1–4. IEEE (2008)
8. Giblin, P., Kimia, B.: On the local form and transitions of symmetry sets, medial axes, and shocks. International Journal of Computer Vision 54(1-3), 143–156 (2003)
9. Huttenlocher, D., Klanderman, D., Rucklige, A.: Comparing images using the Hausdorff distance. IEEE Transactions on Pattern Analysis and Machine Intelligence 15(9), 850–863 (1993),
   `citeseer.ist.psu.edu/huttenlocher93comparing.html`
10. Indyk, P.: Algorithmic aspects of geometric embeddings. In: Proceedings of the 42nd Annual Symposium on Foundations of Computer Science (2001)
11. Jiang, X., Münger, A., Bunke, H.: On median graphs: Properties, algorithms, and applications 23(10), 1144–1151 (2001)
12. Lozano, M., Escolano, F.: Protein classification by matching and clustering surface graphs. Pattern Recognition 39(4), 539–551 (2006)
13. Matousek, J.: On embedding trees into uniformly convex banach spaces. Israel Journal of Mathematics 237, 221–237 (1999)
14. Torsello, A., Hancock, E.: Graph embedding using tree edit-union. Pattern recognition 40(5), 1393–1405 (2007)
15. Torsello, A., Hancock, E.R.: Learning shape-classes using a mixture of tree-unions. IEEE Transactions on Pattern Analysis and Machine Intelligence 28(6), 954–967 (2006)
16. Torsello, A., Robles-Kelly, A., Hancock, E.: Discovering shape classes using tree edit-distance and pairwise clustering. International Journal of Computer Vision 72(3), 259–285 (2007)
17. Weiszfeld, E.: Sur le point pour lequel la somme des distances de n points donns est minimum. Thoku Mathematical Journal 43, 355–386 (1937)
18. Xiao, B., Torsello, A., Hancock, E.: Isotree: Tree clustering via metric embedding. Neurocomputing 71(10), 2029–2036 (2008)

Robust Selective Stereo SLAM
without Loop Closure and Bundle Adjustment

Fabio Bellavia, Marco Fanfani, Fabio Pazzaglia, and Carlo Colombo

Computational Vision Group, University of Florence
Via Santa Marta, 3, 50139, Florence, Italy
{marco.fanfani,carlo.colombo}@unifi.it,
{bellavia.fabio,fabio.pazzaglia}@gmail.com

Abstract. This paper presents a novel stereo SLAM framework, where a robust loop chain matching scheme for tracking keypoints is combined with an effective frame selection strategy. The proposed approach, referred to as *selective SLAM* (SSLAM), relies on the observation that the error in the pose estimation propagates from the uncertainty of the three-dimensional points. This is higher for distant points, corresponding to matches with low temporal flow disparity in the images. Comparative results based on the reference KITTI evaluation framework show that SSLAM is effective and can be implemented efficiently, as it does not require any loop closure or bundle adjustment.

Keywords: SLAM, Structure from Motion, RANSAC, feature matching, frame selection.

1 Introduction

The interest for visual Simultaneous Localization and Mapping (SLAM) has been increasingly growing in the computer vision community during the last few years. The main goal of SLAM is to simultaneously estimate both the camera positions and a geometrical 3D representation of the environment with real-time constraints [3]. Early SLAM implementations were based on probabilistic frameworks [4,16], and employed Bayesian filtering techniques, such as the Extendend Kalman Filter (EKF), to couple together in the same process the 6 DoF camera positions and all the 3D points, incrementally updated. Later, alternative SLAM implementations were proposed [17], influenced by the Structure from Motion (SfM) paradigm. These approaches exploit the epipolar geometry constraints to first estimate the camera positions and the 3D map, in general by using the RANdom SAmple Consensus (RANSAC) paradigm [5,8]. Successive refinement steps by iterative non-linear optimization techniques, such as bundle adjustment [21], over a selected sub-set of frames (keyframes) are used to minimize the global error. This allows separating pose estimation from 3D map computation, thus efficiently decoupling the process flows, as 3D structure needs not be optimized at each pose update but only when needed [11].

A. Petrosino (Ed.): ICIAP 2013, Part I, LNCS 8156, pp. 462–471, 2013.
© Springer-Verlag Berlin Heidelberg 2013

Both kinds of approaches have some drawbacks. In the Bayesian frameworks, points have to be added and discarded as the estimation proceeds, since the 3D map cannot grow excessively for computational limits. On the other hand, keyframe-based approaches, in order to achieve real-time operation, can perform local optimizations only occasionally. According to [19], keyframe based solutions outperform Bayesian approaches, due to their ability to maintain more 3D points in the estimation procedure.

Single camera (or mono) [3, 4, 11], stereo [13, 15, 18] or multiple cameras [22] setups can be used in SLAM systems, and different features and matching strategies can be employed to detect and track keypoints across image frames [3, 7, 11, 22]. In general, stereo or multiple camera configurations provide better solutions, since the rigid calibration of the cameras increases the accuracy in the 3D map computation and provides more robust matching correspondences. Further issues must be taken into account in mono SLAM design, such as the delayed 3D feature initialization [16] (i.e. when a point is seen for the first time) and the scale factor uncertainty [20].

Since SLAM system design is affected by the input scene, different implementation choices can be found for indoor [3, 11], outdoor [13, 15, 20] or even underwater [14] environments. Large scenarios present more challenging tasks since long tracks tend to accumulate an error drift. In order to alleviate this issue and to achieve finer and better estimates, loop closure detection techniques [9] have been developed to enforce pose constraints by recognizing already visited scenes, which obviously requires the camera to perform a looping path.

SLAM systems have to deal with errors mainly introduced during the extraction and the matching of 2D features. This paper implements a stereo SLAM system with a robust loop chain matching scheme [7], where the recent HarrisZ detector [2] and the sGLOH descriptor [1] are used to extract and match keypoints respectively, in order to provide more stable matches. Furthermore, only high temporal flow disparity frames are used to estimate the pose, since the error in the pose estimation is propagated from the uncertainty of the three-dimensional points, which is larger for distant points corresponding to low temporal flow disparity matches in the images. This strategy is effective at detecting and discarding frames with a similar visual content. The proposed system, referred to as *selective SLAM* (SSLAM), does not rely upon loop closure detection or bundle adjustment, and only considers the most reliable data measurements. This proves beneficial to both algorithmic accuracy and efficiency.

The paper is organized as follows. In Sect. 2 details of SSLAM are presented. Comparative results according to the reference KITTI evaluation framework [6] are discussed in Sect. 3. Conclusions and final remarks are offered in Sect. 4.

2   Selective Stereo SLAM

Given a calibrated and rectified stereo sequence $S = \{f_t\}$, where the frame $f_t = (I_t^l, I_t^r)$ is composed by the left I_t^l and right I_t^r input images taken at time $t \in \mathbb{N}$, SSLAM alternates between two main steps. The former step matches keypoints between the previous accepted SLAM frame f_i and the current frame f_j, while the latter estimates the relative camera pose $P_{i,j} = [R_{i,j}|t_{i,j}] \in \mathbb{R}^{3 \times 4}$, where $R_{i,j} \in \mathbb{R}^{3 \times 3}$ is the rotation matrix and $t_{i,j} \in \mathbb{R}^3$ is the translation vector. Note that since $i < j$, frames that can potentially lead to a large error in pose estimation due to high uncertainty in the 3D data measurements can be discarded. In this way, the system is able to keep small errors also for long trajectories relative to the first frame f_0, taken as absolute reference, even without global optimization or loop closure techniques. The absolute pose at time n is defined as $P_n = P_{0,n}$. P_n can be computed by concatenating the poses $P_{0,0}, P_{0,k} \ldots, P_{i,j}P_{j,n}$, where time steps $0 < k < \ldots < j < i < n$ belong to accepted frames.

2.1 Keypoints Detection and Matching

In the matching step the HarrisZ detector [2], which provides results comparable to other state-of-the-art detectors, is used to extract robust and stable corner features in the affine scale-space on the images $I_i^l, I_i^r, I_j^l, I_j^r$. The sGLOH descriptor [1] with the Nearest Neighbour matching is used to obtain the candidate correspondences between image pairs $(I_i^l, I_i^r), (I_i^l, I_j^l), (I_i^r, I_j^r), (I_j^l, I_j^r)$ after spatial and temporal constraints have been imposed to refine the candidates matches.

Let $\mathbf{x}_s^d = [x_s^d, y_s^d]^T \in \mathbb{R}^2$, $d \in \{l, r\}$, $s \in \{i, j\}$ be a point in the image I_s^d, a spatial match $(\mathbf{x}_s^l, \mathbf{x}_s^r)$ between the images on the same frame is computed by the stereo epipolar constraints imposed by the calibration

$$|x_s^l - x_s^r| < \delta_x \tag{1}$$
$$|y_s^l - y_s^r| < \delta_y \tag{2}$$

where δ_y is the error band allowed by epipolar rectification and δ_x is the maximal allowed disparity. In the case of a temporal match $(\mathbf{x}_i^d, \mathbf{x}_j^d)$ between corresponding images at different frames, the flow motion restrictions

$$\| \mathbf{x}_i^d - \mathbf{x}_j^d \| < \delta_r \tag{3}$$

are taken into account, where δ_r is the maximal flow displacement. Matches which form a *loop chain*

$$\mathcal{C} = \left((\mathbf{x}_i^l, \mathbf{x}_i^r), (\mathbf{x}_i^l, \mathbf{x}_j^l), (\mathbf{x}_j^l, \mathbf{x}_j^r), (\mathbf{x}_i^r, \mathbf{x}_j^r) \right) \tag{4}$$

are retained, see Fig. 1. In order to filter the candidate matches, four distinct RANSAC tests are finally run among the four image pairs to refine the epipolar geometry so that only a subset of chain matches $C_{i,j} \subseteq \{\mathcal{C}\}$ is selected. Note that the proposed matching scheme is similar to [7], but achieves longer and more stable keypoint tracks, which are crucial for the pose estimation—see the experimental evaluation.

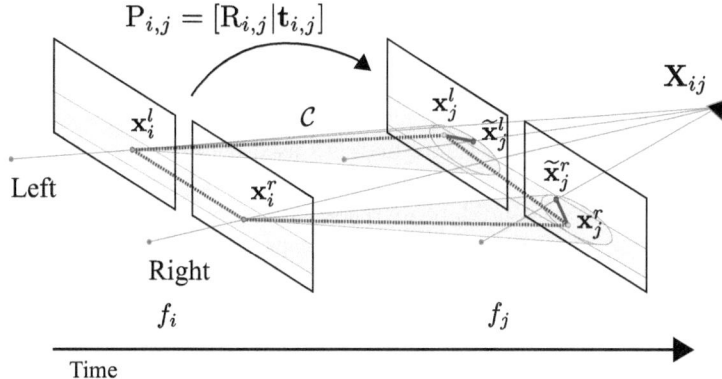

Fig. 1. (Best viewed in color) Keypoint matches between the frame f_i and f_j must satisfy the spatial constraint imposed by the epipolar rectification (yellow band) as well as the temporal flow motion restriction (orange cone). Furthermore, the four matching points must form the loop chain $\mathcal{C}$ (dotted line). In the ideal case, points x_j^l, x_j^r in frame f_j must coincide with the projections $\tilde{x}_j^l$, $\tilde{x}_j^r$ of points x_i^l, x_i^r in f_i obtained by triangulation of $X_{i,j}$ in order for the chain $\mathcal{C}$ to be consistent with the pose $P_{i,j}$. Due to data noise, in the real case the distances $\| \tilde{x}_j^l - x_j^l \|$ and $\| \tilde{x}_j^r - x_j^r \|$ must be minimal.

2.2 Pose Estimation Constrained by Temporal Flow

The relative pose $P_{i,j}$ between frames f_i and f_j is estimated in the second step of our SSLAM approach—see again Fig. 1. The 3D point $X_{i,j}$ corresponding to the match pair (x_i^l, x_i^r) in frame f_i can be estimated by triangulation [8], since the intrinsic and extrinsic calibration parameters of the system are known. Let $\tilde{x}_j^l$ and $\tilde{x}_j^r$ be the projections of $X_{i,j}$ onto frame f_j, according to the estimated relative pose $P_{i,j}$. The distance

$$\mathcal{D}(P_{i,j}) = \sum_{C_{i,j}, d \in \{l,r\}} \| \tilde{x}_j^d - x_j^d \| \tag{5}$$

among the matches of the chain set $C_{i,j}$ must be minimized, in order for the estimated pose $P_{i,j}$ to be consistent with the data. Due to the presence of outliers in $C_{i,j}$, a RANSAC test is run [7], where the number $\mathcal{D}_R(P_{i,j})$ of outliers chain matches over $C_{i,j}$ exceeding a threshold value δ_t is minimized so that pose $P_{i,j}$ be consistent with data:

$$\mathcal{D}_R(P_{i,j}) = \sum_{C_{i,j}} T_d \left(\| \tilde{x}_j^d - x_j^d \| > \delta_t \right) \quad . \tag{6}$$

In Eq. 6, $d \in \{l,r\}$, and the indicator function $T_x(P(x))$ is 1 if the predicate $P(x)$ is true for all the admissible values of x, and 0 otherwise. The final pose estimation $\overline{P}_{i,j}$ between frames f_i and f_j is chosen as

$$\overline{\mathrm{P}}_{i,j} = \underset{\mathrm{P}_{i,j}}{\arg\min}\, \mathcal{D}_R(\mathrm{P}_{i,j}) \quad . \tag{7}$$

In the traditional approach used in [7], at each iteration RANSAC estimates a candidate pose $\mathrm{P}_{i,j}$ using a minimal set of matches, i.e., 3 matches, in order to be robust to outliers [5]. The candidate matches used to build the pose model $\mathrm{P}_{i,j}$ are sampled from the set of candidate matches $C_{i,j}$. The pose $\mathrm{P}_{i,j}$ is validated against the whole set of candidate matches $C_{i,j}$ according to (6) and the best model found so far is retained. The process stops when the probability to get a better model is below some user-defined threshold value, and the final pose $\overline{\mathrm{P}}_{i,j}$ is refined [10] on the set $G_{\overline{\mathrm{P}}_{i,j}}$ of inlier matches where

$$G_{\mathrm{P}_{i,j}} = \left\{ \mathcal{C} \in C_{i,j} \mid T_d \left(\| \, \tilde{\mathbf{x}}_j^d - \mathbf{x}_j^d \, \| < \delta_t \right) \right\} \tag{8}$$

for a generic pose $\mathrm{P}_{i,j}$.

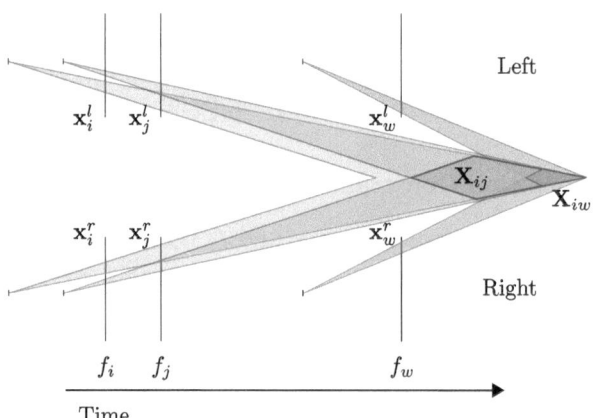

Fig. 2. (Best viewed in color) The uncertainty of matches in the image planes is lower bounded by the image resolution (red) and it is propagated to the 3D points. In order to estimate the 3D point $\mathbf{X}_{i,j}$, by using close frames f_i and f_j, a low temporal disparity flow is present in the image planes, and the 3D point location $\mathbf{X}_{i,j}$ can assume an higher range $\mathbf{X}_{i,j}$ of values (dark gray quadrilateral). In the case of distant frames f_i and f_w, the possible locations $\mathbf{X}_{i,w}$ are more circumscribed (blue quadrilateral), for the same resolution limits.

In our SSLAM approach, pose is estimated in a slightly different way, by taking advantage of the following observation. The image resolution provides a lower bound to the uncertainty in the keypoint match locations, which are triangulated to get the corresponding 3D point, and eventually estimate the relative pose between two temporal frames. Close frame matches have a low temporal flow disparity and the associated 3D point position has a high uncertainty with respect to distant frames, due to the error propagation from the matches on the image planes. Only points with sufficient displacement can give information about the translational and rotational motion, as shown in Fig. 2. According to this observation, SSLAM singles out from the set of chain matches $C_{i,j}$ for frames f_i and f_j the set $F_{i,j}$ containing the fixed points, i.e., points with low temporal flow disparity:

$$F_{i,j} = \{\mathcal{C} \in C_{i,j}, T_d(\| \mathbf{x}_i^d - \mathbf{x}_j^d \| \leq \delta_f)\} \ , \tag{9}$$

for a given threshold δ_f. In order for frame f_j to be accepted, the number of fixed matches between frames f_i and f_j must be low:

$$\frac{|F_{i,j}|}{|C_{i,j}|} < \delta_m \quad . \tag{10}$$

Indeed, if the estimation is severely corrupted by noise and can lead to a very bad estimation, the frame f_j is discarded and the next frame f_{j+1} is tested. This provides an adaptive threshold to discard bad frames containing low motion information—examples are shown in Fig. 3. A similar approach has been recently employed in [12].

Fig. 3. (Best viewed in color) Examples of successive keyframes retained according to the temporal flow for two different sequences of the KITTI dataset. The two temporal keyframes involved are superimposed as for anaglyphs, only images for the left cameras are shown. Good fixed and not fixed matches are shown in blue and light violet, respectively, while wrong correspondences are reported in cyan.

Finally, we add a pose smoothing constraint between frames, so that the current relative pose estimation $P_{i,j}$ cannot abruptly vary from the previous $P_{z,i}$, $z < i < j$. This is achieved by imposing that the relative rotation around the origin between the two incremental rotations $R_{z,i}$ and $R_{i,j}$ is bounded, as well as for the corresponding translation directions $\mathbf{t}_{z,i}$ and $\mathbf{t}_{i,j}$:

$$|\mathbf{r}_{i,j}^{k}{}^{\mathrm{T}}\mathbf{r}_{z,i}^{k}| < \delta_{\theta_1} \tag{11}$$

$$\frac{|\mathbf{t}_{i,j}{}^{\mathrm{T}}\mathbf{t}_{z,i}|}{\|\mathbf{t}_{i,j}\|\|\mathbf{t}_{z,i}\|} < \delta_{\theta_2} \ , \tag{12}$$

where $\mathbf{r}_{a,b}^{k}$ is any k-th column of the rotation matrix $R_{a,b}$. This last constraint can resolve issues in the case of no camera movement or when moving objects crossing the camera path cover the scene.

3   Evaluation

In order to evaluate the proposed system, the odometry dataset and the evaluation protocol of the KITTI vision benchmark suite has been used, which has becoming a reference evaluation framework for SLAM systems in recent years [6]. The dataset provides sequences recorded from car driving sessions on highways and inside cities. In particular we used the first 11 stereo sequences of the dataset, for which ground-truth data obtained by laser and GPS sensors are provided.

We compared different versions of our SSLAM system, corresponding to the successive improvements of the pipeline proposed in Sect. 2, both implemented as non-optimized Matlab code. In particular we indicate by SSLAM* the first version which only includes the loop chain matching described in Sect. 2.1, while the adaptive frame discarding strategy is incorporated in SSLAM.

The freely available SLAM library VISO2-S [7], one of the best performing SLAM in the KITTI ranking, is added in the evaluation as reference, since it uses a loop chain matching scheme similar to ours and a standard RANSAC pose estimation. The default parameter settings provided by the authors have been used for VISO2-S, while additionally for our systems we set $\delta_r = 500$ px, $\delta_x = 300$ px, $\delta_y = 14$ px, $\delta_f = 55$ px, $\delta_m = 0.05$, $\delta_{\theta_1} = 15°$ and $\delta_{\theta_2} = 10°$—see Sect. 2.2.

In order to analyze the robustness and the effectiveness of the proposed method, the SSLAM system was tested with a different number of RANSAC iterations for the pose estimation. In particular, results of SSLAM with 500, 15 and 3 RANSAC iterations, and SSLAM* with 500 iterations are presented, indicated respectively by SSLAM/500, SSLAM/15, SSLAM/3 and SSLAM*/500. The VISO2-S system uses 200 RANSAC iterations by default.

Fig. 4 shows the average translation and rotation errors of the different SLAM methodologies for increasing path length and speed, according to the KITTI evaluation framework [6]. Only frames common to all the methods, i.e. not discarded during the process by any of the proposed implementations, are used. This does not affect the results, since the computed error measures rely on the SLAM absolute positions, which remain the same. Both the different versions of the proposed method provide results better than the VISO2-S system.

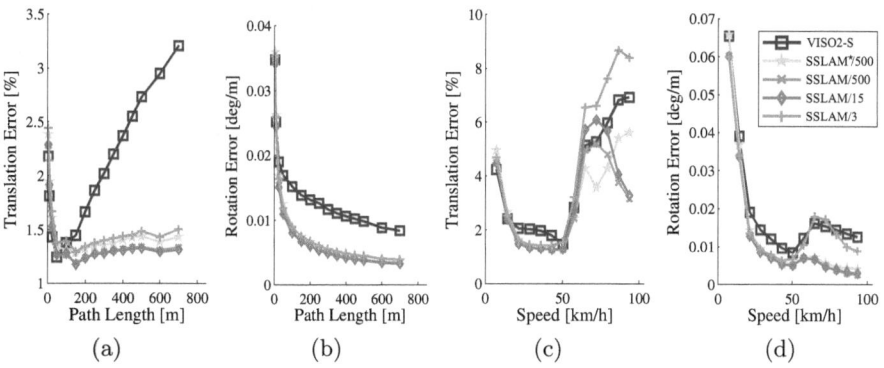

Fig. 4. (Best viewed in color) Average error on the first 11 sequences of the KITTI dataset. Plots (a-b) refer to the average translation and rotation error for increasing path length respectively, while plots (c-d) refer to increasing speed.

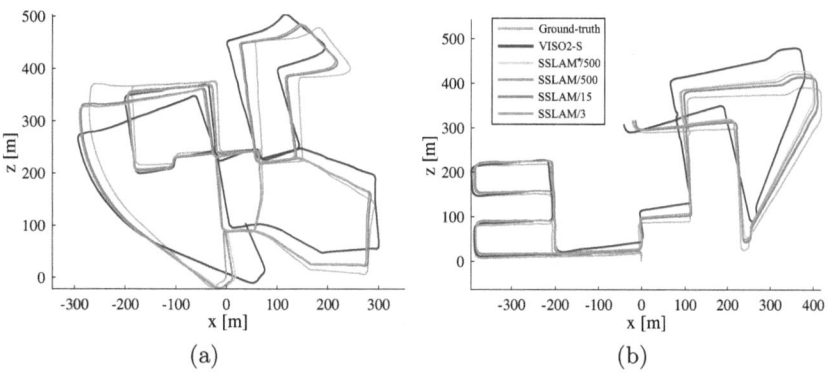

Fig. 5. (Best viewed in color) An example of the final paths computed for Sequence 00 (a) and Sequence 08 (b) of the KITTI dataset

The chain loop matching scheme together with the chosen keypoint detector and descriptor is robust even for long paths, without relying on bundle adjustment or loop closure detection. Furthermore, dropping low temporal flow disparity frames in SSLAM improves on the standard pose estimation used in SSLAM*, allowing the tracking of longer paths. Note that SSLAM drops for each sequence from 35% to 70% of the total frames, which mainly occur on straight paths covered at low and medium speeds.

Moreover, results for SSLAM/15 and SSLAM/500 are equivalent, while SSLAM/3 obtains inferior results but similar to those obtained by SSLAM*/500, giving an evidence of the robustness of the proposed matching selection strategy and pose estimation, which can also improve the final running time.

Finally, on average the SSLAM and SSLAM* systems extract about 300 matches and VISO2-S around 250. The proposed SSLAM methodologies retain about 95% of extracted matches as inliers after the pose estimation, while only about 50% are instead preserved by VISO2-S, giving a further evidence on the effectiveness of the proposed system.

Fig. 5 shows the estimated paths for two sequences of the KITTI dataset. By inspecting the tracks it can be clearly seen that both SSLAM and SSLAM* (in this order) paths are closer to the ground-truth with respect to VISO2-S. In particular, rotations are the major source of incremental error, but our approach succeeds to better solve this issue.

4   Conclusion

In this paper a new stereo SLAM system was presented. The approach achieves a low drift error even for long paths, without relying on loop closure or bundle adjustment. A robust loop chain matching scheme for tracking keypoints is provided, sided by a frame discarding system to improve pose estimation. According to the experimental results, dropping low temporal flow disparity frames for discarding highly uncertain models is an effective strategy to reduce error propagation from matches. Results validated on the KITTI dataset showed the effectiveness of the system, which is robust even for an extremely small number of RANSAC iterations.

Future work will include the implementation of an efficient optimized code of the system, experimental results on a wider range of sequences and the development of a more advanced adaptive sampling scheme for model estimation.

Acknowledgements. This work has been carried out during the THESAURUS project, founded by Regione Toscana (Italy) in the framework of the "FAS" program 2007-2013 under Deliberation CIPE (Italian government) 166/2007.

References

1. Bellavia, F., Tegolo, D., Trucco, E.: Improving SIFT-based descriptors stability to rotations. In: Proc. of International Conference on Pattern Recognition (2010)
2. Bellavia, F., Tegolo, D., Valenti, C.: Improving Harris corner selection strategy. IET Computer Vision 5(2) (2011)
3. Davison, A.: Real-time simultaneous localization and mapping with a single camera. In: Proc. 9th IEEE International Conference on Computer Vision, pp. 1403–1410 (2003)
4. Davison, A., Reid, I., Molton, N., Stasse, O.: MonoSLAM: Real-time single camera SLAM. IEEE Trans. on Pattern Analysis and Machine Intelligence 29(6), 1052–1067 (2007)
5. Fischler, M.A., Bolles, R.C.: Random sample consensus: a paradigm for model fitting with applications to image analysis and automated cartography. Commun. ACM 24(6), 381–395 (1981)

6. Geiger, A., Lenz, P., Urtasun, R.: Are we ready for autonomous driving? The KITTI vision benchmark suite. In: Proc. of Computer Vision and Pattern Recognition (2012)
7. Geiger, A., Ziegler, J., Stiller, C.: StereoScan: Dense 3D reconstruction in real-time. In: IEEE Intelligent Vehicles Symposium (2011)
8. Hartley, R.I., Zisserman, A.: Multiple View Geometry in Computer Vision, 2nd edn. Cambridge University Press (2004)
9. Ho, K., Newman, P.: Detecting loop closure with scene sequences. International Journal Computer Vision 74(3), 261–286 (2007)
10. Horn, B.K.P.: Closed-form solution of absolute orientation using unit quaternions. Journal of the Optical Society of America A 4(4), 629–642 (1987)
11. Klein, G., Murray, D.: Parallel tracking and mapping for small AR workspaces. In: Proc. IEEE/ACM International Symposium on Mixed and Augmented Reality, pp. 225–234 (2007)
12. Lee, G.H., Fraundorfer, F., Pollefeys, M.: RS-SLAM: RANSAC sampling for visual FastSLAM. In: Proc. of IEEE/RSJ International Conference on Intelligent Robots and Systems, pp. 1655–1660 (2011)
13. Lim, J., Pollefeys, M., Frahm, J.M.: Online environment mapping. In: Proc. IEEE Conference on Computer Vision and Pattern Recognition (2011)
14. Mahon, I., Williams, S., Pizarro, O., Johnson-Roberson, M.: Efficient view-based SLAM using visual loop closures. IEEE Trans. on Robotics 24, 1002–1014 (2008)
15. Mei, C., Sibley, G., Cummins, M., Newman, P., Reid, I.: RSLAM: A system for large-scale mapping in constant-time using stereo. International Journal of Computer Vision 94, 198–214 (2011)
16. Montiel, J., Civera, J., Davison, A.: Unified inverse depth parametrization for monocular SLAM. In: Proc. of Robotics: Science and Systems. IEEE Press (2006)
17. Mouragnon, E., Lhuillier, M., Dhomeand, M., Dekeyserand, F., Sayd, P.: Real time localization and 3D reconstruction. In: Proc. IEEE Conference on Computer Vision and Pattern Recognition, pp. 363–370 (2006)
18. Paz, L., Piniés, P., Tardós, J., Neira, J.: Large-scale 6-DoF SLAM with stereo-in-hand. IEEE Trans. Robotics 24(5), 946–957 (2008)
19. Strasdat, H., Montiel, J., Davison, A.: Visual SLAM: Why filter? Image and Vision Computing 30, 65–77 (2012)
20. Strasdat, H., Montiel, J.M.M., Davison, A.J.: Scale drift-aware large scale monocular SLAM. In: Proc. of Robotics: Science and Systems (2010)
21. Triggs, B., McLauchlan, P.F., Hartley, R.I., Fitzgibbon, A.W.: Bundle adjustment – A modern synthesis. In: Triggs, B., Zisserman, A., Szeliski, R. (eds.) ICCV-WS 1999. LNCS, vol. 1883, pp. 298–372. Springer, Heidelberg (2000)
22. Zou, D., Tan, P.: CoSLAM: Collaborative visual SLAM in dynamic environments. IEEE Trans. on Pattern Analysis and Machine Intelligence 35(2), 354–366 (2013)

Demographics versus Biometric Automatic Interoperability

Maria De Marsico[1], Michele Nappi[2], Daniel Riccio[2], and Harry Wechsler[3]

[1] Department of Computer Science, Sapienza University of Rome, Italy
demarsico@di.uniroma1.it
[2] Biometric and Image Processing Lab, University of Salerno, Italy
{mnappi,driccio}@unisa.it
[3] Department of Computer Science, George Mason University, Fairfax Virginia, USA
wechsler@cs.gmu.edu

Abstract. Humans are naturally experts in recognizing faces. Such skills are enforced through a mix of cultural and cognitive processes. This allows human vision system to be especially efficient and effective in processing faces in a familiar environment. Automatic recognition system are currently not able (will ever be?) to achieve similar performance, especially when cross-demographic features are involved (gender, ethnicity, and age). Recent studies suggest a significant decrease of the number of recognition errors by limiting the search space to faces with the same demographics. This can be obtained by preliminarily annotating faces with a demographic profile, or by using demographic features as soft biometrics to be determined as a support to actual recognition. Especially in the second case, a multi-demographics dataset is needed to appropriately train a recognition system, and/or to test its performance. In this paper we use EGA dataset to test how interoperability relationships between biometric and demographics can be exploited for better recognition, though avoiding human intervention to preventively select appropriate demographic parameters.

1   Introduction

Biometric recognition performance is affected by many factors, among them, protocols ("best practices"), performance evaluation and validation [2]. Protocols should enforce the proper and effective use of biometric systems, while performance evaluation and validation should take into account uncontrolled-settings and interoperability in order to determine the robustness and reliability of the biometric authentication decisions even when datasets used during system implementation are different from those actually used during normal system operation. A robust system should present a low sensitivity to biometric variability, including incomplete data ("occlusion"), corrupted data ("disguise"), as well as pose, illumination and expression (PIE) variations. More and more studies recently focus on further variability elements, namely demographics (gender, ethnicity, and age), since they revealed a strong influence on recognition performance, if the datasets used for implementation do not present the same range of variations as the actual population to be processed. This also affects

A. Petrosino (Ed.): ICIAP 2013, Part I, LNCS 8156, pp. 472–481, 2013.
© Springer-Verlag Berlin Heidelberg 2013

reliability, i.e. the consistency and stability of the predictions made on authentication. Both robustness and reliability are therefore influenced by variability of information due to different factors. Uncontrolled settings are especially widespread in real life scenarios, e.g., mass screening, surveillance using smart camera networks, and tagging within social networks. Performance for such settings is definitely significantly lower than the performance for tightly controlled biometrics evaluations, particularly in large scale authentication scenarios. This paper evaluates different operational biometric scenarios for the purpose of interoperability in different demographic settings, with demographic composition and quality of biometric data the varying factor. In general, interoperability is the ability of diverse systems and/or organizations to work together (inter-operate) through well-defined tasks and protocols. In the biometric context, it refers to suites of sensors, data flows, subjects' condition, and algorithms. Especially when demographic differences are involved, we can consider interoperability as the link between biometrics and forensics, within the context of distributed data collection and associated federated identity management systems [2]. Biometric interoperability is addressed here in terms of diversity of population enrolled.

Human observers of a face can easily make subtle judgments for gender, identity, age and expression. These categorizations reasonably exploit different visual information, though it is not possible, in general, to isolate the specific information they use. Many attempts to investigate such issue mainly pertain to vision research. Human skills are enforced through a mix of cultural and cognitive processes, evolution (phylogeny) within a specific ethnic group as well as development (ontogeny) of a person within a mixed population, including different ages and genders. These processes allow the human vision system to be especially efficient and effective in processing faces in a familiar environment. The implicit, unique combination of such processes may also explain some controversial phenomena known to undermine human face recognition. The most known one is the other-race effect, i.e. the poor ability to recognize a subject from a different race. Though often ascribed to the so-called "contact hypothesis" [1], which refers to the single individual experience, it cannot be completely explained by it [7]. On the other hand, the concept of "familiarity" is often the best explanation for excellent human recognition performance. The investigation of cognitive processes regarding cross-demographic features affecting human face recognition ability, and their relevance to automatic systems, is expected to significantly improve face recognition performance. Even for computer systems specific problems arise when cross-demographic features are involved. This phenomenon was seldom systematically observed before the 2006 NIST Face Recognition Vendor Test (FRVT) [9]. The simple reason was that biometric systems implemented in a specific country, e.g. Asian, would usually rely on a home-made dataset of subjects of the same ethnicity. Therefore the cross-demographics interoperability was hardly a problem. The main obstacle for interoperability is that biometric face templates are mostly obtained by extracting the same features regardless the population at hand. It is, however, clear that different categorizations (age, race, ethnicity gender, etc.) should rely on different features. A comprehensive study of the influence of demographic features on recognition often suffers from the lack of large datasets with a wide, uniform variation of such features and possibly pre-annotated data [11]. If these features are significant categorization criteria, one can conjecture that about the same number of subjects from each group should be represented in the dataset used to train biometric systems, to assure robustness and reliability even in contexts different from the implementation one. The work by Klare *et al.* [4] is the most recent attempt to systematically investigate the role of cross-demographics in automatic face recognition.

Regarding interoperability achieved through the use of different datasets, the experiments in [2] demonstrate a significant dependence between recognition performance and the use of different datasets for training and testing. Experiments in [4] show that such dependence is even heavier if demographics differences are also involved. Along the same line, we present experiments performed using EGA dataset (Ethnicity, Gender and Age face database) [11] to further investigate strategies to avoid human intervention in pre-selection of subsets related to specific demographics.

2   The EGA Face Database

One of the most crucial limits for the growing number of works in the area of face recognition, also including multimodal recognition, face age/gender/ethnicity estimation, and face re-identification, is the lack of a face dataset sufficiently large and variegated. Most studies still refer to FERET dataset [8], which to the best of our knowledge, is still the one with the largest number of subjects including good diversification of features related to ethnicity, gender, illumination, and pose. FERET, however, is lacking in both the total number of subjects, and in their overall demographic heterogeneity. The latter is found in the dataset used in [4], namely the set of one million mug shot face images from the Pinellas County Sheriff's Office (PCSO). However, these lack a sufficient amount of Hispanic subjects, as well as younger (<18) or elder (>50) people. EGA dataset (Ethnicity, Gender and Age face database) [11] which is used here, has the advantage to be flexibly updateable and extendable in time to increase the amount of significant variations. In other words, the set of its images is not fixed once and for all, but can be extended according to the needs of the experiments at hand. EGA integrates into a single dataset more sets of face images from different databases which are available nowadays, and are quite different depending on the place and the context where they were gathered. Templates are further organized according to features such as ethnicity, gender and age.

EGA is gathered through a set of links to files that have been processed in advance by appropriate scripts. Each user can ask and obtain on her own the original datasets with the images needed to build the needed version of EGA, whether they are free or require some form of registration/payment. The scripts reorganize and rename the requested images, according to the structure that was devised for EGA. It is not only possible to easily build the dataset, but even to expand it, as new datasets become available and after they are annotated. Thanks to the annotation of single images, it is also possible at any moment, to discard specific samples. In the present composition, distortions such as illumination, pose, expression and occlusions are minimized, since representative datasets are already available, while demographic aspects are appropriately emphasized. Figure 1 shows the conceptual organization and the current composition of EGA.

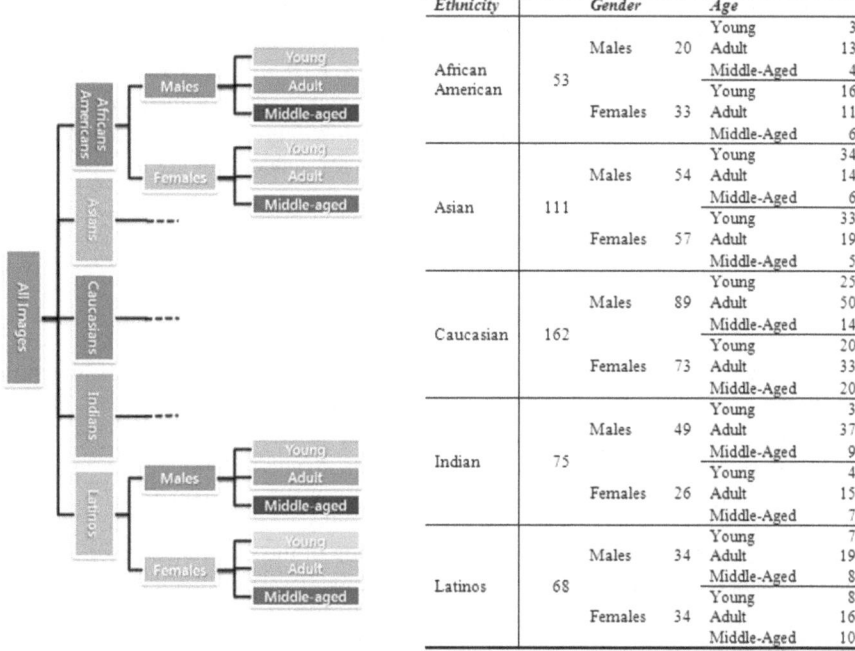

Ethnicity		Gender		Age	
African American	53	Males	20	Young	3
				Adult	13
				Middle-Aged	4
		Females	33	Young	16
				Adult	11
				Middle-Aged	6
Asian	111	Males	54	Young	34
				Adult	14
				Middle-Aged	6
		Females	57	Young	33
				Adult	19
				Middle-Aged	5
Caucasian	162	Males	89	Young	25
				Adult	50
				Middle-Aged	14
		Females	73	Young	20
				Adult	33
				Middle-Aged	20
Indian	75	Males	49	Young	3
				Adult	37
				Middle-Aged	9
		Females	26	Young	4
				Adult	15
				Middle-Aged	7
Latinos	68	Males	34	Young	7
				Adult	19
				Middle-Aged	8
		Females	34	Young	8
				Adult	16
				Middle-Aged	10

Fig. 1. Logical organization and composition of EGA

EGA v1.0 gathers images from six different datasets.

— CASIA-Face V5 [12]: 2,500 images of 500 subjects, captured in a single session with resolution 640×480, 16bit color. Faces possibly present distance, pose, illumination, and expression variations, including wearing eyeglasses. Subjects are mostly young and of Eastern ethnicity.

— FEI [13]: 2,800 images of 200 subjects (100 male and 100 female), with 14 color images each of 640×480 resolution. Faces have been acquired on a white background and belong to subjects from 19 to 40 years old, mostly of Latin ethnicity.

— FERET [8]: 14,126 images of 1,199 subjects of different ethnicities, genders and age, acquired in 15 sessions between August 1993 and July 1996. Resolution is 256×384 for 8bit greyscale. Different face sets (fa, fb, dup I, dup II) differ in pose and acquisition period, and present slight variations in illumination and expression.

— FRGC [9]: 50,000 images from 4,003 subject sessions (for each session four images in controlled conditions, two images in non-controlled conditions, and a 3D model). Images are divided into training and validation partitions. Resolution is 1704×2272 for 24bit color. Most subjects are Caucasians, and are mainly concentrated in a same age range (young/adult). Gender is quite uniformly represented.

— JAFFE [5]: 2130 images of 10 subjects. It was mainly gathered for facial expression analysis. Image resolution 256×256 for 8bit greyscale. Subjects are all female and Japanese of uniform age.

— Indian Face Database [3]: the dataset contains 40 subjects in frontal pose with eleven different looking directions for each individual, plus some additional image when available. Images are captured on a uniform background, and further present four expression variations. The dataset includes one folder for males and one for females. Image resolution is 640×480 pixels, with 256 grey levels. All subjects are of Indian ethnicity, with an adequate distribution for gender, but not for age.

3   Proposed Approach

Experiments by Klare *et al.* in [4] highlighted, among others, two relevant aspects:

- Training a recognition system on a data set of heterogeneous images, does not reduce limitations of its performance with respect to "hardest" ethnic groups, i.e. those which are more difficult to recognize.
- The performance of a recognition system for a specific ethnicity is improved when it is trained on that ethnic group.

Klare et al. suggest, therefore, that the priori selection of the ethnicity (performed by a human operator) of an input image can produce an increased accuracy of the recognition system. However, notwithstanding the precise and accurate analysis of all the assumptions, this hypothesis is not extensively verified in the referenced work. Moreover, the suggested exploitation of the presented findings requires human intervention in the preliminary selection of the ethnicity of each face. The authors assume that "in most operational scenarios, particularly those dealing with forensics and law enforcement, the use of face recognition is not being done in a fully automated, "lights out" mode. Instead, an operator is usually interacting with a face recognition system, performing a one-to-one verification task, or exploring the gallery to group together candidates for further exploitation" [4]. However, in further equally significant scenarios, such as video-surveillance or access control, continuous human interaction might not be feasible. We want to explore viable strategies for this second hypothesis, which allow to avoid the process of human demographics pre-selection. We still rely on multiple classifiers, each trained on a specific demographic feature. However, and this is the core of our work, we rule out any intervention during normal system operations, except for particularly hard cases where an appropriate alert system may signal the need for an operator's check. There are two different ways to achieve this, both relying on a pool of classifiers, each trained for a specific demographic feature:

1. A Priori Demographics Selection (APrDS): thanks to automatic recognition of relevant demographic features (ethnicity, gender, age, or a combination of them), each probe image is submitted to the corresponding classifier; this requires a further training for a first identification of the involved soft biometrics classes, and a further preliminary step during each recognition operation.

2. A Posteriori Demographics Selection (APoDS): it is assumed that the demographic features of the subject are not known a priori, therefore the probe image is inputted to all the classifiers; to complete the recognition process, it is

necessary to adopt a criterion for the selection of   the global best answer; this is the real novelty of our proposal, since we also rule aout any kind of automatic demographics preprocessing.

In the experiments discussed here, we simulated both kinds of system. In particular, we chose ethnicity as the first test-bed, since this feature usually presents the highest number of different significant classes, and therefore is prone to the highest degree of confusion. In the specific case, since demographics (soft biometrics) recognition is not our final goal, the APrDS system relied on using EGA as dataset. In particular, the selection of ethnicity of a probe image was simulated by exploiting the metadata provided by the nomenclature of the dataset. On the other hand, a APoDS system requires more attention. This kind of system is structured in such a way that the same probe image is supplied to the different classifiers, which each perform recognition in relation to their own gallery (consisting of images of a specific ethnic group) and provide as output a similarity score s and a reliability margin φ, in this work   we adopted as φ the function defined as (more details are available in [6]):

$$\varphi(p)=1-|N_b|/|G| \quad \text{with} \quad N_b = \{g_{i_k} \in G \mid F(d(p,g_{i_k})) < 2 \cdot F(d(p,g_{i_1}))\} \quad (1)$$

where $|G|$ is the gallery cardinality, d is a distance measure, g_{i_k} is a gallery element (template), p is a probe template, g_{i_1} is the first element   returned in the list corresponding to the identification of p, and F is a score normalization function (we used here the Quasi-Linear Sigmoidal (QLS) [6]). The architecture is sketched in Figure 2.

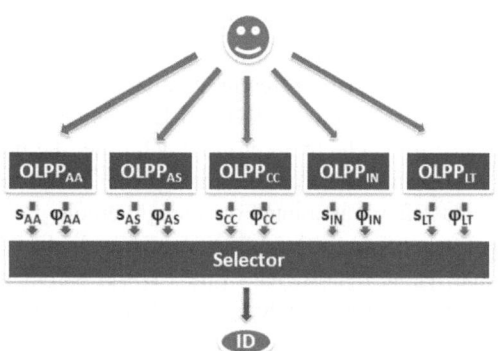

Fig. 2. The architecture of APoDS system

Following EGA composition, we have five classifiers based on ethnicity: Afro-Americans, Asians, Caucasians, Indians   and Latinos. The system includes a Selector module, which collects the scores and reliability margins produced by the involved classifiers and selects the best response (classifier) as follows:

$$cl = \arg\max_j \{(s_j * \varphi_j), j = 1,2,...,5\} \quad (2)$$

In other terms, the Selector selects the response from the classifier which achieves the maximum product of the pair (s_j, φ_j). It is worth noticing that our APoDS approach addresses also the mixed ethnicities, since taking the most reliable response from the classifiers implies to exploit the dominant ethnic features for each probe subject.

4   The Experiments

The experiments were carried out on images from EGA. They were focused on ethnicity, leaving to future work the verification of the same issues regarding gender and age. Ethnicity was preferred, as it provides a greater degree of variation (five possible classes in EGA) than the others (two for gender and three for age). We used widely available classifiers (LDA and OLPP), in order to obtain baseline and well repeatable results about the investigated issues. The general adopted protocol was the following. For each subject, 5 images were considered (1, 2, 3, 4, 5). For each of three experimental rounds three images were used for training, 1 for the gallery, and 1 as the probe. The reported Recognition Rate (RR) represent the average over the three rounds of subject identification:

-     1 round: 1,2,3 (training), 4 gallery, 5 probe
-     2 round: 2,3,4 (training), 5 gallery, 1 probe
-     3 round: 3,4,5 (training), 1 gallery, 2 probe

The first part of the experiments aimed at ensuring that the behavior of the chosen classifier was consistent with the outcomes by Klare *et al.* in [4]. The first experiment aimed at assessing how and to what extent the training set used for a given classifier affects its performance, if it is applied in a different context. In this case, the LDA classifier was trained on the five different ethnicities: African-Americans (AA), Asians (AS), Caucasians (CC), Indians (IN) and Latinos (LT). The experiment was repeated five times, once for each ethnic group. Each time, the probe and gallery images were made up of only members of the ethnic group under consideration. On the other hand, the same classifier was trained on a single ethnic group, either the same or a different one with respect to testing. After it was applied to the pair probe / gallery, RR was computed. Table 1 shows RR of the LDA classifier with single-ethnicity training sets, when applied to probe / gallery sets belonging to a single ethnicity.

Table 1. RR of LDA classifier trained (rows) and tested on single ethnicities

LDA	AA	AS	CC	IN	LT
AA	0.62	0.56	0.66	**0.71**	0.58
AS	0.74	0.74	0.73	**0.77**	0.74
CC	0.70	0.70	**0.71**	0.64	0.65
IN	0.70	0.72	0.72	0.69	**0.73**
LT	0.68	**0.74**	0.72	0.66	0.69

Intuitively, we would expected that a classifier trained on images belonging to a specific ethnic group, gives the best result of the probe / gallery images belonging to the same ethnic group. The results in Table 1 seem to refute this hypothesis. We initially assumed that this phenomenon could depend on the different number of subjects available for each ethnic group. Therefore we chose the ethnicity with the minimum number of subjects, and limited the set of images to this value for all ethnic groups, so as to ensure an even distribution for the different ethnic groups. However, the behavior of the LDA classifier was found to be entirely consistent with that seen before. The second hypothesis focused on a further factor, extremely relevant to take into consideration, which is the sensitivity of the classifier to any distortion in the images. In fact, it is known from the literature that the LDA, although more robust to distortions than other dimensionality reduction techniques, is not particularly accurate, when such distortions are particularly relevant. The images in the EGA database were selected in such a way as to be all in frontal pose and nearly neutral expression. However, coming from different datasets, they show different illuminations (an important aspect in the LDA) and different resolutions. In order to verify whether the inconsistent behavior was actually determined by the lack of robustness of the classifier, the same experiment was performed with a different, more robust classifier, i.e. OLPP. Table 2 shows now a behavior which is fully consistent with the expected outcomes: a classifier which is relatively robust to other sources of distortion, once trained with images from a certain ethnicity, achieves the best RR values with probe/gallery sets which only contain images from the same ethnicity.

Table 2. RR of OLPP classifier trained (rows) and tested on single ethnicities

OLPP	AA	AS	CC	IN	LT
AA	**0.67**	0.62	0.64	0.65	0.62
AS	0.69	**0.73**	0.72	0.71	0.71
CC	0.67	0.69	**0.73**	0.68	0.69
IN	0.72	0.71	0.70	**0.73**	0.71
LT	0.73	0.73	0.76	0.75	**0.77**

Given these results, it is particularly interesting to verify the behavior of the same classifier, OLPP from now on, when it is trained on a single ethnicity but tested on probe/gallery sets with images from all ethnic groups (all five in this case). Table 3 shows results of this experiment.

Table 3. RR of OLPP classifier trained on single ethnicities and tested on probe/galleries with five ethnicities

OLPP	AA	AS	CC	IN	LT
	0.66	0.69	0.70	0.68	0.68

In all cases in Table 3, we observe that RR is lower to that achieved by the same classifier when it is trained and tested on a same specific ethnicity.

In the opposite situation, the classifier was trained on images from all ethnic groups, and tested   on a single one. Table 4 shows the results.

Table 4. RR of OLPP classifier trained on five ethnicities and tested on probe/galleries with single ethnicities

OLPP	AA	AS	CC	IN	LT
	0.64	0.72	0.70	0.73	0.75

It is clear that the classifier in question provided the best performance in terms of RR when working independently on individual ethnic groups.

Generally, in a system functioning in a real world application, it is much more likely that the classifier is trained on a set of images that is less representative and varied than that on which it will then be applied at operating time. This is because companies and laboratories have limited datasets. Furthermore, most existing systems do not make a preliminary judgment about the ethnicity of the acquired probe image, therefore the case more likely to be considered for comparison is one in which the classifier is trained on all ethnic groups and is then tested on all ethnicities. By performing such a test, in this specific case, we obtained a RR of 0.68, which is well below the performance achieved by the same classifier, when applied to the individual ethnic groups. The results suggest, as in [4], that if it was possible to identify a priori the ethnicity of the subject and submit the image to the classifier trained for that specific ethnicity, the accuracy would be greater. By performing this experiment we observed that the RR rises to 0.73. We remind that the correct determination of the ethnicity was simulated in this experiment of APrDS, using the nomenclature of files in EGA.

Finally, we tested the accuracy of a APoDS simulation. We assumed an unknown ethnicity of the input subject. We separately submitted the image to each classifier trained on individual ethnic groups, and verified if it was possible, from time to time, to retrieve the response (possibly correct) of the classifier corresponding to that ethnicity. The RR provided by the APoDS simulation is 0.67, which is comparable with the performance of the system trained and tested on all ethnic groups (0.68). However, this value is still less than what one would get with a APrDS system (RR 0.73). This suggests that, although it is possible to improve the performance of an identification system by considering the different ethnic groups (RR 0.73), it is a complex problem to be able to do this by selecting a posteriori the response of the most suitable classifier, or in other words, inferring ethnicity a posteriori.

5   Conclusions

An increasing number of papers addresses the problem of interoperability of trainable classifiers with respect to demographic variations. However, systematic studies of this phenomenon are still few. This work has the goal to evaluate the weight of ethnicity on the training / testing of a classifier, and to verify the hypothesis that the combination / selection of classifiers, each trained on a separate ethnic group, can produce an increased accuracy in terms of RR. The results obtained show that, once the ethnicity of the subject is known a priori, it is possible to actually increase the RR by about 7%. In contrast, a system that selects the correct answer downstream of the recognition process (a posteriori estimate of ethnicity) is able to achieve performance at least comparable to those of the system trained / tested on all ethnic groups, but still further improvable. A possible topic for future work will be the identification of effective criteria to apply implemented by the module Selector to this aim. Moreover, we will test more demographics and their possibly combined influence on recognition.

References

1. Chiroro, P., Valentine, T.: An investigation of the contact hypothesis of the own-race bias in face recognition. Quarterly J. Experimental Psychol. Human Experimental Psychol. 48A, 879–894 (1995)
2. El Khiyari, H., De Marsico, M., Abate, A.F., Wechsler, H.: Biometric Interoperability Across Training, Enrollment, and Testing for Face Authentication. In: Proceedings of 2012 IEEE Workshop on Biometric Measurements and Systems for Security and Medical Applications (BioMS 2012), Salerno, Italy, pp. 1–8 (2012)
3. Jain, V., Mukherjee, A.: The Indian Face Database (2002), http://vis-www.cs.umass.edu/~vidit/IndianFaceDatabase/
4. Klare, B.F., Burge, M.J., Klontz, J.C., Vorder Bruegge, R.W., Jain, A.K.: Face Recognition Performance: Role of Demographic Information. IEEE Transactions on Information Forensics and Security 7(6), 1789–1801 (2012)
5. Lyons, M.J., Akamatsu, S., Kamachi, M., Gyoba, J.: Coding Facial Expressions with Gabor Wavelets. In: Proceeding of the IEEE International Conference on Automatic Face and Gesture Recognition, pp. 200–205 (1998)
6. De Marsico, M., Nappi, M., Riccio, D., Tortora, G.: NABS: Novel Approaches for Biometric Systems. IEEE Transactions on Systems, Man and Cybernetics—Part C: Applications and Reviews 41(4), 481–493 (2011)
7. Ng, W., Lindsay, R.C.: Cross-race facial recognition: Failure of the contact hypothesis. J. Cross-Cultural Psychol. 25, 217–232 (1994)
8. Phillips, P.J., Wechsler, H., Huang, J., Rauss, P.: The FERET Database and Evaluation Procedure for Face-Recognition Algorithms. Image and Vision Computing J. 16(5), 295–306 (1998)
9. Phillips, P., Flynn, P., Scruggs, T., Bowyer, K.W., Chang, J., Hoffman, K., Marques, J., Min, J., Worek, W.: Overview of the Face Recognition Grand Challenge. In: Proc. of IEEE Conf. on Computer Vision and Pattern Recognition (CVPR), San Diego, CA (2005)
10. Phillips, P., Scruggs, W., O'Toole, A., Flynn, P., Bowyer, K., Schott, C., Sharpe, M.: FRVT 2006 and ICE 2006 large-scale experimental results. IEEE Trans. Pattern Anal. Mach. Intell. 32(5), 831–846 (2010)
11. Riccio, D., Tortora, G., De Marsico, M., Wechsler, H.: EGA - Ethnicity, Gender and Age, a pre-annotated face database. In: Proceedings of 2012 IEEE Workshop on Biometric Measurements and Systems for Security and Medical Applications (BioMS 2012), Salerno, Italy, pp. 38–45 (2012)
12. CASIA-FaceV5, http://biometrics.idealtest.org/
13. The FEI face database (2012), http://www.fei.edu.br/~cet/facedatabase.html

Edge Detection on Polynomial Texture Maps

Cristian Brognara[1], Massimiliano Corsini[2],
Matteo Dellepiane[2], and Andrea Giachetti[1]

[1] Dip. Informatica, Università di Verona, Strada le Grazie 15 - 37134 Verona, Italy
andrea.giachetti@univr.it
[2] Visual Computing Laboratory, ISTI-CNR, 56124 Pisa, Italy

Abstract. In this paper we propose a simple method to extract edges
from Polynomial Texture Maps (PTM) or other kinds of Reflection Transformation Image (RTI) files. It is based on the idea of following 2D lines
where the variation of corresponding 3D normals computed from the
PTM coefficients is maximal. Normals are estimated using a photometric stereo approach, derivatives along image axes directions are computed
in a multiscale framework providing normal discontinuity and orientation maps and lines are finally extracted using non-maxima suppression
and hysteresis thresholds as in Canny's algorithm. In this way it is possible to discover automatically potential structure of interest (inscriptions,
small reliefs) on Cultural Heritage artifacts of interest without the necessity of interactively recreating images using different light directions.
Experimental results obtained on test data and new PTMs acquired in
an archaeological site in the Holy Land with a simple low-end camera,
show that the method provides potentially useful results.

Keywords: Polynomial Texture Maps, Cultural Heritage, Edge Detection.

1   Introduction

Polynomial Texture Maps [7] are an extremely useful tool for the documenta-
tion and the visual analysis of ancient coins, bas-reliefs, paintings and many
other Cultural Heritage objects. They are relightable image, i.e. image where
the user can modify interactively the lighting conditions. Each pixel contains a
bi-quadratic polynomial that encodes an approximate reflectance function of the
scene allowing the possibility modify the image given the illumination direction.
PTMs of an object can be created from multiple images acquired under different incident light directions, without specific hardware (low-end digital camera
provides enough resolution to produce good PTMs, and almost any type of light
source can be used). Several applications of this technique have been proposed,
mainly in the field of Cultural Heritage [9,2,3], where this type of image and
other similar ones [5] have demonstrated to be very useful for analysis purposes.

PTM data can be visualized with ad-hoc software packages, that we describe
in the next, allowing the interactive relighting of the scene that can be used to
find the best way to discover relevant information. Specific methods to enhance
structures using multiple light information have been proposed as well [10].

A. Petrosino (Ed.): ICIAP 2013, Part I, LNCS 8156, pp. 482–491, 2013.

2 Polynomial Texture Maps: Acquisition and Interpretation

A PTM can be generated using different light directions sampling. From this set of photos, the coefficients $(a_0, \ldots, a_5)$ of a bi-quadratic polynomial expressing the reflectance properties of the image can be estimated. This polynomial is in the form:

$$
\begin{aligned}
L(l_u, l_v, x, y) &= a_o(x,y) + a_1(x,y)l_u + a_2(x,y)l_v + \\
&\quad + a_3(x,y)l_u l_v + a_4(x,y)l_u^2 + a_5(x,y)l_v^2 \\
R_f(l_u, l_v, x, y) &= L(l_u, l_v, x, y)R(x, y) \\
G_f(l_u, l_v, x, y) &= L(l_u, l_v, x, y)G(x, y) \\
B_f(l_u, l_v, x, y) &= L(l_u, l_v, x, y)B(x, y)
\end{aligned}
\tag{1}
$$

where (l_u, l_v) is the light direction vector (normalized) projected on the image plane (see Fig. 1). Since this vector is normalized the third component is redundant $(l_z = \sqrt{l_u^2 + l_v^2})$. (R_f, G_f, B_f) is the final color for the given pixel and the given light direction. This type of PTM is called LRGB PTM since the "luminance" term modulates the RGB color channels. In the case of an RGB PTM each channel has its own coefficients, for a total of 18 coefficients per pixel. We refer to $L(l_u, l_v, x, y)$ as "luminance" but, more precisely, this term represents the reflectance functions with the self-shadowing effects embedded. Other basis can be substituted for PTM to better approximate materials characterized by more complex reflectance behavior, such as gold or marble [8].

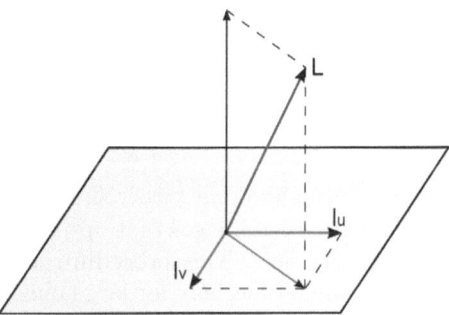

Fig. 1. Projection of light vector on the image plane

Hewlett-Packard labs, where this technology has been developed, freely distribute software tools for the creation and the interactive visualization of PTM data as well as test data, available at the web site *http://www.hpl.hp.com/research/ptm*.

Using the HP tool PTM files can be created from sets of images acquired with standard cameras, provided that the direction of the light is known. A trick to evaluate this direction directly from the acquired images is to include a reflective sphere near the object allowing the direction estimation (l_u, l_v) from the specular highlight. For a correct creation of PTMs of object of interest, therefore, it is not necessary to have an expensive setup: for the acquisition we made in the archaeological site in the Holy Land, we used simply:

- A low-end digital camera (Fuji FINEPIX HS 20 EXR)
- Two tripods, one for the camera and one for the black reflective ball
- One lamp able to generate approximately uniform light on the size of the objects of interest.

Even if the requirements are simple, the acquisition protocol should be, especially for large objects, carefully planned, in order to produce high quality PTMs. In [2] the problem of optimizing acquisition parameters for medium-large objects is discussed. In our work, we acquire a set of large planar stones. For this case we developed and applied the following protocol:

- Take the measures of the object of interest, find its center and its distance from the ground.
- Put the camera on the tripod at a distance of approximately 5/3 of the side of the object from its center
- Measure aperture and shutter speed under the illumination of the central light. Keep these values fixed for all the photos, in order to have a constant exposure.
- Take 40-60 photos moving the light source in a way to attempt to sample uniformly the hemisphere

In order to calculate a precise illumination function, a critical factor is that the digital camera must not move from one photo to the other. Even a misalignment of a few pixel can produce a bad result, with visible aliasing.

3   Edge Detection on PTMs

The analysis of PTM data is usually done with visualization packages allowing the interactive generation of relighted images that optimally put in evidence the structures of interest in the artifacts. This procedure, however, can be sometime difficult and time consuming, especially for large images and when it is not known in advance what should be searched exactly in the images. It could be useful, therefore, to use some automatic processing tools able to extract relevant information and suggest, where structures of interest are located. Filters for the enhanced visualization of this type of images have been proposed in [10] and implemented in a specific tool (*RTIViewer*) available at the web site *http://culturalheritageimaging.org*.

The visualization enhancement is possible thanks to the information provided by the per-pixel bi-quadratic polynomial. Here, we propose an alternative way to help the experts in the artifacts' examination, i.e. not to try to enhance the rendering, but to extract specific information from data.

In ancient tablets or stones, structures of interest are human artifacts and inscriptions that are usually composed by a set of lines. The classical image processing tools used to search lines in images are edge detectors, trying to create lines joining pixels with high values of image gradient. Edges should highlight the important structures in the image rejecting noise and negligible information. We can do something similar for the PTM image data: instead of processing relighted images trying to find optimal visualization, we can simply extract edges from the PTM coefficients and use them to interpret more easily the data acquired.

The detection of depth discontinuities from multi-flash images has been proposed in [11], where authors look for pixels where the difference between baseline and specific illumination is high, and use the results to obtain nice non-photorealistic rendering able to clearly separate objects from background. Our approach is different as well as the application: we used, in fact, classical image processing techniques able to follow continuous lines possibly belonging to artifacts or inscriptions by exploiting PTM coefficients instead of by differentiating the multi-flash images.

Edge detection methods working on standard images first compute a color discontinuity map, e.g. the gradient magnitude, then try to follow its directionally maximal lines to trace edge lines. We propose to replace here the gradient magnitude with a measure of the local variation of the normal vectors that can be estimated at pixel locations from the PTM coefficients. In PTM images, in fact, for each pixel the local normal can be estimated considering the light direction that maximizes the value of the polynomial $L(u, v, l_u, l_v)$. The maximum can be obtained by finding the zero of the spatial derivatives as in [7]. By solving the system:

$$\partial L/\partial u = \partial L/\partial v = 0 \tag{2}$$

it is possible to obtain first the projections of surface normal (n_u, n_v)

$$n_u = \frac{a_2 a_4 - 2a_1 a_3}{4a_0 a_1 - a_2^2} \quad n_v = \frac{a_2 a_3 - 2a_0 a_4}{4a_0 a_1 - a_2^2} \tag{3}$$

and then the full normal as

$$\boldsymbol{N} = (n_u, n_v, \sqrt{1 - n_u^2 - n : v^2}) \tag{4}$$

This normal estimation, however, can be not accurate in certain cases, and tends to be oversmoothed [6]. To reduce these problems we propose to compute normals using a typical photometric stereo [12] approach that consists in assuming the material Lambertian, i.e. a pure diffusive material, and setup a linear system in the following way:

$$\begin{aligned}
N_u l_{u,1} + N_v l_{v,1} + N_z l_{z,1} &= L_1 \\
N_u l_{u,2} + N_v l_{v,2} + N_z l_{z,2} &= L_2 \\
&\cdots \\
N_u l_{u,n} + N_v l_{v,n} + N_z l_{z,n} &= L_n
\end{aligned} \tag{5}$$

where (N_x, N_y, N_z) is the unknown normal for the specific pixel, (l_x, i, l_y, i, l_z, i) is a light direction and L_i is the value of the polynomial for that direction.

The number of directions here used is 32. The set of directions is obtained by sampling the polar coordinates that identify a direction (elevation and azimuth angles) by step of 20 degree starting from 10 to 70, for the elevation, and subdividing the azimuth in 8 parts. In this way the sampling is not uniform (more samples at the pole), but the relatively high number of samples provides a good estimation of the real surface normal. The estimated normal differs from the real one as the material deviates by a Lambertian reflector. The linear system of equation (5) can be solved by using a Singular Value Decomposition (SVD) approach which consists in decomposing the matrix A into three sub-matrix that can be combined to obtain a robust least square solution of the system.

From the normal vector N we can estimate the normal discontinuity along u and v directions (at different scales). If we consider directional derivatives of normal vectors $\partial N/\partial u, \partial N/\partial v$ we can obtain the intensity of the discontinuity as

$$I = \left(\frac{\partial N}{\partial u}\right)^2 + \left(\frac{\partial N}{\partial v}\right)^2 \tag{6}$$

that can be computed using ad hoc masks at different scales. The direction of this edge can be computed as well as

$$\theta = arctan(-\frac{\partial N}{\partial v} \cdot e_2 / \frac{\partial N}{\partial u} \cdot e_1) \tag{7}$$

From the maps I (normalized dividing by its maximum) and θ estimated we can then apply the classical steps of Canny edge detection [1] e.g. non-maxima suppression and hysteresis thresholding. The first step checks if the intensity values are maximal perpendicularly to the edge direction and put to zero all the matrix elements that do not meet this condition. Angles representing edge directions are in the implementation rounded at sampled values of 0, 45, 90 and 135 degrees. The result is a list of candidate edge points that are sorted by decreasing values. Edge lines are finally traced connecting edge candidates with hysteresis thresholding, e.g. using two different thresholds. Edge candidate location with the highest intensity is removed from the list and used to start paths if the corresponding value is larger than the highest threshold, then neigboring candidates (that are as well removed from the list) are joined until their values are higher than the second threshold. The procedure is repeated until all the possible starting points have been removed.

In our implementation we computed normal derivatives using 5-pixel masks, and supported a multiscale estimation of the derivatives/discontinuity as follows: derivatives at different levels of detail are obtained using iterative smoothing and taking derivative masks with increased sampling steps. The final multiscale I and are obtained selecting for each pixel position the value corresponding to the scale maximizing I. Results, as in the usual edge detectors, clearly depend on the parameters chosen (thresholds) and on a possible initial normal smoothing (not necessary however, being the normal maps derived by the PTM coefficients already smoothed). However, for our test a higher threshold of 0.2-0.3 and a smaller one of about 0.05 were able to provide reasonable results on the tested images.

4   Experimental Results

The normal maps computed with the proposed technique appear sufficiently accurate to capture the fine structures of the PTM image.

Fig. 2 shows the results obtained on a sample PTM downloaded from the HP site representing an ancient tablet from the Archæological Research Collection of the University of Southern California. Fig. 2 A shows a single relighting of the data, Fig. 2 B the normal map obtained with our photometric stereo approach and Fig. 2 the normal discontinuity map I computed with two-scale derivatives. The result shows clearly the inscriptions. It is also interesting to compare the edge map obtained applying non maxima suppression and hysteresis thresholding on the previous map, compared with a similar edge detection performed on a single relighted image. It is possible to see that the classical edge detection as well as the gradient map does not highlight correctly the structures of interest, while the PTM based method does (Fig. 3)

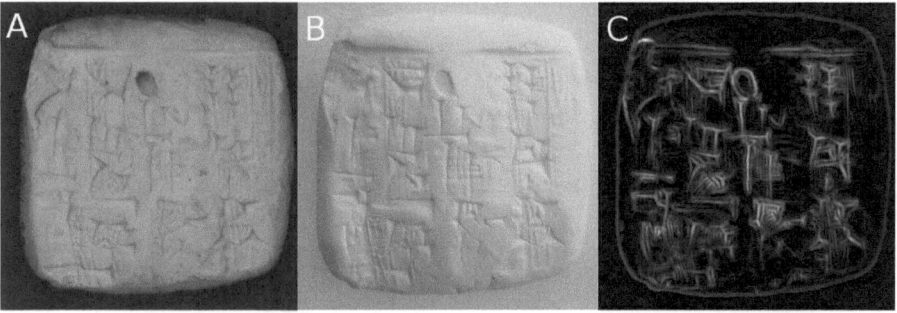

Fig. 2. A: A relighted images from a demo PTM acquisition. B Color coded normal map estimated with the photometric stereo approach. C: Normal discontinuity measure, computed with a 2-scale derivative estimation.

Similar results can be obtained also on our on-field acquisitions on the Holy Land site. Fig. 4 shows an image created by relighting a PTM created with the acquired images. It represents a large stone wall in a tomb, and it is difficult to determine if something interesting can be found in it using the default frontal illumination used.

Looking at small details and changing the light direction something interesting can be, however found. Fig. 5 A shows a central detail on the image of Fig.4, where it is possible to find a written text. To find it automatically, without the necessity of searching manually the optimal light direction, it is possible to try an enhancement filter like those implemented in RTI viewer.

Fig. 5 B shows, for example, the output of the coefficient unsharp masking filter, while Fig.5 C show the result of the static multi-light enhancement, where the inscription is more visible.

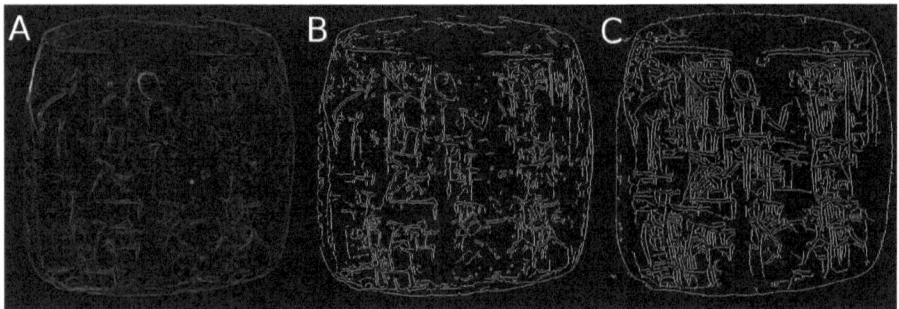

Fig. 3. A: Gradient magnitude computed on a single relighted image. B Corresponding edges detected with the Canny approach. C Edge detection with non maxima suppression and hysteresis thresholding applied to the normal discontinuity map.

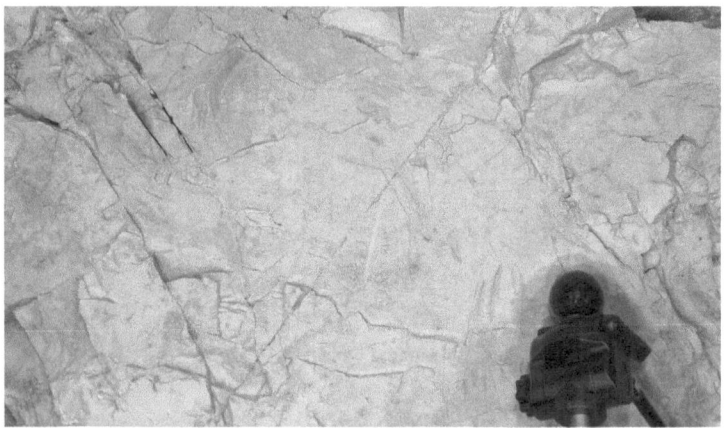

Fig. 4. Relighted PTM acquired on site in an archaeological site in the Holy Land. The quality of the PTMs computed is rather good despite the simple and low-cost acquisition setup.

Extracting normal discontinuities, however, the fact that an inscription is present in that region becomes clearer. Fig. 6 A shows the normal discontinuity map computed at the finest scale, and Fig. 6 B the extracted edges. It is possible to see lines belonging to a text.

Fig. 7 shows another example of edge extraction revealing the location of inscriptions. In Fig. 7 A a frontal relighting shows the lid of a tomb, where it is not easy to see where human artifacts are present. Fig. 7 B shows the normal discontinuity map extracted using derivatives at two scales. Fig. 7 C shows the edges extracted where it is possible to see squares and lines traced in the stone. Fig. 8 shows another part of the archaeological site where the normal discontinuity map and the edges reveal artifacts and text fragments.

Fig. 5. Detail of the previous image where it is possible to find an inscription. It is not visible using a default light direction (A) and still not visible using filters like unsharp masking on coefficients (B). The static multi-light enhancement makes the lines of interest partially visible.

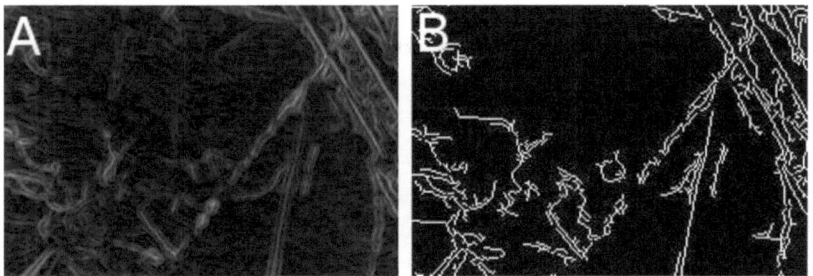

Fig. 6. The normal discontinuity map on the previous detail evidentiates text lines (A). The subsequent edge tracing (B) partially detects the lines of interest.

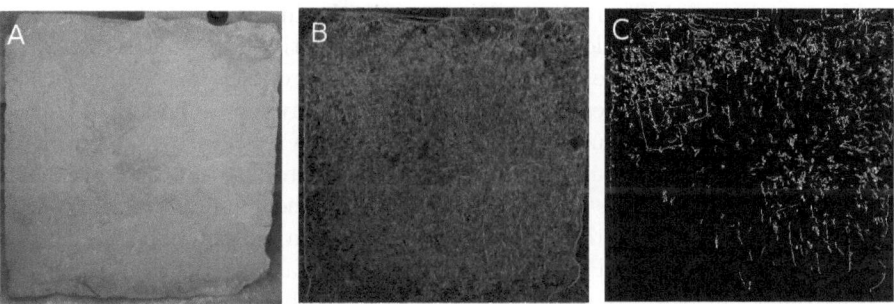

Fig. 7. A: Frontal relighting of a square lid of a tomb. B normal discontinuity map extracted using derivatives at two scales. C: edges extracted after hysteresis thresholding.

A possible improvement of the method could consist in storing local information about the normals and estimating the position of lights enhancing the lines of interest that could be combined creating an adaptively multiple source relighted image.

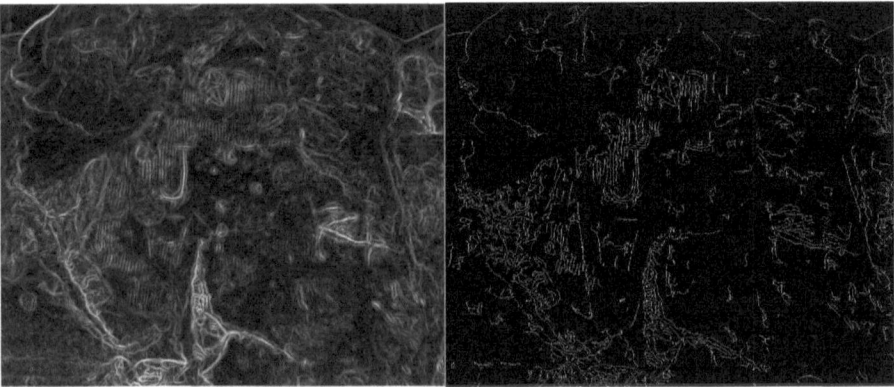

Fig. 8. Left: Normal discontinuity map showing another inscription on a captured stone wall. Right: extracted edge lines.

5   Discussion

Reflection Transform Imaging (RTI) techniques encode a per-pixel parametric approximation of the reflectance behavior of the object depicted, allowing the creation of rendered images with varying lighting of the scene of interest. RTIs are becoming a popular tool to study archaeological sites, stones and artifacts. The analysis of the acquired data is usually done interactively by testing the effects of different light sources on the rendered image. Automatic tools able, for example, to identify automatically regions of interests in the images could be extremely important to help archaeological study of large collections of data. In this paper we proposed a simple edge detector applied to PTM data (a type of RTI) that can be used to enhance lines at different scales, possibly corresponding to human artifacts. Computing 3D surface normals discontinuity and tracing discontinuity profiles on the 2D image plane using a classical edge detection approach, we found that is possible to capture relevant information on the acquired data. Normals are computed using a photometric stereo approach on relighted images and not directly derived from the PTM coefficients, allowing a more precise result.

Lines characterizing inscriptions and artifacts have been successfully extracted on PTM acquired by us on an archaeological site with a simple and cheap setup.

The technique has been here applied to Polynomial Texture Maps, but can be applied as well on different types of RTI [4] which provide a better approximation of the per-pixel reflectance function. We expect to obtain, in this case, improved results. We plan also to extend the idea of automatic processing of RTI data, testing other types of computer vision/pattern recognition methods. For example unsupervised and supervised classification could be applied using the local reflectance coefficients as feature vectors in order to detect regions corresponding to different materials. Pattern recognition tools could be applied as well to recognize characters from lines obtained with the proposed edge detector.

References

1. Canny, J.: A computational approach to edge detection. IEEE Trans. Pattern Anal. Mach. Intell. 8(6), 679–698 (1986), http://dx.doi.org/10.1109/TPAMI.1986.4767851

2. Dellepiane, M., Corsini, M., Callieri, M., Scopigno, R.: High quality ptm acquisition: Reflection transformation imaging for large objects. In: Proceedings of the 7th International Symposium on Virtual Reality, Archaeology and Intelligent Cultural Heritage (VAST 2006), pp. 179–186. The Eurographics Association (2006)

3. Earl, G., Martinez, K., Malzbender, T.: Archaeological applications of polynomial texture mapping: analysis, conservation and representation. Journal of Archaeological Science 37(8), 2040–2050 (2010)

4. Gunawardane, P., Wang, O., Scher, S., Davis, J., Rickard, I., Malzbender, T.: Optimized image sampling for view and light interpolation. In: VAST 2009: 10th International Symposium on Virtual Reality, Archaeology and Cultural Heritage. Faculty of ICT. University of Malta (2009)

5. Happa, J., Mudge, M., Debattista, K., Artusi, A., Gonçalves, A., Chalmers, A.: Illuminating the past: state of the art. Virtual Reality 14(3), 155–182 (2010)

6. MacDonald, L., Robson, S.: Polynomial texture mapping and 3d representations. In: Proc. ISPRS Commission V. Symp. Close Range Image Measurement Techniques (2010)

7. Malzbender, T., Gelb, D., Wolters, H.: Polynomial texture maps. In: Proceedings of the 28th Annual Conference on Computer Graphics and Interactive Techniques, pp. 519–528. ACM (2001)

8. Mudge, M., Malzbender, T., Chalmers, A., Scopigno, R., Davis, J., Wang, O., Gunawardane, P., Ashley, M., Doerr, M., Proenca, A., Barbosa, J.: Image-Based Empirical Information Acquisition. In: Scientific Reliability, and Long-Term Digital Preservation for the Natural Sciences and Cultural Heritage. Eurographics Association, Crete (2008), http://www.eg.org/EG/DL/conf/EG2008/tutorials/T2.pdf

9. Padfield, J., Saunders, D., Malzbender, T.: Polynomial texture mapping: a new tool for examining the surface of paintings. ICOM Committee for Conservation 1, 504–510 (2005)

10. Palma, G., Corsini, M., Cignoni, P., Scopigno, R., Mudge, M.: Dynamic shading enhancement for reflectance transformation imaging. Journal on Computing and Cultural Heritage (JOCCH) 3(2), 6 (2010)

11. Raskar, R., Tan, K.H., Feris, R., Yu, J., Turk, M.: Non-photorealistic camera: depth edge detection and stylized rendering using multi-flash imaging. In: ACM Transactions on Graphics (TOG), vol. 23, pp. 679–688. ACM (2004)

12. Woodham, R.J.: Shape from shading. chap. Photometric method for determining surface orientation from multiple images, pp. 513–531. MIT Press, Cambridge (1989)

A Ripplet Transform Based Statistical Framework for Natural Color Image Retrieval

Manish Chowdhury, Sudeb Das, and Malay K. Kundu

Machine Intelligence Unit, Indian Statistical Institute, Kolkata 700 108, India
{st.manishc,to.sudeb}@gmail.com, malay@isical.ac.in

Abstract. We present a novel Content Based Image Retrieval (CBIR) scheme for natural color images using Multi-scale Geometric Analysis (MGA) of Ripplet Transform (RT) Type-I in the statistical framework based on Generalized Gaussian Density (GGD) model and Kullback-Leibler Distance (KLD). The system is based on modeling the marginal distributions of RT coefficients by GGD framework and computing the similarity between the model parameters using the KLD. Least Square-Support Vector Machine (LS-SVM) classifier is used to classify the images of the database. Extensive experiments were carried out to evaluate the effectiveness of the proposed system on two image databases consisting 1000 (Simplicity) and 2788 (Oliva) images, respectively. Experimental results and comparisons show that the proposed CBIR system performs efficiently in image retrieval field.

Keywords: RT, KLD, LS-SVM, CBIR, MGA, GGD.

1 Introduction

The task of retrieving relevant images from a large image database (DB), by measuring similarities between the query image and the database images has become a potential area of research. High retrieval efficiency and less computational complexity, are the desired characteristics of an efficient CBIR system [3].

A modern CBIR system consists of three main parts: feature extraction (FE), similarity measurement (SM) and relevance feedback (RF), respectively. Various FE schemes have been reported in the literature for finding the significant visual information of the images [4]. Generally, there are two types of image representation: spatial domain representation and frequency domain representation. It has been found that the frequency domain representations are more robust against noise and capable of representing images more efficiently than the spatial domain representation [14].

Among the various Multiresolution Analysis (MRA) tool, Wavelet Transform (WT) and its variants (like M-band WT, WT packets etc.) have been used in CBIR systems extensively [14,10]. But the problem with WT is that it is inherently non-supportive to directionality and anisotropy. To overcome these limitations of WT, recently a theory called Multi-scale Geometric Analysis (MGA) for

A. Petrosino (Ed.): ICIAP 2013, Part I, LNCS 8156, pp. 492–502, 2013.

high-dimensional signals has been proposed and several MGA tools are developed like Curvelet, Contourlet etc., with applications in various problems [6,1]. CBIR systems based on these MGA tools are found to perform better than the system based on WT [15,7]. Recently, Ripplet Transform (RT) Type-I is proposed as an extension to Curvelet Transform (CVT) with alternative scaling law, and possessing higher efficiency for various image processing applications [18,2]. In frequency domain methods, feature vectors are computed from various simple subbands' statistics (such as mean, standard deviation and energy etc.,) [2]. Whereas, in statistical approach, images are modeled in terms of statistical distribution [3,5]. In this article, we have extended this statistical framework to RT for constructing an efficient image representative feature vector.

Many machine learning techniques such as MLP, SVM, etc., have been used to improve the retrieval accuracy of the CBIR system [19,13]. However, the training process of MLP is time consuming and its convergence depends on the initial parameter setting. Similarly, training with SVM is computationally expensive for high dimensional data sets, such as image data. To reduce the computational demand, the least square version of SVM, i.e. least square support vector machine (LS-SVM) was developed, which solves in a set of linear equations instead of quadratic programming and simplifies the training procedure [16]. Therefore, to reduce the time complexity and prolonged training process for parameters optimization, we have used LS-SVM in our proposed CBIR system.

Traditionally, the most commonly used similarity measures in CBIR is Euclidean distance (ED). ED is often used in CBIR, because of its computational simplicity and rotation invariance property [3]. However, the main problem using ED is its scale problem, because features that have an inherently larger scale would be predominant. In contrast to these similarity measures, Kullback Leibler Divergence (KLD), an information-theoretic measure is found to be more effective [5,8].

Even though statistical modeling has been used extensively in texture analysis problem [3,5]. The effectiveness of these approaches in natural color image classification is not explored well. In this article, we propose to apply the statistical framework based on Generalized Gaussian Density (GGD) and KLD on Discrete Ripplet Transform (DRT) coefficients to investigate its performance in natural color image classification problem. To improve the retrieval accuracy as well as to reduce the computational complexity, we have used LS-SVM in the proposed system. Extensive experiments and comparisons with state-of-the-art CBIR systems, shows that the proposed CBIR system works more efficiently .

The paper is organized as follows: Section 2 describes the theoretical preliminary of the RT. The detail descriptions of the feature extraction procedure, classification through LS-SVM and KLD based similarity measure are explained in Section 3, Section 4 and Section 5, respectively. The proposed CBIR system is described in Section 6. Section 7 discusses the experimental system and results. Finally, Section 8 concludes the article.

2   Ripplet Transform Type-I (RT)

Conventional transforms like Fourier Transform (FT) and WT suffer from discontinuities such as edges and contours in images. To address this problem, Jun Xu et al. proposed a new MGA-tool called RT [18]. RT is a higher dimensional generalization of the CVT. RT provides a new tight frame with sparse representation for images with discontinuities along C^d curves. There are two questions regarding the scaling law used in CVT: 1) Is the parabolic scaling law optimal for all types of boundaries? and if not, 2) What scaling law will be optimal? To address these questions, Jun Xu et al., intended to generalize the scaling law, which resulted in RT. RT generalizes CVT by adding two parameters, i.e., support c and degree d. CVT is a special case of RT with $c = 1$ and $d = 2$.

As digital image processing needs discrete transform instead of continuous transform, here we describe the discretization of RT [18]. In the frequency domain, the corresponding frequency response of ripplet function is in the form

$$\hat{\rho}_j(r, \omega) = \frac{1}{\sqrt{c}} a^{\frac{m+n}{2n}} W(2^{-j} \cdot r) V(\frac{1}{c} \cdot 2^{-\lfloor j \frac{m-n}{n} \rfloor} \cdot \omega - l) \tag{1}$$

where W and V are the 'radial window' and 'angular window', respectively and satisfy the following admissibility conditions:

$$\sum_{j=0}^{+\infty} |W(2^{-j} \cdot r)|^2 = 1 \tag{2}$$

$$\sum_{l=-\infty}^{+\infty} |V(\frac{1}{c} \cdot 2^{-\lfloor j(1-1/d) \rfloor} \cdot \omega - l)|^2 = 1 \tag{3}$$

given c, d and j. Here, the scale parameter a is sampled at dyadic intervals. b (position parameter) and θ (rotation parameter) are sampled at equal-spaced intervals. a_j, $\vec{b}_k$ and θ_l substitute a, $\vec{b}$ and θ respectively, and satisfy that $a_j = 2^{-j}$, $\vec{b}_k = [c \cdot 2^{-j} \cdot k_1, 2^{-j/d} \cdot k_2]^T$ and $\theta_l = \frac{2\Pi}{c} \cdot 2^{-\lfloor j(1-1/d) \rfloor} \cdot l$, where $\vec{k} = [k_1, k_2]^T$, and j, k_1, k_2, $l \in \mathbb{Z}$. $(\cdot)^T$ denotes the transpose of a vector. $d \in \mathbb{R}$, since any real number can be approximated by rational numbers, so we can represent d with $d = n/m$, $n, m \neq 0 \in \mathbb{Z}$. Usually, we prefer $n, m \in \mathbf{N}$ and n, m are both primes.

The 'wedge' corresponding to the ripplet function in the frequency domain is

$$H_{j,l}(r, \theta) = \{2^j \leq |r| \leq 2^{2j}, |\theta - \frac{\pi}{c} \cdot 2^{-\lfloor j(1-1/d) \rfloor} \cdot l| \leq \frac{\pi}{2} 2^{-j}\} \tag{4}$$

The DRT of an $M \times N$ image $f(n_1, n_2)$ is in the form of

$$R_{j, \vec{k}, l} = \sum_{n_1=0}^{M-1} \sum_{n_2=0}^{N-1} f(n_1, n_2) \overline{\rho_{j, \vec{k}, l}(n_1, n_2)} \tag{5}$$

where $R_{j, \vec{k}, l}$ are the ripplet coefficients and $\overline{(.)}$ denotes the conjugate operator.

3   Ripplet Coefficient Modeled Using GGD

The images in the DB prior to RT decomposition are transformed from RGB to YCbCr color space. This ensures that the textural characterization of the images are independent of the color characterization. RT decomposition over the intensity plane (Y) characterizes the texture information, while the RT decomposition over chromaticity planes (Cb and Cr) characterizes color. Texture and color information are extracted by using RT on each color plane with 4 level (1, 2, 4, 4) decomposition. This decomposition configuration produces 11 ($= 1+2+4+4$) subbands for each image of the DB for each color plane. As, there are 3 color planes, so altogether we get 33 ($= 3 \times 11$) subbands for each image of the DB. The distribution of ripplet subband coefficients are then modeled with GGD which is defined as,

$$p(x; \alpha, \beta) = \frac{\beta}{2\alpha\Gamma(1/\beta)} e^{-(|x|/\alpha)^{\beta}} \qquad (6)$$

where x is the ripplet subband coefficients and $\Gamma(.)$ is the gamma function, i.e., $\Gamma(z) = \int_0^\infty e^{-t} t^{z-1} dt$, $z > 0$. Here the scale parameter α models the width of the probability distribution function (PDF) peak (standard deviation), while the shape parameter β is inversely proportional to the decreasing rate of the peak. These two parameters need to be estimated for feature vector creation. As Maximum Likelihood (ML) estimator is best suited for estimating heavy-tailed distribution like GGD for both small and large samples, we have used ML estimator in our proposed scheme.

For the sample set $x = (x_1, x_2, x_3, ..., x_k)$, x_i is the ripplet coefficients at the ith subband, and $i \leq L$, the ML estimator is defined as [5].

$$L(x; \alpha, \beta) = log \prod_{i=1}^{L} p(x_i; \alpha, \beta) \qquad (7)$$

GGD parameters are defined with the following equations, which have a unique root in probability

$$\frac{\partial L(x; \alpha, \beta)}{\partial \alpha} = -\frac{L}{\alpha} + \sum_{i=1}^{L} \frac{\beta |x_i|^{\beta} \alpha^{-\beta}}{\alpha} = 0 \qquad (8)$$

$$\frac{\partial L(x; \alpha, \beta)}{\partial \beta} = \frac{L}{\beta} + \frac{L\Psi(1/\beta)}{\beta^2} - \sum_{i=1}^{L} \left(\frac{|x_i|}{\alpha}\right)^{\beta} log\left(\frac{|x_i|}{\alpha}\right) = 0 \qquad (9)$$

where $\Psi(.)$ is the digamma function, i.e., $\Psi(z) = \Gamma'(z)/\Gamma(z)$. α has a unique, real, positive solution and can be obtained from Eq.(8) by fixing $\beta > 0$:

$$\hat{\alpha} = \left(\frac{\beta}{L} \sum_{i=1}^{L} |x_i|^{\beta}\right)^{1/\beta} \qquad (10)$$

Substituting this into (9), the shape parameter β is the solution of the following *transcendental* equation

$$1 + \frac{\Psi(1/\hat{\beta})}{\hat{\beta}} - \frac{\sum_{i=1}^{L} |x_i|^{\hat{\beta}} \log |x_i|}{\sum |x_i|^{\hat{\beta}}} + \frac{\log \left(\frac{\hat{\beta}}{L} \sum_{i=1}^{L} |x_i|^{\hat{\beta}} \right)^{1/\hat{\beta}}}{\hat{\beta}} = 0 \qquad (11)$$

which can be solved numerically using the Newton-Raphson iterative procedure. Therefore, we obtain two features from a ripplet subband (l). Considering s ($= 3l$) as the total number of subbands for an image I, we obtain 2s dimensional ripplet GGD feature vector. The final feature vector of an image I in the DB is as follows:

$$f_{vec}^{I} = [\alpha_1, \beta_1, \alpha_2, \beta_2, ..., \alpha_s, \beta_s,] \qquad (12)$$

where α_l and β_l are the GGD parameters computed from the coefficients of the lth ripplet subband and $1 \le l \le s$.

4   Classification through LS-SVM

Once the ripplet GGD feature vectors are obtained, LS-SVM classifier is used to classify images of the database. LS-SVM avoids solving quadratic programming problem and simplifies the training procedure of conventional SVM [16]. Considering a linearly separable binary classification problem:

$$(x_i, y_i)_{i=1}^{n} \quad and \quad y_i = \{+1, -1\} \qquad (13)$$

where x_i is an n-dimensional vector and y_i is the label of this vector. LS-SVM can be formulated as the optimization problem:

$$\min_{w,b,e} \mathcal{J}(w, b, e) = \frac{1}{2} w'w + \frac{1}{2} C \sum_{i=1}^{n} e_i^2 \qquad (14)$$

subject to the equality constraints

$$y_i[w'\varphi(x_i) + b] = 1 - e_i \qquad (15)$$

where $C > 0$ is a regularization factor, b is a bias term, w is the weights vector, e_i is the difference between the desired output and the actual output and $\varphi(x_i)$ is a mapping function.

The lagrangian for problem of Eq.(14) is defined as follows:

$$\mathcal{L}(w, e_i, b, \alpha_i) = \min_{w,b,e} \mathcal{J}(w, b, e) -$$

$$\sum_{i=1}^{n} \alpha_i \{y_i[w'\varphi(x_i) + b] - 1 + e_i\} \qquad (16)$$

where α_i are Lagrange multipliers. The Karush-Kuhn-Tucker (KKT) conditions for optimality

$\frac{\partial \mathcal{L}}{\partial w} = 0 \rightarrow w = \sum_{i=1}^{n} \alpha_i y_i \varphi(x_i); \frac{\partial \mathcal{L}}{\partial e_i} = 0 \rightarrow \alpha_i = Ce_i; \frac{\partial \mathcal{L}}{\partial b} = 0 \rightarrow \sum_{i=1}^{n} \alpha_i y_i = 0; \frac{\partial \mathcal{L}}{\partial \alpha_i} = 0 \rightarrow y_i[w'\varphi(x_i) + b] - 1 + e_i = 0$,is the solution to the following linear system

$$\begin{bmatrix} 0 & -Y \\ Y & \varphi\varphi' + C^{-1}I \end{bmatrix} \begin{bmatrix} b \\ \alpha \end{bmatrix} = \begin{bmatrix} 0 \\ \bar{1} \end{bmatrix} \tag{17}$$

where $\varphi = [\varphi(x_1)'y_1, ..., \varphi(x_n)'y_n], Y = [y_1, ..., y_n], \bar{1} = [1, ..., 1]$, and $\alpha = [\alpha_1, ..., \alpha_n]$. For a given kernel function K(,) and a new test sample point x, the LS-SVM classifier is given by

$$f(x) = sgn[\sum_{i=1}^{n} \alpha_i y_i K(x, x_i) + b] \tag{18}$$

5   Similarity Measurement Using KLD

After the image is classify using LS-SVM according to the ripplet GGD feature vectors, KLD is used for distance calculation between the query image and the database images.

Let $p_1(x)$ and $p_2(x)$ be two continuous probability distribution functions. KLD distance between these two probability distribution function is defined by [5]

$$D(p_1, p_2) = \int p_1(x) log\left(\frac{p_1(x)}{p_2(x)}\right) dx \tag{19}$$

We obtain the KLD between two GGDs substituting the GGD probability distribution function of Eq.(6) into Eq.(19), which is defined as [5]

$$D(p(,;\alpha_1, \beta_1)||p(,;\alpha_2, \beta_2)) = \log \frac{\beta_1 \alpha_2 \Gamma(1/\beta_2)}{\beta_2 \alpha_1 \Gamma(1/\beta_1)}$$
$$+ \left(\frac{\alpha_1}{\alpha_2}\right)^{\beta_2} \frac{\Gamma((\beta_2 + 1)/\beta_1)}{\Gamma(1/\beta_1)} - \frac{1}{\beta_1} \tag{20}$$

Ripplet coefficients of each subband are assumed to be independent. Therefore, KLD between two images is obtained by calculating the sum of all the Kullback-Leibler distances across all the ripplet subbands of these images. Let, α_j^i and β_j^i be the ripplet GGD features of subband j of image i. Using the definition of KLD between two GGDs shown in Eq. (20), the KLD between two images P and Q is defined by Eq.(21)

$$D(P, Q) = \sum_{j=1}^{s} D(p(.; \alpha_j^{(P)}, \beta_j^{(P)})||p(.; \alpha_j^{(Q)}, \beta_j^{(Q)}))$$
$$= \sum_{j=1}^{s} log \frac{\beta_j^{(P)} \alpha_j^{(Q)} \Gamma(1/\Gamma_j^{(Q)})}{\beta_j^{(Q)} \alpha_j^{(P)} \Gamma(1/\Gamma_j^{(P)})} +$$
$$\sum_{j=1}^{s} \left(\frac{\alpha_j^{(P)}}{\alpha_j^{(Q)}}\right)^{\beta_j^{(Q)}} \frac{\Gamma((\beta_j^{(Q)} + 1)/\beta_j^{(P)})}{\Gamma(1/\beta_j^{(P)})} - \sum_{j=1}^{s} \frac{1}{\beta_j^{(P)}} \tag{21}$$

where s is the total number of ripplet subbands of an image. For a given query image, KLD is measured between the query image and each of the database images using Eq.(21). Database images are ranked according to ascending order of distances and are retrieved from the database.

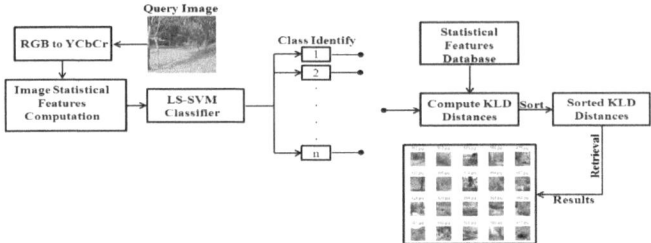

Fig. 1. Block Diagram of the Proposed CBIR System

6   Proposed CBIR System

Here, we outlines the salient steps of the proposed CBIR system, considering the statistical features of the images in the database are already computed and stored in the 'Statistical Features Database'. The Fig. 1, shows the block diagram of the proposed CBIR system.

Steps:

1. Input the query image.
2. Convert the query image from RGB color plane to YCbCr color plane.
3. Compute the GGD based statistical parameters of the query image as described in Section 3.
4. Identify the class of the query image using LS-SVM, depending on the RT based statistical features as described in Section 4.
5. As describe in Section 5, find the KLD distances between the statistical features of the query image and the statistical features of the image DB stored in "Statistical features Database".
6. Sort the distances in ascending order and display 20 images corresponding to the first 20 sorted distances.
7. Stop.

7   Experimental Results

Extensive experiments were carried out to evaluate the performance of the proposed CBIR system, and its performance is compared with several state-of-the-art CBIR systems.

7.1   Experimental Setup

Two publicly available image databases: (1) SIMPLIcity [17] and (2) Oliva [12] were used in the experiments. SIMPLIcity database consists of 1000 images of 10 different categories containing 100 images each. Oliva dataset consists of 2688 images with 8 categories. LS-SVM classifier is trained with 70% and tested with 30% of the leveled data from the image database using stratified random sampling method, respectively. For training, we have used the Radial Basis Function (RBF), as the kernel. There are two tunable parameters while using the RBF kernel in LS-SVM classifier: C and γ. It is not known beforehand which values of C and γ are the best for the classification problem at hand. Hence, a $2X5$ fold cross-validation (CV) is conducted, where various pairs of (C, γ) are tried and the one with the lowest CV error rate is picked. We have achieved the classification accuracy of 80% for SIMPLIcity image database. With similar configuration, 75% the classification accuracy is obtained for Oliva dataset.

Quantitative evaluation of the proposed CBIR system is analyzed using two statistical measures: Mean average precision (MAP) and Mean average recall (MAR). We computed the precisions and recalls considering all the images of the used databases as the query images, and then took the average of the obtained precision and recall values over all the images as the final evaluation result.

7.2   Results and Discussion

The retrieval performance of the proposed CBIR system (RT+GGD+LS-SVM) is shown in the Fig. 2, in terms of precision and recall curve. The performance comparisons of WT and CVT with RT are also shown in Fig. 2 (WT+GGD+LS-SVM and CVT+GGD+LS-SVM, respectively). Here, only the images of the SIMPLIcity database were considered for performance evaluation. It is clear from the Fig. 2, that RT performance much better than WT and CVT in CBIR domain.

Fig. 3(a) and Fig. 3(b), shows two examples of the visual results obtained by the proposed CBIR system on the SIMPLIcity and Oliva dataset using query images from "*Building*" and "*Forest*" classes, respectively. From all the given instances, we can clearly see that all the retrieved images are from the respective classes corresponding to the query images.

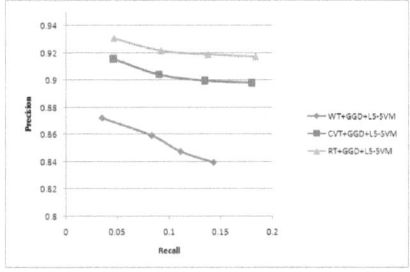

Fig. 2. Performance comparison in terms of precision and recall curves

Fig. 3. Visual results of the proposed CBIR system for SIMPLIcity and Oliva database (top left image is the query image)

Table 1. Comparisons with other existing CBIR systems in terms of average precision on SIMPLicity database

Class	Wang [17]	CTCHIRS [11]	IGA-CBIR [9]	Proposed Method
Africans	48	68.30	61	**89.42**
Sea	32	54	**93**	76.66
Building	35	56.15	85	**86.04**
Buses	36	88.80	71	**97.90**
Dinosaurs	95	99.25	**100**	100
Elephant	38	65.80	80	**93.77**
Flower	42	89.10	82	**98.08**
Horse	72	80.25	79	**100**
Mountain	35	52.15	56	**85.96**
Food	38	73.25	**99**	96.22
Average	47	72.70	80.6	**92.40**

Table 1, shows the results of the performance comparisons between the proposed CBIR system with some of the existing CBIR systems namely, SIMPLIcity [17], CTCHIRS [11] and IGA-CBIR [9]. The images of the SIMPLIcity database are used in this experiment. It can be easily seen from the Table 1 that our proposed CBIR system have overall higher retrieval accuracy than these existing systems. It is also to be noted that for only two categories of images ("Sea" and "Food"), the proposed CBIR system have achieved lower retrieval accuracy than the method described in [9]. The '**bold**' values of the Table 1 denote the highest retrieval accuracy obtained. The CBIR systems described in [11] and [9] used 71 and 86 features, whereas we have used only 66 features to represent every image of the database. It is obvious from the results given in Table 1 that the proposed CBIR system performs much better than the above mentioned existing CBIR schemes. The MAP and MAR of our proposed algorithm is 92.40% and 16.40% for SIMPLIcity dataset and for Oliva dataset, 85.98% and 12.45% respectively.

8   Conclusions

From our experiments, we have noticed that statistical modeling of RT based image coding is suitable for representing low level features of the images. The proposed CBIR system with LS-SVM classifier based on RT based features is able to improve the accuracy of the retrieval performance and to reduce the computational cost. To overcome the problem of misclassification in pre-classification phase, we are trying to implement fuzzy ranking membership function. The proposed mechanism could be tested for video retrieval as future scope of research.

References

1. Candes, E., Donoho, D.: Continuous curvelet transform: I. resolution of the wavefront set. Appl. and Comput. Harmon. Anal. 19, 162–197 (2003)
2. Chowdhury, M., Das, S., Kundu, M.K.: Novel CBIR system based on ripplet transform using interactive neuro-fuzzy technique. Elect. Letters on Comp. Vision and Image Anal. 11, 1–13 (2012)
3. Datta, R., Joshi, D., Li, J., Wang, J.Z.: Image retrieval: Ideas, influences, and trends of the new age. ACM Comput. Surv. 40, 1–60 (2008)
4. Deselaers, T., Keysers, D., Ney, H.: Features for image retrieval: an experimental comparison. Inform. Retri. 11, 77–107 (2008)
5. Do, M.N., Vetterli, M.: Wavelet-based texture retrieval using generalized gaussian density and kullback-leibler distance. IEEE T. Image Process 11(2), 146–158 (2002)
6. Do, M.N., Vetterli, M.: The contourlet transform: an efficient directional multiresolution image representation. IEEE T. Image Process 14, 2091–2106 (2005)
7. Duc, H.N., Tien, T.L., Honl, T.D., Thu, C.B., Xuan, T.N.: Image retrieval using contourlet based interest points. In: Proc. of the 10th ISSPA, pp. 93–96 (2010)
8. Johnson, D.H., Sinanovic, S.: Symmetrizing the kullback-leibler distance. IEEE T. Inf. Theory (2001)
9. Lai, C.C., Chen, Y.C.: A user-oriented image retrieval system based on interactive genetic algorithm. IEEE T. Instrum. Meas. 60, 3318–3325 (2011)
10. Li, L.Y., Yan, Y., Yang, C.: Implementation of texture based image retrieval using M-band wavelet transform. Wuhan University J. of Natural Sciences 8(4), 1107–1110 (2003)
11. Lin, C.H., Chen, R.T., Chan, Y.K.: A smart content-based image retrieval system based on color and texture feature. Image Vision Comput. 27, 658–665 (2009)
12. Oliva, A., Torralba, A.: Modeling the shape of the scene: A holistic representation of the spatial envelope. Int. J. of Comp. Vision 42, 145–175 (2001)
13. Pourghassem, H., Ghassemian, H.: Content-based medical image classification using a new hierarchical merging scheme. Comp. Med. Imaging and Graphics 32, 651–661 (2008)

14. Quellec, G., Lamard, M., Cazuguel, G., Cochener, B., Roux, C.: Adaptive non-separable wavelet transform via lifting and its application to content-based image retrieval. IEEE T. Image Process 19, 25–35 (2010)
15. Sumana, I.J., Islam, M.M., Zhang, D., Lu, G.: Content based image retrieval using curvelet transform. In: Proc. of 10th IEEE MMSP, pp. 11–16 (2008)
16. Suykens, J.A.K., Vandewalle, J.: Least squares support vector machine classifiers. Neural Process. Lett. 9, 293–300 (1999)
17. Wang, J.Z., Li, J., Wiederhold, G.: SIMPLIcity: Semantics-sensitive integrated matching for picture libraries. IEEE T. Pattern Anal. 23, 947–963 (2001)
18. Xu, J., Yang, L., Wu, D.: Ripplet: A new transform for image processing. J. Vis. Commun. Image R. 21, 627–639 (2010)
19. Yang, C.C., Prashar, S.O., Landry, J.A., Perret, J., Ramaswamy, H.S.: Recognition of weeds with image processing and their use with fuzzy logic for precision farming. Can. Agr. Eng. 42, 195–200 (2000)

A Fully Automatic Approach for the Accurate Localization of the Pupils

Marco Leo, Dario Cazzato, Tommaso De Marco, and Cosimo Distante

National Research Council of Italy - Institute of Optics
via della Libertà, 3 73010 Arnesano (Lecce)
marco.leo@cnr.it

Abstract. This paper presents a new method to automatically locate pupils in images (even with low-resolution) containing human faces. In particular pupils are localized by a two steps procedure: at first self-similarity information is extracted by considering the appearance variability of local regions and then they are combined with an estimator of circular shapes based on a modified version of the Circular Hough Transform. Experimental evidence of the effectiveness of the method was achieved on challenging databases containing facial images acquired under different lighting conditions and with different scales and poses.

Keywords: Self-similarity, Saliency, Circularity Analysis, Pupil Localization.

1 Introduction

As one of the most salient features of the human face, eyes and their movements play an important role in expressing a person's desires, needs, cognitive processes, emotional states, and interpersonal relations. For this reason the definition of a robust and non-intrusive eye detection and tracking system is crucial for a large number of applications: advanced interfaces, control of the level of human attention, biometrics, gaze estimation for example for marketing purposes, etc. A detailed review of recent techniques for eye detection and tracking can be found in [1], where it is clear that the most promising solutions use invasive devices (*active eye localization and tracking*). In particular some of them are already available on the market and require the user to be equipped with a head mounted device [2], while others obtain accurate eye location through corneal reflection under active infrared (IR) illumination [3]. Passive eye detection and tracking systems are only recently introduced and they attempt to obtain information about eye location just starting from images acquired from one or more cameras. Most popular approaches in this area use complex shape models of the eye: they work only if the important elements of the eye are visible and then zoomed, or high resolution views are required [4]. Other approaches explore the characteristics of the human eye to identify a set of distinctive features around the eyes and/or to characterize the eye and its surroundings by the color distribution or filter responses. The method proposed by Asteriadis et al. [5] assigns a vector to every pixel in the edge map of the eye area, which points to the closest edge pixel. The length and the slope information of these vectors are consequently used to detect and localize the eyes by matching them with a training

A. Petrosino (Ed.): ICIAP 2013, Part I, LNCS 8156, pp. 503–512, 2013.

set. Timm and al. [6] proposed an approach for accurate and robust eye center local-ization by using image gradients. They derived an objective function whose maximum corresponds to the location where most gradient vectors intersect and thus to the eye's center. A post-processing step is introduced to reduce wrong detection on structures such as hair, eyebrows, or glasses. In [7] the center of (semi)circular patterns is inferred by using isophotes. In a more recent paper by the same authors, additional enhance-ments are proposed (using mean shift for density estimation and machine learning for classification) to overcome problems that arise in certain lighting conditions and occlu-sions from the eyelids [8]. A filter inspired by the Fisher Linear Discriminant classifier is instead proposed in [9] to localize the eyes. A sophisticated training of the left and right eye filters is required. In [10] a cascaded AdaBoost framework is proposed. Two cascade classifiers in two directions are used: the first one is a cascade designed by bootstrapping the positive samples, and the second one, as the component classifiers of the first one, is cascaded by bootstrapping the negative samples. A method for precise eye localization that uses two Support Vector Machines trained on properly selected Haar wavelet coefficients is presented in [11]. In [12] an Active Appearance Model (or AAM) is used to model edge and corner features in order to localize eye regions.

Unfortunately, the analysis of the state of the art reveals that most of the methods uses a supervised training phase for modeling the appearance of the eye or, alternatively, introduces ad-hoc reasoning to filter missing or incorrect detection. For this reason, although leading to excellent performance in specific contexts, they can not be directly used in different contexts (especially in the real world ones) without some adjustments of the models previously learned.

This paper explores the possibility to introduce a pupil detection approach that does not require any training phase (or post filtering strategy). It detects the pupil in low-resolution images by combining self-similarity and circularity information: in other words the total variability of local regions are characterized by saliency maps that are then related with gradient based features specifying point-wise circularity. This way the proposed approach is more suitable to operate in real contexts where it is not generally possible to ensure uniform boundary conditions. Experimental evidence of the effec-tiveness of the method was achieved on benchmark data sets containing facial images acquired under different lighting conditions and with different scales and poses.

2   Overview of the Proposed approach

Figure 1 schematically shows the main steps of the proposed approach. Each input im-age is initially analyzed by using the boosted cascade face detector proposed by Viola and Jones [13]. The rough positions of the left and right eye regions are then estimated using anthropometric relations. In fact, pupils are always contained within two regions starting from 20×30 percent (left eye) and 60×30 percent (right eye) of the detected face region, with dimensions of 25×20 percent of the latter [14]. The innovative pro-cedure based on the combination of self-similarity and circularity information is finally applied to the cropped patches in order to accurately find the pupil location.

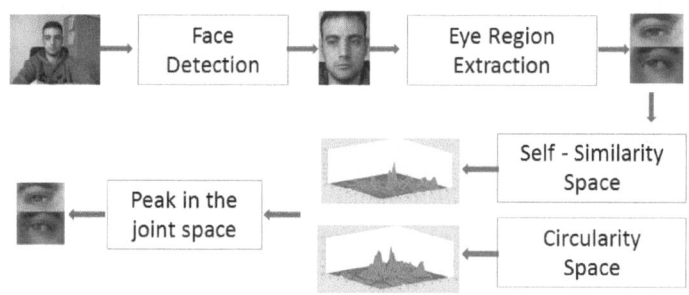

Fig. 1. A quick overview of the proposed multistep approach for pupil localization

2.1 Self-similarity Space Computation

The key idea is to initially find regions with high self-similarity, i.e. regions that retain their peculiar characteristics even under geometric transformations (such as rotations or reflections), changes of scale, viewpoint or lighting conditions and possibly also in the presence of noise. Self-similarity score can be effectively computed as a normalized correlation coefficient between the intensity values of a local region and the intensity values of the same geometrically transformed local region [15]. A local region is self-similar if a linear relationship exists:

$$I(T(x)) = a + bI(x) \qquad \forall x \in P \tag{1}$$

where P is a circular region of radius r and x is a point located in P. $I(x)$ denotes the intensity value of the image I at location x, and T represents a geometric transformation defined on P. For the purposes of the paper, T is limited to a reflection and a rotation. Both reflection and rotation, preserve distances, angles, sizes, and shapes. To better clarify the notions of reflection and rotation into the specific context under consideration, point locations can be represented in polar coordinates, hence $x = (r, \phi)$. Every reflection is associated to a mirror line going through the center of P and having orientation denoted by $\vartheta \in [0; 2\pi]$. Having said that, a reflection is defined as the geometric transformation that maps the location (r, ϕ) to location $(r, 2\vartheta - \phi)$.

Similarly every rotation has a centre and an angle. Let the centre of the rotation be the centre of P and let the rotation angle α be one of the angles $\frac{2\pi}{n}$, where n is an integer. A rotation maps the location (r, ϕ) to location $(r, \phi + \alpha)$.

Given these preliminary concepts, from the operational point of view, the cornerstone of this first phase is the search of the points that are closest to satisfy the condition in equation 1 considering that on real data, it can hardly be fulfilled for all points of P. This way, highlighted points should correspond to the pixels of the eye which has both (almost) radial and rotational symmetry. In particular the strength of the linear relationship in equation 1 can be measured by the normalized correlation coefficient:

$$ncc(P, T) = \frac{\sum_i (I(x_i) - \bar{I})(I(T(x_i)) - \bar{I})}{\sqrt{(\sum_i (I(x_i) - \bar{I}))^2 (I(T(x_i)) - \bar{I})^2}} \tag{2}$$

Here i counts all points of P and $\bar{I}$ represents the average intensity value of points of P.

At a given location, the normalized correlation coefficients in equation 2 can be computed for different mirror line orientations or different angles of rotation. All give information of region self-similarity.

In this paper the average normalized correlation coefficient computed over all orientations of the mirror line (*radial similarity map S*) at a given location is used as a measure of region self-similarity[1].

Let the sampling intervals for θ be $\Delta\theta = \frac{2\pi}{N}$ the similarity measure is then computed as

$$S(P) = \frac{1}{N} \sum_{i=0}^{N-1} ncc(P, T_{\theta_i}) \tag{3}$$

To overcome the problems related to the processing near the borders, for the calculation of the self-similarity scores, a wider area (10 pixels in each direction) is considered and then the portion of the self-similarity space relative to the original size of the eye patches is extracted. At the end a $(m \times n \times M)$ data structure S_r is available where $m \times n$ is the size of the input image and M is the number of sampled radii r (i.e. the number of considered scales). Local maxima and minima are then obtained by comparing each pixel value to its eight neighbors in the current similarity map and nine neighbors in the scale level above and below. A point is selected only if it has an extreme value compared to all its neighbors. The self-similarity map (of size $m \times n$) at the scale corresponding to the region (among the selected ones) with the highest self similarity score is the outcome of this first phase.

2.2 Circularity Measurements

The second phase starts with the estimation of the spatial distribution of circular shapes and this is done by a modified version of the Circular Hough Transform (CHT). A circle detection operator, that is applied over all the image pixels, produces a maximal value when a circle with a radius in the range $[Rmin, Rmax]$ is detected. It is defined as follows:

$$u(x,y) = \frac{\int \int_{D(x,y)} e(\alpha, \beta) \cdot O(\alpha - x, \beta - y) d\alpha d\beta}{2\pi(R_{max} - R_{mim})} \tag{4}$$

The domain D(x,y) is defined as:

$$D(x,y) = \{(\alpha, \beta) \in \Re^2 | R_{min}^2 \le (\alpha - x)^2 + (\beta - y)^2 \le R_{max}^2\} \tag{5}$$

where e is the normalized gradient vector:

$$e(x,y) = [\frac{E_x(x,y)}{|E|}, \frac{E_y(x,y)}{|E|}]^T \tag{6}$$

---

[1] The self-similarity coefficients computed when T is a reflection are equal to those computed when T is a rotation. This has been mathematically proven in [15]

and O is the kernel vector

$$O(x, y) = [\frac{\cos(\tan^{-1}(y/x))}{\sqrt{x^2 + y^2}}, \frac{\sin(\tan^{-1}(y/x))}{\sqrt{x^2 + y^2}}]^T \tag{7}$$

The use of the normalized gradient vector in the equation (4) is necessary in order to have an operator whose results are independent from the intensity of the gradient in each point.

2.3   Pupil Localization

The final step of the proposed approach integrates the selected self-similarity map and the circular shape distribution space. Both data structures are normalized in the range [0,1] and then point-wise added. The peak in the resulting data structure is the point that is finally selected as the pupil center.

Figure 2 shows an example of how the proposed procedure works; in particular figure 2(a) shows the cropped region of the eye whereas figure 2(b) shows the points with highest self similarity values across all the scales. The resulting self-similarity map at the peak scale is then reported in figure 2(c) and figure 2(d) reports the circular shape distribution space built by using the modified version of the Circular Hough Transform. Finally figure 2(e) shows the joint space obtained by appropriately combining the two data structures. Notice how this joined space is the most suited to easily locate the pupil that is highlighted in figure 2(f) by a red circle.

From section 2 it is quite straightforward to derive that the proposed approach is invariant to rotation, illumination and scale changes. Moreover it works without making the assumption that the image sequences contain only face images.

3   Experimental Results

Experimental evidence of the effectiveness of the method was achieved on challenging benchmark data sets. The MATLAB implementation of the boosted cascade face detector proposed by Viola and Jones [13] with default parameters is used in our experiments, discarding false negatives from the test set.

In the first experimental phase the BioID database [16] is used for testing and in particular the accuracy of the approach in the localization of the pupils was evaluated. The BioID database consists of 1.521 gray-scale images of 23 different subjects and has been taken in different locations and at different times of the day under uncontrolled lighting conditions. Besides non-uniform changes in illumination, the positions of the subjects change both in scale and pose. Furthermore in several examples of the database the subjects are wearing glasses. In some instances the eyes are partially closed, turned away from the camera, or completely hidden by strong highlights on the glasses. Due to these conditions, the BioID database is considered a difficult and realistic database. The size of each image is 384×288 pixels. A ground truth of the left and right eye centers is provided with the database. The *normalized error*, indicating the error obtained by the worse eye estimation, is adopted as the accuracy measure for the found eye locations.

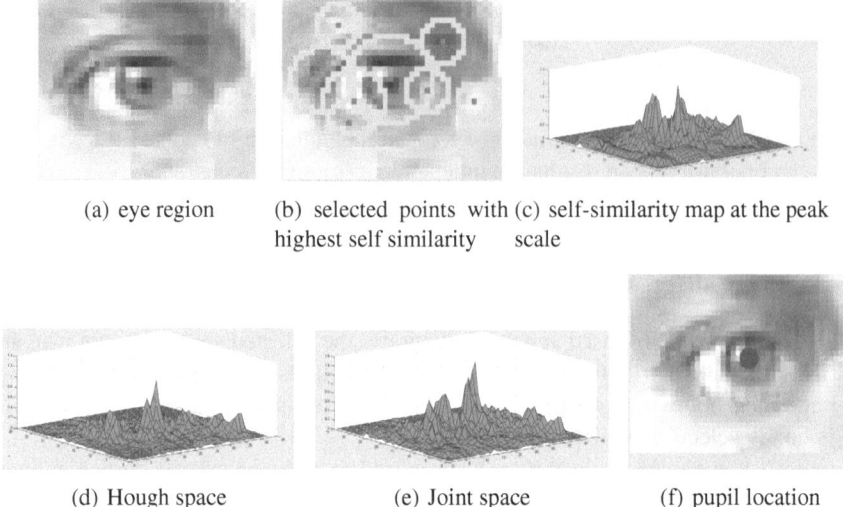

(a) eye region   (b) selected points with (c) self-similarity map at the peak
                  highest self similarity   scale

(d) Hough space   (e) Joint space   (f) pupil location

Fig. 2. The outcomes of the proposed approach

This measure is defined in [17] as $e = \frac{max(d_{left}, d_{right})}{w}$ where d_{left} and d_{right} are the euclidean distances between the found left and right eye centers and the ones in the ground truth and w is the euclidean distance between the eyes in the ground truth. In this measure, $e \leq 0.25$ (a quarter of the interocular distance) roughly corresponds to the distance between the eye center and the eye corners, $e \leq 0.1$ corresponds to the range of the iris, and $e \leq 0.05$ to the range of the pupil.

In figure 3 the performances of the proposed approach on the BioID database are reported (blue line) and compared with those obtained using self-similarity (i.e. the point with the maximum value in the saliency map) or circularity (i.e. the point with the maximum value in the accumulation space) information. The graph shows that the combination of the feature related to appearance (that is to say the self-similarity) and to oriented edge location as feasible centers of circumferences (i.e. Modified Circular Hough Transform) allows to increase the localization performance. These results are very encouraging especially when correlated with state-of-the-art methods in the literature. To this end in figure 4 the comparison with some of the most accurate techniques in the literature which use the same dataset and the same performance metric is shown. Looking at the graph it can be seen that only the methods proposed in [8] and [6] provide slightly better results both for $e \leq 0.1$ and $e \leq 0.05$ measures. In general the proposed approach outperforms most of the related methods, even if it does not make use of supervised training or post processing adjustments.

In figure 5 two images of the BioID database processed by the proposed approach are shown. In the one on the left pupils are correctly detected (normalized error 0.0267), whereas in the one on the right some highlights on the glasses mislead the algorithms that miss the detection of the pupil of the left eye (normalized error 0.1158).

To systematically evaluate the robustness of the proposed pupil locator to lighting and pose changes, one subset of the Extended Yale Face Database B [18] is used.

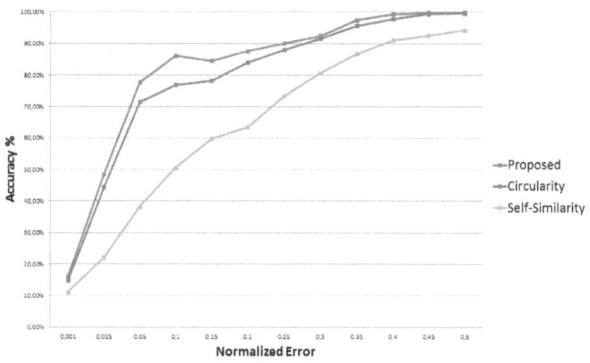

Fig. 3. Results obtained on the BioID database

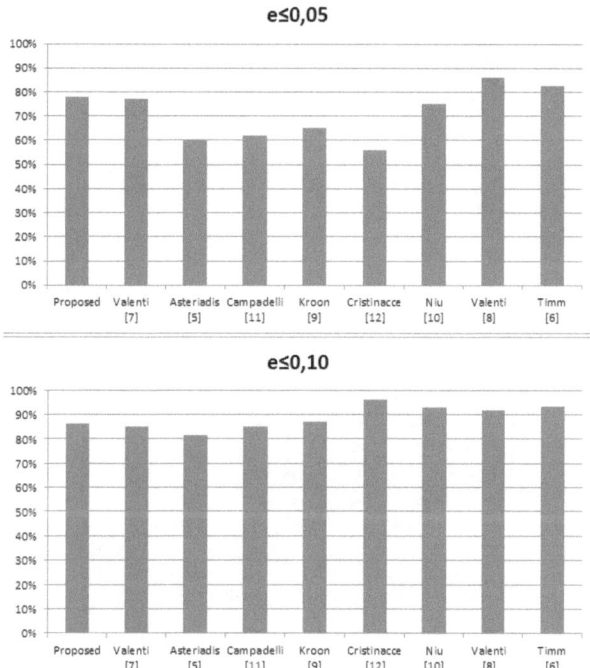

Fig. 4. Comparison with state-of-the-art methods in the literature on the BioID database

The full database contains 16128 images of 28 human subjects under 9 poses and 64 illumination conditions. The size of each image is 640×480 pixels. In particular the system was tested on the 585 images of the subset $\#39B$. The performance in accuracy of the proposed approach on this challenging dataset are $61, 66\%$ ($e \leq 0.05$) and $73, 16\%$ ($e \leq 0.01$). By analyzing the results, it is possible to note that the system is able to deal with light source directions varying from $\pm 35°$ azimuth and from $\pm 40°$ elevation with respect to the camera axis. The results obtained under these conditions

Fig. 5. Two images of the BioID database processed by the proposed approach (first row) and the the corresponding details on the eye regions (second row). In the image on the left both pupils are correctly detected, whereas in the one on the right some highlights on the glasses mislead the algorithms that miss the detection of the pupil in the left eye.

Fig. 6. Some images of the Extended YALE database B in which the approach correctly detects the pupils

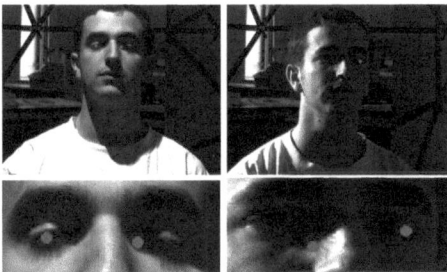

Fig. 7. Some images of the Extended YALE database B in which the detection of the pupils is either less accurate or completely fails

are $77,95\%$ ($e \leq 0.05$) and $84,66\%$($e \leq 0.01$). For higher angles, the method is often successful for the less illuminated eye and sporadically for the most illuminated one: if the eye is uniformly illuminated, the pupil is correctly located, even for low-intensity images. In figure 6 some images of the Extended YALE database B in which the approach correctly detects the pupils even under different lighting conditions and pose changing are shown. In figure 7 some images in which the detection of the pupils is either less accurate or completely fails are instead shown.

4   Conclusions and Future Works

A new method to automatically locate pupils in low-resolution images containing human faces is proposed in this paper. In particular pupils are localized by a two steps procedure: at first self-similarity information is extracted by considering the appearance variability of local regions and then they are combined with an estimator of circular shapes based on a modified version of the Circular Hough Transform. The proposed approach does not require any training phase or decision rules embedding some a priori knowledge about the operating environment. Experimental evidence of the effectiveness of the method was achieved on challenging benchmark data sets. The results obtained are comparable (sometimes outperform) with those obtained by the approaches proposed in literature (making use of training phase and machine learning strategies).

With regard to the computational load, the calculation of the similarity space has a complexity $O(kM^2)$ where k is the number of pixels in the image and M represents the maximal considered scale. The circle detection has instead $O(kn)$ complexity where k is again the number of pixels in the image and n is the dimensionality of the operator used in the convolution implemented by equation 4. However considering that the calculation of the two spaces is embarrassingly parallel (no effort is required to separate the problem into a number of parallel tasks) it is possible to approximate the computational load to the maximum of the two terms above. This therefore leads to a complexity comparable to that of the state of the art methods, however, offering better performance of detection and although not requiring training or other specific post-processing steps that limit their ability to work under various operating conditions.

Future work will address the improvement of the construction of the area of circularity through techniques derived from differential geometry in order to make the system even more accurate.

References

1. Hansen, D.W., Qiang, J.: In the Eye of the Beholder: A Survey of Models for Eyes and Gaze. IEEE Transactions on Pattern Analysis and Machine Intelligence 32(3), 478–500 (2010)
2. SMI, SensoMotoric Instruments, http://smivision.com/en/gaze-and-eye-tracking-systems/products/iview-x-hed.html (retrieved on December 2012)
3. Zhu, Z., Ji, Q.: Robust Real-Time Eye Detection and Tracking under Variable Lighting Conditions and Various face Orientations. Computer Vision and Image Understanding 98(1), 124–154 (2005)
4. Coutinho, F.L., Morimoto, C.H.: Improving Head Movement Tolerance of Cross-Ratio Based Eye Trackers. International Journal of Computer Vision 101(3), 459–481 (2013)
5. Asteriadis, S., Nikolaidis, N., Pitas, I.: Facial feature detection using distance vector fields. Pattern Recognition 42(7), 1388–1398 (2009)
6. Timm, F., Barth, E.: Accurate Eye Centre Localisation by Means of Gradients. In: Proceeding of the International Conference on Computer Vision Theory and Applications, pp. 125–130 (2011)
7. Valenti, R., Gevers, T.: Accurate eye center location and tracking using isophote curvature. In: Proceeding of the IEEE International Conference on Computer Vision and Pattern Recognition, CVPR, pp. 1–8 (2008)

8. Valenti, R., Gevers, T.: Accurate Eye Center Location through Invariant Isocentric Patterns. IEEE Transactions on Pattern Analysis and Machine Intelligence 34(9), 1785–1798 (2012)
9. Kroon, B., Hanjalic, A., Maas, S.: Eye localization for face matching: is it always useful and under what conditions? In: Proceedings of the 2008 International Conference on Content-based Image and Video Retrieval, CIVR, Niagara Falls, Canada, pp. 379–388 (2008)
10. Niu, Z., Shan, S., Yan, S., Chen, X., Gao, W.: 2D Cascaded AdaBoost for Eye Localization. In: Proceeding of the 18th International Conference on Pattern Recognition, ICPR, vol. 2, pp. 1216–1219 (2006)
11. Campadelli, P., Lanzarotti, R., Lipori, G.: Precise Eye and Mouth Localization. International Journal of Pattern Recognition and Artificial Intelligence 23(3), 359–377 (2009)
12. Cristinacce, D., Cootes, T., Scott, I.: A Multi-Stage Approach to Facial Feature Detection. In: Proceedings of the British Machine Conference, pp. 231–240 (2004)
13. Viola, P., Jones, M.: Robust real-time face detection. International Journal of Computer Vision 57, 137–154 (2004)
14. Prendergast, P.M.: Facial Proportions. Advanced Surgical Facial Rejuvenation, 15–22 (2012)
15. Maver, J.: Self-Similarity and Points of Interest. IEEE Transactions on Pattern Analysis and Machine Intelligence 32(7), 1211–1226 (2010)
16. BioID: Technology Research, the BioID Face Database (2001), www.bioid.com
17. Jesorsky, O., Kirchberg, K.J., Frischholz, R.W.: Robust face detection using the hausdorff distance. In: Bigun, J., Smeraldi, F. (eds.) AVBPA 2001. LNCS, vol. 2091, pp. 90–95. Springer, Heidelberg (2001)
18. Georghiades, A., Belhumeur, P., Kriegman, D.: From few to many: Illumination cone models for face recognition under variable lighting and pose. IEEE Trans. Pattern Anal. Mach. Intelligence 23(6), 643–660 (2001)

Problems in Distortion Corrected Texture Classification and the Impact of Scale and Interpolation

Michael Gadermayr[1], Michael Liedlgruber[1], Andreas Uhl[1], and Andreas Vécsei[2]

[1] Department of Computer Sciences, University of Salzburg, Austria
{mgadermayr,mliedl,uhl}@cosy.sbg.ac.at
[2] St. Anna Children's Hospital, Endoscopy Unit, Vienna, Austria

Abstract. In the field of computer aided celiac disease diagnosis, wide-angle endoscopy lenses are employed which introduce significant barrel type distortions. Although the images can be rectified using distortion correction methods, computer based diagnosis suffers from missing information in highly distorted image regions. First, we investigate the impact of simple and advanced interpolation techniques on the classification rates. Furthermore we explore the effect of considering different image resolutions. Whereas in previous studies distortion correction in most cases turned out to be disadvantageous, we show that for certain setups distortion correction definitely is advantageous.

Keywords: Distortion correction, endoscopy, classification, celiac disease.

1 Introduction

Celiac disease [11] is an autoimmune disorder which affects the small intestine in genetically predisposed individuals after introduction of gluten containing nutrient. Characteristic for this disease is an inflammatory reaction in the mucosa of the small bowel caused by a dysregulated immune response triggered by ingested gluten proteins. During the course of celiac disease the mucosa loses its absorptive villi and hyperplasia of the enteric crypts occurs leading to a diminished ability to absorb food. According to a study [2], the overall prevalence in the USA in not-at-risk groups was 1:133 .

Computer aided celiac disease diagnosis relies on images taken during endoscopy. The employed cameras are equipped with wide angle lenses, which suffer from a significant amount of barrel type distortion. Whereas the distortion in central image pixels can be neglected, peripheral regions are highly distorted. Thereby, the feature extraction as well as the following classification is compromised. Based on camera calibration, distortion correction (DC) techniques are able to rectify the images. However, although the barrel type distortion can be undone, especially in peripheral regions there remains a lack of information, as the DC method stretches the image. The lack of information has to be compensated using an interpolation technique.

In recent studies, the impact of barrel type distortion [9] and distortion correction [5] on the classification rate of celiac disease endoscopy images was investigated. The authors have shown that image patches in peripheral regions, which are more strongly affected by the distortion are more likely to be misclassified. However, with distortion

A. Petrosino (Ed.): ICIAP 2013, Part I, LNCS 8156, pp. 513–522, 2013.

correction, the classification rate on average even suffers. In [4], different distortion correction techniques have been investigated.

In this paper, priority is given to the following aspects:

- **Interpolation:** First, different interpolation methods are investigated. Not the visual quality, but only the classification rate is our interest. Apart from the simple and commonly used bilinear and nearest neighbor interpolation, we investigate a Lanczos filter which effectively imitates the perfect low pass filter. Moreover, a (non-linear) edge preserving method is analyzed.
- **Scale:** Furthermore, we investigate the impact of the image scale on the classification rate. For all features a setup is chosen, that only pixels in the 8-pixel neighborhood are considered. These features are adjusted implicitly, by downscaling the input image. We know that distortion correction in a small neighborhoods (achieved with the original image), mostly does not improve the performance. However, we suspect that by downscaling the image, the profit of distortion correction could be increased and thereby a benefit in the classification performance might be achieved.

The paper is organized as follows: In Sect. 2, the problems of DC based feature extraction are explained. In Sect. 3, the utilized DC method, the different interpolation methods and the used features are explained. In Sect. 4, experiments are shown and the results are discussed. Section 5 concludes this paper.

2   Motivation – Problems within Distortion Correction

Distortion correction of barrel type distortions is known to effectively rectify the geometrical properties of images taken with wide angle or fish eye lenses. Visually, the distortion corrected images seem to be better suited for classification of computer based celiac disease diagnosis than the distorted images. However, there are certain inadequacies which arise if DC is applied. Whereas the images are geometrically corrected, in case of considering small details (i.e. texture), new problems are introduced.

Intuitively, the probability of a feature or of single characteristics of a feature should be uniformly distributed over all coordinates of the image. For example this condition could be violated in case of systematic sensor faults. Distortion correction, is another source for a violation of the mentioned condition.

We utilize images taken with the same endoscope and we compute pixel based properties (which are also included in features), for each pixel in all images. Afterwords, the average of the features for each coordinate over all images is calculated, to get the mean for each point. Having a large number of images, the resulting data should be approximately homogeneous. A high degree on inhomogeneity indicates an anomaly.

As an example, we visualize a property which is a part of the well known local ternary patterns (LTP [14]). Especially, we consider the probability that the pixel above the current pixel has approximately (± 2) the same value as the current pixel. In Fig. 1, the distributions of this certain property are given for the original (Fig. 1a) and for the distortion corrected image (with different downscaling factors: Fig. 1b-1d). A similar behavior can be achieved not just with the mentioned property, but with many others too (e.g. LBP [12], ELBP [8]). In the original image, apart from slight stripes (caused by

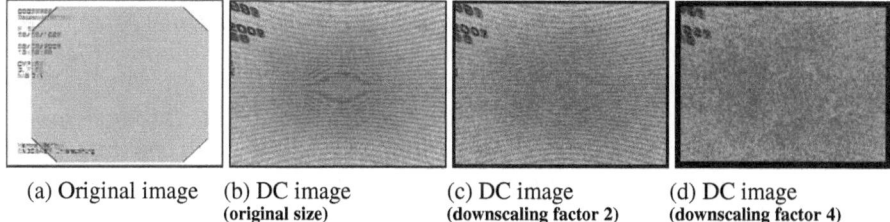

| (a) Original image | (b) DC image (original size) | (c) DC image (downscaling factor 2) | (d) DC image (downscaling factor 4) |

Fig. 1. Distribution of the upper sample of local ternary patterns: Especially we consider the occurrences of the sample value equaling approximately (± 2) the center value

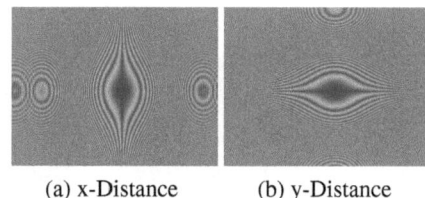

(a) x-Distance   (b) y-Distance

Fig. 2. The distances to the next sample point in interpolation within DC are represented by gray values. A dark region corresponds to a small distance and vice versa.

the sensor), the property is approximately uniformly distributed. In the DC images, the property definitely is not approximately uniformly distributed, however, with downscaling the uniform distribution can be recovered. With blurring after applying the distortion correction, a similar behavior can be achieved. For these experiments, we utilized bilinear interpolation within distortion correction as done in previous studies [9,5,4]. We identified two major issues which occur in distortion correction:

- **Interference patterns**
  Especially in central, but also in peripheral regions strange (interference) patterns can be observed (Fig. 1b).
  Explanation: During interpolation (needed in rasterization), some required pixel values are near to real sample pixel values in the original image. In opposite, others are approximately in the middle of four sample pixels. The problem in distortion correction is the following: Whereas in the first case, edge frequencies are maintained quite effectively, in the second case the edges are blurred during interpolation. In Fig. 2, the distances of interpolation coordinates to the next sample coordinates are visualized. It can be seen, that the pattern shown in Fig. 2b, can also be recognized in the averaged feature images (Fig. 1b).
- **Contrast between center and peripheral regions**
  Apart from the mentioned patterns, a significant difference between central and peripheral regions can be observed with some feature properties.
  Explanation: In central image regions, the image is not changed significantly within DC. But in peripheral image regions, the original image has to be stretched strongly, in order to rectify the barrel type distortion. The stretching results in a decrease of

high frequencies which is visualized in Fig. 3. These figures show the average local frequencies, contained in the distorted (Fig. 3a) and in the distortion corrected images (Fig. 3b) with reference to the distance to the center of distortion. Therefore, we computed the local discrete Fourier transform (with window size 16 × 16) for all of our endoscopy images and for all pixels. Then we averaged all Fourier spectra having the same distance to the center of distortion. In Fig. 3 (a) and (b), for each frequency (corresponds to a ring in the 2-D Fourier domain) on the x-axis and for each distance to the center of distortion (y-axis), the logarithmic average value of the Fourier power spectrum is given. In order to emphasize on the differences, in Fig. 3c, we subtracted the distortion corrected from the distorted frequency image. In this image, it can be seen, that with an increasing distance to the center of distortion, especially high frequencies are decreasing in case of distortion correction.

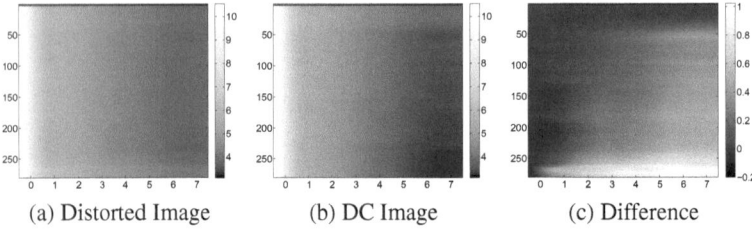

(a) Distorted Image   (b) DC Image   (c) Difference

Fig. 3. Local DFT: These figures show averaged local frequencies. On the x-axis the frequencies (a low value corresponds with a low frequency and vice versa) and on the y-axis the distances from the center of distortion are given.

3   Theory

3.1   Distortion Correction

We utilize the distortion correction method based on the work of Melo et al. [10]. In this approach, the circular barrel type distortion is modeled by the division model [3]. Having the center of distortion $\hat{x}_c$ and the distortion parameter ξ, an undistorted point x_u can be calculated from the distorted point x_d as follows:

$$x_u = \hat{x}_c + \frac{(x_d - \hat{x}_c)}{||x_d - \hat{x}_c||_2} \cdot r_u(||x_d - \hat{x}_c||_2) \,. \tag{1}$$

$||x_d - \hat{x}_c||_2$ (in the following r_d) is the distance (radius) of the distorted point x_d from the center of distortion $\hat{x}_c$. The function r_u defines for a radius r_d in the distorted image, the new radius in the undistorted image:

$$r_u(r_d) = \frac{r_d}{1 + \xi \cdot r_d^2} \,. \tag{2}$$

Figure 4 shows a distorted and the corresponding undistorted image of a checkerboard pattern.

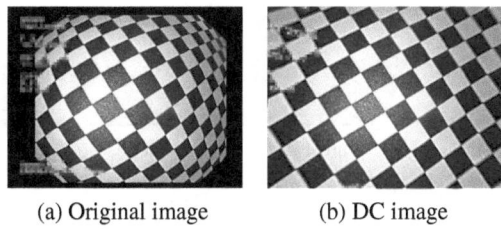

(a) Original image   (b) DC image

Fig. 4. Distorted and undistorted image of a (planar) checkerboard pattern

3.2 Interpolation Techniques in Distortion Correction

In order to get a rasterization after applying a distortion correction method, an interpolation method must be applied. We investigate the following techniques, which are identified to have quite different properties:

- **Bilinear interpolation**
- **Nearest neighbor interpolation**
- **Lanczos interpolation**
  The Lanczos filter [1] with $a = 3$ is known to imitate the perfect low pass filter (sinc) quite effectively and reduces ringing artifacts. With this filter, the available frequencies are retained, whereas aliasing is avoided.
- **Edge preserving interpolation**
  The Lanczos filter is known to retain the available frequencies. However, in this case due to the stretching of the image in peripheral regions, high frequencies (sharp edges) are missing after undistortion. Consequently, maintaining the available frequencies is not enough to reconstruct the real edge information. Therefore, an edge preserving interpolation method [15] has been implemented. Unlike usual (linear) interpolation kernels, with this nonlinear approach the behavior of the interpolation depends on the image properties. Near edges, high frequencies are encouraged (similar to nearest neighbor interpolation), whereas smooth regions are retained smooth (similar to linear interpolation).

In Fig. 5, a perfect (undistorted) checkerboard pattern (Fig. 5a - 5d) and a smooth gray value gradient image (Fig. 5e - 5h) are undistorted (leads to an inverse distortion) and the different interpolation techniques are applied. Whereas from the first images, the ability of edge preservation, from the second image the ability of preserving smooth gradients can be deduced visually. Although undistorting an undistorted image is not a sensible application, the differences of the interpolation methods can be seen well with artificial images. The major positive (+) and negative (–) properties of the interpolation techniques are outlined in Table 1. It can be seen, that the chosen methods have quite different strengths and weaknesses.

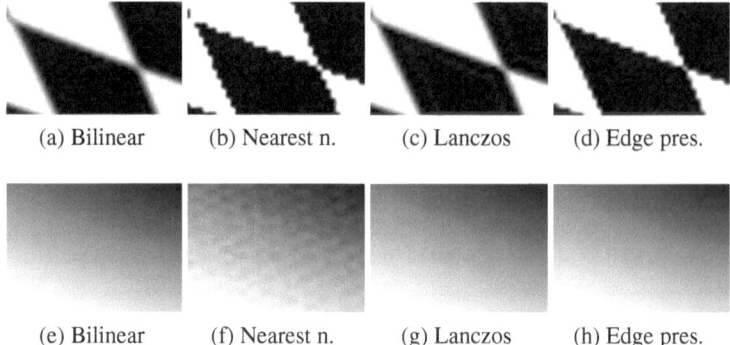

| | | | |
| (a) Bilinear | (b) Nearest n. | (c) Lanczos | (d) Edge pres. |

| | | | |
| (e) Bilinear | (f) Nearest n. | (g) Lanczos | (h) Edge pres. |

Fig. 5. The impact of the different interpolation methods on a distorted checkerboard pattern (top row) and a smooth gray value gradient image (bottom row)

Table 1. Properties of interpolation methods

	Bilinear	Nearest n.	Lanczos	Edge pres.
edge preservation	–	+	o	+
smoothness	+	–	+	o
avoid ringing	+	+	o	+

3.3 Features in Classification

In order to investigate the feature extraction methods with differently downscaled images, all of our features are adjusted, to operate only in the direct neighborhood (only the 8 neighbors are considered). For the experiments, the following features are used:

- **Local binary patterns** [12] (LBP): This features is used with a radius (i.e. the distance to the neighboring samples) of 1 and 4 (LBP$_4$) and 8 (LBP$_8$) neighboring samples, respectively.
- **Local ternary patterns** [14] (LTP): This feature is used with a radius of 1 and with 8 neighbors. The threshold, which has been evaluated during exhaustive search, is set to 1.
- **Extended local binary patterns** [8] (ELBP): As LBP, ELBP is used with a radius of 1 and 8 neighbors.
- **Gray level co-occurrence matrix** [6] (GLCMP): The feature vector consists of the Haralick features [6] contrast, correlation, energy and homogeneity of the gray level co-occurrence matrices with different offset vectors. In order to focus on small neighborhoods (as above), the offset vectors $(0, 1)^T$, $(1, 0)^T$, $(1, 1)^T$, $(1, -1)^T$ have been chosen for creating the matrices.
- **Edge co-occurrence matrix** [13] (ECM): To get the ECM, first the orientations of all local image gradient quantized to 8 directions have to be evaluated. Whereas non-edge pixels (this is determined using the canny-edge detector) are ignored, with the orientation-labels (1 to 8) of edge pixels, the gray level co-occurrence matrix is computed (as above). As ECM feature vector the whole matrix is used.

4 Experiments

4.1 Experimental Setup

The image test set used contains images of the *duodenal bulb* taken during duodeno-scopies at the St. Anna Children's Hospital using pediatric gastroscopes (with resolution 768×576 and 528×522 pixels, respectively). In a preprocessing step, texture patches with a fixed size of 128×128 pixels were manually extracted. The size turned out to be optimally suited in earlier experiments on automated celiac disease diagnosis [7]. In case of distortion correction, the patch position is adjusted according to the distortion function. Downscaling with bicubic interpolation is executed after patch extraction (i.e. the considered regions are always the same, only the resolution differs). To generate the ground truth for the texture patches used, the condition of the mucosal areas covered by the images was determined by histological examination from the corresponding regions. Severity of villous atrophy was classified according to the modified Marsh classification scheme [11]. Although it is possible to distinguish between the different stages of the disease (called Marsh 3A-3C), we only aim in distinguishing between images of patients with (Marsh3A-3C) and without the disease (called Marsh0). We decided for this policy, because the two classes case is more relevant in practice. Our experiments are based on a database containing 163 (Marsh 0) and 124 (Marsh 3A-3C) images, respectively. Example texture patches are shown in Fig. 6. For classification, we use the k-nearest-neighbor classifier. This rather weak classifier has been chosen to emphasize on quantifying the discriminative power of the features proposed in this work. To avoid any bias in the results, leave-one-patient-out cross validation is utilized. We evaluated all different combinations of downscaling factors and interpolation techniques. We also considered previously blurring in combination with downscaling, but thereby the achieve classification results can not be improved. Moreover, different interpolation techniques within image resizing did not lead to significantly different rates.

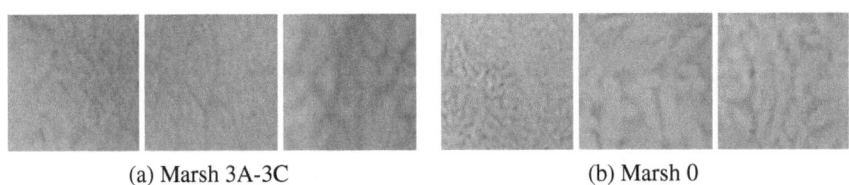

(a) Marsh 3A-3C   (b) Marsh 0

Fig. 6. Example patches of patients with (left) and without the disease (right)

4.2 Results

In Fig. 7a - 7f, the achieved classification rates with the different features and different image resolutions can be observed. Whereas the bold lines represent the approach without distortion correction, the different thin lines stand for different interpolations within distortion correction. In all figures can be seen, that the original approach (i.e. without distortion correction) is beneficial in case of the original image size (left most points). However, with increasing downscaling factor, in general the classification rates

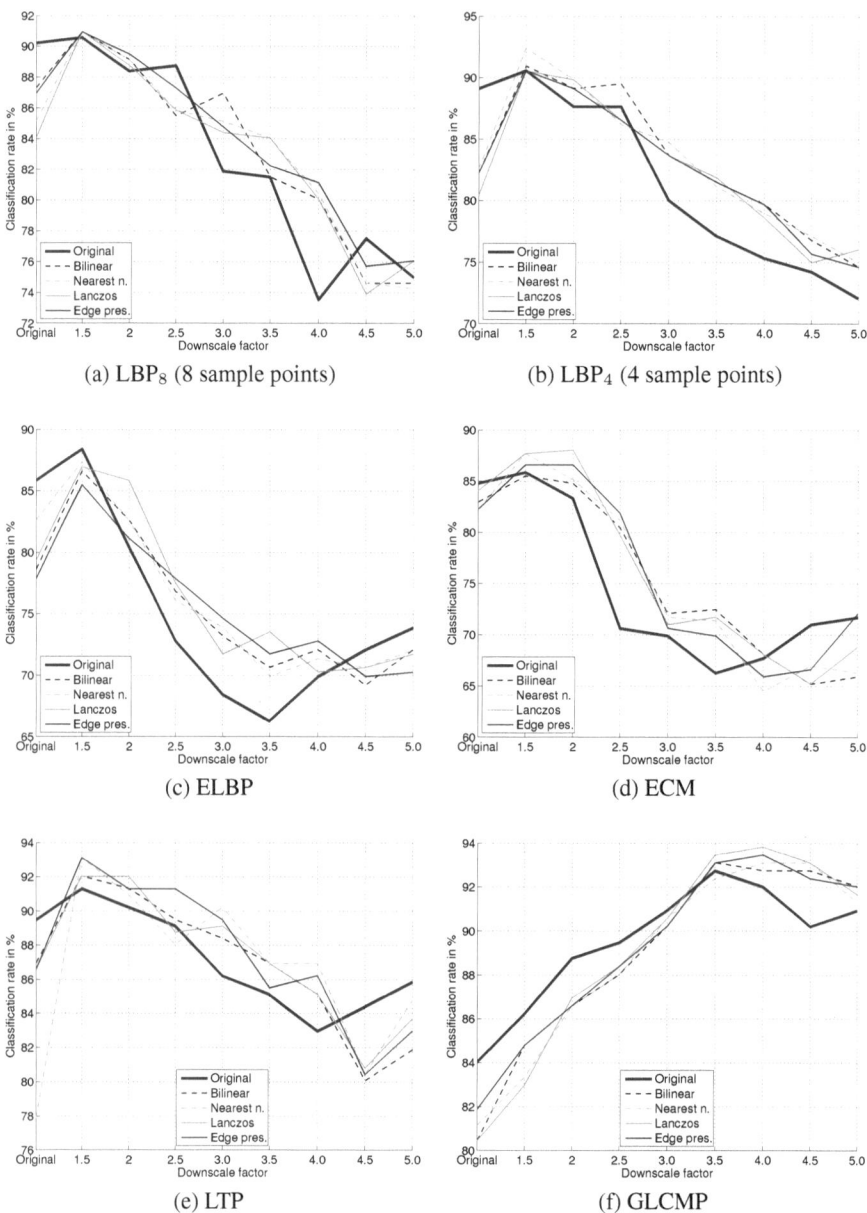

Fig. 7. The plots show classification rates achieved with the considered features. Especially, the dependency of the scaling factor (x-axis) and the different interpolation methods (thin lines) can be seen.

with distortion correction are increasing compared with the approach without DC. For each feature, downscaling factors exist, for which DC is beneficial (points where all thin lines are above the thick line). If the discriminative power of the feature benefits from the slight image downscaling (factors 2 and above), then the distortion correction has a positive effect on the classification rate as shown in Fig. 7b, 7d, 7e and 7f. With the features LBP_4, ECM and LTP the best overall classification rates are achieved with distortion correction and a moderate downscaling factor (between 1.5 and 2.0). With GLCMP, the best rate is achieved with distortion correction and a high downscaling factor (4.0). Especially using the GLCMP feature, a considerable performance increase (considering all downscaling factors) can be observed. Otherwise, if the classification rate falls significantly with the downscaling factor, the best classification rate is achieved without DC (Fig. 7c: With ELBP the classification rate strongly falls between downscaling factors 1.5 to 3).

4.3 Discussion

A positive effect of distortion correction can be observed, especially when the images are downscaled. Downscaling the image implies, that pixels which a higher displacement (in the original image) are considered. Consequently, high frequencies (i.e. sharp edges) in the image are no longer extracted. Instead, with the downscaled images, lower frequencies are extracted by the feature. With the original images, distortion correction in general does not provide better results. A reason for this behavior is, that the regarded high frequencies are destroyed by the image stretching of the DC method in peripheral regions. With downscaled images, the geometric context and low frequencies are getting more important than high frequencies (sharp edges). Therefore, the distortion correction has a beneficial effect on the classification rate. The enhanced interpolation methods do not imply significantly better classification results. As distortion correction on original images is disadvantageous, the missing information cannot be compensated effectively even by the enhanced methods.

5 Conclusion

If images are downscaled, a positive effect of distortion correction can be observed. That means, distortion correction is especially sensible, if the classification rate utilizing a specific feature benefits from the downscaling (e.g. with the gray level co-occurrence matrix). Using feature extraction methods which suffer strongly from downscaling (e.g. ELBP), distortion correction does not improve the best classification result. The difference between the interpolation techniques is quite small. Even with the more sophisticated methods (the Lanczos and the edge preserving method [15]) no significant improvement of classification rates can be achieved (compared to the very simple bilinear and nearest neighbor interpolation). We showed that high frequencies cannot be maintained with any interpolation method, whereas the lower frequencies are maintained with each of the investigated interpolation techniques.

Acknowledgments. This work is partially funded by the Austrian Science Fund (FWF) under Project No. 24366.

References

1. Duchon, C.E.: Lanczos filtering in one and two dimensions. J. Appl. Meteor. 18(8), 1016–1022 (1979)
2. Fasano, A., et al.: Prevalence of celiac disease in at-risk and not-at-risk groups in the united states: A large multicenter study. Archives of Internal Medicine 163(3), 286–292 (2003)
3. Fitzgibbon, A.W.: Simultaneous linear estimation of multiple view geometry and lens distortion. In: CVPR, pp. 125–132 (2001)
4. Gschwandtner, M., Hämmerle-Uhl, J., Höller, Y., Liedlgruber, M., Uhl, A., Vécsei, A.: Improved endoscope distortion correction does not necessarily enhance mucosa-classification based medical decision support systems. In: Proc. of the Intern. Workshop on Multimedia Signal Processing (MMSP 2012), pp. 158–163 (September 2012)
5. Gschwandtner, M., Liedlgruber, M., Uhl, A., Vécsei, A.: Experimental study on the impact of endoscope distortion correction on computer-assisted celiac disease diagnosis. In: Proc. of the 10th Intern. Conf. on Information Technology and Applications in Biomedicine (ITAB 2010) (November 2010)
6. Haralick, R.M., Shanmugam, K., Dinstein, I.: Textural features for image classification. IEEE Trans. on Systems, Man, and Cybernetics 3, 610–621 (1973)
7. Hegenbart, S., Kwitt, R., Liedlgruber, M., Uhl, A., Vécsei, A.: Impact of duodenal image capturing techniques and duodenal regions on the performance of automated diagnosis of celiac disease. In: Proc. of the 6th Intern. Symp. on Image and Signal Proc. and Analysis (ISPA 2009), pp. 718–723 (September 2009)
8. Liao, S., Zhu, X., Lei, Z., Zhang, L., Li, S.Z.: Learning multi-scale block local binary patterns for face recognition. In: Lee, S.-W., Li, S.Z. (eds.) ICB 2007. LNCS, vol. 4642, pp. 828–837. Springer, Heidelberg (2007)
9. Liedlgruber, M., Uhl, A., Vécsei, A.: Statistical analysis of the impact of distortion (correction) on an automated classification of celiac disease. In: Proc. of the 17th Intern. Conf. on Digital Signal Processing (DSP 2011), Corfu, Greece (July 2011)
10. Melo, R., Barreto, J.P., Falcao, G.: A new solution for camera calibration and real-time image distortion correction in medical endoscopy-initial technical evaluation. IEEE Trans. Biomed. Eng. 59(3), 634–644 (2012)
11. Oberhuber, G., Granditsch, G., Vogelsang, H.: The histopathology of celiac disease: time for a standardized report scheme for pathologists. European Journal of Gastroenterology and Hepatology 11, 1185–1194 (1999)
12. Ojala, T., Pietikäinen, M., Harwood, D.: A comparative study of texture measures with classification based on feature distributions. Pattern Recognition 29(1), 51–59 (1996)
13. Rautkorpi, R., Iivarinen, J.: A novel shape feature for image classification and retrieval. In: Proc. of the International Conference on Image Analysis and Recognition (ICIAR), pp. 753–760 (2004)
14. Tan, X., Triggs, B.: Enhanced local texture feature sets for face recognition under difficult lighting conditions. In: Zhou, S.K., Zhao, W., Tang, X., Gong, S. (eds.) AMFG 2007. LNCS, vol. 4778, pp. 168–182. Springer, Heidelberg (2007)
15. Wang, F., Xu, Y., Zhao, Y., Hu, F.: A new nonlinear interpolation algorithm for edge preserving. In: 2010 International Conference on Multimedia Technology (ICMT), pp. 1–4 (October 2010)

On Optimized Color Image Coding
Using Correlation of Primary Colors

Eyal Braunstain[1] and Moshe Porat[2]

[1] Department of Biomedical Engineering
Technion - Israel Institute of Technology, Haifa 32000, Israel
seyalbra@tx.technion.ac.il
[2] Department of Electrical Engineering
Technion - Israel Institute of Technology, Haifa 32000, Israel
mp@ee.technion.ac.il

Abstract. The RGB color primaries in natural images are characterized by a high degree of inter-correlation. Many compression algorithms use this information redundancy to reduce the amount of bits required for coding, by transforming the color information to a decorrelated color space - such as YUV. The human visual system is more sensitive to luminance than chrominance components, so more bits are allocated to luminance. We examine a different approach, by expressing two of the color components (termed subordinate colors) as a functional of the other color component (termed base color). Unlike some compression algorithms (e.g. JPEG) that perform the analysis on NxN blocks in the image, we utilize segmentation by Region Growing in both gray level and color (RGB) images. We suggest a method for selection of optimal base color for each region separately. The proposed approach could be useful for color compression and progressive transmission applications.

Keywords: Color compression, Correlation, Polynomial approximation.

1 Introduction

Natural images are characterized by a high degree of correlation between the color primaries (RGB). This inter-dependence yields the possibility of reducing the amount of data required for representation, e.g. by compression algorithms that transform the color components to a decorrelated color space, e.g. YUV (Y - luminance component, U, V - chrominance), and utilizing the fact that the human visual system is more sensitive to details in luminance than chrominance, and encode the luminance information using more bits. Another approach to utilizing the correlation between the color primaries (RGB) is by expressing two of the color components, termed 'subordinate colors' as a function of a third color component, termed 'base color' [1], [2].

In many compression algorithms, e.g., JPEG, the image is divided into NxN - pixels blocks, without reference to the spatial correlation between adjacent pixels. In the proposed approach, we refer to image segments, acquired by Region Growing

A. Petrosino (Ed.): ICIAP 2013, Part I, LNCS 8156, pp. 523–531, 2013.

segmentation [3], [4]. More advanced segmentation methods can be utilized, e.g. Quick Shift [5]. In methods that divide the image into arbitrary NxN blocks, distorting blockiness artifacts are created in the compressed image, which can be diminished by segmentation into natural regions. Region Growing can be performed on a gray level image, using a simple absolute distance to grow regions from seeds. However, performing the segmentation with full color information, rather than relying on brightness alone, can yield superior segmentation results. Therefore, we examine the compression results achieved when the Region Growing is performed on an RGB color image, using Euclidean metric to grow regions from seeds. This approach generally provides better segmentation, which results in superior reconstruction (decompression) results, both visually and by PSNR (Peak Signal to Noise Ratio) measure.

The image segments (texture) and region boundaries (contours) are encoded separately. The texture information of each segment is encoded using a base color, and by expressing the subordinate colors as functions (0 or 1st order) of this base color. The base color itself is encoded as a 0 / 1st / 2nd order polynomial approximation of the actual base color pixels in the segment. In [1], the authors propose to compress an image by using NxN pixels blocks, and in each block express two subordinate colors as a function of a base color, which can be chosen variably for each block by an optimization method, according to the highest correlation of a base color to the other two colors, or by choosing the base color that leads to the lowest subordinate colors' reconstruction error for the block. We propose to choose the base color that yields the minimal reconstruction error (MSE) of both the base and subordinate approximated colors for a region. Our proposed approach, based on image segmentation by color Region Growing and optimal base color selection for the utilization of inter-correlation of colors, can lead to significant improvement of compression results, compared to the widespread JPEG compression algorithm, especially for high compression ratios, as shown by our experimental results.

2   Methodology

The compression algorithm is comprised of encoder and decoder blocks. The stages of the encoder block are as follows:

1. Construct the luminance image (Y) from the R, G, and B color primaries:
   $Y = 0.3R + 0.59G + 0.11B$.
2. In order to get a smoother image, and suppress false contours in the luminance image, we filter the Y image using the inverse gradient filter [6]. This filter reduces the noise in Y image, while preserving contours (high gradients).
3. Perform Region Growing segmentation on the smoothed image. We examine the compression results for both gray level and color (RGB) Region Growing. In gray level Region Growing, a region is grown from a seed pixel by adding to the region neighboring pixels whose gray level is inside a fixed gray level interval from the region mean gray level. Seed pixels are chosen randomly, from pixels not yet assigned to specific regions. In RGB Region Growing, the above gray level interval is replaced by a Euclidean distance in RGB color space, meaning that a new pixel

will be added to a current region if its (R, G, B) color triplet is distant less than a fixed threshold from the region's current mean (R, G, B) values. In both Region Growing options, a pixel that is rejected from a region is marked as contour pixel.

The RGB Euclidean metric for Region Growing is defined as follows:

$$d_{RGB} = \sqrt{(R_{pixel} - R_{regMean})^2 + (G_{pixel} - G_{regMean})^2 + (B_{pixel} - B_{regMean})^2} < d_{RGB\_Thresh} \quad (1)$$

4. Contour artifacts, e.g. double contours and non-separating contours pixels are removed from the segmentation image. Small regions (having a number of pixels less than threshold size) that are connected through contour with other regions, are removed, i.e., united with a neighbor region with the closest gray level [7].
5. The contour information is encoded by chain coding [8]. The contour encoding is based on Freeman's chain-code, with 8-connected contours. In natural images' contours, sharp turns are less frequent than gradual turns [7]. Thus, a statistical approach is employed, and the difference in direction between sequential contour pixels is encoded using Huffman coding. The most probable direction of movement is the same as the previous; the next probable direction is a 45^0 turn, etc. This enables the reduction of the number of bits required to code the image contours.
6. For each region, a base color is chosen (R, G, or B), and approximated by a $0 / 1^{st} / 2^{nd}$ order two-dimensional polynomial function. For example, a 1st order approximation of some region's color by the spatial axes x and y, will be of the form $Y_{approx} = a_1 + a_2 x + a_3 y$, where a_1, a_2, a_3 are the approximation coefficients, and Y_{approx} is the approximation of the region color surface Y. The solution for a_1, a_2, a_3 is given by the minimization of the square error between Y and Y_{approx}, which leads to the solution of the following system of equations [7]:

$$\begin{pmatrix} n & \sum x_i & \sum y_i \\ \sum x_i & \sum x_i^2 & \sum x_i y_i \\ \sum y_i & \sum x_i y_i & \sum y_i^2 \end{pmatrix} \cdot \begin{pmatrix} a_1 \\ a_2 \\ a_3 \end{pmatrix} = \begin{pmatrix} \sum BC_i \\ \sum x_i BC_i \\ \sum y_i BC_i \end{pmatrix} \quad (2)$$

where x_i, y_i are the image coordinates, and BC_i are the base color values of pixels.

A similar derivation exists for 2nd order approximation of base color, by minimization of a square error. Zero-order approximation is given by the region average.

The other two colors (subordinate colors) are encoded by polynomial expansion of the base color for each region [1]:

$$C_1 = a_k C_0^k + a_{k-1} C_0^{k-1} + ... + a_1 C_0 + a_0$$
$$C_2 = b_k C_0^k + b_{k-1} C_0^{k-1} + ... + b_1 C_0 + b_0 \quad (3)$$

where C_1, C_2 are the interpolated subordinate colors, C_0 is the base color, and a_k, b_k are the polynomial expansion coefficients for the subordinate colors.

For 1st order polynomial expansion of subordinate color R as function of base color G, we have [7]:

$$R_{app} = r_1 G + r_0 \tag{4}$$

The coefficients r_i that minimize the mean square error between the original R component in the region, and the approximated R component, are hence derived:

$$g_0 = \frac{1}{n}\sum_1^n G_i \;\rightarrow\; r_0 = \frac{1}{n}\sum_1^n R_i \;,\quad r_1 = \frac{\sum_1^n (R_i - r_0)\cdot(G_i - g_0)}{\sqrt{\sum_1^n G_i^2}} \tag{5}$$

The above summation is made upon all pixels in region i.

The base color can be fixed - e.g. green, thus determining the subordinate colors to be red and blue. The base color for a region can also be chosen as the color component that has the highest correlation with the other two components, from the color triplet. The approach adopted in this algorithm, is to choose a base color for a region that minimizes the reconstruction error of all three colors in the region - i.e., minimum base and subordinate colors' approximation mean square errors (base color - BC, subordinate color - SC):

$$BC_{Reg\_i} = \underset{BC\in\{R,G,B\}}{\arg\min}\{BC\_RecErr_{Reg\_i} + SC\_RecErr_{Reg\_i}\}$$

$$BC\_RecErr_{Reg\_i} = \underset{p}{mean}(BC_{p\_approx} - BC_p)^2 \tag{6}$$

$$SC\_RecErr_{Reg\_i} = SC_1\_RecErr_{Reg\_i} + SC_2\_RecErr_{Reg\_i}$$

$$SC_k\_RecErr_{Reg\_i} = \underset{p}{mean}(SC_{p\_k\_approx} - SC_{p\_k})^2 \quad k = 1,2$$

The average over p is for pixels in region i. *RecErr* refers to MSE reconstruction error. Utilizing other perceptual measures instead of MSE, might cause preference and bias towards specific colors to serve as base colors, which might result in a smaller compression ratio.

7. The coefficients of the optimal base color for each region, and coefficients of polynomial expansion for subordinate colors are quantized.

The stages of the decoder block are as follows:

1. The segmentation contours are reconstructed from the compressed data.
2. Region color texture is reconstructed from base and subordinate color coefficients.
3. The contours pixels are each approximated by the pixels' color components of the surrounding dominant (maximal size) reconstructed segment. The inclusion of surrounding pixels from only the dominant neighboring region in the interpolation of contour pixels is made to prevent smearing and blurring of contour areas in the final reconstructed image. The transition between neighboring segments is made bit more abrupt [9], but we avoid the distorting smearing of colors in contour areas.

The block diagrams for the encoder and decoder are given in Fig. 1.

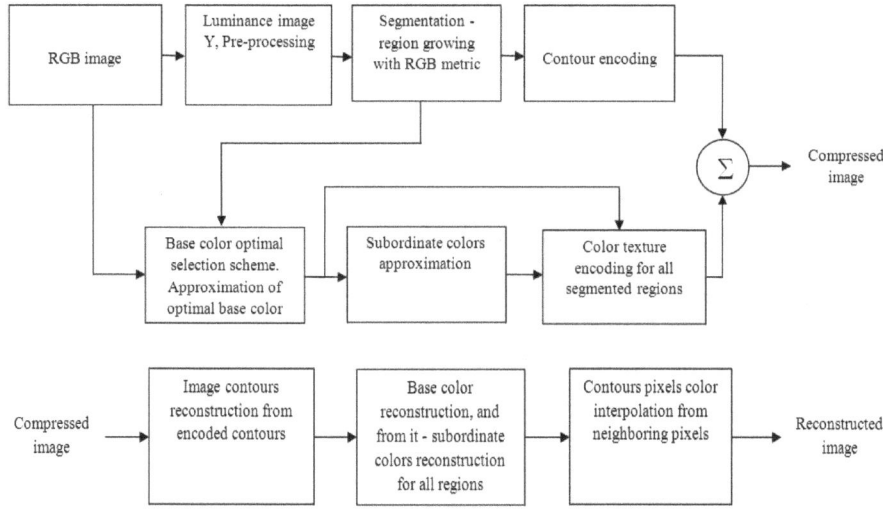

Fig. 1. - Block diagrams of the encoder (top) and decoder (bottom)

3   Application Details and Results

Our main proposed developments to improve the quality of reconstructed images are:

(1)   Region Growing segmentation using full color information (Euclidean metric in RGB color space), rather than absolute distance in gray levels.
(2)   An optimization scheme for the selection of an optimal base color for each segmented region.

To demonstrate the superiority of the reconstructions attained from the compression of color images using the above developments, we examine the compression results on the images "Peppers", "SolidWorks" and "Lena".

The parameters and thresholds are chosen as following: for Region Growing, the gray-level interval threshold [10] for merging a pixel to a region is chosen as 10 (/ 256). In RGB Region Growing, the Euclidean distance threshold is reasonably chosen approximate to $\sqrt{3 \cdot 10^2} = 17.32$. The region size threshold for merging small regions is chosen as 10 pixels. The above parameters can be changed, to compare compressions with same compression ratios by PSNR measure.

In Fig. 2, the reconstructed images of the different compression procedures options are displayed, for the "Peppers" image. Here 'GL' stands for gray level; 'Optimal' refers to optimal base color selection. The areas of the image containing perceptually discernible distortions between the reconstructed images and the original image are mainly the transition areas between the red and green peppers and the green pepper's stalk. This distortion is significantly reduced by the suggested approach, as can be observed from Fig. 2(e), with RGB region growing and optimal base color. It is also noted that in addition to improvement in PSNR, optimal base color selection can lead to an increase in compression ratio (Fig. 2(d) and (e)), due to the possibility of approximation of regions' colors textures by lower degree polynomials, thus leading to fewer coefficients required for the encoding. An improvement in PSNR by more than 1 [dB] is acquired by RGB Region Growing and optimal base color selection. Fig. 2(f) displays JPEG2000 compression, for same ratio as in 2(e). For this image, which exhibits a degree of smoothness, the JPEG2000 is superior to our method by PSNR measure. Visually, however, JPEG2000 produces a result that is somewhat blurred.

The "SolidWorks" image was also examined, and a comparison of reconstruction results to JPEG [11] and JPEG2000 [12] compression algorithms was made.

The difference between Figs. 3(b), 3(c) is very distinct. In Fig. 3(b) the colors of some parts of the solid are distorted, and these artifacts are diminished in Fig. 3(c). Fig. 3(d) displays the JPEG compression result, which produces lower PSNR than our method (5 [dB] difference) and much lower compression ratio than our algorithm.

Fig. 3(f) displays the results of JPEG2000 compression algorithm for the same compression ratio as in Fig. 3(c). Our method produces higher PSNR by more than 1 [dB]. This improvement exists in such structured images, but may be less pronounced for finely detailed, highly-textured images.

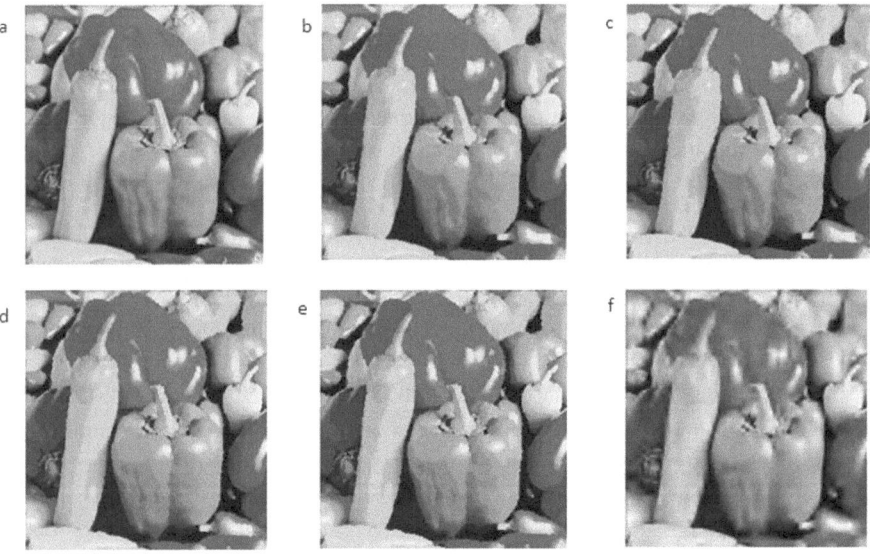

Fig. 2. Example of color image compression reconstruction results. (a) Original image.
(b) Region Growing (GL), base color (G) PSNR = 24.4332 [dB], compression ratio 89.
(c) Region Growing (GL), base color (Optimal) PSNR = 24.7226 [dB], compression ratio 92.
(d) Region Growing (RGB), base color (G) PSNR = 25.3185[dB], compression ratio 92.
(e) Region Growing (RGB), base color (Optimal) PSNR = 25.4374[dB], compression ratio 94.
(f) JPEG2000 compression result, PSNR = 27.6061, compression ratio 94.

Fig. 4 presents the algorithm sample run on the image "Lena", with highly detailed texture [13] in the hair area. The visual improvement between images 4(b) and 4(c) is perceptually noticeable - the smearing effects around the eyes and mouth in Fig. 4(b) are reduced in Fig. 4(c).

Table 1 contains information regarding the segmentation and compression of the presented images, when using RGB region growing and optimal base color selection procedure, e.g. number of segmented regions, contours properties etc.

Table 1. Segmentation and compression information for presented images

image	Number of regions	Avg. num. Coefficients - Green	Avg. num. Coefficients - Red	Avg. num. Coefficients - Blue	Avg. contour length [pixels]
Peppers	1329	2.51	1.97	2.32	18.18
SolidWorks	356	2.14	1.86	2.4	10.91
Lena	1253	2.63	1.85	2.22	21.14

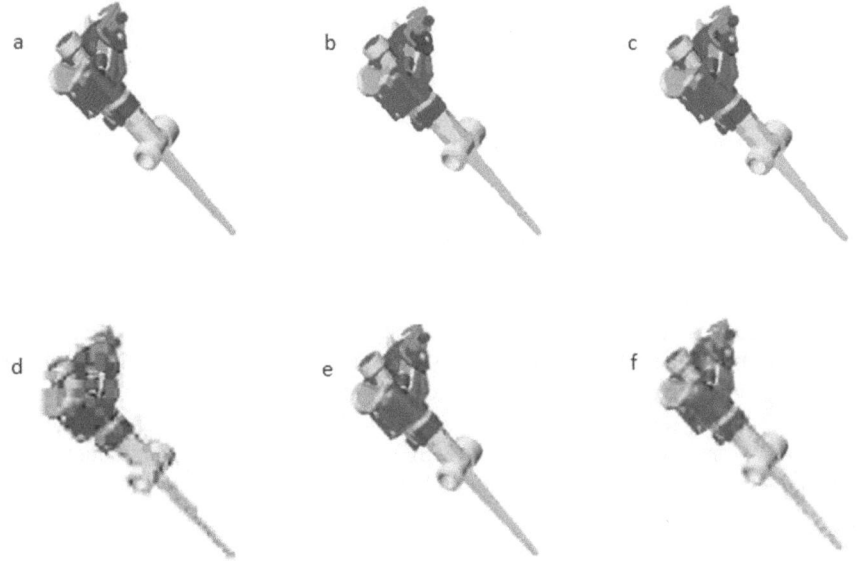

Fig. 3. A second example of color image compression reconstruction. (a) Original image.
(b) Region Growing (GL), base color (G) PSNR = 27.0026 [dB], compression ratio 190.
(c) Region Growing (RGB), base color (Optimal) PSNR = 29.2514[dB], compression ratio 190.
(d) JPEG compression, PSNR = 24.2 [dB], compression ratio 85.
(e) JPEG2000 compression, PSNR = 29.88 [dB], compression ratio 85.
(f) JPEG2000 compression, PSNR = 28.16 [dB], compression ratio 190 (same as 3c).

a   b   c

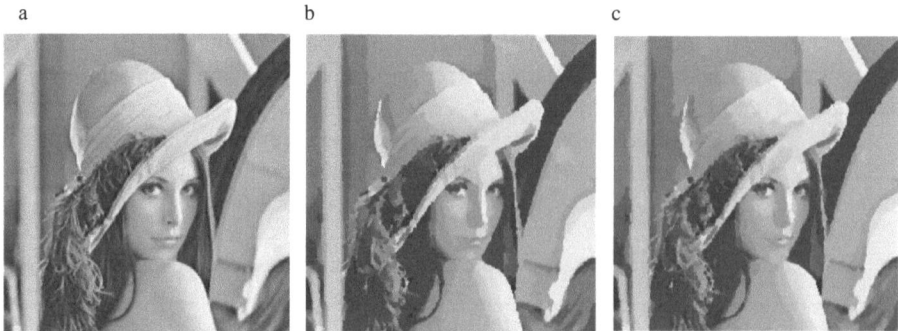

Fig. 4. Another example of color image compression reconstruction. (a) Original image.
(b) Region Growing (GL), base color (G) PSNR = 26.0021 [dB], compression ratio 81.
(c) Region Growing (RGB), base color (Optimal) PSNR = 26.7238 [dB], compression ratio 83.

RGB Region Growing produces segmentation maps that are more detailed and more correlating to the human observer's perception of segmentation. RGB segmentation, compared to gray-level segmentation, is more computationally consuming. The compression ratios of the suggested algorithm vary in the range 30-90 for typical images, and can go up to 200 for structured images with well defined color regions.

4   Summary and Conclusions

In this work, we have expanded and developed existing methodologies of color image compression using inter-color correlation between the color primaries. Past works have dealt with exploiting this inter-correlation by expressing some color components as a functional of a base component, thus reducing the amount of data required for representation. There has also been use of image regions acquired from segmentation, instead of NxN square blocks. We suggested making full use of the color information in the Region Growing segmentation process, and implementing an optimization [14] procedure for the selection of a unique base color for each image region, which minimizes the colors' reconstruction errors (MSE) for that region. The above methodology can produce reconstructed results with higher quality than JPEG, for high compression ratios. It can also produce superior results compared to JPEG2000 compression standard for images of well-defined separate color regions.

These two main developments have been shown to improve compression reconstruction quality, each by its own, and together, both perceptually and by PSNR measure, with an increase in PSNR of more than 1 [dB] for examined images.

Performing RGB Region Growing, instead of gray level, increases the compression PSNR measure significantly. Optimal base color selection can generally lead to an increase in compression ratio, by approximating regions' color textures of base and subordinate colors by lower degree polynomial functions, thus lowering the amount of bits required for encoding the color texture of an image. Our conclusion is that the proposed approach could improve the performance of presently available color compression methods.

Acknowledgment. This research was supported in part by the Ollendorff Minerva Center. Minerva is funded through the BMBF.

References

1. Goffman-Vinopal, L., Porat, M.: Color image compression using inter-color correlation. In: Proceedings of the International Conference on Image Processing 2002, vol. 2, pp. II-353–II-356 (2002)
2. Kotera, H., Kanamori, K.: A novel coding algorithm for representing full color images by a single color image. J. Imag. Technol. 16, 142–152 (1990)
3. Brice, C.R., Fennema, L.: Scene analysis using regions. Artif. Intell. (1), 205–226 (1970)
4. Kunt, M., Ikonomopoulos, A., Kocher, M.: Second-generation image-coding techniques. Proc. IEEE 73(4), 549–573 (1985)
5. Vedaldi, A., Soatto, S.: Quick Shift and Kernel Methods for Mode Seeking. In: Forsyth, D., Torr, P., Zisserman, A. (eds.) ECCV 2008, Part IV. LNCS, vol. 5305, pp. 705–718. Springer, Heidelberg (2008)
6. Wang, D.C.C., Vaganucci, A.H.: Gradient inverse weighted smoothing scheme and evaluation of performance. Comput. Graph. Image Process. 15 (1981)
7. Roterman, Y., Porat, M.: Color image coding using regional correlation of primary colors. Image and Vision Computing 25, 637–651 (2007)
8. Freeman, H.: On the encoding of arbitrary geometric configuration. IRE Trans. Electron. Comput. EC-10, 260–268 (1961)
9. Porat, M., Zeevi, Y.Y.: The generalized Gabor scheme of image representation in biological and machine vision. IEEE Transactions on Pattern Analysis and Machine Intelligence (PAMI) 10(44), 452–468 (1988)
10. Eldar, Y., Lindenbaum M., Porat M., Zeevi Y.Y.: The farthest point strategy for progressive image sampling. IEEE Transactions on Image Processing 6(9), 1305–1315 (1997)
11. Wallace, G.K.: The JPEG still picture compression standard. IEEE Transactions on Consumer Electronics 38(1), xviii-xxxiv (1992)
12. Christopoulos, C., Skodras, A., Ebrahimi, T.: The JPEG 2000 Still Image Coding System: An Overview. IEEE Transactions on Consumer Electronics 46(4), 1103–1127 (2000)
13. Nemirovsky, S., Porat, M.: On texture and image interpolation using Markov models. Signal Processing: Image Communication 24(3), 139–157 (2009)
14. Genossar, T., Porat, M.: Optimal bi-orthonormal approximation of signals. IEEE Transactions on Systems, Man and Cybernetics 22(3), 449–460 (1992)

Towards Learning Hierarchical Compositional Models in the Presence of Clutter[*]

Jan Mačák and Ondřej Drbohlav

Center for Machine Perception (CMP),
Department of Cybernetics, Czech Technical University in Prague
{macakj1,drbohlav}@cmp.felk.cvut.cz

Abstract. Our goal is to identify hierarchical compositional models from highly cluttered data. The data to learn from are assumed to be imperfect in two respects. Firstly, large portion of the data is coming from background clutter. Secondly, data generated by a recursive compositional model are subject to random replacements of correct descendants by randomly chosen ones at every level of the hierarchy. In this paper, we study the limits and capabilities of an approach which is based on likelihood maximization. The algorithm makes explicit probabilistic assignments of individual data to compositional model and background clutter. It uses these assignments to effectively focus on the data coming from the compositional model and iteratively estimate their compositional structure.

1   Introduction

A long-term goal in machine learning field is to learn usable models of objects from observations of the world, in an unsupervised way. It is generally believed that suitable representations of the world should be hierarchical and compositional [11,10]. Such representations are efficient as for the computational resources (due to hierarchy) and in memory requirements (share-ability of parts in compositional structures).

What principles should be used in order to identify structural building blocks in the data? Possibilities are usually centered around the Minimum Description Length (MDL) principle [1], and proceed by looking for what has been termed "suspicious coincidences" in the data (data appearing jointly with probability higher than chance). Notable results have been obtained in specific cases, e.g. in vision (e.g. [14,8,3]). Our primary research interest is also centered in vision, and we will be inclined to motivate the design settings of our experiment, or comment on the results, from point of view of a computer vision researcher, while keeping the analysis general.

---

[*] The authors were supported by the Czech Science Foundation under the Project P103/12/1578 (SeMoA).

A. Petrosino (Ed.): ICIAP 2013, Part I, LNCS 8156, pp. 532–541, 2013.

1.1   Key Features of This Work

Our goal is to identify hierarchical compositional models from the observed data. Consider an example of the visual domain: The data would consist of images, and the hierarchical compositional models would comprise of models for shapes and visual qualities of different objects, their parts, and (recursively) the sub-parts of these parts. In this paper, we however do not wish to be bound to a particular domain (visual, auditory, or other); our aim is to find out whether the proposed algorithm can generally work in such adverse settings. For this reason, we extract two usual aspects of the real-world data which make the task difficult: i) the data are, for a large part, composed of noise, or background clutter; ii) the compositions (or objects) are not present in the data in a clean form, but any part (or its sub-part) in the object's compositional structure may be replaced by a different part. In the visual domain, aspect i) would correspond to the fact that large portions of images are composed of non-object data (textures or uniform areas), while aspect ii) would roughly correspond to occlusion by another object, articulation or changed appearance.

   In this paper, we wish to show general capabilities and limits of an approach based on Maximum Likelihood (ML). The aim of our algorithm is to find a suitable set of models (compositions) on each layer of hierarchy such that the data likelihood is maximized (the model is described in a more detail in Section 2). The algorithm works in iterations. In each iteration, it learns the compositions layer by layer, starting from the bottom one, and estimates soft assignments of data to clutter and non-clutter. This enables it to use only the non-clutter data to refine the library of compositions in the next iteration (full description of the algorithm is in Section 3.)

1.2   Contribution

We design an algorithm for compositional models extraction from data which is suited for the two peculiarities of the data. The algorithm proceeds iteratively in a bottom-up and top-down manner. The essential part of the algorithm is that it makes probabilistic assignments of each data to clutter and non-clutter part of the model. We show the capabilities and limits of this ML-based approach on synthetically generated and real data. The study is made on a problem which is simplified to a maximum (aligned data, known graph structure) while maintaining the essential features of the real-world data (high amount of clutter, and missing parts in compositions).

1.3   Prior Work

Fukushima [5] was one of the first to research the topic of a neural network able to learn without supervision. His approach has been inspired by biological principles. He constructed a hierarchical neural network and demonstrated its performance on digits recognition. Frey et al. [4] also used a neural network for unsupervised learning and used local rules for adjusting weights on connections in it. They were able to adjust weights such that the neural network learns two classes of simple patterns (horizontal and vertical stripes). In effect, they employed the principle of KL divergence minimization. This was followed by the work of Hinton et al. [7] which extended the previous results and proposed efficient algorithm for learning in deep belief nets.

Compositionality has been recently an active topic in the specific area of computer vision. For example, Ommer and Buhmann [9] suggested an algorithm to learn compositional representation of visual objects and used this approach for object recognition. Image patches are used as building blocks for the lowest layer. Leonardis [3,2] uses sparse features (edges) as building blocks, and constructs a hierarchical model which is, notably, generative. The model is constructed intuitively but not cast in probabilistic framework. Zhu et al. [14,13] published works on semi- and unsupervised learning in visual domain combining bottom-up and top-down processes. Short theoretical analysis of this problem has been recently performed by Yuille [12].

2   Model

A tree-shaped graphical model has been employed in this paper (please see Fig. 1 and the notation symbols explained there) to represent the data. The tree has five layers in total: a root layer consisting of a single node σ, three inner layers $\mathbf{a}$, $\mathbf{b}$ and $\mathbf{c}$, and the data layer $\mathbf{D}$. Note that we use the 5-layer model to employ a concrete example but in practice we could use any number of layers.

The root of the graph hosts variable σ which can be thought of as representing different objects. Thus its states are from a set Σ which represents the set of top-level compositions.

The probability of a set of node values is

$$P(\sigma, \mathbf{a}, \mathbf{b}, \mathbf{c}, \mathbf{D}) = P(\sigma) \prod_{i=1}^{5} \left(P(a_i|\sigma) \prod_{j=1}^{5} \left(P(b_{ij}|a_i) \prod_{k=1}^{5} P(c_{ijk}|b_{ij}) P(D_{ijk}|c_{ijk}) \right) \right).$$

(1)

Here, the node names are as described in Fig. 1 (please see this figure for a self-contained explanation of the graph structure and notation conventions), and the bold-faced letters $\mathbf{a}$, $\mathbf{b}$, $\mathbf{c}$ and $\mathbf{D}$ denote a set collecting all the nodes from the respective layers. Note that following (usual) minor abuse of notation, a_i is sometimes used both to identify the node in the graph *and* to denote a specific value $a_i \in \mathcal{A}$ in that node.

A state in any of the inner nodes x is either one of the compositions $x \in \mathcal{X} = \{x^1, x^2, .., x^{|\mathcal{X}|}\}$, or the background state x^{bg}. Having a parent x and its child y (take e.g. $x = b_{25}$ and $y = c_{251}$, say), the conditional probability $P(y|x)$ takes a form according to the type of the state. For compositional states, this probability reflects the preference of the state x to have a specific part y^k as its child:

$$
\begin{array}{c|c|c|c|c|c|c}
y & y^1 & y^2 & \cdots & y^k & \cdots & y^{|\mathcal{Y}|} & y^{bg} \\
\hline
P(y|x) & \frac{\delta}{|\mathcal{Y}|} & \frac{\delta}{|\mathcal{Y}|} & \cdots & \Delta & \cdots & \frac{\delta}{|\mathcal{Y}|} & \frac{\delta}{|\mathcal{Y}|}
\end{array}
\qquad
\begin{array}{c}
x \in \mathcal{X} \\
\Delta \gg \delta\,, \Delta + \delta = 1.
\end{array}
$$

(2)

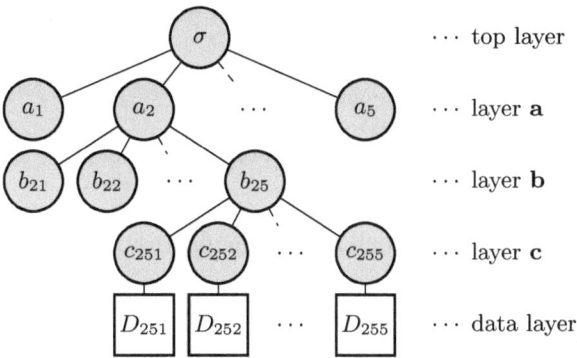

Fig. 1. The structure of the graph. Each node in the root layer and in layer **a** and **b** has five children, of which only some are shown for the sake of clarity. Nodes in layer **c** are connected to leaves by single links. The leaves are the data nodes. The nodes are denoted using indexing where children inherit the parent's indices and get one more own index. Thus children of a_i are denoted b_{ij} where j identifies one of the five children of a_i. Likewise, children of b_{ij} are denoted c_{ijk}. Compositionality is reflected by that any of the nodes in layer **a** has states from the same set $\mathcal{A} = \{A_1, A_2, \ldots, A_{|\mathcal{A}|}\}$, thus $a_i \in \mathcal{A}, \forall i = 1, 2, \ldots, 5$. The same is true about nodes in layers **b** and **c**. For the same reason of compositionality, the edge potentials for a given parent and child at a given layer are also the same, thus e.g. $(b_{ij} = b_{op} \wedge c_{ijk} = c_{opk}) \Rightarrow P(c_{ijk}|b_{ij}) = P(c_{opk}|b_{op})$.

The other possibility is that $x = \mathrm{x}^{bg}$, and in that case the preference for any of its child's identities is described by the relative frequencies $\{\hat{p}_{\mathrm{y}^1}, \hat{p}_{\mathrm{y}^2}, \ldots, \hat{p}_{\mathrm{y}^{|\mathcal{Y}|}}, \hat{p}_{\mathrm{y}^{bg}}\}$ in the data:

$$
\begin{array}{c|c|c|c|c|c}
y & \mathrm{y}^1 & \mathrm{y}^2 & \cdots & \mathrm{y}^{|\mathcal{Y}|} & \mathrm{y}^{bg} \\
\hline
P(y|x) & \hat{p}_{\mathrm{y}^1} & \hat{p}_{\mathrm{y}^2} & \cdots & \hat{p}_{\mathrm{y}^{|\mathcal{Y}|}} & \hat{p}_{\mathrm{y}^{bg}}
\end{array}
\qquad x = \mathrm{x}^{bg} . \qquad (3)
$$

3   Algorithm

The task of the algorithm is the following. The input consists of the observed data D. The required goal is to find the parameters of the model in the Eq. 1 such that the probability of observing the data is maximized. The parameters of the model are: i) prior probabilities over all layers ($P(\sigma)$ resp. $P(x)$), the lower layer priors are used as $\hat{p}$ in Eq. 3, and ii) five-tuple of indices k specifying child components, for each $x \in \mathcal{X}$. The conditional probabilities of the data, that is $P(D_{ijk}|c_{ijk})$, are given.

To learn the model basically means to build up a library of compositions. The library is hierarchically organized and each layer is learned interdependently based on previous layers and has the capability of storing a certain number of compositions. These capacities are set in advance. The learning works in a *Bottom-Up* manner with *Top-Down* adjustment.

for $i \leftarrow 1$ **to** #*Iterations* **do**

    1) learn compositions in the bottom layer **b** — *(details provided in the subsection 3.1)*;

    2) learn compositions in the middle layer **a**;

    3) learn compositions in the top layer σ;

    4) propagate the information from the top-node to lower nodes — *(details provided in the subsection 3.2)*;

end

Algorithm 1. Learning algorithm

3.1 *Bottom-Up* **Learning**

Bottom-up learning is an iterative process performed as the Steps 1)–3) of the Algorithm 1 and deals with learning a single layer. It basically takes the set of all five-tuples in the preceding (lower) layer and comes up with compositions in the current layer which represent them best.

The library is initialized randomly and all compositions are set to be equally likely. The ML estimates for the conditional probabilities $P(y|x)$ and prior $P(x)$ are given by

$$P(y|x) = \frac{1}{\lambda_1} \sum_{n=1}^{N} P(y, x|D_n), \quad P(x) = \frac{1}{\lambda_2} \sum_{n=1}^{N} P(x|D_n), \tag{4}$$

provided that we assume for the moment a reduced graphical model only up to the current layer (x being the top node). In this equation, the summation is over the whole dataset and λ_1, λ_2 are the normalizing factors. The joint probability is estimated as $P(y, x|D_n) = P(y|D_n)P(x|D_n)$, which means that information from the top node x to the lower-level node y does not affect the marginal at the lower-level node. The conditional probabilities used are estimated using standard Bayes formula from $P(D_n|y)$, $P(D_n|x)$ and priors $P(x)$ and $P(y)$. The former is estimated simultaneously with the layer compositions, the latter is known from learning of the previous layer.

The probability in Eq. 4 is then converted to a composition by taking the max-probability item y^k in the distribution $P(y|x)$ and assigning value Δ to it, while assigning values $\delta/|\mathcal{Y}|$ to all other items. When learning the background (cf. Eq. 3), the conditional probability stays as it is.

The process iterates by re-estimation of assignments $P(y|D_n)$, $P(x|D_n)$ using adjusted compositions and re-estimation of conditional and prior probabilities. The number of iterations is fixed.

3.2 *Top-Down* **Adjustment**

Top-down adjustment is performed as the Step 4 of the Algorithm 1. Once the learning process reaches the top layer, it is possible to propagate the top-node information on the soft assignments of individual data vectors to compositional model or to clutter. This essentially helps to focus on non-clutter data in the next algorithm iteration and improve the model.

The assignment of each data to compositional model or background is propagated by a top-down message constructed as

$$
\begin{array}{c|c|c|c|c|c}
 & \mathbf{y}^1 & \mathbf{y}^2 & \cdots & \mathbf{y}^{|\mathcal{Y}|} & \mathbf{y}^{bg} \\
\hline
P(y|x), x \in \mathcal{X} & \frac{\Delta}{|\mathcal{Y}|} & \frac{\Delta}{|\mathcal{Y}|} & \cdots & \frac{\Delta}{|\mathcal{Y}|} & \delta \\
P(y|\mathbf{x}^{bg}) & \frac{\delta}{|\mathcal{Y}|} & \frac{\delta}{|\mathcal{Y}|} & \cdots & \frac{\delta}{|\mathcal{Y}|} & \Delta
\end{array}
\tag{5}
$$

Consequently, in the next iteration, the non-model data are effectively assigned to the clutter model σ^{bg} and do not affect the learning of the compositional model.

Additionally, before propagating the top-down information, models with low significance are removed from library. This is basically a simple implementation of the MDL principle [6]. It means that models which do not contribute much to data log-likelihood are not considered to belong to the model and corresponding data are assigned to be clutter. The contribution of each learned composition is computed as

$$
\Delta L_{\sigma^i} = \sum_{n=1}^{N} \ln \frac{P(D_n|\sigma^i)P(\sigma^i) + P(D_n|\sigma^{bg})(1 - P(\sigma^i))}{P(D_n|\sigma^{bg})},
\tag{6}
$$

where $P(\sigma^i)$ is probability of the compositional model σ^i, the model σ^{bg} is the model for clutter. The composition σ^i is removed if its contribution of the composition is less than threshold value. The Eq. 6 is simply a difference of the baseline data log-likelihood of the model without any compositional model and the data log-likelihood supposing there is a compositional model σ_i which represents well a $P(\sigma^i)$ portion of data. The values of $P(\sigma^i)$ and $P(\sigma^{bg})$ are derived from ML estimates of the distribution of top layer models. Using the value $P(\sigma^i)$ gives a lower bound on ΔL_{σ^i}, because, under the assumption of no other compositional models, the optimal $P^*(\sigma^i)$ must lie within the interval of $\langle P(\sigma^i), P(\sigma^{bg}) \rangle$.

In practice, the learning converges in few steps even for the cases of low concentration of a compositional model data in clutter. As can be seen in the Fig. 2, two iterations are sufficient for learning the model.

4 Experiments

4.1 Artificial Data Experiment

The learning algorithm was tested on artificially generated compositional data. This data was generated from a randomly generated compositional model which consisted of five classes (templates) on the top layer, eight templates on the middle layer and nineteen templates on the bottom layer. Each template was build from five components (templates from the lower layer) randomly. The number of five top classes was chosen small, because the aim was to investigate how the learning worked for highly cluttered data. For high number of classes, the amount of the clutter data would have had to be extremely large.

The compositional data was generated using the templates converted to potentials specified by Eq. 2. Several values for Δ were used, of which we show results for $\Delta = 0.99$ (lower number of random replacements, thus easier case) and $\Delta = 0.75$ (harder case). Background clutter was generated from the uniform distribution over the finite interval of $\langle 1, 25 \rangle$.

In the learning data, there were 5000 data vectors of which only 5 % was generated by the compositional model. The capacities of library while learning were set to accommodate up to 20, 40 and 60 models at the bottom, the middle and the top layer respectively.

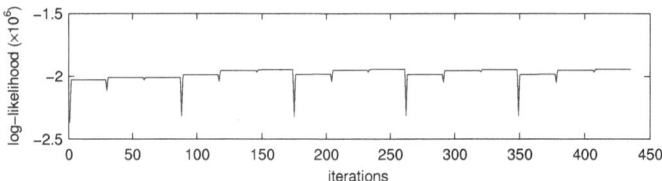

Fig. 2. Development of data log-likelihood. The horizontal axis shows inner cycle iterations. There are 30 iterations at each layer, so 90 iterations are performed in each cycle of *Bottom-Up* learning. The drop of the log-likelihood after the end of the previous cycle is caused by random initialization of clusters.

The Fig. 3 shows results of learning. It shows the euclidean distance of the learned models and the ground truth. The model was learned correctly as there is zero distance between five models from top layer and generating models. The rate was estimated also correctly, assigning 95% of data to clutter, and approx. 1% of data to each of the five models. The true numbers are 95% clutter, 1% each compositional model. This result was obtained after five iterations. The assignment of data to individual top-layer models is shown in the Fig. 5 and as can be seen, the actual number of learned models on the top-most layer is correct, despite the fact that the library size is set excessively.

Another interesting question would be that of what is the break-down point of this method. This was estimated experimentally. The experiment was to keep all the statistics and parameters unchanged and iteratively reduce the number of compositional data instances. The total number of data vectors was 1000 and the concentration was falling from 5% to 0.5%. The algorithm began to fail at around 1.5% - 2%. Success and failure was measured as whether the data was correctly partitioned.

4.2   Controlled Experiment with Real Data

The learning algorithm was tested on a more realistic example of learning an optimal representation for random sequences of characters. The source characters were taken from a scanned letter written on a typewriter, which means that there

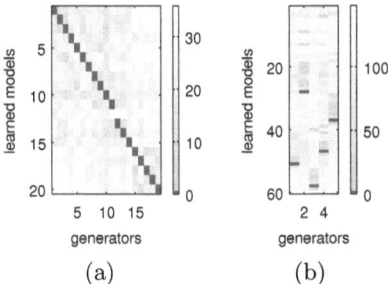

(a)   (b)

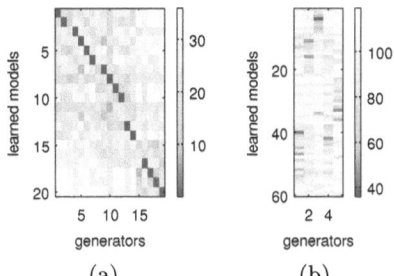

(a)   (b)

Fig. 3. Distances of learned models to ground truth, for $\Delta = 0.99$. (a) bottom layer **b**, (b) top layer σ. Vertical axis shows indices of learned templates, horizontal axis shows indices of ground truth templates. All models have been found.

Fig. 4. Distances of learned models to ground truth, harder data ($\Delta = 0.75$). (a) 1st layer, (b) top layer. It can be seen in the (a) that some compositions were missed (3rd, 8th, 15th).

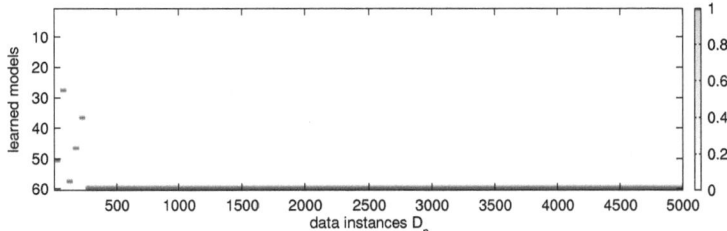

Fig. 5. Top layer model-data assignment. The horizontal axis shows individual data vectors and their assignments to the top layer models (on the vertical axis) the indices correspond to the vertical indices in the Fig. 3(b). The "compositional" data are grouped together and aligned to the left for the sake of visual inspection. The probability of $P(\sigma|D_n)$ is color-coded with the scale shown on the right side.

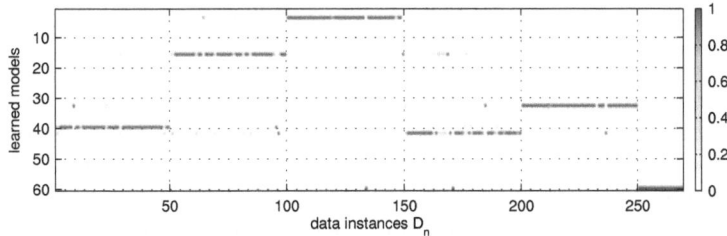

Fig. 6. Top layer model-data assignment for harder data ($\Delta = 0.75$), the meaning is the same as in the Fig. 5. This time, only the detailed picture of the assignment of "compositional" data is shown. It can be seen that majority of data instances are assigned correctly, however there are some wrong assignments.

are various differences within the sets of same characters. The thickness depends on a stroke style, the noise due to scanning is present, etc.

The data itself was generated in the following way. First, the individual characters were cropped from the picture and coarsely aligned. Then, five random sequences were selected and ten instances for each sequence built using randomly selected character instances. Finally, the data was completed by randomly misaligned samples from the same picture as to obtain the background distribution with similar statistical properties. The ratio between compositional and background data was 1:1 (5 × 10 model instances, 50 random instances). The five sequences to be learned and their learned representations are shown in the Fig. 7. The numbers of elements on individual layers were set to 50, 50 and 10 respectively.

The characters were not learned precisely, but are distinguishable. The interesting observation is that there are both multiple words represented by one model and multiple models for one word. This is caused by the fact that the characters were not aligned precisely and the algorithm is not shift invariant and consequently a slightly shifted letter might get completely different representation on the lowest layer.

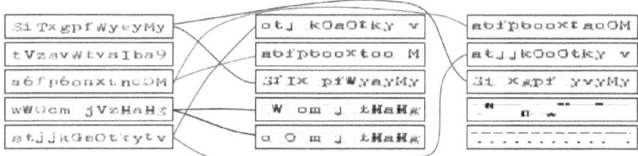

Fig. 7. Ground truth sequences in the left column and their learned representations in the middle and the right column. As can be seen, even for such a small dataset 4 out of 5 sequences were learned.

5   Conclusion

Our main goal in this paper was to see how far we can get with likelihood optimization in learning hierarchical models from heavily cluttered data. We have constructed a simple and very controlled framework, and an experiment which enabled us to study the merits of likelihood maximization in isolation.

We have observed that hierarchical models can be learnt even in a very adverse scenario when 95 % of data were outliers. This was greatly helped by the feedback nature of the algorithm, where the bottom-up stage is followed by a top-down step, thus possibly improving the selection of proper models at the lower levels of hierarchy. We have also observed how the results depend on the rate of the clutter data, as well as on the number of random replacements (parameter Δ). Experiment on scanned characters confirmed the expected limitation of the algorithm for this type of application due to the requirement of data being aligned with data nodes.

References

1. Bienenstock, E., Geman, S., Potter, D.: Compositionality, MDL priors, and object recognition. In: Mozer, M., Jordan, M.I., Petsche, T. (eds.) NIPS, pp. 838–844. MIT Press (1996)
2. Fidler, S., Boben, M., Leonardis, A.: Optimization framework for learning a hierarchical shape vocabulary for object class. In: BMVC. British Machine Vision Association (2009)
3. Fidler, S., Leonardis, A.: Towards scalable representations of object categories: Learning a hierarchy of parts. In: Proc. CVPR (2007)
4. Frey, B., Dayan, P., Hinton, G.E., Jenkin, I.M.: A simple algorithm that discovers efficient perceptual codes. In: Mechanisms, L.R.H., Mechanisms, B. (ed.) Computational and Psychophysical Mechanisms of Visual Coding, pp. 296–315. Cambridge University Press (1997)
5. Fukushima, K.: Neocognitron: A hierarchical neural network capable of visual pattern recognition. Neural Networks 1(2), 119–130 (1988), http://www.sciencedirect.com/science/article/pii/0893608088900147
6. Grünwald, P.D.: The Minimum Description Length Principle. MIT Press (2007)
7. Hinton, G.E., Osindero, S.: A fast learning algorithm for deep belief nets. Neural Computation 18 (2006)
8. Kokkinos, I., Yuille, A.: Inference and learning with hierarchical shape models. International Journal of Computer Vision 93, 201–225 (2011), http://dx.doi.org/10.1007/s11263-010-0398-7
9. Ommer, B., Buhmann, J.: A compositionality architecture for perceptual feature grouping. In: Rangarajan, A., Figueiredo, M.A.T., Zerubia, J. (eds.) EMMCVPR 2003. LNCS, vol. 2683, pp. 275–290. Springer, Heidelberg (2003)
10. Ommer, B., Buhmann, J.: Learning the compositional nature of visual object categories for recognition. IEEE Trans. Pattern Anal. Mach. Intell. 32(3), 501–516 (2010)
11. Tsotsos, J.K.: Analyzing vision at the complexity level. Behavioral and Brain Sciences 13(03), 423–445 (1990), http://dx.doi.org/10.1017/S0140525X00079577
12. Yuille, A.L.: Towards a theory of compositional learning and encoding of objects. In: ICCV Workshops, pp. 1448–1455. IEEE (2011)
13. Zhu, L., Chen, Y., Yuille, A.L.: Learning a hierarchical deformable template for rapid deformable object parsing. IEEE Trans. Pattern Anal. Mach. Intell. 32(6), 1029–1043 (2010)
14. Zhu, L., Chen, Y., Yuille, A.L.: Recursive compositional models for vision: Description and review of recent work. Journal of Mathematical Imaging and Vision 41(1-2), 122–146 (2011)

Social Groups Detection in Crowd through Shape-Augmented Structured Learning

Francesco Solera and Simone Calderara

DIEF University of Modena and Reggio Emilia, Italy
francesco.solera@gmail.com, simone.calderara@unimore.it

Abstract. Most of the behaviors people exhibit while being part of a crowd are social processes that tend to emerge among groups and as a consequence, detecting groups in crowds is becoming an important issue in modern behavior analysis. We propose a supervised correlation clustering technique that employs Structural SVM and a proxemic based feature to learn how to partition people trajectories in groups, by injecting in the model socially plausible shape configurations. By taking into account social groups patterns, the system is able to outperform state of the art methods on two publicly available benchmark sets of videos.

Keywords: group detection, proxemic theory, Structural SVM.

1 Introduction

Group detection in video surveillance systems is profoundly motivated by behavior analysis and security issues in enhancing scene understanding, as the truth is that many interesting behaviors don't occur at an individual level but are the results of complex interactions between individuals in specific subsets of the crowd, namely groups. It is known, in fact, that the existence of groups highly influences the behavior of the individuals as it is at this level that people start to experiment structured interactions [1]. While there isn't a general solution to the problem of locating groups, data driven approaches have recently been obtaining interesting results, mainly motivated by the improvements in tracking performance that can be achieved when considering groups as structured entities of the scene [2]. Group tracking can be partitioned according to the availability of tracklets. In *group-based* approaches [3][4][5] groups are seen as the atomic entities to look for, with the major drawback of creating models that are too simplistic and cannot be used to further infer on groups behavior. On the other hand, *individual-based* tracking algorithms [6][7][8] focus on single pedestrians trajectories, which are the most informative features we can extract from a crowded scene. This latter approach has been gaining momentum only lately as tracking even in dense crowds is becoming everyday a more feasible task [9].

The common drawback of all these approaches lays in their scarce consideration of decades of sociological theories which can though provide many brilliant insights. Methods that rely exclusively on data assume to have at disposal a

A. Petrosino (Ed.): ICIAP 2013, Part I, LNCS 8156, pp. 542–551, 2013.

(a)   (b)   (c)

Fig. 1. Example of small groups in crowd

highly descriptive and complete dataset, which is quite an unrealistic assumption to state when considering dynamic concepts as groups and crowds are. As opposite, we believe social dynamics can help to deeply understand the group formation phenomenon. In particular, despite the well-assessed theory about collective crowd will, new results are underlining that people in groups tend to produce predefined shapes while maintaining their identity and goals [1], as can be observed in Fig. 1b. Ge *et al.* [8] actually present a statistical shape analysis method to analyze the spatial position of all group members jointly and estimate the typical group formations of walking pedestrians; but then they don't take advantage of these configurations to improve their group detection method.

Given the aforementioned sociological breakthrough we devise a new algorithm for group detection based on structured learning, which let us incorporate shapes structure evaluation in the optimization process yielding to more sociologically plausible predictions. We propose to train a supervised hierarchical bottom-up correlation clustering when trajectories of pedestrians are available. As we want to focus on the sociological side of the problem, we employ a feature founded on Hall's *proxemic distance theory* [10] and through Structural SVM we decide, based on previously annotated scenes, how this feature can be significant in explaining the concept of groups in any given scenario.

2   Structured Output Learning for Group Detection

Pedestrian trajectories encode many sociological and physical information about the way people interact. If two pedestrians have diverging trajectories it's very unlikely that they were on the scene together and, at the same time, in a group of friends everyone will likely have very similar and compact trajectories over a generic period of observation. Starting from this consideration we reformulate the problem of finding groups in the scene as the one of clustering trajectories, or partitioning the set of those. The solution of a clustering algorithm is a collection of members' assignments, thus conventional classifiers aren't suitable to deal with the combinatorial size of this problem output space. Building up on the work of Finley and Joachims [11], we propose a structured supervised algorithm able to learn how to partition crowds using sociologically founded features between pairs of trajectories. The method is summarized in the scheme of Fig. 2.

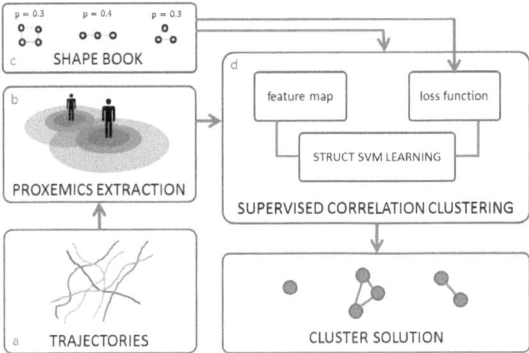

Fig. 2. Block diagram illustrating our group detection algorithm

2.1   Hall's Proxemic Distance: A Sociological Feature

An important relation between space and social interaction has been first formulated in Hall's proxemic theory and has already been applied to people trajectories in order to better understand diadic interactions [12]. Proxemic theory
states that the social distance between people is reliably correlated with their
physical distance, and more important it tells us that this relation is not linear.
Intuitively, the theory defines circles around every individual and the interaction
between pairs of individuals is classified according to which circle they mutually
reside in, as depicted in Fig. 2b. The result is a non linear quantization of their
distance in intimate, personal, social and public space. We generalize the original
quantization by approximating it with an exponential term. The proxemic score
of any two trajectories r and t is computed as

$$f_{rt}^{prox} = \frac{1}{\max\{|r|, |t|\}} \sum_{i \in I_{t,r}} e^{-\sqrt{(t_x^{(i)} - r_x^{(i)})^2 + (t_y^{(i)} - r_y^{(i)})^2}}, \tag{1}$$

where $I_{t,r}$ is the subset of time instances which restricts the summation to temporal intersection only and the coefficient outside the sum is needed in order for
f_{rt}^{prox} to be normalized.

2.2   Supervised Correlation Clustering through Structural SVM

While Hall's proxemic theory is a valuable instrument which can be exploited
in order to better understand group dynamics, it is not sufficient to solve the
problem of their detection. It is possible to grasp the complexity of the task by
considering a highly crowded scene where all the pedestrians are touching each
others, here distance is much less discriminant. Other than crowd density, also
the environment conformation, the local culture and other factors that cannot
be modeled explicitly make groups a challenging concept to define. For this
reason we adopt a supervised clustering approach in order to learn how proxemic

distance can be significant to describe groups in different scenarios. In [13] we prove our framework is also able to balance the contributions of multiple features useful to describe crowded scenarios, while here we only employ Hall's distance as we rather investigate the importance of group patterns in the detection task.

Since information about groups doesn't reside in the trajectories, but in the spatiotemporal relationships they are engaged in, we can't apply standard clustering techniques. **Correlation clustering** [14] operates exactly in this scope, where we don't want to describe the elements themselves but rather their pairwise relationships, computed as in Eq. 1 for the special case of Hall's proxemic distance. Formally, correlation clustering takes as input an affinity matrix $W = \{W_{rt}\}_{rt}$ where for $W_{rt} > 0$ we say that elements r and t are similar with certainty $|W_{rt}|$, and for $W_{rt} < 0$ we say elements r and t belong to different clusters with certainty $|W_{rt}|$. Our model aims to parametrize $W_{rt} = \mathbf{w}^T \phi_{rt}$ in $\mathbf{w}$ so that we can fix the feature but still be able to adjust its importance for as much discriminative information it can provide on the current scenario. By defining our feature vector to be $\phi_{rt} = [f_{rt}^{prox}, 1 - f_{rt}^{prox}]^T$ we can create negative entries in W_{rt} representing unlikely pairs of individuals. The correlation clustering $\bar{\mathbf{y}}$ of a set of trajectories $\mathbf{x}$ is the configuration that maximizes the sum of affinities for item pairs in the same cluster:

$$\bar{\mathbf{y}} = \arg\max_{\mathbf{y}} \sum_{y \in \mathbf{y}} \sum_{r \neq t \in y} W_{rt} = \arg\max_{\mathbf{y}} \sum_{y \in \mathbf{y}} \sum_{r \neq t \in y} \mathbf{w}^T \phi_{rt} \qquad (2)$$

Given the parametric model of the correlation clustering, the weight vector $\mathbf{w}$ can be learned using structured learning. **Structural SVMs** [15] offer a generalized framework to model and solve structured output problems by learning a mapping $f : \mathcal{X} \to \mathcal{Y}$ between input space $\mathcal{X}$ and structured output space $\mathcal{Y}$ given a sample of input-output pairs $S = \{(\mathbf{x}_1, \mathbf{y}_1), \ldots, (\mathbf{x}_n, \mathbf{y}_n)\}$. Recall $\mathbf{x}_i$ is a set of trajectories and $\mathbf{y}_i$ is a clustering solution for $\mathbf{x}_i$. In contrast to standard multiclass classification, where a different prediction function for each class is learned independently, we define a discriminant function $F : \mathcal{X} \times \mathcal{Y} \to \Re$ over the joint input-output space where $F(\mathbf{x}, \mathbf{y})$ can be interpreted as a compatibility measure between $\mathbf{x}$ and $\mathbf{y}$. We remark that $F(\mathbf{x}, \mathbf{y})$ cannot be defined out of the context of the problem, as it is the problem itself that specifies what kind of solution we want given a particular input. As a matter of fact, the parametric formulation of correlation clustering presented in Eq. 2, implicitly defines the compatibility of an input-output pair. We can thus restrict the space of F to linear functions over some combined feature representation $\Psi(\mathbf{x}, \mathbf{y})$, yielding to $F(\mathbf{x}, \mathbf{y}; \mathbf{w}) = \mathbf{w}^T \Psi(\mathbf{x}, \mathbf{y})$. From Eq. 2 it follows

$$\Psi(\mathbf{x}, \mathbf{y}) = \sum_{y \in \mathbf{y}} \sum_{r \neq t \in y} \phi_{rt}. \qquad (3)$$

Given F we define the prediction function $f(\mathbf{x}) = \arg\max_{\mathbf{y} \in \mathcal{Y}} F(\mathbf{x}, \mathbf{y}; \mathbf{w})$. According to Finley and Joachims [11] the problem of finding f can be restated as a n-slack margin rescaling maximum-margin problem:

$$\min_{\mathbf{w},\xi} \quad \frac{1}{2}\|w\|^2 + \frac{C}{n}\sum_{i=1}^{n}\xi_i$$
$$\text{s.t.} \quad \forall i : \xi_i \geq 0,$$
$$\forall i, \forall \mathbf{y} \in \mathcal{Y}\backslash\mathbf{y}_i : \mathbf{w}^T\delta\Psi_i(\mathbf{y}) \geq \Delta(\mathbf{y},\mathbf{y}_i) - \xi_i, \tag{4}$$

where $\delta\Psi_i(\mathbf{y}) = \Psi(\mathbf{x}_i,\mathbf{y}_i) - \Psi(\mathbf{x}_i,\mathbf{y})$, ξ_i are the slack variables introduced in order to accommodate for margin violations and $\Delta(\mathbf{y},\mathbf{y}_i)$ is the loss function.

The optimization problem of Eq. 4 introduces a constraint for every possible wrong clustering of the set. Since the number of wrong clusterings scales more than exponentially with the number of items, we choose to employ the *cutting plane* algorithm [15] that, starting with no constraints, aims at iteratively finding the most violated one $\hat{\mathbf{y}}_i = \arg\max_{\mathbf{y}} \Delta(\mathbf{y}_i,\mathbf{y}) - \mathbf{w}^T\delta\Psi_i(\mathbf{y})$ and re-optimize until convergence. Finding the most violated constraint requires to solve the correlation clustering problem, which we know to be NP-hard [14]. Finley and Joachims [11] propose a greedy approximation which works by initially considering each element in its own cluster, then iteratively merging the two clusters whose union would produce the worst clustering score.

The learning ability of the algorithm strongly depends on the choice of the loss function in Eq. 4, as it has the power to force or relax input margins. Given the analogy between trajectories clustering and the noun-coreference problem [17], we adopt the **MITRE** score [18] from NLP in defining the loss of Eq. 4:

$$\Delta(\mathbf{y}_i,\mathbf{y}) \equiv \Delta_M(\mathbf{y}_i,\mathbf{y}) = 1 - F_1 \tag{5}$$

where F_1 is the harmonic mean of precision and recall computed as in [18].

3   Shape-Augmented Learning

The spatial organization of pedestrians inside groups tends to obey to patterns that facilitate social interactions and verbal communications, while trying to avoid collisions with in-group members and out-group pedestrians. This patterns are highly dependent on the crowd density, the environment conformation and the group speed and it's not completely clear yet how these elements are all correlated together. Nevertheless recent behavior analysis research has been focusing on the formalization of such concepts. Moussaïd *et al.* [1] show from real-world data that up to 70% of observed pedestrians in a commercial street are walking in groups and provide a distribution of those groups size. Despite the different experimental scenario, Bandini *et al.* [16] provide empirical results about the frequency of group patterns which converge accordingly to the work of Moussaïd *et al.* [1]. As a consequence, we think it does make sense to consider some pattern more probable than others while searching the crowd for groups.

A peculiar consideration emerging from these studies [16] states that in dense scenarios, people tend to rearrange in formations which allow to feel compact and protected from out-group individuals, Fig. 3b (case B), while if space is available people prefer to reside in a line-like shape, Fig. 3c (case B). In both

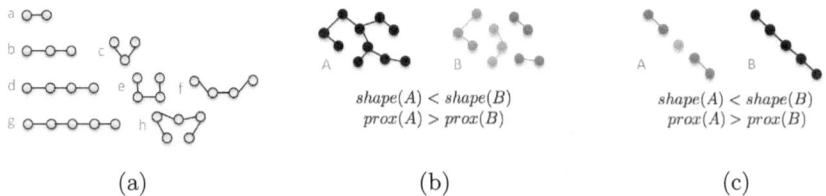

Fig. 3. (a) reports the most frequent group patterns. (b) and (c) show two scenarios where proxemic measure would produces socially implausible grouping predictions.

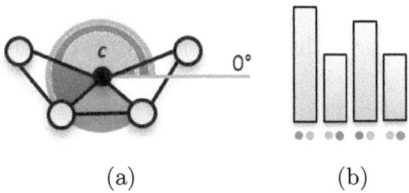

Fig. 4. The process employed in order to obtain a shape histogram (b) from a given group pattern (a)

cases, minimizing mutual distances among members (*i.e.* exploiting proxemics) is not sufficient to produce socially plausible groups, Fig. 3b and 3c (cases A).

3.1   Modeling Shapes

In order for a group-shape descriptor to be considered effective, it should be invariant under translation, rotation and scale, as relative distances are already taken into account by pairwise features. Moreover it should be robust against small variations, *i.e.* small differences in the group configuration should correspond to small differences in the shape description.

We begin by finding the mean point C between group members and starting from the closest one and following a counter clock-wise direction, we keep track of the angular position of each member w.r.t. C, as shown in Fig. 4a. This step guarantees that the scoring measure will be invariant under translation and rotation. In order to ensure scale invariance as well, we choose to neglect the distance between group members focusing only on their sequential angular distance, *i.e.* we only measure the angular distance from one member to the next one, yielding to the histogram reported in Fig. 4b. Note that every histogram obtained as described above is characterized by a number of bins equal to the number of group members and a normalization value of a turn. Shapes can now be compared using a trivial histogram intersection measure.

3.2   Shape-Aware Loss Function

Given the social groups configurations depicted in Fig. 3a and their relative shape histograms obtained as explained in Sec. 3.1, we propose to build a codebook of preferred shapes, *shapebook*, where each configuration is associated with an a-priori probability of occurrence extracted from sociological studies [1][16]. The configurations listed in Fig. 3a are only a subset of all the possible ones, but we remark that a group can still be detected as such, even if it lays in a configuration that we do not model. It will simply be considered less probable.

Given a possible clustering solution $\mathbf{y}$, we can measure how much the detected groups are sociologically conceivable by (i) finding, for each frame f in the time window T and for the configuration $s_{y,f}$ of each group $y \in \mathbf{y}$ the most similar shape $\bar{s}_{y,f}$ among the ones described in the shapebook, (ii) evaluating the similarities of their histograms through histogram intersection and by (iii) assigning a score proportional to the probability of the group pattern $\bar{s}_{y,f}$ and its similarity with pattern $s_{y,f}$. The process is synthesized in the following formula:

$$\Delta_S(\mathbf{y}) = 1 - \frac{1}{|T|} \sum_{y \in \mathbf{y}} \sum_{f \in T} [s_{y,f} \cap \bar{s}_{y,f}] \, p(\bar{s}_{y,f}). \tag{6}$$

As shapes embody our a-priori knowledge of the problem, we want to force them into the learning framework. To our knowledge, there isn't any explicit proposal in literature focused on the idea of encoding a-priori data in Structural SVM. Moreover, since we let correlation clustering define the joint feature representation, we employ pairwise features which cannot capture global level information such as the shape of the group they are in. Due to the impossibility of including shapes at a feature level, we consider the loss function level, where information about all individuals in the scene can be accessed simultaneously. By pursuing this idea, we define the *shape-aware* loss function $\Delta(\mathbf{y}_i, \mathbf{y})$ as

$$\Delta(\mathbf{y}_i, \mathbf{y}) \equiv \lambda \Delta_M(\mathbf{y}_i, \mathbf{y}) + (1 - \lambda) \Delta_S(\mathbf{y}). \tag{7}$$

By replacing the loss function in the constraints of Eq. 4 with the one of Eq. 7, we obtain

$$\begin{aligned}
\min_{\mathbf{w}, \xi} \quad & \frac{1}{2}\|w\|^2 + \frac{C}{n}\sum_{i=1}^{n} \xi_i \\
\text{s.t.} \quad & \forall i : \xi_i \geq 0, \\
& \forall i, \forall \mathbf{y} \in \mathcal{Y} \backslash \mathbf{y}_i : \mathbf{w}^T \delta \Psi_i(\mathbf{y}) - (1 - \lambda)\Delta_S(\mathbf{y}) \geq \lambda \Delta_M(\mathbf{y}, \mathbf{y}_i) - \xi_i.
\end{aligned} \tag{8}$$

The new set of constraints state that it doesn't matter if the margin of the feature space isn't optimally maximized as the classifier can also count on the fact that poorly structured solutions will be highly penalized. This allows $\mathbf{w}$ to be slightly less fit on the data, enhancing the generalization capability of the algorithm.

Given the optimization problem of Eq. 8, questions arise on the correctness and on the convergence of classical algorithms to solve this unconventional SSVM problem. We modified the original cutting plane algorithm [15] and solved the dual form of the quadratic problem of line 11 in Alg. 1 through the Sequential Minimal Optimization (SMO) approach, inspired by the work of Lee and

Jang [19]. Let $\mathcal{L}_P$ be the Lagrangian of the problem in Eq. 8, by differentiating it and back-substituting we obtain its dual counterpart $\mathcal{L}_D$, as in Eq. 9.

$$\max_{\alpha,\beta} \quad \mathcal{L}_D(\alpha,\beta) = -\frac{1}{2} \sum_{i,\mathbf{y}\neq\mathbf{y}_i} \sum_{j,\bar{\mathbf{y}}\neq\mathbf{y}_j} \alpha_{i\mathbf{y}}\alpha_{j\bar{\mathbf{y}}}\delta\Psi_i(\mathbf{x}_i,\mathbf{y})\delta\Psi_j(\mathbf{x}_j,\bar{\mathbf{y}})$$

$$+ \sum_{i,\mathbf{y}\neq\mathbf{y}_i} \alpha_{i\mathbf{y}}\Delta(\mathbf{y}_i,\mathbf{y}) + \sum_{i,\mathbf{y}\neq\mathbf{y}_i} \beta_{i\mathbf{y}}(1-\lambda)\Delta_S(\mathbf{y}) \qquad (9)$$

$$\text{s.t.} \quad \forall i: \sum_{\mathbf{y}\neq\mathbf{y}_i} (\alpha_{i\mathbf{y}} + \beta_{i\mathbf{y}}) = \frac{C}{n}, \quad \forall i, \forall \mathbf{y}: \alpha_{i\mathbf{y}} \geq 0, \beta_{i\mathbf{y}} \geq 0.$$

The maximum of $\mathcal{L}_D$ can be found by differentiating with respect to both α and β, resulting in the identity $\lambda\Delta_M(\mathbf{y}_i,\mathbf{y}) = \mathbf{w}^T\delta\Psi_i(\mathbf{x}_i,\mathbf{y})$. It can be shown that by exploiting this equivalence, the SMO step update results as in line 8 of Alg. 2.

Algorithm 1. Cutting Plane Shape

Input: $\{(\mathbf{x}_i,\mathbf{y}_i)\}_n, C, \epsilon$
Output: w
1: $S_i := 0, \forall i \in \mathbb{N}_n$
2: **repeat**
3: **for** $i := 1 \to n$ **do**
4:    $H(\mathbf{y}) = \Delta(\mathbf{y}_i,\mathbf{y}) - \mathbf{w}^T\delta\Psi_i(\mathbf{x}_i,\mathbf{y})$
5:    $\hat{\mathbf{y}}_i = \arg\max_{\mathbf{y}} H(\mathbf{y})$
6:    $\xi_i = \max\{$
7:       $\max_{\mathbf{y}\in S_i}(1-\lambda)\Delta_S(\mathbf{y}),$
8:       $\max_{\mathbf{y}\in S_i} H(\mathbf{y})\}$
9:    **if** $H(\hat{\mathbf{y}} > \xi_i + \epsilon$ **then**
10:     $S_i \leftarrow S_i \cup \{\hat{\mathbf{y}}\}$
11:     $\alpha \leftarrow$ opt. dual over $S = \bigcup_i S_i$
12:   **end if**
13: **end for**
14: **until** no S_i changes during iteration

Algorithm 2. SMO Shape

Input: $\{(\mathbf{x}_i,\mathbf{y}_i)\}_n, S, \alpha, C$
Output: α
1: $\mathbf{w} := \sum_{i,\mathbf{y}\neq\mathbf{y}_i} \alpha_{i\mathbf{y}}\delta\Psi_i(\mathbf{x}_i,\mathbf{y})$
2: **repeat**
3: **for all** $(\mathbf{x}_i,\hat{\mathbf{y}}) \in S$ **do**
4:    **if** $(\mathbf{x}_i,\hat{\mathbf{y}})$ violates KKT **then**
5:      $s := \frac{\lambda\Delta_M(\mathbf{y}_i,\hat{\mathbf{y}})-\mathbf{w}^T\delta\Psi_i(\mathbf{x}_i,\hat{\mathbf{y}})}{\|\delta\Psi_i(\mathbf{x}_i,\hat{\mathbf{y}})\|^2}$
6:      $s_{\text{clip}} \leftarrow \min\{s, \frac{C}{n} - \sum_{\mathbf{y}\neq\mathbf{y}_i}\alpha_{i\hat{\mathbf{y}}}\}$
7:      $s_{\text{clip}} \leftarrow \max\{s_{\text{clip}}, -\alpha_{i\hat{\mathbf{y}}}\}$
8:      $\alpha \leftarrow \alpha + s_{\text{clip}}$
9:      $\mathbf{w} \leftarrow \mathbf{w} + s_{\text{clip}}\delta\Psi_i(\mathbf{x}_i,\mathbf{y})$
10:   **end if**
11: **end for**
12: **until** no $\alpha_{i\hat{\mathbf{y}}}$ changes during iteration

Eq. 10 provides the KKT conditions of line 4 of Alg. 2 that have to be met in order for the algorithm to converge to an optimal solution.

$$\alpha_{i\mathbf{y}} = 0 \quad \Rightarrow \quad \mathbf{w}^T\delta\Psi_i(\mathbf{x}_i,\mathbf{y}) \geq \lambda\Delta_M(\mathbf{y}_i,\mathbf{y})$$
$$0 < \sum_{\mathbf{y}\neq\mathbf{y}_i}\alpha_{i\mathbf{y}} < \frac{C}{n} \quad \Rightarrow \quad \mathbf{w}^T\delta\Psi_i(\mathbf{x}_i,\mathbf{y}) = \lambda\Delta_M(\mathbf{y}_i,\mathbf{y}) \qquad (10)$$
$$\sum_{\mathbf{y}\neq\mathbf{y}_i}\alpha_{i\mathbf{y}} = \frac{C}{n} \quad \Rightarrow \quad \mathbf{w}^T\delta\Psi_i(\mathbf{x}_i,\mathbf{y}) \leq \lambda\Delta_M(\mathbf{y}_i,\mathbf{y})$$

4   Experimental Results

We tested our system on two publicly available datasets, namely the *BIWI Walking Pedestrians* dataset [6] and the *Crowds-By-Examples (CBE)* dataset [20]. The former records two low crowded scenes, one outside a university, named `eth`, and one, `hotel`, at a bus stop, both shown in Fig. 5, while the latter records a high density crowd video outside another university, `student003` (`stu003`). As it can be seen from Fig. 1, `stu003` dataset provides some real challenge as the

(a) *BIWI* hotel   (b) *BIWI* eth

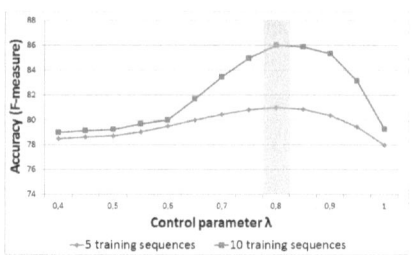

Fig. 6. Test results for different values of λ on stu003

(c) *CBE* student003

Fig. 5. Visual results on test videos[1]

Table 1. Performance comparison in terms of precision and recall computed according to the MITRE [18]

	our $\lambda = 0.8$		[6]		[7]	
	$\mathcal{P}$	$\mathcal{R}$	$\mathcal{P}$	$\mathcal{R}$	$\mathcal{P}$	$\mathcal{R}$
hotel	**93.2**	93.7	-	-	91.3	**95.9**
eth	**88.4**	**91.6**	-	-	83.0	80.2
stu003	**85.9**	**86.3**	46.0	82.0	80.5	77.0

density of the pedestrians is significant as well as for the presence of stairs and multiple entry and exit points. In the test setting we trained the classifier with one minute and two minutes of crowd videos; the trajectories were acquired on a time window of 10 seconds with no overlap. We evaluate the impact on performances of taking into account group shapes. The control parameter λ of Eq. 7 can be seen as a trade-off between how correct the solutions must be (according to the training data) and how valuable are to be considered the structured patterns within those solutions. In particular, Fig. 6 shows the accuracy trend obtained by varying λ in the stu003 dataset, with a significant improvement at $\lambda = 0.8$. Table 1 highlights the performance gain is more relevant in this particular dataset as the density of the crowd doesn't allow the feature alone to easily separate groups, suggesting it is at this level that shapes become actually incisive. Moreover we compare our results with state-of-the-art methods [6][7] and the quantitative results shown in Tab. 1 indicate that our method outperforms all the approaches in most of the proposed videos. This is due to the use of a sociological feature, supervised learning able to better generalize to previously unseen scenarios and a specifically designed loss function. Our method takes about 1 second to cluster 10 seconds of observed trajectories in an averagely crowded scene. Fig. 5 reports a visual example of the classifier solutions.

5   Conclusions

We proposed a method for detecting small groups of pedestrian in crowd by employing supervised structured learning and sociological theories. In particular,

---

[1] See more video examples at http://imagelab.ing.unimore.it.

we devised a method to encode common groups shapes as a-priori knowledge. Results prove the effectiveness of adopting a social perspective on the task as our method outperforms current state of the art work. This project was supported by Modena local police and Softech ICT center of Regione Emilia Romagna.

References

1. Moussaïd, M., Perozo, N., Garnier, S., Helbing, D.: The Walking Behaviour of Pedestrian Social Groups and Its Impact on Crowd Dynamics. In: PLoS ONE, vol. 5 (2010)
2. Pang, S.K., Li, J., Godsill, S.J.: Models and Algorithms for Detection and Tracking of Coordinated Groups. In: Aerospace Conference, pp. 1–17 (2008)
3. Feldmann, M., Fränken, D., Koch, W.: Tracking of Extended Objects and Group Targets Using Random Matrices. IEEE Trans. Signal Processing 59 (2011)
4. Lau, B., Arras, K., Burgard, W.: Multi-Model Hypothesis Group Tracking and Group Size Estimation. I. J. Social Robotics 2, 19–30 (2010)
5. Lin, W.C., Liu, Y.: A Lattice-Based MRF Model for Dynamic Near-Regular Texture Tracking. PAMI 29, 777–792 (2007)
6. Pellegrini, S., Ess, A., Van Gool, L.: Improving Data Association by Joint Modeling of Pedestrian Trajectories and Groupings. In: Daniilidis, K., Maragos, P., Paragios, N. (eds.) ECCV 2010, Part I. LNCS, vol. 6311, pp. 452–465. Springer, Heidelberg (2010)
7. Yamaguchi, K., Berg, A.C., Ortiz, L.E., Berg, T.L.: Who Are You with and Where Are You Going? In: CVPR, pp. 1345–1352 (2011)
8. Ge, W., Collins, R.T., Ruback, R.B.: Vision-Based Analysis of Small Groups in Pedestrian Crowds. PAMI 34, 1003–1016 (2012)
9. Rodriguez, M., Laptev, I., Sivic, J., Audibert, J.-Y.: Density-Aware Person Detection and Tracking in Crowds. In: ICCV, pp. 2423–2430 (2011)
10. Hall, E.T.: The Hidden Dimension. Doubleday (1966)
11. Finley, T., Joachims, T.: Supervised Clustering with Support Vector Machines. In: ICML, pp. 217–224 (2005)
12. Calderara, S., Cucchiara, R.: Understanding Dyadic Interactions Applying Proxemic Theory on Videosurveillance Trajectories. In: CVPRW, pp. 20–27 (2012)
13. Solera, F., Calderara, S., Cucchiara, R.: Structured learning for detection of social groups in crowd. In: Proc. of Advanced Video and Signal-Based Surveillance (2013)
14. Bansal, N., Blum, A., Chawla, S.: Correlation Clustering. Machine Learning 56, 89–113 (2004)
15. Tsochantaridis, I., Hofmann, T., Joachims, T., Altun, Y.: Support Vector Machine Learning for Interdependent and Structured Output Spaces. In: ICML (2004)
16. Bandini, S., Gorrini, A., Manenti, L., Vizzari, G.: Crowd and Pedestrian Dynamics: Empirical Investigation and Simulation. In: Proc. of the Measuring Behavior (2012)
17. Cardie, C., Wagstaff, K.: Noun Phrase Coreference As Clustering. In: Proc. of the Empirical Methods in NLP and Very Large Corpora (1999)
18. Vilain, M., Burger, J., Aberdeen, J., Connolly, D., Hirschman, L.: A Model-Theoretic Coreference Scoring Scheme. In: Conf. on Message Understanding (1995)
19. Lee, C., Jang, M.-G.: Fast Training of Structured SVM Using Fixed-Threshold Sequential Minimal Optimization. Etri Journal 31, 121–128 (2009)
20. Lerner, A., Chrysanthou, Y., Lischinski, D.: Crowds by Example. Computer Graphics Forum 26, 655–664 (2007)

Evaluation of Statistical Features
for Medical Image Retrieval

Cecilia Di Ruberto and Giuseppe Fodde

Department of Mathematics and Computer Science, University of Cagliari,
via Ospedale 72, 09124 Cagliari, Italy
dirubert@unica.it, giufodde@libero.it

Abstract. In this paper we present a complete system allowing the classification of medical images in order to detect possible diseases present in them. The proposed method is developed in two distinct stages: calculation of descriptors and their classification. In the first stage we compute a vector of thirty-three statistical features: seven are related to statistics of the first level order, fifteen to that of second level where thirteen are calculated by means of co-occurrence matrices and two with absolute gradient; finally the last eleven are calculated using run-length matrices. In the second phase, using the descriptors already calculated, there is the actual image classification. Naive Bayes, RBF, Support VectorMachine, K-Nearest Neighbor, Random Forest and Random Tree classifiers are used. The results obtained applying the proposed system both on textured and on medical images show a very high accuracy.

Keywords: texture, feature extraction, feature selection, classification, medical image analysis.

1 Introduction

Texture analysis is a process that allows the characterization of different surfaces and objects through the identification of their specific statistical properties. Through rigorous techniques of image capture, you can get a texture concerning a certain surface that uniquely identifies its structure depending on the lighting and the intensity captured during acquisition. From this you can extract characteristics or features that allow to describe a texture image, through an appropriate mathematical formulation. In this work we propose a complete system that allows the classification of medical images in order to detect possible diseases present in them. The proposed method is developed in two distinct stages: extraction of descriptors and classification. In the first stage we extract thirty-three features from the image: seven are related to statistics of the first level order, fifteen to that of the second level (thirteen are calculated using the co-occurrence matrices and two by the absolute gradient) and, finally, the last eleven are calculated using the run-length matrices. In the second phase, instead, using the descriptors already calculated, there is the actual image classification by using different classifiers. The results obtained from the proposed system are

A. Petrosino (Ed.): ICIAP 2013, Part I, LNCS 8156, pp. 552–561, 2013.

encouraging and show that the analysis carried out both on textured and on medical images lead to have a high accuracy. The rest of paper is organized in the following way. In section 2 there is a texture analysis techniques overview. Section 3 describes the proposed system. In section 4 there is a comparison with other methods and the experimental results. Finally, in section 5 there are the conclusions and possible future works.

2   Related Works

In literature there are several methods for the extraction of features from medical images, each of them based on a different type of texture. In addition, most of them are only concerned with the extraction of the feature of texture. In the field of Image Processing the term texture refers to any and repetitive geometric arrangement of the gray levels of an image. The texture provides important information about the spatial arrangement of the gray levels and the relationship with their surroundings. The human visual system determines and recognizes easily different types of texture characterizing them in a subjective manner and, although for a human observer it is simple and intuitive to associate a surface with a particular texture, to give a rigorous definition for this is very difficult. In fact, there is no general definition of texture and a methodology for measuring the texture acceptable to all. Typically you only use qualitative definitions to describe texture with attributes as coarse, granular, random, ordered, filiform, punctuating, fine-grained, etc. It can easily guess that the quantitative analysis of texture passes through statistical and structural relations among the basic elements (the texels) of what we call just texture.

The analysis of texture has three main aspects: classification, segmentation and shape from texture. Classification concerns the search for particular regions of texture among different predefined classes of texture. The classification of textures is carried out using statistical methods that define the descriptors of the texture, i.e. from the images in gray level, a measure of the characteristics of texture is evaluated through co-occurrence matrices, run-length matrix, contrast, homogeneity, entropy, etc. This type of statistical approach is particularly suitable when the texture is made up of very small and complex elementary primitives, typical of microstructures. Segmentation determines the boundaries between regions with different textures. Segmentation occurs when the previous statistical approaches do not provide accurate measures of the textural characteristics and therefore are insufficient to characterize the texture of a region. In addition, the segmentation is indispensable when you do not have any knowledge nor on the number of classes of texture features nor on a priori characteristics. Shape from texture, instead, is essential for the reconstruction of the surface of objects starting from the information associated with the macrostructure, such as density, size and orientation.

There are various methods for feature extraction and texture analysis and one very important is the one that makes use of the statistics of various orders useful for our purpose. The statistical approach is particularly suitable for the analysis of microstructures. The elementary primitives of the microstructures are determined by analyzing the characteristics of the texture associated with a few pixels of the image. The approaches of this type comprise three different orders: first, second and higher order. *Statistics of first order* measure the likelihood of observing a gray value in random position in the image. The statistics of the first order can be calculated from the histogram of the image gray levels. This depends only on the single gray level of the pixel and not on the interaction co-occurring with the pixels of the surroundings, thus it does not depend on the absolute position and reciprocal pixels. It is ascertained experimentally that this type of statistics is the most strong, stable and representative of an image. *Statistics of second order* are defined as the likelihood of observing a pair of gray levels measured at the ends of a segment randomly positioned in the image with a random orientation. The scientist Julesz [2] proposed the theory of textons to explain the early mechanisms that occur in the brain in discriminating two pairs of regions with different textures. The textons are visual events (i.e. collinearity, terminations, closures, etc..) whose presence is detected and used to discriminate the texture. Terminations are the heads of a segment or an angle. This category includes all those statistics to determine the mutual correlation between the gray levels. *Statistics of higher order* involve the analysis of higher-order statistics information. This type of statistics contains information regarding the runs of gray levels, which make up the image matrix. This information concerns to the size of these runs, i.e. their lengths. It is possible to define also the sets of consecutive pixels that all have the same level of gray and then go to make up a run.

Image analysis techniques have played an important role in various medical applications. In general, applications involve the automatic features extraction from the images which are then used for classification, such as the distinction of normal or healthy tissue from abnormal or sick one. Depending on the particular classification, the features extracted capture morphological, color or texture properties of the image. The properties of calculated texture are closely related to the application domain to be used. For example, Sutton and Hall [3] discussed the classification of lung diseases by examining texture features. Some diseases, such as interstitial fibrosis affecting the lungs, lead to have change of textures in the radiographic images, on the contrary of the lesions that are clearly delineated. In such applications, the methods of textures analysis are ideal for such images. Sutton and Hall [3] proposed three types of texture features to distinguish the sick pulmonary tissue from the healthy one. These features were calculated on the basis of a measure of isotropic contrast, a measure of directional contrast and a sampling of Fourier energy domain. In their experiments of classification, the best results were obtained using the directional contrast measure. Harms [4] used the image texture combined with the features of color to diagnose leukemic malignant tumors in samples of blood cells. He extracted particular texture features, such as the "textons" (regions with nearly uniform color), and analyzed the related features, such as the total number of pixels of textons having a specific color, the average radius and the size for each color and various shape features of textons. The features of texture, combined with the color, have increased the percentage of correct classification of type of cell in the blood compared to the classification done using only color. Landeweerd and Gelsema [5] extracted various statistics of the first order (as the average gray level in a region) and second order (as arrays of co-occurrences of gray levels) to differentiate the different types of white blood cells. Insana [6] used the features of texture of ultrasound images to estimate the parameters of dispersion of the tissue. He made a meaningful use both of knowledge about the physics used in ultrasound images and of characteristics of the tissue to design the model of the texture. Chen [7] used the fractal texture features for classifying the ultrasonic images of the liver and for improving the analysis X-rays of the chest. Lundervold [8] used fractal texture features in combination with other functions to analyze the ultrasound images of the heart (the ultrasound images are temporal sequences of the left ventricle of the heart).

3   The Proposed System

Let us now explain the system designed to perform the texture analysis of medical images. The method implemented is divided into two main parts: in the first we have the extraction of texture features while in the second part we have the image classification using the descriptors obtained in the first phase. After choosing to analyze the entire image or only a part of it, we calculate thirty-three features: run-length matrices (eleven features), absolute gradient (two features), first order statistics (seven features) and co-occurrence matrices (thirteen features). This first part is implemented using the Matlab development tool. In the second step, we classify the data previously obtained: through the Weka environment, in fact, we build the dataset containing the calculated descriptors. These data are classified using various types of classifiers. We also use a feature selection, as sometimes the obtained accuracy results are not satisfactory, while in other cases the feature selection do not improve the results already obtained. We can see graphically the system diagram in Fig. 1.

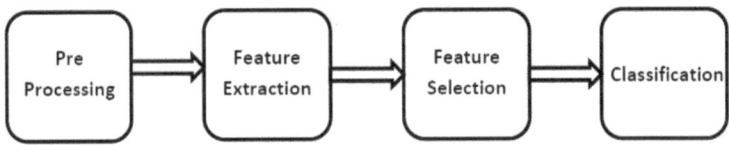

Fig. 1. Proposed system diagram

3.1   Pre-processing

In this first phase, we need sometimes to resize and crop the image manually in order to isolate the regions of interest from medical images and to eliminate portions of them not related to human tissue but to the instruments used during the mammography generation.

3.2   Features Extraction

The first features (higher-order statistics) to be analyzed are related to arrays of run lengths (GLRLM), from which we obtain eleven features: Short Run Emphasis, Long Run Emphasis, Gray Level Nonuniformity, Run Length Nonuniformity, Run Percentage, Low Gray-Level Run Emphasis, High Gray Level Run Emphasis, Short Run Low Gray-Level Emphasis, Short Run High Gray-Level Emphasis, Long Run Low Gray-Level Emphasis and Long Run High Gray-Level Emphasis. After the run-lengths matrices, we calculate other two descriptors, still related to higher order statistics, those of the absolute gradient: mean and variance. Other implemented descriptors are the seven statistics of the first order: mean, standard deviation, entropy, kurtosis, uniformity, smoothness and skewness. The latest features (statistics of the second order) are related to the co-occurrence matrices (GLCM) : contrast, angular second moment, correlation, entropy, variance, homogeneity, sum average, sum variance, sum entropy, difference variance, difference entropy, two information measures of correlation, as proposed by Haralick [9].

3.3   Feature Selection

This phase belongs to the second part of our system when, using the descriptors calculated, we have the step of classification of real images. During this step, we select the most predictive attributes based on their correlation with the attribute class (positive correlation) in descending order. The most frequently features removed from feature selection, both for Brodatz's images and for DDSM database, are: contrast, correlation, homogeneity, difference variance, difference entropy, Info measure correlation 1 and 2 from GLCM and Gray Level non Uniformity and Run Percentage from GLRLM. This part is introduced why during the testing phase we have achieved not very satisfactory accuracy results. It was then decided to try to improve these results adopting this technique that, in some cases, has led to an improvement of the initial accuracy up to 10% more. In few cases, however, the feature selection has not brought improvements to the results already obtained.

3.4   Classification

The phase of classification is the latest one. In fact, after the descriptors extractions and the feature selection, then we have to choose the classifiers that in our approach are six: SVM, RBF, Random Tree, Random Forest, Nave Bayes and k-NN. For this stage we decide to use the technique of ten fold cross-validation where the original dataset is divided into subsets each composed by the same number of samples (in all our cases ten). The data are first classified by analyzing the original dataset and by applying the classifiers listed above, and then by making the feature selection on the original dataset and by proceeding with the application of the various classifiers.

4   Experimental Results and Comparisons

The implemented system is applied to two different types of images. The first type includes all the images taken from Brodatz album [1]. Texture is fundamental for analyzing medical tissue and methods of texture analysis are widely studied, confirming the utility and the goodness in classifying a human tissue as healthy or sick. Therefore, we also apply our system to a medical database, the DDSM (Digital Database for Screening Mammography) [10]. We test the performance of our system by comparing the accuracy value obtained by our classification and that presented by eight alternative methods. We analyze the same subset of images (from Brodatzs album) proposed in each paper. The accuracy values, that is a measure of how many images are correctly classified after feature selection, are obtained using six different classification models, as described in section 3. In all the cases, the feature selection never decreases the accuracy level obtained without it. Finally, the best result is compared with the accuracy value presented in each considered alternative method in classifying the same set of images. Let's show the comparison results for each of them.

4.1   Brodatz Album

In this experiment, we use 37 classes of textures of Brodatz album from which we create non-overlapping sub-images. We compare the performance of our system on the same set of Brodatz textures analized in each considered alternative method, each time creating a sub-set of textures ad hoc for each of them. Naive Bayes, RBF, Random Tree, Random Forest, k-Nearest Neighbors and Support Vector Machine classifiers are used.

In the paper [11] for classification of Brodatz textures five feature sets are calculated: Sobel edge detector, discrete cosine transform (DCT), speeded up robust features (SURF), gray level run length matrix and the eigenvalues. The classification is performed using decision trees, random trees and support vector machines. The first set of images (Set 1) is composed by the following classes: D1, D5, D6, D15, D21, D41, D49, D60, D67, D74, D82, D87, D94, D102, D109. The average accuracy is 68.36% and the best result is 84.44% obtained with the SVM. In our work the best result is 100% obtained with the KNN. Paper [12] presents a new feature selection scheme that automatically determines a reduced subset of methods whose integration produces classification results comparable to those obtained when all the available methods are integrated but with a significantly lower computational cost. The second set of image (Set 2) is composed by: D3, D15, D32, D37, D41, D54, D91, D94. The average classification is 91.13% and the best result is 96.8% obtained with all methods. In our work the best result is 100% obtained with Random Forest and KNN. In the paper [13] a combined statistical and structural approach is used. The spectrum of image is used for unsupervised texture classification. The third set of images (Set 3) is composed by two classes: D84 and D112. An average correct classification of 96% is obtained. In our work the best result is 100% obtained with all classifiers. In the paper [14] features computed as statistics (e.g. histograms) of local filters responses are reported as the most powerful descriptors for texture classification and segmentation. The set of images (Set 4) is composed by: D3, D4, D6, D21, D24, D49, D68, D71, D82, D87. The average classification is 89.86%. In our work the best result is 93.5% obtained with the KNN. In the paper [15] texture images are splitted into 64 non-overlapping sub-images and then these sub-images are decomposed through wavelet transformation to obtain sub-band images. These sub-band images are further used to extract statistical texture features. For images classification neural networks are used. The set of images (Set 5) is composed by: D9, D12, D15, D19, D21, D24, D29, D30, D112. The average classification is 92,58% and the best result is 100% obtained with the MRBF. In our work the best result is 100% obtained with KNN. In the paper [16] the main focus is to do texture segmentation and classification: GLCP method is used to extract features from texture images and GSVM has been proposed to do classification on extracted features. The set of images (Set 6) is composed by: D4, D6, D57, D64. The average classification is 95,13% and the best result obtained is 98,49% with GSVM. In our work the accuracy varies from 74,2% to 100% and the best result is 100% obtained using all classifiers except the Random-Tree. In the paper [17] a set of invariant descriptors of each image is extracted. These descriptors are vector-quantized from key-points. PLSI and NMF to perform unsupervised classification are used. The set of images (Set 7) is composed by: D35, D64, D74, D99. The average classification is 62,92%. In our work the accuracy varies from 95% to 100% where the best result is 100% obtained using all classifiers except the Random-Tree and Random-Forest. The paper [18] describes the usage of wavelet packet neural networks (WPNN) for texture classification problem. The proposed schema is composed of a wavelet packet feature extractor and a multi-layer perceptron classifier. Set 8 is composed by: D9, D12, D15, D16, D19, D28, D54, D68, D94, D112. The overall success rate is about 95%. In our work the success rate varies from 83% to 98,6% obtained with KNN and RBF.

The best accuracy values obtained by our system and those of the methods present in literature are reported in table 1. As it can be seen from the results shown in the table, our method leads to a very high accuracy in the classification of textured images.

4.2   Digital Database for Screening Mammography (DDSM)

This database [10] contains a series of mammography screenings stored in four different categories: Normal, Cancer, Benign, Benign Without Callback. For each of these categories we use 20 different images, for a total of 80 images to classify. The classifiers are the same used with Brodatz images. Again, we first make a classification with the original data, then we refine it by using a feature selection. The process starts downloading from the archive some images for each case contained in DDSM. More specifically, eighty pictures are downloaded in total, twenty for each case. The feature extraction are made both on the entire mammographic image and on a cropped sub-image after eliminating superfluous

Table 1. The best accuracy values on Brodatz textures achieved by our method and others: Set 1,..., Set 8 are the sets of textures analysed by each method.

Images	Our System		Other Methods	
	Classifier	*Accuracy*	*Classifier*	*Accuracy*
Set 1	KNN	100%	SVM[11]	84.44%
Set 2	R-Forest	100%	All Methods[12]	96.8%
Set 3	All Class.	100%	AVG[13]	96%
Set 4	KNN	93.5%	WAVELET[14]	96.9%
Set 5	KNN	100%	MRBF[15]	100%
Set 6	SVM	100%	GSVM[16]	98.49%
Set 7	RBF	100%	PLSI[17]	64.46%
Set 8	KNN	98.6%	WPNN[18]	95.7%

background (see Fig. 2). The first images analyzed are those related to the "normal" case, followed by "cancer", "benign" and finally "benign without callback" cases. Once calculated values for the first two cases, the results obtained are used to create a first dataset consisting of only two classes ("normal" and "cancer"), to see the behavior of the system in the diagnosis of possible diseases. The other dataset is composed proceeding with the calculation of descriptors for the last two cases and integrating them to the previous one, going to get a multi-class classification in which there are all the four categories listed above. So, we have two datasets: one composed by forty images and only two classes and the other one composed of eighty images and four classes.

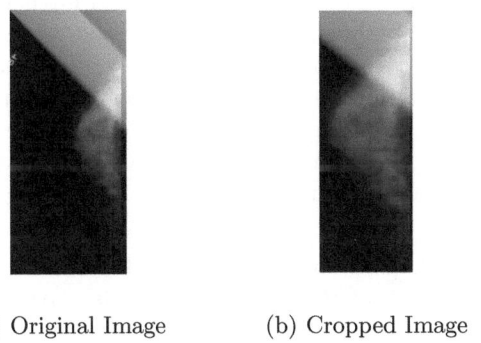

(a) Original Image   (b) Cropped Image

Fig. 2. A mammography image example from DDSM [10]

In [10] the classification is conducted by CAD algorithms where the success rate is about 95%. Our best classification rate is 98,5% obtained by Random-Forest. The best accuracy values obtained by our system are reported in table 2.

Table 2. The best accuracy values (Acc) on mammography images (without Feature selection (FS) and with Feature Selection) for the relative classifier (Class)

	Original Images			Cropped Images			
	Without FS		With FS		Without FS		With FS
Classes	*Class*	*Acc*	*Class*	*Acc*	*Class*	*Acc*	*Class*	*Acc*
2 classes	RBF	92.6%	SVM	92.6%	KNN	93.5%	R-Tree	97.1%
4 classes	KNN	81.4%	KNN	81.4%	KNN	96.2%	R-Forest	98.5%

5 Conclusions and Future Works

The purpose of this work is to combine various statistical methods for features descriptors calculation in order to classify images containing different textural information. We combine together the various statistics of first, second and higher level order to see if a combined application of these methods is more discriminating than the application both of a different approach and similar methods. The main aim is to look for a system that is able to find in medical images various possible diseases through a texture analysis quickly. The results obtained in different experimental phases show how the implemented system is able to massively discriminate the textures of the images analyzed going to track pathologies within a diseased tissue. The rates of accuracy are just fine, but we achieve further refinements through feature selection. So, the implemented method is able to identify various diseases quickly and very satisfactorily. This work could be further developed going to implement and test other methods for texture analysis as the wavelet transforms, the Gabor filters and the Fourier spectrum and to see so if the addition of new descriptors allows a further discrimination compared to that already obtained, thus to increase the values of accuracy already achieved. In addition, it is also necessary to increase the size of the dataset to allow a better classification, without maybe having feature selection. Finally, given the encouraging experimental results, of course our future interests and possible developments will include a detailed analysis of a possible dependance analysis among the chosen features.

Acknowledgments. This work has been funded by Regione Autonoma della Sardegna (R.A.S.) Project CRP-17615 DENIS: Dataspace Enhancing Next Internet in Sardinia.

References

1. Brodatz, P.: Texture: A photographic album for Artists and Designers. Dover Pubblications (1966)
2. Julesz, B.: Textons, the Elements of Texture Perception, and Their Interactions. Nature 290, 91–97 (1981)
3. Sutton, R., Hall, E.L.: Texture Measures for Automatic Classification of Pulmonary Disease. IEEE Transactions on Computers C-21, 667–676 (1972)

4. Harms, H., Gunzer, U., Aus, H.M.: Combined Local Color and Texture Analysis of Stained Cells. Computer Vision, Graphics, and Image Processing 33, 364–376 (1986)
5. Landeweerd, G.H., Gelsema, E.S.: The Use of Nuclear Texture Parameters in the Automatic Analysis of Leukocytes. Pattern Recognition 10, 57–61 (1978)
6. Insana, M.F., Wagner, R.F., Garra, B.S., Brown, D.G., Shawker, T.H.: Analysis of Ultrasound Image Texture via Generalized Rician Statistics. Optical Engineering 25, 743–748 (1986)
7. Chen, C.C., Daponte, J.S., Fox, M.D.: Fractal Feature Analysis and Classification in Medical Imaging. IEEE Transactions on Medical Imaging 8, 133–142 (1989)
8. Lundervold, A.: Ultrasonic Tissue Characterization - A Pattern Recognition Approach, Technical Report. Norwegian Computing Center Oslo Norway (1992)
9. Haralick, R.M.: Statistical and Structural Approaches to Texture. IEEE Proceedings 7, 786–804 (1979)
10. Hearth, M., Bowyer, K., Kopans, D., Moore, R., Kegelmeyer, W.P.: The digital database for Screening Mammography. Medical Physics Publishing, 212–218 (2001)
11. Desager, C., Geerts, S., Van der Schueren, F., Ledda, A.: Texture Analysis and Classification of Brodatz Textures using ESAC Framework. The E-Lab Masters Theses 2010–2011 (2011)
12. Puig, D., Garcia, M.: Automatic texture feature selection for image pixel classification. Pattern Recognition 39, 1996–2009 (2006)
13. Umarani, C., Ganesan, L., Radhakrishnan, S.: Combined Statistical and Structural Approach for Unsupervised Texture Classification. International Journal of Imaging Scienze and Engineering, 162–165 (2008)
14. Karoui, I., Fablet, R., Boucher, J.M., Pieczynski, W., Augustin, J.M.: Fusion of textural statistics using a similarity measure: application to texture recognition and segmentation. Pattern Analysis & Applications 11, 425–434 (2008)
15. Chuan-Yu, C., Shih-Yu, F.: Image Classification using a Module RBF Neural Network. In: Proceedings of the First International Conference on Innovative Computing, Information and Control (2006)
16. Hee-Kooi, K., Hong-Choon, O., Ya-Ping, W.: Image Texture Classification using Combined Grey Level Co-occurrence Probabilities and Support Vector Machines. In: Proceedings of Fifth International Conference on Computer Graphics, Imaging and Visualization, pp. 180–184 (2008)
17. Lei, Q., Qingfang, Z., Shuqiang, J., Qingming, H., Wen, G.: Unsupervised texture classification: Automatically discover and classify texture patterns. Image and Vision Computing 26, 647–656 (2008)
18. Sengur, A., Turkoglu, I., Ince Cevdet, M.: Wavelet packet neural networks for texture classification. Expert Systems with Applications 32, 527–533 (2007)

Hierarchical Image Representation Simplification Driven by Region Complexity

Petra Bosilj[1], Sébastien Lefèvre[1], and Ewa Kijak[2]

[1] Univ. Bretagne-Sud, UMR 6074, IRISA, F-56000 Vannes, France
{Petra.Bosilj,Sebastien.Lefevre}@irisa.fr
[2] Univ. Rennes I, UMR 6074, IRISA, F-35000 Rennes, France
Ewa.Kijak@irisa.fr

Abstract. This article presents a technique that arranges the elements of hierarchical representations of images according to a coarseness attribute. The choice of the attribute can be made according to prior knowledge about the content of the images and the intended application. The transformation is similar to filtering a hierarchy with a non-increasing attribute, and comprises the results of multiple simple filterings with an increasing attribute. The transformed hierarchy can be used for search space reduction prior to the image analysis process because it allows for direct access to the hierarchy elements at the same scale or a narrow range of scales.

Keywords: hierarchical representation, tree structures, image filtering, segmentation and grouping, image region analysis.

1 Introduction

Hierarchical image representations have been used in applications as diverse as object detection [22], video segmentation [11], image segmentation and filtering [1,18], image simplification [16] and image compression [11]. They encode the composition of complex image structures by proposing the unions between simple image regions most likely to correspond to objects [22]. The choice of representation is application specific and is a compromise between *representation size* and *information retained by the representation elements* [10,22]. The organization these hierarchies reflects the inclusion relations between the regions and provides the means to detect and analyze objects at different levels of detail by examining the fine features present in small scale regions [10].

Among many equivalent representations used for such hierarchies (e.g. [1,5,7]), a representation by a tree structure is the direct one. Each node in such a tree corresponds to one of the regions in the representation, while the parent-child relations correspond to inclusions between regions (cf. Fig. 1).

A common property among all tree representations is that the leaf nodes represent the fine image structures, increasing in complexity with proximity to the root. The *coarseness inherent to the representation* could be defined as a distance from the node to the root of the tree. This inherent coarseness *does*

A. Petrosino (Ed.): ICIAP 2013, Part I, LNCS 8156, pp. 562–571, 2013.
© Springer-Verlag Berlin Heidelberg 2013

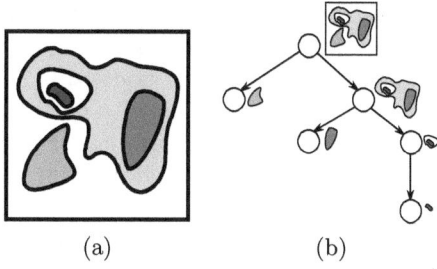

(a)   (b)

Fig. 1. The tree in (b) is a hierarchical representation of (a)

not, in the general case, accurately reflect the region complexity and can not be used to compare any two regions. But, if some coarseness measure for the objects of interest is known prior to main image analysis step, the relevant search space could be limited to structures with a similar level of coarseness.

The transformation presented hereafter assigns an external coarseness measure to all the nodes and rearranges them accordingly while preserving the hierarchical relations. New coarseness measure is chosen among increasing attributes on the tree, reflecting that the complexity of regions increases along each branch even if it can not be directly compared. The nodes of the same coarseness are pruned and at most one region of a certain coarseness per branch is kept. The result is a representation where the node levels correspond to the coarseness of the regions represented by the nodes and every tree level comprises only nodes of the same coarseness. This in turn enables limiting the search space when dealing with objects whose coarseness can be estimated by directly accessing only the regions of the relevant coarseness. Additionally, the search space is reduced even for objects with unknown coarseness, as the number of regions after the transformation can only decrease. This property makes the transformation suited for processing hierarchies that are too fine before the image analysis step.

In the next section, basic notions used throughout the article are introduced. Section 3 explains the effects of the proposed transformation, gives the algorithm and the estimation of the algorithm complexity. The approach is compared to previous works on the topic in Sect. 4, while the advantages and the potential applications of the presented algorithm are summarized in Sect. 5.

2   Properties of Tree Superclasses

In this section, we first offer a summary of some basic definitions about trees before discussing the differences between two superclasses, the *partitioning* and *inclusion* trees. Notations similar to the ones used here are used throughout literature (e.g. [14, 16]), and the reader can refer to these works for more details about directed and undirected graphs associated with image representations.

A tree $\mathcal{T} = (M, P)$ can be defined as a directed graph, where M is the vertex (node) set and P the edge set of the graph. If there exists an edge $e_{m,n}$ between two nodes m and n in $\mathcal{T}$, m is called *the parent of n* and n a *child of m*.

The nodes without children are called *leaf* nodes. The only node in the tree with no parent is called the *root*. A path $\mathcal{P}$ in $\mathcal{T}$ from n_1 to n_k is defined as $(n_1, \ldots, n_k)$ such that for all $1 \leq i < k$, the node n_i is a parent of n_{i+1}. The vertex m is called *an ancestor of n* if there exists a path $\mathcal{P}$ in $\mathcal{T}$ from m to n.

Let $C = \{n_1, \ldots, n_k\}$ be a set of nodes. C is a *cut of the tree* if every path $\mathcal{P}$ from the root to any leaf passes through a single node $n_j \in C$. The *level* of the root is the length of the longest path from the root to any of the leaves. Let n be any node in the tree except the root and l_n the length of the path from the root to n. The level of the node n is equal to the level of the root decreased by l_n. A *level of a tree* is a cut in which all the nodes have the same level.

2.1   Trees as Hierarchical Image Representations

Let I be a monochannel digital image consisting of a set of pixels, f a function that assigns to each pixel p its intensity value, and $\mathcal{G} = (V, E)$ the associated undirected graph. The vertex set V contains all the pixels while the edge set E indicates their adjacency relation (cf. [14] for more details). In a tree $\mathcal{T} = (M, P)$ corresponding to a hierarchical representation of I, every node n corresponds to a connected image region. The underlying graph $\mathcal{R}(n)$ of a connected region represented by the node n consists of a set of region pixels $V(n)$ and the edge set $E(n) = \{e_{p,q} \in E | p, q \in V(n)\}$ by the adjacency relation.

In such a representation the sets of pixels of two nodes are either disjoint, $V(n) \cap V(m) = \emptyset$, or one of the nodes is an ancestor of the other. In case the node m is the ancestor of n, the following relation holds: $V(n) \subseteq V(m)$. If m is a parent node in the tree and $n_1, \ldots, n_k$ are all the children of m, the edge set of m can, in general case, be described as follows:

$$V(m) = V(n_1) \cup \ldots \cup V(n_k) \cup S, \tag{1}$$

where S is a pixel set comprising pixels with the following property:

$$S = \{p_1, \ldots, p_l\}, l \geq 0 \tag{2}$$
$$\text{such that } \forall i \in \{1..l\}, \forall j \in \{1..k\}, \ p_i \notin V(n_j) \ .$$

Additionally, the pixel set S has to be such that the region $\mathcal{R}(m) = (V(m), E(m))$ is a *connected* region in I. Additional constraints will be introduced when describing trees from each of the superclasses. Despite the differences in inclusion order and types of regions dependent of the choice of tree, there are common properties within the superclasses, discussed hereafter.

2.2   Partitioning and Inclusion Trees

The common property among all **partitioning trees** is that the leaves correspond to regions of a fine partition of the image (e.g. initial over-segmentation [22], image flat zones [7, 10], watershed segmentation results [7]). Every cut of the tree is also a partition. Similarity between neighboring regions is either calculated based on a global similarity measure between region models (e.g. binary partition tree [22]) or depends on local measures between neighboring pixels on the region borders (e. g. α-tree [7,9,16], alternate local measures [2,17]). More similar regions are merged earlier in the construction process.

We will presume in this article that the tree is constructed *with no redundancies*, i.e. no two nodes represent the same region of the image I (the term is also used in [9] in the context of α-trees). For partitioning trees, this means that every inner node has strictly more than 1 child. If, due to the construction process, the same region appears multiple times in the hierarchy, we will simply say that the node representing it appears on multiple levels without duplicating the node. Instead of just a level, we assign a *level range* [*lMin*, *lMax*] to the node. *lMin* and *lMax* are then defined as the lowest levels in the tree on which the node does and does not appear on, respectively. The maximal level of a node is equal to the minimal level of its parent. The root node is the exception, existing only on the maximal level present in the tree. Representing a tree with redundancies as a tree without them results in no loss of information (illustrated in Fig. 2).

Since the leaf nodes already form an image partition, no new pixels are added when constructing a parent node. Any inner region is a union of at least two of its children. Partitioning trees with no redundancies can be formalized by adding the constraint $k > 1$ in (1) and $l = 0$ in (2).

For **inclusion trees**, no explicit measure defining the merging order is present. Leaves in these trees are either image intensity maxima, minima, or both, and do not form a partition of the image. *Level sets* [4,19] of an image are obtained by thresholding the image, and looking at connected components formed by pixels p with intensities $f(p)$ above (for *upper* level sets) or below the threshold (for *lower* level sets). Inclusion trees are built based on inclusion relations between the level sets of the same type (min- and max-trees [3,6]) or of both types (level line trees [4,19,21], dual-input max-trees [8]).

We also presume no redundancies in the inclusion trees, which in this context means that a region not represented by a leaf node can only be formed by adding, to an existing region (or regions), the pixels of one or more flat zones not included in the region(s) yet. *Flat zones* [13] of an image I are connected regions of same intensity pixels with maximal size. The set of pixels of an inner node can not be formed by merging multiple child regions without adding new pixels. All the construction algorithms for commonly used inclusion hierarchies produce inclusion trees with no redundancies as their output [4,6,19]. The only constraint to add to the general definition of tree hierarchies is $l > 0$ in (2).

3   The Proposed Transformation

The algorithm for imposing an external coarseness measure on a tree representation of an image without redundancies is presented in this section. The output is also a tree representation with no redundancies, whose levels comprise nodes with the same value of the coarseness measure.

The first assumption of the algorithm, explained already in Sect. 2.2, is that the tree representation is without redundancies. The second condition pertains

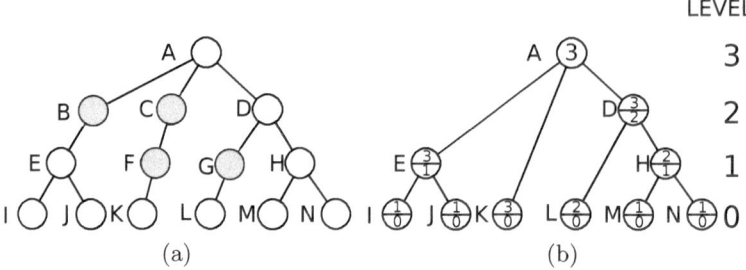

Fig. 2. The tree shown in (a) has redundant nodes, marked in gray. The levels of the nodes are displayed on the right of the trees. By removing the redundancies we get the tree shown in (b) (the level range $[lMin, lMax)$ is displayed inside the nodes).

to the attribute $K(\cdot)$ used as a new coarseness measure for regions. An attribute $K(\cdot)$ is said to be *increasing* [14] on a tree if the attribute value of a parent m node is always greater than or equal to the attribute values for all its children n_i: $K(m) \geq K(n_i)$. The coarseness measure used must be an increasing attribute (discussed in Sect. 1), and the algorithm assumes that the values of this increasing attribute were assigned to the nodes of the tree *before the transformation*. Many interesting attributes (e.g. intensity range, component area) can be assigned to nodes directly during tree construction. More attribute possibilities can be found in [12, 15], but the final choice always depends on the intended application and known properties of object of interest and image domain.

Transformation results can be interpreted as a hierarchy formed by stacking, for threshold values ranging from zero to maximal value of the attribute, the leaf nodes of trees obtained by performing attribute filtering [12] on the original tree with an increasing attribute. The result is a tree representation of this hierarchy with no redundancies. A node present as a leaf in the hierarchy after an attribute filtering with a threshold T will have T included in its level range in the result. After the transformation, the tree cuts stacked to produce a new tree can be directly accessed. This definition extends easily to attributes that take continuous values, where the node can belong to a continuous range of levels.

3.1   The Algorithm

Assuming no redundancies and that the attribute values are already calculated for every node, the algorithm can be described in very simple terms: in a bottom-up traversal of the tree, if we discover a node with an attribute value equal to the attribute value of its parent, we should add all the child-nodes of this node to the children set of its parent, and then delete the node. This is summarized in Algorithm 1. The attribute value assigned to a node in the original tree becomes the minimal level of that node if the node is kept after the transformation. The tree before and after the transformation is shown in Fig. 3(a) and 3(b).

If the tree is stored in the straightforward way, the memory requirements are proportional to number of image pixels (c.f. Sect. 3.2). Highest cut of the

```
1  function rearrangeTree(Node):
2      foreach Child ∈ Node.children do
3      └ rearrangeTree(Child)
4      if Node.attributeValue = Node.parent.attributeValue then
5      │   add Node.children to Node.parent.children
6      └   delete Node
7      else
8      │   Node.minLevel ← Node.attributeValue
9      └   Node.maxLevel ← Node.parent.attributeValue
```

Algorithm 1. The proposed transformation

tree comprising nodes with coarseness lower or equal to the desired level is then selected by performing a top-down traversal of the tree and keeping the first node in each branch with satisfying coarseness level. Memory requirements rise if we want faster access. For each level of the tree we store a pointer to the left-most node and, for each node and each level in the level range of the node, the first next node at that level in the tree. Figures 3(b) and 3(c) illustrate the information which needs to be stored to enable direct access to any tree level. Once the first node in a level of a tree is accessed, the pointers to the next nodes can be followed to access all the nodes of that level.

3.2   Complexity Analysis

From the pseudocode, it is visible that the algorithm is linear in the number of nodes in the tree. In order to estimate the complexity of the algorithm, we must estimate the maximum number of tree nodes in a tree $\mathcal{T} = (M, P)$ with no redundancies, compared to the number of pixels N in the image I with the corresponding graph $\mathcal{G} = (V, E)$, where $N = \mathrm{card}\,(V)$. The estimation is done separately for partitioning and inclusion types of trees.

The maximum number of leaves in a **partitioning tree** is achieved if the elements of the finest partition represented by the tree are pixels. Since the tree has no redundancies, every inner node in the tree has at least 2 distinct child nodes. The number of inner nodes is less or equal than that of a binary tree with the same number of leaves, and never exceeds the number of leaves in the tree. There are at most N leaves, so the maximum number of nodes in any partitioning tree is lower than $2N$.

The maximum number of nodes in an **inclusion tree** is achieved if every node adds the smallest possible number of pixels, i.e. 1 pixel, to the region it represents. As there are no overlaps between the regions of different branches, each pixel in the image can be added to the set of region pixels exactly once. Regardless of the number of leaf nodes in the inclusion tree and given that there are N pixels in the image, there can be at most N nodes in any inclusion tree.

Considering that the number of nodes in any tree image representation without redundancies is linear in the number of image pixels and the transformation

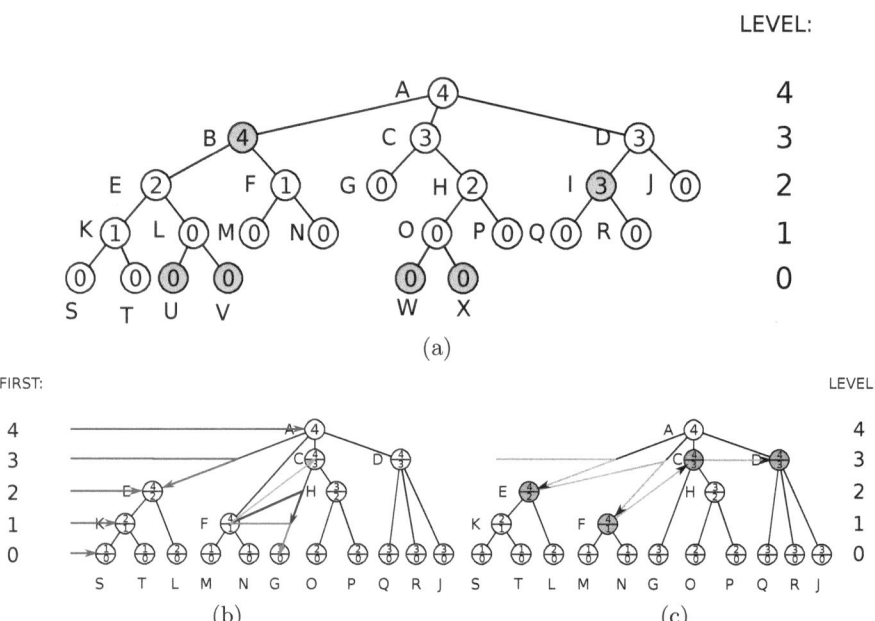

Fig. 3. Subfigure (a) shows the original tree before the transformation, with attribute values displayed in the nodes and the nodes to be removed highlighted in green. In both subfigures (b) and (c), the level range is displayed inside the nodes. Subfigure (b) shows pointers to beginning of every tree level, needed for direct access to levels. Entries like the one shown for node F keep the next node for every level $[1, 4\rangle$ (level 1: G (purple), 2: G (blue), 3: C (green)) and need to be stored for every node. Subfigure (c) shows accessing the third level of the tree using the stored pointers.

algorithm is linear in the number of tree nodes, we can conclude that the complexity of the algorithm is linear in the number of image pixels, $O(N)$. The overall complexity of producing such a transformed tree from the original image depends on the choice of the original underlying tree, the complexity of the construction algorithm for the chosen tree type and additional costs (if any) of calculating node attribute values.

4   Similar Approaches

The search through the literature for alternate approaches to selecting the coarsest tree-cut satisfying some constraint found such attempts only in the domain of partitioning trees. Imposing constraints on components of another partitioning hierarchy in a way that the hierarchical relations between the remaining components are preserved was first explored in [15]. The work in [15] only provides the definitions of constrained components and the potential applications while the algorithm for selecting such constrained components is not proposed.

The approaches presented in [5, 9] were demonstrated on α-trees, but are applicable to all partitioning trees. Due to the definition of α-connectivity, the α-connected components suffer from the chaining effect [17,18]. This effect causes some nodes close to the leaves of the tree to be complex regions instead representing the fine details of the image. Because of this, the concept of constrained connectivity was introduced [16] together with the notion of an (ω)-hierarchy by imposing global component range constraints on α-connected components. This component range constraint can be viewed as an external coarseness measure.

In [9], the approach to extract just one level of the (ω)-hierarchy at a time is presented. Their approach for creating the desired partition is equivalent to the process of attribute filtering (with an increasing attribute) [12] on the hierarchical image representation, but they only use the leaf nodes of the filtered hierarchy. The algorithm presented herein was inspired by the idea from [9] and adopted its bottom-up approach to tree traversal. The difference is that stopping the tree traversal as soon as the attribute values are above the chosen threshold, as in [9], returns only one partition, while continuing the traversal to the root of the tree as proposed here builds a structure containing the information on partitions for every possible threshold value. The algorithm presented here can be compared to the direct rule of simplifying the tree with a non-increasing criterion [11]. The criterion for keeping the node is that the attribute value assigned to it is strictly smaller than that of its parent.

The main contribution of [5] is a representation of partitioning tree hierarchies in the framework of edge-weighted graphs, proving that any partitioning hierarchy can be represented by an *ultrametric watershed*. They present an algorithm to calculate the ultrametric watershed representation for any hierarchy by imposing an increasing constraint on the initial hierarchical segmentation. Applying the constraints on the original hierarchy is linear in the number of image pixels and the complexity, $O(N)$, is commensurable with the complexity of the algorithm presented in this article. Similarly to the algorithm presented here, the overall complexity of calculating a hierarchical segmentation based on constrained connectivity as presented in [5] depends on the complexity of calculating the original hierarchy and calculating the attribute values for each of the original tree components. However, since [5] is operating on ultrametric watersheds, their approach is not applicable to inclusion trees.

5   Discussion and Conclusion

Applying additional constraints on a hierarchical partitioning of an image is already well explored in [5, 9, 15]. The earliest works on the topic [15] suggest varying the constraint threshold parameters to control the degree of image simplification. When the node level does not coincide with the perceived complexity of the represented region (e.g. the chaining effect in α-trees), applying constraints rearranges the hierarchy according to a more precise external coarseness measure [9,15]. The results can also be interpreted as performing an attribute filtering with all threshold values simultaneously and storing all the results within a same structure [9]. In image segmentation, regions of a hierarchical image partition are treated as "puzzle pieces" [15, 16] used to compose a segmentation. The transformation reduces hierarchy size, lowering the number of "puzzle choices" and simplifying the calculation of the segmentation. Binary partition trees used for object detection [22] generate partitions so fine that a second merging criterion is used in order to generate a coarse partition in which the potential detections are marked before the object can be detected in the fine parts of the hierarchy. By reducing the size of the hierarchy, better detection could be achieved by using more complex algorithms or a more exhaustive search.

In the domain of inclusion trees, many applications would benefit of the reduction in the search space. Finding the k most prominent structures in an image (cf. [6]) depends directly on the size of the tree. Transforming an inclusion hierarchy simplifies the image in a way similar to the one presented in [15] for partitioning trees. The image simplification technique using level line trees proposed in [4], based on area size, can also be applied to a hierarchy first simplified using a different coarseness attribute. An image comparison method proposed in [4] relies on assigning attributes to describe the regions of the hierarchy and then checking one of the images for presence of shapes similar to shapes present in the other image. Since the method is already working by finding *similar* (and not the same) shapes, a simplification of hierarchies before image comparison would reduce the overall number of comparisons speed up the process. The beginnings of using hierarchical image representations in content-based image retrieval rely on examining image structural elements corresponding to the nodes in the tree [21] and would also benefit from the reduction in hierarchy size. As the approach to background detection presented in [20] depends on the values of several thresholds, multiple precision results could be obtained simultaneously by applying the presented transformation instead of a simple tree filtering only.

This article presented a technique which imposes an external coarseness measure on hierarchical image representations, applicable to both partitioning and inclusion trees. In addition to solving the problem for a larger class of hierarchies than [5], the proposed approach is intuitive and simple to implement and operates on the direct representations of image hierarchies – the tree representations. Many potential applications listed provide interesting directions for future work, taking advantage of the transformation going a step further in search space size reduction compared to directly using hierarchical image representations.

References

1. Cousty, J., Najman, L.: Incremental Algorithm for Hierarchical Minimum Spanning Forests and Saliency of Watershed Cuts. In: Soille, P., Pesaresi, M., Ouzounis, G.K. (eds.) ISMM 2011. LNCS, vol. 6671, pp. 272–283. Springer, Heidelberg (2011)
2. Gueguen, L., Soille, P.: Frequent and Dependent Connectivities. In: Soille, P., Pesaresi, M., Ouzounis, G.K. (eds.) ISMM 2011. LNCS, vol. 6671, pp. 120–131. Springer, Heidelberg (2011)
3. Jones, R.: Connected Filtering and Segmentation using Component Trees. Computer Vision and Image Understanding 75(3), 215–228 (1999)

4. Monasse, P., Guichard, F.: Fast Computation of a Contrast-Invariant Image Representation. IEEE Transactions on Image Processing 9(5), 860–872 (2000)
5. Najman, L.: On the Equivalence Between Hierarchical Segmentations and Ultrametric Watersheds. JMIV 40(3), 231–247 (2011)
6. Najman, L., Couprie, M.: Building the Component Tree in Quasi-Linear Time. IEEE Transactions on Image Processing 15(11), 3531–3539 (2006)
7. Soille, P., Najman, L.: On Morphological Hierarchical Representations for Image Processing and Spatial Data Clustering. In: Köthe, U., Montanvert, A., Soille, P. (eds.) WADGMM 2010. LNCS, vol. 7346, pp. 43–67. Springer, Heidelberg (2012)
8. Ouzounis, G., Wilkinson, M.: Mask-Based Second-Generation Connectivity and Attribute Filters. IEEE Transactions on Pattern Analysis and Machine Intelligence 29(6), 990–1004 (2007)
9. Ouzounis, G.K., Soille, P.: Pattern Spectra from Partition Pyramids and Hierarchies. In: Soille, P., Pesaresi, M., Ouzounis, G.K. (eds.) ISMM 2011. LNCS, vol. 6671, pp. 108–119. Springer, Heidelberg (2011)
10. Salembier, P., Garrido, L.: Binary Partition Tree as an Efficient Representation for Image Processing, Segmentation, and Information Retrieval. IEEE Transactions on Image Processing 9(4), 561–576 (2000)
11. Salembier, P., Oliveras, A., Garrido, L.: Antiextensive Connected Operators for Image and Sequence Processing. IEEE Transactions on Image Processing 7(4), 555–570 (1998)
12. Salembier, P., Wilkinson, M.H.F.: Connected Operators. IEEE Signal Processing Magazine 26(6), 136–157 (2009)
13. Serra, J.C., Salembier, P.: Connected operators and pyramids. In: Dougherty, E.R., Gader, P.D., Serra, J.C. (eds.) Image Algebra and Morphological Image Processing IV. SPIE, vol. 2030, pp. 65–76. SPIE Press, San Diego (1993)
14. Soille, P.: Morphological Image Analysis: Principles and Applications, 2nd edn. Springer, New York (2003)
15. Soille, P.: On Genuine Connectivity Relations Based on Logical Predicates. In: Proc. of the 14th Int. Conf. on Image Analysis and Processing, pp. 487–492. IEEE Computer Society Press, Los Alamitos (2007)
16. Soille, P.: Constrained Connectivity for Hierarchical Image Partitioning and Simplification. IEEE Transactions on Pattern Analysis and Machine Intelligence 30(7), 1132–1145 (2008)
17. Soille, P.: Preventing Chaining Through Transitions While Favouring It within Homogeneous Regions. In: Soille, P., Pesaresi, M., Ouzounis, G.K. (eds.) ISMM 2011. LNCS, vol. 6671, pp. 96–107. Springer, Heidelberg (2011)
18. Soille, P., Grazzini, J.: Constrained Connectivity and Transition Regions. In: Wilkinson, M.H.F., Roerdink, J.B.T.M. (eds.) ISMM 2009. LNCS, vol. 5720, pp. 59–69. Springer, Heidelberg (2009)
19. Song, Y.: A Topdown Algorithm for Computation of Level Line Trees. IEEE Transactions on Image Processing 16(8), 2107–2116 (2007)
20. Song, Y., Zhang, A.: Locating Image Background by Monotonic Tree. In: Caulfield, H., Chen, S.H., Cheng, H.D., Duro, R., Honovar, V., Kerre, E.E., Lu, M., Romay, M., Shih, T., Ventura, D., Wang, P., Yang, Y. (eds.) 6th Joint Conf. on Information Sciences, pp. 879–884. Association for Intelligent Machinery, Inc., Durham (2002)
21. Song, Y., Zhang, A.: Analyzing scenery images by monotonic tree. ACM Multimedia Systems J. 8(6), 495–511 (2003)
22. Vilaplana, V., Marques, F., Salembier, P.: Binary Partition Trees for Object Detection. IEEE Transactions on Image Processing 17(11), 2201–2216 (2008)

Anisotropic Diffusion and Curve Evolution for Segmentation of Color Images in Cultural Heritage

Luigi Cinque[1] and Rossella Cossu[2]

[1] Dipartimento Informatica, Unversitá degli Studi, Sapienza,
Via Salaria 113, 00185 Roma, Italy
[2] Istituto per le Applicazioni del Calcolo-CNR,
Via dei Taurini 19, 00185 Roma, Italy
r.cossu@iac.cnr.it
http://www.iac.cnr.it

Abstract. We propose an innovative procedure for extracting decay regions from color images of stony materials. The use of appropriate image analysis techniques can offer an important contribution to be used together with the traditional methodologies for studying and diagnosing the decay of stony materials that constitute ancient monuments. The presented approach is constituted by the PDE (Partial Differential Equations) model of anisotropic diffusion and by the level set/fast marching method. The anisotropic diffusion is applied in order to limit the smoothing at the zones of high gradient. In the segmentation process, the numerical technique of the level set/fast marching is applied in order to extract from the image only the color region examined. The study case concerns impressive remains of the city of Aosta (Italy).

Keywords: Color image segmentation, anisotropic diffusion, fast marching, level set.

1 Introduction

The image analysis can represent an important tool for Cultural Heritage scientists to analyze and collect information about the preservation state of historical monuments. In the images of ancient buildings, pixels of different color characterize the decay areas [1]. On the wall facades of the monuments, in fact, we note different types of decay such as oxidation, sediment and/or cavities which are recognizable by the color feature. The acquired images, related to the facades of a monument, show, for example, reddish stains due to oxidation, whitish ones due to the sediment of saline efflorescence or generally dark/black ones due to lack of material as in cavities. For this reason a crucial step to extract decay regions, characterized by chromatic alteration, can be the application of an image segmentation strategy to color images of stony materials.

In this paper we present the procedure developed for extracting decay zones from color images of stony materials. It is based on the PDE model anisotropic diffusion [2], [3] and on the level set/fast marching method [4],[5].

A. Petrosino (Ed.): ICIAP 2013, Part I, LNCS 8156, pp. 572–581, 2013.

To remove the noise, we used the anisotropic diffusion because the decay areas do not have a uniform color. They, in fact, show colored pixels, which are not distributed in a homogeneous way. For example, in an image a reddish stain (oxidation) can contain red pixels or red-yellow pixels together with very dark pixels due to dust. This may make the extraction of the decay from color images very complex. The basic idea is to smooth the regions, but to reduce the smoothing effect near edges. Nonlinear PDE diffusion was first proposed by Perona and Malik [6] applied to gray levels images. We applied an anisotropic diffusion to the color images [3].

The traditional techniques of color segmentation can present advantages and drawbacks so that the best segmentation technique depends on the image and application [7],[8]. The level set is a region-based method, which has the advantage of contextually taking into account both color distribution in color space and its repartition in spatial domain. Then the contours represented by the level set function may break or merge naturally during the evolution, so that complex shaped regions can be detected and handled implicitly [5]. It mainly works to extract two regions, one of foreground and one of background. In the segmentation process we used the level set numerical method just because it works in way that it segments from the image only the region under study. In our application we suppose to have a monotonic evolving front so that the level set equation can be reduced to the eikonal equation, which describes the evolution of a curve. In order to solve this equation we used the fast marching numerical technique. It, starting from seed points located in the region of interest, expands by incorporating step by step an increasing number of pixels as far as the contour of the object. The generated front evolves according to an appropriate speed function. The choice of correct speed function is of fundamental importance to achieve a good segmentation. This technique is also used in medical applications to find boundaries of pathological or anatomical structures. Similarly our purpose is to find the decay regions.

Every pixel of the images has been analyzed in $L*a*b*$ color components. CIEL*a*b*, abbreviated CIELAB, color space has been defined by the CIE (Commission International de l'Eclairage) as a uniform color space [9]. For our application, we have used color images taken from the Research Project between ITABC-CNR and the Superintendence of Aosta [10].

The paper is structured as follows. Sections 2 describes the anisotropic diffusion and its extension to color images. Section 3 contains the description of the segmentation by curves evolution applied to color images. The Section 4 presents the results obtained. A conclusion is drawn in Section 5.

2   Color Anisotropic Diffusion

In image analysis it is fundamental to smooth the regions of the image preserving their boundaries.

To overcome this problem, Perona and Malik [6] proposed a nonlinear adaptive diffusion process, termed anisotropic diffusion, for gray levels images, described by the PDE (Partial Differential Equation) equation

$$\frac{\partial I(x,y,t)}{\partial t} = \nabla \cdot (g(|\nabla I|)\nabla I), \qquad I(x,y,0) = I_0(x,y) \tag{1}$$

where $g(|\nabla I|) > 0$ is a decreasing function of the gradient. We choose as diffusion coefficient function

$$g(|\nabla I|) = \frac{1}{1 + (\frac{|\nabla I|}{k})^2}$$

where the parameter k is a constant and acts as an "edge threshold" [3].

In order to extend this approach to the color images, let $\mathbf{I} : \Omega \to \mathbb{R}^n$ be an intensity image function with the domain $\Omega \subset \mathbb{R}^2$. This definition clearly includes gray level images as in particular case ($n = 1$). A color image ($n = 3$) is constituted by the three monochrome components, that is it may be represented as a vector function $\mathbf{I} = (f_1(x,y), f_2(x,y), f_3(x,y))$. It can be processed in different color spaces as **RGB**, **XYZ** or CIEL*a*b*.

The application of nonlinear diffusion PDEs to color images $\mathbf{I}$ is based on a local vector geometry [2], in this case the natural extension of the equation (1) becomes

$$\frac{\partial \mathbf{I}}{\partial t} = div(g(\|G\|) \cdot \nabla \mathbf{I}) \tag{2}$$

where $\|G\|$ is a norm of a tensor defined as

$$G = \sum_{j=1}^{n} \nabla f_j \nabla f_j^T \tag{3}$$

which is a 2×2 symmetric and semi-positive-definite matrix, called by us *diffusion tensor*, and gives the local variation of the image.

In this way $\|G\|$ fuses the different components of the image. In fact, the components are diffused independently of one another, but they are linked by means of the norm of the tensor G.

In (3), each ∇f_j corresponds to the spatial gradient of the jth vector component of the color image $\mathbf{I}$.

So in analogy with (1)the coefficient function $g(\|G\|)$ in (2) is

$$g(\|G\|) = \frac{1}{1 + (\frac{\|G\|}{k})^2} \tag{4}$$

where $\|G\| = \sqrt{\sum_i |\nabla f_i|^2}$ and the parameter k is based on mean tensor norm $\|G\|$ of the whole image.

3   Segmentation by Curves Evolution

The image segmentation through the level set method has become very popular over the last decades. The basic idea of the level set method is to represent contours as the zero level set of an implicit function defined in a higher dimension, called the level set function, and to evolve it according to a partial differential equation [4],[5].

In our application we suppose to have a monotonic evolving front so that the level set equation can be reduced to the eikonal equation, which describes the evolution of a curve. In order to solve this equation we used the fast marching numerical technique. It, starting from seed points located in the region of interest, expands by incorporating step by step an increasing number of pixels as far as the contour of the region. The generated front evolves according to an appropriate speed function of fundamental importance to achieve a good segmentation.

Let $\mathbf{I} : \Omega \to \mathbb{R}^n$ be an intensity image function with the domain $\Omega \subset \mathbb{R}^2$. The goal of image segmentation is based on partitioning Ω in order to extract disjoint regions from the image $\mathbf{I}$ such that they cover Ω.

The boundaries of these regions may be considered as curves belonging to a family whose time evolution is described by a level set equation. The main advantages of using the level set is that complex shaped regions can be detected and handled implicitly.

In order to obtain the governing equation of a front evolution we consider a family of parameterized closed contours $C(x(t), y(t), t) : [0, \infty) \to R^2$, generated by evolving an initial contour $C_0(x(0), y(0), 0)$ [11]. We underline that in the curves evolution theory the geometric shape of the contour is determined by the normal component of the evolution velocity. If $F(x, y)$ is a scalar function representing the curve speed in the normal direction $\mathbf{n}$, the normal velocity components are

$$\frac{dx}{dt} = F(x, y) \cdot n_1 \qquad \frac{dy}{dt} = F(x, y) \cdot n_2$$

with $\mathbf{n} \equiv (n_1, n_2)$. So that the curve evolves in time according to the following equation [5]

$$\frac{\partial C(x(t), y(t), t)}{\partial t} = F(x, y) \cdot \mathbf{n} \qquad (5)$$

Supposing that $C(x(t), y(t), t)$ is a moving front in the image, if we embed this moving front as the zero level of a smooth continuous scalar 3D function $\phi(x(t), y(t), t)$, known as the level set function, the implicit contour at any time t is given by $C(x(t), y(t), t) \equiv \{(x(t), y(t))/\phi(x(t), y(t), t) = 0\}$. By differentiating respect to t the expression $\phi(x(t), y(t), t) = 0$ and substituting in (5), the equation of motion for the level set function may be derived

$$\frac{\partial \phi(x(t), y(t), t)}{\partial t} + F(x, y)|\nabla \phi(x(t), y(t), t)| = 0 \qquad (6)$$

If we choose as level set function $\phi(x, y, t) = T(x, y) - t$, where $T(x, y)$ is the function representing the arrival time at which the curve crosses a given point at the time t, it follows

$$\frac{\partial \phi(x, y, t)}{\partial t} = -1, \qquad |\nabla \phi(x, y, t)| = |\nabla T(x, y)|$$

and from (3) we obtain a simpler equation

$$F(x,y)|\nabla T(x,y)| = 1 \tag{7}$$

Moreover, we assume the special case of a front moving with speed $F(x,y) > 0$ (the case in which $F(x,y)$ is negative everywhere is also allowed). We then have a monotically-advancing front. This leads to a nonlinear eikonal equation

$$|\nabla T(x,y)| = \frac{1}{F(x,y)} \tag{8}$$

where $\frac{1}{F(x,y)} = f(x,y)$ is the curve slowness.

In order to solve the nonlinear eikonal equation (8), we adopted the fast marching numerical technique because it is more appropriate for image segmentations related to single regions. The algorithm starts from initial point, named seed. The front, evolving from these points, propagates to the normal direction with a opportune speed $F(x,y)$. The known term $F(x,y)$ is an image, that represents the front speed point by point. So, known $F(x,y)$, the solution of differential equation is the minimum time $T(x,y)$ necessary to the initial front to arrive at the pixels located on the final contour.

It is well known that the choice of the speed term in the fast marching method is a fundamental task to achieve a good segmentation. In this paper we propose a speed function constituted by both speed of the image threshold based on thresholding computation in color coordinates and speed based on color image tensor computation [6]. In particular let be I_{thr} the threshold image given

$$I_{thr}(x,y) = \begin{cases} L^*(x,y), & p(x,y) \in [\Delta L^* \wedge \Delta a^* \wedge \Delta b^*] \\ K, & p(x,y) \notin [\Delta L^* \wedge \Delta a^* \wedge \Delta b^*] \end{cases}$$

where $K >> max(L^*(x,y)$ and $\Delta L^* \wedge \Delta a^* \wedge \Delta b^*$ constitute the color range selected, in particular $L^*(x,y)$ is the luminance value at the pixel $p(x,y)$. So let be F_{thr} and F_{ten} given by

$$F_{thr} = \frac{1}{1 + |\nabla I_{thr}(x,y)|} \qquad F_{ten} = \frac{1}{1 + \|G\|}$$

where $\|G\|$ is the tensor obtained by the anisotropic diffusion. Then the new velocity is given by

$$F(x,y) = F_{thr} \bigcup F_{ten} \tag{9}$$

The speed is defined in such a way that the front proceeds rather fast in low gradient, while it slows down in high gradient zones. In the opinion of sector expert scientists this speed has the advantage solving the problem of non-uniform distribution of the colors in the region without smoothing decay boundaries.

4 Experimental Results

In this section we first apply the developed procedure to test images.

According to the traditional methodology used by experts, every single pixel of the images is analyzed and processed in $L^*a^*b^*$ components. In fact, the sector experts, in order to identify colorimetric characteristics of the monument decay, normally use the colorimeter instrument that measures tristimulus values [9] of the examined region in CIE (L^*, a^*, b^*) coordinates. The extraction of the decay regions is difficult because the color pixels are not distributed in a homogeneous way on the image. For instance, an oxidation stain can look compact red, or compact red and yellow or all of those colors with a minimal presence of red-yellow color pixels.

The images of Figures 1 (a) and (c) have been obtained by adding uniform noise to an original synthetic image constituted by gray background and green color of the letters (S,B,A,H). These images are similar to gray levels images used in [5]. The images of Figures 1 (b) and (d) are the segmented regions represented by their edge contours.

The images of Figures 2(a) (b) show the resulting contours from test image of the color peppers. These results are examples of contours extracted from objects of compact red and compact green colors.

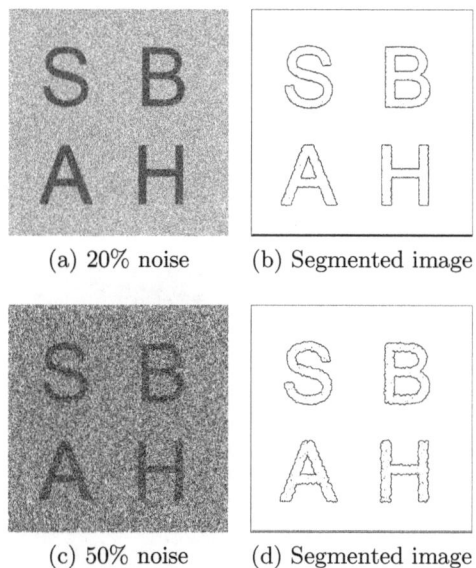

(a) 20% noise   (b) Segmented image

(c) 50% noise   (d) Segmented image

Fig. 1. Segmentation of synthetic images

We show the segmentation obtained by applying the developed procedure to real images of decay. In particular we computed the results from images of the Roman Theater, Roman Walls and Arch of Augustus in the city of Aosta [10].

(a) Green pepper in fore-
ground

(b) Red pepper in fore-
ground

Fig. 2. Segmentation of a green pepper and red pepper

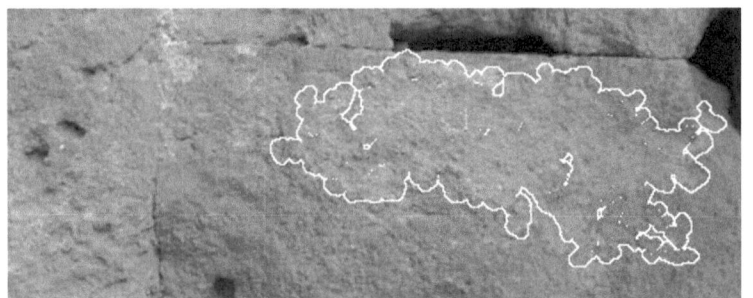

(a) Oxidation

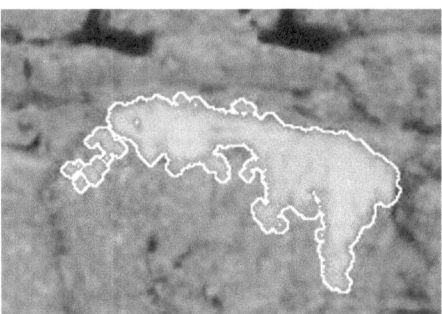

(b) Sediment

Fig. 3. Roman Theater: reddish stain and whitish stain

The acquired images have been geometrically and chromatically corrected by using a colorimetric target under controlled light conditions. An example of the obtained results related to oxidation and sediment decays from images of Roman Theater are shown in the Figures 3 (a) (b) respectively.

In the Figures 4 (a) (b) we can observe the resulting contours of a red stone and a cavity extracted from images of the Roman Walls.

The acquired images of the Augustus Arch have been geometrically corrected and ortho-rectified [10]. The image of Figure 5 (a) shows stains of additional material of black color obtained by a naked eye analysis performed from expert. The boundaries obtained applying our integrated procedure of the same decay are shown in Figure 5 (b). In Figure 6 (a) the selected contours show the stains of reddish decay obtained by naked eye analysis. Analogously in Figure 6 (b) the contours obtained by the procedure application are presented. The results obtained on three different monuments have satisfied the expectation of the sector experts.

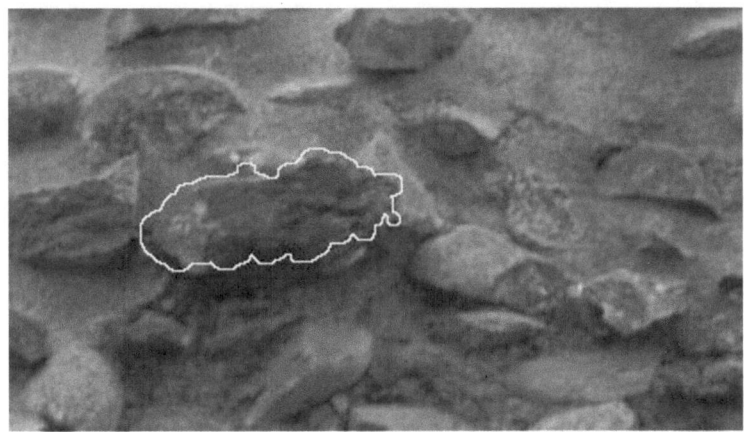

(a) Dark red stone

(b) Cavity

Fig. 4. Roman Walls: reddish stone and cavity

(a) Results of naked eye analysis   (b) Results of segmentation

Fig. 5. Black stains of material addition

(a) Results of naked eye analysis   (b) Results of segmentation

Fig. 6. Reddish stains of biological decay

5   Conclusions

In this paper we have presented a procedure of segmentation applied to color images, based on curve evolution method and an anisotropic diffusion extended to color images. This procedure has been suggested by the type of the real images examined, which present decay regions constituted by pixels of different colors not uniformly distributed. Therefore this application has required decay areas of homogeneous color preserving edges.

Acknowledgements. The authors would like to thank Paolo Salonia (ITABC-CNR) for allowing us to use digital images of Aosta monuments.

References

1. Cerimele, M.M., Cossu, R.: Decay Regions Segmentation from Color Images of Ancient Monuments Using Fast Marching Method. J. Cul. Her. 8, 170–175 (2007)
2. Åström, F., Felsberg, M., Lenz, R.: Color Persistent Anisotropic Diffusion of Images. In: Heyden, A., Kahl, F. (eds.) SCIA 2011. LNCS, vol. 6688, pp. 262–272. Springer, Heidelberg (2011)

3. Tschumperlé, D., Deriche, R.: Vector Valued Image Regularization with PDEs: A Common Framework for Different Applications. IEEE Transaction on Pattern Analysis and Machine Intelligence 27(4), 506–517 (2005)
4. Sapiro, G.: Color Snakes. Computer Vision and Image Understanding 68(2), 247–253 (1997)
5. Sethian, J.: Level Set Methods and Fast Marching Methods. Cambridge University Press (1999)
6. Perona, P., Malik, J.: Scale-Space and Edge Detection Using Anisotropic Diffusion. IEEE Transaction on Pattern Analysis and Machine Intelligence 12(7), 629–639 (1990)
7. Arbelaez, P., Maire, M., Fowlkes, C., Malik, J.: Contour Detection and Hierarchical Image Segmentation. IEEE TPAMI 33(5), 898–916 (2011)
8. Cheng, H.D., Jiang, X.H., Sun, Y., Wang, J.: Color Image Segmentation: Advances and Prospects. Pattern Recognition 34, 2259–2281 (2001)
9. Wyszecki, G., Stiles, W.S.: Color Science: Concepts and Methods, Quantitative Data and Formulae, 2nd edn. John Wiley and Sons (1982)
10. Salonia, P., Bellucci, V., Scolastico, S., Marcolongo, A., Leti Messina, T.: 3D Survey Technologies for Reconstruction, Analysis and Diagnosis in the Conservation Process of Cultural Heritage. In: Atti del XXI CIPA International Symposium, Atene (2007)
11. Cerimele, M.M., Cossu, R.: A Numerical Modeling for the Extraction of Decay Regions from Color Images of Monuments. Mathematics and Computers in Simulation 79, 2334–2344 (2009)

Comparison of Three Approaches for Scenario Classification for the Automotive Field

Nicola Bernini, Massimo Bertozzi, Luca Devincenzi, and Luca Mazzei

Dip. Ing Informazione, Parco Area delle Scienze 181A, 43124 Parma, Italy
bertozzi@vislab.it
www.vislab.it

Abstract. To extend the functionalities of Advanced Driver Assistance Systems (ADAS) and have a more accurate control on the parameters of sensors mounted on an intelligent vehicle, a tool that can classify the scenarios which the vehicle moves in, is needed.

This article presents a comparison of three classification techniques (PCA, ANN and SVM) to obtain a fast and robust scene classifier based only on images. The systems presented in this paper have been trained on three different categories of traffic scenarios: urban, highway, and rural, on a total of more than 23 hours of driving in different countries.

Keywords: scenario classification, intelligent vehicles, automotive.

1 Introduction

During recent years the number of integrated sensors on an intelligent vehicle is progressively increased. In this context ADAS have became significantly relevant allowing more comfortable and safe driving.

The reliability of a sensor is strongly dependent on the scenario in which the vehicle is moving. As an example, if the vehicle is driving in the center of a town among tall buildings or trees, or even inside a tunnel, the positioning system based on GNSS (Global Navigation Satellite System) may have difficulty in giving accurate information. The laserscanner is considered reliable in tight areas or at night, but it is generally negatively affected by the rain, that reflects the light beam, and by dust.

The knowledge of the context in which the vehicle is moving is of primary importance to adjust the sensor configuration in order to improve ADAS reliability and expanding the scope of such systems, freeing them from the restriction to operate only in a restricted number of environments.

Moreover, the knowledge of the scene is valuable to improve high-level applications, allowing fine tuning of planning, such as increasing controls reactivity on the safety distance (on highways, moving at high speed, it should be higher than in city, moving at lower speed) or decide whether or not to rely on GNSS to detect position.

This article presents a comparison of three machine learning techniques for scenario classification in the automotive field. These Systems must to be able

A. Petrosino (Ed.): ICIAP 2013, Part I, LNCS 8156, pp. 582–591, 2013.

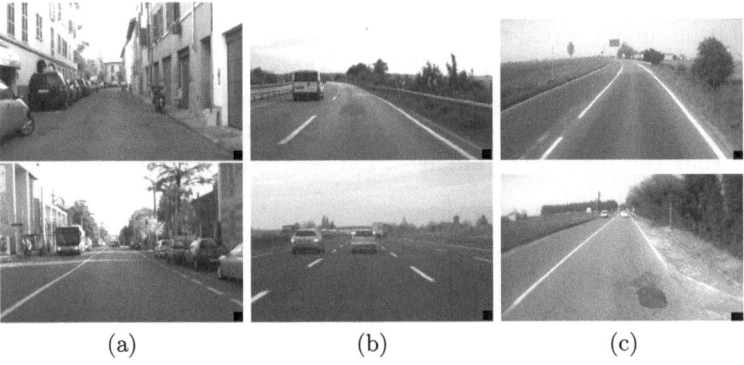

Fig. 1. Examples for different scene classes: (a) City, (b) Highway, and (c) Country Rural Road

to recognize the correct scenario among three possible classes; figure 1 shows examples images of the possible choices: city, highway, and country rural roads.

This paper is organized as follows. The section 2 shows a brief report on the state of art about images segmentation and classification. The section 3 analyzes the whole system structure and shows the developed classifiers in depth. The section 4 presents the results obtained in terms of correct detection and time consumption for the three classification methods, with a brief comparison. Finally, section 5 ends the paper with some final remarks.

2   Related Works

Concerning scenes classification on images, many works are related to image retrieval techniques on image databases. In sever works indoor scenes classification is considered as [5]. In [4] an approach with deformable part model, DPM, is presented. On the other hand, outdoor scenarios classification is performed by low level features and selected categories of scenes, indoor/outdoor, landscape/city, etc. as in [6–8]. Also object bank approach, a high level features method, is used to classify outdoor scenarios, as in [2]. In [3] classification of the scenes is applied to the entire image, with a global scene footprint based on information obtained by the extraction of FFT (Fast Fourier Transform) spectrum from the image.

In [1], he suggested to divide the image into equal parts and, on each of these, to apply a frequency transformation to obtain the features according to the hypothesis that every scenario corresponds to a specific Fourier footprint. These features are reduced using a technique called HPCC (Hierarchical Principal Component Classification), based on PCA, and then LDA and ANNs are used with good results: more than 80% accuracy.

The system detailed in this paper uses a similar approach for the processing, because we believe it's an effective strategy to exploit properly the informative content of the image that's mainly located in subparts of it (the sky and the street in front of the car are generally not very useful to detect the context the vehicle is moving in), and compares additional classifiers for a performance confrontation, under a real-time constraint..

The novelty of the approach is related to the automotive scenario application and real time performance. Generally, scenario classification techniques are performed on really different ones, while, concerning the automotive environment, the fact that the differences amongst the selected contexts are really limited makes the task tougher.

3   Algorithm Description

The system, which is based exclusively on the analysis of a single frame, is composed of a first common layer that is in charge of pre-process frames with the purpose of obtaining features in the frequency domain. These features are then processed using PCA and analyzed with a Gaussian probability distribution function, artificial neural networks (ANN), and Support Vector Machine (SVM) with Gaussian RBF (Radial Basis Function) Kernel to test which of these three systems is the most reliable, fast and robust forthe classification.

Figure 1 shows typical images of cities, highways and rural roads environments which represent the more common scenarios for vehicles to move in, they constitute the three classes used for the classification in this study.

The overall system is divided in three main parts, as shown in figure 2. The first part (preprocessing) is the image pre-processing that reduces the effect of changes in illumination and eliminates small frequency variations, then the image is resized so it can be divided equally into 16 subparts, which are then separately computed in the sampling phase.

The cutting scheme problem has been heuristically addressed, identifying image ROIs suitable to solve the context detection problem. These ROIs are

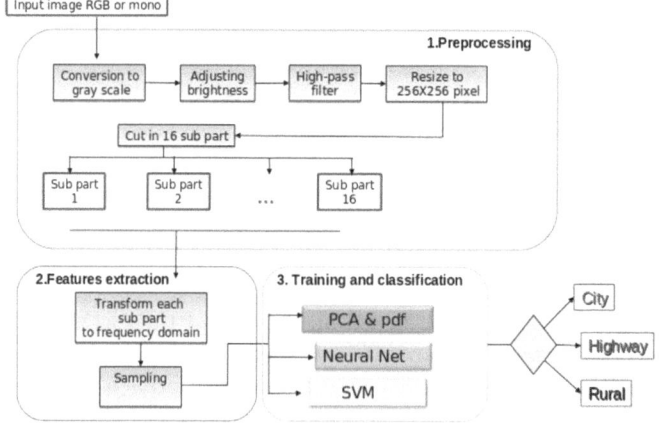

Fig. 2. Algorithm overview flow diagram

square-shaped because FFT will be calculated over them and their dimension depends on the specific image region we want to analyze (sky, front street, sides).

The subdivision phase is a key step in order to be able to exploit the images high similarity in the traffic scenario. Such images have many common parts (presence of road, lines, sky, etc) so dividing them into subparts and treating each of them as a standalone image makes the search for differences a simplified task. This step is similar to the one used in [1] and has been adapted to this specific scenario.

In the second phase (features extraction) an extraction of features to obtain relevant information from the images is performed. The system transforms each subpart of the initial image into frequency domain, using a DFT, so 16 spectra are obtained. Each spectrum is sampled using median filters with different resolutions. The sampled data is then scaled to fit in a predetermined range.

In the third step (training and classification), three methods of classification are applied on the computed samples. In the first approach data is reduced with PCA and a Gaussian probability distribution function for each class is computed. In the second approach an ANN analysis is performed on each subpart of the image, then the each partial result is passed to another ANN in charge of achieving the final classification. In the third approach a SVM is used to map the input space into a new space with higher dimensionality with respect to the previous ones, in order to be able to better discriminate the different classes.

In the following paragraph each algorithm stage is detailed.

3.1 Preprocessing

The preprocessing stage can take as input both gray level or color images; RGB images are converted to grayscale. After that, to expand the dynamics of the

Fig. 3. Image elaboration: (a) input image, (b) histogram normalization, (c) image after suppression of low frequency, (d) rescaled image, (e) subdivision on 16 blocks, (e) DFT results, and (f) 128 masks feature extraction areas

gray levels and to reduce the influence of different illumination conditions, the histogram is normalized. In the following step, a high pass filter is applied, whose frequency response is given by the following equation,

$$H(k,l) = \begin{cases} 1 & se\ k = 0 \cap l = 0 \\ 1 - (1 - 0.9e^{-\frac{k^2+l^2}{156.25}}) & else \end{cases}$$

in order to mitigate the small differences in terms of contrast and frequency that would otherwise produce edges effects in the DFT. The image is then resized to 256×256 pixels in order to have a standard size and it is divided into 16 equal subparts of 64×64 pixels. The division is made to inspect information stored in the amplitude values in the frequency domain in different image regions. This would not be possible if the entire image is directly transformed to the frequency domain. Images 3.b-d show an example of the different preprocessing steps.

3.2   Feature Extraction

The second stage of the system algorithm involves a FFT (Fast Fourier Transform) on each 16 subparts of the image. It is composed of two steps. Initially, FFT is computed on each subpart according to the following formula

$$F_k(i,j) = |F_k(i,j)|e^{\phi(i,j)},\ \text{with}\ k \in \{1...16\}\ \text{subpart}$$

Concerning the purpose of this work, only FFT magnitude is used while the phase information is discarded.

Finally, in the second step of this stage, a sampling of the FFT is performed. Since FFT provides a huge amount of information, a sampling step becomes necessary to enable the following steps to reach real-time performances.

Each spectrum is here filtered by a mask that provides as output 128 values. In the case of road environments, the higher concentration of information is around the origin; while there is little variation at high frequencies. The mask is sized to have low resolution at high frequencies and high resolution at low frequencies, as shown in figure 3.f, Sampling is performed using 2×2 masks in the middle of the subimage, 4×4 mask in the intermediate zones, and a mask of 8×8 pixels for the more external part.

In the same way as a median filter, values of the pixels are used to obtain 128 samples by the 64×64 subpart. The final output of this step of features extraction is 16 vectors of 128 samples. In order to perform a better classification, data used by classifier are scaled in a fixed chosen range $[0, 1]$.

3.3   Training and Classification

Three classifiers are implemented to measure their robustness, performance and execution time.

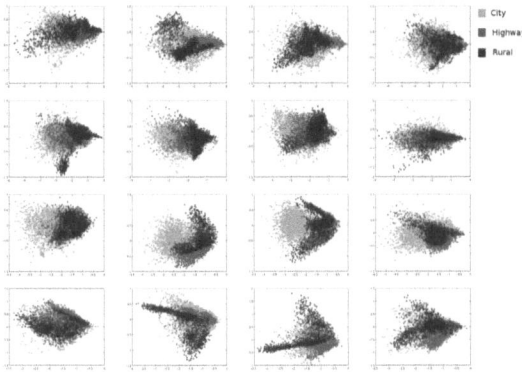

Fig. 4. 2D representation of extracted features

Classification with PCA. This first classifier uses a PCA reduction to re-organize the samples in a small features space, in a second step a vote for each subpart of the image is performed in order to take a classification decision.

The algorithm obtains the 16 matrices for each subpart of the image, as input. Each matrix has a number of row equal to the number of the training set frames, and a number of column equal to the number of the samples.

A reduction in the workspace is performed with a PCA: using the firsts 2 eigenvectors obtained from the SVD (Singular Value Decomposition) to compose a transformation matrix that maps each of the 16 vectors from 128 samples into a 2-dimensional space. Applying this reduction to all of the training set vectors for each subpart, a point cloud is obtained for each class. Figure 4 shows the 2D representation of extracted features: image shows the 16 blocks point clouds. The green points represent the city class, red points the highway class and the blue points to the rural class.

In order to describe statistically these clouds of points, they are characterized by a normal probability density function $\mathcal{N}$ calculating the mean and variance for each class.

During the classification process the same PCA reduction is applied to the image subparts, then the 3 probability density functions are computed to get 3 votes.

To perform the final classification of the entire image, a naïve Bayesian classifier is used to compute hypothesis among classes according to the following formula.

$$\arg\max_k \sum_i \mathcal{N}(\mathbf{x_i}, \mu_{k,i}, \sigma_{k,i})$$

with $k \in \{r, c, h\}$ classes, and $i \in [1...16]$ subparts of the current image.

Classification with Artificial Neural Networks. The architecture overview of the artificial neural networks is shown schematically in figure 5, and features a

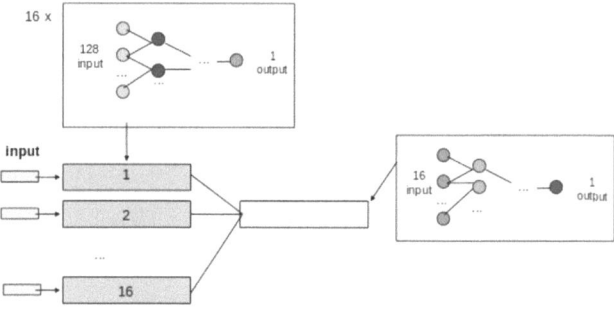

Fig. 5. Artificial neural networks structure

two levels scheme. The first level is based on 16 ANNs, one for each subpart, with 128 inputs and 3 outputs. Each ANN takes decisions using a binary encoding for the 3 class labels. Another ANN, with 48 (16 × 3) inputs, gathers up the results of the previous 16 networks and takes the final decision on the entire image.

In order to improve this system performances, a proper tuning is need in order to be able to find the optimum setup for parameters like the number of hidden layers, neurons, activation functions, average error, and training time. The proposed approach involves a backpropagation training scheme with an incremental algorithm. In the training test phase, the best results in terms of MSE (Mean Squared Error) (less than 0.005) are obtained with 3-layer networks and a sigmoid hidden layer activation function. The output layer has a stepwise linear approximation to sigmoid activation function. The neurons number of the 16 ANNs on the first level are 97, 66, and 35; while the ANN on the second level has 49, 34, and 19 neurons.

Classification with SVM. The structure of the SVM classifier consists of 16 parallel SVMs that process data from the 16 vectors of subimage samples, producing 16 output, and a second layer formed by another SVM that classifies the entire image using the first layer output: a structure very similar to the ANN-based solution one. The SVM classifier uses a gaussian radial basis kernel function (RBF) to map the input space into a high dimensional feature space, looking for a separating hyperplane. The following formula shows the gaussian radial base function: $k(x_i, x_j) = e^{(-\gamma \|x_i - x_j\|^2)}$ for $\gamma > 0$ (SVM classifier uses $\gamma = 1/2\sigma^2$).

RBF kernel is the best choice because linear kernel and sigmoid kernel can be seen as special RBF cases and they have the same performance with specific parameters. Finally the kernel RBF has the lowest number of computing complexity, moreover cross-validation and grid search can be used to determine the best values for its two parameters: C (used to weight the penalty related to misclassification in soft margin SVM) and γ (Gaussian RBF Kernel free parameter).

4 Results

The training set has been built recording 23 hours of driving and manually assigning parts of sequences to the right category: city, highway and rural. Images of the training set has been sampled with a frequency of 1 FPS so they were not too similar. Images have been taken in Italy, Germany, Russia, Austria and U.S. to have a very broad and general set. Samples for the city class consist of up to 2698 frames including images that show narrow streets, no more than two lanes wide, tall buildings, as well as roundabouts and intersections. The training set for highway class includes 2046 frames where the road is wide, with two or more lanes, few or none tall trees on the sides and there are no subways or tunnels which can occlude the sky. The samples for the rural scenarios refer to typical rural landscape, with fields and vegetation on the sides of the road that is at least two lanes wide, with few or no intersections, roundabouts, houses and buildings, for a total of 2205 frames. The test set therefore sums up to 19231 images which were also introduced noisy elements in, such as plants and highway overpass, squares and boulevards in towns.

Table 1 shows the correct detection on test set and figure 6 shows correct detection for the three classes. On the left corner there are three box that encode the vote of PCA, ANNs and SVMs. A vote for city is shown as green, while votes for highway or rural scenarios are rendered in red or blue respectively.

All the three classifiers correctly discriminate urban environment, where the SVM percentage reaches 100%. The best method for the highway classification is the ANN. Concerning the rural context only the SVMs, that work in a larger feature space proved to be able to identify an environment that is someway really similar to the other two ones. The PCA usually mismatches rural environment as urban scenario, while the ANN as highways. SVMs, however, managed to reach a high percentage of correctness of about 95% in all scenarios.

The confusion matrices for all classifiers are detailed in table 1. The ANN confusion matrix clearly shows that ANN were not able to distinguish rural images in the test set, always mismatching them as highways.

Concerning the execution time, a large part of runtime is required by the image pre-processing stage, because of the filters, the DFT, and the sampling. The total

Table 1. Correct Detections and confusion matrix for the 3 classification methods

Method	Correct Detections		
	City	Highway	Rural
PCA	95.6%	96.9%	7.0 %
ANN	91.6%	100%	6.0 %
SVM	100%	84.4%	95.0%

Method	Scenario	City	Highway	Rural
	City	95.6%	4.0%	0.4%
PCA	Highway	2.4%	96.9%	0.7%
	Rural	64.0%	29.0%	7.0%
	City	91.6%	8.1%	0.3%
ANN	Highway	0.2%	93.3%	0.5%
	Rural	0.3%	93.7%	6.0%
	City	100.0%	0.0%	0.0%
SVM	Highway	15.6%	84.4%	0.0%
	Rural	5.0%	0.0%	95.0%

(a)   (b)   (c)

Fig. 6. Results on classification: correct detection of (a) city, (b) highway, (c) and rural class.

execution time of the pre-elaboration stage is 15.2 ms. For the classification stage, PCA and ANN processing algorithms proved to be both quick requiring only 2 ms and 1.3 ms respectively. SVM was the classifier that required longer time with 18.5 ms. Even in the slowest case, the complete algorithm was still below 38 ms and seems therefore suited for the use in the automotive field. The tests were performed on a Intel Core 2 P7450@2.133 GHz Duo, with 2 GBytes of RAM at 1600 MHz.

5   Conclusions

This paper presents the implementation of a classifier for traffic scenarios through the comparison of different machine learning techniques, PCA, ANN, and SVM.

The classification tasks are performed only on images, of any kind, that are pre-processed to extract spectral information. Results obtained demonstrate that it is necessary to work in a larger space in order to be able to discriminate among classes with elements so similar, as shown by the SVM classifier. In all the cases, except for ANN with rural scenario, high correct detection rate was achieved and results were accompanied by excellent execution times well below the human average reaction time.

In city or highway scenarios, either PCA and ANN classifier can be used, due to their high rates of accuracy. Vice-versa, if the purpose is to discriminate amongst all of the three classes, only SVM classifier is reliable.

From the data collected it emerges that SVM is generally the best method and reaches the best classification performance while respecting the time constraint related to the Automotive Environment, that's a mandatory requirement.

SVM outperforming ANN is a fact that has already been observed in other kind of application [9, 10]. The reason is probably related to the Non-Linear SVM ability to discriminate in a non linear way, due to use of the kernel trick that transforms the problem of a non-linear separating curve identification, in a problem of a separating hyperplane identification in high dimensionality. In the present case, a Gaussian RBF Kernel has been used, so the classification has occurred in an Infinite Dimensional Hilber Space.

The proposed system is well suited for a variety of different applications in the automotive fields, as examples: the support to other ADAS for the automatic setting of driving parameters or the environment analysis on completely unmanned intelligent vehicles.

Acknowledgments. The work described in this paper has been developed in the framework of the OFAV Project funded by the European Research Council (ERC) within an Advanced Investigators Grant.

References

1. Kastner, R., Schneider, F., Michalke, T., Fritsch, J., Goerick, C.: Image-based classification of driving scenes by hierarchical principal component classification (hpcc). In: 2009 IEEE Intelligent Vehicles Symposium, pp. 341–346 (June 2009)
2. Li, L.-J., Xing, E.P., Su, H., Fei-Fei, L.: Object bank: A high-level image representation for scene classification & semantic feature sparsification. In: Neural Information Processing Systems (NIPS), Vancouver, Canada, pp. 1378–1386 (December 2010)
3. Oliva, A., Torralba, A., Dugue, A.G., Herault, J.: Global semantic classification of scenes using power spectrum templates (1999)
4. Pandey, M., Lazebnik, S.: Scene recognition and weakly supervised object localization with deformable part-based models. In: 2011 IEEE International Conference on Computer Vision (ICCV), pp. 1307–1314 (November 2011)
5. Quattoni, A., Torralba, A.: Recognizing indoor scenes. In: IEEE Conference on Computer Vision and Pattern Recognition, CVPR 2009, pp. 413–420 (June 2009)
6. Serrano, N., Savakis, A., Luo, A.: A computationally efficient approach to indoor/outdoor scene classification. In: Proceedings of the16th International Conference on Pattern Recognition, vol. 4, pp. 146–149 (2002)
7. Szummer, M., Picard, R.: Indoor-outdoor image classification. In: IEEE International Workshop on Content-Based Access of Image and Video Database, pp. 42–51 (January 1998)
8. Vailaya, A., Jain, A., Zhang, H.J.: On image classification: city vs. landscape. In: IEEE Workshop on Content-Based Access of Image and Video Libraries, pp. 3–8 (June 1998)
9. Dal Moro, F., Abate, A., Lanckriet, G.R.G., Arandjelovic, G., Gasparella, P., Bassi, P., Mancinim, M., Pagano, F.: A novel approach for accurate prediction of spontaneous passage of ureteral stones: Support vector machines. Kidney International 69, 157–160 (2006)
10. Byvatov, E., Fechner, U., Sadowski, J., Schneider, G.: Comparison of support vector machine and artificial neural network systems for drug/nondrug classification. Journal of Chemical Information and Computer Sciences 43(6), 1882–1889 (2003)

Adverse Driving Conditions Alert: Investigations on the SWIR Bandwidth for Road Status Monitoring

Massimo Bertozzi[1], Rean Isabella Fedriga[1], and Carlo D'Ambrosio[2]

[1] Dip. Ing Informazione, Parco Area delle Scienze 181A, 43124 Parma, Italy
www.vislab.it
[2] Centro Ricerche FIAT, Strada Torino 50, 10043 Orbassano (TO), Italy
www.crf.it

Abstract. The 2WIDE_SENSE (*WIDE spectral band & WIDE dynamics multifunctional imaging SENSor Enabling safer car transportation*) EU funded project is aimed at the development of a low-cost camera sensor for automotive applications able to acquire the full *visible* to *Short Wave InfraRed* (SWIR) spectrum, from 400 to 1700 nm.

This paper presents the results obtained using this extended spectral responsivity sensor for a Road Status Monitoring application to inspect the vehicle's frontal area and detect layers of ice or water on the road surface.

Keywords: SWIR, road status monitoring, large bandwidth cameras, icy road, wet road.

1 Introduction

Rain both reduces visibility and makes roadway surfaces dangerous. Wet brakes are less effective too. Snow and ice cause roads to become even more slippery, especially when the temperature is at or below freezing. Slush makes difficult to steer, hard packed snow increases the danger of skidding and black ice makes driving extremely dangerous. Stopping distances on slippery pavement are from two to ten times farther than on dry pavement so that for a vehicle travelling at $30\,km/h$ they can get up to $52\,m$ on black ice. Moreover, usually, Anti Brake Systems (ABS) are tuned for the most slippery scenario and therefore less effective than they can be in normal situations. Therefore the detection of a general road status or of the presence of slippery spots in front of the vehicle can significantly improve driving safety. It can be noticed that in Europe (EU-18) around 3 800 casualties are due to wet, icy or snowy situations [9].

Most of the proposed solutions to this problem are not based on a true prediction but are focused on the estimation of the road friction namely the monitoring of tyres slippering. These approaches are mainly based on the use of inertial sensors or GPS or on the monitoring of the tyres noise [2, 7, 10]. Conversely, different perception approaches have been proposed for a true prediction like the use of radars [15] or lasers [13]. The use of standard cameras have been proposed as well [3, 8, 11, 14] exploiting the different polarization of the light reflected from the road surface.

Anyway, the most promising approach seems the analysis of the different spectral content of the light reflected from the asphalt in dry, wet, icy, or snow conditions [4].

A. Petrosino (Ed.): ICIAP 2013, Part I, LNCS 8156, pp. 592–601, 2013.
© Springer-Verlag Berlin Heidelberg 2013

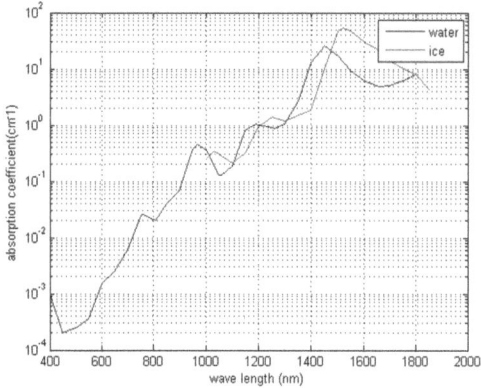

Fig. 1. Water and ice absorption coefficients

More precisely, the ShortWave InfraRed (SWIR, 0.9–1.7 nm) bandwidth shows different light reflection patterns depending on the road status [5, 6, 12]. According to this result, some solutions based on the use of custom spectrometers have been already implemented, as examples, the Volvo's Road eye or the Vaisala's Road Weather Sensors family. While the use of a spectrometer can be effective, the proposed solutions are not suitable for on-board installation on vehicles for costs and size costraints. In addition, the monitored area is too small to allow an exhaustive mapping of slippery road areas when the vehicle is moving.

The European funded 2WIDE_SENSE collaborative project has the aim of developing a low-cost camera able to acquire in the full visible to SWIR bandwidth. This task required the development of a specific sensor and a large bandwidth lens. In addition, the camera will feature a filter on the sensors to enable the independent acquisition of 4 different spectral bandwidths. Each 2×2 pixels window on the sensor in fact will include 4 different pixel-level filters; namely the value of each pixel in the window will be due to a specific bandwidth the pattern of which will be specifically selected according to the automotive needs [1] also allowing to obtain other bandwidths of interest using differences amongst different pixels.

Tests discussed in this article were thus carried out employing a state-of-the-art InGaAs camera module in order to determine the most suitable bandwidths, from visible to SWIR, to recognize the road surface status. The prototypical camera module parameters have been explored (e.g. HDR, integration time, gain) by acquiring images of the asphalt in different conditions (dry, wet, icy, sunny, with shadows).

This paper is organized as follows: section 2 describes the hardware equipments employed. Section 3 describes the experimental measurements carried out both in lab and on-the-field in order to select the most appropriate wavelength bandwidths useful to detect non-dry road condition and to select the optical filter characteristics. Section 4 reports the results obtained and section 5 states our final considerations.

2   Hardware Equipment

The multispectral camera module developed during the project has been available and therefore tested only during the final stage of the 2WIDE_SENSE experiments. A detailed description of the camera sensor, the filter pattern and the large bandwidth lens can be found in [1].

Conversely, most of the tests have been carried out using a state-of-the-art InGaAs camera module equipped with a SWIR high transmission lens. In order to mimic and evaluate the most suitable filters for the final prototype, a number of different filters has been used and tested (see fig. 2).

The camera used for tests is the OWL SW1.7 High sensitivity InGaAs FPA produced by Raptor Photonics and equipped with a sensor developed by Alcatel-Thales III-V Lab, both partners of the project consortium. The camera has a sensitivity bandwidth in the 400–1700 nm interval covering the whole spectrum from visible to the SWIR and acquires 320×256 14 bit images within a 500 ns–500 ms exposure interval.

The lens used is the OB-SWIR25/2 developed and produced by Optec SpA. It is a high transmission lens featuring a transmission rate >94% in the 900–1700 nm interval. The focal length is 25 mm with a 35.5 deg angle of view.

In order to test a number of spectrum bandwidths and to compare the quantity of light reflected by the asphalt for different conditions and wavelengths, several filters have been used. In the preliminary phase of the project tuneable liquid crystal filters have been employed to perform several temporal sequential acquisitions. These tuneable filters allowed to choose different wavelengths with a 20 nm bandwith resolution from 850 nm to 1800 nm and a transmittance around 60%. In the following phase a filter wheel (see fig. 2.b) with 12 filters has been installed between the lens and the camera allowing to select between the available filters. This is a manual operation and therefore limits the use of the filters to still objects.

3   Indoor Tests

The indoor acquisition sessions and measurements for this activity were performed at CRF (Centro Ricerche FIAT) in their electro-optics laboratory using the state-of-the-art InGaAs camera and a set of tunable liquid crystal filters as illustrated in section 2.

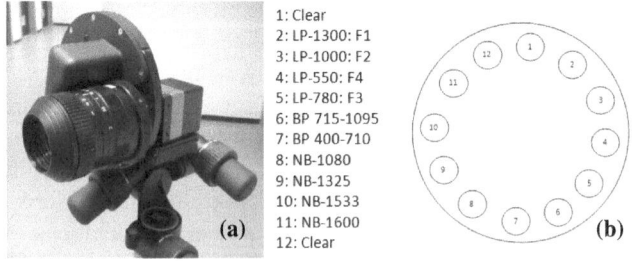

1: Clear
2: LP-1300: F1
3: LP-1000: F2
4: LP-550: F4
5: LP-780: F3
6: BP 715-1095
7: BP 400-710
8: NB-1080
9: NB-1325
10: NB-1533
11: NB-1600
12: Clear

(a)   **(b)**

Fig. 2. The test setup: *(a)* the camera and *(b)* the 12 slots filter wheel outline

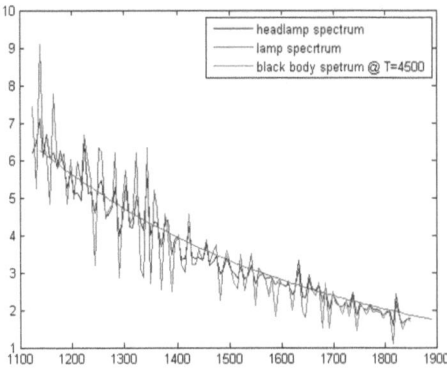

Fig. 3. Laboratory camera and lamp set-up and measures of lamp (green line) and headlamp (blue line) spectra compared to the black body (red line) one

Left image in fig. 3 outlines the laboratory set-up. Asphalt samples were collected and placed in a plastic box covered with commercial aluminum foil to prevent plastic's NIR absorption and avoid interferences in the acquired spectra. The asphalt surface was then illuminated with a stabilized $24\,V$ halogen lamp and the images acquired by the state-of-the-art InGaAs camera described in section 2. The lamp spectrum has been analyzed by a calibrated spectrometer. Graph in fig. 3 reports the spectra comparison between the laboratory acquisition lamp (a vehicle's H7 halogen headlamp) and the blackbody at $4500\,K$.

Several tests have been done in order to set the most appropriate camera control parameters. The camera cooler has been enabled to stabilize the device at $25\,^{\circ}C$ and the NUC (Non-Uniformity Correction) has been applied before each acquisition.

Asphalt samples have been modified in order to reproduce different road states: dry, damp, wet (less than $1\,cm$ water thickness), very wet (about $1\,cm$ water thickness), icy. For each configuration, different illumination conditions were analyzed, from full light to shadow, using metallic stops so as to minimize direct illumination without modifying the lamp spectrum.

4   Outdoor Tests

Outdoor tests have been performed using both the state-of-the-art InGaAs camera with the OB-SWIR25/2 lens and the filter wheel as shown in fig. 2.a.

The acquisition sessions for this activity were done at daytime with sunny and cloudy weather conditions and the road surface both dry and wet or iced in some areas as shown in the examples reported in fig. 4.

All combinations of gain and integration time values were also investigated to find the most suitable acquisition parameters for the RSM function. Some examples of these tests are shown in fig. 5.

Fig. 4. Images acquired in the whole visible to SWIR bandwidth (no filters) with the state of the art camera. Different illumination and road surface conditions are shown.

Fig. 5. Images acquired in the whole visible to SWIR bandwidth (no filters) with the state-of-the-art camera. Low gain and three different integration times have been selected: 1 ms (a), 5 ms (b) and 10 ms (c).

5   Indoor Results

In fig. 6 the results of two different asphalt conditions under different illumination have been reported. 4 ROIs[1] were randomly chosen as test specimens on the asphalt surface (brighter areas of the images belong to the alumunium covered container). Fig. 6 right side reports the corresponding calculated SWIR spectra. For each condition, using the tuneable filters, several images of the samples have been acquired with 50 nm step from 850 nm to 1800 nm. For each ROI, the mean pixel intensity has been computed then the mean intensity of the 1800 nm band subtracted. Since the imager sensitivity in 1800 nm spectral region is negligible, the measure could be used to calculate the pixel background values. This operation allowed to obtain intensity values less affected by noise: $I = mean(I(ROI)) - bkg$ where $bkg = mean(img_{1800})$.

ROI intensity values plotted as a function of wavelength are used to evaluate the spectrum corresponding to each sample in the considered condition. Some comments could be derived:

- the spectrum in dry conditions could be predicted by multiplying the lamp spectrum with the imager spectral responsivity and filter nominal transmittance; the asphalt spectral contribution is practically flat in the considered wavelength range
- there is a noticeable spectral modification from dry to wet conditions around the 1500 nm wavelength

---

[1] Regions Of Interest

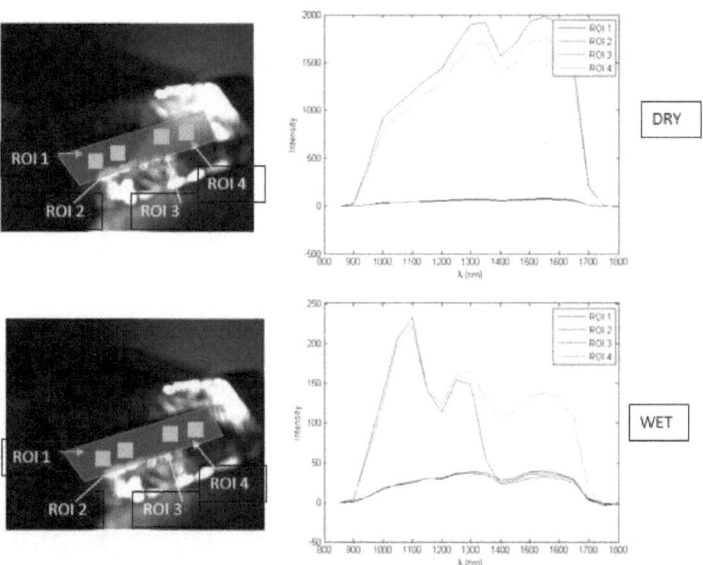

Fig. 6. Laboratory tests with the asphalt sample images (left) acquired in the halogen lamp's full bandwidth light conditions and the 4 ROIs mean pixel intensities graphs (right). The actual size of the asphalt sample is emphasized using the superimposed red bounding box.

Spectrometer filters	Asphalt condition	1500 / 1100	1500 / 1200	1500 / 1300	1500 / 1400	1500 / 1600
ROI_1	dry	1.57	1.30	1.01	1.23	1.02
ROI_2	dry	1.60	1.33	1.02	1.22	1.01
ROI_3	dry	1.63	1.36	1.02	1.23	1.00
ROI_4	dry	1.64	1.34	1.02	1.23	1.01

Spectrometer filters	Asphalt condition	1500 / 1100	1500 / 1200	1500 / 1300	1500 / 1400	1500 / 1600
ROI_1	wet	0.53	0.69	0.53	1.27	1.00
ROI_2	wet	0.13	0.22	0.17	1.50	0.82
ROI_3	wet	0.23	0.43	0.33	1.35	0.99
ROI_4	wet	0.60	1.06	0.82	1.29	1.00

Fig. 7. Ratios relative to fig. 6 asphalt samples

– in non-illuminated ROIs the spectrum is flat and no significant modifications can be observed between the different asphalt conditions

In order to quantify the differences among spectra, these results have been analyzed considering the ratios between some wavelengths. The tables in fig. 7 illustrate the ratios corresponding to the analyzed ROIs. Columns with the 1500/1100 ratios showed the strongest changes in the diverse asphalt conditions.

The most relevant data from these analysis is show how the illumination can be an issue in developing an algorithm for the road monitoring.

6   Outdoor Results

The same methodology described in section 5 has been applied on images acquired in real scenarios. Dry, wet and icy road conditions at daytime have been investigated.

In the following, two scenes showing different illumination and road conditions have been selected (see fig. 8 and fig. 9). The spectral analysis has been done measuring the

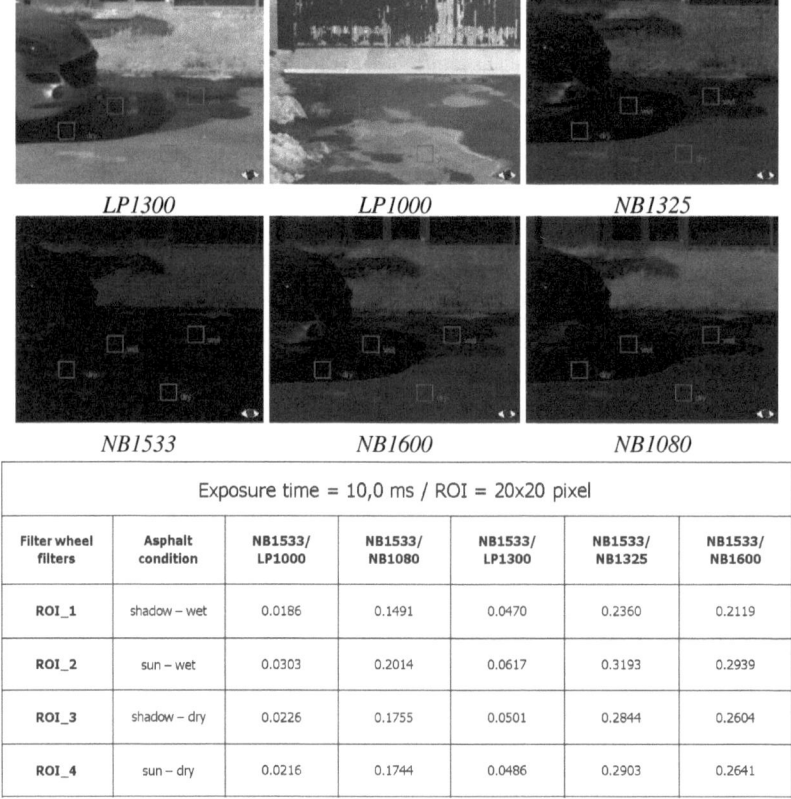

Exposure time = 10,0 ms / ROI = 20x20 pixel						
Filter wheel filters	Asphalt condition	NB1533/ LP1000	NB1533/ NB1080	NB1533/ LP1300	NB1533/ NB1325	NB1533/ NB1600
ROI_1	shadow – wet	0.0186	0.1491	0.0470	0.2360	0.2119
ROI_2	sun – wet	0.0303	0.2014	0.0617	0.3193	0.2939
ROI_3	shadow – dry	0.0226	0.1755	0.0501	0.2844	0.2604
ROI_4	sun – dry	0.0216	0.1744	0.0486	0.2903	0.2641

Fig. 8. Summer scene: dry and wet asphalt areas, illuminated and not, in a clear sky day. The table underneath shows the corresponding ratios.

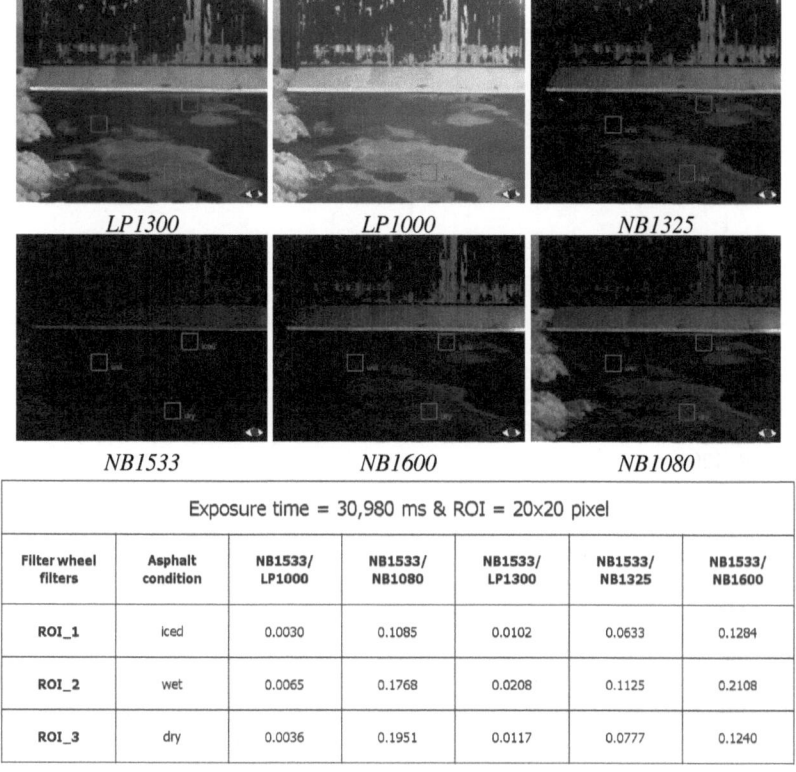

Exposure time = 30,980 ms & ROI = 20x20 pixel						
Filter wheel filters	Asphalt condition	NB1533/ LP1000	NB1533/ NB1080	NB1533/ LP1300	NB1533/ NB1325	NB1533/ NB1600
ROI_1	iced	0.0030	0.1085	0.0102	0.0633	0.1284
ROI_2	wet	0.0065	0.1768	0.0208	0.1125	0.2108
ROI_3	dry	0.0036	0.1951	0.0117	0.0777	0.1240

Fig. 9. Winter scene: dry, wet and icy asphalt areas in a cloudy day. The table underneath shows the corresponding ratios.

intensity values of the selected ROIs by using the filters included in the filter wheel operating in the SWIR bandwidth only.

The resulting ratios, shown in fig. 8 and fig. 9, underline a behavior comparable to the indoor data although some relevant differences are noticeable:

- ratio values are different respect to lab ones due to the different source spectrum, halogen lamp in the lab and sun outdoor
- due to modification in illumination condition (clouds, etc.) during the acquisition (spectra are collected by means of temporal sequential measurements of the filter wheel filters), it is not possible to find a ratio as good indicator for road condition.

Taking into account the previous considerations, several measurements have been done in order to characterize how the presence of clouds could affect the ratios. Spectra in a changeable weather day, initially with clouds and then clear, have been collected using a calibrated spectrometer. In fig. 10.a the temporal spectral evolution is compared to the theoretical solar spectra at sea level (black line).

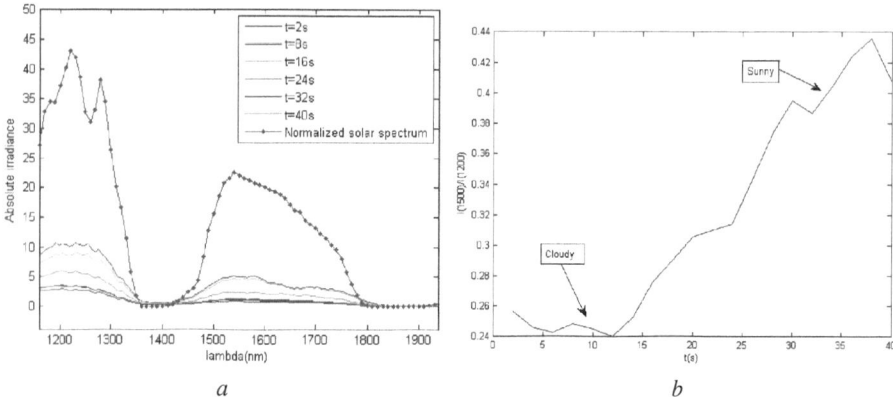

Fig. 10. Dealing with clouds: *(a)* solar spectra collected during a cloudy day and *(b)* variations of the I(1500)/I(1200) ratio

In order to understand the contribution of the clouds, we collected some outdoor spectra during a cloudy day. In fig. 10.b are shown the variations of the I(1500)/I(1200) ratio during the acquisition. It has not been possible to evaluate the I(1500)/I(1100) ratio due to the spectral sensitivity limitation of the spectrometer.

From these measurements it can be noticed that illumination changes not only affect the intensity levels at all wavelengths but, due to the extra absorption of cloud water molecula, some wavelength ranges, i.e. 1500 nm, are more deprived.

7   Conclusions

SWIR is nominally the wavelength range of about 1 to 2 nm that, similar to visible light, responds primarily to light reflected from objects rather than to thermal emissions coming from those objects. Therefore the only applications served by SWIR are those which benefit from the reduced scattering effects of longer wavelengths like illumination from invisible sources as passive illumination due to the upper atmosphere's night glow or active illumination coming from eye-safe lasers or from thermal emissions of objects with temperatures above 150 $°C$. So illumination is an important issue when dealing with SWIR images.

Our tests have shown that where for indoor acquisitions the lamp spectrum affects results only by a multiplicative factor, outdoor the unpredictable changes in illumination not only affect the intensity levels at all wavelengths but, due to the extra absorption of cloud water molecula, different wavelength ranges are also differently affected. Image processing techniques applied to satellite and airborne pictures have also been taken into account looking for a procedure to limit this unwanted behavior, but all spectral analysis techniques are applied to images clear of clouds, a hard restriction which is totally unsuitable for functions to be applied in the automotive field.

Acknowledgments. The work described in this paper has been developed in the framework of the 2WIDE_SENSE (WIDE spectral band & WIDE dynamics multifunctional imaging SENSor Enabling safer car transportation) Project funded by EU within the FP7 Seventh Framework Programme.

References

1. Bertozzi, M., Fedriga, R.I., Miron, A., Reverchon, J.-L.: Pedestrian Detection in Poor Visibility Conditions: Would SWIR Help? In: Procs. of the IAPR Intl. Conf. on Image Analysis and Processing, Naples, Italy (September 2013)

2. Bian, M., Li, K., Jin, D., Lian, X.: Road Condition Estimation for Automotive Anti-Skid Control System Based on BP Neural Network. In: Procs. of the IEEE Conf. on Mechatronics & Automation, pp. 1017–1022. IEEE Computer Society, Niagara Falls (July 2005)

3. Casselgren, J., Jokela, M., Kutila, M.: Slippery road detection by using different methods of polarised light. In: Meyer, G. (ed.) Advanced Microsystems for Automotive Applications 2012, vol. 113, pp. 207–220. Springer, Heidelberg (2012)

4. Casselgren, J., Sjödahl, M., LeBlanc, J.P.: Angular spectral response from covered asphalt. Applied Optics 46(20), 4277–4288 (2007)

5. Casselgren, J., Sjödahl, M., LeBlanc, J.P.: Model–based winter road classification. International Journal of Vehicle Systems Modelling and Testing 7(3), 268–284 (2012)

6. Hansen, M.P., Malchow, D.S.: Overview of SWIR detectors, cameras, and applications. In: Procs. of the SPIE. Thermosense XXX, vol. 6939 (March 2008)

7. Akama, S.I., Tabaru, T., Shin, S.: Bayes Estimation of Road Surface Using Road Noise. In: Procs. of the IEEE Conf. of Industrial Electronics Society, pp. 2923–2928. IEEE Computer Society, Busan (November 2004)

8. Jokela, M., Kutila, M., Lu, L.: Road Condition Monitoring System Based on a Stereo Camera. In: Procs. of the IEEE Conf. on Intelligent Computer Communication and Processing, pp. 423–428 (August 2009)

9. KfV, NTUA, SWOV, and TRL. SafetyNet, Annual Statistical Report 2008. Technical report, European Road Safety Observatory (2008)

10. Lin, P.P., Ye, M., Lee, K.-M.: Intelligent Observer-Based Road Surface Condition Detection and Identification. In: Procs. of the Conf. Systems, Man and Cybernetics, pp. 2465–2470. IEEE Computer Society, Cleveland (October 2008)

11. Lu, Y., Higgins-Luthman, M.J.: Black ice detection and warning system. US Patent App. 11/948, 086 (November 30, 2007)

12. Malchow, D.: NIR Trends: Machine vision in the Short Wave Infrared. In: UTC Aerospace Systems (Sensors Unlimited Products). Goodrich Corporation (March 2009)

13. Mika, M., Hiroyuki, Y., Takao, K., Takeshi, I., Mitsuo, S.: Road Surface Condition Detector Using High Peak Power Fiber Laser. Transactions of the Institute of Electrical Engineers of Japan 10(4), 1198–1204 (2000)

14. Omer, R., Fu, L.: An automatic image recognition system for winter road surface condition classification. In: Procs. of the IEEE Intelligent Transportation Systems 2010, pp. 1375–1379 (September 2010)

15. Viikari, V., Varpula, T., Kantanen, M.: Automotive Radar Technology for Detecting Road Conditions. Backscattering Properties of Dry, Wet, and Icy Asphalt. In: Procs. of the 5th European Radar Conference, pp. 276–279. IEEE Computer Society, Amsterdam (2008)

Simple and Robust Facial Portraits Recognition under Variable Lighting Conditions Based on Two-Dimensional Orthogonal Transformations

Paweł Forczmański[1], Georgy Kukharev[1], and Nadezdha Shchegoleva[2]

[1] Faculty of Computer Science and Information Systems,
West Pomeranian University of Technology, Szczecin, Poland
{pforczmanski,gkukharev}@wi.zut.edu.pl
http://pforczmanski.zut.edu.pl
[2] Saint Petersburg State Electrotechnical University (LETI), Russia
stil_hope@mail.ru

Abstract. The paper addresses the problem of face recognition for images registered in variable lighting, which is common for real-world conditions. Presented algorithm is based on orthogonal transformation preceded by simple transformations comprising of equalization of brightness gradients, removal of spatial low frequency spectral components and fusion of spectral features depending on average pixels intensity. Two types of transformations: 2DDCT (two-dimensional Discrete Cosine Transform) and 2DKLT (two-dimensional Karhunen-Loeve Transform) were investigated in order to find the most optimal algorithm setup. The results of experiments conducted on Yale B and Yale B+ datasets show that a quite simple algorithm is capable of successful recognition without high computing power demand, as opposite to several more sophisticated methods presented recently.

Keywords: face recognition, illumination compensation, dimensionality reduction, DCT, PCA.

1 Introduction

The problem of facial portraits recognition has been attracting scientists interest for several recent years. Although it might has been solved in a satisfactory way, due to its extensive characteristic, it is still interesting. It is because most of the proposed solutions operate mainly in the very limited real-world conditions (i.e. controlled conditions of imaging, very restricted assumptions related to pose, orientation, expression etc.). The continuous research is also driven by practical implementations on low computing power devices like smartphones and tablets. Hence, there is still a need for robust solutions as simple as possible, yet without serious performance degradation.

One of the most vital issues in face recognition practice is a variation of light intensity in input images. In such case, we have to deal with two types of distortions perceived in the face area and its background. The first one is related

A. Petrosino (Ed.): ICIAP 2013, Part I, LNCS 8156, pp. 602–611, 2013.

to the presence of local shadows, while the second one is linked to so called global shadows. Local shadows tend to change not only the form of elementary parts of face (eyes, nose and mouth) but also influence the boundaries of whole face area. On the other hand, global shadows significantly reduce the possibility to distinguish face areas from the background or even hide them completely.

Above problems highly degrade the accuracy of FaReS (Face Recognition System), thus this is one of the main reasons of constant interest of face recognition specialists [1–11]. Literature review shows, that the problem of variable lighting conditions is solved by one of just a few methods, namely 1. extending the FaReS database with a set of new images characterizing all distortions related to lighting problems, 2. transforming the images in order to find features that are independent of light characteristics, 3. reducing local intensity gradients or 4. transforming the images from spatial domain of into spectral one by means of models based on Eigenfaces, wavelets or cosine functions together with elimination of low-frequency components.

It should be emphasized, that in practice we have in our disposal only a limited training database, thus the implementation should employ FaReS base which is not extended with templates having all possible variants of shadows [4, 6, 9] as opposite to setups presented in [1–3, 5, 7, 8].

In order to successfully recognize images in difficult lighting conditions we often need to find their light-invariant representation. There are several approaches that combine different image characteristics, like Gabor features or Local Binary Patterns (LBP). Such an approach presented in [13] leads to the very complex multi-tier algorithm, which also requires a normalization of scales within different features.

One of the most popular approaches related to eigenbasis representation employs Principal Component Analysis (PCA), Linear Discriminant Analysis (LDA) [1–3], and Canonical Correlation Analysis (CCA) [6]. All of them often use Discrete Cosine Transform (DCT) as image preprocessing step.

There are also several approaches to the problem of illumination compensation involving wavelets, e.g. [10] or very complex models consisting of pose estimation and further illumination handling [11]. The latter requires a complex pose estimation (often performed manually), then performs edge filtering and PCA+LDA to estimate light direction. Finally, it compensates the illumination (providing it has properly recognized a direction) and extracts features using another LDA. Above approach is time-consuming and depends on many intermediate results, which in practice may be unpredictable.

The application of 2DDCT has been investigated and proved its high robustness at the stage of facial portraits recognition [1, 14]. In such methods, during feature extraction stage, the low frequency spectral components are often removed, as corresponding to shadows covering faces. The procedure in [2] involves the inverse transformation of images from spectral domain into spatial domain just before PCA step (which, according to authors' claims, increases the recognition accuracy). Similar approach [5], employs calculations related to PCA optimized by the Gramm matrix evaluation. Even though, presented results are not better than average. The typical processing scheme employing DCT was also presented in [15]. The authors performed DCT on the whole image and kept first 49 coefficients later used in a standard PCA recognition scenario. Unfortunately, the authors fail to justify the selection of such number of spectral components. Moreover, the experiments were performed on a few small databases which do not give an objective knowledge about the method's robustness. Some local features derived form DCT were also employed in [16], however the performance of the method was evaluated on a very small number of images.

As it will be shown, above approaches do not give higher recognition accuracy in contrast to the method presented in our work. Although, the PCA used at the final stage of many of them is a bottleneck when it comes to fast database updates, it is often included in the processing pipeline. Hence, in this paper we focus on a a comparison of two methods involving orthogonal transformations (PCA/KLT and DCT). It should be also noted, that most of approaches use simple 1D PCA, which requires eigen-decomposition of very large covariance matrices. In our approach we use 2D method [17] that employs smaller covariance matrices (which is many times faster and requires less memory storage).

This paper presents an algorithm aimed at lighting problems elimination in the aspect of facial portraits recognition, which is very simple in comparison to the earlier presented approaches, yet its accuracy is very similar. Its main features are: template database not containing images characterizing local and/or global shadows, adaptive preprocessing of original images aimed at reduction of intensity gradient (gamma correction and intensity logarithm), orthogonal transformation (2DDCT or 2DPCA/2DKLT) as the only instruments of transformation of original data into the low dimensional space, and finally, the adaptive combination of spectral features at the classification step.

2   Algorithm Details

2.1   Image Preprocessing

It should be noted that presented method is not fully invariant to rotations and shifts. However, there are several methods of face normalization that can be used. Hence, we assume that the input face is geometrically processed. On the other hand, it is evident that without brightness equalization the recognition of images taken under complex lighting conditions would be very difficult.

The methods of brightness enhancing can utilize gamma correction or logarithmic correction which compensate light intensity for properties of human vision, in order to maximize the use of bandwidth relative to how humans perceive light [1–9]. However, it may influence such parameters of input image as mean value, contrast, local brightness and boundaries of shadows. Moreover, it may partially change the form and characteristics of image, which can have a negative influence on the recognition accuracy. Figure 1 presents images taken under variable lighting conditions and the results of applying one of two enhancing procedures: gamma correction (G) and brightness logarithm (Log). After such operations we can still observe some distortions (like shadows, bright spots, low contrast and noise). Despite this fact, the boundaries of all face parts can be easily detected, and its anthropometric parameters may be successfully explored. It can be also seen that above operations unveil different parts of face, which are originally hidden in the shadows, hence it may increase the FaReS accuracy. In contrary to the works presented earlier, in our approach we propose a fusion of features obtained after gamma correction and logarithmic transform for images with extreme lighting conditions. According to our observations, the combination of these two types of preprocessing stages outperforms single one especially when an image is very dark. Hence, the further increase in the recognition accuracy is caused by the adaptation to the image brightness.

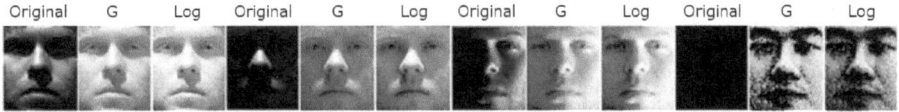

Fig. 1. Gamma correction (G) and intensity logarithm (Log) applied to original images

In many previous works additional stages of processing were based on PCA/KLT approach. Our experiments described recently [18] and in the further part of this paper proved, that using much simpler approach, namely DCT, makes it possible to obtain good recognition results. The basis functions of cosine transform are optimal (in terms of minimal loss of energy) for approximation of eigenfunctions for the whole population (or just large collections) of digital images. As the cosine transformation represents original face images with relatively small number of features, we decided to exclude the entire preprocessing step based on PCA/KLT. For an adequately precise reconstruction of the face (in case of an image with uniform lighting) it is enough to take not more than 20 spectral components, selected from upper left corner of spectrum matrix [14]. However, for the comparison purposes, we show the results of experiments involving KLT.

2.2 Recognition Stage

Developed algorithm for image of size $M \times N$ consists of five following stages, and is common for both 2DDCT and 2DPCA/2DKLT methods (see Fig. 2):

1. Two projection matrices are build: F_1 and F_2 of sizes $d_1 \times M$ and $N \times d_2$, respectively ($d_1 \ll M$ and $d_2 \ll N$; $d_1 \neq d_2$ in general case). In case of 2DDCT this stage is equal to selecting proper functions from the cosine basis, while in case of 2DPCA/2DKLT, such functions are calculated by means of 2DPCA/2DKLT applied for training images only [17]. The step related to 2DPCA/2DKLT requires extra computations, and from the practical point of view, increases the computational complexity of the whole approach.

2. Transformation (2DDCT or 2DPCA/2DKLT, respectively) for templates I with preliminary processing, described above, resulting a database of processed templates $\tilde{I}$. We define both two-dimensional linear transformations in a matrix form as the following multiplication: $Y = F_1 \tilde{I} F_2$, where Y - transformation result; F_1 and F_2 - projection matrices of size $d_1 \times M$ and $N \times d_2$, Note, that the projection of original images of size $M \times N$ into new feature space, results in a new representation with spectral matrices Y of size $d_1 \times d_2 \ll MN$.

3. Gamma correction (G) or gamma correction followed by logarithmic correction (G&Log) performed on input image. The preprocessing type depends on average image brightness J. If values of J is lower than a certain threshold P, then both optional steps are executed, otherwise only one step i.e. gamma correction or logarithmic correction is employed (the performance of both variants is similar).

4. Transformation of preprocessing results into a new feature space based on 2DDCT/2DKLT. Projection result consists of spectral matrices (precisely, one or two matrices) of size $d_1 \times d_2 << MN$. Final projection result (steps 2 and 4) is presented as a vector, read from top left corner of spectral matrices. Number of elements of such vector is equal to $d \times (d + 1)/2$, where $d = \min(d_1, d_2)$.

5. Classification of the result obtained in step 4 through comparison with features of images from all known classes, stored in the training set (results of step 2). A classification criterion is the minimum distance in metric L_1. Test image is assigned to a class (one of known classes) having a minimum distance.

The problem of recognizing extremely dark images is solved trough the following (additional) processing. Before the classification stage, mean brightness J of input image is calculated and if it is lower than a predefined threshold $J < P$, then the projection results (step 4 of the algorithm) are fused in a one common vector, which is compared to respective collection of features stored in the database. In presented experiments, a value of P was set to 110, as corresponding to shadows covering minimum half of face area.

3 Experiments

3.1 Characteristics of Experimental Data

Most of the reported experiments in the area of facial portrait recognition in the presence of variable lighting are conducted on Yale database (including original Yale B images and its extension called Yale B+) [12]. It features a quite wide diversity of illumination variations and the proper registration of images, namely they all are cropped and normalized in such way, that the eyes positions are constant in all images across all subjects. It allows to investigate the illumination influence without the overhead of geometrical transformation.

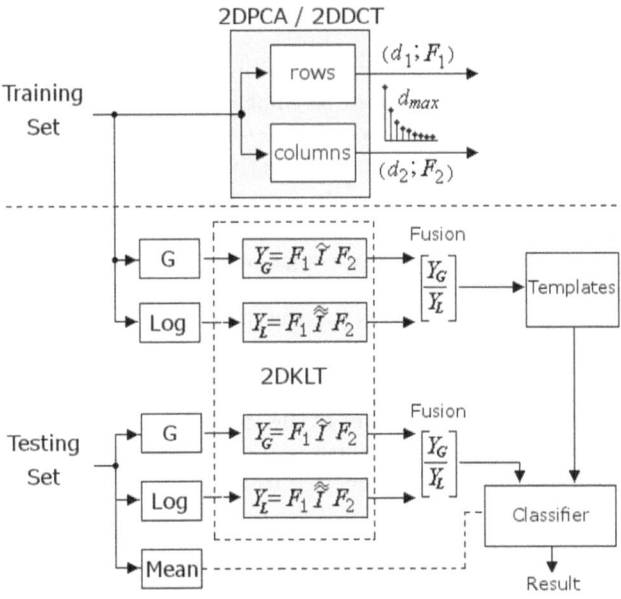

Fig. 2. Structure of a recognition system based on 2DPCA/2DDCT

In our experiments we used 2452 out of 2470 images from Yale B and Yale B+ sets, containing the central part of face area of 38 subjects (18 images were eliminated since they cannot be read from files published on web site [12]). The grayscale images are stored in matrices of 192×168 pixels. They are divided into 6 sets, labeled Subset 0 – Subset 5, respectively. In Fig. 3 one can see examplary images coming from this dataset. Images in Subset 0 have no blinks nor shadows and are evenly illuminated. Images in Subset 1 – Subset 5 were obtained by modeling the spatial movement of a light source, thus they represent various variants of shadows/ flashes (Subset 1 and Subset 2), local shadows (Subset 3) and lateral shadows (Subset 4 and Subset 5), as well as global shadows (Subset 5). The most challenging are images from Subset 3 – Subset 5.

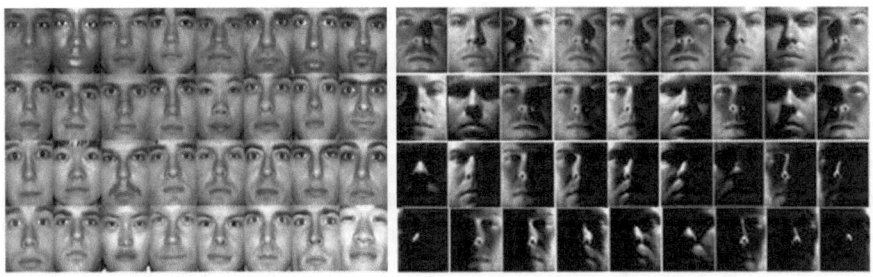

Fig. 3. Selected images from Yale set: Subset 0 (left) and Subsets 1–5 (right)

3.2   Results

The experiments employed two scenarios of dividing the database: the first one employed a training set consisting of 38 images (one image per each class) and 2414 testing images, while the second one consisted of 228 training images (6 images per class) from Subset 1 and 2151 testing images from Subset 2 – Subset 5 (from all 38 classes). All experiments were conducted without extending the database with any other preprocessed images.

We selected G&Log approach as a compromise between complexity/accuracy and investigated the influence of training set selection strategy and transformation type on the recognition accuracy. We investigated the influence of the number of spectral components on the recognition accuracy (results in Tab. 1). In most cases we used 20–40 first spectral components, thus they are strongly related to illumination characteristics of the input image. For images taken from Subset 5, we selected components that are further from the origin, which makes it possible to remove any artifacts related to directional illumination.

Table 1. The results of classification involving 2DDCT and 2DPCA/2DKLT

Number of templates	38 (one image per class)						228 (6 images per class)				
Subset No.	1	2	3	4	5	All	2	3	4	5	All
Number of test images	263	456	455	526	714	2414	456	455	526	714	2151
Transformation	2DDCT										
Misclassified images	3	1	30	27	69	130	0	0	1	16	17
Recognition rate [%]	98.9	99.8	93.4	94.9	90.3	94.6	100	100	99.8	97.8	99.2
Transformation	2DPCA/2DKLT										
Misclassified images	4	1	31	43	91	170	0	0	4	22	26
Recognition rate [%]	98.5	99.8	93.2	91.8	87.2	93.0	100	100	99.2	96.9	98.8

The experiments exposed the problem of recognizing images taken at a very low level of illumination. Several misclassified images are presented in Fig. 4 (most of them belong to Subset 5). As it can be seen, these images have a very low intensity level and a very high contrast. The proposed preprocessing scheme does not increase the visual quality of such images since they contain hardly no information. Mean intensities of presented images vary from 0 to 28. This observation may be a valuable hint for developing a practical implementation in which images with such low brightness level should be rejected as unusable.

3.3   Comparison with State-of-the-Art

It should be noted, that there are many recent methods that use DCT and characterize good recognition accuracy. One of them [1] employs DCT+PCA. The reported accuracy is 94%, which is lower than ours. The most similar approach [2] includes logarithm operator, DCT and low-frequency components extraction. The reported accuracy (98.3%) is a little bit lower than ours. The perfect recognition (100%) is reported in [24]. However, it works on extracted facial parts and

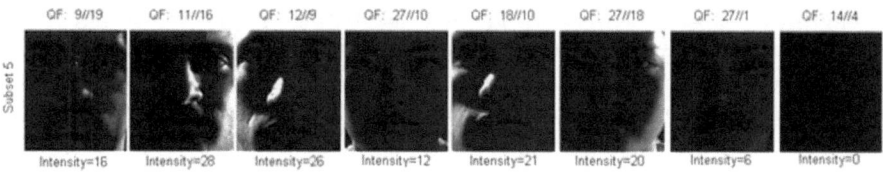

Fig. 4. Selected misclassified images

the increase in the recognition accuracy was achieved through the independent recognition of all facial parts and a highly-complex processing (reduction of DCT components, cross-correlation, PCA, k-NN etc.).

The face recognition with illumination suppression involving other popular techniques is provided below. The experiments were performed on Yale B and Yale B+ datasets. We focus on recognition rates for the following methods: Single Scale Retinex algorithm (SSR) [19], Logarithmic Total Variance technique (LTV) [20], Wavelet-Denoising-based method (WD) [13] and Nuisance Attribute Projection (NAP) [21]. All above techniques were implemented without any pre- or post-processing (e.g., histogram equalization) [21]. The computational expenses comparison of those methods shows that the proposed method is less complex. NAP [21] requires a projection matrix for all the images in the database. When the database changes, the projection should be re-build. Moreover, it requires an eigenvectors decomposition which is itself very time-consuming task. The LTV [20] also is not a trivial one. It features an edge detection stage (convolution filtering) at different scales and again, PCA, at the stage of features reduction. The SR [19] is a global approach and the results of directional lighting compensation are much worse than any other propositions. The last of the methods addressed in the paper [13] requires fast wavelet transform but again, PCA at the final stage slows the whole method down.

The results of the accuracy comparison (see Tab. 2) suggest that our technique gives competitive results when compared to other state-of-the-art methods.

Table 2. Comparison of the performance [%] of other state-of-the-art methods

Subset No.	Our Method	NAP	SSR	LTV	WD
2	100	100	100	100	100
3	100	100	99.2	100	100
4	99.8	99.3	82.9	99.3	98.6
5	97.8	96.8	81.1	99.5	99.5

4   Summary

The article presents a simple, yet effective method of recognizing facial portraits taken under difficult lighting conditions. There is still a gap between simple methods (fast and easy to implement) and sophisticated ones (that require complex computations and tuning). Our approach involves simple preprocessing stages (gamma correction and logarithmic correction) and feature extraction by means of two-dimensional orthogonal transformations (2DDCT and 2DPCA/2DKLT). They are handled by many hardware systems that work on fixed-point arithmetic. Here, we propose an adaptive fusion of features obtained after gamma correction and logarithmic transform for images with extreme lighting conditions which increases overall recognition accuracy. During experiments performed on Yale datasets it was shown that the 2DDCT method together with the brightness correction, fusion of features according to current mean value of brightness as well as removal of low frequency components of spectrum allow to achieve higher efficiency of recognition in case of facial portraits with illumination problems. The accuracy of the developed algorithm is better than ones presented in [1–10, 13, 19–21]. Moreover, it is easier to describe and simple in implementation and does not require any other pre-processing (like downsmpling [4]).

Another interesting observation is related to the differences between 2DDCT and 2DPCA/2DKLT as the main transformation stage. Despite theoretically better adaptation of basis functions (2DPCA), in our experiments, 2DDCT components perform with slightly better efficiency. Its higher efficiency is caused by the fact, that 2DPCA/2DKLT requires much wider representation of training data to form an optimal eigenbasis. In the performed experiments the basis was calculated using quite low number of images which resulted in worse adaptation to the testing set. In contrary, 2DDCT components in such case give a much better basis. This is especially visible when the basis is calculated for images without directional illumination, strong shadows or low brightness. We showed, that Eigenfaces method is not always the best choice when it comes to appearance-based recognition.

References

1. Chen, W., Er, M.J., Wu, S.: PCA and LDA in DCT domain. Pattern Recognition Letters 26, 2474–2482 (2005)
2. Chen, W., Er, M.J., Wu, S.: Illumination compensation and normalization for robust face recognition using discrete cosine transform in logarithm domain. IEEE Trans. Syst. Man Cybern. Part B 36(2), 458–466 (2006)
3. Tan, X., Triggs, B.: Preprocessing and Feature Sets for Robust Face Recognition. In: IEEE Conf. on Computer Vision and Pattern Recognition, CVPR 2007, pp. 1–8 (2007)
4. Xie, X., Zheng, W.-S., Lai, J., Yuen, P.C.: Face Illumination Normalization on Large and Small Scale Features. In: IEEE Conf. on Computer Vision and Pattern Recognition, CVPR 2008 Anchorage, pp. 1–8 (2008)
5. Abbas, A., Khalil, M.I., AbdelHay, S., Fahmy, H.M.A.: Illumination invariant face recognition in logarithm discrete cosine transform domain. In: IEEE Inter. Conf. of Image Processing, ICIP 2009, pp. 4157–4160 (2009)
6. Shao, M., Wang, Y.: Joint Features for Face Recognition under Variable Illuminations. In: Fifth Inter. Conf. on Image and Graphics, ICIG 2009, pp. 922–927 (2009)

7. Liau, H.F., Isa, D.: New Illumination Compensation Method for Face Recognition. Inter. Journal of Computer and Network Security 2(3), 308–321 (2010)
8. Han, H., Shan, S., Qing, L., Chen, X., Gao, W.: Lighting Aware Preprocessing for Face Recognition across Varying Illumination. In: Daniilidis, K., Maragos, P., Paragios, N. (eds.) ECCV 2010, Part II. LNCS, vol. 6312, pp. 308–321. Springer, Heidelberg (2010)
9. Goel, T., Nehra, V., Vishwakarma, V.P.: Comparative Analysis of various Illumination Normalization Techniques for Face Recognition. Inter. Journal of Computer Applications 28(9), 1–7 (2011)
10. Cao, X., Shen, W., Yu, L.G., Wang, W.L., Yang, J.Y., Zhang, Z.W.: Illumination invariant extraction for face recognition using neighboring wavelet coefficients. Pattern Recognition 45, 1299–1305 (2012)
11. Choi, S., Choi, C.-H., Kwak, N.: Face recognition based on 2D images under illumination and pose variations. Pattern Recognition Letters 32, 561–571 (2011)
12. The Extended Yale Face Database B,
    http://vision.ucsd.edu/~leekc/ExtYaleDatabase/ExtYaleB.html
    (accessed May 01, 2012)
13. Zhang, T., Fang, B., Tang, Y.Y., Shang, Z., Li, D., Lang, F.: Multiscale facial structure representation for face recognition under varying illumination. Pattern Recognition 42(2), 251–258 (2009)
14. Forczmański, P., Kukharev, G.: Comparative analysis of simple facial features extractors. Journal of Real Time Image Processing 1(4), 239–255 (2007)
15. Hafed, Z.M., Levine, M.D.: Face Recognition Using the Discrete Cosine Transform. International Journal of Computer Vision 43(2), 167–188 (2001)
16. Schwerin, B., Paliwal, K.: Local-DCT features for facial recognition. In: 2nd Inter. Conf. on Signal Processing and Communication Systems, ICSPCS 2008, pp. 1–6 (2008)
17. Kukharev, G., Forczmański, P.: Facial images dimensionality reduction and recognition by means of 2DKLT. Machine Graphics & Vision 16(3/4), 401–425 (2007)
18. Forczmański, P., Kukharev, G., Shchegoleva, N.: An algorithm of face recognition under difficult lighting conditions. Przeglad Elektrotechniczny (Electrical Review) 10b, 201–205 (2012)
19. Jobson, J., Rahman, Z., Woodell, G.A.: Properties and performance of a Center/Surround Retinex. Trans. on Image Processing 6(3), 451–462 (1997)
20. Chen, T., Yin, W., Zhou, X.S., Comaniciu, D., Huang, T.S.: Total variation models for variable lighting face recognition. TPAMI 28(9), 1519–1524 (2006)
21. Struc, V., Vesnicer, B., Mihelic, F., Pavesic, N.: Removing illumination artifacts from face images using the nuisance attribute projection. In: ICASSP 2010, pp. 846–849 (2010)
22. Zhiming, L., Chengjun, L.: Fusion of color, local spatial and global frequency information for face recognition. Pattern Recognition 43(8), 2882–2890 (2010)
23. Akrouf, S., Sehili, M.A., Chakhchoukh, A., Mostefai, M., Youssef, C.: Face Recognition Using: PCA and DCT. In: Proceedings of the 2009 Fifth International Conference on MEMS NANO, Smart Systems, ICMENS 2009, pp. 15–19 (2009)
24. Vishwakarma, V.P., Pandey, S., Gupta, M.N.: An Illumination Invariant Accurate Face Recognition with Down Scaling of DCT Coefficients. Journal of Computing and Information Technology - CIT 18(1), 53–67 (2010)

Investigation of Different Classification Models to Determine the Presence of Leukemia in Peripheral Blood Image

Lorenzo Putzu and Cecilia Di Ruberto

Department of Mathematics and Computer Science, University of Cagliari,
via Ospedale 72, 09124 Cagliari, Italy
lorenzo.putzu@gmail.com, dirubert@unica.it

Abstract. The counting and classification of blood cells allows the evaluation and diagnosis of a vast number of diseases, such as the ALL - Acute Lymphocytic Leukemia, detected through the analysis of white blood cells (WBCs). Nowadays the morphological analysis of blood cells is performed manually by skilled operators, involving numerous drawbacks, such as slowness of the analysis and a non-standard accuracy, dependent on the operator skills. In literature there are only few examples of automated systems able to process a whole image in order to analyze and classify all the WBCs included. This paper presents a complete and fully automatic method for WBCs identification from microscopic images and an evaluation of different classification model to determine the presence of leukemia. Experimental results show that the proposed method is able to identify the cells carrying leukemia and consequently to determine whether a patient is suffering from this disease.

Keywords: Automatic detection, Classification, Feature selection, Leukemia, Segmentation, White blood cell analysis.

1 Introduction

ALL is a blood cancer that influences a group of leukocytes called lymphocytes, and primarily affects children and adults over 50 years and due to its rapid expansion into the bloodstream and vital organs can be fatal if left untreated [1]. An early diagnosis of the disease is crucial for patients' recovery, especially in the case of children. The observation of blood samples under a microscope is one of the possible procedures for the diagnosis of ALL. This method suffers from slowness and provides a non-standard accuracy dependent on the operator skills. Image processing techniques can help to count the cells in the human blood quickly and, at the same time, provide more accurate information on the cells morphology. Unfortunately the generic term leukocytes refers to a set of cells that are very different between them, in shape and size, which includes neutrophils, basophils, eosinophils, lymphocytes and monocytes, so data extraction from WBCs can present some complications. Furthermore lymphocytes suffering from ALL, called lymphoblasts, have additional morphological changes, like

A. Petrosino (Ed.): ICIAP 2013, Part I, LNCS 8156, pp. 612–621, 2013.

shape and size irregularities, that increase with increasing severity of the disease. Therefore, in this paper we propose a fully automatic procedure to support the medical activity, able to identify all types of WBCs present in the microscopic images, which need various steps to reach the goal, and then classify WBCs as suffering from ALL or not. The identification of the leukocytes is carried out in the first step, described in Section 2. The second step deals with the selection of the nucleus and the cytoplasm of each leukocyte, described in Section 3. The third step deals with the features extraction, described in Section 4, and the last phase proceeds to the classification of WBCs, described in Section 5. Each phase of the method, applied on a sample image, is analyzed in detail and compared with other approaches present in literature. The whole process can be schematized as showed in Figure 1.

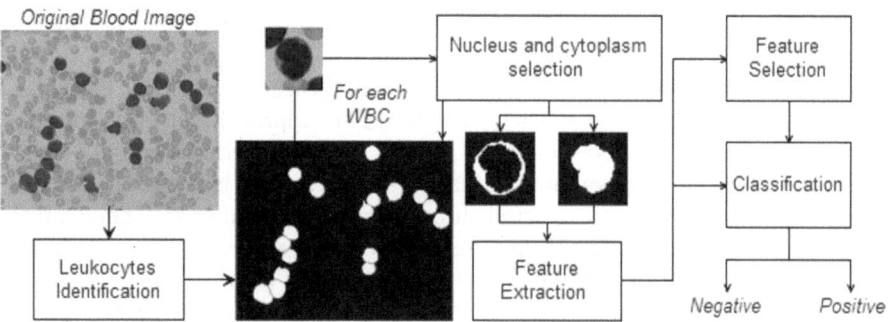

Fig. 1. Proposed method diagram

2   Leukocytes Identification

In many methods present in literature the idea is to identify firstly the nuclei which are more prominent than other components [8] and then to detect the entire membrane, for example by region growing [2], [6]. In the proposed method instead, the membrane is detected firstly thanks to the conversion in the CMYK color model, because the leukocytes are more contrasted in the Y component. After a redistribution of image gray levels by histogram equalization (Fig. 2(b) shows an example), we segment the image using a threshold automatically calculated through the triangle method or Zack algorithm [11] (see Fig. 2(c)), that we consider the best threshold technique available in the literature [5] for this application. To get a better result it is necessary to remove the image background. Some approaches for background extraction are present in literature, such as that showed by Scotti [10] that makes use of a collection of images for the estimation of the background pixels. The proposed approach involves the use of an automatic threshold, calculated again using the triangle method, but this time starting from the G component of the RGB color space (see Fig. 2(e)). Background removal was performed with an arithmetical operation and an area opening operation (see Fig. 2(f)). Another problem to be addressed in analysis of blood image is the presence of agglomerates of leukocytes. Several methods can be used to verify the presence of adjacent leukocytes [5]. In our work each connected component having a roundness value lower than 0.80, identified during our experimentation as the optimal threshold, is classified as grouped leukocytes and so it must be separated. Some approaches to separate the adjacent cells, used by Kovalev [6], work on sub-images extracted from the original image by cutting a square around the nucleus previously segmented. So, assuming that each sub-image has a single WBC, a clustering around the nucleus is performed, by using shape and color information. Our approach works on the whole image, avoiding problems that may arise after nucleus identification (even the nuclei of white blood cells may be in contact) and is based on the method proposed by Lindblad [7] which uses the distance transform. The latter, applied to the binary image, associates to each pixel its distance from the border. A watershed segmentation is then applied to the distance transform to make a first separation between adjacent leukocytes. This approach performs well only in the presence of rounded leukocytes, but it does not perform equally well in the presence of multiple complex forms. For this reason it is necessary a second step to refine the contours extracted through watershed transform. Then, all the pixels of the component under examination, which are located along the border and for which passes a watershed line are considered as a concavity point, for which the line of exact separation will have to pass. Therefore, by exploiting the information of the points of concavity and the information related to the points of maximum image in gray level, it is possible to obtain a cutting line that best fits the contour of the leukocytes, as it can be seen in Fig. 2(g). The last step for the identification of leukocytes consists of a cleaning process of the image, by removing the cells located along the borders and by removing all the components with irregular shape and size, as it can be seen in Fig. 2(h) (for details see [12]). The abnormal components are detected using a solidity value of 0.90, identified during the experiments performed as the most discriminatory value.

3   Nucleus and Cytoplasm Selection

Once the leukocytes have been identified, it is possible to move to the second segmentation level that provides the selection of nucleus and cytoplasm. This step is performed from sub-images, created using the bounding box size and to which is applied an operation of border cleaning with the aim to have a single leukocyte for each sub-image, as it is shown Fig. 3(c). Since by definition, leukocytes nucleus is internal to the membrane, it is possible to crop the entire portion of the image outside the leukocyte in question, in order to excludes artifacts during nucleus selection. Nucleus selection approach takes advantage from Cseke [3] observations, who found that WBCs nuclei are more in contrast on the green component of the RGB colour space. Threshold operation using

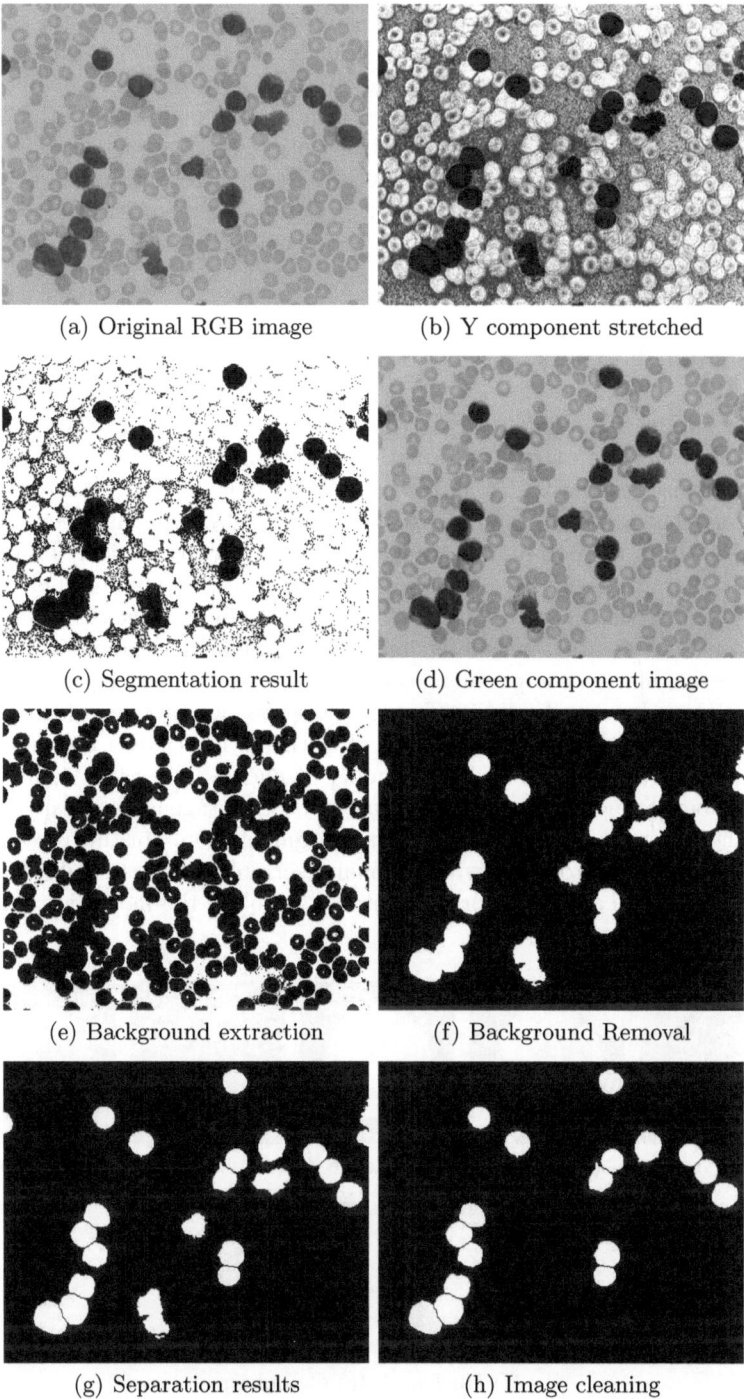

(a) Original RGB image

(b) Y component stretched

(c) Segmentation result

(d) Green component image

(e) Background extraction

(f) Background Removal

(g) Separation results

(h) Image cleaning

Fig. 2. Leukocytes identification process

Otsu [9] in this color space, however, does not produce clean results, especially with the presence of granulocytes, whose granules are selected erroneously as part of the nucleus. For this the binary image obtained from the green component is combined with the binary image, obtained from the a* component of the CIELab color space, again through a threshold operation. The mask obtained allows to extract clearly the leukocytes nucleus. At the end, to obtain the cytoplasm you just have to perform a subtraction operation between the binary image containing the whole leukocyte and the image containing only the nucleus (see Fig. 3(f),3(g)).

4   Feature Extraction

Speak about feature extraction in this context means to transform the images into data, then extract information reflecting the visual patterns which the pathologists refer to, but at the same time it is necessary to extract the descriptors that are most relevant to the subsequent classification process. For this reason, from the sub-images calculated previously are extracted 3 different types of descriptors: shape features, color features and texture features. Starting from the binary sub-images of nucleus and cytoplasm we have extracted shape descriptors such as area, perimeter, major axis, minor axis, orientation, eccentricity, rectangularity, compactness, convex hull, convex area, convex perimeter, convexity, roundness and solidity. To these classical measures we added two specific measures for the analysis of leukocytes, the ratio between the area of the cytoplasm and the nucleus and the number of nucleus lobes (for details about lobes number extraction see [13]). The main disadvantage of the shape features is that they are very susceptible to errors in segmentation. For this reason, these descriptors were used together with regional descriptors less susceptible to errors. Among these there are the color descriptors, which are the most discriminatory

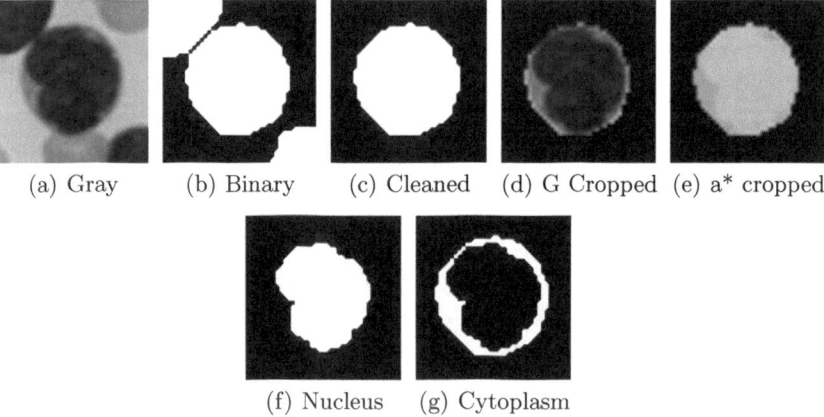

(a) Gray   (b) Binary   (c) Cleaned   (d) G Cropped   (e) a* cropped

(f) Nucleus   (g) Cytoplasm

Fig. 3. Nucleus and cytoplasm selection process

features of the blood cells. The color descriptors used are mean, standard deviation, smoothness, skewness, kurtosis, uniformity and entropy, calculated from the sub-images in shades of gray. Often, however, the descriptors based only on histograms have some drawbacks as they do not give information on the mutual position of the pixels. Some objects have in fact a repeating pattern as the primary visual characteristic and so it is necessary to consider not only the intensity distribution but also the positions of the pixels having a similar gray level. Then we have evaluated the descriptors applied to the matrix of co-occurrence calculated starting from the sub-images in gray level. The descriptors are auto-correlation, contrast, correlation, cluster prominence, cluster shade, dissimilarity, energy, entropy, homogeneity, maximum probability, sum of squares (variance), sum average, sum variance, sum entropy, difference variance, difference entropy, information measure of correlation1, information measure of correlation2, inverse difference normalized and inverse difference moment normalized. These features have been calculated for angles of 0, 45, 90 and 135 degrees. The total number of extracted features is then 117: 30 shape descriptors, 7 color descriptors and 80 texture descriptors.

5   Classification and Experimental Results

The proposed method was finally tested with the database ALL-IDB1 [4] which consists of 108 original blood sample images. Each image has an associated text file containing the coordinates of the centroid of each candidate linfoblast, manually estimated by a skilled operator. The test was carried out with a subset of 33 images acquired from the same camera and under the same lighting conditions. These images were taken with an Olympus C2500L camera and have a resolution of 1712x1368. From this sample of images, in the earlier stages of the analysis process, have been properly extracted 245 sub-images containing individual leukocytes, with an accuracy of 92% (detailed results can be observed in [12]). From these sub-images are then extracted a feature matrix with size 117x245 and a classification vector with size 1x245, which can be used to test the final stage of the process. In our previous work [13] the classification process was carried out using only 50 features and using the SVM classifier, since this model is particularly suitable for binary classification problems, in which the separation between the classes depends on a large number of variables. In this work, since the number of features is even higher we decided to proceed with different approaches. Firstly, the classification was performed using different models, both by exploiting the entire feature vector and by reducing its dimensionality through an operation of feature selection. The models tested for the classification process are still the SVM, K-Nearest Neighbor using different values of K, Naive Bayes by a Gaussian and Kernel data distribution, Decision Trees, Random Forest and different ensemble models such as AdaBoost, RobustBoost, LogitBoost, GentleBoost, Bag and Subspace. For the feature selection, firstly we compared different methods that use the sequential forward feature selection based on K Nearest Neighbor, Naive Bayes, Decision Trees, and Random Forest. In all cases,

Table 1. Experimental results with sequential forward feature selection: for each test accuracy (acc) and standard deviation (SD) are reported

Classifier	No FS acc	No FS S D	k-NN FS acc	k-NN FS S D	NB FS acc	NB FS S D	TREE FS acc	TREE FS S D	RF FS acc	RF FS S D
SVM	0,76	0	0,80	0,009	0,85	0,004	0,87	0,009	0,89	0,006
NN k=1	0,73	0,016	0,90	0,005	0,89	0,003	0,74	0,006	0,83	0,004
NN k=2	0,73	0,015	0,90	0,007	0,81	0,009	0,74	0,006	0,83	0,005
NN k=3	0,75	0,014	0,91	0,005	0,85	0,004	0,79	0,009	0,84	0,003
NN k=4	0,75	0,011	0,91	0,004	0,85	0,007	0,79	0,011	0,84	0,007
NN k=5	0,73	0,011	0,89	0,006	0,85	0,007	0,79	0,008	0,84	0,007
NN k=6	0,74	0,011	0,89	0,007	0,85	0,006	0,79	0,006	0,85	0,003
NN k=7	0,73	0,009	0,88	0,006	0,86	0,007	0,79	0,009	0,85	0,003
NN k=8	0,75	0,014	0,88	0,005	0,86	0,008	0,79	0,009	0,86	0,006
NN k=9	0,72	0,007	0,87	0,006	0,85	0,004	0,82	0,006	0,83	0,005
NN k=10	0,73	0,021	0,88	0,004	0,85	0,004	0,81	0,008	0,84	0,004
NB	0,81	0,006	0,82	0,008	0,89	0,003	0,84	0,007	0,81	0,003
NBK	0,85	0,008	0,84	0,007	0,91	0,006	0,87	0,006	0,86	0,003
tree	0,87	0,016	0,82	0,015	0,86	0,016	0,88	0,014	0,89	0,005
RF	0,89	0,005	0,87	0,004	0,90	0,006	0,90	0,009	0,92	0,006
ADA	0,87	0,008	0,85	0,006	0,87	0,008	0,89	0,006	0,92	0,006
Robust	0,85	0,015	0,86	0,009	0,87	0,012	0,85	0,018	0,88	0,007
Logit	0,87	0,008	0,85	0,005	0,89	0,004	0,89	0,006	0,91	0,009
Gentle	0,87	0,005	0,85	0,005	0,88	0,010	0,89	0,011	0,92	0,008
Bag	0,90	0,006	0,87	0,003	0,90	0,003	0,90	0,012	0,92	0,004
Subspace	0,78	0,008	0,78	0,012	0,83	0,010	0,82	0,005	0,80	0,010
Mean	0,79	0,010	0,86	0,007	0,86	0,007	0,83	0,08	0,87	0,005

given the small size of the dataset used, the performance of the models were then evaluated by a 10-fold Cross-Validation. The experimental results are shown in Table 1.

The results obtained show that in general all the classification models benefit from the process of feature selection. In particular we can observe how the SVM accuracy increments more with more elaborate feature selection processes. It's interesting also to note that all classification models have better performance associated with feature selection based on the same classifiers, for example the k-NN improves more with feature selection based on k-NN, Naive Bayes improves with feature selection based on Naive Bayes and so on. The only classifiers that don't get substantial performance improvements are the ensemble classifiers, which by their nature are generally more robust.

The feature selection has been implemented also with the algorithm of ReliefF, which, unlike the methods of sequential feature selection, does not provides the best feature vector, but it returns a vector containing all the features sorted according to their relevance. The classification was then tested, using the same classification models previously seen, and by using an increasing number of features (from 5 to 25). Also in this case the performances of the models were then

Table 2. Experimental results with ReliefF algorithm: for each test accuracy (acc) and standard deviation (SD) are reported

Classifier	5 feat		10 feat		15 feat		20 feat		25 feat	
	acc	S D	acc	S D	acc	S D	acc	S D	acc	S D
SVM	0,82	0,008	0,85	0,001	0,85	0,002	0,78	0,006	0,8	0,004
NN k=1	0,75	0,006	0,8	0,009	0,87	0,007	0,74	0,002	0,78	0,013
NN k=2	0,75	0,004	0,82	0,008	0,87	0,011	0,74	0,02	0,79	0,002
NN k=3	0,77	0,01	0,83	0,007	0,89	0,009	0,74	0,008	0,79	0,015
NN k=4	0,78	0,004	0,84	0,007	0,9	0,007	0,73	0,007	0,8	0,017
NN k=5	0,78	0,003	0,83	0,008	0,9	0,007	0,73	0,006	0,82	0,012
NN k=6	0,78	0,011	0,84	0,003	0,9	0,004	0,73	0,016	0,83	0,006
NN k=7	0,79	0,011	0,84	0,003	0,9	0,007	0,74	0,009	0,82	0,006
NN k=8	0,80	0,006	0,84	0,007	0,9	0,007	0,74	0,005	0,84	0,011
NN k=9	0,80	0,01	0,83	0,004	0,89	0,005	0,74	0,015	0,82	0,009
NN k=10	0,80	0,008	0,83	0,006	0,9	0,006	0,75	0,008	0,84	0,009
NB	0,81	0,003	0,82	0,003	0,85	0,006	0,82	0,006	0,79	0,002
NBK	0,83	0,006	0,87	0,004	0,88	0,003	0,85	0,005	0,8	0,005
tree	0,80	0,019	0,85	0,007	0,85	0,008	0,77	0,004	0,8	0,002
RF	0,82	0,008	0,9	0,009	0,91	0,009	0,91	0,005	0,87	0,006
ADA	0,83	0,013	0,88	0,008	0,88	0,006	0,89	0,018	0,87	0,007
Robust	0,81	0,008	0,87	0,007	0,89	0,008	0,89	0,008	0,87	0,014
Logit	0,82	0,005	0,87	0,011	0,89	0,01	0,89	0,009	0,87	0,016
Gentle	0,81	0,01	0,87	0,011	0,88	0,016	0,88	0,016	0,87	0,016
Bag	0,83	0,012	0,91	0,005	0,90	0,01	0,91	0,013	0,88	0,012
Subspace	0,82	0,007	0,82	0,006	0,82	0,004	0,83	0,01	0,80	0,003
Mean	0,80	0,008	0,85	0,006	0,88	0,007	0,80	0,009	0,80	0,009

evaluated by a 10-fold Cross-Validation. The experimental results are shown in Table 2.

Even in this case the performances of classification models benefit from feature selection process, but unlike the previous case where the results of a single classification model were more or less good according to the feature selection technique used, in this case all classification models have a similar behavior. In fact for all of them the performance increases with the increase of the features taken into account, up to the maximum value obtained with a number of features equal to 15, then decreased again. From the data collected it was possible to trace the best features selected from various methods of feature selection during the test phase. These features are shown in Table 3, sorted by relevance.

To verify if the selected features are effectively the best we have tested again all classification models seen previously and each time using an increasing number of features from 3 to 15. The best results were obtained using the first 11 features, by which was obtained a value of average accuracy (for all classifiers) equal to 0.88. The value of maximum accuracy has been obtained also in this case with the Random Forest classifier with an accuracy equal to 0.92.

Table 3. Most used descriptors after feature selection

Relevance	Type	Feature
1	Shape	Nucleus Area
2	Shape	Area ratio
3	Texture	Inverse Difference
4	Shape	Minor Axis Length
5	Shape	Lobes number
6	Shape	WBC Area
7	Color	Skewness
8	Color	Kurtosis
9	Shape	WBC Compactness
10	Shape	Nucleus Convex Area
11	Texture	Autocorrelation
12	Shape	WBC Perimeter
13	Shape	WBC Convexity
14	Shape	WBC Convex Area
15	Shape	Nucleus Perimeter

6   Conclusions

In this work we have proposed an innovative method for the completely automatic identification and classification of leukocytes by microscopic images, in order to provide an automated procedure as support medical activity, in recognition of acute lymphocytic leukemia. So, starting from an original image we can process it entirely, without performing manual crops or without identifying a working area, and we are able to process it in a fully automatic way without need for manual intervention by the user. The results obtained show that the proposed method is able to identify in a robust way the WBCs present in the image, being able to properly classify all leukocytes suffering from disease and offering a good level of overall accuracy. We have also obtained a list of features which can be proposed as single set of features to be used for the classification of leukemic images. In fact with this set of features the level of accuracy is comparable to that one obtained with the best methods of feature selection and classification, but with the advantage of completely overthrowing the computation time that are dilated for the extraction of a large number of feature and for the features selection. Further developments could affect the classification phase, in fact to increase the level of overall accuracy it is required the use of a multiclass classification model for the identification of various types of leukocytes and finally of the lymphoblasts. It will also be necessary to expand the size of the dataset in order to provide to the classification model a greater number of useful examples in the training phase and to allow us the use of a validation method different from 10-fold Cross-Validation.

Acknowledgments. This work has been funded by Regione Autonoma della Sardegna (R.A.S.) Project CRP-17615 DENIS: Dataspace Enhancing Next

Internet in Sardinia. Lorenzo Putzu gratefully acknowledges Sardinia Regional Government for the financial support of his PhD scholarship (P.O.R. Sardegna F.S.E. Operational Programme of the Autonomous Region of Sardinia, European Social Fund 2007-2013 - Axis IV Human Resources, Objective 1.3, Line of Activity 1.3.1.). We wish to thank Dr. Scotti for having made available the Dataset on which we could test our method.

References

1. Biondi, A., Cimino, G., Pieters, R., Pui, C.H.: Biological and Therapeutic Aspects of Infant Leukemia. Blood 96(1), 24–33 (2000)
2. Cheewatanon, J., Leauhatong, T., Airpaiboon, S., Sangwarasilp, M.: A New White Blood Cell Segmentation Using Mean Shift Filter and Region Growing Algorithm. International Journal of Applied Biomedical Engineering 4, 30–35 (2011)
3. Cseke, I.: A Fast Segmentation Scheme for White Blood Cell Images. In: Proceedings of the IAPR International Conference on Image, Speech and Signal Analysis, vol. 3, pp. 530–533 (1992)
4. Donida Labati, R., Piuri, V., Scotti, F.: ALL-IDB: the Acute Lymphoblastic Leukemia Image DataBase for Image Processing. In: Proceedings of the ICIP International Conference on Image Processing, pp. 2045–2048 (2011)
5. Gonzalez, R.C., Woods, R.E., Eddins, S.L.: Digital Image Processing Using MATLAB. Pearson Prentice Hall Pearson Education, Inc., New Jersey (2004)
6. Kovalev, V.A., Grigoriev, A.Y., Ahn, H.: Robust Recognition of White Blood Cell Images. In: Proceedings of the 13th International Conference on Pattern Recognition, pp. 371–375 (1996)
7. Lindblad, J.: Development of algorithms for digital image cytometry. Uppsala University, Faculty of Science and Technology (2002)
8. Madhloom, H.T., Kareem, S.A., Ariffin, H., Zaidan, A.A., Alanazi, H.O., Zaidan, B.B.: An Automated White Blood Cell Nucleus Localization and Segmentation using Image Arithmetic and Automated Threshold. Journal of Applied Sciences 10(11), 959–966 (2010)
9. Otsu, N.: A Threshold Selection Method from Gray-Level Histograms. IEEE Transactions on Systems, Man, and Cybernetics 9(1), 62–66 (1979)
10. Scotti, F.: Robust Segmentation and Measurements Techniques of White Cells in Blood Microscope Images. In: Proceedings of the IEEE Instrumentation and Measurement Technology Conference, pp. 43–48 (April 2006)
11. Zack, G., Rogers, W., Latt, S.: Automatic Measurement of Sister Chromatid Exchange Frequency. Journal of Histochemistry and Cytochemistry 25, 741–753 (1977)
12. Putzu, L., Di Ruberto, C.: White Blood Cells Identification and Counting from Microscopic Blood Images. In: Proceedings of the WASET International Conference on Bioinformatics, Computational Biology and Biomedical Engineering, vol. 73, pp. 268–275 (January 2013)
13. Putzu, L., Di Ruberto, C.: White Blood Cells Identification and Classification from Leukemic Blood Image. In: Proceedings of the IWBBIO International Work-Conference on Bioinformatics and Biomedical Engineering, pp. 99–106 (March 2013)

Estimating the Serial Combination's Performance from That of Individual Base Classifiers

Gian Luca Marcialis, Luca Didaci, and Fabio Roli

Dept. of Electrical and Electronic Engineering, University of Cagliari
Piazza d'Armi, 09123 Cagliari, Italy
{marcialis,didaci,roli}@diee.unica.it
http:\\prag.diee.unica.it\en

Abstract. Although the large number of MCS topics, serial fusion of multiple classifiers has been poorly investigated so far. In this paper, we propose a model which, starting from the performance of individual classifiers and the traditional hypothesis of decision independence given the class, is able to estimate the performance, in terms of error rates, of the whole serial classification scheme. The model is tested on a large set of data sets and classifiers, and the importance of the basis hypothesis is evaluated under different scenarios, which can be in agreement or not with such hypothesis.

1 Introduction

Fusion of multiple classifiers can be performed in parallel or serially [1, 2]. Although the most of works relies on parallel fusion at several levels (feature-level, measurement-level, decision-level) [3, 4], very few papers deal with sequential, or serial, fusion of multiple classifiers [5–8, 16–18].

To the best of our knowledge, the only papers which try to analyze analytically serial fusion is the paper by Pudil et al. [5] and Trapeznikov et al. [18], where the problem is dealt by the point of view of the risk assessment. Other works analyze specifically a specific Unfortunately, Pudil's modelling does not allow to point out any specific pros and cons of serial fusion of multiple classifiers. In other cases, we have practical applications of the serial fusion with an experimental assessment of pros and cons [6, 7]. Theoretical and experimental approaches have been proposed for biometric applications [9–11]. In this paper, as in [9], we deal with the serial fusion of multiple classifiers from the point of view of the performance prediction: in other words, we do not try to find the optimal parameters allowing the minimization of a risk function as in [18], but tries to model the sequential fusion in order to have a prediction of the overall error rate, given the error rates of individual classifiers.

In this preliminary work, we derive the general expression of the error rate of a serial fusion of two classifiers for a two-classes classification problem, under the simple hypothesis of decisions independence among classifiers, and propose a

A. Petrosino (Ed.): ICIAP 2013, Part I, LNCS 8156, pp. 622–631, 2013.
© Springer-Verlag Berlin Heidelberg 2013

model which allows deriving further information on the performance achievable by the serial system with respect to the best individual classifier. The model is then tested on a battery of several UCI data sets, and on classifiers whose set up confirms or not the independence hypothesis, in order to see at which extent the proposed model is valid.

The rest of the paper is as follows. The model is described in Section 2. Section 3 report experiments. Conclusions are drawn in Section 4.

2   The Proposed Model

Fig. 1 describes the serial model used in this paper. Observation x_1, a statistical or structural representation of the pattern at hand, is input to the C_1 classifier, whose output is given in terms of d^1, where $d^1 \in \Omega$, being $\Omega = \{w_1, w_2\}$, the state of nature. On the basis of a reject option [12, 16, 18], we add to Ω a third class, namely, w_0. In this case, the final decision is left to the second classifier, named C_2. Thus, for sake of clarity, three decisions can be taken by the first classifier and we will indicate these with d_0^1, d_1^1, and d_2^1, corresponding to the three classes w_0, w_1, w_2 . In the case that C_1 decides for w_0, observation x_2, related to the same pattern, is taken as input by C_2, whose decision can be d_1^2 or d_2^2. In other words, no reject option is given for the second classifier.

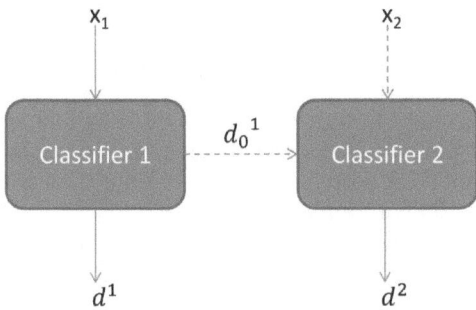

Fig. 1. The proposed model for serial fusion of multiple classifiers

Let d_i^{12} be the decision of the sequence $C_1 \rightarrow C_2$ for class w_i. The probability of error $P(d_i^{12}|w_j)$, $i \neq j$, is given as follows:

$$P(d_i^{12}|w_j) = P(d_i^1|w_j) + P(d_0^1 \bigcap d_i^2|w_j).$$

By hypothesizing the independence of d^1 and d^2, given w_j, we obtain:

$$P(d_i^{12}|w_j) = P(d_i^1|w_j) + P(d_0^1|w_j) \cdot P(d_i^2|w_j). \tag{1}$$

It is easy to see that, given a rejection region for C_1, the error rate is a linear function of the error rate of C_2, where errors of the first classifier represent the bias and the slope.

Aim of this paper is to verify this formula under different practical scenarios. In order to perform a complete investigation, we should perform experiments on all possible rejection regions.

However, in this paper, we will focus only on a particular rejection region, for sake of simplicity. The choice is not arbitrary, because this rejection region is acceptable in many cases.

We investigated the rejection region such that no classification errors are done by C_1. In other words:

$$P(d_i^1|w_j) = 0 \tag{2}$$

Actually, several applications, especially related to the information security, require that, on a chain of possible classifiers, no errors are made by the stages preceding the last one. For example, this occurs in personal identity verification through biometric systems, where rejecting an authorized user at first stage, as well as accepting an impostor at the intermediate stages of a serial system, is a serious drawback [9, 11].

Accordingly, the bias of the error rate indicated in Eq. 3 can be removed, thus obtaining:

$$P(d_i^{12}|w_j) = P(d_0^1|w_j) \cdot P(d_i^2|w_j). \tag{3}$$

This simple expressions leads us to an important, but general, property of serial fusion of multiple classifiers: if $P(d_i^2|w_j) < P(d_i^1|w_j)$, then the sequence $C_1 \rightarrow C_2$ allows to obtain a better performance than that of the best individual classifier, namely, C_2. This is evident from Eq. 3.

Thus, the *best* sequence is that which considers the best classifier at the second stage. It is worth to point out that this is not the "optimal" sequence in terms of error rate minimization, but only with respect to the best individual classifier. On the other hand, nothing can be said about the sequence $C_2 \rightarrow C_1$, under the hypothesis above.

In order to set the rejection region, we opt for a simple strategy suggested in [13], which is an extension of the so-called Chow's rule for rejection option [12]. In particular, this rule takes into account that in practical applications, individual classifiers do not provide the "real" posterior probability of each class but an estimation. In this case, Ref. [13] showed that this rule is able to derive the best trade-off between error rate and rejection rate with respect to the Chow's rule.

On the basis of the above observation, let's consider the output of the classifier C_k, indicated with p_i^k, as an estimation of the probability of w_i, given the pattern. Since we are considering only two-class classification problems, $p_2^k = 1 - p_1^k$.

According to the Chow's rule, we can set a rejection region as function of the interval $[\gamma_2^1, \gamma_1^1]$, and assigning decisions d_0^1, d_1^1, and d_2^1 as follows:

1. d_1^1 if $p_1^1 > \gamma_1^1$;
2. d_2^1 if $p_1^1 < \gamma_2^1 \rightarrow p_2^1 > 1 - \gamma_2^1$;
3. d_0^1 if $\gamma_2^1 \leqslant p_1^1 \leqslant \gamma_1^1$.

Worth noting, $\gamma_1^1 \in (0.5, 1]$ and $\gamma_2^1 \in [0, 0.5)$. In the case that $\gamma_1^1 = \gamma_2^1 = 0.5$, the second stage classifier is required only for the uncertainty value of posterior probability such that $p_1^1 = p_2^1 = 0.5$.

With regard to C_2:

1. d_1^2 if $p_1^2 > \gamma^2$;
2. d_2^2 if $p_1^2 \leqslant \gamma^2$;

Where $\gamma^2 \in [0, 1]$.

Thus, Eq. 3 becomes:

$$P(d_1^{12}|w_2) = P(\gamma_2^1 \leqslant p_1^1 \leqslant \gamma_1^1|w_2) \cdot P(p_1^2 > \gamma^2|w_2) \qquad (4)$$
$$P(d_2^{12}|w_1) = P(\gamma_2^1 \leqslant p_1^1 \leqslant \gamma_1^1|w_1) \cdot P(p_1^2 \leqslant \gamma^2|w_1) \qquad (5)$$

According to Eq. 2, Eqs. 4-5 can be simplified as:

$$P(d_1^{12}|w_2) = P(p_1^1 > \gamma_2^1|w_2) \cdot P(p_1^2 > \gamma^2|w_2) \qquad (6)$$
$$P(d_2^{12}|w_1] = P(p_1^1 < \gamma_1^1|w_1) \cdot P(p_1^2 \leqslant \gamma^2|w_1) \qquad (7)$$

This is the final scheme we followed in our experiments. This model allows to predict the performance of the serial system by starting from errors of the individual classifiers, under the assumption of decisions independence. In other words, a designer can draw the error rate curves in order to find the best sequence on the basis of simple evaluations derived from the classifiers taken individually. No complex estimation of joint probabilities is necessary, as well as further experiments on the overall system.

3   Experimental Results

Aim of this Section is to validate the model given by Eqs. 6-7, by assessing the prediction against the real error rate obtained by experiments.

3.1   Data Sets and Protocol

The experiments have been carried out on 32 data sets form the UCI repository [14] (see Table 1). The main assumption of our model is the conditional independence (given the class) between set of features used to represents the classification task at hand, as illustrated in Section 2. In order to create a setting that meets this assumption, we subdivide the original feature space into two views using the random-restart hill climbing method proposed in [15], which aims at maximizing their independence. Basically, this method splits features set into two views in a random manner, then each feature is switched to the other view once a time, thus obtaining a group of new generated split. All these new splits are evaluated to check their independence, and the one that yields maximum independence is selected. In order to check the robustness of the proposed method when the independence assumption doesnt hold, we performed a second series of experiments in which datasets have been subdivided into two views in a random manner, each view being composed of half of the features, without forcing the independence between views.

In Table 1 we report the characteristics of the data sets used in the experiments. Columns #patterns indicate the number of patterns (total, class A and class B) whilst columns #features indicate the number of features for each view

The ample battery of data sets explored in this paper shows several and different experimental and practical conditions which may affect the design of multiple classifiers systems combined serially. First of all, we may see that some data sets exhibit a strongly unbalanced number of patterns for the involved classes (see for example, Audiology and Solar flare data sets). In other cases, the number of patterns is not balanced with respect to the size of feature set (see for example, Sonar and Breast Cancer data sets). Finally, we may see unbalanced performances among classifiers where the sizes of each feature set strongly differs. Therefore, the experimental results we report will show the reliability of our model as it may be expected in general.

Three different classifiers are adopted for exploring the generality of our observations: K-Nearest Neighbour, Naive Bayes, and Linear Log. We have arbitrarily chosen parameters for the classifiers (k=3 for the K-NN classifier and N=10 for the Nave Bayes classifier) because the aim of the paper is to calculate the final performance of the system from the performance of the individual classifiers; it is not essential to maximize the performance of individual classifiers.

Data sets were splitted training and test sets. The former is made of 70% of available patterns, whilst the latter of the remaining 30%.

3.2   Results

In Figs. 2 we report some ROC curves predicted on the test set by the model according to Eqs. 4-5, and the ones reported on the same data but during system's operations. The thresholds γ have been tuned on the training set in order to obtain zero error for the classifier at the first stage.

Four cases were chosen for showing the effectiveness of our model:

1. imbalanced distributions of patterns over classes and maximum independence among views;
2. imbalanced distributions of patterns over classes and no independence among views;
3. balanced distributions of patterns over classes and maximum independence among views;
4. balanced distributions of patterns over classes and no independence among views.

Table 1. Characteristics of UCI data sets adopted for experiments in terms of number of patterns for the two-class classification problem, and number of features per view (classifier)

Dataset	#patterns			#features - maxInd split		#features random split	
	tot	class A	class B	view 1	view 2	view 1	view 2
1-Audiology	200	48	152	31	24	28	27
2-Automobile	193	130	63	22	2	12	12
3-Congressional Voting Records	232	124	108	15	1	8	8
4-Contraceptive Method Choice	1473	629	844	2	7	5	4
5-Credit Approval	653	296	357	9	6	8	7
6-Dermatology	366	112	254	17	16	17	16
7-Ecoli	336	143	193	3	4	4	3
8-Flag	194	134	60	15	13	14	14
9-Glass identification	214	138	76	1	8	5	4
10-Heart statlog	270	150	120	4	9	7	6
11-Horse-colic	368	232	136	4	1	3	2
12-Ionosphere	351	225	126	6	27	17	16
13-kr-vs-kp	3196	1669	1527	13	23	18	18
14-Mushroom	8124	4208	3916	19	1	10	10
15-Pima indians diabetes	768	500	268	6	2	4	4
16-Sonar	208	97	111	34	26	30	30
17-Spambase	4601	2788	1813	21	36	29	28
18-Splice	3186	1532	1654	32	28	30	30
19-Tic-tac-toe	958	626	332	3	6	5	4
20-Breast Cancer-BCW	699	458	241	5	3	4	4
21-Breast Cancer-WDBC	569	212	357	11	19	15	15
22-Breast Cancer-WPBC1	194	46	148	11	22	17	16
23-Breast Cancer-WPBC2	198	47	151	13	19	16	16
24-Heart disease-Cleveland	303	164	139	5	6	6	5
25-Heart disease-Hungarian	294	188	106	1	3	2	2
26-Heart disease-LongBeachVA	200	144	56	2	2	2	2
27-Heart disease-Switzerland	123	75	48	1	2	2	1
28-Hepatits1	80	13	67	12	7	10	9
29-Hepatits2	155	32	123	1	3	2	2
30-Solar-flare-1	1389	1171	218	1	9	5	5
31-Solar-flare-2	1389	1321	68	8	2	5	5
32-Solar-flare-3	1389	1377	12	9	1	5	5

For reasons of clarity, Figs. 2 shows only two relevant examples of (almost) balanced and unbalanced datasets (respectively, Splice and Solar Flare 3), but the reported data does not deviate from the results obtained on other datasets.

Items 1-2 are shown by the ROC curves of Solar Flare 3 data set, whilst Splice data set has been chosen for items 3-4.

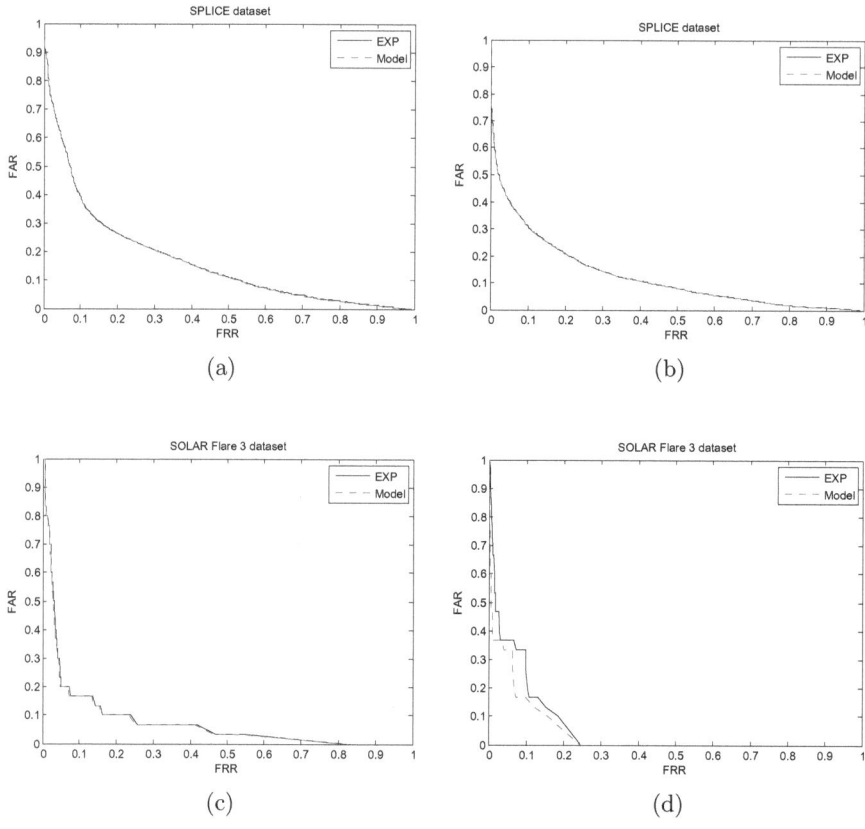

Fig. 2. ROC curves for serial system. Best classifier at second stage. (a) balanced data set, maximum independence split; (b) balanced data set, random split; (c) unbalanced data set, maximum independence split; (d) unbalanced data set, random split.

It is evident that ROC prediction in all cases is very near to the ROC computed by experiments. No appreciable difference is given when random selection of features for views generation is done.

In results reported in Figs. 2, the best classifier is always at the second stage, according to findings reported in Section 2. In the following tables, we explored the difference between ROCs predicted by our model and ROCs computed by experiments, even if the worst classifiers precedes the best one in the serial system.

Table 2 shows results in this last configuration. Reported values are the differences among AUCs for all datasets under consideration. In all cases, the very low values clearly show that our model correctly predict the final performance of the serial system. In fact, a difference of about 10^{-3} on average means that the error rate for one of two classes, in the worst case, is of about 10^{-1}, and on average, of $10^{-1.5}$, which corresponds to an error rate prediction of 3% on both classes.

This is true even when random split is performed for feature selection. In particular, we may see that difference between prediction on maximum independence and prediction on random split is negligible (last column of Table 2), even if, as it may be expected, the AUC difference is slightly higher in the random split case. This could mean that features on UCI data sets investigated are uncorrelated, thus even choosing them randomly does not impact on the model's predictions.

Table 3 confirms what reported in previous Table. For all datasets in 1, a very low AUC difference is reported.

Table 2. Difference of AUC values between the model's prediction and the experimental test. Results refer to the classifiers sequence where the worst classifier is at the first stage.

Classifier	split	AUC difference ($*10^{-3}$)			diff
		min	max	mean ($\pm stddev$)	
KNN- k=3	MAX IND	0	18.9	3.5($\pm$6.0)	0.7
	RANDOM	0	27.7	4.2($\pm$7.8)	
Naive Bayes - N=10	MAX IND	0	52.9	9.9($\pm$11.5)	2.1
	RANDOM	0	42.7	12.0($\pm$11.9)	
Linear Log	MAX IND	0	37.0	8.2($\pm$9.7)	2.7
	RANDOM	0	52.8	10.9($\pm$12.1)	

Table 3. Difference of AUC values between the model's prediction and the experimental test. Results refert to che classifiers sequence where the best classifier is at the first stage.

Classifier	split	AUC difference ($*10^{-3}$)			diff
		min	max	mean ($\pm stddev$)	
KNN- k=3	MAX IND	0	22.7	3.6($\pm$6.1)	1.5
	RANDOM	0	30.4	5.1($\pm$8.0)	
Naive Bayes - N=10	MAX IND	0	39.1	9.9($\pm$10.5)	7.5
	RANDOM	0	65.0	17.4($\pm$15.8)	
Linear Log	MAX IND	0	55.4	12.9($\pm$13.5)	1.1
	RANDOM	0	38.2	14.0($\pm$11.4)	

4   Conclusions

In this preliminary paper, we have shown the general expression of the error rate for a serial system of two classifiers in the case of two-classes classification problem. This expression allows predicting the performance of the system given the performance of the individual classifiers, under the assumption of decision independence.

The model has been validated by a large battery of UCI data sets, which have been set up in order to follow or not the assumption above. Results have shown that the model is highly reliable, even in the case that independence hypothesis is not satisfied in practice.

Validation has been performed only on an operational point, namely, the point for which no errors are allowed for the first stage. Although this choice can be motivated for several practical applications, further works will rely on extending this investigation for any size of the rejection region. An ample set of experiments on realistic use-cases and scenarios will be also taken into account.

Finally, the case of more than two classifiers and the multiclass problem will be also considered in future works.

References

1. El Gayar, N., Kittler, J., Roli, F. (eds.): MCS 2010. LNCS, vol. 5997. Springer, Heidelberg (2010)
2. Sansone, C., Kittler, J., Roli, F. (eds.): MCS 2011. LNCS, vol. 6713. Springer, Heidelberg (2011)
3. Suen, C.Y., Lam, L.: Multiple Classifier Combination Methodologies for Different Output Levels. In: Kittler, J., Roli, F. (eds.) MCS 2000. LNCS, vol. 1857, pp. 52–66. Springer, Heidelberg (2000)
4. Kittler, J., Hatef, M., Duin, R.P.W., Matas, J.: On Combining Classifiers. IEEE Transactions on Pattern Analysis and Machine Intelligence 20(3), 226–239 (1998)
5. Pudil, P., Novovicova, J., Blaha, S., Kittler, J.: Multistage Pattern Recognition with Reject Option. In: Proc. 11th IAPR-ICPR International Conference, vol. 2, pp. 92–95 (1992)
6. Last, M., Bunke, H., Kandel, A.: A feature-based serial approach to classifier combination. Pattern Analysis and Applications 5(4), 385–398 (2002)
7. Sansone, C., Vento, M.: Signature verification: increasing performance by a multi-stage system. Pattern Analysis and Applications 3, 169–181 (2000)
8. Pao, L., Trailovic, L.: The optimal order of processing sensor information in sequential multisensor fusion algorithms. IEEE Transactions on Automatic Control 45(8), 1532–1536 (2000)
9. Marcialis, G.L., Roli, F., Didaci, L.: Personal identity verification by serial fusion of fingerprint and face matchers. Pattern Recognition 42(11), 2807–2817 (2009)
10. Marcialis, G.L., Mastinu, P., Roli, F.: Serial fusion of multi-modal biometric systems. In: Proc. of IEEE International Workshop on Biometric Measurements and Systems for Security and Medical Applications, BioMS 2010 (2010), doi:10.1109/BIOMS.2010.5610438
11. Allano, L., Dorizzi, B., Garcia-Salicetti, S.: Tuning cost and performance in multimodal biometric systems: a novel and consistent view of fusion strategies based on the Sequential Probability Ratio Test (SPRT). Pattern Recognition Letters 31, 884–890 (2010)
12. Chow, C.K.: On optimum rejection error trade-off. IEEE Transactions on Information Theory 16(1), 41–46 (1970)
13. Fumera, G., Roli, F., Giacinto, G.: Reject Option with Multiple Thresholds. Pattern Recognition 33(12), 2099–2101 (2000)
14. Frank, A., Asuncion, A.: UCI Machine Learning Repository. University of California, School of Information and Computer Science, Irvine (2010), http://archive.ics.uci.edu/ml

15. Du, J., Ling, C.X., Zhou, Z.-H.: When Does Co-Training Work in Real Data? IEEE Transactions on Knowledge and Data Engineering 23(35), 788–799 (2011)
16. Giusti, N., Masulli, F., Sperduti, A.: Theoretical and Experimental Analysis of a Two-Stage System for Classification. IEEE Transactions on Pattern Analysis and Machine Intelligence 24(7), 893–904 (2002)
17. Senator, T.: Multi-stage classification. In: Proc. of IEEE 5th Int. Conf. on Data Mining, ICDM 2005 (2005)
18. Trapeznikov, K., Saligrama, V., Castanon, D.: Multi-stage classifier design. In: JMLR Asian Conference on Machine Learning (ACML 2012), pp. 1–16 (2012)

3D Interest Points Detection Using Symmetric Surround-Based Surface Saliency

Yitian Zhao and Yonghuai Liu

Department of Computer Science, Aberystwyth University
yyz10@aber.ac.uk

Abstract. The detection of interest points is an important pre-processing step for the analysis of mesh surfaces. This paper proposes a new method for the detection of the interest points for 3D surface. Our method incorporates the Relative Distance-based Laplacian and *Retinex* to generate the symmetric surround-based surface saliency. The effectiveness of this method is demonstrated by studying the repeatability of the detected interest points under different perceptual conditions, such as different viewpoints and noise corruption. The results show that the proposed method achieves better performance than competitors.

Keywords: salient, surface, bilateral, region, repeatability.

1 Introduction

Many applications have benefited from the wide diffusion of 3D models. In recent years, there has been a growing interest in the use of detailed models for better representation. Consequently, most of the latest 3D scanners can generate huge quantities of data points within a limited time. This creates a number of challenges for storage, editing and transmission. These challenges place a greater burden on feature detection tasks. 3D interest points detection is helpful in capturing the property of a point or region on a surface. Interest points, also referred to as feature points, salient points, or keypoints, are those points which are distinctive in their locality and stable at all instances of an object, or of its category of objects [1]. Therefore, interest points detection has a wide range of applications in the fields of computer vision and graphics, such as mesh simplification, view point selection, point cloud matching, and object recognition. Use of interest points has the advantage of providing local features that are semantically significant and also invariant to rotation, scaling, noise, deformation, and articulation.

Due to the efficiency of visual persuasion in traditional art and technical illustrations, visual saliency has now been widely used in computer vision applications. Recent years have witnessed the rapid development of methods for saliency and interest points detection on 3D surfaces. Most of the current 3D interest point detection methods have been developed over the last decade, as have defined functions summarizing the geometrical content of localities on a

A. Petrosino (Ed.): ICIAP 2013, Part I, LNCS 8156, pp. 632–641, 2013.

3D model in multiple scales, and selected local extrema of those functions as interest points. Gelfand et al. [2] proposed a method that is related to the surface curvature, and which is invariant to rotation and translation. Lee et al. [3] produced an approach that used a center-surround operator on the local curvature as the discriminative feature and generated the mesh saliency. Castellani et al. [4] applied Difference-of-Gaussian (DoG) on various scales, and vertices that are highly displaced after the filtering are marked as interest points. Sun et al.al [5] proposed a Heat Kernel Signature (HKS) feature point detection and the maxima of the kernel were chosen as keypoints. Mian et al. [6] proposed a keypoint detection algorithm along with an automatic scale selection technique for subsequent feature extraction. The keypoints are detected by calculating the ratio of the covariance matrices. The 3D-Harris [7] approach calculated the derivatives by fitting a quadratic surface to a neighborhood of the vertex, then the interest points are located by the first order derivatives along two orthogonal directions on the 3D surface. A 3D-SIFT technique was proposed which constructed a scale space by applying 3D Gaussian filters with increasingly large scales to the voxelized model [8]. The extreme points are detected by searching the DoG space in both spatial and scale dimensions.

Some techniques were inspired by corresponding 2D approaches. However, such extensions are not always straightforward: for example, the keypoints represent interesting information at fine scales and thus, may be sensitive to noise and other transformations. Therefore, it is necessary to find larger and interesting structures to overcome the problems at fine scales. In this paper, we propose an algorithm to select interest points from salient regions. As we mentioned above, the sensitivity to noise affects the accuracy of the interest points detection. In order to overcome this problem, we employ the following three measures:

1. A **Relative Distance-Based Laplacian (RDL)** is used to characterize the local details of the geometry, since it uses relative distance rather than the absolute Euclidean distance. The local scale changes of features will not affect its value: as a result, it is able to capture both large and small scale features.
2. **Retinex** is a theory of color constancy that is usually applied for purposes of image enhancement. This method has been adapted for estimation of viewpoint invariant information in the saliency detection stage, so that the same salient regions will be independent of changes of viewpoint.

Our method proposes the RDL and *Retinex* as the relative importance values to define a new 'color' space. Based on the fused values from RDL and Retinex, the saliency is estimated for stability and continuity.

2   Relative Distance-Based Laplacian

In this paper we propose to use the RDL, which is learned from the data, and better reflects the relative geometry. For two sets of points with similar neighboring relationships but different densities, the absolute distances between corresponding points differ dramatically from each other, but the relative distances are in general similar. This is an advantage of the relative distance over the Euclidean distance in reflecting the relative density. The relative distance is used in case the distribution of the data is not uniform: the distance metric mainly focuses on the representation of the neighboring relationships between points [11]. We use the RDL for mesh processing. It represents each point of a mesh as the difference between the point and its neighborhood. The structure of a shape, either in terms of topology or geometry, can be modelled by matrices. We would then expect its set of eigenvalues to provide an adequate characterization of the shape [9]. The eigenvalues serve as compact global shape descriptors. They are sorted by their magnitudes so as to establish a correspondence for computing the similarity distance between two shapes. Given a point set M containing m points $x_1, x_2, ...x_m$: we begin by defining two kinds of relative distances between a pair of points x_i, $x_j \in M$ as follows, according to [11]:

1: Relative maximum distance: $rd\_max(x_i, x_j) = \frac{\|x_i - x_j\|^2}{max_{x_k \subset M}(\|x_i - x_k\|^2)}$

2: Relative average distance: $rd\_ave(x_i, x_j) = \frac{\|x_i - x_j\|^2}{ave_{x_k \subset M}(\|x_i - x_k\|^2)}$

The $max_{x_k \subset M}(\|x_i - x_k\|^2)$ and $ave_{x_k \subset M}(\|x_i - x_k\|^2)$ are the maximum and average Euclidean distance between x_i and other points belonging to M respectively. We note that the algorithm for computing the surface Laplacian is related to the set of algorithms for computing the Laplacian of point clouds in data analysis and machine learning by using the heat equation. We construct a weighted graph with n nodes, each corresponding to a point in the observed space. If x_i is one of the K nearest neighbors of x_j, or x_j is one of the K nearest neighbors of x_i, we construct an edge between points i and j. The heat kernel is used to weight the adjacency graph. Corresponding to the definition of the relative distance, there are two ways to calculate the heat kernels W:

Definition 1. *If node j is one of the K nearest neighbors of i, or vice versa, rd_max distance-based W_{ij} is defined as:*

$$W_{ij} = \frac{1}{\sqrt{4\pi t}} exp\{-\frac{(rd\_max(x_i, x_j))^2}{t}\} \tag{1}$$

and otherwise, $W_{ij} = 0$. t is the parameter of the heat kernel.

Definition 2. *If node j is one of the K nearest neighbors of i, or vice versa, rd_ave distance-based W_{ij} is defined as:*

$$W_{ij} = \frac{1}{\sqrt{4\pi t}} exp\{-\frac{(rd\_ave(x_i, x_j))^2}{t}\} \tag{2}$$

and otherwise, $W_{ij} = 0$. Both of the relative distances are defined on the K nearest neighbors of point x_i.

Eigenmaps. We compute the eigenvectors and eigenvalues for the generalized eigenvector problem by: $Ly = \lambda Dy$, where D is a diagonal weight matrix, and its entries are column or row sums of W, since W is symmetric. $D_{ii} = \sum_j W_{ji}$. This leads to the geometric Laplacian, which we use in the remainder of the paper:

(a)   (b)   (c)   (d)

Fig. 1. Example of two kinds of RDL on model *lobster*. (a) Maximum distance based RDL, t=1. (b) Maximum distance based RDL, t=3.(a) Average distance based RDL, t=1.(a) Average distance based RDL, t=3.

$L = D - W$. Let $y_0, y_1..., y_d$ be the solutions of $Ly = \lambda Dy$, ordered according to their eigenvalues, $0 = \lambda_0 \leq \lambda_1 \leq, ... \leq \lambda_d$. We leave out the eigenvector y_0 corresponding to the eigenvalue 0 and use the remaining eigenvectors for embedding in d dimensional Euclidean space,

$$x_i \rightarrow (y_1(i), y_2(i), ...y_d(i))^T \tag{3}$$

Fig. 1 shows RDL surfaces with two different kinds of relative distances. Both of them have the ability to represent the shape.

3   Symmetric Surround Saliency

In this section we propose an algorithm that aims to detect salient region and select interest points from a 3D surface. [14] treats the entire 2D image as the common surround for any given pixel, and then the saliency map is obtained by computing the Euclidean distance of the average CIELAB vector of all pixels of an input image with each pixel of a Gaussian blurred version of the same image. LAB color space is designed to approximate human vision. In this paper, we extend this scheme to generate 3D surface saliency. Initially, we employ the *Retinex* [10] for estimation of viewpoint invariant properties of the given surfaces. Normally, human perception and objective information with respect to vision are not in agreement, especially with respect to the regions that have been captured by camera or scanner but to which human eyes are relatively insensitive. The human brain interprets an image of a 3D shape differently from how photosensors or scanners may sense it by consciously correcting brightness, removing noise, shadows, glare, or reflections. After the application of the *Retinex*, the 3D surfaces are represented more faithfully by comparison with the original, simulating the human visual system. The *Retinex* algorithm for 3D surfaces has been proposed in our previous papers [12] [13] for surface details enhancement. Then Relative average distance-based Laplacian (RDL) and *Retinex* are used instead as the dimensions of LAB to estimate the saliency of the surfaces. For a given vertex p_i, where $p_i = (x_i, y_i, z_i)$, we estimate the attention value $\mathscr{E}$ in both RDL (L) and *Retinex* (R) frame:

$$\mathscr{E}(G_i, G_j) = \frac{\|G_i - G_j\|}{\sqrt{(x_i - x_j)^2 + (y_i - y_j)^2 + (z_i - z_j)^2}} \tag{4}$$

where G is either of the geometric measures of a point: L and R, $\|\cdot\|$ calculates the Euclidean distance of vertices p_i and p_j in the mesh. The relative importance value $\mathscr{C}$ for each vertex is estimated as:

$$\mathscr{C}(p_i) = \sqrt{\sum_{i,j \in p} \mathscr{E}(R_i, R_j)^2 + \sum_{i,j \in p} \mathscr{E}(L_i, L_j)^2} \tag{5}$$

where $\mathscr{E}(\cdot, \cdot)$ indicates the visual attention values of the p_i due to p_j in terms of R and L, respectively.

We voxelize the surface that has been mapped with $\mathscr{C}$ by the bounding box, which it is divided into sub-boxes. The scale r of the voxelization is specified by the user. Let A and B be the faces that are located in two voxels. $\mathscr{C}'_A$ and ρ_A denote the mean relative importance value and mean position of the patch A. We define dissimilarity between faces A and B as:

$$diss(\mathscr{C}_A, \mathscr{C}_B) = \frac{\|\mathscr{C}'_A - \mathscr{C}'_B\|}{1 + c \cdot d(\rho_A, \rho_B)} \tag{6}$$

where $d(\rho_A, \rho_B)$ is the Euclidean distance between the mean positions of patches A and B. We set $c = 3$ in this paper as it achieved best performance. The dissimilarity measure is proportional to the difference in shape index and inversely proportional to the positional distance. We then initialize all elements of the dissimilarity map (DM) to 0, and start updating dissimilarity values as follows:

$$DM_A = DM_A + diss(\mathscr{C}_A, \mathscr{C}_B) \tag{7}$$

In practice, if the most similar faces (low dissimilarity faces) are significantly different from face A, then clearly all the other patches are also highly different from face A. Accordingly, considering the Q most similar faces is sufficient. Faces $\{q_k\}_{k=1}^{Q}$ can be identified from Eq. 6. The multi-scale saliency value of face A at scale r is defined as:

$$SM_A^r = 1 - exp\{-\frac{1}{Q} \sum_{k=1}^{Q} DM(A^r, q_k^r)\} \tag{8}$$

where $r \in R$. In this paper, the voxelisation scales $R = 1, 2, 3mm \cdots$, (if size $r = 3mm$, the size of each sub-voxel is 3*3*3). The scanner has a resolution of $0.7mm$ for the dataset we used in this paper. If $r = 1mm$, which is larger than $0.7mm$, one voxel could include more than one point. The smaller the scale r, the more accurate the saliency of the faces.

Finally, the symmetric surround saliency value at the point (v) is obtained as:

$$S(v) = \|SM'(v) - \mathscr{B}(v)\| \tag{9}$$

where $\mathscr{B}(v)$ is the corresponding vertex vector in the bilateral filtered version. Lee et al. [3] introduced the mesh saliency by using a center-surround operator on Gaussian-weighted mean curvature. The disadvantage of taking a Gaussian-weighted average is that it might make two opposite and symmetric vertices have

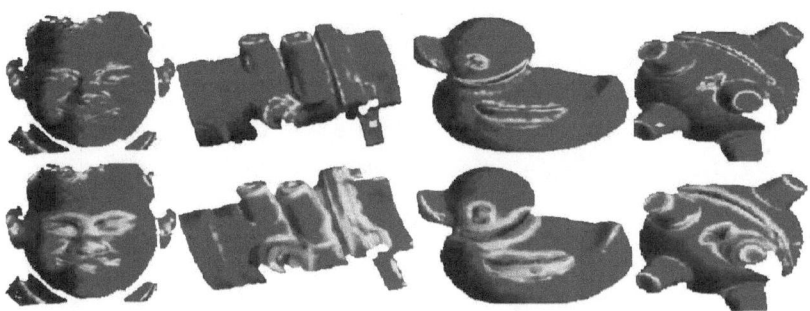

Fig. 2. Example of detected saliency on the model *par-face*, *valve*, *duck* and *frame*, respectively. Top row: Mesh saliency [3]. Bottom row: Our method.

the same saliency. To avoid this, we employ the bilateral filter operation instead of the Gaussian filter. The bilateral filtering for real function $f(v)$ on each vertex v as follows:

$$\mathscr{B}(f(v),\sigma) = \frac{\displaystyle\sum_{u\in N(v,2\sigma)} f(u)W_c(\|u-v\|)W_s(|f(u)-f(v)|)}{\displaystyle\sum_{u\in N(v,2\sigma)} W_c(\|u-v\|)W_s(|f(u)-f(v)|)} \tag{10}$$

where $N(v,\sigma) = \{u\|\|u-v\| < \sigma, u \text{ is a vertex}\}$, denotes a neighborhood for a vertex v, the closeness spatial smoothing is the standard Gaussian filter with parameter σ_c: $W_c = exp[-u^2/(2\sigma_c^2)]$, the feature preserving weight function is $W_s(u) = exp[-u^2/(2\sigma_s^2)]$, where σ_s penalizes large variation in the scalar function $|f(u)-f(v)|$. Compared with the Gaussian operation, the output of the bilateral operation on a vertex v is also a weighted average of the surrounding vertices. However, the weight depends not only on the spatial distance, but also on the scalar function difference. For the purposes of this paper, we chose $\sigma = 8.0\varepsilon$, where ε is 0.3% of the length of the diagonal of the bounding box of the model [3]. Based on a number of experiments, we chose $\sigma_c = \sigma_s = 1.5\sigma$ in our implementation. Fig. 2 shows the results of saliency detection by mesh saliency [3] and the proposed method. Both of the methods capture the visually important features and locate the large curvature regions. However, our method detects the salient region more completely and faithfully than [3].

Interest Points Selection. The first step is the voxelization of the saliency mapped model. Voxels without saliency content are removed. The entropy of saliency values of points in a voxel is calculated based on the saliency of the histogram map as $H(X) = \sum_{i=1}^{n} - P(S_i)*log_2(P(S_i))$, where $P(S_i)$ is the probability of saliency value S_i in a local voxel and n denotes the total number of the saliency values. The entropy measures how the saliency of vertices in a voxel varies. The larger the variation, the larger the entropy, the more detail the voxel contains. Thus the entropy can be used to guide the points sampling. Normally, the maximum entropy will be chosen by default. The minimum distances between the samples are used to address the possibility that the generated samples

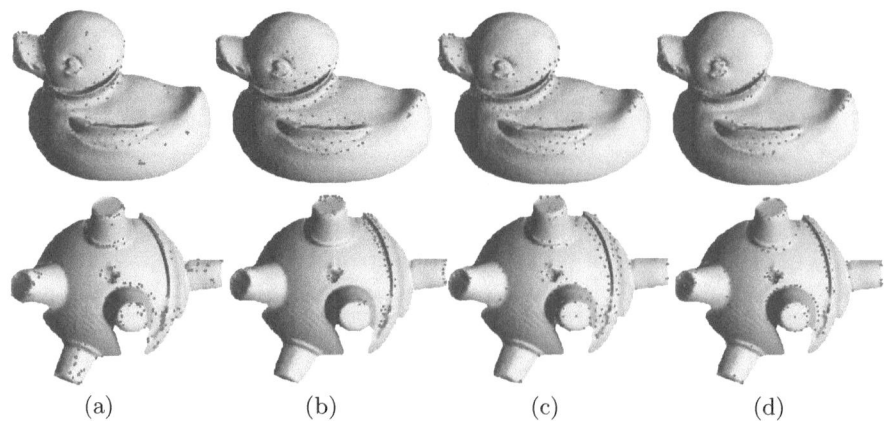

(a)   (b)   (c)   (d)

Fig. 3. Examples of the results of interest points detectors on *duck* and *frame*. (a) 3D-SIFT [8]. (b) 3D-Harris [7]. (c) MBO [6]. (d) Our method.

are close to the boundary between two or more adjacent voxels which might be too close to each other. Fig. 3(d) shows the detected interest points on the *duck* and *frame* by applying the proposed method.

4   Experimental Evaluation and Discussion

In this section we perform a comprehensive evaluation of our interest point detector, investigating its performance under different variations of input data: viewpoints changes and noise corruption. The performance of the proposed method is also compared with existing state-of-the-art 3D surface point detectors: 3D-Harris [7], 3D-SIFT [8] , and MBO [6]. Fig. 3(a)-(c) show the outcomes of the different 3D interest point detectors on models *duck* and *pat-face*.

To evaluate the proposed and other interest point detectors, we study the repeatability of their detection of interest points. This characteristic measures the ability of the detector to find the same set of keypoints on different instances of a given model, where the differences may vary under conditions of changes viewpoint or noise corruption. Once the correspondence of the points under these different conditions has been obtained, we define the repeatability rate of interest points:

$$rep = \frac{|np_i \cap np_r|}{|np_r|} \tag{11}$$

where np_i is the number of interest points found under one of the instances and np_r is the number of interest points detected from the original given model (reference model). For a perfect detector, it detects the same interest points in the first and last frame, i.e. $rep = 1$. In this paper, we select only 1% of the total number of the vertices as interest points, as too few keypoints may not be

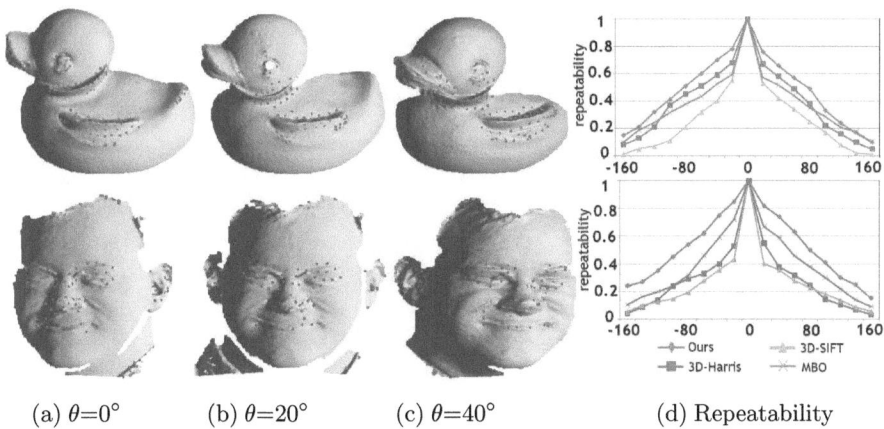

(a) θ=0°   (b) θ=20°   (c) θ=40°   (d) Repeatability

Fig. 4. (a)-(c) Model *duck* and *pat-face* subject to a rotation of θ at intervals of 20° around an unknown rotation axis are used to test the robustness of the proposed method. (d) Repeatability of interest points on the *duck* and *pat-face* with different detectors under different rotation of θ from 0° to ±160°.

enough to represent the global shape or supply further geometrical verification, while too many points leads to unnecessary waste of computational resources.

4.1 Changes of Viewpoint

This experiment evaluates the susceptibility of the proposed detector and its competitors to changing viewpoints. We set the models *buddha* and *lobster* at rotation angle 0° as the reference viewpoint and calculated the repeatability of interest points from 16 alternative viewpoints, of which 8 viewpoints were anti-clockwise rotated (20°,40°,...,160°) and 8 viewpoints were clockwise rotated (-20°,-40°,...,-160°). The effect of these rotations is shown in Fig. 4(a-c).

Fig. 4(d) shows the repeatability score of correspondences. The best results are obtained by our method for both *duck* and *pat-face* surfaces. It shows excellent tolerance with small rotations, such as ±20°. This is due to the high detection accuracy on the salient region, and selection of distinctive points from the salient region as the interest points. The repeatability score for a viewpoint change of 20 degrees on model *duck* achieves repeatability rates around 0.8 by our method, the 3D-Harris achieves a 0.68 repeatability rates and MBO gives a repeatability rate around 0.6, while 3D-SIFT only achieves 0.55. With increasing rotation angles, repeatability rates fall, since the larger the rotation angles the larger the difference between the points distributions. For instance, our detector is able to find approximately 85% of the overlapping points in the image with a rotation angle difference of 20° on model *pat-face*. However, this falls to 60% after the rotation angle increases to 60°. The second best performer is MBO, which achieves an accuracy of 50% at a rotation angle of 60°, as much as 10% lower than our method. The worst approach in this case is 3D-SIFT, which detected

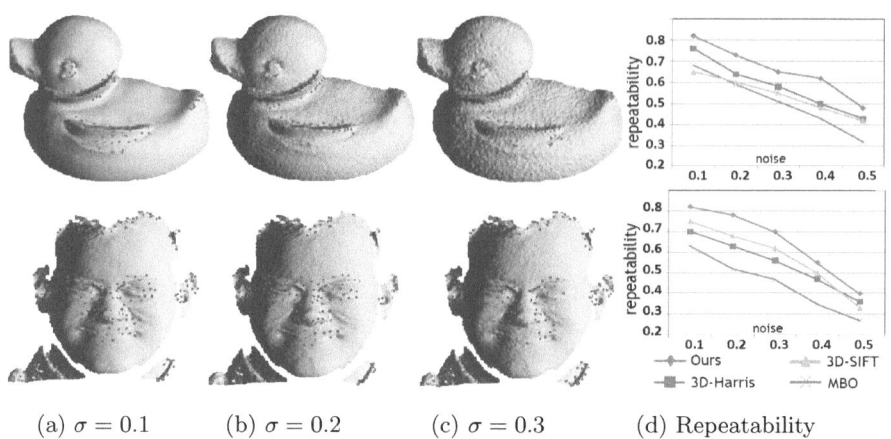

(a) $\sigma = 0.1$ (b) $\sigma = 0.2$ (c) $\sigma = 0.3$ (d) Repeatability

Fig. 5. Interest points detected with different noise levels.(a) $\sigma = 0.1$. (b) $\sigma = 0.2$.(c) $\sigma = 0.3$. (d) Repeatability of interest points on the *duck* (top) and *pat-face* (bottom) with different detectors applied under different noise level.

only 30% of overlapping points; a reduction of almost 30% compared with our detector. In summary, the proposed method achieves the best performance.

4.2 Noise Corruption

To demonstrate the robustness of our method and others, we also added different levels of random Gaussian white noise with a standard deviation of σ to the original data. However, the distribution of detected interest points on the surface remains, as shown in Fig 5(a)-(c). These outcomes show the adaptive neighborhood estimation of the proposed method. After the surface was noise corrupted, the local tessellations around a vertex changed considerably. In this case, our proposal estimated good neighborhoods to mitigate the noise aspects.

The resulting interest points repeatability rates for models *duck* and *pat-face* before and after noise addition are shown in Fig. 5(d). The results were obtained by adding different levels of Gaussian noise. As Gaussian noise is generated randomly, we ran the experiments multiple times at each noise level and report the average repeatability rate. Fig. 5(d) illustrates the repeatability of interest points under different levels of noise. All detectors have nearly linear repeatability curves. However, it can be observed that MBO is clearly more sensitive to this type of transformation, since MBO is relies on the principal curvature to measure the keypoints and the curvature is known to be sensitive to noise. As we expected, the noise level affects the calculation of repeatability: as σ increases, repeatability rates decrease for all the detectors. 3D-Harris achieves a high repeatability rate in the *duck* case. 3D-SIFT presents a relatively stronger tolerance than MBO. The proposed method demonstrates the highest robustness. By contrast with the alternative methods, the proposed method has the most stable repeatability results under noise corruption.

5   Conclusions

In this paper, we propose and demonstrate a novel saliency detection method for mesh surfaces. We define a relative distance-based Laplacian, which can be used in the event that the distribution of the data is not uniform and the distance metric in the algorithm mainly focuses on the representation of neighboring relationship between points. The generated relative distance-based mesh Laplacian is fused with the *Retinex* algorithm in order to estimate a new importance value for the surface. Detection of salient regions or interest points of generic objects is one of those computer vision problems without ground-truth. In addition, we tested the repeatability of interest points of our and three other state-of-the-art detectors. The results show that the proposed method yielded the best scores in response to viewpoints changing and noise corruption. Future research will include applying the detected saliency for segmentation and automatically fine-tuning parameters in the proposed method.

References

1. Dutagaci, H., Godil, A.: Evaluation of 3D interest point detection techniques via human-generated ground truth. Vis. Comput. 28(9), 901–917 (2012)
2. Gelfand, N., Ikemoto, L.: Geometrically Stable Sampling for the ICP Algorithm. In: Proc. of 3D Digital Imaging and Modeling (2003)
3. Lee, C., Varshney, A., Jacobs, D.: Mesh saliency. ACM Transactions on Graphics 24(3), 659–666 (2005)
4. Castellani, U., Cristani, M., Fantoni, A., Murino, V.: Sparse points matching by combining 3D mesh saliency with statistical descriptors. Comput. Graph. Forum 27(2), 643–652 (2008)
5. Sun, J., Ovsjanikov, M., Guibas, L.: A Concise and Provably Informative Multi-Scale Signature Based on Heat Diffusion. Comput. Graph. Forum 28(5), 1383–1392 (2009)
6. Mian, A., Bennamoun, M., Owens, R.: On the repeatability and quality of keypoints for local feature-based 3D object retrieval from cluttered scenes. Int. J. Comput. Vis. 89(2-3), 348–361 (2010)
7. Sipiran, I., Bustos, B.: Harris 3D: a robust extension of the Harris operator for interest point detection on 3D meshes. Vis. Comput. 27(11), 963–976 (2010)
8. Godil, A., Wagan, A.I.: Salient local 3D features for 3D shape retrieval. In: Proc. of 3DIP (2011)
9. Taubin, G.: A signal processing approach to fair surface design. In: 22nd Annual Conference on Computer Graphics and Interactive Techniques, pp. 351–358. ACM (1995)
10. Land, E., McCann, J.: Lightness and Retinex Theory. JOSA 61(1), 1–11 (1971)
11. Zhong, G., Hou, X., Cheng, L.: Relative Distance-based Laplacian Eigenmaps. In: CJKPR, Nanjing, China (2009)
12. Zhao, Y., Liu, Y., Song, R., Zhang, M.: Extended non-local means filter for surface saliency detection. In: Proc. of ICIP, pp. 633–636 (2012)
13. Zhao, Y., Liu, Y.: Patch based saliency detection method for 3D surface simplification. In: Proc. of ICPR, pp. 845–848 (2012)
14. Achanta, R., Estrada, F., Wils, P., Strunk, S.: Frequency tuned Salient Region Detection. In: Proc. of CVPR (2009)

Robust Silhouette Extraction from Kinect Data

Michele Pirovano[1,2], Carl Yuheng Ren[3], Iuri Frosio[1], Pier Luca Lanzi[2],
Victor Prisacariu[3], David W. Murray[3], and N. Alberto Borghese[1]

[1] Department of Computer Science, University of Milano, Milano, Italy
{pirovano,frosio,borghese}@di.unimi.it
[2] Dipartimento di Elettronica, Informazione e Bioingegneria, Politecnico di Milano, Italy
pierluca.lanzi@polimi.it
[3] Department of Engineering Science, University of Oxford, Oxford, England
{carl,victor,david}@robots.ox.ac.uk

Abstract. Natural User Interfaces allow users to interact with virtual environments with little intermediation. Immersion becomes a vital need for such interfaces to be successful and it is achieved by making the interface invisible to the user. For cognitive rehabilitation, a *mirror view* is a good interface to the virtual world, but obtaining immersion is not straightforward. An accurate player profile, or silhouette, accurately extracted from the real-world background, increases both the visual quality and the immersion of the player in the virtual environment. The Kinect SDK provides raw data that can be used to extract a simple player profile. In this paper, we present our method for obtaining a smooth player profile extraction from the Kinect image streams.

1   Introduction

Natural user interfaces based on devices such as the Microsoft Kinect have revolutionized video-gaming: the player is represented by an avatar inside the game and interacts with virtual objects, resulting in a more realistic and intuitive control experience. Furthermore, Microsoft has released a SDK [1] that allows easily integrating Kinect as an input device that provides the motion of the player, defined as a set of segments connected by joints. The pose is described by a set of quaternions, one for each joint, along with the translation of the root joint located at the hip level.

A player tracking modality that is much less explored, is that of tracking the player's *profile*. This has been shown to be particularly effective in cognitive rehabilitation games as it allows the patient to see himself mirrored inside the virtual environment while performing exercises. It has been hypothesized that such modality activates *mirror neurons* [2, 3] that might drive a cortical reorganization to regain the lost function. Moreover, such a *mirror view* largely increases the sense of immersion inside the game. The DuckNeglect platform [4] has been taking advantage of this to design exploratory mini-games that, from preliminary results, have been shown to be

A. Petrosino (Ed.): ICIAP 2013, Part I, LNCS 8156, pp. 642–651, 2013.

effective in treating neglect patients[1]. DuckNeglect forces the patient to explore his neglected hemi-field requiring him to reach targets inside a virtual scenario and to avoid distractors. The patient receives feed-back on his movement in real-time through his profile moving inside the scenario. The limitation of DuckNeglect is the use of a 2D camera that requires uniform lighting and background to achieve a robust and accurate extraction of the patient profile: the gaming space has to be specially set-up and this prevents the patient from exercising at home where he could exercise more intensively and feel more comfortable.

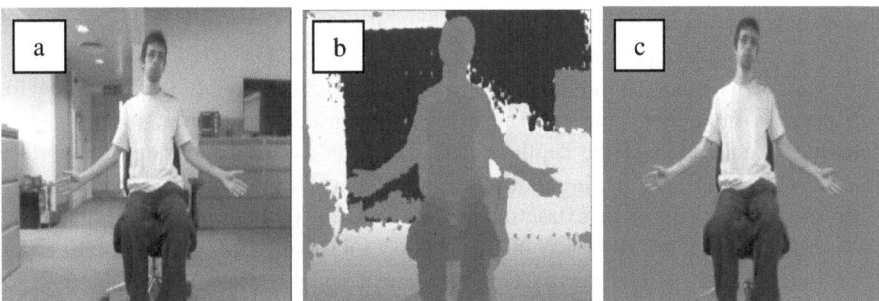

Fig. 1. - The input data for the sitting player. (a) The color image. (b) The depth image. Magenta points in the depth map have an unknown value. (c) our extracted player profile.

The Kinect has the potential to make silhouette extraction more robust through the integration of depth and color data. However, its SDK does not produce a good enough profile. This is due to two main reasons: the depth-to-color calibration parameters provided internally in the SDK are not sufficiently accurate [6], so that the alignment of the two images is inaccurate (Fig. 3b). Moreover, once the two images are aligned, a sharp identification of the profile is not produced due both to the low resolution of the depth map and to noise. The result may be sufficient for entertainment games such as *Get Fit With Mel B*[2], but it can be annoying and distracting to the rehabilitating patient who generally moves slowly and therefore is aware of all the little inaccuracies on the screen, resulting in a decreased sense of presence.

We show here how we can improve the quality of the player profile through different steps. We first achieve a better alignment of the color and depth images through an innovative on-the-field calibration of the Kinect sensor that only requires moving a ball in front of the device. With the computed parameters, we extract in real time a smooth and robust player silhouette by filtering the player color image identified through the depth map.

The paper is structured as follows. In section 2, we present the materials and the proposed methods for extracting the silhouette. Section 3 presents the application of our methods to a serious game. We draw conclusions in Section 4.

---

[1] Neglect is the inability of a person to process and perceive stimuli on one side of the body or environment. Such inability is not due to a lack of sensation [5].

[2] uk.gamespot.com/get-fit-with-mel-b

2   Materials and Methods

The Kinect sensor integrates a color and a depth camera. The color camera outputs an RGB image I_c at a standard resolution of 640x480 pixels at 30 frames per second (fps) (Fig. 1a). The depth camera consists of an infrared sensor paired with an infrared emitter that projects onto the scene an irregular circular pattern (Fig. 2a). The depth map image I_d (Fig. 1b) is recovered through triangulation and speckle pattern decorrelation [7]; it has a maximum resolution of 640x480 pixels, which is actually up-sampled from a resolution of 320x240 pixels at 30 fps. The depth map returned by the Kinect sensor is bound to be noisy, thus a bilateral filter is internally applied to it. The result is a smoother depth map that preserves sharp edges but is still noisy and holds many unknown values. In addition, even with a static scene, the depth reading for each pixel is often not consistent between two consecutive frames, resulting in a flickering visualization of the depth stream. The SDK isolates and identifies up to 6 different players in I_d assigning a different index to each player. This can be used to create an effective silhouette extraction procedure (cf. [4]): an *alpha bitmask* image I_b is created independently for each player where each pixel is set to 1 (visible) if the player index for the pixel corresponds to the player and to 0 (invisible) otherwise. An RGBA player image I_p is created by combining the RGB color of each pixel in I_c with the corresponding alpha value in I_b.

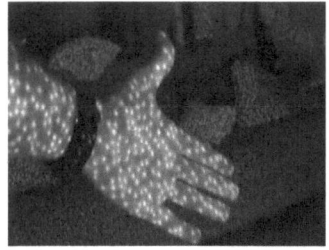

Fig. 2. On the left the typical pattern projected by the infrared emitter[3]. On the right, calibrating Kinect with a LED ball.

However, the direct overlap of I_b with I_c produces an incorrectly aligned I_p as seen in Fig. 3a. This happens because the Kinect's color and depth sensors are not aligned, either in time (the acquisitions from the two sensors, even at the same frequency, are up to 7ms apart), or in space, as the images are acquired from different viewpoints. If time difference can be dealt directly with the function *AllFramesReady* of the SDK that guarantees a maximum of 3ms delay between the frames of different cameras[4], the spatial alignment obtained through the SDK can instead be improved (Fig. 3b-c).

---

[3] from www.youtube.com/watch?v=nvvQJxgykcU

[4] from http://social.msdn.microsoft.com/Forums/sl-SI/
kinectsdknuiapi/thread/2e172449-3d18-4914-9370-fbc1cfa31aa6

2.1   Depth to Color Camera Calibration

Methods for a complete photogrammetric calibration of the Kinect camera have been proposed (e.g. [7-8]), but they require grids and structured markers and are not suitable for the home environment. We propose here an efficient on-the-field calibration method for any Kinect device that can accurately recover the geometrical parameters of the stereo-system constituted of the color and depth camera by tracking a moving ball. It is based on an adapted version of bundle adjustment [14].

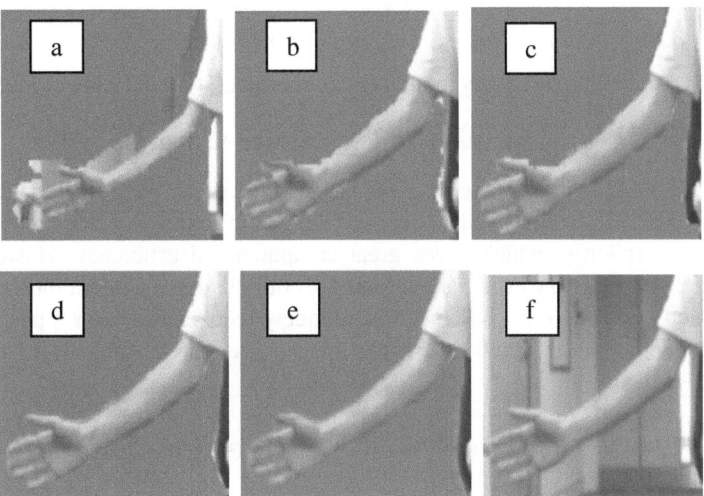

Fig. 3. In panel (a) a close-up of the silhouette produced by directly overlapping depth and color data: the offset is evident. A residual offset is still present when using SDK to extract silhouette (b): the background is visible below the arm and on the left of the chair, whereas the upper part of the forearm has been partially eroded. A better result has been obtained with the method proposed here (c). Notice that in all cases a jagged border is obtained. The silhouette is then smoothed with the moving average method (d) or with the frequency filtering method (e). Panel (f) shows the original color image.

Given I_d, we define each depth image pixel x_d at depth images coordinates (u_d, v_d) having depth d. The corresponding color camera pixel x_c on I_c has coordinates (u_c, v_c). The homogeneous 3D coordinates of x_d and x_c are respectively

$$x_d = [u_d d \quad v_d d \quad d]^T \tag{1}$$

$$x_c = [u_c \quad v_c \quad 1]^T \tag{2}$$

With the world coordinate system centered on the depth camera, we represent a 3D point in the world coordinate as

$$X = [x \quad y \quad z]^T. \tag{3}$$

Both cameras are modeled using the pinhole model and distortions are not considered. The relationship between the pixels and the 3D points is thus

$$x_d = A_d X \tag{4}$$

$$sx_c = A_c(RX + T) \tag{5}$$

where R is the 3x3 relative orientation matrix between the two cameras, T the relative position and s is an arbitrary scale factor. A_d and A_c are the intrinsic matrices of the two cameras in the form:

$$A = \begin{bmatrix} f_x & 0 & c_x \\ 0 & f_y & c_y \\ 0 & 0 & 1 \end{bmatrix}. \tag{6}$$

Given the depth pixel x_d as input, we can reproject it on the camera image as

$$sx_c = A_c(RX + T) = A_c(RA_d^{-1}x_d + T) = A_c RA_d^{-1}x_d + A_c T = Hx_d + M, \tag{7}$$

where H is a 3x3 matrix and M is a 3x1 translation vector. Note that using this reprojection model we do not need to recover the actual depth or color camera intrinsic parameters explicitly, which gives great computational efficiency. However, given known color camera intrinsic parameters, A_c, both the depth camera intrinsic parameters A_d and the relative pose R,T can be uniquely recovered. Given n pairs of corresponding points on the depth and color images, we find H and M through a least squares estimate

$$[M,H] = \arg_{H,M} \min \left(\sum_{i=1}^{n} \left\| x_{c_i} - x_c \left(M, H, x_{d_i} \right) \right\|^2 \right). \tag{8}$$

The n pairs of corresponding points are obtained by tracking a known object both in the depth and color camera. Our object of choice is a LED ball, which proves useful due to its bright color and simple shape. We use color detection in the HSV space on I_c to find the 2D center of the LED ball x_c (Fig. 2b). We use the 3D level-set based tracker described in [9] to track the position of the LED ball in I_d and we find its 3D center x_d. Each tracking frame provides a corresponding ball-center pair (x_c, x_d). Eq. (7) can then be used to find in real-time for each depth pixel x_d the corresponding pixel x_c.

2.2   Silhouette Extraction and Smoothing

The silhouette of the player is jagged (Fig. 3b-c) due to the low resolution of the depth stream and to the irregular pattern used to estimate depth (Fig. 2a). The flickering depth stream, in turn, produces a flickering silhouette as well.

We have tackled this problem through adequate filtering of the player segmentation data. We have developed two alternative solutions: one in the space and one in the frequency domain. Both methods take I_b as input. The first method is inspired by [10] and has two steps. In the first step, we apply a moving average low-pass filter to I_b with a sliding window of 5x5 pixels. To obtain a graded transition, the alpha values are converted from a binary value to a byte value (0 to 255), obtaining I_b'. In the second step, we perform a weighted average over the pixel values in the last N frames $I_b'(t - i)$ with $i = 0 \dots N$, assigning a weight $N - i$ to each frame to avoid adding a significant delay. As a result, we achieve a smoother player profile I_p, reducing the flickering effect of the silhouette with a smooth transition between different frames (Fig. 3d). The method produces a smooth silhouette, but its drawbacks are the local smoothing, that doesn't allow an accurate identification of the silhouette, and the trailing effect that can be noticed especially during fast motion. This can be acceptable and desirable for some implementations but it can be annoying for rehabilitation. For this reason, we have developed a second method that is based on frequency filtering. We first compute the Discrete Cosine Transform (DCT) of the input bitmask $I_b(u, v)$ where $u \in [0, U - 1]$ and $v \in [0, V - 1]$. We then eliminate the higher frequency components, thus realizing a box low-pass filter. Finally, we perform an inverted DCT (IDCT) to obtain a smoother player profile I_p (Fig. 3e). The DCT of I_b, $C(k_u, k_v)$, is a sum of cosine functions computed as

$$C(k_u, k_v) = \alpha_u(k_u)\, \alpha_v(k_v) \sum_{u=0}^{U} \sum_{v=0}^{V} \beta_u(k_u, u)\, \beta_v(k_v, v)\, I_b(u, v) \tag{9}$$

with $k_u \in [0, K_U]$ and $k_v \in [0, K_V]$, K_U and K_V being the number of harmonics on the u and v coordinate that we want to keep in the DCT. Usually, K_U and K_V are set to $U - 1$ and $V - 1$ respectively and cannot be larger.

The coefficients $\alpha_n(k_n)$ assume the form

$$\alpha_n(k_n) = \begin{cases} \frac{1}{\sqrt{K_n}} & for\ k_n = 0 \\ \sqrt{\frac{2}{K_n}} & otherwise \end{cases}. \tag{10}$$

The functions $\beta_n(k_n, n)$ represent the cosine basis functions on the u axis:

$$\beta_n(k_n, n) = \cos\left(\frac{\pi\, k_n\, (2n-1)}{2N}\right). \tag{11}$$

Similarly, $\alpha_v(k_v)$ and $\beta_v(k_v, v)$ can be derived for the v axis. One important property of the DCT is its separability, due to which it can be rewritten as

$$\begin{aligned} C(k_u, k_v) &= \alpha_v(k_v)\, \alpha_u(k_u) \sum_{v=0}^{V} \beta_v(k_v, v) \sum_{u=0}^{U} \beta_u(k_u, u)\, I_b(u, v) \\ &= \alpha_v(k_v)\, \alpha_u(k_u) \sum_{v=0}^{V} \beta_v(k_v, v)\, D_u(k_u, v) \end{aligned}. \tag{12}$$

We can thus separate the computation in a first matrix multiplication that computes $D_u(k_u, v)$ and a second matrix multiplication that computes $C(k_u, k_v)$. The procedure is illustrated in Fig. 4a. A similar separation can be performed for the IDCT (Fig. 4b), that can be written as

$$\begin{aligned} I_p(u, v) &= \sum_{k_v=0}^{K_v} \alpha_v(k_v)\beta_v(k_v, v) \sum_{k_u=0}^{K_u} \alpha_u(k_u)\beta_u(k_u, u)\, C(k_u, k_v) \\ &= \sum_{k_v=0}^{K_v} \alpha_v(k_v)\beta_v(k_v, v)\, D_v(u, k_v) \end{aligned} \tag{13}$$

In our implementation, we observe that we do not need to compute the full DCT of the input bitmask, since we are going to filter out most of the higher frequency data. As such, we compute the DCT only for the first M harmonics, from 20 to 30 in the present case, that we wish to keep in the final image. As a result, we can work with smaller DCT images and thus decrease both computation and memory load. We then apply a circular window in the frequency domain setting to zero the coefficients of the DCT components for which $\sqrt{k_u{}^2 + k_v{}^2} > M$. To increase the speed, we also pre-compute the required coefficients and samples of the cosine functions: $\alpha_n(k_n)$ and $\beta_n(k_n, n)$.

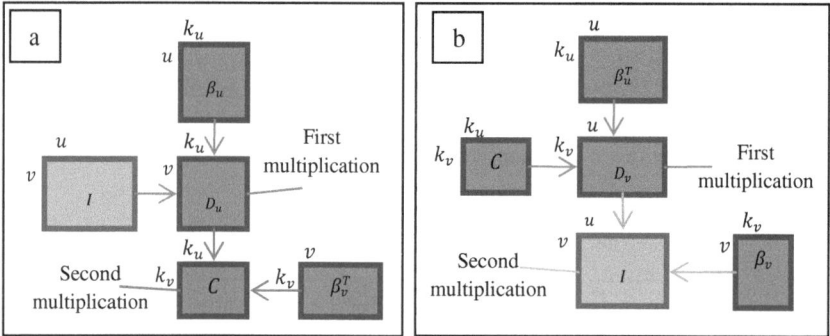

Fig. 4. Visual representation of DCT (a) and IDCT (b) computation using two matrix multiplications

We implemented the algorithm in CUDA for an additional speed-up using a parallel method to compute the DCT and its inverse [11]. We first load the input bitmask I_p into device global memory. We then perform two subsequent matrix multiplications using parallel CUDA Kernels to compute C, using one parallel Kernel thread for each matrix coefficient, thanks to the separability of the DCT. After the DCT is computed, we apply a step filter with a circular shape to remove the high frequencies, then we compute the IDCT with the same parallel multiplication procedure, obtaining I_p.

However, the sharp transition of the step filter introduces oscillations (Gibb's artifacts) on the final image, especially close to the border of the patient profile where I_p goes from one to zero in a few pixels. We thus apply a sigmoid function to each pixel value of I_p to map the resulting pixel values in a 0-255 range and drastically reduce the oscillations. The filter is heavily parameterized in order to easily tune the shape of the outline. A flowchart of the different steps is reported in Fig. 5.

3   Results

These algorithms have been integrated inside the Intelligent Game Engine for Rehabilitation (IGER), developed for the Rewire project [12], to guide and monitor patients during rehabilitation at home and have been extensively tested. In the experiment presented here, the player sits at about $1.5m$ from the Kinect device that is installed above a large TV screen (50 inches) connected to the PC on which the Game Engine is run. The calibration procedure required a maximum of $15s$ to acquire the data and calibrate: 300 frames were typically sufficient to guarantee a good calibration with residuals lower than 2 pixels.

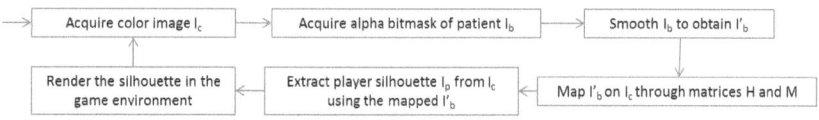

Fig. 5. The block diagram associated to processing each Kinect frame

The player profile is used during a cognitive rehabilitation game called *Balloon Popper*, where the player has to reach and touch a set of balloons that appear in the virtual scene, avoiding the bombs. We played the game using several settings for the silhouette extraction and obtained different visual results, as shown in Fig. 6 and in the related series of videos[5]. The background was cluttered with office objects such as chairs, desks, computers and their displays. Fig. 6a shows the raw misaligned silhouette. The player's profile is jagged and noisy and a large part of the background can be seen. It is also hard to understand when the player is actually touching a game object. In Fig. 6b, we applied our calibrated depth-to-color mapping to the silhouette. The background is correctly subtracted and the player sees only himself inside the profile area. In Fig. 6c, we apply moving average smoothing. The jagged silhouette disappears and there is a less clear separation between the real player and the virtual scene, as if the two were merging together. However, the smoothing is only local and thus the resulting silhouette has a "wavy" look and there is a noticeable trailing. In Fig. 6d, we apply DCT smoothing instead. We obtain a smooth contour around the player, and the player's silhouette is merging into the virtual background as if the player was actually part of the virtual scene, increasing immersion.

4   Discussion

Rehabilitation exercises can be dull and boring due to their repetitive and slow movements. Accordingly, consciousness has raised in the last years towards the need for more compelling experiences, under the form of virtual rehabilitation games, to hold the focus of patients for longer periods and thus address compliance. Such rehabilitation games are often set in a virtual fictitious environment, such as the Rewire farm, in order to create a compelling fantasy setting.

Moreover, a mirrored view of oneself allows a greater immersion inside the virtual world. This increased sense of presence can be a useful way to promote neural reorganization [4]. However, to achieve this, the patient must not be reminded of the difference between his own realistic image, reflected in the mirrored view, and the virtual environment. For this reason, the player's mirrored image must be of high quality and it must be segmented with accuracy and in real-time, in order to let it blend in the virtual environment. Otherwise, the result can easily fall inside the uncanny valley [13] (Fig. 6b) and the display will alienate the patient from exercising.

Using as a base Kinect's official SDK, we obtained a robust and smooth extraction of the player's silhouette in real-time. We improved the depth-to-color mapping

---

[5]   homes.di.unimi.it/~pirovano/rewire/iciap2013

through an efficient and robust calibration. The accuracy of the algorithm can be further improved using circle fitting insensitive to occlusions [6]. We explicitly remark that the projection of the sphere on the image plane is, in general, an ellipse whose center is not the projection of the sphere center. However, under very general assumptions, such displacement can be safely neglected [15]. The resulting procedure is particularly suitable for the home environment and it allows to easily recalibrate the device, whenever required (e.g. if the device falls to the ground).

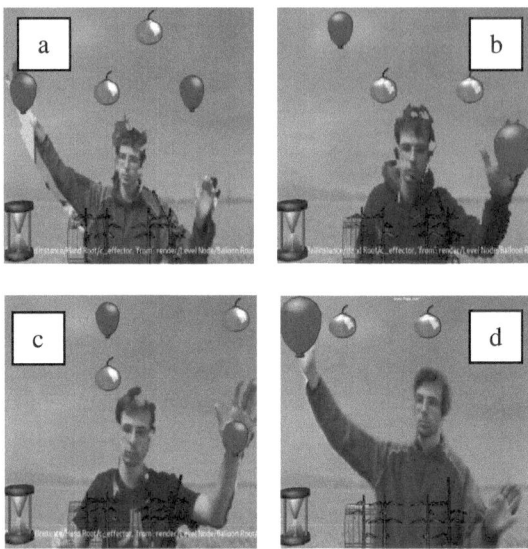

Fig. 6. Playing the Balloon Popper game using different silhouette settings. (a) No mapping nor smoothing. (b) Calibrated mapping, with no smoothing. (c) Calibrated mapping with moving average smoothing. (d) Calibrated mapping with DCT smoothing

Thanks to CUDA and the specific DCT implementation, the silhouette is extracted and smoothed efficiently, freeing the CPU for game engine processing. The resulting silhouette lets the player immerse into the game without being distracted or mislead by inaccuracies or the bad quality of the silhouette. Thanks to the depth camera of Kinect, this result can be achieved even in bad light conditions and with a really noisy background, making this method preferable in respect to color-based methods. As future work, the authors aim to assess quantitatively the quality of our results.

The method depends on the number of DCT components, M, that are retained. The actual choice is a trade-off between the amount of details retained and noise suppression, due to the frequency of the noise being similar to that of the silhouette detail (i.e. the fingers). A similar trade-off on the border of the silhouette between showing the background beneath and the eroding part of the player profile is achieved by tuning the steepness of the sigmoidal function. Still, improvements are required to stabilize the obtained silhouette: unknown depth data and hair, that do not produce a clear reflection of the infrared pattern, can contribute to a variation over time of the shape of the input bitmask. To allow a different cut-off frequency in the different regions of the body, one possibility is to substitute DCT with wavelet decomposition and retain a different number of components in different areas. This approach can be combined with a local color image processing with the aim of an identification of the hair region that is more stable over time.

Acknowledgements. This work was partially supported by the FP7 Project REWIRE, grant 287713 of the European Union.

References

1. Kinect for Windows SDK,
   http://www.microsoft.com/en-us/kinectforwindows/
2. Rizzolatti, G., Craighero, L.: The mirror neuron system. Annual Review of Neuroscience 27, 169–192 (2004)
3. Cameirão, M.S., Badia, S.B.I., Oller, E.D., Verschure, P.F.M.J.: Neurorehabilitation using the virtual reality based Rehabilitation Gaming System: methodology, design, psychometrics, usability and validation. Journal of Neuroengineering and Rehabilitation 7, 48 (2010)
4. Mainetti, R., Sedda, A., Ronchetti, M., Bottini, G., Borghese, N.A.: Duckneglect: videogames based neglect rehabilitation. Technology and Health care (in press, 2013)
5. Unsworth, C.A.: Cognitive and Perceptual Dysfunction. In: Schmitz, T.J., O'Sullivan, S.B. (eds.) Physical Rehabilitation, pp. 1149–1185. Davis Company, Philadelphia (2007)
6. Frosio, I., Borghese, N.A.: Real-time accurate circle fitting with occlusion. Pattern Recognition 41(3), 1041–1055 (2008)
7. Menna, F., Remondino, F., Battisti, R., Nocerino, E.: Geometric investigation of a gaming active device. In: Proc. of Videometrics, Range Imaging and Applications XI, SPIE Optical Metrology (2011)
8. Khoshelham, K., Elberink, S.O.: Accuracy and Resolution of Kinect Depth Data for Indoor Mapping Applications. In: Sensors 2012, vol. 12, pp. 1437–1454 (2012)
9. Yuheng Ren, C., Reid, I.: A unified energy minimization frameworkfor model fitting in depth. In: 2nd Workshop on Consumer Depth Cameras for Computer Vision, CDC4CV (2012)
10. Sanford, K.: Smoothing Kinect Depth Frames in Real-Time (2012)
11. Obukhov, A., Kharlamov, A.: 8x8 Discrete Cosine Transform with CUDA, Whitepaper,
    http://docs.nvidia.com/cuda/cuda-samples/index.html
12. Pirovano, M., Mainetti, R., Baud-Bovy, G., Lanzi, P.L., Borghese, N.A.: Self-Adaptive Games for Rehabilitation at Home. In: Proc. IEEE Conference on Computational Intelligence and Games, pp. 179–186 (2012)
13. Mori, M.: The Uncanny Valley. Energy 7(4), 33–35 (1970)
14. Hartley, R., Zisserman, A.: Multiple View Geometry, Cambridge University Press (2000)
15. Frosio, I., Borghese, N.A.: 6dof tracking, with axial markers. Medical & Biological Engineering & Computing (in press, 2013), doi:10.1007/s11517-013-1052-7

A Modified SIFT Descriptor for Image Matching under Spectral Variations

Sajid Saleem and Robert Sablatnig

Computer Vision Lab, Institute of Computer Aided Automation,
Vienna University of Technology, 1040 Vienna, Austria
{ssaleem,sab}@caa.tuwien.ac.at

Abstract. In multispectral imaging multiple discrete wavelength bands are used to image a scene. The imaging process maps the scene contents to different intensity levels and varies the scene appearance from band to band. This induces intensity variations among the spectral images and effects the performance of SIFT for cross spectral image matching. This paper proposes modifications to the SIFT descriptor in order to improve its robustness against spectral variations. The proposed modifications are based on fact, that edges remain well preserved in multispectral imaging and we can achieve better image matching results by boosting the contribution of local edges in the SIFT descriptor construction process. Therefore, we propose a Local Contrast (Δ) and a Differential Excitation (ξ) function for the construction of SIFT descriptors. The experimental results show, that the performance of Δ-SIFT and ξ-SIFT is superior to standard SIFT for image matching under spectral variations.

Keywords: SIFT, spectral images, interest regions, image matching.

1 Introduction

Multispectral imaging decomposes a scene into multi-band images [2]. The decomposition process generates useful information about the scene contents to solve the visual computing problems related to scene recognition [1], remote sensing [4] and visual surveillance [6] efficiently. The Scale Invariant Feature Transform (SIFT) [7] has been widely used for these applications [1,4]. It extracts keypoints from the images and constructs keypoint descriptors using image gradients. Recent studies show, that intensity differences among the spectral images effect the performance of SIFT [10,12].

To overcome the effects of such spectral variations Yi et al. propose scale restricted SIFT [13]. They use similar scale SIFT as candidates for descriptor matching. The scale restriction is found efficient in reducing the number of outliers and improves the SIFT performance [13]. In the orientation restricted SIFT (π-SIFT) approach [12] the gradient orientations are mapped to the $[0,\pi]$ range to overcome the intensity reversal problem in the images which in turn improves the SIFT performance against spectral variations. However, it has been found that the performance of SIFT remains low for images where spectral variations are high inspite of such modifications [10].

A. Petrosino (Ed.): ICIAP 2013, Part I, LNCS 8156, pp. 652–661, 2013.

In multispectral images we observe that, edges remain well preserved due to their sharp changes in intensity nature. This motivates us to boost their contributions in the descriptor construction process in order to achieve better robustness for SIFT against spectral variations. Therefore, we propose a Local Contrast (Δ) [11] and a Differential Excitation (ξ) [3] function in this paper to construct modified SIFT descriptors [7]. Each function has a spectral invariant response to local edges and improves the performance of SIFT for spectral images as compared to gradient magnitude (Ω) based SIFT descriptors [7].

The Δ function [11] uses minimum and maximum gray levels to estimate the edge magnitude in a local window, whereas ξ estimates the edge magnitude via the ratio between the local sum of gray level differences to the gray level of the pixel under study [3]. Each function assigns high magnitude weights to edge pixels which in turn cast spectral invariant votes for their corresponding gradient orientation feature histogram bins in the Δ-SIFT and ξ-SIFT descriptors and improves the performance of SIFT for image matching under spectral variations.

To illustrate the significance of Δ and ξ functions two regions of interest from 460nm and 720nm wavelength bands are shown in Figure 1. It can be seen that, only sharp edges are visible in the Ω images whereas the edges in the low contrast regions are suppressed. Also the corresponding edge magnitudes are different which lead to less correlated Ω based SIFT descriptors. In the Δ images the edges are equally enhanced irrespectively to image contrast which makes the Δ-SIFT descriptors more spectral invariant as compared to SIFT. In the ξ images the edge magnitudes are more spectral invariant as compared to Ω and Δ functions which increases the correlation between ξ-SIFT descriptors as suggested by the table which summarizes the cross spectral descriptor matching scores. Our experiments on spectral images of three different scenes suggest that, the performance of ξ-SIFT and Δ-SIFT is superior to Ω based SIFT [7] for image matching under spectral variations.

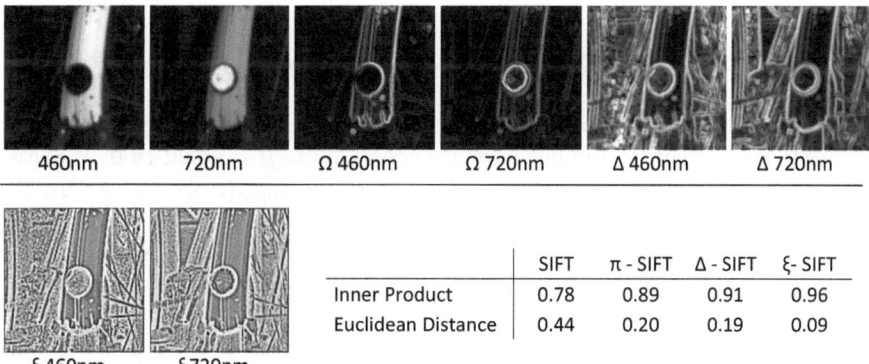

	SIFT	π - SIFT	Δ - SIFT	ξ- SIFT
Inner Product	0.78	0.89	0.91	0.96
Euclidean Distance	0.44	0.20	0.19	0.09

Fig. 1. Two regions of interest from 460nm and 720nm wavelength bands. The edge magnitude images are obtained via gradient magnitude (Ω), local contrast (Δ) and differential excitation (ξ) functions. The table summarizes the cross spectral descriptor matching score for each SIFT type.

The rest of the paper is organized as follows. In Section 2 we discuss the Ω, Δ and ξ functions for edge magnitude estimation. In Section 3 we describe the experimental setup in detail. Section 4 presents the experimental results and finally, we conclude the paper in Section 5.

2   Edge Magnitude Estimation

In this section we briefly describe the Ω, Δ and ξ functions for edge magnitude estimation. We also discuss their significance in the SIFT descriptor construction process.

2.1   Gradient Magnitude (Ω)

The gradient magnitude function uses local gray level changes to estimate the edge magnitudes i.e, $\Omega = (G_x^2 + G_y^2)^{1/2}$ where G_x and G_y are image gradients along x and y directions respectively using $[-1, 0, 1]$ and $[-1, 0, 1]^t$ filter kernels, where t stands for matrix transpose operation. The Ω based descriptor in this paper is referred to as SIFT [7]. In Figure 1 we can see, that gradient magnitude function is sensitive to local gray level changes and produces high magnitude edges in the high contrast regions whereas the edges in the low contrast regions are suppressed due to small changes in the gray levels.

2.2   Local Contrast (Δ)

The Δ function [11] uses minimum $I_{min}(x, y)$ and maximum $I_{max}(x, y)$ gray levels in a local window around each image pixel $I(x, y)$ to estimate the edge magnitude according to (1).

$$\Delta(x, y) = \frac{I_{max}(x, y) - I_{min}(x, y)}{I_{max}(x, y) + I_{min}(x, y) + \epsilon} \tag{1}$$

where (x, y) represents spatial location of the pixel under study and ϵ is an infinitely small positive number added when $I_{max}(x, y)$ is equal to 0. We use a local window of 3×3 size to compute $I_{min}(x, y)$ and $I_{max}(x, y)$ gray levels. The Δ function produces high magnitude edges irrespectively to image contrast as shown in Figure 1 which suggest that, Δ based edge magnitudes are superior to Ω and result in superior performance for Δ-SIFT under spectral variations.

2.3   Differential Excitation (ξ)

Differential excitation is an edge operator [3]. It estimates edge magnitude in a 3×3 region around every image pixel as described in (2) where $d(x, y)$ is a local sum of gray level differences and $I(x, y)$ is the gray level of the pixel under study. The ξ function simulates the local salient variations similar to human perception [3]. Therefore, the edge pixels receive high magnitude weights via ξ and the non edge pixels are suppressed.

$$\xi(x,y) = \arctan\left(\frac{d(x,y)}{I(x,y)}\right); \quad d(x,y) = -9I(x,y) + \sum_{i=-1}^{i=1} \sum_{j=-1}^{j=1} I(x+i, y+j) \quad (2)$$

The domain range for ξ is $[-\pi/2,\ \pi/2]$. We map ξ to the $[0,\pi]$ range via $\xi :=$ $\pi/2 + \xi$ in order to avoid negative values in the descriptor construction process. The descriptor constructed from ξ in this paper is referred to as ξ-SIFT. It can be seen from Figure 1 that, the ξ function boosts the local edges irrespectively to the wavelength band, which increases the correlation between corresponding ξ-SIFT descriptors and results in superior descriptor matching measures as compared to the SIFT descriptor in the presence of spectral variations.

2.4   Statistics of Edge Magnitudes

The histograms of Ω, Δ and ξ based edge magnitudes are shown in Figure 2. Each histogram is constructed from 12,000 different Harris Laplace (HarLap) regions [5,9] of size 41×41 pixels with gray levels in the normalized [0,1] range. We use HarLap regions in this paper to evaluate the performance of Δ-SIFT and ξ-SIFT. Therefore, the edge magnitude statistics in Figure 2 are useful to understand the distribution and the contribution of each function values in the descriptor construction process.

The histogram of Ω exhibits decaying exponential nature with majority samples in the [0, 0.3] range. This range represent homogenous and low contrast regions where the edges are low in magnitude. The Δ distribution is relatively uniform as compared to Ω because it boosts the edges irrespectively to image contrast. In the ξ case, the majority samples lie close to the function boundaries which represent regions of low gray levels. The edge pixels also belong to such regions because the scene edges diffract the incoming light rays and less of them

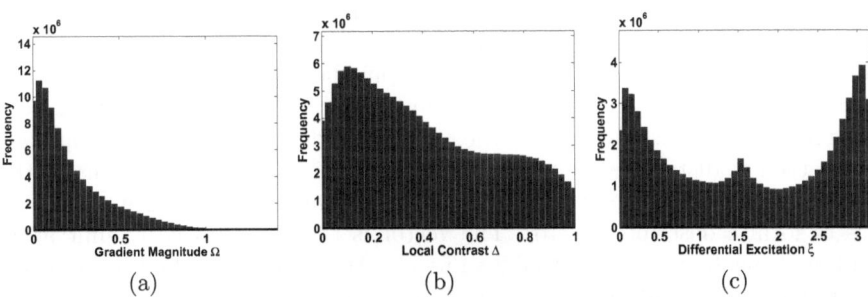

Fig. 2. The histograms of (a) gradient magnitudes (b) local contrast and (c) differential excitations computed from 12,000 different intensity normalized Harris Laplace regions [5,9]

are reflected back towards the camera. The ξ values in the middle represent homogenous regions where the local sum of gray level differences is close to zero in magnitude.

3   Experimental Setup

This section describes the experimental setup in detail. It covers test images, the SIFT descriptor construction process and performance evaluation measures.

3.1   Test Images

We use 460nm and 720nm wavelength band images of three different scenes from Real World Hyperspectral Images (RWHI) dataset[1] as test images. These images are shown in Figure 3. The test images are under different levels of spectral variations which make the cross spectral image matching a challenging problem to solve. Here motivation is to evaluate the performance of SIFT [7] under such intensity variations and then improve its spectral invariant characteristics via Δ and ξ functions.

3.2   Interest Regions

We use HarLap regions [9] for cross spectral image matching. The descriptors constructed from HarLap regions are scale invariant. We can also construct rotation invariant descriptors by rotating the regions in the direction of their dominant gradient orientations according to the application requirement. In the experiments we resize each HarLap region to the size of 41×41 pixels and normalize its gray level to the $[0, 1]$ range for descriptor construction [5,8].

3.3   Descriptor Construction

In the descriptor construction stage every HarLap region is split into 4×4 cells and for each cell a gradient orientation feature histogram is constructed [7]. The gradient samples lying inside the cell region cast votes for their corresponding orientation bins. The votes are computed from the product of gradient magnitudes with a Gaussian window. The window uses high weights for samples near to the region center as compared to the region boundary. A soft binning approach is then used to distribute the votes among the adjacent bins to compensate the effect of region shift. Finally, the feature histograms are concatenated over all cells to form a descriptor vector. The vector is then normalized to unit norm and the elements are limited to 0.2 value [7]. At the end, the descriptor vector is renormalized to unit norm. We use Δ and ξ functions for the construction of Δ-SIFT and ξ-SIFT instead of gradient magnitudes whereas SIFT and π-SIFT descriptors are constructed from gradient magnitudes.

---

[1] http://vision.seas.harvard.edu/hyperspec

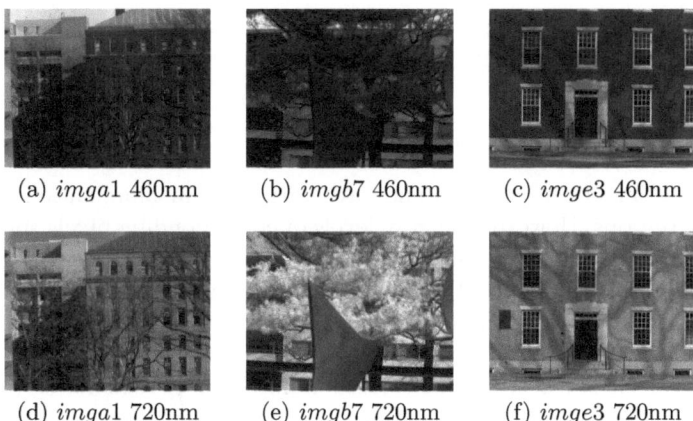

(a) *imga1* 460nm   (b) *imgb7* 460nm   (c) *imge3* 460nm

(d) *imga1* 720nm   (e) *imgb7* 720nm   (f) *imge3* 720nm

Fig. 3. Spectral images of three different test scenes for performance evaluation of SIFT descriptors

3.4   Evaluation Criteria

The performance evaluation criterion is based on the number of correct and false matches. We use the descriptor matching strategies of [8] for cross spectral image matching. A match is declared where the distance between the descriptor vectors of two interest regions is below a threshold. The ground truth for this match is established via the overlap error [9]. This error measures how well two regions A and B correspond under a known homography H and it is computed from the ratio of intersections to union of the regions i.e, $\epsilon_s = 1 - (A \cap H^T BH)/(A \cup H^T BH)$ [9]. A match is declared as a correct match if $\epsilon_s < 0.5$, otherwise it is considered as a false match. At the end, recall and 1-precision scores are computed using (3) for performance comparison. The *num_correspondences* term represents the number of matching regions ($\epsilon_s < 0.5$) between the images and the *num_all* is the sum of correct (*num_correct*) and false (*num_false*) matches. The perfect descriptor would give a recall value equal to 1 for any precision score [8]. We also compute the area under the recall versus 1-precision curve (AUC) as a single valued measure for performance comparison.

$$recall = \frac{num\_correct\ matches}{num\_correspondences}; \quad 1\text{-}precision = \frac{num\_false\ matches}{num\_all\ matches} \quad (3)$$

4   Experimental Results

This section presents the experimental results for image matching between the spectral images of the test scenes shown in Figure 3. The image matching uses HarLap regions with SIFT, π-SIFT, Δ-SIFT and ξ-SIFT based descriptor approaches. We use three different descriptor vector matching strategies for image matching [8] i.e, distance threshold, nearest neighbour and distance ratio.

Each matching strategy computes corresponding descriptor matches between the 460nm and 720nm wavelength band images.

4.1   Discussion on Test Images

The spectral images of *imga*1 scene have variations in illumination as well as spectral variations. However the gray levels of corresponding pixels suggest that illumination variations are dominant. We use normalized HarLap regions (see Section 3.2), therefore, each SIFT approach is illumination invariant and we expect similar performance measures for each of them in this experiment. In the *imgb*7 case, the spectral images appear different due to intensity reversal. This effects the correlation between the corresponding HarLap region descriptors. But there also exists HarLap regions which have only illumination variations, therefore, each SIFT approach also performs better in this experiment. The spectral images of *imge*3 are challenging because most of the corresponding HarLap regions are under spectral variations. This experiment is useful in comparing the evaluation measures for each SIFT type under such spectral variations.

4.2   Distance Threshold Based Matching

In distance threshold (t_d) based matching two interest regions are considered as a match if the distance between their descriptors are less than a threshold [8]. This matching strategy allows several matches for a single query descriptor and several of them may be correct due to $\epsilon_s < 0.5$. The evaluation curves for t_d based matching are shown in Figure 4. The performance measures of each SIFT type are almost similar for *imga*1 due to illumination differences between the spectral images. However, Δ-SIFT and ξ-SIFT perform slightly better than SIFT.

In *imgb*7 the intensity reversal makes the contents of 720nm band image spectrally different from its 460nm band image. This effects the SIFT performance because the corresponding gradient magnitudes are different, however, Δ-SIFT and ξ-SIFT perform relatively better than SIFT especially in the low 1-precision range. In *imge*3 every corresponding HarLap region is under spectral variations

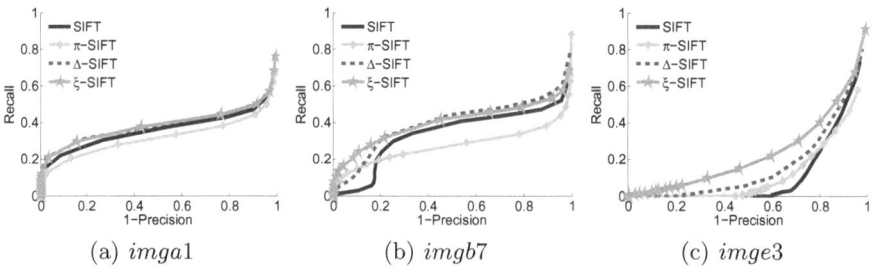

(a) *imga*1   (b) *imgb*7   (c) *imge*3

Fig. 4. Performance comparison of SIFT, π-SIFT, Δ-SIFT and ξ-SIFT using distance threshold (t_d) based matching between the spectral images of Figure 3

which in turn effects the correlation between corresponding descriptors. However, we can see that, the evaluation measures of ξ-SIFT is less effected by such variations as compared to its counterparts. The t_d based image matching suggest, that by boosting the contribution of local edges via Δ and ξ functions we can improve the SIFT robustness against illumination and spectral variations.

4.3 Nearest Neighbour Based Matching

In nearest neighbour (t_n) based descriptor matching, a nearest neighbor descriptor match is searched in the 720nm band image for each query descriptor of the 460nm band image and a match is declared if distance between the query descriptor and its nearest neighbor is found below a threshold. This matching strategy allows only one match for each query descriptor, which in turn results in better evaluation scores as compared to t_d based matching as shown in Figure 5. This is because the nearest neighbour matching ends up with the correct matches [8]. The t_n based matching suggest that, the performance of ξ-SIFT and Δ-SIFT is superior to SIFT and π-SIFT for image matching under spectral variations.

4.4 Distance Ratio Based Matching

In distance ratio (t_r) based descriptor matching, the distance ratio between the nearest and the second nearest neighbour is computed. If the ratio is below a threshold then a match is declared for the query descriptor. The evaluation curves are shown in Figure 6. The curves suggest, that the SIFT performance is superior in the $imgb7$ case but for the other scenes the ξ-SIFT and Δ-SIFT evaluation measures are superior to SIFT for t_r based image matching.

Table 1 summarizes the area under the t_d, t_n and t_r based recall versus 1-precision evaluation curves (AUC%). The AUC% measures suggest that, the performance of ξ-SIFT on average is superior to other SIFT approaches for t_d and t_n based image matching. However, in t_r based matching SIFT performs better than ξ-SIFT. It means that the nearest and the second nearest neighbour SIFT descriptors are less correlated as compared to ξ-SIFT. From AUC measures we

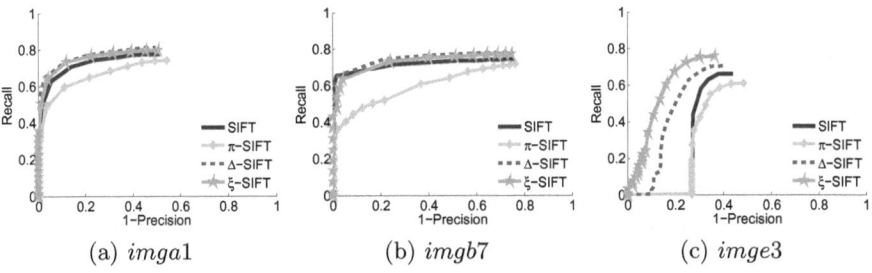

(a) $imga1$ (b) $imgb7$ (c) $imge3$

Fig. 5. Performance comparison of SIFT, π-SIFT, Δ-SIFT and ξ-SIFT using nearest neighbour (t_n) based matching between the spectral images of Figure 3

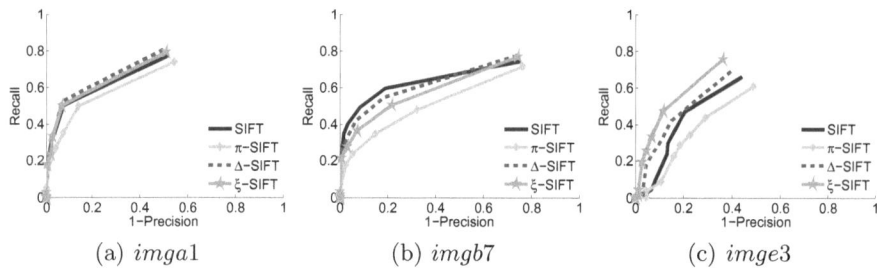

Fig. 6. Performance comparison of SIFT, π-SIFT, Δ-SIFT and ξ-SIFT using distance ratio (t_r) based matching between the spectral images of Figure 3

Table 1. Performance comparison of SIFT, π-SIFT, Δ-SIFT and ξ-SIFT using AUC% measures computed from recall versus 1-precision evaluation curves

(a) imga1					(b) imgb7			
	SIFT	π-SIFT	Δ-SIFT	ξ-SIFT		SIFT	π-SIFT	Δ-SIFT	ξ-SIFT
t_d	35.3	31.7	38.5	<u>38.7</u>	t_d	34.1	28.1	<u>40.1</u>	40.0
t_n	37.4	35.1	<u>37.8</u>	<u>37.8</u>	t_n	54.0	44.8	<u>55.5</u>	54.4
t_r	31.2	29.8	<u>31.3</u>	31.1	t_r	<u>47.2</u>	37.4	45.1	42.2

(c) imge3					(d) mean			
	SIFT	π-SIFT	Δ-SIFT	ξ-SIFT		SIFT	π-SIFT	Δ-SIFT	ξ-SIFT
t_d	10.2	10.0	15.6	<u>23.7</u>	t_d	26.5	23.3	31.4	<u>34.1</u>
t_n	15.4	14.4	16.4	<u>19.2</u>	t_n	35.6	31.4	36.6	<u>37.1</u>
t_r	17.3	16.1	17.8	<u>18.5</u>	t_r	<u>31.9</u>	27.8	31.4	30.6

conclude that ξ and Δ functions improve the SIFT robustness against spectral variations and the idea of boosting local edges in the descriptor construction process produces superior results for image matching.

5   Conclusion

This paper proposes modifications to the SIFT descriptor to improve its performance for image matching under spectral variations. The modifications are based on using the Local Contrast (Δ) and Differential Excitation (ξ) functions instead of gradient magnitudes (Ω) for descriptor construction. Each function produces high magnitude responses to edge pixels and boosts their contributions in the SIFT descriptor construction process. This results in better image matching performance as compared to Ω based SIFT. We validate the proposed Δ-SIFT and ξ-SIFT on the spectral images of three different test scenes.

We use three different descriptor vector matching strategies for image matching i.e, distance threshold (t_t), nearest neighbour (t_n) and distance ratio (t_r). Experimental results show that Δ-SIFT and ξ-SIFT perform better than SIFT for t_d and t_n based image matching whereas the SIFT performance is found superior in t_r based image matching.

Acknowledgments. The first author of this paper was supported by the Vienna PhD School of Informatics, Austria.

References

1. Brown, M., Su, S.: Multi-spectral SIFT for scene category recognition. In: IEEE Conference on Computer Vision and Pattern Recognition, pp. 177–184 (2011)
2. Chakrabarti, A., Zickler, T.: Statistics of Real-World Hyperspectral Images. In: IEEE Conference on Computer Vision and Pattern Recognition, pp. 193–200 (2011)
3. Chen, J., Shan, S., He, C., Zhao, G., Pietikainen, M., Chen, X., Gao, W.: WLD: a robust local image descriptor. IEEE Transactions on Pattern Analysis and Machine Intelligence 32(9), 1705–1720 (2010)
4. Hasan, M., Jia, X., Robles-Kelly, A., Zhou, J., Pickering, M.R.: Multi-spectral remote sensing image registration via spatial relationship analysis on SIFT keypoints. In: IEEE International Geoscience and Remote Sensing Symposium, pp. 1011–1014 (2010)
5. Ke, Y., Sukthankar, R.: PCA-SIFT: A more distinctive representation for local image descriptors. In: IEEE Conference on Computer Vision and Pattern Recognition, vol. 2, pp. 511–517 (2004)
6. Leykin, A., Hammoud, R.: Pedestrian tracking by fusion of thermal-visible surveillance videos. Machine Vision and Applications 21(4), 587–595 (2010)
7. Lowe, D.G.: Distinctive image features from scale-invariant keypoints. International Journal of Computer Vision 60(2), 91–110 (2004)
8. Mikolajczyk, K., Schmid, C.: A performance evaluation of local descriptors. IEEE Transactions on Pattern Analysis and Machine Intelligence 27(10), 1615–1630 (2005)
9. Mikolajczyk, K., Tuytelaars, T., Schmid, C., Zisserman, A., Matas, J., Schaffalitzky, F., Kadir, T., Gool, L.: A comparison of affine region detectors. International Journal of Computer Vision 65(1), 43–72 (2005)
10. Saleem, S., Bais, A., Sablatnig, R.: A performance evaluation of SIFT and SURF for multispectral image matching. In: International Conference on Image Analysis and Recognition, pp. 166–173 (2012)
11. Van Herk, M.: A fast algorithm for local minimum and maximum filters on rectangular and octagonal kernels. Pattern Recognition Letters 13(7), 517–521 (1992)
12. Vural, M., Yardimci, Y., Temizel, A.: Registration of multispectral satellite images with orientation-restricted SIFT. IEEE International Geoscience and Remote Sensing Symposium 3, 243–246 (2009)
13. Yi, Z., Zhiguo, C., Yang, X.: Multi-spectral remote image registration based on SIFT. Electronics Letters 44(2), 107–108 (2008)

A New Fuzzy Skeletonization Algorithm and Its Applications to Medical Imaging

Dakai Jin[1] and Punam K. Saha[1, 2]

[1] Department of Electrical and Computer Engineering, University of Iowa, USA
[2] Department of Radiology, University of Iowa, USA
{dakai-jin,punam-saha}@uiowa.edu

Abstract. Skeletonization provides a simple yet compact representation of an object and is widely used in medical imaging applications including volumetric, structural, and topological analyses, object representation, stenoses detection, path-finding etc. Literature of three-dimensional skeletonization is quite matured for binary digital objects. However, the challenges of skeletonization for fuzzy objects are mostly unanswered. Here, a framework and an algorithm for fuzzy surface skeletonization are developed using a notion of fuzzy grassfire propagation which will minimize binarization related data loss. Several concepts including fuzzy axial voxels, local and global significance factors are introduced. A skeletal noise pruning algorithm using global significance factors as significance measures of individual branches is developed. Results of application of the algorithm on several medical objects have been illustrated. A quantitative comparison with an ideal skeleton has demonstrated that the algorithm can achieve sub-voxel accuracies at various levels of noise and downsampling. The role of fuzzy skeletonization in thickness computation at relatively low resolution has been demonstrated.

1 Introducation

Availability of a wide spectrum of medical imaging techniques together with routine production of large image datasets for both clinical and research purposes have intensified the image processing needs for computerized extraction of knowledge from acquired images. A common objective of medical imaging is to extract information of internal human organ or tissue through *in vivo* or *ex vivo* imaging. Often, these images are processed through complex cascades of processing and analysis steps. Skeletonization is a transformation process that reduces a volumetric object into a significantly reduced, simplified and compact representation, referred to as "medial axis" or "skeleton", while preserving the topology [1] and geometric properties of the object.

Skeletonization has been widely used in different medical imaging applications, including, thickness computation [2], topological classification [3,4], path finding [5], and object shape modeling [6]. Many 3D skeletonization algorithms [7-11] have been reported for binary digital objects. But the same is not true for fuzzy skeletonization. Although, a few works on gray scale skeletonization have been presented in literature [12-15], the fundamental challenges related to fuzzy skeletonization in the presence of

A. Petrosino (Ed.): ICIAP 2013, Part I, LNCS 8156, pp. 662–671, 2013.
© Springer-Verlag Berlin Heidelberg 2013

partial voluming and degradation of object are mostly unanswered and a complete skeletonization algorithm for fuzzy digital objects is missing. For example, the fuzzy skeletonization algorithm by Pal [15], first, binarized a fuzzy object under optimization of fuzzy compactness [14] and then a binary skeletonization algorithm is applied. Yim *et al.* [12] used a significant path selection approach in an ordered region growing (ORG) graph; although, this approach does not use binarization, it is not obvious how to generalize this idea to extract the skeleton of a surface-like object. Sanniti di Baja *et al.* [13] used 3D convertion of a 2D grey image by uplifting each 2D voxel by a height equivalent to its grey-value and then followed a 3D binary skeletonization and collapsing from 3D to 2D. Here, a framework and an algorithm for fuzzy surface skeletonization are developed using a new notion of fuzzy grassfire propagation. The process of fuzzy grassfire propagation is simulated using fuzzy distance transform (FDT) [16]. Arcelli and Sanniti di Baja [17] first used DT in skeletonization in 2-D and discussed its advantages. Saito and Torowaki [9] and others [10] have used DT to define the voxel erosion sequence. Here, FDT is used to define the fuzzy grassfire propagation. Several new concepts including fuzzy axial voxels, local and global significance factors are introduced in this paper. Also, a skeletal noise pruning algorithm using significance measures at individual branch level is developed and its effectiveness is experimentally demonstrated.

2   Fuzzy Skeletonization Theory and Algorithm

Blum's pioneering work on grassfire transform [18] led to the notion of skeletonization that converts a volumetric object into a union of surfaces and curves. The process is defined using fire propagation on a grass field, where the field resembles an object. The fire is simultaneously set at all boundary points and it propagates inwardly at a uniform speed. The skeleton is defined as the set of quench points where two or more fire fronts meet. However, the notion of skeletonization for fuzzy objects has not yet been defined. To define a fuzzy skeletonization process, we suggest modifying the Blum's grassfire transform for a fuzzy object where the membership function is interpreted as local material density so that the speed of grassfire at a given point is inversely proportional to its material density. Following this notion, it can be shown that fuzzy distance transform (FDT) [16] value at a point p is proportional to the time when the fire front reaches p. Therefore, during the fuzzy grassfire propagation, the speed of a fire front at a point equates to the inverse of local material density and this equality is violated only at quench points where the propagation process is interrupted. Thus, a voxel $p \in Z^3$, where Z is the set of integers and Z^3 represents a rectangular image grid, is a *fuzzy quench voxel* in a fuzzy digital object $\mathcal{O} = \{(p, f_O(p)) \mid p \in Z^3 \wedge f_O: Z^3 \to [0,1]\}$ if the following inequality holds for every neighbor q of p

$$FDT(q) - FDT(p) < \tfrac{1}{2}(f_O(p) + f_O(q))|p - q|. \tag{1}$$

Saha and Wehrli [19] introduced the above definition of fuzzy quench voxel which was further studied by Svensson [20] where she referred to it as the center of fuzzy maximal ball (CFMB). Also, it may be noted that the definition of fuzzy quench voxel is equivalent to that of center of maximal ball (CMB) [21] for binary digital objects.

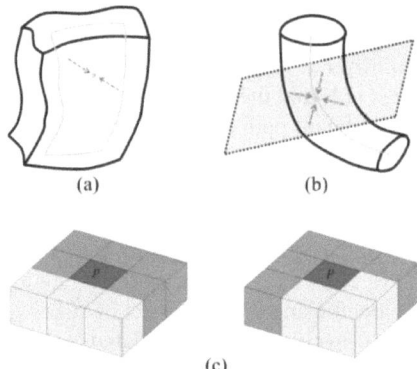

(a)   (b)

(c)

Fig. 1. (a,b) Illustration of independent fire fronts meeting at surface-like (a) and curve-like (b) quench points. (c) Two example masks used to compute the significance of a surface quench voxel. Voxels colored in green are used for average significance computation. Four coplanar geometrically identical masks may be constructed from each of these two examples.

During the process of fuzzy skeletonization, voxels are removed in the increasing order of their FDT values. The overall fuzzy skeletonization process is summarized in the following. Here, $O = \{p| f_O(p) > 0\}$ denotes the support of the fuzzy object O.

Primary skeletonization
```
select voxels p ∈ O in the order of FDT values
   if p is not a fuzzy axial voxel and its deletion pre-
   serves 3D topology and 2D topology on mid-planes
      remove p from O, i.e., set f_O(p) = 0
```

Final skeletonization:
```
select voxels p ∈ O in the order of FDT values
   if p is in two-voxel thick structure and its deletion
   preserves 3D topology and 2D topology on mid-planes
      remove p from O, i.e., set f_O(p) = 0
select voxels p ∈ O in the order of FDT values
   if topologic and geometric features of p fail to agree
   and its deletion preserves 3D topology
      remove p from O, i.e., set f_O(p) = 0
```

This notion of primary and final skeletonization was simultaneously introduced by Saha *et al.* [8] and Arcelli *et al.* [21]. Saha *et al.* described it in 3-D while Arcelli *et al.* presented the idea in 2-D for a DT-based skeletonization algorithm.

Removal of a voxel $p \in O$ preserves the topology of O if and only if p is a (26,6) simple voxel [1] in O. Beside the 3D topology preservation condition, an additional constrain of 2D topology preservation in all three middle planes of the candidate voxel is subjected to ensure continuity of surface-like structures and to avoid undesired drilling effects as illustrated by Saha *et al.* [8].

In the following, several new concepts including fuzzy axial voxels, local and global significance factors and two-voxel thick structures are introduced for fuzzy objects. Also, a condition defining the disagreement between topological and geometric features of a voxel is provided. Finally, a new noisy skeletal branches pruning algorithm based on global significance factor (GSF) is presented.

2.1   Fuzzy Axial Voxel

Two types of quench points may form –surface- and curve-quench points (Fig. 1(a,b)). A surface quench point is formed when two opposite fire fronts meet while a curve quench point is formed when fire fronts meet from all directions on a plane. In a digital space, surface-quench voxel is formed when two opposite fire fronts meet along x-, y- or z-direction and a curve-quench voxel is formed when fire fronts meet from all eight directions in xy-, yz-, or zx-planes. A voxel $p = (p_x, p_y, p_z) \in O$ is an *x-surface-quench voxel* if the following two conditions are satisfied for $p_{x-} = (p_x - 1, p_y, p_z)$, $p_{x+} = (p_x + 1, p_y, p_z)$, and $p_{x++} = (p_x + 2, p_y, p_z)$:

1) $FDT(p) > FDT(p_{x-})$,
2) $FDT(p) > FDT(p_{x+}) \vee (FDT(p) = FDT(p_{x+}) \wedge FDT(p) > FDT(p_{x++}))$.

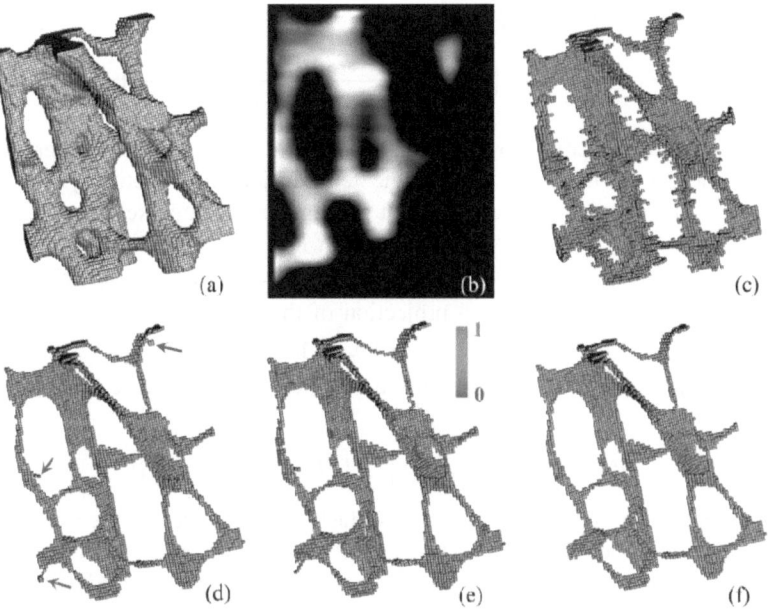

(a)   (b)   (c)

(d)   (e)   (f)

Fig. 2. Results of intermediate steps of fuzzy skeletonization. (a) 3D display of trabecular bone region in a micro-CT image of a cadaveric distal tibia specimen. (b) A sagittal image slice displaying the fuzziness in the image. (c) All quench voxels before filtering with local significance measure. (d) Results of final skeleton; noisy branches are indicated by red arrows. (e) Results of local significance computation. (g) Final results after noisy branch pruning.

To define a curve-quench voxel, let us first consider the formulation of the situation when fire fronts meet in the xy-plane. Curve quench voxels may form a 2x2 clique on the xy-plane.Let $P_{xy} = \{(p_x, p_y, p_z), (p_x + 1, p_y, p_z), (p_x, p_y + 1, p_z), (p_x + 1, p_y + 1, p_z)\}$ denote the 2x2 clique. Let $Q(P_{xy})$ denote the set of voxels within the 2x2 clique P_{xy} with their FDT value identical to that of p, i.e.,

$$Q(P_{xy}) = \{q \mid q \in P_{xy} \wedge FDT(q) = FDT(p)\}.$$

Thus, the fire front reaches simultaneously at every voxel of $Q(P_{xy})$ from all directions on the xy-plane. Therefore, a voxel $p = (p_x, p_y, p_z) \in O$ is an *xy-curve-quench voxel* if the following condition holds for $\forall\, q \in M_{xy}(p) - Q(P_{xy})$

$$q \text{ is 26-adjacent to } Q(P_{xy}) \quad \text{implies} \quad FDT(q) < FDT(p),$$

where, $M_{xy}(p)$ is the set of all voxels constructing the xy-plane through p.

Although the quench voxels captures the notion of fuzzy grassfire transform, it suffers from the fact that a large number of spurious quench voxels are created (Fig. 2(c)). Therefore, it is imperative to filter out some of these quench voxels based on their significance. Here, we introduce a function that resembles the "local significance factor" (LSF) of individual voxels and use LSF measures in the neighborhood to determine the significance of a quench voxel. *Local significance factor* or *LSF* of any voxel $p \in O$, denoted by $LSF(p)$, is defined as follows:

$$LSF(p) = 1 - f_+ \left(\max_{q \in N^*(p)} \frac{FDT(q) - FDT(p)}{\frac{1}{2}(f_o(p) + f_o(q))|p - q|} \right), \tag{2}$$

where the function $f_+(x)$ returns the value of x if $x > 0$ and zero otherwise, and $N^*(p)$ is the set of neighbors of p. The term inside the function f_+ essentially represents the inverse of speed of fire front propagation at the voxel p normalized by local material density. The above formulation of LSF is used to determine the significance of a quench voxel. Let $p = (p_x, p_y, p_z)$ be an x-surface quench voxel. To compute the support for p, first, a projection of three voxels $\{q_{i,j}^+ = (p_x - 1, p_y + i, p_z + j), q_{i,j} = (p_x, p_y + i, p_z + j), q_{i,j}^- = (p_x + 1, p_y + i, p_z + j)\}$, for some $i, j \in \{-1, 0, 1\}$, is computed to generate a 3×3 field of significance map $M_p^x(i, j)$ as follows:

$$M_p^x(i, j) = \max\{LSF(q_{i,j}^+), LSF(q_{i,j}), LSF(q_{i,j}^-)\}.$$

To determine the significance of an x-surface quench voxel p, the average significance value m_i^x over each of eight different masks $D_i \mid i = 1, \cdots, 8$ (see Fig. 1(c)) is computed. An x-surface-quench voxel p is referred to as *x-significant surface-quench* voxel, if any of the average values $m_i^x \mid i = 1, \cdots, 8$ is greater than a preset threshold. It may be noted that each of the mask D_i contains five '1' values (green) which are used for computation of average significance m_i^x and all green voxels in the mask fall on one side of p. A voxel $p \in A$ is referred to as a *significant surface-quench* voxel if it is an x-, y-, or z-significant surface-quench voxel.

Significance of an xy-curve-quench voxel $p = (p_x, p_y, p_z)$ is defined by the maximum LSF value on either of the two 3x3 planar cliques $C_z^+ = \{(p_x + i, p_y + j, p_z + 1) \mid i, j \in \{-1, 0, 1\}\}$ and $C_z^- = \{(p_x + i, p_y + j, p_z - 1) \mid i, j \in \{-1, 0, 1\}\}$.

Specifically, p is an xy-significant curve-quench voxel if the largest LSF value in either of the two cliques C_z^+ and C_z^- is greater than a preset threshold. An xy-, yz-, or zx-significant curve-quench voxel is referred to as a *significant curve-quench* voxel. A significant surface- or curve-quench voxel is referred to as a *fuzzy axial voxel*.

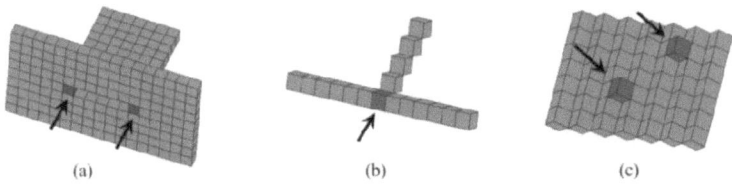

(a)   (b)   (c)

Fig. 3. Examples of voxels with contradiction between their topological and geometric properties. Red voxels are simple voxels which may appear at surface edge (a) or at curve end (b,c). However, these voxels fail to satisfy the geometric properties of a surface edge or a curve end.

2.2   Final Skeletonization

Final skeletonization converts two-voxel thick structures into one-voxel thick structures and it is completed in two steps. Intuitively, a voxel is two-voxel thick if its three non-opposite 6-neighbors are skeletal voxels and forms two-voxel thick surface across x-, y-, and/or z-direction [8]. During the first step, thick voxels are considered for erosion in the order of their FDT values. Specifically, a voxel satisfying two-voxel thickness along all three coordinate directions is deleted if it is a (26,6) simple voxel. A voxel satisfying two-voxel thickness along two directions, say x and y, is deleted if it is a (26,6) simple voxel and it preserves 2D topology in $M_{xy}(p)$. Finally, a voxel satisfying two-voxel thickness along only one directions, say x, is deleted if it is a (26,6) simple voxel and it preserves 2D topology in both $M_{xy}(p)$ and $M_{zx}(p)$.

The second step of final skeletonization brings a new idea of removing voxels with contradicting topological and geometric properties. Specifically, simple voxels must appears at a surface edge or at a curve end. Therefore, a simple voxel is deleted if it fails to satisfy geometric properties of a surface edge or a curve end. For example, for Fig. 3(a), the red voxels do not have two opposite 6-neighbors in background to satisfy the geometry of a digital surface. Similarly, for Fig. 3(b,c), the singleton simple voxels (red) are expected to appear at curve end; but, there is no plane where are isolated failing to satisfy the geometric property of a curve end.

2.3   Skeleton Pruning

The goal of a skeleton pruning algorithm is to discriminate between significant and non-significant branches so that only false branches may be removed. This goal is accomplished by computing LSF-weighted length of an individual branch from its edge to the corresponding junction voxel. This LSF-weighted branch length is used as a global significance factor (GSF) of a specific skeletal branch. This overall process is implemented using the following steps – (1) digital topological analysis (DTA), (2) conversion of two-voxel wide curve-like structures into a true digital curve, (3) computation of GSF for all branches, and (4) removal of non-significant branches.

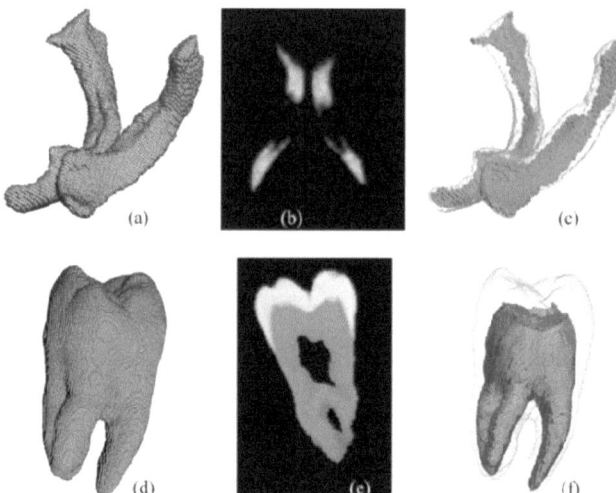

Fig. 4. Results of application of fuzzy skeletonization on two anatomic structures fuzzy segmented from acquired images. (a) A part of cerebrospinal fluid segmented from human brain MR imaging. (b) An axial image slice illustrating the fuzziness and noise. (c) Surface rendering of the fuzzy skeleton. (d-f) Same as (a-c) but for micro CT data of a human tooth.

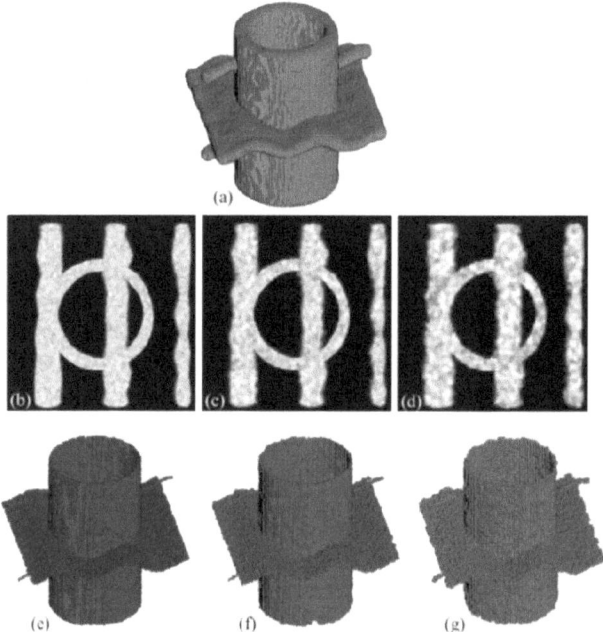

Fig. 5. Results of application of fuzzy skeletonization on a phantom data at different levels of noise and down sampling. (a) 2D display of the original phantom. (b-d) Axial image slices at SNR of 24, 12 and 6 and downsampling of three, four, and five voxels. (e-g) Results of fuzzy skeletonization of the phantom at three levels of noise and downsampling of (b-d).

3   Experiment Results

Results of intermediate steps of steps of skeletonization and pruning on a small region of trabecular bone image are illustrated in Fig. 2. Results of application of skeletonization and pruning of two other medical images are presented in Fig. 4. For all these examples, the results of skeletonization and pruning are visually encouraging.

To quantitatively examine the accuracy of the method, we generated a 3D binary object and its ground truth skeleton at high resolution. Then test phantom images were generated from the high resolution binary image at different downsampling and signal-to-noise ratio (SNR). Error was calculated by comparing computed and true skeletons. The high resolution image and true skeleton was generated by sampling an ideal skeleton in the continuous 3-D space R^3. Let S_T be the set of N_T number of sampled points; S_T is considered as the true skeleton. A Euclidean distance transform $DT_T: Z^3 \rightarrow R^+$ is computed from S_T. To generate a binary object with non-uniform thickness value, a smooth thickness field $f_{thickness}: Z^3 \rightarrow R^+$ is computed. Finally, the volumetric object is defined as the set of all voxels with its DT_T value less than or equal to the local thickness value $f_{thickness}$. Three high resolution binary objects with true skeletons were generated and used for this experiment. Their original image was generated in a 500×500×500 array. An example binary object is shown in Fig. 5. Let $S_C^{l,\rho}$ denote the computed skeleton at the downsampling rate of l and the noise at SNR of ρ and let $N_C^{l,\rho}$ denote the number of voxels in $S_C^{l,\rho}$. The skeletonization error is computed by the following equation:

$$Error_{l,\rho} = \frac{1}{2N_T} \sum_{p \in S_T} \min_{q \in S_C^{l,\rho}} |p - q| + \frac{1}{2N_C^{l,\rho}} \sum_{p \in S_C^{l,\rho}} \min_{q \in S_T} |p - q|. \qquad (3)$$

The average error for phantoms at each level of noise and downsampling is presented in Table 1. As shown in the table, the average error is less than a voxel and as shown by Saha et al. [2], the error of digitization is close to 0.38 voxel. Therefore, after deducting the digitization error, the performance of the fuzzy skeletonization algorithm even at the highest level of noise and downsampling is encouraging.

Table 1. Skeletonization errors at different levels of noise and downsampling

Different downsampling	Different signal to noise ratio			
	noise free	SNR 24	SNR 12	SNR 6
3x3x3	0.486	0.516	0.540	0.576
4x4x4	0.520	0.534	0.545	0.578
5x5x5	0.573	0.577	0.587	0.595

Finally, the application of fuzzy skeletonization in thickness computation is demonstrated in Fig. 6. Here, the basic principle of FDT-based thickness computation by Saha and Wehrli [2] is adopted except the fact that fuzzy skeletonization is used instead of binary skeletonization. Here, FDT-based depth values are sampled along the skeleton of the target object; thus providing the regional thickness distribution over the object. Let $S_{skeleton}$ denote the set voxels in the fuzzy skeleton of a fuzzy object

$\mathcal{O}$ and let *FDT* give the fuzzy distance transform map. At each skeletal voxel $p \in S_{\text{skeleton}}$, its thickness value $\tau(p)$ is computed as twice the FDT value $FDT(p)$. Finally, thickness values at non-skeletal voxels are inherited from the nearest skeletal voxel using the feature propagation algorithm introduced in [4]. Results of thickness distribution shown in Fig. 6 are visually satisfactory.

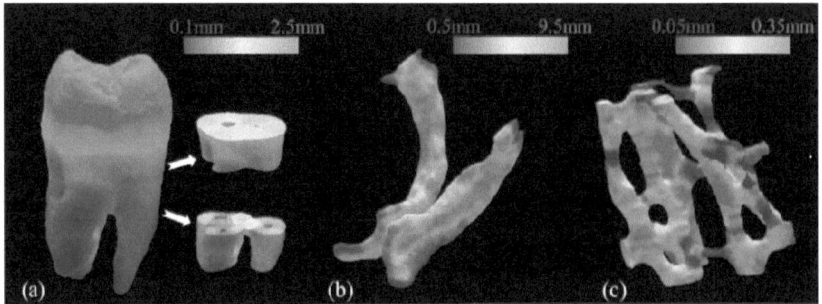

Fig. 6. Results of local thickness computation for three anatomic objects using fuzzy skeletonization, FDT, and feature propagation as described in the text

4   Conclusion

In this paper, a theoretical framework and a robust algorithm for fuzzy surface skeletonization are developed using a new notion of fuzzy grassfire propagation. The process of fuzzy grassfire propagation has been defined and the equation for fuzzy quench voxels has been formulated. Several new concepts including fuzzy axial voxels, local and global significance factors related to fuzzy skeletonization have been introduced and their theoretical formulations have been presented. A skeletal noise pruning algorithm using significance measures of individual branches has been developed and its performance has been demonstrated. Results of application of the algorithm on several medical objects have been illustrated. A quantitative comparison with an ideal skeleton has demonstrated that the algorithm can achieve sub-voxel accuracies at various levels of noise and downsampling rate. The application of fuzzy skeletonization in thickness computation at relatively low resolution has been demonstrated and the results are visually satisfactory.

Acknowledgements. This work was supported by the NIH grant R01-AR054439.

References

1. Saha, P.K., Chaudhuri, B.B.: Detection of 3-D simple points for topology preserving transformation with application to thinning. IEEE Trans.on PAMI 16, 1028–1032 (1994)
2. Saha, P.K., Wehrli, F.W.: Measurement of trabecular bone thickness in the limited resolution regime of in vivo MRI by fuzzy distance transform. IEEE Trans. on MI 23, 53–62 (2004)

3. Saha, P.K., Gomberg, B.R., Wehrli, F.W.: Three-dimensional digital topological characterization of cancellous bone architecture. IJIST 11, 81–90 (2000)
4. Saha, P.K., Xu, Y., Duan, H., Heiner, A., Liang, G.: Volumetric topological analysis: a novel approach for trabecular bone classification on the continuum between plates and rods. IEEE Trans. on MI 29, 1821–1838 (2010)
5. Wan, M., Liang, Z., Ke, Q., Hong, L., Bitter, I., Kaufman, A.: Automatic centerline extraction for virtual colonoscopy. IEEE Trans. on MI 21, 1450–1460 (2002)
6. Pizer, S.M., Fritsch, D.S., Yushkevich, P.A., Johnson, V.E., Chaney, E.L.: Segmentation, registration, and measurement of shape variation via image object shape. IEEE Trans. Med. Imaging 18, 851–865 (1999)
7. Ma, C.M., Sonka, M.: A fully parallel 3D thinning algorithm and Its applications. Computer Vision Image Understanding 64, 420–433 (1996)
8. Saha, P.K., Chaudhuri, B.B., Dutta Majumder, D.: A new shape preserving parallel thinning algorithm for 3D digital images. Pattern Recognition 30, 1939–1955 (1997)
9. Saito, T., Toriwaki, J.-I.: A sequential thinning algorithm for three dimensional digital pictures using the Euclidean distance transformation. In: Proc. 9th SCIA, Uppsala, Sweden, pp. 507–516 (1995)
10. Arcelli, C., Sanniti di Baja, G., Serino, L.: Distance-driven skeletonization in voxel images. IEEE Trans. on PAMI 33, 709–720 (2011)
11. Tsao, Y.F., Fu, K.S.: A parallel thinning algorithm for 3D pictures. Computer Graphics and Image Processing 17, 315–331 (1981)
12. Yim, P.J., Choyke, P.L., Summers, R.M.: Gray-scale skeletonization of small vessels in magnetic resonance angiography. IEEE Trans. on MI 19, 568–576 (2000)
13. di Baja, G.S., Nyström, I., Borgefors, G.: Discrete 3D tools applied to 2D grey-level images. In: Roli, F., Vitulano, S. (eds.) ICIAP 2005. LNCS, vol. 3617, pp. 229–236. Springer, Heidelberg (2005)
14. Pal, S.K., Rosenfeld, A.: Image enhancement and thresholding by optimization of fuzzy compactness. Pat. Rec. Let. 7, 77–86 (1988)
15. Pal, S.K.: Fuzzy skeletonization of an image. Pat. Rec. Let. 10, 17–23 (1989)
16. Saha, P.K., Wehrli, F.W., Gomberg, B.R.: Fuzzy distance transform: theory, algorithms, and applications. Computer Vision and Image Understanding 86, 171–190 (2002)
17. Arcelli, C., di Baja, G.S.: A width-independent fast thinning algorithm. IEEE Trans. on PAMI 7, 463–474 (1985)
18. Blum, H.: A transformation for extracting new descriptors of shape (Models for the Perception of Speech and Visual Form). MIT Press, Cambridge (1967)
19. Saha, P.K., Wehrli, F.W.: Fuzzy distance transform in general digital grids and its applications. In: Proc. 7th JCIS, Research Triangular Park, NC (2003)
20. Svensson, S.: Aspects on the reverse fuzzy distance transform. Pat. Rec. Let. 29, 888–896 (2008)
21. Arcelli, C., di Baja, G.S.: Well-shaped, stable, and reversible skeletons from the (3,4)-distance transform. Journal of Visual Communication and Image Representa

A Subunit-Based Dynamic Time Warping Approach for Hand Movement Recognition

Yanrung Wang[1], Atsushi Shimada[1],
Takayoshi Yamashita[2], and Rin-ichiro Taniguchi[1]

[1] Graduate School of Information Science and Electrical Engineering
Kyushu University, Japan
{kenyou,atsushi,rin}@limu.ait.kyushu-u.ac.jp
[2] OMRON Corporation, Japan
takayosi@omm.ncl.omron.co.jp

Abstract. A subunit-based Dynamic Time Warping (DTW) approach is proposed for hand movement recognition. Two major contributions distinguish the proposed approach from conventional DTW. (1) A set of hand movement subunits is constructed using a data-driven method. The common sub-movements (subunits) are shared across hand gestures to obtain a smaller training data size and search space to improve recognition performance. (2) A similarity measure robust to variability is offered using subunit-to-subunit matching to absorb the difference between two similar sub-sequences belonging to the same subunit, and only keeping the distances between sub-sequences that relate to different subunits. Our experimental results demonstrate the efficiency and accuracy of the proposed approach.

Keywords: hand movement, gesture recognition, subunit.

1 Introduction

Vision-based hand gesture recognition has attracted considerable attention because of its new and fascinating applications such as interactive human-machine interfaces, sign language interpretation, and virtual environments [1]. Features such as hand location, appearance, motion, shape, and orientation often play an important role in hand gesture recognition. In this paper, we consider hand gestures as hand movement trajectories and focus on recognition of the movement trajectories.

Dynamic time warping (DTW) [2] is widely used to recognize movement trajectories, because it simultaneously aligns time-variable data and computes a likelihood of similarity. However, there are two major limitations to the use of DTW in hand movement recognition. (1) DTW matching uses information about individual training examples that it is sensitive to variations in training data. Hence, it is difficult to support efficient personalized gesture recognition. (2) DTW is sensitive to noise and unable to distinguish movement trajectories that have similar sub-sequences, as it requires continuity along the warping path.

A. Petrosino (Ed.): ICIAP 2013, Part I, LNCS 8156, pp. 672–681, 2013.

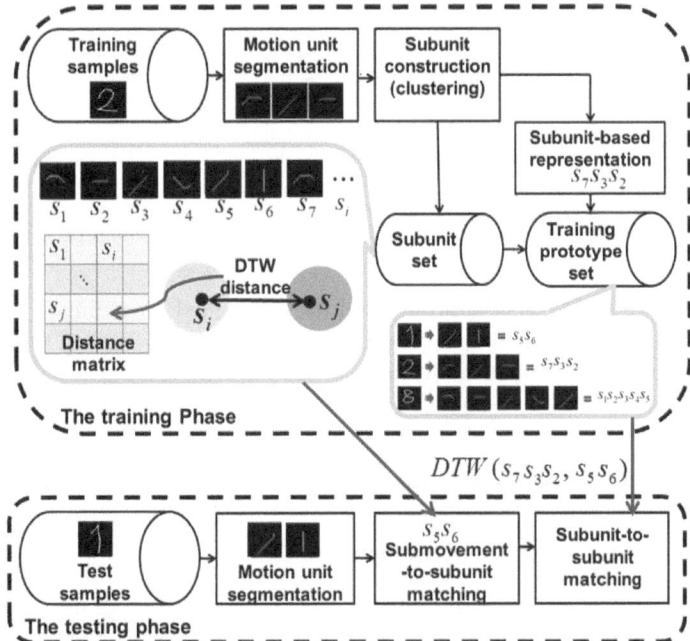

Fig. 1. Flowchart of the proposed approach

The conventional DTW consequently requires the development of many prototypes to achieve proper performance, leading to an expensive computational load.

To address these issues, we develop an effective recognition approach that combines the use of the DTW distance metric and subunits, widely investigated in the field of sign language [3][4]. Subunits are elementary units in a language and there are far fewer subunits than words in the vocabulary of the language, which is expected to lead to smaller data size in training and a smaller search space in recognition.

2   Overview of the Proposed Approach

Our system handles color image sequences in real time to recognize numbers from 0 to 9 by the hand movement trajectories. Figure 1 is a block diagram of our hand movement recognition system. In the training phase, all training data are mapped to sequences of digits between 0 and 7 according to their orientation feature and then segmented into the set of basic motion units according to changes in orientation. Next, subunits are selected via k-medoids clustering and set as the yielded cluster centers. In this case, each training sequence is mapped to a sequence of subunits. In the testing phase, the test sequence is also

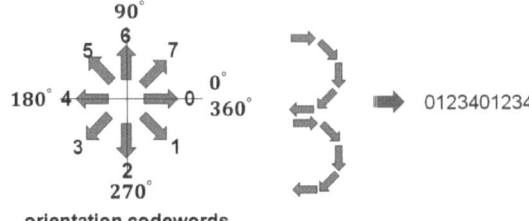

Fig. 2. Orientation codewords and an example of movement representation

represented as a sequence of subunits and then classified according to DP matching between the test sequence and training sequences. Specifically, DTW distance
is measured by subunit-to-subunit matching to improve recognition accuracy and
online learning is used to adapt the training set to the user's individual habits.

3   Hand Movement Representation and Learning

3.1   Hand Movement Representation

Hand movement trajectories are obtained by detecting the top most point of the
hand skin region as the fingertip. To represent and describe these trajectories,
we use the orientation feature, which has been shown to provide high accuracy
and robustness in hand movement recognition in previous work [5]. A hand
movement is a spatio-temporal trajectory that consists of fingertip positions
$(x_i, y_i), t = 1, 2, ..., T - 1$, where T indicates the length of movement trajectory.
Similar to [5], we calculate the orientation feature according to the positions of
fingertips between consecutive frames as follows.

$$\theta_t = arctan(\frac{y_{t+1} - y_t}{x_{t+1} - x_t}); t = 1, 2, ..., T - 1 \tag{1}$$

The orientation θ_t is quantized into a set of codewords from 0 to 7 by dividing
it by 45°. Therefore, a hand movement can be represented by a sequence of digits
between 0 and 7 according to the yielded codewords as shown in Figure 2.

3.2   Hand Movement Subunit Construction

Currently, subunit-based recognition is a main focus of sign language research.
There are two methods with which to perform recognition by means of subunits.
The first method is based on linguistic analysis to determine subunits [6][7]. The
second method segments signs into subunits employing a data-driven process
without any linguistic knowledge about sign languages, and all subunits are
self-organized from the data themselves [4].

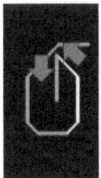

(a) current orientation differing   (b) significant changes between
from the starting orientation   consecutive frames

Fig. 3. Segment motion unit according to changes in orientation

Motivated by [4] and [3], we consider a hand movement as a sequence of basic motion units, referred to as the common pattern of hand movements, and carry out a self-organization process to select a representative set of motion units from the training set as subunits. All training data are segmented into a set of sub-movements that are considered in a common pool of features. Clustering is performed for these sub-movements (basic motion units) to find the common pattern of hand movements.

Motion Unit Segmentation. Motion unit segmentation can be thought of as a boundary detection problem. The use of trajectory discontinuity and motion speed discontinuity has been shown to be effective in detecting the subunit boundary in sign language recognition [8]. We employ changes in orientation as trajectory discontinuity metrics to detect unit boundaries when the current orientation is very different from that in a neighboring frame or that in the starting frame of the motion unit. As shown in Figure 3, all training data can be broken into sub-movements according to the characteristic.

Subunit Clustering. To select a set of representative subunits from all sub-movements of the training set, we perform k-medoids clustering using the DTW distance metric. The k-medoids algorithm, a variant of k-means clustering, computes medoids instead of centroids as cluster centers to minimize the sum of intra-class distances. As opposed to k-means clustering, k-medoids clustering is more robust to outliers and avoids the difficulty of the averaging step of k-means clustering with DTW distance metric [9].

To determine the number of clusters, we employ an iterative k-medoids algorithm that selects all sub-movements of the training set as the initial cluster centers and iteratively merges similar clusters until convergence to obtain the "optimal" number of clusters. Furthermore, we use a k-means-like algorithm for k-medoids clustering [10] to overcome the drawback that partitioning around medoids (PAM, K-medoids) works inefficiently for large data sets because of the complexity.

The yielded cluster centers are then used as subunits to map the training data to subunit sequences. We build a subunit-to-subunit distance matrix computed off-line during construction of subunits and use it as a look-up table to speed up the recognition procedure.

3.3   Subunit-Based Learning

Instead of training entire hand movements composed of orientation codewords, we train each movement as a concatenation of subunits. The advantages are as follows. (1) The amount of training materials needed is reduced as all training data are composed of a limited set of subunits. (2) A simplified enlargement of training data is achieved by composing new training data using the existing subunits.

4   Subunit-Based Recognition

We propose two-step submovement-to-subunit and subunit-to-subunit matching in the recognition process to improve the performance of recognition and employ subunit-based online learning to overcome sensitivity to variations in the training data.

4.1   Submovement-to-Subunit Matching

Let P_x be a testing sequence and $S = \{s_1, s_2, ..., s_{|S|}\}$ be a set of $|S|$ subunits constructed from training data. Similar to training sequences, the test sequence P_x is also mapped to a sequence of digits between 0 and 7 according to changes in orientation and then segmented into m submovements u_{xi} in the same way as described in Section 3.2. We calculate DTW distances between submovement u_{xi} of the index i and all subunits to find the nearest subunit s_{xi} and then use these subunits to recompose the testing sequence P_x. The yielded testing sequence $P_x = \{s_{x1}, s_{x2}, ..., s_{xi}, ..., s_{xm}\}$ is used to perform subunit-to-subunit matching with training data.

4.2   Subunit-to-Subunit Matching

Hand movement trajectories are recognized through dynamic subunit sequence matching. Let $P_y = \{s_{y1}, s_{y2}, ..., s_{yj}, ..., s_{yn}\}$ be a training sequence consisting of n subunits. The DTW distance $DTW(P_x, P_y) = D(s_{xm}, s_{yn})$ is calculated as follows.

$$D(s_{xi}, s_{yj}) = \min \begin{cases} D(s_{xi-1}, s_{yj}) + cost \\ D(s_{xi}, s_{yj-1}) + cost \\ D(s_{xi-1}, s_{yj-1}) + cost \end{cases} \tag{2}$$

$$cost = \begin{cases} 0 & if \ s_{xi} = s_{yj}, \\ dist(s_{xi}, s_{yj}) & if \ s_{xi} \neq s_{yj} \end{cases} \tag{3}$$

Here, $dist(s_{xi}, s_{yj})$ is obtained using the look-up table generated during the construction of subunits.

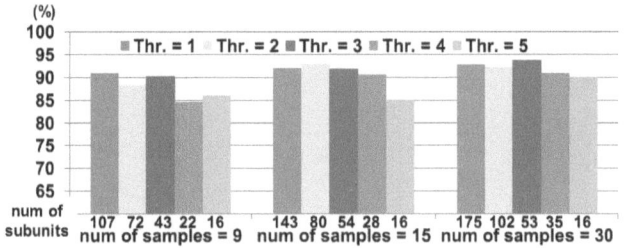

Fig. 4. Average recognition rate using different numbers of subunits

5 Experiments and Discussion

To test the proposed subunit-based DTW approach for hand movement recognition and to compare with conventional DTW, we perform evaluations in terms of the recognition rate and average computational time for two locally collected hand movement corpus. Here, the average computation time is the average time taken to calculate the distance.

5.1 Construction of Hand Movement Subunits

As illustrated in Section 3.2, we merge similar clusters iteratively to select a set of representative subunits. The similarity between clusters is measured as the DTW distance between medoids of clusters. Two clusters are merged if their distance is smaller than a threshold Th.

To evaluate the recognition performance with different thresholds Th, we performed experiments on subunit-based DTW with Th ranging from 1 to 5. The used corpus contains 10 different classes of hand movement trajectories from 0 to 9, performed by seven subjects in our laboratory environment. Each of the 10 classes of trajectories is repeated 25 times by each subject. We randomly select 9, 15, and 30 training samples from each class, performed by three subjects, to construct the training set. The other data corresponding to the other four subjects are used as a test set.

Figures 4 and 5 present the average recognition rate and computational time using different numbers of subunits, with Th ranging from 1 to 5. In general, the greater the number of subunits, the more the different sub-movements can be distinguished, contributing to more discriminative representation of the training data. This yields better recognition results. At the same time, the computational cost is high due to the lager search space as shown in Figure 5.

5.2 Recognition in Different Backgrounds

These experiments are aimed at demonstrating the robustness of our approach with respect to variability. We use a locally collected corpus consisting of 10 different classes of hand movement trajectories from 0 to 9. performed by 16

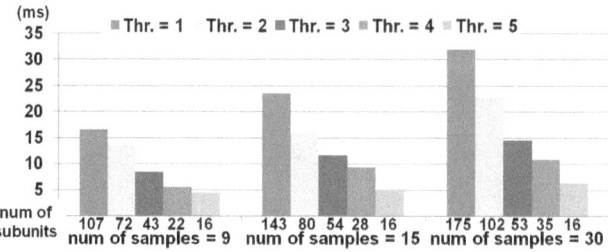

Fig. 5. Computational time using different numbers of subunits, constructed with Th ranging from 1 to 5

Table 1. Comparison of recognition rate in two different backgrounds

method \ background	blue	window
Conventional DTW	93.08%	71.70%
Subunit-based DTW	90%	82.5%

subjects in two different backgrounds: 1) the blue background easy to detect the position of fingertip and 2) the window background hard to to detect the position of fingertip. Each of the 10 classes of trajectories is performed one time in each background. We select 15 samples performed by 15 subjects in the blue background, to construct the training set. The trajectories performed by the other one subject in the blue and window background are used as test data.

Evaluation of the Recognition Rate. Recognition rates classified according to two different background are listed in Table 1. When using test data performed in the blue background, the subunit-based DTW achieves a recognition rate of 90%, which was 3.08% lower than that of conventional DTW. One possible reason could be that feature details is omitted due to the subunit-based movement representation. In contrast to recognition in the blue background, the subunit-based DTW approach shows a significant improvement of 10.8% when using test data performed in the window background. The two main reasons for the improvement are as follows.

A Similarity Measure Robust to Variability: The conventional DTW distance metric is sensitive to noise and unable to find movement trajectories that have similar sub-sequences. Therefore, similar trajectories may be treated as dissimilar, leading to inaccurate recognition. As illustrated in Figure 6, the proposed subunit-based DTW approach offers a more accurate similarity measure because it absorbs the difference between two similar sub-sequences belonging to the same subunit and only keeps the distances between sub-sequences that relate to different subunits.

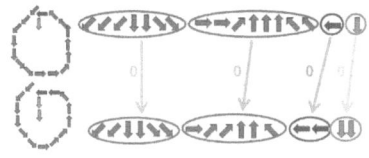

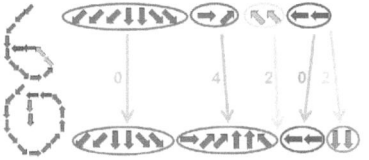

(a) distance between '0' and '0' is zero   (b) distance between '0' and '6' is 8

Fig. 6. Examples of the subunit-based DTW distance between movements

Increase in the Variety of Training Data: To train each movement as a concatenation of subunits increases the variety of training data such that it is possible to recognize new training patterns not seen in training.

For instance, we might have a training set of three training data $P_y = \{u_{y1}, u_{y2}, u_{y3}\}$, $P_z = \{u_{z1}, u_{z2}\}$, and $P_w = \{u_{w1}, u_{w2}\}$, where u_{yj}, u_{zk}, and u_{wl} are segmented submovements and are clustered into three subunits $s_1 = \{u_{y1}\}$, $s_2 = \{u_{y2}, u_{z1}, u_{w1}\}$, and $s_3 = \{u_{z2}, u_{w2}\}$. According to the yielded subunit set, training data are mapped to sequences of subunits $P_y = \{s_1, s_2, s_3\}$, $P_z = \{s_2, s_3\}$, and $P_w = \{s_2, s_3\}$.

In the example, we only train two training prototypes $s_2 s_3$ and $s_1 s_2 s_3$ for three training data because P_z and P_w are mapped to the same prototype $s_2 s_3$. The reduction of training prototypes improves learning efficiency while maintaining the variety of training data to avoid loss of recognition accuracy. In addition, training patterns that can be represented by the training prototype $s_2 s_3$ are not only P_z and P_w but also $u_{w1} u_{z2}$, $u_{w1} u_{y3}$, and so on. That is, the variety of P_z and P_w is increased to $|s_2||s_3|$ training patterns because of the use of existing subunits that include motion units from the other training data. It is thus also possible to recognize new patterns, even though they are not seen in the training. These merits overcome the shortcoming of the expensive computational load resulting from a large number of training data.

These findings support the claim, made above, that the subunit-based DTW distance is able to overcome the sensitivity to training data of conventional DTW. This is expected to solve difficulties in hand movement recognition such as the larger variation within a class due to a userfs individual habits and noises.

Evaluation of Average Computation Time. Figure 7 presents the average computation time and the number of training prototypes when using test data performed in different backgrounds . The results indicate that a significant improvement in computational complexity was obtained. The reduction of the number of training prototypes, due to the fact that multiple training data were mapped to single training prototype, was one of the causes of the improvement in computational complexity. The major reason for the improvement is that the distance between subunits was rapidly obtained using the lookup table in the procedure of subunit-to-subunit matching.

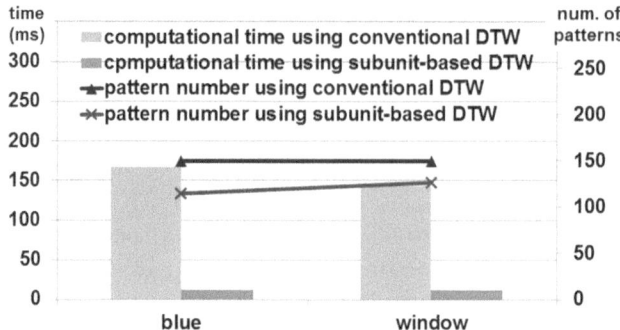

Fig. 7. Average computation time (ms) of recognition using test data performed in different backgrounds

Table 2. Comparison of average recognition rate on different size of training dataset

method \ # training samples	9 samples	15 samples	30 samples
conventional DTW	95.46%	94.43%	95.24%
subunit-based DTW	90.54%	90.98%	93.3%
subunit-based DTW using online learning	95.06%	93.82%	94.5%

5.3   Training Datasets of Different Size

To evaluate performance changes resulting from differing sizes of training data, we use the same corpus described in Section 5.1. To obtain results that are more reliable, the construction of subunits and evaluationof recognition performance were repeated five times using different datasets constructed relating to different subjects.

Table 3. Comparison of average computation time on different size of training dataset

method \ computation time(ms)	9 samples	15 samples	30 samples
conventional DTW	97.7	157.1	304.9
subunit-based DTW	11.6	14.425	16.2
subunit-based DTW using online learning	12.6	14.3	16.9

Recognition rates classified according to three different sizes of training set are compared in Table 2. The subunit-based DTW achieves recognition rates of 90.54%, 90.98%, and 93.3%, which are slightly lower than those of conventional DTW due to the loss of feature details. However, the proposed approach is easily applicable to incremental learning without a time consuming problem. Table 2 and Table 3 prove that a subunit-based online learning is able to address the problem and keep low computation cost.

6   Conclusion

This paper proposes a subunit-based approach to hand movement recognition. In contrast to conventional DTW approaches, we share subunits across hand movements to obtain a smaller training data size and search space to improve recognition performance. In addition, a more robust similarity measure, using subunit-to-subunit matching, is offered. The experimental results demonstrate that the proposed approach is both accurate and efficient for hand movement recognition. Although a similar approach has been also proposed by [11], our proposed method realized much faster matching by introducing two-step DTWs and the efficient lookup table. Our future research will focus on incremental learning for the subunit itself to support efficient personalized recognition.

References

1. Wachs, J.P., Kolsch, M., Stern, H., Edan, Y.: Vision-based hand-gesture applications. Commun. ACM 54(2), 60–71 (2011)
2. Okada, S., Hasegawa, O.: Motion recognition based on dynamic-time warping method with self-organizing incremental neural network. In: 19th International Conference on Pattern Recognition, pp. 1–4 (2008)
3. Roussos, A., Theodorakis, S., Pitsikalis, V., Maragos, P.: Hand tracking and affine shape-appearance handshape subunits in continuous sign language recognition. In: Int. Conf. ECCV Wkshp: SGA (2010)
4. Bauer, B., Kraiss, K.-F.: Towards an automatic sign language recognition system using subunits. In: Wachsmuth, I., Sowa, T. (eds.) GW 2001. LNCS (LNAI), vol. 2298, pp. 64–75. Springer, Heidelberg (2002)
5. Elmezain, M., Al-Hamadi, A., Michaelis, B.: Real-time capable system for hand gesture recognition using hidden markov models in stereo color image sequences. Journal of WSCG, 65–72 (2008)
6. Liddell, S.K., Johnson, R.E.: American sign language: the phonological base. Sign Language Studies 64, 197–277 (1989)
7. Stokoe, W.: Sign language structure: an outline of the visual communication systems of the American deaf. Studies in Linguistics: Occasional Papers 8 (1960)
8. Han, J., Awad, G., Sutherland, A.: Modelling and segmenting subunits for sign language recognition based on hand motion analysis. Pattern Recognition Letters, 623–633 (2009)
9. Niennattrakul, V., Ratanamahatana, C.A.: On clustering multimedia time series data using k-means and dynamic time warping. In: International Conference on Multimedia and Ubiquitous Engineering, pp. 733–738 (2007)
10. Park, H.S., Jun, C.H.: A simple and fast algorithm for k-medoids clustering. Expert Systems with Applications: An International Journal, 3336–3341 (2009)
11. Oszust, M., Wysocki, M.: Modelling and Recognition of Signed Expressions Using Subunits Obtained by Data-Driven Approach. In: Ramsay, A., Agre, G. (eds.) AIMSA 2012. LNCS (LNAI), vol. 7557, pp. 315–324. Springer, Heidelberg (2012)

Softmax Regression for ECOC Reconstruction

Roberto D'Ambrosio, Giulio Iannello, and Paolo Soda

Integrated Research Centre, Universitá Campus Bio-Medico di Roma, Rome, Italy
{r.dambrosio,g.iannello,p.soda}@unicampus.it

Abstract. Classification by binary decomposition is a well-known method to solve multiclass classification tasks since a large number of algorithms were designed for binary classification. Once the polychotomy has been decomposed into several dichotomies, the decisions of binary learners on a test sample are aggregated by a reconstruction rule to set the final multiclass label. In this context, this paper presents a reconstruction rule based on softmax regression which considers the reconstruction task as a new classification problem. To this aim, as second-order features we use both the crisp labels and the reliabilities of binary decisions. Six heterogeneous datasets and three different classification architectures have been used to test our method, whose performance favorably compare with those provided by other three reconstruction rules both in terms of global accuracy and geometric mean of accuracies.

1 Introduction

Classification tasks in general consist in assigning input patterns to a finite set of classes and they should be divided into *dichotomies* and *polychotomies*. The former are classification problems with two classes, and are also referred to as binary learning, that is, positive/negative classification. The latter are also known as multiclass learning and they are recognition tasks with more than two classes.

Although several applications deal with binary classification, there are many others belonging to fields with more than two classes. In this latter case, different classification approaches can be applied. Indeed, we may appeal to typical multiclass techniques, e.g. neural networks and decision tree. Nevertheless, it is well-known that discriminating between two classes is much easier than simultaneously distinguishing among many [17]. This observation has motivated research on decomposition methods, which reduce classification complexity through the decomposition of the polychotomy in less complex binary subtasks. A binary classifier, also referred to as dichotomizer in the following, is then trained on a single binary decomposition of the original polychotomy. To provide the final classification, dichotomizers' crisp or soft labels are combined according to a *reconstruction rule*.

Existing reconstruction rules should appeal to the majority voting [14], minimize the distance from class codewords [6], look at the largest output [6], or minimize the loss on the induced binary problems [1]. Differently, we consider here the reconstruction task as a new classification problem where the second-order features are the crisp labels and the reliabilities of dichotomizers' classifications. Using the latter quantity in the reconstruction rule introduces useful information on the many issues influencing

A. Petrosino (Ed.): ICIAP 2013, Part I, LNCS 8156, pp. 682–691, 2013.

the achievement of a correct classification, providing a deep insight in the classification process. Then, crisps labels and reliabilities are given to the softmax regression to estimate the posterior distribution of classification acts.

The performance of our proposal have been estimated trough a series of experiments on six public and heterogeneous datasets, where the polychotomies were decomposed by the Error-Correcting Output Codes. The proposed reconstruction rule favorably compares with other three reconstruction rules, and we also observe that its performance are promising also to recognize the most under-represented classes.

The rest of the paper is organized as follows: next section briefly revises existing decomposition methods and reconstruction rules, whereas section 3 presents our proposed reconstruction rule. Section 4 introduces the experiments, and section 5 reports and discusses the results. Finally, section 6 provides concluding remarks.

2   Background

Decomposition schemes reduces polycothomies on K classes (with $K > 2$) in binary subtasks that are easier to solve than the original problem. This reduction is usually performed according one of the following three methods [1,6,8,11,22]: One-per-Class (OpC), Pairwise Coupling (PC), and distributed output code.

OpC decomposition method, which is also known as One-against-All, addresses the polycotomy with K binary learning functions, each one separating a single class from all the others.

The second approach, PC decomposition, reduces the polychotomy into $K * (K - 1)/2$ dichotomies, each one addressed by a dichotomizers specialized in discriminating between pair of classes. This decomposition method is also cited as n^2 classifier, One-against-One [9] or even Round Robin [8] classification.

Distributed output code decomposition assigns a unique codeword, i.e. a binary string, to each class. If we assume that the string has L bits, this approach sets a recognition system composed by L binary classification functions. The most known method within this framework is known as *Error-Correcting Output Codes* (ECOC), where the use of error correcting codes as distributed output representation yielded a recognition system less sensitive to noise [6].

It is worth noting that further to these popular approaches, there exist other proposals that do not perfectly fit this categorization, e.g. the hierarchical dichotomies generation [13], but this does not introduce any limitations in the rest of the paper.

When feed with the test sample $x \in \Re^n$, these decompositions produce the vector $\mathbf{M}(x) = \{M_1(x), M_2(x), \ldots, M_L(x)\}$, whose length L depends upon the decomposition approach used. In this notation, $M_j(x)$ is a function classifying x in two separate superclasses, represented by the label set $\Omega_j = \{-1; 1\}$. Decomposition schemes can be unified in a common framework representing the outputs of the dichotomizers by a binary code matrix, named as decomposition matrix $D \in \Re^L$ x $\Re^K$. Its elements d_{ji} are defined as:

$$d_{ji} = \begin{cases} 1 & \text{if class } i \text{ is in the subgroup associated to label 1 of } M_j \\ -1 & \text{if class } i \text{ is in the subgroup associated to label -1 of } M_j \\ 0 & \text{if class } i \text{ is in neither groups associated to label -1 or 1 of } M_j \end{cases} \tag{1}$$

Hence, the dichotomizer M_j is trained to associate patterns belonging to class ω_i with values d_{ji}.

Once a decomposition method has been applied, binary outputs $M_j(x)$ must be combined to predict the final label: therefore, we need a transformation, i.e. the reconstruction rule, associating $\mathbf{M}(x)$ to a label in the finite set $\Omega = \{\omega_1, \omega_2, \ldots, \omega_K\}$.

In the case of OpC, the simplest reconstruction rule looks at the function that returns the highest activation [6], whereas PC dichotomies are aggregated to a final decision by a voting criterion. For example, in [11,22] the Authors proposed a voting scheme adjusted by the credibilities of the base classifiers, which were calculated during the learning phase of the classification.

In the case of ECOC, the input sample is usually assigned to the class with the closest codeword. This approach has derived two popular reconstruction rules. The former is named as *Hamming decoding* (HMD): it uses the crisp labels of dichotomizers, and sets the index s of the final class $\omega_s \in \Omega$ as:

$$s = argmin_i d_H(\mathbf{D}(\omega_i), \mathbf{M}(x)) \tag{2}$$

where $\mathbf{M}(x)$ collects the crisp dichotomizers' outputs and

$$d_H(\mathbf{D}(\omega_i), \mathbf{M}(x)) = \sum_{j=1}^{K} (\frac{1 - sign(D(\omega_i, j)M_j(x))}{2}). \tag{3}$$

It is worth noting that such a reconstruction rule can be used whatever the type of the classifier, i.e. abstract, rank or measurement[1], since it requires the crisp labels, only.

When the classifier outputs a soft label, i.e. a real number representing the degree of support given by classifier to the hypothesis that the test sample comes from the output class, we can exploit this quantity by applying the reconstruction rule introduced in [6], where the assignment of a new input to a certain class is performed looking at the function that returns the highest activation. Allwein et al. [1] extended this rule paying attention for binary learners based on the margin of a training example, e.g. support vector machine and adaboost, proposing a reconstruction rule named as *loss-based decoding* (LBD). The margin of a training example is a number that is positive if and only if the example is correctly classified by a given classifier and whose magnitude is a measure of confidence in the prediction. In this case, the index s of the final class $\omega_s \in \Omega$ is given again by equation 2 where, however, d_H is replaced by d_L that is computed as follows:

$$d_L(\mathbf{D}(\omega_i), \mathbf{M}(x)) = \frac{1}{mK} \sum_{i=1}^{m} (\sum_{j=1}^{K} (\Gamma(D(\omega_i, j)f_j(x)))) \tag{4}$$

where Γ is a loss function and $f_j(x)$ represents the soft label of the j-th classifier.

---

[1] The various classification algorithms can be divided into three categories [23]: type I (abstract), that supplies only the label of the presumed class, type II(rank) that ranks all classes in a queue where the class at the top is the first choice, type III (measurement) that attributes each class a value that measures the degree that the input sample belongs to that class.

Further to these well-known reconstruction rules, others have been recently proposed [4,5,15,18]. While a review of all such approaches is out of the scope of the paper, we observe that Shiraishi et al. in [20] presented a reconstruction rule for OpC and PC decompositions adopting support vector machines and combining their outputs using a logistic regression model, L2 and Group of Lasso penalty.

In this last paragraph of background section, let us to introduce the notion of *classification reliability*, since it is used in the rest of the paper. This quantity is derived from classifier soft label $f(x)$ and permits to properly estimate the reliability of each classification act [2,12]. The reliability takes into account the many issues influencing the achievement of a correct classification, e.g. the noise affecting the samples domain or the difference between the objects to be recognized and those used to train the classifier. Hence, it would provide useful information to derive a reconstruction rule. Without loss of generality, we can assume that the reliability of a sample x varies in $[0, 1]$. In the following, we will denote the reliability as $\psi(f(x))$ or, simply, $\psi(x)$. A low value of $\psi(x)$ suggests that the decision on sample x is not safe since, for example, it can be a borderline instance or it can be affected by noise in the feature space. A large value of $\psi(x)$ suggests that the recognition system is more likely to provide a correct classification [2,12]. Note that, in general, the use of classification reliability does not limit the choice of classifier architecture since it is always possible to obtain a measurement $f(x)$ for each classification act of any kind of classifier [10].

3   Reliability-Based Softmax Reconstruction Rule

In this section we will describe the proposed reconstruction method, which we refer as Reliability-based Softmax reconstruction rule (RBS) in the following.

RBS reconstruction rule considers dichotomizers' outputs as a new feature vector which have to be classified. We assume that the polychotomy is addressed by L binary classifiers providing the binary decision $\mathbf{M}(x)$ and the reliability vector $\psi(x) = \{\psi_1(x), \psi_2(x), \ldots, \psi_L(x)\}$.

On these positions, we introduce the quantity $\chi_j(x)$ that summarizes both the information provided by classifiers. Indeed $\chi_j(x)$ integrates the crisp label and the classification reliability that the j-th binary classifier provides for each sample x by multiplying them. Hence, for the whole decomposition we have: $\chi(x) = \psi(x)^T \odot \mathbf{M}(x) = \{\chi_1(x), \chi_2(x), \ldots, \chi_L(x)\}$.

Considering now $\chi(x)$ as second-order features, we have to face with the classification problem $\{\chi(x), \omega(x)\}$, where each sample $\chi(x)$ is a vectors described by L features with label $\omega(x)$. The classification task consists in predicting the label $y(x) \in \Upsilon$, where $\Upsilon = \{y_1(x), y_2(x), \ldots, y_K(x)\}$, so that $\omega(x) = y(x)$ for each sample. For the sake of clarity we omit in the following the dependence of all symbols from sample x.

We solve this classification task by using the softmax regression to estimate the posterior distribution of classification acts. Softmax regression is a natural choice since multiclass problems show multinomial distribution for the output. Defining a set of $K - 1$ vector of parameters, $\Theta = \{\theta_1, \theta_2, \ldots, \theta_{K-1}\}$, to parameterize the multinomial distribution over K different outputs, the conditional distribution of y given χ is given by:

$$p(\omega = y_i | \chi; \Theta) = \frac{e^{\theta_i^T \chi}}{\sum_{j=1}^{K} e^{\theta_j^T \chi}} \quad i = 1, 2, \ldots, K - 1. \tag{5}$$

It is straightforward observing that $p(\omega = y_K | \chi; \Theta) = 1 - \sum_{i=1}^{K-1} p(\omega = y_i; \Theta)$. The final label is set by:

$$y = argmax_i(p(\omega = y_i | \chi; \Theta)). \tag{6}$$

In order to perform this reconstruction technique we have to estimate Θ. To this aim, consider a training set tr composed of m_{tr} samples. Denoted by χ^{tr} the values of χ of samples belonging to tr, Θ can be estimated maximizing the log-likelihood l:

$$l(\theta) = \sum_{i=1}^{m_{tr}} log \prod_{l=1}^{K} \left(\frac{e^{\theta_l^T \chi_i^{tr}}}{\sum_{j=1}^{k} e^{\theta_j^T \chi_i^{tr}}} \right)^{1\{y_i=l\}} \tag{7}$$

where $1\{-\}$ denotes the index function, which is one if the statement inside the bracket is true, zero otherwise.

To reduce the correlation between classifier outputs when we perform the maximization of eq. 3 we use $L2$ penalty, as suggested in [20]. Note that χ^{tr} is computed performing a stacking procedure, which avoids problem of reusing training samples during parameter estimation. Indeed, we first divide tr into p folds, and then use $p - 1$ folds for training and one to estimate χ_h^{tr}, where $h \in [1; p]$. When all folds were considered as test fold, we compute $\chi^{tr} = \{\chi_h^{tr}\}_{h=1}^{p}$.

Finally, we discuss now the differences between our proposal and the contribution presented in [20], which can be summarized into two main points: (i) in [20] the Authors use raw outputs of the binary classifiers: this choice does not permit to use classifiers which provide crisp labels only, e.g. the k-Nearest Neighbour. Conversely, our contribution uses the quantity χ, which combines the crisps labels with the reliability, i.e. a measure providing us a deep insight in the classification process. Note that this choice permits us to employ any kind of classifiers; (ii) Shiraishi et al. [20] consider OpC and PC decomposition, whereas we focus on the ECOC framework. Hence, our contribution extends the work of Shiraishi et al. [20] since it uses the ECOC decomposition and it permits to apply the softmax reconstruction rule in case of rank and abstract classifiers. Furthermore, the novel use of the reliability in the regression not only extends the work of [20], but provides larger classification performance as will be reported in section 5.

4   Test Configurations

In this section we first describe datasets used in our tests, then we discuss performance metrics and finally we present the experimental protocol.

Datasets. We use six public datasets which provide an heterogeneous set of classification tasks in terms of number of samples, features and classes. Their characteristics are summarized in Table 1.

Datasets shows also different skewness among classes permitting to assess how the classification system performs when a class is under-represented in comparison to others. We deem that such an evaluation is of great extent for several researchers and partitioners in machine learning and pattern recognition since this phenomenon very often occurs in real world classification tasks.

Table 1. Summary of the used datasets

Dataset	Number of samples	Number of classes	Majority class	Minority class	Number of features
BRTISS	106	6	20.8%	13.2%	9
DERM	366	6	30.6%	5.5%	33
ECOLI	327	5	43.7%	6.1%	7
FER	876	6	28.1%	7.5%	50
GLASS	205	5	37.0%	6.3%	9
IIFI	600	3	36.0%	31.5%	14

Performance Metrics. To evaluate the performance of a classification systems the most used performance metrics are the accuracy (acc) and its counterpart, the error rate ($1-acc$). The recognition accuracy is defined as:

$$acc = \frac{\sum_{j=1}^{K} n_{jj}}{m} \tag{8}$$

where n_{jj} is the number of elements of class j correctly labeled and m is the total number of samples in the training set.

Although the accuracy summarizes global information on classifiers performance, it does not provide any data on the accuracy per class. Such an information is very useful to assess classification performance when one or more classes are largely under-represented in comparison to others. This happens in several real world datasets where samples of classes of interest rarely occurs. As an example, consider a dataset with five classes where the prior class probabilities are $56\%, 24\%, 10\%, 7\%$ and 3%. In this case, we should derive a classifier missing all minority class samples, but still achieving an accuracy equal to 97%. Such a failure can be described by the accuracy per class, defined as $acc_j = \frac{n_{jj}}{m_j}$ where m_j is the number of samples in class j. Data on acc_j can be synthetically reported by the geometric mean of accuracies (g), which is given by:

$$g = \left(\prod_{j=1}^{K} acc_j \right)^{\frac{1}{K}} \tag{9}$$

g is a non-linear measure where a change in one of its arguments has a different effect depending on its magnitude; for instance, if a classifier misses the labels of all samples in the jth class, it results in $acc_j = 0$, and $g = 0$.

Experimental Protocol. We test RBS method to solve multiclass tasks in ECOC framework. We apply ECOC using the method proposed by Dietterich et al. [6] for code generation. In particular, if $3 \leq K \leq 7$ we use exhaustive codes; if $8 \leq K \leq 11$ we generate exhaustive codes and then select a good subset of decomposition matrix columns given by the GSAT algorithm [19].

We test three different types of classifiers, belonging to different classification paradigms. Therefore, the binary learners are: Adaboost (ADA) as an ensemble of classifiers, Multilayer Perceptron (MLP) as a neural network and Support Vector Machine (SVM) as as a kernel machine.

We use as the "Adaboost M1" algorithm proposed in [7], where weak learners are decision stumps. The number of iteration is equal to 100. The reliabilities of ADA classifications are estimated using an extension of method [3], where we compute the difference between the outputs related to winning and losing class.

In the case of MLP, we use a number of hidden layers equal to half of the sum of features number plus class number. The number of neurons in the input layer is given by the number of the features. The number of neurons in the output layer is always two when the MLP is employed as dichotomizer. To evaluate the reliability of MLP decisions for multiclass classification problems we adopted a method that estimates the test patterns credibility on the basis of their quality in the feature space [3].

In case of SVM, we use a Gaussian radial basis kernel. Values of regularization parameter C and scaling factor σ are selected within $[1, 10^4]$ and $[10^{-4}, 10]$, adopting a log scale to sample the two intervals. The value of each parameter is selected according to average performance estimated by five fold cross-validation on a validation set. The reliability of a SVM classification is estimated as proposed in [16], where the decision value of the SVM is transformed in a posterior probability.

We compare the proposed reconstruction rule with HMD, LBD, and with the method proposed in [20], which is referred to as SHI in the following. In this latter case, the softmax regression defines $\chi(x)$ as $\psi(x)^T \odot f(x)$, where $f(x) = \{f_j(x)\}_{j=1}^L$ collects the soft labels of the binary classifiers (section 2). Moreover, we use an exponential loss-function for LBD reconstruction, as suggested in [1].

Furthermore, all experiments reported in the following are performed according to a five folds cross validation.

5   Results

This section presents the results achieved by the three classifiers on the tested datasets varying the reconstruction rule used, as reported in section 4.

Tables 2, 3 and 4 report results obtained by ADA, MLP and SVM classifiers. For each table, the left and right side reports performance measured in term of accuracy and geometric mean of accuracies, respectively.

Turning our attention to results achieved using the ADA classifier (Table 2), we observe that RBS globally has larger performance than other methods. Indeed, considering both the accuracy and the geometric mean of accuracies, RBS shows the largest values in five out of six datasets. As an example showing the advantage of using RBS, compare its performance on BRTISS dataset with those achieved by the second best

Table 2. Values of accuracy (%) and geometric mean (%) for AdaBoost classifier, using HMD, LBD, SHI and RBS reconstruction rules

Rule	Accuracy Datasets					Rule	G mean Datasets				
	BRTISS	DERM	ECOLI	FER	GLASS	IIFI		BRTISS	DERM	ECOLI	FER	GLASS	WINE
HMD	68.91	97.54	85.94	53.87	68.77	60.02	HMD	27.95	95.30	74.05	0.52	0.00	54.90
LBD	69.72	98.09	74.05	57.64	71.18	66.34	LBD	44.97	97.08	78.90	11.43	0.00	64.69
SHI	87.22	94.79	82.82	60.59	69.36	56.66	SHI	77.75	94.06	59.78	51.50	37.93	54.25
RBS	94.45	98.08	88.12	64.70	74.68	66.50	RBS	92.14	97.47	80.02	53.65	12.76	65.14

method. The differences with SHI in terms of acc and g are 7.23% and 14.39% respectively. Furthermore, it is worth observing g data on the GLASS dataset: only using a reconstruction rule based on logistic regression will permit to attain a value of g larger than zero. This means that both HMD and LBD misclassifies all samples of one class, at least. On the other side, this does not occur for SHI and RBS methods. We also note that on GLASS dataset SHI recognizes more samples belonging to the minority classes, while RBS correctly classifies more samples belonging to the majority ones, as shown by the pairs of acc and g values (i.e. larger acc for RBS and larger g for SHI).

Table 3. Values of accuracy (%) and geometric mean (%) for Multilayer Perceptron, using HMD, LBD, SHI and RBS reconstruction rules

Rule	Accuracy Datasets					Rule	G mean Datasets				
	BRTISS	DERM	ECOLI	FER	GLASS	IIFI		BRTISS	DERM	ECOLI	FER	GLASS	IIFI
HMD	68.09	98.36	88.06	94.52	71.25	68.84	HMD	13.86	98.40	80.77	92.86	14.94	65.57
LBD	67.17	98.36	88.03	95.32	73.08	70.14	LBD	27.73	98.40	82.02	93.57	14.88	68.67
SHI	92.72	64.19	70.06	45.54	62.27	43.51	SHI	80.00	40.79	53.66	35.15	22.84	35.00
RBS	90.00	99.17	86.88	98.62	76.95	69.31	RBS	80.00	99.10	78.78	98.11	55.68	67.27

Turning our attention to Table 3 reporting results achieved by MLP classifier, we observe that in terms of acc RBS achieves the best results in three out of six datasets. However, there is not another prevalent method, since best performance are attained by SHI and LBD method in one and two datasets, respectively. With respect to g, RBS method shows: (i) larger performance than other methods on three datasets out of six, (ii) best performance on BRTISS dataset which are also equal to those provided by SHI method, (iii) lower performance than LBD method in the other two cases. As an example, we consider the GLASS dataset where RBS shows best performance: in case of acc, the difference with the second best method (LBD) is 3.77%, while in case of g the difference with respect to SHIis 32.84%.

Let us now focus on the results obtained by the SVM classifier (Table 4). On the one side, results measured in terms of accuracy show that RBS performs better than others methods for all datasets. On the other side, results expressed in terms of g show that RBS outperforms other methods for all datasets except of the ECOLI. As an example, we consider the GLASS dataset where we have the largest performance differences between

Table 4. Values of accuracy (%) and geometric mean (%) for Support Vector Machine, using HMD, LBD, SHI and RBS reconstruction rules

	Accuracy						G mean				
Rule	Datasets					Rule	Datasets				
	BRTISS	DERM	ECOLI	FER	GLASS	IIFI		BRTISS	DERM	ECOLI	FER	GLASS	IIFI
HMD	71.67	96.96	87.75	97.48	63.88	62.50	HMD	40.58	96.60	80.40	96.98	23.34	56.12
LBD	71.58	97.23	87.76	97.60	67.87	67.33	LBD	40.73	96.83	87.86	97.24	36.62	64.89
SHI	84.34	82.74	88.06	92.80	62.66	56.00	SHI	80.00	80.00	81.62	80.00	41.18	25.82
RBS	90.43	99.45	88.98	98.97	77.11	68.00	RBS	80.00	99.47	79.06	99.08	73.36	65.40

RBS and other methods. In this case, the second best method is SHI, but its value of g is 32.18% lower than the one provided by RBS.

The results reported so far show that RBS method provides performance that are higher that those provided by other methods in the large majority of tests, regardless of the classifier architecture. Furthermore, if we do not consider RBS, we observe that SHI is not the best performing method. These observations suggest us that the introduction of the reliability in the reconstruction rule has large advantage.

As final issue, we notice that the proposed reconstruction rule provides values of acc and g that are, together, larger than the corresponding ones of other reconstruction rules. We deem that this result is relevant since most of the algorithms coping with class imbalance improve the geometric mean of accuracies harming the global accuracy [21]. Being able to provide larger values of both acc and g implies that RBS improves the recognition ability on the minority class without, or with a small extend, affecting the recognition accuracies on majority classes.

6   Conclusion

In this paper we have presented a reconstruction rule providing the final multiclass label by considering the outputs of binary classifiers addressing the decomposition as a set of second order features of a new classification problem. This task is solved using the softmax regression, presented in [20] in case of One-per-Class and Pairwise coupling decompositions, and here extended to ECOC framework. However, differently from [20], we introduce the classification reliability in the decision making process and test the reconstruction rule using also the Adaboost and Multilayer Perceptron classifiers. Given six heterogeneous datasets, our proposal satisfactory compares with two popular reconstruction rules as well as with the method proposed in [20].

References

1. Allwein, E.L., Schapire, R.E., Singer, Y.: Reducing multiclass to binary: a unifying approach for margin classifiers. Journal of Machine Learning Research 1, 113–141 (2001)
2. Cordella, L.P., Foggia, P.: Reliability parameters to improve combination strategies in multi-expert systems. Pattern Analysis and Applications 2, 205–214 (1999)

3. Cordella, L.P., Foggia, P., et al.: Reliability parameters to improve combination strategies in multi-expert systems. Pattern Analysis & Applications 2(3), 205–214 (1999)
4. D'Ambrosio, R., Iannello, G., Soda, P.: A one-per-class reconstruction rule for class imbalance learning. In: 21st Int. Conf. on Pattern Recognition, pp. 1310–1313. IEEE (2012)
5. D'Ambrosio, R., Soda, P.: Polichotomies on imbalanced domains by one-per-class compensated reconstruction rule. In: Gimel'farb, G., Hancock, E., Imiya, A., Kuijper, A., Kudo, M., Omachi, S., Windeatt, T., Yamada, K. (eds.) SSPR & SPR 2012. LNCS, vol. 7626, pp. 301–309. Springer, Heidelberg (2012)
6. Dietterich, T.G., Bakiri, G.: Solving multiclass learning problems via error-correcting output codes. Journal of Artificial Intelligence Research 2, 263 (1995)
7. Freund, Y., Schapire, R.: Experiments with a new boosting algorithm. In: Machine Learning-International Workshop then Conference, pp. 148–156 (1996)
8. Fürnkranz, J.: Round robin classification. J. of Machine Learning Research 2, 721–747 (2002)
9. Hsu, C.W., Lin, C.J.: A comparison of methods for multi-class support vector machines. IEEE Transactions on Neural Networks 13(2), 415–425 (2002)
10. Iannello, G., Percannella, G., Sansone, C., Soda, P.: On the use of classification reliability for improving performance of the one-per-class decomposition method. Data & Knowledge Engineering 68, 1398–1410 (2009)
11. Jelonek, J., Stefanowski, J.: Experiments on solving multiclass learning problems by n^2 classifier. In: Nédellec, C., Rouveirol, C. (eds.) ECML 1998. LNCS, vol. 1398, pp. 172–177. Springer, Heidelberg (1998)
12. Kittler, J., Hatef, M., Duin, R.P.W., Matas, J.: On combining classifiers. IEEE Transactions on Pattern Analysis and Machine Intelligence 20(3), 226–239 (1998)
13. Kumar, S., Ghosh, J., Crawford, M.M.: Hierarchical fusion of multiple classifiers for hyperspectral data analysis. Pattern Analysis & Applications 5(2), 210–220 (2002)
14. Masulli, F., Valentini, G.: Comparing decomposition methods for classication. In: Fourth International Conference on Knowledge-Based Intelligent Engineering Systems & Allied Technologies, KES 2000, pp. 788–791 (2000)
15. Moreira, M., Mayoraz, E.: Improved pairwise coupling classification with correcting classifiers. In: Nédellec, C., Rouveirol, C. (eds.) ECML 1998. LNCS, vol. 1398, pp. 160–171. Springer, Heidelberg (1998)
16. Platt, J.: Probabilistic output for support vector machines and comparisons to regularize likelihood methods. Advanced in Large Margin Classifiers. MIT Press (2000)
17. Rajan, S., Ghosh, J.: An empirical comparison of hierarchical vs. two-level approaches to multiclass problems. In: Roli, F., Kittler, J., Windeatt, T. (eds.) MCS 2004. LNCS, vol. 3077, pp. 283–292. Springer, Heidelberg (2004)
18. Rtsch, G., Smola, A.J., Mika, S.: Adapting codes and embeddings for polychotomies (2003)
19. Selman, B., Levesque, H., Mitchell, D., et al.: A new method for solving hard satisfiability problems. In: 10th National Conference on Artificial Intelligence, pp. 440–446 (1992)
20. Shiraishi, Y., Fukumizu, K.: Statistical approaches to combining binary classifiers for multiclass classification. Neurocomputing 74(5), 680–688 (2011)
21. Soda, P.: A multi-objective optimisation approach for class-imbalance learning. Pattern Recognition 44, 1801–1810 (2011)
22. Stefanowski, J.: Multiple and hybrid classifiers. pp. 174–188 (2001)
23. Xu, L., Krzyzak, A., Suen, C.: Methods of combining multiple classifiers and their applications to handwriting recognition. IEEE Transactions on Systems, Man and Cybernetics 22(3), 418–435 (1992)

Multisubjects Tracking by Time-of-Flight Camera

Piercarlo Dondi[1], Luca Lombardi[1], and Luigi Cinque[2]

[1] Department of Electrical, Computer and Biomedical Engineering,
University of Pavia, Via Ferrata 1, 27100 Pavia, Italy
[2] Department of Computer Science, Sapienza University of Rome,
Via Salaria 113, Roma, Italy
{piercarlo.dondi,luca.lombardi}@unipv.it,
cinque@di.uniroma1.it

Abstract. Time-of-Flight cameras are the state of art sensors for a fast detection of depth data in a scene. This kind of sensors can be very useful for tracking, in particular in indoor ambient, since, using light in near-infrared spectrum, they are less affected by abrupt change in illumination. In this paper we propose a new method for the tracking of multiple subjects based on Kalman filter. The first step of our solution is a ToF based foreground segmentation, that retrieves all significant clusters in the scene, followed by a robust tracking system able to correctly handle occlusions and possible merging between clusters.

Keywords: Tracking, Time-of-Flight camera.

1 Introduction

From their introduction in 2003 [1], Time-of-Flight (ToF) cameras are quickly become the state of art sensors for achieving a real-time (from 20 fps of the older models to 54 fps of the newer ones) depth measurement of a scene. ToF cameras do not need reference points or external illumination sources, they work emitting light in near-infrared spectrum and distances are estimated according to the time spent by the reflected light to come back from objects to the sensor. This kind of active sensors have trigged the interest of many researchers, in various fields, such as 3D object reconstruction, human-computer interaction, tracking, augmented reality or also medicine and bio-informatics.

In this paper we focus on the problem of real-time tracking of multiple subjects. Our approach is based on the use of a Kalman filter. In particular we extend the standard Kalman filter of the OpenCV library, with some automatic methods specifically designed to associate, as well as possible, the detected clusters with the respective trackers. The tracker works on the clusters retrieved by a fast foreground segmentation that exploit the particular kind of data provided by a ToF camera (depth data and intensity of reflected light).

A preliminary implementation of this solution was presented in [2], then the results achievable by a more complete version was presented in [3]. In this paper

A. Petrosino (Ed.): ICIAP 2013, Part I, LNCS 8156, pp. 692–701, 2013.

we introduce new improvements such as a more precise initial association between clusters and trackers (that guarantees a correct tracking of both near and far clusters) and the handle of clusters merging due to occasional imprecision in the segmentation (e.g. subjects too close). From the computational point of view, the current implementation is more efficient than the original one and the new features does not influences the overall performance of the system.

The paper is organized as follow: section 2 supplies a brief overview of the sensor; section 3 provides the state of the art of tracking method based on ToF cameras; section 4 describes the segmentation method; section 5 analyzes in details the tracking procedure; section 6 shows the experimental results; section 7, at last, draws some conclusions.

2   Time-of-Flight Camera Overview

Time-of-Flight cameras are active imaging sensors that can provide distance measures of an ambient using laser light in near-infrared spectrum. There are two main technologies: pulsed light and modulated light. In the first case a coherent wavefront (similar to a "light wall") hits the targets and then the distances are measured analyzing the deformation in the reflected "wall". In the second case, currently the most widespread technology, the camera emits a modulated light and the depth information are gained by phase delay detection.

Respect to other depth measuring systems, such as stereo cameras or laser scanners, ToF cameras supply some advantages: they can work in real-time, the depth data are directly provided by the sensors without complex additional computations; they do not need external light (the illumination is self-provided); they can operate in any kind of scenario without external reference points or colors contrast; the shape of the objects does not influence the measures.

On the other hand there are also some disadvantages that must be considered for a better use of these devices. ToF cameras have still a limited resolution (e.g. 176x144 of Mesa SR4000 or 200x200 of PMD CamCube 3.0) and they are affected by different kinds of noise: the "flying pixels" due to areas with abrupt changes in depth (e.g. the corners of an object); the "motion artifacts", measurement errors proportional to the speed of moving subjects; the noise cause by sunlight that can significantly alter the result and limit the applicability of these sensors to indoor use. Finally the precision of the measures strictly depends on the reflectivity of the objects, if it is too high it can saturate the sensor, while if it is too low the object can be not correctly detected [4].

All the experiments in this work are been made with a modulated light cam, the Swissranger SR3000. This ToF camera is able to supply simultaneously two images per frame: a distance map and the map of the intensity of the reflected light. Both of them have QCIF resolution (176x144 pixels) and a color depth of 16 bits. The 55 active leds emit in the near infrared around 850nm with a frequency of 20Mhz, a value that guarantees a nominal range without ambiguities of 7.5m. The depth accuracy goes from a few centimeters to millimeter in optimal conditions. The distance accuracy depends on distance range, signal intensity and the background illumination. The field of view (FOV) is about 47.5×39.6 degrees [5]. The camera has been used at 18-20 fps in order to maintain a good ratio between noise and real-time capability: the noise depends directly on the frame rate.

3   State of Art

A lot of approaches have been used for obtaining a good ToF-based tracking. In [6] the retrieved clusters are projected on the ground plane for creating a so called "flat-map", then an Expectation Maximization algorithm has been applied to that map. A method for multiple people tracking based on Shape from Silhouettes (SfS) is proposed in [7]: it appears robust but greatly limited by the high numbers of the cams needed (six in the proposed solution) and by the small dimension of the room. In [8] instead a traditional Kalman filter has been used with the camera placed to provide a top-down view of the scene. This particular position simplifies the tracking problem, but significantly decreases the visible area, moreover almost all the details of the detected subjects are lost.

All these methods segment the scene and retrieve the clusters by a background subtraction algorithm. This solution, even if wide diffused, suffers of known problems such as ghosts appearing at changes in background objects or absorption of still people . The generation of the model can also be computationally expensive, especially if it needs a high resolution or a dynamic adaptation to ambient changes (such as light variations).

An alternative approach, based only on the analysis of depth data, can be found in [9]. The described algorithm allows the detection and the classification of objects in the scene studying the probability density function (pdf) of depth image and its histogram distribution. Then an appropriate distance metric, based on the integrated square error between the pdfs, is used to recognize the clusters through consecutive frames.

A particular category of solutions involves the combination of traditional RGB cameras with a ToF one. In [10] is proposed an approach that exploits two particle filter-based visual trackers, one for each stream data type (RGB and depth). Depending from the scene the system uses the one that guarantees the better performances (generally RGB for outdoor and depth for indoor). In [11], instead, the fusion of color and depth data is adopted for compensating the respective weaknesses of the two type of sensors. More specifically the segmentation and the tracking are achieved using a well designed method based on mean shift algorithm.

4   Segmentation

The proposed ToF-based segmentation method provides different advantages: no need of preprocessing operation (no learning phase); no need of a priori knowledge of the environment (the system is therefore robust to background changes); the shape of objects has no influence on the results.

The main steps are summarized in Fig. 1: firstly a thresholding based on values in the intensity map is applied to the distance map; then a region growing algorithm, that starts from seeds planted in the peaks of the intensity, is executed on the filtered distance map to produce separate clusters corresponding to foreground objects.

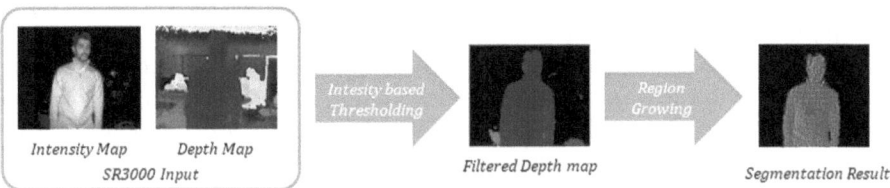

Fig. 1. The main steps of foreground segmentation from the ToF data input to the full body extraction

In the intensity map, foreground objects are brighter than those in background (they received more light), so, the reflectance data can be successfully used as a mask on the depth map to reduce the area of investigation on which the region growing will be applied (Eq. 2). The region growing is applied on the distance map and not directly on the intensity one, due to the greater stability of depth information (intensity is enough precise for a preliminary thresholding, but it varies too much due to different reflection properties of the objects framed). The best seeds for region growing are found applying an opportune intensity threshold (λ_{seed}), estimated using the Otsu's method, a well known thresholding algorithm based on image histogram.

The similarity measure S between a cluster pixel x and a neighboring pixel y is defined as follow:

$$S(x,y) = |\mu_x - D_y| \tag{1}$$

where D_y is the distance value of pixel y and μ_x is a local parameter related to the mean distance value around x (Eq. 3). The lower S is, the more similar pixels are. When a seed is planted, μ_x is initialized to D_x. Considering a 4-connected neighborhood, a pixel x belonging to a cluster C absorbs a neighbor y according to the following conditions:

$$\{x \in C, \ S(x,y) < \theta, \ I_y \geq \lambda, \ D_y < \delta\} \rightarrow \{y \in C\} \tag{2}$$

λ is the intensity threshold, proportional to λ_{seed}, dynamically calculated for each frame (Otsu's threshold λ_{seed} proves to be very effective to find the peak values of the intensity image, but it has turned out to be too strict for the thresholding required in this phase); θ is a constant parameter, experimentally estimated, related to clusters separation; δ is an optional parameter used for excluding a priori all points beyond a fixed distance. It quickly reduces the search area and can be very useful in those applications in which the maximum operating distance is known a priori or if the shot is made too close to a wall.

When a neighbor y of seed x is absorbed, the average distance value μ_y is computed in an incremental manner as follows:

$$\mu_y = \frac{\mu_x * \alpha + D_y}{\alpha + 1} \tag{3}$$

where α is a learning factor of the local mean of D. If pixel y has exactly α neighbors in the cluster, and if the mean of D in this neighbor is exactly μ_x, then μ_y becomes the mean of D when y is added to the cluster.

Every region grows excluding the just analyzed pixels from successive steps. The process is iterated for all seeds in descending intensity order. Regions too small are discarded for removing the possible false positive areas over the threshold (for example little surfaces with high reflectivity).

The performance of the method, varying its parameters, has been analyzed in [12]: all the tests show a correctness (the ratio between true positive and the sum of true positive and false positive) between 94% and 97% and completeness (the ratio between true positive and the sum of true positive and false negative) between 92% and 96%.

5   Tracking

The multi-subjects tracking method presented in this work is based on the use of Kalman filter. As mentioned in section 3, this solution for ToF based tracking was first studied, with good results, by [8]. Respect to that paper the proposed implementation adopts a more general approach, the camera has not been placed to provide a top-down view, but a frontal view. The purpose of this choice is to obtain a more versatile solution, with a major visible area combined with a better angle of view. Our tracking method can also correctly handle the occlusions and the merging between clusters.

5.1   Kalman Tracker

A Kalman filter is characterized by a six dimensions state, (x, y, z, v_x, v_y, v_z) which refers to the position of the assigned cluster: (x, y, z) represent the centroid, while (v_x, v_y, v_z) is the velocity vector. All these parameters are expressed in image coordinates, since the SR3000 supplies data already organized in 3D Cartesian coordinates. When the tracking starts, at each detected cluster is assigned a new Kalman tracker, that it is initialized with the current position of the centroid of the real cluster. At time t, each Kalman predicts the most probable position of its correspondent cluster at the next frame. So, at time $t+1$, the predicted coordinates of each Kalman will be compared with the real positions of the centroids of each clusters retrieved, in order to find again the previous objects. The association between measured clusters and Kalman trackers is evaluated by minimum square euclidean distance between their centroids. In particular each Kalman tracker connects itself with the nearest object that is not yet been assigned to another closer Kalman (Eq. 5). The Kalman and the cluster associated in this way are excluded by the respective set of possible candidates; the system iterates until all the visible clusters are connected with a Kalman. Since sometimes the ToF noises (especially motion artifacts) can generate false small clusters that last less than a second, a Kalman tracker is assigned to a new cluster only if it is in scene at least from 5-7 frames.

According to Kalman filter behavior, after the association, the prediction is corrected with the real measurement in order to refine the state estimation.

For reducing to minimum the errors each Kalman tracker searches correspondences only in a limited spherical area. The set P_c of all the cluster that are enough near at least to one Kalman tracker is defined as follow:

$$\left\{ \sqrt{d_e(c,k)} < \alpha \right\} \rightarrow \{c \in P_c\} \tag{4}$$

where $d_e(c,k)$ is the euclidean distance between the coordinates of retrieved cluster c and of Kalman tracker k; α is an association threshold experimentally defined considering the limited field of view and the limited resolution of the ToF camera.

The assignment procedure can be summarized by the following equation:

$$\left(\sqrt{d_e(c,k)} = \min \left[\sqrt{d_e(x,y)} \right], \forall x \in P_c, \forall y \in K \right) \rightarrow (c \text{ ass. to } k) \tag{5}$$

where K is the set of all active Kalman trackers, and $(c \text{ ass. to } k)$ indicates that cluster c has been recognized as the cluster assigned to Kalman k at previous frame.

After all these steps, if one or more retrieved clusters are not associated to any Kalman, they are considered as new objects entered in scene, so an equivalent number of Kalman trackers are initialized. Otherwise, if one or more active Kalman trackers are not associated, probably there is an occlusion. In this case the Kalman maintains all the precedent data and tries to estimate the most probable path of the disappeared cluster using its last detected movements; the research area is also doubled in order to compensate estimation errors. If the cluster reappears shortly (within 30-40 frames) in a position closed to the predicted one it is immediately recognized and reassigned to its precedent tracker. On the contrary, if the cluster does not reappear in a fixed time, it is considered out of scene, so its Kalman is reset and can be reassigned to a new subject.

The described behaviour of a Kalman tracker can be defined by three possible states: 0, not assigned; 1, assigned to a visible cluster; 2, assigned to a not visible cluster probably occluded. The possible changes between these states are showed in Fig. 2 with a 2D example: the crosses are the retrieved clusters, the arrows are their directions, the points are the Kalman trackers and the circles are their search areas. Figure 2(a) presents a typical situation with two moving clusters assigned to two Kalman trackers. When one cluster is occluded by the other (Fig. 2(b)), its correspondent Kalman goes in state 2 (note the research area increased). From this situation there are two possible exits: the cluster reappears near to the predicted position, so it is associated again to its precedent Kalman that returns in state 1 (Fig. 2(c)); the cluster does not reappear, so the Kalman

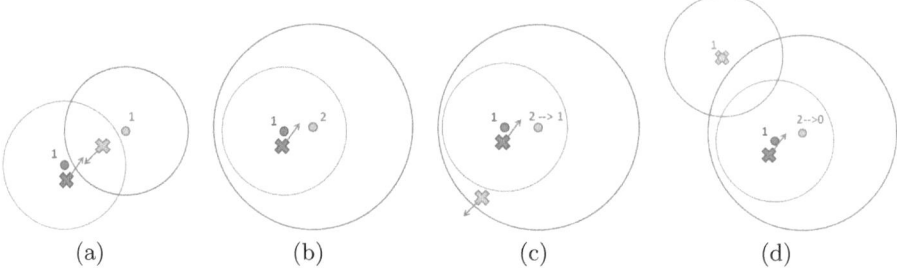

Fig. 2. 2D representation of Kalman trackers behaviour: (a) standard situation, two clusters in movement (the crosses) assigned to two Kalmans (the points), no occlusions; (b) one cluster occludes the other; (c) the occluded cluster reappears and it is assigned again to its precedent Kalman tracker; (d) the occluded cluster does not come back, a new cluster enters in scene and it is assigned to a new Kalman tracker

comes back to state 0 (Fig. 2(d)). At the same time, if a cluster out of all the research areas comes in the scene, it will be assigned to a new Kalman tracker (Fig. 2(d)). Note as in Fig. (Fig. 2(c)) the green tracker is associated to the green cross even if the blue one is nearest; this happens according with Eq. 5 because the blue tracker is closer to blue cluster, so the green tracker associates itself with the second closer object inside its search area.

5.2   Feedback for Smart Seeding

The Kalman predictions can be actively used for increasing the precision of the seeding and so correcting cluster detection mistakes. Considering a case in which there are two subjects. If one of them gets too close to the sensor (Fig. 3, left) its intensity values grow too much so all the seeds will be concentrated on it. As a consequence the faraway cluster is excluded from region growing and disappeared in the final results (Fig. 3, top right), even if it has been correctly included, after thresholding, in the filtered range image (Fig. 3, middle).

This issue can be overcome using the predictions of Kalman trackers as additional input for the seeding phase [3]. At time t new seeds are planted in pixels around to all centroids coordinates predicted at time $t-1$. This smart seeding allows the concurrent detection and tracking of middleground and foreground objects (Fig. 3, bottom right).

However such solution does not work well if a new subject enters in scene when the first one is close to the camera. In this case the new subject is initially undetected because he has not an assigned Kalman tracker. In this situation we plant seeds also in distance peaks of the filtered range image. This approach is not so precise as Kalman seeding, because there is not a direct correlation between distance and presence of a subject, but it is fast and simple and it can correctly handle such kind of issue. Moreover adding the distance seeding is not computationally expensive, so it can be successfully combined with the Kalman seeding for retrieving new entering clusters.

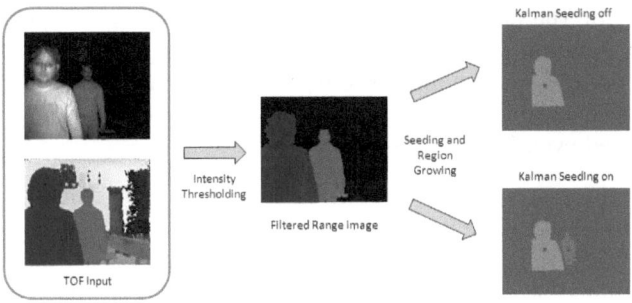

Fig. 3. Segmentation without and with Kalman seeding

5.3   Control for Clusters Merging

A final improvement involves the incorrect clusters merging. The region growing is able to correctly divide near clusters if their Z coordinates are enough different (Fig. 4 left). However, if two clusters have distance values too much similar, they will be wrongly fused in a single cluster (Fig. 4 center). This is not a common situation and, even when occurs, it is usually of short duration (also a little movement can vary enough the Z) – so, most of the times, it can be correctly handled as a traditional occlusion. For those cases in which the fusion takes too much time, and can confuse the tracker, a simple control can be adopted.

Fig. 4. Cluster merging correction: (left) two correctly separated clusters; (center) wrong merging of two clusters in one; (right) clusters subdivision after corretion

At time t, after the segmentation, the system saves the current positions, the size and the number of the current clusters. Different axes of vertical symmetry are traced between each near clusters. At time $t+1$, if there is a reduction in the number of clusters and there is a cluster with a size consistent with the sum of two or more previous clusters, a merging may have occurred (Fig. 5.3). In that case if there is one axis that passes through this cluster, that axis is chosen as lines of separation for splitting the cluster in two parts (Fig. 4 right). When we track people, another practical index of a clusters merging is the presence of a cluster with two heads (that can be recognized with a standard face detection system). The obtained subdivision is only an approximation not much precise, in particular for the shape of the split clusters; however the positions of the centroids are still quite accurate, so they can be used for maintaining a correct tracking. This correction should be used mainly when the moving area is very reduced (i.e. the clusters are very close) and when it is crucial not to confuse the subjects tracked even for a moment.

6   Experimental Results

A series of evaluation tests have been made to prove that the tracking system is able to manage the concurrent movements of multiple subjects and is also robust to the occlusions. The only possible sources of errors are moving objects with abrupt changes of direction or new clusters that appear closed to a just active Kalman. Figure 5 shows some frames of an example sequence. The colored spheres on the top of the three subjects are the markers of Kalman trackers, note how the correspondence between clusters and trackers is always maintained.

Fig. 5. Tracking sequence of three moving subjects with multiple occlusions

The test starts with three people placed at different distances. Then the first one crosses the stage until exiting from the framed area. Clusters temporarily occluded (Fig. 5, second image) correctly maintain their "labels" when they are again visible (Fig. 5, third image). A similar event happens when the second subject occludes the third one (Fig. 5, fourth and fifth images); in this case, also, the first subject re-enters in scene in a position closed to the exit one and after a short time, so it is correctly recognized.

Even if there is not a theoretical limit to the number of clusters that can be followed at the same time, the limited resolution of the SR3000 reduces the useful number for a correct working to a maximum of 3-4 objects. When the number is bigger, the risk to fill all the field of view of the camera with clusters all closed to each other, with a consequent increase of possible sources of errors, is very high.

7   Conclusions

This paper presents a new robust approach to the multi-subjects tracking, based on Kalman filter, that does not need any a-priori information about ambient or clusters. Due to the use of a ToF camera our system can work in any indoor scenario, in particular without controlled illumination sources. The algorithm allows the concurrent tracking of moving subject, correctly handling occlusions or accidentally clusters merging. A flexible seeding system that uses Kalman, depth and intensity data guarantees a fast detection of people placed at difference distance from the camera. Now we are studying how to improve the description of the clusters in order to recognize reentering objects after a longer time than the actual few seconds.

References

1. Oggier, T., Lehmann, M., Kaufmann, R., Schweizer, M., Richter, M., Metzler, P., Lang, G., Lustenberger, F., Blanc, N.: An all-solid-state optical range camera for 3D real-time imaging with sub-centimeter depth resolution (SwissRanger). In: Proceeding of the SPIE, vol. 5249, pp. 634–645 (2003)
2. Bianchi, L., Gatti, R., Lombardi, L., Lombardi, P.: Tracking without Background Model for Time-of-Flight Cameras. In: Wada, T., Huang, F., Lin, S. (eds.) PSIVT 2009. LNCS, vol. 5414, pp. 726–739. Springer, Heidelberg (2009)
3. Dondi, P., Lombardi, L.: Fast Real-Time Segmentation and Tracking of Multiple Subjects by Time-of-Flight Camera. In: 6th International Conference on Computer Vision Theory and Applications (VISAPP 2011), pp. 582–587 (2011)
4. Kolb, A., Barth, E., Koch, R., Larsen, R.: Time-of-Flight Cameras in Computer Graphics. Journal of Computer Graphics Forum 29, 141–159 (2010)
5. CSEM: SwissRanger SR-3000 Manual, Mesa Imaging (2006)
6. Hansen, D.W., Hansen, M.S., Kirschmeyer, M., Larsen, R., Silvestre, D.: Cluster tracking with Time-of-Flight cameras. In: Proceedings of Computer Vision and Pattern Recognition Workshops (CVPRW 2008), pp. 1–6. IEEE Computer Society (2008)
7. Guomundsson, S.A., Larsen, R., Aanaes, H., Pardas, M., Casas, J.R.: TOF imaging in Smart room environments towards improved people tracking. In: Proceedings of Computer Vision and Pattern Recognition Workshops (CVPRW 2008), IEEE Computer Society (2008)
8. Bevilacqua, A., Di Stefano, L., Azzari, P.: People Tracking Using a Time-of-Flight Depth Sensor. In: Proceedings of the AVSS 2006, Video and Signal Based Surveillance, p. 89. IEEE Computer Society (2006)
9. Parvizi, E., Jonathan Wu, Q.M.: Multiple Object Tracking Based on Adaptive Depth Segmentation. In: Proceedings of Canadian Conference of Computer and Robot Vision, pp. 273–277. IEEE Computer Society (2008)
10. Sabeti, L., Parvizi, E., Jonathan Wu, Q.M.: Visual Tracking Using Color Cameras and Time-of-Flight Range Imaging Sensors. Journal of Multimedia 3(2), 28–36 (2008)
11. Bleiweiss, A., Werman, M.: Fusing Time-of-Flight Depth and Color for Real-Time Segmentation and Tracking. In: Kolb, A., Koch, R. (eds.) Dyn3D 2009. LNCS, vol. 5742, pp. 58–69. Springer, Heidelberg (2009)
12. Bianchi, L., Dondi, P., Gatti, R., Lombardi, L., Lombardi, P.: Evaluation of a foreground segmentation algorithm for 3D camera sensors. In: Foggia, P., Sansone, C., Vento, M. (eds.) ICIAP 2009. LNCS, vol. 5716, pp. 797–806. Springer, Heidelberg (2009)

3D Tracking of Honeybees Enhanced by Environmental Context

Guillaume Chiron[1], Petra Gomez-Krämer[1],
Michel Ménard[1], and Fabrice Requier[2,3]

[1] L3I, University of La Rochelle, Avenue M. Crépeau, 17000 La Rochelle, France
[2] INRA, UE 1255, UE Entomologie, 17700 Surgères, France
[3] CEBC, CNRS, UPR 1934, 79360 Beauvoir sur Niort, France
{guillaume.chiron,petra.gomez,michel.menard}@univ-lr.fr,
fabrice.requier@magneraud.inra.fr

Abstract. This paper summarizes an approach based on stereo vision to recover honeybee trajectories in 3D at the beehive entrance. The 3D advantage offered by stereo vision is crucial to overcome the rough constraints of the application (number of bees, target dynamics and light). Biologists have highlighted the close scale influence of the environment on bees dynamics. We propose to transpose this idea to enhance our tracking process based on Global Nearest Neighbors. Our method normalizes track/observation association costs that are originally not uniformly distributed over the scene. Therefore, the structure of the scene is needed in order to compute relative distances with the targets. The beehive and especially the flight board is the referent environment for bees, so we propose a method to reconstruct the flight board surface from the noisy and incomplete disparity maps provided by the stereo camera.

Keywords: stereo vision, honeybees 3D tracking, surface reconstruction, beehive monitoring.

1 Introduction

Forced to observe the worldwide decline of honeybees (Apis mellifera), biologists began to study different hypotheses that could explain the phenomenon. Recently, the authors of [7] highlighted the evidence of behavioral alterations caused by pesticides. In that study, entrances and exits data was collected by a Radio-frequency identification monitoring device placed at the beehive entrance. So far, no biological study has been conducted at a big scale on flight behaviors. In cause, the lack of suitable methods to collect trajectories of honeybees in flight. The only method used by biologists (harmonic radar) is intrusive and suffers from biases. We believe that computer vision can effectively achieve this task with the respect of the application constraints.

Our challenging application of tracking bees in 3D at the beehive entrance has been studied in [4], laying down the application constraints and a detect-before-track approach based on Kalman Filter coupled with Global Nearest Neighbor. In this current paper, we focus our attention on the weakness of our tracking algorithm nearby the flight board. As a main contribution, we propose to enhance

A. Petrosino (Ed.): ICIAP 2013, Part I, LNCS 8156, pp. 702–711, 2013.

the tracking process by taking in consideration the targets relative to environment. This is achieved by a normalization of track/observation association costs relying on the distance of the target from the flight board. Also, as a second contribution, we introduce a method to recover the flight board surface, which is not adequately represented by individual disparity images provided by the stereo camera.

This paper is organized as follows. First, Section 2 summarizes the base of our monitoring system, detailing the acquisition method, the segmentation, and the tracking processes. Then, Section 3 introduces in first, our normalization method of track/observation association costs, and then our method to recover and model the flight board surface in 3D. Results are shown and discussed in Section 4. And finally, Section 5 concludes and opens promising perspectives.

1.1   Related Work

In the following are discussed papers dealing with trajectometry based on video related to insects or animals. The process of recovering trajectories can be split into two distinct parts which are presented below: detection and tracking.

For target detection, several methods have been proposed. Detect-before-track approaches are generally based on a segmentation process, and then the observations are associated in order to know tracks using an assignment method. Many methods based on that approach use a more or less advanced background subtraction (e.g. [1,3]). In [3], potential false alarms are filtered using a shape (ellipsis based) matching process. In contrast, some methods introduced do not require any background subtraction. The authors of [13] detect bees using the well-known Viola-Jones method [18], and the authors of [11] introduce an approach based on vector quantization which is able to detect individual bees among hundreds of walking bees. In [16], flying bats are detected taking the advantage of multiple cameras by directly applying Direct Linear Transform. In case of track-before-detect approaches, the position of the target is first estimated, and a probability for that estimation to correspond to the target drives the next estimation. For that kind of approach, a likehood function based on appearance models (pre-computed "eigenbees" in [10], or adaptive models in [17,12]) are used.

Many methods have been proposed for tracking. When following one clearly detected target moving along a simple trajectory, approaches such as those used in [6], near neighbor or mean shift may be sufficient. But in the case of tracking multiple targets, assignment problems arise due to missed detections and occlusions and thus cause false alarms. The authors of [1,3] use Global Nearest Neighbor (GNN) for track assignment, instanciation and destruction. In [16] the authors track multiple flying targets thanks to a multiple hypotheses tracker (MHT). Compared to GNN, MHT integrates the time in the assignment decision process. In [10,17,12,9], a nonlinear motion model is considered for their targets and they base their tracking on a particle filter [15], which corresponds to a track-before-detect approach. In [9], the authors introduced a MRF-augmented particle filter for multiple targets with reasonable computational cost, contrasting with the three other methods used in [10,17,12] that are less suitable in term of performance when working on multiple interacting targets.

When dealing with many flying targets in natural conditions (uncontrolled light and background), the range of application remains narrow and methods remain to be explored. Compared to most approaches mentioned above, our stereo vision based system brings in this paper the advantage of the the 3rd dimension. This work extends the preliminary study made in [4] by considering the spatial environment into the tracking process.

2   Detect-before-Track Based on Stereo Vision

This section presents our stereo vision acquisition system and summarizes how we recover trajectories from a sequence of couples of intensity/disparity images. More details on the acquisition system and our segmentation are given in [4].

2.1   Acquisition System and Constrains

The following constraints make our application especially challenging: Up to 15 simultaneous flying targets with chaotic dynamics, uncontrolled lighting, and gradual soiling of the background. Among potential suitable cameras (stereo vision and time of flight), we chose the TYZX G3 EV stereo vision camera which seemed to satisfy the best our requirements, with the following configuration: 3 cm baseline and 62 degree HFOV lenses. The camera targets the flight board and is located 50 cm above it. This system acquires a couple of intensity/disparity images of 752×480 px at an average frame rate of 47 fps (unstable rate). Figure 1 shows a capture sample. The quality of the disparity map provided by the camera depends on lighting and texturing conditions. Bees standing on the flight board measure about 10 pixels on those images, and during high activity periods, the flight board can be crowded of bees, making the board difficult to discern.

Fig. 1. Couple of intensity (left) / disparity (right) images provided by the stereo vision camera. The beehive is visible on the top. Bees take off/land on the flight board.

2.2   Segmentation of Flying Targets

Under the constraints listed above, classical motion detection methods based on background modeling fail. Our system detects flying bees thanks to an hybrid segmentation that takes advantage of both intensity and disparity images provided by the stereo camera. Our segmentation relies on an adaptive background model for the intensity, and combined mathematical morphologies and adapted thresholding for the disparity images. As a result, we obtain a mask containing regions corresponding to our targets. A target is defined by (u, v) the center of mass of the region, and d the median value of the depth values in the disparity image. So far, targets are expressed in the image reference coordinates. To ensure later a coherent tracking, targets are rather defined by (x, y, z) in the camera coordinate space (3D Euclidian space) using a projection based on stereo-camera calibration parameters. Our segmentation is globally robust but still returns in rough conditions up to three false alarms per image around the flight board. Also relatively few miss detections occur for too fast, too close or too far targets from the camera.

2.3 Multi-target Tracking in 3D

In our detect-before-track approach, each target is tracked by a Kalman Filter [8]. Despite the apparent rough dynamic of bees, we acquire frames at a sufficiently high frequency (about 47 fps) so that we can assume a constant speed model. Noise matrices (Q for the model, R for the measure) are tuned to prevent the model from derivation. Let us suppose $Y_{1:n}$ the series of observations corresponding to a target from time 1 to n. A Kalman Filter is instantiated with Y_1 and later destroyed when the step $k > n$. For a given step k, an observation is defined by the vector $Y_k = [x, y, z]^T$, and the estimated state of a target is defined by the vector $X_k = [x, y, z, \dot{x}, \dot{y}, \dot{z}]^T$ combining its 3D position and velocity. Figure 2 lays down the recursive mechanism of the Kalman Filter for the estimation of the state vector X_k and the prediction X_{k+1} given an sequence of observation from step 1 to k.

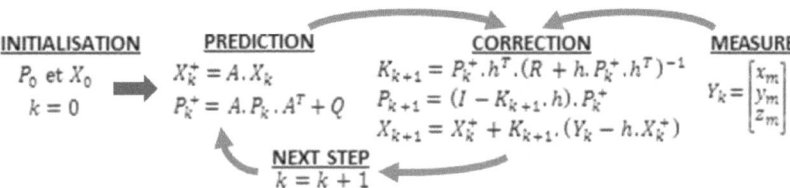

Fig. 2. Recursive mechanism of the Kalman Filter for the estimation of X

In our application, up to 15 targets can be observed at each step, which requires a multi-target tracking approach. The commonly used Global Nearest Neighbor (GNN) [2] method handles observation to track associations, track instantiations and destructions. In GNN, the assignment matrix $A[c_{i,j}]$ represents all the possible associations and the costs generated by those associations. A includes the possibility for each observation to be associated to an existing track, not to be associated to any track or to be associated to a new track. $c_{i,j}$ is the cost for the observation i to be assigned to the possibility j. The best configuration of associations is the solution that minimizes the global cost (e.g. solved by the hungarian method). The association cost between an observation and a track is computed by the Mahalanobis distance given by:

$$d^2 = (Y - MX^+)'S^{-1}(Y - MX^+)$$
$$S = ME^+M' + Em$$
(1)

where Y is the measure vector corresponding to the observation, M the measurement matrix, X^+ the predicted a priori position, S the innovation covariance with Em as the measure noise matrix, and E^+ the predicted noise covariance matrix.

3   Tracking with Environmental Consideration

In GNN, the multi-target assignment process relies on comparisons of association costs d defined by (1). For a track/observation association, if the minimum cost exceeds a fixed threshold later called Association Cost Limitation (ACL), the association is not considered and the observation becomes a false alarm or a new track. The global chaotic dynamics of flying bees forces the adoption of a tolerant threshold. In our application the problem is, in addition to an high frequentation, most of wrong detections (false alarms and miss detections) are located near the flight board. Moreover, targets near the flight board tend to progress nearly on the same plan resulting in the lost of the 3D advantage. So when tracking a bee near the flight board, in case of miss detection the loose threshold allows the track to be associated with a nearby false alarm or other candidate observations, which causes the degeneration of the track. The magnitude of association costs is not uniformly spread over the scene, so adapting the association costs to the situation is needed.

3.1   Association Costs Normalization

As reference, a classic approach consists in adapting the uncertainty at the track initialization. Bees entering the scene from the outside generate tracks with a big initial uncertainty, which has the effect of decreasing d defined by (1). In contrast, bees taking-off from the flight board have a low velocity and thus can generate tracks with a small initial uncertainty, which has the effect of increasing d. In this way, tracks are less vulnerable to wrong associations nearby the flight board and more able to be associated to relatively far observations when entering the scene. The problem is only partially solved because, the convergence of the uncertainty (in the Kalman Filter) is naively driven by the elapsed time, which does not corresponds well to the random chaotic trajectories exhibited by honeybees.

The authors of [14] explain the relation between honeybees' speed and surrounding objects. Using our classic tracker based on [4], we effectively observed lower speeds and also lower association costs nearby the flight board. The lower association costs are explained by the more stable targets' dynamic resulting in easily predictable positions. As a contribution, we propose to normalize association costs all over the scene relying on targets' relative distances from the board. Figure 7 shows an example of association cost distribution, within which we can identify the two following distinct effects. The primary growing effect ($l < 50$) potentially corresponds to the gradual stability in the dynamic adopted by bees when approaching the flight board in the last centimeters. The second effect ($l > 50$) is less perceptible, and potentially corresponds to changes in bees' dynamic at a larger scale (e.g. approach of the beehive). Relying on the modeling (functions α and β) of those effects, we propose the normalized cost c_f based on d^2 that has the advantage to penalize potential wrong detections near the board and thus limits the degeneration of the track. c_f is given by:

$$c_f = \left\{ \begin{array}{l} d^2/\alpha_f \text{ if } f \leq l \\ d^2/\beta_{f-l} \text{ if } f > l \end{array} \right\} \tag{2}$$

where f is the closest distance of the target from the flight board, l the limit of strong influence of the environment, and α and β are respectively the quadratic and affine normalization functions of f relative to the application. Coefficients of α and β can be estimated by quadratic and linear regressions from a set of association costs relative to the board distance taken from well recovered tracks.

The relative distance target/flight board can be computed only with the knowledge of the structure of the flight board surface. Therefore, the following section proposes a method to reconstruct the flight board surface under the constraints of our application.

3.2 Surface Reconstruction in Cluttered Conditions

Individual disparity images provided by the stereo camera are noisy, incomplete and do not represent the flight board surface as needed. Therefore, we need to compute from a sequence of disparity images a sample of pixels that represent the surface as complete as possible and without any bee on it. As a paradox, a medium activity constitutes an optimal condition to retrieve the structure of the flight board. Indeed, zero activity means that untextured parts remain untextured over all the sequence, which is not good for the disparity computation. In contrast, an overcrowded flight board makes it hard to distinguish the surface. In the following we propose a method that iteratively filters pixels from a sequence of disparity images that represent the best the flight board structure.

A region of interest corresponding to the flight board is defined manually, and the flight board is isolated from the rest of the image. A median filter is applied on each disparity image in order to filter inconsistent disparities due to stereo matching errors. A model is initialized with the first image of the sequence. Then image after image, we update each pixel of the model with the greater disparity value between the model and the current image. Figure 3 shows step by step the flight board being extracted. N is the number of disparity images used. It has to be high enough to include a sample of complementary depth maps that could be robust to the clutters by passing through it. But, due to the maximization of the disparity, an overestimation of N increases the chances to add noise (wrong disparity values that have not been filtered by the median filter). An optimal number of iterations (or number of disparity images to use) can be estimated by finding the minimum standard deviation of the depth values on the board over iterations. Figure 5 shows an example with 180 as the optimal number of iterations which corresponds to a flight board approximately located at 380 mm from the camera.

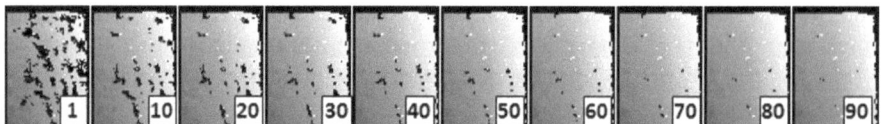

Fig. 3. Step by step flight board reconstruction (90° rotated, partial top view). The numbers are the number of disparity images used. The darker the closer from the camera. Black areas correspond to unavailable disparity information.

As the model still contains holes and inconsistencies, we then apply a fitting based on the locally weighted scatterplot smoothing method (LOWESS) regression modeling method [5] with the Tukey's bisquared function for outliers resistance. Figure 4 shows a mesh reconstructed from the surface model representing a curved flight board.

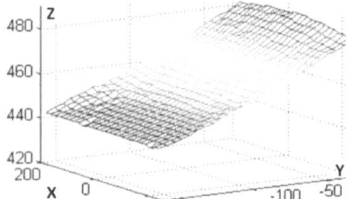

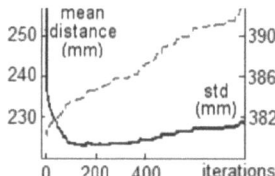

Fig. 4. Reconstructed surface model representing a curved flight board

Fig. 5. Standard deviation (blue line) and mean values (red dashed line) of depth values of the flight board under reconstruction

4   Results and Discussion

Since no trajectory ground truth is available for our application, we based our evaluation on simulated bee trajectories using a multi-agent approach. Our simulator is based on the following assumptions: Bees tend to slow down nearby the flight board and also when changing direction. Bees located at the limit of the field of view can reach a maximum speed of 3 cm/step. We simulated 200 trajectories (100 landings, 100 takeoffs) with a constant number of 15 bees in the field of view. At each step of the simulation (or frame), 3 false alarms are added, and each bee has 10% changes to be undetected. Those conditions of simulation result in similar data obtained from the terrain under challenging conditions. Figure 6 shows a set of simulated/recovered trajectories with respect to a real reconstructed flight board. The shape of the trajectoires are unique but still tend to follow a similar pattern.

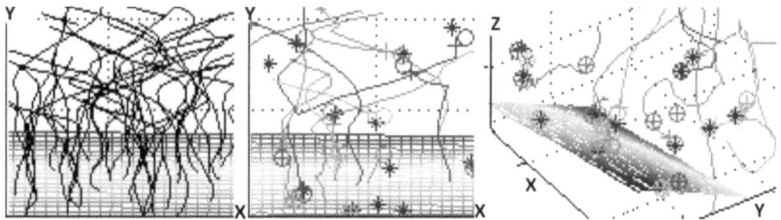

Fig. 6. Top view of simulated trajectories (left) based on a real recovered flight board, and simulated bees tracked with our normalized method (center). 3D view of bees tracked from a real capture with our normalized method (right). Crosses and circles respectively correspond to observations and predicted targets position.

Figure 7 illustrates the effect of our normalization on the relation association cost / distance from the flight board. The normalization model is tuned as followed: Area A with a polynomial function having as coefficients 0.00048, -0.0029 and 0.038 and Area B with an affine function having as coefficients 0.0013 and 1. Therefore, normalized costs can be compared to a constant ACL.

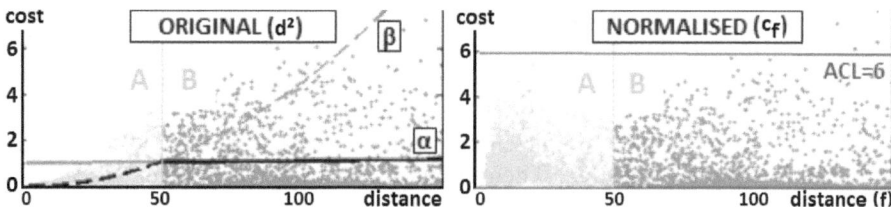

Fig. 7. Association cost / distance to the flight board relation between the original (unnormalized) and our normalized method. The lower original costs nearby the board are explained by easily predictable moves. Areas A and B are separated at the distance of l=50 mm. The quadratic function α (dashed curve) and the affine function β (plain line) are obtained respectively by fitting of points of A and B. As an example, ACL=6.

Figure 8 shows comparative results between the original and our normalized method according to different ACL. The classic method recovers at most 60% of the tracks (ACL=7), where our normalized method recovers 71% (ACL=13). Figure 9 shows the importance of recovering accurately the flight board when using our normalized method. Considering an optimal ACL of 13, the result decreases when adding a positive or negative error to the depth of the flight board surface.

If the flight board surface can not be well recovered (e.g. too much activity), it is preferable to manually lower the flight board estimation (positive depth error in Figure 9) to take in consideration the layer of walking bees. This layer measures up to 15 mm, which does not affect results drastically. Without automatic feedback on the quality of the recovered flight board surface, it would be acceptable to consider by default the 15 mm layer of walking bees (being there or not).

The use of a simulation to evaluate tracking results is debatable. Despite our efforts to create a truthful simulator, there will always be differences between simulated and real data. Nevertheless, we confirmed (without quantification) the benefit of our normalization method on real trajectories (see Figure 6). Further studies involving real data and comparisons of simulation methods have already been started.

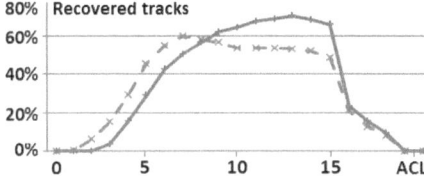

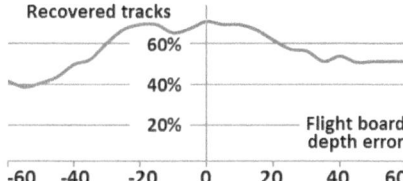

Fig. 8. Comparative results between the unnormalised (dashed line) and our normalized (plain line) methods using different Association Cost Limitation (ACL)

Fig. 9. Effect of a potential wrong flight board recovery (delta depth) on the normalized method based on the optimal ACL, which is 13

5   Conclusion

In addition to our global tracking system based on stereo vision partially presented in [4], we introduced in this article the idea of taking into account targets' environment into the tracking process. The accurate reconstruction of the flight board surface made possible the computation of the relative distance of the bee from its landing/takeoff area. Then our normalization method for association costs made in case of miss detections the tracking process less vulnerable to wrong associations coming from false alarms. Our normalization method provides an improvement of 17.5% on simulated trajectories, which shows the importance taking in account the difference magnitude of association costs when the application allows it. As an example, it works well with our honeybees which tend to follow a typical behavior nearby the flight board.

Comparing costs is a common step in other tracking methods such as MHT, it would be interesting to check the portability of our method. Concerning the flight board reconstruction, our method introduced in Section 3 works with a camera located above the flight board. It could be also interesting to study the possibility of recovering the flight board surface from a non-vertical view.

Concerning long term perspectives, biologists are interested in high level applications such as abnormal behavior detection. On the one hand, behavior models could focus on individual bee trajectories. We can imaging a tracker that takes in consideration more parameters (e.g. activity, weather, time of the day) to adapt motion models. On the other hand, more global models could focus on the colony activity. We can imagine abnormal colony behavior detector based on some basic rules (e.g. low activity during a sunny day).

Acknowledgement. This work has been supported by the European Regional Development Fund (contract : 35053) and Poitou-Charente region. We would like to thanks the Apilab company for making hives available during the system conception.

References

1. Balch, T., Khan, Z., Veloso, M.: Automatically tracking and analyzing the behavior of live insect colonies. In: 5th International Conference on Autonomous Agents, vol. 2001, pp. 521–528. ACM (2001)
2. Blackman, S., Popoli, R.: Design and Analysis of Modern Tracking Systems. Artech House Radar Library, Artech House (1999)
3. Campbell, J., Mummert, L., Sukthankar, R.: Video monitoring of honey bee colonies at the hive entrance. In: VAIB 2008, vol. 8, pp. 1–4 (2008)
4. Chiron, G., Gomez-Krämer, P., Ménard, M.: Outdoor 3d acquisition system for small and fast targets. application to honeybee monitoring at the beehive entrance. In: GEODIFF 2013, pp. 10–19 (2013)
5. Cleveland, W.S.: LOWESS: A Program for Smoothing Scatterplots by Robust Locally Weighted Regression. The American Statistician 35(1), 54–54 (1981)
6. Hendriks, C., Yu, Z., Lecocq, A., Bakker, T., Locke, B., Terenius, O.: Identifying all individuals in a honeybee hive - progress towards mapping all social interactions. In: VAIB 2012 (2012)
7. Henry, M., Beguin, M., Requier, F., Rollin, O., Odoux, J.F., Aupinel, P., Aptel, J., Tchamitchian, S., Decourtye, A.: A common pesticide decreases foraging success and survival in honey bees. Science 336(6079), 348–350 (2012)
8. Kalman, R.E., et al.: A new approach to linear filtering and prediction problems. Journal of Basic Engineering 82(1), 35–45 (1960)
9. Khan, Z., Balch, T., Dellaert, F.: Efficient particle filter-based tracking of multiple interacting targets using an mrf-based motion model. In: IEEE/RSJ International Conference on Intelligent Robots and Systems, vol. 1, pp. 254–259. IEEE (2003)
10. Khan, Z., Balch, T., Dellaert, F.: A rao-blackwellized particle filter for eigentracking. In: Computer Vision and Pattern Recognition, vol. 2, pp. II–980 (2004)
11. Kimura, T., Ohashi, M., Okada, R., et al.: A new approach for the simultaneous tracking of multiple honeybees for analysis of hive behavior. Apidologie 42(5), 607–617 (2011)
12. Maitra, P., Schneider, S., Shin, M.: Robust bee tracking with adaptive appearance template and geometry-constrained resampling. In: 2009 Workshop on Applications of Computer Vision (WACV), pp. 1–6. IEEE (2009)
13. Miranda, B., Salas, J., Vera, P.: Bumblebees detection and tracking. In: VAIB 2012 (2012)
14. Portelli, G., Ruffier, F., Roubieu, F.L., et al.: Honeybees' speed depends on dorsal as well as lateral, ventral and frontal optic flows. PLoS ONE 6(5), e19486 (2011)
15. Ristič, B., Arulampalam, S., Gordon, N.: Beyond the Kalman Filter: Particle Filters for Tracking Applications. Artech House Radar Library, Artech House (2004)
16. Theriault, D., Wu, Z., Hristov, N., Swartz, S., Breuer, K., Kunz, T., Betke, M.: Reconstruction and analysis of 3d trajectories of brazilian free-tailed bats in flight. Tech. rep., CS Department, Boston University (2010)
17. Veeraraghavan, A., Chellappa, R., Srinivasan, M.: Shape-and-behavior encoded tracking of bee dances. IEEE Transactions on PAMI 30(3), 463–476 (2008)
18. Viola, P., Jones, M.: Rapid object detection using a boosted cascade of simple features. In: Computer Vision and Pattern Recognition, vol. 1, pp. I–511 (2001)

Classification of Pollen Apertures
Using Bag of Words

Gildardo Lozano-Vega[1,2], Yannick Benezeth[2],
Franck Marzani[2], and Frank Boochs[1]

[1] i3mainz, Fachhochschule Mainz, Lucy-Hillebrand-Strasse 2, 55128 Mainz, Germany
gildardo.lozano@fh-mainz.de
[2] Le2i, Université de Bourgogne, B.P. 47870, 21078 Dijon Cedex, France

Abstract. The taxonomical recognition of microscopic biological particles such as pollen and spores is relevant for medical and aerobiological
applications. Focusing on an accurate and automatic vision-based pollen
recognition system, we propose a method for classification of pollen apertures based on bag-of-words strategy, with the ability of learning new
types from different taxa without the need of new algorithms. Results
demonstrate suitable performance and ability to add new taxa.

Keywords: pattern recognition, classification, local jets, bag of words,
apertures, palynology.

1 Introduction

The accurate recognition of micro-particles in the field of biology is an important
task for the combat of diseases. Knowledge and forecasting of the concentration
of airborne spores and pollen, for example, is useful for diagnosis, medication and
prevention of diseases. Traditional counting methods consist of trapping airborne
samples and counting particles manually. These methods are time consuming,
involve costly and specialized labor which at the same time is highly susceptible
to human error and inconsistency. An important application of recognition of
pollen taxa is allergenic analysis, where it is desirable to identify pollen at genus
level. The task faces a great complexity due to the high diversity, for example,
pollen of the same taxonomic family usually share many similarities even when
they belong to different genus branches. Their morphological and aerobiological
characteristics are the key for taxa recognition, and are widely described within
the palynological literature [1][2]. The main morphological characteristics include
average size, shape, equatorial outline, aperture type and amount, exine and
intine structure, and type of ornamentation. On the other hand, aerobiological
characteristics provide information about the periods of pollination, specifically
for each taxon. This knowledge is highly related with the levels of concentration
of pollen in the air in a specific time slot.

One important discriminating characteristic of pollen is the type and amount
of apertures. Apertures can be observed externally, and due to their considerable
size with respect to the particle, most can be observed by means of typical

A. Petrosino (Ed.): ICIAP 2013, Part I, LNCS 8156, pp. 712–721, 2013.

brightfield microscopes in typical lens magnifications ranging from 10x to 40x. Nevertheless, the appearance of apertures shows high variability due to both numerous taxa, and to the change of orientation of the particle with respect to the observer. Figure 1 shows examples of the same type of aperture seen from different perspectives.

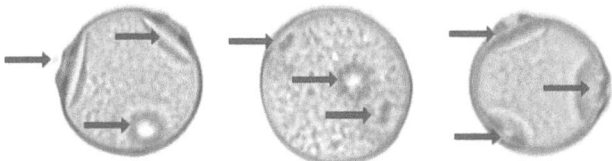

Fig. 1. Examples of aperture of birch pollen marked with arrows. Even though it is the same type of aperture, different appearances can be noticed due to change of the point of view.

The advent of computer vision has brought plausible techniques for fast and robust detection of visual characteristics in order to aid the identification of pollen. Those characteristics, for example shape, texture and color, are employed to describe globally the entire pollen particle in typical recognition systems. Only few works have focused on detecting specific particle structures such as apertures besides the global description of the particle. For instance, the ASTHMA project [3] created individual detectors of aperture, reticulum and cytoplasm. Although no much detail is provided, typical segmentation techniques were employed to detect those regions that were later characterized locally with shape and color features. In the study developed by Chen et al.[4], different methods specific to the pollen type and the point of view were developed in order to detect characteristic patterns formed by the aperture and colpus of birch and mugwort pollen. For circular appearances of pollen apertures, Hough transform and thresholding were used for detection. For the case of apertures located on the boundary of the particle, analysis on the intensity profile of the polar transformation of the particle image, like frequency spectrum and template matching, was conducted. In both cases, measurements on the detected structures were fed together to shape, textures or color features in a larger classifier of pollen taxa. No evaluation was done independently on apertures or other structures. Therefore there is not a clear baseline for comparison. The aforementioned strategies need a particular algorithm for each type of aperture, and in most cases, for different points of view. Consequently, both are too inflexible to learn new type of apertures because completely new algorithms would be needed.

We propose a method for classification of pollen apertures considering intra and inter-class variability. Moreover, it has the ability to add types of apertures from new taxa, without the need of designing a new specific detector. Flexibility is important in order to enable the classifier to learn new varieties of apertures from different points of views with the same method. The method relies on describing apertures in terms of common visual components in a similar way to the Bag of Words approach (BOW)[5]. BOW is a method originating from strategies of text categorization and has been since then widely applied to image processing with several extensions and adaptations [6] [7]. However, to our knowledge, there is no previous application of BOW to microscopic objects. The core concept under the BOW strategy is the creation of a codebook of visual words which are estimated by computing local descriptors at strategic points within the input image. Descriptors with similar values are grouped together in clusters that represent the different visual words. Next, a histogram keeps the frequency of the visual words contained in the image. This particular histogram's bin arrangement characterizes the different regions of interest in the image. It is expected that regions belonging to the same category display a similar pattern in the histogram of words. Consequently, a classifier is employed to learn the different histogram patterns in order to define the classes.

The rest of the paper is organized as follows: In section 2, the method of creating the visual codebook and the differences to typical BOW are explained. In section 3, the setup of the experiments and results are presented. Conclusions about our method and outline of the future work are stated in section 4.

2   Application of Bag of Words to Aperture Description

In this section, we describe how to create the codebook of visual words and how the apertures are represented. The input of the method are image sections of pollen particles, referred in this text simply as regions regardless whether they are apertures or not. The method consists of the following steps: it begins with the subdivision of the regions into a grid of small patches. Next, local descriptors are computed for each patch in order to capture the particular arrangement of pixels, which are subsequently clustered together according to their similarity. These clusters are regarded as the visual words that build the codebook of patterns. Finally, regions are represented by creating their particular histogram of visual words. The overall scheme can be seen in figure 2.

2.1   Computation of Local Descriptors on Region Patches

Our BOW approach is based on the creation of a codebook that keeps the collection of possible visual patterns present in pollen particles. Consistency of the characterization of similar visual patterns is achieved by means of local descriptors. In order to learn these patterns, different regions of the particle must be observed. Contrary to typical BOW, our method divides a region into a grid of small patches of regular size, rather than detecting salient keypoints. Next, local descriptors are computed densely on each of the grid patches. A comprehensive survey of local descriptors can be found in [8]. We have employed the local-jets descriptor, which has demonstrated good performance describing shape, while also being independent from the detection of salient keypoints.

The local-jets descriptor is a differential descriptor designed by Schmid and Mohr [9] from the measures derived by Koenderink and van Doorn [10].

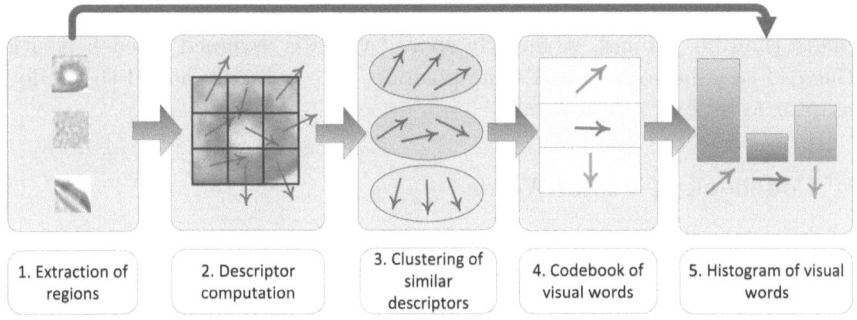

1. Extraction of regions | 2. Descriptor computation | 3. Clustering of similar descriptors | 4. Codebook of visual words | 5. Histogram of visual words

Fig. 2. Overall method of creating the codebook and representing the regions with visual words. 1. Regions are extracted from the particle. 2. Local descriptors are computed on a dense grid. 3. Similar local descriptors are clustered. 4. The codebook is formed with cluster representations. 5. Histograms of visual words is created for all regions.

A local-jet measure L_n is defined as the convolution of the image $I(x, y)$ with a series of Gaussian derivatives $G_n(x, y, \sigma)$ of different scale σ, that is:

$$L_n(x, y, \sigma) := (G_n \star I)(x, y, \sigma). \tag{1}$$

In this equation the Gaussian of order n yields to the local jet L_n. The local-jet measure uses derivatives up to 3rd order. It is also computed at different scales ($\sigma = [0, 0.5, 1]$) in order to provide scale robustness. Finally, all local-jet measures at different scales and orders are concatenated to form a single local-jet descriptor.

Although the local-jets descriptor is not completely rotation invariant because there is no dominant rotation estimation, enough robustness is achieved by training the algorithm with views of the regions at different rotations. Complete scale invariance is not strongly required since the input images are assumed to have been processed with the same scale parameters, and additionally the size of the apertures does not change considerably throughout instances of the same type. However, the local-jets descriptor achieves needed scale robustness since the Gaussian derivatives are computed at several scales.

The training set is formed by grouping together all the local-jets descriptors computed from different regions. At this point, it does not matter whether the local-jets descriptors belong to an aperture or not, since we are interested only in learning all of the different possible patterns.

Because the training image size is such that it contains exclusively the aperture, most local descriptors are computed only on patches of the grid that are strongly informative. This allows to reduce the number of computed local descriptors, unlike typical BOW. A second difference to typical BOW is that our method keeps the position of the relative to the region in order to consider spatial information. Some strategies to achieve this, for example those based on pyramid matching, focus on the computation of abundant descriptors [11][12]. Instead, for our case of fewer regularly spaced local descriptors, we can simply code the relative position of the patches in the grid because the gird has typically no more than 25 patches. A linearly spaced value is assigned to each position of the grid ranging from 0 to 1. The position code is finally added to the local descriptor as an extra dimension.

2.2 Codebook of Visual Words

Because pollen exhibits intra-class variability due to both its biological nature and changes of perspective, similar visual patterns have no identical local descriptor values. In order to identify categorical patterns, local descriptors need to be grouped according to their similarity. For this purpose, k-means clustering is employed where the number of clusters represents the number of learned patterns. This parameter is to be determined experimentally. The collection of clusters, interpreted as visual words, represents the different patterns in the codebook. Subsequently, they will be used as a reference for identifying patterns on unknown instances. Examples of patches with similar patterns for birch pollen are shown in figure 3.

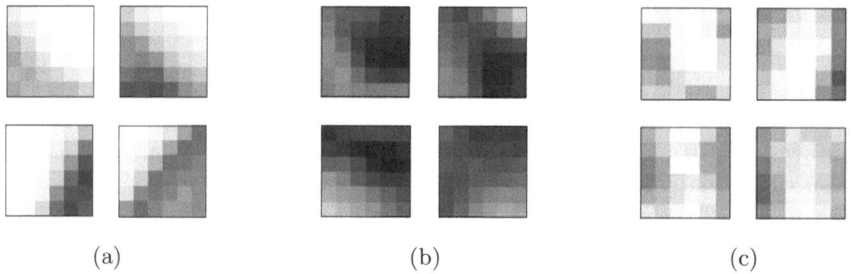

(a)   (b)   (c)

Fig. 3. Example of patches of birch pollen, grouped according to the assigned cluster. Those in the same cluster, show similar patterns.

Apertures and the rest of the regions can now be represented by their content of visual words. During the training phase, words in the codebook are extracted from images of both cases, aperture and not an aperture. A histogram indicating the frequencies of each visual word in the codebook is built for all regions. Given that input regions are initially labeled, a machine learning algorithm is employed to associate similar histogram patterns to the classes "aperture" and "not an aperture".

Different points of view affect directly the look of an aperture. A strategy to solve this problem is to split the task into as many binary classifiers as there are different views of the same type. Later, all classes corresponding to the same aperture are grouped together. A second strategy is to employ more complex algorithms, powerful enough to recognize different views as a single positive class and to differentiate them at the same time from the negative class.

Our method applies Support Vector Machines (SVM) for classifying regional histograms. In recent years, SVM have shown good results in binary classification problems by mapping data into a higher dimension where a linear separation is possible [13]. Later will be experimentally shown that this method is suitable for classification of apertures with multiple points of view. Nevertheless, since the classification stage is independent from the codebook, it is possible to interchange machine learning algorithms.

3   Experiments and Results

We tested our method on the classification of apertures of birch pollen (Betulaceae Betula). Images of pollen were scanned with a brightfield slide scanner Leica SCN400 with magnification 40x. Particles were stained with magenta dye. Images of regions belonging to pollen particles were sampled at specific locations. The birch dataset consisted of 184 regions of which, 92 showed an aperture and 92 did not show an aperture. The classification of an unknown region consists of extracting the visual words contained in the image, based on the previously learnt codebook, and building the corresponding histogram. Finally, the SVM model is applied in order to assign the test region into a class.

The size of the region is a parameter to be evaluated, and it is consistent in all the instances of the set. The size of a region is determined by two factors: the size of the image patch on which an individual local descriptor is computed, and the number of patches in which the grid of image region is split. In all cases, image patches and regions are square. The size of the region is always chosen as a multiple of the image patch size such that the grid fits exactly into the region. Therefore, a square region size is denoted as NxNxP, where N is the number of image patches that fit into any of the region's sides, and P is the number of pixels of the patch size on one side.

Input images were converted to grey levels and the intensity of the average background content was compensated throughout the image by subtraction. The whole set was employed to create the codebook. Then, the local-jets descriptor was computed on the subdivided regions. After computing the word histograms, a binary SVM classification was performed and evaluated using 10-fold cross-validation, labeling the "aperture" class as positive and the "not an aperture" class as negative examples.

The parameters, of both region size and the number of visual words in the codebook were tested for different configurations. The number of clusters, which directly determines the number of visual words, was varied within the range of 3 and 49. Similarly, the region size varied between 20 and 40 pixels for different combinations of NxNxP. Performance was evaluated by means of a Receiver Operating Characteristic (ROC) curve that compared the number of *hits* (recall) versus the number of *false alarms* (fall-out). Figure 4 shows the ROC curve using the best parameters: 4x4x6 and 32 visual words. The best configuration has an accuracy of 95.8% employing a threshold equal to 0.60, having a confidence interval of $\pm$ 2.9 with a 0.95 confidence level. It was found that the variation of

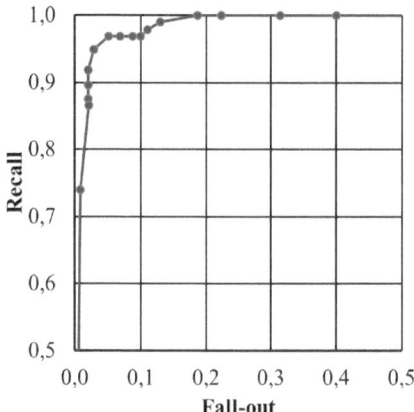

Fig. 4. ROC curve of the BOW performance of classification of birch pollen apertures using best parameters, region size of 4x4x6 and 32 visual words with local-jets descriptor.

the number of visual words does not significantly change the performance. However, the optimal size of the region has to be found such that it fits the typical size of the aperture. Very small or large regions cause a decrease in performance.

We also experimented by substituting the local-jets descriptor. We used a Zernike descriptor, based on Zernike moments, which are particularly useful for describing shape [14][15] and have been applied widely in image retrieval tasks. The descriptor is rotation invariant, and translation invariance can be achieved by image normalization with respect to the centroid. By evaluating both the region size and the number of visual words in the codebook, we found the best performance using a size of 4x4x7 and 5 visual words. The best configuration has an accuracy of 95.4% employing a threshold equal to 0.45, having a confidence interval of ± 3.0 with a 0.95 confidence level. Results confirm the best size for capturing information from apertures and the noncritical number of visual words. Therefore, it is expected that optimal region size changes for different types of apertures. Figure 5 shows the ROC for this configuration.

Tests were also done using the SIFT descriptor in our method [16], but did not produce satisfactory results. We attribute it to the fact that the discriminative power of the SIFT descriptor is effective when working together with the detection of plenty keypoints, rather than in a small grid.

Experimentally, it was confirmed that adding the extra dimension with the spatial information code improves classification results on average 4% compared to the case without it. This difference is statistically significant when compared to the confidence interval of 2.9% and 3.0% of the methods using local-jets and Zernike features respectively.

Finally, in order to test the ability of the method to learn apertures from new taxa, we evaluated our method on the classification of apertures of two types of pollen at the same time. An additional dataset of alder pollen (Betulaceae

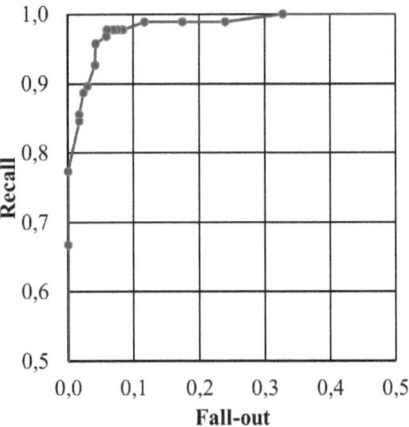

Fig. 5. ROC curve of the BOW performance of classification of birch pollen apertures using best parameters, region size of 4x4x7 and 5 visual words with Zernike descriptor.

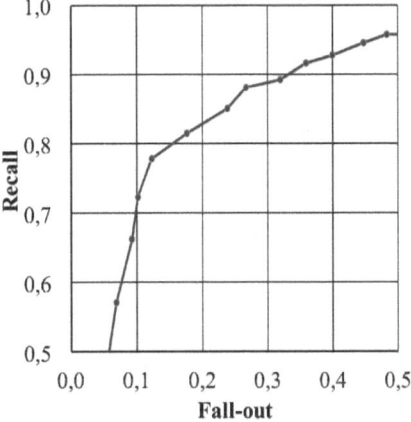

Fig. 6. ROC curve of the BOW performance of classification of alder and birch pollen apertures using 40 visual words with local-jets descriptor.

Alnus) was employed, which consisted of 146 unstained regions of which 73 showed an aperture and 73 did not show an aperture. Optimal region size was experimentally selected for each type, 4x4x6 for betula apertures and 2x2x20 for alder apertures. The performance ROC curve using the local-jet descriptor is shown in figure 6. The method adjusted very well to the newly added taxon, simply through learning the correct new region size. Best results were obtained when using more than 15 visual words. This reflects the need of the codebook for containing more patterns due to the addition of the new taxon. The main reason of the slight performance reduction compared to one-taxon classification is likely that the alder samples were not previously stained with magenta dye. This sort of preparation of the samples is common in order to increase the contrast of the pollen characteristics. Since previous works focus on the final identification of the pollen types, there is no independent evaluation of the aperture detection. Therefore, there is not a clear baseline to compare our results with.

4   Conclusions and Future Work

We have presented a method for detecting apertures in pollen particles. It was tested on the aperture detection of two different pollen types. The results demonstrate that the method is robust and yet flexible enough to learn more types of apertures. Moreover, the presented BOW approach can be extended to process different types of specialized regions of different biological particles with minimal taxa-specific adjustments, for example in spores.

It was shown that local descriptors together with the modified bag-of-words approach are suitable for describing apertures by capturing the most significant pixel patterns, and are also able to identify them against regions of different contents. Despite the random orientation of pollen particles in the sample slide, the method showed robustness to changes of perspective and rotation. Finally, the spatial relationship among the patterns provided useful information with an improved classification performance.

With the goal of an automatic pollen recognition, future work should involve taxon-specific semantic information of pollen such as spatial location and amount of apertures. It is also planned to study additional identifying information present in different focal image planes (layers).

Acknowledgements. The authors are grateful for their financial support to the Bundesministeriums für Wirtschaft und Technologie in Germany through the 3PGM project under the program Zentrales Innovationsprogramm Mittelstand ID KF2848901FR1, to the Conseil Regional de Bourgogne in France and also to the Fond Europeen de Developpement Regional (FEDER). The authors are also grateful to the project partner Bluestone Technology GmbH for providing pollen images and to the Max Plank Institute for Chemistry for permitting the use of their facilities.

References

1. Erdtman, G.: An Introduction To Pollen Analysis. Chronica Botanica Company, USA (1943)
2. Hesse, M., Halbritter, H., Weber, M., Buchner, R., Frosch- Radivo, A., Ulrich, S.: Pollen Terminology. In: An Illustrated Handbook, Springer, Austria (2009)
3. Boucher, A., Hidalgo, P.J., Thonnat, M., Belmonte, J., Galan, C., Bonton, P., Tomczak, R.: Development of a semi-automatic system for pollen recognition. Aerobiologia 18(3), 195–201 (2002)

4. Chen, C., Hendriks, E.A., Duin, R.P., Reiber, J., Hiemstra, P., De Weger, L., Stoel, B.: Feasibility study on automated recognition of allergenic pollen: grass, birch and mugwort. Aerobiologia 22, 275–284 (2006)
5. Csurka, G., Dance, C., Bray, C., Fan, L., Willamowski, J.: Visual categorization with bags of keypoints. In: Pattern Recognition and Machine Learning in Computer Vision Workshop, ECCV Grenoble, France, pp. 1–22 (2004)
6. Wu, J., Tan, W.-C., Rehg, J.M.: Efficient and Effective Visual Codebook Generation Using Additive Kernels. Journal of Machine Learning Research 12, 3097–3118 (2011), Georgia Institute of Technology
7. López-Sastre, R.J., Tuytelaars, T., Acevedo-Rodríguez, F.J., Maldonado-Bascón, S.: Towards a more discriminative and semantic visual vocabulary. Computer Vision and Image Understanding 115, 415–425 (2011)
8. Mikolajczyk, K., Schmid, C.: A performance evaluation of local descriptors. IEEE Transactions on Pattern Analysis and Machine Intelligence 27, 1615–1630 (2005)
9. Schmid, C., Mohr, R.: Local grayvalue invariants for image retrieval. IEEE Transactions on Pattern Analysis and Machine Intelligence 19(5), 530–535 (1997)
10. Koenderink, J.J., Doorn, A.J.: Representation of local geometry in the visual system. Biological Cybernetics 5(6), 367–375 (1987)
11. Grauman, K., Darrell, T.: Pyramid matching kernel: Discriminative classification with sets of image features. In: Tenth IEEE International Conference on Computer Vision, ICCV 2005, vol. 2, pp. 1458–1465 (2005)
12. Lazebnik, S., Schmid, C., Ponce, J.: Beyond Bags of Features: Spatial Pyramid Matching for Recognizing Natural Scene Categories. In: IEEE Computer Society Conference on Computer Vision and Pattern Recognition, CVPR 2006, vol. 2, pp. 2169–2178 (2006)
13. Byun, H.-R., Lee, S.-W.: Applications of support vector machines for pattern recognition: A survey. In: Lee, S.-W., Verri, A. (eds.) SVM 2002. LNCS, vol. 2388, pp. 213–236. Springer, Heidelberg (2002)
14. Teague, M.R.: Image analysis via the general theory of moments. Optical Society of America 70(8), 920–930 (1979)
15. Vorobyov, M.: Shape Classification Using Zernike Moments. Technical Report. iCamp-University of California Irvine (2011)
16. Lowe, D.: Distinctive image features from scale-invariant keypoints. In: IJCV (2004)

Evaluating Temporal Information for Social Image Annotation and Retrieval

Tiberio Uricchio, Lamberto Ballan, Marco Bertini, and Alberto Del Bimbo

Media Integration and Communication Center (MICC)
Università degli Studi di Firenze, Italy

Abstract. Can we use the temporal evolution of annotations in Web images to improve tasks such as annotation, indexing and retrieval? This important question is the main motivation for this work. Typically visual content, text and metadata, are used to improve these tasks. A characteristic that has received less attention, so far, is the temporal aspect of social media production and tagging. The main contribution of this paper is a thorough analysis of the temporal aspects of two popular datasets commonly used for tasks such as tag ranking, tag suggestion and tag refinement, namely NUS-WIDE and MIR-Flickr-1M. The correlation of the time series of the tags with Google searches shows that for certain concepts web information sources may be beneficial to annotate social media.

Keywords: Temporal information, image annotation, image retrieval, image tagging, social media.

1 Introduction

The huge success of sites and applications for creation, sharing and tagging of user-generated media - such as Flickr, Facebook and YouTube - has lead to a strong interest by the multimedia and computer vision communities in researching methods and techniques for annotating and searching social media. Typically visual content, text and metadata, such as geo-tags, are used to improve tasks such as annotation, indexing and retrieval of the huge quantities of media produced every day by the users of such systems. For instance, visual content similarity is used in [15] to perform tag suggestion and image retrieval, tag co-occurrence has been proposed in [19] for tag suggestion, geo-tags have been used in [20] for tag recommendation, content classification and clustering. A recent review of the state-of-the-art in areas related to web-based social communities and social media has been presented in [21], considering in particular the contribution of contextual and social aspects of media semantics to multimedia applications.

A characteristic that has received less attention, so far, is the temporal aspect of social media production. As noted in [2], extracting time information from documents may improve several applications such as hit-list clustering and exploratory search. More recently, several researchers have shown that the temporal information associated to search engine queries (e.g. frequency of query

A. Petrosino (Ed.): ICIAP 2013, Part I, LNCS 8156, pp. 722–732, 2013.
© Springer-Verlag Berlin Heidelberg 2013

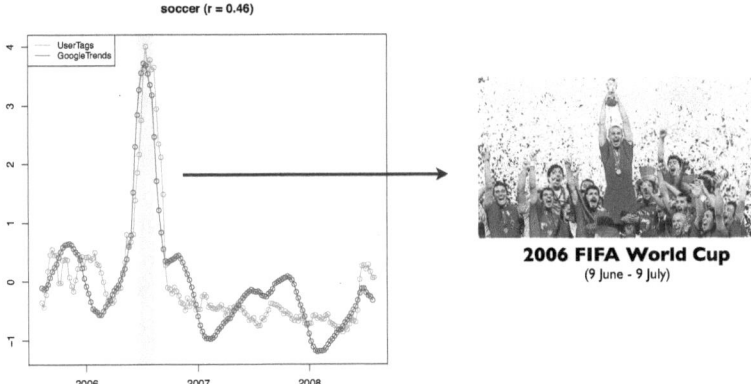

Fig. 1. Time series of user tags and Google searches for "soccer" in NUS-WIDE dataset

keywords over time) can be used to predict trends and behaviors related to economics and medicine, such as claims for unemployment benefits [4], and detection of flu epidemics [7].

In [18] "burst" analysis techniques derived from signal processing are compared against a novel method to identify social events in the associated social media, using the tags and geo-localization information of Flickr images. In [14], the temporal evolution of topics in social image collections is proposed to perform subtopic outbreak detection and to classify noisy social images. The authors used a non-parametric approach in which images are represented using a similarity network, created using Sequential Monte Carlo, where images are the vertices and the edges connect the temporally related an visually similar images. Temporal dynamics of social image collections has been studied in [13] to improve search relevance at query time, addressing both a general case and personalized interest searches. The authors propose a unified statistical model based on regularized multi-task regression on multivariate point process, in which an image stream is considered an instance of a process and a regression problem is formulated to learn the relations between image occurrence probabilities and temporal factors that influence them (e.g. seasons).

Analysis of the temporal evolution of social media collections have been proposed in [10] to predict political success and product sales; regression-based and diffusion-based models have been adapted to account for a Flickr-based index, combining images' metadata and visual similarity, that models the popularity of politicians and products. The work presented in [12] re-casts the problem of image retrieval re-ranking as a prediction of which images will be more likely to appear on the web at a future time point. Both collective group level and individual user level cases are considered, using a multivariate point process to model a stream of input images, and using a stochastic parametric model to solve the relations between the occurrences of the images and factors such as visual clusters, user descriptors and month of the image.

All the datasets used in these works are based on custom selections of user-generated images selected from Flickr, and are not publicly available. The main contribution of this paper is a thorough analysis of the temporal aspects of two "standard" datasets commonly used for tasks such as tag ranking, tag suggestion and tag refinement [16] [15] [23] [17] [3]: NUS-WIDE [5] and MIR-Flickr-1M [9]. These datasets provide images and associated metadata, along with a ground-truth annotation of 81 and 18 tags, respectively. Analysis of the temporal evolution of both user tags and ground-truth tags allows to evaluate the social context (e.g. use of tags related to the semantics associated to social interaction, and not necessarily associated with image content) and visual content (e.g. use of tags that are more strictly related to image content). The correlation of the time series of the tags with Google searches (see Fig. 1) shows that for certain concepts web information sources may be beneficial to annotate social media.

2   Data Analysis Method

2.1   Datasets

To measure the impact of temporal information for image annotation purposes, we performed a quantitative analysis over two image datasets: NUS-WIDE [5] and MIR-Flickr-1M [9].

NUS-WIDE is a large scale dataset collected from Flickr. It contains 269,648 images, provided as multiple visual features and source URLs, with 5,018 tags of which 81 have been manually checked and can be considered ground-truth tags. Tab. 1 reports the classification of these tags according to their main WordNet category. In order to obtain all temporal metadata not contained in the set, we had to download again all the original images from Flickr. Unfortunately, some images are not available anymore, therefore we had to use a subset of 238,251 images that are still present on Flickr. We refer to this subset as NUS-WIDE-240K. Images are unbalanced with respect to time, having very different number of images per date. The time interval goes from year 1900 (old photo scans) to 2009, concentrating most of the images between 2005-2008.

MIR-Flickr-1M is also a large dataset crawled from Flickr which contains 1 million images, selected by their Flickr interestingness score [1] [8]. Every image provided has full *Flickr metadata* which includes *taken* and *posted* timestamps, indicating when a photo was taken and when it was shared on Flickr. However, only about half of the images provide a valid "taken" timestamp, in particular only 584,892 are valid, as 330,454 have no timestamps and 84,654 have an invalid timestamp. Like NUS-WIDE-240K, images are unbalanced with respect to time. Images are concentrated around years 2007-2009. A ground-truth comprised of

Table 1. WordNet categories of NUS-WIDE ground-truth tags

Object	12	Animal	13	Location	2	Substance	2
Action	5	Plant	4	Top	4	Time	2
Artifact	26	Event	4	Phenomenon	4	Person + Groups	3

18 tags is provided for the first 25,000 images only, that compose a subset called MIR-Flickr25K [8].

2.2 Temporal Features

Given a set of images I, all taken in a set of dates D (as a daily interval), we denote as T the set of all tags used and U the set of all users. For every image $i \in I$ we denote tag$(i) \subseteq T$ the set of tags associated, $day(i) \in D$ the timestamp associated and $user(i) \in U$ the user who owns the image. We also consider two other time spans, a set of weeks W and a set of months M, easily computed by integrating over the interval of days considered. These can be thought as time series over the selected index set. For every set considered, we computed a set of features, as proposed in [12]:

- **Images per day**: the number of relevant images which are *taken* in a day. More specifically, given a day $d \in D$, the number of images per day (IMD) is defined as

$$\text{IMD}(d) := |\{i \in I | day(i) = d\}| \tag{1}$$

Similarly we also define a feature for the number of images per week (IMW) and per month (IMM).
- **Images per day for a tag**: the number of relevant images associated with a tag which are *taken* in a day. More specifically, given a tag $t \in T$ and a day $d \in D$, the number of images with t per day (ITD) is defined as

$$\text{ITD}(t, d) := |\{i \in I | day(i) = d \wedge t \in tag(i)\}| \tag{2}$$

Similarly we also define a feature per week (ITW) and per month (ITM).

However, a phenomenon associated with a social source is that of *batch tagging*: a user may decide to upload an entire album of photos and, instead of carefully tagging each photo, he could simply opt to tag each photo with the same tags (e.g. tag the album instead of every single photo). This may result in a kind of noise with respect to the normal use of tags in time. In addition, the features defined above are sensitive to this kind of noise, producing noisy peaks over single days. To produce a more meaningful analysis we decide to collapse all images that are batch tagged into a single entry. A set of images are considered *batch tagged* if they are all uploaded by the same user on the same day and have the same set of tags. More specifically, given a user $\hat{u} \in U$, a day $\hat{d} \in D$ and a set of tags $\hat{t} \subseteq T$, a set of images $I_B = \{i_1, i_2, \ldots, i_k\}$ are considered *batch tagged* if tag$(i) = \hat{t}$, user$(i) = \hat{u}$, day$(i) = \hat{d}$ $\forall i \in I_B$.

2.3 Flickr Popularity Model

As described in [10], available images from the two datasets are only a sample of all images in Flickr. In addition, the number of images over time in Flickr are mostly variable, based on the popularity of the site itself. This slow change over time can be modeled as a trend over all tags, independent from any particular query. Unfortunately, no statistics are released publicly and other sources such as Alexa[1] or Google Trends[2] are affected by the impact of news. Based on this preliminary analysis and supposing an uniform sampling in Flickr searches, we use the feature IMD to remove this background deviation by normalizing the ITD feature.

Given a tag $t \in T$ and a date $d \in D$ we compute:

$$\overline{ITD}(t,d) = \frac{ITD(t,d)}{IMD(d)} \tag{3}$$

This may also be considered as a frequentist probability distribution of tag t in day d with respect to all other tags considered, which is $p(t; d)$. Similarly we also compute $\overline{ITW}$ and $\overline{ITM}$ by considering a week and a month granularity, respectively. After collapsing all batch tagged images, the two datasets retain 179,128 images for NUS-WIDE-240K and 531,670 images for MIRFLICKR-1M respectively.

2.4   Processing

First of all we present a qualitative analysis by measuring the occurrence of tags in time. Given that NUS-WIDE-240K has the biggest ground truth of all datasets considered and that we are looking to discover the relations between tags and image content with respect to time, we choose to use it as the main reference. We use all the 81 manually checked tags as T set and consider four different information sources which are different in the kind of underlining latent process :

- From NUS-WIDE-240K, for all images, we consider the T set of tags using the **manually validated** tags which constitute the entire ground truth; we refer to this source as **NUS-GT**.
- From NUS-WIDE-240K, for all images, we consider the T set of tags using the **user tags** (e.g. the tags provided by the respective Flickr users); we refer to this source as **NUS-TAGS**.
- From MIRFLICKR-1M, for all images, we consider the T set of tags using the **user tags**; we refer to this source as **MIR-TAGS**.
- Beside image datasets, we also consider a source of temporal query information given by Google Trends. From Google Trends, we have downloaded all available query data for the T set of tags considered; we refer to this source as **GOO-TAGS**.

All sources are to be considered subject to different kinds of noise, in particular all images are highly unbalanced over time, resulting in days with hundreds of images and others with at most ten images. To reduce this effect, we choose to

---

[1] Alexa Internet, Inc. http://www.alexa.com
[2] Google Trends. http://www.google.com/trends

consider only the largest time span with at least 350 images per week. In addition
the two image datasets differ in the time interval which has the most images.
This forced us to use a reduced time interval that we choose as starting from
2005-06-01 and ending in 2008-08-01 for NUS-WIDE-240K (retaining 161,176
images from 179,128) and from 2007-01-01 to 2008-08-01 for MIR-Flickr-1M
(retaining 110,064 images from 531,670). Those filters were processed with a
combination of Python scripts and Google Refine[3]. After this we used the R
package [22] to plot and execute any successive analysis. A plotting of features of
this data revealed an insufficient reduction in noise to be able to clearly visualize
most characteristics pattern. To make the time series patterns more clear, we
computed a simple moving average over all time series, varying the windows size
n from 2 to 10 weeks. For a day time series defined over a time span Ψ for a tag
$t \in T$ is defined as:

$$ITD_n(t, d) = \frac{1}{n} \sum_{i=-n}^{n} \overline{ITD}(t, d+i) \quad \forall d \in \Psi \tag{4}$$

This has the effect to smooth the series, letting to visualize more clearly the
trend. On the other hand, tags which have very sparse frequency tends to be
worsened, so we adjusted the window size empirically, based on visualization
clearness. The final time series are composed of 1,158 and 579 week samples
respectively for NUS-WIDE-240K and MIR-Flickr-1M.

2.5   Correlation Analysis

To exploit the underlining time process and to be able to improve image annotation using temporal information, we need a way to evaluate quantitatively the
possible correlation between sources. This allows us to analyze if a series can be
estimated by another one and how a generalized model may describe the original
time series. To this end we compute a correlation measure over two series. First
of all we standardize all time series: given a time series $X = \{x_i : i \in D\}$, we
compute $x_i = \frac{x_i - \overline{X}}{s}$, where $\overline{X}$ is the sample mean and s is the sample standard
deviation. Even if sample mean and sample standard deviation are sensible to
outliers, those are removed thanks to the filtering and smoothing procedure described above. To evaluate the correlation between two time series, we choose
to use the *sample Pearson correlation coefficient*, often denoted as r. Given two
time series X and Y of n samples, r is defined as the ratio between covariance
and the product of X variance and Y variance:

$$r = \frac{\sum_{i=1}^{n}(x_i - \overline{X})(y_i - \overline{Y})}{\sqrt{\sum_{i=1}^{n}(x_i - \overline{X})^2}\sqrt{\sum_{i=1}^{n}(y_i - \overline{Y})^2}} \tag{5}$$

which is defined in $[-1, 1]$. Values towards the positive or negative end reveal
a strong correlation between the two time series, changing only in the sign.

---

[3] Google Refine. `http://code.google.com/p/google-refine`

We can reformulate it as the mean of the products of the standard scores, which permits us to use standardized time series $\hat{x}_i = \frac{x_i - X}{s_X}$ and $\hat{y}_i = \frac{y_i - Y}{s_Y}$:

$$r = \frac{1}{n-1} \sum_{i=1}^{n} \left(\frac{x_i - X}{s_X} \right) \left(\frac{y_i - Y}{s_Y} \right) = \frac{1}{n-1} \sum_{i=1}^{n} \hat{x}_i \hat{y}_i \qquad (6)$$

Given that the strength of correlation is not dependent on the direction or the sign, we also computed r-square. Unfortunately the interpretation of a correlation coefficient depends heavily on the context and purposes that can't be easily defined at this stage of work. However several works like [6] offered some guidelines which can be used to interpret our analysis, that are reported in Tab. 2.

Table 2. Guidelines for sample Pearson correlation coefficient

Correlation	None	Small	Medium	Strong
Positive	0.0 to 0.09	0.1 to 0.3	0.3 to 0.5	0.5 to 1.0
Negative	-0.09 to 0.0	-0.3 to -0.1	-0.5 to -0.3	-1.0 to -0.5

3   Experiments and Discussion

In the following we will consider both the presence of the tags that have been added by the users that uploaded the images to Flickr (referring to them as "user tags") and the tags that have been manually checked by the creators of NUS-WIDE as referring to visual content of images (referring to them as "ground-truth" tags). In fact, several studies have shown that tags are often ambiguous

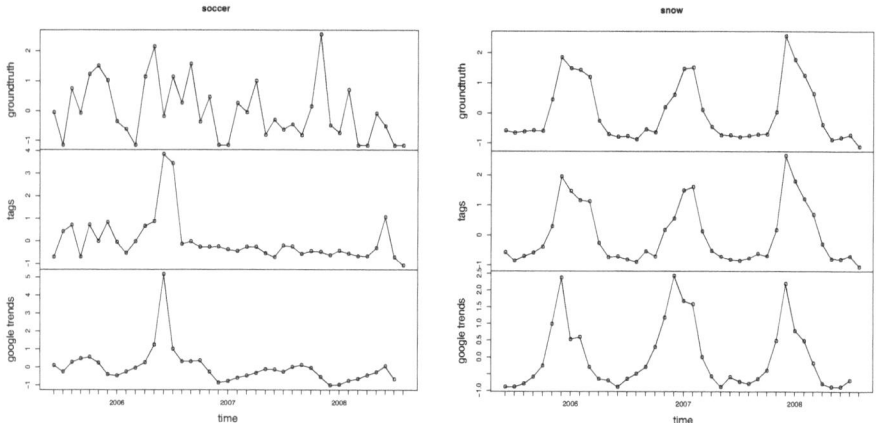

Fig. 2. *left)* frequency of "soccer" in NUS-GT, NUS-TAGS and GOO-TAGS: the peak of Google Trends and user tags in the summer of 2006 are related to the World Soccer Championship; *right)* frequency of "snow" in NUS-GT, NUS-TAGS and GOO-TAGS: the peaks are associated with winter seasons. Tag frequencies have been normalized by the number of images of the same day.

and personalized [11] [19], and do not necessarily reflect the visual content of the image. As an example consider Fig. 2, showing the temporal usage of the tags "snow" and "soccer" in NUS-WIDE, along with the respective Google searches, as obtained from Google Trends. It can be observed that the peak in usage of the "soccer" tag - associated with the 2006 FIFA World Cup - reflects that in Google Trends, but the peak is much less pronounced in the ground truth tags; this indicates that for this tag the relationship between tags and image may exist because of how people react to social events, rather than uploading photos depicting that event on Flickr. On the other hand the peaks of both user and ground truth "snow" tag are corresponding to that of Google Trends: in this case the relationship may exist because it is more likely that people take pictures of snow scenes during winter, and this concept is less related to social aspects than to visual content of these images.

3.1   Temporal Evaluation

Considering time series composed of the frequencies of image tags (either user or ground-truth) and Google searches obtained from Google Trends, it is possible to observe that they exhibit the presence of different components, that may appear mixed together:

trend long term variation, that can be increasing, decreasing or also stable (see Fig. 3 left). Terms such as "computer" or "military" have this pattern;

cyclical variation repeated but not periodic variations. Tags like "sports" or "flags" have this pattern;

seasonal variation periodic variations, e.g. due to concepts associated with some regular event (see Fig. 3 center). Concepts related to seasons show this behavior, like "garden", "snow", "beach" or "frost";

irregular variation random irregular variations, e.g. due to the sudden emergence of a topic (see Fig. 3 right), that appears as a burst of activity. Concepts that exhibit this pattern are related to social or natural events like "soccer", "earthquake" and "protest".

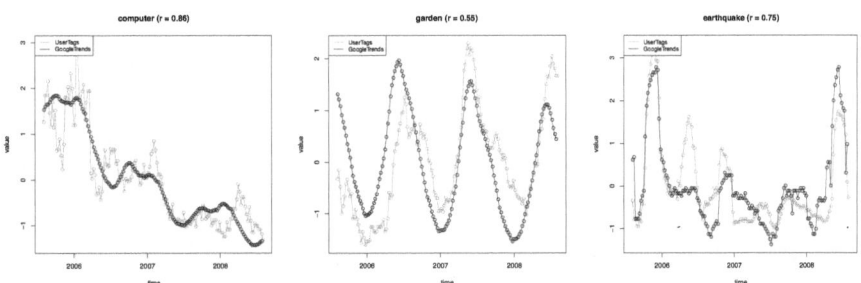

Fig. 3. Time series patterns of NUS-TAGS and GOO-TAGS, averaged over 10 weeks. *left)* trend (computer); *center)* seasonal (garden); *right)* episodic (earthquake: peaks correspond to earthquakes in China and Pakistan).

3.2   Correlation Analysis

Fig. 4 reports the outcome of correlation analysis of NUS-TAGS with NUS-GT, NUS-TAGS with GOO-TAGS and NUS-GT with MIR-TAGS. In particular it can be observed that the correlation of NUS-TAGS and NUS-GT has a vast majority of "Medium" and "Strong" values, while the correlation between user tags and Google searches is overall weaker and can be useful for a selected number of tags. The correlation between NUS-GT and MIR-TAGS has a large number of "Medium" and "Strong" values, suggesting that the temporal information of NUS-WIDE can be used in MIR-Flickr-1M.

Correlation analysis of NUS-TAGS with GOO-TAGS, followed by averaging of r-square values over tags classes (Fig. 5 left) shows that Plant, Event, Phenomenon and Action obtain the higher values. A second group of categories comprises Artifact, Person+Group, Animal, Object and Time. In general, the categories that obtain the best performances are benefitting from tags whose time series show seasonal behaviors (e.g. "snow", "frost", "grass", "leaf") or

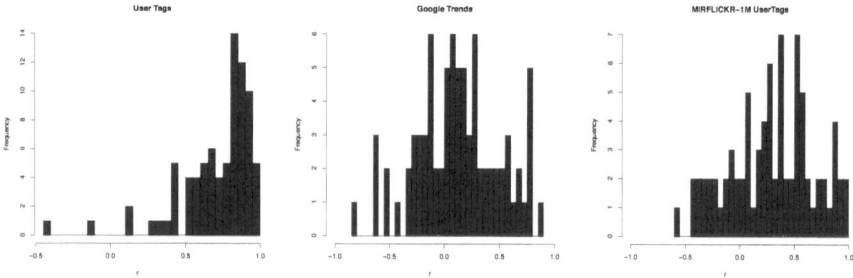

Fig. 4. *left)* r values computed between NUS-TAGS and NUS-GT; *center)* r values computed between NUS-TAGS and GOO-TAGS; *right)* r values computed between NUS-GT and MIR-TAGS

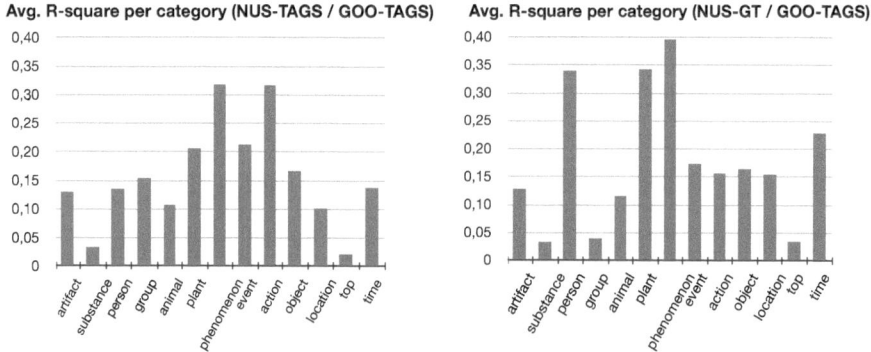

Fig. 5. NUS-WIDE dataset: r-square averages for tags classes. *left)* NUS-TAGS correlation with GOO-TAGS; *right)* NUS-GT correlation with GOO-TAGS.

have a "burst" behavior associated with specific social events (e.g. "soccer", "protest", "earthquake").

Correlation analysis of NUS-GT with GOO-TAGS (Fig. 5 right) shows that Plant and Phenomenon categories maintain their position among the best performing classes, because of the tags that exhibit a seasonal pattern. Instead the correlation of Event and Action categories is lower because the ground-truth tags that have an episodic pattern like "soccer", "protest" and "earthquake" have a lower correlation. This is due to the fact that these tags are employed by users also when the content of the image is not visually related to the described event.

4 Conclusion

This paper presented a thorough analysis of the temporal aspects of user annotations in two popular large-scale datasets. The correlation of the time series of the tags with Google searches showed that for certain concepts web information sources may be beneficial to annotate social media.

References

1. von Ahn, L., Dabbish, L.: Labeling images with a computer game. In: ACM CHI (2004)
2. Alonso, O., Gertz, M., Baeza-Yates, R.: On the value of temporal information in information retrieval. SIGIR Forum 41(2), 35–41 (2007)
3. Uricchio, T., Ballan, L., Bertini, M., Del Bimbo, A.: An evaluation of nearest-neighbor methods for tag refinement. In: IEEE ICME (2013)
4. Choi, H., Varian, H.: Predicting the present with Google Trends. Tech. rep., Google (2011)
5. Chua, T.S., Tang, J., Hong, R., Li, H., Luo, Z., Zheng, Y.: NUS-WIDE: A real-world web image database from National University of Singapore. In: ACM CIVR (2009)
6. Cohen, J.: Statistical power analysis for the behavioral sciences. Routledge Academic (1988)
7. Ginsberg, J., Mohebbi, M.H., Patel, R.S., Brammer, L., Smolinski, M.S., Brilliant, L.: Detecting influenza epidemics using search engine query data. Nature 457(7232), 1012–1014 (2009)
8. Huiskes, M.J., Lew, M.S.: The MIR Flickr retrieval evaluation. In: ACM MIR (2008)
9. Huiskes, M.J., Thomee, B., Lew, M.S.: New trends and ideas in visual concept detection: the MIR Flickr retrieval evaluation initiative. In: ACM MIR, pp. 527–536 (2010)
10. Jin, X., Gallagher, A., Cao, L., Luo, J., Han, J.: The wisdom of social multimedia: using Flickr for prediction and forecast. In: ACM MM, pp. 1235–1244 (2010)
11. Kennedy, L.S., Chang, S.F., Kozintsev, I.V.: To search or to label? Predicting the performance of search-based automatic image classifiers. In: ACM MIR (2006)
12. Kim, G., Fei-Fei, L., Xing, E.P.: Web image prediction using multivariate point processes. In: ACM SIGKDD, pp. 1068–1076 (2012)

13. Kim, G., Xing, E.P.: Time-sensitive web image ranking and retrieval via dynamic multi-task regression. In: ACM WSDM, pp. 163–172 (2013)
14. Kim, G., Xing, E.P., Torralba, A.: Modeling and analysis of dynamic behaviors of web image collections. In: Daniilidis, K., Maragos, P., Paragios, N. (eds.) ECCV 2010, Part V. LNCS, vol. 6315, pp. 85–98. Springer, Heidelberg (2010)
15. Li, X., Snoek, C.G.M., Worring, M.: Learning social tag relevance by neighbor voting. IEEE Transactions on Multimedia 11(7), 1310–1322 (2009)
16. Liu, D., Hua, X.S., Yang, L., Wang, M., Zhang, H.J.: Tag ranking. In: WWW (2009)
17. Liu, D., Yan, S., Hua, X.S., Zhang, H.J.: Image retagging using collaborative tag propagation. IEEE Transactions on Multimedia 13(4), 702–712 (2011)
18. Rattenbury, T., Good, N., Naaman, M.: Towards automatic extraction of event and place semantics from flickr tags. In: ACM SIGIR, pp. 103–110 (2007)
19. Sigurbjörnsson, B., van Zwol, R.: Flickr tag recommendation based on collective knowledge. In: WWW, pp. 327–336 (2008)
20. Sizov, S.: Geofolk: latent spatial semantics in web 2.0 social media. In: ACM WSDM, pp. 281–290 (2010)
21. Sundaram, H., Xie, L., De Choudhury, M., Lin, Y.R., Natsev, A.: Multimedia semantics: Interactions between content and community. Proceedings of the IEEE 100(9), 2737–2758 (2012)
22. Team, R.C.: R: A language and environment for statistical computing. vienna, austria: R foundation for statistical computing; 2008 (2011)
23. Zhu, G., Yan, S., Ma, Y.: Image tag refinement towards low-rank, content-tag prior and error sparsity. In: ACM Multimedia (2010)

VSCAN: An Enhanced Video Summarization Using Density-Based Spatial Clustering

Karim M. Mahmoud, Mohamed A. Ismail, and Nagia M. Ghanem

Computer and Systems Engineering Department
Faculty of Engineering, Alexandria University
Alexandria, Egypt

Abstract. In this paper, we present VSCAN, a novel approach for generating static video summaries. This approach is based on a modified DBSCAN clustering algorithm to summarize the video content utilizing both color and texture features of the video frames. The paper also introduces an enhanced evaluation method that depends on color and texture features. Video Summaries generated by VSCAN are compared with summaries generated by other approaches found in the literature and those created by users. Experimental results indicate that the video summaries generated by VSCAN have a higher quality than those generated by other approaches.

Keywords: Video Summarization, Color and Texture, Clustering, Evaluation Method.

1 Introduction

The revolution in digital video has been driven by the rapid development of computer infrastructure in various areas such as improved processing power, enhanced and cheaper capacity of storage devices, and faster networks. This revolution has brought many new applications and as a consequence research into new technologies that aim at improving the effectiveness and efficiency of video acquisition, archiving, cataloguing and indexing as well as increasing the usability of stored videos. This leads to the requirement of efficient management of video data such as video summarization.

A video summary is defined as a sequence of still pictures that represent the content of a video in such a way that the respective target group is rapidly provided with concise information about the content, while the essential message of the original video is preserved [13].

Over the past years, various approaches and techniques have been proposed towards the summarization of video content. However there are many drawbacks for these approaches. First, most of video summarization approaches that achieved a relatively high quality are based only on a single visual descriptor such as the color of the video frames; while other descriptors like texture is not considered. Second, clustering algorithms used in current video summarization techniques could not detect noise frames automatically; instead some of these techniques have to detect noise frames using separate methods which require

A. Petrosino (Ed.): ICIAP 2013, Part I, LNCS 8156, pp. 733–742, 2013.

additional computation. Third, the current video summarization approaches depend on special input parameters that may not be suitable for all cases. For example, many approaches utilizes k-means partitioning-based clustering algorithm that requires the number of clusters as an input; while the number of clusters is not related to the perceptual content of the automatic video summary. To overcome this problem, additional stage is required to filter key frames which increases the complexity of the video summarization process and makes using the clustering algorithm inefficient. Finally, current evaluation methods depend only on color features for comparing different summaries and do not consider other features like texture; which gives a less perceptual assessment of the quality of video summaries.

In this paper, we present VSCAN, an enhanced approach for generating static video summaries that operates on the whole video clip. It relies on clustering color and texture features extracted from the video frames using a modified DBSCAN [6] algorithm to summarize the video content which overcomes the drawbacks of the other approaches. Also, we introduce an enhanced evaluation method that depends on color and texture features. VSCAN approach is evaluated using the enhanced evaluation method and the experimental results show that VSCAN produces video summaries with higher quality than those generated by other approaches.

The rest of this paper is organized as follows. Section 2 introduces some related work. Section 3 presents VSCAN approach and shows how to apply it to summarize a video sequence. Section 4 illustrates the evaluation method and reports the results of our experiments. Finally, we offer our conclusions and directions for future work in Section 5.

2   Related Work

A comprehensive review of video summarization approaches can be found in [19]. Some of the main approaches and techniques related to static video summarization which can be found in the literature are briefly discussed next.

In [11], an approach based on clustering the video frames using the Delaunay Triangulation (DT) is developed. The first step in this apporach is pre-sampling the frames of the input video. Then, the video frames are represented by a color histogram in the HSV color space and the Principal Component Analysis (PCA) is applied on the color feature matrix to reduce its dimensionality. After that, the Delaunay diagram is built and clusters are formed by separating edges in the Delaunay diagram. Finally, for each cluster, the frame that is closest to its center is selected as the key frame.

In [7], an approach called STIMO (STIll and MOving Video Storyboard) is introduced. This approach is designed to produce on-the-fly video storyboards and it is composed of three phases. In the first phase, the video frames are pre-sampled and then feature vectors are extracted from the selected video frames by computing a color histogram in the HSV color space. In the second phase, a clustering method based on the Furthest-Point-First (FPF) algorithm is applied.

To estimate the number of clusters, the pairwise distance of consecutive frames is computed using Generalized Jaccard Distance (GJD). Finally, a post-processing step is performed for removing noise video frames.

In [3], an approach called VSUMM (Video SUMMarization) is presented. In the first step, the video frames are pre-sampled by selecting one frame per second. In the second step, the color features of video frames are extracted from Hue component only in the HSV color space. In the third step, the meaningless frames are eliminated. In the fourth step, the frames are clustered using k-means algorithm where the number of clusters is estimated by computing the pairwise Euclidean distances between video frames and a key frame is extracted from each cluster. Finally, another extra step occurs in which the key frames are compared among themselves through color histogram to eliminate those similar key frames in the produced summaries.

3   VSCAN Approach

Fig.1 shows the steps of VSCAN approach to produce static video summaries. First, the original video is pre-sampled (step 1). Second, color features are extracted using color histogram in HSV color space (step 2). Third, texture features are extracted using a two-dimensional Haar wavelet transform in HSV color space (step 3). In step 4, video frames are clustered by a modified DBSCAN clustering algorithm. Then, in step 5, the key frames are selected. Finally, the extracted key frames are arranged in the original order of appearance in the video to facilitate the visual understanding of the result. These steps are explained in more details in the following subsections.

3.1   Video Frames Pre-sampling

The first step towards video summarization is pre-sampling the original video which aims to reduce the number of frames to be processed. Choosing a proper sampling rate is very important. A low sampling rate leads to poor video summaries; while a large sampling rate shortens the video summarization time. In VSCAN approach, the sampling rate used is selected to be one frame per second. So, for a video sample of duration one minute, and a frame rate of 30 fps (i.e., 1800 frames); the number of extracted frames is 60 frames.

3.2   Color Features Extraction

In VSCAN, color histogram [18] is applied to describe the visual content of video frames. In video summarization systems, the color space selected for histogram extraction should reflect the way in which humans perceive color. This can be achieved by using user-oriented color spaces as they employ the characteristics used by humans to distinguish one color from another [17,3]. One popular choice is the HSV color space, the HSV color space was developed to provide an intuitive representation of color and to be near to the way in which humans perceive color [3].

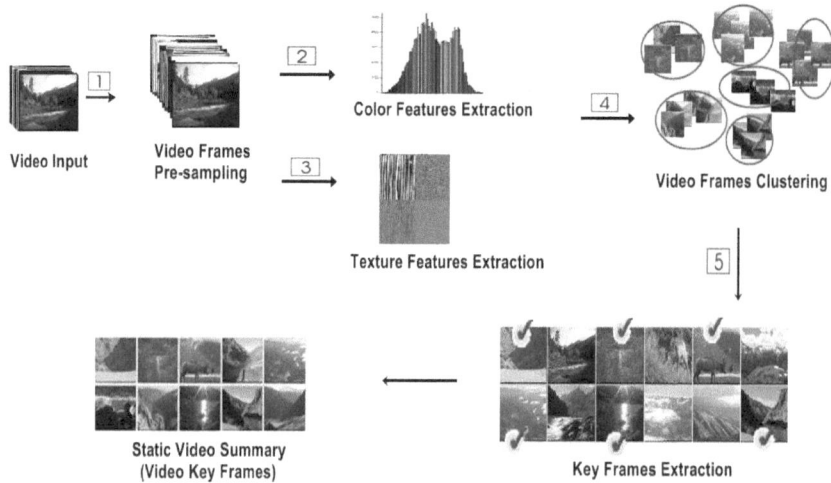

Fig. 1. VSCAN Approach

The color histogram used in VSCAN is computed from the HSV color space using 32 bins of H, 4 bins of S, and 2 bins of V. This quantization of the color histogram is established through experimental tests and aims at reducing the amount of data without losing important information.

3.3 Texture Features Extraction

Texture is a powerful low-level feature for representing images. It can be defined as an attribute representing the spatial arrangement of the pixels in a region or image [1].

Discrete Wavelet Transformation (DWT) is commonly used to extract texture features of an image by transforming it from spatial domain into frequency domain [15]. Wavelet transforms extract information from signal at different scales by passing the signal through low pass and high pass filters. Also, Wavelets provide multi-resolution capability and good energy compaction. In addition, they are robust with respect to color intensity shifts and can capture both texture and shape information efficiently [14].

In VSCAN, Discrete Haar Wavelet Transforms [16] is used to compute feature vectors as a texture representation for video frames, because it is fast to compute and also have been found to perform well in practice. Each video frame is divided into color channels and the Discrete Haar Wavelet Transform is applied on each channel. It is well known that the RGB color space is not suitable to reflect human perception of color [3,8,10]. Instead of using RGB, the video frame image is converted to HSV color space; moreover the video frames size is reduced into 64 X 64 pixels in order to reduce computation without losing significant image information. Next step is applying a two-dimensional Haar Wavelet transform on the reduced HSV image data with decomposition level 3. Finally the texture features of the video frames are extracted from the approximation coefficients of the Haar Wavelet Transforms.

3.4 Video Frames Clustering

DBSCAN (density-based spatial clustering of applications with noise) [6] is a density-based algorithm which discovers clusters with arbitrary shape using minimal number of input parameters. The input parameters required for this algorithm is the radius of the cluster (Eps) and the minimum points required inside the cluster (Minpts) [6].

Using DBSCAN clustering algorithm has many advantages. First, it does not require specifying the number of clusters in the data a priori, as opposed to partitioning algorithms like k-means [12]. Second, DBSCAN can find arbitrarily shaped clusters. Third, it has a notion of noise. Finally, DBSCAN requires minimal number of input parameters.

In VSCAN, we apply a dual feature space DBSCAN algorithm. The proposed clustering algorithm used in VSCAN aims at adapting and modifying DBSCAN to be used by a video summarization system that utilizes both color and texture features. Instead of accepting only one input dataset as in the original DBSCAN, the clustering algorithm in VSCAN accepts both color and texture features of video frames as input datasets, with the Bhattacharya distance [9] as a dissimilarity measure. The Bhattacharyya distance between two discrete distributions P and Q of size n; is defined as:

$$BhattacharyyaDistance = \sum_{i=0}^{n} \sqrt{\sum Pi \bullet \sum Qi} \qquad (1)$$

Selecting the Bhattacharyya distance as dissimilarity measure has many advantages [2]. First, the Bhattacharyya measure has a self-consistency property, as by using the Bhattacharyya measure all Poisson errors are forced to be constant therefore ensuring the minimum distance between two observations points is indeed a straight line [2]. The second advantage is the independence between Bhattacharyya measure and the histogram bin widths, as for the Bhattacharyya metric the contribution to the measure is the same irrespective of how the quantities are divided between bins; therefore it is unaffected by the distribution of data across the histogram [2]. Third advantage is that the Bhattacharyya measure is dimensionless; as it is not affected by the measurement scale used [2].

The values of the Bhattacharyya distance between features vectors of two frames p and q, occurs between 0 and 1; where 0 means that two frames are completely not similar and 1 means that they are exact similar.

Original definitions of DBSCAN Algorithm can be found in [6]. Following are the definitions of the proposed video clustering algorithm used by VSCAN .

Definition 1. (CD- Color database) *a database containing color features extracted from video frames.*

Definition 2. *(TD- Texture database) a database containing texture features extracted from video frames.*

Definition 3. *(EpsColor-color-based similarity of video frame)*
EpsColor-color-based similarity of video frame p, denoted by $S_{EpsColor}(p)$ is defined by: $S_{EpsColor}(p) = \{q \in CD | BhatDist(p, q) \geq EpsColor\}$, where BhatDist is Bhattacharyya distance.

Definition 4. *(EpsTexture-texture-based similarity of video frame)*
EpsTexture-texture-based similarity of video frame p, denoted by $S_{EpsTexture}(p)$ is defined by: $S_{EpsTexture}(p) = \{q \in TD | BhatDist(p, q) \geq EpsTexture\}$, where BhatDist is Bhattacharyya distance.

Definition 5. *(Eps composite similarity score of a video frame)*
Eps composite similarity of a video frame p, denoted by $S_{Eps}(p)$ is defined by: $S_{Eps}(p) = \{q \in CD, TD | score(p, q) = Eps\}$, where Eps & $score(p, q) \in \{0, 1, 2\}$ in which possible values are defined as follows:

> **0** : *if $p \notin S_{EpsColor}(q)$ AND $p \notin S_{EpsTexture}(q)$, in this case p,q are NOT similar.*
> **1** : *if $p \in S_{EpsColor}(q)$ OR $p \in S_{EpsTexture}(q)$, in this case p,q are color-based similar OR texture-based similar.*
> **2** : *if $p \in S_{EpsColor}(q)$ AND $p \in S_{EpsTexture}(q)$, in this case p,q are color-based similar AND texture-based similar.*

Definition 6. *(Directly-similar) A frame p is directly-similar to a frame q wrt. Eps, MinPts if $p \in S_{Eps}(q)$ and $|S_{Eps}(q)| \geq MinPts$ (**core frame condition**).*

Definition 7. *(Indirectly-similar) A frame p is indirectly-similar to a frame q wrt. Eps and MinPts if there is a chain of frames $p_1, ..., p_n$, $p_1 = q$, $p_n = p$ such that p_{i+1} is directly similar to p_i*

Definition 8. *(Connected-similar) A frame p is connected-similar to a frame q wrt. Eps and MinPts, if there is a frame o such that both, p and q are indirectly-similar to o wrt.Eps and MinPts.*

Definition 9. *(Video cluster) Let D be a database of frames. A cluster C wrt. Eps and MinPts is a non-empty subset of D satisfying the following conditions:*

> – $\forall\, p, q$: *if $p \in C$ and q is indirectly-similar to p wrt. Eps and MinPts, then $q \in C$ (**Maximality**).*
> – $\forall\, p, q \in C$: *p is connected-similar to q wrt. Eps and MinPts (**Connectivity**).*

Definition 10. *(Noise) Let $C_1, ..., C_k$* be the video clusters of the database D. The noise is defined as the set of frames in the database D not belonging to any cluster C_i, i=1,..,k. i.e. Noise = $\{p \in D | \forall i : p \notin C_i\}$

The steps involved in VSCAN clustering algorithm are as follows:

1. Select an arbitrary frame p.
2. Retrieve all frames that are indirectly-similar from p w.r.t. Eps and Minpts.
3. If p is a core frame, a video cluster is formed.
4. If p is a border frame, no frames are indirectly-similar from p and the next frame of the database is visited.
5. Continue the process until all the frames have been processed.

For clustering the video frames, we apply the proposed clustering Algorithm on extracted color and texture features of the pre-sampled video frames. According to our experimental tests, we set the input parameters in VSCAN algorithm as follows: EpsColor = 0.97, EpsTexture = 0.97, Eps = 2 and Minpts = 1. As per provided definitions, the EpsColor is the Bhattacharyya distance threshold for grouping frames using color features, this means that only frames with Bhattacharyya distance greater than or equal to 0.97 are color-based similar and eligible to be in the same cluster. While EpsTexture is the Bhattacharyya distance threshold for grouping frames using texture features, this means that only frames with Bhattacharyya distance greater than or equal to 0.97 are texture-based similar and eligible to be in the same cluster.

As per the clustering algorithm definitions, setting Eps value to 2 means that video frames are eligible to belong in same cluster if they are color-based similar and texture-based similar. While Minpts input parameter value is the key value for noise elimination mechanism, Minpts in the algorithm is the minimum number of neighbor frames allowed to create a cluster within the current frame, i.e. setting Minpts to 1, means that minimum cluster size equals to 2 and any cluster of size 1 will be considered as a noise. Since we have selected sampling rate of 1 frame per second, setting Minpts to 1 is equivalent to discarding those video segments of duration less than 2 seconds.

3.5   Key Frames Extraction

After clustering the video frames, the final step is selecting the key frames from the video clusters. In this step the noise frames are discarded and then for each cluster the middle core frame in the ordered frames sequence is selected to construct the video summary. According to our experiments we found that this middle core frame usually is the best representative of the cluster to which it belongs.

4   Experimental Evaluation

In this paper, a modified version of an evaluation method Comparison of User Summaries (CUS) described in [3] is used to evaluate the quality of video summaries. In CUS method, the video summary is built manually by a number of users from the sampled frames and the user summaries are taken as reference (i.e. ground truth) to be compared with the automatic summaries obtained by different methods [3].

The modifications proposed to CUS method aims at providing a more perceptual assessment of the quality of the automatic video summaries. Instead of comparing frames from different summaries using color features only as in CUS method, both color and texture features (as in section 3.2 and section 3.3) are used to detect the similarity of the frames. Once two frames are color-based similar and texture-based similar, they are excluded from the next iteration of the comparison process. In this modified CUS version, the Bhattacharya distance is used to detect both color and texture similarity; in this case the distance threshold value for color and texture similarity is set to 0.97.

In order to evaluate the automatic video summary, the F-measure is used as a metric. The F-measure consolidates both Precision and Recall values into one value using the harmonic mean [4], and it is defined as:

$$F\text{-}measure = \frac{2 \times Precision \times Recall}{Precision + Recall} \qquad (2)$$

The Precision measure of video summary is defined as the ratio of the total number of color-based similar frames and texture-based similar frames to the total number of frames in the automatic summary; and the Recall measure is defined as the ratio of the total number of color-based similar frames and texture-based similar frames to the total number of frames in the user summary

VSCAN approach is evaluated on a set of 50 videos selected from the Open Video Project [1]. All videos are in MPEG-1 format (30 fps, 352   240 pixels). They are distributed among several genres (documentary, historical, lecture, educational) and their duration varies from 1 to 4 min. Also, we use the same user summaries used in [3] as a ground-truth data. These user summaries were created by 50 users, each one dealing with 5 videos, meaning that each video has 5 summaries created by five different users. So, the total number of video summaries created by the users is 250 summaries and each user may create different summary.

For comparing VSCAN approach with other approaches, we used the results reported by three approaches: VSUMM [3], STIMO [7], and DT [11]. In addition to that, the automatic video summaries generated by our approach were compared with the OV summaries generated by the algorithm in [5]. All the videos, user summaries, and automatic summaries are available publicly [2].

In addition to previous approaches, we implemented a video summarization approach called DB-Color using the original DBSCAN algorithm with the color features only as an input. We used the same color features extraction method used in the proposed VSCAN approach as in section 3.2 and also the same input parameters for color similarity and noise detection as in section 3.4, i.e. Eps = 0.97 and Minpts = 1. The reason for implementing DB-Color is to test the effect of using color only instead of combining both color and texture as in VSCAN.

---

[1] Open Video Project. http://www.open-video.org
[2] http://sites.google.com/site/vscansite/

Table 1. Mean F-measure achieved by different approaches

Approach	OV	DT	STIMO	VSUMM	VSCAN	DB-Color
Mean F-Measure	0.67	0.61	0.65	0.72	**0.77**	0.74

Table 1 shows the mean F-measure achieved by the different video summarization approaches. The results indicate that VSCAN performs better than all other approaches. Also, we notice that combining both color and texture features together as done in VSCAN gives better results than using color features only as in DB-Color. However, DB-Color achieved better results if compared to the other four approaches (OV, DT, STIMO, and VSUMM), which indicates that using DBSCAN clustering algorithm is efficient for generating static video summary.

5 Conclusion

In this paper, we presented VSCAN, a novel approach for generating static video summaries. VSCAN utilizes a modified DBSCAN clustering algorithm to summarize the video content using both color and texture features of the video frames. Combining both color and texture features enabled VSCAN to overcome the drawback of using color features only as in other approaches. Also, as an advantage of using a density-based clustering algorithm, VSCAN could overcome the drawback of determining a priori number of clusters; thus, the extra step needed for estimating the number of clusters is avoided. Also, as an advantage of using a modified DBSCAN algorithm, VSCAN can detect noise frames automatically without extra computations.

As an additional contribution, we proposed an enhanced evaluation method based on color and texture matching. The main advantage of this evaluation method is to provide a more perceptual assessment of the quality of automatic video summaries.

Future work includes combining other features to VSCAN approach like edge and motion descriptors. Also, another interesting future work could be generating video skims (dynamic key frames, e.g. movie trailers) from the extracted key frames. Since the video summarization step is usually considered as a prerequisite for video skimming [19], the extracted key frames from VSCAN can be used to develop an enhanced video skimming system.

References

1. IEEE Standard Glossary of Image Processing and Pattern Recognition Terminology, IEEE Std. 610.4-1990 (1990)
2. Aherne, F.J., Thacker, N.A., Rockett, P.I.: The bhattacharyya metric as an absolute similarity measure for frequency coded data. Kybernetika 34(4), 363–368 (1998)

3. de Avila, S.E.F., Lopes, A.P.B., et al.: Vsumm: A mechanism designed to produce static video summaries and a novel evaluation method. Pattern Recognition Letters 32(1), 56–68 (2011)
4. Blanken, H.M., De Vries, A., Blok, H.E., Feng, L.: Multimedia retrieval. Springer, Heidelberg (2007)
5. DeMenthon, D., Kobla, V., Doermann, D.: Video summarization by curve simplification. In: Proceedings of the Sixth ACM International Conference on Multimedia, pp. 211–218. ACM Press (1998)
6. Ester, M., Kriegel, H.P., Sander, J., Xu, X.: A density-based algorithm for discovering clusters in large spatial databases with noise. In: Proceedings of the 2nd International Conference on Knowledge Discovery and Data mining, vol. 1996, pp. 226–231. AAAI Press (1996)
7. Furini, M., Geraci, F., Montangero, M., Pellegrini, M.: Stimo: Still and moving video storyboard for the web scenario. Multimedia Tools and Applications 46(1), 47–69 (2010)
8. Girgensohn, A., Boreczky, J., Wilcox, L.: Keyframe-based user interfaces for digital video. Computer 34(9), 61–67 (2001)
9. Kailath, T.: The divergence and bhattacharyya distance measures in signal selection. IEEE Transactions on Communication Technology 15(1), 52–60 (1967)
10. Liu, T., Zhang, X., Feng, J., Lo, K.T.: Shot reconstruction degree: a novel criterion for key frame selection. Pattern Recognition Letters 25(12), 1451–1457 (2004)
11. Mundur, P., Rao, Y., Yesha, Y.: Keyframe-based video summarization using delaunay clustering. International Journal on Digital Libraries 6(2), 219–232 (2006)
12. Parimala, M., Lopez, D., Senthilkumar, N.: A survey on density based clustering algorithms for mining large spatial databases. International Journal of Advanced Science and Technology 31, 59–66 (2011)
13. Pfeiffer, S., Lienhart, R., Fischer, S., Effelsberg, W.: Abstracting digital movies automatically. Journal of Visual Communication and Image Representation 7(4), 345–353 (1996)
14. Singha, M., Hemachandran, K.: Signal & image processing: An international journal (sipij). Content Based Image Retrieval using Color and Texture 3(1), 39–57 (2012)
15. Smith, J.R., Chang, S.F.: Transform features for texture classification and discrimination in large image databases. In: Proceedings of the IEEE International Conference on Image Processing, ICIP 1994, vol. 3, pp. 407–411 (1994)
16. Stanković, R.S., Falkowski, B.J.: The haar wavelet transform: its status and achievements. Computers & Electrical Engineering 29(1), 25–44 (2003)
17. Stehling, R.O., Nascimento, M.A., Falcao, A.X.: Techniques for color-based image retrieval. Multimedia Mining, 61–82 (2002)
18. Swain, M.J., Ballard, D.H.: Color indexing. International Journal of Computer Vision 7(1), 11–32 (1991)
19. Truong, B.T., Venkatesh, S.: Video abstraction: A systematic review and classification. ACM Transactions on Multimedia Computing, Communications, and Applications (TOMCCAP) 3(1), 3 (2007)

On the Impact of Alterations on Face Photo Recognition Accuracy

Matteo Ferrara, Annalisa Franco, Davide Maltoni, and Yunlian Sun

Department of Computer Science and Engineering, University of Bologna, Cesena, Italy
{matteo.ferrara,annalisa.franco,davide.maltoni,
yunlian.sun2}@unibo.it

Abstract. This work is framed into the context of automatic face recognition in electronic identity documents. In particular we study the impact of digital alteration of the face images used for enrollment on the recognition accuracy. Alterations can be produced both unintentionally (e.g., by the acquisition or printing device) or intentionally (e.g., people modify images to appear more attractive). Our results show that state-of-the-art algorithms are sufficiently robust to deal with some alterations whereas other kinds of degradation can significantly affect the accuracy, thus requiring the adoption of proper detection mechanisms.

Keywords: ICAO, eMRTD, face recognition, image alteration, digital beautification.

1 Introduction

In recent years traditional identity documents have been replaced by electronic documents able to store biometric features to be used for machine-assisted identity verification [1] [2]. With the Berlin resolution (2002), the International Civil Aviation Organization (ICAO) selected the face as the primary globally interoperable biometric characteristic for machine assisted identity confirmation in electronic Machine Readable Travel Documents (eMRTD) [3].

In order to facilitate the automatic identity verification process, the images stored in an ICAO compliant electronic document have to fulfill very restrictive quality standards, i.e. no elements that could compromise the recognition accuracy should be present. A number of indications about the geometric and photometric properties of the face images to be used in e-documents are given in the ISO/IEC 19794-5 standard following the guidelines initially proposed by ICAO. For instance the subject should have a well-controlled pose, a proper lighting, a natural expression, no accessories that could partially occlude some important facial characteristics, etc…

Some of the countries issuing e-documents, acquire the face image of the subject at the enrollment station with a digital camera. Other countries (e.g., Italy) require the user to provide a printed face photo (ID format) to the issuing authority. While in the first case it is enough to verify (manually or in machine-assisted way) ISO/IEC

A. Petrosino (Ed.): ICIAP 2013, Part I, LNCS 8156, pp. 743–751, 2013.

19794-5 compliance [4], in the second case a further validation should be done to ensure that the face in the printed photo has not (intentionally or unintentionally) been altered.

In fact, several problems may raise in different scenarios if an altered face image were included in the document:

- in a verification scenario, for instance in an automatic gate in an airport, the alterations may determine a high rate of false rejections, i.e. the system would not recognize the person thus making necessary the human intervention;
- in a watch-list scenario, where a list of subjects wanted by the police has to be checked in order to raise proper alarms, the presence of altered face images in the documents could imply missing the suspect that, in this case, could intentionally alter the face image to reduce the probability of being identified.

More in detail, possible alterations includes: i) intentional digital alteration of the face image, for instance to obtain a beautification; ii) geometric alterations introduced by the acquisition device or a bad printing process.

This paper presents a study of the effects of digital image alteration on face recognition performance. For the experimentations we used three reference recognition approaches: two commercial systems and one algorithm at the state-of-the-art are used.

The paper is organized as follows: in section 2 the digital alterations considered in the experiments are presented, section 3 details the experiments carried out and finally section 4 draws some conclusions.

2   Digital Face Image Alteration

Digital alterations of face images can be broadly organized in two categories: geometric and appearance-based. The first category includes transformations that are typically introduced by either the acquisition or the printing device (e.g. barrel distortion or change in the image aspect ratio); this kind of alterations are usually unintentional. The second kind of transformation includes all the alterations that can usually be performed by some image processing software. Such alterations are usually intentionally introduced to make the image more attractive; moreover several software applications that allow to simulate plastic surgery interventions are now available. In this section, we will investigate different types of alterations which are likely to be found in practical cases.

2.1   Alterations

For each kind of alteration, we modify the original image at different levels (i.e., with different strength). We use a parameter p to describe the strength. A large value of p denotes a more significant alteration.

Barrel distortion

Barrel distortion is one of the most common types of lens distortions and represents the typical defect that could be introduced by a low quality acquisition device. In this transformation, a barrel distortion with a strength of p is imposed on the original image while preserving the image size. The approach described in [5] has been adopted to implement this transformation. The value of p is increased in steps of 2% from 10% to 20%, i.e. $p \in \{0.10, 0.12, 0.14, 0.16, 0.18, 0.20\}$. An altered image obtained applying the barrel distortion with $p = 0.20$ is shown in Fig. 1 (b).

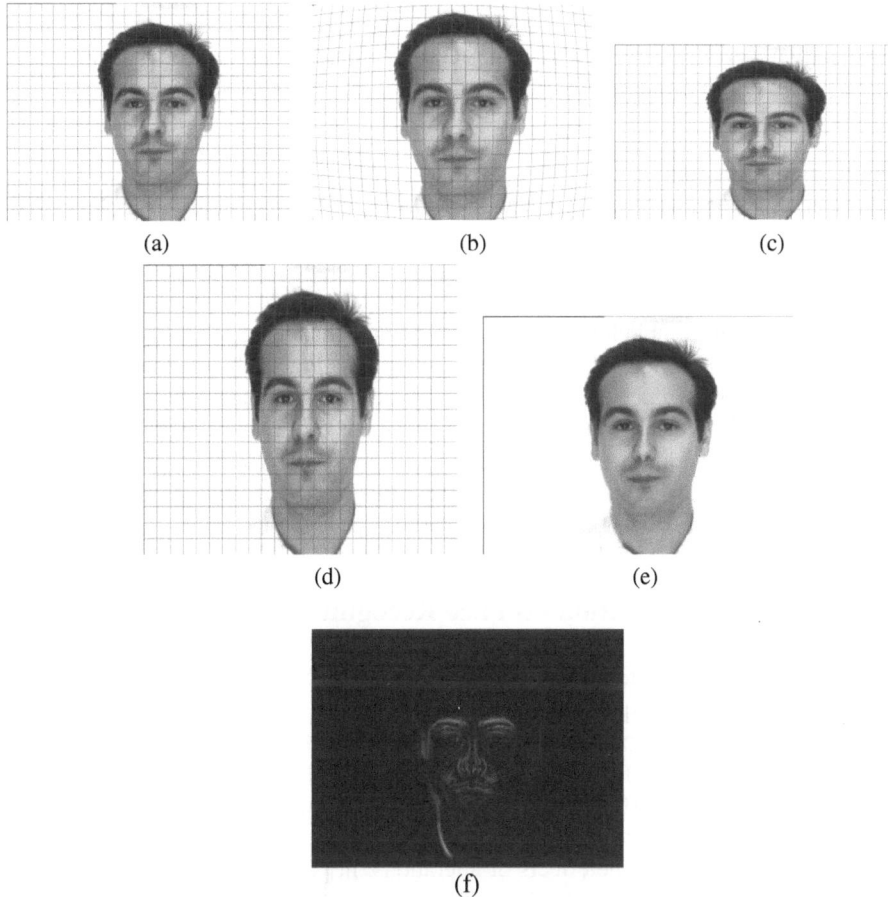

(a)   (b)   (c)

(d)   (e)

(f)

Fig. 1. Example of each alteration. (a) Original image. (b) Altered image with barrel distortion. (c) Altered image with vertical contraction. (d) Altered image with vertical extension. In a), b), c) and d) a squared grid is superimposed to the image to better highlight the effects of geometric alterations. (e) Altered image with digital beatification. (f) Pixel difference between original a) and digitally beautified image e).

Vertical Contraction

In this alteration, we vertically compress the image while keeping the width fixed. In particular we reduce the original height by a multiplying factor of $(1 - p)$. The values of p remain the same to those in the barrel distortion. Fig. 1 (c) shows an altered image after vertical contraction with $p = 0.20$.

Vertical Extension

On the contrary, in vertical extension, the height is increased by a multiplying factor of $(1 + p)$ while keeping the width invariable. Here too we increase the strength of extension from 10% to 20% in a step of 2%. An altered image after vertical extension with $p = 0.20$ is shown in Fig. 1 (d). This alteration (and the previous one) which are essentially a modification of the face aspect ratio, could be unintentionally introduced when processing the image with a photo-editor tool or could be the result of a bad printing.

Digital Beautification

To obtain this alteration, we use LiftMagic [6], an instant cosmetic surgery and anti-aging makeover tool that produces realistic image beautification. The tool presents a very simple web interface that allows to load an image and to simulate different plastic surgery treatments at different levels. It makes available 17 treatments: 16 local treatments (e.g., injectable for forehead, eyelid fold enhancement, lip augmentation, et al.) and one treatment integrating all the local ones. For each treatment, a specific selection bar allows to personalize the strength of the modification. In this alteration, we consider only the integrated treatment and three different strengths obtained by positioning the selection bar at three equidistant positions. We name the three levels 'low', 'medium' and ' high'. In other words, $p \in \{low, medium, high\}$. Fig. 1 (e) presents the altered image after this alteration with $p = high$.

3   Effects of Alteration on Face Recognition Performance

In this section we evaluate the effects of the above described alterations on face recognition accuracy. The experiments have been conducted with three different state-of-the-art face recognition approaches: two commercial software (Neurotechnology VeriLookSDK 2.1 [7] (VL) and Luxand FaceSDK 4.0 [8] (LU)) and a SIFT-based matching algorithm [9] [10] (SI). The performance measured for the three systems on the unaltered database described below are good (see Fig. 2), so they constitute a good test bed to evaluate the effects of alterations: in particular the measured EER is 0.003% (VL), 1.693% (LU) and 2.217% (SI).

3.1   Database

The choice of a proper face database is here an important issue. In fact, in the context of electronic documents, face images are expected to be high quality; hence, variations caused by illumination, expressions, poses, etc. should not be presented in the selected database. The selected database is AR face database [11]. This database consists of 4,000 frontal images taken under different conditions in two sessions, separated by two weeks. The images relevant to our study are well controlled and high quality images (with neutral expressions and good illumination), so the poses 1 and 14 are selected for the tests. We denote them as *No*1 and *No*14 respectively (see Fig. 3 for an example).

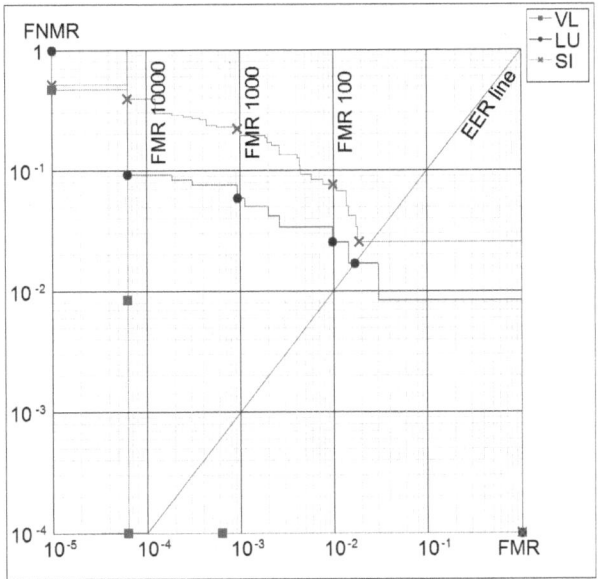

Fig. 2. DET curves of the three reference systems on the unaltered database

Fig. 3. Two unaltered images of the same subject in the AR database (pose 1 on the left, pose 14 on the right)

In our test we assume that the images *No*14 are used during enrollment (i.e., are stored in the e-documents), while the images *No*1 are used as probe (i.e., at the point of verification). The alterations are thus applied to images *No*14 to simulate the inclusion in the document of an altered image.

3.2   Face Recognition Results

To evaluate the effects of the various alterations on face recognition accuracy, a systematic experimentation has been carried out. Starting from the original database, for each alteration, face images with different alteration strength have been generated by modifying the original images with different transformations (see Section 2).

The performance evaluation of face recognition algorithms is based on a set of genuine and impostor recognition attempts. In a genuine recognition attempt, two face images of the same individual are compared, while in an impostor attempt, two images from different persons are compared. In each genuine/impostor attempt, the first image is supposed to have been acquired during an 'enrollment' stage (and included into the document) and the second one during a 'verification' stage. The following performance indicators are used: False Non-Match Rate (FNMR) at a False Matching Rate (FMR) of 1% (FMR_{100}) and 1‰ (FMR_{1000}) [12].

In the following definitions, each database DB consists of two sets of face images: DB_e (acquired during enrollment) and DB_v (acquired during verification).

The original database (without alterations) is denotes as $DB^O = \{DB_e^O, DB_v^O\}$. DB_e^O is made of all the original $No14$s of 120 subjects, while DB_v^O is composed of all the original $No1$s (of 134 subjects). For genuine attempts, each $No14$ is compared against the $No1$ of the same subject; since only 118 subjects have both pose 1 and 14, the number of genuine attempts is 118. For impostor attempts, the $No14$ of one subject and all the $No1$ of the other subjects are compared. Hence, the total number of impostor attempts is 15962.

As to the altered databases, for a given alteration a let $DB_a^p = \{(DB_e)_a^p, DB_v^O\}$ be a database that simulates enrollment face images reporting alteration a with a strength of p. For genuine attempts, the original $No1$ and the altered $No14$ from the same subject are compared. Impostor attempts are the same as in the original database DB^O.

The results of the barrel distortion are reported in Fig. 4. It can be observed that both FMR_{100} and FMR_{1000} change slightly and irregularly as the degree of barrel distortion increases for LU and SI, while there is no significant performance change for VL. Overall this alteration has no noticeable effects on the recognition accuracy.

Fig. 5 and Fig. 6 illustrate the results of the vertical contraction and extension respectively. For both FMR_{100} and FMR_{1000}, as the strength of the alterations increases, the accuracy of LU significantly decreases. SI shows a less noticeable performance drop than LU, while there is no significant performance change for VL.

Finally the results of the digital beautification are reported in Fig. 7. For both FMR_{100} and FMR_{1000}, this alteration produces a performance drop for all the three system (even if LU shows a less noticeable reduction of the recognition accuracy).

Overall the experimental results show that the barrel alteration does not affects significantly the recognition accuracy. This is probably due to the fact that in the central part of the image containing the face, the barrel distortion produces simply a sort of scaling effect, which is well handled by the algorithms analyzed.

Aspect ratio alteration is critical for some approaches (for instance the vertical contraction at the maximum strength causes a performance drop of FMR_{1000} of about 11 times for LU) while it is just slightly disturbing other systems. In particular, we believe that face recognition based on local features only are quite insensitive to global geometric changes.

Finally, alteration such as digital beautification, when applied with high strength, produce marked performance drop to all the system tested.

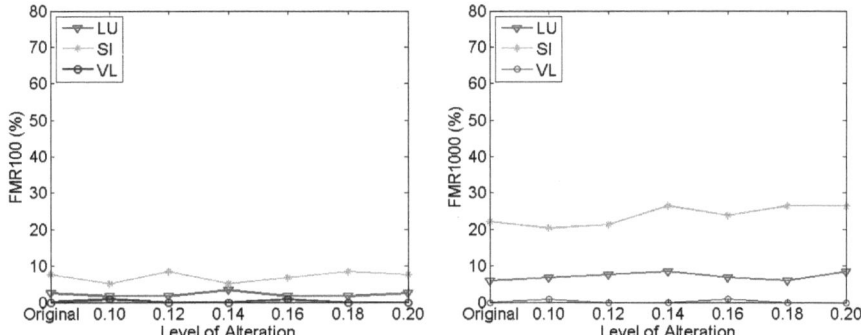

Fig. 4. Performance comparison before and after barrel distortion: FMR_{100} (left) and FMR_{1000} (right)

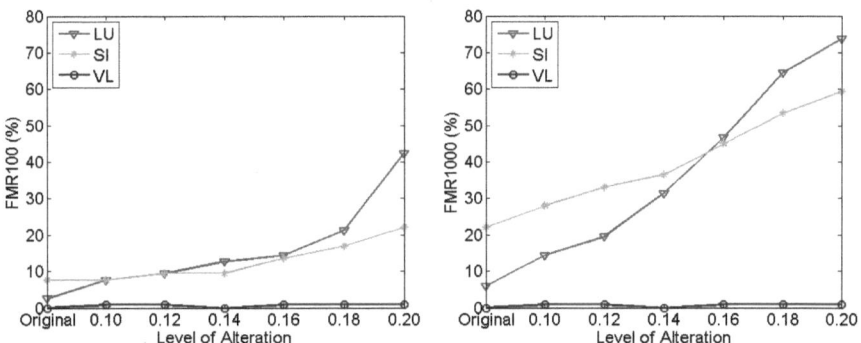

Fig. 5. Performance comparison before and after vertical contraction: FMR_{100} (left) and FMR_{1000} (right)

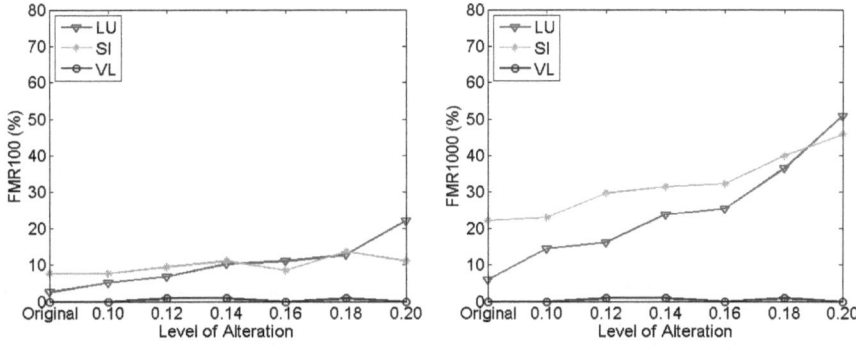

Fig. 6. Performance comparison before and after vertical extension: FMR_{100} (left) and FMR_{1000} (right)

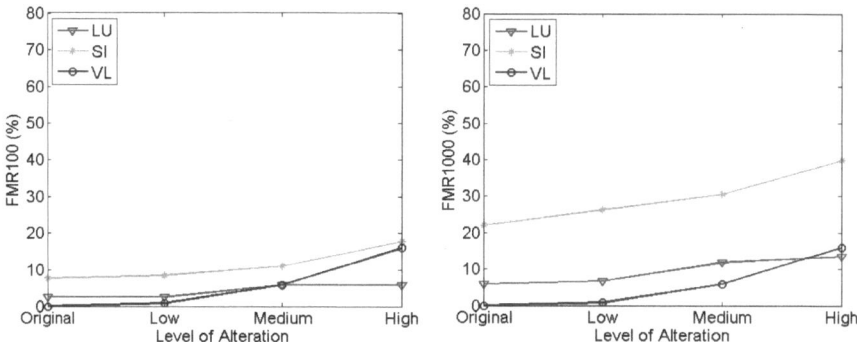

Fig. 7. Performance comparison before and after digital beautification: FMR_{100} (left) and FMR_{1000} (right)

4   Conclusions

The outcome of this study is that state-of-the-art algorithms are able to overcome limited digital alterations but are sensitive to more relevant modifications, thus suggesting that proper countermeasures have to be taken to avoid storing digitally altered photo in eMRTD.

In particular we suggest that authorities issuing e-documents, based on printed ID photos provided by citizens, carefully check these photos to detect intentional or unintentional alterations. To this purpose the officer workstation could be equipped with a software supporting the operator in comparing the scanned version of the ID photo with the live acquired face image. Automatic face recognition system (possibly based on both global and local features) could be used to issue warning in case of low matching scores and a graphical superimposition of the two face image could easily reveal to the officer the presence of alteration.

Acknowledgment. The work leading to these results has received funding from the European Community's Framework Programme (FP7/2007-2013) under grant agreement n° 284862.

References

[1]   Bourlai, T., Ross, A., Jain, A.K.: On matching digital face images against passport photos. In: 2009 International Conference on Biometrics, Identity and Security (BIdS), pp. 1–10 (2009)

[2]   Bourlai, T., Ross, A., Jain, A.K.: Restoring degraded face images for matching faxed or scanned photos. IEEE Transactions on Information Forensics and Security 6(2), 371–384 (2011)

[3]   Biometric Deployment of Machine Readable Travel Documents. ICAO (2004)

[4]   ISO/IEC 19794-5, Information technology - Biometric data interchange formats - Part 5: Face image data (2011)

[5]   Vass, G., Perlaki, T.: Applying and removing lens distortion in post production. In: Proceedings of 2nd Hungarian Conference on Computer Graphics and Geometry (2003)

[6]   LiftMagic - Instant cosmetic surgery and anti-aging makeover tool (2013),
      `http://makeovr.com/liftmagic`

[7]   Neurotechnology Inc. Neurotechnology web site (2013),
      `http://www.neurotechnology.com/`

[8]   Luxand Inc. Luxand Web Site (2013), http://luxand.com

[9]   Lowe, D.G.: Distinctive Image Features from Scale-Invariant Keypoints. International Journal of Computer Vision 60(2), 91–110 (2004)

[10] Bicego, M., Grosso, A., Tistarelli, M.: On the Use of SIFT Features for Face Authentication. In: Conference on Computer Vision and Pattern Recognition Workshop, CVPRW 2006, p. 35 (2006)

[11] Martinez, A.M., Benavente, R.: The AR face database. Computer Vision Center, CVC Technical Report (1998)

[12] Maltoni, D., Maio, D., Jain, A.K., Prabhakar, S.: Handbook of Fingerprint Recognition, 2nd edn. Springer, New York (2009)

An Efficient Indexing Scheme
Based on Linked-Node m-Ary Tree Structure

The-Anh Pham, Sabine Barrat, Mathieu Delalandre, and Jean-Yves Ramel

PolytechTours, 64 Avenue Jean Portalis, 37200 Tours, France
phamtheanh@hdu.edu.vn,
{sabine.barrat,mathieu.delalandre,jean-yves.ramel}@univ-tours.fr

Abstract. Fast nearest neighbor search is a crucial need for many recognition systems. Despite the fact that a large number of indexing algorithms have been proposed in the literature, few of them (e.g., randomized KD-trees, hierarchical K-means tree, randomized clustering trees, and LHS-based schemes) have been well validated on extensive experiments to give satisfactory performance on specific benchmarks. In this work, we propose a linked-node m-ary tree (LM-tree) algorithm, which works really well for both exact and approximate nearest neighbor search. The main contribution of the LM-tree is three-fold. First, a new polar-space-based method of data decomposition is presented to construct the LM-tree. Second, a novel pruning rule is proposed to efficiently narrow down the search space. Finally, a bandwidth search method is introduced to explore the nodes of the LM-tree. Our experiments, applied to one million 128-dimensional SIFT features and 250000 960-dimensional GIST features, showed that the proposed algorithm gives the best search performance, compared to the aforementioned algorithms.

Keywords: Image Indexing, Locality-Sensitive Hashing, Clustering Trees.

1 Introduction

In fact, there has been a great interest of researchers to deal with the fast nearest neighbor search as this task plays a critical role in many computer vision systems such as object matching, object recognition, and CBIR. A comprehensive survey of the indexing algorithms in vector space is presented by Böhm et al. [2]. These algorithms are often categorized into space-partitioning-based, clustering-based, and hashing-based approaches. We will discuss hereafter the most representative indexing algorithms.

For the space-partitioning-based approaches, KD-tree is probably argued as one of the most popular techniques [3]. The basic idea is to iteratively partition the data X into two roughly equal-sized subsets, using a hyperplane perpendicular to a split axis in a D-dimensional vector space. Two new nodes are then created corresponding to these subsets. This process is then repeated for the two subsets until the size of each subset falls below a threshold. Searching for a nearest neighbor of a given query point q is proceeded using a branch-and-bound technique whose the pruning rule works as follows: a node u is selected

A. Petrosino (Ed.): ICIAP 2013, Part I, LNCS 8156, pp. 752–762, 2013.

to explore if its hyper-rectangle *does* intersect the hyper-sphere centered at q with a radius equal to the distance of q to the nearest neighbor found so far. The KD-tree has been shown to work very efficiently for the exact nearest neighbor (ENN) search in low-dimensional space. Several variations of the KD-tree have been investigated to deal with the approximate nearest neighbor (ANN) search. The *Best-Bin-First* search or priority search in [1] is a typical improvement of the KD-tree. The basic idea of the BBF technique is twofold: it limits the maximum number of data points to be searched; and it visits the nodes in the order of increasing distances to the query. The use of priority search was further improved in [12], where the authors proposed to construct multiple randomized KD-trees (RKD-trees) and principal component KD-trees (PKD-trees). The RKD-trees are constructed by selecting the split axis at random from a small set of dimensions having the highest variances. The PKD-trees are constructed in a similar manner but the data are aligned in advance to the principal axes obtained from PCA analysis. A last noticeable improvement of the KD-tree for the ENN search is principal axis tree (PAT-tree) [9]. The PAT-tree extends the KD-tree at twofold. First, it constructs a bigger fanout tree by partitioning the data at each step into m subsets ($m \geq 2$). Second, the split hyper-plane is chosen to be perpendicular to the principal axis of the underlying data. Although the computation of lower bounds in the PAT-tree is quite complicated, the PAT-tree still outperforms many other indexing schemes.

For the clustering-based approaches, Fukunaga et al. [4] proposed to recursively partition the data into smaller regions using the K-means technique. This process terminates when the size of every region falls below a threshold, resulting in a hierarchical K-means tree of the data. Nearest neighbor search is then proceeded using a branch-and-bound algorithm. Muja et al. [10] extended the work of Fukunaga et al. by incorporating the priority search to deal with the ANN task. Particularly, proximity search is proceeded by traversing down the tree and always choosing the node whose cluster center closest to the query. Each time, when we pick up a node for further exploration, the other sibling nodes are inserted to a priority queue, which contains a sequence of nodes stored in the increasing order of the distances to the query. This continues until a leaf node is reached followed by a sequence search for the points contained in this node. Backtracking is then invoked starting from the top node stored in the priority queue. Multiple randomized clustering trees were also explored in [11] by the same authors, where the trees are constructed by selecting the centroids at random. The experiments showed a significant improvement of search performance, compared to other indexing algorithms for many datasets.

For hashing-based approaches, Locality-Sensitive Hashing (LHS) [5] has been known as one of the most popular hashing-based methods, which can perform ANN search with a truly sub-linear time. The key idea of the LHS is to design the hash functions so that the similar points may be hashed with a high probability of collision, while the dissimilar points may likely be hashed with different keys. Given a query, proximity search is proceeded by first projecting the query using the LSH functions. The obtained indices are then used to access the appropriate buckets followed by a sequence search for the data points contained in the buckets. Given a sufficiently large number of hash tables, the LSH can perform ANN search in a truly sub-linear time complexity. Qin Lv et al. [8] introduced *multi-probe* LSH to substantially reduce the number of hash tables, while retaining the same search precision. The basic idea is to search multiple buckets, which probably contain the potential nearest neighbors of the query. In this way, the proposed method reduces the space requirement and increases the chance of finding the true answers. Kulis et al. [7] extended the LSH to the case when the similarity function is an arbitrary kernel function κ: $D(p,q) = \kappa(p,q) = \phi(p)^T\phi(q)$, where $\phi(x)$ is some unknown embedding function. In their work, the LSH hash function is constructed as: $h(\phi(x)) = sign(r^T\phi(x))$, where r is a random hyperplane drawn from $N(0, I)$ and is computed as a weighted sum of a subset of the database feature vectors. As the function $h(\phi(x))$ satisfies the LSH property, the new indexing scheme can perform similarity search in a sub-linear time complexity, while being useful to the cases of kernelized data.

For summary, the hashing-based approaches give a great advantage of search efficiency, but the main drawback is the use of a huge amount of memory to construct the hash tables. In addition, the search precision could be a problem because the true nearest neighbors could be hashed into many adjacent buckets, making the access to a single hash bin insufficient to recover the true answers. The clustering-based approaches have shown quite good performance in a wide range of feature types and dataset sizes [10], [11]. The main disadvantage is the expensive time of the process of tree construction. The space-partitioning-based approaches, particularly the KD-tree-based algorithms, seem to be the most appropriate solutions for all aspects of search precision, search speedup, and tree construction time.

In this work, we propose a linked-node m-ary tree (LM-tree) algorithm, which works really well for both ANN and ENN tasks. Three main contributions are attributed to the proposed LM-tree. First, a new method of data decomposition is presented to construct the LM-tree. Second, a novel pruning rule is proposed to efficiently narrow down the search space. Finally, a bandwidth search method is presented to deal with the ANN task. Our experiments, applied to one million 128-dimensional SIFT features and 250000 960-dimensional GIST features, showed that the proposed algorithm gives a significant improvement of search performance, compared to the baseline indexing algorithms. The rest of this paper is organized as follows: The proposed indexing algorithm is detailed in Section 2. Experimental results are presented in Section 3. We conclude the paper in Section 4.

2   The Proposed Algorithm

2.1   Construction of the LM-Tree

Given a dataset X composing of N feature vectors in a D-dimensional space R^D, we present, in this section, an indexing structure to index the dataset X supporting efficient proximity search. For better presentation of our approach, we use the notation $\mathbf{p}$ as a point in the R^D feature vector space, and p_i as the ith component of $\mathbf{p}$ ($1 \le i \le D$). We also denote $p = (p_{i_1}, p_{i_2})$ as a point in a 2D space. Before constructing the LM-tree, the dataset X is normalized by aligning it to the principal axes obtained from PCA analysis. Note that no dimension reduction is performed in this step. In fact, PCA analysis is only used to align the data. Next, the LM-tree is constructed by recursively partitioning the dataset X into m roughly equal-sized subsets as follows:

- Sort the axes in the decreasing order of variance, and choose randomly two axes, i_1 and i_2, from the first L highest variance axes ($L < D$).
- Project every point $\mathbf{p} \in X$ into the plane $i_1\mathbf{c}i_2$, where $\mathbf{c}$ is the centroid of the set X, and then compute the corresponding angle:

$$\phi = \arctan(p_{i_1} - c_{i_1}, p_{i_2} - c_{i_2})$$

- Sort the angles $\{\phi_t\}_{t=1}^n$ in the increasing order ($n = |X|$), and then divide the angles into m equal-sized sub-partitions: $(0, \phi_{t_1}] \cup (\phi_{t_1}, \phi_{t_2}] \cup \ldots \cup (\phi_{t_m}, 360]$.
- Partition the set X into m subsets $\{X_k\}_{k=1}^m$ corresponding to m angle sub-partitions obtained in the previous step.

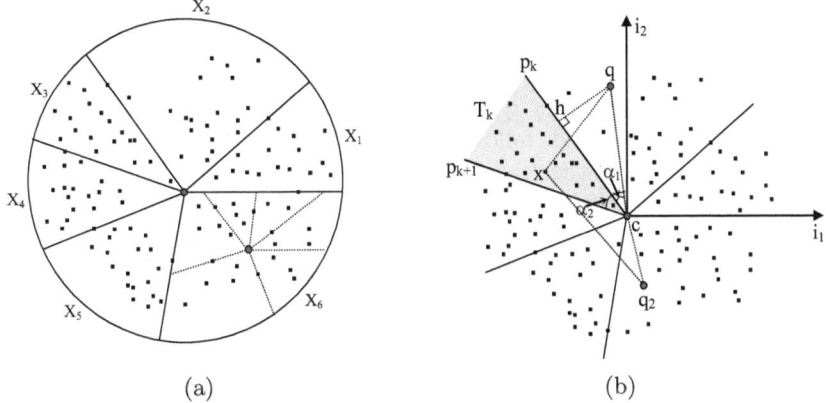

Fig. 1. (a) Illustration of the iterative process of data partitioning in a 2D space: the 1st partitioning applied to the dataset X, and the 2nd partitioning applied to the subset X_6 (the branching factor $m = 6$); (b) Illustration of the lower bound computation

For each subset X_k, a new node T_k is constructed and then attached to its parent node, where we also store the following information: the split axes (i.e., i_1 and i_2), the split centroid (c_{i_1}, c_{i_2}), the split angles $\{\phi_{t_k}\}_{k=1}^m$, and the split projected points $\{(p_{i_1}^k, p_{i_2}^k)\}_{k=1}^m$ where the point $(p_{i_1}^k, p_{i_2}^k)$ corresponds to the split angle ϕ_{t_k}. For efficient access across these child nodes, a direct link is established between two adjacent nodes T_k and T_{k+1} ($1 \le k < m$), and the last one T_m is linked to the first one T_1. Next, we repeat this partitioning process for each subset X_k associated to the child node T_k until the number of data points in each node falls below a pre-defined threshold L_{max}. It is worth pointing that each time, when a partition is proceeded, the two highest variance axes of the corresponding dataset are employed. This is contrast to many existing tree-based indexing algorithms, where only one axis is often employed to partition the data. Consequently, as argued in [12], considering a high-dimensional feature space, such as 128-dimensional SIFT features, the total number of axes involved in the tree construction is rather limited, making any pruning rules inefficient, and the tree less discriminative for later usage of searching. Naturally, the number of principal axes involved in partitioning the data is proportional to both the search efficiency and precision. Figure 1(a) illustrates the first and second levels of the LM-tree construction with a branching factor $m = 6$.

2.2 Exact Nearest Neighbor Search in the LM-Tree

Exact nearest neighbor search in the LM-tree is proceeded using a branch-and-bound algorithm. Given a query point $\mathbf{q}$, we first project $\mathbf{q}$ into a new space using the principal axes as we have processed in the LM-tree construction. Next, starting from the root, we traverse down the tree, and use the split information stored at each node to choose the best child node for further exploration. Particularly, given an internal node u along with the corresponding split information $\{i_1, i_2, c_{i_1}, c_{i_2}, \{\phi_{t_k}\}_{k=1}^m, \{(p_{i_1}^k, p_{i_2}^k)\}_{k=1}^m\}$ which is already stored at u, we first compute an angle: $\phi_{q_u} = \arctan(q_{i_1} - c_{i_1}, q_{i_2} - c_{i_2})$. Next, binary search is applied to the query angle ϕ_{q_u} over the sequence $\{\phi_{t_k}\}_{k=1}^m$ to choose the child node of u closest to the query $\mathbf{q}$ for further exploration. This process continues until a leaf node is reached, following by partial distance search (PDS) [9] to the points contained in the leaf. Backtracking is then invoked to explore the rest of the tree. Each time, when we are positioned at some node u, the lower bound is computed as the distance from the query $\mathbf{q}$ to the node u. If the lower bound exceeds the distance from $\mathbf{q}$ to the nearest point found so far, we can safely avoid exploring this node and proceed with other nodes. In this section, we present a novel rule to compute efficiently the lower bound. Our pruning rule was inspired by the work presented in the principal axis tree (PAT) [9]. PAT is a generalization of the KD-tree, where the page regions are hyper-polygons rather than hyper-rectangles, and the pruning rule is recursively computed based on the law of cosines. The disadvantages of this pruning rule are the computation cost of the complexity (i.e., $O(D)$) and being inefficient when working on a high-dimensional space, due to the fact that only one axis is employed at each partition. As our method of data decomposition (i.e., LM-tree construction) is quite different from that of the KD-tree-based structures, we have developed a significant improvement of the pruning rule used in PAT. Particularly, we have incorporated two following major advantages for the proposed pruning rule:

- The lower bound is computed as simple as in a 2D space, regardless of how large the dimensionality D is. Therefore, the time complexity is just $O(2)$ instead of $O(D)$ as in the case of PAT.
- The magnitude of the proposed lower bound is significantly greater than that in PAT. This enables the proposed pruning rule to work efficiently.

We now come back to the description of computing the lower bound. Let u be the node in the LM-tree at which we are positioned, and T_k be one of the children of u which is going to be searched, and $p_k = (p_{i_1}^k, p_{i_2}^k)$ be the kth split point corresponding to the child node T_k (see Figure 1(b)). The lower bound $LB(\mathbf{q}, T_k)$, from $\mathbf{q}$ to T_k, is recursively computed from $LB(\mathbf{q}, u)$ as follows:

- Compute the angles: $\alpha_1 = \angle qcp_k$ and $\alpha_2 = \angle qcp_{k+1}$, where $q = (q_{i_1}, q_{i_2})$ and $c = (c_{i_1}, c_{i_2})$.
- If one of two angles α_1 and α_2 is smaller than 90^0, we have the following fact due to the rule of cosines [9]:

$$d(q, x)^2 \geq d(q, h)^2 + d(h, x)^2 \tag{1}$$

where x is any point in the region of T_k, and $h = (h_{i_1}, h_{i_2})$ is the projection of q on the line cp_k or cp_{k+1} taking into account that $\alpha_1 \leq \alpha_2$ or $\alpha_1 > \alpha_2$. Then, we applied the rule of lower bound computation in PAT in a 2D space as follows:

$$LB^2(\mathbf{q}, T_k) \leftarrow LB^2(\mathbf{q}, u) + d(q, h)^2 \tag{2}$$

Next, we treat the point $\mathbf{h} = (q_1, q_2, \ldots, h_{i_1}, \ldots, h_{i_2}, \ldots, q_{D-1}, q_D)$ in place of $\mathbf{q}$ in the means of lower bound computation to the descendant of T_k.
- If both angles, α_1 and α_2, are greater than 90^0 (e.g., the point q_2 in Figure 1(b)), we have a more restricted rule as follows:

$$d(q, x)^2 \geq d(q, c)^2 + d(c, x)^2 \tag{3}$$

Therefore, the lower bound is easily computed as:

$$LB^2(\mathbf{q}, T_k) \leftarrow LB^2(\mathbf{q}, u) + d(q, c)^2 \tag{4}$$

Again, we treat the point $\mathbf{c} = (q_1, q_2, \ldots, c_{i_1}, \ldots, c_{i_2}, \ldots, q_{D-1}, q_D)$ in place of $\mathbf{q}$ in the means of lower bound computation to the descendant of T_k.

As the lower bound $LB(\mathbf{q}, T_k)$ is recursively computed from $LB(\mathbf{q}, u)$, an initial value has to be set for the lower bound at the root node. Obviously, we set $LB(\mathbf{q}, root) = 0$. It is also noted that when the point $\mathbf{q}$ is fully contained in the region of T_k, no computation of the lower bound is required.

2.3 Approximate Nearest Neighbor Search in the LM-Tree

Approximate nearest neighbor search is proceeded by constructing multiple randomized LM-trees to account for different viewpoints of the data. The idea of using multiple randomized trees for ANN search was originally presented in [12], where the authors proposed to construct multiple randomized KD-trees. This technique was then incorporated with the priority search and successfully used in many other tree-based structures [10], [11]. Although the priority search was shown to give better search performance, it is certainly subjected to high computation cost because the process of maintaining a priority queue during the online search is rather expensive.

Here, we exploit the advantages of using multiple randomized LM-trees but without using the priority queue. The basic idea is to restrict the search space to the branches not very far from the considering path. In this way, we introduce a specific search procedure, so-called *bandwidth* search, which is proceeded by setting a search bandwidth to every intermediate node of the ongoing path. Particularly, let $P = \{u_1, u_2, \ldots, u_r\}$ be a considering path obtained by traversing on a single LM-tree, where u_1 is the root node, and u_r is the node at which we are positioned. The proposed bandwidth search indicates that for each intermediate node u_i of P $(1 \le i \le r)$, every sibling node of u_i at a distance of $b + 1$ nodes $(1 \le b < m/2)$ on both sides from u_i is no need to be searched. The value b is called search bandwidth.

There is a notable point that when the projected query q is too close to the projected centroid c, all the sibling nodes of u_i should be inspected. In our case, this is happened at the node u_i if $d(q, c) \le \epsilon D_{med}$, where D_{med} is the median value of the distances between c and all projected data points associated to u_i, and ϵ is a tolerate parameter. In addition, in order to obtain a varying range of search precision, we would need a parameter E_{max} of maximum data points to be searched on a single LM-tree. As we are designing an efficient solution dedicated to ANN search, it would make sense to use an approximate pruning rule rather than an exact one. This is not only reduces much of computation cost but also ensures that a larger fraction of nodes will be inspected yet few of them would be actually searched after checking the lower bound. In this way, it increases the chance of reaching the true nodes closest to the query. Particularly, we have used only the formula (4) as an approximate pruning rule as follows:

$$LB^2(\mathbf{q}, T_k) \leftarrow \kappa \cdot (LB^2(\mathbf{q}, u) + d(q, c)^2) \tag{5}$$

where $\kappa \ge 1$ is a pruning factor, which controls the rate of pruning the branches in the tree. This factor can be adaptively estimated during the tree construction given a specific precision and a particular dataset. However, we have fixed this value $\kappa = 2.5$ and shown, in our experiments, that it is possible to achieve satisfactory search performance on many datasets. Recall that the rule (5) requires the computation of $d(q, c)$ in a 2D space, where the point c has been already computed during the offline tree construction. The computation of this rule is thus very efficient.

3   Experimental Results

We evaluated our system versus several representative fast proximity search systems in the literature, including randomized KD-trees (RKD-trees) [10], [12], hierarchical K-means tree (K-means tree) [10], randomized K-medoids clustering trees (RC-trees) [11], and multi-probe LSH algorithm [8]. These indexing systems were well-implemented and widely used in the literature thanks to the open source FLANN library[1]. The source code of our system is also publicly

---

[1] http://www.cs.ubc.ca/~mariusm/index.php/FLANN/FLANN

available at this address[2]. Note that the partial distance search were implemented in these systems in order to improve the efficiency of sequence search at the leaf nodes. Two datasets, ANN_SIFT1M and ANN_GIST1M [6], were used for all the experiments. The ANN_SIFT1M dataset contains a database of one million 128-dimensional SIFT features and a test set of 5000 SIFT features, while the ANN_GIST1M dataset is composed of a database of one million 960-dimensional GIST features and a test set of 1000 GIST features. Since the dimensionality of the GIST feature is very high and our computer configuration is limited (i.e., Windows XP, 2.4G RAM), we were not able to load the full ANN_GIST1M dataset into memory. Consequently, we have used 250000 GIST features for search evaluation. Following the convention of the evaluation protocol used in the literature [1], [10], [11], we computed the search precision and search time as the average measures obtained by running 1000 queries taken from the test sets of the two datasets. To make the results independent on the machine and software configuration, the speedup factor is computed relative to the brute-force search.

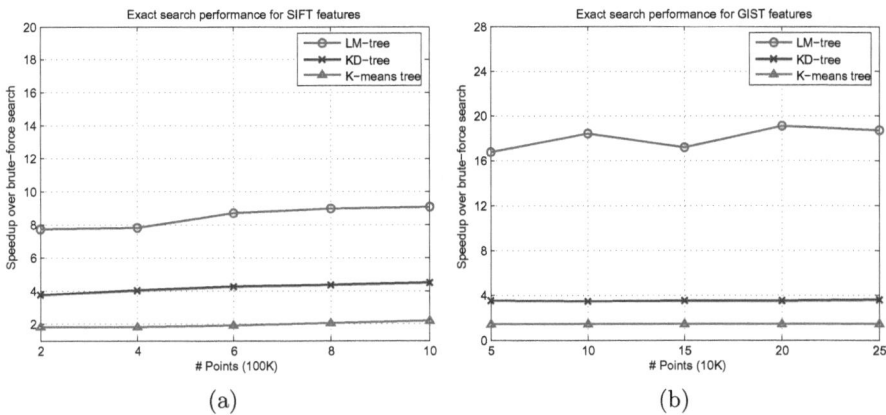

Fig. 2. ENN search performance for the SIFT (a) and GIST (b) features

3.1 ENN Search Evaluation

For ENN search, we set the parameters involved in the LM-tree as follows: $L_{max} = 10$, $L = 2$, and $m \in \{6, 7\}$ with respect to the GIST and SIFT datasets (see Section 2.1). We compared the performance of ENN search of three systems: the proposed LM-tree, the KD-tree, and the hierarchical K-means tree. Figure 2(a) shows the speedup over brute-force search of the three systems, applied to the SIFT datasets with different sizes. We can notice that the LM-tree outperforms the two other systems on all tests. Figure 2(b) presents the search performance of the three systems for the GIST features. The proposed LM-tree again outperforms the others and even performs far better than the SIFT features. Taking the

---

[2] https://sites.google.com/site/ptalmtree/

test where $\#Points = 150000$ on Figure 2(b), for example, the LM-tree gives a speedup of 17.2, the KD-tree gives a speedup of 3.53, and the K-means tree gives a speedup of 1.42 over the brute-force search. These results confirm the efficiency of the LM-tree for ENN search relative to the two baseline systems.

3.2   ANN Search Evaluation

For ANN search, we fixed the following parameters: $L_{max} = 10$, $L = 8$, $b = 1$, and $m \in \{6, 7\}$. Four systems participated in this evaluation, including the proposed LM-trees, RKD-trees, RC-trees, and K-means tree. We used 8 parallel trees in the first three systems, while the last one uses a single tree because it was shown in [10] that the use of multiple K-means trees does not give better search performance. For all the systems, the parameters E_{max} and ϵ (i.e., the LM-tree) are varied to obtain a wide range of search precision. Figure 3(a) shows the search speedup versus the search precision of the four systems for 1 million SIFT features. As we can see, the proposed LM-trees algorithm gives significantly better search performance everywhere than the other systems. Taking the search precision of 95%, for example, the speedups over brute-force search of the LM-trees, RKD-trees, RC-trees, and K-means tree are 167.7, 108.4, 122.4, and 114.5, respectively. To make it comparable with the multi-probe LSH indexing algorithm, we converted the real SIFT features to the binary vectors and tried several parameter settings (i.e., the number of hash tables, the number of multi-probe levels, and the length of the hash key) to obtain the best search performance. However, the result obtained on one million SIFT vectors is rather limited. Taking the search precision of 74.7%, for instance, the speedup over brute-force search (using Hamming distance) is just 1.5.

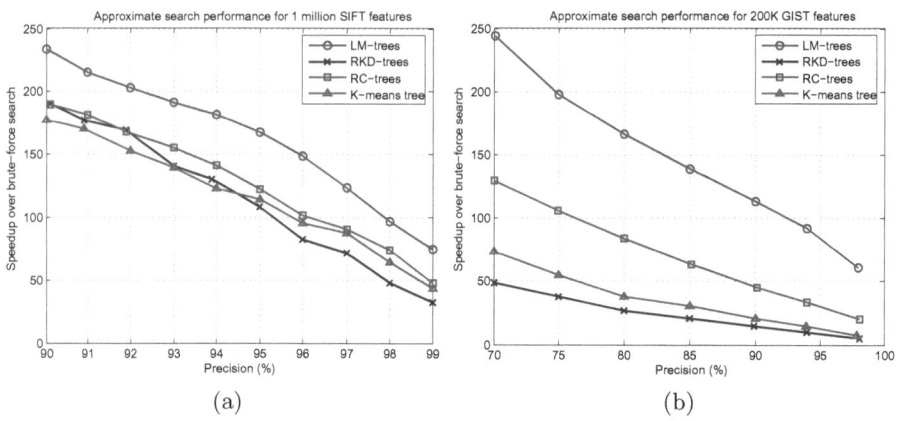

Fig. 3. ANN search performance for the SIFT (a) and GIST (b) features

Figure 3(b) shows the search performance of all the systems for 200000 GIST features. Again, the LM-trees algorithm clearly outperforms the others and tends to perform much better than the SIFT features. The RC-trees algorithm also works reasonably well, while the RKD-trees and K-means tree work poorly for this dataset. Taking the search precision of 90%, for example, the speedups over brute-force search of the LM-trees, RKD-trees, RC-trees, and K-means tree are 113.5, 15.0, 45.2, and 21.2, respectively. Three crucial factors explain these outstanding results of the LM-trees. First, the use of the two highest variance axes for data partitioning in the LM-tree gives more discriminative representation of the data in comparison to the common use of the sole highest variance axis as in the literature. Second, by using the approximate pruning rule, a larger fraction of nodes will be inspected but much of them would be eliminated after checking the lower bound. In this way, the number of data points, which will be actually searched, is retained under the pre-defined threshold E_{max}, while covering a larger number of inspected nodes, and thus increasing the chance of reaching the true nodes closest to the query. Finally, the use of bandwidth search gives much of benefit in terms of computation cost, compared to the priority search used in the baseline indexing systems.

Table 1 summarizes the main results of the proposed indexing algorithm in terms of speedup and absolute query time. The speedup is computed relative to the brute-force search as before, but the absolute query time is averaged over the 1000 queries. We report the results for both ENN and ANN search, where the search precision is set to 95% for ANN search.

Table 1. Summary of our search speedup and mean query time (second)

Results	ENN search		ANN search ($P = 95\%$)	
	SIFT 1M	GIST 250K	SIFT 1M	GIST 200K
Speedup	9.1x	18.7x	168.9x	84.7x
Mean query time	51.2	47.8	2.7	8.1

4 Conclusion and Future Works

In this paper, a new indexing algorithm in feature vector space has been presented. The main contribution of the proposed LM-tree is three-fold. First, a new method of data decomposition has been presented to construct the LM-tree. Second, a novel elimination rule has been proposed to efficiently prune the search space. Finally, a bandwidth search technique has been introduced to deal with the ANN task, in combination with the use of multiple randomized LM-trees. The proposed LM-tree has been validated on 1 million SIFT features and 250000 GIST features, demonstrating that it works really well for both ENN and ANN search, compared to the baseline indexing algorithms. More experiments on different feature types would be performed in the future to study thoroughly the performance of the proposed LM-tree. Dynamic insertion and deletion of data points in the LM-tree would be also investigated in future works.

Acknowledgment. This work has been supported by the Vietnam International Education Development (VIED) project.

References

1. Beis, J.S., Lowe, D.G.: Shape indexing using approximate nearest-neighbour search in high-dimensional spaces. In: Proceedings of the 1997 Conference on Computer Vision and Pattern Recognition, CVPR 1997, pp. 1000–1006 (1997)
2. Böhm, C., Berchtold, S., Keim, D.A.: Searching in high-dimensional spaces: Index structures for improving the performance of multimedia databases. ACM Comput. Surv. 33(3), 322–373 (2001)
3. Friedman, J.H., Bentley, J.L., Finkel, R.A.: An algorithm for finding best matches in logarithmic expected time. ACM Trans. Math. Softw. 3(3), 209–226 (1977)
4. Fukunaga, K., Narendra, M.: A branch and bound algorithm for computing k-nearest neighbors. IEEE Trans. Comput. 24(7), 750–753 (1975)
5. Indyk, P., Motwani, R.: Approximate nearest neighbors: towards removing the curse of dimensionality. In: Proceedings of the Thirtieth Annual ACM Symposium on Theory of Computing, STOC 1998, pp. 604–613 (1998)
6. Jégou, H., Douze, M., Schmid, C.: Product Quantization for Nearest Neighbor Search. IEEE Trans. Pattern Anal. Mach. Intell. 33(1), 117–128 (2011)
7. Kulis, B., Grauman, K.: Kernelized locality-sensitive hashing. IEEE Trans. Pattern Anal. Mach. Intell. 34(6), 1092–1104 (2012)
8. Lv, Q., Josephson, W., Wang, Z., Charikar, M., Li, K.: Multi-probe lsh: efficient indexing for high-dimensional similarity search. In: Proceedings of the 33rd International Conference on Very Large Data Bases, pp. 950–961 (2007)
9. McNames, J.: A fast nearest-neighbor algorithm based on a principal axis search tree. IEEE Trans. Pattern Anal. Mach. Intell. 23(9), 964–976 (2001)
10. Muja, M., Lowe, D.G.: Fast approximate nearest neighbors with automatic algorithm configuration. In: VISAPP International Conference on Computer Vision Theory and Applications, pp. 331–340 (2009)
11. Muja, M., Lowe, D.G.: Fast matching of binary features. In: Proceedings of the Ninth Conference on Computer and Robot Vision, pp. 404–410 (2012)
12. Silpa-Anan, C., Hartley, R.: Optimised kd-trees for fast image descriptor matching. In: IEEE Conference on Computer Vision and Pattern Recognition (CVPR 2008), pp. 1–8 (2008)

A Natural Interface for the Training of Medical Personnel in an Immersive and Virtual Reality System

Alberto Del Bimbo, Andrea Ferracani,
Daniele Pezzatini, and Lorenzo Seidenari

MICC - University of Firenze, Firenze, Tuscany, ITA
delbimbo@dsi.unifi.it,
{andrea.ferracani,daniele.pezzatini,lorenzo.seidenari}@unifi.it
http://www.micc.unifi.it/vim/
Universitá degli Studi di Firenze, Media Integration and Communication Center,
Viale Morgagni 65,
Firenze, Italy

Abstract. In this paper we present an immersive system, developed for the RIMSI project, designed for the training of medical and paramedical personnel in emergency medicine. Virtual reality systems have been used recently in combination with natural interaction systems for patients' rehabilitation but are proving especially useful for educational applications in the field of medical training and teaching. The idea is to mimic real three-dimensional environments and to make them available through natural interaction to enable users to explore, interact, collaborate and assist patients in virtual scenarios that otherwise could not be simulated easily in the real world without an high cost (e.g. to prepare the environment for a plane or a car crash) or possible risks of injuries to actors involved in the simulation (e.g. perilous situations like a gas leak). In addition, these simulation scenarios, using a digital environment, are repeatable and can be recorded, easing the process of errors' highlighting. The RIMSI prototype provides a virtual first aid scenario with interactive 3D graphics which can be controlled and navigated through a natural gesture interface based on Kinect^{TM}.

Keywords: virtual reality, medical training, body tracking.

1 Introduction

Educational virtual environments are used, through computer applications, in a variety of fields of study to facilitate the learning and the acquisition of knowledge and to allow users to feel immersed in a common environment, in specific situations, in order to ease the exchange of skills. Network systems have been designed, in the recent past, to solve issues related to distance learning, mediated by specialists; but often, in such systems, priority has been given to theoretical rather then practical learning. In systems featuring natural interfaces instead

A. Petrosino (Ed.): ICIAP 2013, Part I, LNCS 8156, pp. 763–772, 2013.

medical trainees can realistically manipulate 3D anatomical parts of the body, reproduced in graphics, or assist virtual injured people in dangerous scenarios. This aspect is very important in preparing trainees psychologically to deal with real situations.

In the field of medical training the main problem of digital simulation systems emerged in recent years is, on the one hand, the excessive orientation towards the design of too configurable systems, given the technical difficulties due to the correct interfacing between hardware, software and the often complicated medical procedures, difficulties that usually lead to the development of very simple prototypes. On the other hand, the proliferation of very specialized systems has resulted in applications not sufficiently adaptable to several scenarios. Furthermore, for the most part, all these systems are not fully immersive.

The Cybermed framework [6], for example, is a system for medical training via computer network. It has advanced features for configuring some aspects of the software like the number of participants, the manipulation of objects, the type of device (mouse or haptic systems), the definition of the actors for remote mentoring and distance learning (tutor or participants), but does not really allow an exhaustive characterization of the situation or the definition of complex medical procedures. SOFA [2], ViMet [3] and Gipsy [8] frameworks instead are open-source projects which feature an high level modularity and rely on the capability of a multi-model representation of simulation models (deformable models, collision models, instruments) exposing appropriate APIs. Although they easily could allow to simulate a scenario, e.g. a laparoscopy, they neither provides a real natural and immersive interface nor they go beyond enabling the system to respond to punctual stimuli (e.g., in the laparoscopy use case, deformation of the liver and collisions with the ribs). Finally, Spring [7] is a more specific mouse based desktop framework for real-time surgical simulations which allows a basic configurability of patient-specific anatomy.

As regard to more immersive solutions Honey et al. have created a virtual environment in Second Life for hemorrhage management[13]; Cowan et al. [11] have developed a system for the inter-professional education (IPE) for critical care providers where an immersive 3D situation is accessed across the network in a "multi-player online" environment, allowing trainees to participate as avatars from remote locations. These systems have all the benefits of realism but still lack the naturalness of human gestures and actions that can be performed in solutions exploiting natural interaction.

Several natural interfaces featuring natural interaction systems have been used in medical rehabilitation, although sparsely, leveraging different technologies (e.g. Nintendo Wii, PlayStation EyeToy) or the Microsoft Kinect, as part of physical therapy. Lange et al. [12] of the Institute of Creative Technologies of the University of Southern California, for example, have developed a Kinect-based rehabilitation game "JewelMine" that consists of a set of balance training exercises to stimulate the players to reach out of their base of support. These kind of softwares take advantage from the realism and the naturalness of the situation to encourage physical exercise but are not common in medical operators training. In this sense they are patient oriented and not trainees oriented.

2   RIMSI: Using Natural Interaction for Medical Training

The use of natural interaction combined with an intelligent and realistic virtual environment allows to improve training systems and to overcome the difficulties which arise, for example, in an haptic enabled environment where it is usually only the teacher that shows the type of movements to be performed on a 3D model [1] and there is only one feedback from teacher to learner via network, as providing all students of a haptic tool would be very expensive.

Furthermore for virtual environments and natural interfaces to be effective, it is essential that the scenarios are very realistic [9]. With scenery we mean the places, the events, but also the virtual humans involved in a particular situation. Such humans should be able to give realistic feedback, express emotions, react to natural stimuli. Many medical simulation centers are currently using patient simulator systems provided with mechanical patients: trainee's task is to carry out some emergency maneuvers on a patient whose vital signs are in the meantime monitored and controlled by training doctors present in another room. The trainees have the possibility to evaluate these values through a vital sign monitor. Their behaviors influence the responses and the reactions of the patient, guided interactively by trainers. The so-called debriefing usually follows this phase. During the debriefing the trainers and the team of trainees discuss actions and mistakes often with the aid of a video record of the whole session: this is the stage where trainees learn more. Laboratories equipped with high and low fidelity medical mannequins are expensive to build and to maintain. Moreover, the traditional classrooms or debriefing for learning have a teacher - centered approach. In the training sessions mediated by mannequins users undergo reactions and behaviors whilst in natural interaction systems individual responsibility, represented by the avatar or character in the game, is more relevant. Our prototype (RIMSI) uses a learner-centered teaching approach where he is the user himself, through interaction, to guide the learning process resulting in a better memorability. As showed by Villneuve et al. the generation of digital natives prefer the use of technology in learning [10].

RIMSI has been designed with the aim of becoming a flexible framework in the context of very specific scenarios. As use case we propose a first aid BLSD (Basic Life Support Defibrillator) scenario providing the software with a configuration interface that allows educators to change several environment variables, such as the location of objects, and some events and actors in the scene, etc. (Figure 1).

The core of the system is on the natural way in which users can exploit the simulation environments. Trainees can move in the space like in real life and interact with realistic objects and avatars (Figure 2). This is very important from the point of view of training as a naturally usable environment eases the learnability and the memorability of the procedures.

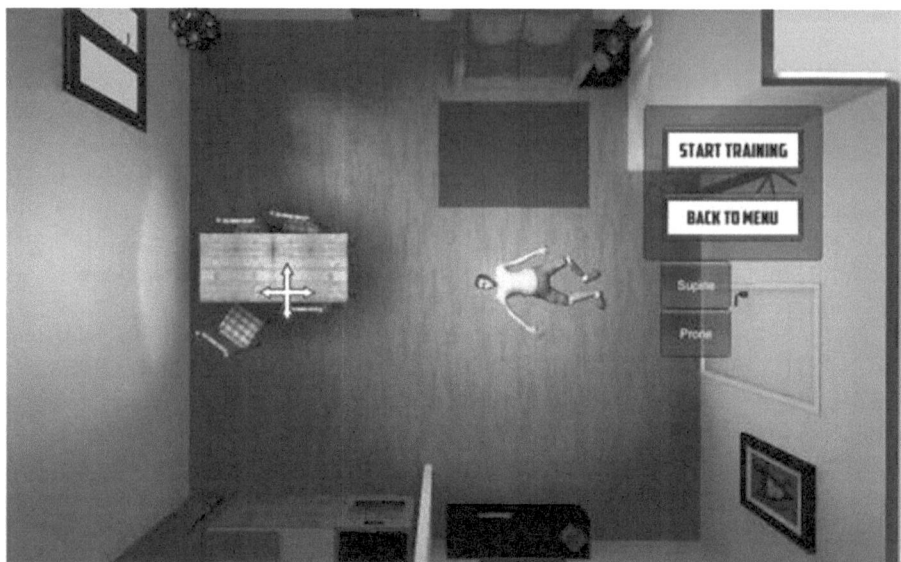

Fig. 1. The graphic configuration interface

Fig. 2. Avatars performing actions in the environment

3 Use Cases and Scenarios

Simulation in medical training can address a wide range of aspects, spacing from the application of emergency procedures to the enhancement of a specific skill. Usually, training of specific manual and practical skills is performed via haptic devices that simulates medical tools in a very accurate way or by using body part replication [4] [5]. Since this kind of training require a very accurate evaluation of the results (e.g. a few millimeters error may results in a critical outcome for the patient) it would be very difficult to obtain a good simulation in a complete virtual environment. Anyway, other aspects of medical training such as the application of procedures, environmental awareness or communication with team members can gain a great improvement from virtual simulations. Some medical scenarios, such as Basic Life Support in emergency medicine or sepsis treatment, present a well-known sequence of actions to be taken in order to accomplish a correct patient assistance.

According to this, the proposed system consists of a virtual reality simulator for emergency situations in which trainees are asked to take some actions on the environment and to apply medical procedures on the patient depending on the proposed situation. The system is configurable and it provides a scenario editor through which the medical instructors can define the type of the simulation. It is possible to place patients in different positions in the environment, arrange the environment itself (e.g. pre-hospital first aid, emergency room) and to add to each scenario several additional threats to the safety of patients and medical operators. Furthermore, its also possible to insert different patients conditions, like his state of consciousness and his body pose, and different responses to therapies that require appropriate standardized procedures. In this way, a different virtual simulation can be generated for each training session.

The simulator allows to use Non-Playing Characters (NPCs) in the scenario. NPCs can have different roles: they may be members of the medical team performing the procedure, but they can also be relatives of the patient or simple by-standers. In both cases they are employed to deliver information about the patient or the environment or to give some hints about the correct procedure to the trainees.

We present a possible use case scenario: a basic life support and defibrillation (BLSD) procedure with a patient inside a house with a gas leak. In this scenario the trainee, associated with the doctor character, can perform the full procedure assisted by a NPC, the medical assistant. As the simulation begin, the assistant deliver informations about the house situation (e.g. she claims she is smelling gas). The player can navigate in the virtual room to check the environment safety or go directly to the patient to check his conditions. In order to perform a correct procedure on the patient, the player should locate the gas source, block the leak and open a window before a certain amount of seconds. When the room is safe, the assistant give feedback to the player, so he can approach the patient in order to check his state. Then, to complete the procedure, the trainee have to choose between different action proposed by the simulator and perform them in the correct order and within a certain time. Errors are reported by the medical assistant with audio e textual feedback. Timers are used in order to assess the player performance and to consider the need of different procedures. To finalize the procedure, the player has to grab the proper tools and activate them in the correct way. As an example, if a defibrillator is needed, the player has to grab it, turn it on and wait until it is completely charged before activating it. The simulation ends successfully if the player completes the procedure correctly or unsuccessfully if he commits too many errors or delays in taking decisions.

4   Navigation and Interaction System

The proposed system is a free-roam game. It means that there is no predefined order in which actions need to be taken in order to obtain accomplishments. This is also a way to improve realism since, as in real life, a user may decide to assist a patient before another or check the environment to assess its safety before beginning the medical procedures. Obviously, some actions need to be taken in the correct order to perform a correct training session. As an example, if multiple patients are in the scene, the more severely injured should be assisted first.

The navigation in the RIMSI virtual environment is made possible by a gesture driven Kinect^{TM}interface, through which the variation of user posture is evaluated in order to interact with the scenario. Simple or complex gestures can be recognized with the use of a depth camera and a skeletal tracking system [14]. Based on information extracted form the depth and RGB cameras, the Microsoft Kinect APIs track the position and orientation parameters of all the joints of the skeleton model in real time. For every gesture that users can perform in the simulator, we defined a simple Finite State Machine (FSM) and implemented the relative FSM recognizer. In general, gesture recognition relies only on some specific body part, so for every gesture we only consider significant skeleton joints.

Defined gestures can be found in Table 1, along with details about considered skeleton joints and actions performed by the virtual character.

Gestures have been dened according to the requirements of the simulation context. Since the training sessions should consist in applying the correct procedures and not in learning the semantic of predefined gestures, users should interact in the virtual environment as they would do in real life. Simplest actions and behaviors can be easily translated in natural body gestures or poses. As an example, the rotation of the the torso will cause a rotation of the character, while the action of bending will be used to check the conditions of the patient.

However, not every phase of the medical procedure can be associated to natural gestures. When the trainee has to face a choice between different options, virtual menus will appear to allow the player to choose the action to perform. We defined a "pointing" gesture that produce a cursor on the virtual environment, allowing user to select one of the options by maintaining the persistence over it. Figure 3 shows some phases of the interaction.

Table 1. Associations between gestures and character actions

Gesture	Body pose	Actions
Take / Grab	One of the hands moves in front of the body (on Z axes)	The character grab objects in front of him, open/closes handles
Rotate	The rotation of the torso is evaluated using x,y values of users shoulders	The camera rotate in order to explore the scene
Walk-in-place	Sequences of x,y values of legs' joints are evaluated to check several phases of the walk	The character moves along the direction he is facing
Bend	Sequences of decreasing y values of chest and knees	The character bends on his knees in order to examine the patient
Point	User's hand in front of his body, x,y values are normalized respect to the user's position	A cursor is visualized in order to allow multiple choices

The only limitation of the standard Kinect SDK has is the lack of a hand pose recognition algorithm. The recognition of the hand pose could improve the system capabilities, allowing to recognize on/off actions such as activating, grabbing or manipulating an object in the 3D world. We propose to employ an improvement to this system based on a robust hand pose classifier we developed [16]. The proposed module is trained on a large dataset (30k+) of hands in open and closed status recorded at several distances, from eight subjects of different genders. This vision module improves the responsiveness of the interface and enables the activation and manipulation of 3D objects using informations about the state of the hands of the player. In Figure 4 an example of the detector output is shown.

As an example, in the BLSD scenario, the player could grab the electrodes one at a time or with both hands by simply pointing at them in the 3D space and closing the hand(s), and then place them on the patient chest by opening his hand(s).

The vision module represents a hand, segmented with a skeletal tracker, with a regular grid of 5 SURF[15] descriptors. This feature is fed to a nonlinear SVM classifier that predicts the state of the hand. This module has a very low overhead in terms of computation, in fact a detection can be obtained in less than a frame, although this detection can be noisy. The classifier is cascaded with a temporal Kalman filter that outputs a smooth estimate of the hand state. The σ parameter can be varied obtaining a trade-off between responsiveness and robustness of the system. Overall the hand status recognition module has an extremely high reliability with an accuracy of to 98.95%.

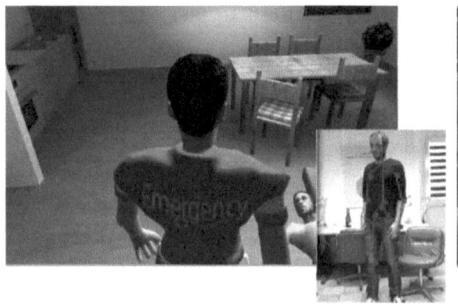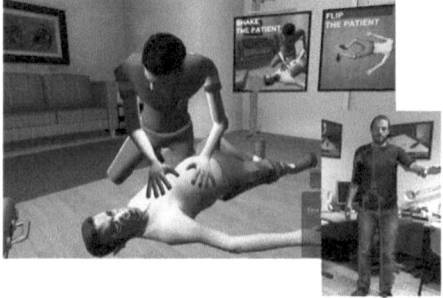

Fig. 3. Avatars performing actions based on user's gestures. On the right, a virtual menu with multiple options is shown.

Fig. 4. Hand status detection: square (open hand), circle (closed hand)

The game is based on the Unity3D game engine while gesture detection module has been developed in C# using the official Microsoft Kinect SDK. Scenes and models have been realized with Maya.

5   Conclusion

The RIMSI prototype (RIMSI) is part of the so-called serious games technologies for the improvement of skills and knowledge in medicine. Serious games provide an interactive environment where students and teachers are represented by avatars. Currently most of the simulated training sessions in medicine occur in real environments such as laboratories but more realistic digital simulations that facilitate learning in a safe environment can be implemented through emerging technologies such as tracking and VR environments.

RIMSI features a 3D virtual environment, high resolution graphics and natural interaction interfaces to improve the effectiveness of the training of medical personnel in emergency situations. It provides an immersive and intelligent digital environment which can respond to natural gestures and actions of more team members in a simulated situation of emergency. The system has been developed to improve the learning methods currently in use and to solve some common issues.

In particular RIMSI allows to reduce costs of simulations, optimize some critical phases,like debriefing, with a better management of the time spent and of the use of personnel in the centers of simulation. Furthermore, compared to simulations with mannequins, the RIMSI system allows: 1) to have an intelligent system that responds automatically to user actions, 2) have continuous feedback, 3) to simulate an immersive environment such as that with the use of mannequins, but at a lower cost, enhancing learnability and memorability of procedures 4) to be corrected on the fly and have system's feedback in realtime (the debriefing phase assumes instead a period of time between the wrong action and the correction).

Interactive sessions with RIMSI can also be recorded and used as training materials. In this way a large database can be build quickly to evaluate errors and the most common difficulties of trainees. This data, in addition to increase the effectiveness of training, can be useful to assess how well these simulation environments work and how they can be improved. In fact, future work will include the testing of the prototype developed by the research unit and subsequent focus groups to discuss critical points.

Acknowledgments. This work was supported in part by the Regione Toscana RIMSI Project (Program POR CREO FESR 2007-2013).

References

1. Gunn, C., Hutchins, M., Stevenson, D., Adcock, M., Youngblood, P.: Using collaborative haptics in remote surgical training. In: Eurohaptics Conference and Symposium on Haptic Interfaces for Virtual Environment and Teleoperator Systems (WHC 2005), Italy (2005)
2. Allard, J., Cotin, S., Faure, F., Bensoussan, P.J., Poyer, F., Du-riez, C., et al.: SOFA - an Open Source Framework for Medical Simulation. Medicine Meets Virtual Reality 15, 1–6 (2007)
3. Oliveira, A.C.M.T.G., Botega, L.C., Pavarini, L., Rossatto, D.J., Nunes, F.L.S., Bezerra, A.: Virtual Reality Framework for Medical Training: Implementation of a deformation class using Java. In: Proceedings of the SIGGRAPH International Conference on Virtual-Reality Continuum and its Applications in Industry (SIGGRAPH 2006), Hong Kong, pp. 347–351 (2006)
4. Basdogan, C., Ho, C.-H., Srinivasan, M.A.: Virtual environments for medical training: graphical and haptic simulation of laparoscopic common bile duct exploration. IEEE/ASME Transactions on Mechatronics 6(3), 269–285 (2001)
5. Coles, T.R., Dwight, M., Nigel, W.J.: The role of haptics in medical training simulators: a survey of the state of the art. IEEE Transactions on Haptics 4(1), 51–66 (2011)
6. Sales, B.R.A., Machado, L.S., Moraes, R.M.: Interactive Collaboration for Virtual Reality Systems related to Medical Education and Training. Technology and Medical Sciences, 157–162 (2011)
7. Montgomery, K., Bruyns, C., Brown, J., Sorkin, S., Mazzella, F., Thonier, G., Tellier, A., Lerman, B., Menon, A.: Spring: A General Framework for Collaborative, Real-time Surgical Simulation. In: Medicine Meets Virtual Reality (MMVR 2002), pp. 23–26 (2002)

8. Goktekin, T.G., Çavuşoğlu, M.C., Tendick, F., Sastry, S.: GiPSi: An Open Source Open Architecture Software Development Framework for Surgical Simulation. In: Cotin, S., Metaxas, D. (eds.) ISMS 2004. LNCS, vol. 3078, pp. 240–248. Springer, Heidelberg (2004)

9. Lok, B., Ferdig, R.E., Raij, A., Johnsen, K., Dickerson, R., Coutts, J., Stevens, A., Lind, D.S.: Applying virtual reality in medical communication education: current findings and potential teaching and learning benefits of immersive virtual patients. Virtual Real. 10(3), 185–195 (2006)

10. Villeneuve, M., MacDonald, J.: Toward 2020: Visions for nursing. Technical Report. Canadian Nurses Association, Ottawa, Ontario, Canada (2006)

11. Cowan, B., Shelley, M., Sabri, H., Kapralos, B., Hogue, A., Hogan, M., Jenkin, M., Goldsworthy, S., Rose, L., Dubrowski, A.: Interprofessional care simulator for critical care education. In: Proceedings of the Conference on Future Play: Research, Play, Share (Future Play 2008), pp. 260–261. ACM, New York (2008)

12. Lange, B., Koenig, S., McConnell, E., Chang, C., Juang, R., Suma, E., Bolas, M., Rizzo, A.: Interactive game-based rehabilitation using the Microsoft Kinect. In: IEEE Virtual Reality Short Papers and Posters, VRW (2012)

13. Honey, M.L.L., Diener, S., Connor, K., Veltman, M., Bodily, D.: Teaching in virtual space: Second Life simulation for haemorrhage management. In: Ascilite Conference, Aukland (2009)

14. Shotton, J., Sharp, T., Kipman, A., Fitzgibbon, A., Finocchio, M., Blake, A., Cook, M., Moore, R.: Real-time human pose recognition in parts from single depth images. Commun. ACM 56(1), 116–124 (2013)

15. Bay, H., Ess, A., Tuytelaars, T., Van Gool, L.: SURF: Speeded up robust features. Computer Vision and Image Understanding 110(3), 346–359 (2008)

16. Bagdanov, A.D., Del Bimbo, A., Seidenari, L., Usai, L.: Real-time hand status recognition from rgb-d imagery. In: International Conference on Pattern Recognition (2012)

Saliency Based Image Cropping

Edoardo Ardizzone, Alessandro Bruno, and Giuseppe Mazzola

Dipartimento di Ingegneria Chimica, Gestionale, Informatica, Meccanica,
Università degli studi di Palermo, Italy
{edoardo.ardizzone,alessandro.bruno15,giuseppe.mazzola}@unipa.it

Abstract. Image cropping is a technique that is used to select the most relevant areas of an image, discarding the useless ones. Handmade selection, especially in case of large photo collections, is a time consuming task. Automatic image cropping techniques may help users, suggesting to them which part of the image is the most relevant, according to specific criteria. We suppose that the most visually salient areas of a photo are also the most relevant ones to the users. In this paper we present an extended version of our previously proposed method, to extract the saliency map of an image, which is based on the analysis of the distribution of the interest points of the image. Three different interest point extraction algorithms are evaluated within an automatic image cropping system, to study the effectiveness of the related saliency maps for this task. We furthermore compared our results with two state of the art saliency detection techniques. Tests have been conducted onto an online available dataset, made of 5000 images which have been manually labeled by 9 users.

Keywords: Image Cropping, Visual Saliency, Visual Perception, Saliency Map.

1 Introduction

The development of network and computer hardware technology, and mobile media devices (smartphones, digital cameras and PDA) has had a large impact in people's everyday life. Today many people use these devices to take a lot of photos, and share them by social networks. Image cropping is a technique that is used to resize an image by selecting its most relevant areas, discarding its useless or redundant parts. People may wish to remove portions of photo background to emphasize the subject (fig.1), to fit an image to fill a frame, to create thumbnails for browsing purposes, or to select the most important (to the observer) parts of the image, to improve photo-composition. In standard photo-editing applications, designers manually crop an area around the important content of the image. Some commercial products allow users to manually crop images to generate thumbnails. Handmade cropping, in case of large photo collections, is an onerous and time-consuming task. Automatic cropping methods suggest to the users which are the most important parts of the image. What does it mean for "the most important parts of a photo" in this context? In our work we suppose that the most salient regions of the image are considered as the most relevant parts of a photo. The aim of visual saliency detection methods is to build a saliency map that replicates the human visual

A. Petrosino (Ed.): ICIAP 2013, Part I, LNCS 8156, pp. 773–782, 2013.

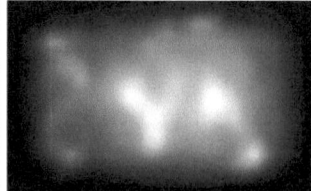

Fig. 1. Input Image (left), saliency map (center) and image crop (right)

system (HVS) behavior in the visual attention process. From a visual perception viewpoint, many images are made of salient regions surrounded by unnecessary background areas. In this paper, we present an extended version of our previous work on visual saliency[1] and we study the effectiveness of saliency maps in image cropping methods. We compare the results obtained by our method with those obtained with two state of the art techinques, within a common automatic image cropping system. We evaluate results using a free available dataset made of manually cropped images (more details about the dataset are given in section 5). The paper is organized as follows: in section 2 we discuss some State of the art methods about Image Cropping; in section 3 we discuss our saliency detection algorithm and the two reference methods; in section 4 we describe the Saliency-Based Image Cropping system; in section 5 we show and discuss the experimental results; section 6 contains conclusive considerations.

2   State of the Art

In the last years, there has been much interest in automated cropping methods, especially using visual attention information. Suh et al. [2] used both saliency maps and automatic face detection to evaluate candidate cropped regions to determine best crop. Ma et al. [3] first segmented the images, then ROI are selected according to the image entropy, the size and the closeness to the center of the image. Zhang et al. [4] formulated automatic cropping as an optimization problem, which consists of three sub models (composition, conservative, and penalty), and employed   a particle swarm optimization (PSO) to obtain the optimal solution by maximizing the objective function. Santella et al. [5] employed an eye tracking system to identify the important content of a photo, for automatic snapshot recomposition, adaptive documents, and thumbnailing. Stentiford [6] proposed a method for automatically cropping photos and camera zooming based upon a new visual attention model. Ciocca et al. [7] proposed a self-adaptive image cropping algorithm where the processing steps are driven by the classification of the images into semantic classes, exploiting both visual and semantic information. In She et al. [8] the authors first classified photos in five categories, and then build a dictionary for each category, extracting the visual saliency maps of these photos. Some more recent works on image cropping emphasize the pleasantness of resulting cropped photos, taking into account photographic composition rules (as the rule of thirds, the rule of filling the frame and leading the lines) or quality scores. In Luo [9] the author performed image cropping method based on subject detection algorithm, using a belief map that probabilistically indicates the subject content. Nishiyama et al. [10] build a quality classifier using a dataset of photos into which users manually insert quality scores to photos. They finally used the classifier to find the cropped region with the highest quality score. Cheng et al. [11] proposed a photo quality evaluation metric for automatic professional view finding, exploiting professional photographers' knowledge and composition rules. Bhattacharya et al. [12] presented an interactive application that helps users to improve the visual aesthetics of their digital photographs using spatial re-composition. Liu et al. [13] applied the rule of thirds, the diagonal dominance, visual balance and sizes of salient regions for equally evaluation. Ahn et al. [14] used crowdsourcing techniques to collect many crops for a set of 68 photographs. They analyze the crops with respect to the composition guidelines recommended in photography and art literature. In McManus et al. [15] the authors analyze the psychic aspects of Image Cropping and the influence of color, semantics and expertise of the users on the resulting crops.

3   Visual Saliency

Visual Saliency deals with identifying the most important regions of an image from a perceptual point of view (Frintrop et al. [16]). In the first three seconds a human observer fixates some particular points inside an image and tends to group them into visual significant areas (Judd et al. [17]). In [18] Achanta et al. exploit features of color and luminance to detect salient objects into the image. In this paper we compared five saliency maps to evaluate the effectiveness these methods, for image cropping applications: Itti et al. [19], Harel et al. [20], and three variations of our previous work [1]. All the analyzed methods are bottom-up, stimulus-driven and unsupervised:

- Itti-Koch model [19] is based on a multi-scale analysis of the image. Multi-scale image features are combined into a single topographical saliency map. A dynamical neural network selects attended locations in descending order of saliency.
- Harel [20] saliency approach is based on a biologically plausible model, and it consists of two steps: activation maps on certain feature channels and normalization, which highlights conspicuity.
- In our method [1] we analyze the distribution of the keypoints onto the image, with different scales of observation. In particular SIFT-point Density Maps (SDM) are built to study the relationship between the keypoints extracted by the SIFT algorithm [21] and real human fixation points. We furthermore extended our previous work by considering also other two types of keypoint extraction algorithms: Harris Corner [22] and SURF[23]. More particularly, our previous method [1] analyzes the distribution of the SIFT interest points along the image, and build the SIFT Density Maps (SDM). The most salient areas are those that maximizes the difference between the SDM value and the most frequent value of the SDM (which we suppose related to the background).

In the remainder of the paper we will refer to them with ITTI, GBVS, SIFT, SURF and HARRIS, respectively.

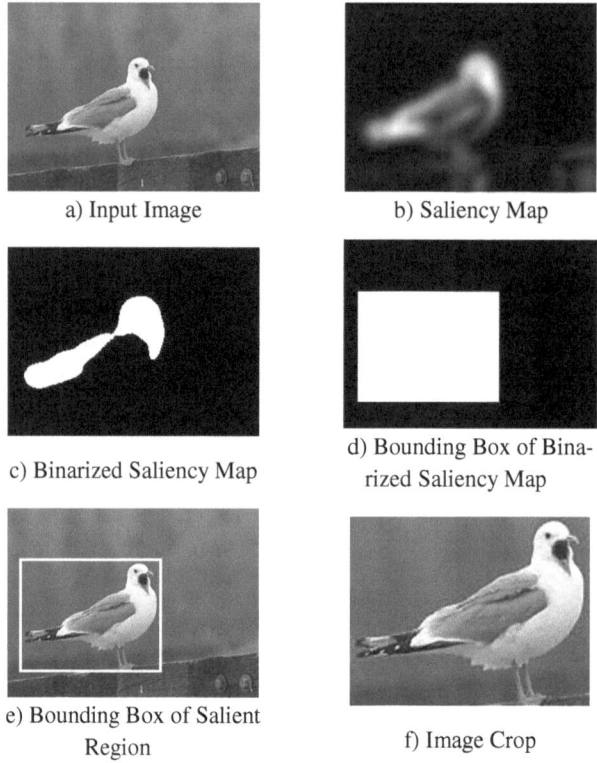

a) Input Image   b) Saliency Map

c) Binarized Saliency Map

d) Bounding Box of Bina-
rized Saliency Map

e) Bounding Box of Salient
Region

f) Image Crop

Fig. 2. The steps of the Image Cropping Method pipeline

4   Saliency Based Image Cropping

The aim of the present work is to evaluate the effectiveness of saliency maps when used to support automatic cropping. Our system is subdivided into (see fig.2): Salien-cy Map Extraction, Saliency Map Binarization (Thresholding), Bounding Box Extrac-tion, Photo Cropping, Evaluation. Given an image, we compute the five saliency maps described in section 3 (ITTI, GBVS, SIFT, SURF, HARRIS). Each saliency map is then binarized using different threshold values (see section 5) and then the bounding box of all the pixels, which values are above the threshold, is selected and used to crop the photo (fig.2). Results are evaluated (in terms of precision, recall and F-measure) comparing the resulting crops with handmade selected crops, from the dataset created by Liu et al. [24], that will be further described in the next section. We selected ITTI and GBVS as reference methods to compare, as they are the most cited methods in literature, and they obtain very good results in any application in which they are used. The center-weighted Gaussian estimator is also a solution but, to our knowledge, both ITTI and GBVS typically achieve much better results. Moreover, the photographic "rule of the thirds" suggests that the subject of the photo should not be placed in center of the image, but along the intersections of a 3x3 grid, superim-posed to the image. Therefore, the "center" method will not work in these cases.

5   Experimental Results

Our experiments has been conducted onto a freely available dataset [23] which consists of 5000 images labeled by 9 users, who have been invited to select the most salient object of the scene represented in the image, by drawing a rectangle. For our purposes, we compute the "average crop" of the handmade crops of this dataset considering, for each image, only the pixels that have been selected at least by N

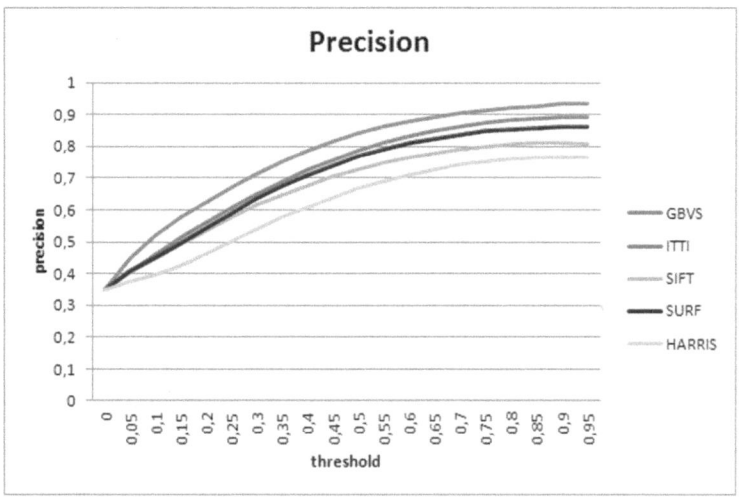

Fig. 3. Experimental Results: Precision

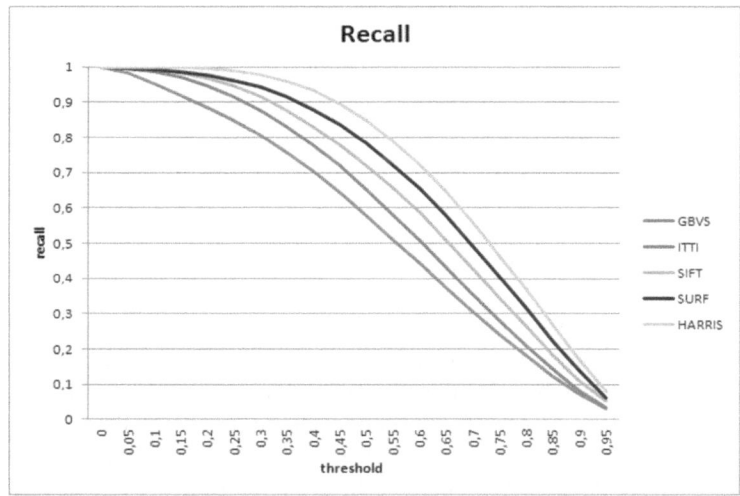

Fig. 4. Experimental Results: Recall

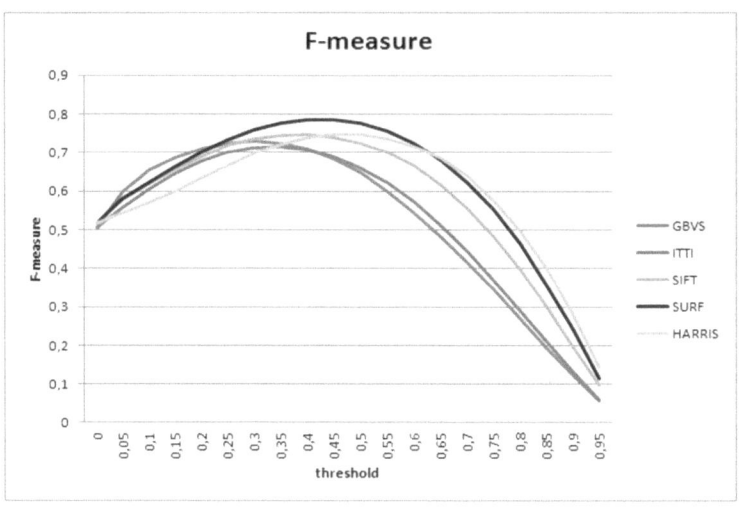

Fig. 5. Experimental Results: F-measure

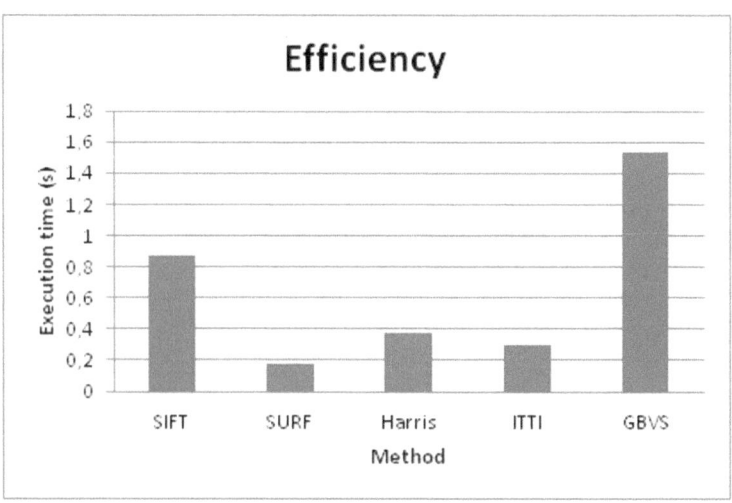

Fig. 6. Experimental Results: Efficiency

(in our experiments N=5) of the 9 users, i.e. the majority of the users. For each of 5000 images of the dataset, we computed the five saliency maps (ITTI, GBVS, SIFT SURF, HARRIS) and we binarize them by using different threshold values T (from 0 to 0.95 with step 0.05). An interesting study could be to analyze the relationship between the threshold and the content of the images. This study could be done only after having labelled the images in the dataset, in terms of some pre-established classes. But this is not the focus of our paper. The accuracy of the results is measured

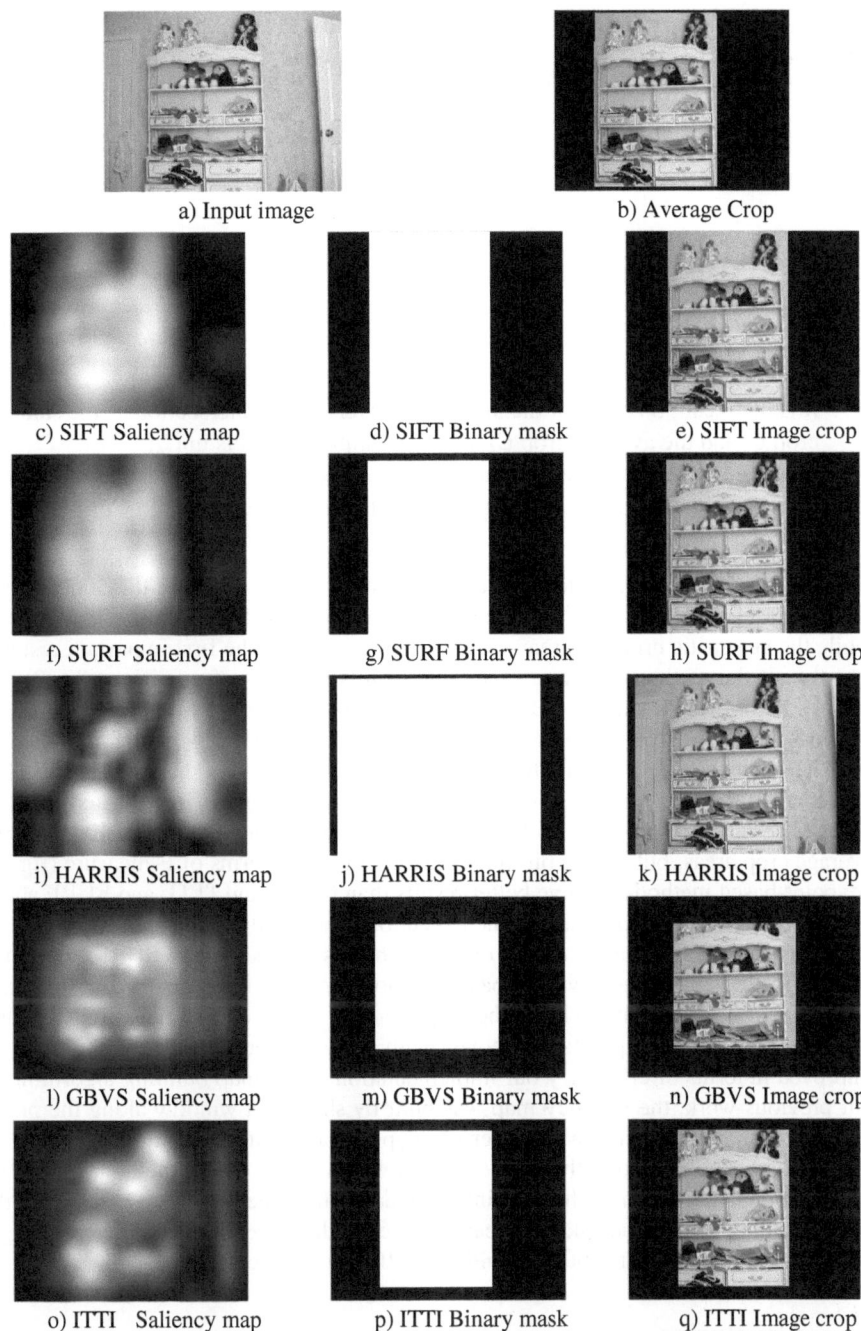

a) Input image

b) Average Crop

c) SIFT Saliency map

d) SIFT Binary mask

e) SIFT Image crop

f) SURF Saliency map

g) SURF Binary mask

h) SURF Image crop

i) HARRIS Saliency map

j) HARRIS Binary mask

k) HARRIS Image crop

l) GBVS Saliency map

m) GBVS Binary mask

n) GBVS Image crop

o) ITTI   Saliency map

p) ITTI Binary mask

q) ITTI Image crop

Fig. 7. A visual example of image crops with the different saliency maps. Saliency maps are binarized with threshold = 0.35, which is a good tradeoff for all the methods (see fig. 6)

comparing the binary mask C_S of the crops (fig. 2.d), for a given threshold, with the binary mask C_A of the "average crop", in terms of recall, precision and F-measure:

$$P = \frac{n(C_s \cap C_A)}{n(C_s)} \tag{1}$$

$$R = \frac{n(C_s \cap C_A)}{n(C_A)} \tag{2}$$

$$F_1 = 2\frac{P \cdot R}{P + R} \tag{3}$$

where:

- R is the recall, the ratio of the number of pixels in the intersection of the saliency crop C_s and the average crop C_A, and the number of pixels in C_A;
- P is the precision, the ratio of the number of pixels in the intersection of the saliency crop C_s and the average crop C_A, and the number of pixels in C_s;
- F_1 is the F-measure.

Figures 3,4,5 show Precision, Recall and F-measure, averaged on the 5000 images in the dataset. Note that the "average crop" is independent from the threshold values, while the automatic cropped areas vary with these values. The first important result is that the all the methods achieve very good results in terms of precision (fig. 3), for most of the threshold values. GBVS and ITTI are the best ones, while keypoint-based methods have slightly worse results. In terms of recall (fig. 4), SIFT, SURF and HARRIS achieve better results than GBVS and ITTI. It means that GBVS and ITTI return smaller crops which, however, includes fewer pixels of the related average crops. The last three methods return larger crops, which include more pixels of the average crop areas, but a little bit more false positives. In terms of F-measure (fig. 5), keypoint-based methods achieve better results than GBVS and ITTI, and SURF above all. About efficiency, fig. 6 shows the execution times of the compared methods, averaged for the 5000 images of the dataset. The fastest method is SURF, while the slowest is GBVS. For the keypoint-based methods, most of the time is spent to extract keypoints, in fact the time to build the map is one of two orders of magnitude smaller than the time to extract points. Moreover, with respect to our previous version[1], we improved in terms of efficiency our implementation of the map building algorithm. In our previous work, the saliency map was built by shifting a window along the pixels of the image and counting the number of keypoints it includes, to study the distribution of the keypoints along the image. The newest version of our method focuses on the keypoints, that is, for each keypoint we update simultaneously the values of all the windows that will include it, drastically reducing the execution time. Finally, fig.7 shows some visual results obtained with the different saliency maps.

6   Conclusions

In this paper we present an extended version of our previous work on visual saliency and we evaluated its effectiveness when used to support automatic image cropping.

Saliency-based crops have been compared to handmade crops, within a standard database. We improved the algorithm implementation, as briefly described in section 5, in terms of efficiency, as our previous version was slower than the reference methods, while the new one is comparable or faster.

Note that a user took typically 10-20 seconds to select the part to crop and to draw a rectangle onto an image, while (bottom-up) saliency detection methods typically aim to reproduce the behavior of the Human Visual System in the very first instants when observing an image. Therefore there is a time "gap" between the moment in which the user recognizes a salient part of an image and the moment in which he manually crops that area. After the first 3 seconds users are guided, when analyzing a scene, also by high-level mental processes (recognizing objects, context, faces), therefore there can be a difference between what is visually salient and what is representative of an image. Results showed that saliency-based approaches are very suitable for automatic image cropping applications. We suppose that results could be further improved if different cropping strategies are adopted for different categories of image, as images can represent scene with any type of visual and semantic content.

Regarding the "multiple objects" question, our work is inspired by the assumption that, according to the photocomposition rules, a "high quality" photo must have only one and distinct subject. This is only a suggestion and not a strict constraint, and it is not true for some categories of photo (e.g. panoramas). We think that saliency based methods do not work as well in case of multiple objects. Probably they will include in the same crop all the salient objects (if they are close enough), or they select only one of the salient object, if they are far. Therefore it strongly depends on the reciprocal distance between the salient objects in the scene. Further experiments are needed to face this specific problem.

In fact we observed that saliency based automatic cropping methods, as expected, give worse results when the background area is very composite, or whenever it is not easy to detect a single foreground object into the scene. In those cases, for automatic cropping an image, saliency detection methods could be supported by segmentation algorithms as a preprocessing step. The method can be further improved when applied to other tasks, e.g. face or object detection, if combined with other types of information, e.g. from face detector or object classifiers. But this is not the focus of the paper, which intend to be general purpose. We intend to further study this problem in our future works.

References

1. Ardizzone, E., Bruno, A., Mazzola, G.: Visual saliency by keypoints distribution analysis. In: Maino, G., Foresti, G.L. (eds.) ICIAP 2011, Part I. LNCS, vol. 6978, pp. 691–699. Springer, Heidelberg (2011)
2. Suh, B., Ling, H., Bederson, B.B., Jacobs, D.W.: Automatic Thumbnail Cropping and its Effectiveness. In: Proc. of the 16th ACM Symposium on User Interface Software and Technology, pp. 95–104 (2003)
3. Ma, M., Guo, J.K.: Automatic Image Cropping for Mobile Devices with Built-in Camera. In: Proc. of the Consumer Communication & Networking Conf., pp. 710–711 (January 2004)

4. Zhang, M., Zhang, L., Sun, Y., Feng, L., Ma, W.: Auto Cropping for Digital Photographs. In: IEEE International Conference on Multimedia and Expo (2005)
5. Santella, A., Agrawala, M., DeCarlo, D., Salesin, D., Cohen, M.F.: Gaze-based interaction for semi-automatic photo cropping. In: CHI 2006, pp. 771–780 (2006)
6. Stentiford, F.: Attention Based Auto Image Cropping. In: ICVS Workshop on Computational Attention & Applications (2007)
7. Ciocca, G., Cusano, C., Gasparini, F., Schettini, R.: Self-Adaptive Image Cropping for Small Displays. IEEE Trans. on Cons. Electronics 53(4), 1622–1627 (2007)
8. She, J., Wang, D., Song, M.: Automatic image cropping using sparse coding. In: 2011 First Asian Conference on Pattern Recognition (ACPR), pp. 490–494 (2011)
9. Luo, J.: Subject Content-Based Intelligent Cropping of Digital Photos. In: IEEE International Conference on Multimedia and Expo, pp. 2218–2221 (2007)
10. Nishiyama, M., Okabe, T., Sato, Y., Sato, I.: Sensation-based photo cropping. In: Proceedings of the 17th International Conference on Multimedia 2009, pp. 669–672. ACM, Vancouver (2009)
11. Cheng, B., Ni, B., Yan, S., Tian, Q.: Learning to photograph. In: Proceedings of the International Conference on Multimedia, Ser. MM 2010, pp. 291–300. ACM (2010)
12. Bhattacharya, S., Sukthankar, R., Shah, M.: A framework for photo-quality assessment and enhancement based on visual aesthetics. In: Proceedings of the International Conference on Multimedia (MM 2010), pp. 271–280. ACM, New York (2010)
13. Liu, L., Chen, R., Wolf, L., Cohen-Or, D.: Optimizing photo composition. Computer Graphic Forum 29(2), 469–478 (2010)
14. Ahn, S., Agrawala, M., Hartmann, B., Barsky, B.A.: Image Cropping: Collection and Analysis of Crowdsourced Data Technical Report No. UCB/EECS-2012-94 (2012)
15. McManus, I.C., Zhou, F.A., l'Anson, S., Waterfield, L., Stöver, K., Cook, R.: The psychometrics of photographic cropping:The influence of colour, meaning, and expertise. Perception 40(3), 332–357 (2011)
16. Frintrop, S., Rome, E., Christensen, H.I.: Computational visual attention systems and their cognitive foundations: A survey. ACM Transactions on Applied Perception (TAP) 7(1), Article 6 (2010)
17. Judd, T., Ehinger, K., Durand, F., Torralba, A.: Learning to predict where humans look. In: 12th International Conference on Computer Vision (2009)
18. Achanta, R., Hemami, S., Estrada, F., Süsstrunk, S.: Frequency-tuned Salient Region Detection. In: IEEE International Conference on Computer Vision and Pattern Recognition (CVPR 2009), pp. 1597–1604 (2009)
19. Itti, L., Koch, C., Niebur, E.: A model of saliency-based visual attention for rapid scene analysis. IEEE Transactions on Pattern Analysis and Machine Intelligence 20(11), 1254 (1998)
20. Harel, J., Koch, C., Perona, P.: Graph-based visual saliency. Advances in neural information processing systems, vol. 19, pp. 545–552. MIT Press, Cambridge (2007)
21. Lowe, D.G.: Distinctive Image Features from Scale-Invariant Keypoints. International Journal of Computer Vision 60(2), 91–110 (2004)
22. Bay, H., Tuytelaars, T., Van Gool, L.: SURF: Speeded up robust features. In: Leonardis, A., Bischof, H., Pinz, A. (eds.) ECCV 2006, Part I. LNCS, vol. 3951, pp. 404–417. Springer, Heidelberg (2006)
23. Harris, C., Stephens, M.: A combined edge corner detector. In: 4th Alvey Vision Conference (1998)
24. Liu, T., Yuan, Z., Sun, J., Wang, J., Zheng, N., Tang, X., Shum, H.Y.: Learning to detect a salient object. IEEE Transactions on Pattern Analysis and Machine Intelligence 33(2), 353–367 (2011)

First Quantization Coefficient Extraction from Double Compressed JPEG Images

Fausto Galvan[1], Giovanni Puglisi[2],
Arcangelo R. Bruna[2], and Sebastiano Battiato[2]

[1] University of Udine, Italy
fausto.galvan@uniud.it
[2] University of Catania, Italy
Dipartimento di Matematica e Informatica - Image Processing Lab
{puglisi,bruna,battiato}@dmi.unict.it

Abstract. In the forensics domain can be useful to recover image history, and in particular whether or not it has been doubly compressed. Clarify this point allows to assess if, in addition to the compression at the time of shooting, the picture was decompressed and then resaved. This is not a clear indication of forgery, but it can justify further investigations. In this paper we propose a novel technique able to retrieve the coefficients of the first compression in a double compressed JPEG image when the second compression is lighter than the first one. The proposed approach exploits the effects of successive quantizations followed by dequantization to recover the original compression parameters. Experimental results and comparisons with a state of the art method confirm the effectiveness of the proposed approach.

Keywords: Double JPEG Compression, Forgery Identification.

1   Introduction

The pipeline which leads to ascertain whether an image has undergone to some kind of forgery leads through the following steps: determine whether the image is "original" and, in the case where the previous step has given negative results, try to understand the past history of the image. To discover that an input image has been manipulated or not is a prelude to any other type of investigation. Regarding the preliminary stage, the EXIF metadata could be examined, but they are not so robust to tampering, so that they can provide indicative but not certain results. To discover image manipulations, many approaches have been proposed in literature. In this scenario, a lot of works as reported in [4], have proved that a very promising way is the analysis of the statistical distribution of the values assumed by the DCT coefficients. In this regard, in [11], [12] and [10] is shown how, by checking the related histogram, it is possible to determine whether the image was or not doubly saved, and also if the quantization coefficient, was greater (or less) than that used in the first compression. Other methods suggest JPEG blocking artefacts analysis [6] and the use of the hash functions [2]. In

A. Petrosino (Ed.): ICIAP 2013, Part I, LNCS 8156, pp. 783–792, 2013.

[9], the authors suggest a method that assess whether an image has been double compressed with the same compression factor. The part of this theory not yet fully developed, at least in our knowledge, concerns the determination of the values of the coefficients of the first quantization. In [10] the authors expose some ideas to retrieve the coefficients of the first quantization: the first and the second one, just using the evaluation of the behavior of normalized histograms, provide unsatisfactory results. They then focused on a method that uses a Neural Network as a classifier. Their approach, however, does not work for medium and high frequencies, and it has been proved only for a specific subset of the AC terms. The works in [14,5] also estimate first quantization coefficients but only to locate forgeries without providing exhaustive results related to its estimation. From here on, q_1 and q_2 will indicate respectively the coefficient of first and second quantization for a generic frequency in the DCT domain. In [8] the author, to estimate q_1, proposes to carry out a third quantization and then calculates the error between the coefficients before and after this step. By varying the coefficient of the third quantization the method is able to detect two minima, one (absolute) in correspondence of q_2, and one (local) in correspondence of q_1. Due to several similarities with the proposed approach, we will discuss in detail the results of this method and its limitations in the remainder of this article.

In this paper we focused on the determination of first quantization coefficients. In particular we demonstrate that, when the second quantization factor is lower than the first one, the accuracy of the proposed approach is pretty high. This is analytically proved by exploiting some interesting properties of integer numbers whenever they are quantized more than once. The proposed method can be used as stand-alone module (just to detect forgeries) or combined with other methods (e.g., in combination with the "Signature detection" algorithm proposed in [7] to retrieve the camera model used to shoot the image). The paper is structured as follows: in Section 2 JPEG compression algorithm together with some properties of double compressed images are reviewed. Section 3 presents the mathematical details of the proposed approach whereas in Section 4 the effectiveness of the proposed solution has been tested considering both synthetic and real data (double JPEG compressed images). Finally, in Section 5 we report conclusions and our prospects for future work in this field.

2   Scientific Background

The JPEG compression engine [13] works just considering a partition of the input image into 8×8 non-overlapping blocks. A DCT transform is then applied to each block; next a proper dead-zone quantization is employed just using for each coefficient a corresponding integer value belonging to a 8×8 quantization matrix [3]. The quantized coefficients, obtained just rounding the results of the ratio between the original DCT coefficients and the corresponding quantization values, are then transformed into a data stream by mean of a classic entropy coding. Coding parameters and other metadata are usually inserted into the header of the JPEG file to allow a proper decoding. If an image has been compressed twice

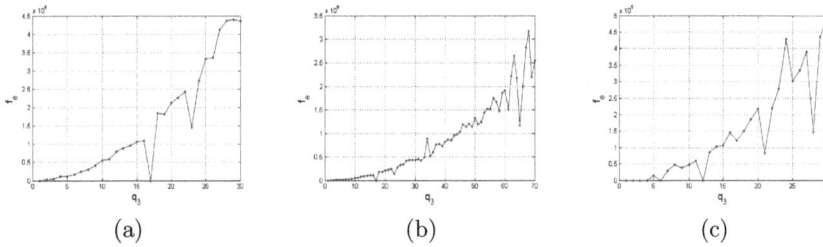

Fig. 1. Examples of the error function values (2) with respect to q_3: (a) $q_1 = 23$, $q_2 = 17$ and q_3 is in the range [1,30]; (b) $q_1 = 23$, $q_2 = 17$ and q_3 is in the range [1,70]; (c) $q_1 = 21$, $q_2 = 12$ and q_3 is in the range [1,30]

(e.g., after malicious manipulation) although, of course, the current quantization values are available, the original quantization factors are lost.

Considering a single DCT coefficient c and the related quantization factors q_1 (first quantization) and q_2 (second quantization), the value of each coefficient is given by [1,2] :

$$\left[\left[\frac{c}{q_1}\right]\frac{q_1}{q_2}\right] \tag{1}$$

To infer the value of q_1, Farid ([8]) suggests to quantize again such value with a novel quantization coefficient (q_3) varying in a proper range and evaluate an error function defined as follows[3] :

$$f_e(c, q_1, q_2, q_3) = \left|\left[\left[\left[\frac{c}{q_1}\right]\frac{q_1}{q_2}\right]\frac{q_2}{q_3}\right]q_3 - \left[\left[\frac{c}{q_1}\right]\frac{q_1}{q_2}\right]q_2\right| \tag{2}$$

For example, the typical outcome of (2) considering the DC term when $q_1 = 23$ and $q_2 = 17$ is reported in Fig. 1(a). In this specific case (both coefficients are prime numbers) we obtain interesting results: both q_1 and q_2 can be easily found, since they correspond to the two evident local minima. Hence, in this case, the first quantization coefficient q_1 can be retrieved (q_2, as mentioned before, is already available). Unfortunately, in real cases, the original quantization coefficient cannot be easily inferred as proved by considering the following cases:

- Taking into account the quantization values used before ($q_1 = 23$, $q_2 = 17$) and the same input image, but varying q_3 in the range [1,70], the outcome is more complex to analyse than before. Several local minima arise and a strong one can be found in 65 (see Fig. 1(b)).

---

[1] For sake of simplicity, truncation and rounding error have not been considered in (1). However, they have been taken into account in the design of the proposed approach (see Section 3)

[2] [.] indicates the *rounding* function

[3] |.|indicates the *abs* function

- By employing the same input image and q_3 in the range [1,30], but with different quantization coefficients $q_1 = 21$ and $q_2 = 12$ (they are both not prime numbers), the outcome reported in Fig.1(c) is obtained. Without additional information about the input image, a wrong estimation could be performed ($q_1 = 28$ in this case).

When $q_3 = q_2$, the error function (2) is equal to 0. This motivates the absolute minima found in the previous examples. Of course, the (2) allows estimating q_2, but this information is not useful since q_2 is already known from the bitstream. Obtaining a reliable q_1 estimation from (2) is a difficult task because too many cases have to be considered. In the following an alternative strategy devoted to increase the reliability of the estimation is then proposed.

3 Proposed Approach

As explained above, (2) allows obtaining q_2 but is not able to provide a reliable estimation of q_1. To overcome this problem we have designed a new error function:

$$f'_e(c, q_1, q_2, q_3) = \left| \left[\left[\left[\left[\frac{c}{q_1} \right] \frac{q_1}{q_2} \right] \frac{q_2}{q_3} \right] \frac{q_3}{q_2} \right] q_2 - \left[\left[\frac{c}{q_1} \right] \frac{q_1}{q_2} \right] q_2 \right| \qquad (3)$$

To properly understand the rationale of (3), especially when $q_3 = q_1$, we have to better analyse the effect of a single quantization and dequantization step. If we examine the behaviour of the following function:

$$\hat{c} = \left[\frac{c}{q} \right] q \qquad (4)$$

we can note that if q is odd, all integer numbers in $[nq - \lfloor \frac{q}{2} \rfloor, nq + \lfloor \frac{q}{2} \rfloor]$[4] will be mapped in nq (with n a generic integer number). If q is even it maps in nq all the integer numbers in $[nq - \frac{q}{2}, nq + \frac{q}{2} - 1]$. From now on we call this range 'space attributable to nq'. Such aspect is synthetically sketched in Fig. 2 (when q is even). As a consequence, we can say that (4) "groups" all integer numbers of its domain in multiples of q. Let observe that the maximum distance between a generic coefficient c and the corresponding $\hat{c}$, obtained by the quantization and dequantization process, is $\frac{q}{2}$ if c is even, $\lfloor \frac{q}{2} \rfloor$ if c is odd. Based on the above observation, we analyse three rounds applied in sequence, when $q_2 < q_1$ and $q_3 = q_1$:

$$\left[\left[\left[\frac{c}{q_1} \right] \frac{q_1}{q_2} \right] \frac{q_2}{q_1} \right] q_1 \qquad (5)$$

---
[4] $\lfloor . \rfloor$ indicates the *floor* function

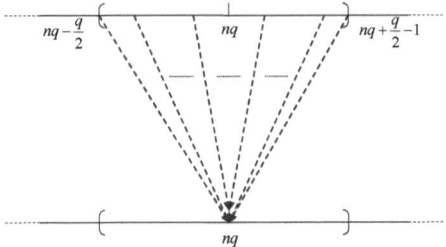

Fig. 2. The effect of quantization and dequantization for a generic term $c \in [nq - \frac{q}{2}, nq + \frac{q}{2} - 1]$ in case of q even

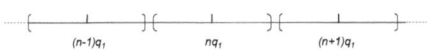

Fig. 3. The effect of quantization and dequantization for coefficient q_1

- $c_1 = \left[\dfrac{c}{q_1}\right] q_1$ for the above observations leads to the situation shown in Fig. 3;
- $c_2 = \left[\dfrac{c_1}{q_2}\right] q_2$ maps multiples of q_1 in multiples of q_2. It is worth noting that, being $q_2 < q_1$, a generic nq_1 will be mapped in a multiple of q_2 (for example mq_2) whose distance from nq_1 will be less than or equal to $\dfrac{q_2}{2}$ (or $\lfloor\dfrac{q_2}{2}\rfloor$ if q_2 is odd), then in the space attributable to nq_1;
- at this point, $\left[\dfrac{c_2}{q_1}\right] q_1$ maps c_2 in nq_1 again, since, as pointed out in the preceding paragraph, c_2 is in the space attributable to nq_1.

With the three steps above, we demonstrated that (see Fig. 4):

$$\left[\left[\left[\frac{c}{q_1}\right]\frac{q_1}{q_2}\right]\frac{q_2}{q_1}\right] q_1 = \left[\frac{c}{q_1}\right] q_1 \tag{6}$$

Therefore the error function in (3) is 0 when $q_3 = q_1$ regardless the c value.

It is worth noting that in real conditions, due to rounding and truncation to eight bit integers, (3) in $q_3 = q_1$ is close to zero but not zero. In presence of multiple minima (e.g., q_1 is a multiple of q_2) this behaviour has to be carefully considered as detailed below. Starting from a double quantized image I_{DQ}, DCT coefficient c_{DQ} are extracted and, for each frequency $f_j \in [1, 2, ..., 64]$, the following algorithm is applied. First, a set of candidate to be the right value of q_1 (C_{f_j}) is collected by simply considering (3) and selecting the strongest minima with the lowest values. In order to select the correct first quantization coefficient, we exploit information coming from the double compressed image I_{DQ}.

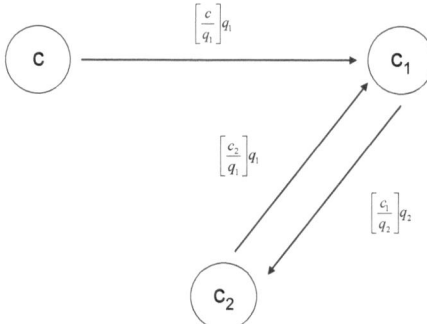

Fig. 4. Scheme describing the effect of three quantization and dequantization with coefficients q_1, q_2 and $q_3 = q_1$ again

Specifically, as already proposed in [10], by performing a proper cropping of the double compressed image, an estimation of the original DCT coefficients can be obtained $(\widehat{c_{f_j}})$. These coefficients are then used as input of a double compression procedure where the first quantization is performed by using a constant matrix with values from C_{f_j} and the second one by simply using the already known values of the second quantization coefficients $(q2_{f_j}$ values are present in the header data). At the end of this step a set of double quantized images are obtained $(I_{DQ_i}, \ i \in [1, ..., |C_{f_j}|])$ related to the different first quantization candidates. Equation (3) is then computed for each candidate image I_{DQ_i} and the output is compared with respect to the one obtained from I_{DQ} by simply using the mean absolute distance. The closest image is then found and the related $\widehat{q1_{f_j}}$ is selected as the correct one.

4   Experimental Results

In order to prove the effectiveness of the proposed approach several tests and comparisons have been performed. A first test has been conducted considering artificial data. Specifically, a random vector of 5000 elements has been built by using a uniform distribution in the range [-1023,1023]. The range corresponds to an input image within the range [0,255] in the spatial domain. In fact, the output of the DCT transform, as it is defined, is three bits wider of the input bit depth and it is centered in zero. These simulated DCT coefficients are then used as input of the error function we proposed (3) by considering several pairs of quantization coefficients with $q_1 < q_2$. As can be easily seen from Figs. 5(a), 5(b) and 5(c), (3) has a global minimum (equal to zero) when $q_3 = q_1$. Moreover, q_1 value can be found in the range $q_3 \in]q_2, +\infty[$. It is worth noting that, sometimes, more than one local minimum can be found in the range $q_3 \in]q_2, +\infty[$. For example, this behaviour can arise when q_1 is a multiple of q_2 (see Fig. 5(d)). However, even in this case, the correct q_1 value can be easily recovered. In fact,

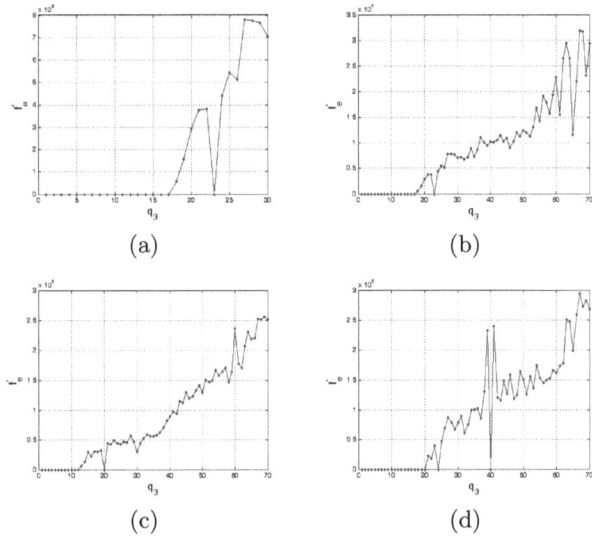

Fig. 5. Examples of error function values vs q_3, computed by using the proposed solution (3): (a) $q_1 = 23$, $q_2 = 17$ and q_3 is in the range [1,30]; (b) $q_1 = 23$, $q_2 = 17$ and q_3 is in the range [1,70]; (c) $q_1 = 20$, $q_2 = 12$ and q_3 is in the range [1,70]; (d) $q_1 = 40$, $q_2 = 20$ and q_3 is in the range [1,70].

the additional minima are always between q_1 and q_2 and their positions depend on the common divisors between q_2 and q_3. Hence, in this case, q_1 is a multiple of q_2, the correct value is the maximum one.

To further assess the performance of our approach, several tests have been conducted considering real double compressed images. Starting from a set of 24 uncompressed images [1], by simple using JPEG encoding provided by Matlab, a dataset of double compressed images have been built just considering quality factors (QF_1, QF_2) in the range 50 to 100 at step of 10. Taking into account the condition $q_1 > q_2$, the final dataset contains 360 images.

Tables 1, 2, 3, 4 and 5 report the average percentage of erroneously estimated q_1 values at varying of quality factor relative to the DC and the first 3, 6, 10 and 15 DCT coefficients. These values have been averaged over all images [1]. The coefficients were considered in zig-zag order. This order, used in JPEG standard, allows sorting the coefficients from the lowest frequency (DC) to the highest frequencies in a 1D vector. As expected, better results are usually obtained for higher QF_1 and QF_2 values corresponding to lower quantization.

Further analyses have been conducted in order to study the performance of the proposed approach with respect to the specific DCT coefficient. In Fig. 6 is reported the average percentage of erroneously estimated q_1 values at varying of the DCT coefficient (from low to high frequencies). These values are obtained averaging over all (QF_1, QF_2). As expected the performance of the proposed solution degrades with DCT coefficients corresponding to high frequencies.

Table 1. Percentage of erroneously estimated q_1 values at varying of quality factor (QF_1, QF_2) relative to the DC coefficient

QF_1 \ QF_2	60	70	80	90	100
50	4.17%	0.00%	0.00%	0.00%	0.00%
60	-	4.17%	0.00%	0.00%	0.00%
70	-	-	0.00%	0.00%	0.00%
80	-	-	-	0.00%	0.00%
90	-	-	-	-	0.00%

Table 2. Percentage of erroneously estimated q_1 values at varying of quality factor (QF_1, QF_2) relative to the first 3 DCT coefficients in zig-zag order

QF_1 \ QF_2	60	70	80	90	100
50	1.39%	0.00%	0.00%	0.00%	0.00%
60	-	1.39%	1.39%	0.00%	0.00%
70	-	-	0.00%	0.00%	0.00%
80	-	-	-	1.39%	0.00%
90	-	-	-	-	0.00%

Table 3. Percentage of erroneously estimated q_1 values at varying of quality factor (QF_1, QF_2) relative to the first 6 DCT coefficients in zig-zag order

QF_1 \ QF_2	60	70	80	90	100
50	2.08%	0.00%	0.00%	0.00%	0.00%
60	-	4.17%	1.39%	0.00%	0.00%
70	-	-	0.00%	0.00%	0.00%
80	-	-	-	0.69%	0.00%
90	-	-	-	-	0.00%

Table 4. Percentage of erroneously estimated q_1 values at varying of quality factor (QF_1, QF_2) relative to the first 10 DCT coefficients in zig-zag order

QF_1 \ QF_2	60	70	80	90	100
50	7.92%	0.00%	0.00%	0.83%	0.00%
60	-	10.42%	10.42%	0.00%	0.00%
70	-	-	0.00%	0.00%	0.00%
80	-	-	-	0.42%	0.00%
90	-	-	-	-	0.00%

To test the effectiveness of the proposed function (3), some comparisons have been performed with the same automatic algorithm described before but considering function (2) proposed in [8] instead of (3). As can be easily seen from

Table 5. Percentage of erroneously estimated q_1 values at varying of quality factor (QF_1, QF_2) relative to the first 15 DCT coefficients in zig-zag order

		QF_2				
		60	70	80	90	100
	50	19.44%	12.78%	0.56%	2.22%	5.83%
QF_1	60	-	24.17%	14.44%	0.00%	0.00%
	70	-	-	1.67%	0.00%	0.00%
	80	-	-	-	0.28%	0.00%
	90	-	-	-	-	0.00%

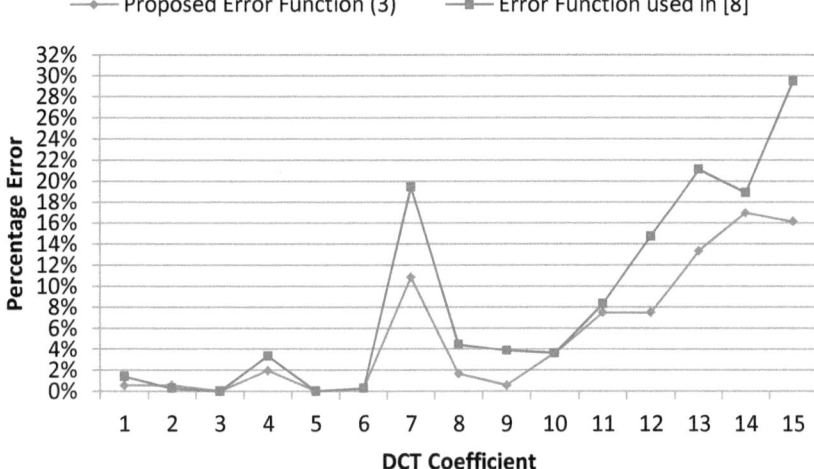

Fig. 6. Percentage of erroneously estimated q_1 values at varying of DCT coefficient position (zig-zag scanning) considering the proposed error function (3) and the one presented in [8]. These values are obtained averaging over all (QF_1, QF_2) and images [1]

Fig. 6, the proposed approach provides satisfactory results outperforming the considered state-of-the-art approach.

5 Conclusion

In this paper we proposed a novel algorithm for the estimation of the first quantization coefficient from double compressed JPEG images. To confirm the effectiveness of the proposed solution, several tests have been conducted both on synthetic and real data. Comparisons have been also performed with the error function proposed in [8]. Future work will be devoted to find smart solutions able to retrieve q_1 also in the case where $q_1 \leq q_2$. Moreover, additional experiments related to the recovering of the overall initial quantization matrix will be performed considering a double compression process achieved by applying actual quantization tables used by camera devices and common photo-retouching software (e.g., Photoshop, Gimp, etc.).

References

1. Dataset Eastman Kodak Company: PhotoCD PCD0992,
   `http://r0k.us/graphics/kodak/`
2. Battiato, S., Farinella, G.M., Messina, E., Puglisi, G.: Robust image alignment for tampering detection. IEEE Transactions on Information Forensics and Security 7(4), 1105–1117 (2012)
3. Battiato, S., Mancuso, M., Bosco, A., Guarnera, M.: Psychovisual and statistical optimization of quantization tables for DCT compression engines. In: Proceedings of the 11th International Conference on Image Analysis and Processing (ICIAP), pp. 602–606 (2001)
4. Battiato, S., Messina, G.: Digital forgery estimation into DCT domain: a critical analysis. In: Proceedings of the First ACM Workshop on Multimedia in Forensics (MiFor), pp. 37–42 (2009)
5. Bianchi, T., Piva, A.: Image forgery localization via block-grained analysis of JPEG artifacts. IEEE Transactions on Information Forensics and Security 7(3), 1003–1017 (2012)
6. Bruna, A.R., Messina, G., Battiato, S.: Crop detection through blocking artefacts analysis. In: Maino, G., Foresti, G.L. (eds.) ICIAP 2011, Part I. LNCS, vol. 6978, pp. 650–659. Springer, Heidelberg (2011)
7. Farid, H.: Digital image ballistics from JPEG quantization: A followup study. Tech. Rep. TR2008-638, Department of Computer Science, Dartmouth College (2008)
8. Farid, H.: Exposing digital forgeries from JPEG ghosts. IEEE Transactions on Information Forensics and Security 4(1), 154–160 (2009)
9. Huang, F., Huang, J., Shi, Y.Q.: Detecting double JPEG compression with the same quantization matrix. IEEE Transactions on Information Forensics and Security 5(4), 848–856 (2010)
10. Lukas, J., Fridrich, J.: Estimation of primary quantization matrix in double compressed JPEG images. In: Proceedings of Digital Forensic Research Workshop, DFRWS (2003)
11. Popescu, A., Farid, H.: Statistical tools for digital forensics. In: Fridrich, J. (ed.) IH 2004. LNCS, vol. 3200, pp. 128–147. Springer, Heidelberg (2004)
12. Redi, J., Taktak, W., Dugelay, J.: Digital image forensics: a booklet for beginners. Multimedia Tools and Applications 51(1), 133–162 (2011)
13. Wallace, G.K.: The JPEG still picture compression standard. Communications of the ACM 34(4), 30–44 (1991)
14. Wang, W., Dong, J., Tan, T.: Exploring DCT coefficient quantization effect for image tampering localization. In: Proceedings of Workshop on Information Forensics and Security (2011)

Soccer Ball Detection
with Isophotes Curvature Analysis

Tommaso De Marco, Marco Leo, and Cosimo Distante

CNR - INO. Istituto Nazionale di Ottica, via della Libertà 3, Arnesano (LE)
{tommaso.demarco,marco.leo,cosimo.distante}@ino.it
http://www.ino.it

Abstract. Circle detection is a critical issue in image analysis: it is
undoubtedly a fundamental step in different application contexts, among
them one of the most challenging is the detection of the ball in soccer
game. Hough Transform based circle detector are largely used but there is
a large open research area that attempt to provide more effective and less
computationally expensive solutions based on randomized approaches,
i.e. based on iterative sampling of the edge pixels. To this end, this
work presents an ad-hoc randomized iterative work-flow, which exploits
geometrical properties of isophotes, the curvature, to identify edge pixels
belonging to the ball boundaries; this allow to consider a large amount of
edge pixels, but limiting most of the time-consuming computation only
on a restricted subset given by pixels with an high probability to lie on a
circular structure. The method, coupled with a background suppression
algorithm, has been applied to a set of real images acquired by fixed
camera providing performances higher than a standard circular Hough
transform solver, with a detection rate $> 86\%$.

Keywords: Ball detection, Isophote, Image sampling.

1 Introduction

Circle detection is a critical issue in pattern recognition and computer vision
[1], as in the automatic sport video analysis. In particular in the soccer game
circle detection is necessary in order to detect and localize the ball on each video
frame. This is a critical task, and, although different techniques have been proposed, further solutions need to be investigated in order to achieve the accuracy
required for a real usage during the matches [2]. The most powerful methods in
literature address this problem dividing the computation in two phases: firstly,
on each frame the regions with an high probability to contain the ball are selected
(*ball candidate extraction*), secondly, selected regions are processed to discover
the real presence of the ball (*ball candidate validation*) [3]. Candidate ball extraction is performed with circle detection methods, as the circular Hough transform
(CHT) or randomized techniques. Randomized approaches are particularly interesting in this context because they reduce the computational power required
by Hough based approaches, and in addition make relatively easy to add further

A. Petrosino (Ed.): ICIAP 2013, Part I, LNCS 8156, pp. 793–802, 2013.

constraints and validity checks to increase the detection performances. In a randomized approach three random edge points are iteratively chosen to determine the parameters of a candidate circle, then the best one(s) is selected according to a voting strategy. However, before the voting process, a series of validation checks are necessary to discard possible invalid circles: in the Randomized Circle Detection (RCD) method [4] at each iteration four edge pixels are selected: three pixels are used to define a possible circle, while the fourth one is used to check if it can be considered a valid candidate or not; then an evidence-collecting process is performed to select best circle(s) among the candidates, without using any accumulator.

In this work we propose an alternative multiple-evidence strategy to discard invalid circles: maintaining the four edge pixels approach, we add further constraints based on the curvature of the isophotes. Isophotes are curves connecting pixels in the image with equal intensity, whose properties make them particularly suitable for objects detection [5]. Analysis of isophote curvature allows to select pixels with a curvature compliant with the ball radius, and consequently limiting the sampling process only to edge pixel with an high probability to lie on the ball circumference. A voting process, based on kernel density estimation, is performed for each candidate circle, determining if it is a real circle or not. Then, results are refined with an error linear compensation algorithm, in order to provide a better fitting with the edge points lying on the recognized circle.

2   Algorithm Overview

The proposed algorithm is based on analysis of the curvature of isophotes, curves connecting pixels in the image with equal intensity. Isophotes curvature is computed for each pixel in the image, then data are restricted to edge pixels extracted using a Canny operator [6]; accuracy of edge extraction is improved performing on each frame an adaptive background suppression [7].

Obtained curvature distribution is analyzed to detect the occurrence of most probable values, providing a clusterization of the edge pixels into subsets of equal curvature; then subset(s) with curvature compliant with the expected ball radius are selected.

Each subset is separately processed by an iterative randomized algorithm; candidate circles parameters are computed by randomly sampling three pixels at each iteration, then invalid circles are discarded by performing a series of evidence checks.

Best circles are then selected between all the candidates by a voting algorithm, and, finally a linear error compensation process is applied to improve accuracy of the recognized circles.

2.1   Adaptive Background Subtraction

At first all the moving regions are detected making use of a background subtraction algorithm. The procedure consists of a number of steps. At the beginning of the image acquisition a background model has to be generated and later continuously updated to include lighting variations in the model. Then, a background subtraction algorithm distinguishes moving points from static ones. Finally, a connected components analysis detects the blobs in the image.

The implemented algorithm uses the mean and standard deviation to give a statistical model of the background. Formally, for each frame $I(x, y)$ at the time t the algorithm evaluates an average $\mu^t(x, y)$ and a standard deviation $\sigma^t(x, y)$ image:

$$\mu^t(x, y) = \alpha I^t(x, y) + (1 - \alpha)\mu^{t-1}(x, y) \tag{1}$$

$$\sigma^t(x, y) = \alpha |I^t(x, y) - \mu^t(x, y)| + (1 - \alpha)\sigma^{t-1}(x, y) \tag{2}$$

It should be noted that (2) is not the correct statistical evaluation of standard deviation, but it represents a good approximation of it, allowing a simpler and faster incremental algorithm that works in real time.

The background model described above is the starting point of the motion detection step. The current image is compared to the reference model, and points that differ from the model by at least two times the correspondent standard deviation are marked. Formally, the resulting motion image can be described as:

$$M(x, y) = \begin{cases} 1 \; if & |I(x, y) - \mu^t(x, y)| > 2 \cdot \sigma^t(x, y) \\ 0 \; otherwise \end{cases} \tag{3}$$

where $M(x, y)$ is the binary output of the subtraction procedure. An updating procedure is necessary to have a consistent reference image at each frame, a requirement of all motion detection approaches based on background. The particular context of application imposed some constraints. First of all, it is necessary to quickly adapt the model to the variations of light conditions, which can rapidly and significantly modify the reference image, especially in cases of natural illumination. In addition, it is necessary to avoid including players who remain in the same position for a certain period of time (goalkeepers are a particular problem for goal detection as they can remain relatively still when play is elsewhere on the pitch) in the background model. To obtain these two opposite requirements, we chose to use two different values for α in the updating equations (1) (2). The binary mask $M(x, y)$ allows us to switch between these two values, and permits us to quickly update static points ($M(x, y) = 0$) and to slowly update moving ones ($M(x, y) = 1$). Let α_S and α_D be the two updating values for static and dynamic points respectively:

$$\alpha(x, y) = \begin{cases} \alpha_S \; if & M(x, y) = 1 \\ \alpha_D \; otherwise \end{cases} \tag{4}$$

In our experiments we used $\alpha_S = 0.02$ and $\alpha_D = 0.5$. The choice of a small value for α_S is owed to the consideration that very sudden changes in light conditions can produce artifacts in binary mask 3: in such cases these artifacts would be slowly absorbed into the background, while they would have remained permanent if we had used $\alpha_S = 0$. The binary image of moving points is the input of the circle detection algorithm based on the analysis of the curvature of isophotes.

2.2   Pixel Subset Detection by Isophotes Curvature

Isophotes properties make them particularly suitable for objects detection and image segmentation, e.g. they are successfully used for accurate eye center location [8]; in particular, it has been demonstrated that their shapes are independent to rotation and varying lighting conditions [5]. Analysis of the curvature can improve the ball detection performance for different reasons. First of all, edge extraction can be performed with a low threshold, in order to detect a large number of edge pixels. This ensures to preserve the ball boundary in the edge map, which otherwise can be lost with severe thresholds in the edge extraction, for example in case of varying environmental conditions (natural or artificial lights, weather changes, etc); this feature is fundamental considering that circle detection implements the *ball candidate extraction* phase. A large edge pixels number does not affect the performances because successively only pixels with a curvature complaint with the ball target radius (with a certain tolerance of few pixels, which takes into account variations in the ball radius given by the non-constant distance between it and the camera) are selected and elaborated. In addition, the sampling process results so limited to a restricted subset given by pixels with an higher probability to lie on the ball boundary; consequently the number of iterations necessary is reduced.

Curvature κ of an isophote, which is the reciprocal of the subtended radius r, can be computed as:

$$\kappa = \frac{1}{r} = -\frac{L_y^2 L_{xx} - 2L_x L_{xy} L_y + L_x^2 L_{yy}}{(L_x^2 + L_y^2)^{3/2}} \qquad (5)$$

where $\{L_x, L_y\}$ and $\{L_{xx}, L_{xy}, L_{yy}\}$ are the first- and second-order derivatives of the luminance function $L(x, y)$ in the x and y dimensions respectively (for further details refer to [8]). Isophotes curvature is restricted to edge pixels V, as shown in the simplistic image in fig. 1 (in order to reduce the aliasing effect due to the image discretization a median filter is applied). Obtained κ distribution is analyzed to detect the occurrence of most probable values, as it was a probabilistic distribution.

The basic idea of this procedure is that, if a circle is in the image, there is an accumulation of edge pixels with the corresponding curvature. We use Mean Shift to detect local maxima in κ distribution, assigning at each edge pixel a

(a)   (b)

Fig. 1. (a) Original image. (b) 3D view of the isophotes curvature at the edges

probability weighted by a 1D gaussian kernel; V is then divided into subsets V_i, given by pixels with the same isophote curvature κ_i. Finally subset(s) V_b with curvature compliant with the target radius of the balloon are selected.

2.3   Iterative Computation of the Candidate Circles

Edge pixels in V_b are processed by an iterative randomized algorithm: at each iteration three pixels (v_j, v_k, v_l) are randomly sampled and used to compute the parameters of a circle C_{jkl} (center coordinates and radius). A series of evidence checks are performed to determine if C_{jkl} can be considered a valid circle or not. Firstly it is verified that C_{jkl} center is located in the image and the radius is compared to the expected $r_i = 1/\kappa_i$; if the difference is too high (i.e. > 1 pixel), C_{jkl} is discarded, because this means that the sampled edge points belong to different circles of equal radius.

A further check is performed applying the curvedness operator to the image [9]. Curvedness is a rotational invariant gradient operator, able to provide a measurement of the level of steepness of the gradient image:

$$curvedness = \sqrt{L_{xx}^2 + 2L_{xy}^2 + L_{yy}^2}. \tag{6}$$

Curvedness is strictly related to the density of the isophotes: in particular it is higher where isphote are denser, e.g. around the boundaries-edge of the objects. Referred to circle detection this means that C_{jkl}, to be a valid closed circle, must necessarily contain at least a local maximum of the curvedness; intuitively, if the circle is relatively small (radius of few pixels), the maximum will be located on circle center (fig. 2). Local maxima are so localized in the curvedness distribution by a non-local maxima suppression filter, with a square window of size equal to the expected ball diameter. When C_{jkl} does not contain any curvedness local maximum is discarded.

In the last check, it is verified that at least an other edge pixel in V_b is close to C_{jkl}, following the four points approach proposed in [10], where it is shown

| (a) Original image | (b) Curvedness in the original image | (c) Curvedness in the scaled image |

Fig. 2. Curvedness values of a real ball

as this allows to reduce the number of necessary iterations until 95%. However, while in [10] the fourth point is randomly sampled from the edge pixel set, in our approach we process all points in V_b, exploiting the reduced edge pixel set.

2.4 Voting Algorithm

For each subset V_b, found candidate circles are analyzed by a voting algorithm, firstly to select the best fitting one, secondly to verify if it is actually in the image or not (false positive). This procedure is based on the analysis of distances $\mathbf{d_0}$ between all edge pixels and center of the candidate C_{jkl}; considering $\mathbf{d_0}$ as a random variable, kernel density estimation $f(d, \mathbf{d_0})$ is performed using Parzen windows with Gaussian kernels $K(\cdot)$:

$$f(d, \mathbf{d_0}) = \frac{1}{2\pi r_i N} \sum_{p=1}^{N} \frac{1}{h_{\mathbf{d_0}}} K\left(\frac{d - d_{0p}}{h_{\mathbf{d_0}}}\right) \qquad (7)$$

with d the generic pixel-center distance, N the number of edge pixels and $h_{\mathbf{d_0}}$ bandwidth of Parzen windows. This procedure does not require parameters, because $h_{\mathbf{d_0}}$ is computed directly from $\mathbf{d_0}$ according to the MAD formulation [11]. Candidate with the highest absolute maximum of f is chosen, and pixels around it are taken as lying along its circumference (namely *inliers*, fig. 3). f is normalized to length $L = 2\pi r_i$ of the expected circumference; this is equivalent to impose that, to obtain the same score, circles with high radius must have more *inliers* than circles with lower radius. Finally, to avoid false positives, detected circle is considered valid only if the ratio of *inliers* over L is higher than a threshold T_{cov}.

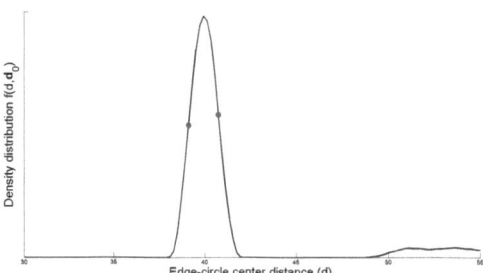

Fig. 3. Kernel density function $f(d, \mathbf{d_0})$ of the distance between edge pixels and the candidate circle center. When estimate is correct, f has a peak around the radius value; points around correspond to edge pixels lying along its circumference (*inliers*). They are selected determining the two points (red dots) around the peak where behaviour of first derivative of f changes.

2.5   Linearization

Detected circle(s) is affected by a bias effect, because parameters have been estimated only from three pixels: consequently if they are not exactly located on the circumference, detected circle does not perfectly match the ideal one. To reduce this effect we use a linear error compensation, adjusting found parameters to have a better fitting with the recognized *inliers*. Given an ideal circle with center $(\bar{x}_c, \bar{y}_c)$ and radius $\bar{R}_c$, it is defined by the following equation:

$$(x - \bar{x}_c)^2 + (y - \bar{y}_c)^2 - \bar{R}_c^2 = 0. \tag{8}$$

Given the detected parameters (x_c, y_c, R_c), affected by errors $(\delta x_c, \delta y_c, \delta R_c)$, (8) becomes:

$$(x - (x_c - \delta x_c))^2 + (y - (y_c - \delta y_c))^2 - (R_c - \delta R_c)^2 = 0. \tag{9}$$

Linearizing (9), and considering the generic edge pixel coordinates (x_i, y_i), we obtain:

$$B\delta x_c + C\delta y_c + D\delta R_c = A \tag{10}$$

where

$$\begin{aligned}
A &= -(x_i - x_c)^2 - (y_i - y_c)^2 + R_c^2 \\
B &= 2(x_i - x_c) \\
C &= 2(y_i - y_c) \\
D &= 2R_c.
\end{aligned} \tag{11}$$

Applying (10) to all N_c *inliers*, we obtain the following over-determined linear system, which is able to provide an evaluation of the parameters errors (to solve this problem we use a Moore-Penrose pseudo-inverse):

$$\begin{pmatrix} B_0 & C_0 & D_0 \\ B_1 & C_1 & D_1 \\ \vdots & \vdots & \vdots \\ B_{N_c} & C_{N_c} & D_{N_c} \end{pmatrix} \begin{pmatrix} \delta x_c \\ \delta y_c \\ \delta R_c \end{pmatrix} = \begin{pmatrix} A_0 \\ A_1 \\ \vdots \\ A_{N_c} \end{pmatrix} \tag{12}$$

3   Experimental Results

The tests are executed on image sequences acquired from fixed cameras placed on the stands of the stadium.

The first experiment is performed on a video sequence acquired during a training session by a camera with a spatial resolution of 1024×768 pixels at 504 fps. In Fig. 4 six frames are shown, with superimposed the detection results of the algorithm. In this experiment ball is always correctly detected, and number of false positives is limited and referred to moving structures in the images. Computation has been performed with a threshold T_{cov} of 90%, and a maximum number of iteration of 1000 for each valid subset V_b.

To evaluate the detection performances of the algorithm, we have extracted from the video under consideration 500 frames where the ball is always present.

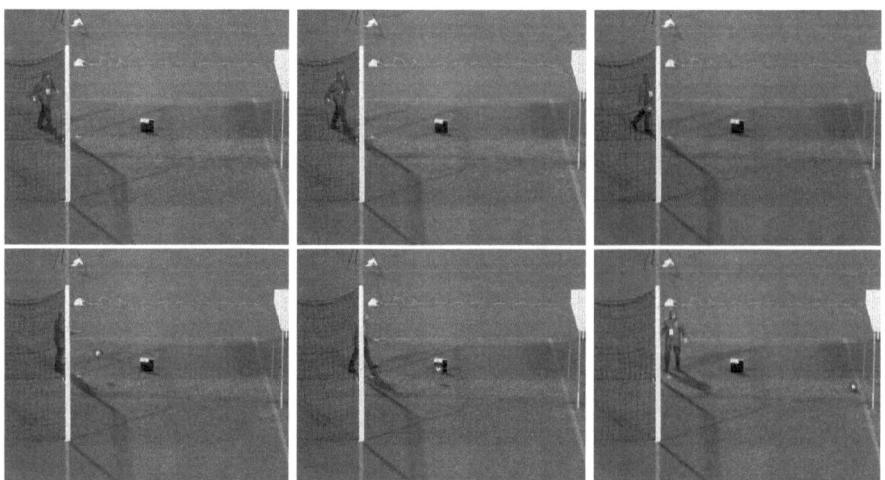

Fig. 4. Ball detection on training video

On these extracted frames we compare the performances of the proposed approach with a reference solver; in particular we use the Matlab implementation of the circular Hough transform, coupled with the same algorithm of background suppression and a level of sensitivity of 0.97; for both solvers we have used a tolerance of two pixels for the target ball radius. Obtained results are shown in table 1. The proposed randomized solver is able to achieve a detection rate higher then circular Hough transform method (more than 90 % versus 64 %), providing in addiction a smaller number of false positives (less than a false positive for frame).

Table 1. Detection performances obtained with 500 frames extracted from the training video

	Randomized solver	Circular Hough
# of detected ball	460	323
Ball detection rate	92%	64%
# of false positives/frame	0.6	3.2

An other series of tests are executed on a set of frames extracted from the *ISSIA-CNR Soccer Dataset* [12]. The dataset contains six synchronized views acquired by six Full-HD *DALSA 25-2M30* cameras, three for each major side of the playing-field, at 25 fps in quasi-stationary playing conditions. Videos are acquired during some matches of the italian "serie A". In Fig. 5 six frames are shown, with superimposed the detection results of the algorithm. In this case we have observed a reduction of the detection rate (however higher than reference solver's one), motivated by the complexity of the scene (tab. 2); in particular

Table 2. Results on videos acquired during a real soccer match

	Randomized solver	Circular Hough
Ball detection rate	86%	79%
# of false positives/frame	0.1	1.2

Fig. 5. Ball detection on real video

cases of missed detection are relative to frames where the ball is stationary, probably because the background suppression algorithm tends to mainly highlight the ball when it is moving. Number of false positive for frame continues to be less than 1.

3.1 Conclusions

In this work a randomized iterative work-flow is presented in order to provide a circle detector able to implement the *ball candidate extraction* stage in a complete ball detector for soccer game. This exploits geometrical properties of the isophote distribution on video frames to cluster a large number of edge pixels into subsets of equal curvature, limiting the successive intensive computation only to the most meaningful ones. The method, coupled with a background suppression algorithm, is applied on real videos and standard database achieving a detection rate higher than standard circular Hough solution, maintaining a lower number of false positives. This can simplify the successive *ball candidate validation* necessary to discard false positives, and eventually, estimate with tracking techniques the positions of the ball in frames where circle detector fails (e.g., in presence of occlusions).

References

1. Davies, E.R.: Machine vision: theory, algorithms, practicalities. Morgan Kaufmann (2004)
2. D'Orazio, T., Leo, M.: A review of vision-based systems for soccer video analysis. Pattern Recognition 43(8), 2911–2926 (2010)
3. Mazzeo, P.L., Leo, M., Spagnolo, P., Nitti, M.: Soccer ball detection by comparing different feature extraction methodologies. Advances in Artificial Intelligence 6 (2012)
4. Chen, T.C., Chung, K.L.: An efficient randomized algorithm for detecting circles. Computer Vision and Image Understanding 83(2), 172–191 (2001)
5. Lichtenauer, J., Hendriks, E., Reinders, M.: Isophote properties as features for object detection. In: IEEE Computer Society Conference on Computer Vision and Pattern Recognition, CVPR, vol. 2, pp. 649–654 (2005)
6. Canny, J.: A computational approach to edge detection. IEEE Transactions on Pattern Analysis and Machine Intelligence PAMI 8(6), 679–698 (1986)
7. Spagnolo, P., D'Orazio, T., Leo, M., Distante, A.: Moving Object Segmentation by Background Subtraction and Temporal Analysis. Image and Vision Computing 24, 411–423 (2006)
8. Valenti, R., Gevers, T.: Accurate eye center location through invariant isocentric patterns. IEEE Transactions on Pattern Analysis and Machine Intelligence 34(9), 1785–1798 (2012)
9. Koenderink, J.J., van Doorn, A.J.: Surface shape and curvature scales. Image and Vision Computing 10(8), 557–564 (1992)
10. Chung, K.L., Huang, Y.H., Shen, S.M., Krylov, A.S., Yurin, D.V., Semeikina, E.V.: Efficient sampling strategy and refinement strategy for randomized circle detection. Pattern Recognition 45(1), 252–263 (2012)
11. Rousseeuw, P.J., Croux, C.: Alternatives to the median absolute deviation. Journal of the American Statistical Association 88(424), 1273–1283 (1993)
12. D'Orazio, T., Leo, M., Mosca, N., Spagnolo, P., Mazzeo, P.L.: A Semi-Automatic System for Ground Truth Generation of Soccer Video Sequences. In: 6th IEEE International Conference on Advanced Video and Signal Surveillance, Genoa, Italy (September 2009)

A Bayesian Approach to Tracking Learning Detection

Giorgio Gemignani[1], Wongun Choi[2], Alessio Ferone[1],
Alfredo Petrosino[1], and Silvio Savarese[2]

[1] DSA, University of Naples "Parthenope", Napoli, Italy
{giorgio.gemignani,alessio.ferone,alfredo.petrosino}@uniparthenope.it
[2] EECS, University of Michigan, Ann Arbor, USA
silvio@eecs.umich.edu, wgchoi@umich.edu

Abstract. Tracking objects of interest in video sequences, referred in computer vision literature as **video tracking** or **visual tracking**, is an essential task for intelligent machines able to understand and react to the surrounding environment. This work investigates the problem of *robust, long-term visual tracking of unknown objects in unconstrained environments*. Such problem is affected by several challenging difficulties arising from fast camera movements, partial or total object occlusions and temporal disappearance. We describe a novel framework based on *Tracking-Learning-Detection* (*TLD*), that combine bayesian optimal filtering with *pn on-line* learning theory [12] to adapt target visual likelihood during tracking. We designed particle filtering algorithm for parameter inference and propose a solution that enables accurate and efficient tracking. The performance and the long-term stability are demonstrated and evaluated on a set of challenging video sequences usually employed to test tracking algorithms.

Keywords: Visual tracking, MCMC particle filter, Adaptive likelihood.

1 Introduction

Single target tracking, defined as the problem of estimating the state X^t of an object of interest at time t in a sequence of images $I_1, \cdots, I_t$, is a fundamental issue in computer vision since it provides low-level information for a wide range of high level analysis applications, such as visual surveillance, activity analysis, vision-based user interfaces, augmented reality, etc.

Visual trackers rely on an appearance model, i.e. an internal representation of the target appearance, learned by extracting from incoming images high discriminative visual features characterizing the target. This model is then evaluated on candidate image regions, through a set of measurements, in order to estimate the most confident target location in the current frame. In [12], *single target* trackers are classified into two broad categories namely, *short term trackers* (*STT*) and *long term trackers* (*LTT*). The former refers to standard tracking approaches such as [15] that try to find frame to frame correspondences

A. Petrosino (Ed.): ICIAP 2013, Part I, LNCS 8156, pp. 803–812, 2013.

assuming no complete occlusion or disappearance of the tracked object between consecutive frames, while the latter refers to sequences of possibly infinite length, affected by frame cuts, fast camera movements and object temporary disappearance from the scene. In [12] single target tracking is defined as the problem of *"long-term on-line tracking with minimum prior information"* where the tracker learns an appearance model by continuously adapting itself to new observed data and exploiting only information from the past. Minimum prior information underlines that object modeling is formulated as semi-supervised problem where labeled data are provided manually by the user *only at the first frame of the sequence*. Such formulation requires a model able to continuously adapt to changes of appearance and at the same time, robust to wrong measurements generated by failures.

Appearance model adaptation introduces several challenges, such as the need for simultaneous fulfillment of the contradicting goals of rapid learning and stable memory referred in [8] as the *stability-plasticity* dilemma. Furthermore, on-line evaluation of new data samples becomes a critical issue in order to detect and learn changes in pose and scale or varying illumination condition. To cope with the challenges of this task, *Adaptive Appearance Trackers (AAT)*, [1,12,10,16,24,6,2,22] rely on models able to learn changing imaging conditions. According to the type of the adopted appearance model, adaptive trackers can be grouped into three classes, namely *generative*, *discriminative* and *hybrid trackers*.

Generative trackers formulates target's appearance modeling as an unsupervised learning problem where model adaption is achieved by re-estimating target appearance distribution with new high likely samples [10,16,17,7]. Such approaches ignore discriminative information coming from the surrounding background, resulting in *high sensitivity to cluttered scenes*. On the other hand, *discriminative trackers*, using a classifier that learns a decision boundary between the appearance of the target and that of the surrounding area, w.r.t. background or other moving objects [1,12,18,24,6,2], are more *robust to clutter or resembling objects* lying in the scene. *Hybrid* trackers combine the aforementioned approaches providing more stable and flexible trackers. Authors in [21], propose to switch between discriminative and generative observation models according to targets proximity in a multi-target scenario; in [23] different generative models are aggregated by means of a weighted combination whose values are learned in each frame, by maximizing the distance to the background appearance; in [3] co-training of a short-term discriminative observation model and long-term generative one is exploited; in [14] two generative non-parametric models of target and background appearance are used to train a discriminative tracker in each frame. Authors in [12] decompose the long-term tracking task into three interacting sub-tasks, *Tracking Learning and Detection (TLD)*, performed by three independent components. The *tracker* is a *STT* component that follows the target exploiting optical flow on local feature points lying on a regular grid generated at each tracking iteration inside the target bounding box. The *detector* localizes all appearances that have been observed so far and if necessary, corrects (re-initialize) the tracker. The *integrator* selects hypothesis coming from the aforementioned components and update the global appearance model defined by a set of patches. During the update stage, it also estimate detector errors and correct it to avoid these errors in the future, by **pn** *on-line learning* paradigm [12].

As stated in [19], many of these techniques ([6,22,2,12,18]) are successful in several scenarios and *TLD* is one of the best performing *AAT* paradigm, however some critical aspects of its *tracker* component have been highlighted by theoretical analysis and verified by experimentation where high sensitivity to strong occlusions and resembling background has been revealed.

Inspired by such analysis, we extensively investigated *TLD* architecture revealing a systematic drifting behavior of the *tracker* component that passes wrong hypothesis to the *integrator* component, causing in the worst case, the learning of wrong examples. We argue that such behavior is due to the design choice of tracking points over a fixed grid that is reinitialized at each tracking iteration on the previous estimated location, assuming complete visibility of the target. As it will be explained in section 2, under occlusions, this strategy drifts the *STT* component, that starts to track the occluding object (see fig. 2) until the *detector* provides more confident hypothesis. If the target have high visual similarity with the occluding object, wrong samples could be injected into the appearance model leading to an inconsistent detector and breaking the overall performance of the tracker. The *reinitialization strategy* is a challenging problem in adaptive visual tracking. In [9] a set of simple features (e.g., optical flow features) is used to track individual parts of the object while distances among features are used to add or remove salient points during tracking. Since the set of features is *geometrically unconstrained*, the tracker is likely to get stuck on the background, losing the target. In [24] *harris corner* is used to detect stable regions for tracking and enforcing a single global affine transformation constraint to avoid drifting. However, authors assume that shape of the object can be approximated with an ellipsoid and that the object does not deform, limiting the generality of the tracker.

In this work we approach the aforementioned problems by focusing short term tracking on high textured regions localized around *harris* local maxima and formulating a novel *reinitialization strategy* that is directly encoded into our probabilistic framework. The novel idea behind our *reinitialization strategy* is to add new regions of interest around high confident points tracked from the previous frame and filter out those regions that are not geometrically consistent with the best explanation of target global appearance during the inference process. Inspired by the outlier filtering scheme proposed in [5] for multiple target tracking, we designed an *Markov Chain Monte Carlo* (*MCMC*) particle filter that automatically rejects local features not consistent with the current estimate of the target location and scale. In this way stable regions are geometrically constrained to the estimated target area *without any assumption on its shape*.

We integrated our method into *TLD* approach, resulting in a new efficient and accurate long term tracker, that we named Bayesian Tracking Learning Detection (*BTLD*).

Our contributions Are Three-Fold: we propose (*i*) a tracker that make *TLD* robust to occlusions and resembling background; (*ii*) a novel bayesian model that jointly estimate target location and select new stable feature points for the next tracking iteration, exploiting adaptive visual likelihood provided by *TLD* learning component; (*iii*) quantitative evaluation on a number of challenging video sequences, verifying how our intuition corrects the baseline method respect to underlined critical conditions.

The paper is organized as follows: section 2 describes *TLD* framework and its weakness; in section 3, the proposed generative tracker and its integration in to the *TLD* framework is presented; in section 4 experimental results are showed comparing the proposed approach with the original *TLD* and other state-of-the arts methods; finally, section 4 summarizes the main contributions and highlights open research challenges.

2   Tracking Learning Detection and Its Limits

As previously introduced, *TLD* is an hybrid long term tracker that performs robust tracking by decoupling object tracking and object detection. It uses a specific object detector, trained on-line with examples found on the trajectory of a short term generative tracker that itself does not depend on the object detector. The system architecture is built on three interacting components namely the *tracker*, *detector* and *integrator*. The *tracker* component is a short term generative tracker that self-learns the appearance model and is based on *median flow tracker* [11] extended with failure detection. At each frame, it tracks by pyramidal Lucas Kanade Tracker (*KLT*) [4], a set of patches centered over points $\mathcal{K}^t = \{K^t_{i,j}\}_{i,j=1...N}$ lying on a regular grid overlapped with the target estimated bounding box. *KLT* failures are controlled by a refinement step, where wrong correspondences are rejected by a median filter computed over measures of visual similarity and motion reversibility (*forward-backward error*) calculated on the set of tracked points $\mathcal{K}^{t+1} = \{K^{t+1}_{i,j}\}_{i,j=1...N}$. Given $\mathcal{K}^t$ and $\mathcal{K}^{t+1}$, visual similarity is computed by normalized cross correlation between patches centered on them. *Forward-backward error* is computed, by backward tracking each point in $\mathcal{K}^{t+1}$ for k frames and measuring the geometric distance between points lying on backward and forward trajectory.

As stated before the adopted *reinitialization strategy* is prone to drift in presence of strong occlusion. In fig. 2 a clarifying example of this critical behavior is verified: the *coke* is characterized by areas of uniform color resembling surrounding background and as it moves behind the leaf the tracker component drifts. Indeed, after reinitialization, stationary local points on the leaf (*blue dots*) are identified by median operator as the correct ones, leading the final solution to drift (*yellow box*). Even if among ensemble's detections (*all other colored boxes*),

Fig. 1. *TLD* failure on *Coke* sequence. On the right column *positive samples*, on the left *negative samples*. In yellow *TLD* final solution corresponding to *median flow tracker* solution, while other colored boxes are detections provided by *detector*

the correct one is present, its confidence is still too low to activate error correction. The time required to correct the tracker depends on leaf similarity respect to target appearance and on the ability of the *detector* component to recover with an higher confidence the correct target hypothesis that enables error correction trigger. Such *detector* is a cascaded object detector, that analyzes the entire image by a scanning window approach, providing new hypothesis to the *integrator* component in order to correct the aforementioned failure. It performs three main stages: at the beginning, it applies a variance filter rejecting all patches with a variance lower than a given threshold; subsequently it classifies remaining patches by an ensemble of random *ferns* [13]; in the last step it evaluates the confidence of the detections by normalized cross correlation, computed with respect to the nearest *positive* and *negative* patches (respectively on the right and left, in fig. 2) defining the appearance model. The *integrator* component is designed to perform *model adaption* by selecting the highest confident results provided by *tracker* and *detector* to estimate current target location X^t and providing new samples to the learning process. It also bootstraps the ensemble classifier building the object detector, by ***pn**-learning* [12]. Such approach is based on structural constraints namely *P-constraint* and *N-constraint* that identify and relabel miss-classified data samples according to the assumption that all patches highly overlapping with the estimated state X^t should be classified as positive while patches far from it should be classified as negative. Retraining the detector with such strategy realizes a feature selection stage where challenging samples, in *fern* feature space, lying near the decision boundary are continuously corrected, providing a strong classifier, able to re-detect the target during drift.

3   Proposed Approach

In this work, we focus *KLT* only on salient regions defined around local maximum of *harris* operator. This strategy as stated in [20], reduces *KLT* failures to resembling background since it restricts the tracking on high textured regions, even if further refinement steps are still necessary to remove the remaining erroneous correspondences. Assuming coherent motion of tracked points, we remove those whose motion does not agree with the motion distribution estimated by kernel density estimation, over the set $\mathcal{K}^{t+1}$ of tracked points. This processes removes KLT serious failures but reduces drastically the number of local feature points. To guarantee a sufficient number of such salient points, new ones, detected near the remaining points are added. This is the most critical stage, since there is no prior knowledge about the "nature" of such new salient regions. The main idea is to allow the selection of new unconstrained salient points followed by a refining step where not consistent elements are rejected. In this way geometrical constraints can be encoded in our $MCMC$ particle filter, where we introduce two competitive likelihood functions: one promotes the maximum number of salient points in $\mathcal{K}^{t+1}$, the other rejects local feature points that are not consistent with the visual model.

3.1   Bayesian Formulation

Following the Sequential Bayesian formulation, the posterior probability of target state X^t a time t is given by

$$\underbrace{p(X^t|\mathcal{O}^t)}_{posterior} \approx \underbrace{p(\mathcal{O}^t|X^t)}_{a} \int \underbrace{p(X^t|X^{t-1})}_{b}\underbrace{p(X^{t-1}|\mathcal{O}^{t-1})}_{c}\,dX^{t-1} \qquad (1)$$

where (a), (b) and (c) in Eq. 1 represent the *observation likelihood*, the *motion model* and the *posterior* from previous time, respectively. The hidden state X^t encodes location and scale of the 2D box enclosing the target, resulting in a 4D state space $\mathbf{X}^t = [\,x\ y\ w\ h\,]$, where x, y, w and h are the coordinates of the center the width and the height of the bounding box, respectively. $\mathcal{O}^t = [\mathcal{K}^t\ \mathcal{A}^t]$ represents the measurement space, where $\mathcal{K}^t = \{\mathbf{K}_i^t \in \mathcal{R}^2\}$ is the set of local points tracked from previous frame and $\mathcal{A}^t$ represent the adaptive global appearance model learned on-line by TLD. Assuming $\mathcal{K}^t$ and $\mathcal{A}^t$ independent, the *observation likelihood* $\mathcal{O}$ is factorized by $p(\mathcal{A}^t, \mathcal{K}^t|X^t) = p(\mathcal{A}^t|X^t)p(\mathcal{K}^t|X^t)$. Observation likelihood of $\mathcal{K}^t$ measures the fraction of local feature points lying inside the candidate target state X^t: $p(\mathcal{K}^t|X^t) = \frac{\mathbf{K}_i^t \in \mathbf{X}^t}{|\mathcal{K}^t|}$. Such distribution promotes candidate states containing the maximum number of tracked local features, assuming that they are free of errors. KLT failures, are automatically rejected by the global appearance likelihood modeled by TLD. It assign low confidence to hypothesis containing local tracked points and not resembling target appearance, assuming an "outliers-rejection" role similar to $RANSAC$. $p(\mathcal{A}^t|X^t)$, measuring the normalized cross-correlation distance respect to target's patches, is given in [12]. We use a linear *dynamic model* defined by a Gaussian distribution over $\mathcal{X}$, centered on previous target X^{t-1} location and scale. Considering the complexity of the given probabilistic formulation, it is extremely challenging to design an analytical inference method for estimating the MAP solution. This challenge is due to the presence of the high nonlinearity of observation likelihood functions. We propose to employ a sampling based sequential filtering technique based on the *MCMC* particle filter. At each time step t given a set of N predictions on hidden variable status X^{t-1}, we propagate samples in the particle filtering framework to get an approximation of the final posterior distribution: $p(X^{t-1}|\mathcal{O}^{t-1}) \approx \{X_s^{t-1}\}_{s=1}^N$. Propagating samples through the *motion model*, we generate particles for the *predictive distribution* and approximate the posterior distribution at time t by Monte Carlo integration:

$$p(X^t|\mathcal{Y}^t,\mathcal{K}^t) \propto p(\mathcal{Y}^t|X^t)p(\mathcal{K}^t|X^t) \sum_{s=1}^N p(X^t|X_s^{t-1})p(X_s^{t-1}|\mathcal{A}^{t-1},\mathcal{K}^{t-1}) \quad (2)$$

Approximation in eq. 2 is achieved by a Markov chain over the joint space of $\mathcal{X}$ that converges over the posterior distribution $p(X^t|\mathcal{Y}^t,\mathcal{K}^t)$. The whole *Metropolis-Hasting* procedure is sketched in algorithm 1. For *MCMC* sampling

Algorithm 1. MCMC Particle Filter

1: **procedure** MCMC PARTICLE FILTER
2:     **Input:** $\mathcal{K}^t$, $\mathcal{Y}^t$, X_i^{t-1}
3:     **Output:** $p(X^t|\mathcal{Y}^t,\mathcal{K}^t)$
4:     Initialize $X_0^t = X^{t-1}$
5:     **while** $i < N_{accept}$ **do**
6:         Propose $X^* \sim N(X^i|X^{i-1})$
7:         Evaluate the acceptance probability $\alpha = min(1, \frac{p(X_s^*|\mathcal{A}^t,\mathcal{K}^t)}{p(X_s^{i-1}|\mathcal{A}^t,\mathcal{K}^t)})$
8:         Accept $X_s^* \rightarrow X_s^{i+1}$ if $\alpha < u \leftarrow$ uniform sample $\in [0\ 1]$
9:     **end while**
10: **end procedure**

to be successful, it is critical to have a good proposal distribution which can explore the hypothesis space efficiently. Our *proposal distribution* generates separate random hypothesis for location and scale subspaces, according to normal deviates from previous accepted hypothesis. Once the sampling method has reached convergence, the maximum a posterior estimate for X^t is analyzed by the *TLD integrator* to establish the final solution.

4   Experimental Results

We evaluate, quantitatively, *BTLD* using challenging sequences from the *MIL-Boost dataset*. Each experiment in this section adopts the evaluation protocol proposed in [12]. The tracker is initialized in the first frame of a sequence and tracks the object of interest up to the end. The performance are evaluated by the average percentage of frames for which the overlap between the identified bounding box and the ground-truth bounding box is at least 50%. Authors in [19], identified in *Coke* and *Faceocc2* the most critical sequences for *TLD*.

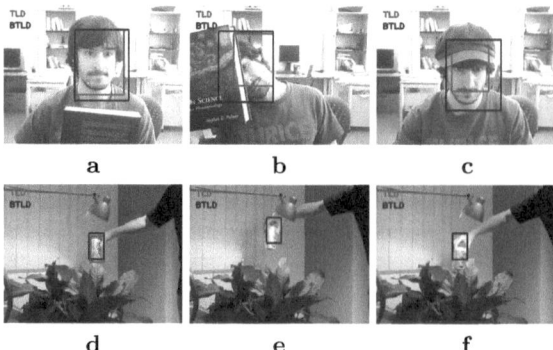

Fig. 2. From top to bottom: *Faceocc2*, *Coke*. In Red *TLD* estimated object state, in blue *BTLD* estimated object state

Fig. 3. Sequences *Sylvester*(a), *David*(b), *Dollar*(c), *Tiger2*(d). In Red *TLD* estimated object state, in blue *BTLD* estimated object state

As stated in Section 2, the *Coke* sequence proves sensitivity to occlusion and resembling background. The target is affected by several occlusions at the beginning of the sequence (fig. 2-**d**) resulting in a not effective object detector. Furthermore, coke continuous motion causes drifting of the *median flow* that in few frames loses the target and is unable to be restarted since the ensemble classifier does not detect the target (fig. 2-**e**,**f**). Our method, by tracking only stable points, does not lose the target (fig. 2-**e**,**f**) outperforming the baseline method and other state of the art approaches. In sequence *Faceocc2* sensitivity to occlusions and changes of appearance is analyzed, since a man is continuously occluding his face behind a book (fig. 2-**b**). Moreover during the sequence the man wears a hat (fig. 2-**c**), so that the adaptivity of the tracker to permanent changes of appearance can be evaluated. Reported frames (**a**,**b**,**c**) highlight how *BTLD* produces more accurate detection results since target state estimation is exploited by temporal consistency that controls variation in position and scale over time. Quantitative results reported in table 4 confirm the improvement in accuracy achieved respect to the baseline method. We evaluated our method also on sequences *Sylvester*, *David*, *Tiger1*, *Tiger2* and *Dollar* in order to verify the ability of our method to improve *TLD* results in other scenarios where the baseline method produces accurate tracking itself. In fig. 3 we show some conditions where our method corrects (fig. 2-**a**,**d**) or produces more accurate results (fig. 2-**b,c**) compared to the baseline method. As expected from the theoretical analysis, by fixing short term tracking instability, we increase tracking performance on all the tested sequences. Results reported in table 4 underline the improvement achieved by integrating our component into the *TLD* method. Furthermore, experiments underline how *BTLD* also affects appearance modeling since it provides more stable hypothesis to the learning component.

Table 1. *recall* measuers. The best performance on each video is boldfaced.

Sequence	frames	MIL [2]	ORF [18]	TLD [12]	BTLD
1. *David*	1200	0.70	0.95	1.00	**1.00**
2. *FaceOcc*	820	0.96	0.70	0.96	**1.00**
3. *Sylvester*	1440	0.93	0.71	0.97	**1.00**
4. *Coke*	292	0.46	0.17	0.60	**0.91**
5. *Tiger1*	353	0.78	0.27	0.88	**0.92**
6. *Tiger2*	364	0.80	0.21	0.85	**0.94**
7. *Dollar*	326	**1.00**	—	0.86	0.93

5   Conclusions

In this paper, we developed *BTLD*, a novel generative tracker that corrects a systematic drifting behavior revealed in the short term tracker of *TLD* approach. We designed a generative model that jointly solve feature selection and resampling exploiting a global adaptive appearance model as outlier removal. A real-time implementation of the *MCMC* particle filter framework has been described in detail and an extensive set of experiments was performed in order to highlight the ability of our approach to increase robustness of *TLD* tracker.

References

1. Avidan, S.: Ensemble tracking. In: CVPR, pp. 494–501 (2005)
2. Babenko, B., Yang, M.H., Belongie, S.: Robust object tracking with online multiple instance learning. IEEE Trans. Pattern Anal. Mach. Intell. 33(8), 1619–1632 (2011)
3. Blum, A., Mitchell, T.: Combining labeled and unlabeled data with co-training. In: Proceedings of the Eleventh Annual Conference on Computational Learning Theory, COLT 1998, pp. 92–100. ACM, New York (1998)
4. Yves Bouguet, J.: Pyramidal implementation of the lucas kanade feature tracker. Intel Corporation, Microprocessor Research Labs (2000)
5. Choi, W., Savarese, S.: Multiple target tracking in world coordinate with single, minimally calibrated camera. In: Daniilidis, K., Maragos, P., Paragios, N. (eds.) ECCV 2010, Part IV. LNCS, vol. 6314, pp. 553–567. Springer, Heidelberg (2010)
6. Tang, F., Brennan, S., Zhao, Q., Tao, H.: Co-tracking using semi-supervised support vector machines. IEEE Trans. Pattern Anal. Mach. Intell., 1–8 (August 2007)
7. Fan, J., Shen, X., Wu, Y.: Closed-loop adaptation for robust tracking. In: Daniilidis, K., Maragos, P., Paragios, N. (eds.) ECCV 2010, Part I. LNCS, vol. 6311, pp. 411–424. Springer, Heidelberg (2010)

8. Grossberg, S.: Competitive learning: From interactive activation to adaptive resonance. Cognitive Science 11(1), 23–63 (1987)
9. Hoey, J.: Tracking using flocks of features, with application to assisted handwash-
   ing. In: British Machine Vision Conference BMVC (2006)
10. Jepson, A.D., Fleet, D.J., El-maraghi, T.F.: Robust online appearance models for
    visual tracking, pp. 415–422 (2001)
11. Kalal, Z., Mikolajczyk, K., Matas, J.: Forward-backward error: Automatic detec-
    tion of tracking failures. In: Proceedings of the 2010 20th International Conference
    on Pattern Recognition, ICPR 2010, pp. 2756–2759 (2010)
12. Kalal, Z., Mikolajczyk, K., Matas, J.: Tracking-learning-detection. IEEE Transac-
    tions on Pattern Analysis and Machine Intelligence 34(7), 1409–1422 (2012)
13. Lepetit, V., Lagger, P., Fua, P.: Randomized trees for real-time keypoint recogni-
    tion. In: CVPR, pp. 775–781 (2005)
14. Lu, L., Hager, G.D.: A nonparametric treatment for location/segmentation based
    visual tracking. In: 2007 IEEE Computer Society Conference on Computer Vision
    and Pattern Recognition (CVPR 2007), Minneapolis, Minnesota, USA, June 18-23,
    IEEE Computer Society (2007)
15. Lucas, B.D., Kanade, T.: An iterative image registration technique with an appli-
    cation to stereo vision. In: Proceedings of the 7th International Joint Conference
    on Artificial Intelligence, IJCAI 1981, vol. 2, pp. 674–679. Morgan Kaufmann Pub-
    lishers Inc., San Francisco (1981)
16. Matthews, I., Ishikawa, T., Baker, S.: The template update problem. IEEE
    PAMI 26, 810–815 (2003)
17. Ross, D.A., Lim, J., Lin, R.S., Yang, M.H.: Incremental learning for robust visual
    tracking (2008)
18. Saffari, A., Leistner, C., Santner, J., Godec, M., Bischof, H.: On-line random forests
19. Salti, S., Cavallaro, A., di Stefano, L.: Adaptive appearance modeling for video
    tracking: Survey and evaluation. IEEE TIP 21(10), 4334–4348 (2012)
20. Shi, J., Tomasi, C.: Good features to track. In: 1994 IEEE Conference on Computer
    Vision and Pattern Recognition (CVPR 1994), pp. 593–600 (1994)
21. Song, X., Cui, J., Zha, H., Zhao, H.: Vision-based multiple interacting targets
    tracking via on-line supervised learning. In: Forsyth, D., Torr, P., Zisserman, A.
    (eds.) ECCV 2008, Part III. LNCS, vol. 5304, pp. 642–655. Springer, Heidelberg
    (2008)
22. Teichman, A., Thrun, S.: Tracking-based semi-supervised learning. Int. J. Rob.
    Res. 31(7), 804–818 (2012)
23. Yang, M., Lv, F., Xu, W., Gong, Y.: Detection driven adaptive multi-cue integra-
    tion for multiple human tracking. In: 2009 IEEE 12th International Conference on
    Computer Vision, pp. 1554–1561. IEEE (2009)
24. Yin, Z., Collins, R.T.: On-the-fly object modeling while tracking. In: CVPR 2007,
    Minneapolis, Minnesota, USA, June 18-23. IEEE Computer Society (2007)

Blind Invisible Watermarking Technique in DT-CWT Domain Using Visual Cryptography

Meryem Benyoussef, Samira Mabtoul,
Mohamed El Marraki, and Driss Aboutajdine

Mohammed V-Agdal University, Faculty of Science, Department of Physics
LRIT Associated Unit to the CNRST-URAC N 29
benyoussef.meryem@yahoo.fr

Abstract. A method for digital image copyright protection is presented in this paper. The proposed method is a blind invisible and robust image watermarking scheme based on Dual Tree Complex Wavelet Transform (DT-CWT) and Visual Cryptography concept (VC). This method does not require that the watermark to be embedded into the original image which leaves the marked image equal to the original one. In the concealing and extracting process, the image is transformed in the complex wavelet domain to generate a secret and a public share respectively, using LL sub-band features and a VC codebook. To extract the watermark from the attacked image, the secret and public shares are stacked together. To improve the visual quality of the extracted watermark, a post process called reduction procedure is also proposed. The experimental results show that the proposed method can withstand several image processing attacks such as cropping, filtering and compression etc...

Keywords: Robust Blind Watermarking, Visual Cryptography, Complex Wavelet Transform, Copyright Protection.

1 Introduction

Visual Cryptography (VC) was first introduced by Moni Noar and Shamir at Eurocrypt'94 [1]. VC is described as a secret sharing scheme of digital images. It involved breaking up the image into n shares using a codebook. Those shares are binary images usually presented in transparencies; so that each participant can hold a transparency (share). The act of decryption is to simply stack shares and view the secret image that appears on the stacked shares. The decoding of the secret image by the Human Visual System (HVS) is the interesting feature that has attracted the researchers in adapting this concept for several applications including watermarking. In accordance with cryptography, the security of a crypto-system does not reside in the algorithm, but resides in the secret key; that is, the security will maintain well even if the algorithm has been published.

Digital image watermarking is the technique of embedding a secret image, called also "watermark pattern", into a cover image, to protect intellectual property. The watermark pattern in the cover image can be either visible [2] or invisible [3]. However the visible watermarking techniques destroy the image quality

A. Petrosino (Ed.): ICIAP 2013, Part I, LNCS 8156, pp. 813–822, 2013.

and are easily attacked through direct image processing, which increase studies on invisible watermarking. By using this later scheme, the owner can prove his copyrights by extracting the watermark pattern from the watermarked image.

Hwang [4] is the first author who proposed a method of how to take benefit of VC to create digital image copyright protection. Since the security characteristics of VC, the watermark pattern is difficult to detect or recover from the marked image in an illegal way. Based in the Hwang idea, others related works have been proposed [5] [6] [7] [8] [9]. In the watermarking schemes using VC, the watermark pattern can be either physically embedded into the cover image or not. The first category schemes which are similar to traditional methods are called watermark embedding schemes. The second category are called watermark concealing schemes, they are particularly useful in protecting highly sensitive images, since the original image is not altered. In [9], a recent VC based watermark concealing scheme in DWT domain is proposed. To improve the security, the authors introduce three new security related performance criteria: column equity, code equity, and color equity. Column equity refers to the probability of selecting each column in the codebook of VC; code equity refers to the similarity of code-block used for coding black and white pixels of the secret image while color equity refers to the distribution of black and white pixels in each code-block.

As traditional techniques, VC based watermarking methods, described in the literature, can be grouped into two main categories. In the first one, the watermark is embedded in the spatial domain, such methods are low computational complexity but vulnerable to attacks [5] [6]. In the second one, the watermark is embedded in the transform domain, especially Discrete Wavelet Transform (DWT) [9]. In general, the DWT produces watermark images with the best visual quality due to the absence of blocking artifacts. However, it has two draw backs [12]:

- Lack of shift invariance, which means that small shifts in the input signal can cause major variations in the distribution of energy between DWT coefficients at different scales.
- Poor directional selectivity for diagonal features, because the wavelet filters are separable and real.

To overcome these problems, Kingsbury introduced the design and implementation of 2-D multi-scale transform, called Complex Dual Tree Wavelet Transform (DT-CWT), that represent edges more efficiently than does the DWT [10] [11], this will be discussed in the next section.

According to the research that we did, we didn't found any work combining VC and DT-CWT. So To benefit from the advantages of DT-CWT transform over DWT transform; we present in this work, a blind, invisible and robust VC-based watermark concealing scheme in DT-CWT domain.

The rest of this paper is organized as follows: the concept of visual secret sharing scheme and DT-CWT are explained briefly in section 2. Section 3 describes the watermark concealing/extracting phases and the reducing procedure. The experimental results and some comparisons are shown in section 4. Finally, section 5 concludes this paper.

2   Techniques Used

2.1   Dual Tree Complex Wavelet Transform (DT-CWT)

Kingsbury [10] [11] introduced a new kind of wavelet transform, called the Dual-Tree Complex Wavelet Transform that exhibits approximate shift in variant property and improved angular resolution. The DT-CWT is an over complete transform with limited redundancy (2^m: 1 for m dimensional signals). This transform has both the advantages of being approximately shift invariant and having additional directionalities ($\pm15,\pm45,\pm75$) compared to three directionalities (H,V,D) for traditional DWT (Fig 1). There are many successful applications

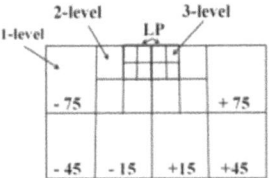

Fig. 1. 3-level DT-CWT decomposition of an image. LP corresponds to low-pass CWT coefficients

on DT-CWT watermarking, for example P. Loo and N. Kingsbury [12] suggested adding the watermark, which is a pseudo-random sequence generated from the valid wavelet coefficients (i.e. coefficients which come from the CWT of an image), to the host image CWT coefficients. Detection involves retrieving the original watermark by correlation between the original watermark CWT coefficients and the CWT of the received image. Y. Hu and al. [13] proposed a new technique of visible watermarking based on the principle of image fusion, and to better protect the host features and increase the robustness of the watermark, the dual-tree complex wavelet transform DT-CWT is used, in [14], the DT-CWT-based watermarking algorithm, robust for affine transformation and time-varying according to the degree of distortion is given.

2.2   Visual Cryptography (VC)

VC, introduced by Naor and Shamir during Eurocrypt in 1994 [1], is a Visual Secret Sharing Scheme (VSS) which uses the Human Visual System to decrypt a secret image without expensive and complicated decoding process. The original problem of VC is the special case of a 2 out of 2 visual secret sharing problem which is the most frequently used. In this scheme the secret image is divided into two shares that consist of random dots. For each pixel P of the secret image, two blocks of 1×2 pixels are generated in the corresponding location for each share. Therefore, the generated shares have a size of $1s \times 2s$ if the original image is of size $1s \times 1s$. If P is white, then the encoder chooses randomly a block of the first two columns in Table 1. If P is black, then the encoder chooses randomly a block of the last two columns in Table 1. Note that if P is white the two blocks generated are identical, but if P is black the two blocks are complementary.

In the decryption process, the two shares are stacked together. Then for a black pixel P, the result is a block with two black subpixels. But for a white pixel, the result is a block with one black subpixel and one white subpixel. By the human vision system, the block with black and white subpixels will be recognized as a white pixel and the block with two black subpixels will be recognized as a black pixel. Therefore, the secret information can be easily detected when these shares are stacked together.

Table 1. Codebook of the basic (2,2) visual cryptography

Pixel	▢		■	
Probability	1/2	1/2	1/2	1/2
Share 1	■▢	▢■	■▢	▢■
Share 2	■▢	▢■	▢■	■▢
Share 1 ⊖ Share 2	■▢	▢■	■■	■■

3   Proposed Watermarking Scheme

The proposed watermarking method consists of three components: watermark concealing, watermark extraction and watermark reduction. In the concealing process, we firstly apply the DT-CWT transform to create a binary matrix B based on LL sub-band features. The binary matrix B is used to generate the secret share from the watermark pattern based on a (2,2) VSS scheme (Table 2). This same process is repeated in the extraction process to generate the public share. This later is superimposed on the secret share to recover the ownership label. Finally, we apply a reduction process to improve the image quality of the recovered watermark (Fig 2).

3.1   Concealing Process

The steps of watermark concealing are given below:

Inputs: Host Image I $(m \times n)$, Watermark Image $(r \times c)$, Secret Key S, Number of decomposition level k
Outputs: Secret Share $(r \times 2c)$
Step 1: Select the number of wavelet decomposition level k.
Step 2: A k-level DT-CWT transform is performed on the image I.

Table 2. Codebook used to generate public and secret shares

Pixel	□		■	
Matrix B	0	1	0	1
Public Share				
Secret Share				
Public Share ⊕ Secret Share				

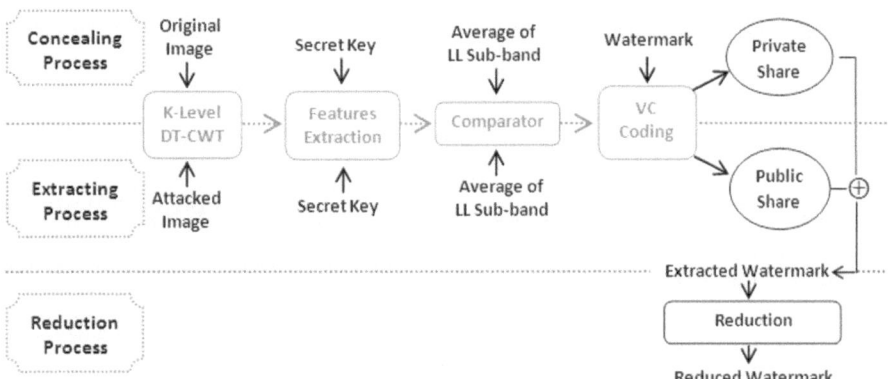

Fig. 2. Proposed watermarking sheme

Step 3: Calculate average gray level LL_{avg} of the LL^k sub-band image.

Step 4: Use S as a seed to select $r \times c$ random pixel locations within LL^k. Let $R_i(x, y)$ be the ith random location. Note that, these positions should not be the last 3 boundary pixels.

Step 5: For each $R_i(x, y)$, select a 7×7 size sub-image area centered at location $R_i(x, y)$, and find its average.

Step 6: Construct a feature image F of size $r \times c$, such that the entries in the matrix are the sample averages obtained in the above step.

Step 7: Construct a binary matrix B:

$$B(x, y) = \begin{cases} 1, & \text{if } F(x, y) \geq LL_{avg} \\ 0, & \text{if } F(x, y) < LL_{avg} \end{cases} \tag{1}$$

Step 8: Use the bits in matrix B to select columns in Table 2 for generating the secret share. Note that, the secret share size is $(r \times 2c)$, due to the codebook used.

3.2   Extracting Process

The steps of watermark extraction are given below:

Inputs: Attacked image I' $(m \times n)$, Secret Share $(r \times 2c)$, Secret Key S, Number of decomposition level k
Outputs: Watermark $(r \times 2c)$
Step 1: A k-level DT-CWT transform is performed on the image I'.
Step 2-6: Same as steps 3-7 of the concealing process.
Step 7: Use the bits in matrix B to select columns in Table 2 for generating a public share. Note that, the code-block assignment for public share corresponding to each secret bit is independent of the pixel pair colors in the watermark image.
Step 8: Perform logical OR operation on the public share and the secret share to extract the watermark.

3.3   Reduction Process

To generate the secret and public shares, the author of the basic system [9] proposed a PWVC (Pair Wise Visual Cryptoraphy). This technique aims at resolving pixels expansion by coding a pair of pixels instead of coding a single pixel each time. However it cannot allow a reduction process, without using the original watermark, to improve the image quality. Due to the proposed codebook that we used to generate the two shares in our method, the extracted watermark has a size of $(r \times 2c)$ compared to the original one. To retrieve the original size and to mitigate the noise effect caused by the watermark extraction, we use a post-process called "reduction process" that can reduce the redundancy data caused by VSS scheme in a blind way. Indeed, this process can perform a function of data reduction (Table 3); that is, a block data with two pixels located in each group will be transferred into a corresponding pixel. As shown in Table 3, if the block is composed of one black and white pixel or two white pixels then the corresponding pixel is white, but if the block is composed of two black pixels then the corresponding pixel is black.

Table 3. The lookup table of reduction process

Block of the extracted watermark	■□ □■ □□	■■
Block of the reduced watermark	□	■

4   Experimental Results

In this section, we present some experimental results concerning the proposed method. To evaluate the effectiveness of the proposed approach, four standard grayscale images are used: F-16, Elain, Boat and Mandrill of size 512×512 pixels. The watermark is a binary image with the size of 100×100 pixels (Fig. 3).

(a) F-16   (b) Elain   (c) Boat   (d) Mandrill   (e) Watermark

Fig. 3. Test images and watermark pattern used

The first type of simulation is done to show the advantage of our method compared the watermarking scheme of [9]. To test the robustness of the algorithm to attacks, our test images are subjected to several common attacks. They are compression, median filer, blurring, salt and pepper noise, histogram equalisation, cropping, resizing, rotation and translation. Table 4 shows the attacked images and the extracted watermark from each one using our method and [9] method.

The second type of simulation is done to show the robustness of the proposed algorithm with any cover image, we give in Tables 5-9 the PSNR (2) values of the attacked images and NC (3) values of the extracted watermark from each one. It can be seen from the high values of NC that our algorithm can successfully resist many common attacks.

$$PSNR = 10 \times log \frac{255^2}{MSE} \tag{2}$$

$$MSE = \frac{1}{m \times n} \sum_{i=1}^{m} \sum_{j=1}^{n} \left(c_{i,j} - c'_{i,j}\right)^2$$

where $c_{i,j}$ and $c'_{i,j}$ denote pixel intensity of the original and attacked images.

$$NC = \frac{\sum_{i=1}^{r} \sum_{j=1}^{c} \overline{\left(w_{i,j} \oplus w'_{i,j}\right)}}{r \times c} \times 100\% \tag{3}$$

where $w_{i,j}$ and $w'_{i,j}$ denote pixel intensity of the original and extracted watermarks.

Table 4. Experimental results outlining the superiority of the proposed algorithm

Attacked Images	[9]'s Results	Our Results	Attacked Images	[9]'s Results	Our Results
Cropping 25%			Cropping 50%		

| Rotation 3 | | | Rotation 10 | | |

| Scale 50% | | | Translate 20 lines | | |

| Compression 10% | | | Compression 40% | | |

| Salt & Pepper Noise 25% | | | Blurring | | |

| Histogram Equalization | | | Median Filter (3x3) | | |

5   Conclusion

The proposed method describes a robust and blind digital image watermarking in frequency domain, which is computationally efficient. This method applies the Dual Tree Complex Wavelet Transform and uses Visual Secret Sharing Scheme to generate a secret share and hide it into the LL sub-band. Thus the proposed

Table 5. PSNR and NC values after adding Salt & Pepper noise

| Images | Salt and Pepper Noise Noise density | | | | | |
| | 0.05 | | 0.15 | | 0.25 | |
	PSNR (dB)	NC (%)	PSNR (dB)	NC (%)	PSNR (dB)	NC (%)
F-16	34.87	99.28	25.45	98.24	21.02	97.41
Elain	38.77	98.54	29.39	97.32	24.99	96.15
Boat	37.55	98.60	28.17	97.19	23.77	95.67
Mandrill	36.23	97.68	26.87	95.90	22.44	93.21

Table 6. PSNR and NC values after image JPEG compression

| Images | JPEG Compression Quality factor | | | | | |
| | 70% | | 40% | | 10% | |
	PSNR (dB)	NC (%)	PSNR (dB)	NC (%)	PSNR (dB)	NC (%)
F-16	46.86	99.93	44.36	99.78	38.01	99.36
Elain	44.07	99.90	40.71	99.79	34.80	99.00
Boat	43.47	99.87	41.40	99.66	36.81	98.76
Mandrill	38.76	99.80	35.76	99.56	31.79	98.52

Table 7. PSNR and NC values after cropping

| Images | Cropping | | | | | |
| | 15% | | 35% | | 50% | |
	PSNR (dB)	NC (%)	PSNR (dB)	NC (%)	PSNR (dB)	NC (%)
F-16	26.37	94.85	16.61	81.89	12.29	75.83
Elain	31.24	96.16	21.60	85.93	16.70	73.40
Boat	38.87	99.12	19.19	83.83	15.24	74.48
Mandrill	30.75	95.58	17.56	69.49	13.25	54.89

Table 8. PSNR and NC values after rotation

| Images | Rotation | | | | | |
| | 3 | | 5 | | 10 | |
	PSNR (dB)	NC (%)	PSNR (dB)	NC (%)	PSNR (dB)	NC (%)
F-16	22.53	88.21	20.16	82.37	16.80	71.37
Elain	23.56	85.52	21.43	78.19	18.79	66.02
Boat	23.67	87.50	21.55	83.83	18.88	77.20
Mandrill	21.90	86.60	20.39	80.78	17.96	70.33

work is the first one that combine VC concept and DT-CWT. The simulation results of this combination, have confirmed that that the proposed method has high fidelity and it is robust against common image processing attacks such as cropping, lossy compression, filtering, etc.

Table 9. PSNR and NC values after Median filter, scale and translation

Images	Median Filter (3x3)		Scale 50%		Translate 20 lines	
	PSNR (dB)	NC (%)	PSNR (dB)	NC (%)	PSNR (dB)	NC (%)
F-16	46.92	99.60	40.23	99.94	15.30	75.08
Elain	35.93	99.68	35.04	99.96	17.34	66.03
Boat	41.70	99.47	39.72	99.87	17.18	72.57
Mandrill	33.15	99.12	32.38	99.84	16.17	69.19

References

1. Naor, M., Shamir, A.: Visual Cryptography. In: De Santis, A. (ed.) EUROCRYPT 1994. LNCS, vol. 950, pp. 1–12. Springer, Heidelberg (1995)
2. Liu, J.C., Chen, S.Y.: Fast two-layer image watermarking without referring to the original image and watermark. Image and Vision Computing 19, 1083–1097 (2001)
3. Chu, S.C., Roddick, J.F., Lu, Z.M., Pan, J.S.: A digital image watermarking method based on labeled bisecting clustering algorithm. IEICE Trans. E87-A(1), 282–285 (2004)
4. Hwang, R.J.: A Digital Copyright Protection Scheme Based on Visual Cryptography. Tamkang Journal of Science and Engineering 3(3), 97–106 (2001)
5. Abusitta, A.H.: A Visual Cryptography Based Digital Image Copyright Protection. Journal of Information Security 3, 96–104 (2012)
6. Wang, C.C., Tai, S.C., Yu, C.S.: Repeating Image Watermarking Technique by the Visual Cryptography. IEICE Trans. E83-A(8), 1589–1598 (2000)
7. Sleit, A., Abusitta, A.: A Visual Cryptography BasedWatermark Technology for Individual and Group Images. Systemics, Cybernetics and Informatics 5(2), 24–32 (2006)
8. Surekha, B., Swamy, G.N.: Multiple Watermarking Scheme for Image Authentication and Copyright Protection using Wavelet based Texture Properties and Visual Cryptography. International Journal of Computer Applications 23(3), 29–36 (2011)
9. Surekha, B., Swamy, G.N.: Sensitive Digital Image Watermarking for Copyright Prottection. International Journal of Network Security 15(1), 95–103 (2013)
10. Kingsbury, N.G.: Complex wavelets for shift invariant analysis and filtering of signals. Journal of Applied and Computational Harmonic 10(3), 234–253 (2001)
11. Selesnick, I.W., Baraniuk, R.G., Kingsbury, N.G.: The Dual-Tree Complex Wavelet Transform. IEEE Signal Processing Magazine 22(6), 123–151 (2005)
12. Loo, P., Kingsbury, N.G.: Watermarking using complex wavelets with resistance to geometric distortion. In: EUSIPCO, pp. 5–8 (2001)
13. Hu, Y., Huang, J., Kwong, S., Chan, Y.K.: Image Fusion Based Visible Watermarking Using Dual-Tree Complex Wavelet Transform. In: Kalker, T., Cox, I., Ro, Y.M. (eds.) IWDW 2003. LNCS, vol. 2939, pp. 86–100. Springer, Heidelberg (2004)
14. Joong-Jae, L., Kim, W., Na-Young, L., Kim, G.: A New Incremental Watermarking Based on Dual-Tree Complex Wavelet Transform. JSC 33(1-2), 133–140 (2005)

An Ensemble Algorithm Framework
for Automated Stereology of Cervical Cancer

Baishali Chaudhury[1], Hady Ahmady Phoulady[1], Dmitry Goldgof[1],
Lawrence O. Hall[1], Peter R. Mouton[2], Ardeshir Hakam[3], and Erin M. Siegel[3]

[1] Computer Science & Engineering, University of South Florida, Tampa, FL 33620
{baishali,hady}@mail.usf.edu, {goldgof,hall}@cse.usf.edu
[2] Dept of Pathology & Cell Biology, University of South Florida School of Medicine,
Tampa, FL 33620
peter@disector.com
[3] H. Lee Moffitt Cancer Center & Research Institute, Tampa, FL 33620
{ardeshir.hakam,erin.siegel}@moffitt.org

Abstract. Stereological procedures to quantify mean nuclear volume are
commonly used to differentiate cancerous from normal tissue. Automatic
quantification of these parameters requires segmentation, which is complicated by the variability in tissue staining and nuclei size. One solution
to deal with such alterations in a robust fashion is to use an ensemble of
segmentation methods. The goal of this work is to demonstrate the use of
an ensemble of simple segmentors in a novel way to improve the performance achieved by the individual segmentors. The contributions of this
paper are three fold: applying an ensemble on the blob level in addition
to the image level, utilizing the image level ensemble to accept or reject
input images based on their segmentation quality and finally applying the
ensembles for discriminating cancer and normal classes. Hematoxylin and
eosin (H&E) stained sections from archival tissues from the normal cervix
and cervical cancer have been used as the dataset. The results presented
here show that both levels of ensembles enable clear class separability as
compared to the individual segmentors, and thus demonstrate the effectiveness of the proposed ensemble framework.

Keywords: Ensemble of segmentations, microscopy images, Otsu,
cervical cancer.

1 Introduction

Subjective examination of tissue and cytology specimens by experts remains
the current approach for the diagnosis, treatment and prognostic assessment of
cervical cancer. However, these approaches suffer from poor inter-rater reliability, rater fatigue, and the morbidity associated with false negatives and false
positives [1]. Unbiased stereological approaches have the potential to strongly
enhance expert-based clinical decisions with accurate assessments of first-order
(number, length, surface area, volume) and second-order (spatial distribution,
clustering, anisotropy) parameters. The application of automatic stereological

A. Petrosino (Ed.): ICIAP 2013, Part I, LNCS 8156, pp. 823–832, 2013.
© Springer-Verlag Berlin Heidelberg 2013

quantification requires an initial first segmentation step, which is complicated by high variability between microscopy images, uneven staining and cell/nuclei population [2]. This initial segmentation when applied to a sufficient number of images from cancer and normal tissue has been shown to reveal biological differences, if present. To provide robust segmentation for catering to the diverse image types, recent developments in segmentation favor an ensemble approach with multiple segmentors rather than a single segmentation algorithm [3] [4][5][6] [7][8] [9][10]. For this study an ensemble of segmentations is applied to microscopy images from cervical cancer and normal tissue. Since staining is used to differentiate the various cell parts, intensity thresholding qualifies as an obvious and a reliable segmentation method [12] [13]. Otsu thresholding [11], an automatic, non parametric and unsupervised method to find the threshold value, is used widely because of its simplicity and stability. Our approach uses a simple ensemble created from applying three-class Otsu followed by a series of morphological operations. Although previous studies apply ensemble of segmentations mainly for parameter tuning [14][15] and optimal algorithm search, here we use the approach for a novel application. Our ensemble of segmentations is used to accept or reject images depending on their segmentation quality. Since the final analysis is based on stereological estimates, it is sufficient to use a subset of images for quantitative analysis. The final goal is to make unbiased estimates of a first-order stereology parameter (mean nuclear volume) for segmented nuclei from normal and cancer tissue. Thus an ensemble of segmentations was applied per segmented blob (nucleus) in addition to an image level ensemble which is the focus of most of the work in this field. The algorithm proposed in this paper is designed to test the hypothesis that cancer nuclei (blobs) will be on average larger than normal nuclei. The methodology (Section 2) begins with an outline of the proposed approach including a detailed description of the algorithm, followed by a description of data acquisition, the experimentation and the results (Section 3) and finally the conclusion (Section 4).

2   Methodology

The ensemble framework has two main parts individual segmentation methods and the application of the ensemble at two different levels: the image level and the blob level. The algorithm starts with an initial screening of input images based on pixel intensity, followed by its segmentation using individual segmentations and then a second level of image screening based on the size of the largest segmented nuclei, which we will call a blob. The algorithm can then be divided into two independent levels of ensemble, the image level and the blob level. The image level ensemble has been utilized here to accept or reject images based on the similarities among the segmentations. For the accepted image, the multiple segmentations are then combined to generate the final segmentation. At the blob level ensemble, the final segmentation is generated by accepting or rejecting the blobs from all the segmentations. The final set of segmented blobs from both the ensemble levels is then used independently for mean nuclear line length calculation to discriminate between cancer and normal cases. The general overview of the proposed ensemble framework is depicted in Figure 1.

2.1 Ensemble Components

Individual Segmentors. The original RGB microscopy images are converted to gray level images using a Karhunen-Loeve transform. This is followed by a three class (nuclei, cytoplasm and backgound) Otsu thresholding, which works to minimize the intra class variance and maximize the inter class variance, to detect objects of interest (nuclei). After applying the three class Otsu threshold algorithm on the gray level image, the class of pixels whose original average color is closest to the target color is selected as the foreground and the rest as background. The target color is determined by the observed stain and must be adjusted if the stain is changed.In addition to Otsu thresholding, morphological operations are applied to enhance the segmentation issues due to overlapped and partial segmented regions. For example erosion might be necessary to separate overlapped segmented regions, dialation and fill hole operations can be used to expand foreground pixels and fill interior holes respectively to rectify the problem of partially segmented regions and the opening operation could be used to remove small foreground regions that may be falsely detected as objects. Different combinations of Otsu and such morphological operations form the four individual segmentors in the ensemble and are listed below:

Segmentor 1: It starts with three class Otsu, followed by morphological opening and then a fill hole operation.

Segmentor 2: For cases where target nuclei is classfied as either of the other two classes due to low intensity variation among the three classes, a second application of three class Otsu is performed.Then a morphological opening operation is done and the interior holes are filled. The top row of Figure 3 demonstrates different steps of segmentor 2.

Segmentor 3: Cancer nuclei stain dark around nuclei edges; however, the area within the nuclei is lightly stained due to the presence of euchromatin which is the result of high cellular activity unlike normal cell nuclei which are darkly stained throughout due to the presence of heterochromatin. Thus in segmentor 3 edge detection is carried out on the original gray scale image which is then combined with the result from segmentor 1.

Segmentor 4: It starts with three class Otsu followed by edge detection, dilation and then a fill hole operation.

In cancer tissues, the cancer cells are actively replicating DNA for cell division as well as actively transcribing DNA for cell growth, processes which are uncommon in normal cells. Typically a cancer biopsy contains a mixture of normal as well as cancer cells. If the smaller normal nuclei (blobs) are retained and classified as cancer nuclei, then the accuracy of the algorithm may be affected. In order to prevent this, a connected component analysis is done after each segmentation and the connected components (nuclei blobs) smaller than *MinBlobSize*

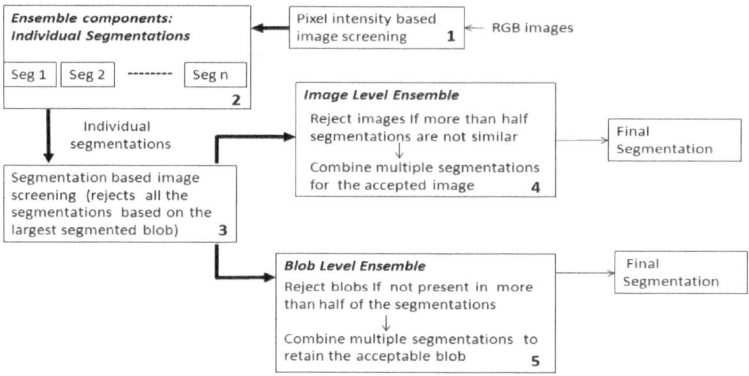

Fig. 1. Block diagram showing the proposed ensemble framework

are removed. It was observed that the nuclei size in both cancer and normal case did not exceed a certain value - *MaxBlobSize*. So if the maximum size of the segmented blob (taken over all of its four segmentation results) is more than the *MaxBlobSize* then it indicates a bad segmentation and the original image is discarded (block 3 of Figure 1). Figures 2 e-f show an example of this segmentation based image screening.

2.2   Different Ensemble Levels

A consensus function is used to combine the results from individual segmentors to get the final result and is the most important step in an ensemble framework. In this paper the ensemble has been applied on both the image and blob level. In the former case the ensemble is also applied for image acceptance/rejection depending on their segmentation quality. While at the blob level, the final segmentation is generated by accepting and rejecting blobs from all the segmentations.

Image Level Ensemble. The first step of the image level ensemble is to accept/reject the images and the second step is to combine the multiple segmentation results based on some consensus function for the accepted images to get the final segmentation. The principle used to achieve the first step is based on the argument that a similarity among at least three of the total four individual segmentations (of a particular image), indicates a good, acceptable segmentation and vice versa. Two approaches are used here to calculate this similarity, accept/reject images based on it, and achieve the final segmentation from the ensemble.

The first step for both the approaches is to consider all the combinations of the four segmentation methods taken three at a time. Let the four individual segmentations be S_1, S_2, S_3 and S_4, and their combinations are C_1 (S_1, S_2, S_3),C_2 (S_1, S_2, S_4),C_3 (S_1, S_3, S_4) and C_4 (S_4, S_2, S_3).

In ensemble approach 1, for each combination a single similarity ratio is calculated taking the three segmentations at a time. While in ensemble approach 2, for each combination three similarity ratios are calculated between the three pairs of segmentations.

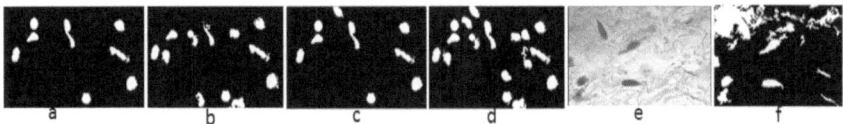

Fig. 2. a-d: Example of a rejected image since the individual segmentations dont satisfy the ImgSimThresh of 55%; e-f: image rejected based on segmentation i.e. the maximum segmented blob size $>$ *MaxBlobSize*

Ensemble approach 1

Step 1: Similarity among three segmentations S_a,S_b and S_c in a combination C_k is calculated by the following formula for all C_1,C_2, C_3 and C_4.
$$ThreeSimilarity(C_k) = \frac{|S_a \cap S_b \cap S_c|}{|S_a \cup S_b \cup S_c|}$$
Step 2: The maximum-*Max* of the four *ThreeSimilarity* values is selected

Step 3: If $Max \geqslant ImgSimThresh$ (pre defined threshold) accept the image, choose the corresponding combination, and conduct step 4 else reject the image. Figures 2 (a-d) shows an example of an image rejected through this criteria.

Step 4: For the selected combination C_k (satisfying the above condition) get the pair-wise similarities between each of the three pairs of segmentations within that combination-*PairSimilarity*.
$$PairSimilarity(S_a, S_b) = \frac{|S_a \cap S_b|}{|S_a \cup S_b|}.$$
Step 5: For each of the segmentations in the combination, get the average of the pair-wise similarities calculated between this and the other two segmentations, *Avg*. Select the segmentation which has the highest *Avg* value as the final segmentation. The middle row of Figure 3 demonstrates this approach.

Ensemble approach 2

Step 1: For a combination C_k get the pair-wise similarities between each of the three pairs of segmentations (as described in step 4 of the previous approach), select the minimum pairwise similarity *Min* and also calculate average of the three pair-wise similarities *AvgSim*

Step 2: Accept the image if *Min* (from all the four combinations) $>$ *ImgSim Thresh* and choose the combination with maximum *AvgSim*, else reject the image.

Step 3: For the selected combination, the final segmentation is generated by taking the intersection of the three segmentations within that combination: $(S_a \cap S_b \cap S_c)$

Blob Level Ensemble. A blob (nuclei) is retained if it appears in at least three segmentations; two blobs are assumed to represent the same nucleus if they pass the following criterion (The bottom row of Figure 3 shows an example of the blob level ensemble).

Co linearity check: If the distance between the blob centroids $< CentDist$, then they are considered to represent the same nucleus.

Area overlap check: If the area overlap between two blobs $\geqslant BlobSimThresh$ then it is considered to represent the same nucleus.

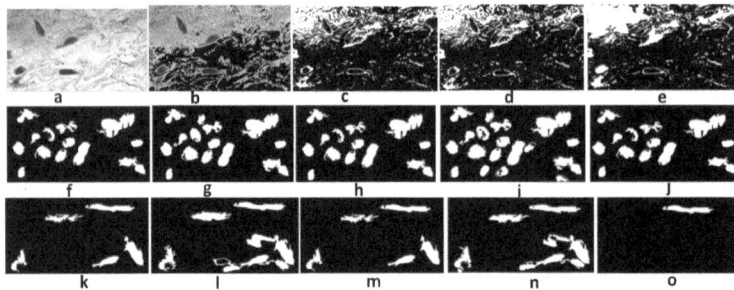

Fig. 3. (a) gray scale image;(b) 1st application of Otsu; (c) 2nd application of Otsu; (d) morphological opening of (c); (e) Fill hole operation on (d); (f)-(i) results from individual segmentors with ImgSimThresh 55% (i) ensemble 1 output;(k)-(o) individual segmentations; (e) blob level ensemble output

2.3   Volume Estimation

Because the nuclei of many cancers are on average larger than nuclei from normal cells, mean nuclear volume (MNV) is often an important feature to discriminate between the cancer and normal classes. Here stereological methods are applied to obtain MNV estimates for the segmented nuclei (blobs). Because a point-sampled intercept is automatically used to sample segmented nuclei for the MNV estimate, the estimator is termed volume-weighted mean nuclear volume (PSI-Vv) [16]. The three main steps involved for estimating PSI-Vv are placement of a point-grid for point sampled intercept (PSI) sampling; collection of line lengths (l) across sampled nuclei (shown in the top-row of Figure 4); and computation of PSI-Vv from the formula given below. The top-row in Figure 4, shows the steps involved for volume estimation. $V_V = \frac{\pi \sum_{i=1}^{N} l_i^3}{3N}$, where *N=total number of nuclei sampled by PSI in the region of interest (ROI)*.

3   Results and Discussion

3.1   Data Acquisition

The data were acquired using an integrated hardware-software-microscope system (*Stereologer*, Stereology Resource Center, Inc. Tampa-St. Petersburg, Fl.).

The source of the input was archived cervical tissue from a cone biopsy or surgery sectioned at 6 um and stained with hematoxylin and eosin (H&E) at the Moffitt cancer center. Data were acquired by placing the tissue sample/ biopsy slide with normal or cancer (squamous cell carcinoma) tissue under the microscope. For data acquisition the automatic XYZ stepping motor and *Stereologer* software were used to manually outline a ROI at low magnification. Within each ROI a minimum of 300 2-D images were captured over a single focal plane at 40x magnification.

3.2   Dataset

Images were acquired from 29 individual biopsy slides/cases, 14 normal and 15 cancerous cases. The tissue was stained to enhance the signal to noise ratio (SNR) of cell nuclei, which are the objects of interest. Some of the cases were removed after a visual screening if they had either poor acquisition quality or are complicated images (images with a large number of overlapped cells, high variation in background intensity). For the final data collection, a total of 13 cases were used, 6 normal and 7 cancer case. There are a total of 4106 and 6145 images of normal and cancer tissues, respectively, and each image has a resolution of 759 X 1138.The middle and bottom row in Figure 4 shows examples of visually acceptable and unacceptable cases respectively.

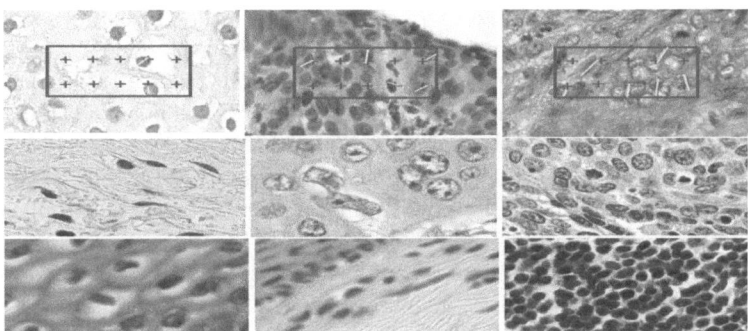

Fig. 4. Top-row: Demonstrates collection of line lengths (yellow lines) across sampled nuclei (red crosses); Middle-row: Examples of visually acceptable cases; Bottom-row: examples of visually unacceptable cases: Left- blurry image, center- background variation within image, right- overcrowded image

3.3   Results

Mean line lengths calculated from the segmented nuclei acquired from the ensembles were used to calculate the volume (PSI-Vv) as discussed in Section 2.3. Through different experiments (parameter settings), the potential of PSI-Vv to discriminate cancer from normal cases was explored. The degree of separability,

S, between the two classes with means μ_1 (for Normal class) and μ_2 (for Cancer Class) and standard deviations $Stddev_1$ (Normal Class) and $Stddev_2$, (Cancer Class) was calculated using the following formula and used to compare different experiments. This degree of separability should be more than 0 for potential separation between two classes. $S = |\mu_2 - \mu_1| - (Stddev_1 + Stddev_2)$

Since in the image level ensemble there is an additional image accept/reject stage unlike the non ensemble case and the blob level ensemble, the total number of final segmented images generated from the image level ensemble (3612 normal and 4440 cancer images) varies from that generated through the latter two scenarios (3910 normal and 4472 cancer images). Also, both image level ensembles accept/reject images in different ways and thus produce slightly different numbers of final segmented images. Results from the image level ensembles, the blob level ensemble and individual segmentators (Seg1, Seg2, Seg3, Seg4) are reported here.

Three parameters- $ImgSimThresh$, $MaxBlobSize$ and $MinBlobSize$ were varied to conduct different sets of experiments. A thorough parameter exploration was not done as the parameter selection was done manually. Though different sets of parameter values were used for experimentation, only two such sets are reported for the image level ensemble, as they provided some significant observations. For the blob level ensemble only one experiment is reported. The performance evaluation of both levels of ensembles was done by comparing their degree of separability with that of the individual segmentations with the same parameter set. Since the objective of this work is to solely demonstrate the effectiveness of an ensemble, its performance has only been compared to the individual segmentors.

For the first set of experiments the parameters $ImgSimThresh$, $MaxBlobSize$ and $MinBlobSize$ have the values of 65% , 80000 and 1000 (Table 1) and, in the second set of experiments their values were 55%, 40000 and 4000 (Table 2). At blob level ensemble the value of $BlobSimThresh$ was set to 75%, while $MaxBlobSize$ and $MinBlobSize$ were set to 40000 and 4000 (Table 2). In all the sets of experiments it should be noted that $ImgSimThresh$ is not applicable for the individual segmentations (it is only applicable for the ensembles).

In the first set of experiments (Table 1) the image level ensemble as well as the individual segmentations generated a negative value for the class Separability measure, thus indicating that they were unable to discriminate cancer from normal cases. However the image level ensembles performed better than the individual segmentations. In the second set of experiments (Table 2) better results were obtained overall by increasing the value for $MinBlobSize$. The algorithm proposed in this paper is designed to test the hypothesis that cancer nuclei will be on average larger than normal nuclei. To avoid diluting this effect in the cancer tissue, which can be a mixture of normal and cancer cells, we recommend primarily sampling the larger sized cells (> 4000) in sections from both normal and cancer tissue. The ensembles not only outperformed the individual segmentors, but were also able to clearly discriminate cancer from normal class.

Table 1. Image level ensemble: Summary table for parameters 65%, 80000, 1000

Experiment	μ_1	μ_2	$Stddev_1$	$Stddev_2$	S
Seg 1	133.96	260.06	48.01	120	-41.9
Seg 2	166.36	357.63	49.3	190.92	-48.96
Seg 3	144.67	274.42	50.01	123.84	-44.1
Seg 4	194.48	400.12	66.22	195.4	-55.98
Ensemble 1	133.79	263.89	43.18	118.85	-31.94
Ensemble 2	131.03	274.5	39.38	138.9	-34.8

Table 2. Image level ensemble: Summary table for parameters 55%, 40000, 4000

Experiment	μ_1	μ_2	$Stddev_1$	$Stddev_2$	S
Seg 1	185.98	309.57	59.68	74.09	-10.19
Seg 2	220.63	388.28	53.85	120.40	-6.61
Seg 3	196.15	321.23	60.33	75.53	-7.78
Seg 4	233.58	382.86	50.94	94.02	4.32
Ensemble 1	187.27	328.82	34.69	86.8377	20.02
Ensemble 2	152.81	301.92	37.29	70.25	41.58
Blob Ensemble	174.43	283.87	34.04	67.29	8.11

4   Conclusions

We have demonstrated the application of an ensemble of segmentations for the analysis of microscopy images from cervical cancer and normal tissue. A simple ensemble of three-class Otsu followed by morphological operations was used in the proposed algorithm. The ensemble approach was used for a novel application to accept and reject images based on their segmentation quality and in turn reduce the computation which may arise due to the automatic acquisition of a large number of images. The ensemble approach was not only applied at the image level but also at a lower - blob level. The final segmentations achieved through both the levels of segmentations were independently used to calculate mean nuclear line length and finally the volume-weighted mean nuclear volume (PSI-Vv) using unbiased stereological rules. A degree of class separability measure was calculated from the PSI-Vv values for each ensemble as well as the individual segmentations, and was used for performance evaluation. Both image and blob level ensembles gave better class separability values than the individual segmentations. By increasing the minimum blob size to 4000 the ensembles were able to separate nuclei of the normal cells from the cancer cells. The best results were given by ensemble approach 2 with the parameter settings for *ImgSimThresh*, *MaxBlobSize* and *MinBlobsize* of 55%, 40000 and 4000 respectively, which gave a class separability measure of 41.58. Hence, we demonstrated that the ensembles framework improved the results from the individual segmentations.

Acknowledgments. This research has been supported by NIMH/SBIR 2R44M H076541-04 and FHT matching.

References

1. Creagh, T., Bridger, J., Kupek, E.: Pathologist Variation in Reporting Cervical Borderline Epithelial Abnormalities and Cervical Intraepithelial Neoplasia. J. Clin. Pathol. 48, 59–60 (1995)
2. Wahlby, C., et al.: Combining Intensity, Edge and Shape Information for 2D and 3D Segmentation of Cell Nuclei in Tissue Sections. J. Micro. 215, 67–76 (2004)
3. Lewin, S., Jiang, X., Clausing, A.: A Clustering-Based Ensemble Technique for Shape Decomposition. In: Gimel'farb, G., Hancock, E., Imiya, A., Kuijper, A., Kudo, M., Omachi, S., Windeatt, T., Yamada, K. (eds.) SSPR&SPR 2012. LNCS, vol. 7626, pp. 153–161. Springer, Heidelberg (2012)
4. Vega-Pons, S., Shullcloper, J.: A Survey of Clustering Ensemble Algorithms. Int. J. Pattern Recogn. 25, 337–372 (2011)
5. Caruana, R., et al.: Ensemble Selection from Libraries of Models. In: Proceeding of 21st International Conference on Machine Learning. ACM (2004)
6. Gu, Y.: Automated Delineation of Lung Tumors from CT Images Using a Single Click Ensemble Segmentation Approach. Int J. Pattern Recogn. 46, 692–702 (2013)
7. Wattuya, P.: Combination of Multiple Segmentations. Diss. Phd. thesis, University of Munster, Germany (2010)
8. Simsek, A.C., Tosun, A.B.: Multilevel Segmentation of Histopathological Images Using Cooccurrence of Tissue Objects. IEEE Transactions on Biomedical Engineering 59, 1681–1689 (2012)
9. Rafiee, G., Dlay, S., Woo, W.: Automatic Segmentation of Interest Regions in Low Depth of Field Images Using Ensemble Clustering and Graph Cut Optimization Approaches. In: IEEE International Symposium on Multimedia, vol. 59, pp. 161–164 (2012)
10. Pantofaru, C., Schmid, C., Hebert, M.: Object Recognition by Integrating Multiple Image Segmentations. In: Forsyth, D., Torr, P., Zisserman, A. (eds.) ECCV 2008, Part III. LNCS, vol. 5304, pp. 481–494. Springer, Heidelberg (2008)
11. Otsu, N.: A Threshold Selection Method from Gray-Level Histograms. IEEE Transactions on Systems, Man and Cybernetics 9, 62–66 (1979)
12. Dufour, A., et al.: 3-D Active Meshes: Fast Discrete Deformable Models for Cell Tracking in 3-D Time-Lapse Microscopy. IEEE Transactions on Image Processing 20, 1925–1937 (2011)
13. Xiong, W., Chia, S., Lim, J.H.: Automated Nuclei ClumpDecomposition for Image Analysis in Neuronal Cell Fluoroscent Microscopy. In: 18th IEEE International Conference on Image Processing, pp. 1577–1580 (2011)
14. Wattuya, P., et al.: A Random Walker Based Approach to Combining Multiple Segmentations. In: International Conference on Pattern Recognition, pp. 1–4 (2008)
15. Franek, L., Abdala, D.D., Vega-Pons, S., Jiang, X.: Image Segmentation Fusion Using General Ensemble Clustering Methods. In: Kimmel, R., Klette, R., Sugimoto, A. (eds.) ACCV 2010, Part IV. LNCS, vol. 6495, pp. 373–384. Springer, Heidelberg (2011)
16. Sorensen, F.B., et al.: Stereological Estimates of Nuclear Volume in Squamous Cell Carcinoma of the Uterine Cervix and its Precursors. Virchows Archive 418, 225–233 (1991)

Attributed Relational SIFT-Based Regions Graph for Art Painting Retrieval

Mario Manzo and Alfredo Petrosino

Department of Science and Technology
University of Naples "Parthenope",
Isola C4, Centro Direzionale, Napoli, Italy
{mario.manzo,alfredo.petrosino}@uniparthenope.it

Abstract. Recently, image retrieval and analysis algorithms have been extensively applied to art related domains. In this field, state-of-the-art approaches mainly focus on feature extraction with the aim of improving reliability of authentication, classification and retrieval of art paintings. In this paper we propose an effective modeling, based on a graph structure, and a retrieval strategy, based on a graph matching algorithm, for art paintings. The proposed approach has been tested on different datasets with high quality results allowing an user to run effective content-based queries on painting records.

Keywords: CBIR systems, graph matching, graph based image representation, local invariant features extraction.

1   Introduction

During last decade, computer graphics and vision experts have focused their attention on the problem of cultural heritage preservation. In this field, effective techniques have been proposed concerning classification [11] and retrieval [20]. In [4] a graph-based method is described for automatic annotation and retrieval of digital art print images. This method has been proved to be particularly useful for art historians to annotate database of digital art print images. In [8] a colorimetric visualization method is proposed based on a spatial organization of colors within the painting. The effectiveness of the method is evaluated on Italian Renaissance images. Other approaches exploits Local Invariant Features Extraction (LIFE) methods [18] for image representation and similarity measurements. Authors in [11] present a novel approach for painting classification based on image segmentation and SIFT [16]/SURF [2] features extraction. In [20] a system for retrieving information about paintings using mobile devices is presented. An augmented reality system based on SIFT[16] is described in [24] to retrieve information about artist and historical context of paintings.

In this paper we propose a novel graph-based image representation along with a graph matching algorithm to effectively tackle the art painting retrieval task. A segmented digital image can be seen as a set of regions, each carrying two types of information: local visual information (color, shape or texture)

A. Petrosino (Ed.): ICIAP 2013, Part I, LNCS 8156, pp. 833–842, 2013.
© Springer-Verlag Berlin Heidelberg 2013

and spatial global information (topological configuration of regions located in a neighborhood). Indeed, relations between local and global information play a key role in human recognition task [13]. Many approaches [4] represent images using a graph structure considering, in this way, the image matching problem as a graph matching problem. In this context, the aim of our paper is threefold: first, a graph structure for image representation called Attributed Relational SIFT-based Regions Graph (ARSRG) is introduced to reduce the gap between local and global features; second a graph matching algorithm is presented to measure regions similarity exploiting information about topological relations; last, the LIFE method is applied in order to extract stable descriptors starting from a given set of image features.

The paper is organized as follows. Section 2 describes the graph based image representation, while Section 3 describes the graph matching algorithm. Relevant results are discussed in Section 4 and conclusions are drawn in Section 5.

2   Graph Based Image Representation

In this section we introduce a novel graph based image representation, composed by two main steps: features extraction and graph construction.

The first step consists in the extraction of the regions of interest (ROIs) from an image, by means of a segmentation technique, and the construction of a *Region Adjacency Graph (RAG)*[23] to encode spatial relations between extracted regions.

The second step consists of the construction of a graph, named by us **Attributed Relational SIFT-based Regions Graph (ARSRG)**, composed by three levels: *Root node, RAG Nodes* and *Leaf nodes*. The *Root node* represents the whole image and is linked to all the *RAG Nodes* at second level. *RAG Nodes* encode adjacency relationships between different image regions. Thus, adjacent regions in the image are represented by connected nodes. Finally, *Leaf nodes* represent the set of SIFT descriptors extracted from the image, in order to tackle invariance to view-point, illumination and scale.

Two types of configurations are provided at this level: *Region based* and *Region graph based* (Figure 1). In *Region based* a keypoint is associated to a region based on its spatial coordinates, whereas, *Region graph based* contains keypoints belonging to the same region connected by edges, which encode spatial adjacency. ARSRG can be defined by structures based on two different *Leaf nodes* configurations.

Definition 1. *An **ARSRG$_{1^{st}}$** (first leaf nodes configuration), G is defined as a tuple $G = (V_{regions}, E_{regions}, VF_{SIFT}, E_{regions-SIFT})$, where:*

- $V_{regions}$, *the set of regions-nodes.*
- $E_{regions} \subseteq V_{regions} \times V_{regions}$, *the set of undirected edges, where $e \in E_{regions}$ and $e = (v_i, v_j)$ is an edge between nodes $v_i, v_j \in V_{regions}$.*
- VF_{SIFT}, *the set of SIFT-nodes.*

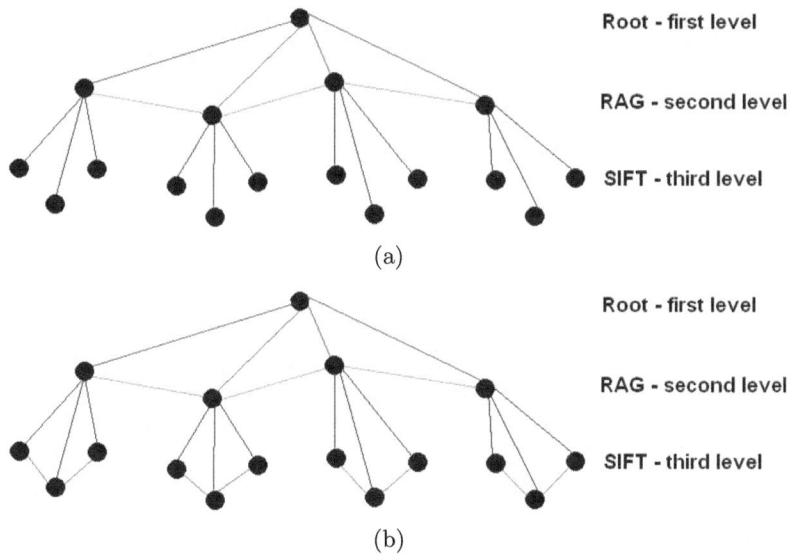

Fig. 1. *Region based* (a) and *Region graph based* (b) Configurations

- $E_{regions-SIFT} \subseteq V_{regions} \times VF_{SIFT}$, *the set of directed edges, where* $e \in E_{regions-SIFT}$ *and* $e = (v_i, vf_j)$ *is an edge between source node* $v_i \in V_{regions}$ *and destination node* $vf_j \in VF_{SIFT}$.

Definition 2. *An* **ARSRG**$_{2nd}$ *(second leaf nodes configuration), G is defined as a tuple* $G = (V_{regions}, E_{regions}, VF_{SIFT}, E_{regions-SIFT}, E_{SIFT})$, *where:*

- $V_{regions}$, *the set of regions-nodes.*
- $E_{regions} \subseteq V_{regions} \times V_{regions}$, *the set of undirected edges, where* $e \in E_{regions}$ *and* $e = (v_i, v_j)$ *is an edge between nodes* $v_i, v_j \in V_{regions}$
- VF_{SIFT}, *the set of SIFT-nodes.*
- $E_{regions-SIFT} \subseteq V_{regions} \times VF_{SIFT}$, *the set of directed edges, where* $e \in E_{regions-SIFT}$ *and* $e = (v_i, vf_j)$ *is an edge between source node* $v_i \in V_{regions}$ *and destination node* $vf_j \in VF_{SIFT}$.
- $E_{SIFT} \subseteq VF_{SIFT} \times VF_{SIFT}$, *the set of undirected edges, where* $e \in E_{SIFT}$ *and* $e = (vf_i, vf_j)$ *is an edge between nodes* $vf_i, vf_j \in V_{SIFT}$

The ARSRG structure has a set of properties arising from two building blocks: features extraction and matching. The first building block includes relations among local features and structural information of image encoded into the RAG configuration located at second level. It has been demonstrated that global configuration and local information of scene play a key role in the human recognition task [13]. Relations can be distinguished in: horizontal and vertical. Horizontal relations provide information about spatial closeness between ROIs (level two) or SIFT features (level three). Vertical relations concern connections among ROIs (level two) and SIFT features (level three). The second building block includes an algorithm for optimal matching and false positive reduction, to refine results. The matching phase is handled through a hierarchical exploration of ARSRG, that can be roughly divided in two steps: filtering of regions based on their size; subgraph matching performed by matching features belonging to single regions located at the third level of of ARSRG.

3   Graph Matching

Given two ARSRGs, the goal is to find best matches among their nodes and to determine a mapping set M containing associated nodes between the two structures. This is done by iterative exploration of best possible nodes mapping and selecting the best pairs at each iteration, adopting two approaches to measure dissimilarity among node pairs: ratio test[16] and graph matching[21].

The first step of the algorithm is the construction of a $n \times m$ matrix, called $Dist\_matrix$, where n and m are the numbers of regions-nodes at the second level of the two ARSRGs respectively. The matrix contains the distances between each node of the first ARSRG and all the nodes of the second ARSRG. In order to find the most promising mapping, a second matrix B, of dimension $n \times m$, stores the mapping corresponding to the minimum value of rows in $Dist\_matrix$. For each possible nodes mapping extracted from B, the algorithm computes matches generated by SIFT descriptors associated to the nodes. Nodes pairs that present a number of matches greater than a given threshold are saved.

Next, the algorithm analyzes the second-smallest elements at each row of matrix $Dist\_matrix$ extracting, from B, the correspondences that contain at least one node-to-node matching and so on, until it reaches the final iteration.

3.1   Regions Matching with Ratio Test

Different approaches can be employed to find the best match for each region. For instance, given two regions with associated SIFT keypoints, the naive approach consists in searching for the best candidate match for each keypoint in the first region by identifying its nearest neighbor in the second region, using a global threshold. This approach produces many false matches, i.e. many keypoints do not match correctly due to the global threshold. Therefore, a different measure is adopted comparing the closest to the second-closest neighbor of each keypoint [16].

3.2   Regions Matching with Graph Matching

Differently from the previous solution, the problem of regions comparison can be reformulated in terms of graph matching [21], with the goal of improving the quality of the matches. We consider SIFT features organized in the form of SIFT Nearest Neighbor Graph (SNNG) according to the following definition:

Definition 3. *A $SNNG = (VF_{SIFT}, E_{SIFT})$ is defined as*

- *VF_{SIFT}: the set of nodes associated to SIFT keypoints*
- *E_{SIFT}: the set of edges*

An edge $e = (v_i, v_p)$ exists, for $v_i, v_p \in VF_{SIFT}$, if $dist(v_i, v_p) < \tau$, where $dist(v_i, v_p)$ is the Euclidean distance, τ is a threshold value and p stems from 1 to k, k being the size of VF_{SIFT}.

SNNG represents SIFT keypoints belonging to image region located at the third level of ARSRG structure according to Definition 2. Matches among SNNGs are described through a matrix S that defines an injective mapping between two SNNGs: $SNNG_1 = (VF_{SIFT1}, E_{SIFT1})$ and $SNNG_2 = (VF_{SIFT2}, E_{SIFT2})$. In particular, if an element $s_{ij} \in S$ is assigned to 1 then the node $v_i \in VF_{SIFT1}$ matches with node $v_j \in VF_{SIFT2}$, otherwise 0. In this context, the goal of algorithm is to initially estimate best matrix S, starting from the initial guess $S^{(1)}$ through the space of matching configurations. We use a combined measure of structural consistency and similarity called W, to compare SNNGs during the matching. Given two nodes $v_a \in VF_{SIFT1}$ and $v_\alpha \in VF_{SIFT2}$, we define

$$W_{a\alpha} = Q_{a\alpha} R_{a\alpha} \tag{1}$$

where

$$Q_{a\alpha} = exp\left[\mu \sum_{b \in V_1} \sum_{\beta \in V_2} D_{ab} M_{\alpha\beta} s_{b\beta}\right] \quad and \quad R_{a\alpha} = \frac{1}{dist(z_a^1, z_\alpha^2)} \tag{2}$$

$Q_{a\alpha}$ is the structural consistency coefficient, D and M are the adjacency matrices of G_1 and G_2, $s_{b\beta}$ is an element of matrix S and $\mu > 0$ is a control parameter. $R_{a\alpha}$ is a similarity nodes matching function, where $dist(z_a^1, z_\alpha^2)$ is the Euclidean distance between SIFT descriptors z_a^1 and z_α^2 corresponding to nodes v_a and v_α.

Moreover, in order to describe the matching node-by-node between two SNNGs, an additional matrix Ω is adopted

$$\Omega = \begin{bmatrix} W_{11} & \cdots & W_{1m} \\ \vdots & W_{a\alpha} & \vdots \\ W_{n1} & \cdots & W_{nm} \end{bmatrix} \tag{3}$$

A cleaning heuristic approach to extract best matches is applied on Ω with the purpose of building matrix S. The iterative procedure is composed by three steps:

1. at the first step, $W_{a,k} = max(W_{a,\alpha})$ is selected at each row a of Ω, $\alpha = 1, ..., m$, such that $W_{a,k}/W_{a,k2} > \frac{1}{\rho}$, where $W_{a,k2}$ is the second greatest element in the a-th row of Ω;

2. the second step finds the maximum element $W_{a,\alpha} \in \Omega$ and activates the corresponding match $s_{a\alpha} \in S$;
3. at the third step, the rows and columns of Ω containing $W_{a,\alpha}$ are sets to zero.

The three steps are repeated until Ω does not contain any other element to analyze, i.e. $W_{ij} = 0$, $\forall i, j$ $i = 1, \ldots, n$ and $j = 1, \ldots, m$.

4 Experimental Results

The proposed approach has been tested on three datasets and compared with other LIFE methods, graph matching algorithms and CBIR system reported in the literature. The first dataset, described in [11], is composed by two sets of images obtained from Olga's gallery[1] and Travel Webshots[2]. The second dataset, described in[10], is composed by painting photos taken from the Cantor Arts Center[3]. The third dataset, described in [20], is composed by 1002 images. Figure 2 shows some examples.

4.1 LIFE Methods Comparison

A first evaluation is performed for dataset used in [11] and through comparisons with LIFE methods. Results are reported in terms of Mean Reciprocal Rank (MRR). As in [11,16], a tuning procedure is applied to ρ parameter that controls tolerance of false matches both in graph matching and ratio test. We used the values of ρ as suggested in [11] and [16]. In particular, ρ values of 0.6 and 0.7 are used in [11] and values greater than 0.8 are rejected as in [16].

Table 1. Quantitative comparison using MRR measure among SIFT[16], SURF[2], ORB[19], FREAK[1], BRIEF[3] and ARSRG matching on dataset in[11]

ρ	$SIFT[16]$	$SURF[2]$	$ORB[19]$	$FREAK[1]$	$BRIEF[3]$	$ARSRG_{1st}$	$ARSRG_{2nd}$
0.6	0.7485	0.8400	0.6500	0.3558	0.4300	0.6700	0.6750
0.7	0.7051	0.6800	0.6116	0.3360	0.3995	0.7133	0.7500
0.8	0.6963	0.5997	0.5651	0.2645	0.4227	0.6115	0.8000

Table 1 shows that graph based approach provides best performance. ρ values of 0.7 and 0.8 give optimal results for ARSRG matching. Graph based image representation clearly captures the topological relationships among features and acts as a filter over the complete set of SIFT features extracted from the image. Indeed, the comparison was performed among descriptors belonging to regions

---

[1] http://www.abcgallery.com/index.html
[2] http://travel.webshots.com
[3] http://museum.stanford.edu/

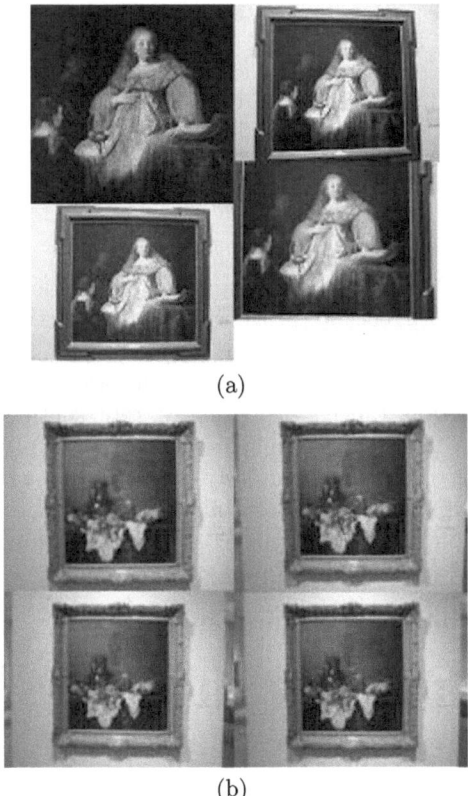

(a)

(b)

Fig. 2. Some examples of art painting images

instead of entire image as proposed in standard approaches. In this way, many false matches are discarded and effectiveness is greatly improved.

A second test has been performed on the dataset adopted in [10], computing performance in terms of Precision and Recall. Values of ρ parameter are the same as in the previous test.

Table 2. Quantitative comparison, using *Recall* measure, among SIFT[16], SURF[2], ORB[19], FREAK[1], BRIEF[3] and ARSRG matching on dataset in[10]

ρ	$SIFT[16]$	$SURF[2]$	$ORB[19]$	$FREAK[1]$	$BRIEF[3]$	$ARSRG_{1st}$	$ARSRG_{2nd}$
0.6	1.0	0.8666	0.8000	0.7333	0.7666	0.7333	0.7333
0.7	1.0	0.9000	0.8666	0.7333	0.8666	0.7666	0.7333
0.8	1.0	1.0	1.0	0.8333	1.0000	0.8000	0.8000

Table 2 shows that SIFT based approach performs better in terms of Recall. In case of ρ equal to 0.8, our approach yields comparable results.

Table 3. Quantitative comparison using *Precision* measure, among SIFT[16], SURF[2], ORB[19], FREAK[1], BRIEF[3] and ARSRG matching on dataset in[10]

ρ	$SIFT[16]$	$SURF[2]$	$ORB[19]$	$FREAK[1]$	$BRIEF[3]$	$ARSRG_{1st}$	$ARSRG_{2nd}$
0.6	0.0674	0.0820	0.2051	0.05584	0.10689	1.0	1.0
0.7	0.0401	0.0441	0.0742	0.04671	0.05664	0.6571	1.0
0.8	0.0312	0.0338	0.0348	0.04072	0.03452	0.1428	0.6666

In contrast, Table 3 shows that our approach, clearly outperforming the other approaches in terms of Precision, proves to be very effective for image retrieval problem. The best results by graph matching algorithm for Precision are provided with ρ equal to 0.6 and 0.7. These results are due to the use of image structural representation. Indeed, graph nodes, representing different image regions, provide a partitioning rule applied on entire set of SIFT. In this way, the subsets obtained are considered separately during matching step. This strategy removes most of false matches that normally belongs to accepted matches. As a consequence, several images are discarded as candidates for final ranking.

4.2   Graph Matching Algorithms Comparison

This section describes performance comparison with graph SIFT-based matching algorithms. Experiments are performed on datasets presented in [11,20] and are evaluated through MRR measure. Results are reported in tables 4 and 5.

Table 4. Quantitative comparison, using *MRR* measure, among HGM[14], RRWGM[15], TM[9] algorithms and ARSRG matching on dataset in[11]

$HGM[14]$	$RRWGM[15]$	$TM[9]$	$ARSRG_{1st}$	$ARSRG_{2nd}$
0.2600	0.1322	0.1348	0.6115	0.8000

Table 5. Quantitative comparison, using *MRR* measure, among HGM[14], RRWGM[15], TM[9] algorithms and ARSRG matching on dataset in[20]

$HGM[14]$	$RRWGM[15]$	$TM[9]$	$ARSRG_{1st}$	$ARSRG_{2nd}$
0.1000	0.0545	0.0545	0.20961	0.39803

In particular, Tables 4 and 5 show comparison, in terms of MMR values, with HGM [14], RRWGM [15], TM [9] algorithms. Also in this case, ARSRG leads to better results compared to those obtained by the other graph SIFT-based matching algorithms. Similarly in this case, the region matching approach, by providing local information about spatial distribution of the features, leads to false matches removal and hence improves final results.

4.3 CBIR System Comparison

This section describe the performance comparison with Lucene Image Retrieval (LIRe) [17] system. Experiments are performed on dataset presented in [11], considering different features implemented in LIRe, and evaluated through MRR measure. Results are reported in table 6.

Table 6. Quantitative comparison using MRR measure, among some features available in (LIRe)[17] system and ARSRG matching on dataset in[11]

$MPEG7[5]$	$Tamura[22]$	$CEDD[6]$	$FCTH[7]$	$ACC[12]$	$ARSRG_{1st}$	$ARSRG_{2nd}$
0.2645	0.1885	0.2329	0.1924	0.1879	0.7133	0.7500

From the reported results, it is clear that LIRe system is not very suitable for art paint retrieval, due to its low performing features, which results in wrong discrimination of relevant and irrelevant images. Consequently, the achieved ranking contains inadequate results, with respect to user's request, which affects heavily its final performance. In contrast, results obtained by ARSRG algorithm, demonstrates once more that the proposed approach is very effective for this application.

5 Conclusion Remarks

In this paper a novel way to capture visual and structural information from digital art paintings has been proposed. The resulting ARSRG structure has proved to be a valid alternative to standard techniques which use color, shape and texture to describe image content. Robustness and effectiveness of the proposed graph matching algorithm have been extensively tested on different public data repositories for the art painting retrieval task. The proposed approach is robust to changes in scale and lighting conditions, and allows to effectively retrieve objects based on the user preferences.

References

1. Alahi, A., Ortiz, R., Vandergheynst, P.: FREAK: Fast Retina Keypoint. In: CVPR, pp. 510–517 (2012)
2. Bay, H., Tuytelaars, T., Van Gool, L.: SURF: Speeded up robust features. In: Leonardis, A., Bischof, H., Pinz, A. (eds.) ECCV 2006, Part I. LNCS, vol. 3951, pp. 404–417. Springer, Heidelberg (2006)
3. Calonder, M., Lepetit, V., Strecha, C., Fua, P.: BRIEF: Binary Robust Independent Elementary Features. In: Daniilidis, K., Maragos, P., Paragios, N. (eds.) ECCV 2010, Part IV. LNCS, vol. 6314, pp. 778–792. Springer, Heidelberg (2010)
4. Carneiro, G.: Graph-based methods for the automatic annotation and retrieval of art prints. In: ICMR, p. 32 (2011)

5. Chang, S.F., Sikora, T., Puri, A.: Overview of the mpeg-7 standard. Circuits and Systems for Video Technology 11(6), 688–695 (2001)
6. Chatzichristofis, S.A., Boutalis, Y.S.: CEDD: Color and Edge Directivity Descriptor. A Compact Descriptor for Image Indexing and Retrieval. In: Gasteratos, A., Vincze, M., Tsotsos, J.K. (eds.) ICVS 2008. LNCS, vol. 5008, pp. 312–322. Springer, Heidelberg (2008)
7. Chatzichristofis, S.A., Boutalis, Y.S.: FCTH: Fuzzy Color And Texture Histogram A Low Level Feature For Accurate Image Retrieval. In: 9th International Workshop on Image Analysis for Multimedia Interactive Services, pp. 191–196 (2008)
8. Colantoni, P., Jean-Baptiste, T., Ruven, P.: Graph-based 3d visualization of color content in paintings. In: VAST, pp. 25–30 (2010)
9. Duchenne, O., Bach, F., Kweon, I.S., Ponce, J.: A tensor-based algorithm for high-order graph matching. PAMI 33(12), 2383–2395 (2011)
10. Etezadi-Amoli, M., Chang, C., Hewlett, M.: A day at the museum (2009)
11. Haladová, Z., Šikudová, E.: Limitations of the SIFT/SURF based methods in the classifications of fine art paintings. Computer Graphics and Geometry 12(1), 40–50 (2010)
12. Huang, J., Kumar, S.R., Mitra, M., Zhu, W.J., Zabih, R.: Image indexing using color correlograms. In: CVPR, pp. 762–768 (1997)
13. Koffka, K.: Principles of Gestalt Psychology. Harcourt, New York (1935)
14. Lee, J., Cho, M., Lee, K.M.: Hyper-graph matching via reweighted random walks. In: CVPR, pp. 1633–1640 (2011)
15. Cho, M., Lee, J., Lee, K.M.: Reweighted random walks for graph matching. In: Daniilidis, K., Maragos, P., Paragios, N. (eds.) ECCV 2010, Part V. LNCS, vol. 6315, pp. 492–505. Springer, Heidelberg (2010)
16. Lowe, D.G.: Distinctive image features from scale-invariant keypoints. International Journal of Computer Vision 60(2), 91–110 (2004)
17. Lux, M., Chatzichristofis, S.A.: Lire: Lucene image retrieval: an extensible Java CBIR library. In: 16th ACM Multimedia, pp. 1085–1088 (2008)
18. Mikolajczyk, K., Schmid, C.: A performance evaluation of local descriptors. PAMI 27(10), 1615–1630 (2005)
19. Rublee, E., Rabaud, V., Konolige, K., Bradski, G.R.: ORB: An efficient alternative to SIFT or SURF. In: ICCV, pp. 2564–2571 (2011)
20. Ruf, B., Kokiopoulou, E., Detyniecki, M.: Mobile museum guide based on fast SIFT recognition. In: Detyniecki, M., Leiner, U., Nürnberger, A. (eds.) AMR 2010. LNCS, vol. 5811, pp. 170–183. Springer, Heidelberg (2010)
21. Sanroma, G., Alquézar Mancho, R., Serratosa, I., Casanelles, F.: Graph matching using SIFT descriptors - An application to pose recovery of a mobile robot. In: 5th International Conference on Computer Vision Theory and Applications, pp. 249–254 (2010)
22. Tamura, H., Mori, S., Yamawak, T.: Textural features corresponding to visual perception. Systems, Man, and Cybernetics 8(6), 460–472 (1978)
23. Tremeau, A., Colantoni, P.: Regions adjacency graph applied to color image segmentation. Trans. on Image Processing 9(4), 735–744 (2000)
24. You, S., Neumann, U.: Mobile augmented reality for enhancing e-learning and e-business. In: International Conference on Internet Technology and Applications, pp. 1–4 (2010)

Video Segmentation Framework by Dynamic Background Modelling

Santiago Molina-Giraldo, Andres M. Álvarez-Meza,
Julio C. García-Álvarez, and Cesar G. Castellanos-Domínguez

Signal Processing and Recognition Group
Universidad Nacional de Colombia - sede Manizales
Campus La Nubia, km 7 via al Magdalena, Manizales-Colombia
{smolinag,amalvarezme,jcgarciaa,cgcastellanosd}@unal.edu.co

Abstract. Detecting moving objects in video streams is the first relevant step of information extraction in many computer vision applications, e.g. video surveillance systems. In this work, a video segmentation framework by dynamic background modelling is presented. Our approach aims to update suitably the background model of a scene that is recorded by a static camera. For such purpose, we develop an optical flow based methodology to suitable track moving objects, which can stop or change smoothly their movement along the video. Moreover, a light variations identification stage, is employed to avoid possible confusions between illumination changes and objects in movement. Regarding this, our approach is able to ensure a suitable background modelling in real world scenarios. Attained results show that our framework outperforms, in well-known datasets, state of the art methodologies.

Keywords: background subtraction, optical flow, tracking.

1 Introduction

A system that monitors an area by camera and is able to detect and track moving objects is called a surveillance system. Intelligent video surveillance systems can achieve unsupervised results using video segmentation, where the moving objects are extracted from video sequences. Nevertheless, the challenge is to devise and implement automatic systems able to perform both activities, detection and tracking. Afterwards, the main goal is to interpret their activities and behaviours to support computer vision analysis (e.g. object classification, tracking, activity understanding, among others) [8,9]. In this regard, it is necessary to segment (subtract) each object from the scene to facilitate such kind of analysis.

Our main goal is on the detection phase of a general visual surveillance system using static cameras. Aside from the intrinsic usefulness of being able to segment video streams into foreground and background components, detecting moving objects provides a focus of attention for recognition, classification, and activity analysis, making these later steps more efficient [3]. Many segmentation algorithms have been proposed, among them, algorithms with background modelling usually show superior performance, by exploiting the prior information about the process, specially, for static

A. Petrosino (Ed.): ICIAP 2013, Part I, LNCS 8156, pp. 843–852, 2013.

cameras environments [2]. Thus, background subtraction, that consists in maintaining an updated model of the background and detecting (segmenting) moving objects as those that deviate from such a model. Given a suitable background model, the system is able to subtract it, as a typical and crucial process for a surveillance system to detect moving objects that may enter, leave, move or left unattended in the surveillance region.

Compared to other approaches, such as optical flow, background modelling is computationally affordable for real-time applications [1]. The main problem is its sensitivity to dynamic scene changes, and the consequent need for the background model adaptation via background maintenance. Such problem is known to be significant and difficult [3]. Furthermore, image sequences with dynamic backgrounds often cause false classification of pixels (e.g. light variations). One common solution is to map the input samples into alternate color spaces, however, it has failed to solve this problem and an enhanced solution is the use of image features, where the distributions at each pixel may be modelled in a parametric manner using a mixture of Gaussians or by using non-parametric kernel density estimation [5, 7]. The self organizing maps have been also explored as an alternative for the background subtraction task, because of their nature to learn by means of local variations [8]. However, these techniques have some drawbacks when dealing with the actualization of the background model, when in the video sequence there are objects which after moving become static.

In this work, a video segmentation framework by dynamic background modelling is presented. Our approach aims to update the model of a scene that is recorded by a static camera. In this regard, an optical flow based methodology is proposed to suitable track moving objects, which can stop or change smoothly their movement along the video. Objects detected as static are stored into a memory, with the purpose of not updating the model in these regions. Moreover, a light variations identification stage, is proposed to avoid possible confusions between illumination changes and objects in movement. Thus, our framework is able to ensure a suitable background modelling in real-world scenarios. The remainder of this work is organized as follows. In section 2, the proposed framework is described. In section 3, the experiments and results are presented. Finally, in sections 4 and 5 we discuss and conclude about the attained results.

2 Theoretical Background

2.1 Region Change Detection

Let $\boldsymbol{\Phi} = \{\boldsymbol{F}^{(t-T)}, \boldsymbol{F}^{(t-T+1)}, \ldots, \boldsymbol{F}^{(t)}\}$ be a set of images of a given video, with $t = T + 1, \ldots, \infty$ and $\boldsymbol{F}^{(t)} \in \mathbb{R}^{h \times w \times N_c}$. Here, h is the number of rows of the frame, w the number of columns, and N_c the number of color channels. Given the frame $\boldsymbol{F}^{(t)}$, the mean intensity matrix $\boldsymbol{M}^{(t)} \in \mathbb{R}^{h \times w}$, with elements $M_{i,j}^{(t)} = \boldsymbol{E}\left\{\boldsymbol{f}_{(i,j)}^{(t)}\right\}$, where $\boldsymbol{f}_{(i,j)}^{(t)} \in \mathbb{R}^{1 \times N_c}$ is a row vector containing the N_c color channels of pixel (i, j) and notation $\boldsymbol{E}\{\cdot\}$ stands for expectation operator. The purpose is to detect changes between the frame t and $t - 1$, then, $\boldsymbol{M}^{(t)}$ is divided into $n_p = h \times w$ patches, and the *Sum of Absolute Differences* (SAD) matrix $\boldsymbol{H}^{(t)} \in \mathbb{R}^{h \times w}$ is calculated as

$$H_{i,j}^{(t)} = \begin{cases} 1 & \left\| \Omega_{(i,j)}^{(t)} - \Omega_{(i,j)}^{(t-1)} \right\|_F^2 < \xi_H ; \\ 0 & \text{Otherwise} \end{cases} \tag{1}$$

where the patch $\Omega_{(i,j)}^{(t)} \in \mathbb{R}^{h_p \times w_p}$ contains the spatial neighbourhood of the pixel (i,j) in $M^{(t)}$, $\xi_H \in \mathbb{R}^+$ is a given threshold parameter, and $\|\cdot\|$ is the Frobenius norm operator. Thus, matrix $H^{(t)}$ in (1) highlights pixels detected as moving from frame $t-1$ to t.

2.2   Motion Modeling by Optical Flow

Now the challenge is to detect the apparent movement of the regions detected as moving in matrix $H^{(t)}$, in this regard, an optical flow procedure is employed. By definition, an optical flow field is a vector field that describes the velocity of pixels in an image sequence. Due to the complexity of most of the optical flow approaches proposed in the state of the art [4, 10], we propose to use a selective optical flow approach based on block matching, where only for the patches highlighted as moving in $H^{(t)}$ the optical flow is computed. Given the patch $\Omega_{(i,j)}^{(t-1)}$ in $M^{(t-1)}$ marked as moving, $M^{(t)}$ is used to compute the optical flow by searching in a neighbourhood of size κ around (i,j) the patch that best matches $\Omega_{(i,j)}^{(t-1)}$ by using the SAD. Then, the relative coordinates $(i + vi, j + vj)$ of such patch are stored in matrices $X^{(t)} \in \mathbb{R}^{h \times w}$ and $Y^{(t)} \in \mathbb{R}^{h \times w}$ respectively. Hence, the apparent movement from frame $t-1$ to t, is described by matrices $X^{(t)}$ for x coordinates and $Y^{(t)}$ for y coordinates.

2.3   Object Movement Identification and Static Object Memory Computation

Given the motion matrix $H^{(t-1)}$, some morphological filters are applied to achieve a smooth representation of the movement. Furthermore, above procedure allows to enhance each object shape representation. Afterwards, the filtered version of $H^{(t-1)}$ is processed to highlight connected components, thus is, a set of objects $\Xi^{(t-1)} = \{\Theta_1^{(t-1)}, \Theta_2^{(t-1)}, \dots, \Theta_{l_{t-1}}^{(t-1)}\}$ is inferred by an 8-connected based operation, where $\Theta_{l_{t-1}}^{(t-1)}$ is a real-valued matrix of size $h_{l_{t-1}}^{(t-1)} \times w_{l_{t-1}}^{(t-1)}$. For each detected object, the mode direction vector $\mu_{l_{t-1}}^{(t-1)} \in \mathbb{R}^{2 \times 1}$ is estimated, where the first and second components of $\mu_{l_{t-1}}^{(t-1)}$ correspond to the mode of pixels (i,j) in $X^{(t-1)}$ and $Y^{(t-1)}$ that are contained in object $\Theta_{l_{t-1}}^{(t-1)}$. In addition, the magnitude of $\mu_{l_{t-1}}^{(t-1)}$ is computed and stored into $v_{l_{t-1}}^{(t-1)} \in \mathbb{R}^+$. So, we are looking for the trend of motion direction for each detected object. Then, each object $\Theta_{l_{t-1}}^{(t-1)} \subset H^{(t-1)}$ is enclosed by a bounding box. Such box is moved according to $\mu_{l_{t-1}}^{(t-1)}$ and $v_{l_{t-1}}^{(t-1)}$, and it is mapped into $H^{(t)}$. After that, the number of projected pixels $\lambda_{l_{t-1}}^{(t)} \in \mathbb{N}$ at instant t, which are contained into the mapped box are calculated. Object $\Theta_{l_{t-1}}^{(t-1)}$ is detected as moving from $t-1$ to t if

$$\frac{\lambda_{l_t}^{(t)}}{\lambda_{l_{t-1}}^{(t)}} > \xi_\lambda, \tag{2}$$

with $\xi_\lambda \in \mathbb{N}$. Otherwise, if (2) is not accomplished, object $\boldsymbol{\Theta}_{l_{t-1}}^{(t-1)}$ is labelled as static. Hence, the static objects set $\boldsymbol{\Psi}^{(t)}$ is composed by the $\boldsymbol{\Theta}_{l_{t-1}}^{(t-1)}$ objects that are labelled as static at instant t. So, a static memory matrix $\boldsymbol{B}^{(t)} \in \mathbb{R}^{h \times w}$ can be written as

$$B_{i,j}^{(t)} = \begin{cases} 1 & (i,j) \in \bigcup_t^{t_i=1} \boldsymbol{\Psi}^{(t_i)} \\ 0 & \text{Otherwise} \end{cases}. \tag{3}$$

Note that, a static object in $\boldsymbol{\Psi}^{(t)}$ only can be considered as moving, if at a posterior time instant the condition (2) is accomplished, then, removing it from the set.

2.4   Background Modeling Updating

The main goal of our approach is to deal with real-world environment conditions, looking for a background model that allows to identify, as well as possible, smooth movements, which are inherent to the scene. Hence, we propose to define a Gaussian-based background model $\zeta(\boldsymbol{\Gamma}^{(t-1)}, \boldsymbol{\Sigma}^{(t-1)})$, with $\boldsymbol{\Gamma}^{(t-1)} \in \mathbb{R}^{h \times w \times N_c}$ and $\boldsymbol{\Sigma}^{(t-1)} \in \mathbb{R}^{h \times w \times N_c}$, as in (4)

$$p(i,j,z) = \exp\left(-\frac{\|f_{i,j,z}^{(t-1)} - \Gamma_{i,j,z}^{(t-1)}\|_2^2}{2\Sigma_{i,j,z}^{(t-1)^2}}\right), \tag{4}$$

being $\Gamma_{i,j,z}^{(t-1)} \in \mathbb{R}^+$ the background model value of pixel (i,j) at time instant $t-1$ in channel z, and $\Sigma_{i,j,z}^{(t-1)} \in \mathbb{R}^+$ a color band-width value.

Thus, in order to update the parameters of the model, the mean color value matrix $\boldsymbol{M}^{(t)} \in \mathbb{R}^{h \times w \times N_c}$ is defined as

$$M_{i,j,z}^{(t)} = \boldsymbol{E}\left\{v_{i,j,z}^{(t)}\right\}, \tag{5}$$

being $v_{i,j,z}^{(t)} = \{\boldsymbol{F}_{i,j,z}^{(t)}, \boldsymbol{F}_{i,j,z}^{(t-1)}, \dots, \boldsymbol{F}_{i,j,z}^{(t-T)}\}$. Similarly, a deviation matrix $\boldsymbol{D}^{(t)} \in \mathbb{R}^{h \times w \times N_c}$ is computed as

$$D_{i,j,z}^{(t)} = \sigma\{v_{i,j,z}^{(t)}\}, \tag{6}$$

where $\sigma\{\cdot\}$ is the deviation operator. Finally, the Gaussian-based background model $\zeta(\boldsymbol{\Gamma}^{(t)}, \boldsymbol{\Sigma}^{(t)})$ is updated as in (7) and (8)

$$\Gamma_{i,j,z}^{(t)} = \begin{cases} M_{i,j,z}^{(t)} & \left(D_{i,j,z}^{(t)} < \Sigma_{i,j,z}^{(t-1)}\right) \cap \left(B_{i,j}^{(t)} = 0\right) \\ \Gamma_{i,j,z}^{(t-1)} & \text{Otherwise} \end{cases}. \tag{7}$$

$$\Sigma_{i,j,z}^{(t)} = \begin{cases} D_{i,j,z}^{(t)} & \left(D_{i,j,z}^{(t)} < \Sigma_{i,j,z}^{(t-1)}\right) \cap \left(B_{i,j}^{(t)} = 0\right) \\ \Sigma_{i,j,z}^{(t-1)} & \text{Otherwise} \end{cases}. \tag{8}$$

It is important to note that the matrices $M^{(t)}$ and $D^{(t)}$ are computed only by using the last T frames, then it is only needed to store these frames for such computation. Given the provided background model $\zeta(\Gamma^{(t)}, \Sigma^{(t)})$, a clustering based segmentation algorithm as described in [6] is used to extract the moving objects from the scene. This approach propose to use kernel representations and a relevance analysis in order to conform an enhanced feature space, then a tuned kmeans algorithm with 2 groups is performed over the enhanced space, aiming to group static pixels in one cluster and moving pixels in the other. Thence, a segmented binary matrix S^t is attained.

2.5 Illumination Change Detection

Nonetheless, some scene conditions (e.g. light variations, casted shadows) perturb the quality of the segmented image. Therefore, it is proposed to debug the matrix $S^{(t)}$ by means of an illumination change detection stage. For such purpose, we propose to use the normalized RGB representation as studied in [5]. Then a normalized channel difference matrix $A_n^{(t)} \in \mathbb{R}^{h \times w}$ is calculated as

$$A_{n_{i,j}}^{(t)} = \begin{cases} 1 & \prod_{z=1}^{N_c} \eta\left(f_{n_{i,j,z}}^{(t)} - \Gamma_{n_{i,j,z}}^{(t-1)}, \xi_{A_n}\right) = 1 \\ 0 & \text{Otherwise} \end{cases}; \tag{9}$$

where

$$f_{n_{i,j,z}}^{(t)} = \frac{f_{i,j,z}^{(t)}}{\sum_{r=1}^{N_c} f_{i,j,r}^{(t)}} \quad \Gamma_{n_{i,j,z}}^{(t-1)} = \frac{\Gamma_{i,j,z}^{(t-1)}}{\sum_{r=1}^{N_c} \Gamma_{i,j,r}^{(t-1)}}; \tag{10}$$

are the normalized color channel z of pixel (i,j) in $F^{(t)}$ and $\Gamma^{(t-1)}$, respectively. And $\eta(f_{n_{i,j,z}}^{(t)} - \Gamma_{n_{i,j,z}}^{(t-1)}, \xi_A) = 1$, if $\|f_{n_{i,j,z}}^{(t)} - \Gamma_{n_{i,j,z}}^{(t-1)}\|_2^2 < \xi_A$. Similarly, a difference matrix $A_n^{(t)} \in \mathbb{R}^{h \times w}$ is calculated as

$$A_{i,j}^{(t)} = \begin{cases} 1 & \prod_{z=1}^{N_c} \eta\left(f_{i,j,z}^{(t)} - \Gamma_{i,j,z}^{(t-1)}, \xi_A\right) = 1 \\ 0 & \text{Otherwise} \end{cases}; \tag{11}$$

Thus, $A_n^{(t)}$ in (9) looks for chromaticity changes in pixels by using a normalized color space computed according to (10), this help to suppress the shadows casted by moving objects. Nonetheless, the lightness representation is lost by using only this information, in this regard, $A^{(t)}$ in (11) is employed. Then, combining $A^{(t)}$, $A_n^{(t)}$, and $S^{(t)}$, it is possible to attain a correct segmented matrix $\hat{S}^{(t)} \in \mathbb{R}^{h \times w}$ avoiding illumination changes as

$$\hat{S}_{i,j}^{(t)} = S_{i,j}^{(t)} A_{i,j}^{(t)} A_{n_{i,j}}^{(t)}. \tag{12}$$

3 Experiments

The proposed methodology for automatic video segmentation is tested using some well known databases of the state of the art. Moreover, a traditional video segmentation algorithm named Self-Organizing Approach to Background Subtraction (SOBS), which builds a background model by learning in a self-organizing manner the scene variations, is employed as benchmark [8]. Each database includes image sequences that represent typical situations for testing video surveillance systems.

3.1 Databases

A-Star-Perception[1]: This database contains 9 image sequences recorded in different scenes. Hand-segmented ground truths are available for each sequence, thus, supervised measures can be used. For concrete testing, the sequences: Water-Surface, Fountain, Shopping-Mall, and Hall are used. The first two sequences are recorded in outdoor scenarios which present high variations due to their nature. Hence, the segmentation process poses a considerable challenge. The other two sequences are recorded in public halls, in which are present many moving objects casting strong shadows crossing each other, being difficult to segment.

Left-Packages[2]: It contains 5 different image sequences recorded at an interior scenario, which has several illumination changes. The main purpose of this database is the identification of abandoned objects (a box and a bag). For testing, hand-segmented ground truths from randomly selected frames are made.

MSA[3]: It contains a single indoor sequence with stable lighting conditions. Nonetheless, strong shadows are casted by the moving objects. The purpose of this sequence is the detection of an abandoned briefcase. Hand-segmented ground truths from randomly selected frames are made in order to give a quantitative measure.

3.2 Experimental Set-Up and Results

For all the experiments, the free parameters above mentioned (section 2) are set as: $T = 15, \xi_H = (h_p \times w_p * 15)^2, \xi_A = 75, \xi_{A_n} = 0.05, \xi_\lambda = 0.5$. The SOBS algorithm is free available[4], and for testing the parameters were left as default. For measuring the accuracy of the methodologies, three different pixel-based measures have been adopted, as described in (13), (14), and (15)

$$Recall = t_p / (t_p + f_n) \tag{13}$$

$$Precision = t_p / (t_p + f_p) \tag{14}$$

$$Similarity = t_p / (t_p + f_n + f_p), \tag{15}$$

---

[1] http://perception.i2r.a-star.edu.sg
[2] http://homepages.inf.ed.ac.uk/rbf/CAVIARDATA1
[3] http://cvprlab.uniparthenope
[4] http://www.na.icar.cnr.it/ maddalena.l/
MODLab/SoftwareSC-SOBS.html

where t_p (true positives), f_p (false positives) and f_n (false negatives) are obtained while comparing against a hand-segmented ground truth [8]. A method is considered good if it reaches high *Recall* measures, without sacrificing *Precision*. *Similarity* is adopted aiming to further compare the results achieved by other proposed algorithms.

In Table 1 are presented the computed measures for the proposed approach and the SOBS algorithm [8]. Additionally, in Fig. 1 an example of the proposed motion direction calculation stage is presented. Fig. 2 shows an example of the proposed stage to detect static objects. Finally, Fig. 3 and Fig. 4 present some relevant segmentation results related to the proposed algorithm and the SOBS method.

Table 1. Segmentation performance for the proposed approach and SOBS

Video	Proposed approach			SOBS		
	Recall	Precision	Similarity	Recall	Precision	Similarity
Water-Surface	0.968 ± 0.027	0.863 ± 0.067	0.837 ± 0.054	0.709 ± 0.103	0.998 ± 0.143	0.708 ± 0.129
Shopping-Mall	0.771 ± 0.144	0.632 ± 0.070	0.543 ± 0.123	0.522 ± 0.192	0.861 ± 0.126	0.482 ± 0.143
MSA	0.989 ± 0.005	0.762 ± 0.026	0.756 ± 0.028	0.778 ± 0.053	0.788 ± 0.042	0.748 ± 0.045
Left-Bag	0.735 ± 0.147	0.863 ± 0.056	0.657 ± 0.126	0.472 ± 0.111	0.642 ± 0.122	0.373 ± 0.116
Left-Box	0.914 ± 0.066	0.891 ± 0.039	0.823 ± 0.075	0.746 ± 0.108	0.806 ± 0.093	0.634 ± 0.104

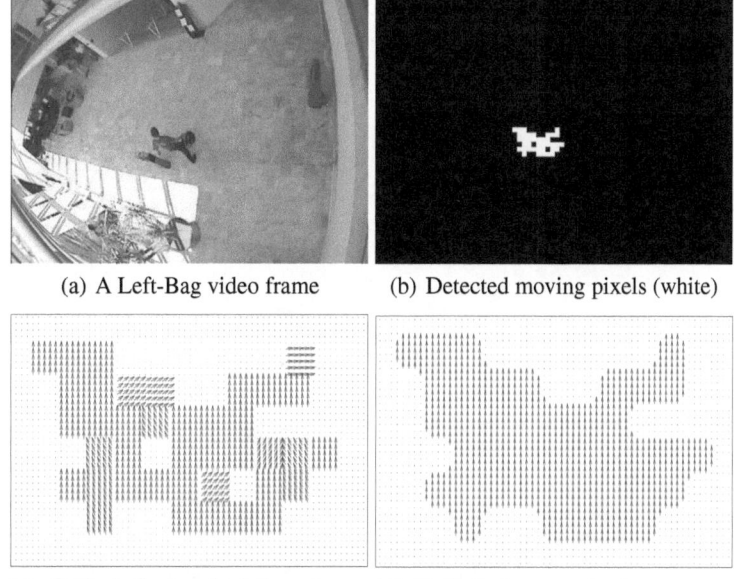

(a) A Left-Bag video frame   (b) Detected moving pixels (white)

(c) Optical flow calculation (zoom over moving pixels)   (d) Optical flow mode (zoom over moving pixels)

Fig. 1. Motion detection results over Left-Bag video

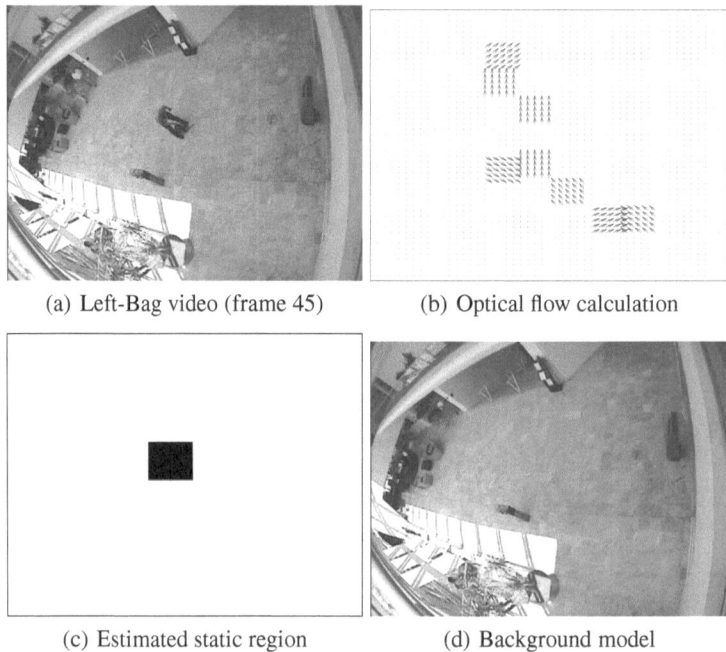

(a) Left-Bag video (frame 45)   (b) Optical flow calculation

(c) Estimated static region   (d) Background model

Fig. 2. Detection of static object after movement (Left-bag video example)

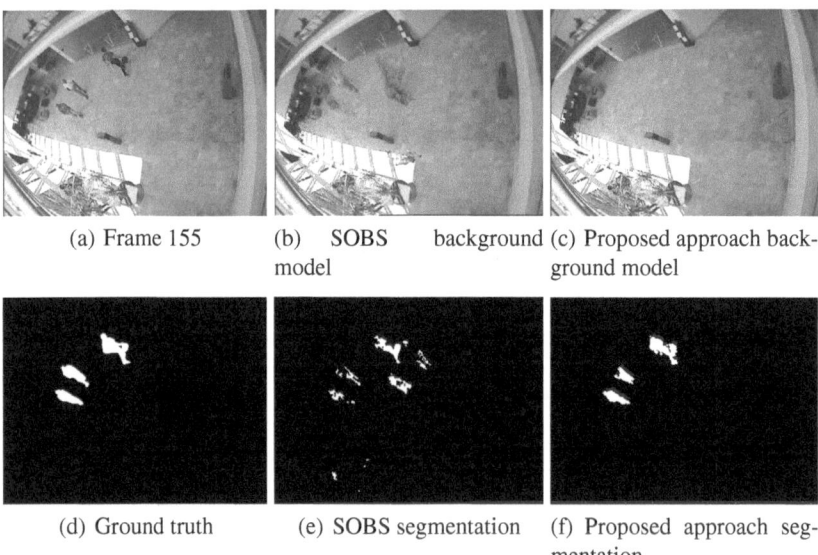

(a) Frame 155   (b)   SOBS   background (c) Proposed approach back-
                     model   ground model

(d) Ground truth   (e) SOBS segmentation   (f) Proposed approach seg-
                                                mentation

Fig. 3. Segmentation results for SOBS and the proposed approach

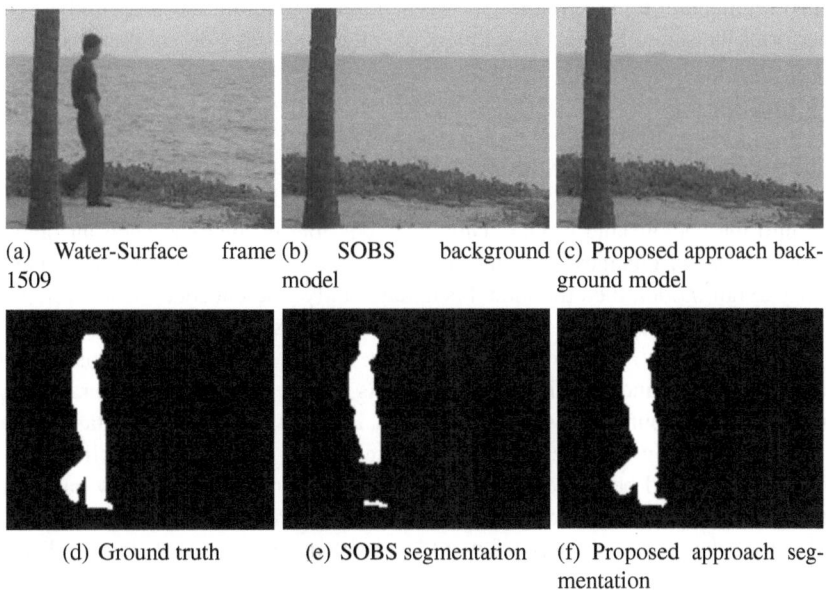

(a) Water-Surface frame
1509

(b) SOBS background
model

(c) Proposed approach background model

(d) Ground truth

(e) SOBS segmentation

(f) Proposed approach segmentation

Fig. 4. Segmentation results for SOBS and the proposed approach

4 Discussion

From Fig.1 it can be noticed the attained results for background modeling of frame 17 in video Left-Bag. Moreover, in Fig. 1(b) is showed that our framework is able to identify pixels that are moved from frame 16 to 17. Given the optical flow calculation (movement direction), a zoom over the pixels that are detecting as movement, and the optical flow direction as computed as shown in Fig. 1(c), where it can be perceived a motion direction trend. Above mentioned trend is suitable softened by our approach (see 1(d)). Thus, our framework is able to track the motion direction of a given object.

Now, from Fig. 2, it is presented the background modelling of video Left-Bag, where a person who had been previously walking stops at frame 45. In this regard, Fig. 2(b) describes the detected motion pixels, however, these pixels do not exhibit a clear direction of movement. Consequently, our approach identifies such region as an static object (see Fig. 2(c)). So, the background model at frame 45 is suitable inferred, as can be visually corroborated in Fig. 2(d).

In addition, in Figs. 3 and 4 some relevant segmentation results are showed. As can be seen from Figs. 3(f) and 4(f) proposed framework obtains suitable segmentation results, while the SOBS approach is not able to deal with light changes, moving/stopping objects, difficult moving trajectories, and/or camouflages, as seen in 3(d) and 4(d). Furthermore, Figs. 3(e) and 4(e) prove how our framework suitable updates the background modelling, even against real-world video conditions (as mentioned above). Otherwise, SOBS approach do not obtain an appropriated background modelling for all the provided cases (see Figs. 3(c) and 4(c)). The above statements are corroborated by the attained segmentation performance shown in Table 1. Overall, the proposed approach outperforms the state of the art benchmark over the analysed video sequences.

5   Conclusions

A video segmentation framework by dynamic background modelling was presented. We developed an optical flow based methodology to take into account object trajectories into the background model. Thus, it is possible to identify when an object in movement stops or changes smoothly their motion along the video, giving a memory property to our model, then the model is updated only when needed. Moreover, a light variations identification stage is employed, which allows to avoid confusions between illumination changes and objects in movement. Hence, our approach aims to update suitable the background model of an scene that is recorded by a static camera. Attained results showed that our framework outperforms, in well-known datasets, state of the art methodologies, being suitable for supporting real video surveillance applications. As future work, it would be interesting to consider other kind of optical flow based models to deal with occluded objects and the inclusion of a multimodal background model. Furthermore, the proposed approach would be implemented as a real-time application.

References

1. Barron, J.L., Fleet, D.J., Beauchemin, S.: Performance of optical flow techniques. International Journal of Computer Vision 12(1), 43–77 (1994)
2. Chen, T.W., Hsu, S.C., Chien, S.Y.: Robust video object segmentation based on k-means background clustering and watershed in ill-conditioned surveillance systems. In: 2007 IEEE International Conference on Multimedia and Expo, pp. 787–790 (July 2007)
3. Collins, R.T., Lipton, A., Kanade, T., Fujiyoshi, H., Duggins, D., Tsin, Y., Tolliver, D., Enomoto, N., Hasegawa, O., Burt, P., et al.: A system for video surveillance and monitoring, vol. 102. Carnegie Mellon University, the Robotics Institute Pittsburg (2000)
4. Correia, M.V., Campilho, A.C.: Real-time implementation of an optical flow algorithm. In: Proceedings of the Pattern Recognition, vol. 4, pp. 247–250. IEEE (2002)
5. Elgammal, A., Duraiswami, R., Harwood, D., Davis, L.: Background and foreground modeling using nonparametric kernel density estimation for visual surveillance. Proceedings of the IEEE 90(7), 1151–1163 (2002)
6. Molina-Giraldo, S., Carvajal-Gonzales, J., Álvarez-Meza, A., Castellanos-Domínguez, G.: Video segmentation based on multi-kernel learning and feature relevance analysis for object classification. In: ICPRAM (2013)
7. Klare, B., Sarkar, S.: Background subtraction in varying illuminations using an ensemble based on an enlarged feature set. In: IEEE Computer Society Conference on CVPR Workshops 2009, pp. 66–73 (June 2009)
8. Maddalena, L., Petrosino, A.: A self-organizing approach to background subtraction for visual surveillance applications. IEEE Transactions on Image Processing 17(7), 1168–1177 (2008)
9. Raty, T.: Survey on contemporary remote surveillance systems for public safety. IEEE Transactions on Systems, Man, and Cybernetics 40(5), 493–515 (2010)
10. Zhou, D., Zhang, H.: Modified gmm background modeling and optical flow for detection of moving objects. In: 2005 IEEE International Conference on Systems, Man and Cybernetics, vol. 3, pp. 2224–2229. IEEE (2005)

Author Index

Abaza, Ayman   I-260
Abdelwahab, Moataz M.   II-601
Abduljalil Abdulhak, Sami   II-191
Abe, Yuichi   II-459
Aboutajdine, Driss   I-813
Achard, Catherine   I-51
Adamo, Alessandro   II-31
Ahmady Phoulady, Hady   I-823
Akhoury, Sharat Saurabh   II-288
Akhtar, Zahid   II-309
Akkoul, Sonia   II-91
Alimi, Adel M.   II-439
Álvarez-Meza, Andres M.   I-843
Alzati, Alberto   I-361
Amelio, Alessia   I-170
Andrade Jr., Aniceto C.   I-151
Andrés-Ferrer, Jesús   I-330
Angelino, C.V.   II-749
Arcelli, Carlo   I-111
Ardizzone, Edoardo   I-773, II-21
Avola, Danilo   II-181
Aydos, Fahri   I-452

Baccaglini, E.   II-749
Bagdanov, Andrew D.   II-239
Bai, Lu   I-181
Balduzzi, Luigi   I-442
Ballan, Lamberto   I-722
Banerjee, Abhirup   II-542
Barbuzzi, Donato   I-121
Barrat, Sabine   I-752
Battiato, Sebastiano   I-381, I-391, I-420, I-783
Bellavia, Fabio   I-270, I-462
Ben Amar, Chokri   II-591, II-611
Benezeth, Yannick   I-712
Benítez-Restrepo, Hern D.   II-121
Bensebaa, Amina   I-340
Benvegna, Francesco   II-500
Benyoussef, Meryem   I-813
Bernard, Guillaume   II-81
Bernini, Nicola   I-582
Berthold, Michael R.   I-131
Bertini, Marco   I-722

Bertozzi, Massimo   I-582, I-592, II-229
Bianco, Simone   II-631, II-652
Bloch, Isabelle   II-562
Boldyš, Jiří   II-369
Boochs, Frank   I-712
Borghese, N. Alberto   I-642
Borgi, Mohamed Anouar   II-611
Bosilj, Petra   I-562
Boughrara, Hayet   II-591
Bourbakis, Nikolaos   II-469
Braunstain, Eyal   I-351, I-523
Bres, Stephane   II-581
Brognara, Cristian   I-482
Bruna, Arcangelo R.   I-783
Bruno, Alessandro   I-773
Buemi, F.   II-731
Burns, Trudy L.   II-349
Bustard, John   II-1

Cafiso, Stefano   I-381
Caglayan, Ali   II-161
Calderara, Simone   I-542
Campadelli, Paola   I-41
Can, Ahmet Burak   II-161
Cancela, Brais   I-400
Capra, Alessandro   II-721
Carullo, Mariarosaria   II-740
Casanova, Andrea   II-721
Casiraghi, Elena   I-41
Castaldo, Francesco   II-552
Castellanos-Domínguez, Cesar G.   I-843, II-121
Castiglioni, Isabella   II-711
Cavaliere, Gianluca   II-740
Cazzato, Dario   I-503
Cerri, Gregorio   II-299
Ceruti, Claudio   I-41
Chan-Hon-Tong, Adrien   I-51
Chao, Yu-Wei   II-489
Chaudhury, Baishali   I-823
Chen, Liming   II-581, II-591
Cheng, Dong-Seon   II-191
Chimienti, Michela   I-61
Chiron, Guillaume   I-702

Choi, Wongun   I-803, II-489
Chowdhury, Manish   I-492
Christersson, Albert   II-479
Chtourou, Mohamed   II-591
Cicala, Luca   II-749
Cinnirella, Alessandro   II-409
Cinque, Luigi   I-572, I-692, II-181
Ciocca, Gianluigi   II-631
Císař, Petr   II-71
Coleman, Sonya   II-532
Colombo, Carlo   I-462
Cordella, Luigi P.   II-219
Cornelis, Jan   II-141
Corsini, Massimiliano   I-482
Cossu, Rossella   I-572
Cozzolino, Davide   II-259
Cristani, Marco   II-131, II-191
Cucchiara, Rita   II-111
Cusano, Claudio   II-652

D'Ambrosio, Carlo   I-592
D'Ambrosio, Roberto   I-682
Das, Sudeb   I-492
Dassisti, Michele   I-61
De Floriani, Leila   II-339
Delalandre, Mathieu   I-752
Del Bimbo, Alberto   I-722, I-763, II-239
Dellepiane, Matteo   I-482
De Marco, Tommaso   I-503, I-793
De Marsico, Maria   I-472
Demirci, M. Fatih   I-452
De Mizio, Marco   II-721, II-749
De Muro, Stefano   II-721
De Prisco, Aniello   II-740
De Stefano, Claudio   II-219
Devincenzi, Luca   I-582
Di Capua, Michele   II-721, II-740
Didaci, Luca   I-622
Di Fina, Dario   II-239
Di Giore, Giuseppe   I-391
Di Graziano, Alessandro   I-381
Di Leo, Giuseppe   II-721, II-753
Di Martino, Gerardo   II-11
Di Ruberto, Cecilia   I-552, I-612
Distante, Cosimo   I-503, I-793
Di Stefano, Luigi   II-299
Dondi, Piercarlo   I-692
Drbohlav, Ondřej   I-532
Drira, Fadoua   II-439
D'Urso, Michele   II-721

El Abed, Haikal   I-251
El'Arbi, Maher   II-611
El-Din, Yomna Safaa   II-329
El Marraki, Mohamed   I-813
Elouedi, Ines   II-249
El Saadany, Omnia S.   II-601
Eramian, Mark G.   I-191
Esposito, Luca Giangiuseppe   II-389
Esposito, Mariana   II-721, II-731

Fadda, Gianluca   I-280
Fakhrzadeh, Azadeh   II-201
Fanfani, Marco   I-462
Farinella, Giovanni M.   I-381, I-420
Fedriga, Rean Isabella   I-592, II-229
Fefilatyev, Sergiy   I-161
Fenu, Gianni   II-721
Fernández, Alba   I-400
Fernández-Beltran, Rubén   I-290
Ferone, Alessio   I-803, II-721
Ferracani, Andrea   I-763
Ferrara, Matteo   I-743
Ferrario, Roberta   II-191
Fioraio, Nicola   II-299
Flammini, Francesco   II-721, II-731
Flusser, Jan   II-369
Fodde, Giuseppe   I-552
Fontanella, Francesco   II-219
Forczmański, Paweł   I-602
Foresti, Gian Luca   II-279
Foti, Enrico   I-420
Fournier, Régis   II-249
Franco, Annalisa   I-743
Frontoni, Emanuele   II-409
Frosio, Iuri   I-361, I-642
Frucci, Maria   II-269
Fusco, Giovanni   I-410
Fusco, Roberta   II-359
Fusiello, Andrea   I-320

Gadermayr, Michael   I-513
Gaetano, Raffaele   I-241, I-371
Galassi, Maurizio   I-320
Galiano, Angelo   I-61
Gallea, Roberto   II-21
Gallivanone, Francesca   II-711
Galvan, Fausto   I-783
Garcia, Christophe   II-439
García-Álvarez, Julio C.   I-843, II-121
Gardiner, Bryan   II-532

Gargiulo, Francesco   II-259
Garibotto, Giovanni   II-721
Garro, Valeria   I-320
Gath, Isak   I-351
Gavelli, M.   II-749
Gemignani, Giorgio   I-803
Ghahramani, Mohammad   II-1
Ghanem, Nagia M.   I-733
Ghiani, Luca   I-280
Giachetti, Andrea   I-482
Giacinto, Giorgio   II-399
Gilardi, Maria C.   II-711
Giménez, Adrià   I-330
Giudice, Oliver   I-381
Goldgof, Dmitry   I-161, I-823
Gomez-Krämer, Petra   I-702
Gori, Marco   II-101
Goto, Hideaki   II-459
Grana, Costantino   II-111
Greco, Luca   II-151
Grossi, Giuliano   II-31
Guclu, Oguzhan   II-161
Guimarães, Silvio Jamil F.   I-11, I-151

Hadid, Abdenour   I-141, II-1, II-309
Hafiane, Adel   II-91
Hakam, Ardeshir   I-823
Hall, Lawrence O.   I-823
Hamouda, Atef   II-249
Hancock, Edwin R.   I-181, I-201, II-41
Haque, S.M. Rafizul   I-191
Haseeb, Muhammad   II-41
Herumurti, Darlis   II-209
Holm, Lena   II-201
Hong, Xiaopeng   I-141
Horn, Martin   I-131
Hou, A'lin   II-449
Hu, Tao   II-683, II-693

Iannello, Giulio   I-682, II-319
Impedovo, Donato   I-61, I-121
Impedovo, Sebastiano   I-91
Ionescu, Radu Tudor   I-1, I-81
Ippolito, Massimo   II-711
Ismail, Mohamed A.   I-733
Iuricich, Federico   II-339

Jaballah, Kabil   II-621
Jaiem, Faten Kallel   I-251
Jansen, Bart   II-141

Jemni, Mohamed   II-621
Jennane, Rachid   II-91
Jin, Dakai   I-662, II-349
Juan, Alfons   I-330

Kampel, Martin   I-71
Kanoun, Slim   I-251
Karaman, Svebor   II-239
Kardoun, Jihain   I-251
Khellah, Fakhry   I-21
Khemakhem, Maher   I-251
Khoury, Ihab   I-330
Kijak, Ewa   I-562
Koutaki, Gou   II-209
Kuijper, Arjan   I-211
Kukharev, Georgy   I-602
Kundu, Malay K.   I-492
Kutics, Andrea   I-310
Kyrgyzov, Olexiy   II-562

Labate, Demetrio   II-611
La Cascia, Marco   II-151
Laganière, Robert   II-288
Lanz, Oswald   II-683, II-693
Lanzarotti, Raffaella   II-31
Lanzi, Pier Luca   I-642
Larabi, Slimane   I-340
Larsson, Sune   II-479
Lebourgeois, Franck   II-439
Lee, John A.   II-81
Lefèvre, Sébastien   I-562
Lengu, Roald   II-721, II-753
Leo, Marco   I-503, I-793
Leoncini, P.   II-749
Lespessailles, Eric   II-91
Levy, Steven M.   II-349
Li, Cheng   II-349
Li, Yang   II-449
Li, You   II-379
Liedlgruber, Michael   I-513
Lippi, Marco   II-101
Lippiello, Vincenzo   II-552
Lisanti, Giuseppe   II-239
Liu, Yonghuai   I-632
Lo Bosco, Giosué   II-500
Lombardi, Gabriele   I-41
Lombardi, Luca   I-692
Lozano-Vega, Gildardo   I-712
Lucat, Laurent   I-51

Luengo Hendriks, Cris L.   II-201
Lupascu, Carmen Alina   I-270

Mabtoul, Samira   I-813
Mačák, Jan   I-532
Maggini, Marco   II-101
Magillo, Paola   II-339
Mahdi, Hani   II-329
Mahmoud, Karim M.   I-733
Maji, Pradipta   II-542
Malmberg, Filip   II-479
Maltoni, Davide   I-743
Mancini, Adriano   II-409
Manfredi, Marco   II-111
Mangini, Francesco Maurizio   I-91, I-121
Manzo, Mario   I-833
Marasco, Emanuela   I-260
Marcelli, Angelo   II-673
Marcialis, Gian Luca   I-280, I-622
Maresca, Mario Edoardo   II-419
Margherita, Roberto   II-631
Marini, Gianluca   II-631
Marrocco, Claudio   II-572
Marrone, Stefano   II-359
Martinel, Niki   II-279
Marzani, Franck   I-712
Masi, Giuseppe   I-241, I-371
Masi, Iacopo   II-239
Mazzei, Luca   I-582
Mazzino, Nadia   II-721, II-753
Mazzocca, Nicola   II-731
Mazzola, Giuseppe   I-773
Melacci, Stefano   II-101
Ménard, Michel   I-702
Micheloni, Christian   II-279
Miron, Alina   II-229
Mironică, Ionuţ   I-431
Mizukami, Yoshiki   I-300
Molina-Giraldo, Santiago   I-843
Molinara, Mario   II-572
Moustafa, Mohamed N.   II-329
Mouton, Peter R.   I-823
Murino, Vittorio   II-131
Murray, David W.   I-642
Murrieri, Pierpaolo   II-721
Musumeci, Rosaria E.   I-420
Mutlu, Sinan   II-683, II-693

Naït-Ali, Amine   II-249
Nakagawa, Akihiko   I-310

Nakanishi, Shinya   I-300
Napoletano, Paolo   II-631, II-652
Nappi, Michele   I-472, II-269
Narducci, Fabio   II-721
Nave, Gennaro   II-740
Nixon, Mark   II-1
Noceti, Nicoletta   I-410, I-442
Nysjö, Johan   II-479
Nyström, Ingela   II-479

O'Connell, Christian   I-310
Odone, Francesca   I-410, I-442
Omelina, Lubos   II-141
Onofri, Leonardo   II-319
Oravec, Milos   II-141
Ortega, Marcos   I-400
Ortis, Alessandro   I-391
Öztimur Karadağ, Özge   II-61

Padovano, Donatella   II-740
Palmieri, Francesco A.N.   II-552
Pantaleo, Giuseppe   II-631
Pantofaru, Caroline   II-489
Paolillo, Alfredo   II-721, II-753
Parziale, Antonio   II-673
Patrocínio Jr., Zenilton K.G.   I-11, I-151
Pazzaglia, Fabio   I-462
Pecoraro, Giovanni   II-11
Pelillo, Marcello   II-131
Penedo, Manuel G.   I-400
Petersen, Thor U.   I-420
Petillo, Ugo   II-721
Petrillo, Antonella   II-359
Petrosino, Alfredo   I-803, I-833, II-419, II-740
Pezzatini, Daniele   I-763
Pham, The-Anh   I-752
Piantadosi, Gabriele   II-359
Piciarelli, Claudio   II-279
Pietikäinen, Matti   I-141
Ping, Bo   II-703
Piras, Luca   II-399
Pirlo, Giuseppe   I-61, I-91, I-121
Pirovano, Michele   I-642
Pirrone, Roberto   II-21
Pizzuti, Clara   I-170
Pla, Filiberto   I-290
Placidi, Giuseppe   II-181
Poggi, Giovanni   I-241, I-371, II-11

Popescu, Marius   I-1, I-81
Porat, Moshe   I-523
Pragliola, Concetta   II-721, II-731
Prisacariu, Victor   I-642
Puglisi, Giovanni   I-783
Puppo, Enrico   II-522
Putzu, Lorenzo   I-612

Raimondo, N.   II-749
Ramel, Jean-Yves   I-752
Ramella, Giuliana   II-642
Ramírez-Rozo, Tomas J.   II-121
Rattani, Ajita   II-309
Ren, Carl Yuheng   I-642
Requier, Fabrice   I-702
Reverchon, Jean-Luc   II-229
Ricciardi, Stefano   II-721
Riccio, Daniel   I-472, II-269
Riccio, Daniele   II-11
Roberto, Vito   II-663
Robertson, Neil M.   I-340
Rocca, Luigi   II-522
Roli, Fabio   I-280, I-622
Rostamzadeh, Negar   I-431
Rozza, Alessandro   I-41
Ruggeri, D.   II-740
Ruichek, Yassine   II-379
Rundo, Francesco   I-391
Russo, Giorgio   II-711

Sabini, Maria G.   II-711
Sablatnig, Robert   I-652
Saha, Punam K.   I-662, II-349
Saleem, Sajid   I-652
Saman, Gul e   I-201
Sanniti di Baja, Gabriella   I-111, II-269, II-642
Sansone, Carlo   II-259, II-359, II-389
Sansone, Mario   II-359
Santoro, Adolfo   II-673
Sao, Anil Kumar   I-221
Sardina, Daniele   II-711
Sasaki, Takahiro   II-459
Savarese, Silvio   I-803, II-489
Savastano, Mario   II-721
Sbihi, Abderrahmane   II-510
Scarpa, Giuseppe   I-241, I-371
Schettini, Raimondo   II-631, II-652
Schlosser, Markus   I-31

Schneider, Kevin A.   I-191
Scopigno, R.   II-749
Scotney, Bryan   II-532
Scotto di Freca, Alessandra   II-219
Sebe, Nicu   I-431
Seidenari, Lorenzo   I-763
Serino, Luca   I-111
Serra, Giuseppe   II-111
Setti, Francesco   II-191
Shchegoleva, Nadezdha   I-602
Shimada, Atsushi   I-101, I-672
Shreve, Matthew   I-161
Siciliano, Bruno   II-552
Siegel, Erin M.   I-823
Simari, Patricio   II-339
Sintorn, Ida-Maria   II-479
Soda, Paolo   I-682, II-319
Solera, Francesco   I-542
Soran, Ahmet   I-452
Sorgi, Lorenzo   I-31
Soukup, Jindřich   II-71
Souloumiac, Antoine   II-562
Spirito, M.   II-731
Spörndly-Nees, Ellinor   II-201
Šroubek, Filip   II-71
Stefano, Alessandro   II-711
Su, Fenzhen   II-703
Su, Weiguang   II-703
Sumer, B. Mutlu   I-420
Sun, Yunlian   I-743, II-171
Svoboda, David   II-429

Tadamura, Katsumi   I-300
Talibi Alaoui, Mohammed   II-510
Taniguchi, Rin-ichiro   I-101, I-672
Tassetti, Anna Nora   II-409
Tegolo, Domenico   I-270, II-500
Thavalengal, Shejin   I-221
Thuerck, Daniel   I-211
Tistarelli, Massimo   II-171, II-309
Toppano, Elio   II-663
Torner, James C.   II-349
Tortorella, Francesco   II-572
Tronci, Roberto   II-399
Tsitsoulis, Athanasios   II-469
Turrini, Cristina   I-361

Uchimura, Keiichi   II-209
Uemura, Takumi   II-209

Uhl, Andreas   I-513
Uijlings, Jasper   I-431
Ulman, Vladimír   II-429
Uricchio, Tiberio   I-722

Valenti, Cesare   I-270
Vascon, Sebastiano   II-131
Vécsei, Andreas   I-513
Verdoliva, Luisa   I-371, II-11, II-259
Verleysen, Michel   II-81
Vernier, Marco   II-279
Vertucci, Raffaele   II-721
Vitabile, Salvatore   II-711

Walha, Rim   II-439
Wang, Ke   II-449
Wang, Yanrung   I-672
Wechsler, Harry   I-472
Wen, Dunwei   II-449

Wiart, Joe   II-562
Wu, Yun   I-231

Xu, Xing   I-101

Yamashita, Takayoshi   I-672
Yang, Yuan   II-562
Yarman Vural, Fatoş T.   II-61
Ylioinas, Juha   I-141
Yonemoto, Satoshi   II-51
Yuan, Chun   I-231

Zambanini, Sebastian   I-71
Zen, Gloria   I-431
Zhang, Yu   II-581
Zhao, Yitian   I-632
Zingaretti, Primo   II-409
Zini, Luca   I-410